Methods in Cell Biology

VOLUME 78

Intermediate Filament Cytoskeleton

Series Editors

Leslie Wilson

Department of Molecular, Cellular and Developmental Biology
University of California
Santa Barbara, California

Paul Matsudaira

Whitehead Institute for Biomedical Research
Department of Biology
Division of Biological Engineering
Massachusetts Institute of Technology
Cambridge, Massachusetts

Methods in Cell Biology

VOLUME 78

Intermediate Filament Cytoskeleton

Edited by

M. Bishr Omary
Department of Medicine
Palo Alto VA Medical Center and
Stanford University
Palo Alto, California

Pierre A. Coulombe
Department of Biological Chemistry
The Johns Hopkins University School of Medicine
Baltimore, Maryland

ELSEVIER
ACADEMIC
PRESS

AMSTERDAM • BOSTON • HEIDELBERG • LONDON
NEW YORK • OXFORD • PARIS • SAN DIEGO
SAN FRANCISCO • SINGAPORE • SYDNEY • TOKYO

Elsevier Academic Press
525 B Street, Suite 1900, San Diego, California 92101-4495, USA
84 Theobald's Road, London WC1X 8RR, UK

This book is printed on acid-free paper.

Permissions may be sought directly from Elsevier's Science & Technology Rights
Department in Oxford, UK: phone: (+44) 1865 843830, fax: (+44) 1865 853333,
E-mail: permissions@elsevier.com.uk. You may also complete your request on-line
via the Elsevier homepage (http://elsevier.com), by selecting
"Customer Support" and then "Obtaining Permissions."

For all information on all Academic Press publications
visit our Web site at www.books.elsevier.com

ISBN: 0-12-564173-7

PRINTED IN THE UNITED STATES OF AMERICA
04 05 06 07 08 9 8 7 6 5 4 3 2 1

CONTENTS

PART I   General Methods to Study Intermediate Filament Proteins — From Molecules to Disease

8. Intermediate Filament Protein Inclusions

Kurt Zatloukal, Conny Stumptner, Andrea Fuchsbichler, Elke Janig, and Helmut Denk

9. Stress Models for the Study of Intermediate Filament Function

E. Birgitte Lane and Milos Pekny

PART II General Methods to Study Intermediate Filament Proteins and Gene Regulation

10. Regulation of Intermediate Filament Gene Expression

Satrajit Sinha

PART III   Methods to Study Specific Intermediate Filament Proteins in Mammalian Systems

19. [^{35}S]Methionine Metabolic Labeling to Study Axonal Transport of Neuronal
Intermediate Filament Proteins *In Vivo*

Stéphanie Millecamps and Jean-Pierre Julien

20. Lamins

Georg Krohne

21. The Intermediate Filament Systems in the Eye Lens

*Ming Der Perng, Aileen Sandilands, Jer Kuszak, Ralf Dahm, Alfred Wegener,
Alan R. Prescott, and Roy A. Quinlan*

PART V   Methods to Study Intermediate Filament Associated Proteins

CONTRIBUTORS

Numbers in parentheses indicate the pages on which the authors' contributions begin.

Ueli Aebi (3, 25), Maurice E. Müller Institute for Structural Biology, Biozentrum, University of Basel, CH-4056 Basel, Switzerland

Robert M. Bellin (519), Department of Biology, College of the Holy Cross, Worcester, Massachusetts 01610

Kelsie M. Bernot (453), Department of Biological Chemistry, The Johns Hopkins University School of Medicine, Baltimore, Maryland 21205

Luca Borradori (757), Department of Dermatology, University Hospital, Geneva, Switzerland CH-1211

Hee-Jung Choi (757), Departments of Structural Biology and Molecular and Cellular Physiology, Stanford University School of Medicine, Stanford, California 94305

Pierre A. Coulombe (xxi, 453), Departments of Biological Chemistry and Dermatology, The Johns Hopkins University School of Medicine, Baltimore, Maryland 21205

Ralf Dahm (597), Project Manager ZF-MODELS IP, Max Planck Institute for Developmental Biology, D-72076 Tübingen, Germany

Helmut Denk (205), Institute of Pathology, Medical University of Graz, A-8036 Graz, Austria

John E. Eriksson (373), Turku Centre for Biotechnology, University of Turku and Åbo Akademi University, FIN-20521 Turku, Finland and Department of Biology, University of Turku, FIN-20014 Turku, Finland

Eric W. Flitney (297), Department of Cell and Molecular Biology, Feinberg School of Medicine, Northwestern University, Chicago, Illinois 60611

Lionel Fontao (757), Department of Dermatology, University Hospital, Geneva, Switzerland CH-1211

Alexandra Fridkin (703), Department of Genetics, The Institute of Life Sciences, The Hebrew University of Jerusalem, Jerusalem 91904, Israel

Andrea Fuchsbichler (205), Institute of Pathology, Medical University of Graz, A-8036 Graz, Austria

Christine Gervasi (673), Department of Biological Sciences and the Center for Neuroscience Research, University at Albany, State University of New York, Albany, New York 12222

Stéphane Gilbert (95), Centre de recherche en cancérologie et Départment de médecine, Université Laval and Centre de recherche de L'Hôtel-Dieu de Québec (CHUQ), GIR 2J6 QC, Canada

Lisa M. Godsel (757), Departments of Pathology and Dermatology, Northwestern University Feinberg School of Medicine, Chicago, Illinois 60611

Robert D. Goldman (297), Department of Cell and Molecular Biology, Feinberg School of Medicine, Northwestern University, Chicago, Illinois 60611

Dmitry Goryunov (787), Columbia University College of Physicians and Surgeons, New York, New York 10032

Kathleen J. Green (757), Departments of Pathology and Dermatology, Northwestern University Feinberg School of Medicine, Chicago, Illinois 60611

Yosef Gruenbaum (703), Department of Genetics, The Institute of Life Sciences, The Hebrew University of Jerusalem, Jerusalem 91904, Israel

Harald Herrmann (3, 25), Department of Cell Biology, TP3, German Cancer Research Center (DKFZ), D-69120 Heidelberg, Germany

Michael Hesse (65), Institut für Physiologische Chemie, Abteilung für Zellbiochemie, Bonner Forum Biomedizin and LIMES, Universitätsklinikum Bonn, 53115 Bonn, Germany

Tracie Y. Hudson (757), Departments of Pathology and Dermatology, Northwestern University Feinberg School of Medicine, Chicago, Illinois 60611

Arthur C. Huen (757), Departments of Pathology and Dermatology, Northwestern University Feinberg School of Medicine, Chicago, Illinois 60611

Ted W. Huiatt (519), Muscle Biology Group, Departments of Biochemistry, Biophysics, and Molecular Biology and of Animal Science, Iowa State University, Ames, Iowa 50011

Masaki Inagaki (353), Division of Biochemistry, Aichi Cancer Center Research Institute, Nagoya, Aichi 464–8681, Japan

Lubomír Janda (721), Institute of Biochemistry and Molecular Cell Biology, University of Vienna, Max F. Perutz Laboratories, Vienna Biocenter, A-1030 Vienna, Austria

Elke Janig (205), Institute of Pathology, Medical University of Graz, A-8036 Graz, Austria

Jean-Pierre Julien (555), Research Center of CHUL and Department of Anatomy and Physiology, Laval University, Quebec, G1V 4G2 QC, Canada

Anton Karabinos (703), Max Planck Institute for Biophysical Chemistry, Department of Biochemistry, 37077 Goettingen, Germany

Aie Kawajiri (353), Division of Biochemistry, Aichi Cancer Center Research Institute, Nagoya, Aichi 464–8681, Japan and Department of Pathology, Nagoya University School of Medicine, Nagoya, Aichi 466–8550, Japan

Vitaly Kochin (373), Turku Centre for Biotechnology, University of Turku and Åbo Akademi University, FIN-20521 Turku, Finland and Department of Biology, University of Turku, FIN-20014 Turku, Finland

Thomas P. Kole (45), Department of Chemical and Biomolecular Engineering, The Johns Hopkins University, Baltimore, Maryland 21218

Laurent Kreplak (3, 25), Maurice E. Müller Institute for Structural Biology, Biozentrum, University of Basel, CH-4056 Basel, Switzerland

Georg Krohne (573), Division of Electron Microscopy, Biocenter of the University of Würzburg, Am Hubland, D-97074 Würzburg, Germany

Nam-On Ku (489), Department of Medicine, Palo Alto VA Medical Center and Stanford University, Palo Alto, California 94304

Jer Kuszak (597), Departments of Ophthalmology and Pathology, Rush-Presbyterian-St Luke's Medical Centre, Chicago, Illinois 60612

E. Birgitte Lane (229), Cancer Research UK, Cell Structure Research Group, University of Dundee, School of Life Sciences, Dundee DD1 5EH, Scotland, United Kingdom

Lutz Langbein (413), Division of Cell Biology, German Cancer Research Center, 69120 Heidelberg, Germany

Math P. G. Leers (163), Department of Clinical Chemistry & Hematology, Atrium Medical Center Heerlen, Heerlen, The Netherlands

Rudolf E. Leube (321), Department of Anatomy, Johannes Gutenberg-University, 55128 Mainz, Germany

Conrad L. Leung (787), Columbia University College of Physicians and Surgeons, New York, New York 10032

Ronald K. H. Liem (787), Columbia University College of Physicians and Surgeons, New York, New York 10032

Pawel Listwan (817), Department of Biochemistry and Molecular Biology and The Centre for Functional and Applied Genomics, University of Queensland, Brisbane, Queensland 4072, Australia

Anne Loranger (95), Centre de recherche de L' Hôtel-Dieu de Québec (CHUQ), G1R 2J6 QC, Canada

Thomas M. Magin (65), Institut für Physiologische Chemie, Abteilung für Zellbiochemie, Bonner Forum Biomedizin and LIMES, Universitätsklinikum Bonn, 53115 Bonn, Germany

Normand Marceau (95), Centre de recherche en cancérologie et Départment de médecine, Université Laval and Centre de recherche de L'Hôtel-Dieu de Québec (CHUQ), GIR 2J6 QC, Canada

Jürgen Markl (627), Institute of Zoology, Johannes Gutenberg University, 55099 Mainz, Germany

W. H. Irwin McLean (131), Epithelial Genetics Group, Human Genetics Unit, Ninewells Medical School, University of Dundee, Dundee, Scotland, United Kingdom

Roland Meier-Bornheim (65), Institut für Physiologische Chemie, Abteilung für Zellbiochemie, Bonner Forum Biomedizin and LIMES, Universitätsklinikum Bonn, 53115 Bonn, Germany

Stéphanie Millecamps (555), Research Center of CHUL and Department of Anatomy and Physiology, Laval University, Quebec, G1V 4G2 QC, Canada

M. Bishr Omary (xxi, 489), Department of Medicine, Palo Alto VA Medical Center and Stanford University, Palo Alto, California 94304

Cecilia Östlund (829), Departments of Medicine and of Anatomy and Cell Biology, College of Physicians and Surgeons, Columbia University, New York, New York 10032

Hanna-Mari Pallari (373), Turku Centre for Biotechnology, University of Turku and Åbo Akademi University, FIN-20521 Turku, Finland and Department of Biology, University of Turku, FIN-20014 Turku, Finland

Harish Pant (373), Laboratory of Neurochemistry, NIH, NINDS, Bethesda, Maryland 20892

Milos Pekny (229), Department of Medical Biochemistry, Sahlgrenska Academy at Göteborg University, 405 30 Göteborg, Sweden

Ming Der Perng (597), School of Biological and Biomedical Sciences, The University of Durham, Durham DH1 3LE, United Kingdom

Silke Praetzel (413), Division of Cell Biology, German Cancer Research Center, 69120 Heidelberg, Germany

Alan R. Prescott (597), CHIPs, School of Life Sciences, The University of Dundee, Dundee DD1 5EH, United Kingdom

Roy A. Quinlan (597), School of Biological and Biomedical Sciences, The University of Durham, Durham DH1 3LE, United Kingdom

Julia Reichelt (65), Institut für Physiologische Chemie, Abteilung für Zellbiochemie, Bonner Forum Biomedizin and LIMES, Universitätsklinikum Bonn, 53115 Bonn, Germany

Günther A. Rezniczek (721), Institute of Biochemistry and Molecular Cell Biology, University of Vienna, Max F. Perutz Laboratories, Vienna Biocenter, A-1030 Vienna, Austria

Richard M. Robson (519), Muscle Biology Group, Departments of Biochemistry, Biophysics, and Molecular Biology and of Animal Science, Iowa State University Ames, Iowa 50011

Michael A. Rogers (413), Division of Normal and Neoplastic Epidermal Differentiation, German Cancer Research Center, 69120 Heidelberg, Germany

Joseph A. Rothnagel (817) Department of Biochemistry and Molecular Biology and The Centre for Functional and Applied Genomics, University of Queensland, Brisbane, Queensland 4072, Australia

Aileen Sandilands (131, 597), Epithelial Genetics Group, Human Genetics Unit, Ninewells Medical School and Department of Molecular and Cellular Pathology, The University of Dundee, Dundee DDI 9SY, Scotland, United Kingdom

Michael Schaffeld (627), Institute of Zoology, Johannes Gutenberg University, 55099 Mainz, Germany

Juergen Schweizer (413), Division of Normal and Neoplastic Epidermal Differentiation, German Cancer Research Center, 69120 Heidelberg, Germany

Satrajit Sinha (267), Department of Biochemistry, State University of New York at Buffalo, Buffalo, New York 14214

Frances J. D. Smith (131), Epithelial Genetics Group, Human Genetics Unit, Ninewells Medical School, University of Dundee, Dundee, Scotland, United Kingdom

Herbert Spring (413), Division of Analytical Microscopy and Microinjection, German Cancer Research Center, 69120 Heidelberg, Germany

Sergei V. Strelkov (25), Maurice E. Müller Institute for Structural Biology, Biozentrum, University of Basel, CH-4056 Basel, Switzerland

Conny Stumptner (205), Institute of Pathology, Medical University of Graz, A-8036 Graz, Austria

Ben G. Szaro (673), Department of Biological Sciences and the Center for Neuroscience Research, University at Albany, State University of New York, Albany, New York 12222

Guo-Zhong Tao (489), Department of Medicine, Palo Alto VA Medical Center and Stanford University, Palo Alto, California 94304

Diana M. Toivola (489), Department of Medicine, Palo Alto VA Medical Center and Stanford University, Palo Alto, California 94304

Yiider Tseng (45), Department of Chemical and Biomolecular Engineering, The Johns Hopkins University, Baltimore, Maryland 21218

Alfred Wegener (597), Experimental Ophthalomology, University of Bonn, Bonn 53105, Germany

William I. Weis (757), Departments of Structural Biology and Molecular and Cellular Physiology, Stanford University School of Medicine, Stanford, California 94305

Gerhard Wiche (721), Institute of Biochemistry and Molecular Cell Biology, University of Vienna, Max F. Perutz Laboratories, Vienna Biocenter, A-1030 Vienna, Austria

Reinhard Windoffer (321), Department of Anatomy, Johannes Gutenberg-University, 55128 Mainz, Germany

Denis Wirtz (45), Department of Chemical and Biomolecular Engineering and Graduate Program in Molecular Biophysics, The Johns Hopkins University, Baltimore, Maryland 21218

Pauline Wong (453), Department of Biological Chemistry, The Johns Hopkins University School of Medicine, Baltimore, Maryland 21205

Howard J. Worman (829), Departments of Medicine and of Anatomy and Cell Biology, College of Physicians and Surgeons, Columbia University, New York 10032

Kurt Zatloukal (205), Institute of Pathology, Medical University of Graz, A-8036 Graz, Austria

Bihui Zhong (489), Division of Gastroenterology, The First Affiliated Hospital of Sun Yet-sen University, Guangzhou 510080, China

Qin Zhou (489), Department of Medicine, Palo Alto VA Medical Center and Stanford University, Palo Alto, California 94304

PREFACE

The intermediate filament (IF) protein field has witnessed a remarkable growth since the initial description of filamentous arrays that were seen by electron microscopy in muscle cells. These initially described arrays were intermediate in size as compared to actin and myosin filaments or to microfilaments and microtubules (Ishikawa et al., 1968). Hence the name that has ignited a field no one could have envisioned would evolve into its present and continuously growing status. Since those pioneering years of early characterization (Fig. 1), the complexity of this large and heterogeneous family of proteins and their interaction with other cytoskeletal elements continues to unfold and surprise present and past investigators (Helfrand et al., 2003; Herrmann et al., 2003; Hesse et al., 2001; Lariviere and Julien, 2004; Lazarides, 1980; Leung et al., 2002; Moll et al., 1982; Oshima, 1992; Steinert and Parry, 1985). Among the three major categories of cytoskeletal proteins, intermediate filaments are fascinating in part because of their cell type-specific distribution (Table I) and their clearly defined function of protection from mechanical and environmental forms of stress (Fuchs and

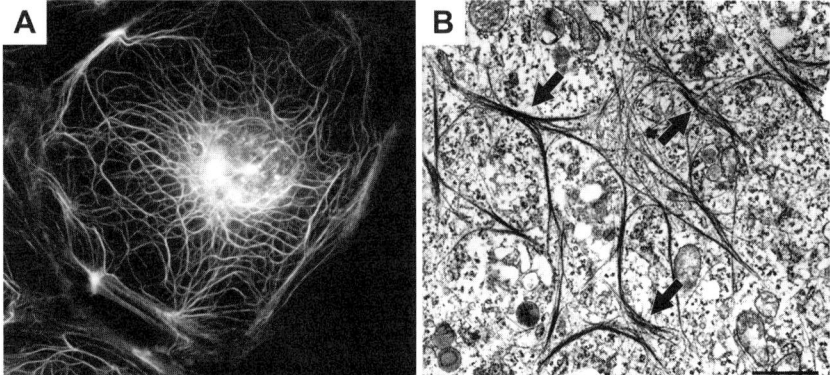

Fig. 1 Intermediate filaments-the early days. Two images reproduced from the pioneering work of Osborn, Franke and Weber (Osborn et al., 1977), when intermediate filament proteins were just being recognized and tools developed for their study. Panel A: Intermediate filaments of methanol/acetone-fixed Pt K2 cells (derived from rat kangaroo kidney epithelium) as visualized by immunofluorescence staining using non-immune normal rabbit serum at a high concentration (1:6 dilution). The reactivity of the antibody was abolished after formaldehyde fixation (not shown). Panel B: Transmission electron microscopy of Pt K2 cells, with arrows highlighting the typically-noted bundles of intermediate filaments (in this case keratin and vimentin filaments). Bar = 1 μm. The images were generously provided by Dr. Mary Osborn.

Table I
Intermediate Filament Proteins

Name	Type[a]	Chromosome[b]	Size (kDa)[b]	No. genes[b]	Assembly group[c]	Cell/tissue distribution
Cytoplasmic						
Keratins	I	17	40–64	>25	A	K9-K20, K23 (Epithelia) Ha1-Ha8 (Hair, nail) Irs1-4 (Hair follicles)
Keratins	II	12	52–70	>25	A	K1-K8 (Epithelia) Hb1-Hb6 (Hair, nail) K6irs1-4 (Hair follicles)
Vimentin	III	10	55	1	B	Mesenchymal
Desmin	III	2	53	1	B	All muscle
GFAP	III	17	52	1	B	Astrocytes
Peripherin	III	12	54–62 [three splice variants]	1	B	Peripheral neurons
Neurofilaments (L,M,H chains)	IV	8 8 22	61 (NF-L), 90 (NF-M), 110 (NF-H)	1	B	Central neurons
α-Internexin	IV	10	61	1	B	Central neurons
Nestin	III, IV, VI	1	240	1	B	Neuroepithelial
Synemin	III, IV, VI	15	180 (α) & 150 (β) [two splice variants]	1	B	All muscle (β isoform mainly in striated muscle)
Syncoilin	III, IV	1	54	1	B	Muscle (mainly skeletal/cardiac)
Phakinin (CP49)	Orphan	3	46	1	D	Lens
Filensin (CP115)	Orphan	20	83	1	D	Lens
Nuclear						
Lamins A, C	V	1	62–78 [four splice variants]	1	C	Nuclear lamina
Lamins B1, B2	V	5, 19	62–78	1 each	C	Nuclear lamina

[a]IF sequence type is defined by gene substructure (position of introns) and nucleotide sequence homology for the region coding for the α-helical rod domain shared by all IF proteins. Some genes (nestin, synemin, syncoilin) can be ascribed to more than one sub-groupings depending on the criteria used.
[b]Chromosome assignment and protein size are given for human IF genes and proteins. The number of IF genes is nearly perfectly conserved in mammalian genomes that have been thoroughly investigated (human, mouse, rat).
[c]Assembly group refers to the ability of IF roteins to copolymerize to form 10–12 nm filaments. Keratins (group A) are strict heteropolymers involving type I and type II chains in a 1:1 molar ratio. Several proteins in group B can polymerize on their own to form IFs; however, they can copolymerize with one another *in vitro*. The nuclear lamins and lens-specific filensin and phakinin represent distinct groups.

Cleveland, 1998; Ku *et al.*, 1999) that is becoming understood in biophysical, structural and biochemical terms (Coulombe and Wong, 2004; Strelkov *et al.*, 2003). Intermediate filament proteins also take center stage when it comes to human disease, given the complex array of human diseases that they either cause or predispose to, which is a reflection of their nearly ubiquitous yet tissue specific expression (Omary *et al.*, 2004).

Our goal in assembling this volume was to provide a comprehensive treatise of methodology and systems analysis essentials for any investigator, actively working- or wishing to work- on IF genes and their gene products and associated proteins. Such a methodology resource is presently not available and the series editors felt, and we wholeheartedly agreed, that covering the topic of IF genes and proteins is timely. Our strategy in seeking expert investigators in the field was to cover state-of-the-art methodologies pertaining to 5 major aspects of IF proteins:

i. IF proteins from molecules to human disease: Here, the Herrmann and Aebi laboratories cover structural aspects of IF proteins in *Chapters 1 and 2*; Kole, Tseng and Wirtz cover, in *Chapter 3*, microrheology as a tool to study the mechanical properties of the cytoskeleton in live cells; in *Chapter 4* the Magin laboratory describe state-of-the-art methods to manipulate IF genes in the mouse genome, and comment on the large impact that such models have had on our understanding of IF-based disease and IF function; Marceau and colleagues cover the cleavage of IF proteins during apoptosis in *Chapter 5*; the McLean laboratory covers, in *Chapter 6*, methodologies and practical aspects pertaining to the genetic assessment of IF proteins in human diseases; Leers presents in *Chapter 7* how the tissue specificity and abundance of IF proteins, coupled with flow cytometry and other techniques, led to the use of IF proteins as tumor makers and a tool to assess tumor spread; in *Chapter 8* Zatloukal and colleagues discuss the ability of IF proteins to form cytoplasmic inclusions and how such inclusions can be studied; and *Chapter 9* by Lane and Pekny closes this section by describing various stress models that can be exploited to study IF function.

ii. IF proteins and gene regulation: Strategies that relate to the study of transcriptional control of IF gene expression are discussed by Sinha in *Chapter 10*. Fluorescence-based imaging modalities to study IF protein organization are covered by Flitney and Goldman in *Chapter 11* and by Windoffer and Leube in *Chaper 12*. Regulation of IF protein function and dynamics by posttranslational modifications, with an emphasis on the characterization of IF phosphorylation and the development of phospho-epitope-specific reagents, are discussed in *Chapter 13* by Kawajiri and Inagaki and in *Chapter 14* by Pant, Eriksson and colleagues.

iii. IF proteins in specific mammalian systems: This section of the volume focuses on hair and hair follicle epithelial keratins in *Chapter 15* by Langbein, Schweizer and colleagues; skin and simple epithelial keratins in *Chapter 16* and *Chapter 17* by the Coulombe and Omary laboratories, respectively; muscle IF proteins in *Chapter 18* by Robson and colleagues; axonal transport of neuronal IF in *Chapter 19* by Millecamps and Julien; lamins in *Chapter 20* by Krohne; and the lens proteins vimentin, phakinin, and filensin in *Chapter 21* by the Quinlan laboratory.

iv. IF proteins in non-mammalian systems: This section recognizes the increasing importance of non-mammalian systems in providing insight into the properties and function of IF proteins, and discusses methods and assays that allow their study. As such, fish IF proteins are extensively discussed by Schaffeld and Markl in *Chapter 22*; Xenopus IF proteins are covered in *Chapter 23* by Gervasi and Szaro; and methodologies to study Caenorhabditis elegans IF proteins are discussed in *Chapter 24* by Fridkin, Karabinos and Gruenbaum.

v. IF-associated proteins: The last section of this volume covers the important topic of IF-associated proteins, an area that is likely to grow as functional and regulatory aspects of IF proteins unfold. Plectin is discussed in *Chapter 25* by the Wiche laboratory; Green, Borradori, Fontao, Weis and colleagues discuss methods to study desmoplakin-IF protein interactions in *Chapter 26*; in *Chapter 27* the Liem laboratory discusses cytolinker proteins and assays to study proteins that regulate IF-microfilament and IF-microtubule interactions; Listwan and Rothnagel discuss IF-binding proteins in the skin in *Chapter 28*; and Ostlund and Worman end this section with *Chapter 29* by covering lamin-associated proteins.

Not all IF proteins could be covered in this monograph due to space constraints, but every effort was made to cover broad categories and, when possible, emphasize a "systems"-oriented approach. We also opted to cover, in the first book of this kind in the field, the most practical and useful techniques to "old" and "new" hands alike. Most chapters include a "Pearls and Pitfalls" section that will hopefully meet its intended goal of providing helpful hints and advice that would otherwise not be routinely seen in the experimental section of manuscripts or in laboratory manuals. In addition, the authors included many useful and detailed tables with relevant reagents, vendors, and important resource websites.

Most if not all of us in this field believe that this is an exciting time for the study of IF genes and proteins. This is certainly reflected in the primary literature, in that the number of research articles dealing with intermediate filaments continues to increase steadily and to unfold their dynamic nature and nonstructural functions. Areas in which important revelations are likely to surface include and are not limited to IF structure, functions pertaining to signaling and other regulatory processes, interactions with other cytoskeletal elements and organelles, characterization of new IF genes and their products, and IF-related disease pathogenesis and treatment. This will be made possible in part through attracting new investigators to this field. We hope that this volume will serve its important purpose of helping new investigators as they establish themselves in the field, assist established investigators wishing to expand their studies on this topic, and disseminate and possibly standardize the practice of many of the methods and assays that are commonly used when studying IF genes and proteins. In closing, we are most grateful to the series editors, all the contributing authors, and the editorial assistance provided by Elsevier to make this project possible.

M. Bishr Omary
Pierre A. Coulombe
Klaus Weber

References

Coulombe, P. A., and Wong, P. (2004). Cytoplasmic intermediate filaments revealed as dynamic and multipurpose scaffolds. *Nat. Cell Biol.* **6,** 699–706.

Fuchs, E., and Cleveland, D. W. (1998). A structural scaffolding of intermediate filaments in health and disease. *Science.* **279,** 514–519.

Helfrand, B. T., Chang, L., and Goldman, R. D. (2003). The dynamic and motile properties of intermediate filaments. *Annu. Rev. Cell Dev. Biol.* **19,** 445–467.

Herrmann, H., Hesse, M., Reichenzeller, M., Aebi, U., and Magin, T. M. (2003). Functional complexity of intermediate filament cytoskeletons: From structure to assembly to gene ablation. *Int. Rev. Cytol.* **223,** 83–175.

Hesse, M., Magin, T. M., and Weber, K. (2001). Genes for intermediate filament proteins and the draft sequence of the human genome: Novel keratin genes and a surprisingly high number of pseudogenes related to keratin genes 8 and 18. *J. Cell Sci.* **114,** 2569–2575.

Ishikawa, H., Bischoff, R., and Holtzer, H. (1968). Mitosis and intermediate-sized filaments in developing skeletal muscle. *J. Cell Biol.* **38,** 538–555.

Ku, N-O., Zhou, X., Toivola, D. M., and Omary, M. B. (1999). The cytoskeleton of digestive epithelia in health and disease. *Am. J. Physiol.* **277,** G1108–G1137.

Lariviere, R. C., and Julien, J. P. (2004). Functions of intermediate filaments in neuronal development and disease. *J. Neurobiol.* **58,** 131–148.

Lazarides, E. (1980). Intermediate filaments as mechanical integrators of cellular space. *Nature* **283,** 249–256.

Leung, C. L., Green, K. J., and Liem, R. K. (2002). Plakins: A family of versatile cytolinker proteins. *Trends Cell Biol.* **12,** 37–45.

Moll, R., Franke, W. W., Schiller, D. L., Geiger, B., and Krepler, R. (1982). The catalog of human cytokeratins: Patterns of expression in normal epithelia, tumors and cultured cells. *Cell* **31,** 11–24.

Omary, M. B., Coulombe, P. A., and McLean, W. H. I. (2004). Intermediate filament proteins and their related diseases. *N. Engl. J. Med.,* in press.

Osborn, M., Franke, W. W., and Weber, K. (1977). Visualization of a system of filaments 7–10 nm thick in cultured cells of an epithelioid line (Pt K2) by immunofluorescence microscopy. *Proc. Natl. Acad. Sci. USA* **74,** 2490–2494.

Oshima, R. G. (1992). Intermediate filament molecular biology. *Curr. Opin. Cell. Biol.* **4,** 110–116.

Steinert, P. M., and Parry, D. A. D. (1985). Intermediate filaments: Conformity and diversity of expression and structure. *Ann. Rev. Cell Biol.* **1,** 41–65.

Strelkov, S. V., Herrmann, H., and Aebi, U. (2003). Molecular architecture of intermediate filaments. *Bioassays* **25,** 243–251.

As co-editors, we wish to take the opportunity of editing this volume to include a dedication.

Bishr Omary dedicates this book to his parents Ibrahim and Dalal; and to Nashwa, Reed, and Isabella for their love and support.

Pierre Coulombe dedicates this book to Carole and Félix, and to adoption, which bridges compatible needs in the interest of children, and in doing so brings out the best in human nature

ACKNOWLEDGMENT

We thank Paul Matsudaira and Leslie Wilson for their invitation, and the opportunity they gave us, to edit this series on the *Intermediate Filament Cytoskeleton*. Paul and Leslie recognized the importance of this component of the cytoskeleton and the lack of any *Methods* volume that covers these proteins. Mica Haley provided editorial support during the initial development of this volume, which was taken over by Kristi Savino, Tari Paschall and Tracy Grace. Kristi, Tari and Tracy worked passionately and relentlessly with us, and always with a smile, on all aspects of this project and were a major force in helping insure its timely completion. We also thank Rich Knauel for his assistance with the Volume Cover. Last and certainly not least, we are very grateful to all the chapter authors for their outstanding and timely contributions that helped us meet our initial publication target date without delay. It was indeed a pleasure and a privilege to have the opportunity to undertake this project, and to work on this effort with our international and distinguished panel of expert colleagues who made this Volume a reality.

PART I

General Methods to Study
Intermediate Filament
Proteins – From Molecules
to Disease

CHAPTER 1

Isolation, Characterization, and *In Vitro* Assembly of Intermediate Filaments

Harald Herrmann,★ **Laurent Kreplak,**† **and Ueli Aebi**†

★Division of Cell Biology, TP 3
German Cancer Research Center (DKFZ)
D-69120 Heidelberg, Germany

†Maurice E. Müller Institute for Structural Biology
Biozentrum, University of Basel
CH-4056 Basel, Switzerland

I. Introduction

Intermediate filaments (IF) are principal structural elements of metazoan cells (Herrmann and Aebi, 2004). In the nucleus, IF proteins constitute a major part of a proteinaceous network attached to the inner nuclear membrane, the nuclear lamina, and for this reason they have been named lamins (Aebi *et al.*, 1986; Gerace and Blobel, 1980). In addition, lamins are also present within the nucleus in structures not characterized in full detail yet (Goldman *et al.*, 1992). In the cytoplasm, most, although not all, metazoan cells contain IF proteins in abundance. In higher verte-brates such as man, more than 60 genes for cytoplasmic IF proteins have been identified, most of which are coding for keratins (Herrmann *et al.*, 2003; Hesse

METHODS IN CELL BIOLOGY, VOL. 78
Copyright 2004, Elsevier Inc. All rights reserved.
0091-679X/04 $35.00

et al., 2001). This corresponds to the execution of complex expression programs during embryogenesis, leading to the generation of the approximately 200 different cell types in man. Despite this tremendous complexity, IF proteins can generally be isolated in a similar way from these various cell types, and only with some tissues special procedures may have to be used.

When cells are extracted with buffers containing moderate levels of non-ionic detergents and high concentrations of monovalent ions, an almost complete solubilization of the cellular material is achieved, except for a remnant often discarded in cell fractionation experiments. In the high-speed centrifugation pellets of such extracts, only a few proteins are found compared with the complex mixtures present in the supernatant fractions. Various procedures to isolate IF proteins from the insoluble residues of different types of cells and tissues have been described, which follow, however, very similar principles (Achstaetter *et al.*, 1986; Aebi *et al.*, 1986; Huiatt *et al.*, 1980; Renner *et al.*, 1981; Starger *et al.*, 1978; Steinert *et al.*, 1982; Vorgias and Traub, 1983). The pelleted material is highly insoluble in physiological buffers but can be dissolved, contrary to many text-books views, under conditions of very low ionic strength or, alternatively, with the help of chaotropic substances such as urea or guanidinium hydrochloride (Steinert *et al.*, 1976; Stromer *et al.*, 1981). In addition to detergent and high salt, the inclusion of nucleases into the extraction buffer is of high importance, because otherwise a compact pellet will not be obtained even after prolonged centrifugation, but instead a highly viscous gel that is hard to handle. In the classical protocols, the extraction procedure is carried out on ice with conventional Dounce homogenizers. Thereby, proteolysis is reduced, and the genomic DNA is sheared to help the action of the nucleases. Nevertheless, it was found to be of utmost importance to suppress proteolysis further by the inclusion of protease inhibitors such as phenylmethyl sulfonyl fluoride (PMSF) conventionally used at 0.1 mM. This proved, however, to be not entirely sufficient. Only by raising the PMSF concentration to 3 mM, which is actually above the saturation concentration for this lipophilic compound in water, was it possible to reproducibly obtain significant amounts of the high molecular weight protein plectin and to keep vimentin degradation low (Pytela and Wiche, 1980). A pronounced difficulty with this toxic compound is its very rapid, pH-dependent hydrolysis. Hence, the half-life time of PMSF at pH 8 is in the second range. Another problem is the low activity of pancreatic DNase I under high-salt conditions and low temperature. Although after prolonged incubation, the viscosity of the solutions is such that it can be taken up with a pipette, a considerable amount of DNA—and RNA—remains in the resulting centrifugation pellets. Remnant nucleic acids can be removed by resuspending insoluble residues in phosphate-buffered saline (PBS) and further incubation with nucleases followed by centrifugation. However, because it is difficult to suppress proteolysis completely, these lengthy manipulations result in the (often complete) loss of protease-sensitive proteins such as the IF-associated protein plectin or high molecular weight IF proteins such as nestin. In addition, the non-α-helical head and tail domains of IF proteins that are not protected through coiled-coil formation are easily degraded (see later; Nelson and Traub, 1983; Quax-Jeuken *et al.*, 1983).

To circumvent these difficulties, a fast cell extraction procedure has been developed and particularly optimized for [^{32}P] phosphate-labeling experiments of cells (Herrmann and Wiche, 1983). Thereby, the large amounts of radioactively labeled lipids and nucleic acids, which are generated in addition to phosphorylated proteins, are quantitatively removed. Lipids and nucleic acids, when present in samples analyzed on standard sodium dodecyl sulfate (SDS) polyacrylamide gels, disturb the documentation of radioactively labeled protein patterns by autoradiography tremendously. With some practice, IFs from two 10-cm Petri culture dishes can be isolated in 20 minutes. This system has further been exploited to investigate the action of cytoskeleton-associated kinases by gentle permeabilization of the plasma membrane first with the help of very low concentrations of digitonin (Fiskum *et al.*, 1980) followed by administration of [γ-^{32}P] adenosine triphosphate (ATP) (Herrmann and Wiche, 1983). After removal of the radioactive compound, IFs can be isolated quickly in the presence of phosphatase inhibitors. In these experiments, it is essential that no significant degradation occurs during the comparatively lengthy incubation and washing steps. Therefore, the first low-detergent extraction was introduced to remove most of the endogenous soluble and membrane-bound proteolytic enzymes followed by further steps to remove chromatin and cytoskeletal proteins. An example of such a differential extraction is shown in Fig. 1. Here it is demonstrated by immunoprecipitation of plectin that, under the conditions described later, proteolysis did not take place to any significant degree (Fig. 1B, lanes 2 and 4, and Fig. 1C, lane 3).

Another important aspect of this procedure is that cells are actually extracted *in situ* and that interference with dynamic cellular processes and cellular structure as encountered during the harvest of cells by scraping or by trypsin-digestion and prolonged storage on ice are essentially avoided. Similarly, when the cleavage activities of distinct proteases in response to extracellular signals are investigated, it is essential that proteins are not degraded as a result of the isolation procedure. Indeed, it was possible to demonstrate with this fast-extraction protocol that the stimulation of the death receptor in MCF 7 cells activates plectin cleavage by caspase 8 (Stegh *et al.*, 2000). Because only one site in the middle of this huge protein is actually targeted by caspase 8, it was obviously essential that no cleavage took place during the isolation and lengthy immunoprecipitation procedure of the factor. Finally, an adaptation of this protocol made it possible to isolate the large, proteolysis-sensitive nuclear protein TPR in significant amounts (Cordes *et al.*, 1997), as well as to investigate bacterial cytoskeletons (Regula *et al.*, 2001).

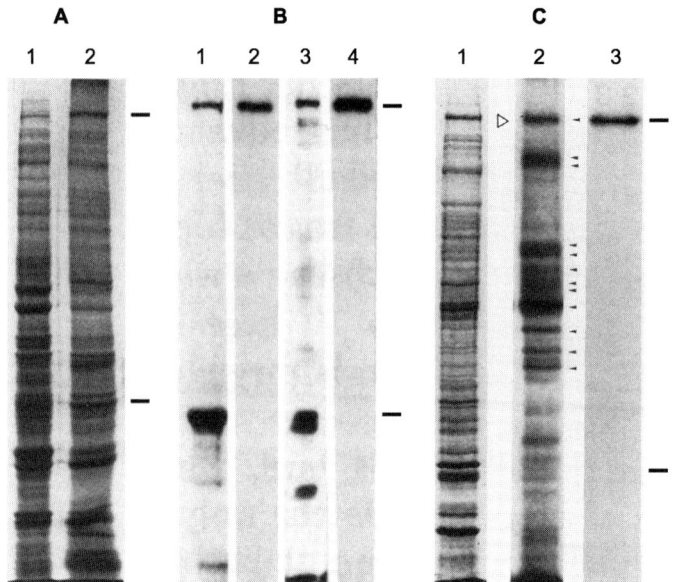

Fig. 1 Gel electrophoretic analysis of protein fractions obtained by differential extraction of monolayers from CHO and BHK 21 cells. (A) Extraction with a low-detergent buffer (0.2% Triton X-100) yields a soluble (lane 1) and a cytoskeltal fraction that stays bound to the substratum but is solubilized with high salt and detergent buffer as described in the text (lane 2). (B) Centrifugation of the high-salt detergent solubilized cytoskeltal fraction of CHO cells yields a pellet that contains nearly exclusively vimentin and plectin (lane 1). With BHK 21 cells, a band migrating slightly faster than vimentin is obtained, which is desmin (lane 3). The high molecular weight compound represents plectin as demonstrated by immunoprecipitation with plectin-specific antibodies (Herrmann and Wiche, 1987). (C) Incubation of cytoskeletons adhering to the substratum with $[\gamma\text{-}^{32}P]$ adenosine triphosphate yields intensive phosphorylation of several proteins, including plectin by cytoskeleton-associated kinases. After labeling, cytoskeletons were solubilized by sodium dodecylsulfate (SDS) sample buffer and, after electrophoresis, stained with Coomassie Brilliant Blue (lane 1). Dried gels were exposed to x-ray film (lane 2). In addition, plectin was immunoprecipitated from SDS-boiled samples after dilution with SDS-free RIPA, subjected to electrophoresis, and the dried gel was exposed to x-ray film (lane 3). Note that plectin is essentially not degraded (Herrmann and Wiche, 1983).

II. Materials and Methods

A. Isolation and Purification

1. Cell Monolayers

Monolayers of mammalian cultured cells grown in 10-cm Petri dishes are first washed three times with PBS containing 2 mM MgCl₂ and 0.05 mM Pefabloc at 37 °C ("cell wash"). PBS is made from a 10-fold concentrated stock solution (10 × PBS), which consists of 1.44 M NaCl (84 g/L), 30 mM KCl (2 g/L), 65 mM Na₂HPO₄ · H₂O (11.5 g/L), and 15 mM KH₂PO₄ (2 g/L). The following operations should be conducted under a fume hood! After removal of all of the PBS, the monolayer is incubated with 1 ml of "low-detergent buffer" (LOW) at room temperature. LOW is made easily from stock solutions: To 5 ml of water in a graded 10-ml plastic centrifugation tube, add 0.5 ml of $10 \times$ PBS, 0.5 ml of 1 M MOPS (pH 7.0), 0.1 ml of 1 M $MgCl_2$, 0.1 ml Pefabloc, 0.1 ml 0.1 M EGTA, 0.15 ml of 10% Triton X-100 in water (v/v) and then fill up to 10 ml. Immediately before use add $7.5 \mu l$ of saturated solution of PMSF in ethanol to 1 ml of extraction buffer. To prepare the PMSF solution, add some PMSF crystals into a weighed Eppendorf vial. Close the cap and weigh the vial again, then add $15 \mu l$ absolute ethanol per milligram of PMSF. Wear gloves and perform this step in a fume hood. PMSF is very toxic! Do not try to weigh a certain amount of PMSF; you will certainly contaminate the scales. The PMSF-ethanol solution is agitated on an Eppendorf shaker for at least 10 minutes. Prepare a fresh solution each time when needed.

The extraction of membranes from cells may be followed on an inverted light microscope positioned next to the hood. About 20 seconds after the beginning of the extraction, the appearance of the cells changes dramatically from one moment to the other as now the outline of the cell nucleus becomes clearly visible, comparable to an image obtained with phase-contrast microscopy. This visual inspection will also give information about how tight cytoskeletons stick to the substratum. After 60–90 seconds of extraction, during which the Petri dish is only gently agitated, the extraction buffer is carefully removed and transferred into an Eppendorf tube. This yields the membrane-bound and the soluble protein fraction of a cell. At this point, an optional second incubation with 1 ml LOW may be carried out.

The Petri dish with the adhering monolayer is then transferred onto ice, and immediately 1 ml of ice-cold "high detergent buffer" (HIGH) is added. HIGH is identical to LOW except that the EGTA is left out and that 1 ml of 10% Triton X-100 (v/v) instead of only 0.15 ml is used per 10 ml. Immediately before use, $50 \mu l$ of DNase I (Roche, Mannheim, Germany) and $75 \mu l$ of the ethanolic PMSF solution are added. It is important to preincubate the extraction buffer after adding the nuclease for at least 5 minutes on ice to inactivate proteases that may be still present in the DNase I preparation. This buffer does not contain high salt, thereby facilitating the rapid digestion of genomic DNA. The reaction may be enhanced by gentle resuspension of the buffer with a Pasteur pipette. In addition, RNase may be added to destroy nuclear RNA–protein complexes, which may form some kind of matrix-type accumulation and resist extraction.

In the meantime, the low-detergent extract is briefly spun in a benchtop centrifuge at full speed for 30 seconds, and the supernatant is saved. If only a small amount is needed for gel electrophoretic analysis, one may only centrifuge $100 \mu l$ and boil the supernatant of the centrifugation for 3 minutes after adding it into a new Eppendorf vial already containing $50 \mu l$ of threefold concentrated SDS sample buffer (Laemmli, 1970). This will give *sample 1*, which represents all the soluble and extractable membrane proteins.

After the cytoskeletons have been incubated with the nuclease/high-detergent buffer for 3 minutes on ice, 0.25 ml of an ice-cold 5 M NaCl solution is added. This will lead, supported by gentle pipetting, to the detachment of the cytoskeletons from the Petri dish. For some cells, such as human epithelial MCF 7 cell, one should remove cytoskeletons from the support with a sterile, disposable plastic cell scraper. At this point, a small amount of the solution may be removed into an Eppendorf vial precooled on ice for later gel electrophoretic analysis. We routinely add 100 μl of the high-salt extract to this Eppendorf vial already containing 100 μl both of water and threefold concentrated SDS sample buffer. It is important to dilute the sample, because otherwise the gel run will be severely disturbed. Note, the sample applied to the gel will still contain more than 350 mM NaCl. However, we still find that 20 μl of sample may be applied to a minigel. This will give *sample 2*, which represents the entire cytoskeletal fraction, including the nucleoskeleton and membrane proteins tightly bound to the cytoskeleton (see Fig. 1A, lane 2). This cytoskeletal fraction still contains proteases, and they may even be active after resuspension of pelleted cytoskeletons in 9 M urea as used with sample buffers for two-dimensional isoelectric focusing (IEF)/SDS-polyacrylamide gel electrophoresis (Fig. 2A).

The rest of the high-salt extract is carefully collected and transferred into an Eppendorf vial sitting on ice and then centrifuged in a benchtop centrifuge at full speed for 10 minutes at 4°C. The supernatant from this centrifugation step is carefully removed from the tight pellet, and a 100 μl sample is processed for gel electrophoretic analysis as described for sample 2. This will give *sample 3*, which represents the high-salt soluble fraction of the cytoskeleton.

For gel electrophoretic analysis, one pellet is taken up in 300 μl of urea-SDS sample buffer (to 600 mg of ultra pure urea in an Eppendorf vial, add 0.5 ml water

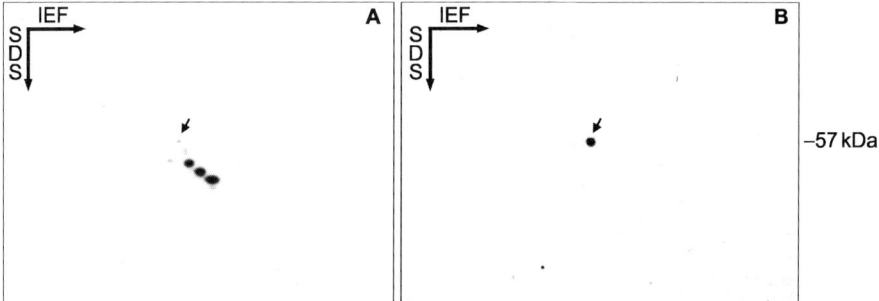

Fig. 2 Immunoblot analysis of cytoskeletal proteins from the rat glioma cell line C6 separated by two-dimensional gel electrophoresis. (A) Cytoskeletal proteins were either directly resuspended in IEF-lysis buffer containing 9 M urea or (B) first resuspended in sodium dodecylsulfate (SDS) sample buffer containing 6 M urea, boiled for 3 minutes, and then diluted with IEF-lysis such that the SDS concentration was <0.2% (for details of IEF in the presence of SDS, see Garrels, 1979). Vimentin was detected by a vimentin-specific antibodies followed by reaction with ^{125}I-protein A and autoradiography (Herrmann *et al.*, 1984).

and 0.5 ml three-fold concentrated SDS sample buffer; this will yield approximately 1.5 ml of sample buffer). Resuspend the pellet gently with the help of an automatic pipette and boil immediately for 3 minutes. This will give *sample 4*, which represents the cytoskeleton proper. In the case of fibroblasts, it consists of vimentin and plectin (Fig. 1B, lane 1). Longer heating is not needed and also not recommended, because it may eventually produce carbamoylated by-products. After 3 minutes of heating, however, no additional isoelectric variants were detected by two-dimensional gel electrophoresis (Fig. 2B). The pelleted material may be washed further by resuspension into 0.5 ml LOW followed by centrifugation at full speed for 10 minutes at 4 °C to remove salt and lipids. In addition, nucleases may be included into the washing buffer to get rid of remnants of nucleic acids binding to highly basic segments of IF proteins as found in vimentin or lamins. This interaction is so strong that it can withstand resuspension in 5.7 M CsCl as used for density gradient centrifugation or the afore-mentioned processing for IEF/SDS-polyacrylamide gel electrophoresis (PAGE) (Cress and Kurath, 1988). Moreover, if actin contamination is to be removed completely, a washing step with 0.6 M potassium iodide may be performed in addition. However, for many purposes, the pellets may be used directly for further purification steps.

If one wishes to isolate IFs only and is not interested in the characterization and keeping of the other fractions, the procedure is much faster, and one may harvest four Petri dishes at a time without losing speed. Moreover, if the next round of extraction is performed when the last set is in the centrifuge, 50 dishes may be easily harvested in 2 hours by one person.

2. Suspension Cells

For suspension cells such as those from the human promyelocytic line HL 60, we follow exactly the protocol outlined previously. With suspension cells, it is even faster, because all steps can be performed in Eppendorf vials, and incubation times may be reduced further. Cells are harvested into Eppendorf vials by centrifugation at room temperature (300 g, 1 minute) and gently resuspended into LOW (1 ml for up to 100 mg of cells). After 1 minute, cytoskeletons are harvested (300 g, 1 minute, or by just turning on the benchtop centrifuge at full speed and immediately turning it off again). The supernatant is removed, the pellet is transferred to ice, and extraction is continued as described in (1_1) for the monolayers.

3. Tissues

Procedures for the extraction of tissues have been described extensively in the articles cited in the "Introduction." As mentioned previously, other protease inhibitors and different combinations of such inhibitors may be used and also different concentrations may be tested (e.g., Wiche *et al.*, 1983). The isolation of desmin from muscular tissue is described in Chapter 19 (this book).

4. Isolation of Intermediate Filament Proteins from Transformed Bacteria

For the expression of IF proteins in TG1 bacteria grown in "terrific broth" (TB) medium, we routinely use a T5 promoter containing plasmid originally described by Bujard and colleagues (Bujard *et al.*, 1987). The isolation of inclusion bodies is performed according to the protocol of Nagai and Thogerson (1987) with the following important modifications. After resuspension of cells in the Dounce homogenizer, we add Triton X-100 to 0.2%, followed by $MgCl_2$ to 10 mM, DNase I to 50 μg/ml and PMSF to 2 mM. The permeabilized bacterial cells are gently agitated with the tight-fitting pistil until the viscosity of the solution is low. We then add the detergent buffer and homogenize the suspension accordingly. The inclusion bodies are harvested by centrifugation, resuspended in the corresponding washing buffer, and are successively transferred back to the Dounce homogenizer precooled on ice. At first, they are homogenized with 10 mM Tris-HCl buffer, pH 7.5, containing 1 mM EDTA, 1 mM DTT, 2 mM PMSF, and 1.5 M KCl. After recentrifugation, this procedure is repeated with the same buffer without the KCl. Finally, the pellets are homogenized with 10 mM Tris-HCl, pH 7.5, 0.1 mM EDTA, to remove salt, which may disturb the binding of the protein to the anion-exchanger (see later). The pellet from the last centrifugation is dissolved in 16 ml of 9.5 M urea in 10 mM Tris-HCl, pH 7.5. We prepare this solution from stock solutions of 10 M urea and 0.2 M Tris-HCl, pH 7.0. The final pH has to be adjusted with a pH meter, because urea increases the pH at this low concentration of buffer considerably! The solubilized IF proteins are freed from debris by centrifugation at $100,000\,g$ for 1 hour at 20 °C in a standard preparative ultracentrifuge. After the run, the supernatant is either directly transferred to an anion exchange column or stored frozen at −80 °C.

5. Column Purification

We found it optimal to start the purification of IF proteins with a Fast Flow DEAE-Sepharose or Q-Sepharose-matrix for keratin (Pharmacia, Freiburg, Germany), equilibrated in 10 mM Tris-HCl, pH 7.5, 1 mM DTT with 8 M urea ("column buffer"). We find Econo columns from Bio-Rad (München) very useful and operate these columns without pumps. Bound proteins are eluted with a salt gradient (0–0.3 M NaCl) in this buffer, and each fraction is analyzed by SDS-PAGE. Peak fractions are pooled and directly applied to a CM-Sepharose or S-Sepharose column made in column buffer. Without any special equipment, the purification in the range from 2 mg–1 g is performed within 5 hours. The procedure is so simple that purification of two proteins in parallel is achieved without problems. As a precaution against carbamoylation of proteins, the 10 M urea solution is stirred in the presence of a mixed-bed ion exchanger resin (TMD-8, M-8157, Sigma). We did not find it necessary to do the purification at 4 °C, although in some instances, it may help to stop unwanted degradation. Again, samples for SDS gel electrophoresis should be taken before loading the column to compare the state of the applied protein with that of the product obtained after column purification.

Importantly, mass spectrometric analysis did not reveal any carbamoylation products. Interestingly, the recombinant human vimentin rod exhibited two acetylation reactions. In addition, both amino-terminal fragments, with and without N-formyl methionine, were detected, indicating the bacteria did not completely cleave off the initiating methionine (H. Herrmann, unpublished results). Correspondingly, recombinant IF proteins often exhibit two isoelectric variants on two-dimensional IEF/SDS-PAGE, the more basic N-formyl methionine-free protein and the N-formyl methionine-containing, full-length protein (see e.g., Gieffers and Krohne, 1991).

B. Characterization of Soluble Complexes

It is a hallmark feature of IF proteins that they can be renatured after solubilization in 8 M urea or in 3 M guanidinium hydrochloride. However, if the latter chemical is used for denaturation, extreme care should be taken that it is completely removed before assembly studies are undertaken. Moreover, guanidinium hydrochloride gradients (0–0.3 M) are also frequently used to dissociate urea-solubilized keratins from ion exchanger matrices, and, even with such relatively low concentrations of this strong denaturant, extreme care should be taken that it is removed completely.

Early on, it was demonstrated that IF proteins are soluble in low ionic strength buffers (see references earlier). Nevertheless, no general rule applies for this property with respect to individual IF proteins. Whereas vimentin forms rather homogeneous complexes when dialyzed into 5 mM Tris-HCl, pH 8.4, or even 2 mM sodium phosphate, pH 7.5, keratins from simple epithelial tissues may require 2 mM Tris-HCl, pH 9 (Herrmann et al., 1996, 1999; Mücke et al., 2004a). Under these conditions, however, epidermal keratins such as human K5 and K14 form extended filaments (Herrmann et al., 2002). Nevertheless, stable and reliable systems for the investigation of the assembly of keratins and their mutated forms, as found in patients or as designed to explain the contribution of various domains to assembly, have been described (Coulombe and Fuchs, 1990; Coulombe et al., 1990). Moreover, nuclear lamins are routinely transferred to buffers of high pH and high ionic strength to keep them soluble. Nevertheless, rather homogeneous complexes are obtained with recombinant human lamin B_1 in 5 mM Tris-HCl, pH 8.4 (D. Lotsch, N. Muecke and H. Herrmann, unpublished observations).

To obtain homogeneous soluble complexes, we routinely dialyze IF proteins from 8 M urea in steps of decreasing urea concentrations (6 M, 4 M, 2 M) into buffer without urea. Our basic dialysis buffer is 5 mM Tris-HCl, pH 8.4, 1 mM EDTA, 0.1 mM EGTA, and 1 mM DTT. These dialysis steps are performed at room temperature over a 1-hour period with dialysis tubing with a molecular weight cut off of 12,000–14,000 (SERVAPOR, 44145, SERVA, Germany). Here we use volumes of 200 to 500 ml of the respective dialysis buffer. To remove traces of urea, a further dialysis against the solubilization buffer, 0.8 L, of at least 1 hour with one buffer change is necessary. We routinely dialyze the samples overnight at 4 °C. It is our impression that IFs look more homogeneous, are longer, and tend much less to aggregate during assembly in a capillary viscometer when this additional dialysis is performed. We assume that tightly binding compounds (see earlier) are removed by this step.

We routinely take and boil in SDS sample buffer aliquots of each protein that is to be investigated before and after dialysis. In addition, we take probes of each protein after incubation in dialysis buffer for 1 week at room temperature. This will give essential information as to how stable the corresponding protein is, which is of high importance for analytical ultracentrifugation experiments and for extended assembly trials with mammalian IF proteins performed at 37 °C.

The biophysical and biochemical properties of these complexes can be investigated by chemical cross-linking (Steinert *et al.*, 1993), glycerol spraying/rotary metal shadowing electron microscopy (Aebi *et al.*, 1983; Herrmann *et al.*, 1996), transient electric birefringence measurements (van Amerongen *et al.*, 1998), and analytical ultracentrifugation (e.g. Herrmann *et al.*, 1996; Huiatt *et al.*, 1980; Quinlan *et al.*, 1984, 1986). To be able to carry out chemical cross-linking experiments of endogenous lysines, for which Tris-buffer cannot be used, we introduced a new assembly procedure that uses 2 mM sodium phosphate buffer, pH 7.5. Under these conditions, vimentin is essentially forming soluble tetrameric complexes as revealed by analytical ultracentrifugation, and these complexes form bona fide IFs when the ionic strength is raised. A rigorous analysis of the behavior of vimentin and several mutated versions of the protein in analytical ultracentrifugation that uses powerful computational programs for the behavior of these highly extended and negatively charged molecules has been recently reported in detail by Muecke *et al.* (2004a), and the reader is referred to this article for further information.

C. *In Vitro* Assembly

To investigate IF protein assembly, one has to consider several parameters that will influence the experimental strategy to choose. First, we have to discriminate three principal assembly groups within the IF protein family:

1. Nuclear IF proteins (i.e., A-type and B-type lamins)

2. Keratins, in particular "soft" and "hard" keratins, (i.e., those from epithelial and those from trichocytic tissues, respectively)

3. Desmin-like proteins (i.e., sequence-homology class (SHC) type III proteins, including vimentin and glial fibrillar acidic protein, and, to a certain extent, SHC proteins such as NF-L and internexin)

Second, whatever type of IF protein is considered to be investigated, an important decision will be whether recombinant or material isolated from cultured cells or tissues is to be used.

1. Cytoplasmic Intermediate Filament Proteins

If only filaments are to be produced for documentation by electron microscopy, IF proteins may directly be dialyzed from 8 M urea into filament assembly buffer. We start, however, from solubilized complexes as described previously, because in this way kinetic measurements are feasible. In a standard regimen, we add 1 volume of 10× concentrated filament buffer to 9 volumes of solubilized IF protein. For vimentin—renatured into 5 mM Tris-HCl, pH 8.4, 1 mM EDTA, 0.1 mM EGTA, 1 mM DTT (Tris-buffer)—this buffer is 0.2 M Tris-HCl, pH 7.0, 0.5 M NaCl. Depending on the protein concentration, the assembly proceeds very fast. Quantitative kinetic measurements are performed in an Ostwald viscometer at 0.5 mg/ml (Herrmann et al., 1992; Hofmann, 1998; Hofmann et al., 1991). We obtain glass capillary instruments designed for 1 ml of protein solution, Cannon-Manning Semi-Micro Viscometer (CANNON Instrument Company, State College PA). They are best cleaned by incubation with 3 M guanidinium hydrochloride followed by at least 10 rinses with distilled water and drying. After assembly has been started in an Eppendorf vial, the protein solution is transferred right away into the viscometer with a Pasteur pipette. The first measurement is done 1 minute after initiation of assembly, followed by measurements every 5 minutes.

To visualize structures formed immediately after initiation of assembly, we developed the following procedure: The protein is solubilized at 0.2 or 0.4 mg in 5 mM Tris-buffer and assembly is started by the addition of an equal volume of filament buffer consisting of 45 mM Tris-HCl, pH 7.0, and 100 mM NaCl in an Eppendorf vial. This yields a protein concentration of 0.1 or 0.2 mg/ml, respectively, in 25 mM Tris-HCl and 50 mM NaCl. Assembly is stopped by the addition of the same volume of assembly buffer (i.e., 25 mM Tris-HCl, pH 7.5, 50 mM NaCl) containing 0.2% glutaraldehyde. A glutaraldehyde-stock solution is prepared freshly from a 25% glutaraldehyde solution delivered in glass ampules (G-5882, SIGMA, Germany). This will stop any protein interaction immediately (Herrmann et al., 1999). The filament suspension is then delivered to a glow-discharged, carbon-coated copper electron microscopy grid (Aebi and Pollard, 1987). The structures formed are allowed to bind the support for 15 seconds followed by removal of water with the help of a filter paper applied to the side of the grid. The sample is then washed with distilled water by placing the grid upside down onto a drop of water residing on a piece of Parafilm. The protein is then stained by incubation with 2% uranyl acetate for 15 seconds, the staining solution is removed, and the grid is air-dried. A representative series of structures formed by recombinant human vimentin at 10 seconds, 1 minute, 5 minutes, and 1 hour after initiation of assembly is shown in Fig. 3. A detailed description of how to use the "negative staining" technique has been described in detail by Bremer et al. (1998).

More insight into the fine structure of filaments can be obtained by subjecting them to glycerol spraying/rotary metal shadowing electron microscopy as shown here for reconstituted neurofilament triplet proteins (Fig. 4A) compared with filaments formed from the low molecular weight neurofilament protein NF-L

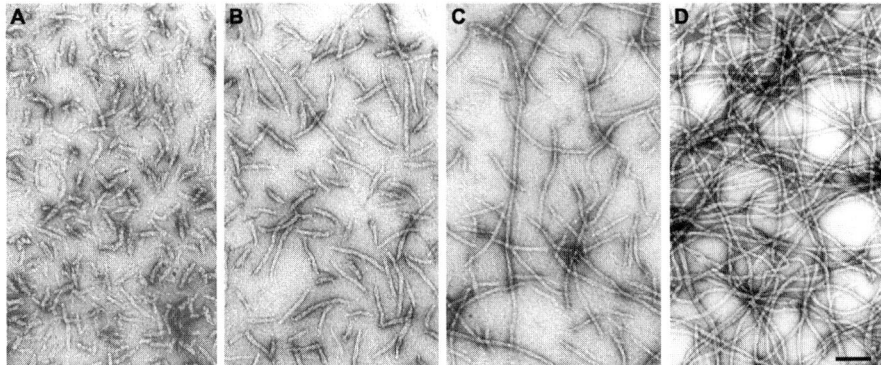

Fig. 3 Time-dependent assembly of recombinant human vimentin as followed by electron microscopy of negatively stained samples taken at (A) 10 seconds; (B) 1 minute; (C) 5 minutes; and (D) 1 hour. Bar, 100 nm.

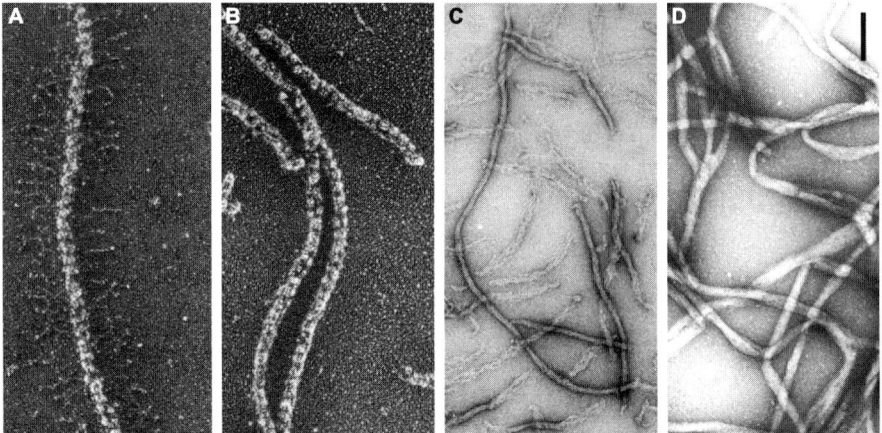

Fig. 4 Electron microscopy of native and *in vitro* reconstituted intermediate filaments (IFs). (A) Native neurofilaments and (B) urea-dissolved, purified, and reconstituted NF-L (in 50 mM Mes, pH 6.25, 170 mM NaCl, 1 mM DTT) were glycerol sprayed/low-angle rotary metal shadowed on freshly cleaved mica and processed for conventional transmission electron microscopy. (C) Recombinant mouse NF-L after reconstitution into "vimentin assembly buffer" (25 mM NaCl, pH 7.5, 50 mM NaCl) and (D) recombinant human vimentin reconstituted into "NF-L assembly buffer" (see B) were visualized by negative staining with uranyl acetate. Note the many "unraveled" filaments in (C) and the general "non-IF" appearance of vimentin in (D). Bar, 100 nm.

alone (Fig. 4B). Here, assembly is performed by dialysis of urea-dissolved protein against 50 mM morpholino ethane sulfonic acid (Mes), pH 6.25, 170 mM NaCl, and 1 mM DTT at 37 °C. In contrast, NF-L assembled under vimentin conditions just described forms only short and often unraveled filaments (Fig. 4C). Vimentin, under the low pH conditions used for NF-L, forms rather irregular and flat non-IF structures (Fig. 4D). Without going into further detail and as alluded to, keratins need much more distinct conditions. Moreover, as soon as divalent cations such as Ca^{2+}, Mg^{2+}, or Zn^{2+} are included into the buffers, matters get even more complex as fibrillar structures with increasing diameter are obtained (Herrmann et al., 1999; Hofmann et al., 1991; Mack et al., 1993). This may engage charge interactions not taking place in this manner under normal conditions within cells. Support for this assumption comes from experiments that use head-truncated IF proteins that do not assemble at all on raising the ionic strength to physiological values. However, in the presence of Mg^{2+}, these proteins form extended non-IF fibrils (Geisler et al., 1982). Again, a thorough presentation of the method has recently been published (Häner et al., 1998).

In principle, it is not easy to decide whether a certain filament obtained under specific conditions is a bona fide IF. One way to approach this problem in a quantitative way is to determine the mass-per-length value of the filament in question, which will permit the calculation of the number of molecules per cross-section for a given fibril. A major problem is, however, that in most cases it is not known how many molecules constitute a specific IF in vivo. Scanning transmission electron microscopy (STEM) measurements of in vitro assembled, freeze-dried, unstained filaments have indicated that significant differences exist between an IF made from keratins, vimentin, desmin, or NF-L (Herrmann et al., 1999). Nevertheless, such measurements facilitate the evaluation of how many molecules assemble, depending on the assembly conditions from one and the same protein, and these values may be compared with authentic IFs isolated from tissues. For example, vimentin IFs have approximately 30 molecules per cross-section after assembly under standard conditions. In contrast, when assembly takes place in 0.7 mM sodium phosphate, 2.5 mM $MgCl_2$, pH 7.5, filaments of highly varying diameter are obtained, with peak values corresponding to 56 and 84 molecules per cross-section (Herrmann et al., 1999). Moreover, with this method it was possible to demonstrate that unit-length filaments formed by all different types of cytoplasmic IF proteins exhibit similar numbers per cross-section immediately after initiation of assembly compared with mature filaments. For example, NF-L at 2 seconds after initiation of assembly consists of a rather homogenous species (Fig. 5), and they exhibit a very similar mass like mature filaments obtained after 1 hour of incubation (Herrmann et al., 1999). This method is, however, not easily established in a standard cell biology laboratory, although it is a must if assembly is to be studied seriously.

2. Nuclear Lamins

If one compares the amino acid sequence and the structural organization of lamins with that of cytoplasmic IF proteins, several significant differences become immediately apparent (Herrmann and Aebi, 2004). Besides the longer helix 1B and the non-α-helical linker L1, the non-α-helical amino-terminal "head" domain is of

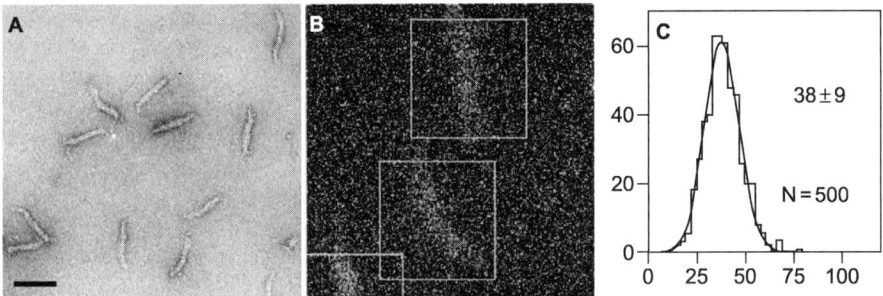

Fig. 5 Quantification of the number of molecules per filament cross-section for NF-L reconstituted into "vimentin assembly buffer." (A) Unit-length filaments (ULFs) negatively stained 10 seconds after initiation of assembly; (B) dark-field picture of ULFs fixed with 0.1% glutaraldehyde in filament buffer 2 seconds after initiation of assembly followed by freeze-drying of unstained filaments; (C) histogram of mass-per-length measurements of ULFs as shown in (B). N = 500, segments were measured and Gaussian curves fitted into the histograms to identify presumptive major peaks. Bar, 100 nm.

particular importance, because it is much shorter in lamins and contains significantly lower numbers of basic amino acids. Clearly, a vimentin with a correspondingly short head will not assemble at all. Moreover, the lamin non-α-helical tail domain is rather long and contains in its center a globular domain. Cytoplasmic IF proteins with similarly long tails do not assemble on their own (Herrmann and Aebi, 2004).

B-type lamins have been demonstrated to yield soluble dimeric complexes when dialyzed into "equilibration buffer" (25 mM Tris, pH 8.5, 150 mM NaCL, 1 mM EGTA, 1 mM DTT, 1 mM PMSF) by Heitlinger and colleagues (1991). It should be mentioned, however, that PMSF at this high pH is hydrolyzed within seconds and should be replaced by more stable compounds such as Pefabloc. Preferably, the protein preparation should be protease-free. These dimers associate longitudinally into polar head-to-tail polymers when dialyzed into "polymerization buffer" (25 mM Mes, pH 8.5, 150 mM NaCl, 1 mM EGTA, 1 mM DTT, 1 mM PMSF) as visible both in negatively stained (Gieffers and Krohne, 1991) and, much more clearly, in glycerol sprayed/low-angle rotary metal shadowed electron microscopic preparations (Fig. 6A; Heitlinger *et al.*, 1991). On further incubation, but especially in the presence of high concentration of divalent cations such as 10–25 mM $CaCl_2$, lateral association of head-to-tail polymers takes place, leading to the formation of beaded filaments and, eventually, paracrystal-like structures (Fig. 6B, C and Aebi *et al.*, 1986; Gieffers and Krohne, 1991; Heitlinger *et al.*, 1991).

More IF-like filaments, which are somehow more stable with respect to their number of molecules per cross-section, have been obtained with the recombinant lamin from *Caenorhabditis elegans* (see Chapter 24, this book).

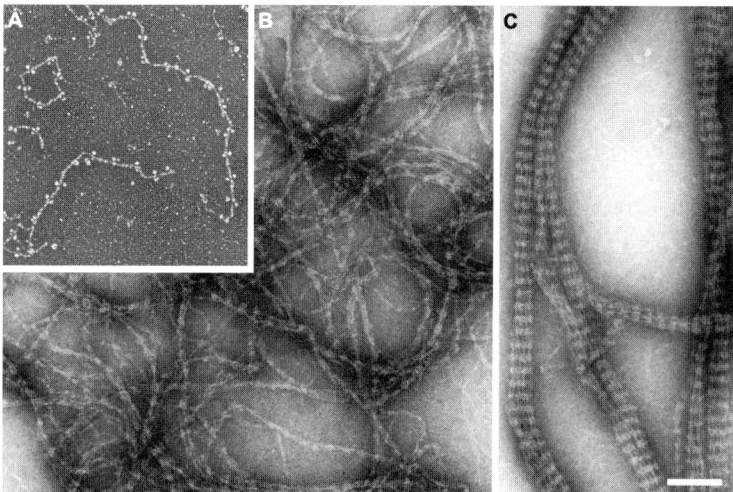

Fig. 6 Buffer-dependent structures formed from recombinant chicken lamin B as visualized (A) by glycerol spraying/low-angle rotary metal shadowing and (B, C) negative staining followed by transmission electron microscopy. (A) Assembly at high pH yields head-to-tail polymers; (B) assembly at low pH results in the formation of beaded filaments with relaxed width control; (C) in the presence of 25 mM $CaCl_2$ paracrystal-like fibers are observed that can grow significantly in width.

D. Atomic Force Microscopy of Filament Assembly

The atomic force microscope (AFM) has been originally developed to image the atomic structure of solid surfaces in air or in liquid. The principle of this technique is to raster scan the surface of interest with a very fine stylus (tip radius, 1–10 nm). The stylus is located at the end of a cantilever, the bending of which is monitored by the deflection of a laser spot focused on its back. In the contact mode, the tip is in permanent contact with the sample, and the force is kept constant. This mode has been used to obtain high-resolution topographs of two-dimensional protein crystals of, for example, bacteriorhodopsin (Müller *et al.*, 1997). However, weakly immobilized structures, like single macromolecules or linear assemblies, are often pushed away by the AFM stylus during scanning in contact mode (Karrasch *et al.*, 1993). To overcome this disadvantage, the tapping mode has been developed (Hansma *et al.*, 1994). In the tapping mode, the stylus oscillates at a given frequency, touching the sample only at the end of its downward movement, which reduces the contact time and the friction forces compared with the contact mode. Thus, a variety of macromolecules have been observed that could not be imaged in contact mode (Fritz *et al.*, 1995). In the case of IFs, filaments covalently cross-linked to a support can be imaged in contact mode, whereas filaments only adsorbed to a support have to be imaged in tapping mode.

1. Imaging Vimentin Intermediate Filaments Adsorbed to Mica

A commercial AFM (Nanoscope IIIa, Digital Instruments Inc., Santa Barbara, CA), equipped with a 120-μm scanner (j-scanner) and a liquid cell, is used. Before use, the liquid cell has to be cleaned with normal dish cleaner (Pril, Henkel Hygiene AG, Pratteln, Switzerland), gently rinsed with ultrapure water ($\approx$18 MΩ/cm; Branstead, Boston, MA), sonicated in ethanol (50 kHz), and sonicated in ultrapure water (50 kHz).

The sample support is a mica sheet (Mica House, Calcutta, India), which is punched to a diameter of about 5 mm and glued onto a Teflon-covered steel disk with water-insoluble epoxy glue (Araldit, Ciba Geigy AG, Basel, Switzerland). The steel disk is required to magnetically mount the mica disk onto the piezoelectric scanner. Notice that other solid supports can be easily glued to the Teflon disklike glass coverslips or graphite sheets (highly oriented pyrolitic graphite); any atomic flat solid substrate can be used. As a routine, we use 100-μm long cantilevers with oxide-sharpened silicon nitride tips, which have a nominal spring constant of 0.38 N/m (type NP-S from Digital Instruments, Santa Barbara, CA). In tapping mode, the cantilever is forced to oscillate at a certain drive amplitude and has a characteristic frequency peak between 7.5 and 9.5 kHz. The images are 512×512 pixels and are recorded with a scan frequency <2 Hz. Human recombinant vimentin at a concentration of 0.1 mg/ml is polymerized for 1 hour at 37 °C in 2 mM phosphate, 100 mM KCl, pH 7.5. The solution is diluted 20 times in the same buffer, and 20 μl is applied to mica for 5 minutes. The drop is then removed with a pipette, and the mica is washed three times with the same buffer. Other assembly conditions can be used, but relatively high ionic strengths have to be used to obtain a good physical adsorption to mica. This is because both the vimentin IFs and the mica surface are negatively charged at neutral pH. As far as we have observed, diluting the filaments to concentrations of a few micrograms per milliliter does not destabilize them (Mücke *et al.*, 2004b).

In a typical overview image of vimentin IFs, individual long filaments are found. They generally have a wavy appearance and are frequently organized into tightly packed spiral structures, often at the ends of the filaments (Fig. 7A). This is not observed with negatively stained specimens. Note that the apparent diameter of the filaments is approximately 40–70 nm compared with 10 nm as seen by EM. This lateral broadening is because the image you see is a convolution between the shape of the tip and the real dimensions of the sample.

2. Imaging Vimentin Intermediate Filaments Assembly Intermediates Absorbed to Mica

In the section concerning the electron microscopy analysis of IF assembly, we have given the protocol to glutaraldehyde-fix filaments along their assembly pathway in the ULF stage. With AFM, it is possible to observe ULFs adsorbed to mica without fixation. Human recombinant vimentin at a concentration of 0.1 mg/ml is polymerized for 10 seconds at 37 °C in 2 mM sodium phosphate, 100 mM

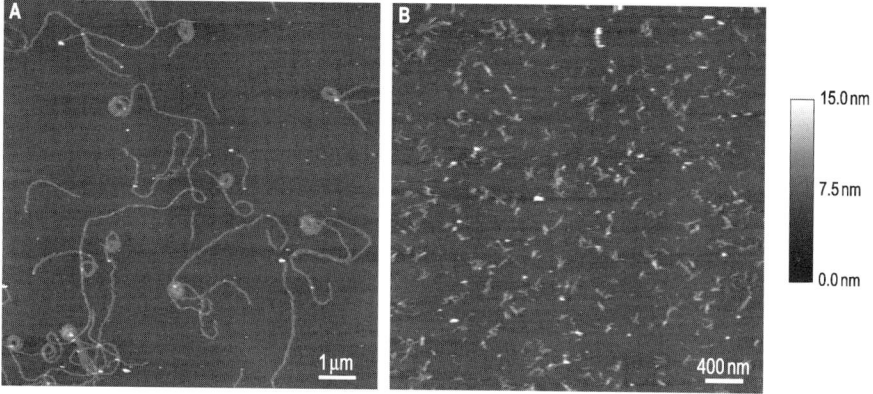

Fig. 7 Tapping mode atomic force microscopy images in 2 mM sodium phosphate, 100 mM KCl, pH 7.5, of recombinant human vimentin intermediate filaments adsorbed to mica after: (A) 1 hour polymerization at 37 °C; (B) 10 seconds polymerization at 37 °C.

KCl, pH 7.5. The solution is diluted 20 times in the same buffer, and 20 μl is applied to mica for 5 minutes. The drop is then removed with a pipette, and the mica is washed three times with the same buffer. On imaging in AFM, short (80–160 nm long) filaments are visible, which frequently interact with each other by means of their ends. However, because they are apparently very tightly bound to the support, they cannot undergo proper fusion as seen with ULFs fixed and negatively stained in solution (Fig. 7B). Obviously, dilution of the sample does apparently disassemble the ULFs.

III. Pearls and Pitfalls

The mechanical properties of assembled proteins, in particular their dependence on associated proteins, can be evaluated by rheometry. Originally, Paul Janmey and colleagues compared microtubules, microfilaments, and IFs in a self-made torsion pendulum (Janmey, 1991; Janmey *et al.*, 1991). We have used this device to demonstrate that the successive deletion of amino acid segments from the non-α-helical carboxyterminal domain of *Xenopus* vimentin did not significantly influence the viscoelastic properties, in particular the dynamic storage modulus G', of the seven mutants compared with the wild-type protein (Rogers *et al.*, 1995). At present, the instrument is used to determine the effect of single point mutations found in desmin of patients with various muscle disorders. The system is most useful when cross-linking factors of filaments are tested for their bridging potential (Janmey *et al.*, 1998). Of course, fancier, but also much more expensive, instruments may be used, which will then allow for the investigation of more specific questions (Yamada *et al.*, 2002).

Although the expression of recombinant IF proteins and fragments thereof in bacteria is a convenient method to get large amounts of unmodified protein, in some cases it turns out to be extremely difficult to yield meaningful amounts or to avoid proteolysis. For example, Heitlinger *et al.* (1991) encountered severe problems with the production of chicken lamin B_2 in BL21(DE3) cells. Here, even transformation of bacteria was not possible at meaningful rates. We encountered the same problem with human lamin B_1. Heitlinger *et al.* solved the problem in part by use of a different expression system (i.e., the bacterial strain JM109), and they obtained about 20 mgL of lamin B_2 of bacterial suspension (Heitlinger *et al.*, 1991). Moreover, they found that most of the lamin was in the supernatant when the bacteria were lysed in a high pH buffer (pH 9). When lysis was performed at pH 6, the expressed protein was pelletable, probably because of an isoelectric precipitation effect. The situation with lamin expression is complex, because some lamins such as the *Xenopus laevis* lamin A and the *Caenorhabditis elegans* lamin are generated easily in standard expression systems (Gieffers and Krohne, 1991; Karabinos *et al.*, 2003). The best thing one can try when expression is a problem is to try out different expression plasmids (e.g., the PET series of Novagen in combination with different cell strains, including new strains such as Tuner [DE3]). We have found that the Shine-Dalgarno sequence of the respective expression plasmid is of considerable importance for the expression of IF proteins. Hence, if one does not want to mutate this side considerably, it seems easier to clone the corresponding cDNA into several different expression plasmids and optimize expression by trial and error. By serendipity, we experienced that protein expression from the multicopy plasmid pBluescript KS (Stratagene), mostly used for cloning, *in vitro* mRNA synthesis, and sequencing purposes, can be very robust. We cloned the *Xenopus* vimentin cDNA in the orientation of the alpha-fragment transcription and yielded very high amounts of protein. A "stop" in the 5'-UTR of the vimentin cDNA terminated the expression of the galactosidase peptide, and a sequence AGGA motif nine nucleotides upstream of the initiation codon ATG within the 5'-UTR obviously served as Shine-Dalgarno element. Therefore, with minimal effort one can modify the pBluescript polylinker such that even a His-tag and a protease cleavage site are introduced in addition to the desired expression unit. Finally, we noted that even a bacterial strain, *E. coli* SURE (Stratagene), conventionally used only as a recombination-proof strain for cloning complex constructs, may produce large amounts of human lamin B_2 when transfected with the human lamin B_2 in pDS5.

IV. Concluding Remarks

Although useful general procedures for the isolation of IF proteins from various tissues and organisms are available, the individual assembly properties vary considerably from one IF protein to another within one species, as well as between orthologous proteins between different species. A striking example is the different behavior as exhibited by vimentins from such distinct organisms as shark, frog, and man. Although the primary sequence is indeed similar, conditions for their optimal assembly differ quite a lot, for instance, with respect to the temperature. If one moves on to keratins or lamins, matters become even more complex. More strikingly, even for one and the same IF protein, quite different polymers may be obtained, depending on the assembly conditions used. This polymorphism may be an insult to structural biologists but may prove to be of high physiological importance. Finally, the fast isolation procedure for isolation of IF proteins from cytoskeletons *in situ* presented here may allow investigation of posttranslational modifications with a new reliability. Labile groups are known to be easily lost in conventional SDS-PAGE, especially during the boiling process. Direct immuno-isolation from solubilized cytoskeletons that use magnetic beads and subsequent analysis of the immune complexes by MALDI-TOF mass spectroscopy may be a way to find new modifications not thought of before.

Acknowledgments

We thank Tatjana Wedig for excellent technical assistance and Michaela Reichenzeller for help with the figures. Eva Gundel helped with the finishing of the manuscript. H. H. was supported by the Deutsche Forschungsgemeinschaft (DFG, He 1853). L. K. was supported by a postdoctoral fellowship from the Swiss Foundation for research in muscular dystrophy. U. A. received support from an NCCR program grant on "Nanoscale Science" by the Swiss National Science Foundation, the M. E. Müller Foundation of Switzerland, and the Canton Basel Stadt.

References

Achtstaetter, T., Hatzfeld, M., Quinlan, R. A., Parmelee, D. C., and Franke, W. W. (1986). Separation of cytokeratin polypeptides by gel electrophoretic and chromatographic tecniques and their identification by immunoblotting. *Methods Enzymol.* **134,** 355–371.

Aebi, U., and Pollard, T. D. (1987). A glow discharge unit to render electron microscope grids and other surfaces hydrophilic. *J. Electron Microsc. Tech.* **7,** 29–33.

Aebi, U., Cohn, J., Buhle, L., and Gerace, L. (1986). The nuclear lamina is a meshwork of intermediate-type filaments. *Nature* **323,** 560–564.

Aebi, U., Fowler, W. E., Rew, P., and Sun, T. T. (1983). The fibrillar substructure of keratin filaments unraveled. *J. Cell Biol.* **97,** 1131–1143.

Bremer, A., Häner, M., and Aebi, U. (1998). Negative staining. *In* "Cell Biology. A Laboratory Handbook," 2nd ed., Vol. III, pp. 277–284. Academic Press, San Diego.

Bujard, H., Gentz, R., Lanzer, M., Stueber, D., Mueller, M., Ibrahimi, I., Haeuptle, M. T., and Dobberstein, B. (1987). A T5 promoter-based transcription-translation system for the analysis of proteins *in vitro* and *in vivo*. *Methods Enzymol.* **155,** 416–433.

Cordes, V. C., Reidenbach, S., Rackwitz, H.-R., and Franke, W. W. (1997). Identification of protein p270/Tpr as a constitutive component of the nuclear pore complex-attached intranuclear filaments. *J. Cell Biol.* **136,** 515–529.

Cress, A. E., and Kurath, K. M. (1988). Identification of attachment proteins for DNA in Chinese hamster ovary cells. *J. Biol. Chem.* **263,** 19678–19683.

Coulombe, P. A., and Fuchs, E. (1990). Elucidating the early stages of keratin filament assembly. *J. Cell Biol.* **111,** 153–169.

Coulombe, P. A., Chan, Y.-M., Albers, K., and Fuchs, E. (1990). Deletions in epidermal keratins leading to alterations in filament organization *in vivo* and in intermediate filament assembly *in vitro*. *J. Cell Biol.* **111,** 3049–3064.

Fiskum, G., Craig, S. W., Decker, G. L., and Lehninger, A. L. (1980). The cytoskeleton of digitonin-treated rat hepatocytes. *Proc. Natl. Acad. Sci. USA* **77**, 3430–3434.

Fritz, M., Radmacher, M., Cleveland, J. P., Gieselmann, R., Schmidt, C. F., Allersma, M. W., Stewart, R. J., Morese, D. E., and Hansma, P. K. (1995). Imaging globular and filamentous proteins in physiological buffer solution in tapping mode atomic force microscopy. *Langmuir* **11**, 3529–3535.

Garrels, J. I. (1979). Two-dimensional gel electrophoresis and computer analysis of proteins synthesized by clonal cell lines. *J. Biol. Chem.* **254**, 7961–7977.

Geisler, N., Kaufmann, E., and Weber, K. (1982). Proteinchemical characterization of three structurally distinct domains along the protofilament unit of desmin 10 nm filaments. *Cell* **30**, 277–286.

Gerace, L., and Blobel, G. (1980). The nuclear envelope lamina is reversibly depolymerized during mitosis. *Cell* **19**, 277–287.

Gieffers, C., and Krohne, G. (1991). *In vitro* reconstitution of recombinant lamin A and a lamin A mutant lacking the carboxy-terminal tail. *Eur. J. Cell Biol.* **55**, 191–199.

Goldman, A. E., Moir, R. D., Montag-Lowy, M., Stewart, M., and Goldman, R. D. (1992). Pathway of incorporation of microinjected lamin A into the nuclear envelope. *J. Cell Biol.* **119**, 725–735.

Häner, M., Bremer, A., and Aebi, U. (1998). Glycerol spraying/low-angle rotary metal shadowing. *In* "Cell Biology. A Laboratory Handbook," 2nd ed., Vol. III, pp. 292–298. Academic Press, San Diego.

Hansma, P. K., Cleveland, J. P., Radmacher, M., Walters, D. A., Hillner, P. E., Bezanilla, M., Fritz, M., Vie, D., Hansma, H. G., Prater, C. B., Massie, J., Fukunaga, L., Gurley, J., and Elings, V. (1994). Tapping mode atomic force microscopy in liquids. *Appl. Phys. Lett.* **64**, 1738–1740.

Heitlinger, E., Peter, M., Häner, M., Lustig, A., Aebi, U., and Nigg, E. A. (1991). Expression of chicken lamin B$_2$ in *Eschericia coli*: Characterization of ist structure, assembly, and molecular interactions. *J. Cell Biol.* **113**, 485–495.

Herrmann, H., and Aebi, U. (2004). Intermediate filaments: Molecular structure, assembly mechanism, and integration into functionally distinct intracellular scaffolds. *Annu. Rev. Biochem.* **73**, 749–789.

Herrmann, H., and Wiche, G. (1983). Specific *in situ* phosphorylation of plectin in detergent-resistant cytoskeletons from cultured Chinese hamster ovary cells. *J. Biol. Chem.* **258**, 14610–14618.

Herrmann, H., and Wiche, G. (1987). Plectin and IFAP-300K are homologous proteins binding to microtubule-associated proteins 1 and 2 and to the 240-kilodalton subunit of spectrin. *J. Biol. Chem.* **262**, 1320–1325.

Herrmann, H., Aberer, W., Majdic, O., Schuler, G., and Wiche, G. (1984). Monoclonal antibody to a 43 000 M$_r$ surface protein of a human leukaemia cell line (THP-1) crossreacts with the fibroblast intermediate filament protein vimentin. *J. Cell Sci.* **73**, 87–103.

Herrmann, H., Hofmann, I., and Franke, W. W. (1992). Identification of a nonapeptide motif in the vimentin head domain involved in intermediate filament assembly. *J. Mol. Biol.* **223**, 637–650.

Herrmann, H., Häner, M., Brettel, M., Müller, S. A., Goldie, K. N., Fedtke, B., Lustig, A., Franke, W. W., and Aebi, U. (1996). Structure and assembly properties of the intermediate filament protein vimentin: The role of its head, rod and tail domains. *J. Mol. Biol.* **264**, 933–953.

Herrmann, H., Häner, M., Brettel, M., Ku, N. O., and Aebi, U. (1999). Characterization of distinct early assembly units of different intermediate filament proteins. *J. Mol. Biol.* **286**, 1403–1420.

Herrmann, H., Wedig, T., Porter, R. M., Lane, E. B., and Aebi, U. (2002). Characterization of early assembly intermediates of recombinant human keratins. *J. Struct. Biol.* **137**, 82–96.

Herrmann, H., Hesse, M., Reichenzeller, M., Aebi, U., and Magin, T. M. (2003). Functional complexity of intermediate filament cytoskeletons: From structure to assembly to gene ablation. *Int. Rev. Cytol.* **223**, 83–175.

Hesse, M., Magin, T. M., and Weber, K. (2001). Genes for intermediate filament proteins and the draft sequence of the human genome: Novel keratin genes and a surprisingly high number of pseudogenes related to keratin genes 8 and 18. *J. Cell Sci.* **114**, 2569–2575.

Hofmann, I. (1998). Measuring the assembly kinetics and binding properties of intermediate filament proteins. *Subcell. Biochem.* **31**, 363–380.

Hofmann, I., Herrmann, H., and Franke, W. W. (1991). Assembly and structure of calcium-induced thick vimentin filaments. *Eur. J. Cell Biol.* **56**, 328–341.

Huiatt, T. W., Robson, R. M., Arakawa, N., and Stromer, M. H. (1980). Desmin from avian smooth muscle. Purification and partial characterization. *J. Biol. Chem.* **255**, 6981–6989.

Janmey, P. A. (1991). A torsion pendulum for measurement of the viscoelasticity of biopolymers and its application to actin networks. *J. Biophys. Biochem. Methods* **22**, 41–53.

Janmey, P. A., Euteneuer, U., Traub, P., and Schliwa, M. (1991). Viscoeleastic properties of vimentin compared with other filamentous biopolymer networks. *J. Cell Biol.* **113**, 155–160.

Janmey, P. A., Shah, J. V., Janssen, K. P., and Schliwa, M. (1998). Viscoelasticity of intermediate filament networks. *Subcell. Biochem.* **31**, 381–397.

Karabinos, A., Schünemann, J., Meyer, M., Aebi, U., and Weber, K. (2003). The single nuclear lamin of *Caenorhabditis elegans* for *in vitro* stable intermediate filaments and paracrystals with a reduced axial periodicity. *J. Mol. Biol.* **325**, 241–247.

Karrasch, S., Dolder, M., Schabert, F., Ramsden, J., and Engel, A. (1993). Covalent binding of biological samples to solid supports for scanning probe microscopy in buffer solution. *Biophys. J.* **65**, 2437–2446.

Laemmli, U. K. (1970). Cleavage of structural proteins during the assembly of the head of bacteriophage T4. *Nature* **227**, 680–685.

Mack, J. W., Steven, A. C., and Steinert, P. M. (1993). The mechanism of interaction of filaggrin with intermediate filaments. The ionic zipper hypothesis. *J. Mol. Biol.* **232**, 50–66.

Mücke, N., Wedig, T., Bürer, A., Marekov, L. N., Steinert, P. M., Langowski, J., Aebi, U., and Herrmann, H. (2004a). Molecular and biophysical characterization of assembly-starter units of human vimentin. *J. Mol. Biol.* **340**, 97–114.

Mücke, N., Kreplak, L., Kirmse, R., Wedig, T., Herrmann, H., Aebi, U., and Langowski, J. (2004b). Assessing the flexibility of intermediate filaments by atomic force microscopy. *J. Mol. Biol.* **335**, 1241–1250.

Müller, D. J., Schoenenberger, C.-A., Schabert, F., and Engel, A. (1997). Structural changes in native membrane proteins monitored at subnanometer resolution with the atomic force microscope: A review. *J. Struct. Biol.* **119**, 149–157.

Nagai, K., and Thogersen, H.-C. (1987). Synthesis and sequence-specific proteolysis of hybrid proteins produced in Escherichia coli. *Methods Enzymol.* **153**, 461–481.

Nelson, W. J., and Traub, P. (1983). Proteolysis of vimentin and desmin by the Ca^{2+}-activated proteinase specific for these intermediate filament proteins. *Mol. Cell. Biol.* **3**, 1146–1156.

Pytela, R., and Wiche, G. (1980). High molecular weight polypeptides (270,000-340,000) from cultured cells are related to hog brain microtubule-associated proteins but copurify with intermediate filaments. *Proc. Natl. Acad. Sci. USA* **77**, 4808–4812.

Quax-Jeuken, Y. E. F. M., Quax, W. J., and Bloemendal, H. (1983). Primary and secondary structure of hamster vimentin predicted from the nucleotide sequence. *Proc. Natl. Acad. Sci. USA* **80**, 3548–3552.

Quinlan, R. A., Cohlberg, J. A., Schiller, D. L., Hatzfeld, M., and Franke, W. W. (1984). Heterotypic tetramer (A2D2) complexes of non-epidermal keratins isolated from cytoskeletons of rat hepatocytes and hepatoma cells. *J. Mol. Biol.* **178**, 365–388.

Quinlan, R. A., Hatzfeld, M., Franke, W. W., Lustig, A., Schulthess, T., and Engel, J. (1986). Characterization of dimer subunits of intermediate filament proteins. *J. Mol. Biol.* **192**, 337–349.

Regula, J. T., Boguth, G., Gorg, A., Hegermann, J., Mayer, F., Frank, R., and Herrmann, R. (2001). Defining the mycoplasma 'cytoskelton': The protein composition of the Triton X-100 insoluble fraction of the bacterium Mycoplasma pneumoniae determined by 2-D gel electrophoresis and mass spectrometry. *Microbiology* **147**, 1045–1057.

Renner, W., Franke, W. W., Schmidt, E., Geisler, N., Weber, K., and Mandelkow, E. (1981). Reconstitution of intermediate-sized filaments from denatured monomeric vimentin. *J. Mol. Biol.* **149**, 285–306.

Rogers, K. R., Eckelt, A., Nimmrich, V., Janssen, K. P., Schliwa, M., Herrmann, H., and Franke, W. W. (1995). Truncation mutagenesis of the non-α-helical carboxyterminal tail domain of vimentin reveals contributions to cellular localization but not to filament assembly. *Eur. J. Cell Biol.* **66**, 136–150.

Starger, J. M., Brown, W. E., Goldman, A. E., and Goldman, R. D. (1978). Biochemical and immunological analysis of rapidly purified 10-nm filaments from baby hamster kidney (BHK-21) cells. *J. Cell Biol.* **78**, 93–109.

Stegh, A. H., Herrmann, H., Lampel, S., Weisenberger, D., Andrä, K., Seper, M., Wiche, G., Krammer, P. H., and Peter, M. E. (2000). Identification of the cytolinker plectin as a major early *in vivo* substrate for caspase 8 during CD95- and tumor necrosis factor receptor-mediated apoptosis. *Mol. Cell. Biol.* **20**, 5665–5679.

Steinert, P. M., Idler, W. W., and Zimmermann, S. B. (1976). Self-assembly of bovine epidermal keratin filaments *in vitro*. *J. Mol. Biol.* **108**, 547–567.

Steinert, P., Zackroff, Aynardi-Whitman, M., and Goldman, R. D. (1982). Isolation and characterization of intermediate filaments. *Methods Cell Biol.* **24**, 399–419.

Steinert, P. M., Marekov, L. N., and Parry, D. A. (1993). Diversity of intermediate filament structure. Evidence that the alignment of coiled-coil molecules in vimentin is different from that in keratin intermediate filaments. *J. Biol. Chem.* **268**, 24916–24925.

Stromer, M. H., Huiatt, T. W., Richardson, R. L., and Robson, R. M. (1981). Disassembly of synthetic 10-nm desmin filaments from smooth muscle into protofilaments. *Eur. J. Cell Biol.* **25**, 136–143.

van Amerongen, H., Kooijman, M., and Bloemendal, M. (1998). Transient electric birefringence in the study of intermediate filament assembly. *Subcell. Biochem.* **31**, 399–421.

Vorgias, C. E., and Traub, P. (1983). Isolation, purification and characterization of the intermediate filament protein desmin from porcine smooth muscle. *Prep. Biochem.* **13**, 227–243.

Wiche, G., Krepler, R., Artlieb, U., Pytela, R., and Denk, H. (1983). Occurrence and immunolocalization of plectin in tissues. *J. Cell Biol.* **97**, 887–901.

Yamada, S., Wirtz, D., and Coulombe, P. A. (2002). Pairwise assembly determines the intrinsic potential for self-organization and mechanical properties of keratin filaments. *Mol. Biol. Cell.* **13**, 382–392.

CHAPTER 2

Intermediate Filament Protein Structure Determination

Sergei V. Strelkov,★ **Laurent Kreplak,**★ **Harald Herrmann,**[†] **and Ueli Aebi**★

★Maurice E. Müller Institute for Structural Biology
Biozentrum, University of Basel
CH-4056 Basel, Switzerland

[†]Division of Cell Biology, TP3
German Cancer Research Center (DKFZ)
D-69120 Heidelberg, Germany

I. Introduction

X-ray crystallography is one of the few experimental methods that makes it possible to study intermediate filament (IF) structure at atomic resolution. In contrast, electron microscopy (EM) has been widely used to examine IF structure and assembly for decades, but it is normally only able to resolve details larger than 2–3 nm. However, the prerequisite for a crystallographic analysis is the ability to produce macroscopic, well-ordered crystals. It was not until several years ago that this was shown to be feasible for several fragments of human vimentin (Strelkov *et al.*, 2001). Since then, several atomic structures of the vimentin rod (Strelkov *et al.*, 2002) and, more recently, of nuclear lamin fragments including the protease-resistant part of the tail (Dhe-Paganon *et al.*, 2002; Krimm *et al.*, 2002), as well as parts of the rod domain (Strelkov *et al.*, 2004) have been determined, revealing essential atomic detail on the structure of the elementary IF dimer. In addition, another diffraction method, small angle x-ray scattering (SAXS), has also been used to investigate IF structure at higher assembly levels than the dimer. Although the structural resolution provided by SAXS is less than that of x-ray crystallography, the advantage of SAXS is that it works on protein solutions and hence does not depend on the availability of crystals. In addition, SAXS allows the monitoring of the consequent steps of IF assembly by varying the solution conditions, and hence provides additional information on the assembly pathway with respect to the approaches discussed in the previous chapter. In fact, both x-ray crystallography and SAXS have a large future potential toward getting more structural insight into the interactions of the elementary IF dimers during assembly.

This chapter will focus on the current experience with x-ray crystallography and SAXS toward studying IF protein structure. The (rather complicated) standard procedures pertinent to either method will not be covered, because they are well described in detail elsewhere. See, for example, (Ducruix and Giege, 1992; McPherson, 1982) (protein crystallization), (Blundell and Johnson, 1976; Cantor and Schirmer, 1980; Drenth, 1995) (x-ray diffraction analysis), and (Svergun and Koch, 2002; Svergun *et al.*, 2001; Vachette *et al.*, 2003) (SAXS data collection and processing). In addition, we will outline the approaches that bring together data on the IF protein structure obtained by various methods, aiming at constructing the three-dimensional model of complete IFs—a goal that has not been quite achieved yet.

II. Approaches and Methods

A. Crystallization and X-Ray Crystallography of Intermediate Filament Proteins

1. "Divide-and-Conquer" Approach to Intermediate Filament Protein Crystallization

For an x-ray crystallographic study, well-ordered three-dimensional crystals are required. However, many full-length wild-type IF proteins are known to be capable of a spontaneous self-assembly into filaments under physiological conditions. In addition, although the self-assembly could, in principle, be inhibited by some means (e.g., by introducing specific mutations or deletions, such as found in vimentin (Herrmann *et al.*, 1996)), the highly elongated, relatively flexible coiled-coil dimer together with its head and tail domains would still represent a difficult subject for crystallization (Brown *et al.*, 2001; Phillips *et al.*, 1986; Strelkov *et al.*, 2001).

A few years ago, we suggested a way to circumvent both these problems. Our method, dubbed the "divide-and-conquer approach," is based on creating a series of overlapping, sufficiently short fragments of the IF protein of interest (see Fig. 1 for examples). The fragments are prepared by recombinant expression in an *Escherichia coli* system, which provides sufficient quantities of the protein

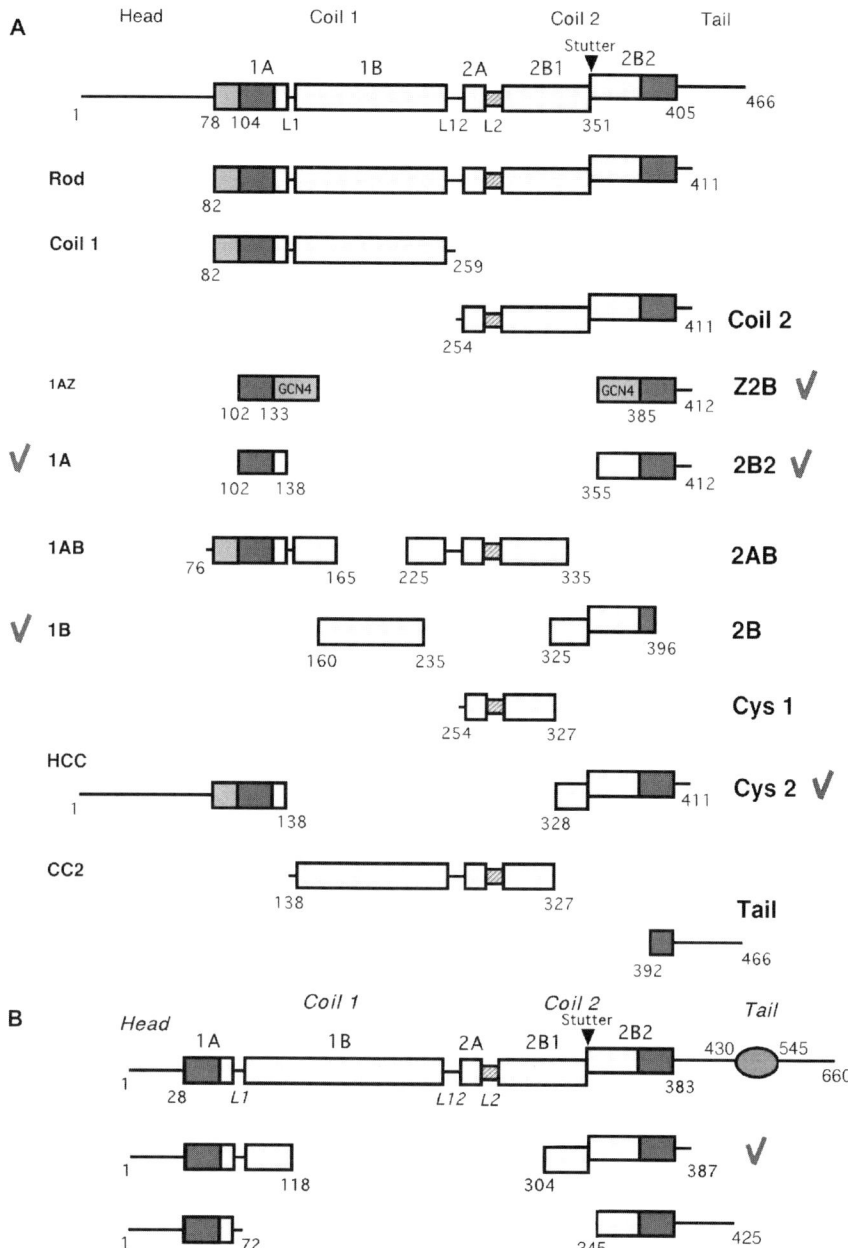

Fig. 1 (A) Schematic diagram of human vimentin and its fragments. The predicted α-helical segments are depicted by rectangles; the highly conserved ends on the rod domain are shaded. The ticks denote the fragments that have been crystallized to date. (B) Schematic diagram of human lamin A and its fragments. The globular part of the tail domain is shown as ovals.

($\geq$10 mg) needed to pursue an extensive search for crystallization conditions. The obtained fragments undergo characterization by biophysical methods such as analytical ultracentrifugation and circular dichroism spectroscopy. In particular, for the fragments of the rod domain, one should investigate whether they (1) are forming proper dimers and (2) are highly α-helical. These are two necessary conditions suggesting the formation of a coiled-coil structure. Indeed, results with vimentin rod fragments (Strelkov *et al.*, 2001), as well as lamin fragments (Strelkov *et al.*, 2004), document that this is often the case.

As we will discuss later, experience shows that IF fragments containing coiled coils up to $\sim$100 residues (corresponding to rodlike particles with a length of $\sim$150 Å) can be successfully crystallized by use of standard methods. Furthermore, it is the elongated, "linear" structure of the IF dimer that justifies its splitting into shorter fragments. Indeed, the fragments of the coiled-coil rod should be expected to form a structure that is close to their native structure within the full dimer (for more discussion on the justification and limitations of the divide-and-conquer approach see Strelkov *et al.*, 2001). Once the crystal structures of individual fragments have been determined, they can be patched together into an atomic model of the full-length dimer.

2. Preparation of Recombinant Intermediate Filament Fragments

During the past few years, we have designed and produced a series of human vimentin fragments (about 25 fragments in total, some of these are shown in Fig. 1A), a series of human desmin fragments and two parallel series of human lamins A and B$_1$, respectively (Fig. 1B). The length of the fragments ranges from 39–330 residues. In addition, Dhe-Paganon and colleagues (Dhe-Paganon *et al.*, 2002) have prepared (and successfully crystallized) the protease-resistant part of the lamin A tail domain (117 residues).

The most successful approach to produce these fragments was by expression of recombinant proteins in *E. coli*. In particular, we have extensively used the prokaryotic expression vector pPEP-T (Brandenberger *et al.*, 1996; Kammerer *et al.*, 1995). This vector (2968 bp) is a pUC19/pBR322 derivative carrying a T7 promoter as found in the pET vector family. At the N-terminus, there is the start Met, a glycine residue and six consecutive histidine residues. This His-tag is followed by a 35-amino acid oligomerization domain, a spacer of five glycines and a thrombin cleavage site (LeuValProArgGlySer). Immediately next to the arginine codon is a *Bam*HI site followed by codons that can be cleaved by *Xma*I, *Sal*I, *Xho*I, and *Eco*RI. Hence, if a cDNA is cloned into the vector by the *Bam*HI site, the peptide obtained after thrombin cleavage will start with residues GlySer. The expression vector pPEP-T may be obtained on request from Dr. Richard A. Kammerer, Wellcome Trust Centre for Cell Matrix Research, School of Biological Sciences, University of Manchester, The Michael Smith Building, Room B.3011, Oxford Road, Manchester, M13 9PT; fax: +44 161 275 1505; e-mail: richard.kammerer@man.ac.uk. To generate the recombinant fragments

(Fig. 1), we have used the full-length cDNA of the IF protein of interest as a template for polymerase chain reaction amplification. Forward primers included a *Bam*HI recognition site and the reverse primers contained an *Eco*RI recognition site together with a stop codon. The amplified products were ligated into the *Bam*HI and *Eco*RI sites of the pPEP-T vector, and the inserted sequences were verified by DNA sequencing.

By use of the pPEP-T vector, the presence of the additional (oligomerization) domain upstream of the sequence of interest provides for high expression yields even for short sequences. The BL21(DE3) *E. coli* cells (Novagen, Madison, WI) were grown in LB medium, and the recombinant expression is induced by IPTG addition at the optical density (600 nm) of 0.6. To optimize the expression of specific constructs, we initially experimented with 2-ml cultures with different amounts of IPTG (0.5–10 mM), different temperatures (most typically 37 °C and 30 °C), as well as different incubation times after the induction. At several time points, 100-μl aliquots of the culture were taken. The aliquots were complemented with 50 μl of the threefold concentrated sodium dodecyl sulfate (SDS) sample buffer (Gallagher, 1995), heated at 95 °C for 15 minutes and subjected to electrophoresis on 15% polyacrylamide/Tricine gels (SDS polyacrylamide gel electrophoresis [PAGE]) (Gallagher, 1995). This treatment disintegrates bacterial DNA into fragments so that the gels can be loaded with standard pipette tips. Thereafter, the gels were analyzed by Western blotting with an anti-His-tag antibody. After expression in a 200-ml culture, the fusion product was purified with a Ni-NTA agarose matrix column (Qiagen, Hilden, Germany) according to the manufacturer's protocol. The IF fragment was liberated from the fusion part by thrombin cleavage and again applied on the Ni-NTA column for the final purification step (Strelkov *et al.*, 2001). A purity of 90%–95% was achieved for all IF fragments obtained (see Fig. 1) as judged by Coomassie Brilliant Blue staining of the gels (Gallagher, 1995).

In addition, larger vimentin fragments, such as the rod, coil 1, and coil 2 (Fig. 1A), were expressed on their own with the pDS5 vector (Bujard *et al.*, 1987; Herrmann *et al.*, 1993). This vector has been shown to be highly efficient for the full-length vimentin, readily yielding as much as 100 mg of protein per 1 L of culture after all purification steps. The transformed BL21 codon plus cells (Stratagene, La Jolla, CA) were grown overnight at 37 °C in "terrific broth" (12 g tryptone, 24 g yeast extract, 4 ml glycerol, 2.31 g KH_2PO_4, and 12.54 g K_2HPO_4 per liter) containing 100 μg/ml ampicillin, the expressed product isolated as inclusion bodies and purified as described in the previous Chapter (see also [Strelkov *et al.*, 2001]).

Furthermore, to produce vimentin fragments HCC, Cys 1 and Cys 2 (Fig. 1A), we have used chemical cleavage of the wild-type protein or a larger fragment. In particular, to produce the HCC fragment, the full-length human vimentin cDNA carrying the point mutation Lys139→Cys was cloned into and overexpressed in BL21 cells. The purified polypeptide contained, in addition to the introduced Cys139, the Cys328, which is the only cysteine present in the wild-type sequence.

The subsequent cleavage was done at cysteines with 2-nitro-5-thiocyanobenzoic acid (NTCB) (Geisler and Weber, 1981), and the resulting fragments, including HCC, were isolated by high-performance liquid chromatography (HPLC) as described (Geisler *et al.*, 1992). The cleavage with NTCB is quite practical, because it generates a normal residue at the C-terminus of the upstream fragment and because the reaction produces few side products (Geisler and Weber, 1981). The fragments Cys 1 and Cys 2 were generated in a similar fashion. To achieve this, the vimentin coil 2 fragment (see Fig. 1) was overexpressed, purified, and subsequently cleaved with NTCB at its only endogenous cysteine. In addition, we have engineered a vimentin with cysteine 328 replaced by alanine. By inserting cysteines into this protein, it is possible to obtain any desired vimentin fragments (Rogers *et al.*, 1996).

Finally, several fragments were prepared by chemical synthesis. Some of them could be crystallized successfully and revealed the proper coiled-coil structure, such as the 59-residue long chimerical protein including the 31-residue-long oligomerization domain (leucine zipper) followed by 28 residues corresponding to the C-terminal part of the vimentin rod domain (see Fig. 1A). The advantage of using chemically synthesized peptides is that they can be prepared on order by several companies, which can save time. However, chemical synthesis is expensive, and because there is often the need to produce a number of fragments, *E. coli* expression is clearly a more economical solution. Furthermore, the most critical limitation of the chemical synthesis is the achievable maximal peptide length. At present, only fragments not exceeding ~50 residues in length can be synthesized as continuous peptides with a sufficient yield of the chemically correct product. At the same time, the availability of milligram quantities of at least 95%–98% pure product is a prerequisite for successful crystallization screening. In summary, although we have found nothing principally wrong with the use of chemically synthesized peptides toward crystallization and three-dimensional structure determination of IF fragments, we believe that the use of recombinant techniques is generally a better approach.

3. Biophysical Characterization of Intermediate Filament Fragments

After purification, the recombinantly expressed IF fragments were collected at a concentration of ~0.3–1 mg/ml in 8 M urea solution. Subsequently, a stepwise dialysis to 6, 4, 2 M urea (every 0.5 hour at 22 °C), and finally to 0 M urea (overnight at 4 °C) was done, yielding, in most cases, a fully soluble fragment in 10 mM Tris-HCl buffer, pH 8.4, with 1 mM ethylenediamine tetraacetic acid (EDTA) and 0.02% NaN_3. In addition, 1 mM DTT was added in cases in which the particular fragment contained Cys residues. Protein concentrations were routinely determined by absorbance at 280 nm with a standard spectrophotometer. The extinction coefficients were normally calculated from the amino acid sequence using the *Protparam* web-based tool (http://www.expasy.org/tools/protparam.html). When the fragment contained no Trp or Tyr residues, its concentration was estimated with the Bio-Rad dye assay (Bio-Rad, Hercules, CA) by absorbance at 595 nm. Most IF fragments could be stored at 4 °C for at least several months, exhibiting no signs of degradation by SDS-PAGE. Furthermore, for the fragments for which crystallization conditions could be found, prolonged storage at 4 °C in the aforementioned buffer did not affect the crystallizability.

Sedimentation velocity and sedimentation equilibrium runs were performed at 20 °C in a Beckman analytical ultracentrifuge, model Optima XLA, equipped with an ultraviolet optical system (see e.g. [Herrmann et al., 1996]). Typically, sedimentation equilibrium runs with several protein concentrations ranging from 0.2–2 mg/ml were performed to explore the concentration dependence of oligomerization. Most samples were in 10 mM Tris-HCl buffer, pH 8.4, without salt or with 150 mM NaCl. Circular dichroism analysis was carried out on a Jasco J-720 spectropolarimeter (Jasco, Japan) at 20 °C. Protein samples at ∼0.1 mg/ml and a cell with a 1-mm path length were used.

4. Crystallization and X–ray Structure Determination

Our experience suggests that the standard methods for growing three-dimensional crystals of soluble proteins can also be successfully used for IF fragments. In particular, the initial screening for crystallization conditions was effectively done by use of commercially available kits of precipitant solutions formulated according to the so-called random sparse matrix approach (Jancarik and Kim, 1991). Normally, we have used the Crystal Screen (50 precipitant solutions) and the Crystal Screen 2 (48 solutions) from Hampton Research (Celiso Viejo, CA), and Wizard I and II kits (48 solutions each) from Emerald Biostructures (Bainbridge Island, WA). Hence, for each new IF fragment, the initial screening was done with a total of 194 precipitant solutions.

Before crystallization, the samples were concentrated to ∼10–20 mg/ml with the Vivaspin-20 (Vivascience, Hannover, Germany) or Amicon Ultra-15 (Millipore, Billerica, MA) ultrafiltration devices and a standard centrifuge. Crystallization setups were done by the hanging drop technique (Ducruix and Giege, 1992) with the 24-well VDX plates. Crystallization drops were typically prepared by mixing 1.2 μl of protein solution with 1.2 μl of reservoir solution. Initial crystallizations were set up and stored at 20 °C and observed under a low-magnification microscope on a regular basis over a period of several months. If no crystals were obtained, the setups were repeated at 4 °C. Crystallization screening solutions, plates, coverslips, and other supplies were purchased from Hampton Research (34 Journey, Aliso Viejo, CA 92656, http://www.hamptonresearch. com) and from Emerald Biostructures (7869 NE Day Road, Bainbridge Island, WA 98110, http://www.decode.com/emeraldbiostructures).

In cases in which the initial screening revealed any kind of crystals or crystalline precipitate, the crystallization conditions were refined toward obtaining single crystals with at least 0.1 mm in each dimension. Therefore, approximately 250 μl of the concentrated protein (10 mg/ml or more) solution was required for the initial screen with 194 precipitant solutions, and additional quantities of the protein were needed for the refinement. Hence, even in the most favorable crystallization scenario, the supply of the purified IF fragment needs to be at least 5 mg. Examples of the obtained crystals are shown in Fig. 2; Table I shows the refined crystallization conditions for several IF fragments. Many of these fragments are forming predominantly α-helical coiled-coil structures. As seen in Table I, only coiled-coil fragments up to 84 amino acid residues could be crystallized thus far, despite extensive screening also done for longer fragments. Furthermore, the successful crystallization conditions for the different fragments are mostly unrelated; in other words, there does not seem to be any preferred reagent or condition for crystallizing IF fragments.

The obtained crystals were used to collect diffraction data with an in-house x-ray source equipped with a copper rotating anode or synchrotron radiation at EMBL/DESY (Hamburg, Germany), ESRF (Grenoble, France), or SLS (Villingen, Switzerland). For the data collection, crystals were flash-frozen to

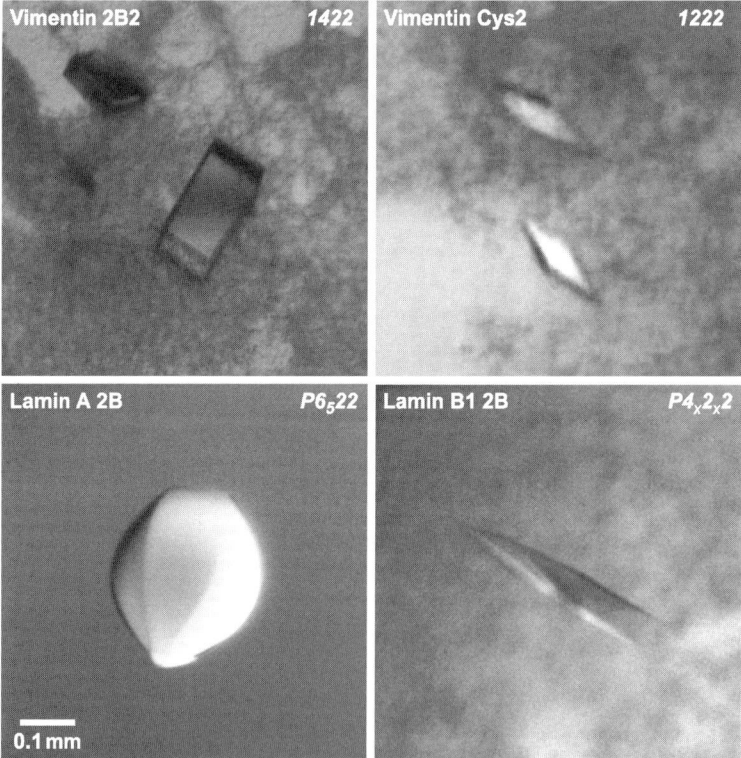

Fig. 2 Photographs of crystals of vimentin and lamin fragments (see Fig. 1 and Table I). The space groups of the crystals are indicated in the upper right-hand corner of each panel.

Table I
Crystals of Intermediate Filament Fragments

Fragment	Residues*	Crystallization conditions	Diffraction limit (Å)	Reference
Vimentin				
1A	102–138 (39)	2.0 M $(NH_4)_2SO_4$, 10% (v/v) dioxane, 0.1 M MES/Na buffer, pH 6.5	1.4	Strelkov *et al.*, 2002
1B	160–235 (76)	0.2 M NH_4 acetate, 30% (w/v) PEG 4k, 0.1 M citrate buffer, pH 5.6	2.1	Strelkov *et al.*, 2001
Cys2	328–411 (84)	0.17 M Na acetate, 25.5% (w/v) PEG 8k, 0.1 M cacodylate buffer, pH 6.5	2.3	Strelkov *et al.*, 2002
2B2	355–412 (58)	39% MPD, 25% (v/v) PEG 400, 0.1 M Tris/HCl buffer, pH 8.5	3.0	Strelkov *et al.*, 2001
Z2B	385–412 (59)	0.55 M $(NH_4)H_2PO_4$, pH 9.0 adjusted with NaOH	1.9	Strelkov *et al.*, 2002
Lamin A				
2B_lamA	305–387 (83)	0.7 M Na/K tartrate, 0.1 M Tris/HCl, pH7.5	2.2	Strelkov *et al.*, 2004
Tail domain	436–552 (117)	0.2 M NH_4 acetate, 25% PEG 4k, 0.1 M Tris/HCl buffer, pH 8.5, 10 mM DTT	1.4	Dhe-Paganon *et al.*, 2002
Lamin B1				
2B_lamB1	307–389 (83)	1.8 M $(NH_4)_2SO_4$, 10 mM $CoCl_2$, 0.1 M MES/Na buffer, pH 6.5	3.3	Strelkov *et al.*, 2004

*The number in parentheses indicates the total number of amino acid residues in the fragment. For the *Z2B* fragment, the total number of residues includes the 31 residues of the leucine zipper.

100 K in cryoloops (Hampton Research). As seen from the examples in Fig. 2, the crystal packing arrangements (i.e., the crystallographic space group) for different fragments are rather unrelated. Subsequent crystallographic computing was performed with standard software (Brünger *et al.*, 1998; CCP4, 1994). The phasing of the diffraction data was done, when possible, by molecular replacement or otherwise with heavy-atom isomorphous derivatization (Blundell and Johnson, 1976; Drenth, 1995). A typical section of the final electron density map for the coil 2B fragment of lamin A is shown in Fig. 3.

B. Small-Angle X-Ray Scattering from Intermediate Filament Proteins in Solution

To our knowledge, the first attempt to use SAXS toward studying IF structure and assembly was done with tissue-purified keratin filaments (Sayers *et al.*, 1990). Recently, we performed a series of SAXS measurements on solutions of recombinantly expressed wild-type vimentin and its mutants. Here we will only briefly outline these experiments; the full account will appear in a forthcoming publication.

Generally, SAXS measurements require a dedicated experimental setup on a synchrotron x-ray source. Our experiments with vimentin solutions were carried

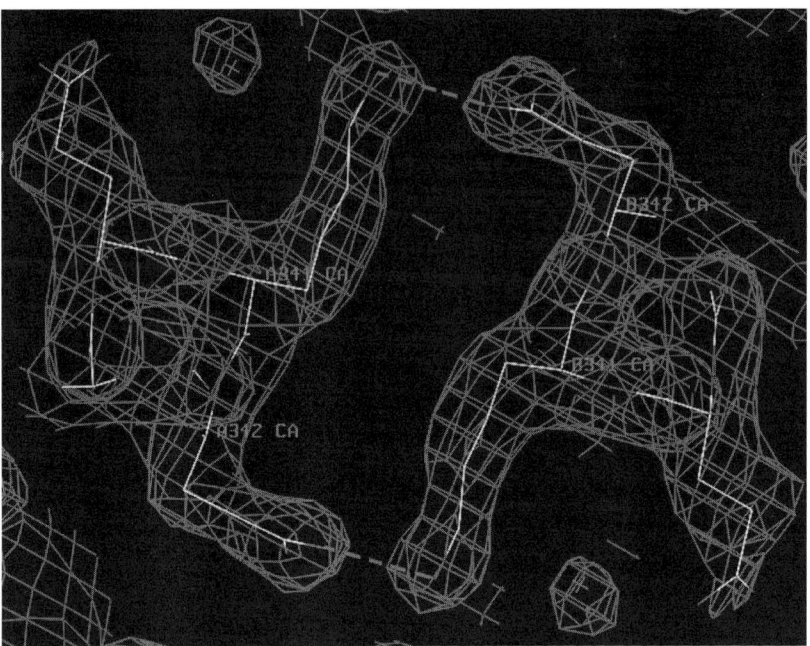

Fig. 3 Part of the atomic structure of the lamin A coil 2B fragment overlaid with the 2.3 Å-resolution electron density map ($2F_{obs}$-F_{calc}, 1.5σ level). The interchain salt bridges formed between residues Lys341 and Glu342 are shown with dashed lines. (See Color Insert.)

out at the NCS beamline of EMBL/DESY in Hamburg. A series of protein samples with varying buffer conditions, including pH (5 mM Tris-HCl buffer, pH 8.4, 7.5, or 7.0) and ionic strength (0–160 mM NaCl), were used. For each pH/ionic strength combination, two to three samples with increasing protein concentrations from approximately 2–10 mg/ml, as well as the reference buffer solution, were prepared to determine whether the oligomerization was concentration dependent. Careful determination of the concentration of each sample (preferably by absorption at 280 nm) was important for the subsequent evaluation of the SAXS data. As a reference, a solution of bovine serum albumin was used. The samples were exposed to x-rays with a wavelength of 1.5 Å for 15 minutes. The sample volume for each experiment was 100 μl, and each sample could be used only once because of radiation damage. Hence, for each pH/ionic strength combination, approximately 1.5–2 mg of pure IF protein was required.

The processing of the collected SAXS curves was done with the ATSAS 2.0 program package (see http://www.embl-hamburg.de/ExternalInfo/Research/Sax/). The obtained data revealed the gradual assembly of vimentin dimers occurring when the ionic strength was increased and the pH was lowered. Although tetramers, octamers, and, eventually, the unit-length filaments (32mers) could be clearly detected, many samples represented mixtures of these oligomers. Few character-istics of the oligomers contained in a particular sample could be determined from the SAXS data directly. These include the average molecular mass and the distribution of the intramolecular distances (Vachette *et al.*, 2003). At the same time, because of the relatively low information content of the SAXS data, its more detailed evaluation required additional constraints. Such constraints were provided, on one hand, by the knowledge of the atomic structure of the elementa-ry dimer (for a discussion on how to assemble an atomic model of the IF dimer from the known crystal structures of fragments see the following) and, on the other hand, by the existence of three lateral dimer–dimer association modes (A_{11}, A_{22} and A_{12}) as detected by cross-linking experiments. Test models of the tetramer and higher oligomers were first assembled from the required number of dimer structures, and then the relative positions of the dimers as rigid bodies were refined to minimize the discrepancy between the observed data and the SAXS curves calculated from the models (Svergun and Koch, 2002; see Fig. 4C for an example).

C. How to Create an Atomic Model of Intermediate Filaments

The principal goal of structural studies of IFs is to obtain a three-dimensional model of the filament at atomic detail. For microtubules and actin filaments, this goal has already been largely achieved. In the following we outline the prerequi-sites for building a structural model of IF architecture, give a short overview of the currently available experimental data, and discuss the missing links.

1. Atomic Structure of the Elementary Dimer

Although laborious, the divide-and-conquer crystallographic approach allows us to obtain atomic detail on various parts of the elementary IF dimer. Our knowledge of the dimer structure is therefore fairly precise already. The exceptions are the head domain and nonglobular portions of the tail domain. These regions appear to be poorly ordered on their own and get structured only on the formation of higher order oligomers (Herrmann and Aebi, 2004). This is the reason that these parts of the dimer cannot be resolved by the "divide-and-conquer" approach.

The obtained structures of the coiled-coil fragments can be readily patched together into a three-dimensional model of the full dimer (Fig. 4A,B). In addition, if the crystal data are not (yet) available, the structure of the coiled-coil segments can be modeled. Indeed, the coiled-coil segments can be reasonably well delineated on the basis of the primary sequence of IFs, as seen from the comparison of the earlier predictions (see e.g., Parry and Steinert, 1999) with the corresponding crystal structures. Furthermore, the coiled-coil structure can be modeled with a higher degree of precision than most other protein folding motifs (Cohen and Parry, 1990). To do such modeling, it is sufficient to use some crystal structure of a two-stranded coiled coil as a backbone template, for instance, the GCN4 leucine

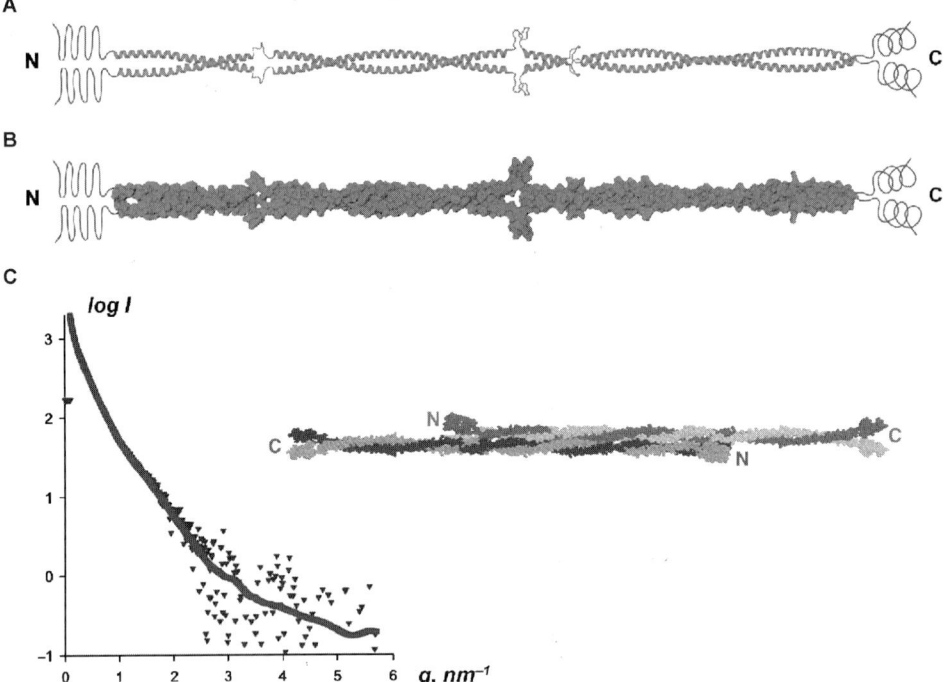

Fig. 4 (A) Model of the vimentin dimer shown as a ribbon diagram. The coiled-coil rod and the terminal domains are shown in red and blue, respectively. (B) Same model with the coiled-coil rod shown as surface. (C) Model of the vimentin tetramer based on SAXS data. (C) Experimental SAXS data obtained from the solution of vimentin tetramers (dark blue triangles), as well as the scattering curve computed from the above tetramer model (light blue line). Shown is the dependence of the scattered intensity (arbitrary units, log scale) on $q = 4\pi \sin \theta / \lambda$, where 2θ is the scattering angle and λ is the wavelength. (See Color Insert.)

zipper (O'Shea *et al.*, 1991) or another IF fragment that has been determined crystallographically (Strelkov *et al.*, 2002). Thereafter, the correct side chains of the sequence of interest can be built in, and the model can be locally optimized by use of standard energy minimizers such as those implemented in the CNS program (Brünger *et al.*, 1998). This approach does not require the use of more sophisticated, computationally expensive minimization approaches such as molecular dynamics with explicit solvent. However, the modeled structure does not substitute for a crystallographic analysis. Although the overall backbone trace of a modeled region is likely to be qualitatively correct, the more subtle details of the structure (such as conformation of particular side chains, exact geometrical parameters of the coiled coil, or the formation of salt bridges) should be taken with a sufficient degree of reservation.

2. Structural Detail of the Dimer–Dimer Interactions

The existence of three distinct modes of lateral dimer–dimer contacts within the filament (A_{11}, A_{12}, and A_{22}) have been established by chemical cross-linking experiments for a number of IF proteins (Steinert *et al.*, 1993b). In addition, there is a single mode of longitudinal ("head-to-tail") contact (Steinert *et al.*, 1993a). A recent ultracentrifugation study has shown that vimentin solutions in low ionic strength buffers contain predominantly A_{11}-type tetramers, whereas the other dimer–dimer contacts take effect only at later stages of assembly (Mucke *et al.*, 2004b).

The SAXS experiments described previously provided a basis for constructing a three-dimensional model of the A_{11} tetramer and higher order oligomers (Fig. 4C). As a first approximation, the dimers within such oligomers can be considered rigid bodies. In fact, each of the four coiled-coil segments of the rod domain can be considered a straight rodlet, because their lengths are smaller than the persistence length of a coiled coil (~ 250 Å) (Schwaiger *et al.*, 2002). At the same time, the linkers connecting the coiled-coil segments, especially L12, may serve as highly flexible "hinges," because EM images of invertebrate IF tetramers appear considerably bent (Geisler *et al.*, 1998). A further refinement of the tetramer and higher order oligomer models may involve further adjustment of the dimers' conformations by energy minimization.

As mentioned previously, the N-terminal head and possibly the C-terminal tail domains are likely to become more ordered once in a tetramer or higher order oligomer (Herrmann and Aebi, 2004). Cross-linking data suggest that the head domain folds back onto the coiled-coil rod (Herrmann and Aebi, 2004; Parry *et al.*, 2002). Moreover, electrostatic attraction between the positively charged head domains and the negatively charged rod domains seem to play a crucial role in IF assembly (Strelkov *et al.*, 2004), but solid data on this are still scarce. In particular, "headless" IF proteins hardly assemble above the tetrameric state in the presence of monovalent ions (Herrmann *et al.*, 1996; Mucke *et al.*, 2004b). Only by inclusion of divalent ions such as Mg^{2+} is the repulsion of the highly negatively charged rods overcome and massive "precipitation" into non-IF fibers induced (Geisler and Weber, 1982).

3. Molecular Architecture of Mature Intermediate Filaments

For many other filament systems, such as actin, EM images of either negatively stained or frozen hydrated samples revealed distinct substructures and well-defined (mostly helical) symmetry (Bremer *et al.*, 1994). Unfortunately, in the case of IFs, EM images of negatively stained specimens mostly reveal smooth filaments with no apparent symmetry elements. At the same time, EM observations of IFs have nevertheless yielded some important structural detail. First, glycerol spraying followed by rotary metal shadowing has revealed a regular 21-nm beading of IFs (Heins *et al.*, 1993), as well as suggested the existence of protofilamentous

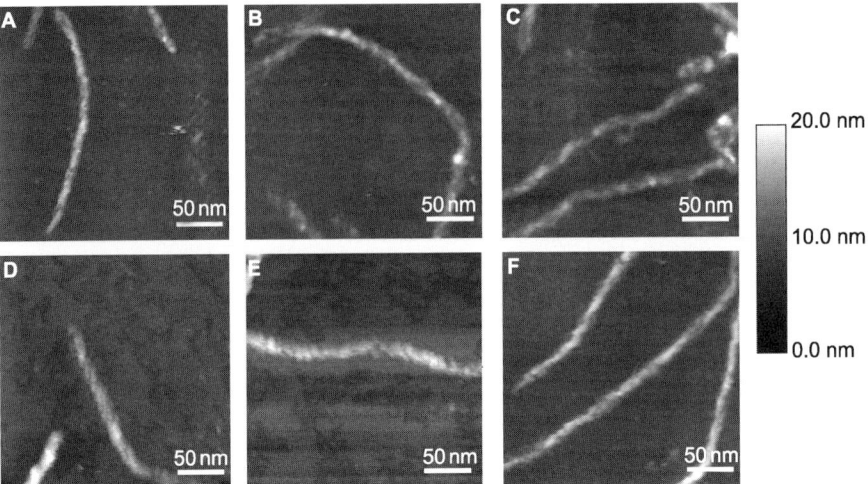

Fig. 5 *In vitro* assembled IFs of human vimentin (A–C) and desmin (D–F) as revealed by tapping mode atomic force microscopy in assembly buffer. The filaments were polymerized at 37 °C in 2 mM phosphate buffer, 100 mM KCl, pH 7.5, for 30 minutes and fixed with 0.1% glutaraldehyde before absorption on a graphite support.

substructures (Aebi *et al.*, 1983). Furthermore, the "sidearm structure" of neurofilament IFs was observed (Mulligan *et al.*, 1991). Second, the scanning transmission EM (STEM) measurements yielded the mass/length ratios for various IFs (Engel *et al.*, 1985; Herrmann *et al.*, 1996) that allow us to calculate the number of dimers per filament cross section. Third, EM observations of lamin assembly have revealed a short "head-to-tail" overlap of the elementary dimers (Aebi *et al.*, 1986), as well as formation of lamin sheets (Goldman *et al.*, 1986).

Furthermore, by use of atomic force microscopy to image glutaraldehyde-fixed vimentin and desmin filaments on a graphite support (Mucke *et al.*, 2004a), we have recently observed protofilamentous substructures coiling around each other in an irregular fashion (Fig. 5). These observations suggest the existence of at least a short range order within the filaments. Moreover, by analyzing the cross-linking products of various IFs, the lateral staggers of the individual dimers could be calculated (Steinert *et al.*, 1993b), and this information provided the basis for the surface lattice models of IFs (Parry *et al.*, 2001). In addition, some information on the density distribution on the IF cross section has been obtained by diffraction studies on hair keratin fibers (Briki *et al.*, 1998; Watts *et al.*, 2002), but until now these data could not provide an unambiguous assignment of dimers and higher order oligomers on cross section.

======   **III. Pearls and Pitfalls**

Thus far, the "divide-and-conquer" approach has been the only method capable of delivering atomic resolution data on IF proteins. We have found that the most effective way to obtain the needed IF fragments is the expression in *E. coli*. As was described in this and the previous chapter, the fragments can be either over-expressed on their own (with the pDS5 vector) or as a His-tagged fusion peptide (with the pPEP-T vector). Occasionally, however, the level of expression of a particular construct turns out to be too low to allow isolation of the needed milligram quantities of the fragment from reasonable volumes of cell culture (most likely because of the toxicity of the expression product for the bacterium). For instance, during the initial attempts to express the HCC fragment of human vimentin (residues 1–138, see Fig. 1A) on its own, the level of expression was low, and only 2–3 mg of pure protein could be obtained from 1 L of bacterial culture. The solution to the problem was to obtain the HCC fragment by chemical cleavage of the full-length protein. The latter can be readily overexpressed in high amounts; moreover, the purification procedure for the full-length vimentin is well established (Herrmann *et al.*, 1993). As a result, approximately 30 mg of the pure HCC fragment could be obtained from 1 L of culture. Finally, in case none of the overexpression strategies work for a particular fragment, the last resort is to try to obtain the fragment by chemical synthesis, which is, however, limited to small fragments up to approximately 50 residues and is expensive.

At the same time, the key "bottleneck" step is the crystallization of IF frag-ments. As explained earlier, thus far the search for crystallization conditions for IF fragments relied on standard crystallization screening techniques. Hence, in case the latter yield no crystals, the only productive solution is to generate additional fragments and try crystallizing these. Another caveat is that the partic-ular IF fragment may not fold in the same way on its own as in the full-length native protein. For example, the recombinantly expressed coil 1A fragment of human vimentin was found to be monomeric in solution and not forming proper dimers (Strelkov *et al.*, 2001). Fortunately, this problem is not as severe for coiled-coil fragments as for fragments of globular proteins (Strelkov *et al.*, 2001). Nevertheless, there is a serious conflict between crystallizability and correct fold-ing of coiled-coil fragments depending on their length. On one hand, the wish to obtain a possibly nativelike conformation speaks in favor of working with the longest constructs possible; on the other hand, the probability of obtaining crystals rapidly decreases with the fragment size, as discussed.

A final remark has to be made regarding the SAXS experiments described previously and also EM analyses of IFs assembled in a test tube. Importantly, it was demonstrated that *in vitro* assembly of IF proteins, such as vimentin or cytokeratins (for an overview, see Herrmann and Aebi, 2004; Parry and Steinert, 1995), may be carried out in such a way that the resulting filaments will be indistinguishable from the corresponding native IFs. The factors promoting the correct assembly include changing the ionic strength and pH of the solution, addition of specific cations, and so on (Herrmann *et al.*, 1996). At the same time, several examples point to the dependence of the result of *in vitro* assembly on the protocol used (Herrmann *et al.*, 1996, 1999). Frequently, polymorphism manifesting itself, (e.g., in the variability of the filament diameter) is observed. Hence, the issue as to whether the *in vitro* assembled filaments or assembly intermediates reflect the natural situation should be always looked at in principle. The main problem here is that the structural data on the native IFs are scarce at present. Unfortunately, fluorescent microscopy, often used to study the IF assembly within cells, does not provide high resolution structural detail Much more promising in this regard is the recently devised cryoelectron tomography of intact cells (Medalia *et al.*, 2002).

IV. Concluding Remarks

As a further extension of the divide-and-conquer approach, x-ray structure determination of stable complexes formed by various IF fragments, including those of the rod and the terminal domains, should be attempted. Crystallization of such complexes is a difficult task; however, this should provide the essential atomic detail on the four dimer–dimer contact types occurring in mature IFs. Furthermore, we believe that the ultimate solution of the IF structure "enigma" will require the determination of the three-dimensional envelope of the filament with cryoelectron microscopy or other imaging techniques. In particular, such an envelope should be accurate enough to allow the unambiguous localization of the dimers or higher order substructural elements.

Acknowledgments

This work was supported by a research grant and the NCCR program grant on "Nanoscale Science" from the Swiss National Science Foundation, the Swiss Society for Research on Muscular Diseases, the M. E. Müller Foundation of Switzerland, and the Canton Basel-Stadt.

References

Aebi, U., Cohn, J., Buhle, L., and Gerace, L. (1986). The nuclear lamina is a meshwork of intermediate-type filaments. *Nature* **323,** 560–564.

Aebi, U., Fowler, W. E., Rew, P., and Sun, T. T. (1983). The fibrillar substructure of keratin filaments unraveled. *J. Cell Biol.* **97,** 1131–1143.

Blundell, T. L., and Johnson, L. N. (1976). *Protein Crystallography.* Academic Press, New York.

Brandenberger, R., Kammerer, R. A., Engel, J., and Chiquet, M. (1996). Native chick laminin-4 containing the beta 2 chain (s-laminin) promotes motor axon growth. *J. Cell Biol.* **135,** 1583–1592.

Bremer, A., Henn, C., Goldie, K. N., Engel, A., Smith, P. R., and Aebi, U. (1994). Towards atomic interpretation of F-actin filament three-dimensional reconstructions. *J. Mol. Biol.* **242,** 683–700.

Briki, F., Busson, B., and Doucet, J. (1998). Organization of microfibrils in keratin fibers studied by X-ray scattering modelling using the paracrystal concept. *Biochim. Biophys. Acta* **1429,** 57–68.

Brown, J. H., Kim, K. H., Jun, G., Greenfield, N. J., Dominguez, R., Volkmann, N., Hitchcock-DeGregori, S. E., and Cohen, C. (2001). Deciphering the design of the tropomyosin molecule. *Proc. Natl. Acad. Sci. USA* **98,** 8496–8501.

Brünger, A. T., Adams, P. D., Clore, G. M., DeLano, W. L., Gros, P., Grosse-Kunstleve, R. W., Jiang, J. S., Kuszewski, J., Nilges, M., Pannu, N. S., Read, R. J., Rice, L. M., Simonson, T., and Warren, G. L. (1998). Crystallography & NMR system: A new software suite for macromolecular structure determination. *Acta Crystallogr. Dev. Biol. Crystallogr.* **54,** 905–921.

Bujard, H., Gentz, R., Lanzer, M., Stueber, D., Mueller, M., Ibrahimi, I., Haeuptle, M. T., and Dobberstein, B. (1987). A T5 promoter-based transcription-translation system for the analysis of proteins *in vitro* and *in vivo. Methods Enzymol.* **155,** 416–433.

Cantor, C. R., and Schirmer, P. R. (1980). *Biophysical Chemistry.* Freeman, New York.

CCP4. (1994). The CCP4 suite: Programs for protein crystallography. *Acta Cryst.* **D50,** 760–763.

Cohen, C., and Parry, D. A. (1990). Alpha-helical coiled coils and bundles: How to design an alpha-helical protein. *Proteins* **7,** 1–15.

Dhe-Paganon, S., Werner, E. D., Chi, Y. I., and Shoelson, S. E. (2002). Structure of the globular tail of nuclear lamin. *J. Biol. Chem.* **18,** 18.

Drenth, J. (1995). *Principles of Protein X-Ray Crystallography.* Springer, New York.

Ducruix, A., and Giege, R. (1992). *Crystallisation of Nucleic Acids and Proteins: A Practical Approach.* Oxford University Press, Oxford, U.K.

Engel, A., Eichner, R., and Aebi, U. (1985). Polymorphism of reconstituted human epidermal keratin filaments: Determination of their mass-per-length and width by scanning transmission electron microscopy (STEM). *J. Ultrastruct. Res.* **90,** 323–335.

Gallagher, S. R. (1995). One-dimensional SDS gel electrophoresis of proteins. *In* "Current Protocols in Protein Science." (J. E. Coligan, B. M. Dunn, D. W. Speicher, and P. T. Wingfield, eds.), Vol. 2, pp. 10.11.11–10.11.34. John Wiley & Sons, New York.

Geisler, N., Schunemann, J., and Weber, K. (1992). Chemical cross-linking indicates a staggered and antiparallel protofilament of desmin intermediate filaments and characterizes one higher-level complex between protofilaments. *Eur. J. Biochem.* **206,** 841–852.

Geisler, N., Schunemann, J., Weber, K., Haner, M., and Aebi, U. (1998). Assembly and architecture of invertebrate cytoplasmic intermediate filaments reconcile features of vertebrate cytoplasmic and nuclear lamin-type intermediate filaments. *J. Mol. Biol.* **282,** 601–617.

Geisler, N., and Weber, K. (1981). Comparison of the proteins of two immunologically distinct intermediate-sized filaments by amino acid sequence analysis: Desmin and vimentin. *Proc. Natl. Acad. Sci. USA* **78,** 4120–4123.

Geisler, N., and Weber, K. (1982). The amino acid sequence of chicken muscle desmin provides a common structural model for intermediate filament proteins. *EMBO J.* **1,** 1649–1656.

Goldman, A. E., Maul, G., Steinert, P. M., Yang, H. Y., and Goldman, R. D. (1986). Keratin-like proteins that coisolate with intermediate filaments of BHK-21 cells are nuclear lamins. *Proc. Natl. Acad. Sci. USA* **83,** 3839–3843.

Heins, S., Wong, P. C., Muller, S., Goldie, K., Cleveland, D. W., and Aebi, U. (1993). The rod domain of NF-L determines neurofilament architecture, whereas the end domains specify filament assembly and network formation. *J. Cell. Biol.* **123,** 1517–1533.

Herrmann, H., and Aebi, U. (2004). Intermediate filaments: Molecular structure, assembly mechanism, and integration into functionally distinct intracellular scaffolds. *Ann. Rev. Biochem.* **73,** 749–789.

Herrmann, H., Eckelt, A., Brettel, M., Grund, C., and Franke, W. W. (1993). Temperature-sensitive intermediate filament assembly. Alternative structures of Xenopus laevis vimentin *in vitro* and *in vivo. J. Mol. Biol.* **234,** 99–113.

Herrmann, H., Haner, M., Brettel, M., Ku, N. O., and Aebi, U. (1999). Characterization of distinct early assembly units of different intermediate filament proteins. *J. Mol. Biol.* **286,** 1403–1420.

Herrmann, H., Haner, M., Brettel, M., Muller, S. A., Goldie, K. N., Fedtke, B., Lustig, A., Franke, W. W., and Aebi, U. (1996). Structure and assembly properties of the intermediate filament protein vimentin: The role of its head, rod and tail domains. *J. Mol. Biol.* **264,** 933–953.

Jancarik, J., and Kim, S. H. (1991). Sparse matrix sampling: A screening method for crystallization of proteins. *J. Appl. Cryst.* **24,** 409–411.

Kammerer, R. A., Antonsson, P., Schulthess, T., Fauser, C., and Engel, J. (1995). Selective chain recognition in the C-terminal alpha-helical coiled-coil region of laminin. *J. Mol. Biol.* **250,** 64–73.

Krimm, I., Ostlund, C., Gilquin, B., Couprie, J., Hossenlopp, P., Mornon, J. P., Bonne, G., Courvalin, J. C., Worman, H. J., and Zinn-Justin, S. (2002). The Ig-like structure of the C-terminal domain of lamin a/c, mutated in muscular dystrophies, cardiomyopathy, and partial lipodystrophy. *Structure (Camb).* **10,** 811–823.

McPherson, A. (1982). *Preparation and Analysis of Protein Crystals.* John Wiley & Sons Inc., New York.

Medalia, O., Weber, I., Frangakis, A. S., Nicastro, D., Gerisch, G., and Baumeister, W. (2002). Macromolecular architecture in eukaryotic cells visualized by cryoelectron tomography. *Science* **298,** 1209–1213.

Mucke, N., Kreplak, L., Kirmse, R., Wedig, T., Herrmann, H., Aebi, U., and Langowski, J. (2004a). Assessing the flexibility of intermediate filaments by atomic force microscopy. *J. Mol. Biol.* **335,** 1241–1250.

Mucke, N., Wedig, T., Burer, A., Marekov, L. N., Steinert, P. M., Langowski, J., Aebi, U., and Herrmann, H. (2004b). Molecular and biophysical characterization of assembly-starter units of human vimentin. *J. Mol. Biol.* **340,** 97–114.

Mulligan, L., Balin, B. J., Lee, V. M., and Ip, W. (1991). Antibody labeling of bovine neurofilaments: Implications on the structure of neurofilament sidearms. *J. Struct. Biol.* **106,** 145–160.

O'Shea, E. K., Klemm, J. D., Kim, P. S., and Alber, T. (1991). X-ray structure of the GCN4 leucine zipper, a two-stranded, parallel coiled coil. *Science* **254,** 539–544.

Parry, D. A., Marekov, L. N., and Steinert, P. M. (2001). Subfilamentous protofibril structures in fibrous proteins: Cross-linking evidence for protofibrils in intermediate filaments. *J. Biol. Chem.* **276,** 39253–39258.

Parry, D. A., Marekov, L. N., Steinert, P. M., and Smith, T. A. (2002). A role for the 1A and L1 rod domain segments in head domain organization and function of intermediate filaments: Structural analysis of trichocyte keratin. *J. Struct. Biol.* **137,** 97–108.

Parry, D. A. D., and Steinert, P. M. (1995). *Intermediate Filament Structure.* Springer Verlag, Heidelberg.

Parry, D. A. D., and Steinert, P. M. (1999). Intermediate filaments: Molecular architecture, assembly, dynamics and polymorphism. *Q. Rev. Biophys.* **32,** 99–187.

Phillips, G. N., Jr., Fillers, J. P., and Cohen, C. (1986). Tropomyosin crystal structure and muscle regulation. *J. Mol. Biol.* **192,** 111–131.

Rogers, K. R., Herrmann, H., and Franke, W. W. (1996). Characterization of disulfide crosslink formation of human vimentin at the dimer, tetramer, and intermediate filament levels. *J. Struct. Biol.* **117,** 55–69.

Sayers, Z., Michon, A. M., Sicre, P., and Koch, M. H. (1990). Structure and assembly of calf hoof keratin filaments. *J. Struct. Biol.* **103,** 212–224.

Schwaiger, I., Sattler, C., Hostetter, D. R., and Rief, M. (2002). The myosin coiled-coil is a truly elastic protein structure. *Nat. Mater.* **1,** 232–235.

Steinert, P. M., Marekov, L. N., Fraser, R. D., and Parry, D. A. (1993a). Keratin intermediate filament structure. Crosslinking studies yield quantitative information on molecular dimensions and mechanism of assembly. *J. Mol. Biol.* **230,** 436–452.

Steinert, P. M., Marekov, L. N., and Parry, D. A. (1993b). Diversity of intermediate filament structure. Evidence that the alignment of coiled-coil molecules in vimentin is different from that in keratin intermediate filaments. *J. Biol. Chem.* **268,** 24916–24925.

Strelkov, S. V., Herrmann, H., Geisler, N., Lustig, A., Ivaninskii, S., Zimbelmann, R., Burkhard, P., and Aebi, U. (2001). Divide-and-conquer crystallographic approach towards an atomic structure of intermediate filaments. *J. Mol. Biol.* **306,** 773–781.

Strelkov, S. V., Herrmann, H., Geisler, N., Wedig, T., Zimbelmann, R., Aebi, U., and Burkhard, P. (2002). Conserved segments 1A and 2B of the intermediate filament dimer: Their atomic structures and role in filament assembly. *EMBO J.* **21,** 1255–1266.

Strelkov, S. V., Schumacher, J., Burkhard, P., Aebi, U., and Herrmann, H. (2004). Crystal structure of the human lamin A coil 2B dimer: Implications for the head-to-tail association of nuclear lamins. *J. Mol. Biol.* In press.

Svergun, D. I., and Koch, M. H. (2002). Advances in structure analysis using small-angle scattering in solution. *Curr. Opin. Struct. Biol.* **12,** 654–660.

Svergun, D. I., Petoukhov, M. V., and Koch, M. H. (2001). Determination of domain structure of proteins from X-ray solution scattering. *Biophys. J.* **80,** 2946–2953.

Vachette, P., Koch, M. H., and Svergun, D. I. (2003). Looking behind the beamstop: X-ray solution scattering studies of structure and conformational changes of biological macromolecules. *Methods Enzymol.* **374,** 584–615.

Watts, N. R., Jones, L. N., Cheng, N., Wall, J. S., Parry, D. A., and Steven, A. C. (2002). Cryo-electron microscopy of trichocyte (hard alpha-keratin) intermediate filaments reveals a low-density core. *J. Struct. Biol.* **137,** 109–118.

CHAPTER 3

Intracellular Microrheology as a Tool for the Measurement of the Local Mechanical Properties of Live Cells

Thomas P. Kole,★ Yiider Tseng,★ and Denis Wirtz★,†

★Department of Chemical and Biomolecular Engineering
The Johns Hopkins University
Baltimore, Maryland 21218

†Graduate Program in Molecular Biophysics
The Johns Hopkins University
Baltimore, Maryland 21218

METHODS IN CELL BIOLOGY, VOL. 78
Copyright 2004, Elsevier Inc. All rights reserved.
0091-679X/04 $35.00

═══════════ **I. Introduction**

A. Background

   The cytoskeleton is an extremely dynamic and highly interconnected network of filamentous proteins that extend throughout the cytoplasm and provide cells with a structural framework. The cytoskeleton is composed of three major classes of structural filamentous proteins (microtubules, microfilaments, and intermediate filaments), which are categorized according to their constituent protein subunits and filament diameters. Accessory proteins that bind, bundle, and/or cross-link these proteins in globular or filamentous form control the dynamic mechanical properties and organization of the cytoskeleton and determine the morphology, intracellular architecture, and mechanics of the cell (Coulombe *et al.*, 2000; Heidemann and Wirtz, 2004; Howard, 2001).

   Because of their aspect ratio and internal architecture (Steinmetz *et al.*, 1997), actin filaments behave as semiflexible polymers (Gittes *et al.*, 1993), which facilitate their entanglements and the formation of highly elastic networks even at low polymer concentrations (Morse, 1998). The complex viscoelastic properties of actin filament networks (Hinner *et al.*, 1998; Palmer *et al.*, 1999) are believed to be required to orchestrate the mechanical behavior that shapes cells and provides them with the ability to move and resist external stresses. Much like conventional polymer networks (Ferry, 1980), cross-linking and/or bundling of actin filaments has profound effects on the organization and mechanical properties of the actin cytoskeleton (Borisy and Svitkina, 2000; Pollard *et al.*, 1994). Actin networks containing actin-binding proteins respond to chemical and physical changes in the cytoplasmic environment, including changes in pH, ionic strength, concentration of regulatory proteins, as well as rates of mechanical stress to locally regulate the mechanical properties of the cell (Sato *et al.*, 1985, 1987; Tseng *et al.*, 2004). Like F-actin, one of the primary functions of intermediate filaments (IFs) is to provide mechanical strength to cells (Coulombe *et al.*, 2000). Most of what we know about the physical properties of IFs has been gathered by use of purified proteins. IFs also form semiflexible polymers *in vitro*. This means that the persistence length of IFs ($\sim$1 μm) is on the same order of magnitude as their contour length ($\sim$2.5 μm) (Mucke *et al.*, 2004). Unlike F-actin, IFs have a natural propensity to self-interact and form bundled/cross-linked structures, a property that depends on pH, ionic strength, and pairwise interactions (Yamada *et al.*, 2002). A mechanical signature of this behavior is that the stiffness of IF networks is mostly independent of the rate of deformation (Janmey *et al.*, 1991; Ma *et al.*, 1999; Yamada *et al.*, 2003): IFs in entangled networks cannot diffuse readily and therefore cannot relax mechanical stresses. IFs behave like cross-linked networks in standard assembly conditions *in vitro*.

   In vitro rheological studies that use purified cytoskeletal proteins suggest molecular mechanisms of regulation of cytoskeleton mechanics (Heidemann and Wirtz, 2004). The mechanical behavior of F-actin and IFs and their tight regulation by auxiliary proteins and environmental conditions seem to be closely correlated with key cellular processes that are mechanical in nature, including cell shape changes during mitosis, the separation of daughter cells by the contractile ring during cytokinesis, cell–cell and cell–matrix interactions, transmembrane signaling, endo-cytosis, secretion, and motility (Schmidt and Hall, 1998). However, the mechani-cal behavior of living cells and the molecular signaling pathways by which it is regulated remain largely unknown. In particular, how the unique physical proper-ties of IFs are exploited by cells to shape their plasma membrane, promote cell–cell interactions, and participate to produce the propulsive forces required for wound healing remains largely unknown. This is largely due to the absence of noninvasive methods that measure unambiguously the local mechanical properties of the intracellular milieu. Here we detail a new method to measure cytoskeleton networks *in vitro* and *in vivo*, which can be implemented within an existing light microscope relatively easily.

B. Methods for the Measurement of the Viscoelastic Properties of Living Cells

There have been numerous attempts to quantify the mechanical properties of the cytoskeleton of living cells; however, most current methods cannot readily measure local mechanical parameters (Heidemann and Wirtz, 2004). Light and electron microscopy have been used successfully to gain insight into the spatial distribution of cytoskeleton arrays and their auxiliary proteins in cells (Schoenenberger *et al.*, 1999; Svitkina and Borisy, 1998, 1999). However, micros-copy does not constitute a functional assay as such and does not measure the physical properties of cells such as cytoplasmic viscosity and elasticity. Magnetic tweezers and magnetic microspheres embedded within the cytoplasm of live cells have been used to probe the frequency-dependent viscoelastic moduli of live cells (Bausch *et al.*, 1999; Crick and Hughes, 1950). However, this approach typically requires the use of large magnetic beads (>1 μm), which can distort their local subcellular environment to reach sufficiently high probing forces. Moreover, specific and nonspecific interactions between the (phagocytosed) beads and sub-cellular structures are difficult to assess. Finally, the slow response time of the probe renders the approach inapplicable to examine the temporal response of a cell to extracellular stimuli. By monitoring the diffusion of small, inert tracer particles by means of fluorescence microscopy, Ragsdale and colleagues (1997) probed the micromechanical properties of fibroblasts. However, in its present form, many hypotheses are required to compute the cytoplasmic elasticity, and no global cellular response is obtained. Scanning probe microscopy (SPM), a technique derived from atomic force microscopy (AFM), has been used to quantify the local mechanical properties of stationary and motile fibroblasts (Haga *et al.*, 2000; Nagayama *et al.*, 2001). By measuring local cell properties, AFM acknowledges the regional variations of the cytoskeleton architectures observed under light microscopy. But AFM presents the central problem of measuring cytoplasmic mechanics by probing the cell from the outside. Therefore, mechanical parameters reported by AFM are ill-defined composites of the plasma membrane tension and the viscoelasticity of the underlying cytoskeleton. Moreover, the type of deformation caused by the AFM tip (shear, stretch, or both) is unknown. This is a serious problem, because complex viscoelastic fluids can respond differently when sheared or stretched. Magnetocytometry uses magnetic microspheres coated with extracellular matrix (ECM) components such as fibronectin (FN) or arginine-glycine-aspartic acid (RGD) peptides to apply calibrated torques to the surface of live cells (Cai *et al.*, 1998; Wang *et al.*, 1993). Although local in nature, this approach does not provide local information, because measurements are averaged over many cells in culture and hundreds of probes. Moreover, the ECM-coated beads promote actin polymerization and the formation of actin-rich focal adhesion complexes underneath the plasma membrane. Hence major reorganization of the cytoskeleton is caused by the probing beads, a problem that makes magnetocytometry highly invasive. Furthermore, the surface area of contact between the cell surface and the beads is not controlled. This makes magnetocytometry ambiguous, because an apparent increase or decrease in cytoplasmic stiffness may be due to a change in plasma membrane tension or a change in the avidity/affinity of the cell surface receptors for the ECM-coated beads. Finally, this approach has not been tested with standard fluids; positive controls are missing.

Other methods have been used to assess cell mechanics. These include parallel microplates (Beil *et al.*, 2003; Thoumine and Ott, 1997), micropipette manipulation (Merkel *et al.*, 2000; Paulitschke *et al.*, 1995) and calibrated microneedles (Heidemann *et al.*, 1999; Rahman *et al.*, 2002). All these methods suffer from some of the same drawbacks detailed previously. These methods attempt to measure cytoplasmic mechanics from the cell exterior, they apply ill-defined mechanical deformations, they do not measure frequency-dependent viscoelastic parameters, and they often cannot directly distinguish viscosity from elasticity. Most importantly, all these methods overlook the central fact that the cytoplasm is highly heterogeneous and warrants a local physical probe.

C. Intracellular Microrheology

To address these shortcomings, we have introduced the method of intracellular microrheology (ICM) to extract the local viscoelastic properties of live cells (Tseng *et al.*, 2002). This work extends previous work from our group on the microrheology of reconstituted cytoskeletal filament networks (Apgar *et al.*, 2000; Ma *et al.*, 2001; Palmer *et al.*, 1998; Tseng *et al.*, 2002; Xu *et al.*, 1998); DNA solutions (Goodman *et al.*, 2002; Mason *et al.*, 1997a), engineered protein polymers (Petka *et al.*, 1998; Xu *et al.*, 2002), and live cell microrheology with organelles used as local cytoplasmic probes (Yamada *et al.*, 2000). ICM consists of introducing fluorescent nanoprobes (<0.2 μm) into the cytoplasm of living cell by use of microinjection and then statistically analyzing their thermally excited motion to extract the local mechanical properties of the cytoskeletal network surrounding the particle. Particles embedded in the cytoplasm of living cells are equivalent to nanoscale rheometers that impose a time-averaged constant stress in the surrounding fluid on the order of $k_B T/a^3$ where k_B is the Boltzmann constant, T is the absolute temperature, and a is the particle radius. The resulting deformation is measured as the particle displacement and can be directly related to the viscoelastic properties of the surrounding fluid (Fig. 1).

Intracellular microrheology avoids most of the limitations of current cell-mechanics methods. ICM measures intracellular mechanical properties, both locally and globally. It can readily be combined with fluorescence/bright field microscopy to correlate cytoskeleton architecture and cell organization with local intracellular mechanics in single live cells. ICM measures both elasticity and viscosity and has been tested against standard fluids of known viscoelasticity with traditional mechanical rheometers (Mason *et al.*, 1997b; Xu *et al.*, 1998). It measures frequency-dependent viscoelastic parameters; moreover, it measures viscoelastic parameters of cytoplasm in the linear regime of small deformations (Tseng *et al.*, 2002). These two qualities allow for a direct, comprehensive comparison of live-cell viscoelastic parameters with those displayed by reconstituted cytoskeleton networks *in vitro*. ICM measurements are rapid and can probe the spatiotemporal mechanical response of a cell a stimulus in real time (Kole *et al.*, 2004). Moreover, because ICM probes mechanics from the inside of the cell, for the first time it allows measurement of the mechanics of cells embedded in three-dimensional networks.

Fluorescent nanospheres are prepared by dialysis against injection buffer and then microinjected into the cytoplasm of live cells (Fig. 2A–C). Microinjected cells containing fluorescent particles are placed on the stage of a microscope at 37 °C. Movies of the fluctuating fluorescent nanospheres are then recorded onto the

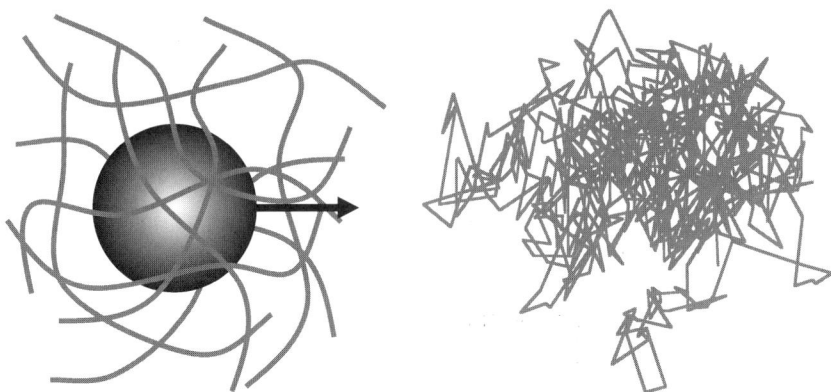

Fig. 1 Brownian motion of a submicron particle entangled in a filamentous protein network. Thermal energy creates a random force on each probe nanosphere of order-of-magnitude $k_B T/a^3$, where a is the radius of the nanosphere. This force creates a local deformation of the viscoelastic medium in the vicinity of the particle, which is observed and measured as the particle's displacement by video-based particle nanotracking.

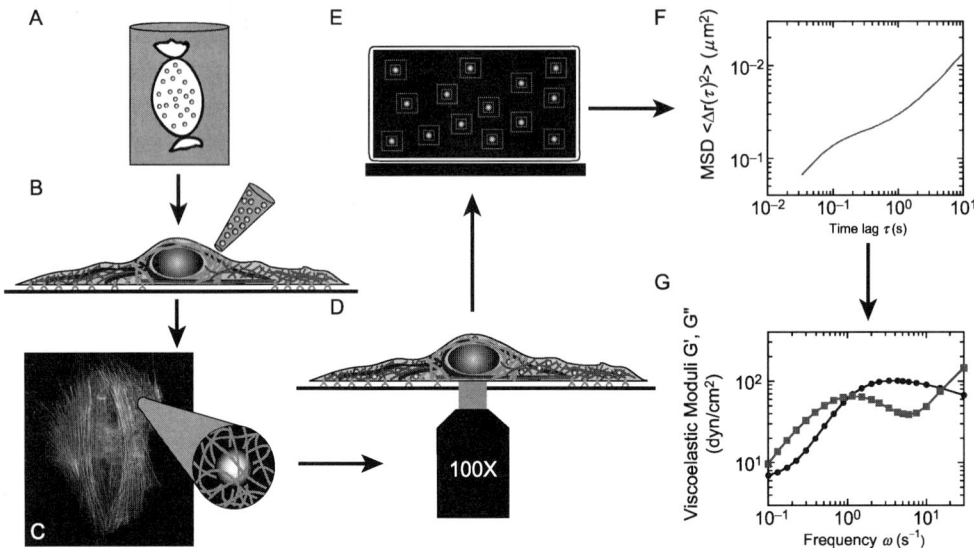

Fig. 2 Schematic of intracellular microrheology (ICM). (A) Particles are first prepared by dialysis against microinjection buffer. (B) Dialyzed particles are then microinjected into the cytoplasm of adherent cells and become embedded in the cytoskeletal network (C). (D) The Brownian displacements of particles injected into the cell are monitored by video-based particle tracking with high-magnification fluorescence microscopy. (E) The time-dependent x and y coordinates of recorded particles are tracked by measurement of intensity-weighted centroid displacements. (F) Mean-squared displacements are calculated for each particle and used to evaluate the frequency-dependent viscoelastic moduli of the cell (G). (See Color Insert.)

random-access memory of a personal computer by a silicon-intensifier target (SIT) camera (VE-100 Dage-MTI, Michigan City, IN) mounted on an inverted epifluorescence microscope (Eclipse TE300, Nikon, Melville, NY) at a frame rate of 30 Hz by use of the Metavue software (Universal Imaging Corp., West Chester, PA). A high magnification 100× Plan Fluor oil-immersion objective (N.A. 1.3) is used for particle tracking, which permits approximately 5-nm spatial resolution over a 120 μm × 120 μm field of view, as assessed by monitoring the apparent displacement of nanospheres firmly attached to a glass coverslip with the same microscope and camera settings as used during live-cell experiments. The displacements $[x(t),y(t)]$, where t is the elapsed time of the particles centroids, are simultaneously monitored in the focal plane of the microscope for 20 seconds (Fig. 2D,E). The same multiple-particle tracking approach has recently been used to track the multistep transport process of gene carriers from the plasma membrane to the nucleus of live cells (Suh *et al.*, 2003) and the micromechanical properties of the interphase nucleus (Tseng *et al.*, 2004).

Movies of fluctuating nanospheres are analyzed by a custom particle tracking routine incorporated into the Metamorph imaging suite (Universal Imaging Corp.) as described (Tseng and Wirtz, 2001). Individual time-averaged mean square displacements (MSDs), $\langle \Delta r^2(\tau) \rangle = \langle [x(t+\tau) - x(t)]^2 + [y(t+\tau) - y(t)]^2 \rangle$, where τ is the time scale, are calculated from the two-dimensional trajectories of the centroids of the nanospheres (Fig. 2F). All control experiments are described in Tseng and Colleagues (2002), including effects of particle size and surface chemistry. For a particle in a perfectly viscous fluid undergoing diffusive motion, the power law slope of the MSD will approach a value of one, whereas a particle embedded in a perfectly elastic solid will yield an MSD with a slope of zero. This implies that a complex viscoelastic fluid will result in power law slopes of the MSD between zero and one. The shear creep compliance can be related to the MSD through the following relationship (Xu et al., 1998):

$$\Gamma(\tau) = \frac{3k_B T}{2\pi a} \langle \Delta r^2(\tau) \rangle \tag{1}$$

The creep compliance is a measure of the deformability of the cell and shares all of the same features as the MSD (Xu et al., 1998).

All the mechanical information is contained in the amplitude and the time scale dependence of the creep compliance. However, by use of the generalized form of the Stokes Einstein relationship (Mason et al., 1997b)

$$\tilde{G}(s) = \frac{2k_B T}{3\pi a s \Delta \langle \tilde{r}^2(s) \rangle} \tag{2}$$

we can approximate the viscoelastic spectrum $\tilde{G}(s)$, where s is the Laplace frequency, from $\langle \Delta \tilde{r}^2(s) \rangle$, the unilateral Laplace transform of $\langle \Delta r^2(\tau) \rangle$. With this expression, we can calculate the traditional frequency-dependent elastic modulus $G'(\omega)$ and loss modulus $G''(\omega)$ from time scale-dependent MSDs as described (Mason, 2000; Mason et al., 1997b) (Fig. 2G). The viscoelastic moduli $G'(\omega)$ and $G''(\omega)$ are the real and imaginary parts, respectively, of the complex modulus $G^*(\omega)$, which is the projection of $\tilde{G}(s)$ in Fourier space, and they obey Kramers–Kronig relationships (Mason et al., 1997b).

In addition, the diffusion coefficient, D, of a nanosphere of radius a can be calculated from the Stokes–Einstein relationship (Berg, 1993; Chandrasekhar, 1943; Einstein, 1905; Qian et al., 1991):

$$D = \frac{k_B T}{6\pi a \eta} \tag{3}$$

where η is the viscosity of the fluid surrounding the particle. In the case of a viscoelastic fluid, such as the cytoplasm of a living cell, η is not a constant and is time scale–dependent, therefore giving rise to a time scale–dependent diffusion coefficient. We can, however, instead approximate η as the shear viscosity η_s, which is the product of the relaxation time (the time scale at which the viscous to elastic crossover occurs, see Fig. 2G) and the plateau value of the elastic modulus (Eckstein *et al.*, 1998). It is important to note that the diffusion coefficient calculated from two-dimensional particle trajectories can be approximated as the three-dimensional diffusion coefficient, assuming that the local environment surrounding each nanosphere is isotropic in three dimensions. This is a valid approximation, even in regions of the cell where long-range interactions between nanospheres and the cell membrane could occur by way of hydrodynamic interactions, because those interactions are screened to within a mesh size of the surrounding network, which is approximately 50 nm. If the cell thickness were similar or smaller than the particle diameter, they would be mostly excluded from those (too thin) areas.

Through the use of ICM, we have found that the mechanical properties of the cell are highly heterogeneous and spatially coordinated (Fig. 3A-C) (Kole *et al.*, 2004; Tseng *et al.*, 2002, 2004). Serum-starved Swiss 3T3 fibroblasts treated with LPA exhibit significant increases in both cytoplasmic elasticity and viscosity. Our results showed that that the mechanical response of Swiss 3T3 cells to Rho activation by lysophosphatidic acid (LPA) is time-dependent and follows the time-dependent profile of Rho activation. This mechanical response is not instantaneous, and the delay between Rho activation and stiffening of the cell may be attributed to the slow gelation kinetics of the actin cytoskeleton. Comparison of the observed micromechanical trends in control cells treated with LPA with fluorescent micrographs of the actin cytoskeleton showed that there was little correlation between F-actin structures and their mechanical function. Although LPA elicited rapid formation of actin stress fibers and global stiffening of the cell, micromechanical relaxation of the cell was not accompanied by disassembly of stress fibers and other organized F-actin structures. We also observed a key distinction between intracellular stiffness and intracellular tension that is generated by myosin-based contraction of the actin cytoskeleton. Treatment of Swiss 3T3 cells with the contractile inhibitor Y-27632 resulted in the increase of intracellular stiffness to levels that are almost twice that of untreated cells.

Recently, using wounded fibroblast monolayers, we have demonstrated that the leading edge of motile cells is much stiffer than the perinuclear region. Particles toward the leading edge exhibit more confined motions than those closer to the nucleus (Fig. 3A,B). Examination of the ensemble-averaged creep compliances of both migrating and quiescent 3T3 fibroblasts revealed that the overall mechanical properties of the cell dramatically changed on the initiation of migration. Cells along the edge of a wounded fibroblast monolayer were significantly less deformable than quiescent fibroblasts and possessed an increased frequency range of elastically dominant mechanical responses to strain. This more solidlike behavior makes migrating cells better able to elastically rebound under mechanical stress and, therefore, potentially less prone to mechanical failure, which may be important *in vivo* where cells are migrating through dense tissue networks and/or against venus/arterial shear flow.

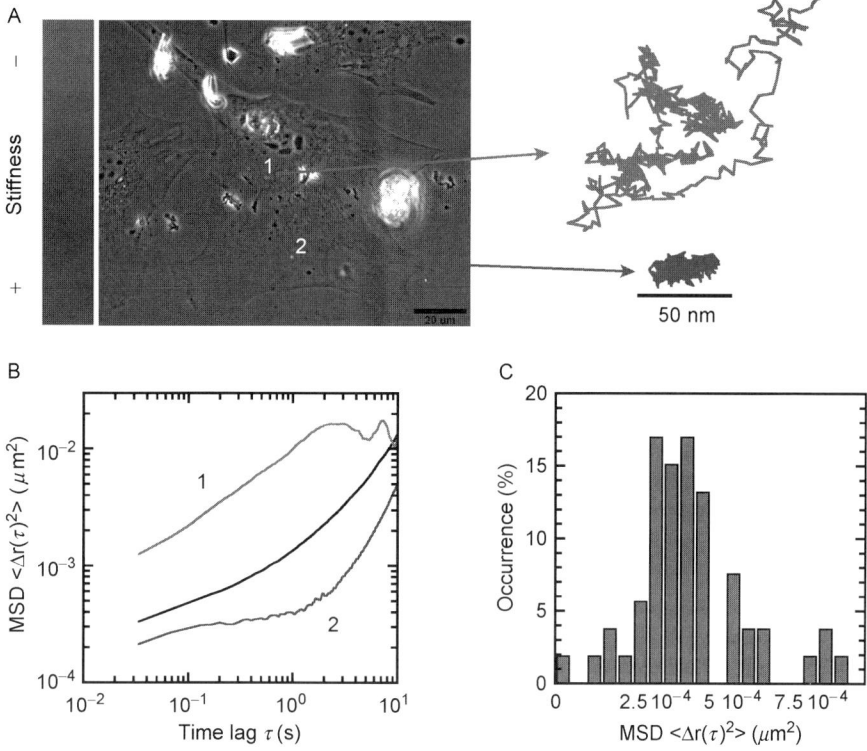

Fig. 3 Application of intracellular microrheology. (A) Local mapping of the mechanical properties of a migrating cell at the edge of a wounded fibroblast monolayer. Here we only show tracked particles and have neglected aggregates that violate assumptions made in the generalized Stokes–Einstein relationship. Each particle position was color coded, corresponding to the local value of the creep compliance at that position in the cell. The color indicators at each particle position do not reflect the size of the particle (100 nm). Indicator size was increased to aid visual presentation. The time-dependent trajectory of a particle close to the nucleus (1) is much more confined than a particle toward the leading edge (2). (B) The mean-squared displacements (MSDs) of particles depicted in (A) together with the ensemble averaged MSD of all particles within the cell. The MSD of the particle close to the nucleus (1) is much larger than a particle toward the leading edge (2), suggesting that the leading edge of migrating cells is much stiffer than the perinuclear region. (C) Distribution of the MSDs of particles embedded within the cytoplasm of the cell shown in (A). The mechanical properties of the cell are highly heterogeneous, thus illustrating the necessity of local measurements within the cell. (See Color Insert.)

II. Materials and Instrumentation

A. Preparation of Probe Nanospheres for Intracellular Microrheology

One hundred nanometer–diameter fluorescent carboxylate modified nanospheres (Cat. No. F8803) from Molecular Probes, Inc. were used as local probes of the mechanical properties of live cells. Particles were prepared by dialysis against

Dulbecco's phosphate-buffered saline (DPBS, Cat. No. 14040-133) from Invitrogen with 300,000 MW Spectra/Por Cellulose Ester dialysis membrane tubing (Cat. No. 131447) from Spectrum Laboratories, Inc. Polyethylene glycol (PEG)–coated particles were prepared with 3.4-kDa amine-terminated PEG (Cat. No. 2V2V0F22) from Shearwater Corp dissolved in 2-[*N*-morpholino]ethanesulfonic acid (MES, Cat. No. M-8250) from Sigma-Aldrich. Sodium hydroxide (NaOH, Cat. No. 3722-05) was purchased from J. T. Baker. 1-ethyl-3-(3-dimethylaminopropyl)-carbodiimide (EDAC, Cat. No. E-2247) was purchased from Molecular Probes, Inc.

B. Cell Culture and Microinjection of Probe Nanospheres

Swiss 3T3 mouse fibroblasts (Cat. No. CCL-92), Dulbecco's modified eagle medium (DMEM, Cat. No. 30-2002), and bovine calf serum (BCS, Cat. No. 30-2030) were from American Type Tissue Culture (ATCC). Fibronectin (Cat. No. 341631) was from Calbiochem. Hank's balanced salt solution (HBSS, Cat. No. 14170-112) and 0.25% trypsin–1 mM ethylenediamine tetraacetic acid (EDTA) (Cat. No. 25200-056) were from Invitrogen; 35-mm Poly-D-lysine–coated glass bottom dishes (Cat. No. P35GC-0-14-C) were from MatTek Corp., and 500-nm inner diameter prepulled glass capillary microneedles (Cat. No. TIP05TW1) were from World Precision Instruments. Microneedle back-loading pipette tips (Cat. No. 930001007), Transjector 5246 microinjection device (Cat. No. 5246000.010), and Micromanipulator 5171 (Cat. No. 5171000.019) were from Brinkmann Instruments; 10X Tris-buffered saline (TBS), pH. 7.4 (Cat. No. 351-086-101), was from Quality Biological Inc. Alexa Fluor 488 10,000-MW fluorescent dextran (Cat. No. D22910) is from Molecular Probes, Inc.

C. Video-Based Live-Cell Intracellular Microrheology

Live cell ICM experiments were conducted on a Nikon TE300 or Nikon TE2000E inverted epifluorescent microscope with a Nikon PlanFluor 100× oil immersion lens (N.A. 1.3). Movies of fluctuating fluorescent nanospheres were recorded onto the random-access memory of a PC computer with a silicon-intensifier target (SIT) camera (VE-100 Dage-MTI, Michigan City, IN) at a frame rate of 30 Hz with the software Metavue (Universal Imaging Corp., West Chester, PA). The time-dependent spatial coordinates of the fluorescent particles were obtained by use of particle tracking routines built into Metamorph Imaging Suite (Universal Imaging Corp.).

III. Procedures

A. Preparation of Probe Nanospheres for Intracellular Microrheology

Carboxylate modified polystyrene nanospheres (100 nm) were purchased as a 2% (w/v) suspension containing 0.1% sodium azide. Fluorescent nanospheres that are to be used for microinjection must first be dialyzed against DPBS injection buffer and adjusted to a suitable particle concentration. Optionally, particles can be passively or covalently surface coated to specifically control particle interactions within the cell. Protein-resistant PEG-coated particles are obtained by covalently coupling amine-terminated PEG to the surface of carboxylate-coated nanospheres by means of the carbodiimide method.

1. Preparation of Carboxylate–Modified Nanospheres

a. Steps

1. Stock nanospheres (1 ml) are pipetted into a 2-cm piece of dialysis tubing sealed at one end and prewet with DPBS.
2. The preceding is dialyzed against 4 L of DPBS with gentle stirring at 4 °C for 12–16 hours. This is repeated three times with fresh DBPS.
3. The particle solution from step 2 is carefully transferred into a sterile 50-ml conical tube.
4. DPBS is added to a final volume of 20 ml and stored at 4 °C.

2. Preparation of Polyethylene Glycol–Coated Nanospheres

a. Solutions

1. 50 mM MES buffer, pH 6.0
2. 1 M NaOH

b. Steps

1. Five milligrams of amine-terminated PEG is dissolved in 1 ml of MES buffer in a glass vial.
2. One milliliter of particle solution is added from step 3 earlier and oral incubated at room temperature for 15 minutes.
3. EDAC (8 mg) is added and mixed thoroughly.
4. The pH of the reaction mixture is adjusted to 6.5 with NaOH and incubated with gentle rocking at room temperature for 2 hours.
5. Glycine (15 mg) is added to quench the reaction and incubated at room temperature for 30 minutes.
6. The reaction mixture is carefully pipetted into a 3-cm piece of dialysis tubing sealed at one end and prewet with DPBS and dialyzed against 4 L of DPBS at 4 °C for 12–16 hours. This is repeated five times with fresh DPBS.
7. The particle solution from step 6 is carefully transferred into a sterile 50-ml conical tube. DPBS is added to a final volume of 20 ml and it is stored at 4 °C.

B. Cell Culture and Microinjection of Probe Nanospheres

Particles are now ready to be microinjected into the cytoplasm of live cells and used as local probes of the intracellular mechanical properties of the cell. Immediately before injection, cells are diluted in injection buffer and centrifuged to remove aggregates that may clog the injection needle. Adherent cells are plated on extracellular matrix (ECM)-coated glass-bottom dishes marked with an x at the center with a carbide tipped pen. Cultures are then serum-starved overnight, microinjected with fluorescent particles dispersed in DPBS, and subsequently used for live cell ICM.

1. Plating of Cells for Microinjection

a. Solutions

1. cDMEM: Dulbecco's modified Eagle's medium containing 10% bovine calf serum.
2. Fifty micrograms per milliliter fibronectin in DPBS.

b. Steps

1. With a carbide-tipped pencil, an x is inscribed on the bottom center of sterile 35-mm poly-D-lysine–coated glass-bottom culture dishes. The dishes are coated with bovine plasma fibronectin by incubating with a 50 μg/ml fibronectin solution in DPBS for 45–60 minutes at room temperature.
2. Medium from a T-75 tissue culture flask containing 70%–80% confluent layer of Swiss 3T3 fibroblasts or another type of adherent cell is aspirated.
3. The mixture is washed with 10 ml of HBSS.
4. Trypsin-EDTA (2 ml) is added and the flask is carefully rotated to cover the entire bottom surface of the flask. This is incubated at 37 °C for 5 minutes or until cells are visually detached from the surface of the flask.
5. cDMEM (8 ml) is added and pipetted up and down to mix thoroughly.
6. The cell suspension is transferred to a 15-ml conical centrifuge tube and pellet cells at 1100 rpm for 5 minutes. The supernatant is aspirated and the pellet is carefully resuspended in 10 ml of cDMEM.
7. The cell density is measured, and 1×10^4 cells/ml suspension in cDMEM is prepared.
8. The fibronectin solution is gently aspirated from the 35-mm glass-bottom dishes and washed once with HBSS. Two milliliters of suspension is added from step 7 to each dish. Cells are allowed to adhere overnight at 37 °C and 5% CO_2.

2. Microinjection of Fluorescent Nanospheres

a. Solutions

1. Particle solution prepared in step 3 (preparation of carboxylate-modified nanospheres) or step 7 (preparation of polyethylene glycol–coated nanospheres).
2. $1\times$ TBS.

b. Steps

1. Previously prepared 35-mm culture dishes are serum starved by washing three times with HBSS and replacing complete media (cDMEM) with DMEM. Cells are allowed to incubate for 24 hours.
2. Twenty microliters of solution 1 are pipetted into 980 μl of DPBS in a sterile 1.5-ml centrifuge tube and spun at $16,000g$ for 15 minutes. Optionally, particle solution can be briefly sonicated for 5 minutes before centrifugation.
3. The top 100 μl of the centrifuged particle solution is carefully pipetted off and transferred to a new sterile 1.5-ml centrifuge tube. This is the microinjection suspension containing approximately 3.6×10^{10} particles/ml. Optionally, fluorescently labeled dextran can be added to the microinjection suspension to final concentration of 2 mg/ml to facilitate visualization of microinjected cells.
4. Previously serum-starved cells are placed on the stage of a microscope equipped with a 37 °C and 5% CO_2 incubator and a suitable microinjection system. Four microliters (4 μl) of microinjection suspension are backloaded from step (3) into a microinjection needle and immediately mounted onto micromanipulator; the tip is immersed into the media of the 35-mm culture dish.
5. The center of the dish marked by an x is located at $10\times$ magnification and then the tip of the microinjection needle is centered over the x. A higher power objective ($60\times$) is used, and the tip of the microneedle is positioned over the perinuclear region of a cell near the x on the bottom of the coverslip.
6. The microneedle is lowered until the tip begins to deform the cell, and this height is set as the injection plane. The microneedle is raised 5–10 μm above the cell and the cell is injected. This procedure is repeated for every cell within a 1-mm radius of the x.
7. The cells are immediately washed three times with TBS and the media are replaced with DMEM. Cells are allowed to incubate for 12–24 hours to facilitate spreading of particles throughout the cytoplasm.

C. Video–Based Intracellular Microrheology

After a 12–24-hour incubation period, microinjected cells are ready to be used in live cell ICM experiments. Experiments are performed with high-magnification objectives in a climate-controlled environment. After initial particle tracking, the procedure can be repeated for more cells, the culture can be fixed for subsequent labeling with fluorescent antibodies, or the culture may be stimulated with an exogenous stimulus and then remeasured to examine the intracellular mechanical effects of the applied stimulus. Movies of fluctuating nanospheres are analyzed by a custom particle tracking routine incorporated into the Metamorph imaging suite (Universal Imaging Corp.) as described (Tseng and Wirtz, 2001). Individual time-averaged MSDs, $\langle \Delta r^2(\tau) \rangle = \langle [x(t+\tau) - x(t)]^2 + [y(t+\tau) - y(t)]^2 \rangle$, where τ is the time scale, are calculated from the two-dimensional trajectories of the centroids of the nanopheres.

1. Acquisition of Moving Particle Video

a. Steps

1. The previously microinjected cell culture is placed on the stage of an epifluorescent microscope surrounded by an air-curtain incubator maintained at $37\,^{\circ}C$ and 5% CO_2.

2. The center of the dish is located and marked by an x at low ($10\times$) magnification and then switched to a high N.A. objective. With a combination of fluorescence and bright field illumination, the microinjected cells are identified within a 1-mm radius of the x.

3. One cell is chosen and with fluorescence microscopy, 20–100 seconds of streaming video are obtained through a SIT camera controlled by Metavue acquisition software. High-resolution bright-field and fluorescence still images of the cell are immediately captured with a CCD camera also controlled by Metavue acquisition software. The movie of the fluctuating particles is saved as an STK file and the still images as TIFF files.

2. Analysis of Movies of Fluctuating Particles

a. Steps

1. An STK file of fluctuating nanospheres is opened in the Metamorph image analysis software. Pixel distances are calibrated with calibration files previously made with a stage micrometer.

2. By use of the track objects command, regions around each particle are created. Regions around aggregated particles are not created. These particles violate assumptions made in our constitutive viscoelastic equations and therefore cannot be used in our analysis.

3. For each particle, the inner region is adjusted so that it extends just beyond the edge of the particle. Similarly, the outer region is adjusted so that it

encompasses an area large enough so that in the subsequent frame no part of the particle will be outside this area. An image of the initial frame with labeled regions visible is duplicated and saved as a TIFF file.

4. The particles are tracked, and the particle number, frame number, and the time-dependent coordinates, $[x(t),y(t)]$, are logged for each frame. These data are saved as an Excel spreadsheet.

3. Calculation of Mean Squared Displacement, Creep Compliance, and Viscoelastic Moduli

From the time-dependent coordinates, $[x(t),y(t)]$, the MSD for each particle is calculated with the following formulas (Qian et al., 1991):

$$MSD_X(\tau) = \sum_{i=1}^{N}(x(t_i + \tau) - x(t_i))^2/(N+1)$$

$$MSD_Y(\tau) = \sum_{i=1}^{N}(y(t_i + \tau) - y(t_i))^2/(N+1) \tag{4}$$

and

$$MSD(\tau) = MSD_X(\tau) + MSD_Y(\tau) \tag{5}$$

The creep compliance is then calculated using Eq. (1) and the results from Eq. (5). Individual particle MSDs and/or creep compliances can be ensemble averaged to obtain a "bulk" mechanical measurement and then used to calculate viscoelastic moduli.

To obtain classical frequency-dependent viscoelastic moduli as measured by conventional mechanical rheometers, the complex shear modulus $\tilde{G}(s)$ from Eq. (2) must first be transformed into the time domain and then Fourier transformed into frequency space. However, because data for $\langle \Delta r^2(\tau) \rangle$ are only measured over a limited time range at discrete times, this method can introduce significant numerical errors at extreme frequencies and requires numerically intensive computations. Alternatively, Eq. (2) can be continued into the Fourier domain by a simple substitution of $s = i\,\omega$ to obtain:

$$G^*(\omega) = \frac{k_B T}{\pi a i\omega \Im_u\{\langle \Delta r^2(\tau) \rangle\}} \tag{6}$$

where $\Im_u\{\langle \Delta r^2(t) \rangle\}$ is the Fourier transform of the time-dependent MSD $\Im_u\{\langle \Delta r^2(t) \rangle\}$ can be estimated algebraically using a wedge assumption (Mason, 2000; Mason et al., 1997b):

$$i\omega \Im_u\{\langle \Delta r^2(\tau) \rangle\} \approx \langle \Delta r^2(1/\omega) \rangle \Gamma[1 + \alpha(\omega)]i^{-\alpha(\omega)} \tag{7}$$

where $\alpha(\omega) = d\ln\langle \Delta r^2(\tau) \rangle/d\ln\tau|_{\tau=1/\omega}$ is the local logarithmic slope of $\langle \Delta r^2(\tau) \rangle$ at the frequency of interest $\omega = 1/\tau$. The frequency-dependent elastic and viscous moduli, G' and G'', respectively, can then be calculated algebraically by use of the following relationships:

$$G'(\omega) = |G^*(\omega)|\cos(\pi\alpha(\omega)/2) \tag{8}$$

$$G''(\omega) = |G^*(\omega)|\sin(\pi\alpha(\omega)/2) \tag{9}$$

where

$$|G^*(\omega)| = \frac{2k_{\mathrm{B}}T}{3\pi a\langle\Delta r^2(1/\omega)\rangle\Gamma(1+\alpha(\omega))} \tag{10}$$

IV. Pearls and Pitfalls

The data generated from ICM are critically dependent on the ability to accurately measure the spatial fluctuations of Brownian particles; therefore, it is critical to maintain a precisely aligned microscope. Misalignments may cause particle displacements along the optical axis to be mistaken as lateral displacements, which will greatly diminish the spatial resolution of particle displacements. It is also important to note that each instrument will have its own resolution, which depends not only on the quality of the optics (lens, N.A., and magnification) and quality of microscope alignment, but also on the mechanical stability of the microscope. The very small displacements such as those measured with microinjected particles in crowded areas of the cytoplasm require the highest possible resolution. Therefore, placing the microscope on an air-floated optical table is highly recommended.

The resolution of the ICM optical setup can be obtained by measuring the root mean square diameter (r.m.s.d) of particles firmly attached to the surface of a glass coverslip. In our system, the r.m.s.d. of a fixed 0.1-μm particle is approximately 5 nm. This implies that the upper limit for elasticity measurements with our system is approximately 1.1 kPa. The derivation of Eq. (2) neglects an inertial term that becomes significant only at high (1 MHz) frequencies, so it is important to avoid frame acquisition rates higher than 10^6 frames/sec. However, this is currently outside the range available to video microscopy.

Because of the relatively similar sizes of particles microinjected into the cell and the inner diameter of microinjection needles, needles can easily become clogged. It is therefore extremely important to apply a slight compensation pressure to the microinjection needle so that there is always a constant flow out of the needle. In the event of a clogged needle, apply a large cleansing pressure to try and pass the clog, or gently chip the tip of the needle on the coverslip until particles begin to flow again.

It is extremely important to wash cells immediately after microinjection to remove any particles that may be floating in the media, which on contact with a cell may be endocytosed. Endocytosed particles spend most of their lifetime within the cell under the influence of directed motor proteins. The nonrandom component of this motion cannot be accounted for by the equations of motion derived here, and, therefore, endocytosed particles may not be used for the measurement of the local viscoelastic properties of the cell by use of ICM. Particles moving under the influence of nonrandom forces will yield MSDs with power-law slopes greater than unity.

Vertical drifting of the culture dish on the microscope stage during particle tracking can occur from heating the oil in between the coverslip and the objective. To prevent this, we suggest placing a weight on top of the dish. Similarly, inverted microscope objective assemblies have a tendency to sink over time. Even in the course of a 20-second experiment, vertical drifting of the objective assembly may be significant. Therefore, it is critical to tighten the focus control to its maximum setting or to implement a laser-monitored control system to maintain a constant objective position.

Our particle tracking technique measures the two-dimensional displacements of an essentially three-dimensional motion. But this has little consequence on the evaluation of the MSD and cell mechanics parameters. Here, we probe the passive motion of inert particles in a locally isotropic medium. Therefore, the MSD in two dimensions, Eq. (5), is simply two thirds of the MSD in three dimensions, because the projections of the random movements along the x-, y-, and z-axes are uncorrelated. This would not be true if the purpose were to track directed motion.

V. Concluding Remarks

Intracellular microrheology is a powerful technique that allows us for the first time to probe the mechanical properties of the cytoskeleton in its natural environment. ICM offers numerous advantages over other techniques that have been developed to quantify the mechanical properties of living cells (Gittes *et al.*, 1997; Mason *et al.*, 1997b; Yamada *et al.*, 2000). ICM is able to provide simultaneous local measurements as well as a global picture of the local stiffness of a single cell (Fig. 3), a feat that cannot be achieved by any other method. Furthermore, ICM measures frequency-dependent mechanical properties and, as such, has revealed the highly dynamic nature of cytoskeletal filament organization and cross-linking.

Through the use of ICM, we have shown that the mechanical properties of the cell are highly heterogeneous and spatially coordinated (Kole *et al.*, 2004; Tseng *et al.*, 2002, 2004). Moreover, we have identified key molecules and molecular mechanisms that are involved in the regulation of the mechanical properties of the cell (Kole *et al.*, 2004). Combined with existing methods such as transmission electron microscopy (TEM), ICM will be used to provide greater insight into the correlation of cytoskeleton structure with mechanical function. It has been widely suggested that the mechanical properties of cytoskeleton filaments play a critical role in many cellular processes (Heidemann and Wirtz, 2004). We are now in a position to test these hypotheses with careful studies that use a multidisciplinary approach combining ICM, molecular biology, and cell biology.

Acknowledgments

This work was supported by a National Institutes of Health and National Aeronautics and Space Administration grant (NAG9-1563, D. Wirtz and Y. Tseng) and a National Science Foundation grant (NES/NIRT CTS0210718, D. Wirtz). T. P. Kole was supported by a National Aeronautics and Space Administration training grant (NGT965).

References

Apgar, J., Tseng, Y., Federov, E., Herwig, M. B., Almo, S. C., and Wirtz, D. (2000). Multiple-particle tracking measurements of heterogeneities in solutions of actin filaments and actin bundles. *Biophys. J.* **79**, 1095–1106.

Bausch, A. R., Môller, W., and Sackmann, E. (1999). Measurement of local viscoelasticity and forces in living cells by magnetic tweezers. *Biophys. J.* **76**, 573–579.

Beil, M., Micoulet, A., von Wichert, G., Paschke, S., Walther, P., Omary, M. B., Van Veldhoven, P. P., Gern, U., Wolff-Hieber, E., Eggermann, J., Waltenberger, J., Adler, G., Spatz, J., and Seufferlein, T. (2003). Sphingosylphosphorylcholine regulates keratin network architecture and visco-elastic properties of human cancer cells. *Nat. Cell. Biol.* **5**, 803–811.

Berg, H. C. (1993). "Random Walks in Biology." Princeton University Press, Princeton, N.J..

Borisy, G. G., and Svitkina, T. M. (2000). Actin machinery: Pushing the envelope. *Cur. Opin. Cell Biol.* **12**, 104–112.

Cai, S., Pestic-Dragovich, L., O'Donnell, M. E., Wang, N., Ingber, D., Elson, E., and De Lanerolle, R. (1998). Regulation of cytoskeletal mechanics and cell growth by myosin light chain phosphorylation. *Am. J. Physiol.* **275**, C1349–C1356.

Chandrasekhar, S. (1943). Stochastic problems in physics and astronomy. *Rev. Mod. Phys.* **15**, 1–89.

Coulombe, P. A., Bousquet, O., Ma, L., Yamada, S., and Wirtz, D. (2000). The 'ins' and 'outs' of intermediate filament organization. *Trends Cell. Biol.* **10**, 420–428.

Crick, F. H. C., and Hughes, A. F. W. (1950). The physical properties of cytoplasm: A study by means of the magnetic particle method part I. *Exp. Cell Res.* **1**, 37–80.

Eckstein, A., Suhm, J., Friedrich, C., Maier, R. D., Sassmannshausen, J., Bochmann, M., and Mulhaupt, R. (1998). Determination of plateau moduli and entanglement molecular weights of isotactic, syndiotactic, and atactic polypropylenes synthesized with metallocene catalysts. *Macromolecules* **31**, 1335–1340.

Einstein, A. (1905). Über die von der molekularkinetischen Theorie der Wärme geforderte Bewegung von in ruhenden Flüssigkeiten suspendierten Teilchen. *Ann. Physik.* **17**, 549.

Ferry, J. D. (1980). "Viscoelastic Properties of Polymers." John Wiley & Sons, New York.

Gittes, F., Mickey, B., Nettleton, J., and Howard, J. (1993). Flexural rigidity of microtubules and Actin filaments measured from thermal fluctuations in shape. *J. Cell Biol.* **120**, 923–934.

Gittes, F., Schnurr, B., Olmsted, P. D., MacKintosh, F. C., and Schmidt, C. F. (1997). Microscopic viscoelasticity: Shear moduli of soft materials determined from thermal fluctuations. *Phys. Rev. Lett.* **79**, 3286–3289.

Goodman, A., Tseng, Y., and Wirtz, D. (2002). Effect of length, topology, and concentration on the microviscosity and microheterogeneity of DNA solutions. *J. Mol. Biol.* **323**, 199–215.

Haga, H., Nagayama, M., Kawabata, K., Ito, E., Ushiki, T., and Sambongi, T. (2000). Time-lapse viscoelastic imaging of living fibroblasts using force modulation mode in AFM. *J. Electron. Microsc. (Tokyo)* **49**, 473–481.

Heidemann, S. R., Kaech, S., Buxbaum, R. E., and Matus, A. (1999). Direct observations of the mechanical behaviors of the cytoskeleton in living fibroblasts. *J. Cell Biol.* **145**, 109–122.

Heidemann, S. R., and Wirtz, D. (2004). Towards a regional approach to cell mechanics. *Trends Cell Biol.* **14,** 160–166.

Hinner, B., Tempel, M., Sackmann, E., Kroy, K., and Frey, E. (1998). Entanglement, elasticity, and viscous relaxation of actin solutions. *Phys. Rev. Lett.* **81,** 2614–2617.

Howard, J. (2001). "Mechanics of Motor Proteins and the Cytoskeleton." Sinauer Associates, Sunderland, Mass.

Janmey, P. A., Euteneuer, U., Traub, P., and Schliwa, M. (1991). Viscoelastic properties of vimentin compared with other filamentous biopolymer networks. *J. Cell Biol.* **113,** 155–160.

Kole, T. P., Tseng, Y., Huang, L., Katz, J. L., and Wirtz, D. (2004). Rho kinase regulates the intracellular micromechanical response of adherent cells to Rho activation. *Mol. Cell Biol.* **15,** 3475–3484.

Ma, L., Xu, J., Coulombe, P. A., and Wirtz, D. (1999). Epidermal keratin suspensions have unique micromechanical properties. *J. Biol. Chem.* **274,** 19145–19151.

Ma, L., Yamada, S., Wirtz, D., and Coulombe, P. A. (2001). A hot-spot mutation alters the mechanical properties of keratin filament networks. *Nat. Cell Biol.* **3,** 503–506.

Mason, T. G. (2000). Estimating the viscoelastic moduli of complex fluids using the generalized Stokes-Einstein equation. *Rheologica Acta.* **39,** 371–378.

Mason, T. G., Dhople, A., and Wirtz, D. (1997a). Concentrated DNA rheology and microrheology. *In* "Statistical Mechanics in Physics and Biology," (D. Wirtz and T. C. Halsey, eds.), Vol. 463, pp. 153–158. Warrendale, Penn.

Mason, T. G., Ganesan, K., van Zanten, J. V., Wirtz, D., and Kuo, S. C. (1997b). Particle-tracking microrheology of complex fluids. *Phys. Rev. Lett.* **79,** 3282–3285.

Merkel, R., Simson, R., Simson, D. A., Hohenadl, M., Boulbitch, A., Wallraff, E., and Sackmann, E. (2000). A micromechanic study of cell polarity and plasma membrane cell body coupling in Dictyostelium. *Biophys. J.* **79,** 707–719.

Morse, D. C. (1998). Viscoelasticity of concentrated isotropic solutions of semiflexible polymers. 2. Linear response. *Macromolecules* **31,** 7044–7067.

Mucke, N., Kreplak, L., Kirmse, R., Wedig, T., Herrmann, H., Aebi, U., and Langowski, J. (2004). Assessing the flexibility of intermediate filaments by atomic force microscopy. *J. Mol. Biol.* **335,** 1241–1250.

Nagayama, M., Haga, H., and Kawabata, K. (2001). Drastic change of local stiffness distribution correlating to cell migration in living fibroblasts. *Cell Motil. Cytoskeleton* **50,** 173–179.

Palmer, A., Xu, J., Kuo, S. C., and Wirtz, D. (1999). Diffusing wave spectroscopy microrheology of actin filament networks. *Biophys. J.* **76,** 1063–1071.

Palmer, A., Xu, J., and Wirtz, D. (1998). High-frequency rheology of crosslinked actin networks measured by diffusing wave spectroscopy. *Rheologica Acta.* **37,** 97–108.

Paulitschke, M., Nash, G. B., Anstee, D. J., Tanner, M. J., and Gratzer, W. B. (1995). Perturbation of red blood cell membrane rigidity by extracellular ligands. *Blood* **86,** 342–348.

Petka, W. A., Harden, J., McGrath, K. P., Wirtz, D., and Tirrell, D. A. (1998). Reversible hydrogels from self-assembling artificial proteins. *Science* **281,** 389–393.

Pollard, T. D., Almo, S., Quirk, S., Vinson, V., and Lattman, E. E. (1994). Structure of actin binding proteins: Insights about function at atomic resolution. *Annu. Rev. Cell Biol.* **10,** 207–249.

Qian, H., Sheetz, M. P., and Elson, E. L. (1991). Single particle tracking. Analysis of diffusion and flow in two-dimensional systems. *Biophys. J.* **60,** 910–921.

Ragsdale, G. K., Phelps, J., and Luby-Phelps, K. (1997). Viscoelastic response of fibroblasts to tension transmitted through adherens junctions. *Biophys. J.* **73,** 2798–2808.

Rahman, A., Tseng, Y., and Wirtz, D. (2002). Micromechanical coupling between cell surface receptors and RGD peptides. *Biochem. Biophys. Res. Commun.* **296,** 771–778.

Sato, M., Leimbach, G., Schwarz, W. H., and Pollard, T. D. (1985). Mechanical properties of actin. *J. Biol. Chem.* **260,** 8585–8592.

Sato, M., Schwarz, W. H., and Pollard, T. D. (1987). Dependence of the mechanical properties of actin/α-actinin gels on deformation rate. *Nature* **325,** 828–830.

Schmidt, A., and Hall, M. N. (1998). Signaling to the actin cytoskeleton. *Annu. Rev. Cell Dev. Biol.* **14,** 305–338.

Schoenenberger, C. A., Steinmetz, M. O., Stoffler, D., Mandinova, A., and Aebi, U. (1999). Structure, assembly, and dynamics of actin filaments *in situ* and *in vitro*. *Microsc. Res. Tech.* **47,** 38–50.

Steinmetz, M. O., Stoffler, D., Hoenger, A., Bremer, A., and Aebi, U. (1997). Actin: From cell biology to atomic detail. *J. Struct. Biol.* **119,** 295–320.

Suh, J., Wirtz, D., and Hanes, J. (2003). Efficient active transport of gene nanocarriers to the cell nucleus. *Proc. Natl. Acad. Sci. U.S.A* **100,** 3878–3882.

Svitkina, T. M., and Borisy, G. G. (1998). Correlative light and electron microscopy of the cytoskeleton of cultured cells. *Methods Enzymol.* **298,** 570–592.

Svitkina, T. M., and Borisy, G. G. (1999). Arp2/3 complex and actin depolymerizing factor cofilin in dendritic organization and treadmilling of actin filament array in lamellipodia. *J. Cell Biol.* **145,** 1009–1026.

Thoumine, O., and Ott, A. (1997). Time scale dependent viscoelastic and contractile regimes in fibroblasts probed by microplate manipulation. *J. Cell Sci.* **110,** 2109–2116.

Tseng, Y., Kole, T. P., and Wirtz, D. (2002). Micromechanical mapping of live cells by multiple-particle-tracking microrheology. *Biophys. J.* **83,** 3162–3176.

Tseng, Y., Lee, J. S., Kole, T. P., Jiang, I., and Wirtz, D. (2004). Micro-organization and visco-elasticity of the interphase nucleus revealed by particle nanotracking. *J. Cell Sci.* **117,** 2159–2167.

Tseng, Y., and Wirtz, D. (2001). Mechanics and multiple-particle tracking microhetrogeneity of alpha-actinin-cross-linked actin filament networks. *Biophys. J.* **81,** 1643–1656.

Wang, N., Butler, J. P., and Ingber, D. E. (1993). Mechanotransduction across the cell surface and through the cytoskeleton. *Science* **260,** 1124–1127.

Xu, J., Tseng, Y., Carriere, C. J., and Wirtz, D. (2002). Microrheology and microheterogeneity of wheat gliadin suspensions studied by multiple particle tracking. *Biomacromolecules* **3,** 92–99.

Xu, J., Viasnoff, V., and Wirtz, D. (1998). Compliance of actin filament networks measured by particle-tracking microrheology and diffusing wave spectroscopy. *Rheologica Acta.* **37,** 387–398.

Yamada, S., Wirtz, D., and Coulombe, P. A. (2002). Pairwise assembly determines the intrinsic potential for self-organization and mechanical properties of keratin filaments. *Mol. Biol. Cell.* **13,** 382–391.

Yamada, S., Wirtz, D., and Coulombe, P. A. (2003). The mechanical properties of simple epithelial keratins 8 and 18: Discriminating between interfacial and bulk elasticities. *J. Struct. Biol.* **143,** 45–55.

Yamada, S., Wirtz, D., and Kuo, S. C. (2000). Mechanics of living cells measured by laser tracking microrheology. *Biophys. J.* **78,** 1736–1747.

CHAPTER 4

Developing Mouse Models to Study Intermediate Filament Function

Thomas M. Magin, Michael Hesse, Roland Meier-Bornheim, and Julia Reichelt

Institut für Physiologische Chemie
Abteilung für Zellbiochemie
Bonner Forum Biomedizin and LIMES
Universitätsklinikum Bonn
53115 Bonn, Germany

I. Introduction

Since the first description of the keratin (K) 8 null mouse in 1993 by Oshima and colleagues (Baribault *et al.*, 1993), more than 20 of the 70 intermediate filament (IF) genes have been altered by gene targeting to gain insights into the function of the corresponding proteins. Most of the targeted alterations represent null mutations, but a few have introduced either point mutations or deletions. If one accepts an oversimplification, the resulting phenotypes fall into two categories: severe cell/tissue fragility syndromes and very mild alterations that require experimental manipulation to become manifest (Coulombe and Omary, 2002; Herrmann *et al.*, 2003). The molecular basis for the latter observation seems to be that other IF proteins are able to compensate for the one deleted. This is most convincingly demonstrated by producing double-deficient mouse lines (Hesse *et al.*, 2000; Tamai *et al.*, 2000).

In addition to gene targeting experiments, conventional transgenic approaches have in some cases provided exciting and unexpected information on protein function. Among the well-established examples are the mutant K14 transgenic mouse, which revealed the causal role of keratins in epidermolysis bullosa simplex and the GFAP overexpressing mouse, which linked GFAP to the fatal Alexander's disease (Brenner *et al.*, 2001; Coulombe *et al.*, 1991; Messing *et al.*, 1998). Both have revealed that gain of toxic function mutations can be very informative because of their dominant nature. Also, mice carrying point mutations in known phosphorylation sites (e.g., for K8 and K18) have been very useful in revealing the role of protein modification (Omary *et al.*, 2002).

One major challenge to produce mice with genetically altered IF genes comes from the range of mutations in IF genes leading to disease in humans. In addition to mutations in at least 15 keratin genes, mutations in several other IF genes, including desmin, GFAP, NF-L, and lamin A, lead to disease (Coulombe and Omary, 2002). For most, if not all, the same mutation can give rise to a different phenotype, reflecting the significance of genetic background and epigenetic variation. Genetically altered mice should bring us closer to understanding the molecular basis of this variation and hence to an understanding of IF function. At present, rather little is known about the binding sites of IF-associated proteins on IF subunits; therefore, it is difficult to predict the outcome of a given mutation. The complexity of this problem is highlighted by the lamin A gene. In this ubiquitously expressed gene, more than 120 different point mutations lead to at least nine human disorders with distinct phenotypes, ranging from muscle degeneration to progeria (Mounkes *et al.*, 2003). Although the binding sites for some lamin-associated proteins are known, it is not yet known why mutations cause apparently different disorders.

To meet the challenge to build appropriate mouse models of IF disease, to understand IF gene regulation, protein function, and, not the least, to develop therapy approaches for IF diseases, conventional gene targeting and transgenic approaches require modifications. Indeed, some recent developments should add to the success of generating genetically altered mice: (1) On the basis of the available mouse genome sequence, vector design and construction can be based on polymerase chain reaction (PCR) instead of library screening. (2) For large-scale genomic alterations, a comprehensive genomic library of loxP sequences is available. (3) Also, a range of ES cell lines with good germline transmission potential, including outbred lines with added capabilities, is available. Their culture conditions (which in the past required extensive testing of sera, growth factors, and personal experience) have recently been much better defined. (4) For conditional transgenesis, a considerable number of mice expressing Cre-recombinase under the control of constitutive or regulatable promoters are accessible. (5) Finally, chimeric embryos to study IF function in distinct lineages during embryonic development has become a reliable technique.

II. Essential Materials and Methods

A. Design of Targeting Vectors

Targeted gene manipulation can be accomplished by the use of replacement or insertion vectors (Melton, 1994). Here we focus on replacement vectors and describe approaches to modify a gene or gene fragment constitutively or conditionally. Then, we describe how large-scale genomic deletions are carried out. All gene targeting vectors should be constructed from isogenic DNA (i.e., from the same mouse strain as the ES cell line to be used). Deviation from this rule will decrease targeting frequencies 10–1000-fold.

1. Constitutive and Conditional Alterations

Most IF genes have a gene structure well suited for gene targeting, because they span only around 10 kb of genomic DNA. If the generation of a null allele or the consecutive introduction of several independent mutations is intended, it is desirable to delete the coding sequences completely from the genome. The efficiency of gene targeting remains high as long as no more than 20 kb of genomic DNA is replaced. Other options to create a functional null allele include deletion of the core promoter and the first coding exon. In this case, alternative start codons should be considered, which can be removed by site-directed mutagenesis.

Replacement-type targeting vectors carry two segments of sequences homologous to the gene of interest (Fig. 1). The short arm should be ~1.5 kb long to allow for easy PCR detection, and the long arm should be 4–6 kb. One PCR primer resides in the selectable marker gene, the other one outside the region of homology. Vector arms can either be isolated from genomic libraries or, more conveniently, using long-range PCR from genomic DNA preparations or BAC clones carrying the gene of interest.

In our hands, this has resulted in targeting frequencies between 1:5 and 1:35 positive ES cell clones (Magin et al., 1998; Peters et al., 2001; Plum et al., 2000; Porter et al., 1996; Reichelt et al., 2001). Vector arms flank the part of the gene to be manipulated and carry between them a positive selectable marker gene. On homologous recombination between vector arms and the corresponding genomic DNA, the selectable marker gene replaces the genomic sequences and creates a modified gene. Available selectable marker genes code for resistance against neomycin, hygromycin, puromycin, or hypoxanthine phosphoribosyl transferase (HPRT). In the latter case, HPRT-deficient ES cells are required. Only ES cells that harbor an HPRT-minigene can grow in hypoxanthine, aminopterin, thymidine (HAT)-containing selective medium. The major advantage of the HPRT system lies in the fact that it allows for negative selection in medium containing 6-thioguanine (Melton, 1994). In our experience, the use of negative selectable marker genes outside the homology arms offers no advantage.

In most cases, selectable marker genes with their own promoters, usually PGK-1, are used. In rare instances, the PGK-1 promoter becomes silenced as it carries a

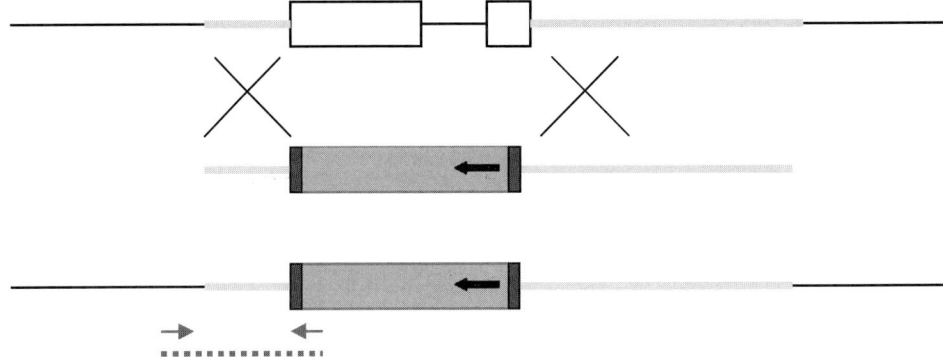

Fig. 1 Conventional gene targeting experiment for creation of null allele. Top, Genomic locus with two exemplary exons (open boxes). Sequences homologous to replacement-type gene targeting vector are in green. Middle, Targeting vector with short and long arms of homology, flanking a selectable marker gene (cross-striped) that carries FRT recognition sites (black) for Flpe-mediated deletion. The arrow indicates the marker gene promoter and its transcription direction. X indicates regions of crossing-over. Bottom, Targeted locus at which the marker gene has replaced two exons. Arrows, Primers used for PCR detection of homologous recombinants. Dotted line; predicted PCR product.

partial pCpG island and may have to be replaced by other promoters (Melton *et al.*, 1997). To minimize transcriptional interference between the gene's own and the minigene promoter, we recommend placing the selectable marker gene opposite the target gene's transcriptional orientation. To minimize potential effects of the selectable marker gene on neighboring genes and on the phenotype of mice, it is desirable to remove it. This can be done by placing loxP or FRT recognition sequences at both ends of the selectable marker cassette. After transient expression of Cre or FLP-recombinase in ES cells, the marker gene is removed, and ES cells become sensitive to the selective agent. This allows the use of the original targeting vector to create ES cells homozygous at the targeting locus (Abuin and Bradley, 1996; Rossant and Nagy, 1995). If conditional gene targeting is planned, it is advisable to use FRT sites to tag the selectable marker gene (see later).

For those IF genes that are transcribed in ES cells, very high gene targeting frequencies can be achieved with promoterless selectable marker genes that are placed in the first coding exon, either in frame with the endogenous ATG codon or by positioning the initiation codon of the marker gene downstream from the major transcription start site. If insertion of a promoterless marker gene is required in an exon further 3′ into the gene, an internal ribosomal entry site (IRES; Jeannotte *et al.*, 1991) can be incorporated to enable translation of the marker gene. A slight modification of the preceding design can be used to introduce a promoterless reporter gene along with the selectable marker to follow the cells carrying the targeted allele. The EGFP-reporter genes or the more sensitive lacZ-reporter genes are convenient but do not necessarily provide quantitative readouts (Godwin *et al.*, 1998; Le Mouellic *et al.*, 1990). There are several

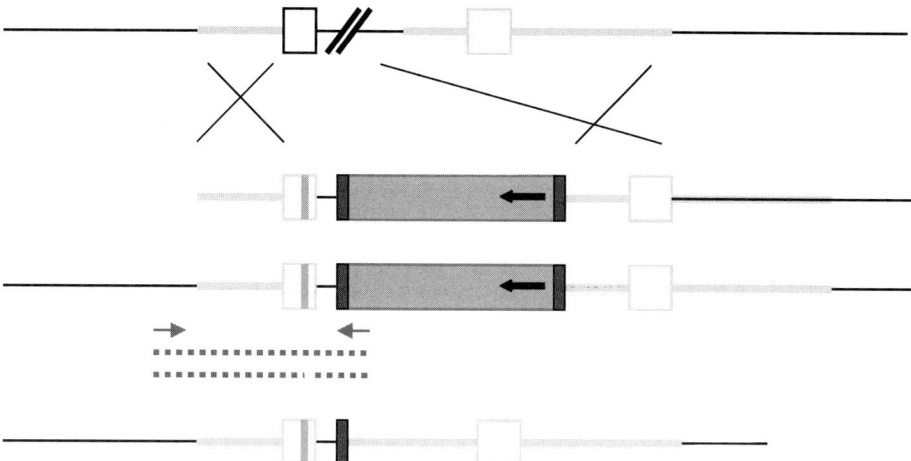

Fig. 2 Introduction of subtle alteration by cotransfer. Top, Genomic locus. Below, Targeting vector. The short arm of homology carries the desired alteration (bar), created by mutagenesis. If, in addition, a new restriction site has been generated without altering the properties of this sequence, the presence of the desired mutation can be directly screened by restriction analysis of PCR products. Below, Target locus after homologous recombination. Presence or absence of desired alteration is indicated by dotted lines before and after restriction of PCR products. Bottom, Target locus after deletion of marker gene by site-specific recombinase. One recognition site resides in the genome.

ways to introduce subtle alterations (e.g., point mutations) into genes. The mutation can be positioned in one of the homology arms of the targeting vector and becomes cotransferred along with the selectable marker, provided recombination does not separate the mutation from the selectable marker (Fig. 2). The selectable marker gene has to be positioned either in a noncoding exon or in an intron, from where it can be removed with Flp or Cre-recombinases, leaving one recombinase recognition site in the genome. In general, cotransfer of mutations works well, provided the distance between the marker gene and the alteration is shorter than between the alteration and the end of the homology arm. One potential drawback of this strategy is that the expression of the target gene can be severely altered by placing the marker gene in regulatory introns (Arin *et al.*, 2001; Meyers *et al.*, 1998; Nagy *et al.*, 1998).

An alternative strategy follows the so-called double replacement procedure (Fig. 3; Askew *et al.*, 1993; Stacey *et al.*, 1994; Wu *et al.*, 1994). In the first round of gene targeting, an HPRT-minigene replaces the (part of the) gene to be modified. After the first round of correct targeting, the minigene is replaced by a modified version of the original gene or by another sequence. In the second step, selection with 6-thioguanine allows cell survival only on loss of the HPRT-minigene. In case HPRT-deficient ES cells are not available, a combination of two selectable marker genes for positive and negative selection has been described (Askew *et al.*, 1993; Wu *et al.*, 1994). In principle, this procedure allows the

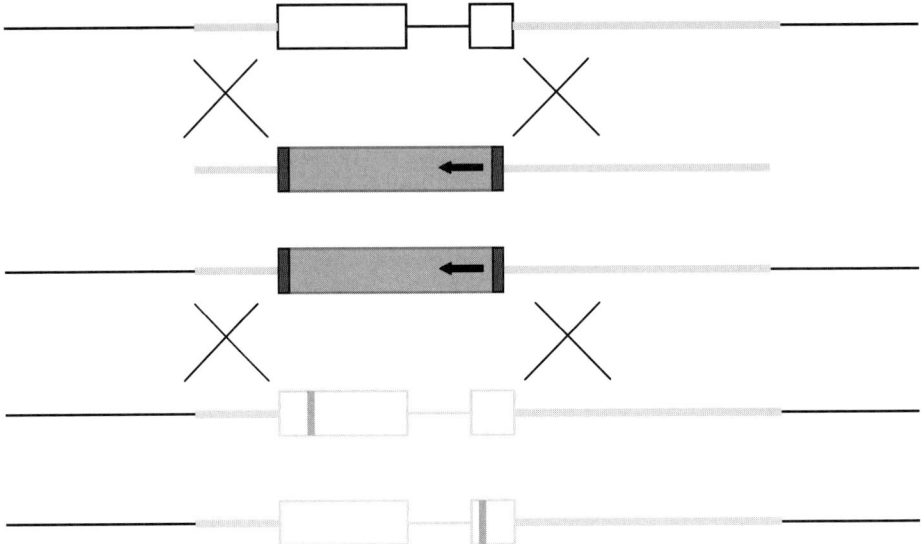

Fig. 3 Double replacement gene targeting procedure that uses positive/negative selection. Top, Target locus. Below, Gene targeting vector. Below, After the first gene targeting step, the gene or gene fragment has been replaced by an HPRT-minigene, providing resistance against HAT-containing media. The target locus can be replaced by modified gene sequences. Below, Second-step gene targeting to introduce a subtle alteration (bar) in the left exon. Below, Alternative second-step gene targeting to introduce a subtle alteration in the right exon. In both cases, selection in media containing 6-TG indicates loss of HPRT-marker gene.

introduction of many point mutations by use of the same second round gene targeting vector. In contrast to Cre/lox or FRT/Flp procedures, no foreign DNA remains in the target locus.

With the use of the site-specific recombinases, conditional, cell-type and temporal gene alterations have become feasible (Danielian *et al.*, 1998; Gu *et al.*, 1994; Guo *et al.*, 2002; Utomo *et al.*, 1999; Wunderlich *et al.*, 2001). The original vector design used a loxP-flanked marker gene and a third loxP site placed in an intron such that the loxP sites flank an exon to be deleted (Gu *et al.*, 1994). After Cre-recombinase activity, three types of excision can occur (Fig. 4), of which complete and type II excisions are the desired ones and are selectable by the loss of the marker gene. The type II excision provides a conditional allele that can be deleted either in ES cells or *in vivo* after mating with a strain of mice expressing the Cre-gene under the control of an appropriate promoter. Because type II excisions can be difficult to obtain, an improved strategy relies on the combined use of loxP and FRT sites. FRT sites are used to flank the selectable marker gene and allow its removal by the Flpe gene (Schaft *et al.*, 2001), either in ES cells or *in vivo*. Two loxP sites are placed at intronic or noncoding exon positions to delete (parts of) the gene of interest (Fig. 5). Conditional alleles created in this way can be

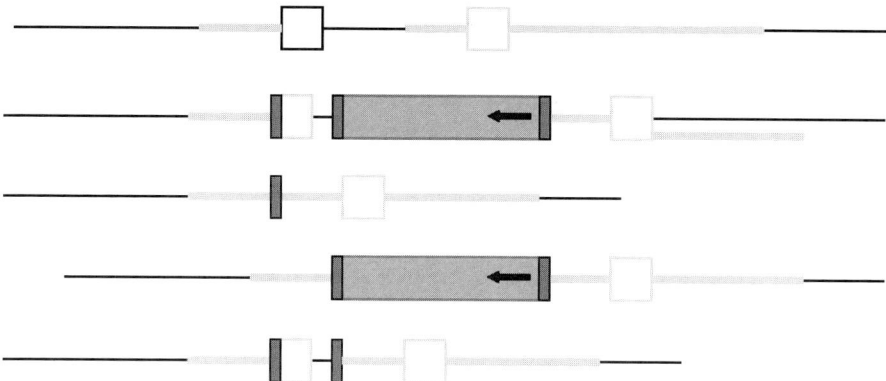

Fig. 4 Conditional gene targeting that uses three loxP recognition sites. Top, Target locus. Below, Gene targeting vector with loxP site flanking the exon to be removed on Cre activation (cross-hatched boxes; loxP recognition sites). The loxP-flanked marker gene is placed in an intron. After Cre activity, several types of deletions can occur. Below, Complete excision creating a null allele (type I), type II (undesired), and type III excision creating a conditional allele. The outcome is determined by various parameters, (e.g., the relative position of the loxP sites and the extent of Cre activity).

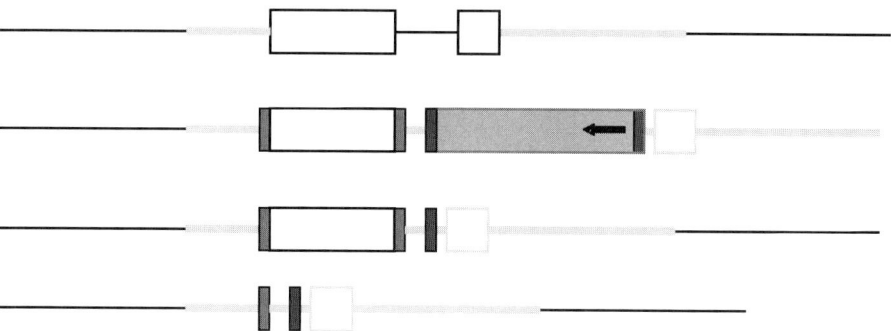

Fig. 5 Conditional gene targeting that uses FRT and loxP recognition sites. Top, Target locus. Gene targeting vector with FRT sites (black) flanking the marker gene (striped) and 2 loxP sites (cross-hatched boxes) flanking the exon to be deleted. After homologous recombination and FLPe activity, the marker gene has been removed with one FRT site left behind. On Cre activity, the "floxed" exon is deleted. At present, there are far more Cre than FLPe transgenic mouse lines available to create spatially or temporally altered mice.

controlled in a temporal manner by use of an inducible Cre-recombinase. At present, tamoxifen, RU 486, and tetracycline-controlled Cre-recombinases, in combination with various promoters, are available. At present, there are two limitations of these inducible transgenes, namely leakiness at the uninduced stage and the time to remove the target gene product after Cre expression. It can take several days for this process, especially for IF proteins with their long half-life

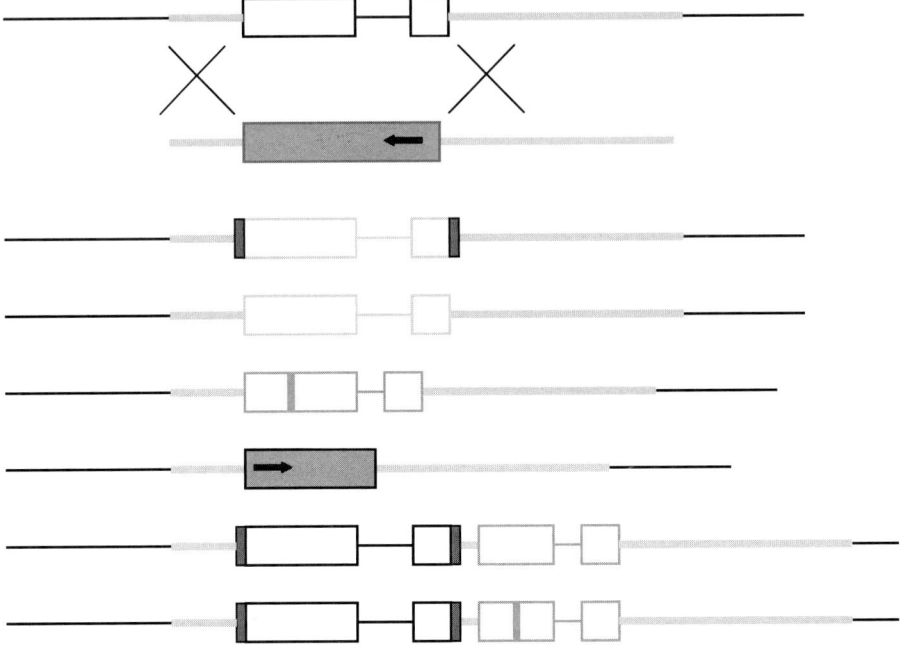

Fig. 6 Gene modifications that use double-replacement gene targeting. Top, Target locus. Below, Gene targeting vector. Below, After the first gene targeting step, the gene or gene fragment has been replaced by an HPRT-minigene, providing resistance against HAT-containing media. The target locus can be replaced by modified gene sequences as indicated below. 1, Replacement with a conditional ("floxed") allele. 2, Replacement with a related gene (open box). 3, Replacement with a mutant gene copy (grey box). 4, Replacement with an inducible, promoterless Cre or FLPe recombinase to be driven by the target gene promoter. 5, Replacement with a conditional, silent gene copy, followed by a wild-type copy of a related gene (grey box). 6, As before, but followed by a mutant (bar) copy of a related gene.

times (Denk *et al.*, 1987). The use of Cre-recombinase and FLP-recombinase can be combined with double replacement and allows a variety of options (Fig. 6).

2. Large-Scale Genome Engineering

If it becomes necessary to delete larger genes or gene fragments from the genome, the approaches described previously become inefficient as the distance between loxP sites increases. To create deletions of a large segment of chromosomal DNA, Bradley and colleagues developed the following scheme based on two consecutive homologous recombination events, one at each end of the target locus. The 5′ vector consists of a neomycin resistance gene for gene targeting, a loxP site, a 5′ half HPRT minigene (nonfunctional HPRT fragment), and approximately 5 kb genomic DNA homologous to the 5′ end of the target locus. Similarly, the 3′ vector contains a puromycin resistance gene, a loxP site, a

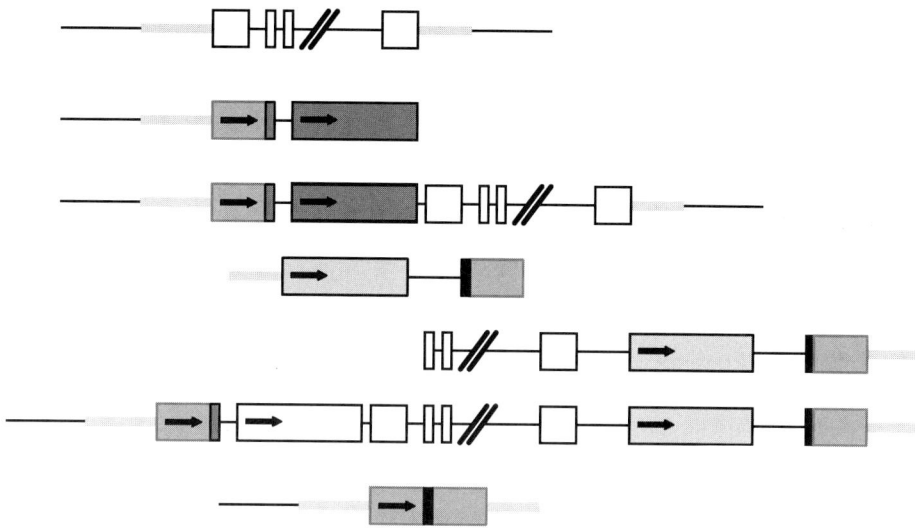

Fig. 7 Generation of large-scale genomic deletions based on an HPRT-deficient ES cell line. Top, Target locus. Below, Targeting vector for 5′ flank of targeting vector, consisting of a nonfunctional half-HPRT-minigene (red arrow indicates minigene promoter), followed by a loxP site (red box) and a neomycin resistance gene (blue box, arrow marking promoter). Below, After homologous recombination, G418-resistant ES cell clones are identified by PCR. Below, Targeting vector for 3′ end, consisting of a puromycin resistance gene (yellow box) and the 3′ half of a nonfunctional HPRT-minigene (red), preceded by a loxP site. Below, After homologous recombination, puromycin-resistant ES cell clones are identified by PCR. Complex rearrangements can occur after Cre activity, depending on the cell cycle phase in which Cre is active and the chromosomal integration of the two targeting vectors (refer to Yu and Bradley, 2001). Below, After Cre activity, the marker genes and the genomic sequences between the two loxP sites are deleted, and surviving correctly targeted ES cell clones are resistant against HAT and sensitive against G418 and puromycin. (See Color Insert.)

3′HPRT-minigene (nonfunctional HPRT fragment), and approximately 5 kb genomic DNA homologous to the 3′ end of the target locus (Yu and Bradley, 2001). The loxP site has been placed in an intron of the HPRT-minigene. These vectors can be generated in a conventional way by sequentially inserting various genetic components into a plasmid construct (Yu and Bradley, 2001), or they can be isolated directly from genomic libraries of premade targeting vectors (http://www.ensembl.org/Mus_musculus/). LoxP sites must be in the same orientation (Fig. 7). At least 50% of the recombination events will occur on the same chromosome (cis). A cis recombination event between two loxP sites in the same orientation, in the presence of Cre-recombinase, will lead to the deletion of the loxP-flanked DNA sequence in a circular molecule, whereas a transrecombination event will lead to deletions and duplication events. Complex recombination products can arise when recombination takes place between sister chromatids and between the nonsister chromatids in the G2 phase of cell cycle (see Yu and Bradley, 2001). On Cre activity, recombination between the loxP sites reconstitutes a functional HPRT-minigene. This provides a positive selection for this event when the ES cells are cultured in HAT medium. These clones carrying the functional HPRT gene are a mixture of those with the desired chromosomal deletion and with the undesired chromosomal duplication, which have been produced because of recombination events in the G1 and G2 phases of the cell cycle.

B. Culture of ES Cells

ES cells are available from several laboratories and from commercial suppliers. Most presently available lines have been derived from one of the several mouse 129 inbred strains, and they maintain good germline transmission capabilities. C57B1/6 ES lines are also available but seem to be more variable with respect to germline transmission. Most recently, ES cells from F1 or mixed hybrid genetic background strains were isolated and show very high germline transmission frequencies caused by hybrid vigor (Eggan *et al.*, 2001, 2002; Seibler *et al.*, 2003). These were developed to shorten the time required for the generation of gene-targeted mice considerably. The concept is based on the use of tetraploid embryos, generated by electrofusion of ES cells, for injection of genetically engineered diploid ES cells.

Tetraploid cells have been shown to give rise exclusively to extraembryonic tissues, whereas diploid ES cells contribute to the embryo proper (Nagy *et al.*, 1993). In consequence, this procedure leads to germline chimeras, which are completely derived from ES cells, except for one breeding step (equaling 3 months) along the route of establishing a line of genetically altered mice. This procedure can be shortened even further by carrying out two rounds of gene targeting to conditionally target both alleles of the gene of interest (Seibler *et al.*, 2003). If this is performed in ES cells that express an inducible Cre gene, it is possible to generate mice with two conditionally modified alleles in 6 months compared with 14 months. Many ES cell lines are grown on feeder cells, either primary embryonic (MEF) or immortalized STO cells (available from the Jackson Laboratory). If feeder cells are used during drug selection, they must express the appropriate resistance gene. Alternatively, ES cells can be grown on gelatinized dishes. Several ES cell lines are available that maintain an excellent germline potential under these conditions, including the HPRT-deficient line HM-1, which is in extensive use in our laboratory (Magin *et al.*, 1992a,b; Ying and Smith, 2003; Ying *et al.*, 2003; Yu and Bradley, 2001).

To maintain the germline transmission potential of ES cells, great care has to be taken with respect to their culture conditions. This includes rigorous testing of serum and growth factors and the use of reagents tested for ES cell work. Most recently, the major signaling pathways responsible for the maintenance of ES pluripotency have been identified and led to the formulation of serum-free media with defined growth factors (Ying and Smith, 2003; Ying *et al.*, 2003). In our hands, these conditions are favorable.

It is important to keep ES cell passage number as low as possible to maximize germline competence and to minimize genetic and epigenetic alterations. It is recommended that a sufficiently large stock of frozen vials is prepared and a vial is exponded every time for a new experiment. In general, ES cells should be split at rates between 1:6 and 1:12. High-density and low-density plating carry the risk of differentiating them. Their growth rate (typical doubling time is approximately 20–22 hours) should be carefully monitored; significant changes in growth rate and morphological appearance signal chromosomal alterations and preclude further use of such cells for gene targeting experiments. In the following section, culture of HM-1 cells in standard and serum-free media is described, both on gelatin-coated dishes and on feeder cells. Culture conditions for other ES cell lines will be similar; however, they may react differently on the same serum batches. It is recommended that one follows the advice of the donating laboratory on the source of serum. In our hands, ES cell–tested serum batches of American, Australian, and New Zealand origin, offered by several companies, are preferable. Changing to serum-free conditions puts cells through a crisis, which lasts three to four passages, after which the line should be expanded and frozen in aliquots [see also (Ying and Smith, 2003; Ying *et al.*, 2003)].

1. Materials

a. Serum–Containing Culture Medium
Glasgow modification of Eagles medium (BHK-21 medium)
Without tryptose phosphate broth
With L-glutamine
Gibco BRL 041-01710
Sodium pyruvate 100 mM Gibco BRL
Nonessential amino acids 100× Gibco BRL
15% fetal calf serum

Note that all batches of sera should be tested for ES cell viability and germline competence compared with existing good batches.

Preparation of complete HM-1 culture medium:
Always prepare appropriate amount of medium (i.e., sufficient for up to 1 week).
Add the following to a 500-ml bottle of Glasgow medium:
  8 ml sodium pyruvate
  8 ml nonessential amino acids
  8 ml 200 mM glutamine (equals 100× concentration)
  75 ml fetal calf serum
  910 μl monothioglycerol

600 μ LIF (this amount will vary from batch to batch) if self-prepared LIF is used. For commercially available LIF, follow manufacturer's recommendations.

Avoid unnecessary warm-up cycles in waterbath.

Store complete medium at 4 °C and use for no more than 1 week.

Preparation of monothioglycerol:

Add 1 ml of Mtg (Sigma M6145, cell culture grade) to 99 ml of sterile water (stock is 9.24 M) and freeze in small aliquots at −20 °C. This gives a concentration of 92.4 mM. The final concentration in the medium is 0.15 mM.

Alternative to Mtg: preparation of beta-mercaptoethanol:

Add 0.2 ml of beta-MeOH to 28.2 ml of water and filter through 0.1-μm pore filter. Store at −20 °C. Avoid more than 20 freezing/thawing cycles. The final concentration in the medium is 10^{-4} M.

10× Hepes-buffered-saline (HBS) for electroporation

16 g NaCl
0.74 g KCl
0.252 g Na_2HPO_4
2 g d-glucose (dextrose)
10 g HEPES

Dissolve all in 180-ml cell culture grade water, adjust pH to 7.2, bring up to 200 ml, and filter through 0.1-μm pore filter. Store at −20 °C. To prepare 1× HBS, dilute with sterile water, which brings pH to 7.05.

Preparation of gelatin solution:

Use Sigma G2500 swine skin type I gelatin. Prepare a 1% solution in cell culture grade water. Autoclave. Leave overnight and autoclave again the following day. Store at 4 °C. Warm up before preparing 0.1% working dilution.

Preparation of trypsin:

For 500 ml of trypsin solution, use
400 ml cell culture grade sterile PBS
100 ml sterile EDTA (1.85 g/L)
5 ml trypsin (Gibco BRL 25090)
5 ml chicken serum (Gibco BRL 16110)
Mix and store as 100-ml aliquots at −20 °C. Once frozen, keep aliquots at 4 °C.

Preparation of ES cell freezing medium:

1 × freezing medium contains complete ES medium plus 10% serum (20% final concentration) and 10% DMSO. To freeze ES cells, prepare 2× freezing medium as follows:

For 20 ml 2× freezing medium, mix

10 ml complete ES medium

4 ml DMSO (ultrapure)

6 ml FCS

Store at −20 °C.

HAT and HT media are bought as 50× stocks from Sigma.

To prepare 100 × HAT yourself, mix

272.2 mg hypoxanthine

48.4 mg thymidine

4 ml of 1 mM aminopterin

Dissolve hypoxanthine in 2 ml 1 N NaOH. Add 8 ml cell culture grade water and mix well. Add another 10 ml of water and mix until most of the hypoxanthine is dissolved. Add 1 ml of 1 N NaOH and mix until hypoxanthine is completely dissolved. Add 75 ml of water, then add thymidine, followed by a further 100 ml of water. Then add aminopterin and mix again (aminopterin is a poison, watch material safety sheet).

Filter through 0.1-μm filter and store as 20 ml aliquots at −20 °C.

To prepare 1 mM aminopterin, weigh out 100 mg and dissolve in 45 ml of cell culture grade water. Add approximately 1 ml 1 N NaOH to dissolve aminopterin. Sterilize as previously and store at −20 °C.

Additional compounds:

200× Pen/Strep is bought from Gibco BRL and stored as 1-ml aliquots at −20 °C.

G 418 is bought from Gibco BRL or any other manufacturer. It is prepared at 40 mg/ml in PBS and filter sterilized. Store at −20 °C. Final concentration is 350–400 μg/ml.

Tylosin is used against mycoplasms as recommended by supplier.

6-Thioguanine (6-TG) is prepared as a 1000× stock (i.e., 2 mM). Dissolve 50 mg in small volume of 1 N NaOH. Add water up to 25 ml, dissolve completely, and filter sterilize.

Store as 5-ml aliquots at −20 °C.

LIF is prepared according to protocol.

b. Serum-Free Culture Medium
Preparation of (50×) stock solutions:

Supplement B27 is 50× concentrated, aliquots of 2 ml, store at −20 °C

BSA Fraktion V, 13.3 ml/100 ml medium (1:1), aliquots of 10 ml, $-20\,°C$

Apo-transferrin, 1 g/10 ml medium (1:1), aliquots of 0.5 ml, $-20\,°C$

Human insulin stock, 4 mg/ml, aliquots of 2 ml, $-20\,°C$

Progesterone, dissolve 1 mg in 1 ml EtOH abs., add 49 ml medium (1:1), filter sterile (filter 0.1 μm), aliquots of 1 ml, $-20\,°C$

Putrescin, dissolve in 6.2 ml medium (1:1)

Na-Selenite stock I 10 mM; 0.865 g/50 ml cell culture grade H_2O Stock II 100 μM = 1 ml stock I/100 ml medium (1:1) Filter sterile (filter 0.1 μm), aliquots of 2 ml, $-20\,°C$

Bone morphogenetic protein (BMP)-4, 23.29 μl 25% HCl plus cell culture grade H_2O up to 100 ml (equals 4 mM), filter sterile (filter 0.1 μm), of this solution, add 1 ml per vial BMP-4, then add 1 μl BSA fra. V, aliquots of 100 μl, store at $80\,°C$.

Components (500 ml)
DMEM-F12 (225 ml)
Neurobasal med. (225 ml)
Supplement B27 (10 ml)
BSA Fr. V (5 ml)
APO-transferrin (500 μl)
Human insulin (3.125 ml)
Progesterone (300 μl)
Putrescin (50 ml)
Na-selenite (300 μl)
Glutamine (5 ml)
Monothioglyerol (910 μl)
BMP-4 (500 μl)
LIF (575 μl)

Components:

DMEM-F12 (Invitrogen 11320-033)

Neurobasal medium (Invitrogen 21103-049)

Supplement B27 (Invitrogen 17504-044) 50×

BSA fr. V (Invitrogen 15260-037)

Apo-transferrin bovine (Invitrogen 11108-016)

Recombinant human insulin (Invitrogen 0030110 SA) 5 ml

Progesterone (Sigma P6149)

Putrescin (Sigma P6024)

Na-selenite (Sigma S 5261)

BMP-4 (R&D Systems 314-BP)

Prepare medium as follows: Store complete medium at $4\,°C$ and use for no more than 1 week. Warm up quickly in water bath. Remove quickly.

2. Methods

a. Routine Culture, Generation, and Characterization of Genetically Altered ES Cells

Here, we describe culture of HM-1 cells, which are derived from strain 129/Ola and are HPRT deficient. They are grown in gelatinized dishes and do not require feeder cells. Other cells are grown essentially in the same way. For routine culture, cells should be passaged twice a week (ideally on a Monday–Friday rota), splitting them between 1:6 and 1:12, depending on conditions. Medium change should take place every other day as long as the culture is thin; later on medium should be changed daily. They should not be plated too thin, because this encourages excessive differentiation. Cultures should not be allowed to become overconfluent for the same reason. Any drastic change in cell doubling times and/or cell morphology is indicative of unwanted damage to the cells (e.g., chromosomal aberrations). Typical doubling time is approximately 22–24 hours.

To retain their excellent germline transmission frequency, it is essential to keep the passage number of HM-1 cells as low as possible. Good germline transmission rates depend on (1) tested sera, media, plasticware, etc., (2) limiting passage number, and (3) experimental skills. We use cell culture plastic from Becton Dickinson, cell culture grade water, and reagents, and we test all batches of sera for germline transmission.

To carry out an electroporation experiment, 1 aliquot of cells is thawed on a single 25-cm^2 flask (day 1). This becomes confluent 2–3 days later. These are trypsinized and plated on a single 75-cm^2 flask (if necessary, two flasks are possible). These should be confluent 3–4 days later (day 6–8), trypsinized and used for experiment. Spare cells should not be used for another electroporation experiment, because their passage number has increased.

b. Thawing and Plating Cells

The flask is gelatinized with 3–5 ml of gelatin per 25-cm^2 and 7–10 ml per 75-cm^2 flask (for any other vessel, the surface should be completely covered with gelatin) and left at room temperature for at least 10 minutes. Before plating cells, these should be aspirated off. (If feeders are used, the appropriate amount of cells are plated together with ES cells or plated 1–2 hours before.)

The centrifuge tube is prepared with 10 ml of prewarmed medium. The cells are thawed quickly in 37°C bath and transfered into prewarmed medium. They are spun for 5 minutes at 1000 rpm. The medium is removed and replaced with 5 ml of fresh medium containing Pen/Strep, if necessary, and all cells are replated on a 25-cm^2 flask. Next day, fresh medium without antibiotics is used. Routine culture without antibiotics is strongly recommended.

To subculture, the medium is aspirated and the cells are washed with 1 ml ES cell–trypsin (prewarmed) per 25-cm^2 flask (2 ml per 75 cm^2). This is removed and replaced with fresh trypsin and left in a 37°C incubator until cells are dislodged by gentle agitation. It is a good sign if this happens after about 2 minutes but may take up to 5 minutes. Overtrypsinization can damage the cells. Four milliliters medium is added per 1 ml of trypsin and pipetted up and down against flask wall about 10–20 times (the distance between pipet tip and flask wall may be varied between 0.5 and 2.5 cm). It is not essential to obtain a single cell suspension for routine culture, but large cell clumps must be avoided, because those will lead to differentiation. Suspension should be checked under the microscope before transferring cells into a centrifuge tube. There should be a uniform size distribution of cells/cell aggregates.

In the meantime, new flask(s) are gelatinized and spun for 5 minutes at 1000 rpm at room temperature. The supernatant is aspirated off and the cell pellet is resuspended gently in a few milliliters of medium, then replated at desired density; 6–8 ml of medium is added to small and 25–30 ml is added for a large flask. A 25-cm^2 flask yields about 7–10 million cells, and a 75-cm^2 flask yields about 25–30 million cells.

c. Electroporation of ES Cells (First–Step Gene Targeting)

The conditions recommended will work with a Biorad gene pulser and cuvettes of 0.4-cm electrode gap.

The setting should be 800 V, 3 microfarads (standard). This setting will work with any ES line and is relatively insensitive to changes in cell number and amount of DNA. It does not yield the highest possible number of transformed cells but is gentle to cells.

Other settings that yield higher efficiencies are 250 V, 500 microfarads (optional). Under these conditions, 25 μg of DNA and 10 mio. cells per electroporation are used. Changes in DNA amounts and cell number alter efficiency and cell viability.

Optimal results are currently achieved if the DNA is purified with Nucleobond or Qiagen columns. EndoFree purified DNA is not necessary but may improve transfection efficiency. For a standard targeting experiment, approximately 20–30 million cells are trypsinized. Cell number is determined in a counting chamber. ES cells are resuspended in 0.8 ml of 1 × HBS buffer. Cells are added to 200-μg linearized DNA in 100-μl sterile TE in an Eppendorf tube (Phenol/ Chloroform-extracted, ethanol-precipitated, 70% ethanol removed in clean bench) and mixed by pipetting up and down with a 1-ml cell culture pipette and transferred into an electroporation cuvette. One pulse is applied 10 minutes at room temperature followed by not touching the cells. Cells are added to the appropriate amount of medium in a 500-ml bottle, depending on the number of dishes to be used altogether: Approximately 1–2 million cells are plated per 10-cm dish in 10 ml of medium. Positive selection is started the next day (G 418 is 350 μg/ml; hygromycin is 150 μg/ml; puromycin is 1 μg/ml; for HAT, see later). One plate is left without selection to check cell recovery. After 3–5 days, medium is changed again (electroporation is day 1). Colonies will appear after 8–10 days. Colonies are isolated as described in ES cell colony isolation protocol, half is used for PCR, and the remaining cells are left in 24-well plate with approximately 1 ml of medium. After a few days, multiple colonies should have appeared. The medium is aspirated off, and it is washed with 0.2 ml of trypsin and again aspirated. Two to three drops of trypsin are added and it is incubated for 3–5 minutes at 37 °C; 0.5 ml of medium is added, and cell clumps are broken up with a blue-tip Gilson pipet. Cells are transferred to a 6-well plate and 5 ml of medium is added. It is grown to near-confluency and trypsinized and plated onto two 25-cm^2 flasks adding two thirds on one and one third onto the other flask. When confluent, one flask is used for freezing cells (2–3 vials) and the other one for preparation of genomic DNA.

Selection is maintained until targeted ES cells are expanded and characterized by Southern blotting. To remove cells from HAT selection, HAT with HT is replaced for at least 2 days before growing cells in normal medium. Cells will grow slightly faster in the absence of selection.

For HPRT negative selection, the final concentration of 6-thioguanine is 5 μg/ml. For correctly targeted ES cells, an aliquot is thawed on a 25-cm^2 flask, grown until confluency, and passaged on one 75-cm^2 flask; 6–8 freezing aliquots are then prepared for blastocyst injections.

d. Second–Step Gene Targeting (Double Replacement with HPRT–Minigenes)

The passage number is kept as low as possible between the first and second gene targeting step to minimize the accumulation of 6-TG–sensitive cells before electroporation. Cells are transferred in normal medium as described. Approximately 20 million cells are electroporated as before and plated onto six 75-cm^2 flasks. 6-TG selection is applied at day 6. Most likely, it will be necessary to split cells first. In this case, they should be split in such a way that 6 nearconfluent flasks remain at day 6. The cells are then trypsinized. From each flask, six × 1–1.5 million cells are plated onto 10-cm dishes in medium containing 6-TG; remaining cells are discarded. Medium is changed after 2 and 4 days, respectively. Colonies will appear around day 10 (there will only be 1–5 colonies per dish).

e. Further Handling of Resistant Colonies

Colonies are passaged when they are approximately 2–3 mm in size with yellow Gilson tips and pipet adjusted at 150 μl. The colony is transferred to gelatin-coated 24-well plate and pipetted up and down a few times, removing a cell aliquot for PCR. When cells are becoming confluent, they are trypsinized and transferred to a 6-well plate. From there, they are passaged onto two 25-cm^2 flasks as described previously. The passage number must be kept as low as possible and the cells should be trypsinized well. Prolonged growth of clumpy cell aggregates should be avoided.

f. Freezing ES Cells

Cells are frozen from near-confluent flasks. One aliquot should contain 2–4 million cells (i.e., 2–3 per small and 5–8 per large flask). It is beneficial to freeze cells as concentrated as possible, (e.g., in 0.5–1 ml per vial). They are frozen at approximately 1 °C/minute (e.g., by use of a Nalgene-freezing container).

The cells are trypsinized as usual once resuspended gently in 1 ml of normal medium. Add 1 ml of 2× freezing medium (dropwise). They are gently mixed and transferred instantly to a Nalgene freezing container and placed into −80 °C overnight (or at least 4 hours). Thereafter, cells are transferred to a liquid nitrogen container.

g. Preparation of HM-1 Cells for Blastocyst Injections

A subconfluent 25-cm^2 flask of cells is sufficient. Ideally, cells are thawed 4 days before injection. Approximately 24 hours before microinjection, they should be near-confluent and 70%–100% of cells should be trypsinized and replated.

The medium should be changed early in the morning injection. Trypsinization should start only when everything at the microinjection facility is ready to go. Cells become fragile after 1.5–2 hours. If microinjection takes a long time, a second flask of cells should be available. Trypsinization: medium is aspirated, 2 ml of trypsin is added and mixed gently and removed. Three milliliters of trypsin is added and incubated 5–10 minutes at 37 °C to receive single cell suspension (check under microscope and continue incubation if necessary). Five milliliters of medium is added and pipetted up and down gently; this is spun 2 minutes at 1200 rpm, the medium is aspirated and cells are resuspended in 5 ml of injection medium (GMEM, 10% FCS, 20 mM HEPES, 1× Pen/Strep.).

Before adding cells to microinjection droplet, 1 μl of ultrapure DNase I is added (20–40 u/μl). The cells are kept on ice during injection.

h. Culture of ES Cells with Serum-Free ES-Medium

Handling essentially follows the description given in the previous protocol, with the following modifications: (1) For trypsinization, the medium is aspirated up and the cells are washed with 1 ml ES cell–trypsin (prewarmed) per 25-cm^2 flask (2 ml per 75 cm^2). This is removed and replaced with fresh trypsin, and the cells are dislodged by gentle agitation. This happens instantly. (2) One milliliter of soybean trypsin inhibitor and 4 ml medium per 1 ml of trypsin are added and are pipetted up and down against the flask wall a few times. The suspension is checked under the microscope before transferring the cells into a centrifuge tube. There should be a uniform size distribution of cells/cell aggregates. In the meantime, new flask(s) are gelatinized. This is spun for 5 minutes at 1000 rpm at room temperature, the supernatant is aspirated off and the cell pellet is resuspended gently in a few milliliters of medium. This is replated at the desired density and 6–8 ml of medium is added to a small flask, and 25–30 ml is added to a large flask.

i. Isolation of ES Cell DNA for PCR-Based Genotyping

With good vector design and the use of isogenic DNA for vector construction, high gene targeting frequencies can be expected. For a range of different genes, we noted frequencies of 1:5–1:35 PCR-positive colonies. Therefore, it should be sufficient to analyze between 100 and 200 colonies per experiment. The procedure detailed in the following is based on the use of approximately half of a colony for DNA isolation and the other half for further growth. In this way, positive colonies can be scored the next day. For analysis, small, slower growing ES cell colonies with a good morphology should be used. Their position should be labeled with a pencil, and they should be isolated with a yellow Gilson tip under the clean bench.

A 24-well plate is coated with gelatin and removed by aspiration after 10–20 minutes.

Colonies are picked along with approximately 150 μl of medium with a P 200 Gilson pipette.

The colony is transferred to a microtiter plate and dispersed by pipetting up and down 3–4 times.

Half of the cells are left in the well, and other half are transferred into an Eppendorf tube. 1 ml of selective medium is added, and cells are grown until 80% confluency.

Cells are spun down in an Eppendorf tube for 2 minutes at full speed, and the supernatant is aspirated off.

Fifty microliters of 1× PCR buffer are added. The buffer corresponding to the Taq polymerase being used is chosen. It must contain approximately 0.5% of N-P 40 or Triton X-100 for cell lysis, including 200 μg/ml proteinase K. This is vortexed briefly and incubated 1 hour at 65 °C, preferably at slow motion (rotating wheel).

Lysate is heated 10 minutes at 95 °C to inactivate proteinase K, and spun briefly, 2–5 μl of cell lysate is used for 25-μl PCR reaction. PCR must be carried out immediately after having the extract ready.

j. Karyotyping of ES Cells

Before using a clone for a second electroporation or for blastocyst injections, the number of chromosomes should be determined by karyotype analysis. A high degree of euploidy (greater than 70%) is necessary but does not guarantee germline competence. Therefore, it is preferable to characterize a sufficient number of euploid clones to increase the probability of germline transmission. Starting with confluent 25-cm^2 flasks, the cells are arrested in metaphase by incubating them for 1 hour at 37 °C, 5% CO_2 with 1 ml fresh medium containing 10 μl demecolcine solution (10 μg/ml in HBBS; Sigma D-1925; toxic-handle and dispose according to safety guidelines). The medium is removed and the cells are washed twice with PBS. One milliliter of trypsin solution is added and incubated for 4 minutes at 37 °C, trypsinization is stopped by adding 5 ml of medium, cells are spun down, medium is removed, and 1 ml of 0.56% KCl is added dropwise to the cell pellet. Cells are resuspended by flicking, then 1 ml of 0.56% KCl is added and mixed by flicking. This is incubated at room temperature for 8–10 minutes and spun down gently (700 rpm, 5 minutes) and the supernatant removed. A chilled fixation solution (2–3 ml) (methanol/acetic acid 3:1, freshly prepared) is added dropwise onto the pellet, it is resuspended by flicking, incubated 5 minutes at room temperature, spun down gently for 5 minutes at 700 rpm, and the supernatant is removed. The fixation procedure is repeated once or twice. Finally, the pellet is resuspended in 0.5–1 ml fixation solution, the solution is dropped onto clean slides from a distance of 15–30 cm, and the slides are air dried. Use microscope to see whether the broken nuclei are sufficiently separated from each other to ensure that every chromosome can be assigned to its metaphase spread. The slides are incubated for 1 minute in Giemsa staining solution (Giemsa stain, modified, 0.4%, Sigma). This is washed twice with distilled water, air-dried, and the slides are mounted. The number of chromosomes in metaphase spreads of the nuclei is counted. Because direct counting under the microscope is tedious and error prone, it is recommended that images be taken on which the counted chromosomes can be marked.

k. Preparation of Feeder Cells from STO or MEF Cells

Here we describe the routine culture and inactivation of feeder cells. Feeder cells should be grown in the same medium as ES cells, except for LIF and monothioglycerol. For serum-free ES cell culture, feeders are grown in serum-containing medium. After trypsinization, they are washed once in serum-free medium and used.

There are two ways to inactivate feeder cells, mitomycin C-treatment and γ-irradiation.

Mitomycin C treatment:

1. Ten micrograms per milliliter of mitomycin C is added to a confluent culture of feeder cells and placed in an incubator for 2–3 hours.

2. The dishes are washed 5× with PBS, and cells are collected by trypsinization (1–2 minutes). Cells are spun 3 minutes at 1000 rpm at room temperature, the supernatant is discarded and resuspend 1× in medium and respun.

3. Cells are resuspended in ES cell medium and counted. At appropriate density they are adjusted (e.g., 10^6/ml) and used. Mitomycin C–treated feeder cells can be frozen, following the same regimen as described for ES cell freezing.

γ-Irradiation:

1. Cells are grown to near confluency, trypsinized from 10–15 75-cm^2 flasks and spun down 3 minutes at 1000 rpm. The supernatant is discarded, and the cells are resuspended in 10–12 ml of culture medium in a 50-ml plastic tube.

2. The cell suspension is irradiated at 15 Gray for 7–8 minutes.

3. The irradiated cells are replated on the same number (as before) of gelatinized flasks overnight, and they are trypsinized and frozen in aliquots or used in experiment.

For use in ES cell growth, feeders are plated on gelatin-coated dishes 1–2 hours before plating ES cells. A density of 50,000 cells/cm^2 is required. Feeder layers last about 1 week after treatment.

l. Superovulating of Females for Preparation of Donor Blastocysts

The consecutive action of two hormones is used to superovulate females. Gonad-otropin from pregnant mare serum (PMS, used to mimic FSH) is the first hormone; human chorionic gonadotropin (hCG, used to mimic LH) is the second. Strain response to such induction is widely variable. The doses used depend on the particular strain and age (or weight) of female and are worked out empirically.

For isolation of blastocysts, we used 4- to 5-week-old C57Bl/6J females; for collecting oocytes, we used 8- to 10-week-old FVB females. The times that the PMS and the hCG are administered are relative to each other and to the light–dark cycle of the mouse room. For a light–dark cycle of 12 hours (6 am–6 pm) the optimal has been found to be:

Injection of the first hormone PMS at 12:00 on day 1; 5 U for 4- to 5-week-old C57Bl/6J, 7.5 U for 8- to 10-week-old FVB. Injection of the second hormone hCG 48 hours later at 12:00 on day 3; 5 U for 4- to 5-week-old C57Bl/6J, 7.5 U for 8- to 10-week-old FVB strain mice.

Females are mated with normal males after the second injection, one female per male. A period of 1 or 2 hours is normally given for the females to recover from the stress of injection before mating. Plugs are checked next morning.

Hormones:

A stock solution is prepared in sterile 0.9% NaCl, as described later. They are kept frozen in convenient aliquots at $-20\,^\circ$C until used up (hormone stocks are stable in the freezer for more than 1 year). They are thawed before use, and should not be refrozen. If aliquots are not used up, they are discarded after injection.

Gonadotropin:

Gonadotropin from pregnant mare's serum (Calbiochem, Cat. No. 367222, 1000 U or 5000 U/vial) is used. The powder is stored at $+4\,^\circ$C. For administra-tion, the PMS is resuspended at 50 U/ml in sterile 0.9% NaCl and divided into aliquots.

Human chorionic gonadotropin:

Human chorionic gonadotropin is supplied from pharmacies (e.g., as Predalon 500 I.E., Organon GmbH; Germany). HCG is resuspended at a final concentra-tion of 50 U/ml in sterile 0.9% NaCl and divided into aliquots.

m. Making and Testing LIF

For a large batch of LIF, 15 flasks (75 ml) of COS 7 cells grown to confluency are used. This makes 500 ml of LIF. It requires approximately 1200 μg of pC10 DNA (no need for linearization; prepare by NucleoBond, Qiagen, or any other high-purity column purification method). Grow COS 7 cells in ES cell medium in the absence of LIF and reducing agent.

1. Trypsinize flasks in batches of three. There is no need to count cells. The cells are pooled from three flasks, spun down (3 minutes at 1000 rpm), and resuspended in 1.6 ml of HBS buffer.

2. Cells 0.8 ml are electroporated with ~120 μg of pC10 DNA; 200 V, 500 μFarad.

3. Cells are left for 10 minutes, then each 0.8 ml of electroporated cells is transfered into a separate 50-ml blue cap tube with 10 ml of medium and plated onto a 140-mm dish. A further 20 ml of medium without antibiotics is added to each dish. This yields 10 dishes in total.

4. The next day, the dishes should be nearly confluent. The medium is aspirated and 25 ml of fresh medium is added; this is left for 3 days.

5. The medium is collected and another 25 ml is added. This is left for further 2 days and collected again.

6. As a precautionary measure, the medium is collected into three separate bottles. One should know which bottle is being used to collect medium from a particular dish. After collection is completed, and if there is no evidence of any culture contamination, the medium is pooled and filtered. A prefilter (0.45-μm and 0.1-μm filters) is used, and 5-ml aliquots are pipetted into tubes and stored at $-20\,°C$. A sterility test is set up on the first and last aliquots.

In our experience, a preparation can be used for up to 5 years without a drop in quality.

Testing:

1. A healthy culture of HM-1 cells is trypsinized in the morning, and 1000 cells/well are plated into a gelatinized 24-well microtiter plate. Also, 250 cells/well are plated into a separate plate.

2. After ~4–5 hours, the medium is aspirated and replaced with 1 ml/well of complete medium containing the new batch of LIF at dilutions of 1000-, 2000-, 5000- and 10,000-fold. Also, a row of four wells with no LIF is set up with the current batch of LIF at the appropriate dilution.

3. After 7 days, these are fixed and stained. Plating efficiency, colony size and, most importantly, level of differentiation are compared in the dilutions of the new batch of LIF against the standard, which should show only a low level (no more than 5%–10%) of differentiation.

C. Production of Chimeric Mice

The setup for the injection of blastocysts or the production of aggregation chimeras has been extensively described (Nagy *et al.*, 2003).

1. Materials

a. Media

KSOM-medium complemented with amino acids is recommended to overcome the inability of two-cell embryos to divide further in culture (two-cell block).

95 mmol/l NaCl
2.5 mmol/l KCl
0.35 mmol/l KH_2PO_4
0.2 mmol/l $MgSO_4$
1.71 mmol/l $CaCl_2 \times 2H_2O$
25 mmol/l $NaHCO_3$
10 mmol/l lactate
0.2 mmol/l pyruvate
0.2 mmol/l glucose
1 mmol/l L-glutamine
0.01 mmol/l EDTA
1 mg ml^{-1} BSA
Phenol red
$100\times$ NEAA (nonessential amino acids)   1:50
$50\times$ EAA (essential amino acids)   1:100

2. Methods

a. Production of Chimeras by Aggregation of Diploid with Tetraploid Embryos

The generation of aggregation chimeras has proven to be a powerful tool for the analysis of genes essential for embryonic development. This technique can be used to produce a mosaic of two different mice or to restrict the contribution of one mouse to the extraembryonic or embryonic compartments. Because of the limited developmental potential of tetraploid embryos, their cells contribute preferentially to the extraembryonic tissues of the embryo, where as ES cells are unable to form those compartments. By aggregation of ES cells and tetraploid embryos, it is possible to generate embryos that consist of a diploid embryo and tetraploid extraembryonic tissues. This system has the power to overcome an embryonic lethal phenotype that is caused by an extraembryonic defect, if tetraploid wild-type embryos are aggregated with diploid knockout ES cells (Jaquemar et al., 2003). If a homozygous ES cell line of the knockout of interest is not available, knockout embryos of the eight-cell stage can be used for aggregation. The efficiency is significantly lower compared with aggregation with ES cells, because the embryos can still contribute to the extraembryonic compartments.

b. Recovery of 2-Cell and 8-Cell Stage Embryos

Two-cell stage embryos are flushed out on day 1.5 post-coitum (p.c.), whereas 8-cell stage embryos are flushed out one day later at day 2.5 p.c.

Supplemental materials:

Dissecting microscope

Flushing needle (The sharp tip of No. 30 G 1/2 needle is cut off and then rounded with a sharpening stone.)

1-ml syringe

Dissecting instruments (fine-pointed scissors, fine forceps)

No. 5 forceps (Dumont)

Mouth pipette (aspirator mouth piece, latex tubing, blue tip, micropipette)

70% ethanol

Sterile tissue culture dishes (35 × 10 mm)

M2 and KSOM-AA mledia.

1. Superovulate an appropriate number of CD-1 females (see III.B.3.), mate with CD-1 males, and check for copulation plugs.

2. On day 1.5 p.c. or 2.5 p.c., the plugged females are killed by cervical dislocation to obtain two-cell or eight-cell stage embryos, respectively. The oviducts are dissected by cutting the upper part of the uterus and underneath the ovary. The oviducts are placed in a 35-mm tissue culture dish filled with M2 medium at RT.

3. An oviduct is transferred into a tissue culture dish with M2 medium under a dissecting microscope. The flushing needle is connected to a 1-ml syringe and filled with M2 medium. The infundibulum is grasped with a fine forceps, and flushing needle is inserted. After successful insertion, the needle is held with the forceps.

4. The oviduct is flushed out with 100–150 μl M2 medium by applying constant pressure. Steps 3 and 4 are repeated for each oviduct dissected.

5. The tissue culture dish containing the embryos is spun manually with fast circular movements. This will bring the embryos to the middle of the dish. All embryos are collected with the mouth pipette rinsed with three drops of M2 medium and three drops of equilibrated KSOM-AA medium.

6. The embryos are stored in KSOM-AA drop under mineral oil in an incubator (37 °C, 5% CO_2).

c. Generation of Tetraploid Embryos

Fusion of the blastomeres of two-cell stage embryos occurs when a square pulse is applied perpendicular to the plane of contact of the two cells.

Adjustable AC field is applied to allow the correct orientation of embryos (enable 1 or 2 V on the display). Too high an AC field can cause lysis.

Supplemental materials:

Cell fusion instrument

Two dissecting microscopes

M2 and KSOM-AA media

0.3 M mannitol (Sigma M4125) (dissolved in ultra pure water with added 0.3% BSA (Sigma A4378) and filtered through 0.22-μm Millipore filter, stored in aliquots at $-20\,°C$); tissue culture dish with KSOM-AA microdrops covered with oil (Sigma M8410); mouth- or finger-controlled pipette

1. Slide with electrodes is placed onto a 100-mm petri-dish by "fixing" it with 2 water drops. A drop of M2-medium and mannitol is dispersed over the slide-with the electrodes with 1 drop of M2 medium under it. A large drop of mannitol is placed between the electrodes. The electrodes are connected with the pulse generator, and all settings are checked for correctness according to the manufacturer.

2. All recovered two-cell stage embryos are placed into the M2 microdrop. A group of ∼30 embryos are washed in the mannitol drop and transferred between the electrodes into the mannitol.

3. The AC current is switched on; the embryos are watched through the microscope for orientation until the contact area of the two blastomeres is parallel to the electrodes.

4. When all embryos are in the proper orientation, a mouth pipet is used to abut nonorientated embryos.

5. A pulse is applied to the embryos, and AC field is immediately switched off.

6. The embryos are washed twice in KMSO media and transferred into a KMSO drop overlayed by light mineral oil and put into the incubator at $37\,°C$, 5% CO_2.

7. The mannitol solution is exchanged to prevent water evaporation and mannitol crystallization.

8. Forty-five minutes after applying the pulse, the embryos are checked for perfect fusing and put into a drop of KSOM-AA with oil overlay. The pulse with unfused two-cell embryos is repeated. Tetraploid embryos are cultured overnight at $37\,°C$, 5% CO_2.

9. After culturing for 24 hours, four-cell stage tetraploid embryos are used for the aggregation experiment; all other embryos are discarded.

d. Preparation of Aggregation Plate
Supplemental materials:

Dissecting microscope

Sterile tissue culture dishes (Easy Grip 35 × 10 mm, Falcon 3001-3)

1-ml syringe with 26 G 1/2 needle of KSOM-AA medium

Light mineral oil (embryo tested) (e.g., Sigma: M8410)

Aggregation (darning) needle (DN-09)

70% ethanol

1. Small drops of KSOM-AA medium (approximately 15 μl each) are dispensed on a 100 mm $\times$ 20 mm tissue culture dish. The dish is flooded with light mineral oil until all drops are covered completely.

2. The aggregation needle is washed briefly in ethanol.

3. Approximately six aggregation wells are made into each of the drops by pressing and turning the needle into the ground of the tissue culture dish.

4. The aggregation plate is placed in an incubator at 37 °C, 5% CO_2 for at least 2 hours.

e. Removal of Zona Pellucida with Tyrode's Solution

Supplemental materials:

Dissecting microscope

Acid Tyrode's solution (Sigma: T1788)

Tissue culture dishes (100 $\times$ 15 mm)

M2 and KSOM-AA medium

Mouth pipette

1. Embryos are added to the first M2 drop. Twenty embryos are rinsed in the first drop of Tyrode's solution and transferred into the second one.

2. The dissolving of the zona pellucida is observed with the microscope and the embryos are instantly transferred to the M2-drop containing 0.3% BSA to prevent embryos from sticking to the plate or to each other.

3. The embryos are washed by transferring them through 3 drops of M2 and 3 drops of KSOM-AA.

4. The embryos are transferred into a KSOM-AA drop without depressions on the aggregation plate.

5. ES cells are trypsinized for aggregation if ES-cell/tetraploid embryo aggregation is planned.

f. ES cells or Diploid Embryo/Tetraploid Embryo "Sandwich" Aggregation (Table I)

Supplemental materials:

Dissecting microscope

Prepared aggregation plate with depressions and embryos with removed zona

Trypsinized ES-cells or eight-cell stage embryos

Mouth-controlled pipette

Table I
Time Schedule for Aggregation of Tetraploid–Diploid Embryos

Day: 1	2	3	4	5	6	7
Superovulation of CD-1 and knockout		Superovulation 5U hCG	Check for copulation plugs	Flushing of 2-cell stage embryos	Flushing of 8-cell stage embryos	Transfer into uterus of day 2.5 pseudopregnant recipients
5U PMSG		Setting up matings of CD-1 and knockout 1:1	Mating of recipients	Electrofusion Setting up aggregation plate Incubation o/n, 5% CO_2 at 37 °C	Setting up aggregation Incubation o/n, 5% CO_2 at 37 °C	

1. Tetraploid embryos are transferred to one microdrop of the aggregation plate without depressions. Diploid eight-cell stage knockout embryos or trypsinized ES cells are transferred into another microdrop.

2. A tetraploid embryo is placed in each of the aggregation wells.

3. An eight-cell stage embryo or a clump of ~10 ES-cells is placed on the tetraploid embryo in each depression.

4. A second tetraploid embryo is placed on top of the eight-cell stage embryo or ES-cell clump, respectively, thus forming a sandwich.

5. Steps 2 to 4 are repeated for all aggregations, and the plate is cultured overnight in an incubator at 37 °C and 5% CO_2.

6. On the afternoon of the next day, aggregates are checked for blastocyst formation. The blastocysts are transferred into the uteri of day 2.5 p.c. pseudopregnant foster mice.

III. Websites, Tables, Reagents, and Vendors of Interest

Online Databases:

http://www.mshri.on.ca/nagy/ Cre-pub.html
http://jaxmice.jax.org/models/ cre_intro.html
http://genetics.med.harvard.edu/ ~dymecki/
http://www.ensembl.org/Mus_musculus

Tables:

Conditional alleles in mice: tissue-specific knockouts, kkwan@mail.mdanderson.org

Vendors:

http://www.invitrogen.com/
http://www.sigmaaldrich.com
http://www.rndsystems.com/
www.calbiochem.com

Vectors:

http://www.aldevron.com/vectors.php?vid=5
http://www.stratagene.com/lit/vector.aspx
http://www.genome.gov/10001852
Mutant loxP vectors: hiroshi@genetics.hpi-uni-hamburg.de

Additional references for materials, equipment and procedures are described in Nagy, A., Gertsenstein, M., Vintersten, K., and Behringer, R., (2003). "Manipulating the Mouse Embryo." Cold Spring Harbor Laboratory, Cold Spring Harbor, New York.

References

Abuin, A., and Bradley, A. (1996). Recycling selectable markers in mouse embryonic stem cells. *Mol. Cell. Biol.* **16,** 1851–1856.

Arin, M. J., Longley, M. A., Wang, X. J., and Roop, D. R. (2001). Focal activation of a mutant allele defines the role of stem cells in mosaic skin disorders. *J. Cell Biol.* **152,** 645–649.

Askew, G. R., Doetschman, T., and Lingrel, J. B. (1993). Site-directed point mutations in embryonic stem cells: A gene-targeting tag-and-exchange strategy. *Mol. Cell. Biol.* **13,** 4115–4124.

Baribault, H., Price, J., Miyai, K., and Oshima, R. G. (1993). Mid-gestational lethality in mice lacking keratin 8. *Genes Dev.* **7,** 1191–1202.

Brenner, M., Johnson, A. B., Boespflug-Tanguy, O., Rodriguez, D., Goldman, J. E., and Messing, A. (2001). Mutations in GFAP, encoding glial fibrillary acidic protein, are associated with Alexander disease. *Nat. Genet.* **27,** 117–120.

Coulombe, P. A., Hutton, M. E., Letai, A., Hebert, A., Paller, A. S., and Fuchs, E. (1991). Point mutations in human keratin 14 genes of epidermolysis bullosa simplex patients: Genetic and functional analyses. *Cell* **66,** 1301–1311.

Coulombe, P. A., and Omary, M. B. (2002). 'Hard' and 'soft' principles defining the structure, function and regulation of keratin intermediate filaments. *Curr. Opin. Cell Biol.* **14,** 110–122.

Danielian, P. S., Muccino, D., Rowitch, D. H., Michael, S. K., and McMahon, A. P. (1998). Modification of gene activity in mouse embryos in utero by a tamoxifen-inducible form of Cre recombinase. *Curr. Biol.* **8,** 1323–1326.

Denk, H., Lackinger, E., Zatloukal, K., and Franke, W. W. (1987). Turnover of cytokeratin polypeptides in mouse hepatocytes. *Exp. Cell Res.* **173,** 137–143.

Eggan, K., Akutsu, H., Loring, J., Jackson-Grusby, L., Klemm, M., Rideout, W. M., III, Yanagimachi, R., and Jaenisch, R. (2001). Hybrid vigor, fetal overgrowth, and viability of mice derived by nuclear cloning and tetraploid embryo complementation. *Proc. Natl. Acad. Sci. USA* **98**, 6209–6214.

Eggan, K., Rode, A., Jentsch, I., Samuel, C., Hennek, T., Tintrup, H., Zevnik, B., Erwin, J., Loring, J., Jackson-Grusby, L., Speicher, M. R., Kuehn, R., and Jaenisch, R. (2002). Male and female mice derived from the same embryonic stem cell clone by tetraploid embryo complementation. *Nat. Biotechnol.* **20**, 455–459.

Godwin, A. R., Stadler, H. S., Nakamura, K., and Capecchi, M. R. (1998). Detection of targeted GFP-Hox gene fusions during mouse embryogenesis. *Proc. Natl. Acad. Sci. USA* **95**, 13042–13047.

Gu, H., Marth, J. D., Orban, P. C., Mossmann, H., and Rajewsky, K. (1994). Deletion of a DNA polymerase beta gene segment in T cells using cell type-specific gene targeting. *Science* **265**, 103–106.

Guo, C., Yang, W., and Lobe, C. G. (2002). A Cre recombinase transgene with mosaic, widespread tamoxifen-inducible action. *Genesis* **32**, 8–18.

Herrmann, H., Hesse, M., Reichenzeller, M., Aebi, U., and Magin, T. M. (2003). Functional complexity of intermediate filament cytoskeletons: From structure to assembly to gene ablation. *Int. Rev. Cytol.* **223**, 83–175.

Hesse, M., Franz, T., Tamai, Y., Taketo, M. M., and Magin, T. M. (2000). Targeted deletion of keratins 18 and 19 leads to trophoblast fragility and early embryonic lethality. *EMBO J.* **19**, 5060–5070.

Jaquemar, D., Kupriyanov, S., Wankell, M., Avis, J., Benirschke, K., Baribault, H., and Oshima, R. G. (2003). Keratin 8 protection of placental barrier function. *J. Cell Biol.* **161**, 749–756.

Jeannotte, L., Ruiz, J. C., and Robertson, E. J. (1991). Low level of Hox1.3 gene expression does not preclude the use of promoterless vectors to generate a targeted gene disruption. *Mol. Cell. Biol.* **11**, 5578–5585.

Le Mouellic, H., Lallemand, Y., and Brulet, P. (1990). Targeted replacement of the homeobox gene Hox-3.1 by the *Escherichia coli* lacZ in mouse chimeric embryos. *Proc. Natl. Acad. Sci. USA* **87**, 4712–4716.

Magin, T. M., McEwan, C., Milne, M., Pow, A. M., Selfridge, J., and Melton, D. W. (1992a). A position- and orientation-dependent element in the first intron is required for expression of the mouse hprt gene in embryonic stem cells. *Gene* **122**, 289–296.

Magin, T. M., McWhir, J., and Melton, D. W. (1992b). A new mouse embryonic stem cell line with good germ line contribution and gene targeting frequency. *Nucleic Acids Res.* **20**, 3795–3796.

Magin, T. M., Schroder, R., Leitgeb, S., Wanninger, F., Zatloukal, K., Grund, C., and Melton, D. W. (1998). Lessons from keratin 18 knockout mice: Formation of novel keratin filaments, secondary loss of keratin 7 and accumulation of liver-specific keratin 8-positive aggregates. *J. Cell Biol.* **140**, 1441–1451.

Melton, D. W. (1994). Gene targeting in the mouse. *Bioessays* **16**, 633–638.

Melton, D. W., Ketchen, A. M., and Selfridge, J. (1997). Stability of HPRT marker gene expression at different gene-targeted loci: Observing and overcoming a position effect. *Nucleic Acids Res.* **25**, 3937–3943.

Messing, A., Head, M. W., Galles, K., Galbreath, E. J., Goldman, J. E., and Brenner, M. (1998). Fatal encephalopathy with astrocyte inclusions in GFAP transgenic mice. *Am. J. Pathol.* **152**, 391–398.

Meyers, E. N., Lewandoski, M., and Martin, G. R. (1998). An Fgf8 mutant allelic series generated by Cre- and Flp-mediated recombination. *Nat. Genet.* **18**, 136–141.

Mounkes, L. C., Kozlov, S., Hernandez, L., Sullivan, T., and Stewart, C. L. (2003). A progeroid syndrome in mice is caused by defects in A-type lamins. *Nature* **423**, 298–301.

Nagy, A., Gertsenstein, M., Vintersten, K., and Behringer, R. (2003). "Manipulating the Mouse Embryo: Laboratory Manual," 3rd Ed. Cold Spring Harbor Laboratory, Cold Spring Harbor, New York.

Nagy, A., Moens, C., Ivanyi, E., Pawling, J., Gertsenstein, M., Hadjantonakis, A. K., Pirity, M., and Rossant, J. (1998). Dissecting the role of N-myc in development using a single targeting vector to generate a series of alleles. *Curr. Biol.* **8,** 661–664.

Nagy, A., Rossant, J., Nagy, R., Abramow-Newerly, W., and Roder, J. C. (1993). Derivation of completely cell culture-derived mice from early-passage embryonic stem cells. *Proc. Natl. Acad. Sci. USA* **90,** 8424–8428.

Omary, M. B., Ku, N. O., and Toivola, D. M. (2002). Keratins: Guardians of the liver. *Hepatology* **35,** 251–257.

Peters, B., Kirfel, J., Bussow, H., Vidal, M., and Magin, T. M. (2001). Complete cytolysis and neonatal lethality in keratin 5 knockout mice reveal its fundamental role in skin integrity and in epidermolysis bullosa simplex. *Mol. Biol. Cell* **12,** 1775–1789.

Plum, A., Hallas, G., Magin, T., Dombrowski, F., Hagendorff, A., Schumacher, B., Wolpert, C., Kim, J., Lamers, W. H., Evert, M., Meda, P., Traub, O., and Willecke, K. (2000). Unique and shared functions of different connexins in mice. *Curr. Biol.* **10,** 1083–1091.

Porter, R. M., Leitgeb, S., Melton, D. W., Swensson, O., Eady, R. A., and Magin, T. M. (1996). Gene targeting at the mouse cytokeratin 10 locus: Severe skin fragility and changes of cytokeratin expression in the epidermis. *J. Cell Biol.* **132,** 925–936.

Reichelt, J., Bussow, H., Grund, C., and Magin, T. M. (2001). Formation of a normal epidermis supported by increased stability of keratins 5 and 14 in keratin 10 null mice. *Mol. Biol. Cell* **12,** 1557–1568.

Rossant, J., and Nagy, A. (1995). Genome engineering: The new mouse genetics. *Nat. Med.* **1,** 592–594.

Schaft, J., Ashery-Padan, R., van der, H. F., Gruss, P., and Stewart, A. F. (2001). Efficient FLP recombination in mouse ES cells and oocytes. *Genesis* **31,** 6–10.

Seibler, J., Zevnik, B., Kuter-Luks, B., Andreas, S., Kern, H., Hennek, T., Rode, A., Heimann, C., Faust, N., Kauselmann, G., Schoor, M., Jaenisch, R., Rajewsky, K., Kuhn, R., and Schwenk, F. (2003). Rapid generation of inducible mouse mutants. *Nucleic Acids Res.* **31,** e12.

Stacey, A., Schnieke, A., McWhir, J., Cooper, J., Colman, A., and Melton, D. W. (1994). Use of double-replacement gene targeting to replace the murine alpha-lactalbumin gene with its human counterpart in embryonic stem cells and mice. *Mol. Cell. Biol.* **14,** 1009–1016.

Tamai, Y., Ishikawa, T., Bosl, M. R., Mori, M., Nozaki, M., Baribault, H., Oshima, R. G., and Taketo, M. M. (2000). Cytokeratins 8 and 19 in the mouse placental development. *J. Cell Biol.* **151,** 563–572.

Utomo, A. R., Nikitin, A. Y., and Lee, W. H. (1999). Temporal, spatial, and cell type-specific control of Cre-mediated DNA recombination in transgenic mice. *Nat. Biotechnol.* **17,** 1091–1096.

Wu, H., Liu, X., and Jaenisch, R. (1994). Double replacement: Strategy for efficient introduction of subtle mutations into the murine Col1a-1 gene by homologous recombination in embryonic stem cells. *Proc. Natl. Acad. Sci. USA* **91,** 2819–2823.

Wunderlich, F. T., Wildner, H., Rajewsky, K., and Edenhofer, F. (2001). New variants of inducible Cre recombinase: A novel mutant of Cre-PR fusion protein exhibits enhanced sensitivity and an expanded range of inducibility. *Nucleic Acids Res.* **29,** E47.

Ying, Q. L., and Smith, A. G. (2003). Defined conditions for neural-commitment and differentiation. *Methods Enzymol.* **365,** 327–341.

Ying, Q. L., Nichols, J., Chambers, I., and Smith, A. (2003). BMP induction of Id proteins suppresses differentiation and sustains embryonic stem cell self-renewal in collaboration with STAT3. *Cell* **115,** 281–292.

Yu, Y., and Bradley, A. (2001). Engineering chromosomal rearrangements in mice. *Nat. Rev. Genet.* **2,** 780–790.

CHAPTER 5

Uncovering the Roles of Intermediate Filaments in Apoptosis

Normand Marceau,[*,†] Stéphane Gilbert,[*,†] and Anne Loranger[†]

*Centre de recherche en cancérologie et Département de médecine
Université Laval
GIR 2J6 QC, Canada

†Centre de recherche de L'Hôtel-Dieu de Québec (CHUQ)
GIR 2J6 QC, Canada

I. Introduction
    A. Intermediate Filament Involvement in Cell Differentiation and Proliferation
    B. Intermediate Filament Protein Involvement in Cell Stress and Apoptosis
II. Materials and Instrumentation
    A. Equipment
    B. Chemicals and Protein Factors
    C. Supplies
    D. Solutions, including Culture Media
III. Procedures
    A. Liver Perfusion and Hepatocyte Isolation
    B. Preparations of Coated Dishes
    C. Primary Culture
    D. Transfection
    E. Induction of DR-Receptor–Mediated Apoptosis
    F. Analysis of Intermediate Filament Protein Involvement Upstream of Caspase Activation
    G. *In Vivo* Analysis of DR-Mediated Hepatocyte Apoptosis
    H. Analysis of Intermediate Filament Protein Involvement Downstream of Caspase Activation
IV. Pearls and Pitfalls
V. Conclusion and Perspectives
    A. Intermediate Filaments as Markers of Differentiation and Apoptosis during Tumor Formation and Progression

I. Introduction

Apoptosis or active cell death is a process whereby the cell dies in response to a wide range of internal and external signals (Kerr *et al.*, 1972; Wyllie *et al.*, 1980). This highly conserved process enables multicellular organisms to eliminate cells that are damaged or mislocated or have become superfluous (Ellis *et al.*, 1991, 1996). The sequence of morphological changes inherent to cell apoptosis, first described by Kerr *et al.* (1972), seems to be common to most cell types and is characterized by cell shrinking to the extent that the cell pulls away from neighboring cells, undergoing then both nuclear and cytoplasmic condensation. The latter leads to the formation of dense inclusions devoid of membrane, and ultimately to membrane-enclosed apoptotic bodies, which are normally taken up by neighboring cells (Wyllie *et al.*, 1980). Although apoptosis can be triggered by diverse stimuli, the central execution machinery is evolutionarily conserved (Zheng and Flavell, 1999) and is based on biochemical changes that include the cleavage of key cellular proteins by caspases, a family of cysteine proteases functioning as a cascade in the death process (Kumar, 1999; Martin and Green, 1995; Zheng and Flavell, 1999). Together with actin microfilaments (MFs) and microtubules (MTs), intermediate filaments (IFs) form the cytoskeleton, and each of them requires interactions with associated proteins to exert their functions properly (Wiche, 1998; Fuchs and Karakesisoglou, 20^1). Although a prominent role for cytoskeleton is to maintain cellular integrity, it seems that IFs constitute a protein scaffold that is largely responsible for the capacity of cells to sustain mechanical and nonmechanical stresses, including those that can lead to apoptosis.

Several excellent reviews describing our current knowledge of the assembly, structure, dynamics, and functions of IFs have been published (Coulombe and Omary, 2002; Coulombe *et al.*, 2000; Fuchs and Cleveland, 1998; Marceau *et al.*, 2001; Oshima, 2002; Owens and Lane, 2003; Paramio and Jorcano, 2002); the description of experimental approaches and methods available for the detailed analyses of these cell biology aspects constitutes the goal of this volume. This chapter focuses on the analysis of the roles fulfilled by events that provide cell resistance to apoptosis stimulation before commitment or those that are part of the execution phase in the committed cells.

A. Intermediate Filament Involvement in Cell Differentiation and Proliferation

The role fulfilled by IFs in apoptosis cannot be addressed without first considering the association of IFs with cell differentiation and proliferation. Indeed, in a tissue, cells can be in a dividing state, in a quiescent state where they undergo differentiation, or in a state into which they can exit the cell cycle and undergo apoptosis or enter the cell death pathway directly from the differentiated state (Walker *et al.*, 1995). Cell differentiation is the result of differential gene expression, and cytoskeletal components and their associated proteins are instrumental in the appearance of the cellular phenotype. Although both actins and tubulins are ubiquitous and encoded by few genes, IF proteins are encoded by more than 60 different genes (Fuchs and Weber, 1994; Hesse *et al.*, 2001), which can be classified into five types and are characteristic of the tissue they are expressed in. The first four types are components of the cytoplasm and are recognized as acid keratins (type I) and basic keratins (type II) present in all epithelia; vimentin, desmin, glial fibrillary acidic protein (GFAP), and peripherin (type III) in mesenchymal, muscle, glial, and neuronal cells, respectively; and neurofilaments in neurons and cells of the peripheral neuroendocrine system (type IV). The type V is provided by the lamins, which are components of the nuclear matrix of all cells and are expressed as three isoforms only (Weber *et al.*, 1989). Although it seems that the differential lamin gene expressions are not particularly associated with cell differentiation, they probably played a central role in IF gene evolution, because phylogenetic analyses suggest that they are the ancestors of the cytoplasmic IF genes. Type I and type II keratins (Ks) constitute the largest and most complex class of IFs and the genes, which exist as various isoforms in mammalian genomes (Hesse, 2004) and are expressed as pairs, exhibit epithelium-specific and cell differentiation–specific regulation (see Coulombe and Omary, 2002, for extensive review). For instance, the terminal differentiation of keratinocytes in epidermis and other cornifying epithelia requires a progressive switch in pair expression from K5/K14 in the basal layer to K1/K10 in the differentiating layers, and the partners exist as very few isoforms (Fuchs, 1996; Hesse, 2004). In comparison, the case of K6, which partners with K16 and K17, is rather unique, given that it exists as multiple isoforms that are expressed as distinctive pairs and at various levels in different epithelial appendages and are induced in several complex epithelia responding to a variety of insults; these and other special features of keratin IFs in cornifying epithelia are addressed in Chapter 18. In simple epithelia, all cells normally contain the K8/K18 pair, and some of them express two to three other keratins (e.g., K7, K19, and K20) in addition (Marceau and Loranger, 1995; Marceau *et al.*, 1989; Omary and Ku, 1997; Oshima *et al.*, 1996). Notably, K8 and K18 constitute the first cytoplasmic IF genes to be expressed in the embryo (Lane *et al.*, 1983; Oshima and Baribault, 1992). Although they progressively mature during development, hepatocytes maintain solely K8 and K18, and from a methodological standpoint, hepatocytes provide a model of choice to assess the contribution of IFs to the maintenance of cellular integrity, particularly in the context of cell resistance to various forms of stress or apoptosis.

Through evolution, individual IFs have diverged and specialized the N-terminal and C-terminal domains that flank their central α-helical (rod) domain required for IF assembly (Herrmann and Aebi, 2000; Herrmann *et al.*, 2003). The rod domain consists of four segments (IA, IB, IIA, and IIB) that are separated by three linkers (L1, L1–2, L2), which act as modulators of the assembly. Although type III proteins, like vimentin and desmin, can assemble as homopolymers or heteropolymers, Ks assemble as obligate heteropolymers of individual type I and type II proteins. The beginning and end portions of the rod domain are conserved among all IF proteins, and point mutations occurring naturally in these rod portions lead to aberrant IF organization and dynamics (Fuchs and Weber, 1994; McLean and Lane, 1995; Omary and Ku, 1997), which in turn result in dramatic loss of cellular integrity and severe pathologies. In contrast, the N-terminal and C-terminal domains contribute to most of the structural heterogeneity of IF proteins and contain the sites that can undergo modifications, like the phosphorylation of particular serine residues by means of various signaling kinase cascades (Ku *et al.*, 1996a, 2002a; Omary *et al.*, 1998), suggesting therefore that they are most important from functional and regulatory standpoints. These terminal domains are also responsible for the immunogenic properties of the individual IF proteins, and immuno-based assays of their differential expression, tissue specificity, and developmental regulation have provided useful tools for analyses of development, differentiation, and neoplastic transformation (Moll *et al.*, 1982; Prasad *et al.*, 1999); as cell-specific markers, IF proteins are also of immense value in tumor diagnosis.

Aside from the overall correlation between differential IF gene expression and cell differentiation, there are proliferative conditions, such as those occurring during tissue regeneration in response to injuries, in which the early repair events include a transient switch in IF gene expression. This is the case for wound repair in the epidermis and other hyperproliferative skin disorders, in which the K6/K16 pair is expressed in the suprabasal layers substituting, therefore, the K1/K10 pair during the repair process (McGowan and Coulombe, 1998; Paladini *et al.*, 1996). Notably, the forced expression of human K16 in the skin of transgenic mice induced a transient hyperproliferation of keratinocytes (Paladini and Coulombe, 1998), whereas the forced expression of K10 in the basal cell layer leads to cell growth arrest (Santos *et al.*, 2002). However, during the tissue regeneration that follows partial hepatectomy, the hepatocytes exhibit a rapid proliferation without any changes in keratin expression (i.e., the K8/K18 pair is maintained throughout the repair process) (Marceau and Loranger, 1995). Nevertheless, the loss of K8/K18 in K8-null or K18-null mice perturbs the progression of hepatocytes through the G2/M phase transition of the cell cycle, and the anomaly is more frequent in regenerating liver (Toivola *et al.*, 2001). Mechanistically, K18 is known to bind 14-3-3 proteins in a phosphorylation-dependent manner, which in turn bind to Cdc25, a key regulator of mitosis (Ku *et al.*, 1998, 2002b), implying, therefore, a perturbed 14-3-3–cdc25 association in K-null hepatocytes (see Chapter 17 for further details on this important cell cycle issue).

B. Intermediate Filament Protein Involvement in Cell Stress and Apoptosis

Because of their accessibility and their specialization, the availability of both tissue and primary culture based approaches, and the severe IF-dependent diseases associated with them, keratinocytes and hepatocytes are the two cell models that are most frequently used, particularly when the central objective of the research is to investigate IF involvement in cell stress and apoptosis under culture conditions can be verified *in vivo*.

Maintenance of the surface membrane mechanical integrity constitutes a prominent role of K IFs in keratinocytes (Coulombe and Omary, 2002; Coulombe *et al.*, 2000). Indeed, point or null mutations in epidermal keratins lead to skin fragility and keratinocyte cytolysis, particularly under mechanical stress (Fuchs and Cleveland, 1998; McLean and Lane, 1995). Although similar mutations in K8 or K18 induce little hepatocyte cytolysis under normal conditions (Baribault *et al.*, 1994), this K pair is required for the maintenance of hepatocyte integrity under mechanical stress conditions (Loranger *et al.*, 1997). For instance, after partial hepatectomy, many damaged hepatocytes are observed in the K8-null mouse liver as a result of a rupture of the surface membrane. A similar rupture of the K8-null hepatocyte surface can be obtained on liver perfusion at a flow rate, which has no effect on wild-type mouse hepatocytes (discussed in detail in "Procedures").

Although resilience to mechanical stress is most likely a common functional feature of IFs, different IF proteins display structural differences, implying that they can also play some nonmechanical tissue-specific protective functions. For instance, we have observed a dramatic decrease in the survival of K8-null mice after partial hepatectomy under pentobarbital (rather than ether) anesthesia, suggesting that K8-null mouse hepatocytes are much more sensitive to toxic stress (Loranger *et al.*, 1997). This is in direct line with other findings that K8-null and Arg89 → Cys human K18 mutant mice are highly sensitive to hepatotoxins like griseofulvin and acetaminophen (Cadrin *et al.*, 1996; Ku *et al.*, 1996b), both of which are generators of oxidative stress, which in turn can result in apoptosis.

Apoptosis induction involves either the stimulation of death receptors at the surface membrane by members of the tumor necrosis factor (TNF) family, TNF-α, Fas ligand (FasL), and TNF-related apoptosis-inducing ligand (TRAIL) or the perturbation of mitochondria by toxic agents or excessive oxidative stress (Ashkenazi and Dixit, 1998; Kaufmann and Gores, 2000; Krammer, 1999). Both mechanisms result in the activation of caspases by other caspases to generate the active enzyme. The initial signaling events leading to caspase activation are well established. For instance, FasL stimulation of the Fas receptor induces trimerization of the latter at the surface membrane, which allows the recruitment of Fas-associated death domain (FADD) protein, which, with the initiator caspase-8, forms the death-inducing signaling complex (DISC) (Chinnaiyan *et al.*, 1996; Kischkel *et al.*, 1995). This triggers caspase-8 activation, which in turn activates effector caspases like caspases-3, 6, and 7 (Hengartner, 2000; Oshima, 2002). In certain cell types, the apoptotic response can be amplified by the release of cytochrome *c* from mitochondria and the subsequent activation of the initiator caspase-9, which in turn enables activation of further caspase 3 (Hengartner, 2000; Kaufmann and Gores, 2000; Krammer, 1999). The activation of effector caspases triggers the execution phase of apoptosis, which includes cleavage of numerous substrates. In contrast to FasL, TNF-α stimulates two receptors, TNF-R1 and TNF-R2 (Ashkenazi and Dixit, 1998; Baker and Reddy, 1996), which normally promote cell proliferation after TNF receptor–associated death domain (TRADD) protein-mediated binding to an appropriate member of the TRAF family. However, in cells sensitized with cycloheximide or actinomycin D, stimulation of TNF-R1, but not TNF-R2 (which lacks the death domain), can lead to apoptosis after binding to TRADD, which recruits FADD to activate the caspase cascade (Chinnaiyan *et al.*, 1996; Hsu *et al.*, 1996). Stimulation of the TRAIL receptors (i.e., TRAIL-R1 and TRAIL-R2) can lead to apoptosis by a sequence of events that correspond to those triggered by Fas (Kischkel *et al.*, 2000; Sprick *et al.*, 2000). In various cell lines, cycloheximide or actinomycin D sensitizes for apoptosis by down-regulating the synthesis of c-Flip, a labile protein homologous to caspase-8 presenting an inactive catalytic site, so that the balance between cell survival and death can be modulated by the relative concentration of the death-receptor and c-Flip (Fulda *et al.*, 2000; Tschopp *et al.*, 1998).

IF proteins seem to be involved upstream and downstream of the caspase machinery in cells responding to apoptotic stimuli. In the first functional trait, IFs are part of the events that provide cell resistance to apoptosis before commitment. In the second aspect, cells undergo an irreversible process that includes the ordered cellular dismantlement, mainly by means of the activation of the caspase cascade in interplay with IFs acting alternatively as targets to and scaffolds for caspases.

1. Intermediate Filament Protein Involvement Upstream of Caspase Activation

Much of our knowledge on the IF modulation of cell apoptosis is based on work performed on the simple epithelium keratins K8/K18. The increased resistance to chemotherapeutic agents of fibroblasts forced to express K8 and K18 has been the first evidence suggesting that this keratin pair could exert a role in apoptosis modulation (Bauman *et al.*, 1994). A key player in the regulation of hepatocyte apoptosis is Fas (Galle *et al.*, 1995; Kaufmann and Earnshaw, 2000; Nagata, 1999). For instance, binding of Fas by FasL or an agonistic anti-Fas antibody (Jo2) induces rapid, massive apoptosis of mouse hepatocytes *in vivo* and less severely in primary culture (Gill *et al.*, 1998; Ni *et al.*, 1994; Ogasawara *et al.*, 1993). Another line of work has shown that the Fas density at the hepatocyte surface is controlled by means of an MT-dependent regulation of trafficking between Golgi and plasma membrane (Feng and Kaplowitz, 2000). In line with these observations, we have shown that K8/K18-null mouse hepatocytes are much less resistant than wild-type mouse hepatocytes to Fas-mediated apoptosis (Gilbert *et al.*, 2001) and that K8/K18 act as a modulator of Fas density at the cell surface. Notably, this increased sensitivity is associated with a higher and more rapid activation of caspase-3, implying that K8/K18 act in events upstream of the caspase execution machinery. In contrast, the loss of K8/K18 does not modulate the hepatocyte response to TNF-α or TRAIL, thus implying a preferential association between Ks and Fas-mediated death. This could have an important application in human liver disorders, considering that the overexpression of a dominant-negative K18 human mutant (K18 Arg89 $\rightarrow$ Cys) in transgenic mouse liver predisposes to Fas but not TNF-mediated apoptosis (Ku *et al.*, 2003).

In a complementary line of work, the Fas system has been shown to contribute massively to the keratinocyte apoptosis occurring during toxic epidermal necrolysis or in response to ultraviolet radiation (Viard-Leveugle *et al.*, 2003). Under normal conditions, keratinocyte apoptosis is a rare event, despite the fact that *in vivo* Fas and FasL are coexpressed in basal and suprabasal keratinocytes and that cell death does occur in cultured keratinocytes on Fas stimulation by exogenous recombinant FasL. The evidence provided by these findings indicates that in nonstimulated epidermal cells FasL remains intracellular and is predominantly associated with keratin IFs, thus suggesting an IF involvement in the regulation of FasL intracellular localization.

However, the K8/K18 modulation of Fas versus TNF receptor–mediated apoptosis may vary with the cell types. Indeed, normal and malignant epithelial cells exhibiting various K8 or K18 perturbations are more sensitive to TNF-α–induced death in the presence of cycloheximide (Caulin *et al.*, 2000), and in these cells K8 is capable of binding TNFR2 and thus modulating the signaling to the caspase cascade by means of a mechanism that may involve a TNFR2 induction of TNF-α at the cell surface (Caulin *et al.*, 2000; Oshima, 2002). Alternatively, from work performed on several cell lines, it seems that K18 attaches to TRADD and that on TNF-α stimulation, there is a competition of activated TNFR1 and K18 for TRADD, in which case TRADD dissociates from K18 and associates with activated TNFR1 to form the signaling complex that activates caspase 8 (Inada *et al.*, 2001). In addition, this K8/K18 modulation of TNF-mediated apoptosis seems to be of prime importance in the context of the maintenance of the trophoblast tissue integrity during mouse development, the evidence being that K8-null trophoblast giant cells are particularly sensitive to maternal-generated TNF, inducing a failure of the placental barrier function (Jaquemar *et al.*, 2003). This provides an explanation for the embryonic lethality observed in K8-null mice of certain genetic backgrounds (Baribault *et al.*, 1993; Hesse *et al.*, 2000). It is worth noting that K14 also binds to TRADD, an association that seems to link the K14-containing aggregates to the TNF-mediated cytolysis in epidermolysis bullosa simplex (EBS) keratinocytes (Yoneda *et al.*, 2004). Whether a similar TNF-dependent mechanism can explain how a K17-null mutation leads to massive apoptosis in all major cell compartments of the hair follicle remains an open question (McGowan *et al.*, 2002).

Moreover, recent studies have shown that death-receptor (DR) receptor stimulation can activate protein kinase signaling pathways that normally regulate cell growth or at least cell survival (Holmstrom *et al.*, 2000; Tran *et al.*, 2001), thus generating events that deviate from the caspase cascade and trigger an antiapop-totic signaling cascade. Phosphoserine sites in the N-terminal and C-terminal domains of IF proteins exhibit motifs recognized by protein kinases, which include mitogenic activated protein (MAP) kinases and cell cycle–dependent kinases that are normally activated in growth-stimulated and mitotic cells (Baribault *et al.*, 1989; Ku *et al.*, 1996a; Marceau *et al.*, 2001; Omary *et al.*, 1998; Paramio and Jorcano, 2002). In simple epithelial cells, phosphorylation also modulates K18 ubiquitination and thus its turnover by proteasomes (Ku and Omary, 2000), and in the same way, phosphorylation affects the caspase cleavage of Ks type I (Ku and Omary, 2001). In addition, Fas stimulation activates the antiapoptotic ERK1/ 2 signaling pathway at the early period and the stress-associated kinase JNK at a later period, and this results in phosphorylation of specific serine residues on K8 and K18 (He *et al.*, 2002). In the same way, activation of the other stress-associated kinase p38 by various insults also leads to phosphorylation of a specific serine site on K8 (Ku *et al.*, 2002a). Of particular interest here, we have direct evidence showing that Fas stimulation leads to a much lower activation of the ERK1/2 signaling pathway in K8-null mouse than in wild-type hepatocytes and that inhibition of this pathway largely increases the sensitivity of wild type (WT) but not of K8-null hepatocytes to induced apoptosis (Gilbert *et al.*, 2004). In addition, the loss of K8/K18 is associated with a marked reduction of c-Flip, a down-regulation that also takes place in a K8-null mouse mammary cell line. Because c-Flip normally competes for caspase 8 activation (Tschopp *et al.*, 1998), the increased caspase signaling observed in K8-null hepatocytes (Gilbert *et al.*, 2001) (Fig. 5) could in part be due to the c-Flip deficit. It seems, therefore, that the K8/K18-dependent resistance to Fas-mediated apoptosis occurs through a regula-tion of Fas density at the surface and a concerted c-Flip–dependent modulation of caspase activation and ERK1/2 activation (Gilbert *et al.*, 2004).

2. Intermediate Filament Protein Involvement Downstream of Caspase Activation

A strong apoptotic challenge inevitably activates the caspase cascade, whereby effector caspases triggered the execution phase. To ensure that the cell dismantle-ment has minimal effects on neighboring cells, the process mainly relies on the ordered cleavage of caspase substrates. Although our knowledge of molecular mechanisms is still at its infancy, recent data have indicated that this cellular disintegration requires a regulated IF reorganization by means of caspase cleavage of IF proteins at specific sites.

In this regard, the L1–2 linker region in the rod domain of IF proteins is known to contain specific consensus sequences for effector caspases (e.g., VEMD and VEVD), where the aspartate is the cleavage site (Ku *et al.*, 1999; Oshima, 2002). The first target protein to be identified was lamin A (Lazebnik *et al.*, 1995), and its cleavage occurs before packaging of the condensed chromatin into apoptotic bodies; an uncleavable lamin A or B mutant leads to delayed cell death and DNA fragmentation. With regard to cytoplasmic IFs, K18 was the first to be reported as IF caspase targets in cells responding to apoptotic agents (Caulin *et al.*, 1997). In fact, cleavage at the L1–2 linker region has been confirmed so far for lamins; type I K14, K15, K17, K18, and K19; vimentin; and desmin (Table I). With regard to the caspase-mediated cleavage of IF proteins other than lamins and Ks, the case of vimentin is particularly intriguing, because its cleavage has been shown to disrupt the IF network and in fact to promote apoptosis (Byun

Table I
Summary of Intermediate Filament Protein Cleavage in Apoptosis

Intermediate filaments	Cut sites (Asp)	Weight (kDa)	Fragment weight (kDa)	Caspase #	References
Lam in A	230	72	47	*f*	*a–f*
Lam in B1	230	76	46, 21, 28	*f, g*	*e–i*
Lam in B2	141	66	45	*f, g*	*f, i, j*
Lam in C	230	62	37	*f*	*e, f*
K14	273	52	29, 23	n.a.	*k*
K15	264, 445	50	28, 25, 22	*c, f, g*	*l*
K17	241, 416	46	26, 23, 20	n.a.	*l*
K18	237, 396	48	45, 26, 22, 19	*c, f, g*	*h, k, m–q*
K19	237	40	38, 28, 20	n.a.	*k, l, p, q*
K8	n.a.	52,5	48, 46	n.a.	*p–r*
Vimentin	85, 259, 429	58	48, 44, 36, 29, 25, 15	*c, f, g, h, l*	*g, r–u*
Desmin	263	53	29, 27	*f*	*v*

[a] Lazebnik *et al.*, 1952.
[b] Oberhammer *et al.*, 1994.
[c] Takahashi *et al.*, 1996.
[d] Orth *et al.*, 1996.
[e] Rao *et al.*, 1996.
[f] Ruchaud *et al.*, 2002.
[g] Morishima, 1999.
[h] Stegh *et al.*, 2000.
[i] Korfali *et al.*, 2004.
[j] Oshima, 2002.
[k] Ku and Omary, 2001.
[l] Badock *et al.*, 2001.
[m] Caulin *et al.*, 1997.
[n] Ku *et al.*, 1997.
[o] Leers *et al.*, 1999.
[p] MacFarlane *et al.*, 2000.
[q] Prasad *et al.*, 1998.
[r] van Engeland *et al.*, 1997.
[s] Hashimoto *et al.*, 1998.
[t] Suarez-Huerta *et al.*, 2000.
[u] Byun *et al.*, 2001.
[v] Chen *et al.*, 2003.
n.a., not available.

et al., 2001). Desmin is also a caspase target, and in this case the stable expression of a caspase cleavage-resistant fragment partially protects cells from TNF-mediated apoptosis (Chen *et al.*, 2003). None of the type II keratins contain similar cleavage sites (Ku *et al.*, 1999; Oshima, 2002). Table I also provides a summary of what is known about the proteolytic fragments that are generated by caspases for lamins, keratins, vimentin, and desmin. To our knowledge, there are no equivalent data on the caspase-mediated cleavage of GFAP, nestin, or neurofilaments, although sequence alignments have suggested that they are likely caspase substrates (Oshima, 2002).

Although a type II keratin, like K8, remains intact in most cases, the two caspase-generated fragments of its type I keratin partner K18 remain attached to K8, and this aberrant association destabilizes the IF network (Caulin *et al.*, 1997). In fact, the IF stability is most likely modulated by successive cleavages, considering that the DALD sequence in the C-terminal domain of K18 is cleaved very early in the process, and this is followed by a cleavage at L1–2 VEVD sequence (Oshima, 2002). Notably, K18 hyperphosphorylation inhibits caspase-3 cleavage at the VEVD site but not the cleavage at DALD (Ku and Omary, 2001; Ku *et al.*, 1997). One can assume that the successive K18 cleavages facilitate the dismantlement of the K–IF network and the disposal of the apoptotic cell content.

In addition, the ordered caspase-dependent dismantlement of apoptotic cells seems to involve an interplay between caspases and IF proteins or some associated (e.g., scaffold) proteins. One of the associated proteins is plectin, a prominent member of the multigene plakin family, that cross-links the MT, MF, and IF cytoskeletal networks in essentially all cell types (Leung *et al.*, 2002; Ruhrberg and Watt, 1997; Wiche, 1998). Although the caspase cascade in DR-mediated apoptosis is initiated by the activation of caspase-8 at the FADD-containing complex, the monitoring of caspase 8 distribution has revealed that it is located mainly at the mitochondria and that on DR-mediated apoptosis, most of the active caspase 8 translocates to plectin and cleaves most of it (Stegh *et al.*, 2000). Notably, this plectin cleavage precedes that of other caspase substrates such as lamin B and keratins and leads to a reorganization of the actin-MFs, without appreciable immediate effects on the IF network (Stegh *et al.*, 2000). From work performed with cultured fibroblasts from plectin-deficient mice, it seems that plectin is, in fact, required for actin-MF reorganization (Andra *et al.*, 1998). Thus, the caspase 8-mediated cleavage of plectin plus the associated actin-MF rearrangement constitute very early events in the dismantlement of apoptotic cells. On these grounds, caspase 8 not only initiates the activation of the caspase cascade, but it triggers in parallel the sequential reorganization of the cytoskeletal networks, starting with actin-MFs.

A second IF-associated protein of interest in epithelial cells and cardiac muscle cells is desmoplakin, another member of the plakin family that stabilizes the desmosomes through keratin–IF and desmin–IF interactions, respectively (Green and Gaudry, 2000; Green *et al.*, 1998). In epithelial cells, the activation of effector caspases has been shown to result in the cleavage of the desmosomal proteins, including desmoplakin (Bojarski *et al.*, 2004; Weiske *et al.*, 2001), and this largely contributes to cell rounding and disintegration of cell–cell contacts before the formation of apoptotic bodies. In terms of the sequential events that lead to the dismantlement of apoptotic cells, it seems that in contrast to plectin cleavage, the desmosomal-protein cleavage constitutes a late event (i.e., certainly after that of type I keratins).

To go a step further in key protein cleavage, IFs are not only caspase targets but they also act as scaffolds in sequential caspase activation, a regulatory process that involves the contribution of a central signaling molecule originally identified as a "death effector domain containing DNA binding domain" (DEDD) (Lee *et al.*, 2002). Although it seems that DEDD functions as a "communicator" between the cytoplasm and the nucleus in apoptotic cells, the experimental evidence accumulated so far indicates that in nonapoptotic simple epithelial cells, a certain amount of DEDD is constitutively associated with K8/K18 IFs and caspase 3, implying that DEDD can direct caspase 3 to the IFs (Dinsdale *et al.*, 2004). Caspase 9 is activated at an early stage of apoptosis, and its active components are then found also on K8/K18 IFs, thus providing an ordered way to activate caspase 3, which in turn cleaves K18 (Dinsdale *et al.*, 2004). Moreover, caspase 3 can activate more caspase 9, thus generating an amplification loop that facilitates the K18 cleavage. This results in the disruption of the IF network and the formation of cytoplasmic cleaved-K18–containing aggregates at the late stage of apoptosis. However, the early sequential events in committed cells also include K8/K18 hyperphosphorylation (Liao *et al.*, 1997; Omary *et al.*, 1998), a prominent molecular modification that seems to render tetramers unsuitable for reassembly into IFs (Strnad *et al.*, 2001, 2002) and to promote the formation of caspase-cleaved K18–containing inclusions. These aggregates not only contain K8/K18 but also DEDD and active caspases, ubiquitin, ubiquitinated proteins, HSP-72, and TRADD (Dinsdale *et al.*, 2004; MacFarlane *et al.*, 2000). Such inclusions are reminiscent of the cytoplasmic structures, known as sequestosomes (Stumptner *et al.*, 2002), that seem to retain potentially noxious proteins for the healthy adjacent cells.

Four main points come out of this background: (1) IF proteins can readily be used as markers of cell lineage establishment and cell differentiation. However, an IF loss does not change the differentiation status of the cells but leads instead to a loss of cell integrity or at least to an increased cell susceptibility to various forms of stress, including those that may result in apoptosis. This provides a dramatic example of cells that may enter the cell death pathway directly from the differentiated state. (2) Apoptosis can occur by stimulation of DR at the surface membrane or perturbation of mitochondria by toxic agents or oxidative stress. (3) The strongest experimental evidence for IF involvement in apoptosis has come from work performed on simple epithelial cells, particularly for the K8/K18 pair. (4) K8/K18 have been shown to be modulators of apoptosis by acting both upstream and downstream of the caspase machinery activation. Despite the strong evidence derived from K8/K18-null mice indicating that liver is a model of choice to study the role of K8/K18 in apoptosis, few data have come out on the use of cultured hepatocytes to address the underlying molecular mechanisms. The rest of the chapter describes the procedures developed in our laboratory for the high yield isolation of viable WT versus K8/K18-null mouse hepatocytes and their setup in primary culture and the main advantages of using these *in vitro* simple epithelial cell models for the study of IF involvement in apoptosis, both as regulators and markers of apoptosis. A change in K8/K18 phosphorylation at specific Ser sites is an early sign of the cell response to apoptosis, whereas cleavage of K18 by caspase is a marker of cells committed to apoptosis. Whether the same *in vitro* approach can be transposed to the analysis of the involvement of other IF proteins in apoptosis remains an open issue.

II. Materials and Instrumentation

A. Equipment

Incubator, model 3158, CO_2, water-jacketed, Forma Scientific

Centrifuge, model TJ-6, Beckman

Laminar-flow hood, model 860, CCI

pH-Meter, Accumet 915, Fisher Scientific

Surgical scissor, straight sharp/sharp, 14002-12, FST

Fine scissor, straight 12 cm, 14068-12, FST

Forceps Semken, straight, 13 cm, 11008-13, FST

Forceps, slim 1×2, 14.5 cm, 11023-14, FST

Hemostatic forceps ($2\times$), curved 12.5 cm, 130096-12, FST

Vascular clamp, bulldog type serrefines, straight, 28 mm, 18050-28, FST

Perfusion apparatus (see Fig. 1):

Hood class 100, model VLF/2/4, Microzone

Peristaltic pump, model no. 7014, Cole Parmer

Water bath, type FJ, HAAKE

Laser scanning confocal microscope:

MRC-1024, Bio-Rad

Krypton/argon 15 mW laser, model 5400B-115-00-2, Ion Laser Technology

Diaphot-TMD inverted microscope equipped with epifluorescence, Nikon

Digital imaging fluorescence system:

Inverted TE2000 microscope equipped with epifluorescence, Nikon

Proscan X,Y,Z stage, Prior

Shutter VMM-DI, Uniblitz

Filter wheel Lambda 10-C, Sutter

Micromax CCD monochrome camera, Princeton

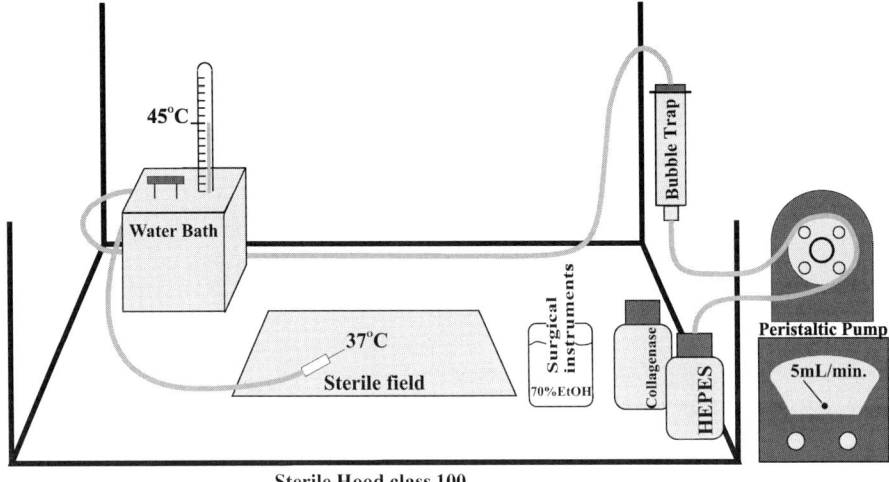

Fig. 1 Schematic representation of the perfusion system. Propelled by the peristalic pump, appropriate solutions pass through the tubing across the bubble trap and enter a water bath to gain the proper temperature at 37 °C at the needle.

Heated stage MC-6, Linkam

Environmental chamber

MetaMorph Imaging System, Universal Imaging Corporation

B. Chemicals and Protein Factors

Albumin bovine fraction V, fatty acid free, 152401, ICN

ALT colorimetric kit, no. AL 1205, Randox

Colchicine, no. C-9754, SIGMA

Collagenase type I, no. CLS-1, Worthington

Cycloheximide and actinomycine D, Sigma

Dexamethasone sodium phosphate, SABEX

Dulbecco's modified Eagle medium, 12800-017, GIBCO

Effectene transfection reagent no. 301427, Qiagen

F-12 nutrient mixture (Ham), 21700-075, GIBCO

Fibronectin, purified from human serum as described before (Engvall and Ruoslahti, 1977)

Holo-transferrin, T 4132, SIGMA

Insulin from bovine pancreas, I 6634, SIGMA

Isoflurane (Baxter)

Matrigel matrix, no. 354234, BD Biosciences
PD 98059, SB 203580 and SP 600125, Sigma
Penicillin/streptomycin, 10000 IU/10000 μg/ml, 30-0020CI, Wisent
Proviodine (Rougier)
Purified anti-mouse Fas (Jo2) NA/LE, no. 554254, BD Pharmingen
R-phycoerythrin–labeled (PE-labeled) anti-Fas (Jo2), no. 554258, BD Pharmigen
Soluble trail mouse recombinant, no. se722-100, Biomol
Tumor necrosis factor alpha (TNFα) from mouse, no. T7539, Sigma

C. Supplies

Cover slips no. 1, 22-mm square, no.48366-067, VWR
Filters 0.45 μm, 25 mm (HAWP02500), and 47 mm (HAWP04700), Millipore
Hypodermic needles 25G 5/8″ and 27G 1/2″
Microhematocrite capillary tubes, plain (blue), no. 22-362-574, Fisher Scientific
Sterile cotton-tipped applicators
Sterile pipettes, 5 ml and 10 ml
Sterile syringes, 1 ml and 60 ml
Sterile tubes, 50 ml
Tissue culture dishes, 35 mm, no. 430165, Corning
Wathman filter paper no.1

D. Solutions, including Culture Media

Ca^2-free HEPES Buffer:

1. Prepare HEPES buffer by adding 6.7 mM KCl, 142 mM NaCl, and 10 mM HEPES
2. Adjust to pH 7.5 with 1.0 N NaOH or HCl
3. Filtrate with 0.45-μm filter
4. Store at 4 °C.

Collagenase Solution:

1. Prepare a 10× stock solution of collagenase. Dissolve 0.25% of collagenase in Ω pure H_2O with 0.1 M $CaCl_2 \cdot 2H_2O$, 0.24 M NaCl, and 0.02 M HEPES. Adjust to pH 7.5 by adding 1.0 N NaOH or HCl. Complete the volume and sterilize by two successive filtrations using 0.8-μm and 0.45-μm filters. Store at -20 °C in 2.5-ml aliquots.
2. At the perfusion day, thaw 1 aliquot and dilute in a 25-ml final volume with sterile basal medium. Calculate 25 ml of 1× collagenase solution/mouse.

Basal Medium:

1. Prepare 2 L of basal medium as follows: in 1.8 L of Ω pure H_2O add $1\times$ powder Dulbecco's modified Eagle medium, $1\times$ powder F-12 nutrient mixture (Ham), 0.2% albumin bovine fraction V fatty acid free, 5.35 μg/ml linoleic acid, 0.03 g/L proline, 2.0 g/L galactose, 6.0 g/L HEPES, 1.68 g/L NaHCO$_3$, and 0.055 g/L Na pyruvate.

2. Adjust to pH 7.5 with NaOH 1 N and complete to 2 L. Sterilize by filtration using a 0.45 μm filter and store at 4 °C.

Attachment Medium:

1. Measure 100 ml of basal medium and add 5 μg/L selenium, 5.0 mg/L insulin, and 5.0 mg/L holo-transferrin.

2. Sterilize by filtration using a 0.45-μm filter and store at 4 °C.

Maintenance Medium:

1. To 100 ml of basal medium add 5 μg/L selenium, 5.0 mg/L insulin, 5.0 mg/L holo-transferrin, 10^{-7} M dexamethasone, and 20 ng/ml EGF.

2. Sterilize the medium by filtration through a 0.45-μm filter and store at 4 °C.

Apoptotic Medium:

1. The apoptotic medium is the maintenance medium without EGF and insulin.

Saline Solution:

1. In Ω pure H_2O, add 0.9% NaCl.

2. If required, sterilize with a 0.4 5-μm filter.

Trypan Blue Solution:

1. In free $Ca^{++}Mg^{++}$ phosphate buffer (PBS), pH 7.4, add 0.6% of Trypan blue.

2. Filtrate through a Wathman filter paper No. 1.

Mounting Medium; pH, 8.6:

1. Prepare glycine buffer 0.2 M (glycine, 15 mg/ml; NaOH, 7 mg/ml; NaCl, 17 mg/ml; NaN$_3$, 1 mg/ml) and adjust the pH at 8.6.

2. Mix 1:1 glycine buffer/glycerol.

III. Procedures

Much of the information on the involvement of IFs in apoptosis has come out of experiments performed with IF-null mice generated by gene targeting homologous recombination. Details on the establishment of the K8-deficient FVB/N and

K18-null S129 mouse lines have been reported by Baribault *et al.* (1994) and Magin *et al.* (1998), respectively. Wild-type and K-null of each mouse line are obtained by regular sibmating. The mice are housed in a specific pathogen-free animal facility; they are routinely monitored for *Helicobacter hepaticus* and *H. bilis* by polymerase chain reaction (PCR)–based screening of the mouse feces (Fox *et al.*, 1996; Shames *et al.*, 1995) to ensure that the pathogen is absent. The animals have access to water and food ad libitum. Experiments are performed with 4-week-old mice, according to the rules of the Animal Care Committee.

A. Liver Perfusion and Hepatocyte Isolation

The livers of WT or K8/K18-null mice are perfused according to a modified version of the two-step method with collagenase originally developed for rats (Deschênes *et al.*, 1979; Loranger *et al.*, 1997). This cell isolation procedure provides high yields of viable hepatocytes, despite the fact that K8/K18-null hepatocytes are sensitive to mechanical stress (Loranger *et al.*, 1997).

1. Prepare fresh $1\times$ collagenase solution.
2. Sterilize surgical instruments and keep them in a beaker filled with 70% EtOH.
3. Anasthetize the mouse with isoflurane.
4. Under a laminar-flow hood, attach legs with the mouse lying on its back and sprinkle the abdomen with 70% ethanol and 1% povidone.
5. With a sterile surgical scissor make two diagonal incisions from the perito-neum area continuing anteriorly to each side of the thorax in a V form. With a hemostatic forceps, clamp the V point and roll up to the sternum.
6. With sterile cotton-tipped applicators, pull the intestines on the right side, and lean the liver against the diaphragm to expose the vena portalis.
7. Cannulate the vena portalis with a 25 G 5/8″ sterile needle (see Fig. 1 for a schematic representation of the perfusion system); fix the vena and the needle with a sterile vascular clamp.
8. Lower the liver lobes, cut the diaphragm, and section the vena cava with a sterile fine scissor.
9. Start the pump and perfuse the liver at a flow rate of 5 ml/min at 37 °C, first with a Ca^2-free HEPES, pH 7.5, containing insulin (0.5 g/ml) and EGTA (0.5 mM) for 3 minutes.
10. Carry on the perfusion with a DME/F12 modified medium containing collagenase (0.2 U Wünsch/ml) and Ca^2 (5 mM) for 3 minutes. Monitor the liver swelling.
11. Stop the pump and use a sterile hemostatic forceps to clamp the dia-phragm near the liver. With a sterile scissor detach the liver by cutting diaphragm around the clamp.

12. Collect the hepatocytes by shaking the clamped liver into a 50-ml sterile tube containing 40 ml of DME/F12 modified medium.

13. Centrifuge isolated hepatocytes for 3 minutes at 50g. Remove the supernatant and add 10 ml of basal medium with sterile pipet to disperse cells, and then complete at 40 ml. Mix gently.

14. Repeat step 10.

15. Centrifuge the cell suspension for 3 minutes at 50g. Remove the supernatant and add 10 ml of attachment medium with a sterile pipet to disperse cells, then complete at 30 ml with the attachment medium and finally mix.

16. Determine the yield of isolated hepatocytes with a hemocytometer and evaluate their viability with a standard Trypan blue exclusion assay. This isolation procedure yields on the average 6×10^7 hepatocytes/liver, with a viability of 90–95%.

B. Preparations of Coated Dishes

1. Dilute the fibronectin stock solution to 20 μg/ml in Ca^2-free HEPES buffer, pH 7.5.

2. At least 30 minutes before hepatocyte seeding, pipette 1.5 ml of the fibronectin solution into 35-mm dishes.

3. Withdraw the fibronectin solution and plate the cells.

C. Primary Culture

1. Plate at a density of 1.8×10^5 hepatocytes/cm^2 in fibronectin-coated dishes in attachment medium (for microscopy analysis, cells are plated onto a sterile fibronectin-precoated glass coverslip placed at the bottom of dishes).

2. After a 4-hour attachment, wash the cells vigorously with sterile HEPES buffer, pH 7.5, and add fresh maintenance medium.

3. Twenty-four hours after seeding, change with fresh maintenance medium and add soluble Matrigel. (Note: With the transfection protocol, the Matrigel is added 18 hours after transfection.)

4. Thereafter, change the complete medium every 2 days.

As shown in Fig. 2, both the WT and K8/K18-null hepatocytes form a full monolayer. Note that although typical bile canaliculi uniformly re-form in WT cultures, those in K8/K18-null hepatocytes exhibit some reassembly variation because the K loss in mouse liver affects the hepatocyte polarity (Ameen *et al.*, 2001). Figure 3 shows the immunolocalization of K8, actin, α-tubulin, desmoplakin, and plectin in WT mouse liver and primary cultured hepatocytes. Note that the distributions observed in culture are much more equivalent to those detected *in situ*, confirming that hepatocytes in primary culture provide reliable *in vitro* cell models.

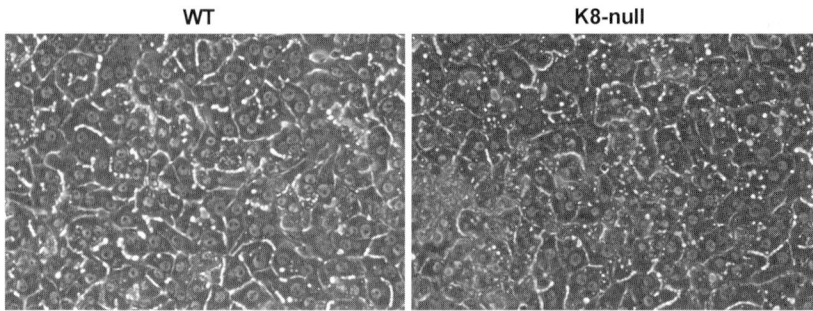

Fig. 2 Phase contrast micrographs of wild-type (WT) and mutated keratin 8 (K8-null) hepatocytes in primary culture 48 hours after seeding.

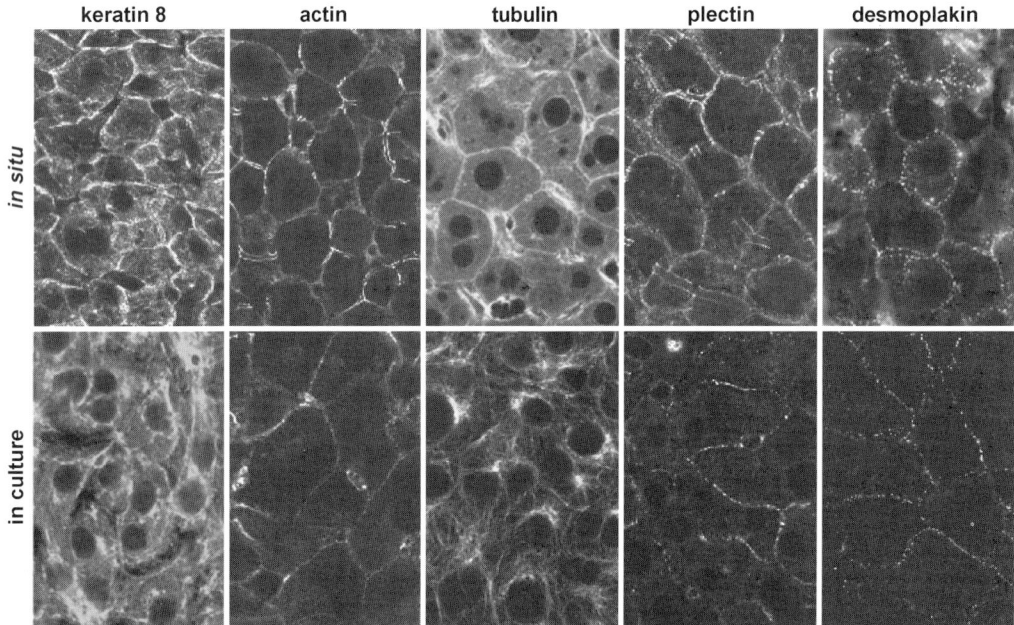

Fig. 3 Immunolocalization of keratin 8, actin, α-tubulin, plectin, and desmoplakin in mouse hepatocytes *in situ* and in monolayer culture 24 hours after seeding. Note that the individual proteins are predominantly localized at the surface membrane. (*Reproduced of BCB* **79**, 543–555, 2001.)

D. Transfection

The K8/K18-null hepatocyte monolayer culture provides a reliable means to assess the contribution of particular K8 and K18 domains, and more specifically the phosphoserine sites, to cell resistance to apoptosis. This involves the transfer of a K8 or K18 mutant cDNA into cultured K8- or K18-null hepatocytes by a transfection protocol that is adequate for this cell type.

1. The transfection is performed with Effectene 52 hours after seeding.
2. According to manufacturer instructions, use a DNA Effectene reagent ratio of 1 μg:10 μl and follow the protocol proposed for adherent cells.
3. Wash the cells twice with sterile HEPES buffer 18 hours after transfection and add the fresh maintenance medium containing soluble Matrigel.

E. Induction of DR-Receptor–Mediated Apoptosis

Hepatocytes are particularly sensitive to apoptosis, and the procedures described in the following focus on their response to FasL (Jo2), TNF-α, and TRAIL stimulation.

1. Fas-Mediated Apoptosis

1. Wash the 24-hour plated mouse hepatocytes with HEPES and add 1.5 ml of apoptotic medium containing Matrigel (0.5 mg/ml). The purpose of this step is to remove the insulin and the EGF, two survival factors, from the medium.
2. After 16 hours, remove the medium (without washing to avoid the loss of the Matrigel) and add the apoptotic medium containing between 0.05 μg/ml and 0.5 μg/ml of Jo2.
3. Incubate the cells in a 37 °C incubator for an appropriate time. To get relevant apoptotic measurements with mouse primary hepatocytes, periods of 6–24 hours are suitable, but to assess MAP kinase activation, short periods starting at 5 minutes to as long as 8 hours are essential to avoid missing some peak of activation.
4. For measurement of apoptosis and MAP kinase activation, see the following procedures.

2. TNF-α and TRAIL-Mediated Apoptosis

1. Wash the 24-hour plated mouse hepatocytes with HEPES and add 1.5 ml of apoptotic medium containing Matrigel (0.5 mg/ml).
2. After 16 hours, remove medium and add apoptotic medium containing between 0.01 μg/ml and 0.1 μg/ml of TNF-α or 0.1–1.0 μg/ml of TRAIL ligand. Hepatocytes are mostly resistant to TNF-α or TRAIL, but the addition of 5 μg/ml–50 μg/ml cycloheximide or 0.5 μg/ml actinomycine D sensitize hepatocytes to both ligand.

3. Incubate cells in a 37 °C incubator for an appropriate time. To get relevant apoptotic measurements with mouse primary hepatocytes, periods of 6–24 hours are suitable.

4. For measurement of apoptosis and MAP kinase activation, see the following procedures.

F. Analysis of Intermediate Filament Protein Involvement Upstream of Caspase Activation

As part of the evaluation of the K8/K18 modulation of DR-mediated apoptosis before commitment, we assess the effect of the K loss on the targeting of Fas to the surface membrane.

1. Assessment of Fas Localization by Fluorescence Microscopy

1. Use a 48-hour coverslip-plated hepatocyte culture.
2. Wash twice with PBS.
3. Put in 1 ml of the fixation solution (2% formaldehyde in PBS), and incubate 10 minutes at room temperature.
4. Wash one time with PBS and put 1 ml 0.1% Triton X-100 in PBS for 10 minutes at room temperature.
5. Wash one time with PBS and put in 1 ml of blocking solution (10% of goat serum in PBS) and incubate 30 minutes at room temperature.
6. Wash one time with PBS, add 60 μl of a R-phycoerythrin–labeled (PE-labeled) anti-Fas Jo2 (5 μg/ml in PBS), and incubate 2 hours at room temperature.
7. Wash twice with PBS, remove the coverslip from the dish, and then mount it on a slide with a drop of mounting medium.
8. Take pictures with a digital imaging fluorescence system or a laser scanning confocal microscope.

Note: with this fixation protocol, double staining with K and Golgi marker antibodies is possible (Gilbert *et al.*, 2001). Jo2-PE provides a red fluorescence that bleaches rapidly, so use a mounting medium and avoid overexposing the slides.

2. Assessment of Fas Density at the Cell Surface by FACS

1. Take 1×10^6 of freshly isolated hepatocytes and centrifuge at 1000 rpm for 2 minutes.
2. Wash twice by resuspending the pellet in 1 ml of the washing buffer (PBS, 0.1% NaN_3 and 2% FBS) and by centrifuging at 1000 rpm for 2 minutes.
3. Resuspend the pellet in washing buffer containing 2 μg/ml of PE-labeled anti-Fas Jo2 and incubate on ice for 20 minutes.
4. Wash three times with the washing buffer, and then analyze with FACS.

Note: Include a control with a nonspecific antibody.

3. Live Cell Imaging of Fas–GFP

1. Use Fas receptor cDNA to construct a fusion protein with the enhanced green fluorescent protein with the pEGFP vector (Clonetech) to generate Fas-GFP–tagged protein.

2. Transfect the resultant plasmid using Effectene (as described earlier) 52 hours after seeding hepatocytes.

3. Then use a digital imaging fluorescence live cell system to take live micrographs of Fas-GFP expressing cells.

As shown in Fig. 4, Fas-GFP is predominantly localized in the Golgi area in WT hepatocytes, whereas it localizes more at the surface membrane in K8-null hepatocytes.

4. K8/K18 Modulation of the Signalization from DR Receptors

The activation of MAP kinase is evaluated in Fas-stimulated WT versus K8/K18-null hepatocytes, as an assay for the contribution of IFs to antiapoptotic signaling pathways, and conversely the effect of MAP kinase inhibitors on Fas-mediated apoptosis is assessed.

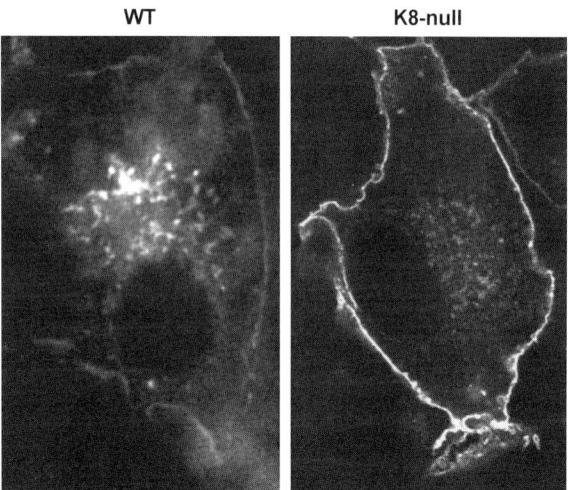

Fig. 4 Fluorescence microscopy of live WT and (K8-null hepatocytes expressing Fas tagged with enhanced green fluorescent protein [Fas-GFP]). Note that although the Fas-GFP is predominantly localized in the Golgi area in WT hepatocytes, it localizes more at the surface membrane in K8-null hepatocytes.

a. Evaluation of MAP Kinase Activation:

Phosphospecific antibodies provide reliable tools for evaluating the activation level of the many MAP kinases, and most of these antibodies are commercially available. The MAP kinase activation can thus be achieved by Western blotting after sodium dodecyl sulfate-polyacrylamide gel electrophoresis (SDS-PAGE) fractionation of protein extracts from stimulated cells. It is important to correlate the activation data with the corresponding nonphosphorylated MAP kinase. Note that these phosphospecific antibodies indicate a phosphorylation level that simply correlates with kinase activity.

b. Inhibition of MAP Kinase Activation:

Several MAP kinase inhibitors are commercially available, and the ones used here are PD 98059 (50 mM) to inhibit MEK1, SB 203580 (10 μM) to block p38, and SP 600125 (20 mM) to inhibit JNK. To efficiently inhibit a MAP kinase, the agent needs to be added for 30 minutes to 1 hour before stimulation. Note that for a prolonged inhibition, the inhibitor efficiency period needs to be evaluated (e.g., PD is active for a 6 to 8-hour period in hepatocytes).

Another way to inhibit MAP kinase activation is to transfect a dominant negative for the targeted pathway (e.g., for ERK1/2 pathway use MEK1 or Raf-1 dominant negative) with the protocol described earlier. The transfection efficiency obtained in primary hepatocytes (around 30%) is usually adequate.

5. Intermediate Filament Hyperphosphorylation and Reorganization

In addition, K8/K18 hyperphosphorylation on specific Ser sites can be monitored with antiphosphoserine-specific antibodies by Western blotting on cell protein lysates or immunocytochemistry on both frozen and formalin-tissue and cell culture (Ku and Omary, 1997; Liao *et al.*, 1997). Note that the same procedure is applicable to other IF proteins like vimentin with antibodies that are available commercially. Details on the protocols are provided in Chapter 17.

G. *In Vivo* Analysis of DR–Mediated Hepatocyte Apoptosis

1. Jo2 Treatment

The sensitivity of mice to a single intraperitoneal (i.p.) injection of Jo2 depends on the mouse strain (Hara *et al.*, 2000; Sarraf *et al.*, 1997).

1. In accordance with the rules provided by the Animal Care Committee and the OCDE, evaluate the response of mice to increasing Jo2 sublethal doses with an i.p. injection.

2. Dilute Jo2 in saline solution for a maximum i.p. injection volume of 0.2 ml. Load a 1-ml sterile syringe fit with a 27-G needle, inject, and monitor animal behavior. Prostration and bristling are restriction points to decide euthanasia.

3. Find the dose at which 90% WT mice survive. An increase of alanine aminotransferase (ALT) serum level by factor 1000 × should be the new restriction point where the animals show no inconvenience.

2. ALT Analysis

1. Just before the i.p. injection of a selected product, collect a blood sample from the saphenous vein with a nonheparinized capillary tube.

2. Twenty-four hours after injection, take a blood sample by cardiac puncture while the mouse is under isoflurane anesthesia.

3. Assess the level of liver damage by the amount of ALT activity released in serum accordingly to the manufacturer instructions of Randox kit.

Evaluation of the ALT level in serum provides a reliable means for the monitoring of the hepatocyte death *in vivo* (Gilbert *et al.*, 2001).

H. Analysis of Intermediate Filament Protein Involvement Downstream of Caspase Activation

Among the various assays that can be used to examine the execution of apoptosis, we propose the caspase cleavage, the nuclear fragmentation, the DNA ladder, and the K18 cleavage.

1. Caspase Cleavage

1. Extract total proteins from the stimulated cells with 300 μl/35-mm plastic dishes of preheated at 90 °C 2× SDS-PAGE sample buffer (4× TrisHCl (0.5 M)/SDS(0.4%), pH 6.8, 20% glycerol, 4% SDS, 2% mercaptoethanol, and 1% bromophenol blue) (Ausubel, 1994).

2. Scrape the cells with a policeman from the dish and transfer into a microcentrifuge tube.

3. Grind the cells by three passages through a 1-ml sterile syringe fit with a 27-G needle.

4. Determine the protein level in each extract and fractionate 10–20 μg of protein with a SDS-PAGE as described (Laemmli, 1970). The protein sample is mixed with a 2× SDS-PAGE sample buffer.

5. Then transfer the proteins electrophoretically from the gel onto a polyvinylidene difluoride (PVDF) membrane to make a standard Western blot, with an antibody recognizing the complete and the cleave portion of the selected caspase.

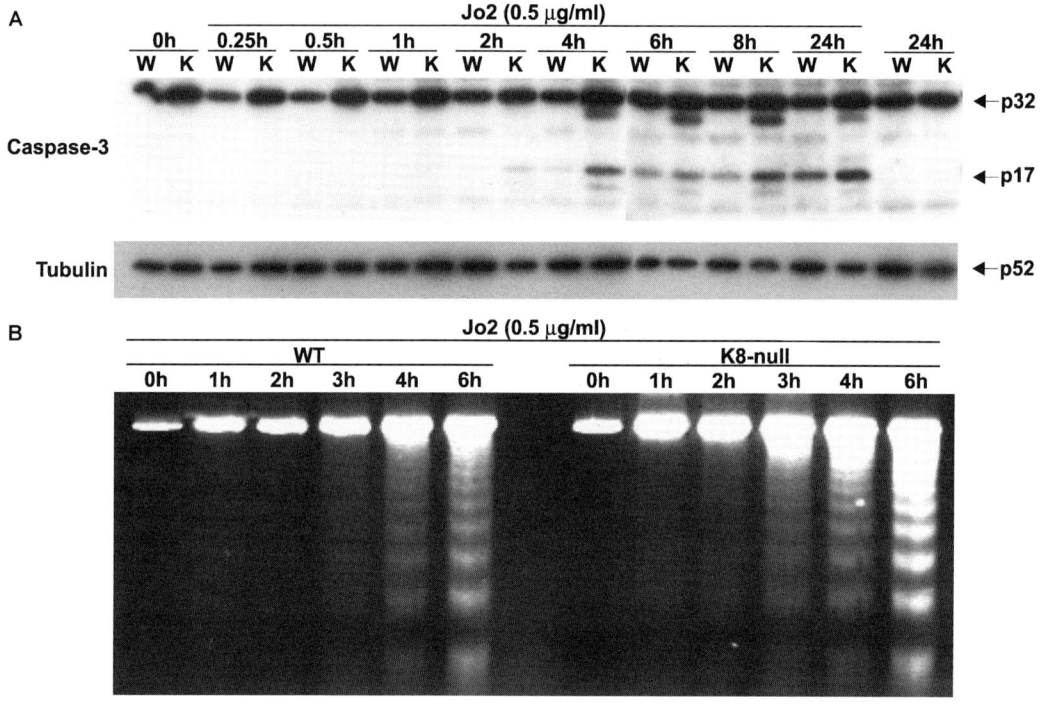

Fig. 5 (A) Western blot analysis of caspase-3 activation after Jo2 induction of Fas showing that procaspase-3 (p32) is cleaved to an active form (p17) more rapidly in K8-null than in WT hepatocytes. The tubulin blot used here as control shows no significant variation in the cellular content of this cytoskeletal protein. (B) The DNA ladder is higher and more rapid in K8-null hepatocytes than in WT hepatocytes after Jo2 stimulation. (*Reproduced of JCB* **154,** 763–773.)

In response to Fas stimulation, procaspase-3 (p32) is cleaved to an active form (p17) more rapidly in K8-null than in WT hepatocytes (Fig. 5A).

2. DNA Fragmentation

Knowing that apoptotic hepatocytes remained attached to the culture substratum, DNA labeling with acridine orange provides an excellent tool to evaluate the percentage of apoptotic cells (Guilhot *et al.*, 1996).

1. Wash the stimulated cells twice with PBS.
2. Add 1 ml of fixation solution (10% acetic acid/90% ethanol) for 1 hour at room temperature.
3. Remove the solution, and add 100% ethanol for 2 hours at room temperature.
4. Wash twice with PBS, and add the acridine orange solution (0.1% in 67 mM PBS, pH 6.0 [67 mM Na_2HPO_4; 67 mM KH_2PO_4]), for 3 minutes.
5. Remove the acridine orange and wash one time with PBS.

6. Add 1 ml of CaCl$_2$ 0.1 M for 1 minute.

7. Remove the CaCl$_2$, add 1 drop of mounting media, and then put on a coverslip.

8. Acridine orange is a fluorochrome able to differentiate between DNA (green fluorescence) and RNA (red fluorescence). With a laser scanning confocal microscope, take enough pictures from optical fields at random to count a significant number of cells (approximately 500 cells).

9. The number of apoptotic cells (as seen by the fragmentation of their nucleus) over the total number of cells provides percentage of apoptosis.

3. DNA Laddering

1. Wash the stimulated cells twice with PBS.

2. Add 500 μl of lysis buffer (Tris-HCl, pH 8.0, 10 mM; EDTA, 10 mM; NP40, 1%; proteinase K, 0.5 mg/ml) for at least 1 hour at room temperature.

3. Put the lysates in microcentrifuge tubes and centrifuge at 20,000g for 10 minutes at 4 °C.

4. Transfer the supernatants in new tubes, add 50 μg/ml RNAse and incubate at 37 °C for 1 hour.

5. Extract the DNA one time with phenol/chloroform/isoamyl alcohol (25:24:1) and precipitate it by adding 0.5 volume of ammonium acetate 7.5 M and 2 volumes of ethanol, followed by a 15-minute incubation on dry ice and a 15-minute centrifugation at 12,000 rpm.

6. Resuspend the precipitate in 20 μl TE buffer, pH 8.0 (Tris, 10 mM; EDTA, 1 mM).

7. Dose DNA at 260–280 nm.

8. Separate 5 μg of the DNA electrophoretically on a 2% agarose gel containing 0.5 μg/ml of ethidium bromide.

9. Destain the gel in water 20 minutes and take a picture.

As shown in Fig. 5B, the DNA laddering is higher and more rapid in K8-null hepatocytes than in WT hepatocytes after Fas-stimulation.

4. K18 Cleavage

As described in the "Introduction," the first step caspase cleavage of K18 at Asp-396 generates a 45-kDa fragment, and the second step cleavage at Asp-237 generates 26/22-kDa fragments (Table I); these fragments can be detected by Western blotting with a standard anti-K18 antibody (Gilbert *et al.*, 2004) after SDS-PAGE fractionation of total cell protein lysates. An excellent tool has been provided by the production of the monoclonal antibody M30 (available commercially) that recognizes a neoepitope created at position 387–396 in the caspase-generated large fragment (Leers *et al.*, 1999). Notably, M30 immunoreactivity precedes annexin V and TUNEL reactivity (Valavanis *et al.*, 2001). It identifies apoptotic epithelial cells, but not viable or necrotic cells, and the neoepitope is lost at the late stage of apoptosis. The detection of the neomarker is independent of the K18 phosphorylation status and can be achieved by Western blotting on protein lysates or immunocytochemistry on both frozen and formalin tissue and cell culture (Valavanis *et al.*, 2001).

IV. Pearls and Pitfalls

Hepatocytes constitute a cell model of choice for the analysis of IF involvement in apoptosis, mainly because they contain only one K pair; they are very sensitive to DR stimulation by various members of the TNF family and to a wide variety of apoptosis-triggering substances, and they are readily accessible for both *in vivo* and primary culture experiments.

K-null mice and several transgenic mouse lines carrying various mutants have been generated, and their use by several laboratories has proven their reliability. Although K8-null females of the FVB/N line could not support pregnancy at the time of establishment, it is now possible to mate homozygote × homozygote, which produces 3–4 pups per litter. Monitoring for the absence of *Helicobacter hepaticus* and *bilis* is important when studying the liver. The use of 3-week-old mice minimizes the contribution of polyploidy at older age.

A low perfusion rate of K-null liver during hepatocyte isolation is crucial for the cell yield and viability. Experiments are performed only with hepatocyte preparations exhibiting at least 90% viability; starting with a lower cell viability may lead to inconsistent results. Although fibronectin, collagen I, and laminin are all excellent substrates for the formation of full hepatocyte monolayer cultures, fibronectin can be routinely purified from human plasma. It is important to use a cell density that favors rapid cell–cell adhesion on seeding to minimize improper spreading and to provide a full monolayer culture.

The addition of soluble Matrigel is important for the maintenance of hepatocyte polarity integrity (e.g., bile canaliculus reformation). It allows efficient DR stimulation and MAP kinase activation under *in vitro* conditions that are much representative of the *in vivo* situation.

Cycloheximide and actinomycin D are general inhibitors of protein and RNA synthesis, and their use as sensitizers of TNF and TRAIL stimulation of cultured cells is not particularly representative of the *in vivo* situation. Care must also be taken when interpreting the data derived from the use of MAP kinase inhibitors, because they may not be entirely signaling-pathway specific, which means that the use of relevant dominant-negative MAP kinase is recommended as a complementary inhibitory tool.

Although the Fas-EGFP construct is most reliable for monitoring Fas trafficking in live cells responding to various apoptotic stimuli, the use of Fas-DsRed may cause death in cells, like hepatocytes, that are highly Fas sensitive because of DsRed-mediated receptor trimerization.

When counting apoptotic hepatocytes with the fragmented nucleus assay, care must be taken to score binucleated hepatocytes as one cell and as one apoptotic cell hepatocytes exhibiting multiple apoptotic bodies.

V. Conclusion and Perspectives

A. Intermediate Filaments as Makers of Differentiation and Apoptosis during Tumor Formation and Progression

Intermediate filament genes are differentially expressed, according to cell lineage establishment and terminal cell differentiation, and in this context IF proteins constitute reliable sets of cell markers in normal tissues. A case of exception is the K8/K18 pair, the first cytoplasmic IF proteins to be expressed in the embryo that remains as the sole IF network in hepatocytes even at the mature stage and remains present in essentially all simple epithelia.

Another interesting feature of K8/K18 expression is their persistence in carcinoma, in which the Ks of origin are sporadically expressed or lost during the tumor progression (Oshima, 2002; Oshima et al., 1996). In addition, K8/K18 reappear in several types of nonepithelial tumors, including melanoma and lymphoma (Hendrix et al., 1996), and their coexpression with vimentin provides a marker for cell aggressiveness (Hendrix et al., 1998).

More than 80% of the human tumors are of epithelial origins, particularly from K8/K18-containing cells, and the versatility of keratins as tumors markers is well recognized (Prasad et al., 1999). Indeed, sera from patients bearing carcinoma contain tissue polypeptide antigen (TPA) that has been identified as a complex of K8, K18, and even K19 fragments, which provide useful serum markers for monitoring the clinical progression of patients with carcinoma (Kramer et al., 2004; Oshima, 2002; Oshima et al., 1996). Both K18 and K19 fragments apparently correspond to those that are generated by caspase cleavage, because the M30 antibody recognizes some of these fragments (Kramer et al., 2004). Considering that these serum K fragments are released by the tumor cells and assuming that the appearance of apoptotic cells is an early event in emerging tumors, the presence of IF protein fragments could offer a relevant tool for the early detection of the related-aberrant process.

B. Intermediate Filaments as Markers for the Monitoring of the Cell Death Efficiency during Cancer Treatment

Increasing the rate of apoptosis offers a means to remove cancer cells, and the benefit from various treatment modalities (e.g., ionizing radiation and chemotherapy) largely depends on the degree to which the agent can favor this cell death process. However, apoptosis is not always the dominating death mode, because significant necrosis can also occur (Kramer *et al.*, 2004; Prasad *et al.*, 1999). For epithelially derived tumors, it seems that the presence of the K18 M30 antigen in serum is associated with apoptosis, whereas uncleaved K18 is released from necrotic cells (Kramer *et al.*, 2004). The same immunochemical procedure can be used for monitoring the treatment efficiency of carcinoma. In principle, the same detection procedure would be applicable to screen the treatment response in patients bearing any other types of tumors, assuming that proper IF proteolytic fragments can be generated by caspases, that these fragments are released in body fluids of patients, and that antibodies are available to these peptides.

C. Intermediate Filaments as Cell Death Markers in Degenerative Tissues

The extent of IF involvement in the emergence of degenerative diseases and associated cell death is largely unsettled. This is particularly the case for neuronal tissues (Julien, 1999; Julien and Beaulieu, 2000), where the presence of caspase-cleaved neurofilament proteins remains to be proven. With regard to muscle degeneration, although fragments can be generated from caspase-cleaved desmin, their use as markers of apoptosis has to be demonstrated. In the case of keratinocyte apoptosis occurring during toxic epidermal necrolysis or in response to ultraviolet radiation, there is a close association between FasL and K IFs before commitment (Viard-Leveugle *et al.*, 2003), but whether the execution of the cell death process is associated with the caspase cleavage of Ks has not been documented. In liver degeneration, such as chronic hepatitis, the detection of the M30 neoepitope in patient serum provides a tool of choice for the routine evaluation of the associated apoptosis (Grassi *et al.*, 2004). By extension, the same procedure could be used to assess whether a down-regulation of apoptosis is beneficial in the treatment of degenerative disorders.

Acknowledgments

We thank R. Kemler for the TROMA-1 hybridoma (anti-mouse K8 rat monoclonal antibody), J. L. Lessard for the anti-actin monoclonal antibody, D. Brown for the anti-tubulin monoclonal antibody, D. R. Garrod for the anti-desmoplakin monoclonal antibody, and M. Vincent for the rabbit anti-plectin polyclonal antibody. We are grateful to S. Nagata for the plasmid-containing Fas cDNA. We also thank L. Galarneau for helpful discussions and critical reading of the manuscript. This work was supported by a grant from Canadian Institutes of Health Research and a grant from The Cancer Research Society.

References

Ameen, N. A., Figueroa, Y., and Salas, P. J. (2001). Anomalous apical plasma membrane phenotype in CK8-deficient mice indicates a novel role for intermediate filaments in the polarization of simple epithelia. *J. Cell Sci.* **114,** 563–575.

Andra, K., Nikolic, B., Stocher, M., Drenckhahn, D., and Wiche, G. (1998). Not just scaffolding: Plectin regulates actin dynamics in cultured cells. *Genes Dev.* **12,** 3442–3451.

Ashkenazi, A., and Dixit, V. M. (1998). Death receptors: Signaling and modulation. *Science* **281,** 1305–1308.

Ausubel, F. M. (1994). "Current Protocols in Molecular Biology." John Wiley & Sons, Inc. New York.

Badock, V., Steinhusen, U., Bommert, K., Wittmann-Liebold, B., and Otto, A. (2001). Apoptosis-induced cleavage of keratin 15 and keratin 17 in a human breast epithelial cell line. *Cell Death Differ.* **8,** 308–315.

Baker, S. J., and Reddy, E. P. (1996). Transducers of life and death: TNF receptor superfamily and associated proteins. *Oncogene* **12,** 1–9.

Baribault, H., Blouin, R., Bourgon, L., and Marceau, N. (1989). Epidermal growth factor-induced selective phosphorylation of cultured rat hepatocyte 55-kD cytokeratin before filament reorganization and DNA synthesis. *J. Cell Biol.* **109,** 1665–1676.

Baribault, H., Penner, J., Iozzo, R. V., and Wilson-Heiner, M. (1994). Colorectal hyperplasia and inflammation in keratin 8-deficient FVB/N mice. *Genes Dev.* **8,** 2964–2973.

Baribault, H., Price, J., Miyai, K., and Oshima, R. G. (1993). Mid-gestational lethality in mice lacking keratin 8. *Genes Dev.* **7,** 1191–1202.

Bauman, P. A., Dalton, W. S., Anderson, J. M., and Cress, A. E. (1994). Expression of cytokeratin confers multiple drug resistance. *Proc. Natl. Acad. Sci. USA* **91,** 5311–5314.

Bojarski, C., Weiske, J., Schoneberg, T., Schroder, W., Mankertz, J., Schulzke, J. D., Florian, P., Fromm, M., Tauber, R., and Huber, O. (2004). The specific fates of tight junction proteins in apoptotic epithelial cells. *J. Cell Sci.* **117,** 2097–2107.

Byun, Y., Chen, F., Chang, R., Trivedi, M., Green, K. J., and Cryns, V. L. (2001). Caspase cleavage of vimentin disrupts intermediate filaments and promotes apoptosis. *Cell Death Differ.* **8,** 443–450.

Cadrin, M., Marceau, N., and Baribault, H. (1996). Griseofulvin hepatotoxicity-related effects in keratin 8 deficient FVB/N mice. *Mol. Biol. Cell* **6S,** 2171.

Caulin, C., Salvesen, G. S., and Oshima, R. G. (1997). Caspase cleavage of keratin 18 and reorganization of intermediate filaments during epithelial cell apoptosis. *J. Cell Biol.* **138,** 1379–1394.

Caulin, C., Ware, C. F., Magin, T. M., and Oshima, R. G. (2000). Keratin-dependent, epithelial resistance to tumor necrosis factor-induced apoptosis. *J. Cell Biol.* **149,** 17–22.

Chen, F., Chang, R., Trivedi, M., Capetanaki, Y., and Cryns, V. L. (2003). Caspase proteolysis of desmin produces a dominant-negative inhibitor of intermediate filaments and promotes apoptosis. *J. Biol. Chem.* **278,** 6848–6853.

Chinnaiyan, A. M., Tepper, C. G., Seldin, M. F., O'Rourke, K., Kischkel, F. C., Hellbardt, S., Krammer, P. H., Peter, M. E., and Dixit, V. M. (1996). FADD/MORT1 is a common mediator of CD95 (Fas/APO-1) and tumor necrosis factor receptor-induced apoptosis. *J. Biol. Chem.* **271,** 4961–4965.

Coulombe, P. A., Bousquet, O., Ma, L., Yamada, S., and Wirtz, D. (2000). The 'ins' and 'outs' of intermediate filament organization. *Trends Cell Biol.* **10,** 420–428.

Coulombe, P. A., and Omary, M. B. (2002). 'Hard' and 'soft' principles defining the structure, function and regulation of keratin intermediate filaments. *Curr. Opin. Cell Biol.* **14,** 110–122.

Deschênes, J., Valet, J. P., and Marceau, N. (1979). Hepatocytes from newborn and weanling rats in monolayer culture: Isolation by perfusion, fibronectin-mediated adhesion, spreading, and functional activities. *In Vitro* **16**(8), 722–730.

Dinsdale, D., Lee, J. C., Dewson, G., Cohen, G. M., and Peter, M. E. (2004). Intermediate filaments control the intracellular distribution of caspases during apoptosis. *Am. J. Pathol.* **164,** 395–407.

Ellis, P. A., Smith, I. E., and Dowsett, M. (1996). Apoptosis—its role in tumour growth and therapy. *Cytopathology* **7,** 201–203.

Ellis, R. E., Yuan, J. Y., and Horvitz, H. R. (1991). Mechanisms and functions of cell death. *Annu. Rev. Cell Biol.* **7,** 663–698.

Engvall, E., and Ruoslahti, E. (1977). Binding of soluble form of fibroblast surface protein, fibronectin, to collagen. *Int. J. Cancer* **20,** 1–5.

Feng, G., and Kaplowitz, N. (2000). Colchicine protects mice from the lethal effect of an agonistic anti-Fas antibody. *J. Clin. Invest.* **105,** 329–339.

Fox, J. G., Yan, L., Shames, B., Campbell, J., Murphy, J. C., and Li, X. (1996). Persistent hepatitis and enterocolitis in germfree mice infected with Helicobacter hepaticus. *Infect. Immun.* **64,** 3673–3681.

Fuchs, E. (1996). The cytoskeleton and disease: Genetic disorders of intermediate filaments. *Annu. Rev. Genet.* **30**, 197–231.

Fuchs, E., and Cleveland, D. W. (1998). A structural scaffolding of intermediate filaments in health and disease. *Science* **279**, 514–518.

Fuchs, E., and Karakesisoglou, I. (2001). Bridging cytoskeletal intersections. *Genes Dev.* **15**, 1–14.

Fuchs, E., and Weber, K. (1994). Intermediate filaments: Structure, dynamics, function, and disease. *Annu. Rev. Biochem.* **63**, 345–382.

Fulda, S., Meyer, E., and Debatin, K. M. (2000). Metabolic inhibitors sensitize for CD95 (APO-1/Fas)-induced apoptosis by down-regulating Fas-associated death domain-like interleukin 1-converting enzyme inhibitory protein expression. *Cancer Res.* **60**, 3947–3956.

Galle, P. R., Hofmann, W. J., Walczak, H., Schaller, H., Otto, G., Stremmel, W., Krammer, P. H., and Runkel, L. (1995). Involvement of the CD95 (APO-1/Fas) receptor and ligand in liver damage. *J. Exp. Med.* **182**, 1223–1230.

Gilbert, S., Loranger, A., Daigle, N., and Marceau, N. (2001). Simple epithelium keratins 8 and 18 provide resistance to Fas-mediated apoptosis. The protection occurs through a receptor-targeting modulation. *J. Cell Biol.* **154**, 763–773.

Gilbert, S., Loranger, A., and Marceau, N. (2004). Keratins modulate c-Flip/ERK1/2 antiapoptotic signaling in simple epithelial cells. *Mol. Cell. Biol.* **24**, 7072–7081.

Gill, J. H., James, N. H., Roberts, R. A., and Dive, C. (1998). The non-genotoxic hepatocarcinogen nafenopin suppresses rodent hepatocyte apoptosis induced by TGFb1, DNA damage and Fas. *Carcinogenesis* **19**(2), 299–304.

Grassi, A., Susca, M., Ferri, S., Gabusi, E., D'Errico, A., Farina, G., Maccariello, S., Zauli, D., Bianchi, F. B., and Ballardini, G. (2004). Detection of the M30 neoepitope as a new tool to quantify liver apoptosis: Timing and patterns of positivity on frozen and paraffin-embedded sections. *Am. J. Clin. Pathol.* **121**, 211–219.

Green, K. J., and Gaudry, C. A. (2000). Are desmosomes more than tethers for intermediate filaments? *Nat. Rev. Mol. Cell. Biol.* **1**, 208–216.

Green, K. J., Kowalczyk, A. P., Bornslaeger, E. A., Palka, H. L., and Norvell, S. M. (1998). Desmosomes: Integrators of mechanical integrity in tissues. *Biol. Bull.* **194**, 374–376; discussion 376–377.

Guilhot, S., Miller, T., Cornman, G., and Isom, H. C. (1996). Apoptosis induced by tumor necrosis factor-alpha in rat hepatocyte cell lines expressing hepatitis B virus. *Am. J. Pathol.* **148**, 801–814.

Hara, A., Yoshimi, N., Yamada, Y., Matsunaga, K., Kawabata, K., Sugie, S., and Mori, H. (2000). Effects of Fas-mediated liver cell apoptosis on diethylnitrosamine-induced hepatocarcinogenesis in mice. *Br. J. Cancer* **82**, 467–471.

Hashimoto, M., Inoue, S., Ogawa, S., Conrad, C., Muramatsu, M., Shackelford, D., and Masliah, E. (1998). Rapid fragmentation of vimentin in human skin fibroblasts exposed to tamoxifen: A possible involvement of caspase-3. *Biochem. Biophys. Res. Commun.* **247**, 401–406.

He, T., Stepulak, A., Holmstrom, T. H., Omary, M. B., and Eriksson, J. E. (2002). The intermediate filament protein keratin 8 is a novel cytoplasmic substrate for c-jun-N-terminal kinase. *J. Biol. Chem.* **277**, 10767–10774.

Hendrix, M. J., Seftor, E. A., Chu, Y. W., Trevor, K. T., and Seftor, R. E. (1996). Role of intermediate filaments in migration, invasion and metastasis. *Cancer Metastasis Rev.* **15**, 507–525.

Hendrix, M. J., Seftor, E. A., Seftor, R. E., Gardner, L. M., Boldt, H. C., Meyer, M., Pe'er, J., and Folberg, R. (1998). Biologic determinants of uveal melanoma metastatic phenotype: Role of intermediate filaments as predictive markers. *Lab. Invest.* **78**, 153–163.

Hengartner, M. O. (2000). The biochemistry of apoptosis. *Nature* **407**, 770–776.

Herrmann, H., and Aebi, U. (2000). Intermediate filaments and their associates: Multi-talented structural elements specifying cytoarchitecture and cytodynamics. *Curr. Opin. Cell Biol.* **12**, 79–90.

Herrmann, H., Hesse, M., Reichenzeller, M., Aebi, U., and Magin, T. M. (2003). Functional complexity of intermediate filament cytoskeletons: From structure to assembly to gene ablation. *Int. Rev. Cytol.* **223**, 83–175.

Hesse, M., Franz, T., Tamai, Y., Taketo, M. M., and Magin, T. M. (2000). Targeted deletion of keratins 18 and 19 leads to trophoblast fragility and early embryonic lethality. *Embo J.* **19,** 5060–5070.

Hesse, M., Magin, T. M., and Weber, K. (2001). Genes for intermediate filament proteins and the draft sequence of the human genome: Novel keratin genes and a surprisingly high number of pseudogenes related to keratin genes 8 and 18. *J. Cell Sci.* **114,** 2569–2575.

Hesse, M., Berg, T., Wiedenmann, B., Spengler, U., Woitas, K. P., and Magin, T. M. (2004). A frequent keratin 8 pL227L polymorphism, but no point mutations in keratin 8 and 18 genes, in pateitns with various liver disorders. *J. Med. Genet.* **41,** e42.

Holmstrom, T. H., Schmitz, I., Soderstrom, T. S., Poukkula, M., Johnson, V. L., Chow, S. C., Krammer, P. H., and Eriksson, J. E. (2000). MAPK/ERK signaling in activated T cells inhibits CD95/Fas-mediated apoptosis downstream of DISC assembly. *EMBO J.* **19,** 5418–5428.

Hsu, H., Shu, H. B., Pan, M. G., and Goeddel, D. V. (1996). TRADD-TRAF2 and TRADD-FADD interactions define two distinct TNF receptor 1 signal transduction pathways. *Cell* **84,** 299–308.

Inada, H., Izawa, I., Nishizawa, M., Fujita, E., Kiyono, T., Takahashi, T., Momoi, T., and Inagaki, M. (2001). Keratin attenuates tumor necrosis factor-induced cytotoxicity through association with TRADD. *J. Cell Biol.* **155,** 415–426.

Jaquemar, D., Kupriyanov, S., Wankell, M., Avis, J., Benirschke, K., Baribault, H., and Oshima, R. G. (2003). Keratin 8 protection of placental barrier function. *J. Cell Biol.* **161,** 749–756.

Julien, J. P. (1999). Neurofilament functions in health and disease. *Curr. Opin. Neurobiol.* **9,** 554–560.

Julien, J. P., and Beaulieu, J. M. (2000). Cytoskeletal abnormalities in amyotrophic lateral sclerosis: Beneficial or detrimental effects? *J. Neurol. Sci.* **180,** 7–14.

Kaufmann, S. H., and Earnshaw, W. C. (2000). Induction of apoptosis by cancer chemotherapy. *Exp. Cell Res.* **256,** 42–49.

Kaufmann, S. H., and Gores, G. J. (2000). Apoptosis in cancer: Cause and cure. *Bioessays* **22,** 1007–1017.

Kerr, J. F., Wyllie, A. H., and Currie, A. R. (1972). Apoptosis: A basic biological phenomenon with wide-ranging implications in tissue kinetics. *Br. J. Cancer* **26,** 239–257.

Kischkel, F. C., Hellbardt, S., Behrmann, I., Germer, M., Pawlita, M., Krammer, P. H., and Peter, M. E. (1995). Cytotoxicity-dependent APO-1 (Fas/CD95)-associated proteins form a death-inducing signaling complex (DISC) with the receptor. *EMBO J.* **14,** 5579–5588.

Kischkel, F. C., Lawrence, D. A., Chuntharapai, A., Schow, P., Kim, K. J., and Ashkenazi, A. (2000). Apo2L/TRAIL-dependent recruitment of endogenous FADD and caspase-8 to death receptors 4 and 5. *Immunity* **12,** 611–620.

Korfali, N., Ruchaud, S., Loegering, D., Bernard, D., Dingwall, C., Kaufmann, S. H., and Earnshaw, W. C. (2004). Caspase-7 gene disruption reveals an involvement of the enzyme during the early stages of apoptosis. *J. Biol. Chem.* **279,** 1030–1039.

Kramer, G., Erdal, H., Mertens, H. J., Nap, M., Mauermann, J., Steiner, G., Marberger, M., Biven, K., Shoshan, M. C., and Linder, S. (2004). Differentiation between cell death modes using measurements of different soluble forms of extracellular cytokeratin 18. *Cancer Res.* **64,** 1751–1756.

Krammer, P. H. (1999). CD95(APO-1/Fas)-mediated apoptosis: Live and let die. *Adv. Immunol.* **71,** 163–210.

Ku, N. O., and Omary, M. B. (1997). Phosphorylation of human keratin 8 *in vivo* at conserved head domain serine 23 and at epidermal growth factor-stimulated tail domain serine 431. *J. Biol. Chem.* **272,** 7556–7564.

Ku, N. O., and Omary, M. B. (2000). Keratins turn over by ubiquitination in a phosphorylation-modulated fashion. *J. Cell Biol.* **149,** 547–552.

Ku, N. O., and Omary, M. B. (2001). Effect of mutation and phosphorylation of type I keratins on their caspase-mediated degradation. *J. Biol. Chem.* **276,** 26792–26798.

Ku, N. O., Azhar, S., and Omary, M. B. (2002a). Keratin 8 phosphorylation by p38 kinase regulates cellular keratin filament reorganization: Modulation by a keratin 1-like disease causing mutation. *J. Biol. Chem.* **277,** 10775–10782.

Ku, N. O., Liao, J., Chou, C. F., and Omary, M. B. (1996a). Implications of intermediate filament protein phosphorylation. *Cancer Metastasis Rev.* **15,** 429–444.

Ku, N. O., Liao, J., and Omary, M. B. (1997). Apoptosis generates stable fragments of human type I keratins. *J. Biol. Chem.* **272,** 33197–33203.

Ku, N. O., Liao, J., and Omary, M. B. (1998). Phosphorylation of human keratin 18 serine 33 regulates binding to 14-3-3 proteins. *EMBO J.* **17,** 1892–1906.

Ku, N. O., Michie, S., Resurreccion, E. Z., Broome, R. L., and Omary, M. B. (2002b). Keratin binding to 14-3-3 proteins modulates keratin filaments and hepatocyte mitotic progression. *Proc. Natl. Acad. Sci. USA* **99,** 4373–4378.

Ku, N. O., Michie, S. A., Soetikno, R. M., Resurreccion, E. Z., Broome, R. L., Oshima, R. G., and Omary, M. B. (1996b). Susceptibility to hepatotoxicity in transgenic mice that express a dominant-negative human keratin 18 mutant. *J. Clin. Invest.* **98,** 1034–1046.

Ku, N. O., Soetikno, R. M., and Omary, M. B. (2003). Keratin mutation in transgenic mice predisposes to Fas but not TNF-induced apoptosis and massive liver injury. *Hepatology* **37,** 1006–1014.

Ku, N. O., Zhou, X., Toivola, D. M., and Omary, M. B. (1999). The cytoskeleton of digestive epithelia in health and disease. *Am. J. Physiol.* **277,** G1108–G1137.

Kumar, S. (1999). Mechanisms mediating caspase activation in cell death. *Cell Death Differ.* **6,** 1060–1066.

Laemmli, U. K. (1970). Cleavage of structural proteins during the assembly of the head of bacteriophage T4. *Nature* **227,** 680–685.

Lane, E. B., Hogan, B. L., Kurkinen, M., and Garrels, J. I. (1983). Co-expression of vimentin and cytokeratins in parietal endoderm cells of early mouse embryo. *Nature* **303,** 701–704.

Lazebnik, Y. A., Takahashi, A., Moir, R. D., Goldman, R. D., Poirier, G. G., Kaufmann, S. H., and Earnshaw, W. C. (1995). Studies of the lamin proteinase reveal multiple parallel biochemical pathways during apoptotic execution. *Proc. Natl. Acad. Sci. USA* **92,** 9042–9046.

Lee, J. C., Schickling, O., Stegh, A. H., Oshima, R. G., Dinsdale, D., Cohen, G. M., and Peter, M. E. (2002). DEDD regulates degradation of intermediate filaments during apoptosis. *J. Cell Biol.* **158,** 1051–1066.

Leers, M. P., Kolgen, W., Bjorklund, V., Bergman, T., Tribbick, G., Persson, B., Bjorklund, P., Ramaekers, F. C., Bjorklund, B., Nap, M., Jornvall, H., and Schutte, B. (1999). Immunocytochemical detection and mapping of a cytokeratin 18 neo-epitope exposed during early apoptosis. *J. Pathol.* **187,** 567–572.

Leung, C. L., Green, K. J., and Liem, R. K. (2002). Plakins: A family of versatile cytolinker proteins. *Trends Cell Biol.* **12,** 37–45.

Liao, J., Ku, N. O., and Omary, M. B. (1997). Stress, apoptosis, and mitosis induce phosphorylation of human keratin 8 at Ser-73 in tissues and cultured cells. *J. Biol. Chem.* **272,** 7565–7573.

Loranger, A., Duclos, S., Grenier, A., Price, J., Wilson-Heiner, M., Baribault, H., and Marceau, N. (1997). Simple epithelium keratins are required for maintenance of hepatocyte integrity. *Am. J. Pathol.* **151,** 1673–1683.

MacFarlane, M., Merrison, W., Dinsdale, D., and Cohen, G. M. (2000). Active caspases and cleaved cytokeratins are sequestered into cytoplasmic inclusions in TRAIL-induced apoptosis. *J. Cell Biol.* **148,** 1239–1254.

Magin, T. M., Schroeder, R., Leitgeb, S., Wanninger, F., Zatloukal, K., Grund, C., and Melton, D. W. (1998). Lessons from keratin 18 knockout mice: Formation of novel keratin filaments, secondary loss of keratin 7 and accumulation of liver-specific keratin 8-positive aggregates. *J. Cell Biol.* **140,** 1441–1451.

Marceau, N., and Loranger, A. (1995). Cytokeratin expression, fibrillar organization, and subtle function in liver cells. *Biochem. Cell Biol.* **73,** 619–625.

Marceau, N., Blouin, M. J., Germain, L., and Noïl, M. (1989). Role of different epithelial cell types in liver ontogenesis, regeneration and neoplasia. *In Vitro Cell. Dev. Biol.* **25**(4), 336–341.

Marceau, N., Loranger, A., Gilbert, S., Daigle, N., and Champetier, S. (2001). Keratin-mediated resistance to stress and apoptosis in simple epithelial cells in relation to health and disease. *Biochem. Cell Biol.* **79,** 543–555.

Martin, S. J., and Green, D. R. (1995). Protease activation during apoptosis: death by a thousand cuts? *Cell* **82,** 349–352.

McGowan, K., and Coulombe, P. A. (1998). The wound repair-associated keratins 6, 16, and 17. Insights into the role of intermediate filaments in specifying keratinocyte cytoarchitecture. *Subcell. Biochem.* **31,** 173–204.

McGowan, K. M., Tong, X., Colucci-Guyon, E., Langa, F., Babinet, C., and Coulombe, P. A. (2002). Keratin 17 null mice exhibit age- and strain-dependent alopecia. *Genes Dev.* **16,** 1412–1422.

McLean, W. H. I., and Lane, E. B. (1995). Intermediate filaments in disease. *Curr. Opin. Cell Biol.* **7,** 118–125.

Moll, R., Franke, W. W., and Schiller, D. L. (1982). The catalog of human cytokeratins; patterns of expression in normal epithelia, tumors and cultured cells. *Cell* **31,** 11–24.

Morishima, N. (1999). Changes in nuclear morphology during apoptosis correlate with vimentin cleavage by different caspases located either upstream or downstream of Bcl-2 action. *Genes Cells* **4,** 401–414.

Nagata, S. (1999). Fas ligand-induced apoptosis. *Annu. Rev. Genet.* **33,** 29–55.

Ni, R., Tomita, Y., Matsuda, K., Ichihara, A., Ishimura, K., Ogasawara, J., and Nagata, S. (1994). Fas-mediated apoptosis in primary cultured mouse hepatocytes. *Exp. Cell Res.* **215,** 332–337.

Oberhammer, F. A., Hochegger, K., Froschl, G., Tiefenbacher, R., and Pavelka, M. (1994). Chromatin condensation during apoptosis is accompanied by degradation of lamin A + B, without enhanced activation of cdc2 kinase. *J. Cell Biol.* **126,** 827–837.

Ogasawara, J., Watanabe-Fukunaga, R., Adachi, M., Matsuzawa, A., Kasugai, T., Kitamura, Y., Itoh, N., Suda, T., and Nagata, S. (1993). Lethal effect of the anti-fas antibody in mice. *Nature* **364,** 806–808.

Omary, M. B., and Ku, N. O. (1997). Intermediate filament proteins of the liver: Emerging disease association and functions. *Hepatology* **25,** 1043–1048.

Omary, M. B., Ku, N. O., Liao, J., and Price, D. (1998). Keratin modifications and solubility properties in epithelial cells and *in vitro*. *Subcell. Biochem.* **31,** 105–140.

Orth, K., Chinnaiyan, A. M., Garg, M., Froelich, C. J., and Dixit, V. M. (1996). The CED-3/ICE-like protease Mch2 is activated during apoptosis and cleaves the death substrate lamin A. *J. Biol. Chem.* **271,** 16443–16446.

Oshima, R. G. (2002). Apoptosis and keratin intermediate filaments. *Cell Death Differ.* **9,** 486–492.

Oshima, R. G., and Baribault, H. (1992). Inactivation of keratin genes by gene targeting: A perspective. *J. Dermatol.* **19,** 786–789.

Oshima, R. G., Baribault, H., and Caulin, C. (1996). Oncogenic regulation and function of keratins 8 and 18. *Cancer Metastasis Rev.* **15,** 445–471.

Owens, D. W., and Lane, E. B. (2003). The quest for the function of simple epithelial keratins. *Bioessays* **25,** 748–758.

Paladini, R. D., and Coulombe, P. A. (1998). Directed expression of keratin 16 to the progenitor basal cells of transgenic mouse skin delays skin maturation. *J. Cell Biol.* **142,** 1035–1051.

Paladini, R. D., Takahashi, K., Bravo, N. S., and Coulombe, P. A. (1996). Onset of re-epithelialization after skin injury correlates with a reorganization of keratin filaments in wound edge keratinocytes: Defining a potential role for keratin 16. *J. Cell Biol.* **132,** 381–397.

Paramio, J. M., and Jorcano, J. L. (2002). Beyond structure: Do intermediate filaments modulate cell signalling? *Bioessays* **24,** 836–844.

Prasad, S., Soldatenkov, V. A., Srinivasarao, G., and Dritschilo, A. (1998). Identification of keratins 18, 19 and heat-shock protein 90 beta as candidate substrates of proteolysis during ionizing radiation-induced apoptosis of estrogen-receptor negative breast tumor cells. *Int. J. Oncol.* **13,** 757–764.

Prasad, S., Soldatenkov, V. A., Srinivasarao, G., and Dritschilo, A. (1999). Intermediate filament proteins during carcinogenesis and apoptosis [review]. *Int. J. Oncol.* **14,** 563–570.

Rao, L., Perez, D., and White, E. (1996). Lamin proteolysis facilitates nuclear events during apoptosis. *J. Cell Biol.* **135,** 1441–1455.

Ruchaud, S., Korfali, N., Villa, P., Kottke, T. J., Dingwall, C., Kaufmann, S. H., and Earnshaw, W. C. (2002). Caspase-6 gene disruption reveals a requirement for lamin A cleavage in apoptotic chromatin condensation. *EMBO J.* **21,** 1967–1977.

Ruhrberg, C., and Watt, F. M. (1997). The plakin family: Versatile organizers of cytoskeletal architecture. *Curr. Opin. Genet. Dev.* **7,** 392–397.

Santos, M., Paramio, J. M., Bravo, A., Ramirez, A., and Jorcano, J. L. (2002). The expression of keratin k10 in the basal layer of the epidermis inhibits cell proliferation and prevents skin tumorigenesis. *J. Biol. Chem.* **277,** 19122–19130.

Sarraf, C. E., Horgan, M., Edwards, R. J., and Alison, M. R. (1997). Reversal of phenobarbital-induced hyperplasia and hypertrophy in the livers of lpr mice. *Int. J. Exp. Pathol.* **78,** 49–56.

Shames, B., Fox, J. G., Dewhirst, F., Yan, L., Shen, Z., and Taylor, N. S. (1995). Identification of widespread Helicobacter hepaticus infection in feces in commercial mouse colonies by culture and PCR assay. *J. Clin. Microbiol.* **33,** 2968–2972.

Sprick, M. R., Weigand, M. A., Rieser, E., Rauch, C. T., Juo, P., Blenis, J., Krammer, P. H., and Walczak, H. (2000). FADD/MORT1 and caspase-8 are recruited to TRAIL receptors 1 and 2 and are essential for apoptosis mediated by TRAIL receptor 2. *Immunity* **12,** 599–609.

Stegh, A. H., Herrmann, H., Lampel, S., Weisenberger, D., Andra, K., Seper, M., Wiche, G., Krammer, P. H., and Peter, M. E. (2000). Identification of the cytolinker plectin as a major early in vivo substrate for caspase 8 during CD95- and tumor necrosis factor receptor-mediated apoptosis. *Mol. Cell. Biol.* **20,** 5665–5679.

Strnad, P., Windoffer, R., and Leube, R. E. (2001). *In vivo* detection of cytokeratin filament network breakdown in cells treated with the phosphatase inhibitor okadaic acid. *Cell Tissue Res.* **306,** 277–293.

Strnad, P., Windoffer, R., and Leube, R. E. (2002). Induction of rapid and reversible cytokeratin filament network remodeling by inhibition of tyrosine phosphatases. *J. Cell Sci.* **115,** 4133–4148.

Stumptner, C., Fuchsbichler, A., Heid, H., Zatloukal, K., and Denk, H. (2002). Mallory body—a disease-associated type of sequestosome. *Hepatology* **35,** 1053–1062.

Suarez-Huerta, N., Lecocq, R., Mosselmans, R., Galand, P., Dumont, J. E., and Robaye, B. (2000). Myosin heavy chain degradation during apoptosis in endothelial cells. *Cell Prolif.* **33,** 101–114.

Takahashi, A., Alnemri, E. S., Lazebnik, Y. A., Fernandes-Alnemri, T., Litwack, G., Moir, R. D., Goldman, R. D., Poirier, G. G., Kaufmann, S. H., and Earnshaw, W. C. (1996). Cleavage of lamin A by Mch2 alpha but not CPP32: multiple interleukin 1 beta-converting enzyme-related proteases with distinct substrate recognition properties are active in apoptosis. *Proc. Natl. Acad. Sci. USA* **93,** 8395–8400.

Toivola, D. M., Nieminen, M. I., Hesse, M., He, T., Baribault, H., Magin, T. M., Omary, M. B., and Eriksson, J. E. (2001). Disturbances in hepatic cell-cycle regulation in mice with assembly-deficient keratins 8/18. *Hepatology* **34,** 1174–1183.

Tran, S. E., Holmstrom, T. H., Ahonen, M., Kahari, V. M., and Eriksson, J. E. (2001). MAPK/ERK overrides the apoptotic signaling from Fas, TNF, and TRAIL receptors. *J. Biol. Chem.* **276,** 16484–16490.

Tschopp, J., Irmler, M., and Thome, M. (1998). Inhibition of fas death signals by FLIPs. *Curr. Opin. Immunol.* **10,** 552–558.

Valavanis, C., Naber, S., and Schwartz, L. M. (2001). In situ detection of dying cells in normal and pathological tissues. *Methods Cell Biol.* **66,** 393–415.

van Engeland, M., Kuijpers, H. J., Ramaekers, F. C., Reutelingsperger, C. P., and Schutte, B. (1997). Plasma membrane alterations and cytoskeletal changes in apoptosis. *Exp. Cell Res.* **235,** 421–430.

Viard-Leveugle, I., Bullani, R. R., Meda, P., Micheau, O., Limat, A., Saurat, J. H., Tschopp, J., and French, L. E. (2003). Intracellular localization of keratinocyte Fas ligand explains lack of cytolytic activity under physiological conditions. *J. Biol. Chem.* **278,** 6183–6188.

Walker, P. R., Testolin, L., Armato, U., Marceau, N., Gourdeau, H., and Sikorska, M. (1995). Modulation of apoptosis by oncogenes. *In* "Apoptosis in Hormone-Dependent Cancers" (M. Tenniswood and H. Michna, eds.), Vol. 14, p. 248. Springer-Verglag, New York.

Weber, K., Plessmann, U., and Ulrich, W. (1989). Cytoplasmic intermediate filament proteins of invertebrates are closer to nuclear lamins than are vertebrate intermediate filament proteins: Sequence characterization of two muscle proteins of a nematode. *EMBO J.* **8,** 3221–3227.

Weiske, J., Schoneberg, T., Schroder, W., Hatzfeld, M., Tauber, R., and Huber, O. (2001). The fate of desmosomal proteins in apoptotic cells. *J. Biol. Chem.* **276,** 41175–41181.

Wiche, G. (1998). Role of plectin in cytoskeleton organization and dynamics. *J. Cell Sci.* **111,** 2477–2486.

Wyllie, A. H., Kerr, J. F., and Currie, A. R. (1980). Cell death: The significance of apoptosis. *Int. Rev. Cytol.* **68,** 251–306.

Yoneda, K., Furukawa, T., Zheng, Y. J., Momoi, T., Izawa, I., Inagaki, M., Manabe, M., and Inagaki, N. (2004). An autocrine/paracrine loop linking keratin 14 aggregates to tumor necrosis factor alpha-mediated cytotoxicity in a keratinocyte model of epidermolysis bullosa simplex. *J. Biol. Chem.* **279,** 7296–7303.

Zheng, T. S., and Flavell, R. A. (1999). Apoptosis. All's well that ends dead. *Nature* **400,** 410–411.

CHAPTER 6

Molecular Genetics Methods for Human Intermediate Filament Diseases

Frances J. D. Smith, Aileen Sandilands, and W. H. Irwin McLean

Epithelial Genetics Group
Human Genetics Unit
Ninewells Medical School
University of Dundee
Dundee, Scotland, UK

I. Introduction

A. Early Intermediate Filament Genetic Disease Studies

In 1991, the intermediate filament (IF) field underwent a dramatic change with the discovery of the first genetic mutations in the keratin 14 (K14) gene causing the hereditary skin blistering disease epidermolysis bullosa simplex (EBS) (Bonifas *et al.*, 1991b; Coulombe *et al.*, 1991). EBS is characterized by cell fragility and cytolysis of the basal cell layer of the epidermis, leading to blisters in response to mild mechanical trauma to the skin (Irvine and McLean, 1999), and so these discoveries revealed that a major function of keratins, and possibly other types of IF proteins, is to provide cells with mechanical strength. Mutations in K5, the type II expression partner of type I keratin K14, were also found to produce the EBS skin fragility phenotype (i.e., mutations in K5 phenocopy those in K14) (Lane *et al.*, 1992). Interestingly, these groundbreaking studies from different laboratories used three differing approaches to establish the role of the K5 and K14 genes in EBS. The Fuchs group used the very powerful approach of transgenics to study the effect of expressing a dominant-negative version of K14 in the epidermis of mice (Vassar *et al.*, 1991). The phenotypical, histological, and ultrastructural analysis of these mice revealed tonofilament aggregation and basal keratinocyte cytolysis remarkably similar to that observed in the skin of patients with EBS (Anton-Lamprecht and Schnyder, 1982). Subsequently, DNA sequence analysis of patients with EBS revealed point mutations in the K14 gene (Coulombe *et al.*, 1991). This method can be viewed as a candidate gene approach to the study of a genetic disorder, in this case informed by the phenotype of transgenic mice. The Epstein group used a genetic linkage approach, specifically what is now commonly known in human genetics as the positional-candidate approach (i.e., limited genetic linkage analysis at one or a small number of chromosomal loci that harbor strong candidate genes). They found that certain alleles of polymorphic genetic markers in the vicinity of the K14 gene showed complete cosegregation with the disease phenotype in families with autosomal-dominant EBS (Bonifas *et al.*, 1991a). This analysis yielded a statistically significant log-of-the-odds (LOD) score of >3.0, which by convention is taken as firm evidence for a gene location. These compelling linkage data were then followed up by point mutations in the K14 gene, confirming the disease association (Bonifas *et al.*, 1991b). The third research group, led by Lane and colleagues, built on their previous work showing that the electron-dense aggregates observed in EBS keratinocytes label with keratin antibodies and were, therefore, likely to be abnormally folded keratin (Ishida-Yamamoto *et al.*, 1991). Immunoblot analysis fortuitously revealed that an epitope on the K5 protein was altered in one EBS family and led to the identification of the causative mutation in the K5 gene (Lane *et al.*, 1992). Again, this is an example of a candidate gene approach.

In the years since these initial studies, there has been a great deal of activity in identifying further human inherited conditions involving genetic defects in keratins and, more recently, nonkeratin IF proteins, as reviewed by Coulombe and Omary (2002), Fuchs and Cleveland (1998), Irvine and McLean (1999), and Porter and Lane (2003). The Human Intermediate Filament Mutation Database has recently been established, cataloging all the published mutations in IF genes (www. interfil.org). At the time of writing, the database had recorded 682 independently occurring mutations in 25 IF genes. This consisted of 419 mutations in 19 keratin genes, 177 mutations in lamin A/C, and 86 mutations in other intermediate filament genes (desmin, glial fibrillary acidic protein [GFAP], neurofilament light polypeptide [NF-L], neurofilament heavy polypeptide [NF-H], and phakinin/CP49). Here, we will briefly review the known IF disease associations, with particular emphasis on the methods involved in linking these genes to their respective diseases and useful trends for designing mutation screening strategies, rather than the clinical details of the diseases. More detailed information on the specific disease phenotypes can be found in the reviews cited earlier in the Human Intermediate Filament Mutation Database; and in the On-line Mendelian Inheritance in Man (OMIM) human genetics database available through the National Center for Bioinformatics (NCBI).

B. Human Keratin Disorders

There are now 19 keratin genes associated with human genetic diseases (Coulombe and Omary, 2002; Fuchs and Cleveland, 1998; Irvine and McLean, 1999; Porter and Lane, 2003; Winter et al., 2004). Of these, 16 are involved in single gene Mendelian traits that mainly exhibit autosomal-dominant transmission. The other three are involved in genetic susceptibility to complex traits. The functional genes are all located in two compact gene clusters on chromosomes 12q13.13 and 17q21.2. The most recent survey of the human genome data, which is now complete across the two gene clusters, reports 27 functional genes in each cluster with four type I and five type II pseudogenes (Hesse et al., 2004). Many pseudogenes, both conventional (intron containing) and processed (intronless), exist elsewhere in the genome (Hesse et al., 2001). The type I/type II clustering of the keratin genes means that genetic linkage analysis can readily be used to eliminate one gene cluster in families with a keratin gene defect before commencing mutation analysis, a fact that has been exploited a number of times in the study of keratin diseases (Irvine et al., 1997a; Smith et al., 1998).

The prototypic keratin disorder EBS comes in various clinical subtypes and is caused by mutations in K5 or K14 (Fine et al., 2000; Irvine and McLean, 1999). The study of this disorder established a number of important paradigms that apply to other IF disorders. The Dowling-Meara form of EBS (EBS-DM) is the most severe variant in which clustered blisters occur all over the body and which can be life threatening in infants. This subtype is associated mainly with missense and small inframe insertion/deletion mutations affecting the helix boundary motifs of the K5 and K14 polypeptides. Mutation detection for patients diagnosed with EBS-DM can therefore focus initially on the exons encoding these protein motifs (exons 1 and 7 of K5; exons 1 and 6 of K14). If this does not yield mutations, the analysis can be extended to cover all remaining exons. The milder Köbner and Weber-Cockayne forms of EBS are characterized by milder widespread blistering and blistering restricted to hands and feet, respectively. These less severe variants have mutations outside of the helix boundary motifs, with clusters in the H1 subdomain of K5, the distal 1A domain, L12 domain, and central 2B domain of both keratins. A specific missense mutation in the V1 domain of K5 has consistently produced a mild form of EBS with pigmentation changes in several unrelated families (Irvine *et al.*, 1997b; Uttam *et al.*, 1996). Thus, once some phenotype–genotype correlation is established for a given gene(s), such as the case in EBS, the mutation detection method can be streamlined. A specific problem with analysis of patients with EBS is the presence of two intron-containing K14 pseudogenes (Wood *et al.*, 2003). Early studies used a reverse transcriptase polymerase chain reaction (RT-PCR) approach to avoid pseudogene contamination or used a published pseudogene sequence to design intronic primers specific for the exons of the functional gene. However, mRNA is not always available from the patients, and the exon-by-exon genomic approach necessitates careful calibration of several sets of PCR primers. Recently, with the complete genome sequence and building on lessons learned in studying the K6, K16, and K17 genes, all of which have multiple pseudogenes (Smith *et al.*, 1999a,b; Terrinoni *et al.*, 2001), a long-range PCR method has been developed to amplify the entire functional K14 gene without pseudogene contamination (Wood *et al.*, 2003). This large PCR fragment can be directly sequenced or used as a template for smaller internal nested PCR fragments. Thus, only one set of specific primers and conditions is needed, streamlining the mutation screening process. An additional benefit of this approach is that it may identify some larger heterozygous genomic deletions that would be missed by single exon PCR methods. There is only one copy of the K5 gene in the human genome, so there are no major problems in analysis of this gene. DNA-based prenatal testing has been carried out for the severe forms of EBS (Rugg *et al.*, 2000).

Keratins K1, K2e, K9, and K10 are involved in phenotypes involving epidermolytic hyperkeratosis—thickening and scaling of the epidermis or specific regions thereof (McLean, 2003; Table I). Only very few mutations occur outside of the helix boundary motifs, and there are no pseudogenes of these keratins, so mutation detection is straightforward and focuses on the four hotspot regions. Recently, some cases clinically diagnosed as epidermolytic palmar plantar keratoderma (EPPK), the K9 disorder (Table I), have been found to carry mutations in K1 rather than K9. It seems that a subset of mutations in K1 can phenocopy K9 mutations, even though K1 is much more widely expressed in the epidermis (Hatsell *et al.*, 2001; Terron-Kwiatkowski *et al.*, 2002, 2004). This is something to be taken into consideration when expected mutations cannot be found in a given intermediate filament gene; they may be in another related gene with an overlapping expression pattern.

Four keratin genes have been linked to the genetic skin condition pachyonychia congenita (PC) (Bowden *et al.*, 1995; McLean *et al.*, 1995; Smith *et al.*, 1998). This

Table I

Disorders Associated with Genetic Defects in Keratin Genes

Intermediate filament protein	Disease	OMIM No.	Main clinical features(s)
Epithelial keratins			
K1, K10	Bullous congenital ichthyosiform erythroderma	113800, 600648 607602	Generalized epidermolytic hyperkeratosis
K1, K16	Nonepidermolytic palmoplantar keratoderma	600962	Focal hyperkeratosis of palm/sole
K1	Striate keratoderma	607654	Hyperkeratosis: linear on palms, focal on soles
K1	Ichthyosis hystrix, Curth-Macklin type	146590	Severe "spiny" epidermolytic hyperkeratosis
K2e	Ichthyosis bullosa of Siemens	146800	Epidermolytic ichthyosis
K3, K12	Meesmann epithelial corneal dystrophy	122100	Fine punctate opacities in cornea
K4, K13	White sponge nevus	193900	Thickened white mucosae
K5, K14	Epidermolysis bullosa simplex	131760, 131800 131900, 131960	Bullous skin lesions, generalized or localized
K6a, K6b, K16, K17	Pachyonychia congenita types 1 and 2	167200, 167210	Nail dystrophy and epithelial dysplasia features
K6hf*	Pseudofolliculitis barbae	Not assigned	Ingrown hairs, pustules after shaving
K8*, K18*	Cryptogenic cirrhosis, hepatitis	215600	Bleeding, ascites, confusion
K8*	Chronic pancreatitis	Not assigned	Malabsorption, pain, weight loss
K8*, K18*	Inflammatory bowel disease	601458, 266600	Ulcerative colitis and Crohn disease
K9, (K1)	Epidermolytic palmoplantar keratoderma	144200	Diffuse hyperkeratosis of palm and sole
K17	Steatocystoma multiplex	184500, 184510	Multiple pilosebaceous cysts
Hair keratins			
Hb1, Hb6	Monilethrix	158000	Variable alopecia, beaded hair

*Not solely causative, predisposition factor in disease.

group of autosomal-dominant ectodermal dysplasias is characterized by hypertrophic nail dystrophy accompanied by hyperkeratosis of specific epidermal appendages and mucosal tissues. Mutation analysis in PC is hampered by the fact that all four genes have multiple pseudogenes and/or isogenes. Initially, RT-PCR was used to circumvent this (McLean *et al.*, 1995), but later, availability of more genomic sequences led to the development of long-range PCR conditions that are able to specifically amplify each of these genes in single PCR fragments, as described previously (Smith *et al.*, 1999a,b; Terrinoni *et al.*, 2001). This allows rapid mutation detection on the basis of genomic DNA and has facilitated prenatal diagnosis with chorionic villus biopsy material (Smith *et al.*, 1999b). Again, the hotspots for mutations in these genes are the helix boundary motifs, with only a very small number falling outside these areas in association with milder phenotypes.

Nonepidermal keratins have also been linked to genetic diseases. K4 and K13 are specifically expressed in mucosal tissue, and mutations in these genes cause white sponge nevus (WSN) or hyperkeratosis of this tissue. The first mutations were identified by candidate gene sequencing (Richard *et al.*, 1995; Rugg *et al.*, 1995). Similarly, K3 and K12 are expressed only in the anterior corneal epithelium, and mutations in these genes cause Meesmann epithelial corneal dystrophy (MECD), a mild ocular disease caused by cytolysis within this tissue. The MECD genes were identified by the positional candidate method–linkage analysis in affected families with polymorphic markers within the two keratin gene clusters to eliminate one gene, followed by mutation screening of the remaining gene (Irvine *et al.*, 1997a). Two of the several hair-specific high-sulfur keratins, hHb1 and hHb6, have so far been linked to a human genetic hair disease, monilethrix (Winter *et al.*, 1997a,b). Again, linkage analysis first established that a type II keratin gene was responsible, essentially halving the required mutation analysis (Healy *et al.*, 1995). Because there are so many hair keratin genes, all of which could be considered candidates for monilethrix, this demonstrates the value of genetic linkage. All the mutations reported in WSN, MECD, and monilethrix fall in the helix boundary motifs, and no pseudogenes have been identified for these genes.

Three keratin genes have been associated with genetic predisposition to disease, rather than being the sole cause of the disorder in question. Sequence variants in the simple epithelial keratins K8 and K18 have been shown to be risk factors in cryptogenic cirrhosis and hepatitis by one laboratory (Ku *et al.*, 1997, 2001), and associated with inflammatory bowel disease by another group (Owens *et al.*, 2004). There are ∼100 copies of pseudogenes for both of these keratins scattered around the human genome (Hesse *et al.*, 2001), which has obvious implications for PCR-based mutation detection. However, these are mainly processed, intronless pseudogenes, so intronic placement of primers can largely avoid any problems. Recently, a common sequence variant in the K6hf gene, expressed specifically in the companion cell layer of hair follicles, has been implicated as a risk factor in pseudofolliculitis barbae, a common disorder in certain populations in which there is ingrowth of hairs in response to shaving, leading to epidermal cysts (Winter *et al.*, 2004). This brings the total number of keratins involved in genetic diseases to 19.

C. The Laminopathies

One of the real surprises from the study of IF diseases is the still growing group of genetic disorders caused by mutations in the lamin A/C gene (*LMNA*), the laminopathies (Table II), reviewed by Herrmann and Foisner (2003), Maidment and Ellis (2002), and Worman and Courvalin (2004). First, it was surprising that mutations affecting a subcellular structure as fundamentally important to all eukaryotic cells as the nuclear lamina could result in viable human phenotypes, especially because a P-element insertion mutation in a *Drosophila* lamin gene had been shown to lead to a number of severe developmental abnormalities (Lenz-Bohme *et al.*, 1997). In fact, the laminopathies are generally late-onset disorders, rather than abnormalities of early development. Second, it turns out that despite the widespread expression of lamins in human cells, lamin A/C mutations lead to a great variety of highly tissue-specific phenotypes, with a total of 10 laminopathies having been identified at the time of writing, as detailed in Table II. Genetic linkage and positional cloning methods have played a large part in unraveling the genetic disorders of lamins. Most of the laminopathy phenotypes would not

Table II

Laminopathies: Disorders Associated with Genetic Defects in Lamin A/C

Disease	OMIM No.	Main clinical features(s)
Atrial fibrillation, early onset	607554	Atrioventricular conduction defect
Cardiomyopathy, dilated, type 1A	115200	Cardiomyopathy, dilated LMNA
Charcot-Marie-Tooth disease	605558	Symmetrical muscle weakness
Familial partial lipodystrophy	151660	Major redistribution of adipose tissue
Hutchinson-Gilford progeria	176670	Premature aging/senescence
Lipoatrophy with diabetes, hepatic steatosis, hypertrophic cardiomyopathy, and leukomelanodermic papules	608056	General lipoatrophy, liver steatosis
Mandibuloacral dysplasia	248370	Mandibular/clavicular hypoplasia acroosteolysis, joint contractures, lipodystrophy
Emery-Dreifuss muscular dystrophy	181350, 604929	Contractures elbows/Achilles tendons-muscle weakness, cardiomyopathy with conduction blocks
Limb girdle muscular dystrophy 1B	159001	Progressive pelvic girdle weakness, atrioventricular cardiac conduction defects, dilated cardiomyopathy
Werner syndrome	277700	Premature aging/senescence

easily have been predicted from the preexisting knowledge of lamin function, and these associations came to light through genetic linkage to the 1q22 locus by a number of independent research groups.

The first lamin A/C disorders to be identified were the autosomal dominant and recessive forms of Emery-Dreifuss muscular dystrophy, which were mapped to 1q22 after genome-wide linkage analysis (Bonne *et al.*, 1999). The protein mutated in the X-linked form of the disease, emerin, was known to be associated with the nuclear lamina (Manilal *et al.*, 1998, 1999), so it followed that lamin mutations could lead to the autosomal-dominant and recessive phenocopies. An independent genome screen in a large family with familial partial lipodystrophy also resulted in linkage to 1q22 and the discovery of further mutations in the *LNMA* gene (Shackleton *et al.*, 2000). After this initial surprising discovery of two tissue-specific lamin A/C phenotypes, a whole range of other diseases, including neuropathies, premature aging syndromes, and developmental defects, have been linked to mutations in this gene (Table II). In view of the phenotypic diversity arising from these mutations, it seems quite likely that further laminopathies will emerge in the future.

D. Other Intermediate Filament Diseases

The diseases associated with mutations in other IF genes are listed in Table III. The first nonkeratin IF gene to be linked to a human disease was the muscle-specific type III protein desmin, mutations that cause desmin-related myopathies and cardiomyopathies (Dalakas *et al.*, 2000; Goldfarb *et al.*, 1998; Munoz-Marmol

Table III
Disorders Associated with Genetic Defects in Other Intermediate Filament Proteins

Intermediate filament protein(s)	Disease	OMIM No.	Main clinical features(s)
Desmin	Desmin-related myopathy	601419	Peripheral/distal muscle weakness arrythmias, restrictive heart failure
Desmin	Cardiomyopathy, dilated, type II	604765	CMDII; pure cardiomyopathy
GFAP	Alexander disease	203450	Early-onset megalencephaly, progressive spasticity, dementia
NF-L	Charcot-Marie-Tooth disease	607684, 607734	Motor/sensory neuropathy (demyelinating) types 2E and type 1F
NF-H[*]	Amyotrophic lateral sclerosis	105400	Rapid loss of motor function
Phakinin	Cataracts, juvenile-onset	604219	Congenital and juvenile cataracts

[*]Not solely causative, predisposition factor in disease.

et al., 1998). GFAP mutations have more recently been found in Alexander disease (Brenner *et al.*, 2001), and two of the neurofilament proteins have also now been connected to inherited neuropathies (Figlewicz *et al.*, 1994; Mersiyanova *et al.*, 2000). Abnormalities in the beaded filament protein CP49/phakinin of the eye lens have now been demonstrated in one type of early-onset cataract (Conley *et al.*, 2000), and it is likely, on the basis of the phenocopy phenomenon seen in the keratin diseases, that mutations in its polymerization partner, filensin, will also play a role in inherited cataract.

Despite the enormous progress in identifying genetic pathologies caused by IF protein defects, there is still much to be done, with only 25 of the 65 or so functional intermediate filament genes linked to disease at this time (Hesse *et al.*, 2001, 2004). This chapter outlines some methods that we have found particularly useful in our studies of the keratin diseases over the past decade or so, many of which have been refined in light of new developments in molecular genetics and the human genome project. We illustrate these techniques with examples based on our previous studies of keratin disorders, from genetic linkage analysis, through mutation scanning and sequencing, to population screening (Figs. 1–5). Human genetics is a fast moving field, with new methods for typing polymorphisms and identifying mutations emerging on a regular basis. The methods given here are the ones used in our laboratory currently and are those that we have found to be particularly robust. Thus, we have not included very new techniques that are as yet unproven in our hands. We hope that these methods are useful to others already in the field, as well as to newcomers to IF biology to study their favorite IF gene/ protein in a human genetics context.

II. Materials and Instrumentation

Websites for clinical and molecular genetics databases, major equipment and consumables suppliers, and other useful resources for human genetics studies are listed in Table IV.

III. Procedures

A. Basic Molecular Genetics Techniques

1. DNA Extraction from Blood, Mouthwash, or Tissue

Genomic DNA is extracted from blood, mouthwash, or tissues samples by use of standard procedures. Those currently used in our laboratory are FlexiGene DNA kit (Qiagen, Crawley, UK) and Nucleon extraction kits (Amersham Life Science, Little Chalfont, Bucks, UK). Five to 10-ml blood samples are collected into sodium ethylene diamine tetraacetic acid (EDTA) tubes and kept at 4 °C until extraction. Mouthwash samples or buccal scrapes are collected in water or saline

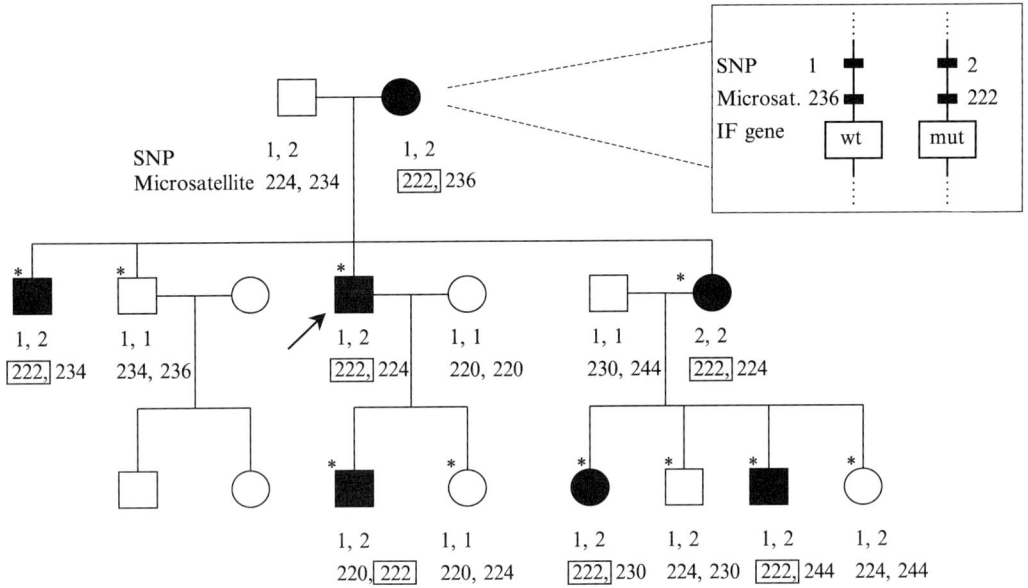

Fig. 1 This pedigree shows classic autosomal dominant inheritance: ~50% of offspring affected; both sexes affected; male-to-male transmission. Asterisk indicates that the person represents a meiosis. These persons, plus the original case, are required for linkage. Spouses and offspring of unaffected persons are irrelevant for linkage purposes. The spouses of affected persons are useful for linkage but not absolutely essential. Genotypes are shown for a single nucleotide polymorphism (SNP) and a microsatellite marker, close to an intermediate filament gene (IF) of interest. The inset schematic shows the arrangement of the marker alleles in relation to the wild-type and mutated copies of the IF gene on the grandmother's two chromosomes. Thus, allele 2 of the SNP and the 222-bp allele of the microsatellite are in linkage with the mutant gene. Following the inheritance of the markers through the family shows that these alleles cosegregate fully with the disease phenotype (i.e., these markers show genetic linkage indicating that the defective gene is nearby). Note that the highly variable microsatellite allows tracking of the gene in all individuals, with allele 222 (boxed) being passed on to all affected offspring. In contrast, the poorly informative SNP only gives linkage information in the index case (arrow) and his family, where the affected parent is heterozygous and his spouse is homozygous. This illustrates the power of microsatellites in genetic linkage analysis. This family is the absolute minimum size to detect a new disease gene (i.e., 10 meioses, giving a statistical log-of-the-odds score of 3.0). For genome screening, a kindred two or more times as large is required.

and should be extracted as soon as possible. *(*NOTE*: certain foods such as apples should not be eaten before collecting mouthwash samples, because compounds present inhibit PCR)*. DNA from cultured cells can also be extracted with these kits; the cells are pelleted before extraction. Both Qiagen and Nucleon provide specialized kits for DNA extraction from tissues. DNA can be extracted from soft tissues either by first homogenizing fresh tissue or by grinding frozen tissue. In the case of hard tissues that do not homogenize easily in lysis buffer, and for paraffin sections, the Nucleon HT kit is used. Samples are treated with proteinase K before extraction.

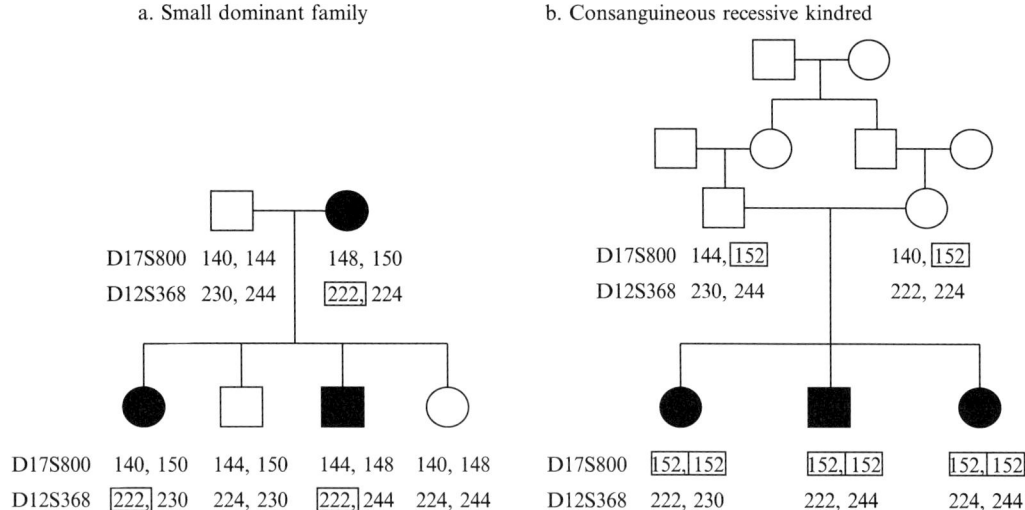

Fig. 2 These small pedigrees with keratin disorders show the value of exclusion analysis. Similar approaches to these are possible for any intermediate filament gene loci. (a) Even though the family is too small for statistically significant linkage (maximum possible log-of-the-odds [LOD] score would be only 1.2), it is possible with a marker close to or within the gene to exclude one of the two keratin loci, in this case the type I locus in which marker D17S800 is located. There is no consistent inheritance of an allele for D17S800 with the phenotype; however, allele 222 of marker D12S368 is consistent with linkage. (b) This recessive consanguineous kindred is very useful for mapping. The type II keratin locus is excluded with D12S368—the parents do not share an allele, and all the three affected offspring are heterozygous. However, with D17S800, a marker in the type I keratin locus, the 152-bp allele is fully linked. Just the genotype data from the three affected persons alone would give a significant LOD score of 3.3, such is the statistical power of inbred families.

2. mRNA Extraction Protocols

mRNA is extracted from cultured cells ($1\times$ 3-cm plate; T25 flask) or small amounts of tissue using the QuickPrep micro mRNA Purification kit (Amersham Biosciences, Little Chalfont, Bucks, UK). Cultured cells can be scraped off tissue culture plates or trypsinized and resuspended in medium containing serum to inactivate trypsin. Cells are then washed with phosphate-buffered saline (PBS) before mRNA extraction. Small tissue samples are either directly homogenized in extraction buffer or cryostat sections cut from a block, 10–15 sections, >10-μm thick, per extraction.

3. cDNA Synthesis

Each mRNA extraction (5 μl) is incubated with 1 μl oligo (dT)$_{15}$ (0.5 mg/ml; Roche) and 1 μl RNAsin (40 U/μl; Promega) for 10 minutes at 70 °C and then cooled to 42 °C; 13 μl of reaction mixture containing $1\times$ AMV buffer, 1 μl of 20 mM dNTPs, and 1 μl (10 U/μl) avian myeloblastosis virus (AMV) reverse

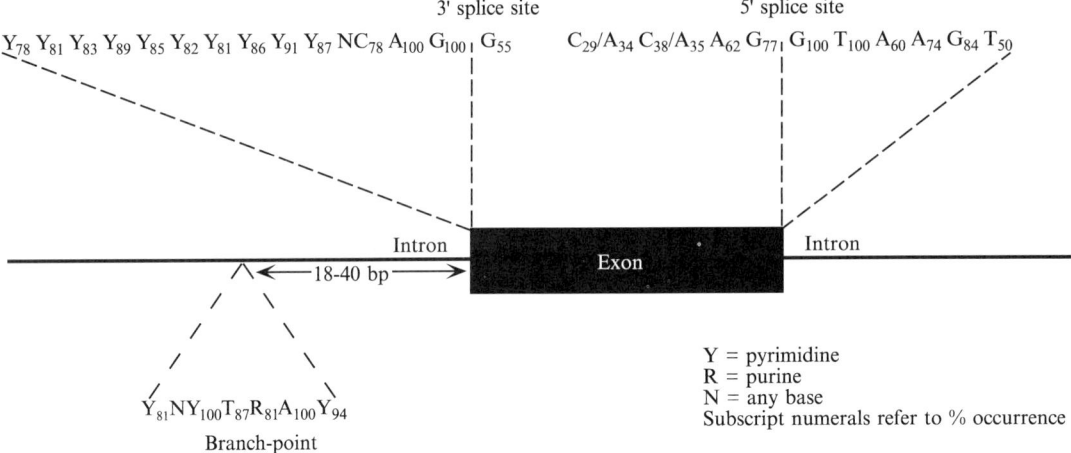

Fig. 3 Typical internal exon of a human gene showing the intronic splice site consensus sequences. Note that the first and last two residues of the intron and one residue in the branch point are invariable. Mutations in these bases will almost certainly abolish normal splicing. Other mutations in these regions may also produce an effect and should be examined further. Figure adapted from Cooper and Krawczak (1993).

transcriptase (Promega) are added, and the reaction is incubated for 1 hour at 42 °C. For difficult transcripts (>2 kb, high GC content, secondary structure, or repetitive sequences), the reaction can be repeated one or more times: heat to 70 °C for 5 minutes, cool to 42 °C, add a further 1 μl of AMV reverse transcriptase, and incubate for an additional hour. The cDNA is then ready to use as a template for PCR. Try using 0.5–1 μl per PCR reaction; it can often be diluted further. Genomic contamination may be a problem, particularly when short RT-PCR fragments do not span an intron, or if investigating splicing mutations. In this case, mRNA can be treated with RNAse-free DNAse I (Promega) before reverse transcription. An mRNA pellet is resuspended in 100 μl DEPC water/DNAse reaction buffer and incubated with 1 U DNAse at 37 °C for 1 hour. The RNA is extracted by use of a standard phenol/chloroform and ethanol precipitation protocol and then reverse transcribed. This treatment is not normally used unless required, because the additional extraction results in some loss of mRNA.

4. Standard Polymerase Chain Reaction Protocols

A reliable PCR reaction that we routinely use to check the quality of genomic DNA is keratin K2e, exon 2. Primers K2e.2L 5′ GCA TCC AGC CTC AGA CAC TT 3′ and K2e.2R 5′ CCA TTA CCA CCA CAA ATG CA 3′ are used in standard PCR buffer (Roche) containing 1.5 mM MgCl$_2$, 0.25 mM dNTPs, 4% (v/v) DMSO, and 1 U *Taq* polymerase (Promega) to amplify a 444-bp fragment spanning exon 2. Amplification conditions are 94 °C for 5 minutes ×1 followed by 30–35 cycles

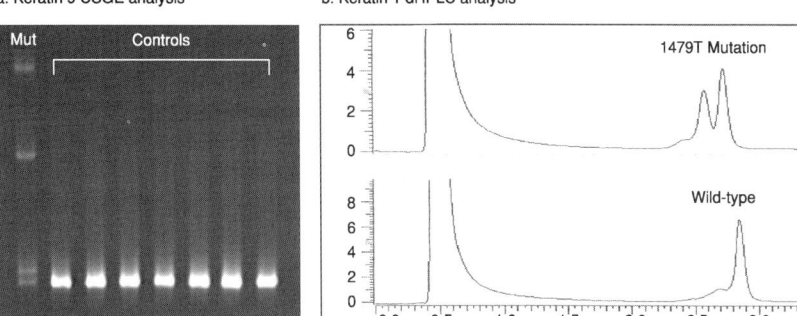

a. Keratin 9 CSGE analysis   b. Keratin 1 dHPLC analysis

Fig. 4 Examples of mutation scanning methods based on the aberrant migration of DNA fragments containing heteroduplexes because of small insertion/deletion or point mutations. (a) Conformation-sensitive gel electrophoresis analysis of polymerase chain reaction (PCR) fragments derived from exon 6 of the keratin 9 gene. The sample in lane 1 (marked "Mut") is from a patient heterozygous for a 3-bp insertion mutation 1362ins3, leading to a duplication of histidine 454 in the helix termination motif of the K9 polypeptide. Four bands can be seen in the mutant sample. The lower two bands are homoduplexes corresponding to the wild-type and mutant alleles. The upper two bands are the two species of heteroduplexes formed. The other lanes represent control samples, showing only single homoduplex bands. Results are obtained after an overnight gel run, and, thus, this is not a very useful technique for large projects. (b) Denaturing high-performance liquid chromatography (dHPLC) analysis of a keratin 1 exon 7 fragments. The upper trace shows dHPLC analysis of a sample from a patient heterozygous for point mutation 1479T, where two peaks are clearly seen. The lower trace is from a normal control. The x-axis shows HPLC elution time; the y-axis is DNA concentration. In dHPLC, heteroduplexes are more rapidly eluted from the column. The Transgenomic Wave system is highly automated and can process unpurified PCR samples in 5–6 minutes from 96-well plates. Thus, this method is suitable for large-scale routine screening.

of 94 °C for 30 seconds, 55 °C for 45 seconds, 72 °C for 45 seconds, 72 °C for 5 minutes ×1. After PCR, 5 μl of product is resolved on a 2% agarose gel. This protocol can be modified as required for other applications.

B. Genetic Linkage Analysis

1. Microsatellites and Single Nucleotide Polymorphisms for Intermediate Filament Genes

The first step in finding markers for a gene of interest is to look for any existing microsatellite markers in the vicinity, using one of the human genome browsers. The University of California Santa Cruz (UCSC) browser is currently the system preferred by most human geneticists, but the NCBI, ENSEMBL, and other systems have similar features. This browser can be set to display known markers identified and characterized by Généthon and Cooperative Human Linkage Center (CHLC), and can now also display uncharacterized "perfect microsatellites" with suggested primers for amplification. As an example, if one searches out the K12 gene in the UCSC browser (gene symbol *KRT12*) and zooms out to show

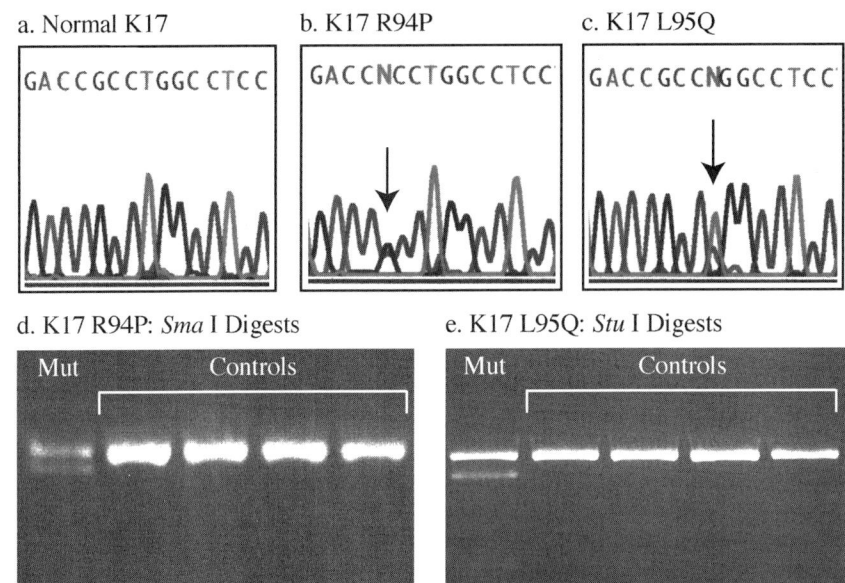

a. Normal K17   b. K17 R94P   c. K17 L95Q

GA CC GC CTGGC CTCC   GACCNCCTGGCCTCC   GACCGCCNGGCCTCC

d. K17 R94P: *Sma* I Digests   e. K17 L95Q: *Stu* I Digests

Mut   Controls   Mut   Controls

Fig. 5 (a–c) Mutations in K17 revealed by direct DNA sequencing; and (d and e), confirmation of mutations by restriction digestion. (a) Normal K17 sequence in exon 1, corresponding to codons 93–97, base numbers 277–291, inclusive. (b) The equivalent sequence shown in (a) from an affected individual with pachyonychia congenita type 2 (PC-2), showing heterozygous missense mutation 281G > C (arrow), which predicts the amino acid substitution R94P in K17. (c) The equivalent sequence shown in (a) derived from another PC-2 patient showing missense mutation 284T > A (arrow) leading to amino acid change L95Q. (d) Confirmation of mutation R94P by *Sma*I digestion. Lane 1; digested polymerase chain reaction (PCR) product from the affected individual (marked "Mut") gives an additional band because of the introduction of a *Sma*I site in one primer, which depends on the mutation. Lanes 2–5 are digested PCR products from normal unrelated controls. (e) Confirmation of mutation L95Q by *Stu*I digestion. Lane 1; digestion of PCR product from the affected individual (marked "Mut") shows an additional band caused by a new *Stu*I site created by the mutation. Lanes 2–5 show digested PCR products from normal unrelated controls. (See Color Insert.)

the flanking keratin genes and displays "STS Markers" (sequence tag sites), one should see nearby known microsatellites colored blue in the STS window closely flanking the K12 gene. The two nearest markers are GATA25A04 and AFM200ZF4. Clicking on these features will give details of primers, PCR conditions, H values, etc. If one displays the "perfect microsatellite" features, there is a CA repeat actually located within the K12 gene, and primers are suggested. In fact, we have already characterized this marker and know that it is, indeed, highly polymorphic (Corden *et al.*, 2000). With these *in silico* tools, it is possible to find microsatellites close to or within any IF gene of interest for genetic linkage. If no known or "perfect" microsatellites are displayed, it is possible to browse the "simple repeats" and choose other short tandem repeats for amplification and analysis. In our experience, di-, tri-, and tetra-nucleotide repeats are most likely to

Table IV
Web Resources and Suppliers

Website	URL
On-line Mendelian Inheritance in Man (OMIM): Clinical and molecular data on all genetic disorders	http://www.ncbi.nlm.nih.gov/Omim/
Intermediate filament mutation database: Clinical and mutational data on keratin disorders and other intermediate filament diseases	http://www.interfil.org/
Lamin mutation website, Leiden	http://www.dmd.nl/lmna_home.html
Génethon	http://www.genethon.fr/
CHLC: Cooperative Human Linkage Center	http://www.gai.nci.nih.gov/
Marshfield	http://research.marshfieldclinic.org/genetics/Default.htm
UCSC: Genome Bioinformatics Group of University of California Santa Cruz	http://www.genome.ucsc.edu/
NCBI: National Center for Biotechnology Information	http://www.ncbi.nlm.nih.gov/
ABI: Applied Biosystems	http://www.appliedbiosystems.com/
MWG	http://www.mwg-biotech.com/
Transgenomics	http://www.transgenomic.com/
Pedigree software	http://www.cyrillicsoftware.com/
Genetic testing	
GeneDx: US-based company offering genetic testing for keratin disorders, genetic skin diseases and other disorders.	http://www.genedx.com/
Human Genetics Unit, Dundee: UK center for molecular diagnosis of keratin disorders.	http://www.humangenetics.org.uk/

be polymorphic. The longer and less interrupted the stretch of repeat is, the more likely it will be polymorphic. Mononucleotide repeats are highly variable but difficult to score. Single nucleotide polymorphisms (SNPs) can be displayed on the UCSC and most other genome browsers. SNPs can be typed by sequencing (expensive) or by restriction enzyme digestion, allele-specific PCR, Affymetrix gene chips, or other high-throughput methods; however, for genetic linkage in families with single-gene traits at the current time, microsatellites are the method of choice.

2. Microsatellite Genotyping Methods

One PCR primer is synthesized with a suitable fluorochrome, and the resulting fluorescent PCR products are analyzed on an automated DNA sequencer. Many of the known microsatellite primer pairs are available from ABI and other suppliers. For custom synthesis, we use the FAM and HEX fluorochromes, which are cheap and readily available from most primer synthesis services, such as MWG Biotech, Ebersberg, Germany. A short additional sequence (GTTTCTT) is added to the 5' end of the unlabeled primer. This promotes addition of the plus-A overhang to the labeled strand by *Taq* polymerase. Without this tail, plus-A addition varies in a sequence-dependent fashion and makes calling of alleles difficult. 7.5 μl PCR reactions are set up for all family members for each primer set (according to the ABI Linkage Mapping Set version 2 protocol). AmpliTaq Gold DNA Polymerase (ABI), a modified heat-activated form of *Taq* polymerase, is used to eliminate nonspecific priming at room temperature before PCR. PCR reactions are heated to 94 °C for 12 minutes to activate the polymerase followed by 10 cycles of 94 °C for 15 seconds, 55 °C for 15 seconds, 72 °C for 30 seconds; then 20 cycles of 89 °C for 15 seconds, 55 °C for 15 seconds, 72 °C for 30 seconds, and a final step of 72 °C for 10 minutes. We have found many markers work better at lower annealing temperatures. PCR products are diluted (1:10–1:40), and markers with nonoverlapping size ranges or differing dyes may be pooled before analysis. Genescan 400HD ROX size standard (ABI) is diluted 1:20 in formamide (Hi-Di grade, ABI); 2 μl of diluted PCR product is mixed with 10 μl of diluted ROX400. The resultant mix is denatured at 95 °C for 2 minutes and chilled on ice before analysis on an ABI 3100 capillary electrophoresis DNA analyzer (or similar). Results are analyzed with Genemapper software (ABI) and are plotted as peaks corresponding to the PCR fragment sizes. The marker alleles can be marked on pedigrees and analyzed "by eye" or used in a linkage analysis program within Genemapper. Alleles can also be exported as Excel spreadsheets.

3. Genome-Wide Screening

Genome-wide linkage is performed with a range of microsatellites covering the entire genome. A well-established system is the ABI PRISM Linkage Mapping Set v2.5 covering the genome at low density (400 markers, <10% recombination between markers [i.e., 10 centiMorgans]) or high density (800 markers, 5 cM). Low-density screening should be adequate for recessive conditions with inbred families. With dominant conditions, especially with smaller families, high-density screening may be required. In both cases, low-density screening is performed first and only extended if linkage is not detected. By use of multiple dyes and different-sized PCR product ranges, marker panels are designed to run up to 20 different PCR products together in one lane/capillary of the sequencer. All the markers from the ABI sets have been previously run on CEPH DNA (individual 1347-02) and the base pair sizes published in the ABI literature. This DNA can be purchased from ABI and should be included as a control in every screen to help identify the correct alleles.

C. Mutation Detection Strategies

1. Polymerase Chain Reaction Design

In 2004, with the human genome project more or less complete and fairly well annotated, the full gamut of human IF genes and their numerous pseudogenes is now emerging (Hesse *et al.*, 2001, 2004). The availability of the full genomic sequence for IF genes, including all intronic and flanking sequences, means that it is now relatively straightforward to develop a comprehensive mutation detection strategy for each member of this large gene family. The excellent annotation and search tools now available through the UCSC genome browser and similar sites also allow avoidance of the main pitfalls of genomic PCR for IF genes. Primers should be checked for cross-reactivity problems caused by pseudogenes or closely related functional genes (isogenes); repetitive DNA, such as *Alu* or other interspersed repeats, should be avoided along with SNPs that could result in failure to amplify both alleles in heterozygous individuals. Repeat sequences and SNPs can be displayed on the UCSC genome browser. Cross-reactivity with other parts of the genome can be checked by the BLAT search tool (UCSC). We routinely use the Oligo 4.0 program for primer design (MBI Inc., Cascade, CO; current version is 6.0). This allows analysis of primer homoduplexes and primer dimer formation. Primer melting temperature (T_m) values should be matched within 1 °C if possible. We normally use 20-mers, aim for Tm values of 62–65 °C, and only vary primer length within the range 18–24 bp if the GC content of the region is unsuitable to get the desired T_m. Primers designed in this way normally work with an annealing temperature of 55 °C and require minimal optimization. Exons are normally amplified with primers placed in introns, allowing ~150 bp upstream to cover the splice site and branch point within the preceding intron and ~100 bp downstream to cover the splice site in the following intron (Fig. 3). Noncoding exons should also be analyzed, because sequence variants therein can affect gene expression. The UCSC browser should be used to scan expressed sequence tags (ESTs) corresponding to the gene, because this can reveal unknown splice variants that will also require screening.

2. Mutation Scanning Techniques

We have used two methods to scan genes for sequence changes by heteroduplex analysis. They can be used to look for new mutations and/or to screen control samples to confirm or exclude a known mutation. When screening for a new mutation, any region showing a heteroduplex is sequenced to identify the exact change. These methods can pick up any sequence change and will therefore detect polymorphisms, as well as pathogenic mutations. dHPLC is the method of choice for high throughput analysis (Cobb *et al.*, 2002) but requires purchase of a wave analyzer. Conformation sensitive gel electrophoresis (CSGE) is based on standard laboratory equipment and is suitable for smaller scale projects (Ganguly *et al.*, 1993).

To screen a gene, PCR primers are designed to cover all exons in overlapping fragments of 200–600 bp. PCR products are first checked on agarose gels before analysis by CSGE or Transgenomic Wave dHPLC System. For CSGE, large manual sequencing gels are poured and polymerized for at least 1 hour. Gels are made up of 40 ml of 99:1 acrylamide:BAP (1,4 bis(acryoyl)piperazine) mixed with 5 ml of 20× GT buffer (US Biochemicals), 30 ml formamide, 20 ml ethylene glycol, 103 ml water, 2 ml of 10% ammonium persulfate, and 138 μl TEMED. Running buffer for both tanks is 0.5× GT buffer. Gels are prerun for 1 hour at 300 V, wells rinsed out, and samples loaded; 5 μl of average-strength PCR products are diluted to 8 μl with water, denatured at 98 °C for 5 minutes, followed by 68 °C for 30–60 minutes and 2 μl loading dye added. Ten microliter samples are run at 300–400 V overnight. Gel plates are then separated, and the one with the gel attached is soaked in 0.5× GT buffer with 0.05 μg/ml ethidium bromide. Bands are visualized on an ultraviolet transilluminator; two or three bands indicate a heteroduplex, and therefore these PCR fragments should be sequenced (Fig. 4a). Depending on the sizes of the PCR fragments, it is possible to double load samples by running in the first loading and then loading a second set.

For dHPLC with the wave analyzer (Transgenomic, Omaha, NE), PCR products are denatured at 95 °C for 5 minutes and slowly renatured using 70 cycles of 22 seconds starting at 95 °C and cooling by 1 °C/cycle to allow heteroduplex formation. Depending on the strength of the PCR products, 5–10 μl is injected according to the manufacturer's protocol. The temperature that the analysis is run at is determined for each PCR fragment as predicted by the WaveMaker program (Transgenomic). Results from the wave system are represented as an elution profile plot (Fig. 4b). Samples showing a different elution profile compared with a normal control, most often two peaks instead of one, indicate a sequence change, and these PCR products require sequencing. One important point for dHPLC analysis is that some PCR buffers contain additives that can damage HPLC columns. One should use a buffer without detergent such as HotStar *Taq* system (Qiagen) or purify PCR products before analysis (QIAquick PCR purification kit). Another general problem of both these techniques is that only heterozygous changes are detected. If a recessive family is being studied and a homozygous change is predicted, the samples can still be screened by mixing equimolar amounts of normal PCR products with the homozygous sample before denaturation.

3. Direct DNA Sequencing

Central sequencing services that use the ABI 377 gel-based system and/or ABI 3100-, 3700, or 3730 capillary systems allow rapid, high-throughput sequencing in most research institutions. Before direct sequencing, PCR products are purified with the QIAquick PCR Purification Kit to remove unwanted primers. In some cases, for example, if two bands are detected because of a large insertion/deletion mutation, the QIAquick Gel Extraction Kit can be used. The current sequencing chemistry in common use is the ABI PRISM Big-Dye system, and it is used according to the manufacturer's straightforward PCR-based protocol. Results from the automated sequencer are plotted as traces in four colors representing each base, A, C, G, and T. Sequence traces of affected individuals should always be compared with those of a normal control and checked by eye, because the base-calling software, particularly for heterozygous point mutations, is not reliable. Gene scanning techniques such as CSGE or dHPLC obviously greatly reduce the amount of sequencing and manual checking required. Examples of these methods are shown in Fig. 5 and 6.

4. Southern Analysis

Some classes of mutations, particularly larger genomic deletions or rearrangements, can be missed by PCR-based techniques, and therefore Southern blot analysis of genomic DNA is required. For example, a 6-Mb genomic deletion identified in the *LMNA* gene would not have been detected by standard exonic PCR methods. Genomic DNA is digested with restriction enzymes and size-fractionated by agarose gel electrophoresis. The gel is then transferred to a solid support such as nitrocellulose or nylon membranes. A labeled DNA probe (either radioactively or nonradioactively labeled) is hybridized and will only bind to complementary DNA sequences in the target DNA. Its subsequent position on the membrane can be compared with the original gel to give an estimation of size. Southern blotting is used to establish a restriction map for the gene of interest, which can then be used to determine the presence of deletions, insertions, or rearrangements in disease situations. For example, a large deletion in a gene may abolish restriction sites, which will consequently alter the pattern of hybridizing fragments on the blot.

In our laboratory, we typically digest 10 μg of genomic DNA overnight with appropriate restriction enzymes. The digested DNA is run on a 1% agarose gel, and before transfer, the gel is photographed to provide a permanent copy. The DNA within the gel is first depurinated with 250 mM HCl for 10 minutes (to aid transfer of large >10 kb fragments), then denatured in 1.5 M NaCl/0.5 M NaOH for 25 minutes before a final neutralization stage in 1.5 M NaCl/0.5 M Tris-HCl,

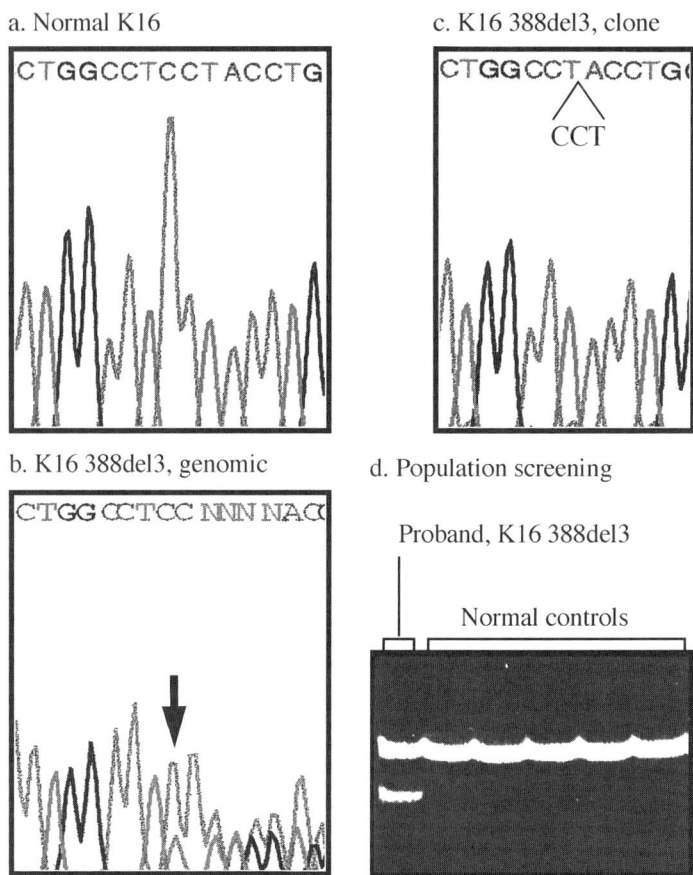

Fig. 6 K16 mutation detection and confirmation. (a) K16 genomic sequence derived from a normal individual by direct sequencing of polymerase chain reaction (PCR) products. Coding sequence base numbers 382–396, encompassing codons 128–132, are shown. (b) The equivalent K16 sequence as shown in (a), derived from an individual with pachyonychia congenita type 1 (PC-1). Arrow indicates the start of sequence overlap caused by the heterozygous 3-bp deletion, 388del3. (c) The equivalent K16 sequence as shown in (a) derived from the mutant allele cloned into pCR2.1 vector. This clarifies that the mutation is a 3-bp deletion: removal of a CCT repeat, as indicated. (d) Confirmation of mutation 388del3 and exclusion from normal control individuals. A 140-bp PCR fragment spanning the mutation was amplified from genomic DNA from the proband and 50 unaffected, unrelated individuals (five of which are shown here). PCR products were separated on 6% sequencing gels. Note that the smaller mutant band is much weaker than the normal one, because the PCR used here amplifies both the functional K16 gene and its pseudogenes. (See Color Insert.)

pH 7.5, for 30 minutes. The gel is then transferred overnight to Hybond N+ membrane (Amersham) with 20× standard saline citrate (SSC) as the transfer buffer. The next day, the position of the gel wells are marked on the membrane with a needle. The transferred DNA is cross-linked to the membrane with an ultraviolet cross-linker. We routinely use a nonradioactive system for labeling DNA probes (Gene Images kit, Amersham). This system uses fluorescein-11-deoxyuridine triphosphate (dUTP) in a random prime-labeling reaction. DNA probes can either be purified from plasmid DNA (after appropriate digestion with restriction enzymes) or can be generated by PCR. The labeled probes, at a concentration of ~10 μg/ml, are hybridized overnight at 60 °C. The next day the membrane is washed in 1× SSC/0.1% (w/v) sodium dodecyl sulfate (SDS) for 15 minutes at 60 °C followed by a second wash in 0.5× SSC/0.1% (w/v) SDS (also at 60 °C). For immunological detection of the bound probe, the membrane is blocked and then incubated with an antifluorescein antibody–alkaline phosphatase conjugate. After antibody incubation, the membrane is washed, and the fluorescein–antibody complex is detected by chemilluminescence with CDP-Star detection reagent (Amersham).

5. Population Screening Techniques

By convention, pathogenic mutations, particularly missense changes, must be excluded from 100 normal chromosomes from the same ethnic background to show that they are not a common polymorphism. There are several ways to do this, depending on the mutation, time available, and cost. In some cases, depending on the gene of interest and the number of samples to be analyzed, it may be easiest to sequence 50 control samples for the specific target regions to prevent the design of individual tests for each new mutation identified. This is the technique we currently use for the hotspot exons of keratin genes. Alternatively, a mutation may either create or delete a restriction enzyme site. PCR primers can be designed around this site to make a suitably sized PCR product. If there is one other site in control samples that is not affected by the mutation and that allows a clear restriction enzyme pattern to be achieved, this extra site acts as an internal control (Fig. 5). However, many mutations will not create or destroy an enzyme site. Here, one can design a mismatch primer, which in combination with the mutation will create a new restriction enzyme site. In these cases, the PCR products need to be kept fairly small (150–200 bp), because it is only the length of the primer that is cut from the PCR fragment (Fig. 5). Insertion or deletion mutations can be screened directly by PCR. If the insertion/deletion is more than approximately 10 bp and primers are designed to amplify a small PCR fragment (~200–300 bp), these can be resolved on a 3%–4% agarose gel or 6% acrylamide Tris-borate-EDTA (TBE) gels (Fig. 6). If the sequence change is only one or a few bp, PCR primers are designed again to make a small fragment, ~200 bp, and one primer is fluorescently tagged and the products analyzed as described for microsatellites. The Transgenomic Wave dHPLC system is ideal for screening of known mutations.

D. Prenatal Testing

In general, many of the disorders caused by keratin mutations are not severe enough to warrant prenatal diagnosis. However, in some of the more severe phenotypes such as epidermolysis bullosa simplex–Dowling-Meara (EBS-DM), bullous congenital ichthyosiform erythroderma (BCIE), or severe pachyonychia congenita (PC), it has been carried out (Rothnagel *et al.*, 1994; Rugg *et al.*, 2000; Smith *et al.*, 1999b), subject to local law and appropriate institutional review. With reliable detection methods available from genomic DNA, prenatal diagnosis can be performed from chorionic villus samples (CVS) when the familial mutation is known. CVS biopsies are taken at an early stage of the pregnancy, normally 11–12 weeks' gestation. Multiple villi are dissected out from the CVS, and DNA is isolated by standard DNA extraction protocols. CVS DNA is screened for the known familial mutation with the PCR and sequencing protocols described previously. Care is needed to prevent contamination of the CVS DNA with maternal cells, because this can lead to problems of diagnosis. A panel of highly polymorphic microsatellite markers can be used to check maternal contamination in the CVS DNA by DNA fingerprinting (e.g., Profiler kit, ABI). Genetic diagnosis can only be performed by accredited National Health Service laboratories in the United Kingdom. In the United States, these services are provided by board-certified genetics laboratories and private companies.

IV. Pearls and Pitfalls

A. Polymerase Chain Reaction Failure

One problem frequently encountered in molecular genetics is a region of a gene that is difficult or impossible to amplify by PCR. There are a number of probable causes, including high GC content, secondary structure caused by inverted repeats, and primers lying within repetitive elements. Primers that inadvertently cover or end on SNPs is another possibility. Even taking these factors into account when designing PCRs, there are still regions that fail to amplify for some unknown reason, even after the standard technique of varying annealing temperature and magnesium concentration. The following course of action is recommended in these instances. First, try a variety of different PCR buffers and enzyme variants from different companies. These vary widely, and often a PCR reaction seems to work well with only one buffer/enzyme combination. The systems we find most useful for difficult templates are the Expand High-Fidelity and Expand Long-Range enzyme mixes and buffers from Roche. Although designed for other applications, these systems work well for short PCR fragments in difficult regions, because they consist of a mixture of thermostable polymerases, one of which may get through the problem sequence. We also recommend a particularly good "home-made" ammonium sulfate bovine serum albumin (BSA) PCR buffer for difficult PCRs, with standard *Taq* or other polymerases (67 mM Tris-HCl, pH 8.8, 16.6 mM $(NH_4)_2SO_4$, 1.5 mM $MgCl_2$, 0.17 mg/ml BSA, and 10 mM 2-mercaptoethanol). Failing that, make a second set of primers and try all four primer combinations. We have encountered exons that will only amplify with one set of primers out of several tested, using only one type of enzyme and buffer.

B. Failure to Detect Mutations

Another problem in human genetics are "missing mutations." In most genetic diseases, there is a subset of patients in whom the diagnosis is good but no mutation can be found. One reason may be that the single base pair mutation has simply been missed by whoever checked the sequence. Independent checking is therefore a good practice. Also, no mutation detection technique is 100% reliable, even sequencing. Heterozygous sequence changes are notoriously easy to miss, so a two-pronged approach of CSGE/dHPLC and sequencing is recommended. Furthermore, some PCR primers may fall on unknown SNPs or other polymorphisms that result in a failure to amplify both alleles. Making a second set of primers for the gene is one way to combat this effect. Checking primers against SNP databases is another.

Some patients may have larger genomic deletions, insertions, or rearrangements, so Southern analysis should be considered. Others may have cryptic splicing mutations within large introns, some of which may be detected by RT-PCR approaches if patient mRNA from relevant tissue or cultured cells can be obtained. Another possibility is that mutations in another related gene produce a copy of this phenotype. A good example is that of patients with EPPK carrying K1 rather than K9 mutations, described previously (Terron-Kwiatkowski *et al.*, 2002, 2004). If it is possible in these unsolved cases to obtain a big enough family, then genetic linkage can be used to exclude loci or to scan the genome for new loci. This type of approach led to the discovery of the first lamin mutations. Certain lamin defects phenocopy the classic form of Emery-Dreifuss muscular dystrophy caused by emerin defects (Bonne *et al.*, 1999).

C. Consequences of Stop Codon Mutations

An important pitfall in human genetics is the issue of premature termination codon (PTC) mutations. This is a problem often not grasped by cell biologists, who frequently introduce PTCs into expression constructs to make deletions. It is a well-established fact that, *in vivo*, most PTC mutations result in destabilization of the mRNA by several orders of magnitude through a mechanism known as nonsense-mediated mRNA decay (Cooper and Krawczak, 1993). Thus, a PTC in an expression plasmid that gives a truncated protein when overexpressed in cultured cells will, *in vivo*, lead to almost complete loss of expression. This was illustrated in one of the original recessive EBS cases, where a frameshift mutation in K14 led to this normally abundant mRNA being undetectable by *in situ* hybridization (Rugg *et al.*, 1994). Thus, RT-PCR of heterozygotes, or Northern blots, and/or *in situ* hybridization of homozygotes, should be pursued in these cases, especially if one is trying to draw important conclusions based on the presence of truncated polypeptides. A related issue is that of splice site mutations. It is difficult to make firm conclusions about their effects on mRNA processing without checking this experimentally by RT-PCR.

V. Discussion and Concluding Remarks

A. Clinical Genetics Considerations

Obtaining families with a relevant human genetic disease is an obvious first step in linking an IF gene to a disease. In the case of a candidate gene approach, where one is seeking a disease that fits with a particular IF protein of perceived function and/or expression pattern, the first step is to look in the medical genetics database, OMIM. This highly searchable and cross-referenced on-line resource is a distillation of essentially all published reports of recognized genetic disorders. It is a valuable short-cut to finding out what phenotypes are already linked to genes, as well as key literature references. The database has more than 10,000 disorders catalogued, most of which still lack causative genes, despite the recent advances in genetics. One example where we used OMIM to good effect was to find the disease for K3 and K12 (Irvine *et al.*, 1997a). We knew from the literature that these keratins were specifically expressed in the anterior corneal epithelium (Chaloin-Dufau *et al.*, 1990). Searching the database yielded approximately 60 corneal disorders, one of which was Meesmann epithelial corneal dystrophy (OMIM 122100). The description in the database and references therein were sufficient to indicate that this was the likely candidate disease.

Another prerequisite for human genetics studies, especially for nonclinicians, is to establish good clinical collaborations to make contact with families and, with suitable ethical permission, to obtain DNA and other materials required for study. Some of the known IF diseases are fairly rare, so this may involve a network of clinicians nationally or internationally. Searching out published case reports and contacting the clinicians involved has been a valuable approach in our experience, and by this means, we have even made contact with descendants of families with keratin disorders that were published as far back as the 1930s (Irvine *et al.*, 1997a). The Internet obviously greatly facilitates tracking authors of papers and establishing correspondence.

The value of good histopathology and ultrastructural analysis of the disease in question, in parallel with molecular genetics studies, cannot be understated. These morphological clues can greatly limit the range of candidate genes that need to be analyzed. An example of this comes from our studies on the role of plectin in EBS with muscular dystrophy. Ultrastructure showed cytolysis low in basal keratinocytes, suggestive of a defect in keratin-hemidesmosome adhesion (Smith *et al.*, 1996). Subsequently, immunohistochemistry showed that plectin epitopes were absent in these patients. Genetic linkage and mutation analysis led to the discovery of loss-of-function mutations in plectin (McLean *et al.*, 1996; Smith *et al.*, 1996).

DNA sample handling and tracking quickly becomes an issue in genetic studies. A good package for handling pedigree information is Cyrillic 2.1, (Cyrillic Software, Reading, UK), and a major advantage of this system is that it allows addition of genetic marker information directly onto the pedigree and contains within it the Linkage 5.1 program required for statistical analysis of genetic linkage data. In terms of patient material, if genome-wide linkage analysis is a possibility, then it is advisable to take 10 ml of blood for DNA extraction, because a genome scan may use several micrograms of DNA per patient. For routine mutation detection of a known gene, a less invasive mouthwash sample may suffice. It is often very valuable to have a source of mRNA, particularly when investigating a new disease, either frozen tissue or cultured cells, to check for expression of both alleles or to check splicing defects (Jonkman *et al.*, 1996). This obviously depends on the tissue involved in terms of accessibility and ethical considerations. Only one or two patients with the condition need to be sampled, and this can often be combined with biopsies for histological examination and/or electron microscopy.

B. Genetic Linkage

1. Basic Concepts

The basic concepts behind genetic linkage analysis using polymorphic markers to track disease genes are thoroughly explained elsewhere (Ott, 1991; Strachan and Read, 2003). The best polymorphic markers are undoubtedly microsatellites— variable number tandem repeat sequences that occur every few kilobases in the human genome, commonly CA repeats (Weber and May, 1989). These markers are highly informative, because they exist as many different allele sizes in the population and consequently have high heterozygosity values (H). For genetic linkage, one needs to distinguish the two copies of a given locus in affected individuals to see whether the same copy is passed on to all the affected descendants. If a marker is often homozygous, which is the case for many SNPs, it seldom allows tracking of alleles through the generations of a family. An example is given in Fig. 1. This family with an autosomal-dominant disorder is fully informative for a microsatellite marker near one of the IF loci. With SNPs, the H value is usually about 0.30 (i.e., only 30% of people are heterozygous and so only some parts of a family may yield linkage information) (Fig. 1). Most microsatellites have H values of 0.85 or greater.

2. Statistical Analysis

The statistics of genetic linkage analysis, with worked examples using the LINKAGE software package, are dealt with by Ott (Ott, 1991). For a new single gene–disease linkage result to be publishable, it requires generation of a statistically significant LOD score. By convention, an LOD score of >3 is accepted as evidence of linkage. This in effect means that there is only a 1 in 1000 (or 10^3) odds of the markers observed to cosegregate with the disease gene in a given family being coincidental. The higher the LOD score, the more likely that this is the correct gene locus. In a dominant condition, a minimum of 10 meioses must show linkage to give an LOD score of 3 (i.e., as a rule of thumb, each linked meiosis contributes about 0.3 to the LOD score). This means finding a family with a founder individual and 10 affected and/or unaffected descendants who could have inherited the disease. In fact, for genome-wide linkage, a much larger family is required, because many markers will be uninformative. The smallest dominant family on which we have successfully performed a genome-wide screen had 22 meioses. We have failed to get a linked locus using families with 14 meioses (McLean, unpublished data). Thus, for a dominant condition, using 22 meioses, 1 CEPH control DNA, a negative control, and 400 microsatellite markers, it will require analysis of 9600 fluorescent PCR products to find linkage. Fortunately, multichannel pipettes, 96-well plate PCR, and mixing of marker panels makes lighter work of this task. In recessive conditions, with material from inbred families, each affected individual represents several meioses, because the same gene can be traced back through the generations on both the paternal and maternal sides of the family, each inheritance step counting as one meiosis. As a rule of thumb, a linked affected person in a consanguineous family contributes ~1.1 to the LOD score. Thus, as few as three affected persons in such a family can yield significant linkage of 3.3, with perhaps as few as 1600 PCR reactions (Fig. 2). Genome-wide linkage is labor-intensive and repetitive in nature, but given the required number of meioses and reliable clinical ascertainment at the outset, it essentially guarantees discovery of a new disease gene.

C. Positional Candidate Approach to Intermediate Filament Genetic Diseases

Once microsatellites are available for a given gene or genes, families with a suspected genetic disease can be genotyped and the alleles tracked through the pedigree. An example for a dominant disorder is given in Fig. 2. Here, a family with a dominant keratin disease was analyzed with a marker within each of the type I and type II keratin loci. One can see that in the case of the type I marker, D17S800, the inheritance of alleles does not follow the disease and so this locus is excluded. However, the allele sized at 222 bp of the type II marker D12S368 is inherited by all the affected individuals throughout the pedigree, consistent with the defective gene residing in this chromosomal location. By this means, 50% of keratin genes can be excluded in families of sufficient size. Similarly, if a family had a suspected neurological disorder, the NF-L, NF-M, NF-H, lamin A/C, and other loci could be examined without resorting to sequencing. Although relatively large families are required for statistically significant linkage, even small families with a few affected and unaffected individuals analyzed for markers close to or within the candidate gene can allow exclusion of loci.

In terms of recessive inheritance, the most useful families for genetic linkage are consanguineous ones, in which the parents are cousins or similar, allowing homozygosity mapping (Fig. 2). In these inbred families, one can fairly safely assume that the parents carry the same copy of a mutant allele inherited from one of their shared ancestors. The affected persons are therefore homozygous at causative gene locus and the parents are heterozygous. The unaffected siblings are either heterozygous or can carry different alleles altogether. Statistically, these families are more powerful for gene mapping than dominant kindreds. An example of an inbred family with a recessive keratin disorder is shown in Fig. 2. The type II keratin locus is excluded using marker D12S368, but the inheritance of markers consistent with linkage is seen with type I keratin marker D17S800.

D. Common Classes of Human Mutation

Point mutations are by far the most common gene defect detected. These, and insertion/deletion mutations, are readily detected by mutation scanning techniques such as CSGE or dHPLC prior to sequencing (Fig. 4). Many point mutations result in synonymous codon changes and so do not affect translation. Nonsynonymous codon substitutions result in an amino acid change or missense mutation, and a change to a stop codon is classed as a nonsense mutation. Premature termination codons can also result from out-of-frame insertion/deletions, including splicing defects. Many IF mutations are dominant, so affected individuals are heterozygous for a missense change seen as a single overlapping peak in an otherwise clean sequence (Fig. 5). Compared with the sequence trace of a normal individual, there is usually a reduction in peak height of the mutant base. Deletion/insertion mutations in a heterozygous individual result in normal sequence traces until the point of the deletion/insertion mutation, after which overlapping sequences are seen (Fig. 6). If only a few bases have been deleted/inserted, it is often possible to work out the mutation; however, to confirm this or determine larger mutations, the PCR fragment should be cloned into a plasmid (for example pCR2.1 TA cloning vector, Invitrogen) and several clones sequenced. Clones containing either the normal or the mutant allele should be obtained. If, however, an individual is expected to be homozygous for a mutation, because of consanguinity, no overlaps are detected, and so careful comparison with a normal sequence is required. Splice site mutations are less common than missense or small insertion/deletion mutations but have been detected in a number of IF diseases (www.interfil.org). Intronic mutations that affect the invariant residues of splice sites or the branch point site will result in aberrant splicing (Fig. 3). Exon skipping is a common effect, and if this involves an out-of-frame exon, a frameshift and premature termination of translation will result. In practice, the effect of these mutations on mRNA processing is difficult or impossible to reliably predict, because splicing to cryptic sites nearby often occurs. As a result, it may require RT-PCR to resolve this issue.

E. Concluding Remarks

The discovery of human genetic diseases of IF proteins has yielded a remarkable amount of information on the *in vivo* functions of this protein family. Who would have guessed 10 years ago that a mutation in a nuclear lamin could lead to the specific loss of fat cells? With the rapid advances that have taken place in human genetics and genomics over the past few years, coupled with new high-throughput techniques such as chip-based resequencing, it is very likely that genetic diseases for most, if not all, members of this protein family will come to light, undoubtedly with many other surprises along the way.

References

Anton-Lamprecht, I., and Schnyder, U. W. (1982). Epidermolysis bullosa herpetiformis Dowling-Meara. Report of a case and pathomorphogenesis. *Dermatologica* **164,** 221–235.

Bonifas, J. M., Rothman, A. L., and Epstein, E. (1991a). Linkage of epidermolysis bullosa simplex to probes in the region of keratin gene clusters on chromosomes 12q and 17q. *J. Invest. Dermatol.* **96,** 550a.

Bonifas, J. M., Rothman, A. L., and Epstein, E. H. (1991b). Epidermolysis bullosa simplex: Evidence in two families for keratin gene abnormalities. *Science* **254,** 1202–1205.

Bonne, G., Di Barletta, M. R., Varnous, S., Becane, H. M., Hammouda, E. H., Merlini, L., Muntoni, F., Greenberg, C. R., Gary, F., Urtizberea, J. A., Duboc, D., Fardeau, M., Toniolo, D., and Schwartz, K. (1999). Mutations in the gene encoding lamin A/C cause autosomal dominant Emery-Dreifuss muscular dystrophy. *Nat. Genet.* **21,** 285–288.

Bowden, P. E., Haley, J. L., Kansky, A., Rothnagel, J. A., Jones, D. O., and Turner, R. J. (1995). Mutation of a type II keratin gene (K6a) in pachyonychia congenita. *Nat. Genet.* **10,** 363–365.

Brenner, M., Johnson, A. B., Boespflug-Tanguy, O., Rodriguez, D., Goldman, J. E., and Messing, A. (2001). Mutations in GFAP, encoding glial fibrillary acidic protein, are associated with Alexander disease. *Nat. Genet.* **27,** 117–120.

Chaloin-Dufau, C., Sun, T.-T., and Dhouailly, D. (1990). Appearance of the keratin pair K3/K12 during embryonic and adult corneal epithelial differentiation in the chick and in the rabbit. *Cell Different. Dev.* **32,** 97–108.

Cobb, C. J., Scott, G., Swingler, R. J., Wilson, S., Ellis, J., MacEwen, C. J., and McLean, W. H. I. (2002). Rapid mutation detection by the transgenomic wave analyser DHPLC identifies MYOC mutations in patients with ocular hypertension and/or open angle glaucoma. *Br. J. Ophthalmol.* **86,** 191–195.

Conley, Y. P., Erturk, D., Keverline, A., Mah, T. S., Keravala, A., Barnes, L. R., Bruchis, A., Hess, J. F., FitzGerald, P. G., Weeks, D. E., Ferrell, R. E., and Gorin, M. B. (2000). A juvenile-onset, progressive cataract locus on chromosome 3q21-q22 is associated with a missense mutation in the beaded filament structural protein-2. *Am. J. Hum. Genet.* **66,** 1426–1431.

Cooper, D. N., and Krawczak, M. (1993). "Human Gene Mutation." BIOS Scientific Publishers Ltd., Oxford.

Corden, L. D., Swensson, O., Swensson, B., Smith, F. J. D., Rochels, R., Uitto, J., and McLean, W. H. I. (2000). Molecular genetics of Meesmann's corneal dystrophy: Ancestral and novel mutations in keratin 12 (K12) and complete sequence of the human KRT12 gene. *Exp. Eye Res.* **70,** 41–49.

Coulombe, P. A., Hutton, M. E., Letai, A., Hebert, A., Paller, A. S., and Fuchs, E. (1991). Point mutations in human keratin 14 genes of epidermolysis bullosa simplex patients: Genetic and functional analysis. *Cell* **66,** 1301–1311.

Coulombe, P. A., and Omary, M. B. (2002). 'Hard' and 'soft' principles defining the structure, function and regulation of keratin intermediate filaments. *Curr. Opin. Cell Biol.* **14,** 110–122.

Dalakas, M. C., Park, K. Y., Semino-Mora, C., Lee, H. S., Sivakumar, K., and Goldfarb, L. G. (2000). Desmin myopathy, a skeletal myopathy with cardiomyopathy caused by mutations in the desmin gene. *N. Engl. J. Med.* **342,** 770–780.

Figlewicz, D. A., Krizus, A., Martinoli, M. G., Meininger, V., Dib, M., Rouleau, G. A., and Julien, J. P. (1994). Variants of the heavy neurofilament subunit are associated with the development of amyotrophic lateral sclerosis. *Hum. Mol. Genet.* **3,** 1757–1761.

Fine, J. D., Eady, R. A., Bauer, E. A., Briggaman, R. A., Bruckner-Tuderman, L., Christiano, A., Heagerty, A., Hintner, H., Jonkman, M. F., McGrath, J., McGuire, J., Moshell, A., Shimizu, H., Tadini, G., and Uitto, J. (2000). Revised classification system for inherited epidermolysis bullosa: Report of the Second International Consensus Meeting on diagnosis and classification of epidermolysis bullosa. *J. Am. Acad. Dermatol.* **42**, 1051–1066.

Fuchs, E., and Cleveland, D. W. (1998). A structural scaffolding of intermediate filaments in health and disease. *Science* **279**, 514–519.

Ganguly, A., Rock, M. J., and Prockop, D. J. (1993). Conformation-sensitive gel electrophoresis for rapid detection of single-base differences in double-stranded PCR products and DNA fragments: Evidence for solvent-induced bends in DNA heteroduplexes. *Proc. Natl. Acad. Sci. USA* **90**, 10325–10329.

Goldfarb, L. G., Park, K. Y., Cervenakova, L., Gorokhova, S., Lee, H. S., Vasconcelos, O., Nagle, J. W., Semino-Mora, C., Sivakumar, K., and Dalakas, M. C. (1998). Missense mutations in desmin associated with familial cardiac and skeletal myopathy. *Nat. Genet.* **19**, 402–403.

Hatsell, S. J., Eady, R. A., Wennerstrand, L., Dopping-Hepenstal, P., Leigh, I. M., Munro, C., and Kelsell, D. P. (2001). Novel splice site mutation in keratin 1 underlies mild epidermolytic palmplantar keratoderma in three kindreds. *J. Invest. Dermatol.* **116**, 606–609.

Healy, E., Holmes, S. C., Belgade, C., Stephenson, A. M., McLean, W. H. I., Rees, J. L., and Munro, C. S. (1995). A gene for monilethrix is closely linked to the keratin gene cluster on chromosome 12q. *Hum. Mol. Genet.* **4**, 2399–2402.

Herrmann, H., and Foisner, R. (2003). Intermediate filaments: Novel assembly models and exciting new functions for nuclear lamins. *Cell Mol. Life Sci.* **60**, 1607–1612.

Hesse, M., Magin, T. M., and Weber, K. (2001). Genes for intermediate filament proteins and the draft sequence of the human genome: Novel keratin genes and a surprisingly high number of pseudogenes related to keratin genes 8 and 18. *J. Cell Sci.* **114**, 2569–2575.

Hesse, M., Zimek, A., Weber, K., and Magin, T. M. (2004). Comprehensive analysis of keratin gene clusters in humans and rodents. *Eur. J. Cell Biol.* **83**, 19–26.

Irvine, A. D., Corden, L. D., Swensson, O., Swensson, B., Moore, J. E., Frazer, D. G., Smith, F. J. D., Knowlton, R. G., Christophers, E., Rochels, R., Uitto, J., and McLean, W. H. I. (1997a). Mutations in cornea-specific keratins K3 or K12 cause Meesmann's corneal dystrophy. *Nat. Genet.* **16**, 184–187.

Irvine, A. D., McKenna, K. E., Jenkinson, H., and Hughes, A. E. (1997b). A mutation in the V1 domain of keratin 5 causes epidermolysis bullosa simplex with mottled pigmentation. *J. Invest. Dermatol.* **108**, 809–810.

Irvine, A. D., and McLean, W. H. I. (1999). Human keratin diseases: The increasing spectrum of disease and subtlety of the phenotype-genotype correlation. *Br. J. Dermatol.* **140**, 815–828.

Ishida-Yamamoto, A., McGrath, J. A., Chapman, S. J., Leigh, I. M., Lane, E. B., and Eady, R. A. J. (1991). Epidermolysis bullosa simplex (Dowling-Meara type) is a genetic disease characterized by an abnormal keratin filament network involving keratins K5 and K14. *J. Invest. Dermatol.* **97**, 959–968.

Jonkman, M. F., Heeres, K., Pas, H. H., van Luyn, M. J. A., Elema, J. D., Corden, L. D., Smith, F. J. D., McLean, W. H. I., Raemakers, F. C. S., Burton, M., and Scheffer, H. (1996). Effects of keratin 14 ablation on the clinical and cellular phenotype in a kindred with recessive epidermolysis bullosa simplex. *J. Invest. Dermatol.* **107**, 764–769.

Ku, N. O., Gish, R., Wright, T. L., and Omary, M. B. (2001). Keratin 8 mutations in patients with cryptogenic liver disease. *N. Engl. J. Med.* **344**, 1580–1587.

Ku, N. O., Wright, T. L., Terrault, N. A., Gish, R., and Omary, M. B. (1997). Mutation of human keratin 18 in association with cryptogenic cirrhosis. *J. Clin. Invest.* **99**, 19–23.

Lane, E. B., Rugg, E. L., Navsaria, H., Leigh, I. M., Heagerty, A. H. M., Ishida-Yamamoto, A., and Eady, R. A. J. (1992). A mutation in the conserved helix termination peptide of keratin 5 in hereditary skin blistering. *Nature* **356**, 244–246.

Lenz-Bohme, B., Wismar, J., Fuchs, S., Reifegerste, R., Buchner, E., Betz, H., and Schmitt, B. (1997). Insertional mutation of the Drosophila nuclear lamin Dm0 gene results in defective nuclear envelopes, clustering of nuclear pore complexes, and accumulation of annulate lamellae. *J. Cell Biol.* **137,** 1001–1016.

Maidment, S. L., and Ellis, J. A. (2002). Muscular dystrophies, dilated cardiomyopathy, lipodystrophy and neuropathy: The nuclear connection. *Exp. Rev. Mol. Med.* **2002,** 1–21.

Manilal, S., Nguyen, T. M., and Morris, G. E. (1998). Colocalization of emerin and lamins in interphase nuclei and changes during mitosis. *Biochem. Biophys. Res. Commun.* **249,** 643–647.

Manilal, S., Sewry, C. A., Pereboev, A., Man, N., Gobbi, P., Hawkes, S., Love, D. R., and Morris, G. E. (1999). Distribution of emerin and lamins in the heart and implications for Emery-Dreifuss muscular dystrophy. *Hum. Mol. Genet.* **8,** 353–359.

McLean, W. H. I. (2003). Genetic disorders of palm skin and nail. *J. Anat.* **202,** 133–141.

McLean, W. H. I., Pulkkinen, L., Smith, F. J. D., Rugg, E. L., Lane, E. B., Bullrich, F., Burgeson, R. E., Amano, S., Hudson, D. L., Owaribe, K., McGrath, J. A., McMillan, J. R., Eady, R. A. J., Leigh, I. M., Christiano, A. M., and Uitto, J. (1996). Loss of plectin causes epidermolysis-bullosa with muscular-dystrophy-cDNA cloning and genomic organization. *Genes Dev.* **10,** 1724–1735.

McLean, W. H. I., Rugg, E. L., Lunny, D. P., Morley, S. M., Lane, E. B., Swensson, O., Dopping-Hepenstal, P. J. C., Griffiths, W. A. D., Eady, R. A. J., Higgins, C., Navsaria, H., Leigh, I. M., Strachan, T., Kunkeler, L., and Munro, C. S. (1995). Keratin 16 and keratin 17 mutations cause pachyonychia congenita. *Nat. Genet.* **9,** 273–278.

Mersiyanova, I. V., Perepelov, A. V., Polyakov, A. V., Sitnikov, V. F., Dadali, E. L., Oparin, R. B., Petrin, A. N., and Evgrafov, O. V. (2000). A new variant of Charcot-Marie-Tooth disease type 2 is probably the result of a mutation in the neurofilament-light gene. *Am. J. Hum. Genet.* **67,** 37–46.

Munoz-Marmol, A. M., Strasser, G., Isamat, M., Coulombe, P. A., Yang, Y., Roca, X., Vela, E., Mate, J. L., Coll, J., Fernandez-Figueras, M. T., Navas-Palacios, J. J., Ariza, A., and Fuchs, E. (1998). A dysfunctional desmin mutation in a patient with severe generalized myopathy. *Proc. Natl. Acad. Sci. U.S.A* **95,** 11312–11317.

Ott, J. (1991). "Analysis of Human Genetic Linkage." Johns Hopkins University Press, Baltimore.

Owens, D. W., Wilson, N. J., Hill, A. J., Rugg, E. L., Porter, R. M., Hutcheson, A. M., Quinlan, R. A., Van Heel, D., Parkes, M., Jewell, D. P., Campbell, S. S., Ghosh, S., Satsangi, J., and Lane, E. B. (2004). Human keratin 8 mutations that disturb filament assembly observed in inflammatory bowel disease patients. *J. Cell Sci.* **117,** 1989–1999.

Porter, R. M., and Lane, E. B. (2003). Phenotypes, genotypes and their contribution to understanding keratin function. *Trends Genet.* **19,** 278–285.

Richard, G., DeLaurenzi, V., DiDona, B., Bale, S. J., and Compton, J. G. (1995). Keratin-13 point mutation underlies the hereditary mucosal epithelial disorder white sponge nevus. *Nat. Genet.* **11,** 453–455.

Rothnagel, J. A., Longley, M. A., Holder, R. A., Kuster, W., and Roop, D. R. (1994). Prenatal diagnosis of epidermolytic hyperkeratosis by direct gene sequencing. *J. Invest. Dermatol.* **102,** 13–16.

Rugg, E. L., Baty, D., Shemanko, C. S., Magee, G., Polak, S., Bergman, R., Kadar, T., Boxer, M., Falik-Zaccai, T., Borochowitz, Z., and Lane, E. B. (2000). DNA based prenatal testing for the skin blistering disorder epidermolysis bullosa simplex. *Prenat. Diagn.* **20,** 371–377.

Rugg, E. L., McLean, W. H. I., Allison, W. E., Lunny, D. P., Macleod, R. I., Felix, D. H., Lane, E. B., and Munro, C. S. (1995). A mutation in the mucosal keratin K4 is associated with oral white sponge nevus. *Nat. Genet.* **11,** 450–452.

Rugg, E. L., McLean, W. H. I., Lane, E. B., Pitera, R., McMillan, J. R., Dopping-Hepenstal, P. J. C., Navsaria, H. A., Leigh, I. M., and Eady, R. A. J. (1994). A functional "knock-out" for human keratin 14. *Genes Dev.* **8,** 2563–2573.

Shackleton, S., Lloyd, D. J., Jackson, S. N., Evans, R., Niermeijer, M. F., Singh, B. M., Schmidt, H., Brabant, G., Kumar, S., Durrington, P. N., Gregory, S., O'Rahilly, S., and Trembath, R. C. (2000). LMNA, encoding lamin A/C, is mutated in partial lipodystrophy. *Nat. Genet.* **24,** 153–156.

Smith, F. J. D., Eady, R. A. J., Leigh, I. M., McMillan, J. R., Rugg, E. L., Kelsell, D. P., Bryant, S. P., Spurr, N. K., Geddes, J. F., Kirtschig, G., Milana, G., de Bono, A. G., Owaribe, K., Wiche, G.,

Pulkkinen, L., Uitto, J., McLean, W. H. I., and Lane, E. B. (1996). Plectin deficiency results in muscular dystrophy with epidermolysis bullosa. *Nat. Genet.* **13**, 450–457.

Smith, F. J. D., Jonkman, M. F., van Goor, H., Coleman, C., Covello, S. P., Uitto, J., and McLean, W. H. I. (1998). A mutation in human keratin K6b produces a phenocopy of the K17 disorder pachyonychia congenita type 2. *Hum. Mol. Genet.* **7**, 1143–1148.

Smith, F. J. D., McKenna, K. E., Irvine, A. D., Bingham, E. A., Coleman, C. M., Uitto, J., and McLean, W. H. I. (1999a). A mutation detection strategy for the human K6A gene and novel mutations in two cases of pachyonychia congenita type 1. *Exp. Dermatol.* **8**, 109–114.

Smith, F. J. D., McKusick, V. A., Nielsen, K., Pfendner, E., Uitto, J., and McLean, W. H. I. (1999b). Cloning of multiple keratin 16 genes facilitates prenatal diagnosis of pachyonychia congenita type 1. *Prenat. Diagn.* **19**, 941–946.

Strachan, T., and Read, A. P. (2003). "Human Molecular Genetics 3." Bios Scientific Publishers Ltd., Oxford.

Terrinoni, A., Smith, F. J. D., Didona, B., Canzona, F., Paradisi, M., Huber, M., Hohl, D., David, A., Verloes, A., Leigh, I. M., Munro, C. S., Melino, G., and McLean, W. H. I. (2001). Novel and recurrent mutations in the genes encoding keratins K6a, K16 and K17 in 13 cases of pachyonychia congenita. *J. Invest. Dermatol.* **117**, 1391–1396.

Terron-Kwiatkowski, A., Paller, A. S., Compton, J., Atherton, D. J., McLean, W. H. I., and Irvine, A. D. (2002). Two cases of primarily palmoplantar keratoderma associated with novel mutations in keratin 1. *J. Invest. Dermatol.* **119**, 966–971.

Terron-Kwiatkowski, A., Terrinoni, A., Didona, B., Melino, G., Atherton, D. J., Irvine, A. D., and McLean, W. H. I. (2004). Atypical epidermolytic palmoplantar keratoderma presentation associated with a mutation in the keratin 1 gene. *Br. J. Dermatol.* **150**, 1096–1103.

Uttam, J., Hutton, E., Coulombe, P. A., Anton-Lamprecht, I., Yu, Q.-C., Gedde-Dahl, T., Fine, J.-D., and Fuchs, E. (1996). The genetic basis of epidermolysis bullosa simplex with mottled pigmentation. *Proc. Natl. Acad. Sci. USA* **93**, 9079–9084.

Vassar, R., Coulombe, P. A., Degenstein, L., Albers, K., and Fuchs, E. (1991). Mutant keratin expression in transgenic mice causes marked abnormalities resembling a human genetic skin disease. *Cell* **64**, 365–380.

Weber, J. L., and May, P. E. (1989). Abundant class of human DNA polymorphisms which can be typed using the polymerase chain reaction. *Am. J. Hum. Genet.* **44**, 338–396.

Winter, H., Rogers, M. A., Gebhardt, M., Wollina, U., Boxall, L., Chitayat, D., Babul-Hirji, R., Stevens, H. P., Zlotogorski, A., and Schweizer, J. (1997a). A new mutation in the type II hair cortex keratin hHb1 involved in the inherited hair disorder monilethrix. *Hum. Genet.* **101**, 165–169.

Winter, H., Rogers, M. A., Langbein, L., Stevens, H. P., Leigh, I. M., Labreze, C., Roul, S., Taieb, A., Krieg, T., and Schweizer, J. (1997b). Mutations in the hair cortex keratin hHb6 cause the inherited hair disease monilethrix. *Nat. Genet.* **16**, 372–374.

Winter, H., Schissel, D., Parry, D. A., Smith, T. A., Liovic, M., Birgitte Lane, E., Edler, L., Langbein, L., Jave-Suarez, L. F., Rogers, M. A., Wilde, J., Peters, G., and Schweizer, J. (2004). An unusual Ala12Thr polymorphism in the 1A alpha-helical segment of the companion layer-specific keratin K6hf: Evidence for a risk factor in the etiology of the common hair disorder pseudofolliculitis barbae. *J. Invest. Dermatol.* **122**, 652–657.

Wood, P., Baty, D. U., Lane, E. B., and McLean, W. H. I. (2003). Long-range PCR for specific full-length amplification of the human keratin 14 gene and novel K14 mutations in epidermolysis bullosa simplex patients. *J. Invest. Dermatol.* **120**, 495–497.

Wood, P., Baty, D. U., Lane, E. B., and McLean, W. H. I. (2003). Long-range polymerase chain reaction for specific full-length amplification of the human keratin 14 gene and novel keratin 14 mutations in epidermolysis bullosa simplex patients. *J. Invest. Dermatol.* **120**, 495–497.

Worman, H. J., and Courvalin, J. C. (2004). How do mutations in lamins A and C cause disease? *J. Clin. Invest.* **113**, 349–351.

CHAPTER 7

Intermediate Filaments and Multiparameter Flow Cytometry for the Study of Solid Tumors

Math P. G. Leers

Department of Clinical Chemistry & Hematology
Atrium Medical Center Heerlen
Heerlen, The Netherlands

I. Introduction

A. Cell Biological Aspects of Intermediate Filaments

Three filamentous systems make up the cytoskeleton of cells. These are actin-containing microfilaments, tubulin-containing microtubules, and intermediate filaments (IFs). The name "intermediate filaments" comes from their diameter (10–12 nm), being intermediate between that of microtubules (25 nm) and microfilaments (7–10 nm). In contrast to microfilaments and microtubules, whose components are highly evolutionarily conserved and very similar within cells of a particular species, IFs display much diversity in their numbers, sequences, and abundance (Fuchs and Weber, 1994). One can define six major types of IFs (I–VI). Type I and type II IF genes encode the "acidic" and "basic" keratins, respectively, which give rise to IFs present in the cytoplasm of all epithelial cells. The prototypic type III gene encodes for:

- Vimentin, which is expressed in a plethora of nonepithelial cells
- Glial fibrillar acid protein (GFAP), expressed in astrocytes and glia
- Desmin, expressed in muscle cells

Type IV sequences, which consist of the neurofilament triplet protein and α-internexin, are expressed in neurons. Type V IF genes encode the lamins, which form the meshwork of filaments located in the nuclear lamina. Finally, the type VI genes encode for nestin, synemin, and desmuslin. Many of the type III and type VI genes are expressed in muscle cells.

In contrast to the other IFs, keratins are the most complex. The family of keratins constitutes a total of 20 different subunits in each mammalian species, with molecular weights varying within the range 40–70 kDa. The classification and numbering system of the keratins (except those of hair and nail) are based on the catalog of Moll *et al.* (1982). In contrast to the homopolymeric vimentin and desmin, keratin filaments contain at least one member from the type I subfamily and one member from the type II subfamily. Pairs of keratins seem to be consistently coexpressed in different types of epithelial cells, so that certain keratin pairs are found only in simple epithelia (type I K18 or K19 and type II K8), whereas others are found in stratified epithelia (type I K14 and type II K4) (Sun and Green, 1978; Sun *et al.*, 1983).

Despite their diversity, members of the IFs share a common structure: a centrally located domain of fixed length is dominated by α-helical segments featuring long-range repeats of hydrophobic/apolar residues (subdomains 1A, 1B, 2A, and 2B). The IFs are generally arranged as a dimer composed of two α-helical chains oriented in parallel and intertwined in a coiled-coil rod. The highly conserved ends of the IF rod associate in a head-to-tail fashion, and mutations in these rod ends have deleterious consequences for the assembly process of most, if not all, IF proteins (Albers and Fuchs, 1987; Letai *et al.*, 1992).

B. Intermediate Filaments and Tumors of Unknown Primary Origin

The demonstration of certain types of IF proteins in a solid tumor can help to identify and to define the organ of origin of an unknown primary tumor or a metastatic tumor. In the diagnosis of tumors, it is the task of the pathologist to decide whether a received biopsy specimen is involved by a true neoplastic process and whether the neoplasia is benign or malignant, as well as to classify the tumor process histogenetically. Although the first two questions generally can be answered on the basis of light microscopic examination of a hematoxylin-eosin–stained section, perhaps supplemented with a few other staining methods, immunohisto-chemical methods are able to give important information regarding the histogenetic classification of tumors.

Histologically, most malignant tumors of unknown origin are adenocarcinomas, often poorly differentiated, whereas "undifferentiated" tumors (i.e., tumors without morphological features in conventionally stained sections to allow a reliable classification) comprise a smaller part. The latter group includes poorly differentiated carcinomas, neuroendocrine tumors, lymphomas, germ cell tumors, melanomas, sarcomas, and embryonal malignancies (Pavlidis et al., 2003).

A correct histopathological classification is important, because specific treatment regimens exist for some conditions that improve the patient's chance of survival (e.g., malignant lymphomas, carcinomas of breast, ovary, thyroid, liver, kidney, and prostate, and neuroendocrine and germ cell tumors, as well as certain mesenchymal tumors (Haskell et al., 1988). Immunohistochemical stainings are very cost-effective compared with other diagnostic methods to classify and determine the origin of tumors. When properly planned, controlled, and used, ideally in an algorithmic approach (DeYoung and Wick, 2000), they can save patients and clinicians inconvenient, time-consuming, and expensive diagnostic procedures in the search for a primary tumor site.

The development of monoclonal antibodies directed against IFs has opened new avenues in investigating the normal and cancerous cells. Because IFs are part of the cytoskeleton, the pattern of expression of these IFs may be used as a device to aid in the classification of undifferentiated neoplasms. The five most used antigenic epitopes are the keratins, vimentin, desmin, neurofilaments (NF), and GFAP. Most neoplasms show predominant expression of one or more of these intermediate filaments (see also Table I). In general, carcinomas usually express

Table I
General Overview of Intermediate Filament Expression in Solid Tumors

Type of tumor	Keratin	Vimentin	Glial fibrillary acid protein	Neurofilaments	Desmin
Carcinoma	+++	±	−	−	−
Lymphoma of leukemia	±*	++	−	−	−
Mesothelioma	++	++	−	−	−
Mesenchymal tumor	+	++	−	−	++[†]
Neural tumor	±	++	−	++	−
Melanoma	±	++	++	−	−
Astrocytoma	±	++	++	−	−
Giloma	±	++	++	+	−
Germ cell tumor	±*	++	−	−	−

*Reported in rare occasions.
[†]In mesenchymal tumors with myomatous differentiation.

cytokeratins; sarcomas, melanomas, and lymphomas typically contain vimentin; myogenic tumors are characteristically positive for desmin and vimentin; and glial tumors are predominantly positive for GFAP. Some tumors characteristically coexpress more than one IF (e.g., renal and thyroid carcinomas contain keratin and often vimentin), and others show aberrant or no intermediate filament expression. These discrepancies serve to reemphasize the value of an approach that incorporates a panel of antibodies in the evaluation of an undifferentiated neoplasm.

Virtually all carcinomas, regardless of their tissue origin, are immunoreactive to keratin antibodies. Keratin typing of carcinomas, particularly metastases, has gained increasing attention in diagnostic oncology in recent years. At first glance, the complexity of the cytokeratin (CK) expression pattern may be confusing. Also, CK patterns usually are not specific for one single carcinoma entity. However, for diagnostic purposes, a small panel of selected CK antibodies may be sufficient to provide in cases of unclear carcinomas—helpful signposts that, in conjunction with clinicopathological data, can guide the pathologist toward the identification of the primary tumor or to narrow down the range of possibilities. In the early days of immunohistochemistry, most antibodies directed against keratin used in diagnostic oncology were not reactive to a specific keratin; rather, they were mixed monoclonal antibodies or polyclonal antibodies that reacted to several keratins. Currently, high-quality antikeratin monoclonal antibodies to all of the 20 different keratins are commercially available. Immunohistochemical detection of keratins has been used in diagnostic oncology for the differential diagnosis of undifferentiated malignancies for more than 20 years. During this period, keratin immunohistochemistry has been used in the differential diagnosis of epithelial-derived tumors from neoplasms of mesenchymal, hematolymphoid, or neural crest origin (Espinoza and Aza, 1998; Gown and Vogel, 1985; Nagle *et al.*, 1983; Osborn and Weber, 1983; Ramaekers *et al.*, 1983; Schlegel *et al.*, 1980; Sun *et al.*, 1979). More recently, keratin immunohistochemistry has been used to differentiate carcinomas of one origin from another (Bartek *et al.*, 1991; Bruderman *et al.*, 1990; Cooper *et al.*, 1985; Eichner *et al.*, 1984; Gown and Vogel, 1985; Lundquist *et al.*, 1999; Miettinen, 1993; Moll *et al.*, 1982, 1993; Osborn and Weber, 1983; Ramaekers *et al.*, 1983; Sun *et al.*, 1979) (see also Table II). Sarcomas rarely express keratins other than CK8 and CK18. The expression of cytokeratins in sarcoma is generally, but not always, accompanied by evidence of epithelial differentiation at the light and electron microscopic level. Unlike in epithelial tumors or the epithelial component of mixed neoplasms, which are usually diffusely cytoplasmic positive for keratins and express a complex set of keratins, the keratin expression in sarcomatous elements is usually anomalous, characterized by immunostaining (even under optimal technical conditions) in a subset of the target cell population (with the exception of epithelial sarcoma).

Neoplasms derived from hematolymphoid tissues virtually never express keratins, which has been adopted as one of principles of diagnostic immunohistochemistry. However, rare cases of hematolymphoid neoplasm may focally express

Table II
Cytokeratin Phenotype of Normal and Neoplastic Tissue

Cytokeratin-number ⇒	1	2	3	4	5	6	7	8	9	10	11	12	13	14	15	16	17	18	19	20
A. Cytokeratin Phenotype of Normal Tissue																				
Adrenal gland																				
Cortex								•										•	•	
Endometrium																				
Normal								•										•	•	
Secretion phase							•	•										•	•	
Proliferation phase							•	•										•	•	
Esophagus																				
Basal cell-layer								•										•	•	
Suprabasal cell-layer				•									•						•	
Gallbladder																				•
Stomach																				
Foveolar cells								•										•	•	•
Corpus mucosa								•										•	•	•
Small intestines								•										•	•	•
Colon								•										•	•	•
Kidneys																				
Glomeruli							•	•										•	•	
Tubuli							•	•										•	•	
Liver																				
Hepatocytes								•										•		
Ductuli							•	•										•	•	
Lung																				
Bronchial epithelium							•	•										•	•	*
Basal cells					•									•	•		•			
Mamma																				
Luminal cells					*		•	•										•	•	
Basal cells					•									•			•			
Merckel cells							•	•										•	•	•
Mesothelium					•		•	•										•	•	
Oral cavity																				
Squamous epithelium				•	•									•	•	•	•			
Ovary																				
Surface epithelium							•	•										•	•	
Granulosa cells								•										•		
Pancreas																				
Exocrine pancreas								•										•		
Endocrine pancreas							•	•										•	•	
Pancreatic ducts				•	•		•	•						•			•	•	•
Pituitary								•										•		
Prostate																				
Basal cells					•	•		•		•				•	•		•	•	•
Luminal cells								•										•		
Salivary gland																				
Myoepithelium					•									•			•			
Luminal cells								•										•		
Salivary ducts								•											•	
Skin																				
Epidermis basal layer					•									•	•					

(continues)

Table II (*continued*)

Cytokeratin-number ⇒	1	2	3	4	5	6	7	8	9	10	11	12	13	14	15	16	17	18	19	20
Suprabasal layer	•	•								•	•									
Hair follicle externa					•	•								•	•	•	•		•	
interna					•											•				
Sebaceous gland				•	•		•	•						•	•	•		•		
Sweat gland							•	•										•	•		
ductuli	•							•	•										•	
Testis																				
Sertoli cells								•										•		
Thymus				•				•										•		*
Thyroid gland								•										•	·	
Urinary bladder																				
Normal							•	•										•	•	
Transitional epithelium			•	•			·						•					•	·	
Umbrella cells								•										•		•
Uterine cervix																				
Reserve cells			·	·	·	•								•	•	•	·	•	•	•
Endocervical cells					·	•												•	·	
B. Cytokeratin Phenotype of Neoplastic Tissue																				
Adrenal gland																				
Carcinoma																		·		
Endometrium																				
Carcinoma							·	•										•	•	
Squamous area			·	·			·											·		
Esophagus																				
Dysplasia				•			•						•						•	
Squamous cell carcinoma				•									•						•	
Gallbladder																				
Adenocarcinoma							•	•										•	•	•
Stomach																				
Metaplasia							•	•										•	•	•
Adenocarcinoma							○	•										•	•	•
Small intestines																				
Adenocarcinoma								•										•	•	•
Colon																				
Adenocarcinoma								•										•	•	•
Kidneys																				
Atrophy prox. tubuli							·	•									·	•	·	
Atrophy dist. tubuli							•	•										•	•	
Renal cell tumor																				
Chromophobe							○	•										•	·	
Eosino-/basophilic							·	•										•	•	
Liver																				
Cirrhosis, hyperplasia								•	•									•	•	
Hepatocellular carcinoma								·	•									•	·	
Fibrolamellar type								•	•									•	•	
Hepatoblastoma								○	•									•	•	
Lung																				
Squamous cell carcinoma				·	•	•		•						·	•	•	·	•	•	
Adenocarcinoma							•	•							○		○		•	•

(*continues*)

Table II (*continued*)

Cytokeratin-number ⇒	1	2	3	4	5	6	7	8	9	10	11	12	13	14	15	16	17	18	19	20
Large cell undifferentiated								•	•									•		
Small cell undifferentiated								•										•		
Mamma																				
Adenocarcinoma								•	•					○		○	•	•	•	
Merckel cell tumor								•	•									•	•	•
Mesothelioma	○	•	○	•	•									○		◑	•	•	•	
Ovary																				
Carcinoma								•	•								•		•	
Mucinous type	○	○		·	•		○					○	○				•	•		•
Leydig cell tumor				•																
Granulosa cell					•												•			
Pancreas																				
Adenocarcinoma	○	○			•	•		○				○	○			○	•	•		○
Pituitary																				
Adenoma					•												•			
Prostate																				
Basal cell hyperplasia					•									•					•	
Carcinoma *in situ*								•										•	•	
Adenocarcinoma							○	•										•	•	
Salivary gland																				
Pleomorphic																				
Myoepithelial part					•									•			•			
Luminal cells								•										•		
Duct structures							•												•	
Skin																				
Basalioma	·				•	•				·	·			•		·	•			
Spinalioma	•	•			•	•		·		•	•			•		•		·	·	
Basal cell carcinoma				·		•	•											·	·	
Condyloma accuminata				•									•							
Psoriasis	•	•								•	•						•			
Testis																				
Germ cell tumor								•	•									•	·	
Thyroid																				
Papillary adenocarcinoma								•										•	•	
Folliculair adenocarcinoma								•										•	○	
Urinary bladder																				
Transitional cell carcinoma							•	•					•				•			
High grade							•	•				·	·				•			•
Uterine cervix																				
Dysplasia CIN I					•			•					•	•	•	•	•			
CIN III	·				·			·	•				·	•	•	•	•	•	•	
Squamous cell carcinoma					•			•					•	•	•	•	•	•	•	
Adenocarcinoma	○	○	○	•	•		○						○	•		○	•	•	•	○

•, Strong expression; ·, mild expression; *, focal expression; ○, some reported cases.

keratins. Reactivity with keratin antibodies seems to be most common in plasma cell neoplasms (Sewell *et al.*, 1986; Wotherspoon *et al.*, 1989). Anaplastic large cell lymphoma (ALCL) is another potentially keratin-positive hematolymphoid neoplasm (Frierson *et al.*, 1994; Gustmann *et al.*, 1991; Ross *et al.*, 1992).

In tumors of the central nervous system, keratin immunohistochemistry was initially regarded potentially as an aid in the recognition of metastatically poorly differentiated carcinoma, the detection of epithelial elements present in intracranial and spinal teratoma, and in the diagnosis of craniopharyngioma (Asa *et al.*, 1981). However, recent studies have indicated the presence of an epitope in primary brain tumors recognized by antibodies directed against cytokeratins that does not necessarily imply an epithelial origin.

C. Intermediate Filaments and Flow Cytometry

In diagnostic oncology, cytological or histological examinations are the most essential parts of daily practice to discriminate among benign, premalignant, and malignant cell proliferations. Although the clinical value of these examinations is not in doubt and despite the fact that in most cases the diagnosis can be reliably established in this way, several attempts have been made to improve the cytological and histological diagnosis over the past 30 years. There is a continuous search for additional parameters, such as "predictors," for response to therapy or prognosis of the disease. In the past three decades, it has been shown that immunophenotyping of tumor cells is useful for the diagnosis, classification, prognostic evaluation, and detection of residual disease in patients with certain malignancies. Quantitative analysis of stained cells has played an important role in this respect. However, immunohistochemistry as a additional tool by the surgical pathologist is not always satisfactory. Solid tumors are of heterogeneous composition. They consist of a mixture of normal stromal, inflammatory, and malignant cells. This heterogeneous cell composition is one of the limiting factors for reproducible quantification of, for example, steroid hormone receptor expression in breast carcinomas. In addition, immunohistochemistry is sensitive to many external factors interfering with accurate quantification of receptor content. For instance, staining intensity is influenced by the kind and duration of fixation, thickness of the tissue section, incubation conditions of the primary antibodies, choice and concentration of the chromogens, and often subjective scoring by the investigators.

Next to this, despite the high technical level of these tools, their application leads to qualitative results. The determination of a quantitative interrelationship of the various cell constituents making up the tumor tissue still remains a task with a high level of subjective influences. Yet, accurate quantification of certain cell biological parameters can be of importance to improve diagnosis and predict therapy response (e.g., the assessment of steroid hormone receptors for patients with breast cancer). The three most important keystones of tissue homeostasis, controlling expansion or regression in tumors, are cell proliferation, differentiation, and cell death (apoptosis). Carcinogenesis can be viewed as a process of cellular evolution in which individual cells acquire mutations that increase survival and/or proliferative capacity or decrease the apoptotic activity. The growth potential and behavior of human neoplasms is the net result of an imbalance between cell proliferation and cell death. Methods that allow the specific quantification of such cell biological parameters may contribute to our understanding of the mutual relationship between these factors in tumor growth and aid in the management of malignancies. Next to this, tumors in general consist of more than one cell population, each with their own characteristics and behavior. It is very difficult to investigate the different tumor cell populations by light microscopy, because of the admixture with nonrelevant normal cells (e.g., inflammatory cells, stromal component). The aforementioned aspects lead us to the conclusion that, in a situation in which quantitative interpretation of cell characteristics is important, a different approach should be looked for. Flow cytometry is a technique that can tackle these problems. One of the first applications in flow cytometry was the cellular detection of DNA (Hudson *et al.*, 1969). A few years later, the first multiparameter measurements were published demonstrating the simultaneous detection of DNA and protein on a cell-by-cell basis (Crissman and Steinkamp, 1973). However, the presence of normal cells can impair flow cytometric analysis of proteins, the expression of which may not be limited to tumor cells alone or is heterogeneous in tumor cells. A way to overcome this problem is the use of a multiparameter approach applying specific monoclonal antibodies direct against IFs of different origin (e.g., monoclonal antibodies against keratin, vimentin, or GFAP), in combination with antibodies directed against the protein of interest, for example, the steroid hormone receptors. In 1984, Ramaekers and coworkers demonstrated in a tumor model the use of monoclonal antibodies directed against keratin in combination with DNA content measurements with a fluorescein isothiocyanate (FITC)-conjugated secondary reagent and propidium iodide as DNA stain. This approach allowed discrimination between cultured bladder carcinoma cells and Molt-4 leukemia cells on the basis of protein expression and DNA content (Ramaekers *et al.*, 1984). The same procedure was later successfully applied to mechanically dissociated human endometrial carcinomas (Oud *et al.*, 1985) and enzymatically dissociated human tumors from the head and neck region (Bijman *et al.*, 1986). This technique was further improved by Zarbo and coworkers (1989). Next to staining for cytokeratin, they introduced labeling of the leukocytes with a monoclonal antibody against leukocyte-common antigen (LCA). This sample served as a patient-specific intrinsic DNA content standard (DNA diploid reference). In this way, no external DNA standard was required for calculating the relative DNA content or DNA index (DI) of the keratin-positive fraction(s). This increased the sensitivity of detecting tumor cells with a DNA content close to that of normal cells (LCA-positive cells) (Corver, 2001).

Later Frei and Martinez (1993) and Nylander and Colleagues (1994) demonstrated that the dual keratin-DNA labeling technique could also be applied to cells isolated from formalin-fixed, paraffin-embedded material. Our group extended these experiments to other tissue types (Leers *et al.*, 1995, 1997a,b) and improved the technique by introducing heating in citrate buffer (Leers *et al.*, 1999b), which was already an accepted technique in immunohistochemistry for unmasking protein molecules in formalin-fixed, paraffin-embedded tissue sections (Cattoretti *et al.*, 1992). This heating step precedes the short enzymatic dissociation step. This strongly improves DNA histogram quality (better CV's of the G_0/G_1-peaks of the first peak in the DNA histogram, reduction in background and debris), release of more single cells, while expression of many epitopes (among which keratin) is retained or restored for identification of tumor cell subpopulations (Leers *et al.*, 1999b), resulting in higher fluorescence signals. In later studies, we demonstrated the usefulness of three-parameter analysis in which cytokeratin was used for labeling the epithelial compartment (conjugated with r-PE) and antibodies against steroid hormone receptor c-erbB2, or apoptosis-related proteins (conjugated with FITC) and DNA content with propidium iodide in several types of cancer specimens (Leers and Nap, 2001; Leers *et al.*, 2000, 2003; Morsi *et al.*, 2000; Nap *et al.*, 2001; Rupa *et al.*, 2003). Direct conjugated primary antibodies with appropriate label/protein ratios will enable the most efficient and economical application of true multiparameter flow cytometric analysis of solid tumors, next to the availability of a straightforward relation of different aspects within the same cell suspension. Indirect, two-step procedures may offer more signal; however, they limit the possibility of the use of certain combinations of antibodies (species differences, isotype-differences, etc.).

In the following sections, some flow cytometric applications for the use of immunochemical detection of intermediate filaments will be highlighted.

1. Flow Cytometric Immunophenotyping of Effusions

Body cavities are frequent sites of tumor metastasis, and they are the site of origin of several tumors. Microscopic examination provides the important diagnostic service of assessing the potential malignant nature of cells from a variety of fluids and aspirations in which tissue architecture has been retained. These examinations are characterized by a high level of both sensitivity and specificity; however, there is still room for improvement. The cytologic distinction between carcinoma cells, inflammatory cells, and reactive or malignant mesothelial cells can be especially challenging. Multiparameter flow cytometry that uses antibodies directed against certain IFs is a valuable adjunctive technology for the evaluation of these effusions. Although relatively low, the number of reported studies is still growing. Risberg and coworkers (2000) showed in peritoneal washings a high concordance with immunohistochemistry when Ber-EP4 (a monoclonal antibody directed against the protein moiety of two glycopeptides in human epithelial cells) was used. Next to this, the evaluation of the fraction of immunoreactive cells was by far easier with flow cytometry (Risberg *et al.*, 2000). Davidson *et al.* (2002) also found good agreement between multiparameter flow cytometry and immunocytochemistry for Ber-EP4. They studied 92 serous effusions for the presence of malignant cells. By use of a panel of four epithelial markers, among which was Ber-EP4, they could discriminate between reactive effusions and malignant specimens (Davidson *et al.*, 2002). The last two studies made no use of DNA staining, which can be an additional tool to study tumor heterogeneity in cytopathology. The group of Sanchez-Carbayo described a multiparameter DNA/cytokeratin flow cytometric assay for the diagnostic workup of urine voided samples from patients with bladder cancer. They stated that this multiparameter DNA-flow cytometric technique combined with a serum assay for soluble cytokeratin fragments can be of great clinical use for screening for tumor recurrence in patients with bladder cancer (Sanchey-Carbayo *et al.*, 2001).

There are enough data that support the view that multiparameter flow cytometry can be a valuable adjunctive technology to the cytopathologist in the evaluation of a variety of cytology specimens. This application is not restricted to effusions but can also be applied to fine-needle aspirations. It is important to restate, however, that a final diagnosis can never be based on flow cytometry results. The flow cytometry results are provided to the cytopathologist who incorporates them into the overall interpretation.

2. Flow Cytometric Detection of Occult Metastatic Tumor Cells

Another approach for the use of IFs in flow cytometry is for the enrichment of tumor cells; enrichment is any procedure that isolates or increases the proportion of a specific cell type or molecule from a heterogeneous population of cells. Enrichment for tumor cells can improve sensitivity in several flow cytometric assays, especially when looking for differences between tumor cells and normal tissue. Neoplastic and nonneoplastic cells are present in varying proportions in solid tumors, and the presence of stromal or inflammatory cells in a sample may obscure these differences or render them inapparent (Yaremko *et al.*, 1996). A possible method for enrichment is immunochemical separation by labeling the tumor cells with the aid of antibodies directed against IFs followed by multiparameter DNA flow cytometric separation, analysis, and eventually cell sorting. Corver and coworkers used this application to perform molecular genetic analysis (Jordanova *et al.*, 2003).

However, another application in which enrichment of cells could be valuable is for the detection of occult, potentially metastatic, tumor cells in an enormous pool of normal or nonrelevant cells. It has already been known for more than a century that tumor cells that are shed from the solid tumor can be detected in the blood circulation of patients with cancer (Ashworth, 1869). Occult dissemination of tumor cells in patients with operable cancer can subsequently lead to formation of metastasis, yet it is usually missed by conventional tumor staging. Several groups of investigators have designed molecular or multiparameter flow cytometric approaches to identify such minimal amounts of occult tumor cells in peripheral blood, bone marrow, or lymph nodes (Hayes *et al.*, 2002; Leers *et al.*, 2002; Lugo *et al.*, 2003; Pantel and Braun, 2001; Racila *et al.*, 1998). In most studies, an immunochemical procedure for the detection of IFs is used. When looking for epithelial tumors, an antibody directed against pan-cytokeratin can be used. The group of Terstappen and coworkers made use of an immunomagnetic labeling of the tumor cells in which the occult epithelial cells are labeled with monoclonal antibodies against epithelial cell adhesion molecule (EpCAM) coupled to magnetic nanoparticles (Hayes *et al.*, 2002; Moreno *et al.*, 2001; Racila *et al.*, 1998). The immunomagnetic-labeled cells can be visualized after magnetic separation by bright field microscopy and fluorescence microscopy (Fig. 1).

Our group recently described a method for the detection of occult metastatic tumor cells in sentinel lymph nodes of patients with breast cancer (Leers *et al.*, 2002). The sentinel lymph node (SLN) is the first lymph node in a nodal basin to drain the primary tumor (Giuliano *et al.*, 1994). This lymph node could be removed by limited surgery and examined to determine whether more extensive lymph node excision should be performed. Conceptually, because the SLN is the first lymph node to receive lymphatic drainage from a tumor, it should be the first to show metastatic tumor. The sensitivity of the procedure for detection of occult micrometastatic tumor cells in the SLN has been variable. The methods most often used for the identification of micrometastases in lymph nodes are hematoxylin–eosin (H&E) serial sectioning and immunohistochemical staining (IHC) of the SLN for a pan-cytokeratin epitope. More recently, single-mRNA marker reverse transcriptase-polymerase chain reaction (RT-PCR) assays have been tested for the detection of micrometastatic cells in SLN. A problem of this technique is the relatively high level of false-positive reactions caused by the presence in

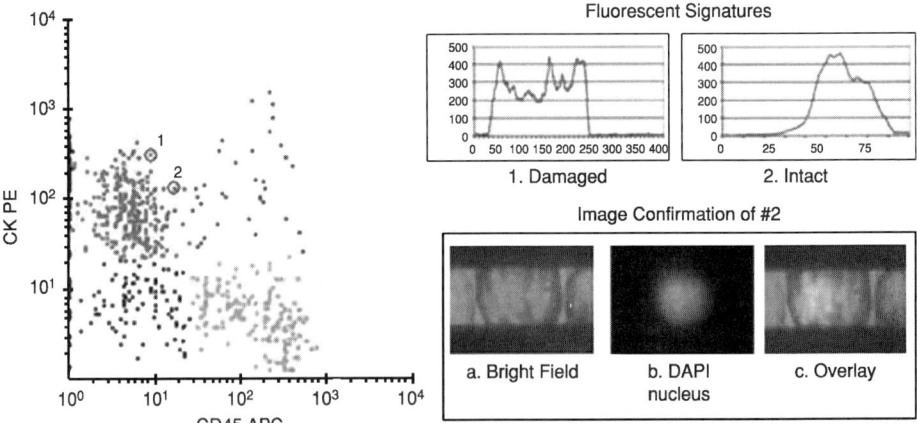

Fig. 1 The dotplot on the left shows the leukocyte staining (CD45-APC) and target cell staining (CK-rPE) of the separated immunomagnetic labeled cells in the CellTracks system. The fluorescent signature of the target candidate no. 1 is displayed in the histograms (upper right), whereas the bright field image of the cytoplasm and the fluorescence signal of DAPI (DNA staining of the nucleus) are displayed in the figure right below. (Source: http://www.immunicon.com) (See Color Insert.)

lymph nodes of nonepithelial cytokeratin-positive elements (Bostick *et al.*, 1998). Multiparameter flow cytometry was investigated by us as an alternative approach for the identification of these occult metastatic tumor cells by labeling them with a fluorescence label. It seemed that multiparameter cytokeratin/DNA flow cytometry was more sensitive than both multilevel histology and immunohistochemistry in the analysis of 238 lymph nodes. In addition, it seemed that most micrometastasis (<2 mm) presented themselves with a diploid DNA content, irrespective of the DNA profile of the primary tumor (Fig. 2).

Approximately 30% of SLN micrometastases (<2 mm) proved to be accompanied by additional non-SLN metastasis. The size of the aneuploid fraction (>60%) in the primary tumor may influence the risk of having both SLN and non-SLN metastases. Next to this, primary tumors in which the aneuploid fraction was larger than 60% showed aneuploid macrometastasis more often (Fig. 2). Furthermore, in this group, additional metastases were more often found in non-SLNs. It seems that the chance of finding aneuploid lymph node metastases and/or macrometastases is directly related to the size of the aneuploid fraction in the primary tumor; when this fraction increases, the chance of aneuploid and/or

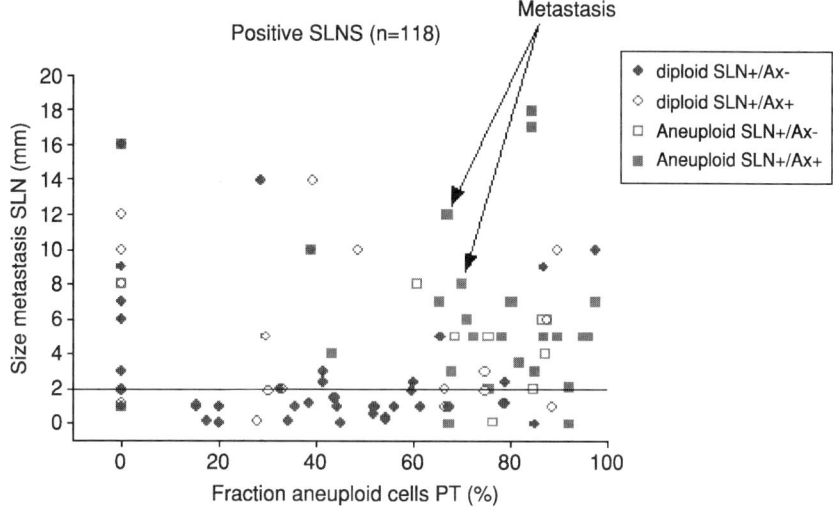

X Labels	SLN+	non-SLN+	nonSLN+(%)
<=2 cm, micrometastasis	35.0	8.0	23.0
<=2 cm, macrometastasis	41.0	21.0	51.0
>2 cm, micrometastasis	8.0	4.0	50.0
>2 cm, macrometastasis	27.0	20.0	74.0

Fig. 2 Comparison of the size and ploidy of the sentinel lymph node (SLN) metastasis with the fraction of aneuploid tumor cells and the N stage of the primary tumor (Ax, axillary lymph node dissection; PT, primary tumor).

lymph node metastasis developing also increases. By applying this technique for more than 3 years in our institute, some follow-up data for these patients become available. It seemed that only two patients had distant metastases develop after a positive SLN procedure and an axillary lymph node dissection. Both patients had an aneuploid primary breast tumor in which the aneuploid fraction was larger than 60% (Fig. 2). Further studies and more patient follow-up are necessary to clarify the relation found between the size of the aneuploid fraction in the primary tumor and the size and the possible prediction for having the chance of non-SLN involvement.

This technique can also be extrapolated to detect isolated tumor cells in bone marrow samples of patients with, for example, breast cancer. Cabioglu and co-workers (2002) removed the epithelial cells in a bone marrow sample with the aid of magnetic microbeads conjugated with a monoclonal antibody directed against cytokeratin to enrich tumor cells in bone marrow samples. With gating strategies, they could demonstrate isolated tumor cells in 53% of patients with stage I/II breast cancer. Although this technique needs some optimization and further investigation, this approach might increase its reliability in the detection of occult metastatic tumor cells and might be useful in selecting patients with a higher risk of relapse regarding tumor load to determine the need for adjuvant therapies (Cabioglu et al., 2002). Zhang and coworkers described a protocol for immuno-phenotyping disseminated breast tumor cells and their microenvironment in fresh bone marrow samples (Zhang et al., 2003).

3. The Use of Flow Cytometry to Study the Fate of Intermediate Filaments during Apoptosis

During apoptosis, cytokeratin and vimentin filaments reorganize into granular structures (Caulin et al., 1997; Leers et al., 1999a; Tinnemans et al., 1995; van Engeland et al., 1997). This process seems to be governed mainly by (hyper)phosphorylation of the cytokeratins during apoptosis (Liao et al., 1997). Furthermore, it has been shown that phosphorylation and dephosphorylation events also critically control cell junction assembly and stability and also regulate the formation of the cadherin–cytoskeleton complex (Aberle et al., 1997), thus influencing the adhesive properties of cells.

Recently, it was shown that cytokeratin IF proteins are cleaved by caspases during apoptosis (Caulin et al., 1997; Leers et al., 1999a). A conserved caspase–cleavage site is found in the linker region of all intermediate filament proteins, except for the type II cytokeratins. Caspase 3, 6, and 7 were all capable of cleaving after the aspartate 238 (D^{238}) and mutation of the $DEVD^{238}$ sequence rendered this site resistant to cleavage both *in vitro* and *in vivo*. A second cleavage site closer to the C-terminal end was deduced on the basis of a 10-residue, monoclonal antibody epitope terminating with a potential caspase cleavage site, $DALD^{397}$, that rendered antibody binding dependent on apoptotic cleavage of CK18 (Leers et al., 1999a). The identification of $DALD^{397}$ as the second caspase cleavage site of

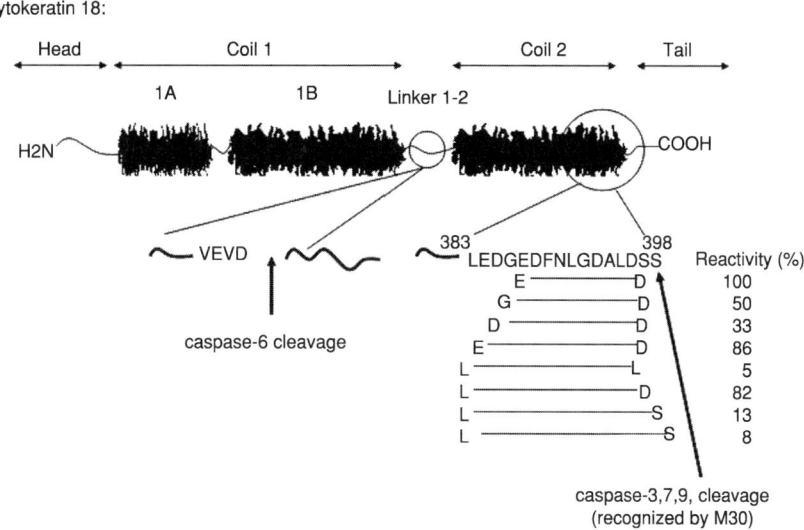

Fig. 3 Schematic representation of CK18. The head, tail, and rod domains are indicated by cylindrical structures and linker region domains as interconnecting solid lines. Caspase cleavage sites, as well as the location of the epitope recognized by M30, are indicated.

CK18 was confirmed by mutagenesis (Ku and Omary, 2001). The monoclonal antibody M30-cytodeath recognizes this neoepitope on CK18, which is not detectable in viable cells (Fig. 3). The immunoreactivity of this antibody is confined to the cytoplasm of apoptotic cells; consequently, it is a tool to demonstrate apoptosis in cell suspension by flow cytometry (Leers *et al.*, 1999a; Mertens *et al.*, 2003; Morsi *et al.*, 2000; Rupa *et al.*, 2003) (Fig. 4).

II. Materials and Instrumentation

A. Principles of Flow Cytometry: The Hardware

Whereas immunohistochemistry and related techniques primarily deal with qualitative identification of phenotypic characteristics of cell components, flow cytometry is focused on the simultaneous detection, measurement, and registration of multiple parameters of thousands of cells that pass the laser beam.

Flow cytometers are instruments constructed to measure and record fluorescence intensity. The basic components of a flow cytometer include a light source, a flow chamber, and an optical assembly (Fig. 5). The measurements are usually performed on cells stained with an appropriate fluorochrome, flowing past an excitation source. The emitted fluorescence level of the stained cell is captured by a photomultiplier tube and digitally converted to an electronic pulse.

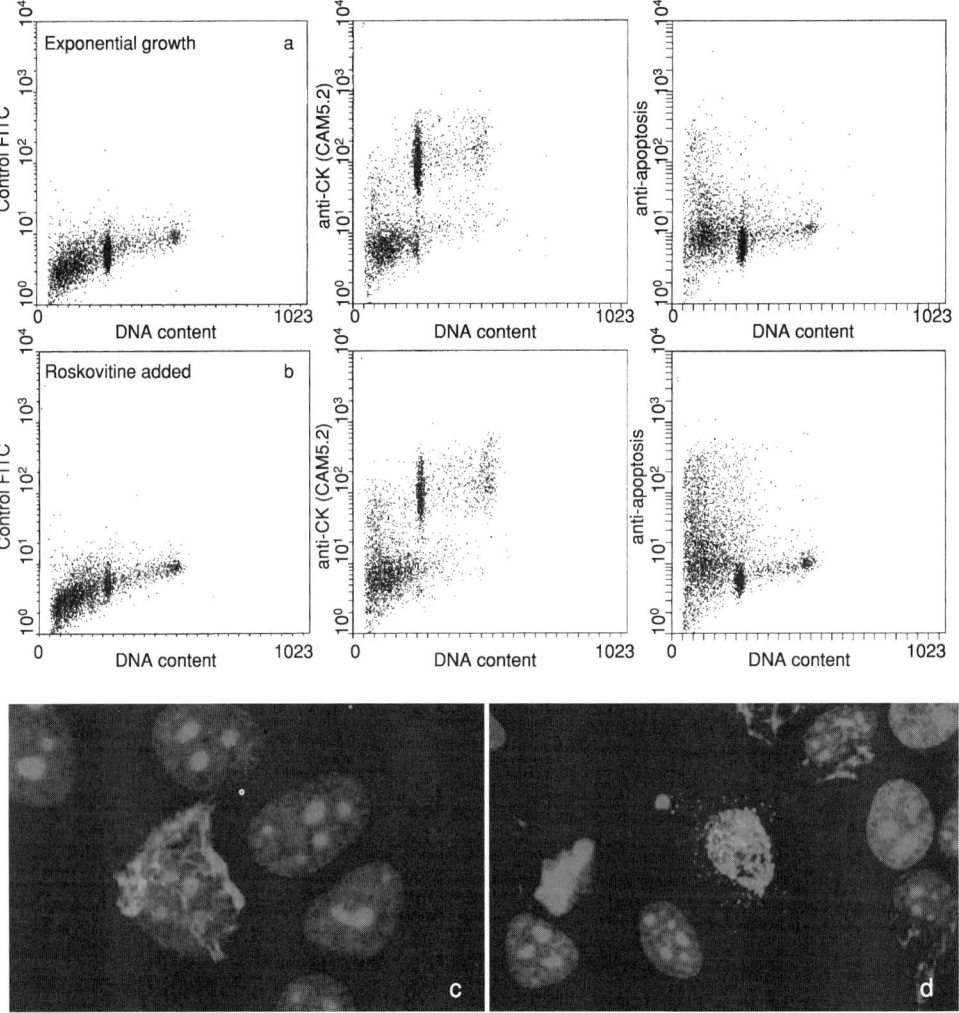

Fig. 4 Multiparameter flow cytometric analysis of apoptosis in exponentially growing MR65 cells (a) and after 8 hours of treatment with roskovitine (b). In the left dotplots, the negative control is depicted (DNA (x-axis) vs. mouse-Ig-FITC); the middle dotplots show the cytokeratin expression (labeled by the pan-CK antibody (CAM5.2) vs. DNA content), whereas in the right dotplots the expression of caspase cleaved CK18 (y-axis) is depicted vs. DNA content (x-axis). Panels (c) and (d) show the expression of this caspase-cleaved CK18 as visualized by confocal scanning laser microscopy after immunostaining with the M30-cytodeath antibody (green) combined with DNA staining with propidium iodide (red). (See Color Insert.)

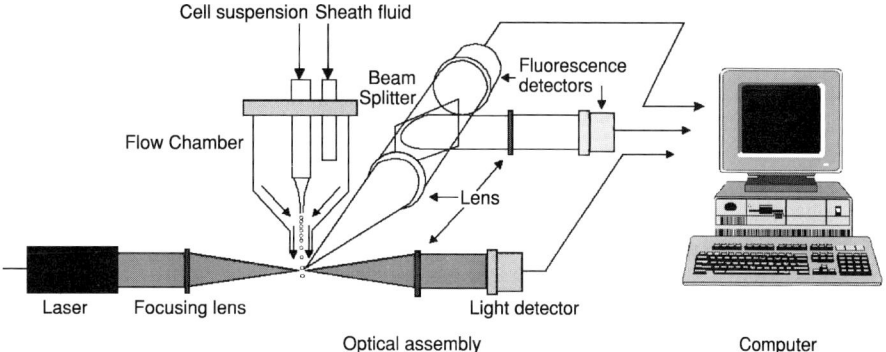

Fig. 5   Schematic representation of a flow cytometer.

1. The Light Source

A variety of light sources have been used in clinical flow cytometers. The most used are arc sources (mercury compact arc lamp) and laser sources (continuous wave, argon-ion gas laser [UV, blue and green light], krypton ion gas laser [yellow and red light], helium–neon gas laser [red light], and the diode [red light] laser). Laser sources are principally used in modern flow cytometers. The advantages of a laser source are that they consist of a single color of light or an extremely narrow range of wavelengths. In addition, the waves comprising a laser beam are in "phase." As a result, the beam of laser light is much more intense than that produced from incoherent light sources. The third advantage of laser light is directionality. The beam of light emerging from a laser is narrow and highly arranged in one direction, while light from ordinary sources is emitted in all directions.

2. The Flow Chamber

The cells to be measured flow in a laminar nonturbulent fluid stream through the flow cell or flow chamber. This laminar flow is achieved by injecting the core fluid containing the sample particles into the center of another smoothly flowing stream (i.e., sheath stream); the two streams will maintain their relative positions and not mix much, a condition called laminar flow. This enables a sequential flow of primarily single cells, sufficiently separated to measure one cell at a time. The sheath stream rate is constant, increasing the pressure or pump speed for the core fluid (sample) results in larger core diameter; more cells can be measured in a given time. However, precision is likely to be decreased, because the illumination from a Gaussian laser beam is less uniform over a larger diameter core. Less speed for the core stream gives a smaller core diameter and a slower measurement, but precision is higher. When measuring DNA content, precision is important; in immunofluorescence measurement, precision is usually of much less concern. The cells within the stream pass next to a measurement station, where they are illuminated by a light source. The point at which the laser beam and the cell stream meet is called the laser interrogation point (Fig. 5). Alignment is critical to successful operation of flow cytometers; suboptimal alignment can result in erroneous data collection, presentation, and interpretation. The newer clinical flow cytometers are constructed to reduce or eliminate the need for daily alignment.

3. Optical Assembly

Once a cell passes the light beam, two events will occur, assuming that fluorochromes are in or on the cell. The first event is that cells will scatter light from the beam at the incident wavelength in 360°. If one collects light scattered along the axis of the laser beam, a parameter known as forward angle light scatter, the quantity of the light is proportional to the size of the cell. If the scattered light is collected orthogonal at right angles to the light beam, the parameter is named 90° light scatter, or side scatter. The side scatter has been shown to be composed primarily of light reflected by internal structures or membrane undulations. Therefore, this parameter correlates with cell granularity. The properties of forward and side scatter are called intrinsic properties, because they can be measured by the flow cytometer without the use of exogenous reagents. Those properties requiring additional reagents for analysis are called extrinsic properties.

The second event that occurs at the laser interrogation point is that fluorochromes present on or in the cell absorb the laser light and reemit the light at a lower energy and a longer wavelength. This property is known as fluorescence. Each fluorochrome possesses a distinctive spectral pattern of excitation and emission. Typically, with argon-ion lasers, the excitation wavelength used is 488 nm, a blue to blue-green light. The fluorochrome must also emit light at a wavelength sufficiently longer than the excitation wavelength so that the two colors of light may be optically separated with selective filters. The difference between the peak wavelength of the excitation light and the peak wavelength of the emitted fluorescence light is a constant factor referred to as the "Stokes shift" (Fig. 6). The most popular fluorochromes used in immunofluorescence analysis are FITC and RPE. If multiple fluorochromes are used, their emission spectra must have minimal overlap so as to be separately quantitated. By use of appropriate excitation and emission filters and dichroic mirrors, the emission spectra of distinct fluorochromes can be separated, and thereby simultaneous analysis of different stainings can be performed. In Fig. 7, the emission spectra of three commonly used fluorochromes are shown. As can be seen, some spectral overlap may not be prevented.

4. Signal Detection and Amplification

The detection system of a flow cytometer consists of a variety of photocells, which collect light and convert it into integrated pulses. Because the burst of light from the particle lasts for only microseconds, the detector must be capable of

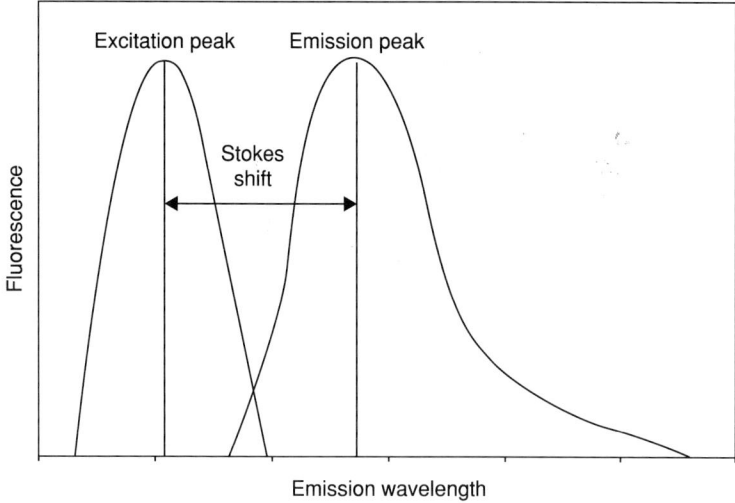

Fig. 6 Excitation spectrum and emission spectrum of a typical fluorochrome. Indicated in this figure is the "Stokes shift".

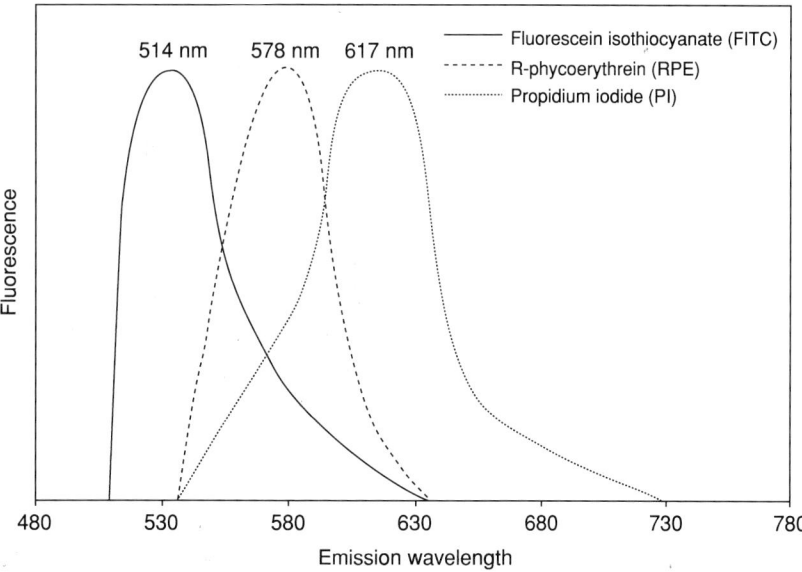

Fig. 7 Emission spectra of FITC, r-PE, and PI. From this figure it becomes clear that with this combination in a three-parameter flow cytometric assay setting, some spectral overlap may not be prevented, and therefore compensation is required.

rapidly processing signals from the detection zone, usually at a rate exceeding 10,000 pulses per second. Because the intensity of fluorescent light emitted by a cell is much less than that of light scatter signals, different types of photocells are used for each parameter. Photodiodes are used as detectors for light scatter signals, whereas photomultiplier tubes are used to detect fluorescent light. The photomultiplier tube (PMT) both detects fluorescent signals and amplifies the weak signals to a useful level. Amplification of the peak or integrated pulses can be used to accentuate the differences between the peaks. Logarithmic amplification increases the difference between small pulses much more than that between larger peaks and is ideal for differentiating between events with similar but slightly different fluorescence signals. Linear amplification accentuates all peaks by the same amount and is preferable when examining events with large fluorescence differences. As a result of overlap between the emission spectra, a correction step called compensation is necessary.

5. Compensation

The goal of compensation is to remove the spillover fluorescence of a particular fluorochrome from the "wrong" channel (Baumgarth and Roederer, 2000). For example, FITC emits mainly green light, which is usually measured in the FL1 (FITC)-channel. However, FITC also emits a significant portion of light with a yellow component that will appear in the FL2(r-PE)-channel (Fig. 8). The appropriate choice of optical filters (band pass and long pass filters) can greatly reduce collection of light from other fluorochromes. Because of this spectral overlap, each fluorochrome will contribute a signal to more than one detector; therefore, the contribution of signals in detectors not assigned to that fluorochrome must be subtracted from the total signal in those detectors. Compensation between detectors can be performed either by hardware (electronic) detection but before logarithmic conversion and/or digitization or afterwards by software. Although compensation is one of the most important steps required for proper data analysis in flow cytometry, it is also perhaps the least well understood. Proper compensation is absolutely necessary for antigen density measurements and to distinguish very weak (dim) positive populations from negative populations. Undercompensation will result in overestimating the frequency of the dim cells; overcompensation will result in underestimating this frequency (Fig. 8; Roederer, 1996). As the number of parameters increases, the complexity and costs of electronic or "hardware" compensation increases; therefore, software compensation becomes a more attractive alternative. For a proper compensation, control stains are very important. The higher the number of fluorochromes and antibodies used in each assay, the greater the risk for artifacts introduced by compensation errors and/or reagent interactions. In general, two types of controls should be included and data collected with every experiment irrespective of the kind of compensation (electronic or software based), compensation controls and staining controls (Baumgarth and Roederer, 2000).

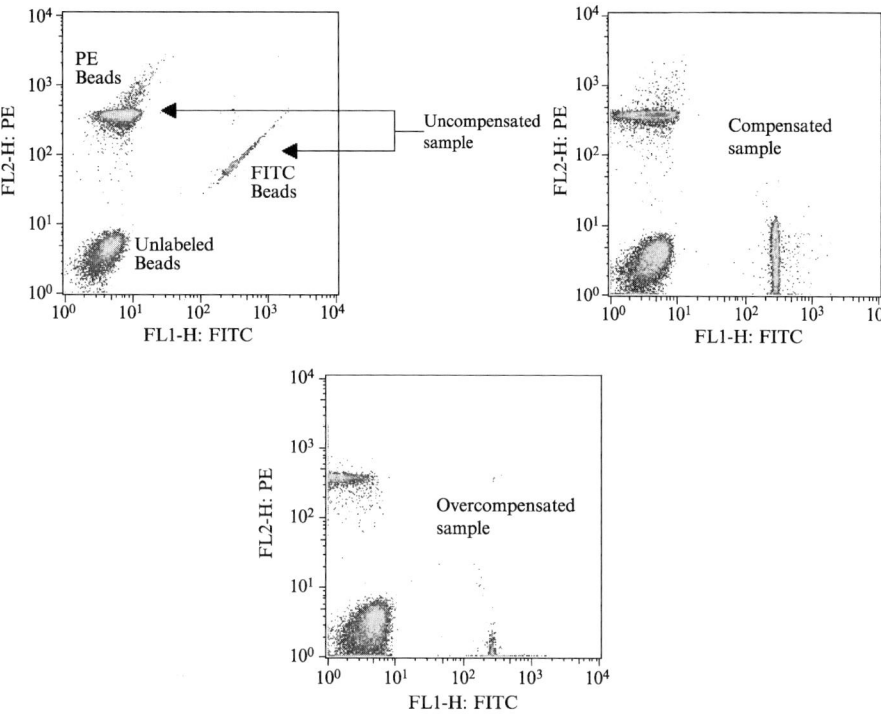

Fig. 8 Three cell populations can be observed in the FITC vs. r-PE dotplot. The unlabeled cells are at the lower left corner. The FITC-positive cells and the r-PE–positive cells can be observed as a diagonal flattened distribution. The contribution of FITC fluorescence signal into the FL2 detector (r-PE channel) is significant. There is also contribution of r-PE fluorescence into the FL1 detector (FITC channel). After compensation, the unwanted contribution of FITC and r-PE fluorescence in the r-PE and FITC-channel, respectively, is corrected. The median FL2 of the FITC-labeled cells is comparable to that of the FL2 of the unlabeled population. Furthermore, the median FL1 of the r-PE positive population is comparable to that of the FL1 of the unlabeled population. This is not the case when the signal is overcompensated as shown in the lower dotplot.

For each fluorochrome used in an assay, one should include a compensation control (i.e., a single color stain) for which data are collected. For example, in a two-parameter flow cytometric assay using FITC and r-PE as fluorochromes, one should include two compensation controls: a tube of cells labeled only with the FITC label and another tube of cells labeled with r-PE. Ideally, the reagents used for the compensation sample should be the same as that used in the two-colored sample.

Next to this, one must include staining control tubes in which the cells are labeled as in the double-staining assay. However, one of the two primary antibodies must be replaced by an isotype-matched and species-matched negative control immunoglobulin (staining control). In this way, one can control for background staining of the "unstained" cells (which can be at very different levels). By use of the aforementioned controls, the spillover of the fluorescence signals in the different channels can be determined and can be compensated (before actual acquisition [electronically or software-based] or afterward [software-based]).

6. Calibration and Maintenance

A good quality control scheme for flow cytometric analysis must be designed to asses the major instrument parameters that affect the reliability and reproducibility of data and must consist of two groups of procedures. The first group of procedures is carried out at relatively large intervals (one or two times a year) by qualified service personnel and includes examination of the efficiency and performance of the laser tube, optical filters, logarithmic and linear amplifiers, and PMT. The second group of procedures consists of frequent (every new use of the machine, e.g., daily) monitoring of instrument performance by the flow cytometer operator to identify both immediate and potential problems. The operator can use labeled beads for this purpose. These beads can be classified into three major categories: alignment beads, reference beads, and calibration beads.

Alignment beads include particles used to align the optics of the flow cytometer. Although many of the modern flow cytometers do not require daily optical alignment because they have fixed optical systems, it is strongly recommended to regularly check instrument alignment when samples seem to be shifted or when peaks are broader than normal. Proper use of the alignment beads ensures the highest resolution between sample populations by allowing the adjustment of the positions of the flow cell and optical components. Alignment maximizes the fluorescence and scatter signal (maximum mean channel number) while minimizing signal variability (minimum CV) (Schwartz et al., 1998).

Reference beads refer to particles with a given fluorescence intensity that are used to set up the instrument. A recommended approach for achieving a unified instrument setup is the establishment of a common window of analysis or analysis region. This is accomplished by use of beads with a high stability of fluorescence signals to prevent drift of the position of the fluorescence window of analysis over time, caused by decay of the fluorochrome on the bead. Daily monitoring with these reference beads ensures reproducibility of the analysis range and also standardizes the position of cell clusters in histograms obtained from different instruments.

Calibration beads encompass the fluorescence particles with multiple populations used to calibrate the response of the instrument and to quantify the fluorescence signal of samples. The ability to resolve or distinguish fluorescence signals of different intensities is the basis of determining negative from positive cell populations. These differences in resolution across the intensity range can be conveniently monitored with calibration beads (Schwartz et al., 1998). At minimum, these calibration beads consist of four types of beads with different fluorescence intensity and one group of nonfluorescent beads. The fluorescence intensities of these beads should cover that part of the fluorescence scale in which a linear response of the instrument to fluorescence signals can be expected.

B. Principles of Flow Cytometry: The Technique

1. Cell Cycle Analysis

Development, growth, renewal, and maintenance of organisms depend on the formation of new cells from parent cells. In other words, the cells need to be copied. This takes place through a process known as the cell cycle. The normal cell cycle is divided into four phases (see also Fig. 9A):

1. G_1-phase: the cells have a diploid or 2N DNA content (equivalent to 46 chromosomes in humans).
2. S-phase: the cycling cells replicate their DNA and have an amount of DNA varying between 2N and 4N. The fraction of cells in the S-phase (SPF) is often used as indication of proliferative status of a tissue.
3. G_2-phase: the cells have a double (4N) or tetraploid DNA content.
4. M-phase: mitosis; the cell divides, thus forming the two daughter cells.

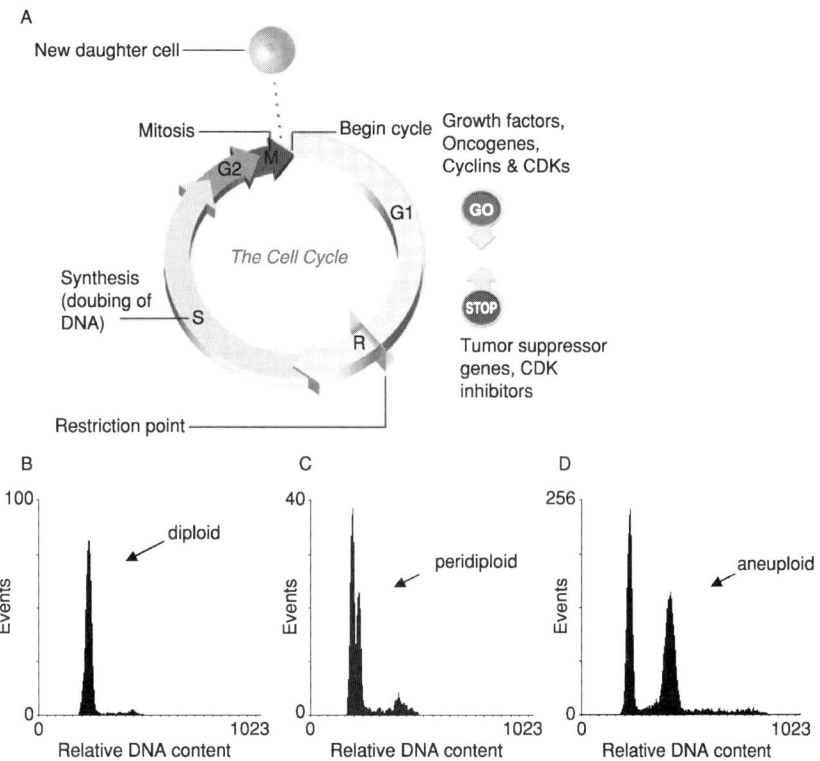

Fig. 9 Cell cycle analysis. Schematic representation of the different stages of the cell cycle (A). Examples of diploid (B), peridiploid (C), and real aneuploid (D) DNA histograms.

The only stage recognizable to the microscopist is the mitotic phase. Cells produced at mitosis reenter the G_1-phase, which is the most variable in duration, and a number of biochemical events occur during this phase and regulate exit from this phase. A separate G_0-phase was proposed by Lajtha to account for cells that do not divide unless stimulated to do so (Lajtha, 1963). It is currently not possible to separate a very long G_1 from G_0. At a certain point after entering the G_1 phase, the cells begin to duplicate their DNA. This phase, during which DNA is synthesized, is termed the S-phase. This phase has duration in the order of 6–16 hours. When the cells have completely doubled their DNA content, they enter a second phase—G_2. This phase typically lasts 4–8 hours. After this phase, the cells enter mitosis.

In any tissue, there are proliferating and nonproliferating cells; the latter are either end-stage, differentiated, or resting. The cells, which are actively involved in the cell cycle, make up the proliferation fraction. The proliferation fraction and the cell cycle time determine the growth activity of any tissue. The discovery of the existence of fluorochromes that bind to DNA in a stoichiometric manner was an important development in quantitative flow cytometric DNA analysis. This type of DNA analysis is the study of the distribution of cells into the different phases of the cell cycle on the basis of the DNA staining among the cells of a population (Gray *et al.*, 1986). These analyses provide clinicians with two potentially important cellular parameters of information. First, it gives information about the size of the fraction of cells that are in the S-phase of the cell cycle. Second, it provides information about the presence and degree of abnormal DNA content in the investigated cell population. DNA histogram analysis requires mathematical analysis to extract the underlying G1, S, and G2 + M phase distribution; methods for this analysis have been developed and refined over the past two decades. Methods to derive cell cycle information from DNA histograms range from simple graphical approaches to more complex deconvolution methods using curve-fitting. The two software programs most widely used for DNA histogram analysis are ModFit (from Bruce Bagwell; Verity Software House) and MultiCycle (from Peter Rabinovitch; Phoenix Flow Systems).

In normal tissue or a DNA diploid tumor, most of the cells are in the G_0/G_1 phase and have a diploid DNA content. This is reflected in the DNA histogram, which shows a single large peak of cells, the G_0/G_1 peak, with 2N DNA content. Normally, a smaller peak of cells, which are in the G_2-phase and M-phase of the cell cycle, is present on the x-axis at twice the distance of the G_0/G_1-peak (4N). The compartment between those two peaks is the S-phase fraction (SPF), with DNA content between 2N and 4N (Fig. 9B). In 1992, a DNA cytometry consensus conference was held, in which a nomenclature for DNA cytometry was recommended (Shankey *et al.*, 1993b). For ploidy, only the terms *DNA diploid* and *DNA aneuploid* should be used, with identification of the degree of DNA content abnormality given by the use of the DNA index (DI). The DI is the ratio of mean or mode of sample G_0/G_1 population divided by mean or mode of diploid reference cells. The definition of DNA aneuploidy includes the requirement that two distinct peaks are present in the DNA histogram (Hiddemann *et al.*, 1984). Furthermore, it was stated that tumors with a DI of 2.0 (DNA "tetraploid") should be recorded separately as a distinct group of DNA aneuploidy, because they may have a distinct prognostic significance in some types of tumors (bladder, prostate). The working definition for DNA tetraploid is DI values between 1.9 and 2.1, with proportions of cells greater than the G_2M fraction of normal tissue samples after correction for aggregates (see also Fig. 6). Many flow cytometric studies have mainly used the calculation of the proportion of cells in the S-phase of the cell cycle to predict the clinical behavior of human tumors. The assumption of these studies is that the fraction of cells, which synthesize DNA (S-phase fraction; SPF), is a direct reflection of tumor proliferation, and hence aggressive behavior (Shankey *et al.*, 1993a). However, the assessment of SPF by DNA flow cytometry encounters some limitations. In many studies, the SPF could not be determined in a considerable number of cases. For technical reasons (e.g., wrong fixatives, prolonged fixation time, delayed time of fixation), only some of the DNA histograms were suitable for analysis by these computer programs. Another possible danger is the underestimation of the SPF in DNA diploid, mixed DNA diploid/aneuploid, or DNA peridiploid tumors. Solid tumors are, in fact a heterogeneous mixture of benign (normal epithelial, stromal, endothelial, and inflammatory cells) and malignant cells (viable, necrotic, and apoptotic cells) (Crissman *et al.*, 1989). Next to this, the malignant cells also show intratumoral heterogeneity in DNA content. All these factors adversely affect DNA ploidy and SPF determinations. Proliferating nonneoplastic cells have in general a lower SPF value than malignant cells, causing an underestimation of SPF in DNA diploid or DNA peridiploid tumors (Dressler *et al.*, 1988; Hedley *et al.*, 1987). Next to this, dilution with nonneoplastic cells impairs sensitivity for detection of minor aneuploid stem lines (Frei *et al.*, 1994; Wingren *et al.*, 1994).

2. Multiparameter Flow Cytometric Analyses

One of the powers of flow cytometry is the ability to simultaneously measure more than one parameter on single cells. Ormerod *et al.* (1995) showed that light scatter/volume measurements could help to distinguish normal from malignant nuclei isolated from paraffin sections, even when both are diploid. To increase the accuracy of ploidy and cell cycle analysis, Ramaekers and coworkers demonstrated in 1984 in a tumor model the use of monoclonal antibodies directed against cytokeratin in combination with DNA content measurements (Ramaekers *et al.*, 1984). Cytokeratins can be considered as epithelium-specific IF proteins expressed in normal and neoplastic conditions (Moll *et al.*, 1982). Exclusion of nonepithelial cells after selecting only the cytokeratin-positive cells in a gated-flow cytometric DNA histogram increases the accuracy of SPF and DI determination. Enkhardt and his coworkers showed, after filtering out the nonrelevant fibroblasts and leukocytes, that the SPF increased significantly in DNA-diploid tumors. Furthermore, they showed that some tumors were aneuploid, whereas they were originally classified as DNA diploid. Multiparameter flow cytometry could offer much more to clinical studies than simply helping to distinguish between normal and malignant cells. DNA content analysis can also be coupled to immunofluorescence staining of a variety of potentially biologically relevant cellular constituents to analyze tissue homeostasis and tumor biodynamics. An example is the combined measurement of the expression of BrdU and DNA (Schutte *et al.*, 1995). This technique can give kinetic information and static indices of proliferation. In Fig. 10, an example of a multiparameter flow cytometric analysis of a breast carcinoma is shown. From this figure, the power of selection of cell populations with different expression of a certain parameter becomes clear.

When performing a DNA flow cytometric analysis, one has to exclude cell aggregates, because these interfere with the accuracy of flow cytometric measurements. For instance, when two G_0/G_1 cells form a doublet, they will yield the same fluorescence signal as a single G_2M cell. A method to reduce the contribution of cell aggregates to the acquired DNA histogram is the use of hardware doublet detection called pulse processing, a kind of electronic gating. In this technique, the signals of the width of the propidium iodide electronic pulses are displayed on the x-axis of a dotplot (e.g., Fig. 10A), and the area of these pulses on the y-axis. Because single cells have a linear relationship between these parameters, cells in the G_0/G_1, S, and G_2M phases of the cell cycle fall on a diagonal line between the axes (Fig. 10A). A gate can be set (R1 in Fig. 10A) along the diagonal to exclude cell doublets, based on the fact that cell doublets show an increased width signal compared with G_2-phase cells. The DNA histogram of this tumor (Fig. 10B) showed two cell populations: a DNA diploid and an aneuploid one. In the next step, the nonrelevant single cells (inflammatory cells, stromal compartment, etc.) are excluded from the analysis. This is performed by selection of cytokeratin-positive cells.

In a dotplot of DNA content (x-axis) vs. cytokeratin signal (y-axis), a region is set around the cytokeratin-positive cells (Fig. 10C; upper parts of the dotplot R2). The cutoff value for determining the threshold for immunoreactivity is based on the signal from the negative control (a single cell suspension incubated with a nonrelevant immunoglobulin). Figure 10D shows a dotplot of DNA content (x-axis) vs. the expression of the progesterone receptor (y-axis) of the total single cell population. Figure 10E and 10F show the dotplot of DNA content vs. PR expression and DNA histogram, respectively, for the cytokeratin-negative cell population (all cells that are excluded from the region from Fig. 10C). From these figures, it becomes clear that this compartment comprises almost only the DNA diploid cell population, whereas the DNA aneuploid population is found in the cytokeratin-positive selection (Figs. 10G and 10H; all cells that are situated in the region of Fig. 10C). Here the highest immunoreactivity for the progesterone receptor is also found. This example shows that by use of more than one parameter simultaneously, plus the possibility of making selections (gates) based

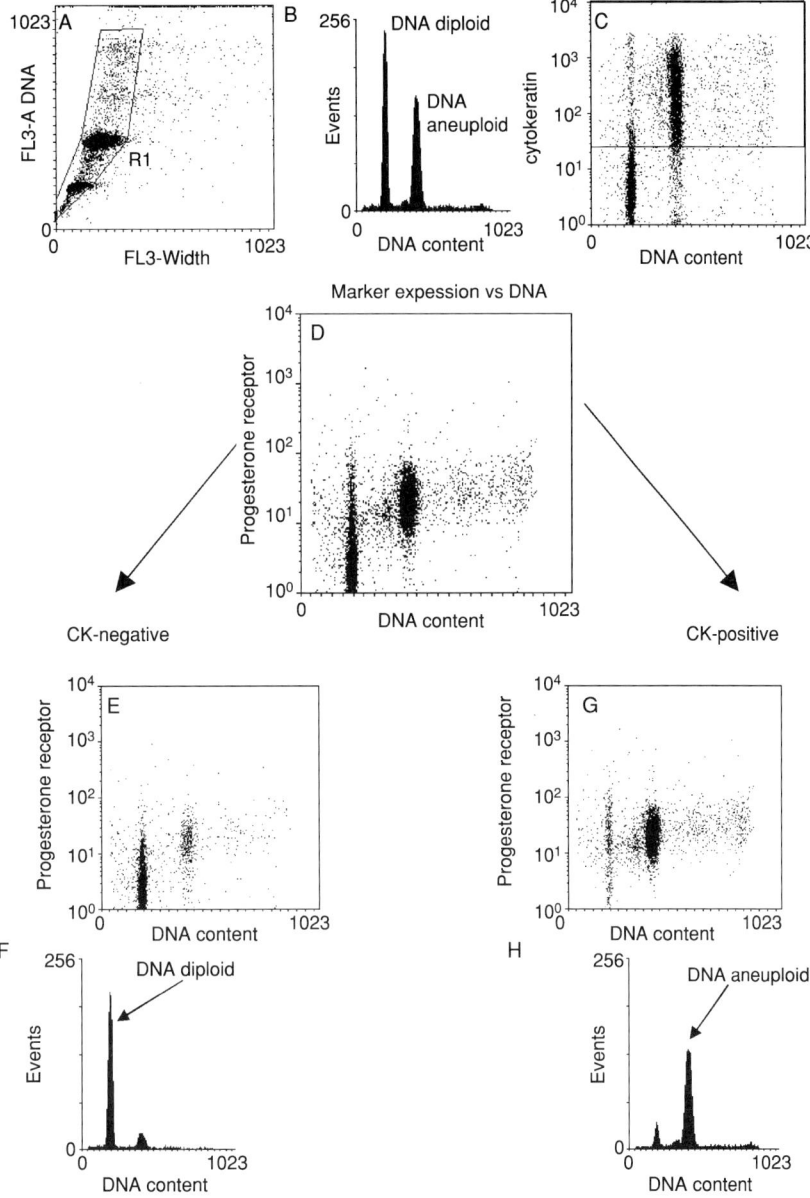

Fig. 10 Example of a multiparameter flow cytometric cytokeratin/PR/DNA analysis of an aneuploid breast carcinoma (see text on page 188 for explanation; CK, cytokeratin). (A) The selection of the single cells (exclusion of doublets, triplets, and aggregates). (B) DNA histogram of the ungated cells. (C and D) The dotplots of DNA (x-axis) vs. cytokeratin (y-axis) and DNA (x-axis) vs. progesterone receptor (PR; y-axis), respectively, for the total cell population. The solid line in (C) denotes the thresholds for immunoreactivity as determined on the negative controls. (E and F) The dotplot (DNA vs. PR) and the DNA histogram, respectively, of the cytokeratin negative cells. Gating for cytokeratin positivity resulted in the dotplot (DNA vs. PR) and DNA histogram as depicted in (G and H).

on the differential expression of markers, a more defined and possibly a more reliable quantification of specific characteristics can be performed in the cell population of interest.

3. Quality of the Analysis

In the ideal case, signals from cells in a certain phase of the cell cycle, with the same DNA content, will accumulate in the same channel of the DNA histogram. However, because of small differences in chromatin compactness between individual cells (which affects the binding of a DNA stain), minor instrumental errors, and preparation artefacts, a Gaussian (normal) distribution is observed (Corver, 2001). A Gaussian distribution is characterized by a mean and a standard deviation (SD). Precision and resolution of DNA histogram analysis is usually monitored by the use of "coefficient of variation" (CV) and gives an impression of the width of the peak. The CV is usually calculated of the DNA diploid G_0/G_1-peak with the following equation:

$$CV = W/(M \times 2.35)$$

where W = full width of the G_0/G_1-peak at the half-maximum height and M = peak channel number of the G_0/G_1-peak. The result is expressed as a percentage. The lower the CV of the peaks in the DNA histogram, the better is the quality. In practice, CVs of less than 1% are exceptional. The quality of the DNA histogram can be affected by, for example, debris and aggregates. The CV significantly affects the accuracy of S-phase calculations. In a DNA cytometry consensus conference, it has been stated that the CV of normal diploid cells in a histogram should be less than 8% (Shankey et al., 1993a,b). If the CV is greater than 8%, the peak may be composed of two or more populations of cells that cannot be resolved by the software program, and estimation of the S-phase fraction should not be attempted. Our experience is that as a rule one should always try to have CVs below 5%.

4. Technical Controls

The fluorescence signal strength of a cell nucleus stained with propidium iodide is many times greater than that of an intact cell stained with a fluorochrome-labeled antibody against a certain cell constituent. In addition, fluorescence intensity of the DNA staining dye must be measured on a linear, rather than a logarithmic, scale because of the direct stoichiometric relationship of DNA content to fluorescence. However, lack of linearity of the amplifier for this signal is a common problem in many instruments. One of the frequent consequences of this nonlinearity is the situation in which G_2/M phase cells, which have twice the amount of DNA in G_0/G_1 phase cells, seem to have substantially more or less DNA. Signal amplifier linearity should be checked on a regular basis with standard particles (standard fluorescent beads, polyploidy liver cells) and/or suitable methods. A common method for fresh or frozen tissue samples enclosed the addition of both chicken erythrocytes (CRBC; with a DNA content of 35% of the human diploid value) and rainbow trout erythrocytes (TRBC; with a DNA content of 80% of the human diploid value) to the sample (Vindelov *et al.*, 1983). The use of two standards eliminates technical errors cause by nonlinearity. Next to this control for linearity, the CRBCs and TRBCs serve as an internal diploid standard to detect staining variations (Vindelov, 1977). In practice, normal human lymphocytes and CRBCs are added to the sample. After flow cytometric analysis, the presence of the G_0/G_1 peak produced by the lymphocytes is a valuable landmark in verifying the position of the diploid G_0/G_1 peak of the clinical sample. The CRBC peak is essential for verifying the linearity of the DNA channel.

For formalin-fixed, paraffin-embedded tissues (variability in fixation [time] and DNA dye accessibility), this control is complicated because of the lack of a known DNA diploid reference population. This makes the identification of the diploid G_0/G_1 peak of the clinical sample difficult. It is possible to add CRBCs to the tissue specimen before fixation and paraffin embedding in the event that future flow cytometric DNA analysis may be required; however, this has several logistical problems. Therefore, the DNA cytometry consensus conference pointed out that because of the rare finding of hypoploid tumors, it is recommended that the leftmost peak in the DNA histogram form paraffin-embedded tissue samples may be assumed to represent the DNA diploid population (Shankey *et al.*, 1993a,b).

III. Procedures

A. Immunohistochemistry

1. Preparation of Tissue Sections for Immunochemistry

   #### a. Cryostat Frozen Sections
   1. Trim tissue to approximately 4 mm³.
   2. Snap-freeze in liquid nitrogen and store at $-70\,°C$. If desired, the tissue may be embedded in a mixture of polyvinyl alcohol/polyethylene glycol or gelatin.
   3. With forceps, slowly immerse the base mold in approximately 2 inches of liquid nitrogen until it is completely frozen.
   4. Frozen blocks are wrapped in aluminium foil and stored at $-70\,°C$.
   5. Blocks to be sectioned are placed in the $-20\,°C$ cryostat for at least 1 hour before sectioning.
   6. Cut sections 4–10 μm thick and thaw mount onto clean glass slides. Use of an adhesive coating or subbing agent, such as poly-L-lysine, charging, or silanization, may improve tissue adherence.
   7. Immediately fix in acetone at room temperature for 10 seconds.

8. Air-dry 12–24 hours. Alternate method: To accelerate the drying process, air-dry for 30 minutes, fix for 10 minutes in fresh acetone, then air-dry again for 30 minutes.

b. Paraffin Sections

1. Fix tissue specimens according to established laboratory protocols. The most commonly used fixatives contain formalin. To determine what fixatives are compatible with an intended marker, check the specification sheet of the primary antibody for further information. In addition to formalin, several other fixatives are commonly used for tissue preservation.

2. Trim tissue to approximately 4 mm^3 and dehydrate in graded alcohols. Typically, the grading is as follows: 60% ethanol, 65% ethanol, 95% ethanol, 95% ethanol, absolute ethanol.

3. Clearing of alcohol from the tissue may be performed with xylene or a xylene substitute.

4. Infiltrate and embed tissue in liquid paraffin. The temperature should not exceed ~58 °C. NOTE: Depending on the molecular weight of paraffin, the melting point (MP) will range from 48–66 °C. Typically, paraffin used in the histology laboratory has a MP ~55–58 °C.

5. Cut sections 4–6 μm thick and mount onto clean glass slides. Use of an adhesive coating or subbing agent, such as poly-L-lysine, charging, or silanization, may improve tissue adherence.

6. Air-dry 12–24 hours at room temperature, overnight at 37 °C or at 60 °C for 1 hour.

NOTE: Use of protein-containing products, such as commercially available compounds that may contain gelatin (Elmer's Glue or Knox gelatin), in the mounting water bath is contraindicated if tissue is being mounted for IHC staining.

Proteins in the water bath will bind with the adhesive coating before the tissue section is mounted on the glass slide. During pretreatment and/or IHC procedures, the tissue may partially lift off the glass slide, allowing IHC reagents to become trapped beneath, or total detachment of the specimen may occur. In addition, IHC reagents may nonspecifically bind to this protein layer, contributing to background staining.

AFTER DRYING: All cryostat section slides that are not to be stained immediately should be placed back-to-back, wrapped in aluminium foil, and stored at −20 °C to −70 °C. Paraffin section slides should be stored at room temperature or 2–8 °C.

BEFORE FIXATION: Allow frozen and refrigerated slides to reach room temperature before unwrapping.

2. Fixation and Dewaxing

a. Cryostat Frozen Sections

1. Fix slide for 10 minutes in cold 100% acetone.

2. Place in phosphate-buffered saline (PBS) for 1–5 minutes.

   3. Do not perform a peroxidase block with H_2O_2 or H_2O_2/methanol until after the primary antibody step.

b. Paraffin Sections
   1. Place slides in a xylene bath and incubate for 5 minutes. Change baths and repeat once.
   2. Tap off excess liquid and place slides in 95%–96% ethanol for 3 minutes. Change baths and repeat once.
   3. Tap off excess liquid and place slides in 70% ethanol for 3 minutes. Change baths and repeat once.
   4. Tap off excess liquid and place slides in DI water for a minimum of 30 seconds.

3. Antigen Retrieval of Formalin-Fixed, Paraffin-Embedded Tissues

a. Proteolytic Method
Materials

   • Trypsin solution: 0.1% trypsin type II, crude from porcine pancreas in 0.1% $CaCl_2$, pH 7.8. Adjust the pH with 0.1% NaOH or 2% HCl and preheat the $CaCl_2$ to 37 °C for at least 15 minutes before adding trypsin.
   • Pepsin solution: 0.4% pepsin in 0.01 M HCl, pH 2.0

   1. Incubate the sections with $200\,\mu l$ enzyme solution (do not preheat the solution) for 15 minutes at 37 °C
   2. Halt digestion by washing in distilled water and then in PBS. The antibody or blocking solution should then be applied to moist but not dry sections.

b. Heat-Induced Epitope Retrieval
   1. Place sections in a Coplin jar with dilute antigen retrieval solution of choice (10 mM citrate acid, pH 6).
   2. Place Coplin jar containing slides in vessel filled with water and microwave on high for 2–3 minutes (700 watt oven).
   3. Check level of retrieval solution; allow cooling for 2–3 minutes, and repeat steps 3 and 4 four times (depending on tissue).
   4. Remove Coplin jar containing sections and allow cooling for 20 minutes at room temperature.
   5. Rinse sections in deionized water, two times for 5 minutes.
   6. Place slides in modified endogenous oxidation blocking solution (PBS + 2% hydrogen peroxide).
   7. Rinse slides once for 5 minutes in PBS.

4. Staining Methods

a. General Avidin Biotin Complex Procedure

Formation of the avidin biotin complex requires that the solutions of the (strept)-avidin and biotinylated enzyme are mixed in an optimal ratio and prepared at least 30 minutes before use. All incubations are carried out at room temperature.

1. Incubate tissue 30 minutes with normal rabbit serum.
2. Tap off serum and wipe away excess. *Do not rinse.*
3. Incubate with primary antibody for 60 minutes at room temperature
4. Rinse sections with PBS for 3–5 minutes.
5. Incubate with biotinylated secondary antibody for 30 minutes at room temperature.
6. Rinse sections with PBS for 3–5 minutes.
7. Incubate with (strept)avidin-biotin complex (prepared at least 30 minutes before use) for 30 minutes at room temperature.
8. Rinse with PBS for 3–5 minutes.

b. Develop and Counterstain

1. Incubate sections for approximately 5–7 minutes in peroxidase substrate solution made up immediately before use as follows:

   - 10 mg diaminobenzidine (DAB) dissolved in 10 ml 50 mM sodium phosphate buffer, pH 7.4
   - 1.25 μl hydrogen peroxide

   CAUTION: Handle DAB carefully because of its possible carcinogenic properties.

2. Rinse slides well three times for 10 minutes in deionized water.
3. Counterstain with hematoxylin for 10 seconds, depending on intensity of counterstain desired.
4. Rinse slides three times for 5 minutes with tap water.
5. Dehydrate two times for 2 minutes in 70% ethanol, 95% ethanol; two times for 2 minutes in 100% ethanol; and two times for 2 minutes in xylene.
6. Mount slides.

B. Multiparameter Flow Cytometry

1. Protocol for Pure Enzymatic Cell Dissociation (Original Hedley Method)

1. Cut 30-μm thick paraffin sections and put them in a glass tube
2. Deparaffinize by rinsing in xylene (2–10 minutes).
3. Rehydrate in decreasing series of ethanol:
   Ethanol 100%, 10 minutes

Ethanol 95%, 10 minutes

Ethanol 70%, 10 minutes

Ethanol 50%, 10 minutes

4. Wash sections twice in distilled water.

5. Resuspend the sections in 1 ml 0.5% pepsin/0.9% NaCl (pH 1.5, adjust with 2 N HCl).

6. Place the tube in a water bath of 37 °C for 30 minutes.

7. Add 2 ml cold PBS to the suspension and vortex thoroughly. Prepare as much as possible single cells by firm mechanical pipetting with a micropipette.

8. Wash the cells with PBS by centrifugation, three times at 400g.

9. Stain DNA with DAPI (4'6'-diamidino-2-phenylindole dihydrochloride: 1 µg/ml RPMI 1640)

10. Allow the cells to incubate for a minimum of 30 minutes at room temperature.

11. Filter cells through nylon gauze.

2. General Protocol for Formalin-Fixed, Paraffin-Embedded Epithelial Tumors (Combined Heat/Enzymatic Digestion)

1. Cut for each assay two 50-µm thick paraffin sections and put them in a glass tube.

2. Deparaffinize by rinsing in xylene (2–30 minutes).

3. Rehydrate in decreasing series of ethanol: ethanol 96%, 2–30 minutes; ethanol 70%, 1–30 minutes; PBS, 1–30 minutes.

4. Add 2 ml cold citrate solution (2 g/l aqua dest, pH = 6.0) to the sections and heat for 120 minutes in a water bath of 80 °C.

5. Allow the sections to cool for 15 minutes at room temperature, tap off the citrate solution.

6. Digest the tissue sections by adding 2 ml *cold* pepsin solution (1 mg/ml 0.1 N HCl) and place the tube for a *maximum* of 10 minutes in an oven of 37 °C.

7. Add 2 ml cold PBS to the suspension and vortex thoroughly. Prepare as much as possible single cells by firm mechanical pipetting with an micropipette.

8. Wash the cells with PBS by centrifugation three times at 400g.

9. Incubate the cells overnight at room temperature with properly diluted primary antibody/ies.

10. Wash the cells with PBS by centrifugation, three times at 400g.

11. Incubate the cells for 90 minutes at room temperature with properly diluted fluorochrome-labeled secondary antibody/ies.

12. Wash the cells with PBS by centrifugation, three times at 400g.

13. Stain DNA by adding 2 ml propidium iodide-solution (1.0 μg PI/ml and 0.1 mg/ml RNAse).

14. Allow the cells to incubate for a minimum of 60 minutes at 4 °C in the dark.

15. Analyze the cells.

Advantages

- High DNA histogram resolution
- Excellent technique for the detection of intermediate filaments (e.g., cytokeratin)
- High yield of isolated cells after pepsin digestion

Disadvantages

- Pepsin digestion destroys most, if not all, cell surface bound epitopes and, for example, the nuclear epitope for Ki-67 (detected by MIB-1)

IV. Comments

Immunophenotyping by multiparameter flow cytometry can provide a rapid, accurate method for both identifying unique cell populations and describing their functional status. This can be achieved because antibodies to cellular proteins can be conjugated or labeled with different-colored fluorochromes and combined in panels for staining heterogeneous cell populations. Traditionally, DNA flow cytometry was one of the earliest applications in flow cytometry (Hudson *et al.*, 1969). However, the value of this technique has often been questioned with regard to the additional information obtained. This doubt is a logical consequence of the existence of several bottlenecks if isolated DNA flow cytometry is applied to fresh tissue or cells. The preparation of single-cell suspensions from solid tumors is complicated by the fact that only limited information is available about the composition of the different contributing populations in the solid tumor. If a frozen section is made to obtain this information, artefacts introduced by freezing and thawing will impair cell and tissue morphology. In case of formalin-fixed paraffin-embedded samples, the morphology is probably better but, even with this knowledge available, single-parameter DNA analysis does not take into account the contribution of inflammatory and other stromal cells to the result of cell cycle parameter calculations. The combination and the use of immunochemical labeling of the IFs and multiparameter DNA flow cytometric analysis allows us to separate the data related to specific immunophenotypes, collect data from a variety of different classes in the same run, and relate these specifically to the population of interest from those in the total population. With this combination, it is virtually possible to visualize the presence and distribution of multiple cell characteristics directly and make calculations with them in relation to each other. By gating and eventually also sorting, it is possible to have an objective impression of the proliferative activity, even in subsets of tumor cells, and relate that to the distribution of other markers of biological activity. Sorting even allows the relative enrichment of rare events to prepare them for further analysis at a molecular level (Bonsing *et al.*, 2000). Another major advantage of applying multiparameter flow cytometry on routinely processed formalin-fixed, paraffin-embedded specimens is that morphological information is still available, because the block from which the 50-μm-thick sections are taken is also used for production of routine H & E staining. The admixture with nontumor elements can thus be anticipated, and, in special cases, a preselection of the area of interest can be done even before the cell suspension preparation step.

Important issues in implementing multiparameter DNA flow cytometry are a solid quality control and management system that have to be set up to guarantee that the data obtained are robust and reliable. A prerequisite for this is that one is not only informed about the specificity of the reagents and machinery, but that one also knows the composition and preservation characteristics of the tissue to be analyzed. This is where the routine skills of the histopathology laboratory are needed. As mentioned already, the use of formalin-fixed, paraffin-embedded tissue blocks allows the introduction of routine H & E-stained sections for evaluation of tissue components and to make a preselection of the area of interest by trimming the paraffin block. Tissue preservation is another important item. Formalin fixation seems to be simple and easy to control. Basically, this is correct, and it is still the most widespread and probably cheapest way of tissue fixation and preservation. At the same time, there are many factors that can interfere with correct fixation, and not all of them are within reach for the pathology staff to monitor. The whole process starts at the moment of tissue removal and the conditions under which the sample is stored and transported to the pathology laboratory. Good and robust logistics are necessary from the start. If no special measurements are taken to ensure complete penetration of the fixative in the tissue, this means that the center of samples greater than 1-cm thick will not easily be preserved properly before the autolysis has done part of its destructive work. Proper fixation for at least 24 hours has been shown to be a minimum time to obtain complete cross-linking of most proteins in the tissue. The second step to preservation of the tissue in paraffin follows a process that will remove all water and, with that, all water-soluble reagents such as formalin. If the fixation process was not completed, there is still a possibility that the degradation of proteins and DNA will slowly continue. Longer fixation is not necessary but, from several experiments with formalin fixation at room temperature up to 72 hours, we have seen no negative effect on DNA histograms (Leers *et al.*, 1999b). Another argument in favor of complete fixation is a bit contradictory, but those familiar with immunohistochemistry will probably recognize it. After incomplete fixation, preservation of optimal morphology will interfere with antigen retrieval methods in a way of destroying cell membranes and blurring nuclear shapes. In addition, endogenous biotin, if available, will become reactive again for avidin or streptavidin, resulting in an unwanted background staining. Because the preparation process of single cells from formalin-fixed paraffin-embedded tissue includes a 2-hour heating step at 80 °C, these negative effects will also influence the quality of the flow cytometric results. Wide CVs for the DNA histogram and increased background and debris will be the result.

Standardization is another problem that needs attention. Whereas results obtained on fresh material often give reproducible information, the results that are obtained in flow cytometry studies on formalin-fixed, paraffin-embedded material, suspended primarily by enzymatic digestion procedures, often show wide variations in CVs and cell cycle characteristics. Before the prolonged heating step was introduced in our routine procedures, the effect of this technical modification on the stability of DNA analysis was also investigated. We could show that the variation in results did not exceed the 5% boundary that is generally accepted as an intrinsic level of variation in biological samples (Leers *et al.*, 1999b). This means that the results from consecutively performed tests have to be monitored, and changes in the mean values, as well as the occurrence of outliers, need to be signaled. Explanations need to be looked for and corrections of procedures or adjustment of calibration values taken care of.

Once quality control and maintenance programs have been established and data analysis rendered less operator dependent, a broad field of possible applications for flow cytometry comes within reach. Preanalytic separation of cells from body fluids, based on size, shape, DNA, and/or immunophenotype, might enhance the possibilities to study the contribution of small numbers of aberrant expression patterns on our understanding of pathogenetic processes or on the course of disease in case of minimal residual pathological elements. Multiparameter flow cytometry based on immunochemical labeling has been applied in hematological routine diagnostics for several years now. This approach has successfully encroached on DNA analysis (Brockhoff *et al.*, 1996, 1998; Corver *et al.*, 1994, 2000; Nowak *et al.*, 1994), most probably enhancing the clinical use. The value of ploidy and SPF calculations under newly optimized conditions in a clinical setting will be unveiled. In this way, multiparameter flow cytometry completed with advanced sorting facilities might become a serious and economically favorable competitor of other more laborious methods such as laser capturing or automated image analysis where only the image is retained and not the actual cell.

V. Pearls and Pitfalls

The most important features in the successful performance of multiparameter flow cytometry on archival tissues are:

- The correct pH of the antigen retrieval buffer (citrate, pH = 6.0, and EDTA, pH = 8.5). An incorrect PH will result in wider CVs in DNA histograms.

- The temperature of the enzyme solution: do not preheat this solution because of excessive and unwanted damage to, or a rapid digestion of, the cells.
- Promoting mechanical dissociation of the tissue section by firmly vortexing the suspension after incubation with the enzyme solution. If this is not done properly, too many aggregates will remain present.
- Correct labeling of the antibodies: for labeling cells derived from archival tissues, FITC-conjugated antibodies with a high F/P ratio (± 5) must be used.
- The use of the right isotype-negative controls for setting up the instrument (compensation and threshold determination).
- When one is interested in a possible weak expression of a protein, it is advisable to visualize the bound antibodies with an FITC-labeled secondary antibody. It is better to use the rPE-labeled secondary antibodies for the visualization of proteins that are abundant in their expression pattern (e.g., cytokeratin, vimentin).
- The rPE-fluorochrome is a large molecule, which penetrates with difficulty into the nucleus. For the visualization of nuclear-bound epitopes, it is advisable to use an immunochemical labeling method based on an FITC label.
- When performing semiquantitative FCM, be aware of the fact that antibodies directed against the same protein but a different epitope can result in different sizes of fractions.

References

Aberle, H., Bauer, A., Stappert, J., Kispert, A., and Kemler, R. (1997). Beta-catenin is a target for the ubiquitonin-proteasome pathway. *EMBO J.* **16,** 3797–3804.

Albers, K., and Fuchs, E. (1987). The expression of mutant epidermal keratin cDNAs transfected in simple epithelial and squamous cell carcinoma lines. *J. Cell Biol.* **105,** 791–806.

Asa, S. L., Kovacs, K., Bilbao, J. M., and Penz, G. (1981). Immunohistochemical localization of keratin in craniopharyngiomas and squamous cell nests of the human pituitary. *Acta Neuropathol. (Berl)* **54,** 257–260.

Ashworth, T. R. (1869). A case of cancer in which cells similar to those in the tumours were seen in the blood after death. *Aust. Med. J.* **14,** 146.

Bartek, J., Vojtesek, B., Staskova, Z., Bartkova, J., Kerekes, Z., Rejthar, A., and Kovarik, J. (1991). A series of 14 new monoclonal antibodies to keratins: Characterization and value in diagnostic histopathology. *J. Pathol.* **164,** 215–224.

Baumgarth, N., and Roederer, M. (2000). A practical approach to multicolor flow cytometry for immunophenotyping. *J. Immunol. Methods* **243,** 77–97.

Bijman, J. T., Wagener, D. J. T., Wessels, J. M. C., Van den Broek, P., and Ramaekers, F. C. S. (1986). Cell size, DNA, and cytokeratin analysis of human head and neck tumours by flow cytometry. *Cytometry* **7,** 76–81.

Bonsing, B. A., Corver, W. E., Fleuren, G. J., Cleton-Jansen, A. M., Devilee, P., and Cornelisse, C. J. (2000). Allelotype analysis of flow-sorted breast cancer cells demonstrates genetically related diploid and aneuploid subpopulations in primary tumours and lymph node metastases. *Genes Chromosomes Cancer* **28,** 173–183.

Bostick, P. J., Chatterjee, S., Chi, D. D., Huynh, K. T., Giuliano, A. E., Cote, R., and Hoon, D. S. (1998). Limitations of specific reverse-transcriptase polymerase chain reaction markers in the detection of metastases in the lymph nodes and blood of breast cancer patients. *J. Clin. Oncol.* **16,** 2632–2640.

Brockhoff, G., Endl, E., Minuth, W., Hofstadter, F., and Knuchel, R. (1996). Options of flow cytometric three-colour DNA measurements to quantitate EGFR in subpopulations of human bladder cancer. *Anal. Cell Pathol.* **11,** 55–70.

Brockhoff, G., Wieland, W., Woelfl, G., Hofstaedter, F., and Knulchel, R. (1998). Evaluation of flow-cytometric three-parameter analysis for EGFR quantification and DNA assessment in human bladder carcinomas. *Virchows Arch.* **432,** 77–84.

Bruderman, I., Cohen, R., Leitner, O., Ronah, R., Guber, A., Griffel, B., and Geiger, B. (1990). Immunocytochemical characterization of lung tumours in fine-needle aspiration. The use of cytokeratin monoclonal antibodies for the differential diagnosis of squamous cell carcinoma and adenocarcinoma. *Cancer* **66,** 1817–1827.

Cabioglu, N., Igci, A., Yildirim, E. O., Aktas, E., Bilgic, S., Yavuz, E., Muslumanoglu, M., Bozfakioglu, Y., Kecer, M., Ozmen, V., and Deniz, G. (2002). An ultrasensitive tumour enriched flow-cytometric assay for detection of isolated tumour cells in bone marrow of patients with breast cancer. *Am. J. Surg.* **184,** 414–417.

Cattoretti, G., Becker, M. H., Key, G., Duchrow, M., Schluter, C., Galle, J., and Gerdes, J. (1992). Monoclonal antibodies against recombinant parts of the Ki-67 antigen (MIB 1 and MIB 3) detect proliferating cells in microwave-processed formalin-fixed paraffin sections. *J. Pathol.* **168,** 357–363.

Caulin, C., Salvesen, G. S., and Oshima, R. G. (1997). Caspase cleavage of keratin 18 and reorganization of intermediate filaments during epithelial cell apoptosis. *J. Cell Biol.* **138,** 1379–1394.

Cooper, D., Schermer, A., and Sun, T. T. (1985). Classification of human epithelia and their neoplasms using monoclonal antibodies to keratins: Strategies, applications, and limitations. *Lab. Invest.* **52,** 243–256.

Corver, W. E. (2001). Multiparameter DNA flow cytometry of human solid tumors: Technical improvements and applications. *In* "Pathology," p. 191. University of Leiden, The Netherlands: Leiden.

Corver, W. E., Cornelisse, C. J., and Fleuren, G. J. (1994). Simultaneous measurement of two cellular antigens and DNA using fluorescein-isothiocyanate, R-phycoerythrin, and propidium iodide on a standard FACScan. *Cytometry* **15,** 117–128.

Corver, W. E., Koopman, L. A., Mulder, A., Cornelisse, C. J., and Fleuren, G. J. (2000). Distinction between HLA class I-positive and- negative cervical tumor subpopulations by multiparameter DNA flow cytometry. *Cytometry* **41,** 73–80.

Crissman, H. A., and Steinkamp, J. A. (1973). Rapid, simultaneous measurement of DNA, protein, and cell volume in single cells from large mammalian cell populations. *J. Cell Biol.* **60,** 523–527.

Crissman, J. D., Zarbo, R. J., Ma, C. K., and Visscher, D. W. (1989). Histopathologic parameters and DNA analysis in colorectal adenocarcinomas. *Pathol. Annu.* **24,** 103–147.

Davidson, B., Dong, H. P., Berner, A., Christensen, J., Nielsen, S., Johansen, P., Brynse, M., Asschenfeldt, P., and Risberg, B. (2002). Detection of malignant epithelial cells in effusions using flow cytometric immunophenotyping. *Am. J. Clin. Pathol.* **118,** 85–92.

DeYoung, B. R., and Wick, M. R. (2000). Immunohistological evaluation of metastatic carcinomas of unknown origin: An algorithmic approach. *Semin. Diagn. Pathol.* **17,** 184–193.

Dressler, L. G., Seamer, L. C., Owens, M. A., Clark, G. M., and McGuire, W. L. (1988). DNA flow cytometry and prognostic factors in 1331 frozen breast cancer specimens. *Cancer* **61,** 420–427.

Eichner, R., Bonitz, P., and Sun, T. T. (1984). Classification of epidermal keratins according to their immunoreactivity isoelectric point, and mode of expression. *J. Cell Biol.* **98,** 1388–1396.

Espinoza, C. G., and Aza, H. A. (1998). Immunohistochemical localization of keratin-type proteins in epithelial neoplasms. Correlation with electron microscopic findings. *Am. J. Clin. Pathol.* **78,** 500–507.

Frei, J. V., and Martinez, V. J. (1993). DNA flow cytometry of fresh and paraffin-embedded tissue using cytokeratin staining. *Mod. Pathol.* **6,** 599–605.

Frei, J. V., Rizkalla, K., and Martinez, V. J. (1994). Proliferative cell indices measured by DNA flow cytometry in node-negative adenocarcinomas of breast: Accuracy and significance in cytokeratin-stained archival specimens. *Mod. Pathol.* **7,** 925–929.

Frierson, H. F., Jr., Bellafiore, F. J., Gaffey, M. J., McCary, W. S., Innes, D. J., Jr., and Williams, M. E. (1994). Cytokeratin in anaplastic large cell lymphoma. *Mod. Pathol.* **7,** 317–321.

Fuchs, E., and Weber, K. (1994). Intermediate filaments: Structure, dynamics, function, and disease. *Annu. Rev. Biochem.* **63,** 345–382.

Giuliano, A. E., Kirgan, D. M., Guenther, J. M., and Morton, D. L. (1994). Lymphatic mapping and sentinel lymphadenectomy for breast cancer [see comments]. *Ann. Surg.* **220,** 391–398; discussion 398–401.

Gown, A. M., and Vogel, A. M. (1985). Monoclonal antibodies to human intermediate filament proteins. *Am. J. Clin. Pathol.* **84,** 413–424.

Gray, J. W., Dolbeare, F., Pallavicini, M. G., Beisker, W., and Waldman, F. (1986). Cell cycle analysis using flow cytometry. *Int. J. Radiat. Biol. Relat. Stud. Phys. Chem. Med.* **49,** 237–255.

Gustmann, C., Altmannsberger, M., Osborn, M., Griesser, H., and Feller, A. C. (1991). Cytokeratin expression and vimentin content in large cell anaplastic lymphomas and other non-Hodgkin's lymphomas. *Am. J. Pathol.* **138,** 1413–1422.

Haskell, C. M., Cochran, A. J., Barsky, C. H., and Steckel, R. J. (1988). Metastasis of unknown origin. *Curr. Probl. Cancer* **12,** 5–58.

Hayes, D. F., Walker, T. M., Singh, B., Vitetta, E. S., Uhr, J. W., Gross, S., Rao, C., Doyle, G. V., and Terstappen, L. W. M. (2002). Monitoring expression of Her-2 on circulating epithelial cells in patient with advanced breast cancer. *Int. J. Oncol.* **21,** 1111–1117.

Hedley, D. W., Rugg, C. A., and Gelber, R. D. (1987). Association of DNA index and S-phase fraction with prognosis of nodes positive early breast cancer. *Cancer Res.* **47,** 4729–4735.

Hiddemann, W., Schumann, J., Andreef, M., Barlogie, B., Herman, C. J., Leif, R. C., Mayall, B. H., Murphy, R. F., and Sandberg, A. A. (1984). Convention on nomenclature for DNA cytometry. Committee on Nomenclature, Society for Analytical Cytology. *Cancer Genet. Cytogenet.* **13,** 181–183.

Hudson, B., Upholt, W. B., Devinny, J., and Vinograd, J. (1969). The use of an ethidium analogue in the dye-buoyant density procedure for the isolation of closed circular DNA: The variation of the superhelix density of mitochondrial DNA. *Proc. Natl. Acad. Sci. USA* **62,** 813–820.

Jordanova, E. S., Corver, W. E., Vonk, M. J., Leers, M. P. G., Riemersma, S. A., Schuuring, E., and Kluin, P. M. (2003). Flow cytometric sorting of paraffin-embedded tumor tissues considerably improves molecular genetic analysis. *Am. J. Clin. Pathol.* **120,** 327–334.

Ku, N. O., and Omary, M. B. (2001). Effect of mutation and phosphorylation of type 1 keratins on their caspase-mediated degradation. *J. Biol. Chem.* **276,** 26792–26798.

Lajtha, L. G. (1963). On the concept of the cell cycle. *J. Cell Comp. Phys.* **62,** 143–145.

Leers, M., Hoop, J., and Nap, M. (2003). Her2/neu analysis in formalin fixed, paraffin embedded breast carcinomas: Comparison of immunohistochemistry and multiparameter DNA flow cytometry. *Anticancer Res.* **23,** 999–1006.

Leers, M., Kolgen, W., Bjorklund, V., Bergman, T., Tribbick, G., Persson, B., Bjorklund, P., Ramaekers, F., Bjorklund, B., Nap, M., Jornvall, H., and Schutte, B. (1999a). Immunocytochemical detection and mapping of a cytokeratin 18 neo-epitope exposed during early apoptosis. *J. Pathol.* **187,** 567–572.

Leers, M., and Nap, M. (2001). Steroid receptor heterogeneity in relation to DNA-index in breast cancer: A multiparameter flow cytometric approach on paraffin embedded tumor samples. *Breast J.* **7,** 249–259.

Leers, M., Schoffelen, R., Theunissen, P., Oosterhuis, J., Bijl, H. V., Rahmy, A., Tan, W., and Nap, M. (2002). Multiparameter flow cytometry as a tool for the detection of micrometastatic tumor cells in the sentinel lymph node procedure. *J. Clin. Pathol.* **55,** 359–366.

Leers, M. P., Schutte, B., Theunissen, P. H., Ramaekers, F. C., and Nap, M. (1999b). Heat pretreatment increases resolution in DNA flow cytometry of paraffin-embedded tumor tissue. *Cytometry* **35,** 260–266.

Leers, M. P., Schutte, B., Theunissen, P. H., Ramaekers, F. C., and Nap, M. (2000). A novel flow cytometric steroid hormone receptor assay for paraffin-embedded breast carcinomas: An objective quantification of the steroid hormone receptors and direct correlation to ploidy status and proliferative capacity in a single-tube assay. *Hum. Pathol.* **31**, 584–592.

Leers, M. P., Theunissen, P. H., Koudstaal, J., Schutte, B., and Ramaekers, F. C. (1997a). Trivariate flow cytometric analysis of paraffin-embedded lung cancer specimens: Application of cytokeratin subtype specific antibodies to distinguish between differentiation pathways. *Cytometry* **27**, 179–188.

Leers, M. P., Theunissen, P. H., Ramaekers, F. C., and Schutte, B. (1997b). Multi-parameter flow cytometric analysis with detection of the Ki67-Ag in paraffin embedded mammary carcinomas. *Cytometry* **27**, 283–289.

Leers, M. P., Theunissen, P. H., Schutte, T. B., and Ramaekers, F. C. (1995). Bivariate cytokeratin/ DNA flow cytometric analysis of paraffin-embedded samples from colorectal carcinomas. *Cytometry* **21**, 101–107.

Letai, A., Coulombe, P. A., and Fuchs, E. (1992). Do the ends justify the mean? Proline mutations at the ends of the keratin coiled-coil rod segments are more disruptive than internal mutations *J. Cell Biol.* **116**, 1181–1195.

Liao, J., Ku, N. O., and Omary, M. B. (1997). Stress, apoptosis and mitosis induce phosphorylation of human keratin 8 at Ser-73 in tissues and cultured cells. *J. Biol. Chem.* **272**, 17565–17573.

Lugo, T. G., Braun, S., Cote, R. J., Pantel, K., and Rusch, V. (2003). Detection and measurement of occult disease for the prognosis of solid tumors. *J. Clin. Oncol.* **21**, 2609–2615.

Lundquist, K., Kohler, S., and Rouse, R. V. (1999). Intraepidermal cytokeratin 7 expression is not restricted to Paget cells but is also seen in Toker cells and Merkel cells. *Am. J. Surg. Pathol.* **23**, 212–219.

Mertens, H. J. M. M., Heineman, M. J., and Evers, J. L. H. (2003). Steroid hormone receptor analysis in human endometrium: Comparison of immunohistochemistry and flow cytometry. *Gynecol. Obstret. Invest.* accepted.

Miettinen, M. (1993). Keratin immunohistochemistry: Update on applications and pitfalls. *Pathol. Annu.* **8**, 113–143.

Moll, R., Franke, W. W., Schiller, D. L., Geiger, B., and Krepler, R. (1982). The catalog of human cytokeratins: Patterns of expression in normal epithelia, tumors and cultured cells. *Cell* **31**, 11–24.

Moll, R., Zimbelmann, R., Goldschmidt, M. D., Keith, M., Laufer, J., Kasper, M., Koch, P. W., and Franke, W. W. (1993). The human gene encoding cytokeratin 20 and its expression during fetal development and in gastrointestinal carcinomas. *Differentiation* **53**, 75–93.

Moreno, J. G., O'Hara, S. M., Gross, S., Doyle, G. V., Fritsche, H., Gomella, L. G., and Terstappen, L. W. M. (2001). Changes in circulating carcinoma cells in patients with metastatic prostate cancer correlate with disease status. *Urology* **58**, 386–392.

Morsi, H. M., Leers, M. P., Radespiel-Troger, M., Bjorklund, V., Kabarity, H. E., Nap, M., and Jager, W. (2000). Apoptosis, bcl-2 expression, and proliferation in benign and malignant endometrial epithelium: An approach using multiparameter flow cytometry. *Gynecol. Oncol.* **77**, 11–17.

Nagle, R. B., McDaniel, K. M., Clark, V. A., and Payne, C. M. (1983). The use of antikeratin antibodies in the diagnosis of human neoplasms. *Am. J. Clin. Pathol.* **79**, 458–466.

Nap, M., Brockhoff, G., Brandt, B., Knuechel, R., Leers, M. P., Schmidt, H., De Angelis, G., Eltze, E., and Semjonow, A. (2001). Flow cytometric DNA and phenotype analysis in pathology. A meeting report of a symposium at the annual conference of the German Society of Pathology, Kiel, Germany, 6–9 June 2000. *Virchows Arch.* **438**, 425–432.

Nowak, R., Oelschlaegel, U., Hofmann, R., Zengler, H., and Huhn, R. (1994). Detection of aneuploid cells in acute lymphoblastic leukemia with flow cytometry before and after therapy. *Leuk. Res.* **18**, 897–901.

Nylander, K., Stenling, R., Gustafsson, H., and Roos, G. (1994). Application of dual parameter analysis in flow cytometric DNA measurements of paraffin-embedded samples. *J. Oral Pathol. Med.* **23**, 190–192.

Ormerod, M. G., Titley, J. C., and Imrie, P. R. (1995). Use of light scatter when recording a DNA histogram from paraffin-embedded tissue. *Cytometry* **21,** 294–299.

Osborn, M., and Weber, K. (1983). Tumor diagnosis by intermediate filament typing: A novel tool for surgical pathology. *Lab. Invest.* **48,** 372–394.

Oud, P. S., Henderik, J. B. J., Beck, H. L. M., Veldhuizen, J. A. M., Vooijs, G. P., Herman, C. J., and Ramaekers, F. C. S. (1985). Flow cytometric analysis and sorting of human endometrial cells after immunocytochemical labeling for cytokeratin using a monoclonal antibody. *Cytometry* **6,** 159–164.

Pantel, K., and Braun, S. (2001). Molecular determinants of occult metastatic tumor cells in bone marrow. *Clin. Breast Cancer* **2,** 222–228.

Pavlidis, N., Briasoulis, E., Hainsworth, J., and Greco, F. A. (2003). Diagnostic and therapeutic management of cancer of an unknown primary. *Eur. J. Cancer* **39,** 1990–2005.

Racila, E., Euhus, D., Weiss, A. J., Rao, C., McConnell, J., Terstappen, L. W. M., and Uhr, J. W. (1998). Detection and characterization of carcinoma cells in the blood. *Proc. Natl. Acad. Sci. USA* **95,** 4589–4594.

Ramaekers, F. C., Beck, H., Vooijs, G. P., and Herman, C. J. (1984). Flow-cytometric analysis of mixed cell populations using intermediate filament antibodies. *Exp. Cell Res.* **153,** 249–253.

Ramaekers, F. C., Huysmans, A., Moesker, O., Kant, A., Jap, P., Herman, C., and Vooijs, P. (1983). Monoclonal antibody to keratin filaments, specific for glandular epithelia and their tumors. Use in surgical pathology. *Lab. Invest.* **49,** 353–361.

Risberg, B., Davidson, B., Dong, H. P., Nesland, J. M., and Berner, A. (2000). Flow cytometric immunophenotyping of serous effusions and peritoneal washings: Comparison with immunohistochemistry and morphological findings. *J. Clin. Pathol.* **53,** 513–517.

Roederer, M. (1996). Compensation (an informal perspective). Available at www.drmr.com/compensation.

Ross, C. W., Hanson, C. A., and Schnitzer, B. (1992). CD30 (Ki-1)-positive, anaplastic large cell lymphoma mimicking gastrointestinal carcinoma. *Cancer* **70,** 2517–2523.

Rupa, J. D., De Bruine, A. P., Gerbers, A. J., Leers, M. P., Nap, M., Kessels, A. G., Schutte, B., and Arends, J. W. (2003). Simultaneous detection of apoptosis and proliferation in colorectal carcinoma by multiparameter flow cytometry allows separation of high and low-turnover tumors with distinct clinical outcome. *Cancer* **97,** 2404–2411.

Sanchey-Carbayo, M., Ciudad, J., and Urrutia, M. (2001). Diagnostic performance of the urinary bladder carcinoma antigen ELISA test and multiparametric DNA cytokeratin flow cytometry in urine voided samples from patients with bladder carcinoma. *Cancer* **92,** 2811–2819.

Schlegel, R., Banks-Schlegel, S., McLeod, J. A., and Pinkus, G. (1980). Immunoperoxidase localization of keratin in human neoplasms. *Am. J. Pathol.* **101,** 41–50.

Schutte, B., Tinnemans, M. M., Pijpers, G. F., Lenders, M. H., and Ramaekers, F. C. (1995). Three parameter flow cytometric analysis for simultaneous detection of cytokeratin, proliferation associated antigens and DNA content. *Cytometry* **21,** 177–186.

Schwartz, A., Marti, G. E., Poon, R., Gratama, J. W., and Fernandez-Repollet, E. (1998). Standardizing flow cytometry: A classification system of fluorescence standards used for flow cytometry. *Cytometry* **33,** 106–114.

Sewell, H. F., Thompson, W. D., and King, D. J. (1986). IgD myeloma/immunoblastic lymphoma cells expressing cytokeratin. *Br. J. Cancer* **53,** 695–696.

Shankey, T. V., Rabinovitch, P. S., Bagwell, B., Bauer, K. D., Duque, R. E., Hedley, D. W., Mayall, B. H., Wheeless, L., and Cox, C. (1993a). Guidelines for implementation of clinical DNA cytometry. International Society for Analytical Cytology. *Cytometry* **14,** 472–477.

Shankey, T. V., Rabinovitch, P. S., Bagwell, B., Bauer, K. D., Duque, R. E., Hedley, D. W., Mayall, B. H., Wheeless, L., and Cox, C. (1993b). Guidelines for implementation of clinical DNA cytometry. International Society for Analytical Cytology [published erratum appears in Cytometry 1993 Oct;14(7):842]. *Cytometry* **14,** 472–477.

Sun, T. T., Eichner, R., Nelson, W. G., Tseng, S. C., Weiss, R. A., Javinen, M., and Woodcock-Mitchell, J. (1983). Keratin classes: Molecular markers for different types of epithelial differentiation. *J. Invest. Dermatol.* **81,** 109s–115s.

Sun, T. T., and Green, H. (1978). Immunfluorescent staining of keratin fibers in cultured cells. *Cell* **14,** 469–476.

Sun, T. T., Skelton, H. G., and Green, H. (1979). Keratin cytoskeletons in epithelial cells of internal organs. *Proc. Natl. Acad. Sci. USA* **76,** 2813–2817.

Tinnemans, M. M., Lenders, M. H., ten Velde, G. P., Ramaekers, F. C., and Schutte, B. (1995). Alterations in cytoskeletal and nuclear matrix-associated proteins during apoptosis. *Eur. J. Cell Biol.* **68,** 35–46.

van Engeland, M., Kuijpers, H. J., Ramaekers, F. C., Reutelingsperger, C. P., and Schutte, B. (1997). Plasma membrane alterations and cytoskeletal changes in apoptosis. *Exp. Cell Res.* **235,** 421–430.

Vindelov, L. L. (1977). Flow microfluorometric analysis of nuclear DNA in cells from solid tumors and cell suspensions. A new method for rapid isolation and straining of nuclei. *Virchows Arch. B Cell Pathol.* **24,** 227–242.

Vindelov, L. L., Christensen, I. J., and Nissen, N. I. (1983). Standardization of high-resolution flow cytometric DNA analysis by the simultaneous use of chicken and trout red blood cells as internal reference standards. *Cytometry* **3,** 328–331.

Wingren, S., Stal, O., Sullivan, S., Brisfors, A., and Nordenskjold, B. (1994). S-phase fraction after gating on epithelial cells predicts recurrence in node-negative breast cancer. *Int. J. Cancer* **59,** 7–10.

Wotherspoon, A. C., Norton, A. J., and Isaacson, P. G. (1989). Immunoreactive cytokeratins in plasmacytomas. *Histopathology* **14,** 141–150.

Yaremko, M. L., Kelemen, P. R., Kutza, C., Barker, D., and Westbrook, C. A. (1996). Immunomagnetic seperation can enrich fixed solid tumors for epithelial cells. *Am. J. Pathol.* **148,** 95–104.

Zarbo, R. J., Visscher, D., and Crissman, J. D. (1989). Two-colour multiparametric method for flow cytometric DNA analysis of carcinomas using staining for cytokeratin and leucocyte-common antigen. *Anal. Quant. Cytol. Histol.* **11,** 391–402.

Zhang, J., Shen, K. W., Liu, G., Zhou, J., Shen, Q., Shen, Z. Z., and Shao, Z. M. (2003). Antigenic profiles of disseminated breast tumour cells and microenvironment in bone marrow. *Eur. J. Surg. Oncol.* **29,** 121–126.

CHAPTER 8

Intermediate Filament Protein Inclusions

Kurt Zatloukal, Conny Stumptner, Andrea Fuchsbichler, Elke Janig, and Helmut Denk

Institute of Pathology
Medical University of Graz
A-8036 Graz, Austria

METHODS IN CELL BIOLOGY, VOL. 78

I. Introduction

A. Overview of Intermediate Filament–Related Inclusions

Cytoplasmic inclusions consisting of abnormally folded proteins are hallmark lesions of a considerable number of human disorders, also designated as protein aggregation or protein misfolding diseases. In general, these inclusions consist of a misfolded and aggregated core protein that is associated with a variety of stress proteins involved in the response to unfolded proteins. Typical examples of cytoplasmic inclusions are neurofibrillary tangles (NFTs) in neurons of patients with Alzheimer's disease, Lewy bodies in neurons of patients with Parkinson disease, Lewy body–like or skeinlike inclusions in amyotrophic lateral sclerosis, Rosenthal fibers in glial cells in Alexander disease, inclusions in skeletal muscle fibers in patients with inclusion body myopathies, Mallory bodies (MBs) in hepatocytes of patients with alcoholic steatohepatitis (ASH), and a variety of nonalcoholic chronic toxic and degenerative liver disorders and hepatocellular neoplasms (for review see Denk *et al.*, 2000; Goebel, 1997; Julien, 1997; Pollanen *et al.*, 1993; Sherman and Goldgerg, 2001).

Intermediate filament (IF) proteins are present in most of these inclusions, and in some types (e.g., MBs), they are the major constituent (Table I). A further common feature of these inclusions is the role of oxidative cell injury in their

Table I

Intermediate Filament Protein Containing Cytoplasmic Inclusions

Inclusion	Intermediate filament	Disease	Cell type
Mallory body	K8, K18	ASH, NASH, Copper toxicosis (including Wilson's disease), HCC, Indian childhood cirrhosis	Hepatocytes
Desmin bodies	Desmin	Desminopathies (desmin storage myopathies)	Skeletal and cardiac muscle fibers
Spheroids, Lewy bodylike inclusions, Bunina bodies	NF	Amyotrophic lateral sclerosis (ALS)	Neurons
Rosenthal fibers	GFAP, vimentin	Alexander's disease, chronic glial scars, and low-grade, fibrillary astrocytomas	Astrocytes
Lewy bodies	NF*	Lewy bodies disease/Parkinson disease	Neurons
NF inclusions	NF	Neuronal intermediate filament inclusion disease (NIFID)	Neurons
Neurofibrillary tangles	NF*	Alzheimer's disease	Neurons

*Neurofilaments contribute to Lewy bodies and neurofibrillary tangles only to a minor extent.

pathogenesis. Oxidative stress is known to induce protein misfolding by modification of amino acid residues. Consequently, proteins either expose hydrophobic domains at their surface, which leads to aggregation by hydrophobic interactions, or form fibrillar polymers by β-sheet structures (Grune *et al.*, 1997). Occurrence of misfolded proteins in cells elicits a response, which involves a variety of stress proteins. These stress proteins bind to misfolded proteins and support refolding (e.g., heat shock protein 70 [HSP70]), prevent aggregation (e.g., HSP25/27, αB-crystallin), mediate their degradation by the proteasome (ubiquitin), or associate with ubiquitinated misfolded proteins as aggregates (p62) (Ehrnsperger *et al.*, 1997; Hartl, 1996; Hershko and Ciechanover, 1998; Shin, 1998; Wisniewski and Goldman, 1998; Zatloukal *et al.*, 2002). In cultured cells, misfolded proteins are transported to the centrosome and deposited as certain structures recently designated as aggresomes (Johnston *et al.*, 1998). Because stress proteins directly interact with and bind to misfolded proteins, various stress proteins are constantly present in most types of cytoplasmic inclusion (Lowe *et al.*, 1988, 1992; Zatloukal *et al.*, 2002).

B. Mallory Body as a Prototype of Intermediate Filament–Related Inclusions

The MB is one of the most common IF-related inclusions in diseases. MB formation is a characteristic feature of liver cell injury in a variety of chronic liver diseases such as ASH and nonalcoholic steatohepatitis (NASH), chronic cholestatic conditions, hepatocellular carcinomas, and certain metabolic disorders, including copper intoxication (for review see Denk *et al.*, 2000). MBs can also be produced in mouse liver by chronic intoxication with the fungistatic antimicrotubular drug griseofulvin (GF) or the porphyrogenic agent 3,5-diethoxycarbonyl-1,4-dihydrocollidine (DDC) (Denk *et al.*, 1975; Tsunoo *et al.*, 1987).

Human and murine MBs are typically seen in enlarged (ballooned) hepatocytes with a deranged or even lost keratin IF network. MBs contain keratins, (predominantly K8), the stress-inducible $M_M 120-1$ antigen, the ubiquitin binding protein p62, HSPs with low and high molecular weights (HSP25/27, αB-crystallin and HSP70), and ubiquitin (Denk *et al.*, 1982; Lowe *et al.*, 1988, 1992; Ohta *et al.*, 1988; Zatloukal *et al.*, 1990, 2002; Fig. 1).

1. Model Systems to Study Mallory Bodies

MB formation can be induced in mouse livers by feeding a GF- or DDC-containing diet. Both substances are metabolized by cytochrome P450–mediated *N*-demethylation, which leads to the generation of methyl radicals. The methyl radicals bind to the heme moiety of cytochromes, which leads to the formation and accumulation *N*-methylprotoporphyrin (Tephly *et al.*, 1981). It is assumed that the oxidative injury induced by the methyl radical is the common pathogenetic principle in GF- or DCC-fed animals and human livers with ASH or NASH, in which free radicals produced by cytochrome P450–mediated oxidation of ethanol and the

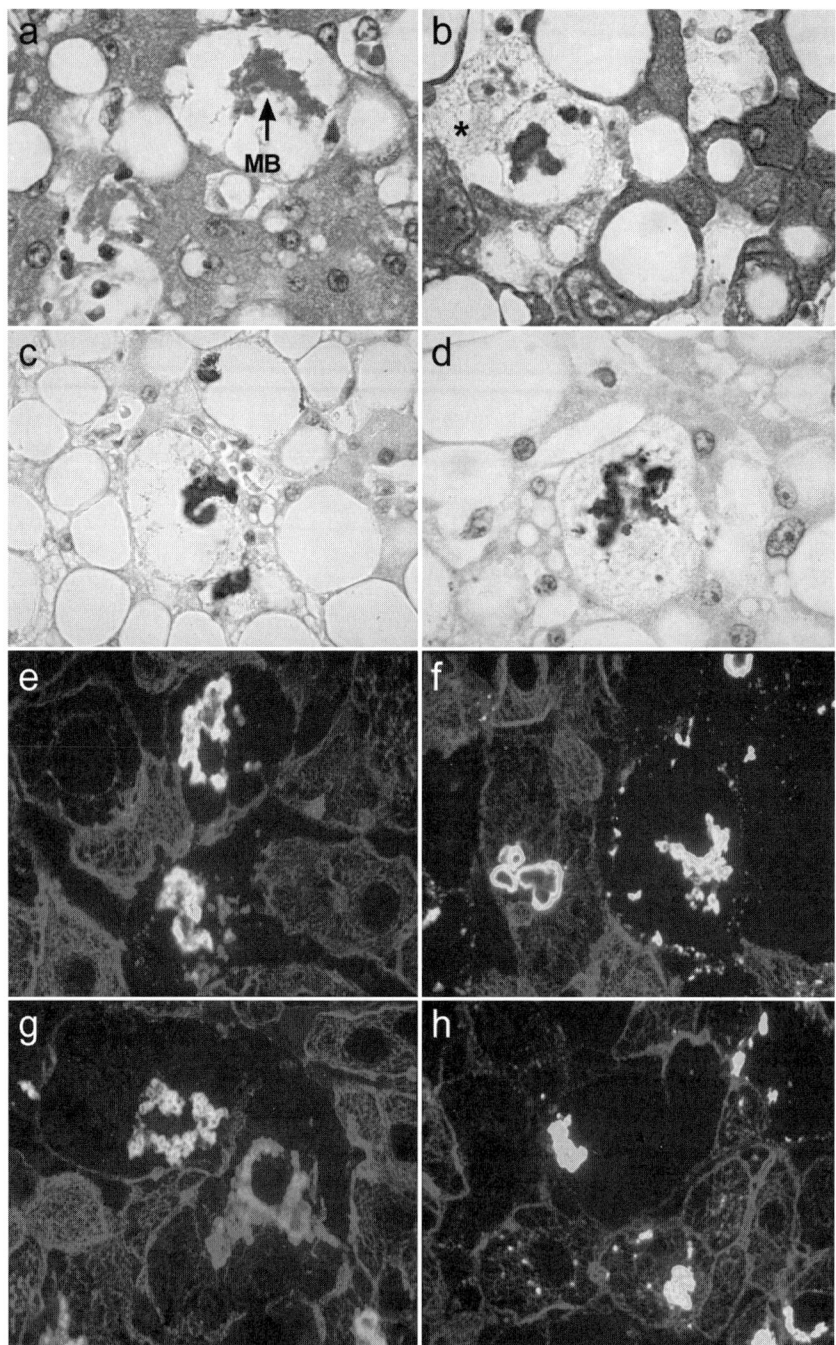

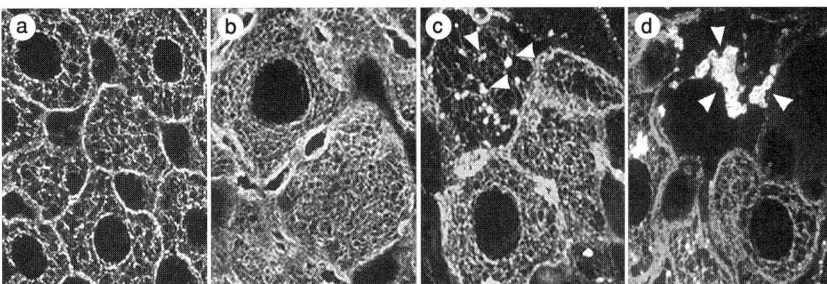

Fig. 2 Immunofluorescence microscopy with the polyclonal antibody 50K160 to K8/K18 on frozen mouse livers at different stages of DDC intoxication. (a) Normal liver; (b) 1 week DDC intoxication; (c) 2 months DDC intoxication; (d) 2.5 months DDC intoxication. Note that at 1 week of DDC intoxication, hepatocytes are enlarged with a denser keratin intermediate filament (IF) network. With prolongation of DDC intoxication, Mallory bodies appear (arrowheads in c and d), and hepatocytes reveal a deranged or even lost keratin IF network.

mitochondrial injury caused by acetaldehyde and free fatty acid overload are central features (Angulo, 2002; Lieber, 2000). Mouse livers respond to GF or DDC intoxication first with ballooning of hepatocytes and formation of a denser keratin IF network (Fig. 2). After approximately 6 weeks of intoxication-ballooned hepatocytes show a reduced density of the keratin IF, and early MBs can be observed as fine granules associated with the keratin IF network. Continuation of intoxication leads to the appearance of large MBs typically located in the perinuclear cytoplasmic region. Most hepatocytes containing large MBs have a markedly reduced or even undetectable cytoplasmic IF keratin network. On cessation of intoxication, MBs disappear within several weeks. At 4 weeks of recovery from intoxication, groups of hepatocytes are devoid of cytoplasmic keratin filaments but still contain small remnants of MBs at the cell periphery in association with desmosomes. If mice are reexposed to GF or DDC, numerous MBs reappear within 24 to 72 hours (Stumptner *et al.*, 2001; Yuan *et al.*, 2000). This enhanced formation of MBs on reintoxication was interpreted—in analogy to allergic reactions—as a toxic memory effect.

Fig. 1 Mallory bodies (MBs) in a human liver with alcoholic steatohepatitis (a–d). (a) Hematoxylin-eosin–stained section (MB in a ballooned hepatocytes is indicated by arrow). (b) Immunohistochemistry on paraffin-embedded tissue with the monoclonal keratin antibodies to K8 and K18 (K8.8 and DC-10, NeoMarkers). (c) Immunohistochemistry with antibody p62CT. (d) Immunohistochemistry with a polyclonal antibody to ubiquitin (Dako). Note that MBs are present in ballooned hepatocytes with a loosened or even undetectable keratin intermediate filament (IF) network (ballooned hepatocyte with no detectable keratin IF is indicated by asterisk in (b). Double-label immunofluorescence microscopy on frozen mouse livers after 2 months of 3,5-diethoxycarbonyl-1,4-dihydrocollidine (DDC) intoxication (e–h) with combinations of the polyclonal antibody to K8/K18 (50K160, in red) and antibodies to nonkeratin MB components (in green). (e) antibody to ubiquitin; (f) antibody to p62; (g) antibody to SMI 31; (h) antibody M$_M$120–1. (See Color Insert.)

In GF- or DDC-fed mice, the alterations of the IF keratin cytoskeleton and structure and chemical composition of MBs are similar, if not identical, to the alterations found in human ASH and NASH (Denk *et al.*, 2000) (Fig. 1). In this context it is noteworthy that other mouse models for alcoholic liver disease on the basis of feeding alcohol-containing diets reproduce the disturbance of fat metabolism and, to some degree, inflammation of human ASH but not the alterations of the keratin IF cytoskeleton and do not lead to MB formation.

Formation of cytoplasmic keratin aggregates can also be achieved in cultured cells by transfection and imbalanced overexpression of type I and II keratins. Overexpressed keratin that cannot be assembled into IF becomes ubiquitinated and accumulates as cytoplasmic aggregates (Ku and Omary, 2000; Zatloukal *et al.*, 2002) (Fig. 3). Occurrence of misfolded keratin also leads to overexpression of p62 and the M_M120-1 antigen, both of which associate with the keratin aggregates. This *in vitro* model is suitable to study the role and fate of improperly assembled keratins in the response to unfolded proteins.

2. Chemical Composition of Mallory Bodies

Biochemical and immunochemical analyses of MBs revealed keratin as the most prominent MB component. In contrast to IF, the equimolar ratio of the type I keratin (K18) and type II keratin (K8) is not maintained in MBs, and K8 prevails over K18 (Zatloukal *et al.*, 1991). Time course studies in DDC-intoxicated mice demonstrated that in hepatocytes keratin expression increases four-fold at 3 days of intoxication and that mRNA concentrations of K8 were higher than that of K18 (Stumptner *et al.*, 2001). This overexpression of keratin is apparently involved in a defense reaction to toxic injury, because hemizygous K8 knock-out mice, which have reduced capacities to react with K8 overexpression, were more sensitive to intoxication than wild-type mice (Zatloukal *et al.*, 2000). Comparative studies in DDC-intoxicated K8 null and K18 null mice demonstrated that K8 is the essential protein for MB formation, because in the absence of K8, no MBs were formed. In contrast, K18 null mice, which still express K8, formed MBs even spontaneously at a high age (Magin *et al.*, 1998). It can be assumed that if K8 exceeds the amounts of its partner K18, prevailing K8 cannot be assembled as IF and is highly sensitive to oxidative stress–induced misfolding. Therefore, situations leading to impaired keratin homeostasis and an excess of K8 predispose to MB formation.

In addition to keratins, a variety of nonkeratin components were identified in MBs by immunohistochemistry, immunoblot, and mass spectrometry analyses (Table II). The nonkeratin MB components include p62, ubiquitin, high and low molecular weight HSPs, and the as yet not further characterized M_M120-1 antigen (Zatloukal *et al.*, 1990, 2002). Furthermore, αB-crystallin, a member of the small HSP family and common component of several other cytoplasmic inclusions, was detected in MBs (Lowe *et al.*, 1992).

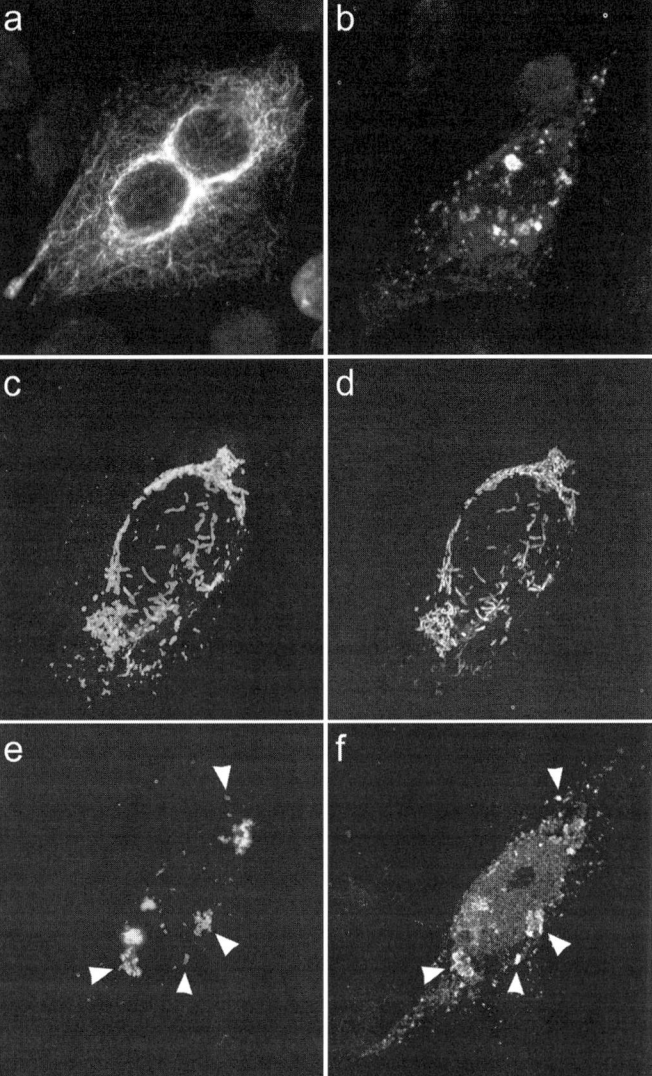

Fig. 3 Double-label immunofluorescence microscopy of transfected CHOK1 cells with the antibody 50K160 to K8/K18 and antibody p62CT. (a) CHOK1 cells cotransfected with K8 and K18 leading to formation of a regular intermediate filament (IF) network. (b) CHOK1 cells transfected with K8 resulting in keratin aggregates. (c and d) CHOK1 cells transfected with N-terminal GFP-tagged K8 and K18. (c) Visualization of GFP-fluorescence; (d) same cell stained with the antibody 50K160 to K8/K18. Note that the GFP-tag influenced IF assembly in that shorter filaments were formed. (e and f) CHOK1 cells cotransfected with K8 and ubiquitin. The same cell is shown in panels e (K8) and f (ubiquitin). Note that most K8 aggregates are heavily ubiquitinated (corresponding aggregates are indicated by arrowheads).

Table II
Antibodies to Mallory Body Components

Mallory body component	Clone	Antibody	Source	Reactivity
Keratin 8	Ks 8.7	mAb	Progen	H, M
Keratin 18	Ks 18.04	mAb	Progen	H, M
Keratin 8	K8.8	mAb	Neomarkers	H, P
Keratin 18	DC-10	mAb	Neomarkers	H, P
Keratin 8+18	50K160	Rabbit	K. Zatloukal *et al.*	H, M, P
Keratin 8 pSer 431	5B3	mAb	Neomarkers	H, M
Keratin 8 pSer 73	LJ4	mAb	Neomarkers	H, M
Keratin 18 pSer 33	IB4	mAb	Neomarkers	H, M
p62 (sequestosome)	p62CT	Guinea pig	K. Zatloukal/H. Heid; Progen	H, M, P
p62 (A170, sequestosome)	3	mAb	BD Biosciences	H, P
Phospho-epitope (on NF, tau)	SMI31	mAb	Sternberger Monoclonal Inc.	H, M, P
Phospho-epitope (on NF, tau)	SMI34	mAb	Sternberger Monoclonal Inc.	H, M, P
Phospho-epitope	MPM-2	mAb	Upstate Biotechnology	H, M, P
To be determined	M_M120-1	mAb	W. W. Franke	H, M
Ubiquitin	—	Rabbit	Dako	H, M, P
Ubiquitin	FPM1		ID Labs Inc.	H, M
hsp70	8B11	mAb	Novocastra	M
hsp25	—	Rabbit	StressGene	M

H, antibody recognizing human Mallory bodies; M, antibody recognizing murine MBs; P, antibody that can be used for paraffin-embedded tissues. Commercial providers of antibodies: Progen, www. progen.com; Neomarkers, www.labvision.com; BD Biosciences, www.bdbiosciences.com; Sternbergen, home.att.net/usternbmonoc/home; Upstate, www.upstatebiotech.com; ID Labs Inc., http://www.idlabs. com/INDXMAIN.HTM; Novocastra, www.novacastra.co.uk, Stress Gene, www.stressgen.com.

P62 has been originally identified as a phosphotyrosine-independent ligand of the SH2 domain of p56[lck] and as a cytoplasmic nonproteasomal ubiquitin-binding protein (Joung *et al.*, 1996; Vadlamudi *et al.*, 1996). It is also involved in tumor necrosis factor (TNF)-α signaling and activation of NF-κB by linking RIP to PKCζ and IKK-β (Sanz *et al.*, 1999). A general role of p62 in the cellular stress response is implied, because p62 expression is increased by a variety of stress stimuli, particularly oxidative stress (Ishii *et al.*, 1996). Because p62 is involved in NF-κB signaling, it could have, on the one hand, an impact on the activation of survival pathways in stress situations. On the other hand, p62 might influence the intracellular fate of misfolded and ubiquitinated proteins. A role of p62 in the cellular response to misfolded proteins is further underlined by the observation that in cells treated with proteasome inhibitors, p62 is a constant component of the aggresome (Stumptner *et al.*, unpublished observation).

In MBs, p62 is also recognized by the antibody SMI 31, which is directed against a phosphorylated epitope also present on neurofilaments and on abnormal tau protein in NFTs in Alzheimer disease (Zatloukal *et al.*, 2002). The common phosphoepitope of p62 in MBs and NFTs implies that similar protein kinases (e.g., proline-directed kinases) are active in both conditions. Another similarity in the phosphorylation state of p62 in MBs and NFTs is the common reactivity with MPM-2 antibodies (Stumptner *et al.*, 2001). The MPM-2 antibody is directed to hyperphosphorylated epitopes generated on diverse proteins during the M-phase of the cell cycle by the action of kinases, like p34 cdc2-, and MAP-kinase (Westendorf *et al.*, 1994).

Ubiquitin is another common denominator of MBs and NFTs and other cytoplasmic protein aggregates, such as Lewy bodies in Parkinson disease, Rosenthal fibers in astrocytomas, and inclusion bodies in motor neuron disease (Lowe *et al.*, 1988).

Besides ubiquitin, other stress proteins, such as αB-crystallin and HSP70, associate with MBs or are induced under conditions leading to MB formation. HSP70 belongs to a family of ubiquitous chaperone proteins that assist in protein folding processes in the course of protein synthesis but also promote refolding of misfolded proteins (Hartl, 1996). HSP70 recognizes, and binds to, exposed hydrophobic surfaces of proteins in an adenosine triphosphate (ATP)–dependent manner. Multiple cycles of binding and detachment of HSP70 facilitate proper folding of proteins.

The M_M120-1 antigen is a MB component with a high molecular weight, which was identified with a monoclonal antibody raised against purified MBs (Zatloukal *et al.*, 1990). *In vitro* studies revealed that M_M120-1-protein can be induced in tissue culture cells by a variety of stress treatments (e.g., Ca-ionophore, sodium arsenite, or heat shock). The biochemical nature of the M_M120-1 antigen has not yet been characterized.

The high molecular weight nature of the M_M120-1 antigen and the occurrence of high molecular weight keratin polypeptides that cannot be dissociated even under denaturing conditions of sodium dodecylsulfate (SDS)-gel electrophoresis suggest modification of MB components by covalent cross-linking (Fig. 4). It has been shown that K8 and K18 are substrates of transglutaminase, a Ca^{++}-dependent cross-linking enzyme, and that MBs contain high amounts of ε-(γ-glutamyl) lysine cross-links resulting from transglutaminase action (Zatloukal *et al.*, 1992). Other mechanisms that could be responsible for cross-linking and stabilization of MB proteins are the formation of aldehyde adducts and the binding of mutated ubiquitin that cannot be cleaved by the proteasome (Bardag-Gorce *et al.*, 2003).

In human ASH and in DDC-fed animals, phosphorylation of keratins at multiple sites and accumulation of phosphorylated keratin in MBs was shown with antibodies that selectively recognize phosphorylated epitopes of K8 or K18. Hyperphosphorylation of keratin occurred within 1 day of DDC intoxication and preceded architectural changes of the IF cytoskeleton. In long-term DDC-intoxicated mice with MB-containing livers, phosphorylated keratins were preferentially associated with MBs but not with the residual keratin IF network adjacent to MBs (Stumptner *et al.*, 2000). Phosphorylation occurred at sites typical for

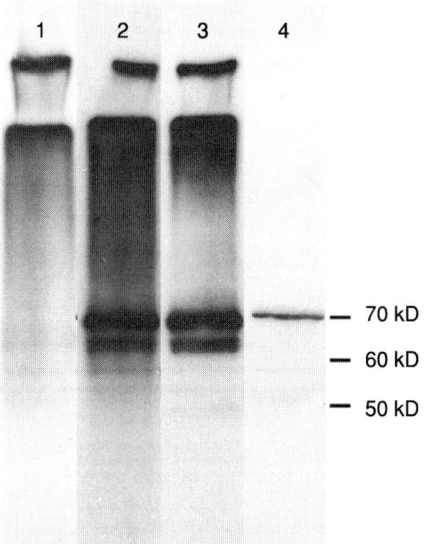

Fig. 4 Western blot analysis of Mallory bodies (MBs) isolated from 3,5-diethoxycarbonyl-1,4-dihydrocollidin–treated mouse liver. Lane 1, antibody to ubiquitin; lane 2, antibody p62CT; lane 3, antibody SMI 31 to phosphorylated p62; lane 4, antibody to HSP70. Note that a major proportion of MB proteins is of high molecular weight and not separated in the sodium dodecylsulfate gel.

stress-activated kinases, such as JNK, and kinases active in mitosis and apoptosis (Liao *et al.*, 1997).

3. Structure of Mallory Bodies

MBs range in size from small granules to large (up to 10 μm in diameter) irregularly shaped dense masses. At the ultrastructural level, MBs consist of 10- to 20-nm thick filaments coated by fuzzy material (Franke *et al.*, 1979). According to their ultrastructural appearance, different types of MBs are distinguished (type I, bundles of filaments in parallel arrays; type II, randomly oriented filaments; type III, granular and amorphous material; Yokoo *et al.*, 1972). The filamentous ultrastructure of MBs led to the hypothesis that MBs originate from a collapse of IF. This hypothesis, however, was refuted by the fact that MB filaments differ in several aspects from IF (Fig. 5) (Fuchs and Weber, 1994; Herrmann *et al.*, 2003): (1) MB filaments and IF differ in diameter, ultrastructure, and arrangement. (2) The equimolar ratio of K8 and K18 is not maintained in MBs. (3) A variety of polyclonal and monoclonal keratin antibodies that recognize conformation-dependent epitopes revealed that keratins in MBs are in a different conformational state than in IF (Hazan *et al.*, 1986). (4) Infrared spectroscopy revealed increased β-sheet conformation in MBs compared with IF (Cadrin *et al.*, 1991).

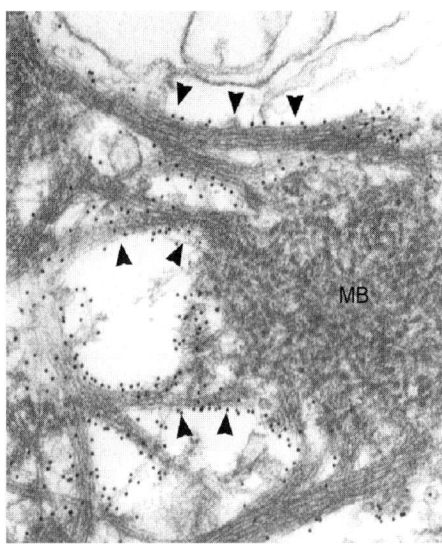

Fig. 5 Immunoelectron microscopy of Mallory bodies (MBs) in 3,5-diethoxycarbonyl-1,4-dihydrocollidine–intoxicated mouse liver. Binding of antibody 50K160 to K8/K18 is visualized by 12-nm gold-conjugated secondary antibodies. Note the different structure and arrangement of intermediate filament (arrowheads) and MB filaments (MB).

4. Biological Relevance of Mallory Bodies

For most diseases, it is still unclear whether cytoplasmic inclusions are instrumental in cell damage. Important clues on the role of MBs came from comparative studies in K8 and K18 null mice. The two types of knock-out mice revealed striking differences in their response to DDC intoxication. K8 null mice were more sensitive to intoxication as reflected in increased lethality and more pronounced alteration of laboratory parameters compared with K18 null or wild-type mice (Zatloukal *et al.*, 2000). Furthermore, no MBs were detectable in DDC-intoxicated K8 null mice, whereas K18 null mice showed enhanced MB formation after short-term DDC feeding (Stumptner *et al.*, manuscript submitted). On the basis of this inverse relationship between MB formation and toxicity, a detrimental effect of MBs on hepatocytes can be excluded.

Rather, formation of MBs can be regarded as a product of the cellular response to the occurrence of abnormally folded proteins. This hypothesis is supported by the results obtained by mass spectrometry analysis of isolated MBs. The analysis revealed that MBs are composed, besides of keratin, predominantly of proteins participating in the response to unfolded proteins (Zatloukal *et al.*, 2002). One may conclude from these data that keratins, particularly K8, are preferred targets for protein unfolding in situations of oxidative stress, and MBs are a certain product of the response to unfolded (misfolded) proteins.

Furthermore, keratins may act as general modulators of toxic liver injury, because transgenic mice harboring specific mutations in keratin genes showed increased sensitivity to various toxic injuries, and mutations in the K8 and K18 genes were recently identified in human livers with cirrhosis, suggesting keratins as important susceptibility genes for chronic liver diseases (for review see Omary *et al.*, 2002).

II. Materials and Instrumentation

A. Experimental Induction of Mallory Bodies in Animals

Instrumentation: Animal isolators (BIO.A.S., Ehret, Emmendingen, Germany)

Animals: Male Swiss Albino mice (strain Him OF1 SPF; Institute of Laboratory Animal Research, University of Vienna, Himberg, Austria), K8 +/− mice in the FVB/N background were obtained from Helene Baribault and Robert Oshima, The Burnham Institute, La Jolla, CA (Baribault *et al.*, 1994), K18 null mice in the 129 ola background were obtained from Thomas Magin, University of Bonn, Germany (Magin *et al.*, 1998).

Reagents: Standard diet (Sniff Spezialdiäten GmbH, Soest, Germany), GF (Griseofulvin, Cat.no. 85,644–4, Sigma-Aldrich, Steinheim, Germany), DDC (3,5-diethoxycarbonyl-1,4-dihydrocollidine see also 1,4-dihydro-2,4,6-trimethyl-pyridin-3,5-dicarbonicaciddiethylester, Cat.no. 13703-0, Sigma-Aldrich). The standard diet containing 2.5% GF or 0.1% DDC was produced as pellets by Sniff (Sniff Spezialdiäten).

B. Induction of Keratin Aggregates in Cultured Cells

Cells: Chinese hamster ovary cells (CHOK1, ATCC No.: CCL 61).

Reagents: Ham's F12 (Cat.no. N-6760, Sigma-Aldrich) with 10% fetal calf serum (FCS) (F-7524, Sigma-Aldrich), trypsin ethylenediamine tetraacetic acid (EDTA) solution (Cat.no. T4049, Sigma-Aldrich), expression vectors for K8 and K18: (hK8-pcDNA3; expression vector pcDNA3 [Invitrogen] harboring the cytomegalo virus [CMV] promotor containing the cDNA of human K8 [NCBI accession number: BC000654; from pos. 65 to pos. 1529], hK18-pcDNA3; expression vector pcDNA3 containing the cDNA sequence of human K18 [NCBI accession: BC020982; from pos. 48 to pos. 1349], pEGFP-C1-hK8; expression vector pEGFP-C1 [Clontech] harboring the CMV promotor and containing the EGFP cDNA followed by the cDNA sequence of human K8 [NCBI accession number: BC000654; from pos. 65 to pos. 1529], pcDNA4/HisMaxC-hUbi; expression vector pcDNA4/HisMaxC [Invitrogen] harboring the CMV promotor containing a cDNA sequence of human ubiquitin [NCBI accession: M26880; from pos. 1960 to pos. 2193]); lipofectamine 2000 (Invitrogen, Groningen, The Netherlands).

Materials: A 75-cm^2 culture flask (Corning, Wiesbaden, Germany), 24-well culture plates (Corning), round coverslips (13 mm in diameter, Assistant, Sondheim/Rhön, Germany).

C. Isolation of Mallory Bodies

Instrumentation: Potter-Elvehjem homogenizator with Teflon pestle (Braun, Melsungen, Germany), ultracentrifuge LE 80 Beckman and SW28 rotor (Beckman, Fullerton, CA), refrigerated laboratory centrifuge (CPKR, Beckman), refrigerated centrifuge for 1.5-ml tubes (Z252MK, Hermle, Wehingen, Germany), refractometer (ABBE refractometer AR10, Schmidt Haensch, Berlin, Germany).

Reagents: Saccharose-tris-EDTA (STE) buffer (250 mM saccharose, 5 mM Tris, 5 mM EDTA (Na$_4$), pH 7.4); high salt buffer (HS buffer; 1.5 M KCl, 1% Triton X100, 10 mM Tris, pH 7.5); Protease Inhibitor Cocktail (for mammalian tissue, Cat. No 8340, Sigma-Aldrich, Austria) was added to all buffers immediately before use; all other buffer components were from Merck (Merck, Darmstadt, Germany).

Sucrose Solutions (w/v; solubilize in 10 mM Tris pH 7.4)

%	g/ml	Refraction index
65%	195 g/300 ml	1.424
70%	210 g/300 ml	1.431
75%	225 g/300 ml	1.438
80%	320 g/400 ml	1.445
85%	255 g/300 ml	1.452

Boil all sucrose solutions to dissolve and filter the solution through folded filter papers into sterilized glass bottles. Adjust refraction index of each solution (with a refractometer).

D. Immunoelectron Microscopy

Instrumentation: Cryocut (Leica CM3050, Leica, Nußloch, Germany), Ultramicrotome (Ultracut; Reichert-Jung, Nußloch, Germany), Electron microscope CM 100 (Philips Electronic Instruments Co., Eindhoven, The Netherlands).

Reagents: Polyclonal rabbit antibody to K8 and K18 (50K160; produced in our laboratory against chromatographically purified mouse liver keratins), swine anti rabbit immunoglobulins, 12-nm gold-conjugated (Jackson Immune Research, West Grove, PA), agar 100 resin (Agar Plano, Essex, U.K.), cacodylate (Fluka, Buchs, CH), glutaraldehyde (Agar Plano), O$_s$O$_4$ (Agar Plano), propylene oxide (Roth, Karlsruhe, Germany), uranyl acetate (Merck), lead citrate (Merck), gelatin capsules (Agar Plano), Formvar-coated grids (Agar Plano), cacodylate-buffer (0.1 M Na-cacodylate, pH 7.3).

E. Immunofluorescence Microscopy

Instrumentation: Cryocut (Leica CM3050, Leica, Nußloch, Germany), incubation chamber (Bioassay plate, Nunc, Roshilde, Denkmark), laser scanning microscope (LSM510, Zeiss, Oberkochen, Germany).

Reagents: Phosphate-buffered saline (PBS), (50 mM potassium phosphate, 150 mM NaCl, pH 8.0–8.5): prepare potassium phosphate by adding solution

B (0.5 M KH$_2$HPO$_4$) to solution A (0.5 M K$_2$HPO$_4$) until pH is 8.0–8.5, then add NaCl to obtain a final concentration of 150 mM NaCl; Mowiol (17% Mowiol 4–88 [Calbiochem Nr. 475904], 34% glycerol in PBS): dissolve 5 g Mowiol in 20 ml PBS and stir overnight at room temperature, add 10 ml of glycerol (100%), stir again overnight at room temperature, centrifuge for 15 minutes at 12,000 rpm (Sorvall, SS-34 Rotor) to clarify solution, transfer supernatant, prepare aliquots, and store mounting medium at 4 °C.

Primary antibodies: SMI 31 (detecting an abnormally phosphorylated epitope on tau protein present in paired helical filaments in Alzheimer disease and hyperphosphorylated neurofilaments; 1:1000; Sternberger Monoclonals Inc., Baltimore, MD), p62CT (polyclonal guinea pig antibody against C-terminal peptide sequence of p62; Zatloukal *et al.*, 2002), antibodies to K8 (Ks 8.7, Progen, Heidelberg, Germany), K18 (Ks 18.04, Progen), K8/18 (50K160), and ubiquitin (ID Labs Inc., London, ON, Canada).

Secondary antibodies: Fluorescein isothiocyanate (FITC)–conjugated goat anti-mouse IgG (Zymed, San Francisco, CA) or Alexa 488-nm)–conjugated goat anti-mouse IgG (Molecular Probes, Leiden, The Netherlands), tetramethylrhodamine isothiocyanate (TRITC)– or FITC-conjugated swine anti-rabbit Ig (Dako, Glostrup, Denmark), and TRITC-conjugated rabbit anti-guinea pig Ig (Dako).

F. Immunohistochemistry

Instrumentation: Microwave oven (conventional household microwave oven with energy control), incubation chamber (Nunc).

Reagents: PBS (see earlier), protease (type XXIV, Sigma Steinheim, Germany), H$_2$O$_2$ (Merck), 3-amino-9-ethylcarbazole (AEC; Dako), Mayer's haemalaun (Merck), secondary antibodies: rabbit anti-guinea pig peroxidase-conjugated (Dako), Multi Link swine anti-goat, mouse, rabbit immunoglobulins (Dako), detection system: Strept ABComplex/HRP (Dako), TSA Biotin System (NEN, Boston, MA). P62CT antibody binding is detected with the TSA Biotin System. Reactivities of SMI 31, ubiquitin, and K8/18 antibodies are detected with the Strept ABComplex system (Dako), Aquatex (Merck).

III. Procedures

A. Experimental Induction of Mallory Bodies in Animals

MBs can be induced in mouse livers by chronic intoxication of various mouse strains with GF or DDC. GF treatment leads to death of approximately 10% of the animals within a 2-month feeding period, whereas DDC is well tolerated. Because the genetic background of mice has potential effects on keratin-associated phenotypes, comparative studies should be performed in the same background.

Therefore, heterozygous K8+/− FVB/N mice were mated with 129 ola wild-type mice over five generations. K18−/− (129 ola) male mice were mated with K18−/− (129 ola) females. Animals are kept in sterile isolators with a 12-hour day-night cycle. Animals receive humane care according to the criteria outlined in the *Guide for the Care and Use of Laboratory Animals* prepared by the National Academy of Sciences and published by the National Institutes of Health (NIH publication 86–23, revised 1985).

Steps:

1. Mice (8 weeks old) are fed a standard diet containing either 0.1% DDC or 2.5% GF for up to 2.5 months.

2. Mice are killed at different time points of intoxication by cervical dislocation, and the livers are either immediately snap-frozen in methylbutane precooled with liquid nitrogen for gel electrophoretic analysis, Western and Northern blotting, immunofluorescence and immunoelectron microscopy, or fixed in 4% buffered formaldehyde for routine histological and immunohistochemical studies.

B. Induction of Keratin Aggregates in Cultured Cells

Keratin aggregates can be induced in cultured cells by transfection with keratin expression gene constructs if this results in imbalanced overexpression of type I and II keratin pairs. Other possibilities are the expression of mutated (assembly incompetent) keratins. Furthermore, certain keratin–GFP fusions tend to aggregate instead of forming IF. Keratin not properly assembled as IF is rapidly degraded, so that only a few of the transfected cells contain keratin aggregates. The number of aggregate-containing cell increases if strong promoters (e.g., CMV promoter) are used or the proteasome is inhibited. Keratin aggregates are targets for the response to unfolded proteins, so that several stress proteins (ubiquitin, HSPs, p62) associate with keratin aggregates (Zatloukal *et al.*, 2002; Fig. 3).

Steps:

1. One day before transfection, trypsinize cells with 0.25% trypsin-EDTA solution and seed 0.5–2 × 10^5 cells in 500 μl culture medium without antibiotics onto sterile round glass coverslips in a 24-well culture plate.

2. For each transfection sample, prepare Lipofectamine 2000 complexes as follows:

   a. Dilute 0.7–2 μg DNA in 37.5 μl medium without serum and mix gently.

   b. Mix 2–3 μl Lipofectamine 2000 gently before use, then dilute in 37.5 μl medium without serum, and mix gently. Incubate for 5 minutes at room temperature and combine the diluted DNA with the diluted Lipofectamine 2000, mix gently. Incubate for 20 minutes at room temperature to allow formation of DNA-Lipofectamine 2000 complexes.

   c. Meanwhile, wash cells with 500 μl medium without serum.

   d. After complex formation, add 112.5 μl medium without serum to the DNA-Lipofectamine 2000 complexes.

   3. Remove the medium from the cells and apply 200 μl of DNA-Lipofectamine 2000 complexes to each well. Mix gently by rocking the plate.

   4. Culture cells at 37 °C in a CO_2 incubator for 3 hours and replace culture medium with medium containing serum. Culture transfected cells for further 18–24 hours.

   5. For immunofluorescence microscopy, wash coverslip onto which cells were grown in PBS and fix in −20 °C acetone-methanol (1:1) for 10 minutes. Rinse in PBS and process coverslips further as described in section E2.

C. Isolation of Mallory Bodies

   The principle of the procedure for isolation and purification of MBs is based on the dense structure of the MBs and poor solubility of MB proteins. The protocol was adapted from the method described by Franke (1979). All steps should be performed on ice with precooled buffers with protease inhibitor added. For one MB preparation (six sucrose gradient tubes), one can use 8–10 g liver tissue (i.e., 1.5 g liver tissue per tube). Do not load more liver tissue for one gradient, because this might exceed the capacity of the sucrose gradient, resulting in impure MB fractions. Time required for the protocol: 2 days.

   Steps:

   1. Kill animals by cervical dislocation (or decapitation).

   2. Remove the liver and determine liver and body weights.

   3. Rinse the whole liver in ice-cold wash buffer for a few seconds.

   4. Transfer liver into a beaker with ice-cold STE buffer, and mechanically mince the liver into small pieces with scissors.

   5. Dilute the tissue with STE buffer to reach a final concentration of 2.5% (w/v) and homogenize in STE buffer (on ice) with a Potter-Elvehjem homogenizator at 600 rpm.

   6. After filtration through four layers of cheesecloth/glass wool, centrifuge the homogenate for 10 minutes at 4000 rpm (in 50-ml tubes) at 4 °C.

   7. Discard the supernatant and resuspend the pellet in ice-cold STE buffer by repeated pipetting. Fill the tubes with STE buffer.

   8. Centrifuge again for 10 minutes at 4000 rpm at 4 °C.

   9. Repeat steps 7 and 8 four times.

   10. Discard the supernatant. Resuspend and pool all pellets in ice-cold STE and fill the tubes with STE buffer.

   11. Centrifuge again for 10 minutes at 4000 rpm at 4 °C.

12. Prepare the sucrose gradient in 36-ml ultracentrifuge tubes (for SW 28 rotor Beckman). Start with 85% sucrose solution and carefully overlay with 80% to 65% sucrose (as in indicated in Fig. 6a).

13. Discard the supernatant from the last centrifugation step (step 11). Resuspend the liver homogenate pellet in 10 mM Tris-HCl, pH 7.4, and 80% sucrose solution by use of a glass Dounce, and adjust the homogenate to a final sucrose concentration of ∼60% sucrose.

14. Overlay the prepared sucrose gradient with this sucrose-homogenate solution.

15. Centrifuge the gradient for 60 minutes at 25,000 rpm in an ultracentrifuge (4 °C). Use a density gradient run modus for low acceleration and deceleration.

16. After centrifugation, collect the material accumulated between 70% and 75% and 75% and 80% sucrose, and pool the fractions on ice (these layers contain MB material as indicated in Fig. 6b). Bring to a final volume of 180 ml with 10 mM Tris-HCl, pH 7.4.

17. Mix and transfer into 36-ml centrifuge tubes. Centrifuge for 40 minutes at 20,000 rpm (4 °C) to pellet MBs.

18. Pool all MB pellets in a 1.5-ml reaction tube with 10 mM Tris-HCl, pH 7.4, and centrifuge for 15 minutes at 13,000 rpm (4 °C). Remove the supernatant and collect pellets for high salt extraction or store pellets at −20 °C until further processing.

19. For high-salt extraction, pool all pellets in 180 ml HS buffer and stir on a magnetic stirrer at 4 °C overnight.

20. Transfer the solution into 36-ml centrifuge tubes, and centrifuge for 90 minutes at 25,000 rpm (4 °C) in a ultracentrifuge to pellet MBs.

21. Pool the MB-pellets into a 1.5-ml reaction tube with 10 mM Tris-HCl, pH 7.4, and centrifuge for 15 minutes at 13,000 rpm (4 °C). Remove the supernatant and store the MB pellets for further analysis at −20 °C.

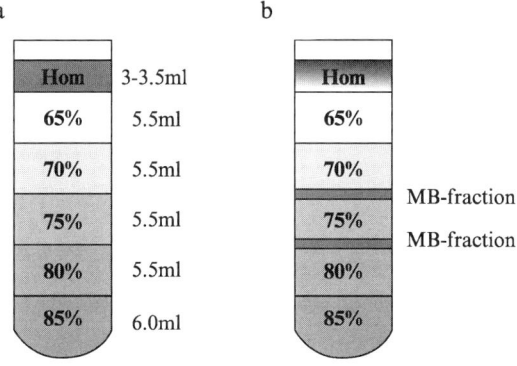

Fig. 6 Sucrose gradients for Mallory body (MB) isolation. (a) Gradient loaded with homogenate; (b) gradient after centrifugation with MB-enriched fractions indicated.

D. Immunoelectron Microscopy

For ultrastructural analysis of MBs, a preembedding immunoelectron micros-copy procedure is recommended. Time required for the protocol: 3 days.

Steps:

1. Prepare frozen liver sections (5-μm thick) and fix in cold acetone ($-20\,^{\circ}$C) for 10 minutes. Transfer slides rapidly to PBS and wash three times with PBS (do not allow the slides to dry).

2. Incubate sections with primary antibody (polyclonal rabbit antibodies to K8/K18; 50K160 diluted 1:50 in PBS) for 60 minutes at room temperature followed by three PBS washings, 5 minutes each. Incubate with secondary anti-body (swine antirabbit immunoglobulins, 12-nm gold-conjugated; 1:30 in 1% bovine serum albumin in PBS) for 60 minutes at room temperature followed by three PBS washings, 5 minutes each.

3. Fix immunolabeled section in 2.5% glutaraldehyde (in cacodylate buffer) for 15 minutes at room temperature. Wash two times in cacodylate buffer for 5 minutes. Contrast with 1% OsO_4 in cacodylatebuffer for 15 minutes. Wash two times in cacodylate buffer for 5 minutes. Remove water in graded ethanol by incubation in 50%, 70%, 80%, and 90% ethanol in cacodylate buffer for 5 minutes each, followed by two steps in 10% ethanol for 5 minutes and two steps in propylene oxide for 5 minutes. Embed in agar 100 resinpropylene oxide (1:1) overnight.

4. Place gelatin capsules filled with freshly prepared agar 100 resin over sections and polymerize at 60 $^{\circ}$C overnight.

5. Detach gelatin capsules by rapid cooling to the temperature of liquid nitro-gen and polymerize for further 2 hours at 60 $^{\circ}$C.

6. Cut ultrathin sections (50-nm thick) with an ultramicrotome, transfer sections to coated grids.

7. Stain with uranyl acetate and lead citrate, examined with a CM 100 electron microscope at 80 kV.

E. Immunofluorescence Microscopy

Protocol for double-labelled immunofluorcence microscopy on frozen section.

All antibodies are diluted in PBS and applied separately in sequential incuba-tions. Fluorochrome-conjugated antibodies should be centrifuged at 16,000g for 5 minutes to remove aggregates before application onto slides. For negative control, first antibodies are replaced by PBS, preimmune serum, or isotype-matched immunoglobulins, respectively.

Steps:

1. Cut cryosections (3-μm thick); air-dry and fix the sections in acetone at $-20\,^{\circ}$C for 10 minutes. Alternatively (particularly if preservation of nuclear

architecture is required), sections are fixed in PBS-buffered 4% formaldehyde for 15 minutes at room temperature, followed by acetone fixation for 5 minutes at $-20\,°C$. Sections can be air-dried after fixation or rinsed in PBS.

2. Apply first primary antibody for 30 minutes at room temperature in a wet chamber. Alternatively, antibodies can be applied overnight at $4\,°C$. Wash three times with PBS for 5 minutes.

3. Apply first secondary antibody for 30 minutes at room temperature in a wet chamber under light protection. Wash three times with PBS for 5 minutes.

4. Apply second primary antibody for 30 minutes at room temperature in a wet chamber under light protection. Wash three times with PBS for 5 minutes.

5. Apply second secondary antibody for 30 minutes at room temperature in a wet chamber under light protection. Wash three times with PBS for 5 minutes.

6. After the last antibody incubation step, rinse slides with distilled water and then with ethanol for a few seconds and let them air-dry. Mount specimens with Mowiol or commercially available mounting medium.

7. Immunofluorescent specimens are analyzed with a LSM510 laser-scanning microscope. For colocalization analyses (dual labeling) images are acquired with the multitrack modus. Merged pictures appear in green/red pseudocolor with yellow color at sites of colocalization.

8. Store slides protected from light at $+4\,°C$. Immunofluorescence samples are stable for several weeks.

F. Immunohistochemistry

Protocol for single-label immunohistochemistry on paraffin-embedded section.

Steps:

1. Sections (4-μm thick) are deparaffinized in xylene and rehydrated in graded ethanol (100%, 90%, 80%, 70%, 50% ethanol) and PBS.

2. For antigen retrieval, incubate rehydrated sections with 0.1% protease type XXIV for 10 minutes at room temperature (for SMI 31 [Sternberger] and ubiquitin [Dako] antibodies) or microwave at 750 W for 10 minutes in 10 mM citrate buffer, pH 6.0 (for the polyclonal K8/18 antibody 50K160, the monoclonal K8 antibody K8.8 [Neomarkers], the monoclonal K18 antibody DC-10 [Neomarkers] and p62CT antibody).

3. After washing in PBS, block endogenous peroxidase by incubation in 1% H_2O_2 in methanol for 10 minutes. Wash in PBS.

4. Incubate with primary antibodies in a wet chamber for 60 minutes at room temperature. Wash three times with PBS.

5. Incubate with Multi Link Swine Anti-Goat, Mouse, Rabbit immunoglobulins (Dako) diluted 1:100 in PBS for 30 minutes at room temperature. Wash three times with PBS and incubate with Strept ABComplex/HRP (Sol A 1:100 and Sol

B 1:100 in PBS) for 30 minutes. Alternatively, incubate with peroxidase-conjugated rabbit anti-guinea pig immunoglobulin secondary antibody (Dako) diluted 1:100 in PBS for 30 minutes, wash three times with PBS, and perform tyramide amplification by applying biotinyl tyramide solution 1:50 in amplification dilutent (TSA, Biotin System) for 5 minutes. Wash three times with PBS and incubate with strepavidin-peroxidase solution (1:100 in PBS) for 30 minutes.

6. For color development, incubate with AEC (Dako) for 5 minutes, wash in PBS, counterstain with Mayr's hemalaun, rinse with tap water, and mount with Aquatex.

IV. Pitfalls

A. Epitope Masking

Keratins are present in MBs in a different structural organization than in IF. Therefore, some antibodies recognizing conformation-dependent epitopes on keratin show different binding properties to MBs and IF (Hazan *et al.*, 1986). We found that several antibodies specific for K8 preferably reacted with MBs (unpublished observation). Furthermore, MBs act as high-affinity and promiscuous binding partners for different keratin polypeptides and stabilize them. In certain liver diseases, when hepatocytes aberrantly express bile duct type keratins, MBs are also positive for K19 or K7 (Dinges *et al.*, 1992; VanEyken *et al.*, 1988).

B. Poor Solubility of Mallory Body Proteins

MBs are very stable structures, and MB components are insoluble in high-salt buffers. Furthermore, because of covalent modifications, including polyubiquitination and cross-linking, a considerable fraction of MB proteins cannot be resolved by SDS-gel electrophoresis and, therefore, escapes further biochemical analysis (Zatloukal *et al.*, 1991; see also Fig. 4).

V. Discussion

A. Not Every Cytoplasmic Inclusion in Hepatocytes Is a Mallory Body

MBs are characteristic features of a variety of liver diseases but have to be distinguished from other cytoplasmic structures. On the basis of biochemical and ultrastructural analyses of human and murine MBs, the best criteria are the positivity for keratin and p62. Although ubiquitin is a common constituent of MBs, some MBs were only variably positive with ubiquitin antibodies. Therefore, ubiquitin is a less reliable marker, particularly to visualize small MBs. In addition to the positivity for keratin and p62, MBs should reveal, at least in part, a filamentous ultrastructure.

MBs have to be discriminated from intracytoplasmic hyaline bodies (IHBs), which occur in hepatocellular carcinoma and copper toxicosis. IHBs have a globular appearance and are constantly positive for p62, positive for ubiquitin to variable extent, and negative for keratin (Stumptner *et al.*, 1999). In some hepatocellular carcinomas, a transition between MBs and IHB was observed, which suggests a possible relationship of these two inclusion types (Stumptner *et al.*, manuscript submitted).

Furthermore MBs have to be distinguished from megamitochondria, alpha$_1$-antitrypsin inclusions, and pale bodies, which consist of fibrinogen.

B. Specific Features of Intermediate Filament–Related Cellular Inclusions

MBs and other IF-related inclusions share several features. For example, all inclusions consist of misfolded IF proteins, ubiquitin, and, to a variable extent, other stress proteins, such as HSPs, αB-crystallin, and p62. Furthermore, abnormal IF protein phosphorylation was found in most types of inclusion. There are also inclusions, such as NFTs or Lewy bodies, in which besides IF proteins other proteins are affected by misfolding and even represent the major aggregate components. In NFTs, there were conflicting results concerning the actual role of neurofilaments and tau protein, because several neurofilament antibodies, which were used to stain NFTs, cross-reacted with epitopes present on abnormally phosphorylated tau protein. Therefore, only a minor, if any, role was attributed to neurofilaments in NFTs. The situation in NFTs became even more complicated when p62, which harbors the same phosphoepitopes as neurofilaments and tau, was identified as an additional NFT component (Zatloukal *et al.*, 2002). However, the constant presence of neurofilaments in NFTs was recently confirmed with a variety of neurofilament-specific antibodies. Furthermore, the complex alterations in neurofilament expression and its correlation with neuronal vulnerability support involvement of neurofilaments in this type of neurofibrillary pathology (Vickers *et al.*, 2000).

In Lewy bodies, besides neurofilaments, α-synuclein was identified as a major constituent. Aggregation of α-synuclein can be caused not only by the action of oxidative stress but also by mutations in the α-synuclein gene. Mutated proteins are particularly prone to misfolding and thus sensitive to oxidative stress. A synergistic role of mutations and oxidative stress is assumed to contribute also to a variety of other IF-related inclusions, such as Rosenthal fibers in Alexander disease, Lewy body–like inclusions in amyotrophic lateral sclerosis, and desmin aggregates in some desmin myopathies.

References

Angulo, P. (2002). Nonalcoholic fatty liver disease. *N. Engl. J. Med.* **16**, 1221–1231.

Bardag-Gorce, F., Riley, N., Nguyen, V., Montgomery, R. O., French, B. A., Li, J., van Leeuwen, F. W., Lungo, W., McPhaul, L. W., and French, S. W. (2003). The mechanism of cytokeratin aggresome formation: The role of mutant ubiquitin (UBB+1). *Exp. Mol. Pathol.* **74**, 160–167.

Baribault, H., Penner, J., Iozzo, R. V., and Wilson-Heiner, M. (1994). Colorectal hyperplasia and inflammation in keratin 8-deficient FVB/N mice. *Genes Dev.* **8**, 2964–2974.

Cadrin, M., French, S. W., and Wong, T. T. (1991). Alternation in molecular structure of cytoskeleton proteins in griseofulvin-treated mouse liver: A pressure tuning infrared spectroscopy study. *Exp. Mol. Pathol.* **55**, 170–179.

Denk, H., Gschnait, F., and Wolff, K. (1975). Hepatocellular hyalin (Mallory bodies) in long term griseofulvin-treated mice: A new experimental model for the study of hyalin formation. *Lab. Invest.* **32**, 773–776.

Denk, H., Krepler, R., Lackinger, E., Artlieb, U., and Franke, W. W. (1982). Immunological and biochemical characterization of the keratin-related component of Mallory bodies: A pathological pattern of hepatocytic cytokeratins. *Liver* **2**, 165–175.

Denk, H., Stumptner, C., and Zatloukal, K. (2000). Mallory body revisited. *J. Hepatol.* **32**, 689–702.

Dinges, H. P., Zatloukal, K., Denk, H., Smolle, J., and Mair, S. (1992). Alcoholic liver disease. Parenchyma to stroma relationship in fibrosis and cirrhosis as revealed by three-dimensional reconstruction and immunohistochemistry. *Am. J. Pathol.* **141**, 69–83.

Ehrnsperger, M., Gräber, S., Gaeste, M., and Buchner, J. (1997). Binding of non-native protein to Hsp25 during heat shock creates a reservoir of folding intermediates for reactivation. *EMBO J.* **16**, 221–229.

Franke, W. W., Denk, H., Schmid, E., Osborn, M., and Weber, K. (1979). Ultrastructural, biochemical and immunologic characterization of Mallory bodies in livers of griseofulvin-treated mice. Fimbriated rods of filaments containing prekeratin-like polypeptides. *Lab. Invest.* **40**, 207–220.

Fuchs, E., and Weber, K. (1994). Intermediate filaments: Structure, dynamics, function, and disease. *Annu. Rev. Biochem.* **63**, 345–382.

Goebel, H. H. (1997). Desmin-related myopathies. *Curr. Opin. Neurol.* **10**, 426–429.

Grune, T., Reinheckel, T., and Davies, K. J. A. (1997). Degradation of oxidized proteins in mammalian cells. *FASEB J.* **11**, 526–534.

Hartl, F. U. (1996). Molecular chaperones in cellular protein folding. *Nature* **381**, 571–580.

Hazan, R., Denk, H., Franke, W. W., Lackinger, E., and Schiller, D. L. (1986). Change of cytokeratin organization during development of Mallory bodies as revealed by a monoclonal antibody. *Lab. Invest.* **54**, 543–553.

Herrmann, H., Hesse, M., Reichenzeller, M., Aebi, U., and Magin, T. M. (2003). Functional complexity of intermediate filament cytoskeletons: From structure to assembly to gene ablation. *Int. Rev. Cytol.* **223**, 83–175.

Hershko, A., and Ciechanover, A. (1998). The ubiquitan system. *Annu. Rev. Biochem.* **67**, 425–479.

Ishii, T., Yanagawa, T., Kawane, T., Yuki, K., Seita, J., Yoshida, H., and Bannai, S. (1996). Murine peritoneal macrophages induce a novel 60-kDa protein with structural similarity to a tyrosine kinase p56[lck]-associated protein in response to oxidative stress. *Biochem. Biophys. Res. Comm.* **226**, 456–460.

Johnston, J. A., Ward, C. L., and Kopito, R. R. (1998). Aggresomes: A cellular response to misfolded proteins. *J. Cell Biol.* **143**, 1883–1898.

Joung, I., Strominger, J. L., and Shin, J. (1996). Molecular cloning of a phosphotyrosine-independent ligand of the p56[lck] SH2 domain. *Proc. Natl. Acad. Sci. USA* **93**, 5991–5995.

Julien, J-P. (1997). Neurofilaments and motor neuron disease. *Trends Cell Biol.* **7**, 243–249.

Ku, N-O., and Omary, M. B. (2000). Keratins turn over by ubiquitination in a phosphorylation-modulated fashion. *J. Cell Biol.* **149**, 547–552.

Liao, J., Ku, N. O., and Omary, M. B. (1997). Stress, apoptosis, and mitosis induce phosphorylation of human keratin 8 at ser-73 in tissues and cultured cells. *J. Biol. Chem.* **272**, 17565–17573.

Lieber, C. S. (2000). Alcoholic liver disease: New insights in pathogenesis lead to new treatments. *J. Hepatol.* **32**, 113–128.

Lowe, J., Blanchard, A., Morell, K., Lennox, G., Reynolds, L., Billet, M., Landon, M., and Mayer, R. J. (1988). Ubiquitin is a common factor in intermediate filament inclusion bodies of diverse type in man, including those of Parkinson's disease, Pick's disease, and Alzheimer's disease, as well as

Rosenthal fibers in cerebellar astrocytomas, cytoplasmic bodies in muscle, and Mallory bodies in alcoholic liver disease. *J. Pathol.* **155,** 9–15.

Lowe, J., McDermott, H., Pike, I., Spendlove, I., Landon, M., and Mayer, R-J. (1992). Alpha B crystallin expression in non-lenticular tissues and selective presence in ubiquitinated inclusion bodies in human disease. *J. Pathol.* **166,** 61–68.

Magin, T. M., Schröder, R., Leitgeb, S., Wanninger, F., Zatloukal, K., Grund, C., and Melton, D. W. (1998). Lessons from keratin 18 knockout mice: Formation of novel keratin filaments, secondary loss of keratin 7 and accumulation of liver-specific keratin 8-positive aggregates. *J. Cell Biol.* **140,** 1441–1451.

Omary, B. M., Ku, N. O., and Toivola, D. M. (2002). Keratins: Guardians of the liver. *Hepatology* **35,** 251–257.

Ohta, M., Marceau, N., Perry, G., Manetto, V., Gambetti, P., Autilio-Gambetti, L., Metuzals, J., Kawahara, H., Cadrin, M., and French, S. W. (1988). Ubiquitin is present on the cytokeratin filaments and Mallory bodies of hepatocytes. *Lab. Invest.* **59,** 848–856.

Pollanen, M. S., Dickson, D. W., and Bergeron, C. (1993). Pathology and biology of the Lewy body. *J. Neuropathol. Exp. Neurol.* **52,** 183–191.

Sanz, L., Sanchez, P., Lallena, M-J., Diaz-Meco, M. T., and Moscat, J. (1999). The interaction of p62 with RIP links the atypical PKCs to NF-κB activation. *EMBO J.* **18,** 3044–3053.

Shin, J. (1998). p62 and the sequestosome, a novel mechanism for protein metabolism. *Arch. Pharm. Res.* **21,** 629–633.

Sherman, M. Y., and Goldberg, A. L. (2001). Cellular defenses against unfolded proteins: A cell biologist thinks about neurodegenerative diseases. *Neuron* **29,** 15–32.

Stumptner, C., Fuchsbichler, A., Lehner, M., Zatloukal, K., and Denk, H. (2001). Sequence of events in the assembly of Mallory body components in mouse liver: Clues to the pathogenesis and significance of Mallory body formation. *J. Hepatol.* **34,** 665–675.

Stumptner, C., Heid, H., Fuchsbichler, A., Hauser, H., Mischinger, H-J., Zatloukal, K., and Denk, H. (1999). Analysis of intracytoplasmatic hyaline bodies in a hepatocellular carcinoma: demonstration of p62 as a major constituent. *Am. J. Pathol.* **154,** 1701–1710.

Stumptner, C., Omary, M. B., Fickert, P., Denk, H., and Zatloukal, K. (2000). Hepatocyte cytokeratins are hyperphosphorylated at multiple sites in human alcoholic hepatitis and in a Mallory body mouse model. *Am. J. Pathol.* **156,** 77–90.

Tephly, T. R., Coffman, B. L., Ingall, G., Abou Zeit-Har, M. S., Goff, H. M., Tabba, H. D., and Smith, K. M. (1981). Identification of N-methylprotophorphyrin IX in lives of untreated mice and mice treated with 3,5-diethoxycarbonyl-1,4-dihydrocollidine: Source of the methyl group. *Arch Biochem. Biophys.* **212,** 120–126.

Tsunoo, C., Harwood, T. R., Arak, S., and Yokoo, H. (1987). Cytoskeletal alterations leading to Mallory body formation in livers of mice fed 3,5-diethoxycarbonyl-1,4-dihydrocollidine. *J. Hepatol.* **5,** 85–97.

Vadlamudi, R. K., Joung, I., Strominger, J. L., and Shin, J. (1996). p62, a phosphotyrosine-independent ligand of the SH2 domain of p56[lck], belongs to a new class of ubiquitin-binding proteins. *J. Biol. Chem.* **271,** 20235–20237.

Van Eyken, P., Sciot, R., and Desmet, V. L. (1988). A cytokeratin immunohistochemical study of alcoholic liver disease: Evidence that hepatocytes can express "bile duct-type" cytokeratins *Histopathology* **13,** 605–617.

Vickers, J. C., Dickson, T. C., Adlard, P. A., Saunders, H. L., King, C. E., and McCormack, G. (2000). The cause of neuronal degeneration in Alzheimer's disease. *Prog. Neurobiol.* **60,** 139–165.

Wisniewski, T., and Goldman, J. E. (1998). αB-crystallin is associated with intermediate filaments in astrocytoma cells. *Neurochem. Res.* **23,** 385–392.

Westendorf, J. M., Rao, P. N., and Gerace, L. (1994). Cloning of cDNAs for M-phase phosphoproteins recognized by the MPM2 monoclonal antibody and determination of the phosphorylated epitope. *Proc. Natl. Acad. Sci. USA* **91,** 714–718.

Yokoo, H., Minick, O. T., Batti, F., and Kent, G. (1972). Morphologic variants of alcoholic hyalin. *Am. J. Pathol.* **69,** 25–40.

Yuan, O. X., Nagao, Y., French, B. A., Wan, Y. J. Y., and French, S. W. (2000). Dexamethasone enhances Mallory body formation in drug-primed mouse liver. *Exp. Mol. Pathol.* **69,** 202–210.

Zatloukal, K., Boeck, G., Rainer, I., Denk, H., and Weber, K. (1991). High molecular weight components are main constituents of Mallory bodies isolated with a fluorescence activated cell sorter. *Lab. Invest.* **64,** 200–206.

Zatloukal, K., Denk, H., Spurej, G., Lackinger, E., Preisegger, K-H., and Franke, W. W. (1990). High molecular weight component of Mallory bodies detected by a monoclonal antibody. *Lab. Invest.* **62,** 427–434.

Zatloukal, K., Fesus, L., Denk, H., Tarcsa, E., Spurej, G., and Böck, G. (1992). High amount of epsilon-(gamma-glutamyl)lysine cross-links in Mallory bodies. *Lab. Invest.* **66,** 774–777.

Zatloukal, K., Stumptner, C., Fuchsbichler, A., Heid, H., Schnoelzer, M., Kenner, L., Kleinert, R., Prinz, M., Aguzzi, A., and Denk, H. (2002). p62 is a common component of cytoplasmic inclusions in protein aggregation diseases. *Am. J. Pathol.* **160,** 255–263.

Zatloukal, K., Stumptner, C., Lehner, M., Denk, H., Baribault, H., Eshkind, L. G., and Franke, W. W. (2000). Cytokeratin 8 protects from hepatotoxicity, and its ratio to cytokeratin 18 determines the ability of hepatocytes to form Mallory bodies. *Am. J. Pathol.* **156,** 1263–1274.

CHAPTER 9

Stress Models for the Study of Intermediate Filament Function

E. Birgitte Lane[*] and Milos Pekny[†]

[*]Cancer Research UK
Cell Structure Research Group
University of Dundee School of Life Sciences
Dundee DD1 5EH, Scotland

[†]Department of Medical Biochemistry
Sahlgrenska Academy at Göteborg University
405 30 Göteborg, Sweden

I. Introduction

A. Intermediate Filaments, Stress, and Pathology

Among the cytoskeleton systems, the intermediate filament (IF) group of cyto-skeleton proteins has always had a close connection with pathology. The tissue specificity of this large multigene family has established IFs as powerful indicators of cell and tissue differentiation for pathologists and basic researchers. After their identification in the mid 1970s, IF proteins and antibodies quickly entered the toolbox of the diagnostic pathologist, and much of the first 10–15 years of research on IFs was taken up with documenting their numbers (currently more than 60 IF genes are in the human genome database) and their tissue expression patterns in the body. With the good analytical tools of monoclonal antibodies, it quickly became clear that there were many known pathological conditions in which IF abnormalities were a characteristic histopathological feature. These include the many degenerative disorders showing cytoplasmic accumulations of IF protein such as Alzheimer's disease, amyotrophic lateral sclerosis, multiple sclerosis, and alcoholic liver cirrhosis. In the past 10 years, a rapidly growing list of pathogenic IF mutations has joined this list (see Chapter 6 by Smith *et al.* in this volume). The mutations, associated with tissue fragility phenotypes, have provided strong evidence of function for these proteins. Although the dogma that IFs show expression patterns that are tightly linked to tissue differentiation is undoubtedly true, it is similarly true that if differentiation is perturbed, IF expression will often change, as shown by examples of altered keratin expression seen in metastatic cancers (Markey *et al.*, 1991; Porter *et al.*, 2000).

The link between IFs and stress is at one level simply intuitive. Intermediate filament proteins are tough, stable proteins that polymerize readily without any extra cofactors, and they are capable of sustaining enormous strains without breaking (Janmey *et al.*, 1998). Even the appearance of IF networks in the cytoplasm of diverse cell types, from nerve cells to keratinocytes, suggests to the observer that these structures might have a physical supporting role in the cell. The epidermal keratino-cytes, with their dense, anastomosing meshwork of IFs, form the outermost physical barrier of the body. Neuronal axons form cell extensions containing long parallel cross-bridged neurofilaments, and these processes can be up to 1 m long but less than a tenth of a millimeter thick, as in the sciatic nerve.

Direct evidence of a stress-resisting role only appeared for keratinocyte fila-ments when the skin fragility mutations in keratins were discovered in the early 1990s; for other IF mutation diseases, the disease mechanism is still mostly speculation. Should further research continue to identify cell reinforcement as the major function of epidermal keratins, this still does not explain the nature of the biological driver for the evolution of the heterogeneity of tissue specificity that exists among IF genes. Do different IFs have different rigidity or plasticity features that are suited to particular tissues situations? Or, maybe there are more subtle and complex roles for these filament proteins still waiting to be discovered.

Stress can be defined as an environmental alteration or potentially detrimental force, acting on something to cause physiological or behavioral changes. Stress is one of the predominant preoccupations of the research interface between medicine and basic science. Research into the nature and treatment of human diseases starts with an analysis of the disease phenotype and typically proceeds to a modeling of the disease process by establishing an experimental system in which the pathogenic stress can be recreated, analyzed, and monitored during the development of a therapeutic intervention. Experimental models of stress could describe a large fraction of the research publications in biomedical science. The question to be answered is rather one of the nature of the stress in any given disease. When defects in IFs are involved, there are several reasons to think that a failure in mechanical stress resistance of the cells may underlie the pathology.

1. Intermediate Filaments Behave like Stress Response Proteins

Some of the keratins, the type I and type II IF protein, and some type III proteins clearly behave like classical stress proteins in that their expression is induced rapidly in response to tissue and cell stress. Keratins (K) are the coexpressed pairs of IF proteins found in epithelial cells, and they were the first IF proteins to be clearly implicated in the stress response of cells and tissues. K6 and K16 were observed to be upregulated in many pathological situations from psoriasis to cancer (Weiss *et al.*, 1984), and they were formerly referred to as "hyperproliferative" keratins because of this phenomenon, although it was subsequently shown that some normal tissues (e.g., mucosal stratified squamous epithelia) express these keratins constitutively, and expression of K6/K16 can be separated from hyperproliferation (Schermer *et al.*, 1989). K6/K16 are certainly upregulated in epidermal wound healing (Mansbridge and Knapp, 1987). There are now known to be at least two human K6 genes, K6a and K6b; the partner for K16 is probably K6a, whereas K6b is probably preferentially coexpressed with K17 (Smith *et al.*, 1998), a stress-response keratin induced by interferon gamma and probably some other factors (Freedberg *et al.*, 2001; Yoshikawa, 1995). Thus, there are two pairs of keratinocyte IF proteins that are expressed in "activated" keratinocytes (Freedberg *et al.*, 2001) and are known to be upregulated by a range of stresses in the body, and these two pairs are regulated differently. In simple (one-layered) epithelial tissues such as liver and pancreas, keratins K8 and K18 are also upregulated by wounding (Loranger *et al.*, 1997).

Nonkeratin IFs can also change their expression in response to stress. The type III IF proteins glial fibrillary acidic protein (GFAP), vimentin, and nestin are all expressed in astrocytes, and their expression in these cells changes both during development and in many pathological states. Through several analyses of knock-out mice, IFs in astrocytes have been functionally linked to the stress response after brain and spinal cord trauma (Nawashiro *et al.*, 1998; Pekny *et al.*, 1999) and osmotic stress (Anderova *et al.*, 2001; Ding *et al.*, 1998; see later for further discussion of these). Reactive (activated) astrocytes upregulate GFAP and vimentin and reexpress their fetal protein nestin in response to trauma such as ischemia, physical injury, and neurodegeneration (Clarke *et al.*, 1994; Eliasson *et al.*, 1999; Kaya *et al.*, 1999; Shibuya *et al.*, 2002). In muscle after injury, nestin and vimentin are transiently induced during the early stages of regeneration (Vaittinen *et al.*, 2001). Neurofilament expression also changes in response to injury of neurons, with downregulation of neurofilament triplet proteins and upregulation of α-internexin (Goldstein *et al.*, 1988; McGraw *et al.*, 2002).

2. Many Intermediate Filament Mutations Show Stress-Related Phenotypes

The number of knock-out mouse models, the variety of spontaneous human mutations recorded, and the number of different related genes carrying these mutations and deletions together suggest that it should now be straightforward to describe the mechanisms underlying the tissue fragility associated with IF malfunction. However, the range of associated phenotypes is very broad, as illustrated by the diverse phenotypes of keratin disorders, which have been the subject of many reviews (see e.g., Irvine and McLean, 1999). Although the keratin mutations are similar and the proteins are highly related, the phenotypic outcomes vary greatly because of the variety of differentiation programs exhibited by the cells expressing them. The breakdown of basal keratinocytes in epidermolysis bullosa simplex occurs on mild physical trauma to the skin and is associated with mutations in K5 and K14. In orogenital epithelia, mutations in K4 or K13 lead to sloughing of suprabasal layers of damaged cells. In epidermal appendages and other locations, mutations in K6a/b, K16, and K17 lead to thickened nails with either fragile but thick hyperproliferative stratified squamous epithelium in the mouth or distorted hair and pilosebaceous cysts. In all cases, the mutations can be seen to be associated with fragility of a subpopulation of epithelial cells of which the mutated keratin is normally a major component, and evidence from hundreds of identified keratin mutations strongly suggests that in keratinocytes and related cells, keratins provide the physical strength and resilience needed to withstand the stresses of everyday wear and tear.

Human mutations in nonkeratin proteins can appear very different: desmin mutations can lead to hypertrophic cardiomyopathy (Li *et al.*, 1999; Sjöberg *et al.*, 1999). Mutations in GFAP have been associated with the neurodegenerative condition Alexander disease (Brenner *et al.*, 2001). Nuclear lamin A/C mutations can lead to diverse conditions, including familial partial lipodystrophy, Emery-Dreifuss muscular dystrophy, Charcot-Marie-Tooth disease type 2B1, or the generalized premature aging of Hutchinson-Gilford progeria (reviewed by Östlund and Worman, 2003). It is quite possible that the simplest explanation for all of these diverse phenotypes will come down to a fragility or reduced resilience in cell structures. However, before this conclusion can be drawn, more analytical work is needed to dissect the mechanism(s) that leads to cell and tissue failure in these different cases.

There is now a discernible commitment within the research community to elucidate the function of IF proteins, born of confidence in the increasing awareness of disease associations. Experimental models are now beginning to come forward for analysis of IF resilience, mostly assay systems designed to answer other questions but now being adapted to IF research. The next few years should yield interesting results in this field.

B. Stressing Intermediate Filaments *In Situ*

In vivo stress experiments on IFs have mostly been carried out by histological investigation of tissue morphology or by measuring tissue performance in physiological preparations of tissues *ex vivo* (usual for muscle experiments). Interpretation of these experiments may require specialist expertise beyond that of many cell biologists and has consequently led to many productive interdisciplinary collaborations; the morphological readout of the former is often in the form of subtle histological changes in tissues, and the latter type of experiment is often carried out in collaboration with biophysicists or physiologists. The source of experimental material for these assays is almost always genetically modified mouse strains.

There are few *in vivo* experimental stress models that address the function of keratin IFs in tissues, because most of the pathogenic keratin defects that mimic human disorders produce rapid and severe effects not conducive to experimental analysis (mice die at birth or soon thereafter). The first observations of several keratin ablations in mice revealed gross phenotypes, however, with extensive skin blistering on loss of K14 (Vassar *et al.*, 1991) or epithelial thickening on loss of K10 (Porter *et al.*, 1996) or K6 genes (Wong *et al.*, 2000), all of which were incompatible with survival. There are some wound healing experiments carried out on keratin knock-out skin grafts (Wong and Coulombe, 2003) and some inducible knock-out and transgenic models are being generated that will allow further analysis, but currently the most informative experiments on IFs and mechanical stress *in vivo* have undoubtedly come from the work on type III IFs.

1. *In Vivo* Stress Models for Desmin

a. Skeletal Muscle

The major muscle protein desmin was the first IF protein to be identified, and because of its tissue location, desmin has always been a strong candidate protein for a role in stress resilience. Desmin-null mice have generated models in which to test this hypothesis of function. A large number of physiological experiments have now been carried out on tissues from desmin-null mice (Balogh *et al.*, 2002, 2003; Lacolley *et al.*, 2001; Li *et al.*, 1997; Sam *et al.*, 2000; Wede *et al.*, 2002). Sam and colleagues reported that skeletal muscle preparations from desmin knock-out mice seem to be less vulnerable to stress-induced damage; they are less able to develop isometric stress, because the desmin-null cells seem to be more compliant than wild-type cells (Sam *et al.*, 2000). Morphologically, the Z lines appear less organized than in wild-type cells. This makes sense, because desmin is known to be concentrated at the Z lines, so the sarcomeres may be less firmly held in register across the muscle fiber when there is no desmin.

b. Smooth Muscle

Observations from analysis of artery structure and function at different positions in the branching arterial "tree" have also supported this view of desmin function (Wede et al., 2002). In wild-type mice, there is more desmin in the endothelial cells of the smallest arterioles in the peripheral microvasculature than there is in the endothelial cells of the large central elastic arteries closer to the heart. Passive and active stress development in the peripheral microvasculature (where desmin is usually expressed) was lower in desmin-null mice than in wild-type mice, whereas there was no significant difference in the behavior of the central vessels that have little or no desmin in any case (Wede et al., 2002). When comparing heart muscle function in mice with or without desmin, the desmin was found to be important for active force generation in the tissue but made no difference to the elasticity of the muscle in passive strain (Balogh et al., 2002). It was also shown that this lower active force generation led to increased fatigue resistance in skeletal muscle (Balogh et al., 2003).

c. Cardiac Muscle

Lack of desmin in mouse models leads to hypertrophic cardiomyopathy (Milner et al., 1996; Thornell et al., 1997). Bloom and colleagues, applying shear force to cardiac muscle cells while fixing rat heart by perfusion under different pressures, showed that mechanical shear force on heart muscle led to electron microscopy (EM)–detectable deflection of the angle between myofibrils, desmin bridges, and the nuclear envelope together with a thinning of the intranuclear chromatin layer that lies along the nuclear envelope lamin layer. They proposed that this suggested a theory for how stretch could lead to altered gene expression in muscle (Bloom et al., 1996).

A curious observation was made by Weisleder and colleagues, who demonstrated that by crossing the desmin$^{-/-}$ mice (Milner et al., 1996) to mice over-expressing bcl-2 under the control of cardiac-specific alpha-myosin heavy chain promoter (Subramaniam et al., 1991), cardiomyopathy is ameliorated, and the morphological and functional changes seen in mitochondria of desmin$^{-/-}$ mice are corrected (Weisleder et al., 2004). This work suggests two conclusions: (1) the mitochondrial pathology is the cause and not the consequence of cardiomyopathy in desmin$^{-/-}$ mice, and (2) desmin might have an important role in protection of mitochondria, rather than only be involved with the mechanical stabilization of the contractile actomyosin apparatus per se. The broader consequences of loss of physical stability and physical resilience in cells must be considered if one is to understand the functions of IFs correctly; many consequences of mechanical stress may not have immediate gross morphological effects.

2. The Astrocyte Model

The importance of type III intermediate filaments in the response to injury in various tissues has been analyzed in several systems. Astrocytes have proven to be especially fertile experimental ground for these studies. Astrocytes express two or three types of IFs simultaneously: vimentin, nestin, and GFAP. Astrocytes are the most numerous cells in the central nervous system (CNS), and the astrocyte/neuron ratio increases with the increasing complexity of the organism. Nevertheless, their function is still poorly understood. Understanding of brain function at the cellular and tissue level remains patchy, and astrocytes, although coupled to each other by gap junctions, do not assemble into any detectable multicellular structures that can be manipulated. Until recently, no single molecule defects were known that primarily affected astrocytes.

Like activated keratinocytes during wound healing of the epidermis, astrocytes show characteristic changes in IF expression in many pathological conditions, which have been interpreted as indicative of altered functions. The term "reactive gliosis" has been used to describe this process of astrocyte activation in diseases affecting the central nervous system (i.e., the brain, spinal cord, and neural retina in the eye). In particular, the upregulation of the IF protein GFAP has been considered a hallmark of reactive gliosis, irrespective of its origin (see Fig. 1). Thus, depletion of GFAP from astrocytes was an obvious experimental route toward learning more about astrocyte function in the steady state and in stress

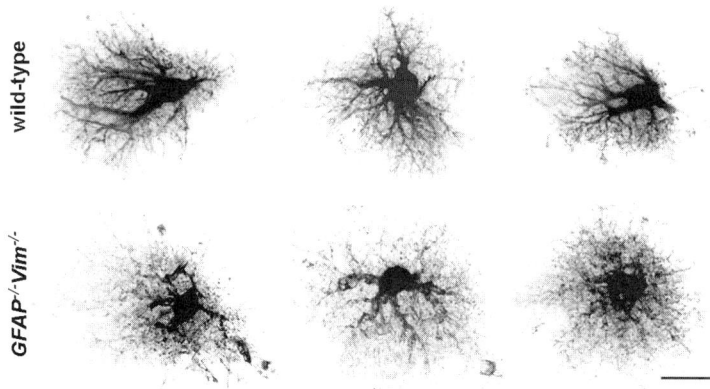

Fig. 1 Morphology of reactive astrocytes as assessed by dye filling of individual cells followed by three-dimensional reconstruction. After central nervous system injury, wild-type reactive astrocytes show prominent hypertrophy of cellular processes, whereas $GFAP^{-/-} Vim^{-/-}$ reactive astrocytes have only a few long and straight processes. Scale bar, 20 μm (modified from Wilhelmsson, 2004; Wilhelmsson et al., 2004).

(Pekny, 2001). What is emerging is a most interesting picture that identifies astrocyte IFs as structures of fundamental importance in different types of stress.

GFAP-deficient ($GFAP^{-/-}$) mice live normal lives and, when unchallenged, showed no obvious phenotype (Gomi *et al.*, 1995; McCall *et al.*, 1996; Pekny *et al.*, 1995). They could therefore be used for addressing the role of astrocyte IFs in reactive gliosis and stress in a number of disease models.

a. Mechanical Stress to the Central Nervous System

The pathology of head injury is complicated and of great clinical importance, yet very difficult to study experimentally. Two models of mechanical stress to the CNS have indicated a significant protective role of astrocyte IFs in this tissue.

Hiroshi Nawashiro and collaborators exposed $GFAP^{-/-}$ mice to percussive head injury by a weight drop device with or without allowing head movement at impact (analogous to whiplash injury). In this model, head movement was accommodated by use of a foam support or prevented by use of a rigid board. When head movement was prevented, $GFAP^{-/-}$ mice survived head impact as did wild-type controls. However, when the foam bed allowed head movement at impact, most $GFAP^{-/-}$ mice, but none of wild-type controls, died after the injury. The GFAP-null mice showed extensive subpial and white matter bleeding in the cervical spinal cord region, probably because of vein rupture (Nawashiro *et al.*, 1998). Although the exact mechanism leading to this differential damage still remains to be fully explained, these experiments suggested that astrocyte IFs may contribute to the resistance of CNS tissue to severe mechanical stress.

The retina is an extension of the CNS into the eye and has been used for studying the impact of stress on the CNS. Using the retina, Lundkvist and colleagues exposed CNS tissue in mice lacking GFAP, vimentin, or both ($GFAP^{-/-}$, $Vim^{-/-}$, and $GFAP^{-/-}Vim^{-/-}$ mice) to severe mechanical stress (Lundkvist *et al.*, 2004). Intramortem application of a severe mechanical stress on the retina caused splitting of the inner limiting membrane with the adjacent tissue from the rest of the retina in $GFAP^{-/-}Vim^{-/-}$ mice and, to a lesser extent, in $Vim^{-/-}$ mice, whereas the same stress left the retinas of wild-type controls intact. Electron microscopy showed that this split in the retina occurred within the end-feet of Müller cells. Müller cells are radial glial-like cells of the retina that normally contain IFs of GFAP and vimentin (Lundkvist *et al.*, 2004).

This decreased mechanical stability of a specific retinal layer was also shown to affect the outcome of another type of stress, retinal hypoxia. Oxygen-induced retinopathy is a widely used model of retinopathy of immaturity, and it also exhibits some features of diabetic retinopathy (Smith *et al.*, 1994). Mice at post-natal day 7 are placed into an environment with increased oxygen concentration. The abundance of oxygen delays the development of the vascular system, and the transfer of mice to a normo-oxygenic environment 5 days later leads to a massive neovascularization triggered by relative hypoxia. The vessels grow from the retina into the vitreous body (as they do in prematurely born babies or patients with diabetes), and their presence there can easily be quantified (Smith *et al.*, 1994). In the oxygen-induced retinopathy model, the efficiency of vessels traversing the presumably weakened region beneath the inner limiting membrane was decreased substantially in $GFAP^{-/-} Vim^{-/-}$ and partially in $Vim^{-/-}$ mice. Thus, the absence of IFs in Müller cells decreases the resistance of their end-feet, and consequently of the corresponding layer of the retina, to mechanical stress, but it also reduces the extent of ischemia-triggered pathological vascularization (Lundkvist *et al.*, 2004).

b. Wound Healing in the Central Nervous System

More than 30 years ago, Larry Eng and coworkers described GFAP in the brain lesions of patients with multiple sclerosis (Eng *et al.*, 1971). It soon became clear that astrocytes in various types of brain or spinal cord lesions—mechanical, ischemic, or neurodegenerative—upregulate IFs, undergo hypertrophy of their cellular processes (Fig. 2), and express a whole range of molecules, including the extracellular matrix proteins that ultimately constitute the glial scar within or around such a lesion. This raises the question of whether IF upregulation by reactive astrocytes is an important step in astrocyte activation, or whether it is just a functionally insignificant consequence of this change in cell behavior. Experiments with mice lacking the IF proteins that are expressed in astrocytes (GFAP, vimentin) were able to provide some of the answers.

Eliasson and coworkers showed that although reactive astrocytes generated in mice with either GFAP or vimentin ablated show a reduced amount of IFs, those

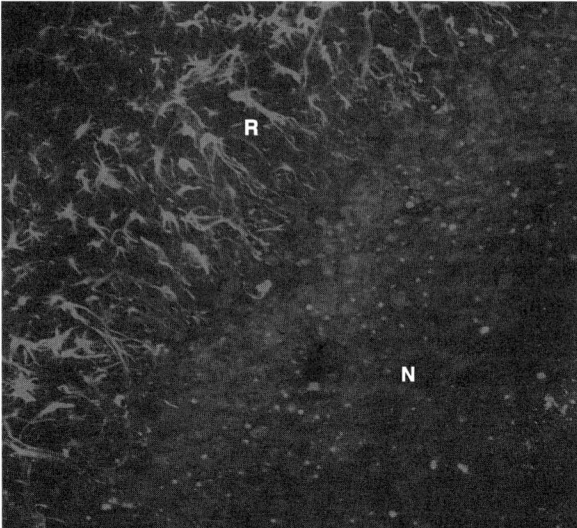

Fig. 2 Reactive gliosis with characteristic hypertrophy of astrocyte processes around an ischemic lesion in the brain. Astrocytes are visualized by use of antibodies against glial fibrillary acidic protein. (See Color Insert.)

of $GFAP^{-/-}Vim^{-/-}$ double-null mice are completely devoid of intermediate filaments (Eliasson *et al.*, 1999). The IFs in wild-type reactive astrocytes consist of GFAP, vimentin, and nestin; in $GFAP^{-/-}$-reactive astrocytes, they contain vimentin and nestin, but $Vim^{-/-}$-reactive astrocytes have IFs made of GFAP only, because nestin can neither self-assemble nor coassemble with GFAP (Eliasson *et al.*, 1999). Two trauma models were applied to all three mutants: (1) a fine-needle injury of the brain cortex and (2) transection of the dorsal funiculus in the upper thoracic spinal cord. Although the responses of wild-type $GFAP^{-/-}$ and $Vim^{-/-}$ mice were indistinguishable from each other, glial scarring was considerably looser and less organized in $GFAP^{-/-}Vim^{-/-}$ mice, suggesting that upregulation of astrocyte IFs is a functional step in astrocyte activation (Pekny *et al.*, 1999). These experiments also showed that these null mutants might help to understand how reactive astrocytes affect the clinical outcome of various pathologies in the CNS, which can one day be expected to open the way for modulation of astrocyte reactivity for the therapeutic benefit to a patient.

Knock-out mouse models have shown that astrocytes actually restrict migration and differentiation of neural precursor cells (Kinouchi *et al.*, 2003) and also the regenerative reponse in the CNS after trauma (Wilhelmsson *et al.*, 2004).

c. Cell Migration after Transplantation

A number of experimental *in vivo* models have been established to test cell capacity for migration in the tissue. Cells can be transplanted to a tissue challenge environment to compare the behavior of cells with and without normal filaments. One of these experiments relates to astrocytes. Because of their morphology and abundance in the adult CNS, astrocytes inevitably come in direct physical contact with any cell that moves from one place to another. Astrocyte migration in tissues is important, for example, in migration of immature neurons born from the endogenous neural stem cells or migration of neuronal precursors from CNS transplants. Chen's and Pekny's groups compared integration efficiency of dissociated retinal cells from 0 to 3-week-old donor mice that ubiquitously express enhanced green fluorescent protein (Okabe *et al.*, 1997) after transplanting them into the retinas of adult wild-type and $GFAP^{-/-}Vim^{-/-}$ recipients (Kinouchi *et al.*, 2003). In the wild-type recipient, very few donor cells migrated out from the transplantation site, and very few integrated into the retina of wild-type hosts. However, in $GFAP^{-/-}Vim^{-/-}$ recipients, transplanted cells moved effectively through the retina, differentiated into neurons, integrated in the ganglion cell layer, and even sent neurites that reached about 1 mm into the optic nerve. They remained alive and well integrated 6 months after transplantation (Kinouchi *et al.*, 2003). Thus, the absence of IFs in astroglial cells of the retina (astrocytes and Müller cells) makes the retinal environment more permissive for integration of neural transplants.

To what extent this is a consequence of increased permissiveness for migration of transplanted cells remains to be established. It is tempting to speculate that depletion of IFs from astroglial cells alters their effective differentiation state so that they functionally resemble more immature astrocytes that are also more supportive of the CNS regeneration (Pekny *et al.*, 2004). This scenario is supported by recent data from Pekny's group showing that $GFAP^{-/-}Vim^{-/-}$ reactive astrocytes exhibit only a very limited hypertrophy of cellular processes (Fig. 2; see Wilhelmsson *et al.*, 2004). The analyses of mice exposed to entorhinal cortex lesions showed that compared with wild-type cells, the $GFAP^{-/-}Vim^{-/-}$ astrocytes were less capable of providing sufficient neuroprotection immediately after the injury. However, later on, these presumably less mature $GFAP^{-/-}Vim^{-/-}$ astrocytes allowed remarkable synaptic regeneration in the region of the hippocampus affected by the lesions (Wilhelmsson *et al.*, 2004). The findings open up interesting therapeutic possibilities for the future (Emsley *et al.*, 2004; Quinlan and Nilsson, 2004).

3. The Epithelial Wound Model

The extensive and complex expression of keratin IF proteins in epidermis and associated structures, plus the increasing number of genetic disorders involving keratin mutations, made skin an early target for functional studies. However, functional experiments on IFs in skin are made difficult by the detrimental phenotypes in mice, which cannot be manipulated directly; there is often a clear phenotype of defective keratins, which is too severe to maintain the animals. The growing trend for developing tissue-specific inducible knock-outs will make more experiments possible (Cao *et al.*, 2001). Caution must be exercised in the interpretation of results from some of these mice (see e.g., Wojcik *et al.*, 2000, 2001; Wong and Coulombe, 2003 for some apparent differences), because the genetic context is critical.

Transplantation experiments have therefore been used to investigate the function of keratins. Ablation of the K6 stress response keratin genes in mice is lethal, because it causes fragility and compensatory hyperplasia of oral epithelia and neonatal mice are unable to suckle (Wong *et al.*, 2000). Wong therefore transplanted K6α/K6β-null epidermal grafts to host animals with wild-type keratins, whereupon they could subsequently wound the transplanted graft and monitor wound healing (Wong and Coulombe, 2003). They reported that in the K6 null mice the newly migrating epidermis was more fragile than in the wild-type graft that expressed normal K6 genes. This suggested that a function of this stress-response keratin might be to strengthen newly forming epidermis. If confirmed, this could be significant, because epidermal wound strength was probably an important feature for the evolution of skin and skin proteins, as suggested by the enhanced scarring that evolution has tolerated in hairless primates as a cost of rapid strong wound healing.

4. Fluid Flow Stress in Blood Vessels

Vimentin is the principal IF protein of endothelial cells, which make up the luminal layer of blood vessels and are the first barrier against the fluid shear stress of blood flow in the vascular system. To investigate the consequences of depleting endothelial cells of vimentin (Colucci-Guyon *et al.*, 1994), experiments have been carried out on blood vessel properties of vimentin −/− mice. Working with isolated segments of mesenteric arteries, Henrion and coworkers tested the effect of fluid flushing on the arterial diameter and pressure differences between the proximal and distal end of the vessel segment. They used an *in vivo* system of an intestinal loop exposed through laparotomy to record blood flow in a segment of the mesenteric artery (Henrion *et al.*, 1997). They observed less flow-induced dilation of the mesenteric artery in $Vim^{-/-}$ mice than in wild-type, although reaction to increasing intraluminal pressure was normal. Their observations implied a role for vimentin in mechanotransduction of shear stress (Henrion *et al.*, 1997).

To further test the physiological relevance of this finding, Terzi and coworkers exposed $Vim^{-/-}$ mice to a complete right nephrectomy combined with a 50% left nephrectomy. $Vim^{-/-}$ mice consistently died within 72 hours postoperatively, whereas all the wild-type controls survived. Isolated second-order renal arteries from $Vim^{-/-}$ mice showed increased active tone in the absence of flow and reduced flow-induced dilation. The kidneys of $Vim^{-/-}$ mice contained more vasoconstriction-promoting endothelin-1 and showed signs of decreased production of nitric oxide, a mediator of vasodilation, both before and after partial nephrectomy (Terzi *et al.*, 1997), indicating that it is a faulty control of the vascular tone mediated by endothelin 1 and/or nitric oxide that leads to renal failure and death. Indeed, the administration by an osmotic pump of bosentan, an endothelin receptor antagonist, completely eliminated the postoperative mortality in $Vim^{-/-}$ mice.

The molecular mechanism linking vimentin IFs with modulation of vascular tone remains to be explained, but it would seem that in both the desmin knock-out and vimentin knock-out mice, changes in cell and tissue response to mechanical force have extensive consequences.

C. Stress Models for Intermediate Filaments in Cell Cultures

Because relatively few experiments have been done on stressing IFs to date, there is no clear consensus yet on the best methods to use. Experiments to stress IFs are still in early developmental stages, and we will need to critically record and assess the outcomes and experiences of different laboratories to find the most effective way to analyze the relationships between IFs and stress. We will review the approaches that seem most promising to date.

1. Mechanical Stress by Fluid Shear

Fluid shear stress is a feature of the cellular environment in many tissues. Blood vessels are subject to variable shear forces, and adhesive behavior of red and white blood cells depends on this. Signaling by osteocytes in bone is triggered by mechanical loading of bone detected in the shear of the fluid pressed through the canaliculi (Mullender *et al.*, 2004). Normal kidney tubule differentiation depends on the epithelial cells' ability to detect flow adequately, because mutations in the PKD1 and PKD2 genes expressed on cilia result in polycystic kidney disease (Nauli *et al.*, 2003; Praetorius and Spring, 2003). In early embryogenesis, loss of function of genes involved in fluid flow detection leads to a breakdown of the mechanisms determining handedness (McGrath *et al.*, 2003; Mercola, 2003).

The resilience of cells in culture to fluid shear stress can be tested by a variety of means, most of which have not yet been used to test IFs. The use of fluid shear stress models on attached cells will primarily target the deforming force to the apical surface domain and its cytoskeleton junctions. Many laboratories have used *perfusion flow chambers* (see e.g., Smith *et al.*, 2004), often of their own design but increasingly using a variety of commercially available models (e.g., manufactured by Bioptech). These are usually chambers of parallel plate configurations, assembled with glass coverslips of high optical quality, on one of which the cells to be tested have been seeded and grown. These chambers are usually kept warm for the duration of the experiment, and culture medium flows between them and across the cells at a controlled rate.

Another popular configuration for fluid shear stress analysis is the *cone and plate viscometer*, in which a shallow cone spins close to the surface on which cells are plated. The angle of the slope of the cone can be engineered such that the decreasing shear force caused by the increasing gap between the cells and the cone (as one moves away from the center of rotation) is balanced by the increasing centrifugal force as the radius of rotation increases, giving even shear force across a large area. This apparatus was designed to measure the viscosity of solutions but has also been used to measure the strength of attachment of cells to a substrate (Skarja *et al.*, 1997); if the cone can be transparent, the cone and plate viscometer can allow monitoring of live cells (Schnittler *et al.*, 1993). Like other mechanical stress stimuli, fluid flow has been shown to stimulate DNA synthesis in endothelial cells with a similar spinning disc apparatus (Ando *et al.*, 1987), although this effect may vary between cell types because fluid shear in a cylindrical growth vessel was also reported to decrease proliferation in arterial smooth muscle cells (Sterpetti *et al.*, 1992).

Experiments targeting "dorsal" stress specifically onto IFs in cells have been carried out in a few cases. In endothelial cells, vimentin distribution changes in response to fluid shear stress. Applying shear stress in a Bioptechs flow chamber, Tsuruta and Jones showed that there was significant association of vimentin with β_3-integrin at forming focal contacts, that the two proteins move together, and that vimentin assembly seems to be concentrated at β_3-integrin sites (Tsuruta and Jones, 2003). Stress induced by fluid shear increased the size of the focal contacts and the thickness of the associated vimentin filament bundles, and knock-down of vimentin reduced the size of focal contacts and reduced cell-substrate attachment.

In another series of experiments, asymmetrical displacement of vimentin filament bundles was observed in endothelial cells in response to fluid shear stress with a heated parallel plate flow chamber (Bioptechs) (Helmke *et al.*, 2001). The significance of this filament displacement is currently unclear.

"Dorsal" shear, to the apical surface of cultured cells, has also been applied to fibroblasts without the use of fluid flow. This was done by attaching arginine-glycine-aspartic acid (RGD) peptide–coated beads to the surface of cells, magnetizing them, and then applying a magnetic torque above the cells. By measuring the resistance to twisting of the cytoplasm, it was concluded that vimentin-null cells show reduced DNA synthesis, less proliferation, and overall were less stiff (Wang and Stamenovic, 2000).

2. Mechanical Stress Through a Deformable Substrate

Stretching of cells by deforming the substrate to which they are attached is perhaps the simplest and most direct sort of model of mechanical stress. This is applicable to cells in tissue culture, and in such experiments the mechanical stress and distortion are targeted at the basal surfaces of the cells. Substrate stretching experiments have been carried out on several different cell types, but limited experimental work has been directed at assessing the role of IFs in stretch resilience.

In the reports of cell stretching to date, there is much variation in the degree to which stretch conditions are defined, and this introduces some ambiguity in analyzing results. There seems to be some cell specificity in the signal transduction pathways that are activated by stretch. Aortic smooth muscle cells (rat) respond to cyclical stretch by activation of p38 MAPK (SAPK2) by means of protein kinase C ras/rac signaling (Li *et al.*, 2000). JNK 2 and ERK 1/2 are also activated. Kippenberger and colleagues applied their own design of linear stretching device to cultured keratinocytes and reported absence of activation of p38 on stretching but found rapid transient activation of β_1-integrin–dependant ERK/MAPK signaling and a slower effect on the JNK/SAPK signaling pathway (Kippenberger *et al.*, 2000). The latter observations were further confirmed in keratinocytes by Russell *et al.* (2004) using an oscillating stretch.

Twenty years ago, Brunette stretched keratinocytes by deforming their plastic substrate and showed that distorting keratinocytes on a plastic substrate increased the proportion of cells in S phase (Brunette, 1984). However, different degrees of stretching can have different effects on cells. Although others have confirmed Brunette's observation and shown that 20% stretch leads to increased DNA synthesis and cell proliferation (hyperplasia) in bladder smooth muscle cells (Orsola *et al.*, 2002) and keratinocytes (Kippenberger *et al.*, 2000; Yano *et al.*, 2004), Orsola also showed that protein synthesis (hypertrophy) was only induced at 6%–12% stretch but was not induced by the 20% stretch (Orsola *et al.*, 2002). It will, therefore, be necessary for experimental reports to specify the degree of stretch involved in some way that can allow comparisons between different laboratories. Yano and colleagues also reported that this 20% stretch induced the stress response keratin K6 and suppressed the secondary differentiation keratin K10, a very rapid response at 24 hours (Yano *et al.*, 2004).

Various designs of commercial purpose–built apparatus can now be used for cell stretching. One of these is Flexcell's Cell Stretcher (Flexcell International, Hillsborough, NC). Russell and colleagues have specifically used this apparatus to look at the resilience of mutant wild-type keratins (Russell, 2004; Russell *et al.*, 2004). They recorded differences in responses to stretching between wild-type keratinocytes and keratinocytes expressing keratin 14 mutation associated with severe epidermolysis bullosa simplex (EBS) skin blistering. By use of stretch at an amplitude of 12% (at each stretch) oscillating at a frequency of 4 Hz, it was observed that in cells expressing mutant keratins, the keratin cytoskeleton was rapidly disrupted and drastically remodeled (Fig. 3), with concomitant disassembly of the cytoplasmic plaques of both desmosomes and hemidesmosomes. This suggests that mechanical tension may be required to maintain these major anchorage junctions in epithelia and that a loss of tension may be an important signal for cytoskeleton remodeling (e.g., in wounding).

Related mechanical stress experiments have been carried out on fibroblasts cultured from lamin A/C-null mice (Lammerding *et al.*, 2004), which demonstrated an increased nuclear deformability in the absence of an efficient nuclear lamina.

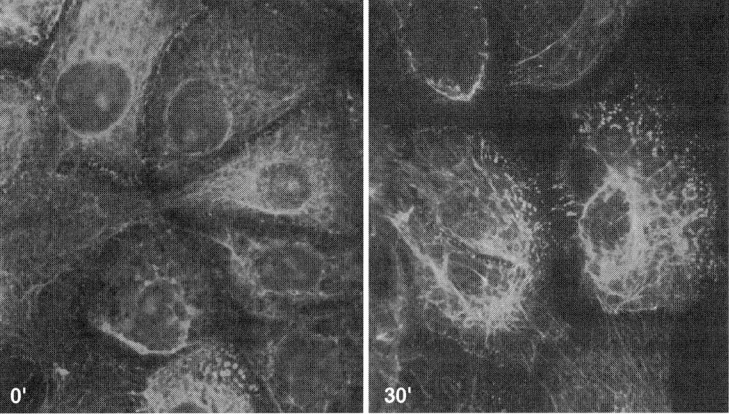

Fig. 3 Mechanical stress can be used to reveal cytoskeleton weaknesses, as demonstrated by the effect of stretch on mutant keratin filaments in cells from patients with EBS (Russell, 2004; Russell *et al.*, 2004). The KEB-7 cells shown here carry the K14 mutation R125P, typical of severe EBS (Morley *et al.*, 2003) and causing basal keratinocyte fragility and epidermal blistering. Before stretch (0') keratin staining reveals occasional aggregates. After 30 minutes of an oscillating stretch (30'), the keratin filament network is severely disrupted. Wild-type cell keratin filaments do not break down under this stress (not shown). Keratin filaments (green) visualized by immunofluorescence with polyclonal antiserum to keratin 5 and fluorescein-conjugated second antibody; desmosomes (red) shown by staining with mouse monoclonal antibody 11-5F to desmoplakin and Texas Red-conjugated second antibody. (Picture courtesy of David Russell.) (See Color Insert.)

3. Osmotic Stress–Induced Swelling as Mechanical Stress

Maintenance of osmotic balance is an energy-requiring process, and any loss of energy to the cell will lead to failure of the membrane's ion pumps and rapidly result in osmotic swelling. At the level of the cytoskeleton, osmotic stress can certainly be classed as a mechanical stress. The rapid cell swelling that accompanies hypoosmotic shock causes dramatic and extensive disruption to the cytoskeleton, not only of actin (which must be wrenched from the plasma membrane as it distends) but also of microtubules. Intermediate filaments are not disrupted so catastrophically, possibly because the network they form is irregular and multistranded and never dependant on single fibers. Another reason may be that IFs do not depend on energy sources for their assembly regulation, whereas actin and tubulin require adenosine triphosphate (ATP) or guanosine triphosphate (GTP) hydrolysis. Thus, IFs are the only filamentous component of the cytoskeleton that remains intact in keratinocytes subjected to hypoosmotic cell swelling, and this by itself may be a powerful reason for their evolution (D'Alessandro et al., 2002).

Various experiments have been carried out on cells with defective, incapacitated, or missing IF proteins to see whether compromising this tissue-specific cytoskeleton component alters the cell's ability to resist cell swelling or its consequences. D'Alessandro and colleagues subjected keratinocyte cell lines to 150 mM urea and showed that this rapidly induced the SAPK1/JNK stress–activated protein kinase signaling pathway as might be expected (D'Alessandro et al., 2002).

The consequences of osmotic stress are particularly harmful in the CNS because of the lack of expansion space. The regulatory volume decrease by astrocytes might be a key mechanism in counteracting the development of brain edema in situations of ischemia or trauma, and it has been proposed that the sensors for cell volume might be cytoskeleton-linked stretch-activated plasma membrane channels (Cantiello, 1997; Cantiello et al., 1993; Moran et al., 1996; Sanchez-Olea et al., 1991). Ding and coworkers subjected primary astrocyte cultures from wild-type, $GFAP^{-/-}$, $Vim^{-/-}$, and $GFAP^{-/-}Vim^{-/-}$ mice to hypoosmotic stress (corresponding to 25 mM reduction in NaCl) in perfusion chambers and assessed subsequent efflux of ^{3}H-taurine. $GFAP^{-/-}Vim^{-/-}$ astrocytes showed up to a 50% reduction in the amount of released taurine, whereas the single mutants exhibited only a slight tendency toward a reduced taurine efflux (Ding et al., 1998). In a later study, Anderova and coworkers found a smaller increase in the potassium concentration around astrocytes in spinal cord slices from $GFAP^{-/-}$ mice compared with wild-type controls after perfusion with either isoosmotic solution with increased concentration of potassium (50 mM) or hypoosmotic solution with reduced sodium concentration (Anderova et al., 2001). These experiments indicate that genetic ablation of astrocyte IFs can compromise the ability of astrocytes to respond to hypoosmotic stress.

Do these findings have any relevance for brain pathologies, in particular those connected with prominent osmotic stress, such as brain ischemia? Nawashiro and coworkers exposed $GFAP^{-/-}$ mice and wild-type mice to brain ischemia induced by middle cerebral artery occlusion for 2 days and reported comparable infarct volumes in the two groups of animals. Interestingly, when the middle cerebral artery occlusion was combined with transient occlusion of the carotid artery, the infarcts in $GFAP^{-/-}$ mice were larger than in the controls (Nawashiro *et al.*, 1998). Thus, the question of a possible protective role of astrocyte IFs remains to be resolved. However, it is known that astrocytes express more IF proteins than just GFAP, and if a clear answer is to be obtained, then $GFAP^{-/-}Vim^{-/-}$ double null mice, in which reactive astrocytes are completely devoid of IFs (Eliasson *et al.*, 1999), are probably required for the study of brain ischemia paradigms.

4. Cell Migration

Cell migration is essential for the organism's survival, and its component features must be the product of acute evolutionary selection pressures. Whatever the tissue context, cell migration leads to enforced cell shape change and always involves significant cytoskeleton reorganization; as such, it constitutes another form of mechanical stress on cells.

Cell migration takes place at several specific times in the lifetime of an organism. It is a critical process for, among other things, CNS development and wound healing. In studying cytoskeleton filament alterations in cell migration, nearly all work has concentrated on actin and tubulin systems. Intermediate filaments are generally thought of as having no role in cell-shape change or cell migration, but some pieces of evidence suggest that this assumption should be reconsidered. In retinal pigment epithelial cells, expression of K18 and K19 was associated with a migratory subpopulation (Robey *et al.*, 1992), while Hendrix and colleagues reported that in melanoma cell lines, there was a correlation between high metastatic behavior and expression of keratins K8 and K18 (Hendrix *et al.*, 1992) and that this phenomenon could be reproduced by experimentally forcing expression of vimentin and K8/K18 (Chu *et al.*, 1996). This group also reported that overexpression of vimentin with keratins in mammary epithelial cells led to increasingly motile and invasive behavior that was reversible by vimentin antisense oligonucleotides (Hendrix *et al.*, 1997).

In tissues, IF protein expression can respond rapidly to the signals that trigger a wound healing response, suggesting that changes in their expression are functionally important in effective wound response. As discussed earlier, IF protein expression changes rapidly on epidermal wounding. Keratinocytes switch from synthesizing K1 and K10 to K6 and K16 within 6 hours of an incisional wound to the stratified squamous epithelium (Paladini *et al.*, 1996). This is concomitant with a change in cell behavior as the cells shift from components of a static tough barrier tissue to a rapidly migrating sheet of highly attenuated epithelial cells. Keratin expression reverts to the steady-state pattern only well after epithelial closure is completed.

It thus seems that appropriate IF expression is probably required for a correct and effective wound response in many tissues. Therefore, a number of assays have been developed to study keratinocyte cell migration in tissue culture. Three strategies are reviewed here.

a. The Transfilter Migration Assay

The many variations of the transfilter migration assay system are based on a modified Boyden chamber, one of the oldest systems for studying cell migration. They are chemotaxis-driven in that the cells are stimulated to migrate up a gradient of growth factor. The output is in the form of cell numbers, migrated through a barrier or not. This type of assay has been mostly used for cells that function as dispersed populations because they can be easily grown as single cells. It has been adapted for use with epithelial cells (Jenner, 2001; Malinda *et al.*, 1999; Robey *et al.*, 1992), but because it requires the cells to be separated from each other, it is not appropriate for all cell types.

In this model, the cells to be tested are plated on the upper surface of a polycarbonate filter with homogeneous channels through it of a predetermined size class. This filter is usually coated with extracellular matrix protein to improve cell adhesion. The cultures are then set up so that the underside of the filter is in contact with growth medium containing a potent chemoattractant as appropriate for the cell type being used, and the upper side of the filter (with cells on) is submerged in normal serum-free medium, establishing a gradient of attractiveness to the cells that crosses the filter at right angles. Cells with then migrate through the pores in the filter to access the more favorable growth environment, and at the end of a given time period, the number of cells on the underside of the filter are counted at low magnification and after staining with, for example, methylene blue. For this assay, cells should be handled and maintained as far as possible as single cells, because the impact of cell growth in colonies on their migration through small channels is not fully defined.

This assay has mostly been used to study the responses of cells to growth factors and cytokines. It has also been used to look at relative migration rates of cells within mixed populations after transient transfection of keratins. Early results suggested that introduction of K18 into a K5/K14-expressing cell line (MCF10A) increased migration, whereas introduction of K10 reduced it (Jenner, 2001). Significant differences may exist *in vivo* in the migration capabilities of epithelial cells expressing different IF proteins.

b. Single Cell Tracking by Microscopy

With recent improvements in microscopy tools, including digital imaging and powerful software, many studies of cell motility and migration now rely on direct monitoring of cell behavior and shape changes over time by directly tracking individual cells or groups of cells microscopically. Cell movement may or may not be stimulated by chemotactic additives. The readout from these assays will consist of individual details from one or more cells as multiple images for comparison or histories of cell movement and images of cell tracks for analysis.

The Dunn chamber (Zicha *et al.*, 1991) is a chemotactic single-cell assay system designed for measuring cell migration by high-resolution monitoring of single cells. It is not suitable for bulk culture counting. Cells are grown in a glass chamber of high optical quality, with the growth zone concentrically flanked by reservoirs of high and low concentrations of growth factors, again establishing a gradient of increasing attractiveness to the cells. The cell substrate can be pre-coated as preferred with extracellular matrix proteins. The chamber is mounted in temperature- and humidity-controlled environment, and the cells are viewed and photographed over time. Cell tracks are recorded and the data processed to evaluate whether their migration patterns deviate from random (i.e., do they respond to the proximity of the growth factor in one of the source wells). This assay can also be used to monitor migration-stimulated changes at specific parts of the cell perimeter by timing the point of addition of the growth factors (Allen *et al.*, 1998; Jones *et al.*, 2002; Webb *et al.*, 1996).

Phagokinetic tracking (Albrecht-Buehler, 1977a) is another method devised by Guenther Albrecht-Buehler to monitor cell movement before and after cell division. Tissue culture plates are coated with gold particles and cells seeded onto this substrate. As the cultured cells migrate, they clear the particles from their track, taking them in by phagocytosis and leaving a clearly marked particle-free trail of exactly where they have been for the last few hours. This technique revealed the striking mirror-image shape changes and migration patterns that are exhibited by fibroblast daughter cells as they move away from each other after cytokinesis (Albrecht-Buehler, 1977b). Similar methods have since been used by others (Chen *et al.*, 1994; Nasca *et al.*, 1999) to monitor keratinocyte migration and to assess the effect of compounds on migration of keratinocytes.

Direct microscopic monitoring and time-lapse filming is now increasingly used for tracking cell migration; as these techniques becomes easier and more accessible, software becomes more powerful and the need for more complex tissue culture models is reducing. The increasing use of green fluorescent protein (GFP) to track individual proteins within living cells has also encouraged this. Direct microscopic monitoring experiments have been carried out to assess the role of IFs on cell migration in a series of experiments on cultured astrocytes lacking IFs. Lepekhin and coworkers assessed the motility of primary cultured astrocytes from $GFAP^{-/-}$, $Vim^{-/-}$, and $GFAP^{-/-}$ $Vim^{-/-}$ mice. They found the fast-moving subpopulation of astrocytes were less well represented among $GFAP^{-/-}$ and $Vim^{-/-}$ astrocytes, and even rarer among $GFAP^{-/-}Vim^{-/-}$ astrocytes (Lepekhin *et al.*, 2001). The relevance of this finding to an *in vivo* wound response remains to be established, but because astrocytes are known to migrate to the injury region over considerable distances (Johansson *et al.*, 1999), slower migration of IF–deficient astrocytes could partially explain the more discrete development of the posttraumatic glial scar seen in $GFAP^{-/-}Vim^{-/-}$ mice (Pekny *et al.*, 1999).

c. Scratch Wound Assays

Wounding is an important biological trigger that induces cell migration. To understand wound healing, the behavior of keratinocytes needs to be well understood; in this cell type, migration is of paramount evolutionary importance because of the need to quickly seal any break in the external barrier. It is also from keratinocytes of the epidermis that the largest single category of human cancers originates. Wounding induces profound remodeling of tissues, and many aspects of a wound response can be reproduced and studied in the much-simplified tissue culture system of the scratch wound. At the leading edge of the migrating epithelial sheet stimulated by scratching, cells become highly active and extend lamellipodia as they migrate across the cleared path (Fig. 4). By marking the culture dish at the start (time of scratching) and end (some arbitrary number of hours afterwards) of the experiment, the area covered by the reepithelializing culture over a set time can be calculated, and the migrating speed of different cell types can be compared. Scratch wound closure assays are now increasingly popular as more experimental cell biology attention turns from solitary cells like fibroblasts to epithelial cells that function in groups and sheets in culture and *in vivo*.

This assay has been previously used to compare the dependence of keratinocyte migration on different substrate components (O'Toole *et al.*, 1997). Recently, we have used this technique to investigate the effect of keratin mutations on cell

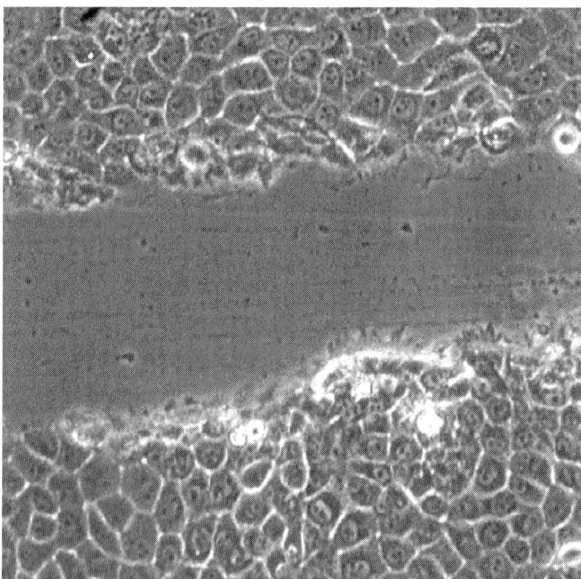

Fig. 4 A culture scratch wound assay can be used to initiate cell migration. This is shown here for epithelial (TR146) cells: a rapid shift from static to migratory behavior is triggered by a scratch across a confluent sheet of cells, initiating a range of cell-shape changes and cytoskeleton alterations. Phase-contrast image; courtesy of Sharon Jenner (Jenner, 2001).

migration efficiency and, interestingly, we observed that cells with keratin mutations covered the scratch area in less time than wild-type cells. At the time, this was attributed to the mutant cells having a constitutively activated stress signaling pathway, probably arising from the defective protein in the cytoplasm, which might have had the same effect as signaling for wound closure—the cells might already have been "primed" for wound healing and might therefore have been able to respond to a migratory signal without delay (Morley *et al.*, 2003).

In vitro studies on the motility of $Vim^{-/-}$ fibroblasts have yielded conflicting data. In one case, scratch wound experiments concluded that vimentin expression did not affect the mobility of polarized cells at the edge of the wound (Holwell *et al.*, 1997), although Eckes and coworkers reported that in comparison to wild-type fibroblasts, $Vim^{-/-}$ fibroblasts exhibited reduced resistance to mechanical stress and reduced migration in both the scratch wound assay and in transfilter migration (Boyden) chambers (Eckes *et al.*, 1998).

II. Materials and Instrumentation

A. Immunoperoxidase Staining of Filaments in Stressed Tissues

- Tissue samples, prepared from experimental systems based on mouse knockout models, by conventional formalin fixation and wax embedding or by snap-freezing in liquid nitrogen.
- Electrostatically charged Superfrost Plus glass microscope slides (VWR, cat. no. 406/0179/00).
- Antibodies as desired.
- Dako Envision system for immunoperoxidase staining (DAKO). This sensitive system is based on a horseradish peroxidase–labeled polymer that is conjugated to secondary antibodies. The labeled polymer does not contain avidin or biotin, and, consequently, nonspecific staining is much reduced.
- Unmasking solution (Vector Labs, H-3300).
- Retriever 2100 histochemical autoclave (Pickcell Laboratories).
- DPX mounting medium for bright-field microscopy.
- Bright field microscope with $16\times$, $25\times$ and $40\times$ objective lenses.

B. Rapid Immunofluorescence of Cultured Cells

- Cultured epithelial cells grown as a monolayer to 50%–80% confluence (e.g., keratinocytes). Culture medium and conditions depend on the cell line.
- Plastic tissue culture grade plates 5 cm in diameter (Nunc); rectangular coverslips to mount over cells. (Alternatively, sterile 13-mm glass coverslips to grow cells on can be submerged inside the culture plates; these are finally mounted onto regular glass microscope slides.)

- Methanol/acetone (1:1) stored at 4 °C.
- If available, unused old tissue culture medium with serum for diluting antibodies. Otherwise, phosphate buffered saline (PBS) with 1% bovine serum albumin can be used as a diluent.
- Primary antibodies as desired. Some we have used are (1) rabbit polyclonal antibody BL18 for visualization of K5 (Purkis *et al.*, 1990), diluted 1:500 in old tissue culture medium with fetal calf serum (FCS); (2) for visualization of K14, mouse monoclonal antibody LL001 (Purkis *et al.*, 1990); (3) for desmoplakin, mouse monoclonal antibody 11-5F, dilution 1:100 in Dulbecco's modified Eagle's medium (DMEM) supplemented with 10% FCS (Parrish *et al.*, 1987).
- Secondary antibodies: Alexa Fluor 488 goat anti-rabbit IgG serum, used at 1:400 dilution; Alexa Fluor 594 goat anti-mouse IgG used at 1:500–1:1000 dilution (as appropriate for species of primary antibody; both from Molecular Probes, Leiden, The Netherlands).
- Fluorescence fade-retarding mounting medium such as Cityfluor (Agar Sciences, Stansted, Essex, UK) or Hydromount (National Diagnostics, Hull, UK) with 2.5% DABCO (Sigma, Poole, UK) to delay bleaching of the fluorochrome.
- Epifluorescence microscope with 25× and 40× objectives.

C. Primary Astrocyte Cultures

- Primary astrocyte cultures dissected from neonatal mouse brains.
- Tissue culture medium. We typically use 90% DMEM (Sigma, D5671), 10% FCS (GIBCO), 2 mM L-glutamine and antibiotics (GIBCO penicillin/streptomycin supplement).
- Bioptech culture chambers for live cell imaging (Bioptech, Butler, PA), if desired.
- Inverted microscope with good phase contrast capabilities (e.g., Zeiss Axiovert 200M) coupled to digital camera.
- For time-lapse microscopy, the microscope should be mounted on an antivibration table and fitted with a maintained environment viewing chamber surrounding the stage to hold the cells at 37 °C and 5% CO_2 during the filming.

D. Scratch Wound–Stimulated Epithelial Cell Migration

- Keratinocyte cells cultured in DMEM with 25% Ham's F12 medium (both from GIBCO BRL, Life Technologies, Paisley, UK) plus 10% FCS (Lab Tech International, Ringmer, Sussex) at 37 °C in 100% humidity, 5% CO_2 atmosphere. Note: Optimal concentration of FCS, and other aspects of optimal growth conditions, must be determined for each cell line by prior experimentation.

- Twenty-four–well plastic tissue culture grade plates for growing cells. Before seeding with cells, these plates can be coated with laminin (from mouse EHS tumor material, Sigma L2020), or collagen IV (Sigma C5533), both at 20 $\mu g/$ml, or other extracellular matrix components as desired.
- Yellow pipette tips (Gilson) for scratching plates.
- Inverted microscope with good phase contrast capabilities with digital camera.
- For time-lapse microscopy of the scratch wound healing event, the microscope should be mounted on an antivibration table and fitted with a maintained environment viewing chamber surrounding the stage to hold the cells at 37 °C and 5% CO_2 during the filming.

E. Osmotic Stress–Induced Cell-Shape Changes

- Cultured cell line to be investigated, such as keratinocytes or astrocytes.
- Solution of 150 mM urea in DMEM at 37 °C.
- Fresh tissue culture medium (DMEM) at 37 °C.
- Phase-contrast inverted microscope for observing shape changes in live cell cultures.

F. Mechanical Stretching

- Epithelial cell cultures as desired.
- BioFlex six-well tissue culture plates, uncoated or coated with extracellular matrix components (Flexcell International, Hillsborough, NC).
- FX-4000T Flexercell Tension Plus system cell stretcher (Flexcell International, Hillsborough, NC), incorporating computer-controlled vacuum system.
- Reagents for fixing and staining cells, as above.

III. Procedures

A. Immunohistochemistry of Filaments in Stressed Tissues

Sources of material for stress assays to study IFs are principally tissues and cells derived from mouse models or from human patient keratinocytes, all used as surrogate models of the relevant disease. Most of these stress models depend on immunohistochemistry or immunocytochemistry to monitor the configuration of the cytoskeleton in tissue cells or cell cultures. Antibodies generally work best on frozen unfixed tissue sections, but the fragility of the tissues in which IFs are defective or altered, plus the small size of mouse organ systems, means that there are few situations in which one can avoid formalin fixation and wax embedding to analyze tissues stressed in organ systems *in situ*. Fortunately, there are now some robust antigen-retrieval methods available, so most antibodies to IFs can now be used on formalin-fixed tissues.

1. Wax sections are cut 7-μm thick and collected onto electrostatically charged slides for improved binding.

2. Sections are dewaxed in 2× 5-minute changes of xylene, followed by 2× 5-minute changes of 100% ethanol. Endogenous peroxidase activity in the sections is then blocked by 30 minutes incubation in 0.3% hydrogen peroxide in methanol. Sections are then rehydrated through 50% ethanol for 5 minutes and washed in running tap water for 5 minutes.

3. Formalin-fixed antigens are then unmasked with either of the following antigen retrieval methods:

    a. Enzyme digestion, in which sections are incubated for 20 minutes in 0.1% trypsin (T-8128: Sigma, Poole, UK), 0.1% $CaCl_2$, and 20 mM Tris, pH 7.8, at 37 °C.

    b. Autoclaving the sections at 120 °C for 20 minutes in antigen unmasking solution (Vector Labs, H-3300) with a Retriever histochemical autoclave (Pickcell Laboratories). Sections are then allowed to cool for at least 2 hours.

4. Sections are then rinsed 3 × 5 minutes in PBS and incubated with primary antibodies (dilution and time depends on the antibodies) in a humidified chamber.

5. Sections are washed for 3 × 3 minutes in PBS. The peroxidase-labeled Envision Dual polymer (anti-mouse and anti-rabbit system, Dakocytomation K-4065) is then applied undiluted to the sections for 30 minutes. After washing sections 3 × 3 minutes in PBS, the DAB + substrate–chromogen system (Dako K3468) is used to visualize the antigen localization.

6. Slides are counterstained in hematoxylin for 2 minutes then dehydrated in ethanol, cleared in xylene, and a coverslip mounted using DPX (VWR, 360292F). Digital images of the stained sections can be captured with Axiovision software on a Zeiss Axiophot 200M microscope or equivalent.

B. Rapid Immunofluorescence of Cultured Cells

This is a quick method for immunofluorescence assessment of cultured cells, which allows cells to be cultured on plastic instead of on glass coverslips. The thickness of the plastic dish does not affect the optics of the preparation when epifluorescence microscopy is used.

1. Cells are cultured to 50%–80% confluence for optimal resolution of cytoskeleton structures. This should give the most cells growing in small colonies, with many cells well spread (growing at the free edges of the colonies) and nearly all cells showing desmosome junctions with their neighbors.

2. Cells are exposed to desired experimental stress.

3. The tissue culture medium is then decanted off and discarded, and 5 ml methanol/acetone (1:1) added, replaced straight away with fresh methanol/ acetone solution, which is then left on for 2–5 minutes to fix and permeabilize the cells. The fixative is then poured off; cells are then rinsed briefly with tap water or PBS.

4. Two milliliters primary antibody is added per dish, diluted as required. Culture dish lid is replaced to prevent dehydration. Antibody is left on for 30–60 minutes at room temperature.

5. Culture plate is then washed with copious tap water (2–4 changes).

6. Two milliliters fluorescent secondary antibody is then added to bind to the primary antibody, diluted as appropriate and left for 30–45 minutes.

7. Secondary antibody is then washed off with three changes of tap water or PBS (3–5 minutes).

8. The side wall of the plastic dish is then cut away with a strong pair of scissors. This allows the plastic plate to be viewed on the microscope stage. The plate is then rinsed with distilled water, flushing any plastic chips away.

9. A rectangular 22 × 13-mm coverslip is then mounted over the cells with antifade fluorescence-compatible mounting reagent. Once the mountant has set, specimens are ready to view by immunofluorescence microscopy.

C. Primary Astrocyte Cultures

In vivo, astrocytes are known to respond to various kind of stress and in this respect can be viewed as guardians of the CNS. Cultured astrocytes are therefore well suited to evaluate responses to diverse challenges, such as osmotic shock or mechanical injury as mimicked, for example, by mechanical stretching or scratch wounding. One advantage of primary astrocyte-enriched cultures over astrocyte cell lines is their closer resemblance to astrocytes *in vivo*, where many different types of astrocytes can be found. An additional stress-inducing factor (apart from the *in vitro* environment itself) is the presence of serum in the culture medium. As other cells in the CNS, astrocytes are normally protected from direct exposure to many molecules from the blood by the blood–brain barrier. The exposure to serum components in culture leads to a certain degree of astrocyte activation and thereby mimics, at least to some extent, conditions associated with broken blood–brain barrier, such as brain trauma or ischemic damage.

In mice, astrocyte-enriched cultures are typically prepared from brains of post-natal day 0–2 (P0–2) mice, either by dissociation of the CNS tissue by trypsin or mechanically. The latter method results in so-called explant cultures and is described in the following (see also Pekny *et al.* [1998]). For a more detailed description, we recommend, for example, the protocol by Hansson and Rönnbäck, (1989) even though it refers to a rat, an animal one order of magnitude larger than a mouse.

1. Brains of 0- to 2-day-old mice are dissected out of the cranium, taking care when exposing the cranium to avoid direct contact with the surface of the skin (to limit the risk of bacterial or fungal infection of cultures).

2. The brain is carefully removed and transferred into sterile PBS in a tissue culture dish. Under the dissection microscope, the meninges are removed as carefully as possible; meningeal mesenchymal cells grow very rapidly and can overgrow astrocytes in culture.

3. The brain tissue is forced through a nylon mesh (80 μm) with a Teflon-coated rod into prewarmed and 5% CO_2-equilibrated complete tissue culture medium in a sterile tissue culture dish. The resulting suspension is distributed between culture dishes, flasks, etc. One mouse brain can be plated on 3–5 10-cm culture dishes.

4. Cultures are maintained at 37 °C and 5% CO_2.

5. Culture medium is changed 3–4 days later, by which time cells are clearly visible migrating out of the brain tissue pieces. These cultures approach confluence after another week.

6. Immunocytochemistry with antibodies to GFAP can be used to assess the proportion of astrocytes in the culture, both in primary cultures and during subsequent passages. Because the number of nonastrocytes in the cultures increases with each passage, primary or passage 1 cultures are most suitable for various stress-inducing, physiological, or pharmacological experiments.

D. Scratch Wound–Stimulated Epithelial Cell Migration

A scratch wound assay is especially useful for analyzing tissues such as skin, where the wound response involves many different cell types interacting in complex ways. A tissue culture scratch wound experiment can be used to look specifically at keratinocyte activation and migration in the absence of dermal responses. Scratch wound assays involve growing cells to confluence and then scraping across the dish to imitate a wound in the monolayer epithelial culture. The clearing of cells by scratching the culture stimulates the cells to become migratory (Fig. 4), and the rate of infilling of the cleared area can then be measured.

1. Keratinocyte cells, or other cells to be investigated, are seeded in a 24-well culture plate (coated if necessary with extracellular matrix material) and grown to confluence. Keratinocyte culture conditions are described elsewhere (Morley *et al.*, 2003).
   Note: Length of time the cells are grown after confluence before the scratch wound is initiated may affect the subsequent behavior of the "wound healing" keratinocytes, because desmosomes continue to mature after confluence.

2. Confluent wells are scratched (through the medium) across their diameter with sterile yellow Gilson tips.

3. For the 0-minute time point, scratched wells are emptied of culture medium by aspiration and immediately fixed with two changes of methanol-acetone (5 minutes; see A previously) at this time. The methanol-acetone is then removed, and the wells are left to air-dry.

4. Remaining wells are fixed in the same way at later time points appropriate to the experiment. For keratinocyte lines, this could be 8 hours after wounding; by this time, significant migration has taken place, but the scratch-wounds are not yet closed. (Note: Rate of closure will depend on the cell line used.)

5. The remaining uncovered area of clear plastic is then measured for each well or time point with (for example) Axiovision 3.0 software (Carl Zeiss Vision GmbH), and the reduction in scratch wound area (i.e. the "wound closure") from time 0 can be calculated.

E. Osmotic Stress–Induced Cell-Shape Changes

Osmotic stress, which causes cells to rapidly shrink (on hyperosmotic stress) or swell (on hypoosmotic stress), is an ever-present hazard for cells, and many cellular strategies have evolved to restrict or rapidly correct for osmotic imbalance. The dangers for unicellular or small multicellular organisms are clear, but even for mammals the risk of osmotic damage is never far away. *In vitro*, cells respond to hypoosmotic environment by sudden rapid cell swelling, but within minutes, they start to return to their original volume (Hoffman, 1991; Kimelberg, 1991). This phenomenon is known as regulatory volume decrease and involves an efflux of osmotically active molecules from cells, among which the amino acid taurine is very important (Hoffman, 1991; Moran *et al.*, 1994; Pasantes-Morales *et al.*, 1990; Vitarella *et al.*, 1994).

1. Cells to be tested are cultured to 80% confluence for maximum cytoskeleton resolution and maximum effect of the osmotic shock.

2. Cells are subjected to hypoosmotic shock by immersion in 150 mM urea for 5 minutes at 37 °C. Monitoring by phase-contrast microscopy will reveal the cell swelling as water passes into the cell, followed by the regulatory volume decrease within 5 minutes of exposure to the 150 mM urea solution. (Water can also be used, but keratinocytes are quite resistant to osmotic swelling by water alone, and the effect takes longer.)

3. Cells are then returned to normal tissue culture medium for varying periods of recovery before further analysis by immunofluorescence microscopy.

F. Mechanical Stretching

Numerous devices are in use for stretching different kinds of cells, but the one that we have found most amenable to mechanical stretching of cultured epithelial cells is the cell stretcher system manufactured by Flexcell International (Hillsborough, NC). This consists of a computer programmable variable vacuum source, which can apply negative pressure to the underside of a deformable membrane on which cells are growing. The membrane forms the base of a six-well plate in which cells can be cultured normally. When the plate of cells is connected to the pump manifold, cells can be cultured while being subjected to a constant or cyclical stretch/relaxation regimen of variable force. The membrane base of each culture well rests on a loading post of nearly the same diameter as the well, leaving a small rim of unsupported membrane around the edge of each well. It is into this space that the membrane is pulled down when the vacuum is applied, resulting in a radial stretch across the surface of the supported cell culture.

1. Cells are seeded onto six-well Bioflex flexible culture plates and grown to 80% confluence (3–4 days).

2. The plates are then connected to the FX-4000T Flexercell Tension Plus cell stretcher (Flexcell International), and the cultured cells are subjected to mechanical stretch.

3. The pattern of stretch can be programmed as required. We have obtained good effects in keratinocytes by use of an oscillating stretch at a frequency of 4 Hz and effective amplitude of 12% for 30 minutes, to which cells are subjected while maintained at 37 °C in 5% CO_2 (Russell *et al.*, 2004). Control wells not being stretched were isolated from the vacuum with FlexStop plugs (Flexcell International).

4. After stretch and recovery as required, the silicone membrane culture well bases from test and control wells are excised with a sharp scalpel from the culture plate and immersed in cold methanol and acetone (1:1) for 5 minutes to fix and permeabilize the cells.

5. Cells are washed twice with cold PBS, then incubated with 5% normal goat serum in PBS for 1 hour at room temperature to block nonspecific antibody binding, and washed again with PBS.

6. Cells are incubated for 1 hour at room temperature with primary antibodies to keratins or associated proteins as desired, washed three times in PBS, then incubated in secondary antibody for 1 hour followed by copious washing in PBS and then distilled water.

7. Silicone membranes bearing the stained cells are then mounted onto glass microscope slides with CitiFluor (Sigma Aldrich, Poole, UK) or equivalent and visualized by fluorescence microscopy (e.g., Fig. 3).

IV. Pitfalls

A. Stress Responses May Depend on Tissue Geometry

Assays to determine cell response to physical deformation have been used to model various human pathologies, predominantly with tissues such as skeletal and cardiac muscle, endothelial cells lining the blood vessels, bladder tissues, bronchial cells lining the respiratory system, and, most recently, skin. The model systems used depend on the geometry of the tissue of interest. Some experiments have used the approach of attaching cells to deformable substrates (thus applying the stretching force to the lower surface of the cell), whereas others use fluid flow to generate shear forces between the tissue culture medium and the cells (i.e., applying a stretching force across the tops of the cultured cells). These two approaches may yield different results. Mechanical stress models that stretch the base of the cells may trigger different signals in a cell from those that impinge primarily on the upper surface of the cell, even if the force of two such stresses is evenly balanced. This is because the first cytoskeleton structures to detect the stress displacement will be different in each case so that different signal transduction cascades may be initiated. The nearest cytoskeleton anchorage junctions to a basal stretch would be the hemidesmosomes, which connect cells through integrin receptors to the extracellular matrix, whereas the nearest anchorage points to an apical shear would be components of the apical junctional complex, (i.e., the tight junctions, adherens junctions, and ultimately desmosomes) connecting cells to one another through cadherins. Each junction type will trigger a slightly different signaling pathway. Although both approaches to mechanical stretching of a cell have the advantage of being applicable to monolayer cultures, they will give information about the properties of the cell in response to different stress signals, so that care will be needed when extrapolating back to the situation in the three-dimensional tissue.

B. The Limitations of Stress Models

When interpreting the results of any stress assay on cells in tissue culture, it must be borne in mind that tissue culture itself imposes a significant metabolic stress on the cells. This is clearly the case for both keratinocytes and astrocytes, in which tissue culture is known to induce altered IF protein expression (compared with cells in tissues). Therefore, the baseline culture effect must be recognized and subtracted from the experimental effects.

The issue of selecting *in vitro* assays or *in vivo* ones merits some forethought. Any assay of biological function that depends on the interaction of multiple differentiated cell types poses a challenge, because it is still unlikely that complex systems can be reproduced to full function in a tissue culture context. Thus, many questions about cell and tissue stress can only ever be answered in a whole animal model. This is particularly true of any functional characteristic that is primarily seen in internal organs or that depends on the interaction of multiple cell types. Most of the intermediate filament defects involving cells in the nervous system would fall into this category; monolayer monocultures of astrocytes, for example, will only ever reveal part of the cells' tissue function.

If cell culture assays are to be informative, the cellular context must be taken into account. An informative assay that can be carried out on a simple monolayer culture has therefore a considerable advantage for many applications. Assays on epithelial cells tend to work well in culture as the original tissue is constructed of a (mostly) homogeneous mass of closely apposed cells from the same tissue (e.g., the epidermis). However, mutations that produce a dramatic *in vivo* phenotype are often less amenable to experimentation *in vivo*, because the phenotype is often too severe or rapid in onset to be useful, as in the case of the EBS keratin mutations that are lethal in mice but not in people (Coulombe *et al.*, 1991; Vassar *et al.*, 1991). It is, therefore, no surprise that tissue culture assays of cell stress in IFs have tended to focus on keratin IFs, whereas functional assays on other IFs, such as GFAP, have usually been carried out in animal models.

V. Discussion/Concluding Remarks

A. Are Stress Model Systems Likely to be Informative?

Although it is early, and superficially it may seem that the functional experiments on defective IFs sometimes give contradictory results, there may be more in common between these results than one first supposes. Most of the experiments carried out so far have shown that cells with abnormal IFs have structural or physical defects: they sustain force less well, they are unable to adopt and maintain complex shapes, they respond less well to transient swelling. It seems that ineffective or inappropriate cell-shape changes trigger stress response signal cascades that are different in nature or in magnitude when cells are not properly reinforced by their IFs. Chemotoxicity (see Chapters 15 and 17) may also lead to cell-shape changes in which a filament-deficient cell may respond aberrantly if the toxicity results in transient metabolic shut down. It seems likely that all the IF defects have generically similar underlying effects on cells which, provided the cells do not actually break, will be translated into stress signals appropriate to that tissue. Thus, the IFs would seem to function not only as mechanoresilient structures but also as a component of the mechanosensory apparatus of the cell, in conjunction with the actin and tubulin systems and probably in response to larger strains and displacements acting on cells (Janmey *et al.*, 1991; Stamenovic *et al.*, 2002). It remains to be determined what exactly those stress signal transduction pathways are that tap into IFs, and what their downstream consequences are in each type of tissue.

Acknowledgments

Birgitte Lane's laboratory is funded by Cancer Research UK, DEBRA, and the Wellcome Trust. Milos Pekny's laboratory is funded by the Swedish Research Council, the Swedish Cancer Foundation, and King Gustav V Foundation. Our thanks to all in our two laboratories, who contributed their own preferred methods and experimental details, and to David Russell and Sharon Jenner for illustrations.

References

Albrecht-Buehler, G. (1977a). The phagokinetic tracks of 3T3 cells. *Cell* **11**, 395–404.
Albrecht-Buehler, G. (1977b). Phagokinetic tracks of 3T3 cells: Parallels between the orientation of track segments and of cellular structures which contain actin or tubulin. *Cell* **12**, 333–339.

Allen, W. E., Zicha, D., Ridley, A. J., and Jones, G. E. (1998). A role for Cdc42 in macrophage chemotaxis. *J. Cell Biol.* **141**, 1147–1157.

Anderova, M., Kubinova, S., Mazel, T., Chvatal, A., Eliasson, C., Pekny, M., and Sykova, E. (2001). Effect of elevated K(+), hypotonic stress, and cortical spreading depression on astrocyte swelling in GFAP-deficient mice. *Glia* **35**, 189–203.

Ando, J., Nomura, H., and Kamiya, A. (1987). The effect of fluid shear stress on the migration and proliferation of cultured endothelial cells. *Microvasc. Res.* **33**, 62–70.

Balogh, J., Li, Z., Paulin, D., and Arner, A. (2003). Lower active force generation and improved fatigue resistance in skeletal muscle from desmin deficient mice. *J. Muscle Res. Cell Motil.* **24**, 453–459.

Balogh, J., Merisckay, M., Li, Z., Paulin, D., and Arner, A. (2002). Hearts from mice lacking desmin have a myopathy with impaired active force generation and unaltered wall compliance. *Cardiovasc. Res.* **53**, 439–450.

Bloom, S., Lockard, V. G., and Bloom, M. (1996). Intermediate filament-mediated stretch-induced changes in chromatin: A hypothesis for growth initiation in cardiac myocytes. *J. Mol. Cell Cardiol.* **28**, 2123–2127.

Brenner, M., Johnson, A. B., Boespflug-Tanguy, O., Rodriguez, D., Goldman, J. E., and Messing, A. (2001). Mutations in GFAP, encoding glial fibrillary acidic protein, are associated with Alexander disease. *Nat. Genet.* **27**, 117–120.

Brunette, D. M. (1984). Mechanical stretching increases the number of epithelial cells synthesizing DNA in culture. *J. Cell Sci.* **69**, 35–45.

Cantiello, H. F. (1997). Role of actin filament organization in cell volume and ion channel regulation. *J. Exp. Zool.* **279**, 425–435.

Cantiello, H. F., Prat, A. G., Bonventre, J. V., Cunningham, C. C., Hartwig, J. H., and Ausiello, D. A. (1993). Actin-binding protein contributes to cell volume regulatory ion channel activation in melanoma cells. *J. Biol. Chem.* **268**, 4596–4599.

Cao, T., Longley, M. A., Wang, X. J., and Roop, D. R. (2001). An inducible mouse model for epidermolysis bullosa simplex: Implications for gene therapy. *J. Cell Biol.* **152**, 651–656.

Chen, J. D., Helmold, M., Kim, J. P., Wynn, K. C., and Woodley, D. T. (1994). Human keratinocytes make uniquely linear phagokinetic tracks. *Dermatology* **188**, 6–12.

Chu, Y. W., Seftor, E. A., Romer, L. H., and Hendrix, M. J. (1996). Experimental coexpression of vimentin and keratin intermediate filaments in human melanoma cells augments motility. *Am. J. Pathol.* **148**, 63–69.

Clarke, S. R., Shetty, A. K., Bradley, J. L., and Turner, D. A. (1994). Reactive astrocytes express the embryonic intermediate neurofilament nestin. *Neuroreport* **5**, 1885–1888.

Colucci-Guyon, E., Portier, M. M., Dunia, I., Paulin, D., Pournin, S., and Babinet, C. (1994). Mice lacking vimentin develop and reproduce without an obvious phenotype. *Cell* **79**, 679–694.

Coulombe, P. A., Hutton, M. E., Letai, A., Hebert, A., Paller, A. S., and Fuchs, E. (1991). Point mutations in human keratin 14 genes of epidermolysis bullosa simplex patients: Genetic and functional analyses. *Cell* **66**, 1301–1311.

D'Alessandro, M., Russell, D., Morley, S. M., Davies, A. M., and Lane, E. B. (2002). Keratin mutations of epidermolysis bullosa simplex alter the kinetics of stress response to osmotic shock. *J. Cell Sci.* **115**, 4341–4351.

Ding, M., Eliasson, C., Betsholtz, C., Hamberger, A., and Pekny, M. (1998). Altered taurine release following hypotonic stress in astrocytes from mice deficient for GFAP and vimentin. *Brain Res. Mol. Brain Res.* **62**, 77–81.

Eckes, B., Dogic, D., Colucci-Guyon, E., Wang, N., Maniotis, A., Ingber, D., Merckling, A., Langa, F., Aumailley, M., Delouvee, A., *et al.* (1998). Impaired mechanical stability, migration and contractile capacity in vimentin-deficient fibroblasts. *J. Cell Sci.* **111(Pt 13)**, 1897–1907.

Eliasson, C., Sahlgren, C., Berthold, C. H., Stakeberg, J., Celis, J. E., Betsholtz, C., Eriksson, J. E., and Pekny, M. (1999). Intermediate filament protein partnership in astrocytes. *J. Biol. Chem.* **274**, 23996–24006.

Emsley, J. G., Arlotta, P., and Macklis, J. D. (2004). Star-cross'd neurons: Astroglial effects on neural repair in the adult mammalian CNS. *Trends Neurosci.* **27,** 238–240.

Eng, L. F., Vanderhaeghen, J. J., Bignami, A., and Gerstl, B. (1971). An acidic protein isolated from fibrous astrocytes. *Brain Res.* **28,** 351–354.

Freedberg, I. M., Tomic-Canic, M., Komine, M., and Blumenberg, M. (2001). Keratins and the keratinocyte activation cycle. *J. Invest. Dermatol.* **116,** 633–640.

Goldstein, M. E., Weiss, S. R., Lazzarini, R. A., Shneidman, P. S., Lees, J. F., and Schlaepfer, W. W. (1988). mRNA levels of all three neurofilament proteins decline following nerve transection. *Brain Res.* **427,** 287–291.

Gomi, H., Yokoyama, T., Fujimoto, K., Ikeda, T., Katoh, A., Itoh, T., and Itohara, S. (1995). Mice devoid of the glial fibrillary acidic protein develop normally and are susceptible to scrapie prions. *Neuron* **14,** 29–41.

Hansson, E., and Rönnbäck, L. (1989). Primary cultures of astroglia and neurons from different brain regions. *In* "A Dissection and Tissue Culture Manual of the Nervous System" (A. Shahar, J. de Vellis, A. Vernadakis, and B. Haber, eds.), pp. 92–104. Allan R. Liss Inc., New York.

Helmke, B. P., Thakker, D. B., Goldman, R. D., and Davies, P. F. (2001). Spatiotemporal analysis of flow-induced intermediate filament displacement in living endothelial cells. *Biophys. J.* **80,** 184–194.

Hendrix, M. J., Seftor, E. A., Chu, Y. W., Seftor, R. E., Nagle, R. B., McDaniel, K. M., Leong, S. P., Yohem, K. H., Leibovitz, A. M., Meyskens, F. L., Jr., *et al.* (1992). Coexpression of vimentin and keratins by human melanoma tumor cells: Correlation with invasive and metastatic potential. *J. Natl. Cancer Inst.* **84,** 165–174.

Hendrix, M. J., Seftor, E. A., Seftor, R. E., and Trevor, K. T. (1997). Experimental co-expression of vimentin and keratin intermediate filaments in human breast cancer cells results in phenotypic interconversion and increased invasive behavior. *Am. J. Pathol.* **150,** 483–495.

Henrion, D., Terzi, F., Matrougui, K., Duriez, M., Boulanger, C. M., Colucci-Guyon, E., Babinet, C., Briand, P., Friedlander, G., Poitevin, P., *et al.* (1997). Impaired flow-induced dilation in mesenteric resistance arteries from mice lacking vimentin. *J. Clin. Invest.* **100,** 2909–2914.

Hoffman, E. (1991). Volume regulation in cultured cells. Academic Press, New York.

Holwell, T. A., Schweitzer, S. C., and Evans, R. M. (1997). Tetracycline regulated expression of vimentin in fibroblasts derived from vimentin null mice. *J. Cell. Sci.* **110(Pt 16),** 1947–1956.

Irvine, A. D., and McLean, W. H. (1999). Human keratin diseases: The increasing spectrum of disease and subtlety of the phenotype-genotype correlation. *Br. J. Dermatol.* **140,** 815–828.

Janmey, P. A., Euteneuer, U., Traub, P., and Schliwa, M. (1991). Viscoelastic properties of vimentin compared with other filamentous biopolymer networks. *J. Cell Biol.* **113,** 155–160.

Janmey, P. A., Shah, J. V., Janssen, K. P., and Schliwa, M. (1998). Viscoelasticity of intermediate filament networks. *Subcell. Biochem.* **31,** 381–397.

Jenner, S. (2001). Stategies to investigate the role of keratin intermediate filaments in epithelial cell migration. *In* Ph.D. thesis, University of Dundee, Dundee, UK.

Johansson, C. B., Momma, S., Clarke, D. L., Risling, M., Lendahl, U., and Frisen, J. (1999). Identification of a neural stem cell in the adult mammalian central nervous system. *Cell* **96,** 25–34.

Jones, G. E., Zicha, D., Dunn, G. A., Blundell, M., and Thrasher, A. (2002). Restoration of podosomes and chemotaxis in Wiskott-Aldrich syndrome macrophages following induced expression of WASp. *Int. J. Biochem. Cell Biol.* **34,** 806–815.

Kaya, S. S., Mahmood, A., Li, Y., Yavuz, E., and Chopp, M. (1999). Expression of cell cycle proteins (cyclin D1 and cdk4) after controlled cortical impact in rat brain. *J. Neurotrauma* **16,** 1187–1196.

Kimelberg, H. K. (1991). "Swelling and Volume Control in Brain Astroglial Cells." Springer, New York.

Kinouchi, R., Takeda, M., Yang, L., Wilhelmsson, U., Lundkvist, A., Pekny, M., and Chen, D. F. (2003). Robust neural integration from retinal transplants in mice deficient in GFAP and vimentin. *Nat. Neurosci.* **6,** 863–868.

Kippenberger, S., Bernd, A., Loitsch, S., Guschel, M., Muller, J., Bereiter-Hahn, J., and Kaufmann, R. (2000). Signaling of mechanical stretch in human keratinocytes via MAP kinases. *J. Invest. Dermatol.* **114,** 408–412.

Lacolley, P., Challande, P., Boumaza, S., Cohuet, G., Laurent, S., Boutouyrie, P., Grimaud, J. A., Paulin, D., Lamaziere, J. M., and Li, Z. (2001). Mechanical properties and structure of carotid arteries in mice lacking desmin. *Cardiovasc. Res.* **51,** 178–187.

Lammerding, J., Schulze, P. C., Takahashi, T., Kozlov, S., Sullivan, T., Kamm, R. D., Stewart, C. L., and Lee, R. T. (2004). Lamin A/C deficiency causes defective nuclear mechanics and mechanotransduction. *J. Clin. Invest.* **113,** 370–378.

Lepekhin, E. A., Eliasson, C., Berthold, C. H., Berezin, V., Bock, E., and Pekny, M. (2001). Intermediate filaments regulate astrocyte motility. *J. Neurochem.* **79,** 617–625.

Li, C., Hu, Y., Sturm, G., Wick, G., and Xu, Q. (2000). Ras/Rac-dependent activation of p38 mitogen-activated protein kinases in smooth muscle cells stimulated by cyclic strain stress. *Arterioscler. Thromb. Vasc. Biol.* **20,** E1–E9.

Li, D., Tapscoft, T., Gonzalez, O., Burch, P. E., Quinones, M. A., Zoghbi, W. A., Hill, R., Bachinski, L. L., Mann, D. L., and Roberts, R. (1999). Desmin mutation responsible for idiopathic dilated cardiomyopathy. *Circulation* **100,** 461–464.

Li, Z., Mericskay, M., Agbulut, O., Butler-Browne, G., Carlsson, L., Thornell, L. E., Babinet, C., and Paulin, D. (1997). Desmin is essential for the tensile strength and integrity of myofibrils but not for myogenic commitment, differentiation, and fusion of skeletal muscle. *J. Cell Biol.* **139,** 129–144.

Loranger, A., Duclos, S., Grenier, A., Price, J., Wilson-Heiner, M., Baribault, H., and Marceau, N. (1997). Simple epithelium keratins are required for maintenance of hepatocyte integrity. *Am. J. Pathol.* **151,** 1673–1683.

Lundkvist, A., Reichenbach, A., Betsholtz, C., Carmeliet, P., Wolburg, H., and Pekny, M. (2004). Under stress, the absence of intermediate filaments in Müller cells in the retina has structural and functional consequences. *J. Cell Sci.* **117,** 3481–3486.

Malinda, K. M., Sidhu, G. S., Mani, H., Banaudha, K., Maheshwari, R. K., Goldstein, A. L., and Kleinman, H. K. (1999). Thymosin beta4 accelerates wound healing. *J. Invest. Dermatol.* **113,** 364–368.

Mansbridge, J. N., and Knapp, A. M. (1987). Changes in keratinocyte maturation during wound healing. *J. Invest. Dermatol.* **89,** 253–263.

Markey, A. C., Lane, E. B., Churchill, L. J., MacDonald, D. M., and Leigh, I. M. (1991). Expression of simple epithelial keratins 8 and 18 in epidermal neoplasia. *J. Invest. Dermatol.* **97,** 763–770.

McCall, M. A., Gregg, R. G., Behringer, R. R., Brenner, M., Delaney, C. L., Galbreath, E. J., Zhang, C. L., Pearce, R. A., Chiu, S. Y., and Messing, A. (1996). Targeted deletion in astrocyte intermediate filament (GFAP) alters neuronal physiology. *Proc. Natl. Acad. Sci. USA* **93,** 6361–6366.

McGrath, J., Somlo, S., Makova, S., Tian, X., and Brueckner, M. (2003). Two populations of node monocilia initiate left-right asymmetry in the mouse. *Cell* **114,** 61–73.

McGraw, T. S., Mickle, J. P., Shaw, G., and Streit, W. J. (2002). Axonally transported peripheral signals regulate alpha-internexin expression in regenerating motoneurons. *J. Neurosci.* **22,** 4955–4963.

Mercola, M. (2003). Left-right asymmetry: Nodal points. *J. Cell Sci.* **116,** 3251–3257.

Milner, D. J., Weitzer, G., Tran, D., Bradley, A., and Capetanaki, Y. (1996). Disruption of muscle architecture and myocardial degeneration in mice lacking desmin. *J. Cell Biol.* **134,** 1255–1270.

Moran, J., Maar, T., and Pasantes-Morales, H. (1994). Cell volume regulation in taurine deficient cultured astrocytes. *Adv. Exp. Med. Biol.* **359,** 361–367.

Moran, J., Sabanero, M., Meza, I., and Pasantes-Morales, H. (1996). Changes of actin cytoskeleton during swelling and regulatory volume decrease in cultured astrocytes. *Am. J. Physiol.* **271,** C1901–C1907.

Morley, S. M., D'Alessandro, M., Sexton, C., Rugg, E. L., Navsaria, H., Shemanko, C. S., Huber, M., Hohl, D., Heagerty, A. I., Leigh, I. M., *et al.* (2003). Generation and characterization of epidermolysis bullosa simplex cell lines: Scratch assays show faster migration with disruptive keratin mutations. *Br. J. Dermatol.* **149**, 46–58.

Mullender, M., El Haj, A. J., Yang, Y., van Duin, M. A., Burger, E. H., and Klein-Nulend, J. (2004). Mechanotransduction of bone cells *in vitro*: Mechanobiology of bone tissue. *Med. Biol. Eng. Comput.* **42**, 14–21.

Nasca, M. R., O'Toole, E. A., Palicharla, P., West, D. P., and Woodley, D. T. (1999). Thalidomide increases human keratinocyte migration and proliferation. *J. Invest. Dermatol.* **113**, 720–724.

Nauli, S. M., Alenghat, F. J., Luo, Y., Williams, E., Vassilev, P., Li, X., Elia, A. E., Lu, W., Brown, E. M., Quinn, S. J., *et al.* (2003). Polycystins 1 and 2 mediate mechanosensation in the primary cilium of kidney cells. *Nat. Genet.* **33**, 129–137.

Nawashiro, H., Messing, A., Azzam, N., and Brenner, M. (1998). Mice lacking GFAP are hypersensitive to traumatic cerebrospinal injury. *Neuroreport* **9**, 1691–1696.

O'Toole, E. A., Marinkovich, M. P., Hoeffler, W. K., Furthmayr, H., and Woodley, D. T. (1997). Laminin-5 inhibits human keratinocyte migration. *Exp. Cell Res.* **233**, 330–339.

Okabe, M., Ikawa, M., Kominami, K., Nakanishi, T., and Nishimune, Y. (1997). 'Green mice' as a source of ubiquitous green cells. *FEBS Lett.* **407**, 313–319.

Orsola, A., Adam, R. M., Peters, C. A., and Freeman, M. R. (2002). The decision to undergo DNA or protein synthesis is determined by the degree of mechanical deformation in human bladder muscle cells. *Urology* **59**, 779–783.

Östlund, C., and Worman, H. (2003). Nuclear envelope proteins and neuromuscular diseases. *Muscle Nerve* **27**, 393–406.

Paladini, R. D., Takahashi, K., Bravo, N. S., and Coulombe, P. A. (1996). Onset of reepithelialization after skin injury correlates with a reorganization of keratin filaments in wound edge keratinocytes: Defining a potential role for keratin 16. *J. Cell Biol.* **132**, 381–397.

Parrish, E. P., Steart, P. V., Garrod, D. R., and Weller, R. O. (1987). Antidesmosomal monoclonal antibody in the diagnosis of intracranial tumours. *J. Pathol.* **153**, 265–273.

Pasantes-Morales, H., Moran, J., and Schousboe, A. (1990). Volume-sensitive release of taurine from cultured astrocytes: Properties and mechanism. *Glia* **3**, 427–432.

Pekny, M. (2001). Astrocytic intermediate filaments: Lessons from GFAP and vimentin knockout mice. *Prog. Brain Res.* **132**, 23–30.

Pekny, M., Eliasson, C., Chien, C. L., Kindblom, L. G., Liem, R., Hamberger, A., and Betsholtz, C. (1998). GFAP-deficient astrocytes are capable of stellation *in vitro* when cocultured with neurons and exhibit a reduced amount of intermediate filaments and an increased cell saturation density. *Exp. Cell Res.* **239**, 332–343.

Pekny, M., Johansson, C. B., Eliasson, C., Stakeberg, J., Wallen, A., Perlmann, T., Lendahl, U., Betsholtz, C., Berthold, C. H., and Frisen, J. (1999). Abnormal reaction to central nervous system injury in mice lacking glial fibrillary acidic protein and vimentin. *J. Cell Biol.* **145**, 503–514.

Pekny, M., Leveen, P., Pekna, M., Eliasson, C., Berthold, C. H., Westermark, B., and Betsholtz, C. (1995). Mice lacking glial fibrillary acidic protein display astrocytes devoid of intermediate filaments but develop and reproduce normally. *EMBO J.* **14**, 1590–1598.

Pekny, M., Pekna, M., Wilhelmsson, U., and Chen, D. F. (2004). Response to Quinlan and Nilsson: Astroglia sitting at the controls? *Trends Neurosci.* **27**, 243–244.

Porter, R. M., Leitgeb, S., Melton, D. W., Swensson, O., Eady, R. A., and Magin, T. M. (1996). Gene targeting at the mouse cytokeratin 10 locus: Severe skin fragility and changes of cytokeratin expression in the epidermis. *J. Cell Biol.* **132**, 925–936.

Porter, R. M., Lunny, D. P., Ogden, P. H., Morley, S. M., McLean, W. H., Evans, A., Harrison, D. L., Rugg, E. L., and Lane, E. B. (2000). K15 expression implies lateral differentiation within stratified epithelial basal cells. *Lab. Invest.* **80**, 1701–1710.

Praetorius, H. A., and Spring, K. R. (2003). The renal cell primary cilium functions as a flow sensor. *Curr. Opin. Nephrol. Hypertens.* **12**, 517–520.

Purkis, P. E., Steel, J. B., Mackenzie, I. C., Nathrath, W. B. J., Leigh, I. M., and Lane, E. B. (1990). Antibody markers of basal cells in complex epithelia. *J. Cell Sci.* **97**, 39–50.

Quinlan, R., and Nilsson, M. (2004). Reloading the retina by modifying the glial matrix. *Trends Neurosci.* **27,** 241–242.

Robey, H. L., Hiscott, P. S., and Grierson, I. (1992). Cytokeratins and retinal epithelial cell behaviour. *J. Cell Sci.* **102,** 329–340.

Russell, D. (2004). The response of keratinocytes to mechanical stretch. Ph.D. thesis, University of Dundee, Dundee, UK.

Russell, D., Andrews, P. D., James, J., and Lane, E. B. (2004). Mechanical stress induces profound remodelling of keratin filaments and cell junctions in epidermolysis bullosa simplex keratinocytes. *J. Cell Sci.* **117,** in press.

Sam, M., Shah, S., Friden, J., Milner, D. J., Capetanaki, Y., and Lieber, R. L. (2000). Desmin knockout muscles generate lower stress and are less vulnerable to injury compared with wild-type muscles. *Am. J. Physiol. Cell Physiol.* **279,** C1116–C1122.

Sanchez-Olea, R., Moran, J., Schousboe, A., and Pasantes-Morales, H. (1991). Hyposmolarity-activated fluxes of taurine in astrocytes are mediated by diffusion. *Neurosci. Lett.* **130,** 233–236.

Schermer, A., Jester, J. V., Hardy, C., Milano, D., and Sun, T. T. (1989). Transient synthesis of K6 and K16 keratins in regenerating rabbit corneal epithelium: Keratin markers for an alternative pathway of keratinocyte differentiation. *Differentiation* **42,** 103–110.

Schnittler, H. J., Franke, R. P., Akbay, U., Mrowietz, C., and Drenckhahn, D. (1993). Improved *in vitro* rheological system for studying the effect of fluid shear stress on cultured cells. *Am. J. Physiol.* **265,** C289–C298.

Shibuya, S., Miyamoto, O., Auer, R. N., Itano, T., Mori, S., and Norimatsu, H. (2002). Embryonic intermediate filament, nestin, expression following traumatic spinal cord injury in adult rats. *Neuroscience* **114,** 905–916.

Sjöberg, G., Saavedra-Matiz, C. A., Rosen, D. R., Wijsman, E. M., Borg, K., Horowitz, S. H., and Sejersen, T. (1999). A missense mutation in the desmin rod domain is associated with autosomal dominant distal myopathy, and exerts a dominant negative effect on filament formation. *Hum. Mol. Genet.* **8,** 2191–2198.

Skarja, G. A., Kinlough-Rathbone, R. L., Perry, D. W., Rubens, F. D., and Brash, J. L. (1997). A cone-and-plate device for the investigation of platelet biomaterial interactions. *J. Biomed. Mater. Res.* **34,** 427–438.

Smith, F. J., Jonkman, M. F., van Goor, H., Coleman, C. M., Covello, S. P., Uitto, J., and McLean, W. H. (1998). A mutation in human keratin K6b produces a phenocopy of the K17 disorder pachyonychia congenita type 2. *Hum. Mol. Genet.* **7,** 1143–1148.

Smith, L. E., Wesolowski, E., McLellan, A., Kostyk, S. K., D'Amato, R., Sullivan, R., and D'Amore, P. A. (1994). Oxygen-induced retinopathy in the mouse. *Invest. Ophthalmol. Vis. Sci.* **35,** 101–111.

Smith, M. L., Sperandio, M., Galkina, E. V., and Ley, K. (2004). Autoperfused mouse flow chamber reveals synergistic neutrophil accumulation through P-selectin and E-selectin. *J. Leukoc. Biol.* **75,** in press.

Stamenovic, D., Liang, Z., Chen, J., and Wang, N. (2002). Effect of the cytoskeletal prestress on the mechanical impedance of cultured airway smooth muscle cells. *J. Appl. Physiol.* **92,** 1443–1450.

Sterpetti, A. V., Cucina, A., Santoro, L., Cardillo, B., and Cavallaro, A. (1992). Modulation of arterial smooth muscle cell growth by haemodynamic forces. *Eur. J. Vasc. Surg.* **6,** 16–20.

Subramaniam, A., Jones, W. K., Gulick, J., Wert, S., Neumann, J., and Robbins, J. (1991). Tissue-specific regulation of the alpha-myosin heavy chain gene promoter in transgenic mice. *J. Biol. Chem.* **266,** 24613–24620.

Terzi, F., Henrion, D., Colucci-Guyon, E., Federici, P., Babinet, C., Levy, B. I., Briand, P., and Friedlander, G. (1997). Reduction of renal mass is lethal in mice lacking vimentin. Role of endothelin-nitric oxide imbalance. *J. Clin. Invest.* **100,** 1520–1528.

Thornell, L., Carlsson, L., Li, Z., Mericskay, M., and Paulin, D. (1997). Null mutation in the desmin gene gives rise to a cardiomyopathy. *J. Mol. Cell Cardiol.* **29,** 2107–2124.

Tsuruta, D., and Jones, J. C. (2003). The vimentin cytoskeleton regulates focal contact size and adhesion of endothelial cells subjected to shear stress. *J. Cell Sci.* **116,** 4977–4984.

Vaittinen, S., Lukka, R., Sahlgren, C., Hurme, T., Rantanen, J., Lendahl, U., Eriksson, J. E., and Kalimo, H. (2001). The expression of intermediate filament protein nestin as related to vimentin and desmin in regenerating skeletal muscle. *J. Neuropathol. Exp. Neurol.* **60,** 588–597.

Vassar, R., Coulombe, P. A., Degenstein, L., Albers, K., and Fuchs, E. (1991). Mutant keratin expression in transgenic mice causes marked abnormalities resembling a human genetic skin disease. *Cell* **64,** 365–380.

Vitarella, D., DiRisio, D. J., Kimelberg, H. K., and Aschner, M. (1994). Potassium and taurine release are highly correlated with regulatory volume decrease in neonatal primary rat astrocyte cultures. *J. Neurochem.* **63,** 1143–1149.

Wang, N., and Stamenovic, D. (2000). Contribution of intermediate filaments to cell stiffness, stiffening, and growth. *Am. J. Physiol. Cell Physiol.* **279,** C188–C194.

Webb, S. E., Pollard, J. W., and Jones, G. E. (1996). Direct observation and quantification of macrophage chemoattraction to the growth factor CSF-1. *J. Cell Sci.* **109,** 793–803.

Wede, O. K., Lofgren, M., Li, Z., Paulin, D., and Arner, A. (2002). Mechanical function of intermediate filaments in arteries of different size examined using desmin deficient mice. *J. Physiol.* **540,** 941–949.

Weisleder, N., Taffet, G. E., and Capetanaki, Y. (2004). Bcl-2 overexpression corrects mitochondrial defects and ameliorates inherited desmin null cardiomyopathy. *Proc. Natl. Acad. Sci. USA* **101,** 769–774.

Weiss, R. A., Eichner, R., and Sun, T. T. (1984). Monoclonal antibody analysis of keratin expression in epidermal diseases: A 48- and 56-kdalton keratin as molecular markers for hyperproliferative keratinocytes. *J. Cell Biol.* **98,** 1397–1406.

Wilhelmsson, U. (2004). Astrocytes, reactive gliosis and CNS regeneration. Ph.D. thesis, Sahlgrenska Academy, Göteborg University, Göteborg, Sweden.

Wilhelmsson, U., Li, L., Pekna, M., Berthold, C.-H., Blum, S., Eliasson, C., Renner, O., Bushong, E., Ellsiman, M., Morgan, T., *et al.* (2004). Absence of glial fibrillary acidic protein and vimentin prevents hypertrophy of astrocytic processes and improves post-traumatic regeneration. *J. Neurosci.* **24,** 5016–5021.

Wojcik, S. M., Bundman, D. S., and Roop, D. R. (2000). Delayed wound healing in keratin 6a knockout mice. *Mol. Cell Biol.* **20,** 5248–5255.

Wojcik, S. M., Longley, M. A., and Roop, D. R. (2001). Discovery of a novel murine keratin 6 (K6) isoform explains the absence of hair and nail defects in mice deficient for K6a and K6b. *J. Cell Biol.* **154,** 619–630.

Wong, P., Colucci-Guyon, E., Takahashi, K., Gu, C., Babinet, C., and Coulombe, P. A. (2000). Introducing a null mutation in the mouse K6alpha and K6beta genes reveals their essential structural role in the oral mucosa. *J. Cell Biol.* **150,** 921–928.

Wong, P., and Coulombe, P. A. (2003). Loss of keratin 6 (K6) proteins reveals a function for intermediate filaments during wound repair. *J. Cell Biol.* **163,** 327–337.

Yano, S., Komine, M., Fujimoto, M., Okochi, H., and Tamaki, K. (2004). Mechanical stretching *in vitro* regulates signal transduction pathways and cellular proliferation in human epidermal keratinocytes. *J. Invest. Dermatol.* **122,** 783–790.

Yoshikawa, K., Katagata, Y., and Kondo, S. (1995). Relative amounts of keratin 17 are higher than those of keratin 16 in hair-follicle-derived tumors in comparison with nonfollicular epithelial skin tumors. *J. Invest. Dermatol.* **104,** 396–400.

Zicha, D., Dunn, G. A., and Brown, A. F. (1991). A new direct-viewing chemotaxis chamber. *J. Cell Sci.* **99(Pt 4),** 769–775.

PART II

General Methods to Study
Intermediate Filament Proteins and
Gene Regulation

Regulation of Intermediate Filament Gene Expression

Satrajit Sinha

Department of Biochemistry
State University of New York at Buffalo
Buffalo, New York 14214

I. Introduction

A. General Background

The intermediate filaments (IF) constitute a large family of proteins that are expressed in multiple tissues and organs. In humans, the family of genes encoding the IF proteins consists of at least 67 members that can be divided into six classes on the basis of their distinctive sequences (Hesse *et al.*, 2001). The recent explosion in genomic sequencing projects has led to the identification of genes encoding IF proteins in a wide variety of organisms that occupy different levels in the evolutionary scale such as nematodes, tunicates, fish, mouse, and others (Karabinos *et al.*, 2001, 2004; Zimek *et al.*, 2003). Considerable variation exists in the number of IF genes present in each species, with higher organisms exhibiting a much larger family size and complexity. The IF family for many of these organisms is also growing as additional genes are identified by data mining and bioinformatics analysis. It is likely that soon these newly discovered genes will be annotated, and a catalog for all IF genes will be available for many species. But for this catalog to be complete and accurate, wet laboratory experiments have to be performed to validate the data obtained from the *in silico* efforts that are in progress. This would, for example, involve the actual cloning of the IF genes, analyzing their expression patterns, and distinguishing them from the large number of potential pseudogenes that are scattered across the genomic land mine.

IF proteins are present in virtually every cell and every tissue of higher organisms, and their expression is highly regulated during different stages of cell growth, development, and differentiation. Indeed, one mode of classification of the IF proteins is based on the cell or tissue type in which they are expressed (Coulombe *et al.*, 2001; Herrmann and Aebi, 2000). Inside these cells, the IF proteins can self-assemble into cytoskeletal filaments that serve primarily as a mechanical scaffold to enable the cells to maintain their shape and sustain physical stress. Interestingly, mutations of IF genes have been associated with a number of human diseases such as those that cause fragility of the skin epidermis and its appendages, thus providing *in vivo* genetic proof of the protective cytoskeletal function of the IF network (Fuchs and Cleveland, 1998). Apart from their obvious mechanical function, IF proteins have also been implicated in other cellular processes such as signaling, gene expression, and apoptosis. Clearly, much work is needed to decipher the emerging role of IF proteins in these critical cellular functions. The large number of IF proteins expressed in diverse cell types is regulated in a context-dependent manner that is tied to the varied and unique needs of individual cell types. Hence, a major research emphasis for many laboratories has been to understand the transcriptional control mechanisms that dictate the expression of the various classes of the IF genes in different cell and tissue populations.

Keratins constitute the largest subgroup of the IF proteins and represent the most abundant proteins in epithelial cells (Coulombe and Omary, 2002). These proteins are encoded by a large multigene family, whose individual members can be divided into type I (acidic) and type II (basic and neutral) classes on the basis of their sequence. The type I and type II genes are regulated in a pair wise, tissue-specific, and differentiation-specific manner in epithelial tissues of various types, including simple epithelia, internal stratified epithelia (such as esophagus), outer stratified epithelia (such as skin), and hair follicles. In the human genome, all the type I keratin genes (except for K18) are clustered on chromosome 17q21, whereas the type II keratin genes are clustered on chromosome 12q13. Interestingly, the gene for K18, a type I keratin, is found adjacent to the gene for its obligatory partner K8 on chromosome 12q13. The expression of the specific pairs of keratins depends largely on the tissue-type, differentiation stage, and the physiological state (Fuchs and Weber, 1994). Thus, basal keratinocytes express K5, K14, K19, and K15 as major keratins in the stratified squamous epithelium (Fuchs and Byrne, 1994; Moll *et al.*, 1982). As these cells differentiate, suprabasal cells turn on a new set of keratins, such as K1/K10/K2 in skin, K4/K13 in buccal epithelium, and K3/K12 in cornea (Moll *et al.*, 1982). The cells of the simple epithelium in portions of the gastrointestinal tract, liver, and pancreas express predominantly K8 and K18 (Owens and Lane, 2003). A different set of keratins—K6, K16, and K17—are associated with activated keratinocytes and with hyperproliferative states, such as wound healing, psoriasis, and various cancers (Machesney *et al.*, 1998; Paladini *et al.*, 1996).

The specific expression of these keratins in distinct cell populations has offered investigators attractive model systems to study the molecular switches that govern tissue-specific and differentiation-specific gene expression. These studies have primarily focused on the identification of the *cis* and *trans* regulatory elements. Indeed, the promoter and enhancer elements for several keratin genes and the transcription factors that bind to these regulatory DNA sequences have been examined. Identification and characterization of these promoters and enhancers have provided major insights into the transcriptional control mechanisms that fine-tune the expression of keratins in various tissues. In addition, these regulatory domains have also proven to be invaluable tools to target transgene expression to specific tissues and organs in mice and to develop model systems to study various normal and aberrant cellular processes such as development, differentiation, and neoplasia.

As discussed in detail later, the transcriptional regulatory mechanisms of the basal-keratinocyte–specific keratins, K5 and K14, have been under intense investigation. Similarly, the differentiation-specific expression of K1 and K10 genes in the suprabasal layers has been exploited to understand what regulatory switches govern differentiation-specific expression (Maytin *et al.*, 1999; Rosenthal *et al.*, 1991; Rothnagel *et al.*, 1993). These studies have led to the identification of control elements that can direct expression of reporter genes in suprabasal keratinocytes in transgenic mice and also respond to differentiation signals in a keratinocyte cell culture system (such as increased Ca^{++} concentration in the medium). Although rapid strides have been made in the delineation of the regulatory *cis* elements for many of the major keratin genes, our understanding of how cell-type and differentiation-specific expression is achieved is limited. This is primarily because the expression of these keratin genes is controlled in a complex manner, and the identity of some of the transcription factors that bind to the regulatory elements *in vivo* remains unknown and awaits further exploration.

The studies on the transcriptional control elements of the K18 gene underscore the complexity of this regulatory process and provide several interesting observations. In this case, identification of the potential regulatory *cis* elements for K18 has been accomplished by DNase I hypersensitive (HS) site mapping, and these elements have been shown to confer position-independent expression on the K18 promoter and heterologous reporter genes in transgenic mice (Neznanov and Oshima, 1993). Interestingly, one regulatory element resides in a protein coding portion (exon 6) of the K18 gene (Neznanov et al., 1997). Furthermore, an Alu-containing element has been shown to be one component of the locus control region associated with the K18 gene and may function to insulate it from transcriptional interference of neighboring genes (Willoughby et al., 2000). Similar analysis of the human K8 gene, a partner for K18, in simple epithelia has demonstrated that important control elements are located both in the body of the gene and in surrounding regions (Casanova et al., 1995). These studies on the regulation of the K18 and K8 gene highlight the importance of analyzing both upstream and downstream regions for potential regulatory elements.

The keratins K6, and its partners K16 and 17, are regulated in both an inducible and constitutive manner with the inducible signals including injury, viral infections, psoriasis, and other challenges to epithelial tissues (Freedberg et al., 2001; Mcgowan, 1998). Hence, the examination of the regulatory regions of these genes, particularly K6, has been vigorously pursued by many laboratories. However, the analysis of K6 gene expression has been complicated by the presence of many isoforms and paralogs and the difficulty in designating the correct ortholog for a specific family member. Indeed, transgenes containing different fragments of the bovine, human, and mouse K6 promoters exhibit different expression characteristics (Mahony et al., 2000; Ramirez et al., 1998; Takahashi and Coulombe, 1997). Whether this disparity is because of differences in the functional elements incorporated into the reporter constructs, species-specific differences, or simply a reflection of the diversity of the K6 isoforms remains to be ascertained. Nevertheless, these studies have led to the identification of critical regions that are important for constitutive, tissue-specific expression or for the inducible expression of K6 in the stratified epithelia of the transgenic mice. Further experiments might provide insight into signaling pathways that lead to the induction of K6 and also important clues into the regulatory processes that govern wound healing and psoriasis. The key transcription factors that bind to the regulatory regions of these keratin genes still remain unidentified, although a recent study of the K16 gene has revealed that Sp1 and AP1 sites in the promoter drive epidermal growth factor (EGF)–induced K16 gene expression (Wang and Chang, 2003). Biochemical analysis of the proximal promoter of the K17 gene has also identified some of the regulatory factors that are involved in constitutive and interferon-induced expression of this gene (Milisavljevic et al., 1996).

Similar studies on transcriptional regulation have also been performed for several other keratins. Thus, transgenic mice containing either the human or mouse K19 promoter has been shown to express faithfully in an epithelial-specific pattern (Bader and Franke, 1990; Brembeck et al., 2001). In addition, it has also been demonstrated that interplay between the transcription factors Sp1 and KLF4 regulates the K19 gene (Brembeck and Rustgi, 2000). The promoter of the rabbit

K3 gene, which is expressed suprabasally in peripheral cornea, has been shown to be regulated by combinatorial effects of multiple motifs and their cognate binding proteins (Wu *et al.*, 1993). A recent study has demonstrated that the K15 promoter can direct reporter gene expression to the cells of the hair follicle bulge in adult transgenic mice and thus provides an important tool to target the epithelial stem cells that are thought to reside in the bulge region (Liu *et al.*, 2003). Another study has identified and characterized the 5′ flanking regions of the K12 gene by use of Gene Gun technology, a particle-mediated gene transfer technique to deliver and analyze K12-promoter reporter genes to rabbit corneal epithelial cells *in vivo* (Wang *et al.*, 2002). This method may be applicable to other epithelial cell types. The hair keratins (hard keratins) constitute a large subclass of keratins whose numbers have steadily grown in the past few years as several new genes have been discovered by database analysis (Langbein *et al.*, 1999, 2001). At present, little is known about the expression patterns and the regulatory domains for many of the hair keratin genes, although recent studies have identified HOX13C and Foxn1 transcription factors as important regulators of some hair keratins (Jave-Suarez *et al.*, 2002; Schlake *et al.*, 2000).

The type III IF genes include desmin, vimentin, glial fibrillary acidic protein (GFAP), and peripherin. Expression of these four proteins is quite widespread in many different kinds of nonepithelial tissues. Desmin is present in striated and smooth muscle and is an early marker of muscle differentiation. The regulatory elements that control the expression of desmin have been examined both *in vitro* and *in vivo*. Mutational analysis has shown that CarG/octamer elements that bind to SRF and OCT-like factors play an important role in the regulation of the desmin gene in arterial smooth muscle (Mericskay *et al.*, 2000). Vimentin exhibits a complex pattern of gene expression during embryonic development and is predominantly expressed in mesenchymal-derived cells. Furthermore, vimentin is often aberrantly expressed during metastasis, making it an important marker for the metastatic potential of many tumor cells. Not surprisingly, the transcriptional regulation of the vimentin gene is complex and involves both positive and negative elements (Zhang *et al.*, 2003). GFAP constitutes the major cytoskeletal protein in astrocytes, and a 1.8-kb promoter fragment is active in the specific regions of the embryonic central nervous system (Andrae *et al.*, 2001). Peripherin is expressed in restricted populations of neurons in the peripheral nervous system. Analysis of LacZ reporter gene expression in transgenic mouse embryos has shown that cell type–specific expression of the mouse peripherin gene requires both upstream and intragenic sequences (Leconte *et al.*, 1996). Thus, the studies of the regulatory regions of the type III IF genes have been useful in providing valuable information about the differentiation processes in many tissues.

Type IV IF genes include the neurofilaments (NF-L, NF-M, and NF-H), which constitute the major cytoskeletal component of differentiated neurons. Each member of the NF gene family is regulated differently during neuronal development and differentiation. Hence, these genes are attractive candidates to identify the molecular mechanisms that control neuronal cell fate and development. The

NF-L gene is so far the best characterized example, because studies on reporter gene expression in both transfected cells and transgenic mice have been performed (Charron *et al.*, 1995). Interestingly, NF-L transcripts are also detected in muscle tissues, and it has been demonstrated that NF-L promoter regions contain distinct regulatory elements for both neuron-specific and muscle-specific gene expression (Yaworsky *et al.*, 1997). The nestin (another neuronal type IV IF gene) promoter has also been used to target reporter genes to neural cells (Lothian *et al.*, 1999). The studies of the regulatory elements of new members of this group, such as syncoilin and synemin, have not yet been reported.

Lamins belong to class V of the IF family and are unique in that they are present in the nucleus, where they are thought to play an important and dynamic role. Perhaps of all the classes of IF proteins, lamins are the least studied in the context of gene regulation. The lamins are divided into two types (A and B), and the A-type lamins are expressed mainly in differentiated tissues and developmentally regulated. Mammalian A-type lamin is encoded by a single gene, lamin A/C, and its promoter has been studied (Lin and Worman, 1997). In one study, the chromatin structure of the mouse lamin A/C gene was examined by DNase I HS site mapping (Nakamachi and Nakajima, 2000). This study also showed that the HS-containing elements possessed transcriptional activity when tested in stably transfected cells.

The last class of the IF family includes the orphan beaded filaments (BF), which consist of two highly divergent members of the IF family, phakinin (CP49) and filensin, whose expression is restricted to the epithelial cells of the eye lens. These genes are markers of the differentiation process that leads to the formation of the primary lens fiber cells and hence have been used as a model system to study this process. Both the phakinin and filensin promoters have been isolated, and the sequences that confer lens-preferred expression of reporter genes have been identified (DePianto *et al.*, 2003; Masaki *et al.*, 2002). The ocular lens of vertebrates has been a wonderful model system to study the processes of development and differentiation, particularly with many IF genes as successive stage-specific markers. Studies of the gene regulatory modules of the IF genes of the eye lens have begun to provide insights into these processes at a molecular level.

B. Studies on K5 and K14 Gene Regulation

Keratins K5 and 14 are highly expressed in the basal keratinocytes of the stratified squamous epithelium. This expression is primarily controlled at the level of transcription, and hence the regulation of these two genes provides a useful model system to understand cell-type and differentiation-specific gene expression (Fuchs and Byrne, 1994). In this segment, the experimental approach to study the transcriptional control mechanisms that govern the expression of K5 and K14 genes and a summary of results will be discussed. It is hoped that these studies may provide a framework for other investigators who wish to pursue the transcriptional regulation studies for any other IF gene. Previous work had demonstrated that large segments of the 5′ regions of K5 and K14 genes could direct reporter gene expression in transgenic mice in a manner that faithfully mimicked the endogenous gene expression (Byrne and Fuchs, 1993; Byrne *et al.*, 1994; Leask *et al.*, 1990; Ramirez *et al.*, 1994). This observation begged the next logical question, what are the *cis*-acting elements that govern this expression, and where are they located? As a first step to answer these questions, we performed DNAse I HS mapping of the K5 and K14 genes. These studies led to the identification of several hypersensitive sites in the 5′ region of the K5 and K14 gene, under specific conditions and in cell types where the genes are actively expressed (Kaufman *et al.*, 2002; Sinha and Fuchs, 2001; Sinha *et al.*, 2000). For example, we showed that two closely spaced HS (HS II and III) between −1700 and −1400 region of the human K14 gene were present in keratinocytes but not in HepG2 and fibroblast cells, where K14 is not expressed (Fig. 1A). The genomic region corresponding to the HS not only showed increased accessibility to DNase I enzyme but also to specific restriction enzymes that cut within this area (Fig. 1B). Interestingly, when the 2300-bp upstream sequences of the human and mouse K14 gene are aligned by the Visualizing Global DNA Sequence Alignments of Arbitrary Length (VISTA) program, the short DNA segments that show higher sequence conservation map to exactly the same region as the HS elements (Fig. 2A). The high level of sequence conservation selectively in the two HS regions between human, rat,

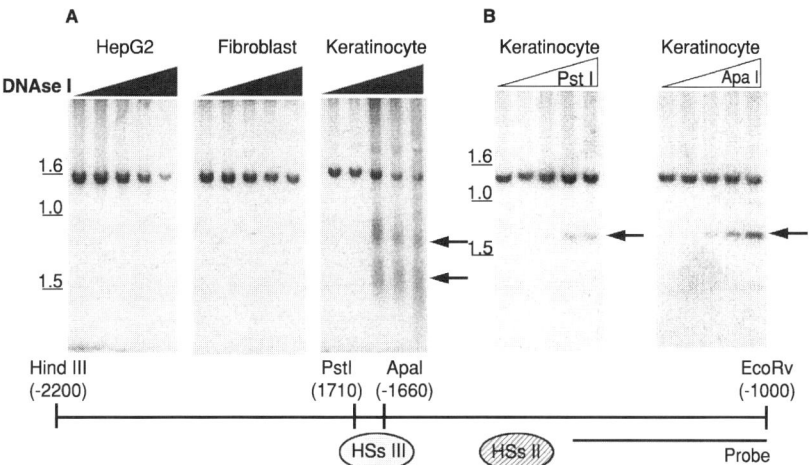

Fig. 1 DNase I hypersensitive (HS) and restriction enzyme accessibility in the 5′ region of the human K14 gene. (A) Nuclei were isolated from the cell types as indicated and treated with increasing amounts of DNase I. Genomic DNA was then subjected to restriction enzyme digestion, and Southern blot analysis was performed with a probe as shown in (B). Nuclei were treated with increasing amounts of restriction enzymes as indicated above. Arrows mark the relative positions of the HS. (From Sinha, S., and Fuchs, E. [2001]. *Proc. Natl. Acad. Sci. USA* **98**, 2455–2460. Copyright [2001] National Academy of Sciences, USA.)

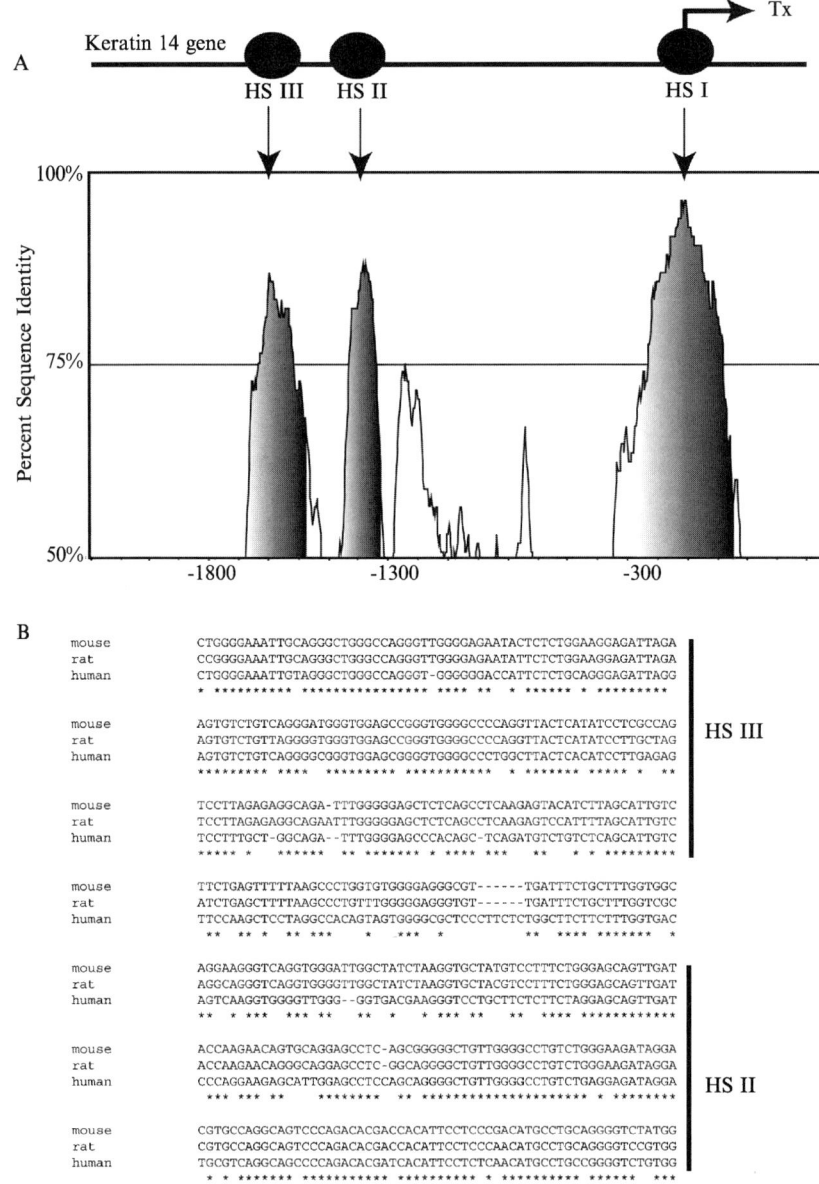

Fig. 2 The hypersensitive (Hs) elements of the K14 gene show a high level of cross-species sequence conservation. (A) VISTA plot of sequence conservation between the upstream sequences of the human and mouse K14 gene. The y-axis represents the percent of sequence conservation at the nucleotide level between 2300 bp of the human and mouse sequences. Regions of at least 75 bp that show >75% identity are shaded: HS III at −1700, HS II at −1400, and HS I corresponding to the proximal promoter region. A schematic diagram of the K14 locus is shown above the VISTA plot to indicate the relative positions of the HS sites. The horizontal arrow marks the direction of transcription (Tx). (B) ClustalW alignment of the sequences corresponding to the HS elements between mouse, rat, and human. The * indicates the nucleotides that are conserved.

and mouse provides additional support to the argument that they are functionally relevant and important for gene expression (Fig. 2B). Similarly, several HS elements present in the 5′ region of the K5 gene were also shown to be conserved in both position and sequence between mouse and human (Kaufman *et al.*, 2002). Our experience thus suggests that it may be feasible to identify potential regulatory segments by performing genome-wide sequence comparison searches across several species. This kind of *in silico* functional genomics approach in combination with DNase I HS mapping offers two parallel and complementary strategies that can lead to the identification of the regulatory elements for any gene of interest.

The next step in our analysis was to examine the transcriptional activity of these *cis* elements. For this purpose, we performed extensive studies of these HS segments by reporter gene assays in both keratinocyte cell culture and transgenic mice. Interestingly, many of the 100 to 200-bp HS elements exhibit robust transcriptional activity when tested individually by transient transfection assays in a keratinocyte cell culture system. However, when examined in transgenic mice, these elements either fail to express in keratinocyte-specific fashion or exhibit altered differentiation-specific expression in sites that normally do not show expression of K5 or K14, such as the inner root sheath of the hair follicles or suprabasal layers of the skin epidermis. On the other hand, proper keratinocyte and differentiation-specific expression of a reporter gene is restored when multiple Hs regions (for example, the K14 HS II and III elements) are tested together in transgenic mice. Our studies lead us to the conclusion that the epidermal and differentiation-specific transcription of the K5 and K14 keratin genes is governed by complex regulatory elements that display remarkable sequence heterogeneity. We hypothesize that these HS elements correspond to unique regulatory modules, and that mixing and matching of these modules generates a code for spatiotemporal specificity of gene expression in keratinocytes.

To gain further insights into the transcriptional regulatory mechanisms that govern the keratin genes, these HS elements were subjected to both database analysis and biochemical analysis. Our studies show that basal-specific and differentiation-specific gene expression in keratinocytes relies on a complex array of regulatory elements that bind diverse families of transcription factors belonging to many families including AP-2, AP-1, ETS, Sp1/Sp3, and as yet unidentified factors. Many of these transcriptional activators and repressors that interact with these HS sites are quite ubiquitous in their expression, and only a few exhibit some level of keratinocyte-restricted expression. This raises an important question of how epidermal-specific gene expression is achieved. Our data strongly argue that gene expression in epidermis is governed by broadly expressed factors that work in a combinatorial fashion or by critical keratinocyte-specific transcription factors yet to be discovered. On the other hand, it is also possible that the transcription factors that confer epidermal-specific gene expression act by keratinocyte-restricted modifications or by recruitment of cofactors that are limited to the epidermis. Many of these alternative possibilities still remain to be sorted out.

In this chapter, an outline of the overall strategy (see Fig. 3) and discussion of various options and tools (both computational and experimental) that may be useful to study the mechanisms that govern the expression of any IF gene of interest are provided. For experimental strategies that are not commonly used in many laboratories, more detailed instructions are provided. It is important to note that many of the experiments described here provide a first step toward the analysis of the regulation of the IF genes. As is the case with most

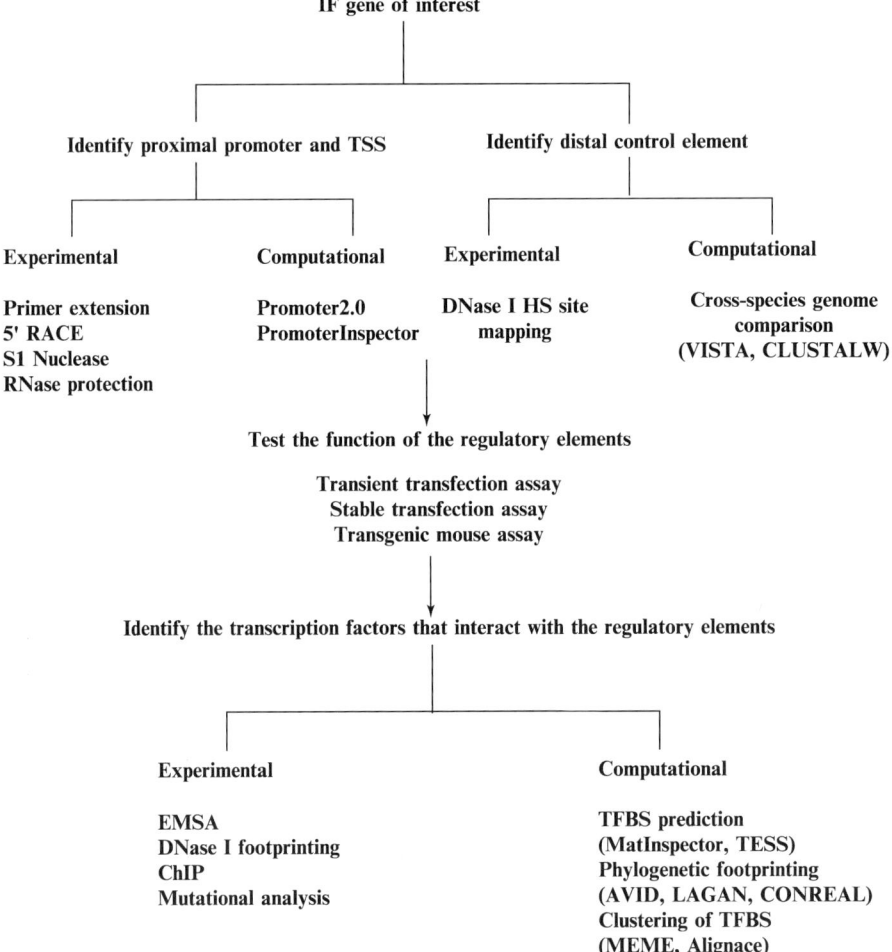

Fig. 3 An outline for a general strategy to study transcriptional regulation of any intermediate filament gene. For computational approaches, only a few examples of the software program are shown. Readers are advised to investigate more programs as they become available. TSS, transcription start site; TFBS, transcription factor binding site; ChIP, chromatin immunoprecipitation.

experimental science, newer technologies and innovations will definitely come along to complement our existing capacities.

II. Materials

A. Experimental Tools and Reagents

Commonly used chemicals and reagents that are described here can be obtained from most vendors such as Fisher Scientific or VWR. Cells for culture can be obtained from ATCC or from investigators who have established them. The list and sources for specialized needs, reagents, and relevant websites are indicated in the following.

1. For DNase I HS mapping and Southern blot analysis: DNase I (Amersham Biosciences, Cat No 27-0514-0), spermidine (Sigma, Cat No S2626), spermine (Sigma, Cat No S3256), proteinase K (Invitrogen, Cat No 25530–049), Zeta-Probe GT membrane (BIO-RAD, Cat No. 162–0196), Klenow Fill-In Kit (Stratagene, Cat No 200410).

2. For transfection and transgenic studies: pGL3 basic vector (Promega, Cat No E1751), PRLTK vector (Promega, Cat No E2241), pCMVLacZ (BD Bioscience Clontech, Cat No 631719), luciferase assay system (Promega, Cat No E1500), EndoFree Plasmid Maxi kit (Qiagen, Cat No 12362), FUGENE 6 (Roche, Cat No 1 814 443), β-Gal Assay kit (Invitrogen, Cat No K1455–01), Dual-Luciferase Reporter Assay System (Promega, Cat No E1910), Dual-Light Luciferase and β-Galactosidase Reporter Gene Assay System (Applied Biosystem, Cat No T1003), Galacto-Light Plus β-Galactosidase Reporter Gene Assay System, Applied Biosystem, Cat No T1007).

3. For EMSA and mutational analysis: Dounce homogenizer (Kontes, VWR Cat No KT 885300–002), Complete Mini EDTA free protease inhibitor (Roche, Cat No 10582900). Poly (dl-dC) (Amersham Biosciences, Cat No 27-7880-01), LightShift Chemiluminescent EMSA kit (Pierce, Cat No 20148), Quick Change Site-directed Mutagenesis kit (Stratagene, Cat No 200519). Purified oligonucleotides can be obtained from many commerical vendors.

Computational Needs

The genomic sequences for any specific IF genes can be obtained by searching the genome databases. The web site of National Center for Biotechnology Information (http://www.ncbi.nlm.nih.gov) has links to the genome resources for human, rat, and mouse. Other web sites of relevance are:

For Promoter Identification:

PromoterInspector (http://www.genomatix.de/)
Promoter2.0 (www.cbs.dtu.dk/services/Promoter/)

For Transcription Factor Binding Site Prediction:

Matinspector (http://www.genomatix.de/)

TESS (www.cbil.upenn.edu/tess/)

Alibaba (www.alibaba2.com/)

For Genome Analysis:

AVID and VISTA (http://www-gsd.lbl.gov/vista/index.shtml)

LAGAN (http://lagan.stanford.edu/lagan_web/index.shtml)

BLAST (http://www.ncbi.nih.gov/BLAST/)

Clustal W (http://www.ebi.ac.uk/clustalw/)

MEME (http://meme.sdsc.edu/meme/website/intro.html)

Alignace (http://atlas.med.harvard.edu/cgi-bin/alignace.pl)

Footprinter (http://bio.cs.washington.edu/software.html)

Dialign (http://bibiserv.techfak.uni-bielefeld.de/dialign/)

III. Procedures

A. Identification of the Regulatory *cis* Elements by Database Analysis, Comparative Genomics, and DNase I Hypersensitive Site Mapping

As evidenced by a large number of studies, a common mode of regulation of the IF genes is at the level of transcription. Hence, to understand the mechanisms that govern the cell type or differentiation-type specific expression of these genes, the first step is to identify the regulatory *cis* elements. These would include the promoter and distal control elements such as enhancers, silencers, or locus control regions. In most cases, it is likely that the transcriptional regulation of a pertinent IF gene depends on multiple elements, as is the case for K5 and K14 genes and many others. Although the promoter elements for any gene are relatively easy to locate because of their close proximity to the 5′ end of a gene of interest, it is often difficult to establish the positions of the distal control elements, because they may be located many kilobases away in both the 5′ and 3′ direction from the gene or even embedded in the DNA sequences corresponding to the gene itself. For a thorough analysis of the regulatory sequences for any IF gene, it is critical that the promoter and distal control elements are first identified by methods described in the following.

1. Identification of the Promoter Region of an Intermediate Filament Gene

The promoter is generally defined as the region of a few hundred base pairs located directly upstream of the site of initiation of transcription. Therefore, identification of the transcription start site (TSS) leads to the location of the promoter of a gene. The location of the TSS and the promoter can be determined by either computational or experimental methods.

Step 1

With the completion of the mouse, human, and soon the rat genome sequencing projects, the information about the genomic organization for any IF gene from these species is readily accessible. These data can be computationally processed to find the promoter. One approach to define the TSS is to align the full-length mRNA (including the complete 5′ UTR) to its counterpart genomic sequence. The expressed sequence tag (EST) database is a useful source to obtain several overlapping ESTs and create a virtual cDNA that contains the most complete 5′ sequence. This is particularly important, because there may be multiple TSSs and variation in the choice of the first exon for a particular IF gene due to the use of alternate promoters.

Step 2

Newly developed software such as the PromoterInspector, Promoter 2.0, and several others can be used to predict promoter regions in unannotated genomic sequences (Qiu, 2003). The success and accuracy of these programs in predicting promoters varies considerably and, therefore, it is recommended that multiple programs be used. Many proximal promoters contain certain conserved signals such as the TATA box, GC–rich elements, and CCAAT boxes that occupy relatively well-defined positions upstream of the TSS. However, an equally large number of the eukaryotic Pol II promoters lack such distinctive motifs in the proximal promoter region. Although the presence of such elements within the putative promoter raises the confidence that, indeed, the promoter elements of a gene have been located, the lack of such motifs in the DNA sequence should not be interpreted as a failure to do so.

Step 3

It is quite likely that the computational methods described previously will lead to the identification of the TSS and the promoter region for any IF gene. However, before proceeding further with functional studies of the promoter region, it is advised that the TSS be experimentally ascertained by primer extension, S1 nuclease digestion, RNase protection, or 5′ RACE protocols. All these approaches have their limitations and advantages, although primer extension is generally considered to be the method of choice for the precise determination of the TSS. Because of space limitations, these methods are not discussed, and readers are advised to follow standard books on molecular biology for details.

2. Identification of the Distal Control Regions by Comparative Genomics

The identification of the distal control elements for any gene presents more of a challenge, because their location is not known a priori. A number of computational approaches are available that attempt to identify gene regulatory elements on a genome-wide scale (Bulyk, 2003; Ureta-Vidal et al., 2003). It is well known that the bulk of noncoding DNA is not conserved between species, but short stretches of highly conserved noncoding sequences (CNS) exist amid the sea of nonconserved DNA. It can be argued that functional sequences such as the coding region and the regulatory regions are more likely to be evolutionarily conserved than the nonfunctional regions. If that were the case, the CNS regions could potentially be attractive candidates as *cis*-regulatory modules that control the expression of nearby genes. Comparative genomics is well suited for identifying such evolutionarily conserved sequences.

Several algorithms such as ClustalW and VISTA have been developed that allow researchers to align large chunks of the genomic sequences (Qiu, 2003). VISTA, for example, can be used to visualize long sequence alignments of DNA from two or more species with annotation information (Mayor *et al.*, 2000). For example, a large segment of DNA sequence corresponding to a human IF gene and its 5′ and 3′ sequence can be compared with the rat and the mouse genome. These data can be easily configured, and alignments of various lengths can be viewed at different levels of resolution. Although this cross-species search for conserved sequence is quite powerful and can identify segments that show a high level of sequence similarity, addressing the functional significance of these conserved sequences is a formidable task for any one gene, let alone for a large family such as the IF proteins. The major limitation of these programs is that there could potentially be large numbers of such CNS regions. It is likely that because of close evolutionary proximity for any two or three species, only a few of such CNS regions may be biologically relevant in context of being an actual regulatory *cis* element for a nearby gene. Further confirmation of such potentially important regions can be ascertained by DNase I HS mapping.

3. Functional Identification of Regulatory Elements by DNase I Hypersensitive Site Mapping

The identification of the DNase I HS sites can be a powerful tool to identify the precise location of the transcriptional regulatory regions for any gene. The presence of a nuclease hypersensitive region indicates that the corresponding genomic DNA segment is likely to be in an open chromatin conformation *in vivo* and thus expected to be bound by regulatory transcription factors. By use of this approach, a large genomic region (~20–50 kb) can be scanned rapidly, and the location of potential distal regulatory sites can be determined. In addition, these experiments can be performed to detect HS elements that are present in a cell-type specific or in an inducible fashion for any IF gene. The experimental procedure consists of two major parts: (1) isolation and DNAse I treatment of nuclei and (2) Southern blot analysis.

a. Isolation and DNAse I Treatment of Nuclei
Step 1

The first step is to identify a cell-type where the IF gene of interest is expressed (by reverse transcriptase-polymerase chain reaction [RT-PCR] or Northern blot analysis). Although for these experiments, the ideal choice is a primary cell culture, which is more likely to mimic the endogenous gene expression, many of the studies have been performed in immortalized cell lines. Start with approximately 10^8 cells, although the amount of starting material will vary depending on the cell type. The experiments can be scaled up and down if necessary.

Step 2

Harvest the cells and wash thoroughly with 1× phosphate-buffered saline (PBS). Scrape cells from the plates with a cell scraper and collect them in 1× PBS. Spin the cells down by centrifugation at 1500 rpm for 10 minutes in a refrigerated centrifuge in disposable 50-ml conical tubes.

Step 3

Resuspend the cell pellet in 1–5 ml of NP-40 lysis buffer (10 mM Tris-HCl, pH 7.5, 3 mM $MgCl_2$, 10 mM NaCl, 1 mM EDTA [pH 8.0], 0.5% NP40, 0.15 mM spermine, 0.5 mM spermidine).

Step 4

Incubate the cell suspension in ice for 5 minutes to lyse the cells, and spin down the nuclei in a refrigerated centrifuge at 2000 rpm for 10 minutes.

Step 5

The pellet is the nuclei. Resuspend the nuclear pellets in a small volume of (~500 μl) DNase I assay buffer (40 mM Tris HCl [pH 7.5], 6 mM $MgCl_2$). Handle the nuclei carefully to prevent nuclear lysis.

Step 6

Dilute the DNase I (7500 U/ml; Amersham) in buffer containing 10 mM Tris HCl (pH 7.5), 50 mM NaCl, 10 mM $CaCl_2$, and 62.5 mM $MgCl_2$ to a final concentration range of 1–6 U/μl.

Step 7

Add 10-μl aliquots of these DNase I solutions (0, 1, 2, 4, and 6 U/μl) to 90 μl of the resuspended nuclei and mix gently. The concentration of DNase I to be added must be optimized for each cell type and amount of starting material.

Step 8

Incubate at 37 °C for 10–20 minutes. Stop the reactions by adding 200 μl of stop buffer (50 mM Tris HCl [pH 7.5], 100 mM NaCl, 1% sodium dodecyl sulfate, 20 mM EDTA [pH 8.0]).

Step 9

Add 40 μl of proteinase K (20 mg/ml) to each reaction, and incubate the DNase I–treated nuclei overnight at 55 °C to digest proteins.

Step 10

Extract the genomic DNA with phenol-chloroform, being careful not to vortex the samples to avoid any shearing of the genomic DNA. The sample(s) that are treated with no or with low concentrations of DNase I will tend be more viscous than the ones that are treated with high concentrations of DNase I.

Step 11

Precipitate the genomic DNA with three volumes of ethanol. After washing the DNA pellet with 70% ethanol, resuspend the genomic DNA in TE (10 mM Tris-HCl [pH 8.0], 1 mM EDTA).

b. Southern Blot Analysis

Since the Southern blot procedures are quite widely used in most laboratories, only a brief outline is provided. Before proceeding with the Southern blot analysis, it is imperative that a detailed restriction enzyme map of the region of interest is obtained. For practically all IF genes, this can be easily obtained from the genome sequence database. On the basis of this information, the genomic DNA can be digested with several restriction enzymes to yield overlapping fragments of 2–20 kb that surround the IF gene of interest. The probes for the Southern blot analysis should be: (1) a 400–500 bp segment of DNA that contains no or few repetitive elements and (2) is located near the end of the fragments generated by the restriction enzyme digestion to facilitate the mapping of the HS sites. The probe can be generated by performing PCR on genomic DNA templates or by restriction digests of genomic DNA clones of the gene, if available. For a typical experiment, 10–20 μg of the genomic DNA should be digested with the appropriate restriction enzymes followed by Southern blot analysis as per standard protocols. (The Zeta-Probe Blotting membrane and associated protocols from BIO-RAD are highly recommended).

c. Data Interpretation and Alternative Approaches

The Southern blot analysis will lead to the identification of the HS sites for the IF gene of interest. Once an HS site is determined, the same membrane can be stripped and hybridized with a different probe at the opposite end of the restriction enzyme fragment to confirm the position of the HS site (see Fig. 4 for an illustration). It is difficult to predict how large a region needs to be scanned for HS site mapping, because regulatory elements can be located at quite a distance in both upstream and downstream regions of a gene. Even if HS are not detected for a particular gene in distal regions, it is expected to find one HS site corresponding to the proximal promoter region. HS elements that show cell-type specificity are of particular interest, because these elements are likely to be involved in cell-type–specific gene expression. This is evident for the HS elements for the K5 and K14 genes, which are only detected in the keratinocytes but not in fibroblasts, where these genes are not expressed. As an alternative to DNase I, the accessibility of a particular region can be confirmed by digestion of nuclei with restriction enzymes followed by Southern blot (see Fig. 1B).

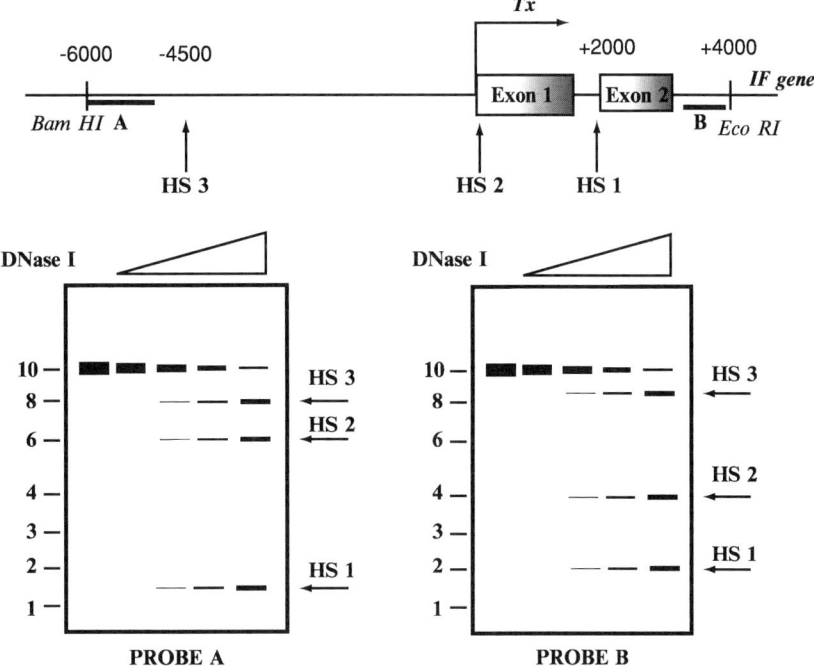

Fig. 4 Results of a DNase I hypersensitive (HS) mapping experiment for a hypothetical intermediate filament gene. The upper panel shows the genomic organization with the transcription start site marked as (Tx). The lower panel shows the results of a Southern blot analysis. Genomic DNA obtained from nuclei treated with varying amounts of DNase I is digested with BamH I and EcoRI. Three HS sites are detected, one in the 5' region (−4500), one at the promoter region, and one in the first intron (+2000). Note that by the use of two different probes, the positions of the HS sites can be confirmed by separate Southern blots. In this hypothetical case, the use of the probe at the 5' end, the positions of the HS sites will be at 1.5 kb, 6 kb, and 8 kb, whereas a probe on the opposite end will detect HS sites at 2 kb, 4 kb, and 8.5 kb, respectively.

B. Testing the Activity of Regulatory *cis* Elements by Functional Analysis in Transient Transfection Assays and/or in Transgenic Mice

As discussed in the previous section, the regulatory regions for any IF gene are either located close to the TSS (promoter) or may be further away (distal control elements). After the identification of these elements, the next logical step is to analyze their transcriptional activity. The most commonly used assay for measuring this activity is the transient transfection assay. In this method, plasmids containing a reporter gene under the control of the relevant *cis* elements of the gene of interest are introduced into cells maintained in culture. The expression of the reporter gene can be measured, which in turn reflects the activity of the control elements.

1. Isolation of the *cis* Elements from a Genomic Library or Bacterial Artificial Chromosome DNA

It is useful to obtain a genomic clone containing the gene of interest and its surrounding sequence by screening a genomic library with a cDNA probe or obtaining a bacterial artificial chromosome (BAC) clone. A BLAST homology search can be performed with any IF cDNA/genomic sequence against the High Throughput Genomic (HTG) sequence database (http://www.ncbi.nlm.nih.gov) to locate a BAC clone that contains large genomic fragments of interest. Once identified, a specific BAC clone can be obtained from BACPAC Resource center (http://bacpac.chori.org/) or commercial sources (Invitrogen) at a nominal price. Alternatively, traditional genomic libraries such as those in lambda phage (commercially available, Stratagene) can also be screened. Once the BAC DNA or a genomic clone is obtained, the relevant areas can be subcloned or amplified by PCR.

2. Cloning of the Regulatory Regions in a Reporter Plasmid and Transient Transfection Experiments

a. Step 1. Choice of Plasmids and Regulatory Elements

The studies of the transcriptional activity can be performed with plasmids containing reporter genes such as those for luciferase, chloramphenicol acetyl transferase (CAT), *LacZ*, and green fluorescent protein (GFP). The most commonly used and perhaps the best choice at present is luciferase. The luciferase gene encodes for an enzyme whose expression can be measured by a quick and easy assay. A variety of luciferase reporter vectors (pGL3-basic is the most commonly used) are commercially available from Promega, which also provides a wide range of technical support and detailed literature. The decision as to how large a DNA fragment needs to be inserted into the luciferase reporter plasmid can be based on information such as the location of the HS sites and the extent of the regions that exhibit high sequence conservation. For most genes, the proximal promoter tends to be located between -400 and $+100$, and hence a fragment of this size can be the starting point and can be placed immediately upstream of the reporter gene. For distal regulatory elements, these regions may be placed upstream or downstream of a constitutive promoter such as those from SV40 or HSV-TK. It should be noted that some enhancers or distal control elements may work better or only with their own natural promoter. To control for transfection efficiency, an additional internal control plasmid such as CMVLacZ (for expression of β-galactosidase) or pRLTK (for expression of renilla luciferase) is usually cotransfected with the promoter/enhancer reporter construct.

b. Step 2. Choice of Cell Line for Transfection

The next step is to choose the cells that express the IF gene of interest and thus provide an optimal model system. Although primary cells are the ideal choice, they are often difficult to grow and passage and to transfect in an efficient manner. Hence, many of the reporter gene assays by transient transfection studies are performed in immortalized cells that also express the gene of interest.

c. Step 3. Preparation of DNA

It is important to begin experiments with good-quality plasmid DNA. Although DNA purified by CsCl probably is the best, commercial kits such as Qiagen Plasmid Maxikit provide sufficiently good quality plasmids for most experimental purposes. To improve the efficiency of transfection, it is advisable to use endotoxin-free DNA, which can be prepared with a commercial kit available from Qiagen.

d. Step 4. Transfection of Reporter Plasmids

The transfection of the promoter constructs containing the regulatory elements can be performed in many different ways. Many reagents/techniques for efficient transfection of mammalian cells have been developed, such as those that are calcium phosphate based, lipid based, or use electroporation. Each cell line has its own characteristics, and, hence, transfection protocols must to be optimized. For example, for transfections in keratinocytes, FUGENE 6 (Roche) seems to work best in our hands. A typical example of a transient transfection experiment is shown as follows:

1. Plate out cells the day before for transfection in 6-well or 12-well plates.
2. Once the cells are 40%–50% confluent, transfect them with the reporter and the internal control plasmids as shown next.

Plasmid: pGL3-basic	1 μg DNA + 0.25 μg pCMVLacZ
Plasmid: pGL3-basic containing promoter 1	1 μg DNA + 0.25 μg pCMVLacZ
Plasmid: pGL3-basic containing promoter 2	1 μg DNA + 0.25 μg pCMVLacZ

For each construct, transfect cells in duplicate by use of a transfection protocol as suggested by the manufacturer. Harvest the cells between 24 and 48 hours after transfection. The cell extracts can be frozen at this stage or processed immediately as discussed in the next step.

e. Step 5. Measuring Luciferase and β-Galactosidase

The activity of luciferase and β-galactosidase can be measured in a variety of ways, and kits are available from many companies. The commonly used luminescent assay (Promega) for the quantitative determination of luciferase activity in transfected cells is optimized for use with microplates, luminometers, or scintillation counters. The Beta-Gal Assay Kit (Invitrogen) provides the reagents required to quickly measure the levels of active β-galactosidase in transfected cells with a colorimetric assay and a spectrophotometer. Newly developed kits such as the Dual-Light luminescent reporter gene assay designed for detection of luciferase and β-galactosidase (if using pCMVLacZ as an internal control) in a single extract aliquot, which simplifies normalization of transfection efficiency. Similarly, Dual-Luciferase reporter assay system (Promega) provides an efficient means to assay the activities of firefly and *Renilla* luciferases (if using pRLTK as an internal control) from a single sample. In both cases, a light signal from each enzymatic reaction can be easily measured sequentially in a luminometer with automatic injectors or other instrumentation in which light emission measurements can be performed within a short period.

f. Data Interpretation and Alternative Approaches

Transient transfection experiments need to be repeated several times to gain statistically significant data. The presence of an internal control allows correction for variations in transfection efficiency. The corrected luciferase values for a control element are represented as fold activation over the empty vector such as pGL3-basic. For a regulatory element with strong transcriptional activity, this value may be 10-fold to 100-fold or higher. The transfection experiments can also be tried out in several cell lines, because in many cases, the regulatory elements may exhibit cell-type–specific activity. Once high levels of activity are detected for any reporter construct, further 5′ and 3′ deletions can be performed to delineate the critical regions of the promoter/enhancer that contain the important regulatory elements.

As an alternate method, stable transfection studies can be performed. This method involves the integration of the reporter plasmid into the host chromosome of a cell and may be used for studying promoter and/or enhancer regions that do not show any significant reporter activity in transient transfection assays. It is thought that the integration of the reporter plasmid in a more natural chromatin configuration mimics the *in vivo* situation. Indeed, some regulatory regions may only demonstrate reporter activity in stable transfections. However, this method requires considerable time and effort and hence is infrequently used by most researchers.

3. Transgenic Mouse Analysis

For any mammalian IF gene, the analysis of the regulatory regions in transgenic mice is the ultimate proof for the *in vivo* functional relevance of the regulatory elements. As discussed for K5 and K14 genes, many regulatory elements that show a high level of reporter activity in transient transfection assays fail to express in transgenic mice or show aberrant expression, such as altered tissue and or differentiation specificity. Hence, testing the regulatory regions in transgenic mice is warranted to demonstrate the functional significance of a regulatory region identified by transient transfection assays. For transgenic analysis, typically multiple lines of mice need to be generated with a construct containing the regulatory fragments and a reporter gene. This is important because some lines may not express the reporter gene or show an aberrant expression pattern because of position effects related to integration site of the transgenic DNA. The *LacZ* reporter is a good choice for these studies, because X-gal staining can be performed to easily follow the expression of the *LacZ* gene in the transgenic mice. Needless to say, analysis of regulatory elements by transgenic mouse studies is prohibitive for many laboratories because of the high cost involved in generating and housing transgenic animals. Although it is understood that it may be impossible to test every reporter construct in a transgenic mouse assay, it is certainly feasible to test a few critical constructs to drive home the most important points such as the functional importance of a key transcription factor binding site. This indeed has been done for many of the regulatory elements for the IF genes. Techniques for generating transgenic mice can be obtained from manuals covering this topic (Nagy *et al.*, 2003).

C. Identification and Analysis of the Transcription Factors Binding to the *cis* Elements by Mutational, Biochemical, and Database Searching

After the regulatory control regions for any gene are identified, the next step is to identify the critical *cis*-acting DNA sequences within these regions that are the important binding sites for transcription factors. Such studies mostly involve three complementary approaches: (1) a comprehensive mutational analysis of the regulatory region in the transient transfection assays, (2) identification of the DNA-binding proteins that bind to the relevant sequences, and (3) computational studies of the *cis* elements.

1. Mutational Analysis of the *cis* Elements

Once a control region is delineated, one strategy is to perform a comprehensive mutational analysis of this entire segment and test the mutant plasmids in a functional assay such as transient transfection assays. However, these experiments can be performed only when a very short segment of DNA is shown to contain important regulatory elements for the transcriptional control of the gene of interest. In such a case, the strategy will involve specific substitution mutations whereby the activity of the region of interest can be scanned by a series of 2- to 3-bp mutations of the regulatory element. Although PCR-based tools such as the Stratagene Quickchange kit have been developed to generate many such mutant constructs in an efficient fashion, such a detailed study undoubtedly involves a significant amount of resources and effort. Hence, an argument can be made for generating targeted mutations on the basis of intelligent guesses and on studies done in parallel as described next.

2. Biochemical Analysis of the Trans-Acting Protein Factors that Bind to the Relevant Elements

The ultimate goal of the studies—to understand the mechanisms that regulate an IF gene expression—often hinges on the identification of critical transcription factors that bind to its regulatory region. Electrophoretic mobility shift assays (EMSAs) provide a quick and efficient method for determining protein/DNA interactions. This is particularly useful when a short DNA segment is identified to possess transcriptional control properties as judged by studies described in prior sections. The steps of EMSA include making nuclear extracts, labeling the DNA fragment, and setting up the reaction and running a gel.

a. Making Nuclear Extracts for Electrophoretic Mobility Shift Assays

The ideal choice for making nuclear extracts will be those cells in which the IF gene of interest is expressed and those that are used for transient transfection experiments. If desired, nuclear extracts can also be prepared from tissues and organs or obtained commercially (Active Motif). The most commonly used method to make nuclear extract is described in the following.

Step 1

Start out with five large plates/flasks of cells. Take off the media, and wash cells with ice cold $1\times$ PBS, repeat once.

Step 2

Scrape cells in PBS and transfer to a 50-ml tube kept in ice.

Step 3

Spin down at 2000 rpm $\times$ 10 minutes at $4\,°C$.

Step 4

Resuspend the cell pellet in at least five $\times$ volume of buffer A (10 mM HEPES [pH 7.9], 1.5 mM $MgCl_2$, 10 mM KCl, 1 mM DTT, protease inhibitors). The volume of the cell pellet will depend on the cell type and starting number of cells.

Step 5

Keep on ice for 10 minutes. Make sure the cells are well resuspended in buffer A.

Step 6

Spin the cells again (2000 rpm for 10 minutes) at $4\,°C$, and resuspend the cells in $2\times$ volume of buffer A.

Step 7

Transfer the cell suspension to a Dounce homogenizer. Apply 10 – 15 strokes of tight pestle to lyse the cells. This process results in lysis of the cells but leaves the nuclei intact.

Step 8

Transfer the lysed cells to a microfuge tube and spin the lysed cells down at 12,000 rpm for 5 minutes at $4\,°C$.

Step 9

The pellet is the nuclei. Discard the supernatant and add two $\times$ volume of buffer B (20 mM HEPES [pH 7.9], 1.5 mM $MgCl_2$, 0.42 M NaCl, 25% glycerol, 0.5 mM EDTA, 1 mM DTT, protease inhibitors). The high salt concentration in buffer B extracts the nuclear proteins, including transcription factors. The concentration of NaCl may be varied, depending on specific needs.

Step 10

Put the microfuge tube containing the nuclei and the extraction buffer in a rotator/shaker at $4\,°C$ for 30 minutes.

Step 11

Spin down the suspension at 12,000 rpm for 15 minutes at 4 °C.

Step 12

The supernatant is nuclear extract (NE); make small aliquots, freeze them in dry ice, and store them at −80 °C. Do not freeze-thaw NE multiple times. If desired, NE can also be dialyzed against a buffer similar to buffer B but containing 100 mM NaCl instead of 0.42 M NaCl.

b. Designing Oligonucleotides and Labeling

The oligonucleotides that are used for EMSAs are usually 20- to 30-bp long. Thus, a 100- to 200-bp region can be easily represented by several overlapping oligonucleotides. Design upper and lower strands of oligonucleotides with overhangs at the end so that they can be end labeled with ^{32}P-dCTP by the klenow enzyme. The use of purified oligonucleotides (by high-performance liquid chromatography [HPLC] or gel purified) and labeling of the double-stranded oligonucleotides to high specific activity is important for the success of EMSA.

c. Setting Up EMSA Reaction and Running the Gel

Step 1

Prepare a 4.5%–5% 0.5× TBE native polyacrylamide gel. The gel can be prepared in advance and stored at 4 °C.

Step 2

Before setting up the EMSA reactions, prerun the gel at 100 V or 10–15 mA. Use 0.5× TBE as running buffer. The prerun should be for at least 1 hour.

Step 3

While the gel is undergoing a prerun, set up the binding reaction of the labeled oligonucleotides with the nuclear extract. An example of a 20-μl binding reaction is as follows:

2 × DNA binding buffer	10 μl
Bovine serum albumin (BSA)	1 μl (final concentration: 1 μg/μl)
Poly (dI:dC): Poly (dI:dC):	1 μl (final concentration: 0.25–1 μg/μl)
Labeled DNA	1 μl (ideally 1 fmole or ~10,000 cpm)
Nuclear extract	2 μl (5–10 μg of total protein)
H$_2$O	5 μl

A typical 2 × DNA binding buffer contains 10% glycerol, 20 mM HEPES, [pH 7.9], 150 mM KCl, 2 mM DTT, 2 mM EDTA, and 5 mM MgCl$_2$. The binding conditions and buffers can be varied on the basis of the requirements for specific transcription factors. Some of the key parameters that can be adjusted and titrated include the concentration of salt, glycerol, and concentration and type of competitor DNA.

Step 4

Incubate the reaction at room temperature or ice for 20 minutes. Stop the gel that was prerunning. Load samples after rinsing out the wells. Use one lane for loading dye-containing bromophenol blue to monitor the progress of the gel. Loading dye should not be added to samples, because it may affect DNA binding of proteins present in the NE.

Step 5

Run gel at 100–150 V for approximately 3 hours or until the dye-front reaches the bottom of the gel.

Step 6

Transfer the gel to a Whatman paper and cover it with plastic wrap. Dry gel on a gel dryer, place the dried gel in a cassette, and expose to autoradiography film or a phosphoimager.

d. Data Interpretation and Alternative Approaches

The DNA–protein complexes detected by EMSA can be further evaluated by various means. For example, the specificity of the complex can be determined by performing competition experiments with oligonucleotides containing mutations that abolish binding. Mutations can be selected on the basis of the potential binding site for known transcription factors or by random choices if there is no match to any known consensus sites. The identity of the transcription factor that binds to a specific oligonucleotide sequence can be determined by use of antibodies, which can give rise to supershifts or abolish the DNA–protein complex. The EMSA studies can also be performed with nonradioactive materials provided in the LightShift Chemiluminescent EMSA Kit (Pierce).

DNase I footprinting is an alternative method to locate the potential sites in a regulatory element that are bound by transcription factors. Although EMSA is more sensitive and often leads to the detection of specific DNA–protein complex even when the binding proteins are present in low amounts in the nuclear extracts, DNase I footprinting has its own advantages. Unlike EMSA, in which the DNA–protein interaction is detected on the gel, the DNase I footprint reaction is carried out in solution and hence may detect complexes that do not survive the unfavorable conditions during electrophoresis in EMSA. For a complete analysis of a regulatory region, both approaches can be used to obtain useful information. Finally, the *in vivo* binding of a transcription factor to a specific regulatory region may be tested by chromatin immunoprecipitation experiments (ChIP). Details about these alternate methods can be found in standard molecular biology manuals.

3. *In Silico* Analysis of the Regulatory *cis* Elements by Computational Methods

The advantage of using database searching as a tool to define and search critical regulatory regions cannot be stressed enough. Once a promoter or a distal control element is identified by the methods described earlier, the search for important regulatory elements and potential transcription factor binding sites within this stretch of DNA can be performed computationally. These searches, when coupled with wet laboratory experiments, can allow investigators to focus on candidate sequences or transcription factors that are more likely to be relevant in controlling the expression of the IF gene of interest. The search can be carried out on three fronts:

1. *Analysis of the potential transcription factor–binding sites by software programs:* It is possible to search for the *cis*-regulatory elements computationally with software. For example, MatInspector is a tool that uses a library of matrix descriptions for transcription factor binding sites to locate matches in any DNA sequence. Because the transcription factor binding sites are often degenerate, short sequences in such searches often lead to a large number of hits, many of which may not be functionally significant. In addition, these methods will fail to recognize binding sites for transcription factors that have not been characterized or are as yet undiscovered. Hence, it is useful to use multiple software programs (such as Alibaba, TESS, etc) to obtain a better idea of the potential transcription factor binding sites. Needless to say, the predicted binding sites need to be tested in functional assays such as EMSA and ChIP.

2. *The use of cross-species genome comparison for identifying critical regulatory regions:* As discussed before, the use of comparative genomics (also referred to as phylogenetic footprinting) can be very useful in identifying the important and relevant transcription factor binding sites within a control element (Qiu, 2003). It is likely that the regulatory sequences for any IF gene are highly conserved between several species. Thus, if the binding site for a specific transcription factor within one or multiple regulatory regions is evolutionarily conserved, it raises the likelihood that the site may be functionally relevant. Consequently, such sites may be chosen as a starting point for further functional and experimental analysis. The programs to perform such studies (BLAST, ClustalW, AVID, LAGAN, VISTA) are readily available on the web and can be used to analyze sequences from a large number of species.

3. *Finding common regulatory modules and transcription factor binding sites by clustering genes on the basis of expression pattern:* This is perhaps the most complex of the computational analyses described here. Although the IF gene family is a small family (at least for this kind of analysis), it is possible to derive a common pattern in terms of coexpression of the IF genes. For example, if a set of IF genes shows a similar cell-type or differentiation-type expression pattern, it is reasonable to assume that the promoters and distal regulatory elements may harbor binding sites for a common set of transcription factors. If that were to be the case, then it would be useful to search for motifs or patterns that are common and overrepresented in the regulatory region of a set of coexpressed IF genes. These kinds of approaches to find regulatory elements have been successful in lower organisms and are now being tested for more complex regulatory networks such as those that exist in mammals. The commonly used programs that can detect such motif overrepresentation include MEME, AlignAce, Dialign, and Footprinter and are available on the web. Genome sequences for several other organisms will be released within the next 2 years, which will provide a substantial stock house of data and information for carrying out these kinds of studies. These genome-wide surveys can therefore provide an important starting point for

understanding the transcriptional regulatory mechanisms for IF genes. They are, however, no substitutes for wet laboratory experiments that need to be performed to validate the *in silico* studies.

IV. Pitfalls

The various approaches that have been described in this chapter for the study of transcriptional control of IF genes are not without shortcomings. One major caveat of the typical experimental setup is the fact that these studies are often carried out in a cell line, which may not fully mimic the differentiation state of cells in their natural tissue environment. Although the use of primary cells provides an improvement from immortalized cell lines, it is quite common even for primary cells to undergo genetic alterations and exhibit variations in the expression profile of many relevant genes when taken out of their natural environment. Thus, when these cells are used as a source of nuclear extracts for EMSAs or for transient transfection experiments to analyze regulatory elements, the results may not be a true indicator of what happens in an organ or tissue. In addition, these experimental protocols also have inherent technical limitations that need to be fine-tuned and adjusted on a case-by-case basis. This is particularly evident, for example, when preparing nuclear extracts for EMSAs where quite often the transcription factors that are more stable and abundant tend to be extracted, whereas relevant transcription factors may not be extracted under the conditions used. Similarly, in transient transfection assays, testing of the regulatory elements are often performed in isolation with small DNA segments. In such a case, the effects of the neighboring regulatory sequences and the natural chromatin structure may be overlooked. Hence, it is important that data obtained from the traditional assays such as EMSAs and transient transfection be verified by more relevant *in vivo* experiments such as ChiP assays and transgenic mice studies.

Although the advent of computational methods, such as cross-species comparison, has added another tool in the arsenal, the bioinformatics approach also has its limitations, the main one being that these analyses are often fraught with high error rates. It should also be stressed that the study of each IF gene needs to be carried out in multiple species. Many of the IF genes exhibit species-specific differences in their expression patterns and response to various growth and differentiation signals. This should be kept in mind when comparing promoter or enhancer elements or testing them in transgenic animals. Finally, it is increasingly apparent that multiple control regions may act in a combinatorial fashion to regulate the precise spatiotemporal expression of IF genes. In view of such complexity, many of the *cis* and *trans* regulatory elements identified previously need to be revisited and reexamined in detail with new tools and techniques to obtain more accurate and pertinent information about their role in the regulation of the IF genes of interest.

V. Discussion and Concluding Remarks

The IF gene family provides an interesting model system to study the diverse processes of development and differentiation in different cell types. In many of these cells, the IF proteins provide not only a mechanical scaffold but are also involved in many dynamic cellular processes. Continuing studies of the regulation of the IF gene expression is hence important and should lead to a better understanding of how important properties of a cell or tissue such as cell fate, shape, development, and differentiation are regulated. Although many of the regulatory regions for IF genes have been identified, most of them remain unexplored. In addition, most of the regulatory studies have been confined to mouse and human species, and few studies have been performed in regulation of IF genes in other model systems. Not surprisingly, our current knowledge of the complex network of transcription factors that govern the expression of IF genes in various cell type is very limited. In this emerging era of system and genome biology, a broad-based multidisciplinary approach to study the process of transcriptional regulation of the IF gene is warranted, and with the advent of new tools and technology, this is within our reach. The future indeed looks promising!

Acknowledgments

I apologize for not referring to many related papers in the field because of lack of space. The work on K5 and K14 described in this book chapter was performed in the laboratory of Dr. Elaine Fuchs. My sincere thanks to Elaine Fuchs for her continued support and to many members of the Fuchs laboratory, particularly Charles Kaufman and Diana Bolotin, for their contribution to these studies.

References

Andrae, J., Bongcam-Rudloff, E., Hansson, I., Lendahl, U., Westermark, B., and Nister, M. (2001). A 1.8 kb GFAP-promoter fragment is active in specific regions of the embryonic CNS. *Mech. Dev.* **107,** 181–185.

Bader, B. L., and Franke, W. W. (1990). Cell type-specific and efficient synthesis of human cytokeratin 19 in transgenic mice. *Differentiation* **45,** 109–118.

Brembeck, F. H., Moffett, J., Wang, T. C., and Rustgi, A. K. (2001). The keratin 19 promoter is potent for cell-specific targeting of genes in transgenic mice. *Gastroenterology* **120,** 1720–1728.

Brembeck, F. H., and Rustgi, A. K. (2000). The tissue-dependent keratin 19 gene transcription is regulated by GKLF/KLF4 and Sp1. *J. Biol. Chem.* **275,** 28230–28239.

Bulyk, M. L. (2003). Computational prediction of transcription-factor binding site locations. *Genome Biol.* **5,** 201.

Byrne, C., and Fuchs, E. (1993). Probing keratinocyte and differentiation specificity of the human K5 promoter *in vitro* and in transgenic mice. *Mol. Cell Biol.* **13,** 3176–3190.

Byrne, C., Tainsky, M., and Fuchs, E. (1994). Programming gene expression in developing epidermis. *Development* **120,** 2369–2383.

Casanova, L., Bravo, A., Were, F., Ramirez, A., Jorcano, J. J., and Vidal, M. (1995). Tissue-specific and efficient expression of the human simple epithelial keratin 8 gene in transgenic mice. *J. Cell Sci.* **108**(Pt. 2)**,** 811–820.

Charron, G., Guy, L. G., Bazinet, M., and Julien, J. P. (1995). Multiple neuron-specific enhancers in the gene coding for the human neurofilament light chain. *J. Biol. Chem.* **270,** 30604–30610.

Coulombe, P. A., Ma, L., Yamada, S., and Wawersik, M. (2001). Intermediate filaments at a glance. *J. Cell Sci.* **114,** 4345–4347.

Coulombe, P. A., and Omary, M. B. (2002). 'Hard' and 'soft' principles defining the structure, function and regulation of keratin intermediate filaments. *Curr. Opin. Cell Biol.* **14,** 110–122.

DePianto, D. J., Hess, J. F., Blankenship, T. N., and FitzGerald, P. G. (2003). Isolation and characterization of the human CP49 gene promoter. *Invest. Ophthalmol. Vis. Sci.* **44,** 235–243.

Freedberg, I. M., Tomic-Canic, M., Komine, M., and Blumenberg, M. (2001). Keratins and the keratinocyte activation cycle. *J. Invest. Dermatol.* **116,** 633–640.

Fuchs, E., and Byrne, C. (1994). The epidermis: Rising to the surface. *Curr. Opin. Genet. Dev.* **4,** 725–736.

Fuchs, E., and Cleveland, D. W. (1998). A structural scaffolding of intermediate filaments in health and disease. *Science* **279,** 514–519.

Fuchs, E., and Weber, K. (1994). Intermediate filaments: Structure, dynamics, function, and disease. *Annu. Rev. Biochem.* **63,** 345–382.

Herrmann, H., and Aebi, U. (2000). Intermediate filaments and their associates: Multi-talented structural elements specifying cytoarchitecture and cytodynamics. *Curr. Opin. Cell Biol.* **12,** 79–90.

Hesse, M., Magin, T. M., and Weber, K. (2001). Genes for intermediate filament proteins and the draft sequence of the human genome: Novel keratin genes and a surprisingly high number of pseudogenes related to keratin genes 8 and 18. *J. Cell Sci.* **114,** 2569–2575.

Jave-Suarez, L. F., Winter, H., Langbein, L., Rogers, M. A., and Schweizer, J. (2002). HOXC13 is involved in the regulation of human hair keratin gene expression. *J. Biol. Chem.* **277,** 3718–3726.

Karabinos, A., Schmidt, H., Harborth, J., Schnabel, R., and Weber, K. (2001). Essential roles for four cytoplasmic intermediate filament protein's in Caenorhabditis elegans development. *Proc. Natl. Acad. Sci. USA* **98,** 7863–7868.

Karabinos, A., Zimek, A., and Weber, K. (2004). The genome of the early chordate Ciona intestinalis encodes only five cytoplasmic intermediate filament proteins including a single type I and type II keratin and a unique IF-annexin fusion protein. *Gene* **326,** 123–129.

Kaufman, C. K., Sinha, S., Bolotin, D., Fan, J., and Fuchs, E. (2002). Dissection of a complex enhancer element: Maintenance of keratinocyte specificity but loss of differentiation specificity. *Mol. Cell Biol.* **22,** 4293–4308.

Langbein, L., Rogers, M. A., Winter, H., Praetzel, S., Beckhaus, U., Rackwitz, H. R., and Schweizer, J. (1999). The catalog of human hair keratins. I. Expression of the nine type I members in the hair follicle. *J. Biol. Chem.* **274,** 19874–19884.

Langbein, L., Rogers, M. A., Winter, H., Praetzel, S., and Schweizer, J. (2001). The catalog of human hair keratins. II. Expression of the six type II members in the hair follicle and the combined catalog of human type I and II keratins. *J. Biol. Chem.* **276,** 35123–35132.

Leask, A., Rosenberg, M., Vassar, R., and Fuchs, E. (1990). Regulation of a human epidermal keratin gene: Sequences and nuclear factors involved in keratinocyte-specific transcription. *Genes Dev.* **4,** 1985–1998.

Leconte, L., Santha, M., Fort, C., Poujeol, C., Portier, M. M., and Simonneau, M. (1996). Cell type-specific expression of the mouse peripherin gene requires both upstream and intragenic sequences in transgenic mouse embryos. *Brain Res. Dev. Brain Res.* **92,** 1–9.

Lin, F., and Worman, H. J. (1997). Expression of nuclear lamins in human tissues and cancer cell lines and transcription from the promoters of the lamin A/C and B1 genes. *Exp. Cell Res.* **236,** 378–384.

Liu, Y., Lyle, S., Yang, Z., and Cotsarelis, G. (2003). Keratin 15 promoter targets putative epithelial stem cells in the hair follicle bulge. *J. Invest. Dermatol.* **121,** 963–968.

Lothian, C., Prakash, N., Lendahl, U., and Wahlstrom, G. M. (1999). Identification of both general and region-specific embryonic CNS enhancer elements in the nestin promoter. *Exp. Cell Res.* **248,** 509–519.

Machesney, M., Tidman, N., Waseem, A., Kirby, L., and Leigh, I. (1998). Activated keratinocytes in the epidermis of hypertrophic scars. *Am. J. Pathol.* **152,** 1133–1141.

Mahony, D., Karunaratne, S., Cam, G., and Rothnagel, J. A. (2000). Analysis of mouse keratin 6a regulatory sequences in transgenic mice reveals constitutive, tissue-specific expression by a keratin 6a minigene. *J. Invest. Dermatol.* **115,** 795–804.

Masaki, S., Yonezawa, S., and Quinlan, R. (2002). Localization of two conserved cis-acting enhancer regions for the filensin gene promoter that direct lens-specific expression. *Exp. Eye Res.* **75,** 295–305.

Mayor, C., Brudno, M., Schwartz, J. R., Poliakov, A., Rubin, E. M., Frazer, K. A., Pachter, L. S., and Dubchak, I. (2000). VISTA: Visualizing global DNA sequence alignments of arbitrary length. *Bioinformatics* **16,** 1046–1047.

Maytin, E. V., Lin, J. C., Krishnamurthy, R., Batchvarova, N., Ron, D., Mitchell, P. J., and Habener, J. F. (1999). Keratin 10 gene expression during differentiation of mouse epidermis requires transcription factors C/EBP and AP-2. *Dev. Biol.* **216,** 164–181.

Mcgowan, K. C., and Coulombe, P. A. (1998). The wound repair associated keratins 6, 16 and 17: Insights into the role of intermediate filaments is specifying cytoarchitecture. *In* "Subcellular Biochemistry: Intermediate Filaments" (J. R., Harris and H. Hermmann, eds.), pp. 141–165. Plenum, New York.

Mericskay, M., Parlakian, A., Porteu, A., Dandre, F., Bonnet, J., Paulin, D., and Li, Z. (2000). An overlapping CArG/octamer element is required for regulation of desmin gene transcription in arterial smooth muscle cells. *Dev. Biol.* **226,** 192–208.

Milisavljevic, V., Freedberg, I. M., and Blumenberg, M. (1996). Characterization of nuclear protein binding sites in the promoter of keratin K17 gene. *DNA Cell Biol.* **15,** 65–74.

Moll, R., Franke, W. W., Schiller, D. L., Geiger, B., and Krepler, R. (1982). The catalog of human cytokeratins: Patterns of expression in normal epithelia, tumors and cultured cells. *Cell* **31,** 11–24.

Nagy, A., Gertsensrein, M., Vintersten, K., and Behringer, R. (2003). "Manipulating the Mouse Embryo: A Laboratory Manual." Cold Spring Harbor Laboratory Press, Cold Spring Harbor, New York.

Nakamachi, K., and Nakajima, N. (2000). DNase I hypersensitive sites and transcriptional activation of the lamin A/C gene. *Eur. J. Biochem.* **267,** 1416–1422.

Neznanov, N., Umezawa, A., and Oshima, R. G. (1997). A regulatory element within a coding exon modulates keratin 18 gene expression in transgenic mice. *J. Biol. Chem.* **272,** 27549–27557.

Neznanov, N. S., and Oshima, R. G. (1993). cis regulation of the keratin 18 gene in transgenic mice. *Mol. Cell Biol.* **13,** 1815–1823.

Owens, D. W., and Lane, E. B. (2003). The quest for the function of simple epithelial keratins. *Bioessays* **25,** 748–758.

Paladini, R. D., Takahashi, K., Bravo, N. S., and Coulombe, P. A. (1996). Onset of re-epithelialization after skin injury correlates with a reorganization of keratin filaments in wound edge keratinocytes: Defining a potential role for keratin 16. *J. Cell Biol.* **132,** 381–397.

Qiu, P. (2003). Recent advances in computational promoter analysis in understanding the transcriptional regulatory network. *Biochem. Biophys. Res. Commun.* **309,** 495–501.

Ramirez, A., Bravo, A., Jorcano, J. L., and Vidal, M. (1994). Sequences 5′ of the bovine keratin 5 gene direct tissue- and cell-type-specific expression of a lacZ gene in the adult and during development. *Differentiation* **58,** 53–64.

Ramirez, A., Vidal, M., Bravo, A., and Jorcano, J. L. (1998). Analysis of sequences controlling tissue-specific and hyperproliferation-related keratin 6 gene expression in transgenic mice. *DNA Cell Biol.* **17,** 177–185.

Rosenthal, D. S., Steinert, P. M., Chung, S., Huff, C. A., Johnson, J., Yuspa, S. H., and Roop, D. R. (1991). A human epidermal differentiation-specific keratin gene is regulated by calcium but not negative modulators of differentiation in transgenic mouse keratinocytes. *Cell Growth Differ.* **2,** 107–113.

Rothnagel, J. A., Greenhalgh, D. A., Gagne, T. A., Longley, M. A., and Roop, D. R. (1993). Identification of a calcium-inducible, epidermal-specific regulatory element in the 3′-flanking region of the human keratin 1 gene. *J. Invest. Dermatol.* **101,** 506–513.

Schlake, T., Schorpp, M., Maul-Pavicic, A., Malashenko, A. M., and Boehm, T. (2000). Forkhead/ winged-helix transcription factor Whn regulates hair keratin gene expression: Molecular analysis of the nude skin phenotype. *Dev. Dyn.* **217,** 368–376.

Sinha, S., Degenstein, L., Copenhaver, C., and Fuchs, E. (2000). Defining the regulatory factors required for epidermal gene expression. *Mol. Cell Biol.* **20,** 2543–2555.

Sinha, S., and Fuchs, E. (2001). Identification and dissection of an enhancer controlling epithelial gene expression in skin. *Proc. Natl. Acad. Sci. USA* **98,** 2455–2460.

Takahashi, K., and Coulombe, P. A. (1997). Defining a region of the human keratin 6a gene that confers inducible expression in stratified epithelia of transgenic mice. *J. Biol. Chem.* **272,** 11979–11985.

Ureta-Vidal, A., Ettwiller, L., and Birney, E. (2003). Comparative genomics: Genome-wide analysis in metazoan eukaryotes. *Nat. Rev. Genet.* **4,** 251–262.

Wang, I. J., Carlson, E. C., Liu, C. Y., Kao, C. W., Hu, F. R., and Kao, W. W. (2002). Cis-regulatory elements of the mouse Krt1.12 gene. *Mol. Vis.* **8,** 94–101.

Wang, Y. N., and Chang, W. C. (2003). Induction of disease-associated keratin 16 gene expression by epidermal growth factor is regulated through cooperation of transcription factors Sp1 and c-Jun. *J. Biol. Chem.* **278,** 45848–45857.

Willoughby, D. A., Vilalta, A., and Oshima, R. G. (2000). An Alu element from the K18 gene confers position-independent expression in transgenic mice. *J. Biol. Chem.* **275,** 759–768.

Wu, R. L., Galvin, S., Wu, S. K., Xu, C., Blumenberg, M., and Sun, T. T. (1993). A 300 bp 5′-upstream sequence of a differentiation-dependent rabbit K3 keratin gene can serve as a keratinocyte-specific promoter. *J. Cell Sci.* **105**(Pt. 2), 303–316.

Yaworsky, P. J., Gardner, D. P., and Kappen, C. (1997). Transgenic analyses reveal developmentally regulated neuron- and muscle-specific elements in the murine neurofilament light chain gene promoter. *J. Biol. Chem.* **272,** 25112–25120.

Zhang, X., Diab, I. H., and Zehner, Z. E. (2003). ZBP-89 represses vimentin gene transcription by interacting with the transcriptional activator, Sp1. *Nucl. Acids Res.* **31,** 2900–2914.

Zimek, A., Stick, R., and Weber, K. (2003). Genes coding for intermediate filament proteins: Common features and unexpected differences in the genomes of humans and the teleost fish Fugu rubripes. *J. Cell Sci.* **116,** 2295–2302.

CHAPTER 11

Fluorescence-Based Methods for Studying Intermediate Filaments

Eric W. Flitney and Robert D. Goldman

Department of Cell and Molecular Biology
Feinberg School of Medicine
Northwestern University
Chicago, Illinois 60611

METHODS IN CELL BIOLOGY, VOL. 78

I. Introduction

The past decade has witnessed unprecedented progress in our understanding of the structure and function of intermediate filaments (IFs). Studies of fixed cells using immunofluorescence and, more recently, of living cells expressing genetically engineered fluorescent fusion proteins have fundamentally reshaped our perception of IF. Time-lapse studies of IF in living cells have uncovered a fascinating wealth of unexpected motile behaviors. Some of these motile processes are powered by molecular motors, interacting with microtubules (MT) or actin microfilaments (MF), whereas others seem to be inherent properties of IF themselves (Helfand *et al.*, 2004). Thus, the widely held notion that IFs are comparatively inert cytoskeletal elements, merely providing cells with the means to resist mechanical stresses, is no longer tenable.

The aim here is to outline some of the relevant fluorescence-based techniques that have been used to study IFs in both fixed and living cells. We highlight the value of combining conventional immunofluorescence techniques with live cell imaging for studying IF dynamics, and we limit our discussion to type III (vimentin, peripherin) and type V (A and B lamins) IFs. Some background material is included, together with step-by-step details of experimental protocols used in our laboratory, in the hope that these will assist the beginner and IF expert alike.

A. Immunofluorescence of Fixed Cells

Immunofluorescence hinges on the ability of an antibody to bind its antigen with high affinity and specificity. The method offers exquisite sensitivity; concentrations of antigen in the picomolar (pM) to the femtomolar (fM) range can easily be detected, and objects that are below the limit of resolution with refraction-based microscopy (differential interference contrast, phase contrast) can be "seen" in the fluorescence microscope as point sources of light.

Immunofluorescence can be performed on cultured cells grown on plastic dishes. This is convenient if the aim is simply to record the presence or absence of a particular IF protein in a population of cells (e.g., in a routine pathology laboratory screen). However, plastic dishes are not suitable for investigating the detailed morphology of IF networks and their interaction with other cytoskeletal elements. This requires that cells be viewed under the best possible optical conditions with oil immersion objective lenses ($\times 63$ or $\times 100$) with high numerical apertures (N.A., 1.4 or 1.3, respectively) to ensure the highest resolution. Because such lenses have short working distances, cells are generally cultivated and processed on thin (No.1) glass coverslips.

Cells must be fixed before immunostaining. The aim is to preserve the *in vivo* distribution of IF as closely as possible—to effectively "freeze" the cell structure—without sacrificing antigenicity. The preferred choice of fixatives for studying IF has not changed appreciably for more than 30 years. Brief immersion in either cold ($-20\,^{\circ}$C) dry methanol or, alternatively, formaldehyde in phosphate-buffered saline (PBS) at room temperature, remain the most popular methods (see section II.A). Methanol is considered marginally better than formaldehyde. However, neither method preserves all components of the cytoskeleton equally well, and this becomes an important issue when investigating possible structural interactions between IF and MF or IF and MT. For example, we find that MFs are not well preserved by cold methanol or glutaraldehyde, whereas formaldehyde gives good preservation. On the other hand, MTs are inadequately preserved by formaldehyde, which can cause them to fragment into short segments, whereas glutaraldehyde and methanol give excellent preservation.

The outer membrane is generally well preserved in aldehyde-fixed cells, and this constitutes an impenetrable barrier to the entry of large molecules into the interior. The membrane must, therefore, be rendered freely permeable before treating the cells with antibodies. Several detergent-based extraction procedures have been devised for this purpose, some of which result in complete removal of the phospholipid membrane, whereas others are more selective and remove certain membrane components only. A separate permeabilization step is not necessary after methanol fixation, because the outer membrane is extensively permeabilized during the fixation procedure.

Indirect immunofluorescence is the method most widely used for studying IF. Cells are fixed, permeabilized if required, and then treated with unlabeled primary IF antibody. Excess (unbound) antibody is removed afterward by careful washing. The cells are then treated with a secondary antibody conjugated to a suitable fluorochrome to disclose the cellular location of the IF protein. In most cultured cells, this usually reveals extensive arrays of IF bundles of varying sizes, coursing from the perinuclear region to the cell surface (Fig. 1). For double indirect immunoflourescence, combinations of rabbit polyclonal and mouse monoclonal primary antibodies are applied to cells either simultaneously or sequentially. In this type of preparation, commercially available anti-rabbit and anti-mouse secondary antibodies should, of course, be conjugated to different fluorochromes. Alternatively, single antibody labeling can be combined with specific stains or dyes conjugated directly to a fluorophore, as in the case of fluorescent phalloidin, widely used to stain F-actin (Fig. 2A–C).

B. Genetically Engineered Fluorescent Protein–Intermediate Filament Probes for Live Cell Imaging

The cloning of a cDNA encoding green fluorescent protein (GFP; Prasher *et al.*, 1992) and the demonstration by Chalfie *et al.* (1994) that this could be expressed in eukaryotic cells were landmark events that had a profound impact on live cell imaging (Cubitt *et al.*, 1995). GFP is a small protein found in the marine coelenterate, *Aequoria victorea*, that emits a characteristic apple-green fluorescence when irradiated with near ultraviolet (UV) or blue light. Mutant forms of GFP have been engineered by site-directed mutagenesis, creating a range of novel fluorescent variants with altered spectral and protein folding properties (see Table I in

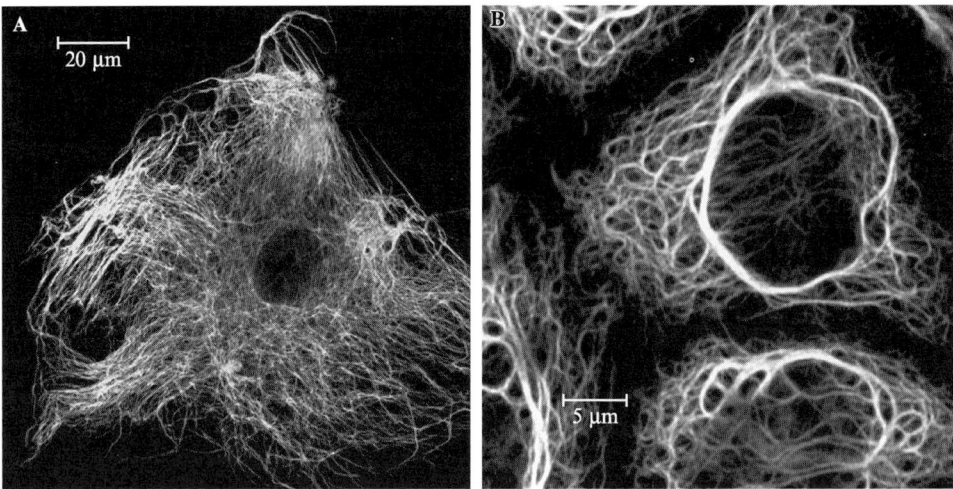

Fig. 1 Actively spreading (A) and confluent (B) human umbilical vein endothelial cells stained with a mouse monoclonal antibody to vimentin (V-9 clone) followed by a goat anti-mouse-Alexa 488 secondary antibody conjugate. Formaldehyde-fixed cells.

Cubbit *et al.*, 1999). The most widely used for live cell work is the double mutant S65T/F64L, called EGFP. This has a red-shifted excitation maximum at 490 nm, conveniently close to the 488-nm line of the Argon ion laser, with a peak emission at 510 nm similar to that of GFP. EGFP is much brighter than the wild-type protein, it oxidizes more rapidly during posttranslational processing, and it is far less prone to photobleaching. The S65G-type mutants produce larger red shifts, generating several bright yellow variants (YFP). These absorb maximally in the range of 512–516 nm and have peak emissions at 518–527 nm. The Y66H group of variants is blue shifted with absorption maxima at 382–384 nm and emission maxima at 446–448 nm.

cDNAs encoding either full-length or truncated IF proteins can be spliced into suitable vectors containing GFP or one if its engineered mutants and then expressed after transfection of cultured cells. In principle, GFP can be attached to either the C- or N-terminus of an IF protein; however, normal function is best preserved when the N-terminus is targeted (Ho *et al.*, 1998; Yoon *et al.*, 1998). Once expressed in a cell, the behavior of the GFP–IF reporter should be identical to that of the endogenous protein if meaningful results are to be obtained. It is critically important to establish this at the outset by performing the appropriate control experiments. Fortunately, GFP retains its fluorescence after fixation. This feature is very useful, because one can easily establish whether the reporter has been fully integrated into the endogenous IF network simply by comparing the pattern of GFP fluorescence with that obtained after immunostaining the IF protein, choosing a secondary antibody conjugated to a red fluorochrome (e.g., rhodamine, Alexa 568). The staining patterns should coalign if the GFP–IF chimera is fully incorporated into the

vimentin   actin   merge

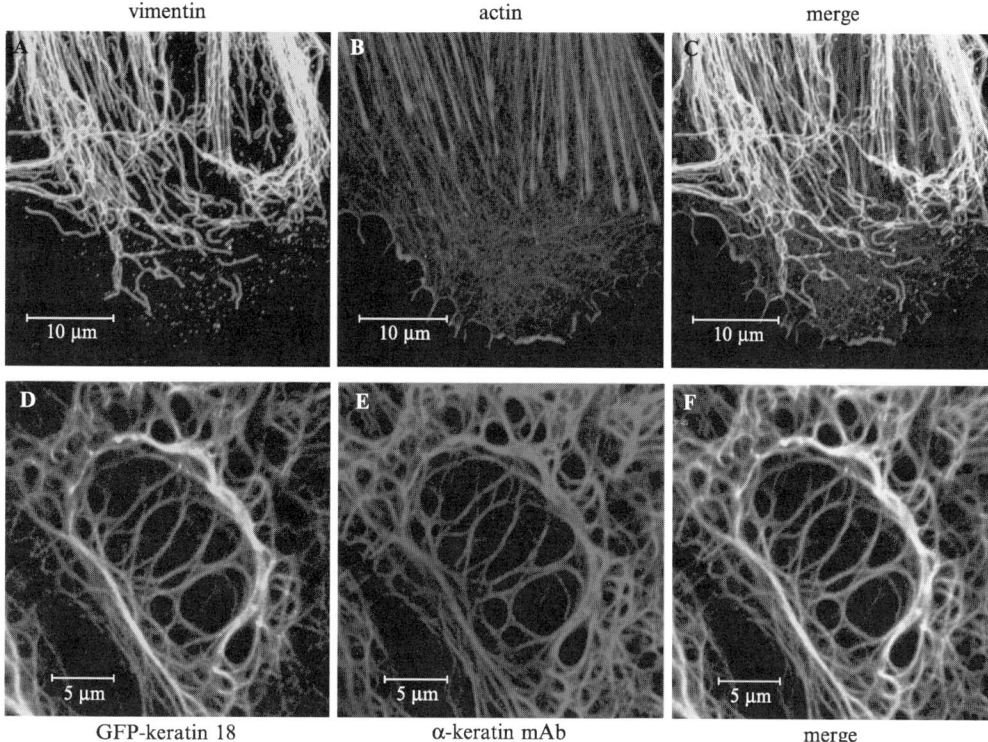

GFP-keratin 18   α-keratin mAb   merge

Fig. 2 Human endothelial cells fixed in formaldehyde and stained with a mouse monoclonal antibody to vimentin (V-9 clone) followed by a goat anti-mouse-Alexa 488 secondary antibody, (A) together with phalloidin-Alexa 568 to show F-actin (B). Overlay (C) reveals colocalization (yellow) between vimentin IF and the tips of actin stress fibers. Transformed lung alveolar epithelial cell (A549 line) expressing green fluovescent protein (GFP)-tagged keratin 18 (D) immunostained with a mouse monoclonal antibody to keratin 18 followed by a goat anti-mouse-Alexa 568 (red) conjugated secondary antibody (E). Overlay shows that the staining patterns are coincident and that GFP-keratin is fully incorporated into the endogenous network (F). (See Color Insert.)

endogenous network, as shown in Fig. 2D–F. The behavior of the fluorescent protein during different phases of the cell cycle, (e.g., throughout mitosis and also during daughter cell spreading after cytokinesis) should also be consistent with what we know about IFs from immunofluorescence observations of nontransfected cells.

II. Materials and Methods

The procedures outlined here are those that are currently used to study IFs in our laboratory. Some are necessarily of a generic nature, and modifications may be required for different cell types or for specific applications.

A. Indirect Immunofluorescence

Methanol and formaldehyde are suitable fixatives for studying IFs by this technique.

1. Immunofluorescence with Methanol Fixation

1. Grow cells on No. 1 glass coverslips in 35-mm Petri dishes.

2. Pour off medium, transfer coverslip(s) to a Columbia jar, and quickly (2 × 10 seconds) rinse in PBSa (6 mM sodium/potassium phosphates (pH 7.4), 0.17 M NaCl, 3 mM KCl).

3. Gently touch the edge of each coverslip against absorbent tissue to remove excess PBSa.

4. Immerse coverslip(s) in *dry* methanol precooled to −20 °C for 3–5 minutes. Keep in the −20 °C refrigerator during fixation.

5. Pour off methanol and wash cells in PBSa (2 × 1 minute).

6. At this stage, a protein solution is applied to the cells to block nonspecific adsorption of antibodies. Place two cocktail sticks on a soaked filter paper in the upturned lid of a 100-mm Petri dish. Remove coverslip(s) from Columbia jar and support (cells up) on the cocktail sticks. Overlay cells with 50–100 μl PBSa containing *either* 1%–5% nonimmune serum from the species in which the secondary antibody was raised *or* 1%–5% bovine serum albumin (BSA). Transfer to a 37 °C incubator and leave for 15–20 minutes, using the base of the Petri dish to seal the chamber. This will maintain a moist atmosphere inside and prevent cells from drying out.

7. Pour off the "blocking" solution and cover cells with 50 μl of "blocking" solution containing the primary antibody(ies), diluted as required. Place in a 37 °C incubator and leave for 20–30 minutes.

8. Remove coverslips and return to Columbia jar. Wash off unbound primary antibody with PBSa containing 0.05% Tween (PBST; 2 minutes), followed by PBSa alone (2 × 3 minutes).

9. Transfer coverslips to the Petri dish and cover cells with 50 μl "blocking" solution containing the secondary antibody(ies). Return to incubator and leave for 20–30 minutes.

10. Place coverslips in Columbia jar and wash thoroughly with PBSa (3 × 3 minutes).

11. Drain off excess PBSa and mount coverslip ("cells down") on a drop of Gelvatol (polyvinyl alcohol; grade 205; Air Products Corporation, Allentown, PA) containing 100 mg · ml^{-1} of 1,4-diazabicyclo [2.2.2] octane antifade agent (DABCO; Sigma). Alternatively, for improved phase-contrast imaging, mount coverslips on a drop of PPD mounting medium (20 mg p-phenylenediamine, 1 ml 0.5 M Tris buffer [pH 9], 2 ml distilled water, 7 ml glycerol) and seal the edges of

the coverslip with nail varnish. Cells mounted with the PPD medium can be viewed immediately. Gelvatol should be allowed to set (~30 minutes at room temperature) before examining cells in the microscope.

Note: Care should be taken to ensure that cells are not allowed to dry out at any time during this procedure. When immunostaining nuclear lamins to study nuclear shape, it is advisable to mount the coverslip supported on four glass "feet" (made by breaking a No. 1 or No. 0 coverslip into small chips) to avoid compressing the nuclei and distorting their shape. The edges of the coverslip can be sealed with nail varnish to retain the mountant and prevent the preparation from drying out. Alternatively, a 1:1:1 mixture of lanolin, petroleum jelly, and bees wax can be melted and applied to the edge of the coverslip with a Pasteur pipette.

2. Immunofluorescence with Formaldehyde Fixation

1. Grow cells on No. 1 glass coverslips in 35-mm Petri dishes.
2. Pour off medium, transfer coverslip(s) to a Columbia jar, and quickly (2×10 seconds) rinse in PBSa.
3. Gently touch the edge of each coverslip against absorbent tissue to remove excess liquid.
4. Fix cells by placing coverslips in a Columbia jar containing freshly prepared formaldehyde (2%–4%; Sigma) dissolved in PBSa for 5–10 minutes at room temperature.
5. Wash cells in PBSa to remove fix (2×3 minutes).
6. Permeabilize cells by immersing coverslips in 0.1% NP-40 (Igepal CA-630; Sigma) or 0.1% Triton X-100 (Sigma) in PBSa (3×3 minutes) at room temperature. Brief exposure to acetone (2 minutes at $-20 °C$) is sometimes used instead of detergent.
7. Remove excess detergent/acetone by washing in PBSa (3×2 minutes).
8. Proceed with steps 6–11 as for methanol-fixed cells (previously).

B. Microinjection Techniques

Intermediate filament proteins can be labeled with biotin and their fate followed after microinjection by fixing and immunostaining cells with an anti-biotin antibody; alternatively, cells can be microinjected with fluorescent-tagged IF proteins and used for live imaging studies. The isolation of different IF proteins from animal and human tissue enables one to study the dynamic properties of terminally differentiated proteins; alternatively, one can use bacterially expressed recombinant IF proteins that have not been subjected to posttranslational processing.

1. Preparation of Purified Vimentin from Bovine Lens (after Vikstrom *et al.*, 1989)

The procedure for obtaining bovine lens vimentin is as follows:

1. Prepare the following solutions:
   a. Buffer H: 50 mM Tris (pH 7.4), 5 mM $MgCl_2$, 0.2% 2-mercaptoethanol, 1 mM phenyl methyl sulfonyl fluoride (PMSF).
   b. Buffer E: 8 M urea, 50 mM Tris (pH 7.4), 1 mM EGTA, 0.2% 2-mercaptoethanol, 1 mM PMSF.
   c. Buffer HA: 6 M urea, 8 mM sodium phosphate (pH 7.2), 0.14 M NaCl, 1 mM dithiothreitol, 0.1 mM PMSF.
   d. Buffer A: 6 mM sodium-potassium phosphate (pH 7.4), 0.17 M NaCl, 3 mM KCl, 0.2% mercaptoethanol, 0.2 mM PMSF.
2. Take 30 g fresh or frozen bovine lenses and homogenize in 250 ml buffer H at 4 °C.
3. Centrifuge homogenate for 20 minutes at 25,000g at 4 °C.
4. Wash resulting pellet (2×) by resuspending it in 250 ml buffer H followed by recentrifugation.
5. Extract final pellet for 4 hours at 4 °C with 100 ml buffer E.
6. Clarify the urea extract by centrifugation (100,000g) for 30 minutes at 4 °C.
7. Add solid ammonium sulfate to the supernatant (36 g/100 ml) to precipitate out the protein.
8. Dissolve precipitated protein in 10 ml buffer HA, desalt by passing over a Sephadex G-25 column (2.5 × 18 cm), then apply to hydroxyapatite column equilibrated with buffer HA at room temperature.
9. Wash column with one half column volume of buffer HA and elute the protein with a linear sodium phosphate gradient (8–40 mM) in buffer HA (450 ml) at room temperature.
10. Run immunoblots of collected fractions with antibody to vimentin and pool those containing vimentin.
11. Dialyze pooled fractions against 500 volumes of buffer A to assemble IF.
12. Freeze droplets of polymerized IF in liquid nitrogen and store at −70 °C until required.

2. Preparation of Bacterially Expressed Vimentin with *c-myc* or *FLAG* Tags

Bacterial expression vectors such as the pET series (Novagen) carrying cDNA for vimentin and a suitable tag (e.g., *c-myc* or *FLAG*) can be used to produce large quantities of unmodified vimentin. The recombinant tagged protein can be purified by column chromatography and used to microinject cells instead of biotinylated or fluorochrome-labeled vimentin. Large amounts of protein are obtained relatively quickly, as follows:

1. Prepare the following solutions:
    a. Growth medium: type 2YT or LB.
    b. Stock ampicillin or kanamycin: 25 mg/ml dissolved in sterile water.
    c. Inducer stock: 100 mM isopropyl-β-D-thiogalactopyranoside (IPTG) in sterile water.
    d. TNE buffer: 10 mM Tris-HCl (pH 8), 100 mM NaCl, 1 mM EGTA.

2. Prepare bacterial culture (5 ml) of BL-21 (DE3 strain) carrying pET– vimentin plasmid with or without a *c-myc* or *FLAG* tag. Add appropriate antibiotics and incubate culture overnight.

3. Add overnight culture to 1 L of prewarmed (37 °C) medium and grow for approximately 3–5 hours on a shaker operating at $\sim$275 rpm.

4. Withdraw a sample of the culture medium periodically and measure its optical density at 595 nm. When the density of the bacterial culture reaches 0.–1.0 AU add IPTG to a final concentration of 0.6 mM and grow bacteria for a further 3 hours.

5. Cool cultures on ice and centrifuge at 4000 rpm (JS 4.2 rotor, Beckman J6-HC centrifuge) for 30 minutes at 4 °C. Discard supernatant and resuspend the cells in 20 ml ice-cold TNE buffer in a 50-ml tube. Centrifuge again at 4000 rpm for 10 minutes to repellet the cells.

6. The recombinant vimentin is present in inclusion bodies inside the bacteria. The pelleted cells can be stored at this stage at -80 °C for use at a later time, or they can be processed immediately to isolate the protein.

7. To continue with the protein purification, prepare the following solutions:
    a. Lysis buffer: 50 mM Tris-HCl (pH 8), 25% sucrose, 1 mM EDTA, 1 mM EGTA.
    b. Detergent buffer: 25 mM Tris-HCl (pH 8), 0.2 M NaCl, 1% NP-40, 1% deoxycholate, 1 mM EDTA.
    c. TNE wash buffer: 10 mM Tris-HCl (pH 8), 100 mM NaCl, 1 mM EDTA, 0.5% NP-40.
    d. Column buffer: 10 mM Tris-HCl (pH 8), 8 M urea, 2 mM EDTA, 10 mM dithiothreitol.

Note: Add protease inhibitors to all the preceding solutions *immediately before use*: 1 mM PMSF and 5 μg/ml each of pepstatin A, leupeptin, and aprotinin. Keep all solutions ice cold.

8. Freeze and thaw pelleted cells and homogenize on ice in lysis buffer (20–30 ml/L of bacterial culture) with a motor-driven Teflon plunger.

9. Add 2 vols of detergent buffer and rehomogenize. (The solution will become very viscous initially, but the viscosity will drop during homogenization).

10. Centrifuge homogenate at 20,000g for 30 minutes at 4 °C.

11. Suspend pellet in 10–15 ml wash buffer containing 3 mM MgCl$_2$ and protease inhibitors. Add 1 mg DNAase I and incubate at 37 °C for 20–25 minutes.

12. Centrifuge at 20,000g for 30 minutes at 4 °C. Wash pellet again with TNE buffer and centrifuge as before.

13. Homogenize pellet in column buffer to solubilize inclusion bodies.

14. Centrifuge at 100,000g for 30 minutes at 4 °C to remove insoluble material. Retain the supernatant and check induction of protein by sodium dodecy sulfate-polyacrylamide gel electrophoresis (SDS-PAGE).

15. Fractionate the supernatant using Mono-Q (FPLC) or Q-Sepharose. Elute protein with 0–0.6 M NaCl gradient.

16. Run gels of column fractions to identify those containing vimentin. Pool vimentin fractions and dialyze against 2 mM phosphate buffer (pH 7.2) containing 1 mM dithiothreitol (three changes).

17. Aliquot recombinant vimentin and store in soluble form at −80 °C.

3. Preparation of Biotinylated Vimentin

Tissue-derived vimentin IFs are first thawed and then taken through one cycle of disassembly–reassembly before labeling with biotin.

1. Prepare the following solutions:
   a. Disassembly buffer: 8 M urea, 5 mM sodium phosphate (pH 7.2), 0.2% 2-mercaptoethanol, 1 mM PMSF.
   b. Subunit buffer: 5 mM sodium phosphate (pH 7.4), 0.2% 2-mercaptoethanol, 0.2 mM PMSF.
   c. Assembly buffer (buffer A, above): 6 mM sodium-potassium phosphate (pH 7.4), 0.17 M NaCl, 3 mM KCl, 0.2% mercaptoethanol, 0.2 mM PMSF.

2. Centrifuge freshly thawed IF suspension (100,000g for 30 minutes), then solubilize the pellet in disassembly buffer.

3. Dialyze the solution against subunit buffer to maintain vimentin in its depolymerized state.

4. Measure protein concentration by Bradford method and adjust to 2.5–3.0 mg/ml.

5. Incubate the solubilized protein in a 100:1 molar excess of N-hydroxysucci-nimidobiotin (Molecular Probes, Eugene, OR) dissolved in 10% dimethyl-formamide and 0.17 M NaCl. IFs will repolymerize under these conditions as biotinylation proceeds.

6. Leave reaction mixture for 90 minutes (the timing *is* critical) and centrifuge labeled IFs (100,000g, 30 minutes).

7. Subject the resulting pellet to *two* cycles of disassembly–reassembly by solubilizing in urea buffer followed by dialysis against assembly buffer.

8. Freeze droplets of suspension of biotinylated vimentin IF in liquid nitrogen and store at −70 °C until required.

Note: The yield of the biotinylation process is typically ~60%–70%. To verify that biotin is conjugated to vimentin, run SDS gels and immunoblot with antibodies to biotin and vimentin. Biotinylated vimentin will be recognized by both antibodies. Its electrophoretic mobility should be retarded somewhat compared with unlabeled vimentin. The competence of biotinylated vimentin to form smooth-walled IFs *in vitro* should be checked in the electron microscope by negative staining with 1% aqueous uranyl acetate.

4. Preparation of Rhodamine-Labeled Vimentin (after Vikstrom *et al.*, 1992)

1–3. Proceed as for preparation of biotinylated vimentin (see section II.B.3).

4. Measure protein concentration by the Bradford method and adjust to 2.0–2.5 mg/ml

5. Add a 40:1 molar excess of 5-(and 6-) carboxy-X-rhodamine succinimidyl ester (Molecular Probes, Eugene, OR) dissolved in dimethyl formamide (DMF). This should be made up immediately before adding to the protein in sufficient amount to bring the final concentration of DMF to 10% v/v. Add NaCl to the reaction mixture to bring final concentration to 0.17 M. This will induce filament formation during the conjugation reaction.

6. Incubate the preceding solution for 60 minutes at room temperature.

7. Collect labeled IFs by centrifugation (100,000g, 30 minutes, 4 °C).

8. Subject pelleted IFs to *two* cycles of disassembly–reassembly to ensure that only polymerization-competent protein is used to microinject cells. Use the following procedure. First, solubilize IF in disassembly buffer and pass through a Sephadex G-25 column equilibrated with subunit buffer. Next, add NaCl to a final concentration of 0.17 M and dialyze overnight at room temperature against assembly buffer. Column chromatography combined with overnight dialysis through each cycle of disassembly–reassembly will ensure that all unbound rhodamine is removed from the rhodamine-tagged vimentin solution.

Note: Check that the labeled protein can form filaments in the electron microscope by negative staining with 1% aqueous uranyl acetate. Typically, the preceding procedure incorporates ~0.5 mole rhodamine per mole of vimentin. Generally, this does not impair protein polmerization.

5. Preparation of Biotinylated Type I Keratin (after Miller *et al.*, 1991)

1. Prepare following solutions:
   a. Buffer A (urea extraction): 8 M urea, 50 mM Tris-HCl (pH 9.0), 20 mM β-mercaptoethanol, 1 mM PMSF.
   b. Buffer B (assembly buffer): 10 mM Tris-HCl (pH 7.4), 0.1 mM PMSF.
   c. Buffer C: PBSa (pH 7.4), 20 mM EDTA, 1 mM PMSF.

    d. Buffer D (depolymerizing buffer): 8 M urea, 5 mM $NaPO_4$ (pH 7.4), 0.2% β-mercaptoethanol, 1 mM PMSF.

    e. Buffer E (column buffer): 9.5 M urea, 20 mM Tris-HCl (pH 8.6), 1 mM EDTA, 1 mM dithiothreitol (DTT).

2. Obtain fresh bovine tongues from the slaughterhouse and store in crushed ice.

3. Remove the mucosa and immerse in buffer C for 12 hours at 4°C.

4. Strip mucosa from underlying dermis, mince with scissors, and extract in urea buffer A for 45–60 minutes at 4°C with vigorous stirring.

5. Centrifuge at 200,000g for 30 minutes to remove urea-insoluble material.

6. Measure protein concentration in supernatant by the Bradford method and adjust to 1.4 mg/ml with buffer A.

7. Dialyze protein solution overnight at room temperature against 50 vol assembly buffer B ($\times$2) to repolymerize keratin.

8. Dissolve 200 mg succinimidyl-D-biotin (Molecular Probes, Eugene, OR) in 55 ml dimethylformamide and pipette slowly into 500 ml reconstituted keratin IF in assembly buffer B with stirring. Incubate reaction mixture for 30 minutes at room temperature.

9. Centrifuge at 200,000g for 30 minutes at 10°C and resuspend pelleted IF, in urea buffer D to depolymerize IFs.

10. Repolymerize labeled IFs by dialysis against assembly buffer B.

11. Repeat cycle (steps 9 and 10) of disassembly–reassembly.

12. Centrifuge twice-cycled biotinylated-keratin IF at 200,000g and discard supernatant.

13. Disassemble IF in column buffer E and separate type I and II keratins on an ion exchange column (1 $\times$ 32 cm; model DE-52; Whatman Chemical Separation Inc., Clifton, NJ) preequilibrated in buffer E with a flow rate of 30 ml/hr. Type II keratins will emerge in the flow-through fraction.

14. Elute type I keratins in 200 ml linear salt gradient (0–100 mM NaCl made up in column buffer). Collect 1-ml fractions.

15. Examine every third fraction by immunoblotting and pool those containing biotinylated type I keratin.

16. Dialyze pooled fractions against 10 mM Tris (pH 7.4).

17. Freeze 200-μl aliquots in liquid nitrogen and store at −70°C for microinjection experiments.

6. Microinjecting Cells with Labeled Intermediate Filament Proteins

    1. Grow cells on locator coverslips (Bellco Glass, Inc.) glued over a $^3/_4$ inch hole in the base of a 35-mm Petri dish. Glue the coverslip to the outer surface of the base with Sylgard 184 (Dow Corning). Sterilize the dishes by washing with 70%

alcohol and exposing them to germicidal UV light (30 minutes). Microinject cells when they attain ~70% confluence.

2. Replace growth medium with Leibovitz L15 culture medium and mount dishes on the inverted stage of a fluorescence-phase contrast microscope, maintained at 37 °C by means of an airstream stage incubator. Microinject cells with an automated injection system (InjectMan NI 2, Ependorf Scientific, Inc.) and keep a careful record of their XY coordinates on the locater coverslip. Inject with the micropipette inclined at a shallow angle into the thickest region of the cell, close to the nucleus. Inject each cell with ~0.06 pl of labeled protein (1–2 mg · ml^{-1}) dissolved in 5 mM sodium phosphate (pH 8.5) containing 0.05% β-mercaptoethanol.

3. The dishes containing microinjected cells can now be transferred to the stage of a laser confocal microscope for high-resolution imaging.

C. Engineering Green Fluorescent Protein–Intermediate Filament Constructs and Transfection Methods for Live Cell Imaging

Constructs for pEGFP-vimentin (Yoon et al., 1998) and pEGFP-peripherin (Helfand et al., 2003) were engineered in our laboratory by subcloning BamHI–BamHI restriction fragments encoding the IF proteins into the BamHI site of the pEGFP-C1 expression vector. cDNAs for full-length lamins A, C, and B1 were also tagged with GFP (Moir et al., 2000). A BamHI–EcoRI fragment composed of full-length prelamin A plus the c-myc 9E10 epitope and a four amino acid linker was subcloned into the BglII–EcoRI site of pEGFP-C1 or pYFP-C1. The same procedure was used to create a pEGFP–lamin C construct. For lamin B1, a BamHI–EcoRI sequence containing the full coding region with a four amino acid linker was cloned into the BglII-EcoRI site of either pEGFP-C1 or into vector for the cyan mutant, pECFP-C1. Human vimentin cDNA was also tagged with CFP with the pECFP-C1 vector.

Cells are transiently transfected with pcDNA3 constructs by electroporation (Moir et al., 2000; Helfand et al., 2003; Yoon et al., 2001) or by lipofectamine (Yoon et al., 1998).

1. Electroporation

1. Grow cells in 100-mm diameter Petri dishes. Trypsinize subconfluent (60%–70% coverage) dishes and suspend cells in 10 ml fresh medium containing 35 μl 1 M Hepes buffer (pH 7.1). Gently centrifuge (1000 rpm, 5 minutes) and discard the supernatant. Resuspend pelleted cells in 250 μl of Hepes-buffered medium.

2. Add 7 μg of the construct plus 13 μg sheared salmon sperm (carrier) DNA (Amresco Inc) to the cell suspension. For double transfection experiments, use 5 μg of each construct with 10 μg carrier DNA.

3. Place the mixture into a 4-mm electroporation chamber (Gene Pulser II, Bio Rad Laboratories) and pulse at 200–260 V/950 μF (dependent on cell type).

4. Seed transfected cells onto 22-mm glass coverslips (No. 1) in 35-mm Petri dishes and use within 24–72 hours, depending on cell type and protein to be expressed.

2. Liposomal Transfection

1. Grow cells on glass coverslips.
2. Mix 1–2 μg GFP-IF plasmid with 6 μl lipofectamine reagent (Gibco-BRL) in 200 μl serum-free medium and incubate for 40 minutes at room temperature.
3. Dilute DNA/lipofectamine mixture to a final volume of 1 ml with serum-free medium. Cover cells and incubate for 3 hours at 37 °C.
4. Remove DNA/lipofectamine mix and wash cells in serum-free medium. Leave overnight in complete growth medium and use cells within 24–72 hours, depending on cell type and protein to be expressed.

D. Fluorescence Recovery after Photobleaching (FRAP) Studies

Photobleaching has been used with great effect to study the dynamic properties of fluorescent-tagged IF proteins by a technique known as fluorescence recovery after photobleaching (FRAP) (Lippincott-Schwarz *et al.*, 2003). This method is performed on live cells previously microinjected with IF protein conjugated to a fluorescent probe or, alternatively, on transfected cells expressing a GFP-IF fusion protein. A region of interest (ROI) within the cell is defined and then deliberately photobleached by exposing it to high-intensity light. Immediately after bleaching, the light intensity is readjusted for normal imaging. If all the labeled material is freely mobile, unbleached molecules will diffuse into the ROI from outside, and the fluorescence will recover fully. However, if the material is completely immobile, there will be no recovery; if some fraction of the labeled material is free to move, there will be partial recovery of fluorescence only. Fluorescence recovery within a bar-shaped ROI, traversing several fibrils within the vimentin network of a BHK-21 cell expressing EGFP-vimentin, is shown in Fig.3A–D. The time course and extent of the recovery is measured by monitoring the fluorescence intensity within the ROI at different time intervals after the photobleach. In the experiment of Fig. 3A–D, FRAP was complete within ∼20 minutes.

III. Pearls and Pitfalls

A. Indirect Immunofluorescence

Indirect immunofluorescence is a relatively straightforward technique, requiring only modest laboratory facilities, although careful attention to detail during processing is necessary to obtain high-quality images. The quality of fixation is of paramount importance. Methanol is suitable for immunostaining all types of IF and for MT. For optimal preservation of IF, it is important that the methanol be

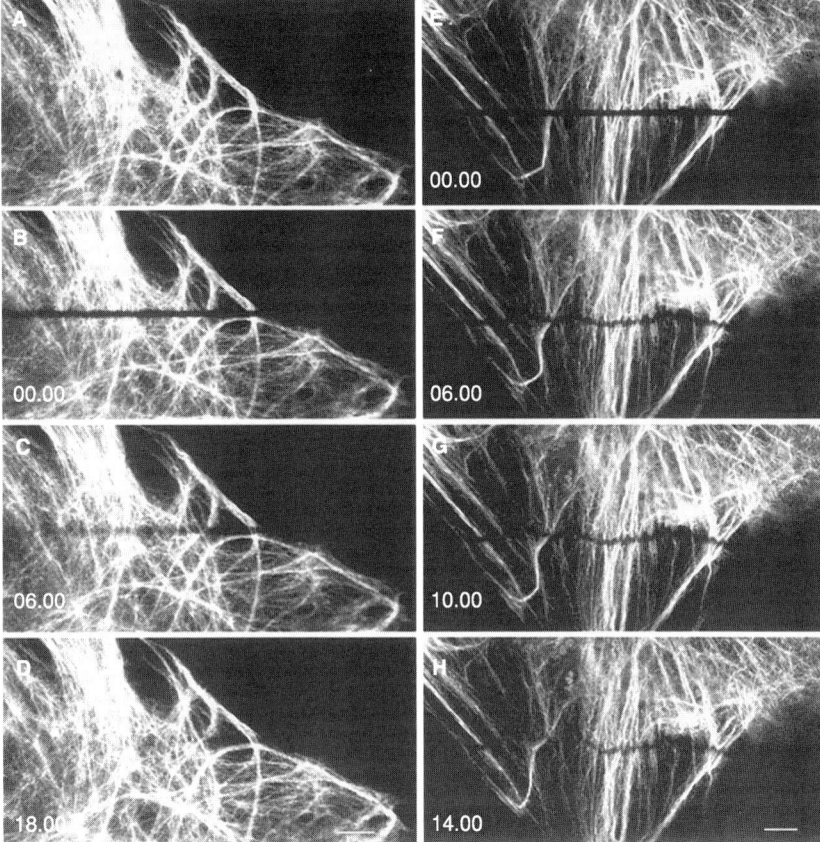

Fig. 3 (A–D) Time-lapse images showing recovery of fluorescence after photobleaching a bar-shaped area of a cell (BHK-21) expressing green fluorescent protein (GFP) vimentin. (A) Shows cell before photobleaching and (D) after the fluorescence has fully recovered. (E–H) Time-lapse images of a BHK-21 cell showing distortion of the original bar-shaped photobleached region caused by differential rates of relative sliding between neighboring GFP-vimentin fibrils. Elapsed times (after bleaching) shown in each panel. Reproduced from the *J. Cell Biol.* 1998, **143**, 152, by copyright permission of the Rockerfeller University Press.

very cold and completely water free. For this reason, we routinely store it over a drying agent (Molecular Sieves, 8–10 mesh beads; Sigma) in a −20 °C refrigerator, and we recommend fixing not more than two coverslips per Columbia jar at any one time. Formaldehyde fixation can also be used for most IFs and for actin MFs. It gives good preservation of IF and also of subfilamentous IF precursors (see section IV.A), although with some loss of antigenicity compared with methanol.

Care must also be taken to eliminate "background" staining by blocking non-specific adsorption of primary and/or secondary antibodies. Nonspecific binding can create a pattern of staining that seems authentic but is entirely artefactual. It is essential, therefore, to conduct appropriate control experiments to assess the extent of any spurious staining. For example, one can omit the primary antibody and immunostain with the secondary antibody only, or one can use a primary antibody to an antigen that is known to be absent from the cells being studied. If the "blocking" procedure is effective, there should be no staining with either of these methods.

The choice of fluorochrome is an important consideration also. Fluorescence "fading," or photobleaching, will cause the microscope image to become progressively degraded under continuous irradiation, and steps must be taken to minimize this as far as possible. The Alex Fluor fluorescent dyes (Molecular Probes, Eugene, OR) are far less susceptible to photobleaching, and they have brighter emissions than traditional fluorochromes (e.g., fluorescein, rhodamine, or Texas Red). Secondary antibodies conjugated to Alexa Fluor dyes have therefore gained in popularity in recent years, and we now use them routinely. Photobleaching can be further minimized by incorporating an antifade agent into the mounting medium and by viewing the specimen with low illumination intensities.

B. Live Cell Imaging

Immunofluorescence studies of cells engaged in different types of physiological activity (e.g., cell spreading and attachment; motility, proliferation, apoptosis) have provided important insights into the dynamic nature of IFs. However, there are several limitations to this approach. Temporal changes must be inferred from relatively few observations made before and either during a cellular event or immediately afterward. The time resolution is therefore poor, and any conclusions drawn are necessarily based on comparisons between different populations of cells.

The situation is greatly improved by studying live cells previously injected with fluorescent IF probes. This is technically demanding, in that IF proteins must first be isolated and purified, labeled with a fluorescent tag, and then microinjected into individual cells. Furthermore, fluorescence fading remains problematic, and this ultimately limits the time available for viewing cells. These difficulties can be alleviated to a large extent by studying living cells expressing GFP–IF fusion proteins. The method is noninvasive, and GFP is more resistant to photobleaching than synthetic fluorochromes, so that cells can be observed for longer periods of time. Of course, it is imperative to show that the fluorescent probe does not compromise the normal behavior of the protein. Several control experiments can be undertaken to examine the *in vivo* behavior of tagged IF proteins. For example, electron microscopy can establish whether a labeled IF can self-assemble into smooth-walled filaments *in vitro*, whereas immunofluorescence can be used to ascertain whether or not the behavior of the fluorescent protein during different phases of the cell cycle is the same as that of the endogenous IF protein.

C. Fluorescence Recovery after Photobleaching

Fluorescence recovery after photobleaching studies can be conducted on cells microinjected with fluorescent IF proteins or in cells expressing GFP–IF chimeras. The technique is simple in principle, although several potential pitfalls must be avoided if it is to give useful information. First, it is important that the fluorescence should not fade appreciably under normal imaging conditions; if it does, the time course of recovery will be prolonged and it will seem to be incomplete, suggestive of an immobile component. The problem can be alleviated to some extent by monitoring fluorescence in regions of the cell outside of (but in close proximity to) the ROI. Once the magnitude of fluorescence fading has been established by this means, the intensities within the bleached zone can be adjusted upward to compensate. Second, translational movement of the cell, and/or of the ROI inside the cell, will distort FRAP measurements. Simultaneous observations with phase-contrast optics should therefore be made, so that cell movement or shape changes can be detected and the relevant images excluded from the analysis. Third, IF must not be damaged by the high illumination intensities needed to achieve rapid photobleaching. This can be evaluated by comparing the photobleached ROI with images obtained after fixing and immunostaining the same cells with antibody to the native IF protein (see e.g., Fig. 2, Vikstrom *et al.*, 1992; also Fig. 6 in Yoon *et al.*, 1998).

IV. Discussion

A. Immunofluorescence Studies of Fixed Cells

As mentioned earlier, IFs are frequently portrayed as static, inert biopolymers that merely aid the cell to resist mechanical deformation. However, immunofluorescence studies of cells undergoing mitosis and during cell spreading reveal a different story. For example, as some cells enter metaphase, their IFs disassemble completely and become reorganized into nonfilamentous particles (Franke *et al.*, 1982; Lane *et al.*, 1982). These particles accumulate near the spindle poles to form brightly staining juxtanuclear "caps" in each daughter cell from which new IF networks originate (Chou *et al.*, 1990; Rosevear *et al.*, 1990). Prahlad *et al.* (1998) found that BHK-21 cells fixed during the early stages of spreading contained many nonmembrane-bound particles of nonfilamentous vimentin, similar to those formed during mitosis, together with numerous short, filamentous structures termed "squiggles" (Fig. 4). The relative abundance of particles and squiggles changed during the first 1.5–2 hours of spreading, favoring fewer particles and more squiggles, and after 3 hours the occurrence of both decreased markedly as cell spreading proceeded and filament length increased. Helfand *et al.* (2003) reported similar structures in neuronal cells (PC12) that express peripherin, a type III IF protein related to vimentin. On the basis of these observations, it was suggested that particles and squiggles might be IF precursors that contribute to the formation of new IFs (see section IV.C).

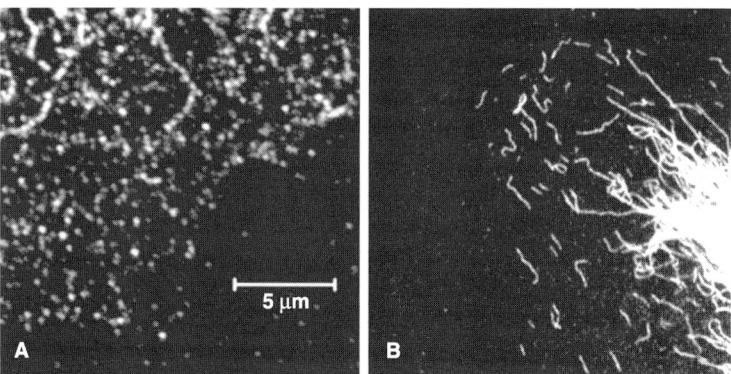

Fig. 4   (A) Peripheral region of cytoplasm in a human endothelial cell immunostained with a mouse monoclonal antibody to vimentin and a goat anti-mouse-Alexa 488 secondary antibody. This image shows several terminals of intermediate filament (IF), some of which appear "beaded," and numerous isolated vimentin particles. Cell fixed in formaldehyde. (B) IF squiggles at the periphery of a transformed human bone marrow endothelial cell fixed in methanol and immunostained with monoclonal antibody to vimentin and goat anti-mouse-Alexa 488 secondary conjugate.

Type V IF, the nuclear lamins, are major components of the lamina, a protein-aceous layer sandwiched between the inner surface of the nuclear membrane and chromatin, and are also found distributed throughout the nucleoplasm. Immunostaining showed that lamin B1 forms nucleoplasmic "spots" during mid-late S1 phase. These brightly staining foci coaligned with monoclonal antibodies to bromodeoxyuridine (BrDU) and proliferating cell nuclear antigen (PCNA), implicating lamin B1 in DNA replication (Moir *et al.*, 1994).

B. Examples of Results Obtained by Microinjecting Cells with Labeled Intermediate Filament Proteins

Experiments that involved microinjecting cells with tagged IF proteins (see section II.B) provided more direct evidence supporting the dynamic nature of IFs. Vikstrom *et al.* (1989) microinjected biotinylated vimentin into BHK-21 cells and monitored its fate by fixing and immunostaining cells at different times after injection. Exogenous vimentin, identified by anti-biotin immunofluorescence, first appeared (<30 minutes) as discrete, brightly stained particles dispersed throughout the cytoplasm. These particles subsequently aggregated near the nucleus to form a "cap" (1–2 hours) that colocalized with accumulations of endogenous vimentin revealed by costaining with an antivimentin antibody. The biotinylated protein then seemed to spread outward from this region to label the entire IF network in ~3–4 hours. In a similar study, Miller *et al.* (1993) microinjected biotinylated type I keratin into primary mouse epithelial (PME) or kangaroo rat kidney epithelial (PtK2) cells. Keratin also formed discrete particles within 20 minutes, resembling those seen after microinjecting vimentin into BHK21 cells. However, keratin particles did not accumulate near the nucleus, but instead formed short filamentous structures that coaligned with endogenous tonofibrils (30–45 minutes). After 1.5–2 hours, virtually all of the anti-biotin immunostaining colocalized with endogenous IF. Similar observations were made by Mittal *et al.* (1989) with fluorescent-tagged desmin and by Weigers *et al.* (1991).

The rapid incorporation of labeled IF protein into endogenous IF networks implies that a pool, albeit small, of soluble protein must exist that can readily exchange with polymerized IF. This was confirmed by FRAP studies of cells microinjected with fluorescent IF protein conjugate. Vikstrom *et al.* (1992) micro-injected 3T3 cells with rhodamine-labeled vimentin and allowed it to integrate into the endogenous IF network. Bar-shaped segments of IF were then photobleached, and the fluorescence was allowed to recover. Full recovery was seen within 30–40 minutes, supporting the notion of a diffusible pool of vimentin subunits (see section II.D) and confirming the existence of a subunit exchange mechanism. The rate of fluorescence recovery was significantly less than anticipated for a diffusion-limited process, suggesting that it might be regulated instead by the rate of dissociation of subunits from polymerized IFs. Finally, the pattern of recovery was uniform along the length of the bleached zone, consistent with the notion that IFs are nonpolar structures made up of antiparallel coiled–coil dimers and so differ fundamentally from MF and MT in this regard.

Microinjection experiments have also been used to track the fate of biotin-labeled lamin A in 3T3 cells (Goldman *et al.*, 1992). Immunofluorescence with anti-biotin antibody showed that the protein rapidly (within a few minutes) accumulated in the nucleoplasm in the form of discrete "spots," then slowly became incorporated into the nuclear lamina. The process culminated in a prominent nuclear "rim" staining pattern with anti-biotin antibody that was indistinguishable from that seen in noninjected cells after immunostaining with anti-lamin A antibody.

Other applications of the microinjection method involved injecting cells with a lamin A mutant lacking its NH_2-terminus (ΔNLA). This mutant had a dominant negative effect; it disrupted the "rim" staining for both A-type and B-type lamins, replacing it with nucleoplasmic aggregates of mutant and wild-type lamin proteins, and simultaneously blocked the elongation phase of DNA replication (Spann *et al.*, 1997). Furthermore, immunofluorescence showed impaired BrDU incorporation into the nuclei of microinjected cells compared with noninjected cells. BrDU staining was least in nuclei where the lamina was most disrupted, implying a role for lamins in transcriptional events (Spann *et al.*, 2002). Significantly, BrDU staining of the nucleoli was normal, an indication that RNA polymerase I activity was not affected. Together, these experiments implicate nuclear lamins in mRNA synthesis and suggest that they may act as a nucleoplasmic "scaffold" for the assembly of factors required for RNA polymerase II transcriptional activity.

C. Correlative Studies of Living Cells Expressing Green Fluorescent Protein–Intermediate Filament Fusion Proteins and Immunofluorescence

The earliest studies of this type, using GFP–vimentin (Ho *et al.*, 1998; Martys *et al.*, 1999; Yoon *et al.*, 1998) confirmed that attaching GFP to vimentin did not introduce any obvious artefactual behavior. Labeled vimentin was readily incorporated into the endogenous IF network, while GFP–IF networks in some cell types disassembled during mitosis, forming nonfilamentous particles that aggregated into juxtanuclear "caps" in each daughter cell. These observations are essentially the same as those described earlier, based on conventional immunofluorescence studies of fixed, nontransfected cells (see Section IV.A) and on microinjection experiments (see Section IV.B).

So what has live cell imaging taught us about IFs? IFs are often regarded as nonmotile components of the cytoskeleton. However, time-lapse observations on cells expressing GFP–IF fusion proteins have shown that this view is erroneous (Ho *et al.*, 1998; Yoon *et al.*, 1998). Many fibrils composed of one or more IF proteins were seen to bend or straighten, to shorten or lengthen, to extend or retract, and even to translocate from one region of the cytoplasm to another. A relative sliding movement between fibrils was observed in FRAP experiments, where bleached zones that started out as perfectly straight lines later became wavy because of differential rates of movement of adjacent fibrils. This effect is illustrated in Fig. 3, E–H. Interestingly, the sliding motion, and the shortening and lengthening of fibrils referred to earlier were impaired by cytochalasin B and nocodazole, implying that MF and MT are involved in both forms of motility. Significantly, however, neither agent affected the bending and straightening movements, showing that these are inherent properties of IFs themselves. FRAP measurements of cells expressing GFP–IF proteins also provided confirmation of the existence of the subunit exchange mechanism between soluble protein and polymerized IF first described by Vikstrom *et al.* (1992), and they estimated the half time for recovery ($t_{1/2}$) to be ~5 minutes (Yoon *et al.*, 1998).

Another revelation from live cell imaging came from observations of vimentin in actively spreading fibroblasts (Prahlad *et al.*, 1998; Yoon *et al.*, 1998). GFP-vimentin particles were seen to move toward the cell periphery in rapid, discontinuous bursts, with a mean velocity of 33 ± 14 μm/min. Short filamentous squiggles were also observed near the cell periphery. Most (>85%) of these were motile and also moved in the anterograde direction, but with a mean velocity of only 3.3 ± 1.9 μm/min. These movements were unaffected by cytochalasin B, but they were almost entirely eradicated in cells lacking intact MT after treatment with nocodazole.

In both of these studies, immunofluorescence observations proved invaluable in helping to explain the mechanisms that underpin these diverse forms of motility. Prahlad *et al.* (1998) showed that subsets of vimentin particles and squiggles colocalized with an antibody to conventional kinesin, a plus-end–directed molecular motor. Helfand *et al.* (2003) later showed that peripherin particles and squiggles in neuronal cells colocalized with kinesin too, and interestingly, most

(>70%) of these structures also colocalized with antibodies to dynein and dynactin, components of a minus-ended motor complex. Furthermore, fully elaborated vimentin and peripherin IF networks were seen to retract toward the nucleus when cells were microinjected with antibody to kinesin (Gyoeva and Gelfand, 1991; Helfand *et al.*, 2003; Prahlad *et al.*, 1998), whereas overexpression of dynamitin, to suppress dynein–dynactin (minus-ended) motor function, caused peripherin IFs to move in the opposite direction and accumulate near the cell periphery (Helfand *et al.*, 2003).

Similar studies of live cells expressing GFP-tagged lamins (A, C, B1) have revealed striking differences in the behavior of A-type and B-type lamins during nuclear envelope formation after mitosis (Moir *et al.*, 2000). FRAP experiments showed that lamins are distributed throughout the cell cytoplasm in a highly diffusible form during early mitosis. Furthermore, lamin B1 was seen to form a shell around decondensing chromosomes during telophase, in advance of the assembly of nuclear pore components. FRAP measurements showed that it was in a more stable polymerized form at this time. On the other hand, lamin A remained in the nuclei of daughter cells into early G1 and only became fully integrated into the lamina after the main components of the nuclear envelope were already in place. These experiments show that lamin B, but not lamin A, participates in the early stages of nuclear envelope assembly.

D. Concluding Remarks

Fluorescence-based methods have made major contributions to our knowledge of the structure and function of IFs, particularly since the introduction of live cell imaging techniques. In the face of mounting evidence to the contrary, the traditional "textbook" view of IFs and their constituent proteins as static, inert elements of cellular architecture is no longer sustainable and should now be laid to rest. Time-lapse observations of cells expressing GFP–IF fusion proteins show unequivocally that IFs and their subfilamentous precursors, so-called particles and squiggles, are in reality extraordinarily dynamic components of the cytoskeleton, continually engaged in a wide spectrum of motilities. We have been able to document some of these movements in considerable detail and in several different cell types. On the basis of these observations, we now know that IF assembly *in vivo* is a modular process, in which several particles merge to form squiggles, and squiggles in turn anneal to form longer IFs. On the other hand, live cell imaging *per se* reveals little about the molecular mechanisms that underly IF dynamics. Here, immunofluorescence studies of cells previously used for live imaging have proved of great value, especially in helping to uncover functional interactions between IF and other cytoskeletal components (Helfand *et al.*, 2003, 2004). Correlative studies of this type show that particles and squiggles are actively transported to sites where new IFs are being assembled by interactions between MTs and both plus-ended and minus-ended molecular motors. There is little doubt that these interactions, and similar ones involving actin-based motors and

MFs, target IFs to regions of the cytoplasm where they are required for numerous physiological functions, ranging from local control of the mechanical properties of the cytoplasm to the construction of elaborate IF networks involved in signal transduction processes.

Acknowledgments

The studies leading to the methods and results described in this chapter have been supported by NIGMS, NIDR, and NHLB.

References

Chalfie, M., Tu, Y., Euskirchen, G., Ward, W. W., and Prasher, D. C. (1994). Green fluorescent protein expression as a marker for gene expression. *Science* **263**, 802–805.

Chou, Y.-H., Bischoff, J. R., Beach, D., and Goldman, R. D. F. (1990). Intermediate filament reorganization during mitosis is mediated by p34cdc2 phosphorylation of vimentin. *Cell* **62**, 1063–1071.

Cubitt, A. B., Heim, R., Adams, S. R., Boyd, A. E., Gross, L. A., and Tsein, R. Y. (1995). Understanding, improving and using green fluorescent proteins. *TIBS* **20**, 448–455.

Cubitt, M., Wollenweber, L. A., and Heim, R. (1999). Understanding structure-function relationships in the Aequoria victorea green fluorescent protein. *Methods Cell Biol.* **58**, 19–30.

Franke, W. W., Schmid, E., Grund, C., and Geiger, B. (1982). Intermediate filament proteins in nonfilamentous structures: Transient disintegration and inclusion of subunit proteins in granular aggregates. *Cell* **30**, 103–113.

Goldman, A. E., Moir, R. D., Montag-Lowy, M., Stewart, M., and Goldman, R. D. (1992). Pathway of incorporation of microinjected lamin A into the nuclear envelope. *J. Cell Biol.* **119**, 725–735.

Gyoeva, F. K., and Gelfand, V. I. (1991). Coalignment of vimentin intermediate filaments and microtubules depends on kinesin. *Nature* **353**, 445–448.

Helfand, B. T., Chang, L., and Goldman, R. D. (2004). Intermediate filaments are dynamic and motile elements of cellular architecture. *J. Cell Sci.* **117**, 133–141.

Helfand, B. T., Loomis, P., Yoon, M., and Goldman, R. D. (2003). Rapid transport of neural intermediate filament. *J. Cell Sci.* **116**, 2345–2359.

Ho, C. L., Martys, J. L., Mikhailov, A., Gundersen, G. G., and Liem, R. K. (1998). Novel features of intermediate dynamics revealed by green fluorescent protein chimeras. *J. Cell Sci.* **111**, 1767–1778.

Lane, E. B., Goodman, S. L., and Trejdosiewicz, L. K. (1982). Disruption of the keratin filament network during epithelial cell division. *Eur. Mol. Biol. Organ. (EMBO) J.* **1**, 1365–1372.

Lippincott-Schwatrz, J., Altan-Bonnet, N., and Patterson, G. H. (2003). Photobleaching and photoactivation: Following protein dynamics in living cells. *Nature Cell Biol.* **5**, S7–S14.

Martys, J. L., Ho., C. L., Liem, R. K., and Gundersen, G. G. (1999). Intermediate filaments in motion: Observations of intermediate filaments in cells using green fluorescent protein-vimentin. *Mol. Biol. Cell* **10**, 1289–1295.

Miller, R. K., Vikstrom, K. L., and Goldman, R. D. (1991). Keratin incorporation into intermediate filament networks is a rapid process. *J. Cell Biol.* **113**, 843–855.

Miller, R. K., Khuon, S., and Goldman, R. D. (1993). Dynamics of keratin assembly: Exogenous type I keratin rapidly associates with type II keratin *in vivo*. *J. Cell Biol.* **122**, 123–135.

Mittal, B., Sanger, J. M., and Sanger, J. W. (1989). Visualisation of intermediate filaments in living cells using fluorescently labeled desmin. *Cell Motil. Cytoskelet.* **12**, 127–138.

Moir, R. D., Montag-Lowy, M., and Goldman, R. D. (1994). Dynamic properties of nuclear lamins: Lamin B is associated with sites of DNA replication. *J. Cell Biol.* **125**, 1201–1212.

Moir, R. D., Yoon, M., Khuon, S., and Goldman, R. D. (2000). Nuclear lamins A and B1: Different pathways of assembly during nuclear envelope formation in living cells. *J. Cell Biol.* **151,** 1155–1168.

Prahlad, V., Yoon, M., Moir, R. D., Vale, R. D., and Goldman, R. D. (1998). Rapid movements of vimentin on microtubules tracks: Kinesin-dependent assembly of intermediate filament networks. *J. Cell Biol.* **143,** 159–170.

Prasher, D. C., Eckenrode, V. K., Ward, W. W., Prendergast, F. G., and Cormier, M. J. (1992). Primary structure of the Aequoria victorea green fluorescent protein. *Gene* **111,** 229–233.

Rosevear, E. R., McReynolds, M., and Goldman, R. D. (1990). Dynamic proerties of intermediate filaments: Disassembly and re-assembly during mitosis in baby hamster kidney cells. *Cell Motil. Cytoskelet.* **17,** 150–166.

Spann, T. P., Moir, R. D., Goldman, A. E., Stick, R., and Goldman, R. D. (1997). Disruption of nuclear lamin organization alters the distribution of replication factors and inhibits DNA synthesis. *J. Cell Biol.* **136,** 1201–1212.

Spann, T. P., Goldman, A. E., Wang, C., Huang, S., and Goldman, R. D. (2002). Alteration of nuclear lamin organization inhibts RNA polymerase II-dependent transcription. *J. Cell Biol.* **156,** 603–608.

Vikstrom, K. L., Borisy, G. G., and Goldman, R. D. (1989). Dynamic aspects of intermediate filament networks in BHK-21 cells. *Proc. Natl. Acad. Sci. USA* **86,** 549–553.

Vikstrom, K. L., Lim, S-S., Goldman, R. D., and Borisy, G. G. (1992). Steady state dynamics of intermediate filament networks. *J. Cell Biol.* **118,** 121–129.

Weigers, W., Honer, B., and Traub, P. (1991). Microinjection of intermediate filament proteins into living cells with and without pre-existing intermediate filament network. *Cell Biol. Int. Rep.* **15,** 287–296.

Yoon, M., Moir, R. D., Prahlad, V., and Goldman, R. D. (1998). Motile properties of intermediate filament networks in living cells. *J. Cell Biol.* **143,** 147–157.

Yoon, K. H., Yoon, M., Moir, R. D., Khuon, S., Flitney, F. W., and Goldman, R. D. (2001). Insights into the dynamic properties of intermediate filaments in living epithelial cells. *J. Cell Biol.* **153,** 503–516.

CHAPTER 12

Imaging of Keratin Dynamics during the Cell Cycle and in Response to Phosphatase Inhibition*

Reinhard Windoffer and Rudolf E. Leube

Department of Anatomy
Johannes Gutenberg-University
55128 Mainz
Germany

* Movies can be viewed http://www.science.direct.com by clicking on 'Books' and going to Volume 78, Chapter 12 in Methods in Cell Biology.

I. Introduction

Among the various intermediate filament (IF) cytoskeletons, the keratin network is considered to be particularly stable, conferring mechanical resilience onto epithelial tissues. Keratin filament (KF) disruption, therefore, results in epithelial weakening as is the case in a number of human diseases (Irvine and McLean, 1999; Porter and Lane, 2003; Smith, 2003). Accordingly, mice synthesizing dominant negative keratin mutants or lacking a keratin cytoskeleton in certain cell populations develop various pathologic conditions in the affected epithelial tissues (Coulombe and Omary, 2002; Herrmann et al., 2003; Porter and Lane, 2003). In vitro analyses provide further evidence for the particular mechanical resilience of KFs (Ma et al., 1999, 2001).

It has turned out, however, that KF networks are not simply rigid scaffoldings that are anchored at specific desmosomal cell–cell adhesion sites as has been portrayed in text books, but they are, instead, highly dynamic cytoskeletal components that are subject to continuous remodeling and rejuvenation (Windoffer and Leube, 1999; Windoffer et al., 2004; Yoon et al., 2001). In addition, considerable motility of KFs and KF precursors has been documented (Helfand et al., 2003b; Liovic et al., 2003; Windoffer and Leube, 1999; Windoffer et al., 2004; Yoon et al., 2001). These properties enable the KF cytoskeleton to respond constantly and quickly to special cellular requirements. The main technology that has helped to identify and analyze these dynamic features is live cell imaging in which cells that synthesize fluorescent keratin polypeptides are monitored by time-lapse fluorescence microscopy. With this technique, key questions that relate to basic organizational principles of the KF system are being addressed, such as the following.

How and where are KFs formed in a living cell? In contrast to the microfilament and microtubule systems, almost nothing is known about the biosynthesis and morphogenesis of KF networks. This lack in understanding is because IFs are intrinsically nonpolar consisting of symmetric tetrameric building blocks (Parry and Steinert, 1999; Strelkov et al., 2003). This absence of directionality thus precludes vectorial growth as is characteristic for the other cytoskeletal filament components. Furthermore, specific initiation sites that could support the formation of the KF network have not been identified on a molecular level. Live imaging of cells that are devoid of an intact KF system but rebuild a new network, as is the case in certain cells during mitosis (e.g., Franke et al., 1982; Horwitz et al., 1981; Lane et al., 1982; Tolle et al., 1987), now affords the examination of de novo KF network formation in situ.

How does KF network turnover occur? Two principal models have been proposed: continuous exchange of subunits throughout the entire filament system (Miller et al., 1991, 1993) and use of certain organizing centers in specific cellular domains (Windoffer et al., 2004 and references therein). Given that the turnover of the KF system exceeds by far the very slow biosynthetic replenishment, live cell analysis is the method of choice to resolve this issue. Photobleach experiments provide powerful means to identify turnover intermediates and to map them to distinct cellular topologies.

How does the keratin cytoskeleton change during the cell cycle? Prominent cell-shape changes are associated with cell division, and different phenotypic altera-tions of the keratin system have been observed in dividing cells (cf. Windoffer and Leube, 2001). Live cell monitoring allows the determination of consecutive steps of the various reorganization events and relates them to specific cell cycle stages.

How are keratin dynamics regulated? KFs form spontaneously and very rapidly *in vitro* without any additional factors (Herrmann *et al.*, 2003). In contrast, living cells regulate this process tightly. They are able to increase the pool of nonfila-mentous keratins to almost 100% (e.g., during mitosis). The balance between the different organizational states of keratins (soluble, granular, filamentous) depends on the state of cellular phosphorylation (e.g., Coulombe and Omary, 2002; Omary *et al.*, 1998; Strnad *et al.*, 2001, 2002). By *in vivo* imaging, one can now directly assess the effects of specific enzyme inhibitors on keratin organiza-tion. Combinatorial application of different drugs furthermore allows us to define complex relationships between morphotype and the action/presence of regulatory factors.

How do diseases affect the dynamic and organizational properties of the keratin cytoskeleton? Synthesis of fluorescent keratin mutants in cDNA-transfected cells (Werner *et al.*, 2004), exposure of cells producing fluorescent keratins to various stress stimuli (Liovic *et al.*, 2003), and synthesis of fluorescent keratins in pathologically altered cells (Riley *et al.*, 2002) facilitate the examination of dynamic alterations of the keratin system in pathophysiologically relevant con-texts. In such model systems, the pathogenesis of genetically induced keratin diseases such as epidermolysis bullosa simplex (Irvine and McLean, 1999; Porter and Lane, 2003; Smith, 2003), of environmentally determined keratin alterations as they occur in toxic liver disease leading to Mallory body formation (Cadrin and Martinoli, 1995), and of various epithelial diseases in the context of cellular stress responses (Coulombe and Omary, 2002) become accessible with close inspection.

These examples shall suffice to point out the potential of live cell imaging to provide novel and unprecedented insights into the dynamic organization of the keratin cytoskeleton. In the following sections, we will outline important aspects of this method. Emphasis will be on the microscope setup, fluorescence recording, and data interpretation that are being used in our laboratory; the reader is referred to excellent reviews that present more general considerations on the topic (e.g., Lippincott-Schwartz *et al.*, 2001; Periasamy and Day, 1999; Rizzuto *et al.*, 1998; Stephens and Allan, 2003).

II. Materials and Instrumentation

A. Special Reagents

- *Okadaic acid (OA)*: The serine-threonine phosphatase inhibitor (Sigma, St. Louis, MO) is dissolved in dimethylsulfoxide (DMSO) at 10 μg/ml and can be stored at $-20\,^\circ$C. It is added to cells at final concentrations between 0.1 μg/ml and 1 μg/ml, which should selectively inhibit protein phosphatases 1 and 2A, although higher concentrations are known to also inhibit phophatase 2B (Cohen *et al.*, 1990; Vandre and Wills, 1992).

- *Sodium orthovanadate (OV)*: The tyrosine phosphatase inhibitor (Aldrich Chemical Corporation; Milwaukee, WI) is freshly dissolved in distilled water at 1 M for each experiment. It is either used directly or incubated before use for up to 1 hour with 50 mM H_2O_2 on ice to generate the more active pervanadate (Feng *et al.*, 1999). Final concentrations range between 2 and 50 mM, depending on the cell type, for pervanadate between 0.2 mM and 1 mM.

- *Hoechst 33342*: The vital DNA stain is from Molecular Probes (Eugene, OR). A stock of 0.733 mg/ml is prepared in H_2O and can be stored at 4 $^\circ$C. Addition of 1 μl stock to 4 ml medium is sufficient to stain chromatin efficiently in living cells in less than 30 minutes.

B. Cell Lines

Preferably, cell lines are used that are well characterized with respect to their IF complement, present an extended KF network, are easy to maintain, and are amenable to standard transfection methods. Among the many cell lines that are available from cell culture banks, several have proven to be useful for live cell imaging. They include those derived from human adenocarcinomas originating either from simple epithelia such as hepatocellular carcinoma PLC cells (ATCC CRL8024), colon carcinoma CaCo-2 cells (ATCC HTB-37), or from complex epithelia such as mammary adenocarcinoma MCF-7 cells (ATCC HTB-22), and those that originate from multilayered epithelia such as vulvar squamous cell carcinoma A-431 cells (ATCC CRL1555). In addition, immortalized HaCaT keratinocytes (Boukamp *et al.*, 1988) are particularly suited for the analysis of the epithelial cytoskeleton. An important and physiologically quite relevant system is provided by primary keratinocytes that are either available commercially (e.g., from Invitrogen GmbH, Karlsruhe, Germany) or prepared from trunks of newborn mice (Hager *et al.*, 1999). Cells are grown in a 5% CO_2 atmosphere at 37 $^\circ$C and should be propagated in growth medium as detailed by the various suppliers.

When selecting a cell type for live cell imaging, various aspects should be considered. For example, a very dense and elaborate network of fine filaments is typical for primary keratinocytes, whereas thick filament bundles are characteristic of PLC cells, which will be easier to image but may differ in their dynamic behavior. Flat and spread-out cells such as PLC and CaCo-2 cells are well suited for high-resolution recordings of the complete cytoplasmic volume, especially in peripheral regions. High efficiency of transfection and stable line formation argue in favor of A-431 cells. Another consideration concerns the different types of keratin cytoskeleton restructuring during the cell cycle. Although the KFs of A-431 cells are almost completely disassembled into granular aggregates during mitosis, most other cells exhibit only filament aggregation. Finally, the responsiveness of the keratin system to phosphatase inhibitors differs significantly between different cell types, both with respect to drug concentration needed to induce alterations and the kinetics of reorganization. These properties have to be determined experimentally in each case.

To examine keratin dynamics in a nonepithelial environment, several cell lines have been used. In some cell types such as 3T3-fibroblasts, only very thick filament bundles/aggregates are formed (e.g., Bader *et al.*, 1991; Domenjoud *et al.*, 1988; Werner *et al.*, 2004). In contrast, an extended KF network is observed in cDNA-transfected human small-cell carcinoma SW13 cells that are derived from the adrenal cortex and do not synthesize any IF protein (ATCC CCL-105; Hedberg and Chen, 1986). A similarly complex KF system is also formed in transgenic human H36CE1 lens cells (Liovic *et al.*, 2003).

For imaging, cells should be transferred to phenol red-free Hanks' medium, because it is superior to other media such as Dulbecco Modified Eagle medium (DMEM) for obtaining high-quality pictures. The medium contains Hanks' salt solution, 25 mM Hepes, MEM nonessential amino acid solution and MEM amino acid solution, 100 U/ml penicillin, 100 μg/ml streptomycin, 5% fetal calf serum (all from Invitrogen), 4.8 mM *N*-acetyl-L-cysteine (Sigma), pH 7.4. The addition of ascorbic acid (0.5 mg/ml; Sigma) may improve fluorescence stability in some cases.

C. Microscopes

The methods described here were developed primarily for the following microscope setups:

• *Olympus IX 70* inverse fluorescence microscope (Hamburg, Germany): The microscope is equipped with a Polychrome IV monochromator (TILL-Photonics, Gräfelfing, Germany). Excitation is at 496 nm for enhanced green fluorescent protein (GFP), 498 nm for enhanced yellow fluorescent protein (YFP), and 436 nm for enhanced cyan fluorescent protein (CFP). Emission filter U-M61008 is used for all colors. The microscope is equipped with a shutter (Uniblitz VMM-D1; Vincent Associates, Rochester, NY) to switch between fluorescence and phase-contrast microscopy. A piezo-driven z-axis stepper (0.1-μm steps; Physik Instrumente, Karlsruhe, Germany) is attached to a 60 × 1.4 N.A. oil immersion objective that is used for most applications. The entire microscope is encased by a Plexiglas chamber and heated to 37 °C. Cells are viewed in γ-irradiated Petri dishes with a glass bottom (Mattek, Ashland, MA). The motorized microscope stage (Märzhäuser, Wetzlar, Germany) is operated with a joystick that is located outside the chamber. Image recording is with an IMAGO slow scan charged-coupled device camera, and the entire system is controlled by TILLvisION software (both from TILL-Photonics). For three-dimensional (3D) delineation of structures, multiple focal planes are recorded at each time point by use of the piezo stepper. The resulting picture stacks are either projected on top of each other or are used to prepare 3D reconstructions with the help of Amira software (TGS; San Diego, CA).

Alternatively, any standard epifluorescence microscope can be used. Such a microscope should be equipped with appropriate filter sets (excitation, dichroic, emission) to image the fluorescent proteins of interest. For multicolor recording, different filter sets are needed. They should be mounted in such a way (e.g., in a filter wheel) that they can be quickly switched with a motorized filter changer. In addition, it is important that the excitation light can be controlled with a shutter to minimize illumination-induced bleaching and phototoxicity. High-sensitivity digital cameras suitable for fluorescence microscopy can be attached to standard microscopes by means of a C-mount. ImagePro Plus software (Media Cybernetics, Silver Spring, CA), MetaMorph (Downingtown, PA), NIH Image (freeware at http://rsb.info.nih.gov), and/or software provided by the various microscope manufacturers can be used to grab images, to operate the filter changer, and to control the shutter. A special culture chamber was designed and constructed by us to examine cells for more than a day without any signs of vitality loss (see following).

• *Leica confocal laser scanning microscope* (model TCS SP2; Leica Microsystems, Wetzlar, Germany): For YFP/GFP detection, the 514-nm line of an argon/krypton laser is used in combination with dichroic DD458,514. For selection of a defined range of emitted light, the monochromator is set to 525–625 nm; 100×1.4 N.A. oil PLAPO objective and a 63×1.4 N.A. oil immersion objective are useful for high-resolution imaging. Other user-friendly confocal laser scan microscopes work equally well for live cell imaging.

D. Culture Chamber

For upright microscopy, a culture chamber was designed that can be mounted directly onto the stage of either an epifluorescence microscope or a confocal laser scan microscope (Fig. 1). The circular chamber is rather small (diameter, 14 mm; height, 1.5 mm) to reduce disturbing vibrations and to allow quick exchange of culture medium. It is embedded in a steel frame with three drill holes (1-mm diameter) for influx and efflux of culture medium and placement of a temperature sensor (Fisher Scientific GmbH, Nidderau, Germany). A peristalsis pump with adjustable flow rates is used to exchange the culture medium either continuously or intermittently. A 16-mm diameter coverslip is placed on a self-made thin silicon ring on top of the culture chamber, and a steel plate with an observation opening is screwed on top of the entire assembly to seal it tightly. Note that cells are growing in an inverted position, which, however, is of no consequence to adhering

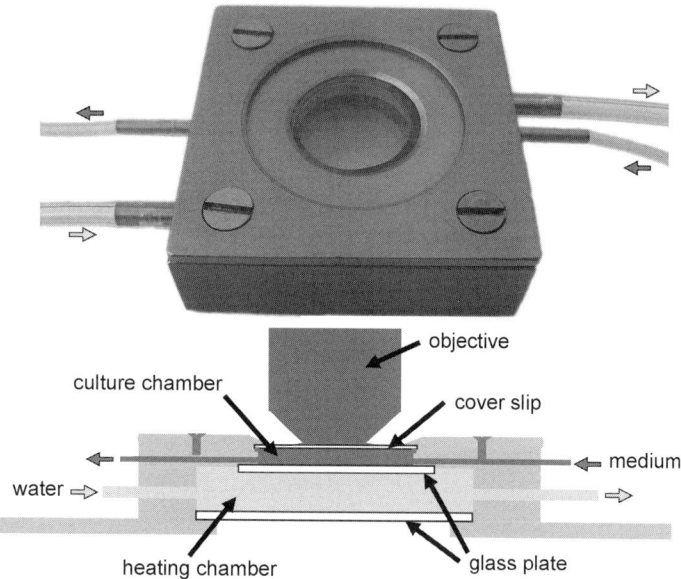

culture chamber

objective

cover slip

medium

water

heating chamber

glass plate

Fig. 1 Photograph (oblique view of top) and schematic drawing (cross-section) of a self-designed culture chamber that is used for live cell imaging. Adhesive cells growing on the inverted coverslip are placed on top of a small culture chamber that is accessible through drill holes for exchange of culture medium and for placement of a temperature sensor. Proper temperature is attained by a heating chamber at the bottom that is connected to tubes containing circulating, preheated water. A clear optical path is maintained by glass plates to enable recording of transmission phase contrast or Nomarski images. The objective is immediately adjacent to the inverse growing cell monolayer for transmission and epifluorescence microscopy. For further details see text.

cells. Underneath the culture chamber is the larger heating chamber (diameter, 20 mm; height, 8 mm) that is encased by 1-mm-thick glass plates at the top and bottom. This chamber is filled with continuously circulating and preheated water to maintain a constant temperature of 37 °C in the culture chamber as measured by the temperature sensor that is positioned next to the observation field.

III. Procedures

A. Generation of Cell Lines Producing Fluorescent Keratin Filaments

1. Preparation of cDNA Constructs

In most instances, strong promoters are selected for high-level expression of fluorescent fusion proteins, and good results are generally obtained with the immediate early promoter elements of the cytomegalovirus (CMV) promoter that is present in many commercially available vectors. The SV40 promoter and the actin promoter work similarly well in many cell types. Preferably, the cDNA coding for the target protein precedes the fluorescent protein-encoding part, although the reverse arrangement works satisfactorily in most, although not all, instances. Comparison of both construct types helps to exclude position-dependent artifacts of the fluorescent hybrid proteins (see, e.g., Fig. 2). Suitable cloning vectors are available from various companies (e.g., Clontech Laboratories [Palo Alto, CA], Qbiogene [Carlsbad, CA]). These vectors also usually include a polyadenylation cassette at the 3'-end of the hybrid cDNA and contain separate selection cassettes. Among the fluorescence tags that excel in terms of fluorescence intensity and low degree of bleaching and that do not influence the distribution patterns of chimeras (e.g., by multimerization) are CFP (433-nm excitation; 475-nm emission), GFP (488-nm excitation; 507-nm emission), and YFP (513-nm excitation; 527-nm emission). The CFP/YFP pair is best suited for colocalization and fluorescence resonance energy transfer experiments (Lippincott-Schwartz *et al.*, 2001; Pollok and Heim, 1999; van Roessel and Brand, 2002). Purification of DNA by Qiagen column chromatography (Qiagen GmbH, Hilden, Germany) is sufficient to obtain high-quality DNA for transfection.

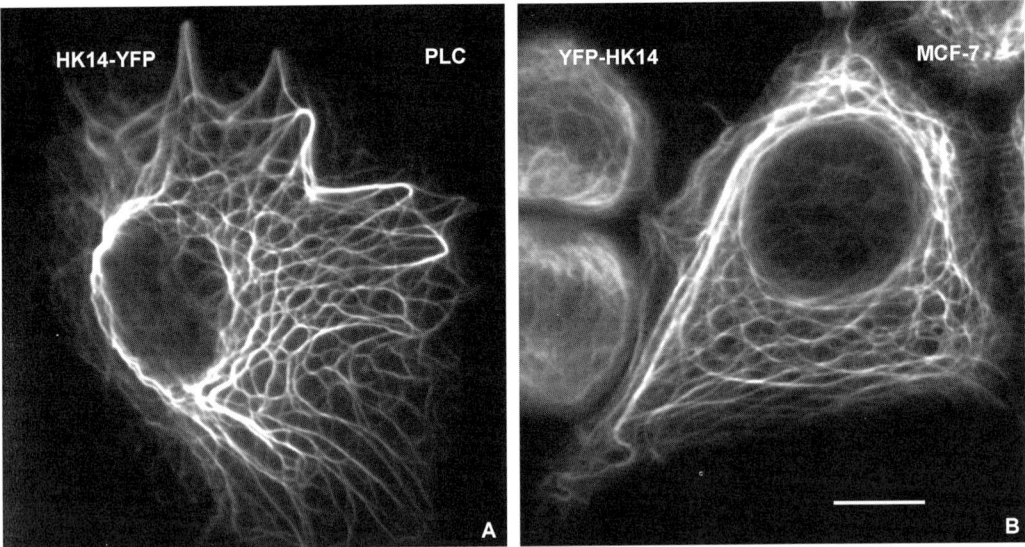

Fig. 2   Epifluorescence microscopy of a methanol–acetone fixed hepatocellular, carcinoma-derived PLC cell (A) and living mammary adenocarcinoma–derived MCF-7 cells (B) expressing human keratin 14 hybrids that contain an enhanced yellow fluorescent protein (YFP) tag either at the carboxyterminus (HK14-YFP; A) or the aminoterminus (YFP-HK14; B). Note the detection of extended filament networks in both instances. Bar, 10 μm.

2. Generation of Stable Cell Lines

Ideally, fluorescent polypeptides should merely act as neutral tags without disturbing the targeted endogenous structures. It is, therefore, important to establish stable cell lines that can be analyzed by biochemical and morphological methods to show that the properties of the mutant polypeptides are not different from their endogenous counterparts and to exclude that the chimeras do not interfere with endogenous functions and morphogenesis. Calcium phosphate precipitation and lipofection have worked well as alternative transfection methods in our hands.

In the case of the calcium phosphate precipitation method, 5 μg purified plasmid DNA is dissolved in 219 μl 10 mM Tris-HCl (pH 7.5). After addition of 31 μl 2 M CaCl$_2$, the mixture is pipetted dropwise into 250 μl 2× HBS (1× HBS is 140 mM NaCl, 0.75 mM Na$_2$HPO$_4$, 50 mM HEPES, adjusted to pH 7.1 with 1 N KOH) while vortexing. The mixture is briefly incubated at room temperature and then added to the medium of a culture of subconfluent cells. Cells are washed three times with DMEM after 5 hours. Depending on the cell type, transfection efficiency is improved considerably by a glycerol shock. In this case, cells are treated with 15% glycerol in 1× HBS for 2–3 minutes after which they are washed three times with DMEM. Finally, growth medium supplemented with 2.5 μg/ml amphotericin B, and 0.1 mg/ml penicillin, 100U/ml streptomycin (all from Invitrogen) is added. After 1–2 days, cells are seeded onto new plates at very low density, and selective medium is added containing up to 1.25 mg/ml G-418 (Geneticin; Sigma) for neomycin resistance–conferring plasmids, up to 300 μg/ml hygromycin (Sigma) for hygromycin-resistance–encoding plasmids, and up to 1 μg/ml puromycin (Sigma) to select for puromycin resistance.

For lipofection, 10 μl serum-free medium is mixed with 10 μl hydrated Gene-PORTER 2 reagent (PEQLAB Biotechnologie GmbH, Erlangen, Germany) before the addition of 2 μg DNA that is freshly diluted to 50 μl with diluent B provided in the transfection kit. The resulting mix is incubated at room temperature for 10 minutes and added afterward drop by drop to subconfluent cells. The medium is replaced with new medium after overnight incubation, and cells can be analyzed after 24–48 hours to evaluate transfection efficiency. Subsequently, selective agents are added (see earlier).

Cultures that are subjected to pharmacological selection are kept in the same dishes with occasional medium changes. Single, drug-resistant colonies are picked and transferred individually into multiwell dishes. Cell clones presenting homogenous patterns of fluorescence are selected for further amplification and analyses.

3. Characterization of Cell Lines

For each construct, several stably transfected cell lines should be subjected to extensive analyses including immunofluorescence microscopy, biochemical assays, and ultrastructural examinations to ensure that the fluorescent polypeptides do not interfere with the normal physiology and structural organization of the keratin cytoskeleton and its associated cellular components. Ideally, only those clones should be selected for this type of time-consuming scrutiny in which all cells are positive for the transgene as determined by direct fluorescence microscopy and in which the transgene products show an identical distribution pattern in all cells, except for the dividing cells that rearrange their keratin system. The overall morphology of selected cell clones should be compared by phase-contrast or Nomarski microscopy to that of the parent cell line to exclude that the transfection and selection procedures singled out cells, with altered properties unrelated to the production of the transgene. Indirect immunofluorescence microscopy that uses standard techniques is performed to determine whether the fluorescent transgenes perturb the keratin cytoskeleton and its associated components in any adverse way. It is necessary to demonstrate that the fluorescent keratin chimeras colocalize completely with the endogenous keratin polypeptides whose overall distribution pattern should reflect that observed in the parent cell line. Further analyses should also include components of the desmosomal keratin adhesion sites and other associated polypeptides such as linkers, bundling factors, or signaling molecules (cf. Coulombe and Omary, 2002).

Biochemical properties of the transfected cell clones should be examined to clarify two major questions:

- *Is the size of the fusion proteins as expected?* In standard immunoblots that use anti-GFP (e.g., from Molecular Probes) and anti-keratin antibodies (e.g., from Progen GmbH [Heidelberg, Germany], Abcam [Cambridge, MA], Sigma, Biomol [Hamburg, Germany]), the size of the fluorescent protein chimeras can be determined. These analyses also provide evidence whether degradation occurs.

- *Are the fusion proteins correctly targeted to the appropriate cell compartment?* To this end, high salt pellets, which usually contain most keratins, are prepared (Achtstaetter *et al.*, 1986). Comparative immunoblotting of the cytoskeletal fraction and the total cell lysate and/or Coomassie Blue staining of the enriched cytoskeleton will give reliable estimates of correct topogenesis (Strnad *et al.*, 2002; Werner *et al.*, 2004).

Ultrastructural properties of the transfected cell clones should be assessed by electron microscopy. It is important to check whether IF bundles are formed properly, whether granule formation is increased, and whether the association of KFs with their desmosomal anchorage sites is inconspicuous (e.g., Windoffer and Leube, 1999). Additional immunoelectron microscopy will provide evidence for correct targeting of keratin chimeras and their even incorporation into the KF network.

B. Imaging of Keratin Filament Dynamics in Interphase

1. Preparation of Cells

Stably transfected cell clones that fulfill the stringent criteria of producing an unperturbed fluorescent keratin cytoskeleton are split and transferred either into culture dishes containing coverslips that can be mounted onto the closed culture chamber or into culture dishes with a glass bottom, depending on which micro-scope setup is used. Cells should be grown under standard conditions to low confluence, which takes ~1–2 days. The culture medium should not contain selective agents to avoid pharmacologically induced stress. Healthy looking cultures are then mounted onto the appropriate microscope stage in Hank's imaging medium. The temperature near the imaged cells should equilibrate to 37 °C as determined by the temperature sensor. CO_2 is not needed, because the Hepes buffering of the medium is usually sufficient for pH maintenance. Suitable cells for image recording are then selected by fluorescence screening and phase-contrast imaging. The "optimal" cell should be flat and should be representative of all cells in a given culture. Giant cells with several nuclei and cells growing in multiple layers should not be used. Short recordings of phase-contrast images will give a fairly good indication of the viability of the selected cells. They should exhibit rapid movements of cytoplasmic structures and abundant motility of their free margins.

2. Image Acquisition

Optimal recording settings have to be determined in each instance. Various aspects are to be considered.

• *Size of imaged area*: Recordings of large areas are helpful in obtaining a general overview of the various fluorescence patterns in a given culture and to assess the variability of reactivities to specific stimuli. Low-magnification imaging will also keep bleaching and phototoxic effects resulting from generation of oxygen radicals at a minimum. Although spatial resolution is rather limited, it will still enable the investigator to decide which cellular domains to analyze further and to determine which recording frequencies are needed to visualize the process of interest. When these parameters are established, high-resolution recordings are carried out, which will also allow small structures in restricted cellular compartments to be resolved and fast reorganization and transport events to be dissected. Major disadvantages of high-magnification recordings, however, are the reduced depth of focus necessitating imaging of multiple focal planes and frequent shifts of the documented area out of view because of cell motility or instability of the imaging system, as well as considerable bleaching and increased phototoxicity. In the case of the keratin system, e.g., low-resolution imaging has provided evidence for oscillating filament motility and overall restructuring of mature filaments (Windoffer and Leube, 1999; Yoon *et al.*, 2001), whereas pictures at the highest magnification were needed to delineate keratin turnover stages in the cell periphery (Fig. 3; movie 1; Windoffer *et al.*, 2004).

• *Speed of dynamic process*: The recording frequency has to be adjusted to the speed of the process of interest. Generally, the recording intervals should be such that the motility of the imaged structures can be traced unequivocally between individual frames, thus resulting in even motions when viewing the assembled

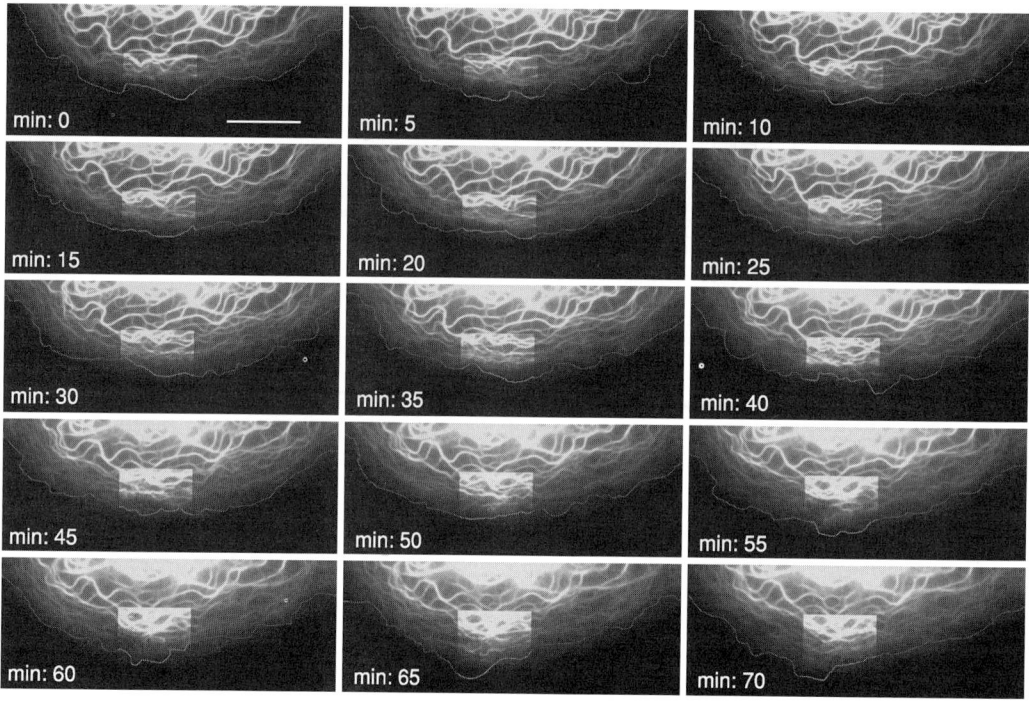

Fig. 3 Epifluorescence images depicting a peripheral region of a living hepatocellular carcinoma–derived PK18-5 cell synthesizing fluorescent keratin hybrid HK18-YFP. Note that changes in keratin distribution are difficult to discern from the series of still images, but that they are easily recognized in the corresponding time-lapse series that is provided as movie 1, revealing a continuous inward flow of fluorescent material. The look-up table was adjusted in the square region in the middle to enhance low fluorescent structures. The cell edge is demarcated by a red line. Its position was determined from phase-contrast images that were recorded in parallel. Bar, 5 μm. (See Color Insert.)

movies. If the imaged fluorescent components undergo rapid shape changes, even higher recording frequencies may be necessary. Other factors to be considered are the density of fluorescent elements and their shape heterogeneity. For most purposes of keratin imaging, intervals ranging between 30 seconds and 2 minutes may provide satisfactory results, although resolution of microtubule-dependent movements require much shorter recording intervals (Liovic *et al.*, 2003; Yoon *et al.*, 2001). Similarly, high-frequency imaging is needed during times of intense keratin restructuring (e.g., during mitosis and in orthovanadate (OV)-treated cells).

• *Number of focal planes*: Recording of multiple focal planes is needed to track and visualize structures that, because of intrinsic motility and/or mechanical shift of the imaging system, migrate in and out of the focal plane. Furthermore, structures that extend beyond the optical focus need to be delineated in this way. Most importantly, recordings in multiple planes can be assembled into 3D reconstructions, thereby providing important information on spatial relationships (see later). Generally, as many confocal planes as possible should be recorded, although fewer may suffice in flat cells (as is the case for most interphase cells) than in round cells, which are generated on entry into mitosis or by treatment with phosphatase inhibitors. Again, bleaching and phototoxicity are limiting factors. The scanning time should be short enough to avoid distortion of structures caused by their movement during the recording. We found that the scanning time should be kept to less than 30 seconds, with recovery intervals from 30 seconds onward.

- *Image resolution*: A color depth of 12 bit is preferred over the standard 8-bit mode to enable look up table (LUT) adjustment after image acquisition and to optimize quantification of gray values when measuring fluorescence intensity for various applications (see later). The image resolution in confocal recordings is 1024 × 1024 pixel for low-frequency recordings and 512 × 512 pixel for high-frequency recordings.

- *Confocal laser scan microscopy*: Settings that have worked well in confocal microscopy are recording intervals between 30 seconds and 2.5 minutes, laser power at 6% of minimum laser power, a medium scan speed, a line average of 4, a photomultiplier gain of 800, a standard pinhole size as given by the software, imaging of up to 10 focal planes, and a resolution of either 1024 × 1024 pixel or 512 × 512 pixel.

As an example, Fig. 3 (and corresponding movie 1) depict time-lapse fluorescence micrographs of the peripheral region of a hepatocellular carcinoma cell of line PK18-5 that was stably transfected with a cDNA construct coding for a human keratin 18-YFP fusion protein. Only a selected region is shown to highlight morphological details of fluorescence patterns in the cell periphery, which contains presumptive KF precursors. To monitor these precursors, which move with a speed of 100–300 nm/min, images were acquired at 30-second intervals. Note that it is difficult to deduce the continuous inward movement which is, however, readily detected in the corresponding movie 1. Because of the thin and extended cell periphery and the high focal depth of epifluorescence microscopy, the entire fluorescence distribution at the cell edge could be recorded in a single focal plane, whereas the fluorescence in the thicker, more central area (top), was not fully resolved. Adjustment of the LUT in the squared middle area helped to visualize further details, although the limits of the current imaging technique is still apparent.

C. Special Aspects of Recording in Mitotic Cells

For time-lapse fluorescence recordings of dividing cells, it is usually sufficient to provide exponentially growing cultures that were seeded at low density. One to two days after plating, mitotic cells can be easily identified as they round up and elongate during this process. To further guide in the identification of dividing cells, the addition of vital DNA stains such as Hoechst 33342 or transfection with fluorescent histone cDNA constructs is of use. The histone H1-EGFP–encoding plasmid (Rolls *et al.*, 1999) works well to monitor chromatin distribution throughout the entire cell cycle (Fig. 4; movie 2). A particularly challenging problem in the high-resolution analysis of keratin alterations is that disassembly of the KF network occurs at the onset of mitosis (i.e., before shape changes of cells, nuclear envelope breakdown, and chromatin condensation are detectable).

The rounding of cells during mitosis necessitates imaging by confocal laser scan microscopy for optimal resolution. Multiple focal planes should be recorded for subsequent 3D reconstruction (Fig. 5; see also later), because this is the best way to identify domain-specific keratin assembly forms and to delineate their true spatial configuration and respective arrangement(s).

A major problem of imaging mitotic cells is their high degree of sensitivity toward intense illumination. Very often, cells arrest in late metaphase without

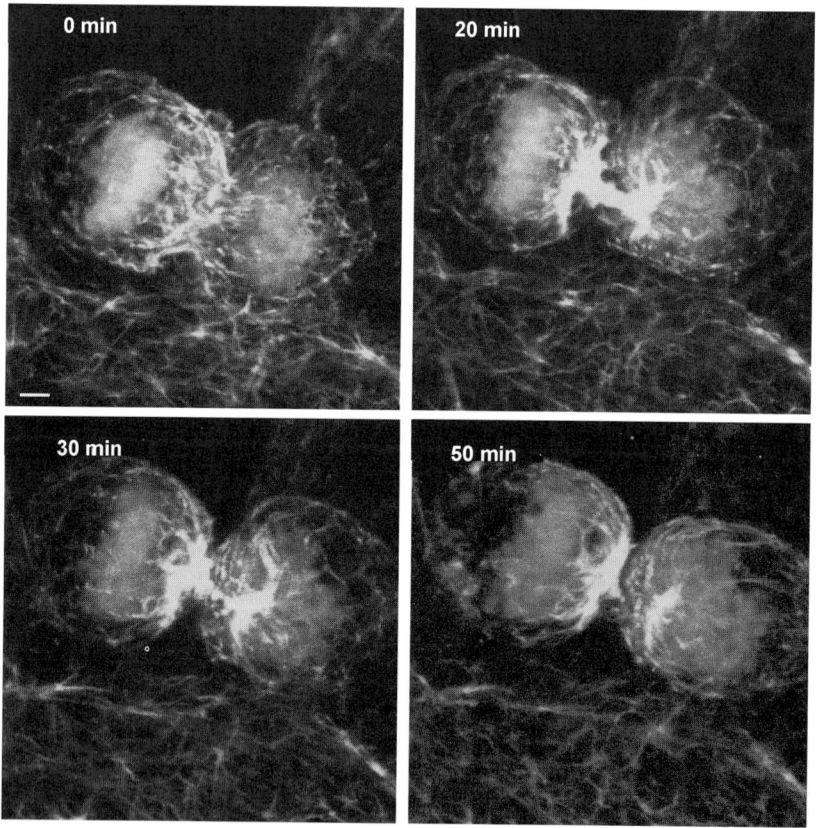

Fig. 4   Projected fluorescence images that were obtained by confocal laser scan microcsopy of vulvar carcinoma–derived AK13-1 cells in anaphase coexpressing keratin chimera HK13-EGFP and a histone H1-EGFP chimera (encoded by plasmid VLP51; Rolls *et al.*, 1999). The complete sequence is provided as movie 2. Note that there is no apparent correlation between keratin dynamics and chromatin movements. Bar, 2.5 μm.

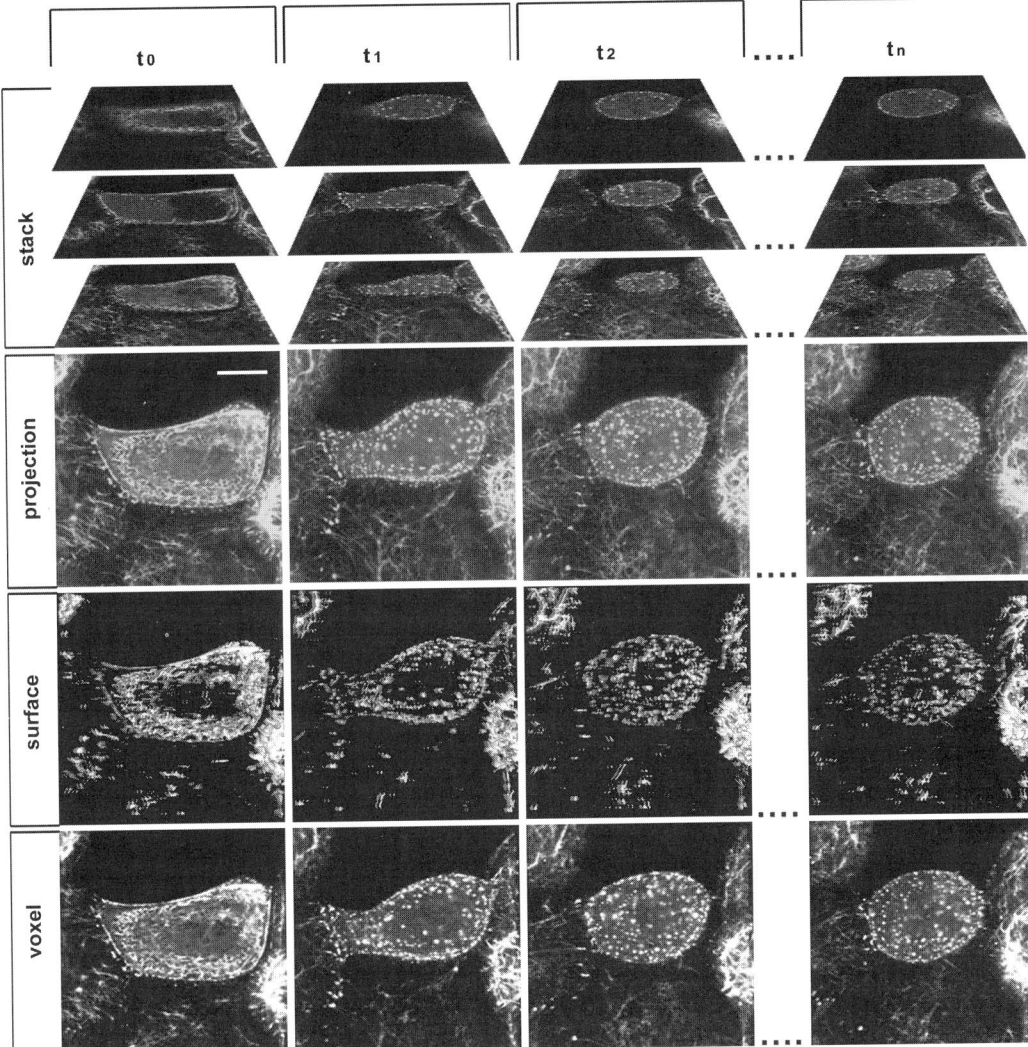

Fig. 5 Different types of presentation of a time series of z-stacks of confocal laser scan fluorescence images that were recorded in a mitotic AK13-1 cell producing HK13-EGFP. The top part of the panel (*stack*) depicts three confocal fluorescence micrographs of a z-stack at four different time points. The corresponding movie 3 was assembled from a single focal plane (recording intervals, 2.5 minutes). The next row of pictures presents projection views of the entire z-stacks at the different time points. The complete series of projection images was combined into movie 4. Below, stacks were reconstructed into 3D anaglyph pictures that should be viewed with red-green glasses for complete 3D visualization. For comparison, surface and voxel reconstructions are shown. The complete time series of the 3D images are provided as movies 5 and 6, respectively. Bar, 5 μm. (See Color Insert.)

reentry into the cell cycle, yet cytoplasmic motility and membrane movements continue for several hours without any further changes in cell shape and keratin morphology. Although the addition of ascorbic acid ameliorates the sensitivity of cells, it is a constant battle to find a compromise between satisfactory temporo-spatial resolution needed for complete monitoring of keratin rearrangement and cell cycle arrest. Therefore, low-resolution images are first taken to obtain a general overview, and high-resolution recordings are done to delineate specific aspects of keratin behavior during individual stages of mitosis.

D. Imaging in the Presence of Phosphatase Inhibitors

An important feature of live cell imaging is the possibility to directly monitor the effects of various drugs on the dynamic behavior of fluorescently labeled structures. In the case of the KF system, substances that alter levels of phosphorylation are of particular interest (Coulombe and Omary, 2002; Omary et al., 1998). Because the phosphatase inhibitors OA and vanadate (orthovanadate as well as pervanadate) are known to induce considerable reorganization of the keratin system in several epithelial cell types (e.g., Blankson et al., 1995; Feng et al., 1999; Kasahara et al., 1993; Strnad et al., 2001, 2002; Yatsunami et al., 1993), we will limit our description to these two agents. Other drugs may be applied and tested in an analogous fashion.

1. Determination of Suitable Drug Concentration

The sensitivity of different cell types to either OA or OV varies greatly. Dilution series should, therefore, be prepared to determine drug levels with the strongest effects on the organization of the keratin system. Precise titration of the lowest possible drug amount needed to elicit these effects is important to avoid alterations of the other cytoskeletal filaments (see, e.g., Strnad et al., 2001, 2002). For the determined "optimal" concentration, time series have to be prepared. Keep in mind that in the case of OA, keratin reorganization takes hours, whereas for vanadate, KF network breakdown occurs within minutes (Figs. 6 and 7).

2. Examination of Reversibility

Ideally, one would like to define conditions under which reorganization of the keratin system can be switched on, arrested, and reversed at will. To find such conditions turns out to be rather tricky, because drug-induced morphological changes often occur only after significant lag periods. In the case of OA treatment (Strnad et al., 2001), a short 30-minute treatment is sufficient to induce the formation of aggregated KF bundles and small, long-lived granular keratin aggregates after ~2 hours. Furthermore, restitution of a normal keratin cytoskeleton is not observed after extended periods. On the other hand, removal of the drug or energy depletion after occurrence of the first morphological signs for keratin reorganization prevent progression of filament network disruption, thus resulting

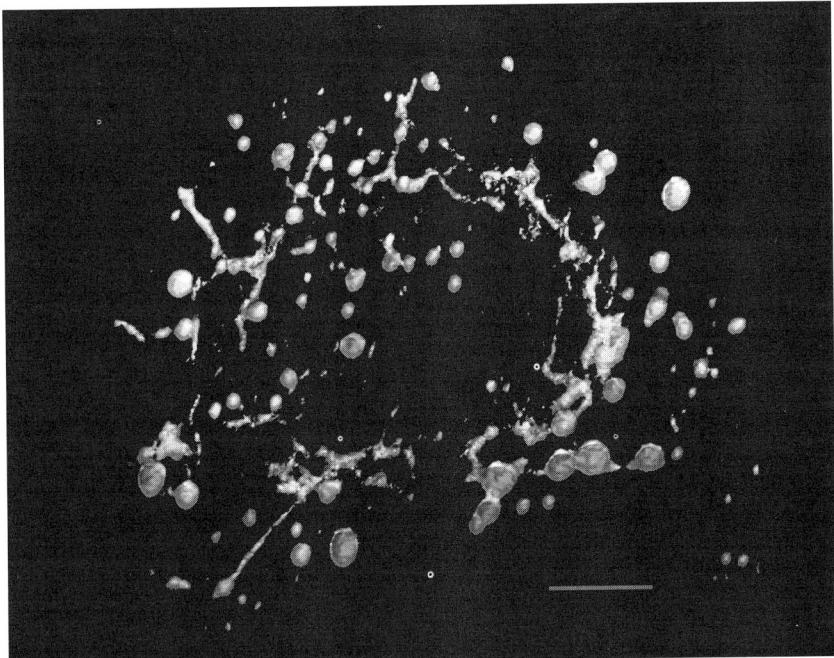

Fig. 6 Three-dimensional reconstruction (surface view) of fluorescence images recorded in 32 confocal planes at 1024 × 1024 pixel of a single AK13-1 cell producing fluorescent HK13-EGFP fusion proteins. The cell was treated with 0.1 μg/ml okadaic acid for 4 hours and was fixed with methanol/acetone before imaging. Keratin granules are seen together with residual perinuclear KF aggregates. Bar, 5 μm. (See Color Insert.)

in a stationary phenotype. In contrast, OV-induced alterations are, at least in part, reversible (Strnad *et al.*, 2002). If the drug is removed after a 10-minute incubation, at which time extensive granular aggregates have formed in most cells, a filamentous keratin network is reestablished in less than an hour (Fig. 7). There is, however, a distinct threshold of drug exposure above which rapid KF reformation is not possible any more. Instead, only a slow reorganization is observed. Both types of reorganization differ not only with respect to their kinetics but are probably dependent on different factors and cellular topologies (Strnad *et al.*, 2002).

3. Light Dependency

Recent findings suggest that the responsiveness of the keratin cytoskeleton can be modulated by light exposition. It was initially observed that, depending on the type of recording protocol and the type of medium, vanadate action differed considerably (Strnad *et al.*, 2002).

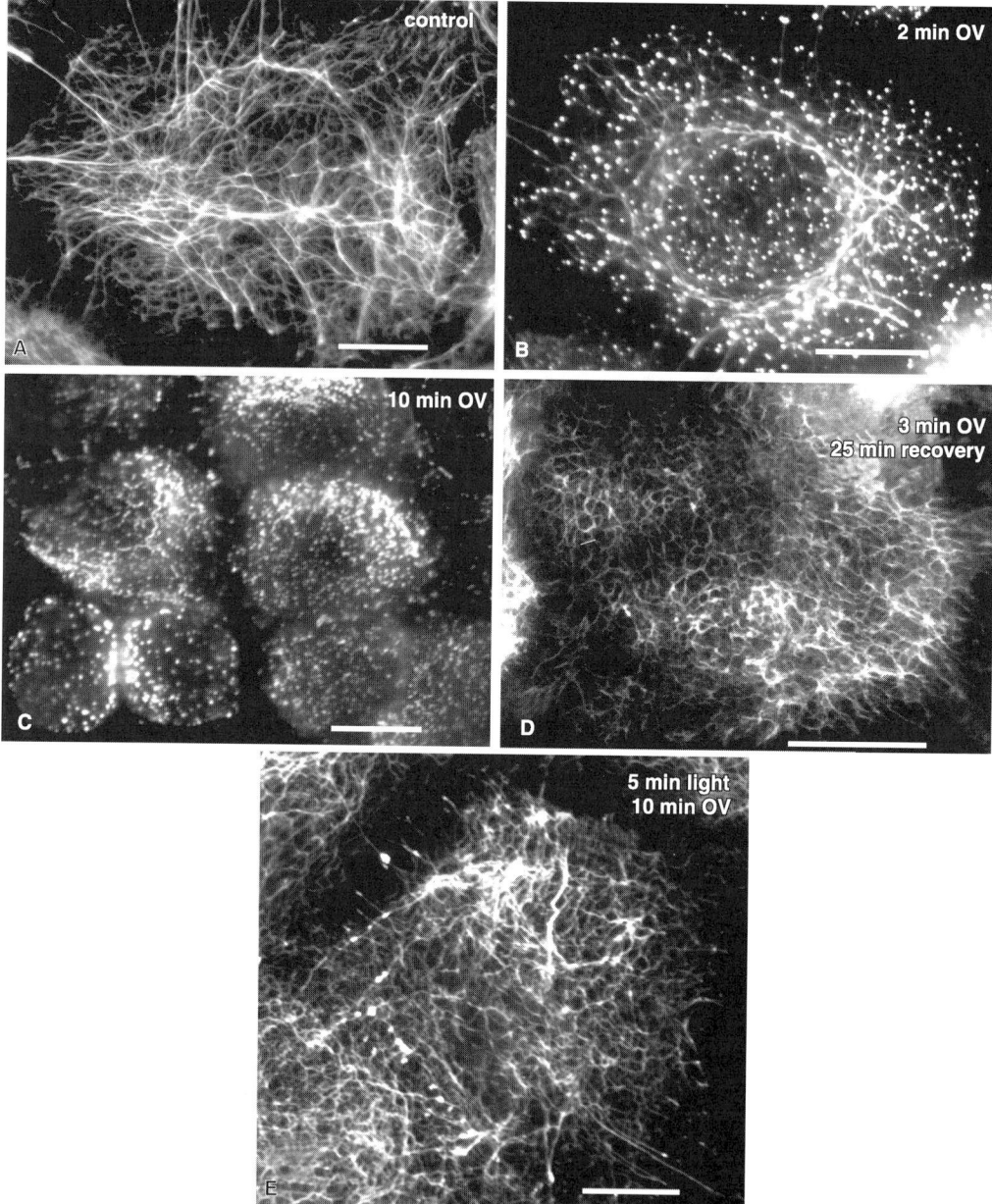

Fig. 7 Reactivity of fluorescent keratins in AK13-1 cells in response to orthovanadate (OV). Epifluorescence micrographs depict control cells in (A), cells after incubation with 10 mM OV for 2 minutes (B) and 10 minutes (C), and 25 minutes after a 3-minute treatment with 10 mM OV in (D). The cells shown in (E) were exposed to monochromatic light (440 nm) for 5 minutes at 4 mW/cm^2 before incubation with 20 mM OV for 10 minutes. Bars, 10 μm.

Systematic investigations (Strnad *et al.*, 2003) revealed that a 1- to 10-minute exposition of cells to normal room light of less than 200 Lux as measured with a digital luxmeter (Mavolux 5032B from Gossen Foto-und Lichtmeβtechnik GmbH, Nürnberg, Germany) is sufficient to inhibit vanadate-induced keratin reorganization (see also Fig. 7E). The protective effect seems to be wavelength-independent and is reversed within 1–2 hours. Interestingly, a similar inhibitory effect on OV action is also observed in cells that are preincubated with the p38 kinase inhibitor SB203580 (Strnad *et al.*, 2003). The maximal protective effect of light is elicited in cells grown in Hanks' medium, whereas DMEM antagonizes light-dependent KF network protection. Although these observations offer a new experimental inroad into the regulation of keratin dynamics, they also add another layer of complexity onto the experimental design of live cell imaging of the keratin system. It is necessary to perform control experiments in the dark for each experimental setting to assess interference of light with a particular dynamic phenomenon under investigation. This may often be a challenging task to accomplish, because manipulations in the dark are rather cumbersome. On occasion, we have even used professional night vision gear.

E. Fluorescence Recovery after Photobleaching

Fluorescence recovery after photobleaching (FRAP) is a powerful technique to determine protein turnover in a given cellular compartment or structure and to visualize motility of fluorescently labeled structures in and out of defined cellular domains (e.g., Lippincott-Schwartz *et al.*, 2001). A confocal microscope has to be used for these experiments, because it allows user-defined bleaching of circumscribed areas of interest with different geometries. A disadvantage of bleaching very small regions is that cell motility often distorts these areas and results in overlap with nonbleached parts of the cell. This may in part explain some of the differences in keratin turnover determinations of the other laboratories that only bleached small bar-shaped segments across filament bundles (Yoon *et al.*, 2001) and our own laboratory that bleached larger, trapezoid cell segments extending from the cell periphery to the perinuclear region (Windoffer *et al.*, 2004). On the other hand, phototoxicity becomes a problem if larger regions are bleached. Therefore, additional transmitted light images should be recorded in this instance to continuously monitor cell viability.

For most experiments, a wide pinhole size (setting of 500) is advantageous to obtain a high focal depth. In this way, a strong fluorescence signal is maintained, enabling short scanning times and minimizing photobleaching, and artificial fluorescence alterations caused by focal shifts are avoided. For higher spatial resolution, the pinhole size can be reduced (a setting of 90 worked well in several experiments), although it is then necessary to increase the time of the recording intervals and to image multiple planes. For bleaching, 100% of medium laser power is applied in the defined area of interest for a total of 20 scans. Immediate postbleach recording is done to confirm complete loss of fluorescence in the area

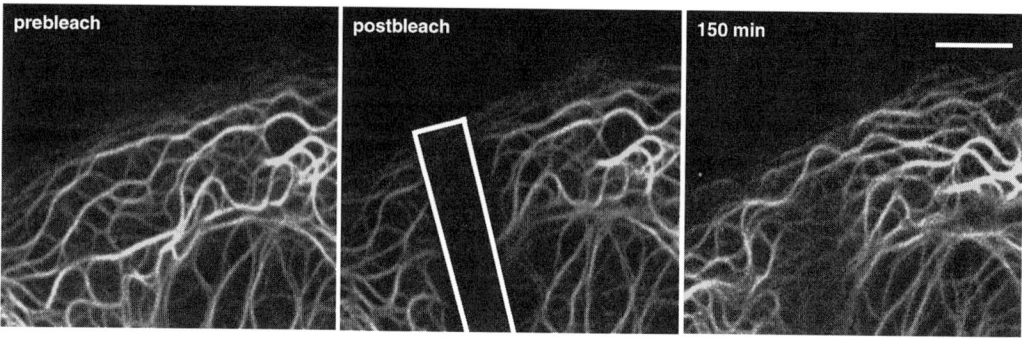

Fig. 8 Fluorescence micrographs taken from a fluorescence recovery after photobleaching experiment showing a section of a PK18-5 cell before bleaching (prebleach), immediately after bleaching (postbleach; box demarcates bleached area), and 150 minutes after bleaching. Note the recurrence of HK18-YFP fluorescence predominantly in the cell periphery. Bar, 5 μm.

of interest. All focal planes should be examined to exclude that focal shift or cell motility contribute erroneously to rapid fluorescence recovery. Further images are recorded at prebleach settings and at low frequency to minimize bleaching to facilitate full fluorescence recovery.

To monitor the net inward-directed motility of keratins (Windoffer and Leube, 1999), cell segments extending from the plasma membrane toward the perinuclear region are bleached. In these bleached areas, not only gradients of recurring fluorescence are revealed, but also keratin conformations are delineated that contribute to the formation of KFs and KF bundles. The method thus allows us to unequivocally define precursor–product relationships. "Young" filaments are detected as a fine mesh in the cell periphery that mature into thick KF bundles that are located in the central cytoplasm (Windoffer *et al.*, 2004). Three time points of a typical FRAP experiment are depicted in Fig. 8, demonstrating that new KFs are formed in the cell periphery, whereas only little turnover of preexisting filaments occurs in the more central cellular domains.

F. Data Analysis

1. Preparation of Movies

The LUT of the recordings should be optimized for black-and-white contrast (see, e.g., Fig. 3 and movie 1 for comparison of different LUT settings in adjacent areas). The correction should be identical for all images of a given series to avoid misrepresentation of fluorescence patterns. In addition, the size of the original recordings should be reduced by altering the color depth from 12 bit to 8 bit. The clarity of visualization can be improved in some instances by image inversion. It is also important to crop the original recordings by use of ImagePro Plus routines to focus on the process of interest and, even more importantly, to reduce the movie size. Self-made ImagePro Plus macros can be prepared to add time stamps and annotations to the movie frames. Finally, the resize routine can be used to shrink the number of pixels per frame to the required final size.

To present movies on the internet, the uncompressed movies are converted into QuickTime movies using "video" compression. Alternatively, movies can be converted into MPEG-1 files that are more compatible with different computer systems and programs.

2. 3D Reconstructions

Recordings of z-stacks as those shown in the top row of Fig. 5 can be processed to gain information on the spatial conformation and arrangement of fluorescent structures.

- *Projection of all images of a z-stack into a single image* (Figs. 4 and 5; movies 2–4): This method is quick and often sufficient for conversion of 3D stacks into an interpretable format. Software of confocal microscopes and ImagePro Plus provide different projection routines. Best results are usually obtained with the maximum projection method. It should be kept in mind, however, that valuable 3D information is lost in this type of presentation.
- *Surface view of 3D reconstructions* (Figs. 5, 6 and 9; movies 5 and 7): Surface visualization provides virtual 3D models of the fluorescent structures that can be displayed from any angle. To enable this type of presentation, a threshold of fluorescence intensity has to be defined for the entire data set. Particular phenomena are highlighted by manually defining the border of the structures of interest, which, however, is extremely time-consuming and can only be applied in specific instances.
- *Voltex representation of 3D data sets* (Fig. 5; movie 6): In this case, each voxel of an image stack is displayed transparently in its 3D position. Adjustment of LUT and of transparency results in a realistic representation of fluorescence in a given volume and may thus be superior to the surface view.

To obtain real 3D images, surface and voltex reconstructions can be presented as anaglyphs, which have to be viewed with special red-green glasses (e.g., Fig. 5; movies 5 and 6). Custom-made Amira scripts are used for automatic generation of the various types of 3D reconstruction from image stacks, and the individual reconstructions from each time point are then assembled into movies.

A major limitation of all 3D reconstructions is their low z-resolution, which is orders of magnitude lower than that in the x–y directions. The technically attainable z-resolution is further compromised by phototoxic effects caused by extensive exposition of cells from repetitive scanning of the same cellular domains. As a consequence, reconstructions often have to be prepared from only a few sections and are therefore prone to many artifacts. This is especially true for surface views, because the manual threshold setting may either result in joining of structures that are actually separate or, conversely, or separation of structures that are actually connected. Careful examination of the unedited images of all focal plane

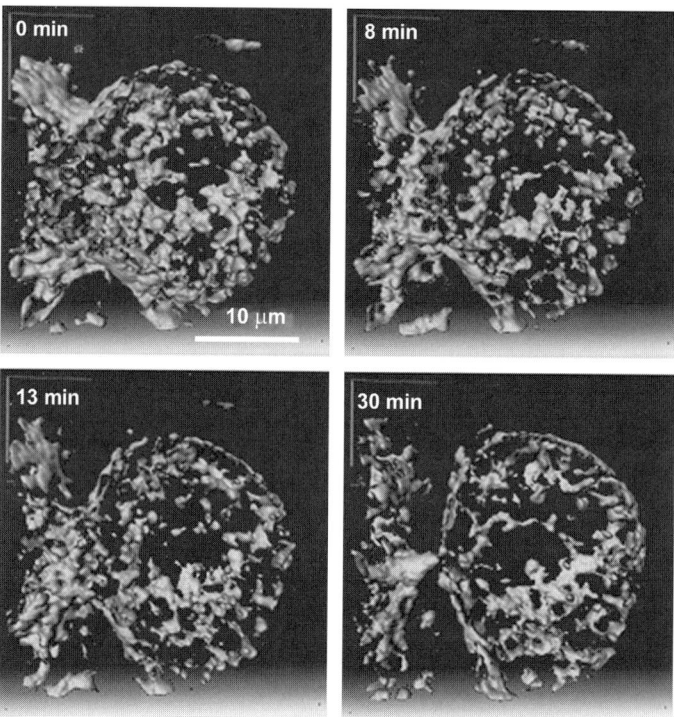

Fig. 9 Four-dimensional analysis of the keratin fluorescence in a dividing AK13-1 cell producing HK13-EGFP. Images were recorded in multiple focal planes with a confocal laser scan microscope every 1.6 minutes and were used to generate the 3D reconstructions (surface view). Four consecutive reconstructions are depicted highlighting redistribution of granular keratin. The complete time series is provided as movie 7. (See Color Insert.)

recordings is therefore a prerequisite before any digital data processing. Yet, after careful and responsible evaluation, the disadvantages are clearly outweighed by the advantages of 3D reconstructions that offer an excellent way to visualize fluorescent structures in 3D space and to delineate the complete keratin cytoskeleton of a given cell (Fig. 6; Windoffer and Leube, 2001). In this way, it was possible, e.g., to decide that the fluorescent cytoplasmic keratin dots in mitotic cells are granules and not filaments/rods in cross section (Windoffer and Leube, 2001). Even more importantly, the transition between granules and filaments can be depicted in 4D movies (Fig. 5; Windoffer and Leube, 2001).

3. Quantification of Soluble Pool

Although the soluble, nonfilamentous keratin pool is very small, it is probably this keratin fraction that is of utmost physiological relevance, because it is highly dynamic by rapid diffusion throughout the entire cytoplasm and is in an exchange equilibrium with the KF network that is most likely subject to precise regulation by phosphorylation (Omary *et al.*, 1998). Most notably, the soluble keratin pool increases significantly during mitosis (e.g., Chou *et al.*, 1993), where it appears as an increased diffuse fluorescence in the cytoplasm of keratin-GFP–labeled cells that can be extracted with Triton X-100 (Windoffer and Leube, 2001). Quantification of the soluble pool works best in mitotic cells with their sparse and/or absent keratin network and the few granular aggregates. By use of ImagePro Plus, gray values are determined for each pixel within a given area. The sum of all gray values of less than 255 is calculated for each time point from the 12-bit images with Excel (Microsoft). White pixels (gray value of 255) are excluded, assuming that they correspond to aggregated and/or filamentous keratins, all of which exhibit very strong fluorescence. The mean gray value is then determined for the entire area of interest at each time point and used for graphical representation of time-dependent changes (see, e.g., Windoffer and Leube, 2001).

4. Quantification of Fluorescence Recovery

A suitable area for measurements of fluorescence recovery has to be defined in FRAP experiments within the bleached area. It should be restricted to the center of the bleached region, because the sharp boundary between bleached and unbleached regions is usually lost within a few minutes, and cells tend to move, thereby distorting and translocating the bleached region within the recording field. In this defined central area, all gray values are summed up with the recorded 12-bit image data. The prebleach value is defined as 100%, the postbleach value as 0%. With the help of Excel spreadsheet routines, diagrams are prepared. From these, the $t_{1/2}$ of fluorescence recovery within the analyzed area is determined. In the case of keratins, one should keep in mind that recovery times differ, depending on intracellular topology, and that other factors such as illumination, phosphorylation, or cell cycle stage may affect turnover rates.

5. Diagrammatic Representations of Time-Dependent Fluorescence Patterns

a. Mobility Diagrams

Given the vectorial movement of keratin fluorescence toward the cell center, a line is selected from fluorescence recordings that is perpendicular to the cell edge and extends from the outside of the cell toward the nucleus to analyze centripetal KF mobility. Along this line, the movement of the cross-sectioned circumferential KFs and peripheral KF precursors are followed between successive images. Amira software is used to depict the fluorescence motility during time, and the results are plotted in diagrams of position along the line versus time. Velocity of the inward-translocation can be easily determined in these diagrams. The motility is greatest in the cell periphery, where KFs are formed and integrated into the preexisting KF network. In this area, the inward-directed keratin movement results in oblique and parallel lines in the diagrams in a Christmas tree–type pattern (Windoffer *et al.*, 2004). The trunk of the tree corresponds to the peripheral KF network, the branches to the inward-moving keratin particles before their integration into the network. Mobility diagrams are also useful for characterization of the dynamics of granules that are present in cells producing epidermolysis bullosa simplex–type keratins or cells that are treated with OV. Such diagrams revealed in both instances that granule production occurs in a submembraneous compartment, followed by vectorial centripetal movement of comparable speed and disassembly at a distinct circumferential zone of the cytoplasm (Werner *et al.*, 2004).

b. Time–Space Diagrams

Time–space diagrams are prepared from movies. Instead of assembling z-stacks of pictures as is done for 3D reconstructions, time-stacks are generated from corresponding regions recorded at consecutive time points. Image processing is analogous to 3D reconsctruction to obtain surface views of time traces (Fig. 10; Windoffer *et al.*, 2002, 2004). Such reconstructions can be used to directly determine the lifetime of fluorescent particles, to depict their direction of movement, to calculate their speed of motility, and to visualize their shape changes during their life cycle. In the example depicted in Fig. 10 for OV-treated cells, the lifetime of the imaged keratin aggregates is ~20 minutes, they move consistently from the cell periphery toward the cell center with a speed of ~250 nm/min, and they first grow to a certain size, which is maintained for most of their lifetime before their rather sudden disassembly.

c. Intensity Diagrams

Diagrams of reappearing fluorescence in relation to time and intracellular topology are prepared from time-lapse fluorescence image series that are recorded in FRAP experiments. In this case, 12-bit gray values are quantified in the central areas of bleached regions with Image-pro Plus. The fluorescence intensity values along this single line are then plotted against time and transformed into a surface view with color-coded intensity peaks with the help of Amira software (Fig. 11; movie 8; see also Windoffer *et al.*, 2004).

IV. Pearls and Pitfalls

A. Transient versus Stable Transfection

A principal problem of any transgenesis is the introduction of mutant polypeptides that may result in nonphysiological perturbations. This is particularly apparent in transient transfection experiments in which various phenotypes are observed using exactly the same construct. Figure 12 presents examples of fluorescent keratin distribution patterns, in which either the characteristic IF-type cytoskeleton (A), an intermediate phenotype with aggregated filament bundles and variously shaped aggregates (B), or a predominantly granular appearance is seen (C). A similar diversity of IF morphotypes was also described for cells producing neurofilament–GFP fusions (Szebenyi *et al.*, 2002). Therefore, transient transfection assays can only provide an overview of the entire spectrum of possible

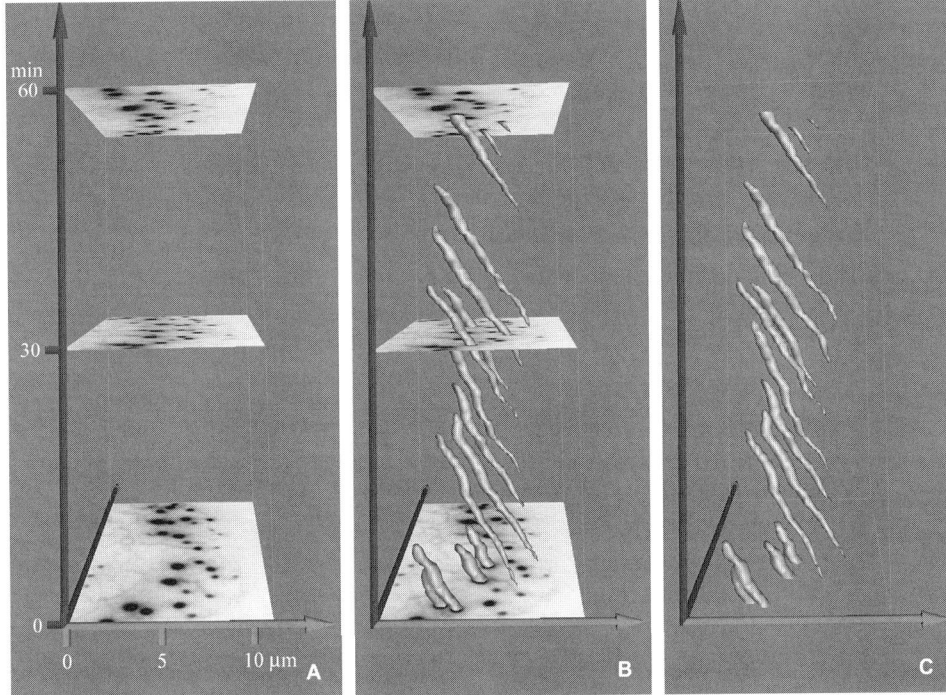

Fig. 10   Preparation of time-space diagrams. (A) Three epifluorescence images (inverse presentation) that are taken from a time-lapse recording of a small region in the cell periphery of an AK13-1 cell synthesizing HK13-EGFP after a short treatment with 10 mM orthovanadate. Surface views were prepared from the time series as indicated in (B). For clarity, only a few representative traces are depicted. These surface views (C) reveal a lifetime of ~20 minutes and a consistent inward-directed movement from the subplasmalemmal region at right to the cell interior at left of ~250 nm/min. (See Color Insert.)

phenotypes but are not useful for analyses at the single cell level, which may lead to misrepresentation of the morphology and dynamics of labeled entities. Thus, stable cell lines are needed as reproducible and reliable model systems that can be examined in detail and in which all cells equal each other. It should be kept in mind, however, that the kinetics of the labeled filaments may still differ considerably from those of the endogenous counterparts. Yet, in the case of keratin polypeptides, half-life determinations suggest that the fluorescent keratin hybrids are, like the endogenous molecules, rather long-lived and may not differ so much in their kinetic behavior (Windoffer *et al.*, 2004).

B. Epifluorescence Microscopy versus Confocal Laser Scan Microscopy

Epifluorescence microscopy and confocal laser scan microscopy complement each other. Epifluorescence microscopy is superior in detecting the curvilinear nature of the extended KFs and KF bundles. In addition, the speed of image

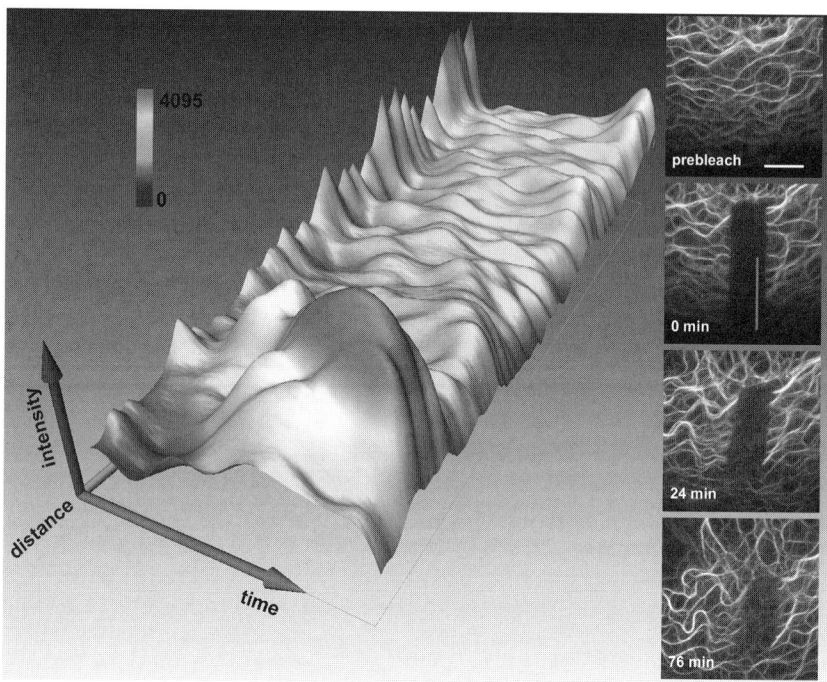

Fig. 11 Time-intensity diagram depicting fluorescence alterations that were observed in a fluorescence recovery after a photobleaching experiment of a PK18-5 cell synthesizing fluorescent HK18-YFP chimeras. The diagram was derived from the image series presented in movie 8. Four images from the movie are shown at right presenting the prebleach keratin cytoskeleton, the location of the bleached area together with the line (in red) that was selected for intensity measurements, and two time points during fluorescence recovery (24 minutes and 76 minutes postbleach). The diagram at left displays the values of fluorescence intensity that were measured along the red line (*distance*) in relation to time. Note that the most fluorescence recurs in the cell periphery originating beneath the plasma membrane that moved gradually outside the imaged area during recording. Bar, 5 μm. (See Color Insert.)

acquisition is unrivaled. The use of a monochromator further increases the versatility of epifluorescence to image structures that are labeled with many different types of fluorophores. To compensate for the lack of confocality, one can record images in different focal planes with the help of a piezo-stepper and use special deconvolution programs (Amira or any other standard program) to obtain 3D space information. Therefore, for many applications of live cell imaging of keratins, epifluorescence microscopy is the method of choice for the recording and preparation of high-quality images and movies. Only in cases in which cells are not sufficiently flat (e.g., during mitosis or after extensive drug treatment) and in bleaching experiments, a confocal laser scan microscope is needed. Furthermore, for optimal 3D resolution confocal recording is clearly the better method.

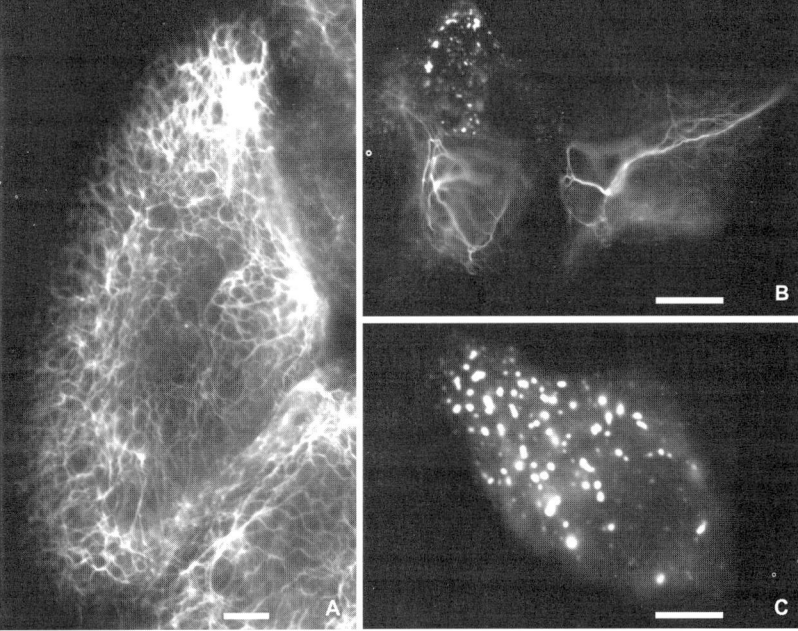

Fig. 12 Fluorescence micrographs of cells shortly after cDNA transfection with expression constructs coding for fluorescent HK13-EGFP fusion proteins. Note the different phenotypes ranging from a typical IF cytoskeleton (A) to a mixed filamentous/granular morphotype (B) and an exclusively granular appearance (C). Bars, 5 μm.

C. Effects of Light

Imaging is intrinsically coupled to light exposure of cells, which, however, interferes not only with the fluorescence signal itself because of photobleaching but has also toxic side effects resulting in altered cell behavior and endows the keratin cytoskeleton with specific properties affecting its responsiveness to certain drugs. Remarkably, KFs are protected against the action of the tyrosine phosphatase inhibitor OV by light. Although OV induces extensive granule formation within minutes in the dark, a brief light exposure is sufficient to prevent keratin disassembly almost completely (Strnad *et al.*, 2003). It is currently unknown how this effect is mediated. But the observation that similar protective mechanisms can be elicited by pretreatment of cells with the p38 inhibitor SB20358 suggests that signaling pathways may be involved. The emerging general caveat for all imaging experiments of the keratin system is that its properties may be altered considerably by light exposure and that possibly other, yet unknown, factors exert similar effects.

V. Discussion and Concluding Remarks

A principal problem of cell biological research has been the interpretation of insights that have been gained from analyses of dead cells, fixed cells, or certain fractions thereof, all of which are prone to multiple types of artifacts. The characterization and development of autofluorescent proteins, most prominently of the green florescent protein that was isolated from the jelly fish *Aequorea victoria*, have provided tools to label cellular structures such that they can be examined in living cells (Tsien, 1998). Analyses of stably transfected cell lines that produce normal-appearing fluorescent keratin hybrids have yielded completely new concepts about the unresolved mysteries of the keratin cytoskeleton as formulated in the Introduction. Thus, we find that *de novo* keratin formation originates predominantly from the cell cortex progressing toward the cell interior at the end of mitosis (Windoffer and Leube, 2001) and that continuous keratin turnover occurs preferentially in a specific submembraneous compartment (Windoffer *et al.*, 2004). Further live cell imaging experiments suggest that the driving force behind the vectorial and dynamic keratin distribution patterns relies both on microtubules and microfilaments and their associated factors (Liovic *et al.*, 2003; Werner *et al.*, 2004; Windoffer and Leube, 1999; Yoon *et al.*, 2001). Finally, it has been demonstrated in keratin-GFP–labeled cells that changes in cellular phosphorylation have a significant impact on keratin organization and dynamics in a temporally and spatially defined manner (Strnad *et al.*, 2001, 2002, 2003). These insights on keratin network modulation have been extended onto disease states by use of cell lines that produce fluorescent dominant negative keratin mutants (Werner *et al.*, 2004).

The studies on the dynamics of the keratin cytoskeleton have been complemented by exciting analyses of the other IF types and systems. In these instances, an astounding variety of dynamic behaviors has been described in living cells producing fluorescent IF protein chimeras. These behaviors include diverse types of motility of individual filaments and filament bundles in the absence of cell shape changes (Ho *et al.*, 1998; Martys *et al.*, 1999; Roy *et al.*, 2000; Wang and Brown, 2001; Wang *et al.*, 2000; Yoon *et al.*, 1998), which may correspond, at least in part, to the reported KF movements (Windoffer and Leube, 1999; Yoon *et al.*, 2001). Furthermore, these analyses have shown that IFs are an integrated part of the cytoskeleton and that their motile properties are predominantly determined by microtubules and their associated motor proteins (Chou and Goldman, 2000; Chou *et al.*, 2001; Helfand *et al.*, 2002, 2003a, 2004; Prahlad *et al.*, 1998). Some of these principles may also be shared by keratins, although clear differences seem to exist (Windoffer and Leube, 1999; Yoon *et al.*, 2001). Interestingly, highly motile, nonfilamentous structures that are referred to as squiggles, IF particles, and/or round/ovoid structures and are transported along microtubules have been identified for vimentin (Helfand *et al.*, 2002; Prahlad *et al.*, 1998; Yoon *et al.*, 2001), peripherin (Helfand *et al.*, 2003b), and neurofilaments (Chan *et al.*, 2003; Prahlad *et al.*, 2000; Yabe *et al.*, 1999, 2001). It is assumed that they are IF precursors (Chan *et al.*, 2003; Chou and Goldman, 2000; Chou *et al.*, 2001; Helfand *et al.*, 2004; Prahlad *et al.*, 1998; Yabe *et al.*, 1999, 2001). In addition, larger filamentous structures have been reported to be transported along axons in living neurons (Roy *et al.*, 2000; Wang *et al.*, 2000). It remains to be shown how and whether these various transport particles are related to keratin squiggles (Yoon *et al.*, 2001), to keratin precursors that were recently identified in the periphery of epithelial cells (Windoffer *et al.*, 2004), and/or to the highly motile particles previously described (Liovic *et al.*, 2003). Finally, it has been shown by imaging of fluorescent IF chimeras that the balance between granular and filamentous IF forms depends on phosphatase activities and the phosphorylation status of IFs (Chan *et al.*, 2003; Yabe *et al.*, 2001), as is the case for the keratin system (Strnad *et al.*, 2001, 2002).

Given the rapid advances in imaging technologies, it can be expected that further exciting findings about the complex dynamics of the keratin cytoskeleton will soon be reported. By use of multicolor imaging, it is possible to simultaneously monitor the relative distribution of various components of the keratin system (e.g., Windoffer *et al.*, 2002). Furthermore, molecular interactions can be visualized by fluorescence resonance energy transfer (FRET; e.g., Lippincott-Schwartz *et al.*, 2001; Periasamy and Day, 1999; Pollok and Heim, 1999; van Roessel and Brand, 2002), and fluorescence correlation spectroscopy (FCS) will afford the investigation of multimerization states of soluble cytoplasmic keratins (e.g., Bacia and Schwille, 2003; Lippincott-Schwartz *et al.*, 2001). This is clearly just the beginning of a new era of *in situ* experimentation, with many more improvements and refinements yet to come to enlarge our knowledge about the dynamic integration of the keratin system into the entire functional and structural cellular assembly.

Acknowledgments

We thank Dr. Pavel Strnad for providing some of the micrographs presented in Fig. 7 and Stefan Wöll for the recording that was used to prepare Fig. 10 (both from this institute). The construct coding for YFP-HK14 was given to us by Drs. Nicola Werner and Thomas Magin (Institut für Physiologische Chemie, Universitätsklinikum Bonn, Bonn, Germany), the histone-EGFP–encoding plasmid is a generous gift of Dr. Melissa Rolls (Department of Cell Biology, Harvard Medical School, Boston, MA). We also thank Dr. Gerd Technau (Institut für Genetik, Universität Mainz, Mainz, Germany) for use of the confocal laser scan microscope. The expert technical assistance of Ursula Wilhelm is gratefully acknowledged. The work was supported by the "Stiftung Rheinland-Pfalz für Innovation" and the German Research Council (LE 566/7).

References

Achtstaetter, T., Hatzfeld, M., Quinlan, R. A., Parmelee, D. C., and Franke, W. W. (1986). Separation of cytokeratin polypeptides by gel electrophoretic and chromatographic techniques and their identification by immunoblotting. *Meth. Enzym.* **134**, 355–371.

Bacia, K., and Schwille, P. (2003). A dynamic view of cellular processes by *in vivo* fluorescence auto- and cross-correlation spectroscopy. *Methods* **29**, 74–85.

Bader, B. L., Magin, T. M., Freudenmann, M., Stumpp, S., and Franke, W. W. (1991). Intermediate filaments formed de novo from tail-less cytokeratins in the cytoplasm and in the nucleus. *J. Cell Biol.* **115**, 1293–1307.

Blankson, H., Holen, I., and Seglen, P. O. (1995). Disruption of the cytokeratin cytoskeleton and inhibition of hepatocytic autophagy by okadaic acid. *Exp. Cell Res.* **218**, 522–530.

Boukamp, P., Petrussevska, R. T., Breitkreutz, D., Hornung, J., Markham, A., and Fusenig, N. E. (1988). Normal keratinization in a spontaneously immortalized aneuploid human keratinocyte cell line. *J. Cell Biol.* **106**, 761–771.

Cadrin, M., and Martinoli, M. G. (1995). Alterations of intermediate filaments in various histopathological conditions. *Biochem. Cell Biol.* **73**, 627–634.

Chan, W. K., Yabe, J. T., Pimenta, A. F., Ortiz, D., and Shea, T. B. (2003). Growth cones contain a dynamic population of neurofilament subunits. *Cell Motil. Cytoskeleton* **54**, 195–207.

Chou, C. F., Riopel, C. L., Rott, L. S., and Omary, M. B. (1993). A significant soluble keratin fraction in 'simple' epithelial cells. Lack of an apparent phosphorylation and glycosylation role in keratin solubility. *J. Cell Sci.* **105**, 433–444.

Chou, Y. H., and Goldman, R. D. (2000). Intermediate filaments on the move. *J. Cell Biol.* **150**, F101–F106.

Chou, Y. H., Helfand, B. T., and Goldman, R. D. (2001). New horizons in cytoskeletal dynamics: Transport of intermediate filaments along microtubule tracks. *Curr. Opin. Cell Biol.* **13**, 106–109.

Cohen, P., Holmes, C. F., and Tsukitani, Y. (1990). Okadaic acid: A new probe for the study of cellular regulation. *Trends Biochem. Sci.* **15**, 98–102.

Coulombe, P. A., and Omary, M. B. (2002). 'Hard' and 'soft' principles defining the structure, function and regulation of keratin intermediate filaments. *Curr. Opin. Cell Biol.* **14**, 110–122.

Domenjoud, L., Jorcano, J. L., Breuer, B., and Alonso, A. (1988). Synthesis and fate of keratins 8 and 18 in nonepithelial cells transfected with cDNA. *Exp. Cell Res.* **179**, 352–361.

Feng, L., Zhou, X., Liao, J., and Omary, M. B. (1999). Pervanadate-mediated tyrosine phosphorylation of keratins 8 and 19 via a p38 mitogen-activated protein kinase-dependent pathway. *J. Cell Sci.* **112**, 2081–2090.

Franke, W. W., Schmid, E., Grund, C., and Geiger, B. (1982). Intermediate filament proteins in nonfilamentous structures: Transient disintegration and inclusion of subunit proteins in granular aggregates. *Cell* **30**, 103–113.

Hager, B., Bickenbach, J. R., and Fleckman, P. (1999). Long-term culture of murine epidermal keratinocytes. *J. Invest. Dermatol.* **112**, 971–976.

Hedberg, K. K., and Chen, L. B. (1986). Absence of intermediate filaments in a human adrenal cortex carcinoma-derived cell line. *Exp. Cell Res.* **163**, 509–517.

Helfand, B. T., Chang, L., and Goldman, R. D. (2003a). The dynamic and motile properties of intermediate filaments. *Annu. Rev. Cell Dev. Biol.* **19**, 445–467.

Helfand, B. T., Chang, L., and Goldman, R. D. (2004). Intermediate filaments are dynamic and motile elements of cellular architecture. *J. Cell Sci.* **117**, 133–141.

Helfand, B. T., Loomis, P., Yoon, M., and Goldman, R. D. (2003b). Rapid transport of neural intermediate filament protein. *J. Cell Sci.* **116**, 2345–2359.

Helfand, B. T., Mikami, A., Vallee, R. B., and Goldman, R. D. (2002). A requirement for cytoplasmic dynein and dynactin in intermediate filament network assembly and organization. *J. Cell Biol.* **157**, 795–806.

Herrmann, H., Hesse, M., Reichenzeller, M., Aebi, U., and Magin, T. M. (2003). Functional complexity of intermediate filament cytoskeletons: From structure to assembly to gene ablation. *Int. Rev. Cytol.* **223**, 83–175.

Ho, C. L., Martys, J. L., Mikhailov, A., Gundersen, G. G., and Liem, R. K. (1998). Novel features of intermediate filament dynamics revealed by green fluorescent protein chimeras. *J. Cell Sci.* **111**, 1767–1778.

Horwitz, B., Kupfer, H., Eshhar, Z., and Geiger, B. (1981). Reorganization of arrays of prekeratin filaments during mitosis. Immunofluorescence microscopy with multiclonal and monoclonal prekeratin antibodies. *Exp. Cell Res.* **134**, 281–290.

Irvine, A. D., and McLean, W. H. (1999). Human keratin diseases: The increasing spectrum of disease and subtlety of the phenotype-genotype correlation. *Br. J. Dermatol.* **140**, 815–828.

Kasahara, K., Kartasova, T., Ren, X. Q., Ikuta, T., Chida, K., and Kuroki, T. (1993). Hyperphosphorylation of keratins by treatment with okadaic acid of BALB/MK-2 mouse keratinocytes. *J. Biol. Chem.* **268**, 23531–23537.

Lane, E. B., Goodman, S. L., and Trejdosiewicz, L. K. (1982). Disruption of the keratin filament network during epithelial cell division. *EMBO J.* **1**, 1365–1372.

Liovic, M., Mogensen, M. M., Prescott, A. R., and Lane, E. B. (2003). Observation of keratin particles showing fast bidirectional movement colocalized with microtubules. *J. Cell Sci.* **116**, 1417–1427.

Lippincott-Schwartz, J., Snapp, E., and Kenworthy, A. (2001). Studying protein dynamics in living cells. *Nat. Rev. Mol. Cell. Biol.* **2**, 444–456.

Ma, L., Xu, J., Coulombe, P. A., and Wirtz, D. (1999). Keratin filament suspensions show unique micromechanical properties. *J. Biol. Chem.* **274**, 19145–19151.

Ma, L., Yamada, S., Wirtz, D., and Coulombe, P. A. (2001). A 'hot-spot' mutation alters the mechanical properties of keratin filament networks. *Nat. Cell Biol.* **3**, 503–506.

Martys, J. L., Ho, C. L., Liem, R. K., and Gundersen, G. G. (1999). Intermediate filaments in motion: Observations of intermediate filaments in cells using green fluorescent protein-vimentin. *Mol. Biol. Cell* **10**, 1289–1295.

Miller, R. K., Khuon, S., and Goldman, R. D. (1993). Dynamics of keratin assembly: Exogenous type I keratin rapidly associates with type II keratin *in vivo*. *J. Cell Biol.* **122**, 123–135.

Miller, R. K., Vikstrom, K., and Goldman, R. D. (1991). Keratin incorporation into intermediate filament networks is a rapid process. *J. Cell Biol.* **113**, 843–855.

Omary, M. B., Ku, N. O., Liao, J., and Price, D. (1998). Keratin modifications and solubility properties in epithelial cells and *in vitro*. *Subcell. Biochem.* **31**, 105–140.

Parry, D. A., and Steinert, P. M. (1999). Intermediate filaments: Molecular architecture, assembly, dynamics and polymorphism. *Q. Rev. Biophys.* **32**, 99–187.

Periasamy, A., and Day, R. N. (1999). Visualizing protein interactions in living cells using digitized GFP imaging and FRET microscopy. *Methods Cell Biol.* **58**, 293–314.

Pollok, B. A., and Heim, R. (1999). Using GFP in FRET-based applications. *Trends Cell Biol.* **9**, 57–60.

Porter, R. M., and Lane, E. B. (2003). Phenotypes, genotypes and their contribution to understanding keratin function. *Trends Genet.* **19**, 278–285.

Prahlad, V., Helfand, B. T., Langford, G. M., Vale, R. D., and Goldman, R. D. (2000). Fast transport of neurofilament protein along microtubules in squid axoplasm. *J. Cell Sci.* **113**, 3939–3946.

Prahlad, V., Yoon, M., Moir, R. D., Vale, R. D., and Goldman, R. D. (1998). Rapid movements of vimentin on microtubule tracks: Kinesin-dependent assembly of intermediate filament networks. *J. Cell Biol.* **143**, 159–170.

Riley, N. E., Li, J., Worrall, S., Rothnagel, J. A., Swagell, C., van Leeuwen, F. W., and French, S. W. (2002). The Mallory body as an aggresome: *In vitro* studies. *Exp. Mol. Pathol.* **72**, 17–23.

Rizzuto, R., Carrington, W., and Tuft, R. A. (1998). Digital imaging microscopy of living cells. *Trends Cell Biol.* **8**, 288–292.

Rolls, M. M., Stein, P. A., Taylor, S. S., Ha, E., McKeon, F., and Rapoport, T. A. (1999). A visual screen of a GFP-fusion library identifies a new type of nuclear envelope membrane protein. *J. Cell Biol.* **146**, 29–44.

Roy, S., Coffee, P., Smith, G., Liem, R. K., Brady, S. T., and Black, M. M. (2000). Neurofilaments are transported rapidly but intermittently in axons: Implications for slow axonal transport. *J. Neurosci.* **20**, 6849–6861.

Smith, F. (2003). The molecular genetics of keratin disorders. *Am. J. Clin. Dermatol.* **4**, 347–364.

Stephens, D. J., and Allan, V. J. (2003). Light microscopy techniques for live cell imaging. *Science* **300**, 82–86.

Strelkov, S. V., Herrmann, H., and Aebi, U. (2003). Molecular architecture of intermediate filaments. *Bioessays* **25**, 243–251.

Strnad, P., Windoffer, R., and Leube, R. E. (2001). *In vivo* detection of cytokeratin filament network breakdown in cells treated with the phosphatase inhibitor okadaic acid. *Cell Tissue Res.* **306**, 277–293.

Strnad, P., Windoffer, R., and Leube, R. E. (2002). Induction of rapid and reversible cytokeratin filament network remodeling by inhibition of tyrosine phosphatases. *J. Cell Sci.* **115**, 4133–4148.

Strnad, P., Windoffer, R., and Leube, R. E. (2003). Light-induced resistance of the keratin network to the filament-disrupting tyrosine phosphatase inhibitor orthovanadate. *J. Invest. Dermatol.* **120**, 198–203.

Szebenyi, G., Smith, G. M., Li, P., and Brady, S. T. (2002). Overexpression of neurofilament H disrupts normal cell structure and function. *J. Neurosci. Res.* **68**, 185–198.

Tolle, H. G., Weber, K., and Osborn, M. (1987). Keratin filament disruption in interphase and mitotic cells–how is it induced? *Eur. J. Cell Biol.* **43**, 35–47.

Tsien, R. Y. (1998). The green fluorescent protein. *Annu. Rev. Biochem.* **67**, 509–544.

van Roessel, P., and Brand, A. H. (2002). Imaging into the future: Visualizing gene expression and protein interactions with fluorescent proteins. *Nat. Cell Biol.* **4**, E15–E20.

Vandre, D. D., and Wills, V. L. (1992). Inhibition of mitosis by okadaic acid: Possible involvement of a protein phosphatase 2A in the transition from metaphase to anaphase. *J. Cell Sci.* **101**, 79–91.

Wang, L., and Brown, A. (2001). Rapid intermittent movement of axonal neurofilaments observed by fluorescence photobleaching. *Mol. Biol. Cell* **12**, 3257–3267.

Wang, L., Ho, C. L., Sun, D., Liem, R. K., and Brown, A. (2000). Rapid movement of axonal neurofilaments interrupted by prolonged pauses. *Nat. Cell Biol.* **2**, 137–141.

Werner, N. S., Windoffer, R., Strnad, P., Grund, C., Leube, R. E., and Magin, T. M. (2004). Epidermolysis bullosa simplex-type mutations alter the dynamics of the keratin cytoskeleton and reveal a contribution of actin to the transport of keratin subunits. *Mol. Biol. Cell* **15**, 990–1002.

Windoffer, R., Borchert-Stuhltrager, M., and Leube, R. E. (2002). Desmosomes: Interconnected calcium-dependent structures of remarkable stability with significant integral membrane protein turnover. *J. Cell Sci.* **115**, 1717–1732.

Windoffer, R., and Leube, R. E. (1999). Detection of cytokeratin dynamics by time-lapse fluorescence microscopy in living cells. *J. Cell Sci.* **112**, 4521–4534.

Windoffer, R., and Leube, R. E. (2001). *De novo* formation of cytokeratin filament networks originates from the cell cortex in A-431 cells. *Cell Motil. Cytoskeleton* **50**, 33–44.

Windoffer, R., Wöll, S., Strnad, P., and Leube, R. E. (2004). Identification of novel principles of keratin filament network turnover in living cells. *Mol. Biol. Cell.* **15**, 2434–2448.

Yabe, J. T., Chan, W. K., Chylinski, T. M., Lee, S., Pimenta, A. F., and Shea, T. B. (2001). The predominant form in which neurofilament subunits undergo axonal transport varies during axonal initiation, elongation, and maturation. *Cell Motil. Cytoskeleton* **48**, 61–83.

Yabe, J. T., Pimenta, A., and Shea, T. B. (1999). Kinesin-mediated transport of neurofilament protein oligomers in growing axons. *J. Cell Sci.* **112**, 3799–3814.

Yatsunami, J., Komori, A., Ohta, T., Suganuma, M., Yuspa, S. H., and Fujiki, H. (1993). Hyperphosphorylation of cytokeratins by okadaic acid class tumor promoters in primary human keratinocytes. *Cancer Res.* **53**, 992–996.

Yoon, K. H., Yoon, M., Moir, R. D., Khuon, S., Flitney, F. W., and Goldman, R. D. (2001). Insights into the dynamic properties of keratin intermediate filaments in living epithelial cells. *J. Cell Biol.* **153**, 503–516.

Yoon, M., Moir, R. D., Prahlad, V., and Goldman, R. D. (1998). Motile properties of vimentin intermediate filament networks in living cells. *J. Cell Biol.* **143**, 147–157.

CHAPTER 13

Approaches to Study Phosphorylation of Intermediate Filament Proteins Using Site-Specific and Phosphorylation State-Specific Antibodies

Aie Kawajiri[*,†] and Masaki Inagaki[*]

[*]Division of Biochemistry
Aichi Cancer Center Research Institute
Nagoya, Aichi 464–8681, Japan

[†]Department of Pathology
Nagoya University School of Medicine
Nagoya, Aichi 466–8550, Japan

I. Introduction

Intermediate filaments (IFs) were long thought to be relatively stable compared with microtubules and actin filaments. However, IFs are reorganized more dynamically than previously considered. We, for the first time, found that site-specific phosphorylation induces dynamic reorganization of vimentin, one of type III IF proteins, *in vitro* (Inagaki *et al.*, 1987). Thereafter, there is increasing evidence that site-specific phosphorylation of various IF proteins modifies their filament structures. Several kinases have been reported to be responsible for the phosphorylation of IF proteins *in vitro* (Inagaki *et al.*, 1996, 1997). Thus, the intracellular organization of the IF network seems to be under the control of protein kinases. These findings suggest that site (domain)-specific phosphorylations of IF proteins are spatiotemporally regulated during cellular events such as cell signaling and cell cycle.

We attempted to visualize the spatiotemporal phosphorylation of IF proteins by protein kinases. There are analytical methods for investigating kinase activity *in vivo*. Immunocytochemical analysis that uses antibodies against protein kinases or biochemical subcellular fractionation analysis has been used to monitor the subcellular distribution of protein kinases. The information obtained by immunocytochemical analysis is useful for examining the distribution of protein kinase. However, the localization of protein kinase is not always equal to that of active kinase. Biochemical subcellular fractionation analysis is also helpful to understand spatial information concerning the localization of protein kinase. It is frequently difficult to define the activity *in vivo* because of technical limitations. To overcome this, we attempted to establish a new method for monitoring *in vivo* phosphorylation states of IF proteins with specific antibodies that recognize the phosphorylated amino acid residues of IF protein (Nishizawa *et al.*, 1990, 1991; Yano *et al.*, 1990, 1991). We developed site-specific and phosphorylation state-specific antibodies against IF proteins, including vimentin (Goto *et al.*, 1998, 2003; Ogawara *et al.*, 1995; Takai *et al.*, 1996; Tsujimura *et al.*, 1994), glial fibrillary acidic protein (GFAP) (Goto *et al.*, 1999; Matsuoka *et al.*, 1992; Sekimata *et al.*, 1996; Yano *et al.*, 1991), and desmin (Inada *et al.*, 1999; Kawajiri *et al.*, 2003) by various kinases such as cyclic adenosine monophosphate (cAMP)–dependent protein kinase (PKA), protein kinase C (PKC), $Ca^{2+}/$ calmodulin-dependent protein kinase II (CaM-K II), Cdc2 kinase, Rho-kinase and Aurora-B kinase (for a review, see Nagata *et al.*, 2001). Analysis with the site-specific and phosphorylation state-specific antibodies made possible the practical visualization of the spatiotemporal phosphorylation states of various IF proteins.

This chapter details the production of site-specific and phosphorylation state-specific antibodies and the methodological approach to visualize the phosphorylation of IF proteins as used in our studies. The flowchart in Fig. 1 shows production of site-specific and phosphorylation state-specific antibodies of IF proteins and study of IF dynamics with antibodies. Although the work in our laboratory concerns the phosphorylation of IF proteins, the method is generally applicable to that of various proteins. Here we present the method to identify the phosphorylation sites of IF

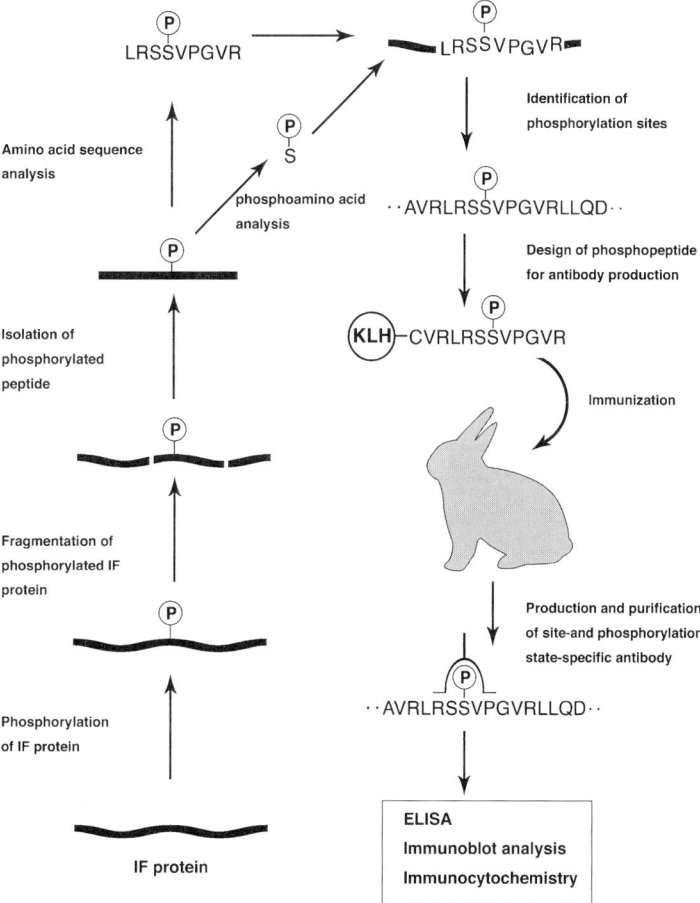

Fig. 1 Strategy for identification of phosphorylation sites on intermediate filament protein and the analysis using site-specific and phosphorylation state-specific antibody.

protein by protein kinase and to visualize the spatiotemporal phosphorylation by developing site-specific and phosphorylation state-specific antibodies.

II. Materials and Instrumentation

A. Detection of Phosphorylation Sites of Intermediate Filament Protein

1. Phosphorylation Assay

Magnesium chloride hexahydrate (MgCl$_2$·6H$_2$O, Cat. No. 135-00165) and 2-amino-2-hydroxymethyl-1,3-propaneidol (Tris, Cat. No. 203-06277) were from Wako Pure Chemical Industries. Calyculin A (Cat. No. C-5552) and adenosine-5′-triphosphate (ATP, Cat. No. 519979) were from Sigma Aldrich and Roche Diagnostics, respectively.

2. Separation of Free Nonreactive [γ-^{32}P] ATP and Phosphorylated Intermediate Filament Protein

Trichloroacetic acid (TCA, Cat. No. 30-3905) was from Katayama Chemical Co., Inc., Japan. Diethyl ether (Cat. No. 32203) was from Riedel-de Haën.

3. Fragmentation of the Phosphorylated Intermediate Filament Protein

Urea (Cat. No. 43009-96) was from Kanto Chemical Co., Inc., Japan. Trifluoroacetic acid (TFA, Cat. No. 206-10731), ammonium hydrogencarbonate (NH$_4$HCO$_3$, Cat. No. 017-02875), calcium chloride dihydrate (CaCl$_2$·2H$_2$O, Cat. No. 031-00435), acetonitrile (Cat. No. 015-08633), and lysyl-endopeptidase (Cat. No. 125-02543) were from Wako Pure Chemical Industries. L-1-tosylamide-2-phenyl-ethyl chloromethyl ketone (TPCK)–treated trypsin (Cat. No. T-1426) was from Sigma Aldrich.

B. Preparation of Site–Specific and Phosphorylation State–Specific Antibodies for the Digested Phosphopeptides

1. Production of Phosphopeptide Antibody

Phosphopeptide and unphosphopeptide were chemically synthesized by Peptide Institute Inc., Japan. Freund's complete adjuvant (Cat. No. 263910) and Freund's incomplete adjuvant (Cat. No. 263910) were from Difco Laboratories, MI. RIBI adjuvant (Cat. No. R-730) was from Corixa Corporation, MT; 0.22-μm Millex-GP filter was from Millipore, MA.

2. Purification of Phosphopeptide Antibody: Preparation of an Affinity Column

Sodium hydrogen carbonate (NaHCO$_3$, Cat. No. 191-01305), sodium carbonate (Na$_2$CO$_3$, Cat. No. 199-01585), Triton X-100 (Cat. No. 203-03215), ethanolamine (Cat. No. 019-12465), and sodium azide (NaN$_3$, Cat. No. 199-11095) were from Wako Pure Chemical Industries. Sodium borohydride (NaBH$_4$, Cat. No. S-9125) and sodium chloride (NaCl, Cat. No. 28-2270-8) were from Sigma Aldrich and Sigma Aldrich Japan, respectively. Disodium β-glycerophosphate pentahydrate (Cat. No. 37177-00) was from Kanto Chemical Co., Inc., Japan. Formyl-cellurofine and Muromac column were from Seikagaku Co., Japan, and Muromachi Kagaku Kogyo Kaisha, Japan., respectively.

3. Purification of Phosphopeptide Antibody: Purification of Antibody

Bovine serum albumin (BSA, Cat. No. 735094) was from Boehringer Mannheim. Glycine (Cat. No. 077-00735) and ethylene glycol (Cat. No. 058-00986) were from Wako Pure Chemical Industries.

C. Specificity Analysis of Site-Specific and Phosphorylation State-Specific Antibody by ELISA

1. Preparation of Plates for ELISA

Sucrose (Cat. No. 28-0010-5) was from Sigma Aldrich Japan. Disodium hydrogenphosphate 12-water ($Na_2HPO_4 \cdot 12H_2O$, Cat. No. 196-02835), sodium dihydrogenphosphate dihydrate ($NaH_2PO_4 \cdot 2H_2O$, Cat. No. 196-02815), and methanol (MeOH, Cat. No. 131-01826) were from Wako Pure Chemical Industries.

2. Examination of Specificity of Antibody with ELISA

Sulfonic acid (H_2SO_4, Cat. No. 192-04696) was from Wako Pure Chemical Industries. *P*-hydroxyphenylacetic acid (Cat. No. H-4377) and thimerosal (Cat. No. T-5125) were from Sigma Aldrich. *O*-phenylenediamin (Cat. No. 32116-30) was from Kanto Chemical Co., Inc., Japan. Horseradish peroxidase (HRP)– conjugated anti-rabbit Ig antibody (Cat. No. NA9340V) was from Amersham Biosciences.

D. Immunocytochemistry Using Site-Specific and Phosphorylation State-Specific Antibody

Formaldehyde (Cat. No. 064-00406) was from Wako Pure Chemical Industries. Alexa Fluor 488 anti-rabbit IgG (Cat. No. A-11034) used as a secondary antibody was from Molecular probes. 4′,6-diamidine-2-phenylindole-dihydrochloride (DAPI, Cat. No. 236276) and propidium iodide (PI, Cat. No. P-4170), which are used for nuclear stain, were from Boehringer Mannheim, Germany and Sigma Aldrich, respectively.

III. Procedures

A. Detection of Phosphorylation Sites of Intermediate Filament Protein

Site-specific phosphorylation induces a dramatic reorganization of IF proteins. The identification of phosphorylated amino acid residue(s) within IF protein is important to understand the mechanisms by which cellular IF reorganization is regulated. As the first step, we tried to identify *in vitro* phosphorylation sites of IF proteins by protein kinases.

1. Phosphorylation Assay

a. Reagents

1. Phosphorylation reaction buffer: 25 mM Tris-HCl (pH 7.5), <3 mM $MgCl_2$, 100 μM ATP, 0.1 μM calyculin A
2. [γ-^{32}P]ATP

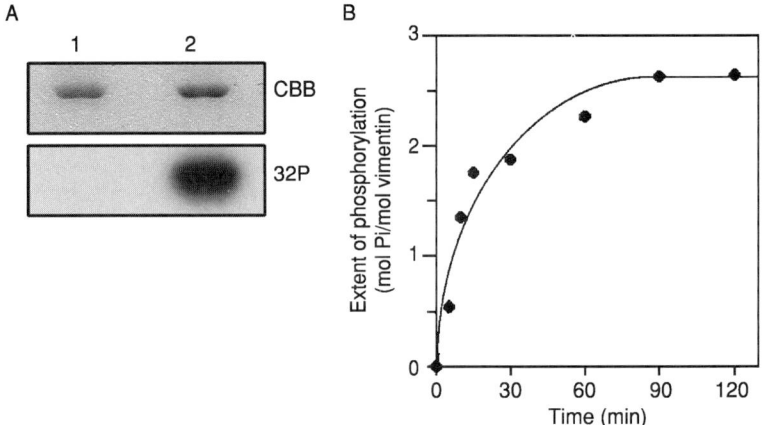

Fig. 2 Phosphorylation of vimentin by Aurora-B. (A) Vimentin was phosphorylated by K/R (kinase dead mutant) (*lane 1*) or WT (*lane 2*) of Aurora-B. After sodium dodecysulfate polyacrylamide gel electrophoresis (SDS-PAGE), the gel was stained with Coomassie Brilliant Blue (*CBB*) then subjected to autoradiography (^{32}P). (B) Time course of vimentin phosphorylation by Aurora-B WT. Modified with permission from Goto *et al.* (2003).

3. IF protein: 100–200 μg/ml

4. Kinase: 5–10 μg/ml

b. Steps

1. Incubate 100–200 μg of IF protein in 1 ml of phosphorylation reaction buffer in the presence of kinase with [γ-^{32}P]ATP at 25 °C for about 60 minutes. Figure 2 shows the phosphorylation of vimentin by Aurora-B *in vitro*.

2. Separation of Free Nonreactive [γ-^{32}P] ATP and Phosphorylated Intermediate Filament Protein

a. Reagents

1. 5% and 30% (w/v) trichloroacetic acid (TCA)

2. Diethyl ether

b. Steps

1. Add 500 μl of 30% ice-cold TCA (final 10% TCA) into 1 ml of the reaction mixture containing radioactive IF protein. Mix immediately and leave the solution on ice for 1–2 hours.

2. Centrifuge the solution at 8000g for 30 minutes at 4 °C. Carefully aspirate the supernatant containing free nonreactive [γ-^{32}P]ATP. The precipitate, which is IF protein precipitated with TCA, could be visible at this stage.

3. Wash the precipitate with 1 ml of 5% ice-cold TCA.

4. Wash the precipitate three times with 1 ml of diethyl ether.

5. Air-dry the precipitate (do not lyophilize). Store at $-20\,^{\circ}$C.

3. Fragmentation of the Phosphorylated Intermediate Filament Protein

a. Reagents

1. Buffer A: 100 mM Tris-HCl, pH 8.0, 8M urea

2. 0.1% trifluoracetic acid (TFA)

3. Lysyl-endopeptidase

4. L-1-tosylamide-2-phenyl-ethyl chloromethyl ketone (TPCK)–treated trypsin

5. 200 mM NH_4HCO_3

6. 100 mM $CaCl_2$

7. Acetonitrile

b. Steps

1. Dissolve the precipitate with 100 μl of buffer A by vortexing for 1–2 hours at room temperature.

2. Add 300 μl of dH_2O to dilute buffer A (total 400 μl, final 25 mM Tris-HCl, pH 8.0, 2 M urea).

3. Incubate with 5 μg of lysyl-endopeptidase for at least 2 hours at $30\,^{\circ}$C.

4. After the incubation, centrifuge at 8000g for 30 minutes and transfer the supernatant to a new tube.

5. Add 1 ml of 0.1% TFA and filtrate the sample through a 0.2-μm filter.

6. Subject the sample to reverse-phase high performance liquid chromatography (HPLC) and collect radioactive fraction(s) (Fig. 3A(a)).

7. Lyophilize the radioactive fraction(s). The lyophilized sample could be stored at $-20\,^{\circ}$C.

8. Resuspend the radioactive fraction with 40 μl of acetonitrile with a vortex mixer for at least 30 minutes.

9. Add 40 μl of 200 mM NH_4HCO_3, 40 μl of 100 mM $CaCl_2$, and 280 μl of dH_2O (total 400 μl, final 10% acetonitrile, 20 mM NH_4HCO_3, 10 mM $CaCl_2$) and suspend gently.

10. Incubate with 1:50 (w/w) TPCK-treated trypsin for 4–8 hours at $37\,^{\circ}$C. Treat with TPCK-treated trypsin for an additional 4–8 hours.

11. After the incubation, centrifuge at 8000g for 30 minutes and transfer the supernatant to a new tube.

12. Add 1 ml of 0.1% TFA and filtrate the sample through a 0.2-μm filter

13. Subject the sample to reverse-phase HPLC and collect radioactive fractions (Fig. 3A(b)).

14. Lyophilize the radioactive fraction(s), and the sample could be stored at $-20\,^{\circ}$C.

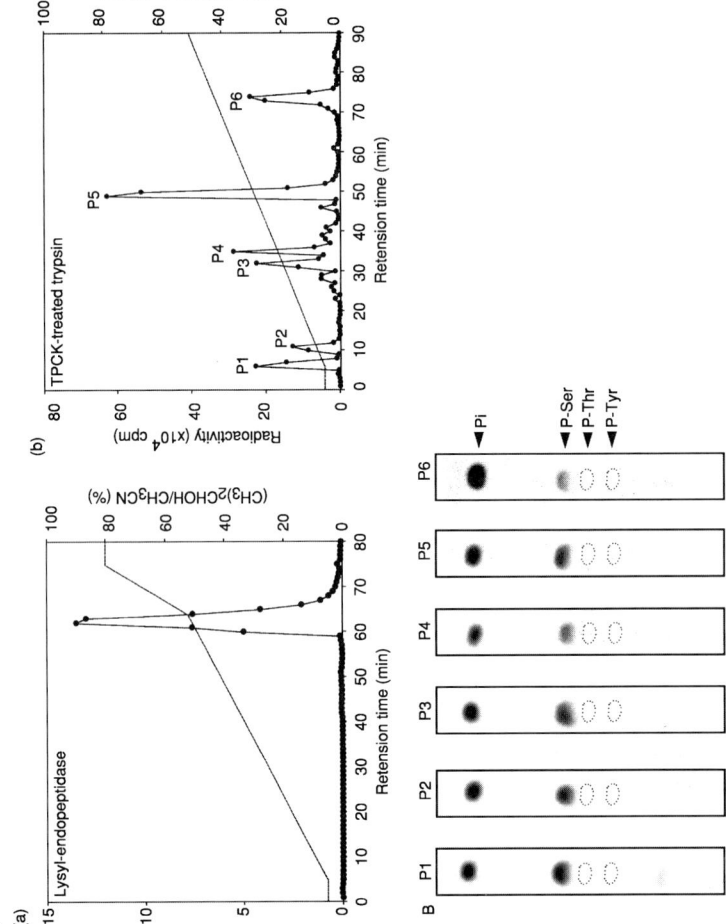

Fig. 3 Identification of Aurora-B phosphorylation sites on vimentin *in vitro*. (A) Vimentin phosphorylated by Aurora-B was treated with lysyl-endopeptidase. Radioactive fraction (amino-terminal head domain of vimentin) was digested with trypsin and fractionated by reverse-phase high-performance liquid chromatography (HPLC) (b). The radioactivity of each fraction (0.8 ml) was measured in a ^{32}P Beckman liquid scintillation counter. (B) Six radioactive peaks (*P1–P6*) were subjected to phosphoamino acid analysis. The positions of phosphoserine (*P-Ser*), phosphothreonine (*P-Thr*), and phosphotyrosine (*P-Tyr*) are indicated. Modified with permission from Goto *et al.* (2003).

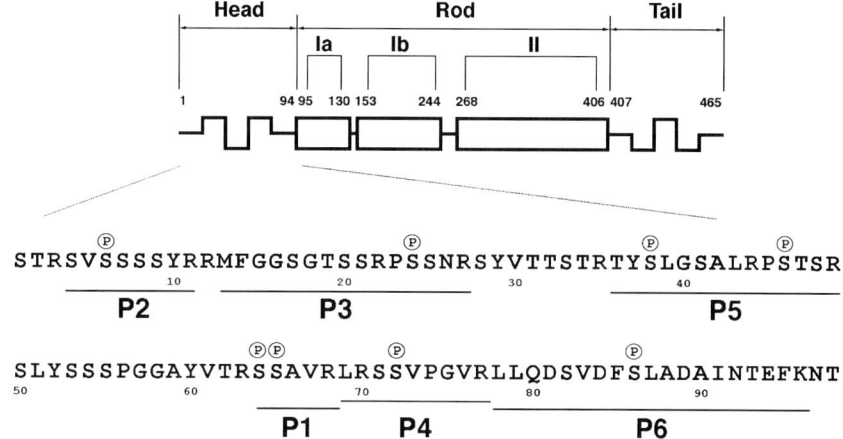

Fig. 4 A map of the vimentin molecule showing phosphorylation sites by Aurora-B. Each determined amino acid sequence of *P1–P6* is also underlined. The phosphorylation sites are indicated by *P* within a circle. Modified with permission from Goto *et al.* (2003).

The phosphorylated amino acid (phospho-Ser, Thr, or Tyr) of each sample is determined by amino acid sequence analysis (Fig. 4) and phosphoamino acid analysis (Fig. 3B).

B. Preparation of Site-Specific and Phosphorylation State-Specific Antibodies for the Digested Phosphopeptides

Next, to monitor the spatiotemporal site-specific phosphorylation of IF proteins, we developed antibodies that recognize the phosphorylation of IF proteins by use of phosphopeptide.

1. Production of Phosphopeptide Antibody

a. Reagents

1. Synthetic phosphopeptide: We introduced a cysteine residue at the N terminal of the synthetic peptide and bound it to the carrier protein, keyhole limpet hemocyanin (KLH), with maleimidobenzoic acid *N*-hydroxysuccinimide ester (MBS). A synthetic peptide design for phosphopeptide antibody against serine 72 of vimentin is shown in Fig. 5.

2. Freund's complete adjuvant and Freund's incomplete adjuvant

3. RIBI adjuvant

4. 0.22-μm Millex-GP filter

PV72 (Phosphopeptide)	CVRLRSSVPGVR
V72 (Peptide)	CVRLRSSVPGVR

Fig. 5 Design of site-specific and phosphorylation state-specific antibody (YG72) for serine 72 of vimentin.

b. Steps

1. Prepare two rabbits 11–12 weeks past birth.

2. Obtain preimmune serum (1 ml) from each rabbit before the primary immunization.

Freund's adjuvant or RIBI adjuvant is used to immunize each rabbit.

Freund's Complete/Incomplete Adjuvant System

3. Immunize rabbit by multiintradermal injections of the peptide (100–200 μg peptide) emulsified with Freund's complete adjuvant.

4. Booster injection is done with the peptide (100–200 μg peptide) emulsified with Freund's incomplete adjuvant 2–4 weeks after the primary immunization.

RIBI Adjuvant System

5. Immunize rabbit with 1 ml of the peptide (100–200 μg peptide) emulsified with RIBI adjuvant (0.05 ml intradermal into six sites, 0.3 ml intramuscular into each hind leg, and 0.1 ml subcutaneous into the neck region).

6. A booster injection of the peptide (100–200 μg peptide)-RIBI adjuvant mixture is given 4 weeks after the primary immunization.

7. After 12–16 days from the first boost, obtain serum samples (about 50 ml) from each rabbit. Incubate samples at 37 °C for 1 hour and stir at 4 °C overnight.

8. Centrifuge the samples at 1000g for 30 minutes at 4 °C and collect the supernatants.

9. Incubate at 56 °C for 30 minutes and filtrate with a 0.22-μm filter. Store in aliquots at −80 °C.

10. After purification of the antibody from the serum (Procedure B. 2 and 3), the specificity of antibody is confirmed by ELISA (Procedure C).

11. The second booster injection of the peptide (100–200 μg)–adjuvant mixture is given, and serum is obtained after another 2 weeks (4 weeks after first boost).

2. Purification of Phosphopeptide Antibody: Preparation of an Affinity Column

a. Reagents

1. Synthetic phosphopeptide
2. Synthetic unphosphopeptide
3. Coupling buffer: 50 mM $NaHCO_3$-Na_2CO_3, pH 8.0–10.0
4. Formyl-cellurofine
5. 70 mg/ml sodium borohydride ($NaBH_4$)
6. Blocking buffer: 50 mM Tris-HCl, pH 8.0, 0.1 M ethanolamine
7. Wash buffer: 20 mM Tris-HCl, pH 7.5, 1 M NaCl, 1% (v/v) Triton X-100
8. TBS: 20 mM Tris-HCl, pH 7.5, 0.15 M NaCl
9. Stock buffer: 20 mM Tris-HCl, pH 7.5, 0.15 M NaCl, 20 mM β-glycerophosphate, 0.2% (w/v) NaN_3

b. Steps

1. Add 1 mg of phosphorylated or unphosphorylated peptide to 1 ml of coupling buffer. Check pH of the solution, which must be 8.0–10.0.

2. Add 1 ml of the formyl-cellurofine equilibrated with coupling buffer.

3. Mix gently for 2 hours at room temperature or overnight at 4 °C on a rotor. Do not use a magnetic stirrer.

4. Add 1/10 gel volume (100 μl/ml of the gel) of 70 mg/ml $NaBH_4$. Rotate the mixture for 2–8 hours at room temperature or overnight at 4 °C on a rotor.

5. Transfer the gel into a suitable column and add 10 gel volumes (10 ml) of blocking buffer. We usually use Muromac column (Muromachi Kagaku Kogyo Kaisha, Japan) for purification of antibody.

6. Rotate the mixture for 2–5 hours at room temperature or overnight at 4 °C on a rotor.

7. Wash away blocking buffer from the column.

8. Wash and equilibrate the column with TBS. The peptide-coupled gel could be stored in stock buffer at 4 °C for several months.

3. Purification of Phosphopeptide Antibody: Purification of Antibody

a. Reagents

1. TBS: 20 mM Tris-HCl, pH 7.5, 0.15 M NaCl
2. Wash buffer: 20 mM Tris-HCl, pH 7.5, 1 M NaCl, 1% (v/v) Triton X-100
3. Elution buffer: 0.1 M glycine-HCl, pH 2.5, 10% (v/v) ethylene glycol
4. Neutralizing buffer: 1 M Tris-HCl, pH 8.5
5. 20 mg/ml BSA in TBS
6. Stock buffer: 20 mM Tris-HCl, pH 7.5, 0.15 M NaCl, 20 mM β-glycerophosphate, 0.2% (w/v) NaN_3

b. Steps

1. Mix 5 ml of the serum and 5 ml of TBS with 1 ml of the phosphorylated peptide-coupled gel in a suitable column (Muromac column). Each step described below should be done in a cold room.

2. Rotate the mixture end-over-end overnight at 4 °C. Do not use a magnetic stirrer.

3. Wash away the mixture of serum and TBS from the column.

4. Wash the column with 10 gel volumes (10 ml) of TBS.

5. Wash the column with 10 gel volumes (10 ml) of wash buffer.

6. Wash the column with 10 gel volumes (10 ml) of TBS.

7. Elute the antibody with elution buffer using a stepwise elution (500 µl/ fraction, about 5–6 times).

8. Collect eluates in each tube. The collection tubes should contain neutralizing buffer (37–39 µl/500 µl of eluate). Mix gently and constantly the eluates and neutralizing buffer.

9. Check the IgG concentration by reading the absorbance at 280 nm (1 mg of IgG/ml, ~1.35) and the purity by sodium dodecylsulfate-polyacrylamide gel electrophoresis (SDS-PAGE). The predominant bands will be IgG light chains (20–25 kDa) and heavy chains (50–55 kDa) under reducing conditions during sample preparation for SDS-PAGE (Fig. 6).

10. Collect two antibody-rich fractions (total 1 ml) and add 50 µl of 20 mg/ml BSA/TBS (final 1 mg/ml BSA).

11. Dialyze for 3–6 hours with two changes of 200 ml TBS at 4 °C.

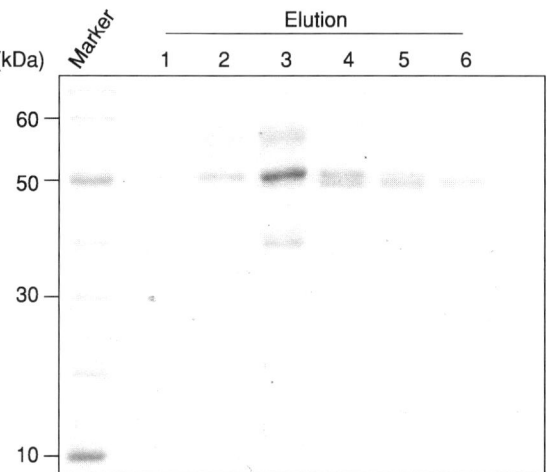

Fig. 6 Purification of site-specific and phosphorylation state-specific antibody (YG72). After SDS-PAGE of each antibody fraction, the gel was stained with Coomassie Brilliant Blue.

12. Mix the antibody and the unphosphorylated peptides–coupled gel in the column.

13. Rotate the mixture gently overnight at 4 °C. The rotation should be sufficient to keep the gel in suspension.

14. Collect the flow-through. Wash away the remaining antibody from the column with 500 μl of TBS and collect the flow-though. In this step, unphosphopeptide antibodies could be removed.

15. Wash columns with excess elution buffer. Wash with TBS and equilibrate with stock buffer. Store at 4 °C.

C. Specificity Analysis of Site-Specific and Phosphorylation State-Specific Antibody by ELISA

We checked specificity of the purified site-specific and phosphorylation state-specific antibody using an ELISA. In Fig. 7A, the specificity of site-specific and phosphorylation state-specific antibody (YG72) against phosphopeptide and non-phosphopeptide was examined with ELISA. Figure 7B also shows the specificity of YG72 determined with an immunoblot analysis.

1. Preparation of Plates for ELISA

a. Reagents

1. Buffer A: 0.1 M $Na_2HPO_4 \cdot NaH_2PO_4$, pH 7.4
2. Blocking buffer: 10 mM $Na_2HPO_4 \cdot NaH_2PO_4$, pH 8.0, 5% (w/v) BSA, 5% (w/v) sucrose, 0.1% (w/v) NaN_3
3. Phosphate buffered saline (PBS)

b. Steps

1. Dilute phosphopeptide or unphosphopeptide to 1 μg/ml with buffer A. The pH of the solution should be 7.0–7.5.
2. Add 60–70 μl of the solution into each well of a 96-well microtiter plate.
3. Incubate for 2 hours at room temperature or overnight at 4 °C.
4. Remove the solution from each well and wash each well three times with 100 μl of PBS.
5. Add 300–350 μl of blocking buffer into each well.
6. Incubate for 2–4 hours at 37 °C or overnight at 4 °C.
7. Remove blocking buffer and store at 4 °C.

2. Examination of Specificity of Antibody Using ELISA

a. Reagents

1. Buffer A: 1% (w/v) BSA, 1% (w/v) sucrose, 0.1% (w/v) NaN_3 in PBS

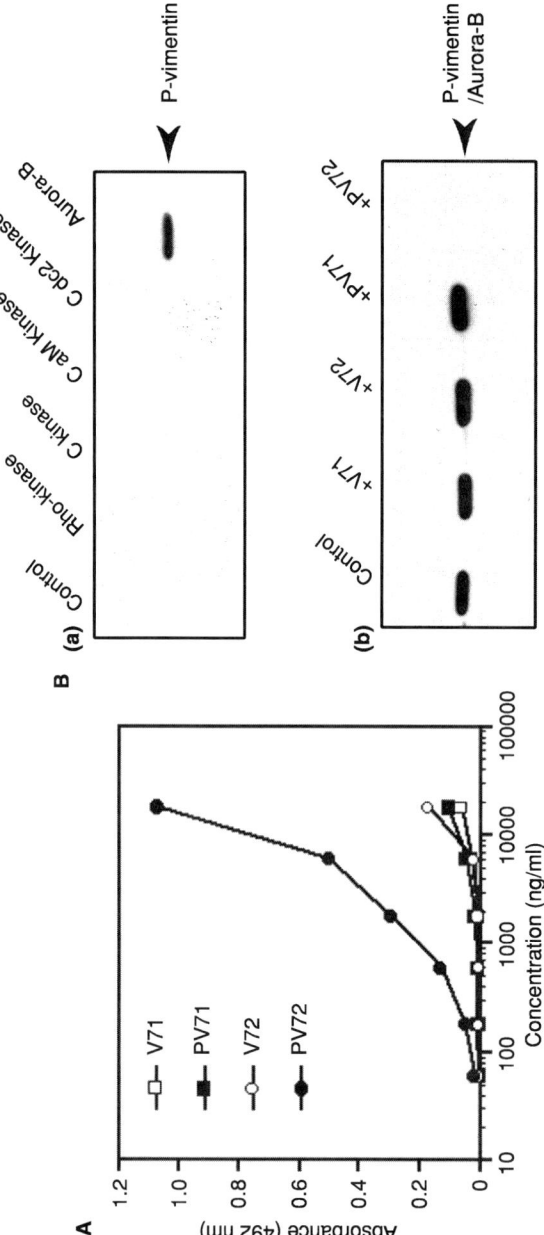

Fig. 7 Specificity analysis of site-specific and phosphorylation state-specific antibody (YG72). (A) Specificity of YG72 for synthetic peptides (*V71*: *Cys-Ala-Val-Arg-Leu-Arg-Ser*[71]*-Ser-Val-Pro-Gly-Val*, *PV71*: *Cys-Ala-Val-Arg-Leu-Arg-phosphoSer*[71]*-Ser-Val-Pro-Gly-Val*, *V72*: *Cys-Val-Arg-Leu-Arg-Ser-Ser*[72]*-Val-Pro-Gly-Val-Arg* or *PV72*: *Cys-Val-Arg-Leu-Arg-Ser-phosphoSer*[72]*-Val-Pro-Gly-Val-Arg*) by enzyme-linked immunosorbent assay. Modified with permission from Yasui *et al.* (2001). (B) Vimentin was unphosphorylated (*control*) or phosphorylated by Rho-kinase, C kinase, CaM kinase II, Cdc2 kinase or Aurora-B, respectively. After SDS-PAGE (100 ng in each lane), samples were transferred onto a poly (vinylidene difluoride) membrane. The membrane was immunoblotted with the antibody YG72. (C) Specificity of YG72 determined by a competition assay. Vimentin phosphorylated by Aurora-B was immunoblotted with YG72 after preincubation with synthetic peptides (50 μg/ml of *V71*, *PV71*, *V72*, or *PV72*) or TBS-T (*control*).

2. Site-specific and phosphorylation state-specific antibody (first antibody, obtained in Procedure B)

3. Buffer B: 10 mM $Na_2HPO_4 \cdot NaH_2PO_4$, pH 8.0, 0.1 M NaCl, 1% (w/v) BSA, 0.1% (w/v) p-hydroxy phenylacetic acid, 0.025% (w/v) thimerosal

4. Reaction buffer: 0.4 mg/ml o-phenylenediamin, 5% (v/v) methanol (MeOH), 0.01% H_2O_2, freshly prepared

5. HRP–conjugated anti-rabbit Ig antibody

6. 2N sulfonic acid (H_2SO_4)

b. Steps

1. Prepare dilutions of first antibody from 1/10–1/10,000 with buffer A.

2. Add 60 μl of each diluted antibody into wells and incubate for 45–60 minutes at 37 °C.

3. Remove the antibody from each well and wash five times with 100 μl of PBS.

4. Add 100 μl of 1/1000 dilutions of HRP-conjugated anti-rabbit Ig antibody (secondary antibody) in buffer B.

5. Incubate for 45–60 minutes at 37 °C.

6. Remove the solution from each well and wash five times with 150–200 μl of PBS.

7. Add 100 μl of reaction buffer into each well and leave for 10–30 minutes at room temperature.

8. Stop the color development by adding 100 μl of 2N H_2SO_4 and measure absorbance at 492 nm with microtiter plate reader.

D. Immunocytochemistry Using Site-Specific and Phosphorylation State-Specific Antibody

After confirmation of the specificity of antibody with ELISA and/or immunoblot analysis, the spatiotemporal distribution of the site-specific phosphorylation in cells was analyzed with a site-specific and phosphorylation state-specific antibody. Figure 8 shows immunofluorescence micrographs of U251 cells stained with site-specific and phosphorylation state-specific antibody (YG72) that recognizes vimentin phosphorylated at serine 72.

a. Reagents

1. 3.7% (v/v) formaldehyde in ice-cold PBS

2. Methanol (MeOH) stored at −20 °C

3. Buffer A: 1% (w/v) BSA, 1% (w/v) sucrose, 0.1% (w/v) NaN_3 in PBS

4. Site-specific and phosphorylation state-specific antibody (first antibody, obtained at Procedure B)

5. Alexa Fluor 488 anti-rabbit IgG

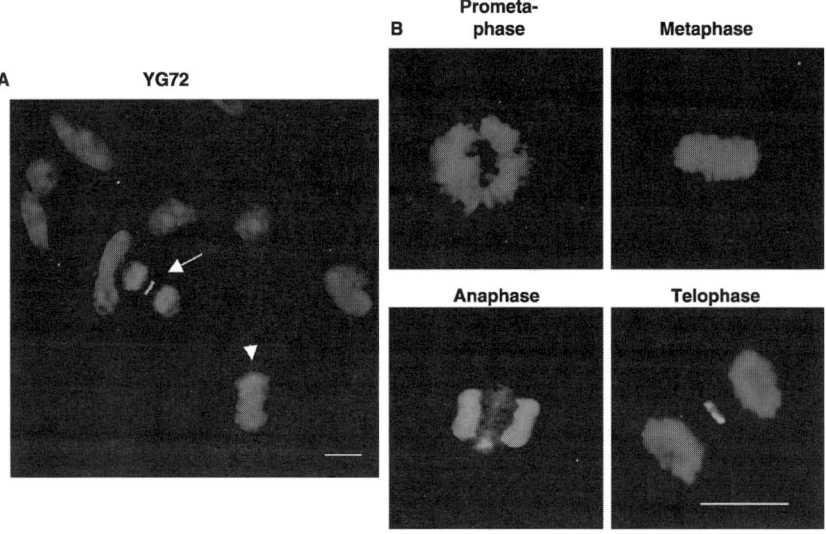

Fig. 8 Immunocytochemistry with YG72. (A) and (B) Fluorescent photomicrographs of U251 glioma cells stained with YG72 (green). Chromosomes were stained with propidium iodide (red). An arrow and an arrowhead in (A) show a telophase and metaphase cell, respectively. Bars represent 10 μm. Modified with permission from Yasui *et al.* (2001). (See Color Insert.)

    6. 0.5 μg/ml of 4′,6-diamidine-2′-phenylindole-dihydrochloride (DAPI)

    7. 0.5 μg/ml of propidium iodide (PI)

b. Steps

    1. Place the sterilized coverslip or slide in a suitable tissue culture dish.

    2. Plate the cell suspension into the dish and culture for at least 24 hours at 37 °C.

    3. Fix the cells with 3.7% ice-cold formaldehyde for 10 minutes.

    4. Wash the coverslip gently three times with ice-cold PBS for 10 minutes.

    5. Treat the fixed cells with MeOH, which is stored at −20 °C, for 10 minutes.

    6. Wash the coverslip gently in three changes of ice-cold PBS for 10 minutes.

    7. Place each coverslip in humidified chamber.

    8. Dilute the first antibody with buffer A.

    9. Apply the first antibody on the coverslip and incubate for 2 hours at 37 °C or overnight at 4 °C.

    10. Wash the coverslip three times with PBS for 10 minutes.

    11. Dilute the Alexa Fluor 488 anti-rabbit IgG with PBS and apply gently on the coverslip.

    12. Incubate for 1 hour at 37 °C in the dark.

13. Wash the coverslip in three changes of ice-cold PBS for 10 minutes.

14. If nuclear stain is necessary, incubate with 0.5 μg/ml DAPI or 0.5 μg/ml PI for 10 minutes at room temperature.

15. Wash the coverslip in two changes of PBS for 10 minutes.

16. Mount the coverslip and analyze using fluorescent microscopy.

IV. Pearls and Pitfalls

The rate of IF phosphorylation will be affected by the concentration of IF protein and Mg^{2+}, pH, ionic strength, and temperature. We recommend the following conditions for phosphorylation assay of IF protein:

1. Concentration of IF protein: 100–200 μg/ml
2. Concentration of Mg^{2+}: <3 mM
3. pH: 7.0–7.5
4. Ionic strength (concentration of NaCl): 0–30 mM
5. Temperature: 25–30 °C

Phosphorylation sites of vimentin, GFAP, desmin, and keratin 8 are mainly located in the amino-terminal head domain. The target sequence for protein digestion by lysyl-endopeptidase is Lys-X. Because the head domain of vimentin and GFAP contains no lysine residues, an intact head domain could be obtained with lysyl-endopeptidase. However, the head domains of desmin and keratin 8 are digested with lysyl-endopeptidase, because they contain some lysine residues. In this case, we use *Staphylococcus aureus* V8 protease (Endoproteinase Glu-C) instead of lysyl-endopeptidase. The target sequence for protein digestion by *S. aureus* V8 protease is Glu-X. Because there are glutamic acid residues in the terminal position of their head domains, *S. aureus* V8 protease is useful for the isolation of their head domains and for identification of their phosphorylation sites.

Two or more rabbits should be used for immunization. We usually use Freund's adjuvant and RIBI adjuvant with the same antigen for immunization of each rabbit. From previous experimental data, the antibodies obtained by Freund's adjuvant system show a high titer by ELISA. On the other hand, the titer of the antibodies obtained with the RIBI adjuvant system tends to be lower than those obtained with the Freund's adjuvant system but shows specific staining by immunocytochemistry. In many cases, the above-mentioned results might be applied to antibodies.

There are various methods used to purify antibodies. Polyclonal antibodies could be purified conventionally using protein A beads, protein G beads, and antigen affinity beads (Formyl-cellurofine, FMP-activated cellurofine, and CNBr-activated

Sepharose 4B [Pharmacia Biotech]). In the case of the above antigen affinity beads, primary amines of ligands are immobilized on beads. Tris or other buffer salts containing amino groups should not be added to coupling buffer, because they will prevent the binding of ligands to the beads. Coupling buffer should be bicarbonate or borate buffer but not Tris-HCl buffer or glycine-HCl buffer.

V. Discussion and Concluding Remarks

We have clarified that the phosphorylation of IF proteins plays an important role in the dynamic IF reorganization. Development of site-specific and phosphorylation state-specific antibodies of IF proteins enables one to monitor the spatio-temporal phosphorylation that occurs during various processes such as cell signaling and cell cycle. These site-specific and phosphorylation state-specific antibodies are available tools to analyze site-specific IF phosphorylation *in vivo*. The antibodies we have shown here are also expected to have a wide application for analysis of phosphorylation of various proteins during cellular events.

Acknowledgments

This work was supported in part by grants-in-aid for scientific research and for cancer research from Ministry of Education, Science, Technology, Sports and Culture of Japan; and by a grant-in-aid for the 2nd-Term Comprehensive 10-year Strategy for Cancer Control from the Ministry of Health and Welfare, Japan. We are grateful to M. Ohara (Fukuoka, Japan) for language assistance and critical comments on the manuscript. A. Kawajiri is a research fellow of the Japan Society for the Promotion of Science.

References

Goto, H., Kosako, H., Tanabe, K., Yanagida, M., Sakurai, M., Amano, M., Kaibuchi, K., and Inagaki, M. (1998). Phosphorylation of vimentin by Rho-associated kinase at a unique amino-terminal site that is specifically phosphorylated during cytokinesis. *J. Biol. Chem.* **273**, 11728–11736.

Goto, H., Tomono, Y., Ajiro, K., Kosako, H., Fujita, M., Sakurai, M., Okawa, K., Iwamatsu, A., Okigaki, T., Takahashi, T., and Inagaki, M. (1999). Identification of a novel phosphorylation site on histone H3 coupled with mitotic chromosome condensation. *J. Biol. Chem.* **274**, 25543–25549.

Goto, H., Yasui, Y., Kawajiri, A., Nigg, E. A., Terada, Y., Tatsuka, M., Nagata, K., and Inagaki, M. (2003). Aurora-B regulates the cleavage furrow-specific vimentin phosphorylation in the cytokinetic process. *J. Biol. Chem.* **278**, 8526–8530.

Inada, H., Togashi, H., Nakamura, Y., Kaibuchi, K., Nagata, K., and Inagaki, M. (1999). Balance between activities of Rho-kinase and Type 1 protein phosphatase modulates turnover of phosphorylation and dynamics of desmin/vimentin filaments. *J. Biol. Chem.* **274**, 34932–34939.

Inagaki, M., Nishi, Y., Nishizawa, K., Matsuyama, M., and Sato, C. (1987). Site-specific phosphorylation induces disassembly of vimentin filaments *in vitro. Nature* **328**, 649–652.

Inagaki, M., Matsuoka, Y., Tsujimura, K., Ando, S., Tokui, T., Takahashi, T., and Inagaki, N. (1996). Dynamic property of intermediate filaments: Regulation by phosphorylation. *BioEssays* **18**, 481–487.

Inagaki, M., Inagaki, N., Takahashi, T., and Takai, Y. (1997). Phosphorylation-dependent control of structures of intermediate filaments; A novel approach using site- and phosphorylation state-specific antibodies. *J. Biochem. Rev.* **121,** 407–414.

Kawajiri, A., Yasui, Y., Goto, H., Tatsuka, M., Takahashi, M., Nagata, K., and Inagaki, M. (2003). Functional significance of the specific sites phosphorylated in desmin at cleavage furrow: Aurora-B may phosphorylate and regulate type III intermediate filaments during cytokinesis coordinatedly with Rho-kinase. *Mol. Biol. Cell* **14,** 1489–1500.

Matsuoka, Y., Nishizawa, K., Yano, T., Shibata, M., Ando, S., Takahashi, T., and Inagaki, M. (1992). Two different protein kinases act on a different time schedule as glial filament kinase during mitosis. *EMBO J.* **11,** 2895–2902.

Nagata, K., Izawa, I., and Inagaki, M. (2001). A decade of site- and phosphorylation state-specific antibodies: Recent advances in studies of spatiotemporal protein phosphorylation. *Genes Cells* **6,** 653–664.

Nishizawa, K., Yano, T., Shibata, M., Takahashi, T., and Inagaki, M. (1990). Distribution of phospho-glial fibrillary acidic protein in astro glial cells. *Cell Struct. Funct.* **15,** 497.

Nishizawa, K., Yano, T., Shibata, M., Ando, S., Saga, S., Takahashi, T., and Inagaki, M. (1991). Specific localization of phospho-intermediate filament protein in the constricted area of dividing cells. *J. Biol. Chem.* **266,** 3074–3079.

Ogawara, M., Inagaki, N., Tsujimura, K., Takai, Y., Sekimata, M., Ha, M. H., Imajoh-Ohmi, S., Hirai, S., Ohno, S., Sugiura, H., Yamauchi, T., and Inagaki, M. (1995). Differential targeting of protein kinase C and CaM kinase II signalings to vimentin. *J. Cell Biol.* **131,** 1055–1066.

Sekimata, M., Tsujimura, K., Tanaka, J., Takeuchi, Y., Inagaki, N., and Inagaki, M. (1996). Detection of protein kinase activity specifically activated at metaphase-anaphase transition. *J. Cell Biol.* **132,** 635–641.

Takai, Y., Ogawara, M., Tomono, Y., Moritoh, C., Imajoh-Ohmi, S., Tsutsumi, O., Taketani, Y., and Inagaki, M. (1996). Mitosis-specific phosphorylation of vimentin by protein kinase C coupled with reorganization of intracellular membranes. *J. Cell Biol.* **133,** 141–149.

Tsujimura, K., Ogawara, M., Takeuchi, Y., Imajoh-Ohmi, S., Ha, M. H., and Inagaki, M. (1994). Visualization and function of vimentin phosphorylation by cdc2 kinase during mitosis. *J. Biol. Chem.* **269,** 31097–31106.

Yano, T., Taura, C., Hirono, Y., Shibata, M., Nishizawa, K., Ando, S., Takahashi, T., and Inagaki, M. (1990). Specific antibodies to phospho-glial fibrillary acidic protein: Production and characterization. *Cell Struct. Funct.* **15,** 497.

Yano, T., Taura, C., Shibata, M., Hirono, Y., Ando, S., Kusubata, M., Takahashi, T., and Inagaki, M. (1991). A monoclonal antibody to the phosphorylated form of glial fibrillary acidic protein; Application to a nonradioactive method for measuring of protein kinase activities. *Biochem. Biophys. Res. Commun.* **175,** 1144–1151.

Yasui, Y., Goto, H., Matsui, S., Manser, E., Lim, L., Nagata, K., and Inagaki, M. (2001). Protein kinases required for segregation of vimentin filaments in mitotic process. *Oncogene* **20,** 2868–2876.

CHAPTER 14

Approaches to Study Posttranslational Regulation of Intermediate Filament Proteins

Vitaly Kochin,[*,†] **Hanna–Mari Pallari,**[*,†] **Harish Pant,**[‡] **and John E. Eriksson**[*,†]

[*]Turku Centre for Biotechnology
University of Turku and Åbo Akademi University
FIN-20521 Turku, Finland

[†]Department of Biology
University of Turku
FIN-20014 Turku, Finland

[‡]Laboratory of Neurochemistry
NIH, NINDS
Bethesda, Maryland 20892

I. Introduction

Intermediate filament (IF) proteins form highly dynamic structures that undergo continuous remodeling, subunit exchange, and shifts in both sequestration and organization (Coulombe and Omary, 2002; Helfand *et al.*, 2003; Helfand *et al.*, 2004). These processes are to a large extent regulated by posttranslational modifications, including phosphorylation, ubiquitylation, glycosylation, and various types of proteolytic cleavage. Of all these modifications, phosphorylation has been established as a key regulator of IF functions, and a large number of protein kinases (PKs) and phosphoprotein phosphatases (PPs) have been implicated in the regulation of overall IF organization (Eriksson *et al.*, 2004; Ku *et al.*, 1996). The current review will focus on how phosphorylation of IF proteins can be approached. Although we will use primarily phosphorylation as an example of how posttranslational IF modifications can be approached, some of the current mass spectrometry (MS)-based techniques could equally well be adopted for studies on other types of posttranslational modifications.

Studies on phosphorylation-based regulation of a given IF protein will require a broad spectrum of techniques, ranging from cell and molecular biology and traditional protein chemistry to advanced MS. Nowadays, peptide and protein identification by MS-based techniques in combination with bioinformatics is performed routinely. However, despite the advances in MS-based phosphoproteomics, the localization and identification of all phosphopeptides in a protein still remain a challenge and, therefore, a combination of different methods, both old and new, will have to be used to maximize the chances of obtaining all phosphorylation sites in a given protein. The aim of this chapter is to give a brief introduction to scientists not familiar with the required method but who would be interested studying whether a given IF is being modified by phosphorylation. A background to all techniques is included, but the section on MS is more thorough, because there are not that many basic-level cell biology–oriented treatises on MS-based phosphoproteomics available. Examples from studies on IF proteins will be used to illustrate the characterization process. We have attempted to include sufficient information so that initial studies can be performed by use of both traditional methods in phosphoproteomics and state-of-the-art MS-based methods. If more extensive studies will be performed, there are a number of excellent in-depth reviews on both traditional methods (Hardie, 1993; Hunter and Sefton, 1991), and MS-based methods (Bennett *et al.*, 2002; Mann *et al.*, 2001, 2002; McLachlin and Chait, 2001).

A. Modification of Kinase and Phosphatase Activities

1. The Use of Pharmacological Agents

One of the initial approaches to test whether a given cytoskeletal component could be regulated by phosphorylation is usually modification of the PK/PP activities suspected to be involved. The complete understanding of the functions of a given phosphorylation-based regulatory function requires the use of a number of different approaches. Nevertheless, by using pharmacological modifiers of PK/PP activities, it is relatively easy and inexpensive to obtain valuable initial information concerning the possible cytoskeletal functions of a number of Ser/Thr and Tyr-directed PKs and PPs. In a previous review, we have listed a number of compounds that could be useful for this approach (Eriksson *et al.*, 1998). Furthermore, many bioscience companies are actively developing new modifier compounds and derivatives of established inhibitors to improve specificity, solubility, stability, and other important properties of the compounds. Hence, current catalogues and/or internet-based product lists of many bioscience companies may often provide useful information on new products and helpful pieces of advice on how to use the listed compounds.

2. Inhibitors of Phosphoprotein Phosphatases

It may often be beneficial to initially use a PP inhibitor rather than a PK activator when the possible roles of phosphorylation are being assessed. There is increasing evidence that IFs are substrates for high constitutive PP activities (Eriksson *et al.*, 1992, 2004; Ku *et al.*, 1996; Toivola *et al.*, 1997). Hence, by preventing dephosphorylation, possible constitutive PK and PP activities can be revealed. Over the past few years, a number of highly specific inhibitors have been established acting directly on Ser/Thr-specific type-1 (PP1) and type-2A (PP2A) PPs, both of which are of major importance in regulating a number of crucial cellular functions and constitute the bulk of the PP activities in any given mammalian cell. These PPs are likely candidates when possible IF-directed PPs are to be considered. Some special consideration should be observed when using inhibitors of PP1 and PP2A (for details, see Eriksson *et al.* [1998]). Depending on the cellular model system and the conditions used, the effect obtained by PP inhibitors may not always stem from direct regulation of a protein by PP1 or PP2A but may also be due to activation of signaling cascades or complexes. Furthermore, when considering these two phosphatase classes, the dose–response curve of their activity is usually extremely steep. Hence, the dose–response of an effect should be titrated carefully by use of multiple inhibitor concentrations. By experimentally using the IC_{50}-ranges of specific PP1 and PP2A inhibitors, it may be possible to determine the involvement of either PP in a given process—okadaic acid (more efficient toward PP2A), tautomycin (more efficient toward PP1), and calyculin-A (equally efficient for both; see Table I). Useful and detailed information on the use of Ser/Thr-specific PP inhibitors is provided in the following references: Cohen *et al.* (1990); Hardie (1993); MacKintosh (1993); and MacKintosh and MacKintosh (1994).

Tyrosine-directed phosphorylation is of special interest with respect to the regulation of matrix-associated proteins, such as integrins or syndecans, and in mediating the signaling going in and out through these attachment complexes (Humphries, 1996). However, recent studies have indicated that IF proteins also may be tyrosine phosphorylated (Angelastro *et al.*, 1998; Feng *et al.*, 1999; Valgeirsdottir *et al.*, 1998). The most useful and specific inhibitor of Tyr PPs is

Table I
Examples of Useful Inhibitors of Protein Phosphatases[†]

PP1 and PP2A

Compound	Structure	Solubility	Potency (IC_{50}) PP1	Potency (IC_{50}) PP2A	Supplier	Reference
Inhibitor-2[1]	Protein	Physiologic buffer	1 nM	—	a	Cohen, 1989
Okadaic acid	Polyether carbocylic acid	Ethanol 10% DMSO	>100 nM	<10 nM	b–h	Fujiki and Suganuma, 1993; Hardie et al., 1991; Ishihara et al., 1989; MacKintosh and MacKintosh, 1994; Matsushima et al., 1990
Tautomycin	Polyketide	DMSO, ethanol	1 nM	10 nM	b, g	Fujiki and Suganuma, 1993; Honkanen et al., 1994; MacKintosh and MacKintosh, 1994
Calyculin A	Phosphorylated polyketide	DMSO, ethanol	1 nM	<1 nM	b, c, e, f, g, h	Fujiki and Suganuma, 1993; Hardie et al., 1991; Ishihara et al., 1989; MacKintosh and MacKintosh, 1994; Matsushima et al., 1990
Cantharidin	Terpenoid	DMSO, ethanol	500 nM	40 nM	b, e, g	Fujiki and Suganuma, 1993; Li and Casida, 1992; MacKintosh and MacKintosh, 1994
Microcystins[2]	Cyclic peptides	DMSO, ethanol, water	1 nM	<1 nM	b, c, e, g	Eriksson et al., 1990; Fujiki and Suganuma, 1993; Hardie et al., 1991; Ishihara et al., 1989; MacKintosh and MacKintosh, 1994; Matsushima et al., 1990

PP2B (calcineurin)

Compound	Comments	Solubility	Potency (IC_{50})	Supplier	Reference
Cyclosporin A	Cyclic undecapeptide	Ethanol	<1 nM	e, g	Liu et al., 1991; MacKintosh and MacKintosh, 1994

[†]Notice that the half-maximal inhibition concentrations are not exact values but rather intended to indicate the approximative specifities against the respective phosphoprotein phosphatase. Useful concentration ranges are likely to be around these values and depend largely on the cell system in question. The supplier list is not comprehensive, but is intended to provide some examples of major suppliers.
[1]Inhibitor 2 is not cell permeable, but intended to be used by transfection or in cell extracts.
[2]Microcystins will enter only into hepatocytes.

a, New England Biolabs, Inc., Beverly, MA; b, Calbiochem, La Jolla, CA; c, Alexis Corporation, San Diego, CA; d, Roche Diagnostics Corporation, Indianapolis, IN; e, Sigma Chemical Company, St. Louis, MO; f, LC Laboratories, Woburn, MA; g, Biomol, Plymouth Meeting, PA; h, Biosource, Camarillo, CA.

vanadate. When using vanadate, the special chemistry of vanadate under physiological conditions should be considered, because vanadate solutions are at physiological pH, and in the presence of reducing agents, vanadate is prone to turn into biologically inactive reduced forms or polymers. An exhaustive presentation on this subject has been published (Gordon, 1991).

3. Inhibitors and Activators of Protein Kinases

A vast number of PKs have been established as being involved in the regulation of IFs (Ku *et al.*, 1996). In Tables II and III, we list examples of a number of useful inhibitor/activator compounds to be used for modulation of kinase activities in cell cultures. Further information on these compounds and protocols on their uses can be found in references listed in the previous review. However, the number of useful kinase-modifying compounds is growing, and, therefore, the list referred to is by no means comprehensive but is meant to illustrate some of useful pharmacological approaches to kinase activation/inhibition. In many cases, whole classes of derivatives of the mentioned compounds have evolved, and the reader is referred to methodological reviews on specific kinase families and/or compounds if more detailed information is required.

4. Transfection of Modified Protein Kinases and Phosphoprotein Phosphatases

As a number of key elements of various signaling cascades or complexes have been cloned, there is an increasing number of PK or PP constructs available with mutations in the regulatory or catalytic domains, enabling overexpression of constitutively active or dominant negative alleles. The use of this approach is an especially good way to explain the functions of a given PK. Many of the available constructs have been tagged with sequences encoding for segments of marker proteins, for example, hemagglutinin, myc, or green fluorescent protein (GFP), enabling identification of transfected cells or levels of exogenously expressed protein by immunofluorescence techniques or Western blotting, respectively. It is also possible to enrich transfected cells by fluorescence-activated cell sorting of cells cotransfected with GFP or by using bidirectional expression vectors that yield both GFP and the protein of interest. It should be noticed that the choice of method for transfection and the specific conditions will vary significantly depending on the cellular model system. For these aspects, the reader is referred to suitable methodological textbooks in molecular biology (for example, Sambrook and Russell [2001]).

5. Cell Cycle Analysis and Synchronization of Cell Cultures

When working with PK or PP whose activities are specific for a certain phase of the cell cycle, it is of interest to obtain cell populations enriched at specific stages of the cell cycle. This can be achieved by use of different drugs that synchronize a cell population at a specific stage of the cell cycle. Nocodazole, which inhibits polymerization of microtubuli, and aphidicolin, a DNA-polymerase inhibitor, are

Table II
Examples of Selective Protein Kinase Inhibitors

Kinase	Compound	Solubility	Potency (IC$_{50}$ μM)	Supplier	Reference
				Selective PK inhibitors	
CaMKII:	KN-62	DMSO	1	a–d, f, g	Hidaka et al., 1991
	KN-93	Methanol, DMSO	0.5	a, b, c, d, g	Sumi et al., 1991
Cdks	Olomoucine[4]	DMSO	1–20	a, b, f	Alessi et al., 1998
	Purvalanols[4]	DMSO	1–20	a, b	Villerbu et al., 2002
	Roscovitine[4]	DMSO	1–20	a, b, f	Alessi et al., 1998; Meijer and Raymond, 2003; Sahlgren et al., 2003
CKI and CKII	CKI-7	DMSO	10/CKI, 100/CKII	d	Hidaka and Kobayashi, 1993; Tamaoki, 1991
MLCK	ML-7	DMSO, 50% ethanol	0.3	a, b, d, g	Saitoh et al., 1987
	ML-9	50% ethanol	4	a, b, d, g	Hidaka and Kobayashi, 1993; Saitoh et al., 1987
MAPK/ERK	PD 98059[1]	DMSO	2–50	a, b, f, g, h	Alessi et al., 1995; Dudley et al., 1995
p38 PK	SB 203580[2]	DMSO	0.6–5	a, b, f, g, h	Badger et al., 1996
JNK	SP600125	DMSO	<0.1	a, b, g, h	Bogoyevitch et al., 2004
p70 S6K	Rapamycin[3]	DMSO, methanol	<0.0002	a, b, f, h	Dumont et al., 1990
PI 3-K	Wortmannin	DMSO ethanol	<0.1	a, b, h	Ui et al., 1995
	LY 284002	DMSO ethanol	<0.01	a, b, f, h	Djordjevic and Driscoll, 2002
PKA	KT5720	DMSO	0.056	a, b, g	Kase et al., 1987
	(Rp)-8-Br-CAMPS	Water	100	e	Gjertsen et al., 1995
	(Rp)-8-Cl-CAMPS	Water	100	e	Gjertsen et al., 1995
PKC	Bisindolylmaleimides	DMSO methanol, water	0.01–0.1	a, b, f	Kiss et al., 1995; Muid et al., 1991
	Calphostin C	DMSO, DMF, ethanol	0.05	a, b, g	Kobayashi et al., 1989; Tamaoki, 1991
	Chelerythrine chloride	DMSO, water	0.7	a, b, c, f, g	Kobayashi et al., 1989; Tamaoki, 1991
PKG	KT5823	DMSO, DMF	0.2–0.3	a, b, g	Grider, 1993

PTK				
Erbstatin	DMSO	0.8	a	Umezawa and Imoto, 1991
Genistein	DMSO	2.0–30	a, b, f, g, h	Akiyama and Ogawara, 1991
Herbimycin A	DMSO	1	a, b	Uehara and Fukazawa, 1991
Lavendustin A	Acetone, DMSO, ethanol	0.01–0.5	a, b	Simon et al., 1998
Tyrphostins	DMSO, ethanol	0.003–45	a, b, c, f	Levitzki et al., 1991

a, Calbiochem, La Jolla, CA; b, Sigma-Aldrich, St. Louis, MO; c, Alexis Corporation, San Diego, CA; d, Seikagaku America Inc., Rockville, MD; e, BIOLOG Life Science Institute, Bremen, Germany; f, LC Laboratories, Woburn, MA; g, Biomol, Plymouth Meeting, PA; h, Biosource, Camarillo, CA.

[1]Inhibits activation of MAPKK by Raf.

[2]Inhibits p38 K.

[3]Inhibits S6 K activation by inhibiting FKBP12 and rapamycin-binding PK, which activates S6 kinase through mechanisms not yet fully understood.

[4]The different Cdk inhibitors show some degree of specificity to different Cdks. By using different conditions (for example, dividing cells vs. nondividing cells), the relative roles of different Cdks in a given process can be inferred. It should be noted that many of the described Cdk inhibitors also inhibit GSK3.

Abbreviations used in Tables II and III. CaMKII, calcium/calmodulin-dependent protein kinase; CKI/CKII, casein kinase I/II; GSK3, glycogen synthase kinase 3; MAPK/ERK, mitogen-activated protein kinase/extracellular signal-regulated kinase; MLCK, myosin light chain kinase; p38 PK, p38 protein kinase; p60 S6K, p60 S6 kinase; PI 3-K, phosphatidylinositol 3-kinase; PKA, cAMP-dependent protein kinase (protein kinase A); PKC, protein kinase C; PKG, cGMP-dependent protein kinase (protein kinase G); PTK, protein tyrosine kinase. The IC$_{50}$ values are approximative values of useful concentrations.

Table III
Examples of Selective Protein Kinase Activators

	Compound	Comments	Solubility	Potency (IC_{50} μM)	Supplier	References
				PK activators		
PKA	Br-cAMP	Cell-permeable cAMP-analog	Water	0.1–1	a, c	Hei et al., 1991
	Forskolin	Activates cAMP	DMSO	5–10	a, c, e, f	Bhat, 1993; Galli et al., 1995; Mokhtari et al., 1985; Seamon and Daly, 1981
PKC	Mezerein		Ethanol	0.01–0.1	b, f	Nishio et al., 1994
	PMA (TPA)	Phorbol-myristate-acetate	DMSO, ethanol	0.01–0.1	a, b, c	Tepper et al., 1995
	SC-9	Ca^{2+}-dependent activation	DMSO	5–10	d	Nishino et al., 1986
	Thymeleatoxin	Selective for α, $\beta 1$ and γ isoforms	DMSO	0.01–0.1	a	Roivainen and Messing, 1993

a, Calbiochem, La Jolla, CA; b, Sigma-Aldrich, St. Louis, MO; c, Alexis Corporation, San Diego, CA; d, Seikagaku America Inc., Rockville, MD; e, LC Laboratories, Woburn, MA; f, Biomol, Plymouth Meeting, PA.

examples of drugs that are commonly used for synchronization of cell growth. A simple example is included on how to enrich mitotic cells during *in vivo* labeling by use of nocodazole. There are a vast number of other approaches, and the reader should consult Fantes and Brooks (1993) for further information on this subject. It should be stressed that the concentration of drugs that interfere with the normal cell cycle process, and the incubation time needed for successful synchronization, varies and often has to be determined separately for each different cell line.

B. Phosphorylation of Intermediate Filament Proteins

1. Metabolic *In Vivo* Labeling

An initial approach to confirm that a given cytoskeletal protein or protein complex is phosphorylated on a given treatment is metabolic *in vivo* labeling of cells with ^{32}P-orthophosphate. Furthermore, because PKs *in vitro* (see below) often phosphorylate substrates that are not their natural substrates *in vivo*, possible results obtained by *in vitro* labeling have to be confirmed by *in vivo* labeling studies. Before treatment with activators, inhibitors or other substances (e.g., receptor ligands, kinase activators, kinase inhibitors, phosphatase inhibitors), the cells' adenosine triphosphate (ATP) pools need to be equilibriated with ^{32}P. The membranes of primary cultures (e.g., hepatocytes, adipocytes, thymocytes) are easily permeable to ^{32}PO$_4$, and the ATP pools may be equilibriated within 1–2 hours. Established cell cultures often require longer preincubation (1–6 hours). For some purposes, it may be useful to determine the rate of ATP pool equilibration, as previously described (Garrison, 1978). Excessive preincubation should be avoided, because this results in higher backgrounds on protein gels as a result of extensive ^{32}P incorporation into phospholipids and nucleic acids. It should also be considered whether the purpose is to determine constitutive or inducible phosphorylation. If the level of constitutive phosphorylation of a given protein is to be determined, then sufficient incubation times are required to allow for equilibration of the phosphate pools of the protein in question. However, IFs contain multiple phosphorylation sites that equilibrate with ^{32}P at different rates. If a certain stimulus-mediated phosphorylation is to be determined, then short equilibration times should be used to avoid masking the stimulus-induced phosphorylation by a high level of basal phosphorylation. To obtain more efficient labeling, phosphate-free cell culture media are often used. Completely phosphate-free media can be used for maximal ^{32}P incorporation, but many cell lines may show signs of detrimental effects if they are subjected to prolonged incubations in fully phosphate-free environment. If 10% fetal calf serum (FCS) is added to a phosphate-depleted medium, cells stay intact for a longer time. In this case, the phosphate contribution of FCS gives a final phosphate concentration of approximately 0.1 mM, which is a sufficient degree of phosphate depletion to allow efficient labeling of most phosphoproteins. When using any of the commercial orthophosphate preparations suitable for metabolic labeling, 100–300 μCi/ml ^{32}PO$_4$ in the medium is usually sufficient.

After the cells have been labeled, they are lysed in a buffer suitable for sodium dodecy sulfate-polyacrylamide gel electrophoresis (SDS-PAGE) two-dimensional (2D) electrophoresis, or immunoprecipitation. Whole cell extracts or immunoprecipitated proteins are then separated by SDS-PAGE or 2D gel electrophoresis, followed by quantitative phosphorimaging analysis and/or quantitative autoradiography. If suitable antibodies are available, the protein(s) in question can be immunoprecipitated before quantification of the ^{32}P labeling. Quantitative immunoprecipitation of IF protein requires denaturing lysis buffers before they are dissolved in radioimmunoprecipitation assay (RIPA) buffer (see enclosed protocol). The specific ^{32}P labeling can be determined by densitometric (Coomassie blue or silver-stained gels) or fluorometric quantification (gels stained with fluorescent protein labels, such as Sypro Orange or Sypro Red from Molecular Probes, Eugene, OR) of the amount of a given phosphoprotein. The specific labeling can also be determined by performing the *in vivo* ^{32}P labeling together with ^{35}S labeling. In this case, autoradiography or phosphorimaging should be carried out with and without four layers of aluminum foil between the gel and the film or intensifying screen. The difference represents the amount of ^{35}S labeling (Boyle *et al.*, 1991), which can be used as a reference value for the amount of protein. However, for this approach to be useful, one has to ensure that the used treatment does not affect the synthesis rate of proteins. Attempts have been made to determine the stoichiometry of phosphoproteins labeled *in vivo*. However, this approach to determine the exact amount of phosphate on a protein is questionable, because so many different factors affect the incorporation rates of ^{32}P-labeled orthophosphate versus unlabeled orthophosphate.

2. *In Vitro* Labeling of Proteins

In vitro labeling of proteins is a rather easy approach to test whether a given PK may have a role in the regulation of any given IF protein. However, possible positive results have to be verified by *in vivo* labeling, because there are proteins that act as common phosphoryl acceptors without necessarily having a physiological relevance (good examples are casein and histones). There are some general considerations when proteins are phosphorylated *in vitro*: (1) the kinase preparation should be highly purified and not contain contaminants of other kinases; and (2) the IF substrate does not have to be a purified protein sample, but it could also be a crude IF preparation. In fact, it may be advantageous to use IF protein complexes for *in vitro* phosphorylation experiments. Crude IF preparations, with all the characteristic associated proteins, are often useful because they reflect the conformations and interactions occurring in these protein assemblies *in vivo*. Hence, the sites available for phosphorylation may be more similar to those occurring *in vivo* than in, for example, bacterially expressed, highly purified proteins. On the other hand, proteins prepared from eukaryotic cells may have some phosphorylation sites partly or completely occupied by phosphorylation and/or glycosylation, which may inhibit attempts to phosphorylate these sites.

There are two additional considerations. ATP is usually the phosphoryl donor, and it is used at concentrations near saturation. Because the K_m values of ATP for most kinases are far below 100 μM, 100–200 μM is usually sufficient. The γ^{32}-P–labeled ATP is added to unlabeled ATP so that the specific activity is 100–200 μCi/1 μmole. Also, the PK/substrate ratio should be high enough so that the amount of kinase is not the limiting factor for obtaining a representative stoichiometry for a given protein substrate. PK to protein substrate ratios of 1:20–1:50 are usually sufficient, although significantly lower ratios may be used if the availability of the kinase in question is at premium.

Many members of the major kinase families have become commercially available over the past few years. Kinases can often also be immunoprecipitated, and the kinase assay can be performed with the immunocomplexed kinase. Good immunoprecipitating antibodies are available for many of the key kinases that are currently the focus of intensive research. When assessing the phosphopeptide maps of substrates phosphorylated *in vitro*, one should bear in mind that this approach may generate artificially high labeling compared with the situation after *in vivo* phosphorylation by the same kinase.

C. Characterization of Intermediate Filament Protein Phosphorylation

1. Phosphoamino Acid Analysis and Phosphopeptide Mapping

Proteins showing alterations in their phosphorylation state during *in vivo* labeling are often initially analyzed to reveal their phosphoamino acid composition. This can be carried out, for example, after electrotransfer of SDS-PAGE–separated proteins to Immobilon-membranes (polyvinylidene difluoride; Millipore Corporation, Bedford, MA), followed by acid hydrolysis of a cut-out protein band with 6 M HCl as described (Boyle *et al.*, 1991; Hardie, 1993). The number of sites that are involved in the phosphorylation of a particular protein can be estimated by tryptic phosphopeptide mapping (also other proteases can be used). The *in vivo* ^{32}P-labeled protein of interest is, after electrophoretic separation, digested with trypsin in the gel, and the resulting peptides are separated on microcrystalline cellulose plates by electrophoresis followed by ascending chromatography. The obtained phosphopeptide map is also a fingerprint of the kinase(s) involved in phosphorylation of a given protein and often allows identification of the kinases regulating the protein (Boyle *et al.*, 1991). Comparing the phosphopeptide maps obtained *in vivo* and *in vitro* is an essential way of determining whether a given PK or PP could be involved in the regulation of a protein. If the pattern from *in vitro* and *in vivo* labeled material is relatively similar, there is a good chance that the tested kinase is responsible for the phosphorylation.

2. Manual Edman Degradation

Manual Edman degradation is a sensitive, simple, and inexpensive (and often overlooked) way to attempt to determine the location of phosphoamino acids in an isolated phosphopeptide. Phosphopeptides are immobilized on arylamine membrane disks with water-soluble carbodiimide, and the immobilized peptides are subjected to manual Edman degradation as described in the enclosed protocol. This approach obviously requires that the sequence of the studied protein is known. The method is especially useful in combination with tryptic phosphopeptide maps, and if the protein to be studied does not contain too many potential phosphoserines, phosphorylation site determination could be obtained by a combination of phosphoamino acid analysis, phosphopeptide mapping, and manual Edman degradation.

3. Purification or Enrichment of Tryptic Phosphopeptides

Obtaining sequence information about the tryptic peptides usually requires a purification step or at least enrichment of the phosphopeptides to reduce suppression effects in MS-based techniques and to enhance the detection sensitivity of individual phosphopeptides. Determination of specific phosphorylation sites is carried out after digestion (e.g., with trypsin) of the *in vivo* or *in vitro* ^{32}P-labeled protein of interest from SDS-PAGE slices. It is crucial that the protein to be digested has been purified to very high purity to avoid contamination of peptides from other proteins. The tryptic peptides can be purified by phosphopeptide mapping, which will yield a sufficient quantity for manual Edman degradation and even for MS-based analysis. A commonly used method to yield purified peptides from tryptic digests is separation on C-18 or C-8 reversed-phase high-performance liquid chromatography (HPLC) analytical columns (preferably microbore scale) followed by collection of the ^{32}P-labeled peptides. C-18 and C-8 columns can be used in sequence to obtain higher purity of peptides (Eriksson *et al.*, 2004). A recent method for enriching phosphoproteins that has become increasingly popular is based on the use of immobilized metal ions or immobilized metal affinity chromatography (IMAC). Immobilized metal ions, such as Fe^{3+}, Ga^{3+}, and $A1^{3+}$ bind with high specificity to phosphoproteins and peptides (Zhou *et al.*, 2000). The method yields sufficient enrichment to significantly reduce the suppression effects by other peptides and will thereby enhance the detection of phosphopeptides. It is useful in both matrix-assisted laser desorption/ionization (MALDI)-MS and electrospray ionization (ESI)-MS analyses (see later). The technique has been successfully used with both on-line and off-line coupling to MS analysis (Nuwaysir and Stults, 1993).

4. Mass Spectrometry–Based Phosphopeptide Identification

Previously, the primary method to obtain the identity of a given peptide was sequencing by an automatic amino acid sequencer. Nowadays, peptide identification by MS-based techniques in combination with bioinformatics is usually the method of choice, because the sensitivity of MS-based techniques is significantly higher than that of traditional sequencing techniques. Numerous MS-based approaches have been applied to phosphoprotein analysis. However, phosphopeptides are difficult to analyze by MS-based methods because of a number of reasons: (1) they are negatively charged, whereas the ionization of phosphopeptides preceding the MS analysis is usually done in the positive mode; (2) in the presence of nonphosphorylated peptides, there is ionic suppression of the phosphopeptides, leading to weakening of phosphopeptide peaks; and (3) very small peptides or very large peptides may not be observed at all. Because of the preceding reasons, the protocols used for proteomic identification of unknown proteins by peptide "fingerprinting" of tryptic protein digests using matrix-assisted laser desorption ionization time-of-flight mass spectrometry (MALDI-TOF MS) are usually not possible to adapt directly for phosphopeptide identification in a peptide mixture. Therefore, either individual phosphopeptides have to be purified by HPLC, as indicated earlier, by use of the ^{32}P label as a marker to follow the elution of phosphopeptides, or the population of phosphopeptides have to be enriched by the afore-described IMAC-based affinity purification procedures.

Until recently, MALDI MS sequencing of selected individual peptide ions was mainly achieved by so-called post-source decay (PSD). Fragment ions are formed as a result of molecular ion decay. PSD happens after acceleration of a molecular (precursor) ion out of the MALDI ion source and before reflection (Fig. 1) (Falick, 1995; Kaufmann et al., 1996; Spengler et al., 1992). The fragmented ions represent the sequenced peptide ion. However, PSD-based analysis is hampered by the fact that it shows quite complex patterns of fragmentation along the peptide backbone, low yield of these fragment ions, limited mass accuracy, and poor sensitivity (Bennett et al., 2002). PSD can be used in some cases to confirm the phosphopeptide presence in MALDI MS spectrum. On PSD of a selected peptide ion, the loss of HPO_3 (-80 Da) from *p-Thr, p-Ser, p-Tyr* and/or H_3PO_4 (-98 Da) from *p-Thr, p-Ser* can occur. Consequently, the peak(s) of dephosphorylated form(s) can be detected at 80 and/or 98 m/z lower than the selected peptide ion (Annan and Carr, 1996; McLachlin and Chait, 2001; Oda et al., 2001).

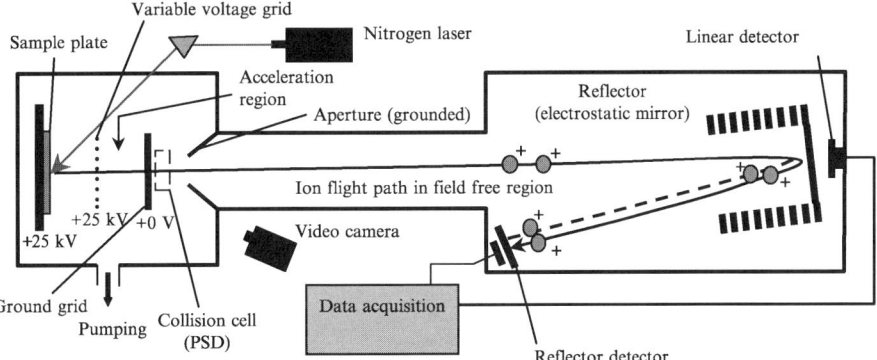

Fig. 1   The principle of MALDI-TOF MS (see text for details). (See Color Insert.)

Phosphopeptides can be investigated by the tandem MS method, which implies that an additional MS run is performed during the exchange (i.e., MS/MS or "mass spectrometry of MS-produced ions"). Sequencing of phosphopeptides by this technique requires prepurification of the phosphopeptides, either off-line or on-line. HPLC fractions of phosphopeptides may be collected and each subjected to MS/MS, or the peptides can be analyzed on-line by use of nanoelectrospray ionization MS/MS (LC-nanoESI-MS/MS); in other words, there the MS unit is directly coupled to a chromatographic separation unit that performs the separation of individual peptides before the MS-based analysis takes place (Asara and Allison, 1999; Jaffe *et al.*, 1998; Kinter and Sherman, 2000).

The combination of MALDI-MS peptide mapping with nanoESI-MS/MS peptide sequencing should in principle be sufficient to both identify most proteins and often also to localize the precise phosphorylation sites of the identified proteins. The efficiency of such MS analysis is much higher if both MS and MS/MS are performed on the same MALDI mass spectrometer. This idea was realized by interfacing a MALDI ion source to an orthogonal injection TOF MS and more recently to a hybrid quadrupole time-of-flight (QqTOF) mass spectrometer (Bennett *et al.*, 2002; Loboda *et al.*, 2000). Bennett *et al.* (2002) investigated application of the MALDI-QqTOF to phosphopeptide analysis. The goal was to sequence phosphopeptides present in a crude digest mixture and to localize the phosphorylation sites. The loss of phosphoric acid, −98 Da (*neutral loss experiment*, see later) from the precursor ion for peptides containing phosphoserine and phosphothreonine, and the presence of the immonium ion at m/z 216.04 (*precursor ion experiment*, see later) generated from peptides containing phosphotyrosine are both diagnostic for determining the presence of these types of phosphorylated amino acids in peptides. MALDI-QqTOF analysis can be combined with IMAC purification to enrich phosphopeptides from digest mixtures, thereby enhancing ion response.

Tandem MS analysis requires the use of two stages of mass separation. It is usually performed with triple quadrupole instruments. Quadropoles are instruments containing ion traps, in which all ions created over a given time can be trapped and then sequentially ejected from the ion trap into a conventional electron multiplier detector. One of these three quadropoles, a collision cell filled with an inert gas such as helium, is placed between the first and third quadrupole instruments (Fig. 2). Selected ions pass from the first quadrupole into the collision cell (i.e., quadrupole 2) and collide with gas molecules. This results in decomposition (fragmentation) of some ions because of their collisionally gained internal energy (*collisionally induced dissociation, CID*). The fragments that are formed are mass-analyzed in the third quadrupole instrument. The resulting mass spectrum, often called a CID spectrum, yields a significant amount of information about the starting compound (Falick, 1995).

The standard (nontandem MS) experiment—*mass spectrum scan*—is used to acquire a general mass spectrum of an analyte. There are three types of tandem mass spectrometry experiments ("scan modes") that are used in parallel for phosphopeptides analysis: *product ion scan, precursor ion scan* and *neutral loss*

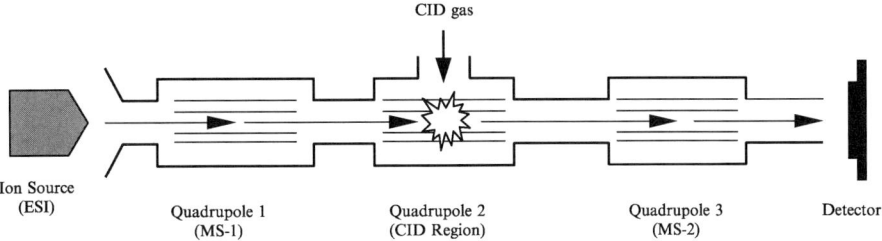

Fig. 2   The principle of tandem mass spectrometry (MS/MS) analysis (see text for details).

scan. All these scan modes are initially based on standard mass spectrum scan experiment.

The *product ion experiment* (Fig. 3A) records the fragment ions formed from a given precursor ion. This precursor ion (which corresponds to one of the peptide ions) is obtained and selected from the standard mass spectrum scan in the first quadrupole instrument (or the first mass spectrometer, MS-1). Then it is passed into the collision cell (quadrupole 2), where its fragmentation ("sequencing") occurs. After the collision cell, the fragmented ions pass into the third quadrupole (or the second mass spectrometer, MS-2), where they are successively detected. Analyzing the mass series of the fragmented ions, one can deduce the sequence of the originally selected peptide ion (precursor ion) and modifications (such as phosphorylation) on it. Each m/z value in the mass series corresponds to the mass of one peptide ion fragment (which can contain one to several amino acid residues with or without modifications). Phosphorylation is usually confirmed by detection of +80 Da shift at the *Tyr, Ser*, and *Thr* residues or by the β-elimination of phosphoric acid (−98 Da) from *p-Thr* and *p-Ser* residues with concomitant conversion of phosphoserine to dehydroalanine and phosphothreonine to β-methyldehydroalanine or 2-aminodehydrobutyric acid (Bennett *et al.*, 2002; Kinter and Sherman, 2000).

The *precursor ion experiment* at the MS-1 stage records m/z of all peptide ions that on fragmentation in the collision cell produce a specific ion (detected at MS-2 stage) with known m/z (Fig. 3B). This scan mode is used to detect peptides with phosphoamino acids. The formation of PO_3^- ion at m/z 79 in the negative ion experiment is diagnostic. Therefore, the m/z of all peptides that produce this specific diagnostic PO_3^- ion on CID will be detected. After identification, the peptides of interest are sequenced by a product ion experiment (see earlier) in the positive ion mode (Bennett *et al.*, 2002; Kinter and Sherman, 2000; Neubauer and Mann, 1999).

In the *neutral loss experiment* (Fig. 3C), the two stages (MS-1 and MS-2) of mass analysis are coordinated with a specific m/z difference. The instrument records the m/z of all ions that fragment (in the collision cell) to form a neutral fragment of a specific mass. These neutrals are not detected (because they are uncharged). Their formation is detected by observation of a specific neutral

A. Product ion scan experiment

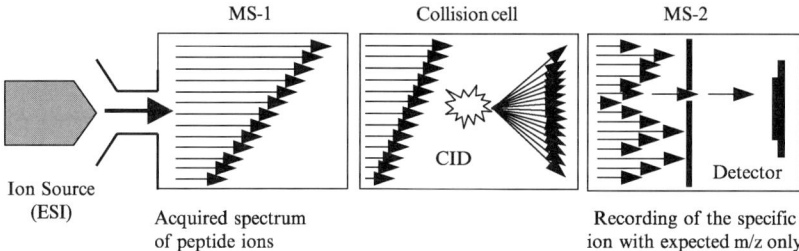

Ion Source
(ESI)

Mass spectrum scan and
precise single m/z selection

Acquired spectrum of
single m/z CID

B. The precursor ion scan experiment

Ion Source
(ESI)

Acquired spectrum
of peptide ions

Recording of the specific
ion with expected m/z only

C. The neutral loss scan experiment

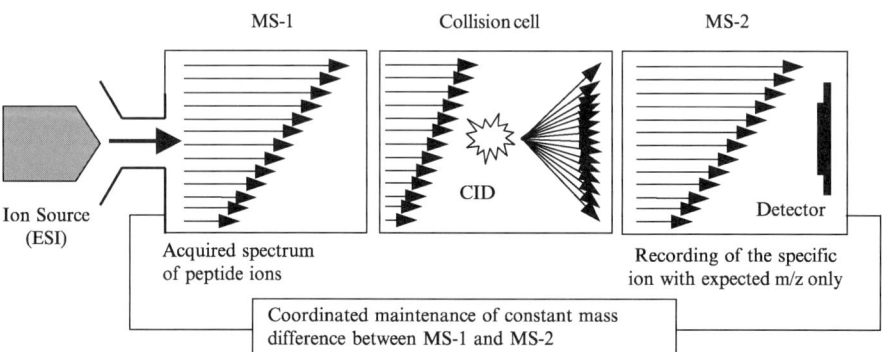

Ion Source
(ESI)

Acquired spectrum
of peptide ions

Recording of the specific
ion with expected m/z only

Coordinated maintenance of constant mass
difference between MS-1 and MS-2

Fig. 3   The principles of (A) a product ion scan experiment, (B) a precursor ion scan experiment, and (C) a neutral loss scan experiment (see text for details).

loss between peptide ions detected before and after CID. Peptides containing phosphorylated amino acids (*p-Ser* and *p-Thr*) lose H_3PO_4 as an uncharged 98 Da mass species. For instance, if the m/z difference between both stages of mass analysis is set as 98, all singly charged peptide ions $[M + H]^+$ losing 98 amu (i.e., H_3PO_4) will be detected. If the m/z difference between both stages of mass analysis is set as 49, only doubly charged phosphopeptides $[M + 2H]^{2+}$ that fragment to lose 98 will be registered. These phosphoserine(s)- or phosphothreonine(s)-containing peptides are then "sequenced" by a product ion experiment (see earlier) (Bennett et al., 2002; Kinter and Sherman, 2000).

Despite many of the advances just described, the MS/MS of phosphopeptides still remains quite challenging because of signal suppression of phosphopeptides in the positive detection mode, inherent lability of the phosphate group on CID, difficulty of achieving full sequence coverage for long peptides, and the presence of low-abundance phosphopeptides or phosphopeptides phosphorylated at substoichiometric levels (Knight et al., 2003).

The direct MALDI MS analysis of a digestion mixture of several phosphopeptides and nonphosphopeptides usually does not represent peaks for all peptides that theoretically should be detected. Two major reasons are: (1) low ionization efficiency; and (2) the suppression effect (i.e., when one peptide suppresses the signal of another). The detection sensitivity of phosphopeptides in MALDI MS is an order of magnitude lower than that for the nonphosphorylated forms. The detection is much worse when several phosphate groups are present on a peptide. For instance, if a peptide containing three phosphate groups is analyzed, it may be difficult to get a signal out of it. After phosphatase treatment, the dephosphorylated form may be detected, but without knowing the number of cleaved phosphate groups. These facts limit the usefulness of MALDI MS for the identification of phosphorylation sites (Asara and Allison, 1999; Liao et al., 1994; Zhou et al., 2000).

5. Enhancing the Ionization of Phosphopeptides

The phosphate group exists in solution in an anionic form with two negative charges. If an analyte with multiple negative charges is trapped in the MALDI matrix crystals, there is not sufficient energy available in the MALDI experiment for the direct desorption of such species. For most analytes, singly charged ions $[M + H]^+$ are predominantly formed. For instance, a tetraphosphorylated peptide carrying a charge of -8 will not be generated in the gas phase on MALDI, and no signal will be registered. To overcome this problem, ammonium salts of acetate or citrate can be included in the MALDI analysis to neutralize the negative charge on the analyte phosphogroups. NH_4^+ ions complex with negatively charged phosphate groups to form $(NH_4^+)_n(\text{phosphate}^{-n})$. Either before or during the MALDI process, ammonia is lost, leaving the phosphate group in a neutral form, $(H^+)_n(\text{phosphate}^{-n})$. These additives enhance the detection signal of phosphopeptides in MALDI MS as a consequence of the improved desorption/ionization efficiency (Asara and Allison, 1999).

Low ionization efficiency of the phosphorylated peptides in the positive ion mode can also be overcome with the negative ion mode experiment, because the MS-signal intensities of phosphorylated peptides are often increased in the negative ion mode compared with their nonphosphorylated analogs (Janek et al., 2001).

6. Selective Chemical Modification of Phosphoamino Acids Residues

To overcome some of the problems just described with MS-based detection of phosphopeptides, different strategies have been developed to yield nonphosphorylated peptides that would be a specific product of a phosphopeptide. One of these approaches has been designing chemical modifications that yield specific proteolysis at the sites of phosphoserine and phosphothreonine residues. The phosphate group of *p-Ser* and *p-Thr* residues can be eliminated under basic conditions, producing dehydroalanine and β-methyldehydroalanine residues. In the next step, dehydroalanine and β-methyldehydroalanine react as Michael acceptors for cysteamine, generating aminoethylcysteine and β-methylaminoethylcysteine residues, respectively. These residues are isosteric with lysine. Proteases that recognize lysine (e.g., trypsin, Lys-C, and lysyl endopeptidase) will cleave proteins at these modified amino acids residues. Moreover, on cleavage, all (β-methyl)aminoethylcysteine residues will be at the C-termini of peptides (Knight *et al.*, 2003). The produced peptides will have highly specific masses that are easily distinguishable from the peptides generated from cleavage of the unphosphorylated protein, thereby exposing the phosphorylation sites in the protein. Cleavage may be also obtained exclusively at the positions of phosphorylation (aminoethylcysteine and β-methylaminoethylcysteine residues), excluding cleavage at the normal lysines. To do this, lysine residues are converted to homoarginine using *O*-methylisourea to block digestion at the lysine residues (Knight *et al.*, 2003).

The aforementioned derivatization chemistry of phosphorylated residues, followed by reversible biotinylation of the modified phosphoresidues, has been used for biotin-based affinity purification of phosphoproteins from normal tryptic digests. The purified peptides can then be analyzed and quantified by MS/MS sequencing (Adamczyk *et al.*, 2001; Oda *et al.*, 2001).

D. Assessing the Functions of Phosphorylation

1. Site-Directed Mutagenesis

Once the specific phosphorylation site has been identified, a common approach to establish its supposed function is to mutate the site so that it cannot function as a phosphate acceptor. This requires that cDNA is available for the given protein. The correctness of the assumed phosphorylation site can be established by tryptic phosphopeptide mapping of the mutant protein expressed in a suitable cell line, which is subjected to metabolic ^{32}P *in vivo* labeling. The previously established phosphorylation site should be absent on the phosphopeptide maps derived from the mutated protein. The mutation can be designed so that it mimics the nonphosphorylated state (Ala for Ser and Thr, and Phe for Tyr), or the residue can be mutated so that it simulates the phosphorylated state (Glu or Asp for phospho-Ser and phospho-Thr). Site-directed mutagenesis also enables examination of the possible effects on protein functions and consequent effects on cellular phenotype.

2. Phosphorylation State-Specific Antibodies

When the identities of target phosphoproteins are known and their specific phosphorylation sites have been determined, it is possible to quantify and localize the phosphorylation in cells or tissues by producing phosphorylation state-specific antibodies. Polyclonal antibodies against phosphorylated peptides, which contain the phosphorylation site and which are conjugated to keyhole limpet hemocyanin (KLH) to enhance the immune response, are produced and further isolated by affinity chromatography. These antibodies can be used both for localization of a specific phosphorylation in a cell population, by use of immunohistochemical techniques, and for quantification of phosphorylation on the site recognized by the antibody, by use of Western blotting. This approach has been successful in the context of IF proteins, because it has been used for determining the spatiotemporal distribution of site-specific phosphorylation on vimentin (Nishizawa et al., 1991; Ogawara et al., 1995; Takai et al., 1996; Tsujimura et al., 1994) gline fibrillatory acidic protein (GFAP) (Marin et al., 1997; Nishizawa et al., 1991; and keratin 8 (Liao et al., 1997), keratin 18 (Chou and Omary, 1994; Ku and Omary, 1995; Ku et al., 1995, 1996; Liao and Omary, 1996; Liao et al., 1997), and nestin (Sahlgren et al., 2001, 2003) during different phases of the cell cycle (Ku and Omary, 2000; Ku et al., 1998), as well as under variable physiological and stressed conditions. Generation of phosphospecific antibodies is described in detail in Chapter 13.

II. Materials and Methods: Characterization of Intermediate Filament Phosphorylation

A. Example 1: Characterization of cAMP–Dependent Protein Kinase–Specific Phosphorylation on Vimentin

It has been established that vimentin is phosphorylated on Ser-38 by cAMP-dependent kinase (PKA; Eriksson et al., 2004). We use the cAMP-dependent protein kinase PKA–specific phosphorylation of vimentin as an example of how a kinase-specific site can be determined *in vitro* and *in vivo*. The different aspects of the methods used at each step are discussed. When the site has been determined, phosphospecific antibodies against this site can be generated as previously described, or the site can be mutated into a phosphorylation-deficient (Ser/Thr $\geq$ Ala) site or a site mimicking constitutive phosphorylation (Ser/Thr $\geq$ Asp) (Eriksson et al., 1998, 2004).

1. Metabolic *In Vivo* ^{32}[P] Labeling of Cultured Cells

To study whether vimentin is phosphorylated on PKA-specific sites *in vivo*, baby hamster kidney (BHK) cells are prelabeled at 37 °C for 3 hours with ^{32}P orthophosphate (see later), and the cultures are then treated for 30 minutes with 50 nM calyculin-A. Cells are lysed in SDS buffer, and the induction of elevated phosphorylation is confirmed by immunoprecipitation (Eriksson *et al.*, 2004).

2. Protocol 1: Procedure for *In Vivo* Labeling

1. For maximal labeling, phosphate-depleted media should be used, although sufficient labeling of major phosphoproteins can usually also be obtained in undepleted media. Wash cells two to three times with the phosphate-depleted media and add media to the cell culture dish, just enough to cover the cells (3 ml to a 100-mm diameter Petri dish is sufficient to safely cover the cells if conservation of radioactive reagents is needed, otherwise 5 ml is recommended).

2. Add ^{32}[P]-orthophosphate to the medium at a concentration of 100–300 μCi/ml and incubate at the appropriate conditions for the given cell line. For cells with fibroblast characteristics, 1–2 hours incubation is sufficient if inducible phosphorylation is to be studied. If constitutive phosphorylation is to be studied, 4–6 hours is usually sufficient. Incubation times up to 12 hours and ^{32}[P]-orthophosphate concentrations up to 1 mCi/ml may be required to study low levels of constitutive phosphorylation. Long incubation times may also be required if inducible dephosphorylation is to be studied.

3. Stimulate or treat cells in desired ways after the prelabeling (e.g., by adding compounds or receptor ligands from stock solutions or by subjecting cells to a given physicochemical stimulus).

4. After treatment, transfer the media to a 15-ml centrifuge tube with cap and collect potentially detached cells by centrifugation (1–2 minutes at 3000g). Remove supernatant and wash the cells on the cell culture dish with ice-cold phosphate–buffed saline (PBS), and transfer this PBS to the collected cell pellet. Resuspend and collect cells by centrifugation as described previously. The washing step can be repeated one to two times to yield a lower background of free label.

5. Prepare the cells for protein extraction by adding appropriate buffers and scraping off cells with a disposable cell scraper. The detached cell pellet is pooled with the cells from the cell culture dish by extracting the cells on the dish first with a suitable extracting buffer (see later) and then transferring the cell extract including debris to the cell pellet in the centrifuge tube.

6. The continued treatment depends on whether a denaturing extraction buffer is used or whether soluble and particulate proteins are collected by some method (see section on buffer systems to be used).

Comments

1. It is important to collect detached cells, because elevated phosphorylation of cytoskeletal phosphorylation is likely to lead to a rounding up and detachment of cells from the cell culture dish.

2. *In vivo* labeling is potentially hazardous and can be a source of major contamination. The person conducting these experiments should therefore be experienced in

the use of radioisotopes. There should preferably be a separate room for the labeling, and stringent routines should be established for radiation safety and waste disposal. For an extended discussion on how to arrange laboratory routines and safety to accommodate *in vivo* labeling procedures, see Garrison (1993).

Protein Extraction Buffers

For quantitative immunoprecipitation of IF proteins, it is often useful to denature all cellular proteins by use of an SDS-based buffer. Extraction of some solubilized IF proteins can be obtained by homogenization or extraction in buffers with Triton X-100 or other detergents. For cell culture 0.5 ml of buffer for a 100-mm Petri dish is suitable.

a. Whole–Cell Protein Extraction of Cultured Cells

Cells are lysed in SDS-buffer: 20 mM Tris-HCl (pH 7.2), 5 mM ethyleneglycol-bis-(2-aminoethyl)-N, N, N′, N′-tetraacetic acid (EGTA), 5 mM ethylene diamine tetraacetic acid (EDTA) 0.4% SDS, 10 mM sodium pyrophosphate, 1 mM phenylmethylsulfonyl fluoride (PMSF), 10 μg/ml antipain, 10 μg/ml leupeptin, 10 μg/ml pepstatin. Cells are detached from the culture dish with a cell scraper, suspended, boiled for 5–10 minutes, and sonicated for 20 seconds with a probe sonicator. This protein extract is suitable for immunoprecipitation when diluted 1:10 with RIPA buffer (for composition see section C).

b. Extraction of Soluble Intermediate Filament Proteins

Cells are homogenized in 20 mM Hepes, (pH 7.6), 100 mM NaCl, 5 mM MgCl$_2$, 5 mM EGTA, 1% Triton X-100, 1 mM PMSF, 10 μg/ml leupeptin, 10 μg/ml antipain, 10 μg/ml pepstatin. Detergent-soluble IF proteins can be obtained by centrifuging these extracts at 15,000g for 15 minutes at +4°C. The proteins in the supernatants and pellets can be analyzed after addition of Laemmli sample buffer, followed by sonication and boiling for 5 minutes. Alternatively, proteins can be immunoprecipitated from the supernatants suitably diluted with RIPA buffer.

c. Buffer for Immunoprecipitation

For immunoprecipitation, cell extracts can be diluted with RIPA-buffer: 20 mM Hepes, (pH 7.4), 140 mM NaCl, 10 mM pyrophosphate, 5 mM EDTA, 0.4% Nonident P-40, 1 mM PMSF, 10 μg/ml leupeptin, 10 μg/ml antipain, 10 μg/ml pepstatin. SDS-based extracts are diluted 1:10 with the RIPA buffer. The NP-40 in the buffer will quench the effect of SDS. Immunoprecipitations can be performed as described (Harlow and Lane, 1988). If cell extracts of soluble proteins in buffers with detergents are used, the NP-40 is omitted from the RIPA buffer.

3. Protocol 2: *In Vitro* Phosphorylation

To study whether the *in vivo* phosphorylated mitosis-specific site of vimentin corresponds to the PKA-specific site of vimentin phosphorylated *in vitro*, vimentin is phosphorylated by purified PKA according to the protocol presented for PKA in the following. Comparative phosphopeptide mapping is performed on *in vivo* and *in vitro* phosphorylated material (Fig. 4A,B) exactly according to the protocol presented in Boyle *et al.* (1991).

The conditions for *in vitro* phosphorylation with specific PKs are different, because the various PKs have different prerequisites for their activities. Some kinases require stimulatory cofactors, whereas some kinases can be used in a simple buffer system. The standard protocol for PKA included later should work for many kinases that do not require cofactors (see comments).

1. Activate freeze-dried PKA, 1 mg (containing 1 μg of the catalytic subunit corresponding to approximately 40 U; Sigma P-2645), by addition of 6 μl of 6 mg/ml dithiothreitol (DTT) (in water) immediately before use. Keep on ice.

2. Combine 50 μl 2× kinase buffer, 50 μl substrate protein (approximately 1 mg/ml), and transfer the mixture to a tube containing 5 μl of ATP-mix (3 μCi ^{32}[P]γ-ATP in 5 μl of 2 mM ATP A-5394, Sigma). The final concentration of ATP in the reaction mixture is 100 μM.

3. Transfer the reaction mixture with the ATP-mix to the tube containing the activated PKA (6 μl), vortex the tube gently, and then incubate at 30 °C. Stop the reaction after 30 minutes by addition of 100 μl 3× Laemmli sample buffer. Boil samples for 5 minutes.

2× kinase buffer: 20 mM Hepes (pH 7.2), 120 mM NaCl, 5 mM EGTA, and 4 mM MgCl$_2$.

Comments

1. Because the kinases are fairly unstable at assay conditions, keep the kinases on ice and add as a last component to the reaction mixture.

2. This same simple buffer system can be used for many common kinases that do not require cofactors, for example, proline-directed kinases such as cdc2, mitogen-activated protein kinase (MAPK), stress-activated protein kinase (SAPK), and p38 protein kinase C (PKC) can be used with this buffer if EGTA is omitted, and 0.5 mM CaCl$_2$, 2 μg diacylglycerol, and 5 μg phosphatidylserine are added to the buffer.

4. Protocol 3: Determination of a Phosphorylation Site by Manual Edman Degradation

In our example, the identity of the assumed PKA-specific phosphopeptide localized on the peptide maps is confirmed by manual Edman degradation (Fig. 4C). This method is a simple and inexpensive way to determine the location of phosphoamino acids in an isolated phosphopeptide. In the procedure suggested here, largely based on previous protocols (Eriksson *et al.*, 1998, 2004), phosphopeptides are immobilized on arylamine membrane disks with water-soluble carbodiimide, and the immobilized peptides are subjected to manual Edman degradation. For immobilization of peptides, Sequelon-AA membranes (Applied Biosystems, Foster City, CA) can be used. Sequelon-AA membranes consist of a polyvinylidene difluoride (PVDF) matrix that has been derivatized with arylamine groups. The C-terminal and side chain carboxyl groups of peptides react with the arylamine groups of the membrane by means of carbodiimide activation.

1. The peptide sample is dissolved in an aqueous acetonitrile solution. Generally, a 30% acetonitrile solution is adequate for dissolving most peptides. The acetonitrile concentration may be varied or 0.1% trifluoroacetic acid (TFA) may be added to aid solubilization of the samples.

2. Apply a small amount of the sample at a time on an arylamine membrane disk that is placed on a Mylar sheet on top of a heating block set at 50 °C. Allow the disk to dry between each application. When the whole sample has been applied, allow the solvent to evaporate for 10–15 minutes before removing the disk from the heating block.

3. Covalent linkage of the peptides to the arylamine membrane disk is accomplished by adding 5 μl of freshly prepared carbodimiide solution (1 mg of water-soluble carbodimiide in 100 μl of 0.1 M maintenance eletrolyte solution [provided by the supplier of the Sequelon membranes], pH 5.0) to the sample disk. After 20 minutes at room temperature (RT), the disk is washed extensively with water and then extracted five times with 0.5 ml trifluoroacetic acid (TFA) to remove unbound peptides. The disk is then extracted three times with 1 ml methanol and subjected to Edman degradation. Alternatively, the disks can be stored in methanol at −20 °C.

4. Edman degradation of immobilized peptides is carried out in 1.5 ml Eppendorf tubes. Extraction and washing can be accomplished by gentle vortexing. Each cycle of degradation is carried out according to the following protocol.

5. Add 0.5 ml of fresh coupling reagent (methanol, water, triethylamine, phenylisothiocyanate (PITC); 7:1:1:1, v/v) to the disk and incubate at 50 °C for 6–10 minutes. Remove the reagent and wash the disk five times with 1 ml methanol. Speed-vacuum dry the disks for approximately 10 minutes.

6. Add 0.5 ml TFA, incubate at 50 °C for 6 minutes, and remove the TFA. Save the TFA wash, and extract the disk with another 0.5 ml TFA. Combine the two TFA washes and dry in oven at 60 °C. It is possible to leave the sample at RT to evaporate, and then dry the samples from each cycle at 60 °C at the same time. Wash the disk six times with 1 ml methanol before beginning a new degradation cycle.

7. Measure the amount of radioactivity released from each cycle after the sample is neutralized by adding 0.5 ml unadjusted 0.5 M Tris and determine the amount of radioactivity that remains bound to the disk.

5. Protocol 4: Matrix-Assisted Laser Desorption/Ionization/Mass Spectrometry Analysis of a Single Phosphopeptide: Overcoming Suppression Effects by Addition of an Ammonium Salt to the Sample Matrix

In MALDI-MS, a peptide M being analyzed ("M" represents the molecular weight of the peptide) forms usually a singly charged ion $[M + H]^+$. If a MALDI mass spectrum peak is 80 Da higher than a predicted $[M + H]^+$, it usually

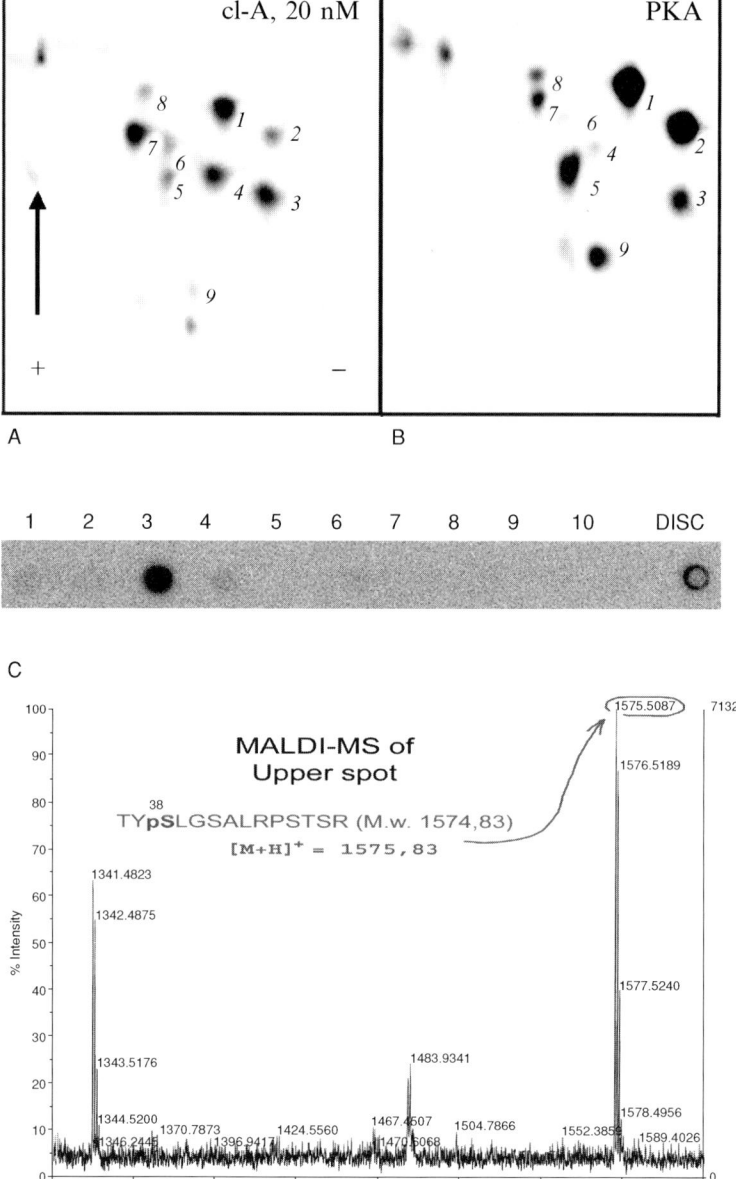

Fig. 4 A major interphase-specific phosphorylation site on vimentin is the same as a major PKA–specific site *in vitro*. Tryptic phosphopeptide maps of (A) vimentin immunoprecipitated from mitotic *in vivo* ^{32}P-labeled cells treated with 20 nM calyculin-A (cl-A) and (B) of vimentin phosphorylated *in vitro* with PKA. The autoradiographies of the TLC plates, with the tryptic peptides separated in two

indicates phosphorylation. This mass shift is the result of HPO_3 addition (80 Da) to the HO– group of a serine, threonine, or tyrosine residues. The presence of this modification can be confirmed by a –80 Da loss of the peptide mass before and after phosphatase treatment, because treatment with phosphatase removes the phosphogroup. Multiple phosphorylations will give a mass shift of 80 Da for each phosphate group (Asara and Allison, 1999; Zhang et al., 1998).

1. A major PKA-specific vimentin ^{32}P-phosphopeptide (peptide 1; Fig. 4B) was accurately scraped off the thin-layer chromatography (TLC) plate, based on the phosphorylated spots obtained by autoradiography (to enable localization) into an Eppendorf tube and then eluted from the cellulose powder twice in 100 μl of 30% acetonitrile, 0.1% TFA solution. First elution: at 55 °C, 50 minutes with occasional vortexing. Sample is then centrifuged at 10,000g for 2 minutes, and the supernatant is collected. Second elution: just short vortexing with 100 μl of the eluate, centrifuging, and supernatants are pooled together.

2. Collected supernatant (about 200 μl) is dried in vacuum centrifuge. Phosphopeptides are dissolved by adding 25–30 μl of 0.1% TFA.

3. Tipping the sample for MALDI analysis: twist the head of a gel loader-tip with forceps to prevent escape of the matrix slurry from the tip. Add 20 μl of methyl alcohol (MeOH) into the tip, then 0.5–1 μl of Poros R3-slurry material (suspension in MeOH), push MeOH through the tip by applying pressure with syringe so that R3-slurry is collected at the twisted part of the tip. Equilibrate the column by washing it through with 2 × 20 μl of 0.1% TFA solution, pipet the sample into the tip, and push it slowly through; wash the column 2 × 10 μl of 0.1% TFA. Cut the tip just after the twisted part with scissors, add 1.5 μl of α-cyano-4-hydroxycinnamic acid saturated solution (AcN; in 60% AcN, 0.1% TFA) into the tip. Slowly push out the peptides with the matrix (by applying syringe pressure) straight to the MALDI plate, let the matrix dry completely. The matrix solution to be used should be fresh for each analysis.

---

dimensions, demonstrates that the in vivo ^{32}P-labeled material has one major phosphopeptide (1) that corresponds to one of the major phosphopeptides generated by PKA-kinase in vitro (directions of electrophoresis (+ and –) and ascending chromatography (arrow) are indicated. Adapted from Eriksson et al., 2004 with permission. (C) The identity of the assumed PKA-specific phosphopeptide localized on the peptide maps is checked by manual Edman degradation as described in protocol 3. Ten Edman degradation cycles were performed, the label at each cycle was spotted on Whatman paper, and the paper was measured on a phosphor imager. Numbers on top of figure indicate cycle number. A significant release occurred at the third cycle. Very little label remains unreleased on the small sequelon disc (DISC), indicating that all label from this particular peptide was released on the third cycle. Judging from the vimentin sequence, there are a few N- and C-terminal tryptic peptides with Ser/Thr at the third position and with a putative consensus site for PKA (R-X_{1-2}-S/T-X). One of them is Ser-38. (D) The correct identity of this site is confirmed by simple MALDI-TOF analysis of the peptide that was eluted from the major PKA-specific spot on the TLC plate. When ammonium citrate is added to the sample matrix used in the MALDI-TOF analysis to enhance the detection of phosphopeptides, there is one significant mass that corresponds to the mass of the phosphorylated form of the tryptic peptide containing Ser-38 (i.e., TY**p**SLGSLRPSTSR).

4. The MS analysis is performed on the Voyager DE PRO MALDI-TOF instrument. The MS analysis is set to detect ions in the range 700–4000 m/z. MS of any calibration mixture should be performed along with the sample MS analysis to calibrate the measurement.

5. To enhance the detection of phosphopeptides in the samples, ammonium citrate can be added to the matrix (Asara and Allison, 1999). If expected peaks do not appear in a spectrum during data acquisition, the MALDI plate can be taken out of the instrument, and 0.5–1 μl of 25 mM ammonium citrate solution (in 50% AcN) can be added to the sample spot. Let the matrix dry and recrystallize. Perform the MALDIMS again.

6. The m/z value of 1575.51 corresponds to the phosphorylated form of the tryptic peptide TYSLGSALRPSTSR (Fig. 4D). In this case, trypsin generates a peptide with an arginine in the middle (trypsin normally cleaves at arginine and lysine residues) because of the presence of a proline C-terminal to arginine. XRPX sequences are cleaved by trypsin with rather low probability. Combined with the data from the manual Edman degradation, the phosphorylation site can be assigned to serine at the third amino acid residue, and the final composition of the phosphopeptide deduced to TYpSLGSALRPSTSR.

Comments

One should be aware of possible chemical modifications of peptides that may occur during sample preparation and MS analysis. One commonly observed modification is oxidation of tryptophan (results in an increase of 32 Da) and methionine (+16 Da) residues. Oxidation of methionine to methionine sulfone (+32 Da) is generally not observed. Other common modifications are disulfide bond cleavage, formylation of amino groups (+28 Da) when formic acid is used as a solvent, cyclization of N-terminal glutamine to pyroglutamic acid under certain aqueous conditions, modification of cysteins by free acrylic acid (+71 Da) and N-terminal and lysyl carbamylation (+43 Da) if proteins were purified in the presence of urea. In addition, there are a number of other possible modifications described (for details see Coligan *et al.* [1995]).

B. Example 2: Analyzing Neurofilament Phosphorylation States in Normal and Diseased Tissue

One of the most distinguishing features of the neuronal phenotype is its morphology—its highly asymmetrical shape with cell body, elongated axon, and branching dendrites, each cellular compartment defined by a unique cytoskeleton. A thorough inventory at the molecular level of the properties of neuron-specific molecules that determine cell size, shape, and stability is pivotal to understanding neuronal cytoskeletal structure/function relationships. The development and maturation of neuronal phenotypes depends on the factors that regulate the cell sorting, processing, and posttranslational modifications of cytoskeletal molecules. Among the many cytoskeletal molecules in nerve cells, the neurofilaments (NFs), the neuron-specific IFs, are exclusively neuronal and serve as phenotypic markers. NFs, neuron-specific class IV IFs, are the major cytoskeletal elements of large axons and, together with microtubules, they determine the size and shape of the neuron. Their expression and phosphorylation is developmentally and topographically regulated.

In a mature nervous system, NFs consist of three subunit proteins (NFPs): low (NF-L), medium (NF-M), and high (NF-H). NF-M and NF-H are extensively phosphorylated, with up to 100 or more potential phosphorylation sites (e.g., NF-H) in their tail domains. After their synthesis in cell bodies, they are transiently phosphorylated at head domain sites by various kinases (e.g., PKA, PKC) and are transported in the axon where the hypervariable lys-ser-pro (KSP) repeat sequences in the C-terminal tail domains of NF-M and NF-H are phosphorylated by proline-directed kinases. Phosphorylated tail domains protrude outward from the 10-nm filament core and form NF-side arms/cross-bridges, which interact with one another and with other cytoskeletal molecules to stabilize the axonal cytoskeleton, increase axon caliber, and increase conduction velocity. During NF processing from cell body to axon terminal, phosphorylation is topographically regulated by a dynamic equilibrium between the activities of kinases and phosphatases within cellular compartments. Although all kinases, phosphatases, regulators, and substrates are synthesized in cell bodies, the extensive stable phosphorylation of the NFP tail domains occurs primarily in the axon during axonal transport. Inasmuch as NFs require multisite phosphorylation, it is possible that during NF processing, a sequence of multisite phosphorylations and dephosphorylations are required before the Lys-Ser-Pro (KSP) repeat sites become accessible for axonal phosphorylation. These processes are not well understood. In the following example, the characterization of neurofilament phosphorylation by MS-based techniques is described.

1. Protocol 5: Characterization of Phosphorylation Sites in Neurofilament Proteins from Mammalian Nervous Tissue by Liquid Chromatography/Mass Spectrometry/Mass Spectrometry

Similar methods are used in characterization and identification of phosphorylated residues in the NFPs as those described previously for other IF proteins. A short description and relevant references are given in the following.

1. Neurofilaments from bovine spinal cords, human, rat/mouse, or other mammalian nervous tissues were obtained and purified as described (Carden *et al.*, 1985; Shetty *et al.*, 1993; Tokutake *et al.*, 1983) (Fig. 5).

2. Human NF-H (50 μg) was dried in a Speed Vac (Savant) and taken up and heated in 25 μl of 8 M urea/0.4 M NH_4HCO_3 at 50 °C for 1 hour to affect solubilization and denaturation of the protein, essentially according to Veeranna *et al.* (1995) but without reduction or alkylation.

3. The resulting solution was diluted with 75 μl of H_2O to 2 M urea/0.1 M NH_4HCO_3 before digestion overnight with 3 μg of modified trypsin (Promega) at

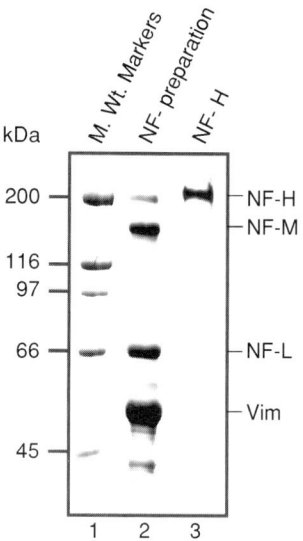

Fig. 5 SDS-PAGE (10% acrylamide, silver staining) analysis of isolation and purification of human NF-H. Lane 1, molecular mass markers; lane 2, human neurofilament preparation; lane 3, human NF-H (From Jaffe *et al.*, 1998 with permission).

37 °C or diluted with 175 μl of H_2O to 1 M urea/0.05 M NH_4-HCO_3 before overnight digestion with 3 μg of endoproteinase Glu-C (Boehringer Mannheim) at ambient temperature.

4. The resulting digests (one third of each) were desalted online by use of a peptide trap cartridge (see Jaffe *et al.* [1998] for details) before separation by microbore RP-HPLC on a 1.0 × 150 mm Monitor C18 column (Column Engineering) and the prefilter eluted at 50 μl/min at 40 °C by use of: (1) a two-step gradient of 2%–10% solvent B over 20 minutes followed by 10%–65% solvent B over 20 minutes (2–10–65 gradient); (2) a two-step gradient of 2%–15% solvent B over 20 minutes followed by 15%–65% solvent B over 20 minutes (2–15–65 gradient); or (3) a one-step linear gradient of 2%–65% solvent B over 40 minutes (2–65 gradient) on a Magic 2002 model microbore HPLC (Michrom Bioresources) equipped with a Model 718 refrigerated autosampler (Alcott).

5. The trypsin digest was analyzed twice by the 2-10-65 gradient. Solvent A was 10/10/980/1/0.2 CH_3-CN/1-PrOH/H_2O/AcOOH/TFA (v/v/v/v/v), and solvent B was 700/200/100/0.9/0.2 CH_3CN/1-PrOH/H_2O/AcOOH/TFA (v/v/v/v/v). Column effluent was monitored at 215 nm (Fig. 6). The HPLC system was coupled to a Model LCQ mass spectrometer (Finnigan) equipped with an ESI.

6. The mass spectrometer was operated in the "triple-play" mode (Fig. 7), in which the instrument was set up to automatically acquire: (1) a full scan, (2) a

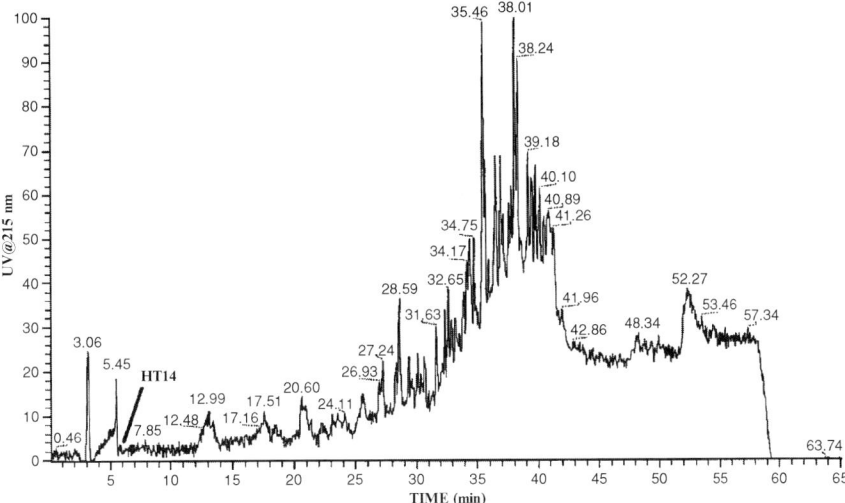

Fig. 6 Microbore RP-HPLC analysis of tryptic digest of human NF-H; 17 mg of human high molecular weight neurofilament protein was digested with trypsin, and the resultant peptides were purified by a two-step 2-10-65 gradient (buffer A/buffer B). UV absorbance was monitored at 215 nm. Phosphopeptides eluted between 5.8 and 25.5 minutes. Phosphopeptide HT14 eluting at 6.1 minutes is indicated (From Jaffe *et al.*, 1998 with permission).

ZoomScan (higher resolution, lower mass range scan) of the $(M + nH)nz$ ion above a preset threshold, and (3) a tandem MS/MS spectrum (relative collision energy 35%) from that ion as shown in Jaffe *et al.* (1998). In this way, the three mass spectra 1–3 are automatically acquired from ions resulting from all major peptide peaks eluting from the HPLC column. Source conditions were as follows: capillary temperature, 220 °C; sheath gas flow, 80 units; auxiliary gas flow, 20 units; ESI spray voltage, 4.2 kV. MS data were acquired on a Gateway 2000 computer (Gateway) and analyzed with the BioExplore software package (Finnigan) or converted to Unix format for analysis on an AlphaStation 200 (Digital) with the Bioworks software package (Finnigan). MS/MS spectra ascertained to be from the same precursor ion were added. HPLC operation and MS calibration were checked daily by injection of synthetic peptides, substance P, or kassinin (Peninsula).

7. Uninterpreted MS/MS spectra were searched with the PEPSEARCH (Finnigan) or SEQUEST (version B22 or C1, by J. Eng and J. Yates, Department of Molecular Biotechnology, University of Washington, Box 357730, Seattle, WA 98195–7730) programs against a database constructed from the published human NF-H sequence. Search parameters were set to reflect the proteolytic enzyme used and to consider possible phosphorylation (+80) at all serine, threonine, or tyrosine

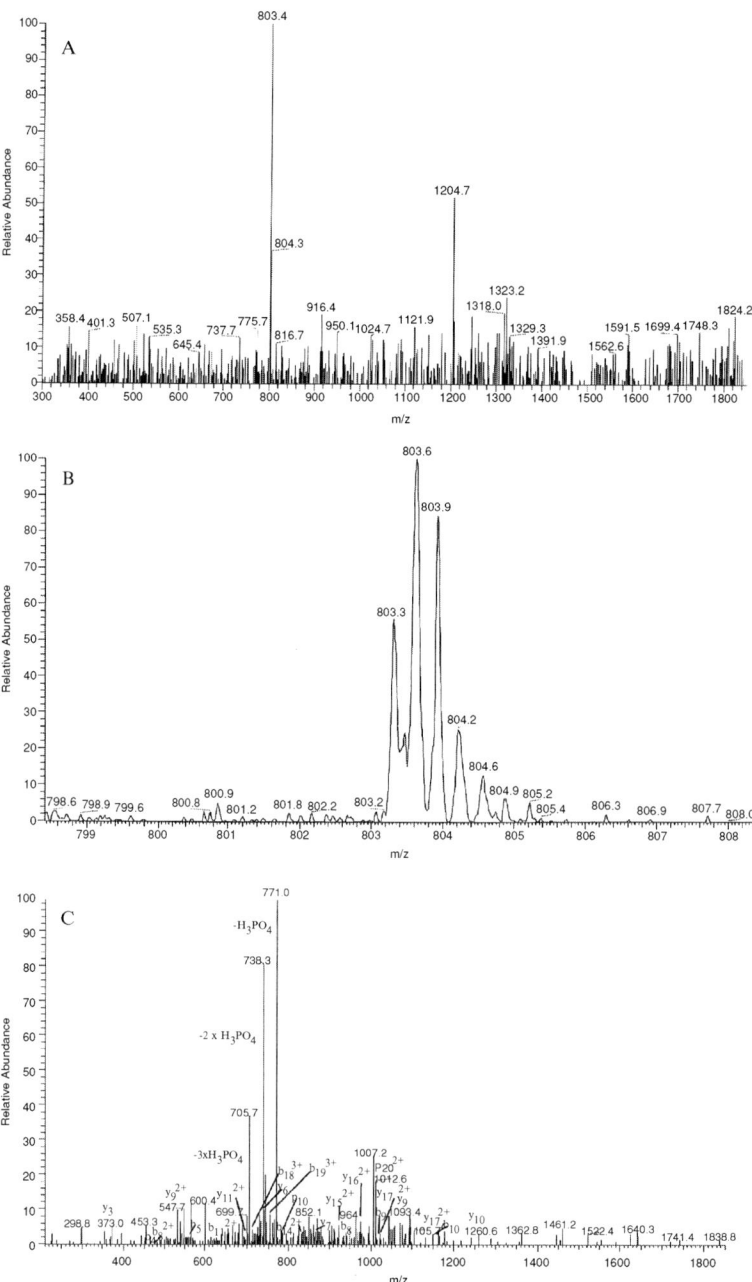

Fig. 7 "Triple play" analysis. The triply charged ion at m/z 803.4, eluting at 6.1 minutes, from human NF-H tryptic was analyzed by triple play analysis. The peak corresponds to the triphosphopeptide, AKS*PVKEEAKS*PEKAKS*PEK. (A) Full-scale scan MS spectrum of the triply

residues. Fragment ions were interpreted and labeled with the PEPMATCH program (Finnigan).

Comments

In addition to the use of human or animal NFs as a source for experiments, *in vitro* or *in vivo* phosphorylation of NFPs can be achieved by use of purified *in vitro* phosphorylated NF preparations, *in vivo* labeled neuronal cultures, or labeled proteins isolated from nervous tissues after injecting the animals with desired labels, as described in earlier publications (Jaffe *et al.*, 1998; Shetty *et al.*, 1993; Veeranna *et al.*, 1995). After a suitable labeling procedure, immunoprecipitation (IP) of individual NF-subunits can then be achieved with commercially available specific antibodies, and the phosphorylated residues can be analyzed.

The medium and high molecular weight neurofilament proteins are stably and extensively phosphorylated *in vivo* in the axonal compartments. The phosphorylated residues in these proteins can be evaluated on isolating these proteins with LC/MS/MS as described earlier or with phosphoproteomics currently used in many laboratories.

III. Pearls and Pitfalls

Although phosphorylation analysis is getting easier, with better established traditional methods and introduction of new and more sensitive methods, there is still no single method that on its own would be likely to yield all the phosphorylation sites on a given IF protein. Eventually, as the new MS-based methods are getting progressively more powerful, reliable, and easily available, they may render traditional methods redundant and unnecessary. However, we are not at this stage yet, but many of the current MS-based methods have a number of general caveats and specific limitations. Therefore, phosphorylation analysis would ideally include a number of parallel methods to maximize the chance of identifying all important sites. When it comes to biochemical tools to identify phosphorylation events, phosphopeptide-specific antibodies are getting increasingly useful in the spatiotemporal phosphorylation analysis of specific phosphorylation sites on individual proteins, especially because there is an increasingly broad range of antibodies available for different IF proteins (although not all of them are commercially

---

charged ion at m/z 803.4. (B) Zoom scan of the triply charged ion at m/z 803.4. Smoothing algorithm (7 points) applied. The one-third m/z unit spacing of the isotope cluster in the ZoomScan high-resolution spectrum identifies the parent ion as triply charged. (C) Full-scan MS/MS spectrum of the triply charged ion at m/z 803.4. The spectrum is dominated by three ions at m/z 771.0, 738.3, and 705.7, consistent with the loss of $1-3 \times 32.7$ ($H_3PO_4/3$) from the triply charged parent ion at m/z 803.4, which flags this spectrum as resulting from a triply charged triphosphopeptide. Observed a,b, b^*, b_o, y, y^*, y_o, and p ions (52–54) consistent with those predicted for this peptide and at least above a threshold of 4% of the base peak counts (140,681 counts) are labeled (From Jaffe *et al.*, 1998 with permission).

available). In contrast, when it comes to detection of phosphoamino acid specificity of a given phosphorylation event, with the exception some relatively well-established phospho-tyrosine–specific antibodies that will detect a rather broad range of phosphotyrosine motifs, there are no phosphoserine or phosphothreonine-specific antibodies available that would be useful as general tools.

IV. Conclusions

IFs are among the most abundant phosphoproteins in the cell. An active phosphorylation-based regulation seems to be a key common regulatory denominator for the members of this large and heterogeneous protein family. The IFs whose phosphorylation has been characterized reveal surprisingly complicated phosphorylation patterns, characteristically with numerous phosphorylation sites and a complex interrelationship between the degrees of phosphorylation on individual sites. Establishing the phosphorylation sites of all individual IFs and, moreover, characterizing the biological functions of individual phosphorylation sites on all different IF proteins will be a major task. Fortunately, the methods for studying phosphoproteins are getting more efficient, and new MS-based techniques will pave the way toward faster characterization and quantification of individual phosphorylation sites. When more sites are known and when individual kinases regulating these sites have been established, IFs will, because of their abundant expression and their tissue-specific expression patterns, provide intriguing possibilities to study the interrelationship between the multitude of signaling pathways that are active at any given moment in a cell. The IFs will, therefore, be an interesting model system for studying posttranslational modification networks, and, also because of their abundance and active phosphorylation, a useful and convenient model system for developing existing and new tools for phosphoproteomics.

Acknowledgments

We extend our apologies to the numerous authors and research groups whose articles we, because of limited space available, could not include among the referred articles. The references on signaling modifying agents especially include only a few useful references on the use and mechanisms of action of theses compounds and do not do justice to the original publications describing the discoveries of these compounds. This work was supported by the Academy of Finland.

References

Adamczyk, M., Gebler, J. C., and Wu, J. (2001). Selective analysis of phosphopeptides within a protein mixture by chemical modification, reversible biotinylation and mass spectrometry. *Rapid Commun. Mass Spectrom.* **15,** 1481–1488.

Akiyama, T., and Ogawara, H. (1991). Use and specificity of genistein as inhibitor of protein-tyrosine kinases. *Methods Enzymol.* **201,** 362–370.

Alessi, F., Quarta, S., Savio, M., Riva, F., Rossi, L., Stivala, L. A., Scovassi, A. I., Meijer, L., and Prosperi, E. (1998). The cyclin-dependent kinase inhibitors olomoucine and roscovitine arrest human fibroblasts in G1 phase by specific inhibition of CDK2 kinase activity. *Exp. Cell Res.* **245**, 8–18.

Angelastro, J. M., Ho, C. L., Frappier, T., Liem, R. K., and Greene, L. A. (1998). Peripherin is tyrosine-phosphorylated at its carboxyl-terminal tyrosine. *J. Neurochem.* **70**, 540–549.

Annan, R. S., and Carr, S. A. (1996). Phosphopeptide analysis by matrix-assisted laser desorption time-of-flight mass spectrometry. *Anal. Chem.* **68**, 3413–3421.

Asara, J. M., and Allison, J. (1999). Enhanced detection of phosphopeptides in matrix-assisted laser desorption/ionization mass spectrometry using ammonium salts. *J. Am. Soc. Mass. Spectrom.* **10**, 35–44.

Badger, A. M., Bradbeer, J. N., Votta, B., Lee, J. C., Adams, J. L., and Griswold, D. E. (1996). Pharmacological profile of SB 203580, a selective inhibitor of cytokine suppressive binding protein/p38 kinase, in animal models of arthritis, bone resorption, endotoxin shock and immune function. *J. Pharmacol. Exp. Ther.* **279**, 1453–1461.

Bennett, K. L., Stensballe, A., Podtelejnikov, A. V., Moniatte, M., and Jensen, O. N. (2002). Phosphopeptide detection and sequencing by matrix-assisted laser desorption/ionization quadrupole time-of-flight tandem mass spectrometry. *J. Mass. Spectrom.* **37**, 179–190.

Bhat, S. V. (1993). Forskolin and congeners. *Fortschr. Chem. Org. Naturst.* **62**, 1–74.

Bogoyevitch, M. A., Boehm, I., Oakley, A., Ketterman, A. J., and Barr, R. K. (2004). Targeting the JNK MAPK cascade for inhibition: Basic science and therapeutic potential. *Biochim. Biophys. Acta* **1697**, 89–101.

Boyle, W. J., van der Geer P., and Hunter T. (1991). Phosphopeptide mapping and phosphoamino acid analysis by two-dimensional separation on thin-layer cellulose plates. *Methods Enzymol.* **201**, 110–149.

Carden, M. J., Schlaepfer, W. W., and Lee, V. M. (1985). The structure, biochemical properties, and immunogenicity of Neurofilament peripheral regions are determined by phosphorylation state. *J. Biol. Chem.* **260**, 9805–9817.

Chou, C. F., and Omary, M. B. (1994). Mitotic arrest with anti-microtubule agents or okadaic acid is associated with increased glycoprotein terminal GlcNAc's. *J. Cell Sci.* **107(Pt 7)**, 1833–1843.

Cohen, P. (1989). The structure and regulation of protein phosphatases. *Annu. Rev. Biochem.* **58**, 453–508.

Cohen, P., Holmes, C. F., and Tsukitani, Y. (1990). Okadaic acid: A new probe for the study of cellular regulation. *Trends Biochem. Sci.* **15**, 98–102.

Coligan, J. E., Dunn, B. M., Ploegh, H. L., Speicher, D. W., and Wingfield, P. T. (1995). "Current Protocols in Protein Science." John Wiley & Sons Inc., New York.

Coulombe, P. A., and Omary, M. B. (2002). 'Hard' and 'soft' principles defining the structure, function and regulation of keratin intermediate filaments. *Curr. Opin. Cell Biol.* **14**, 10–22.

Djordjevic, S., and Driscoll, P. C. (2002). Structural insight into substrate specificity and regulatory mechanisms of phosphoinositide 3-kinases. *Trends Biochem. Sci.* **27**, 426–432.

Dudley, D. T., Pang, L., Decker, S. J., Bridges, A. J., and Saltiel, A. R. (1995). A synthetic inhibitor of the mitogen-activated protein kinase cascade. *Proc. Natl. Acad. Sci. USA* **92**, 7686–7689.

Dumont, F. J., Staruch, M. J., Koprak, S. L., Melino, M. R., and Sigal, N. H. (1990). Distinct mechanisms of suppression of murine T cell activation by the related macrolides FK-506 and rapamycin. *J. Immunol.* **144**, 251–258.

Eriksson, J. E., He, T., Trejo-Skalli, A. V., Harmala-Brasken, A. S., Hellman, J., Chou, Y. H., and Goldman, R. D. (2004). Specific *in vivo* phosphorylation sites determine the assembly dynamics of vimentin intermediate filaments. *J. Cell Sci.* **117**, 919–932.

Eriksson, J. E., Opal, P., and Goldman, R. D. (1992). Intermediate filament dynamics. *Curr. Opin. Cell Biol.* **4**, 99–104.

Eriksson, J. E., Toivola, D., Meriluoto, J. A., Karaki, H., Han, Y. G., and Hartshorne, D. (1990). Hepatocyte deformation induced by cyanobacterial toxins reflects inhibition of protein phosphatases. *Biochem. Biophys. Res. Commun.* **173**, 1347–1353.

Eriksson, J. E., Toivola, D. M., Sahlgren, C., Mikhailov, A., and Harmala-Brasken, A. S. (1998). Strategies to assess phosphoprotein phosphatase and protein kinase-mediated regulation of the cytoskeleton. *Methods Enzymol.* **298**, 542–569.

Falick, A. M. (1995). Matrix-assisted laser desorption ionization-time of flight (MALDI-TOF) analysis of peptides using post-source decay. *In* "PerSeptive Biosystems Technical Bulletin." PerSeptive Biosystems, Foster City, CA.

Fantes, P., and Brooks, R. (1993). "The Cell Cycle, A Practical Approach." IRL Press, Oxford.

Feng, L., Zhou, X., Liao, J., and Omary, M. B. (1999). Pervanadate-mediated tyrosine phosphorylation of keratins 8 and 19 via a p38 mitogen-activated protein kinase-dependent pathway. *J. Cell Sci.* **112(Pt 13),** 2081–2090.

Fujiki, H., and Suganuma, M. (1993). Tumor promotion by inhibitors of protein phosphatases 1 and 2A: the okadaic acid class of compounds. *Adv. Cancer Res.* **61,** 43–94.

Galli, C., Meucci, O., Scorziello, A., Werge, T. M., Calissano, P., and Schettini, G. (1995). Apoptosis in cerebellar granule cells is blocked by high KCl, forskolin, and IGF-1 through distinct mechanisms of action: The involvement of intracellular calcium and RNA synthesis. *J. Neurosci.* **15,** 1172–1179.

Garrison, J. C. (1993). *In* "Protein Phosphorylation, A Practical Approach" (D. G. Hardie, ed.). Oxford University Press, Inc., New York.

Gjertsen, B. T., Mellgren, G., Otten, A., Maronde, E., Genieser, H. G., Jastorff, B., Vintermyr, O. K., McKnight, G. S., and Doskeland, S. O. (1995). Novel (Rp)-cAMPS analogs as tools for inhibition of cAMP-kinase in cell culture. Basal cAMP-kinase activity modulates interleukin-1 beta action. *J. Biol. Chem.* **270,** 20599–20607.

Gordon, J. A. (1991). Use of vanadate as protein-phosphotyrosine phosphatase inhibitor. *Methods Enzymol.* **201,** 110–149.

Grider, J. R. (1993). Interplay of VIP and nitric oxide in regulation of the descending relaxation phase of peristalsis. *Am. J. Physiol.* **264,** G334–G340.

Hardie, D. G. (1993). "Protein Phosphorylation, A Practical Approach." Oxford University Press Inc, New York.

Hardie, D. G., Haystead, T. A., and Sim, A. T. (1991). Use of okadaic acid to inhibit protein phosphatases in intact cells. *Methods Enzymol.* **201,** 469–476.

Harlow, E., and Lane, D. (1988). "Antibodies—A Laboratory Manual." Cold Spring Harbour Laboratory, New York.

Hei, Y. J., MacDonell, K. L., McNeill, J. H., and Diamond, J. (1991). Lack of correlation between activation of cyclic AMP-dependent protein kinase and inhibition of contraction of rat vas deferens by cyclic AMP analogs. *Mol. Pharmacol.* **39,** 233–238.

Helfand, B. T., Chang, L., and Goldman, R. D. (2003). The dynamic and motile properties of intermediate filaments. *Annu. Rev. Cell Dev. Biol.* **19,** 445–467.

Helfand, B. T., Chang, L., and Goldman, R. D. (2004). Intermediate filaments are dynamic and motile elements of cellular architecture. *J. Cell Sci.* **117,** 133–141.

Hidaka, H., and Kobayashi, R. (1993). *In* "Protein Phosphorylation a Practical Approach" (D. G. Hardie, ed.). IRL Press, Oxford, England.

Hidaka, H., Watanabe, M., and Kobayashi, R. (1991). Properties and use of H-series compounds as protein kinase inhibitors. *Methods Enzymol.* **201,** 328–339.

Honkanen, R. E., Codispoti, B. A., Tse, K., Boynton, A. L., and Honkanan, R. E. (1994). Characterization of natural toxins with inhibitory activity against serine/threonine protein phosphatases. *Toxicon* **32,** 339–350.

Humphries, M. J. (1996). Integrin activation: The link between ligand binding and signal transduction. *Curr. Opin. Cell Biol.* **8,** 632–640.

Hunter, T., and Sefton, B. M. (1991). "Methods in Enzymology." **201** (whole volume). Academic Press, New York.

Ishihara, H., Martin, B. L., Brautigan, D. L., Karaki, H., Ozaki, H., Kato, Y., Fusetani, N., Watabe, S., Hashimoto, K., Uemura, D., *et al.* (1989). Calyculin A and okadaic acid: Inhibitors of protein phosphatase activity. *Biochem. Biophys. Res. Commun.* **159,** 871–877.

Jaffe, H., Veeranna, Shetty, K. T., and Pant, H. C. (1998). Characterization of the phosphorylation sites of human high molecular weight neurofilament protein by electrospray ionization tandem mass spectrometry and database searching. *Biochemistry* **37,** 3931–3940.

Janek, K., Wenschuh, H., Bienert, M., and Krause, E. (2001). Phosphopeptide analysis by positive and negative ion matrix assisted laser desorption/ionization mass spectrometry. *Rapid Commun. Mass Spectrom.* **15,** 1593–1599.

Kase, H., Iwahashi, K., Nakanishi, S., Matsuda, Y., Yamada, K., Takahashi, M., Murakata, C., Sato, A., and Kaneko, M. (1987). K-252 compounds, novel and potent inhibitors of protein kinase C and cyclic nucleotide-dependent protein kinases. *Biochem. Biophys. Res. Commun.* **142,** 436–440.

Kaufmann, R., Chaurand, P., Kirsch, D., and Spengler, B. (1996). Post-source decay and delayed extraction in matrix-assisted laser desorption/ionization-reflectron time-of-flight mass spectrometry. Are there trade-offs? *Rapid Commun. Mass Spectrom* **10,** 1199–1208.

Kinter, M., and Sherman, N. (2000). "Protein Sequencing and Identification Using Tandem Mass Spectrometry." John Wiley & Sons Inc., New York.

Kiss, Z., Phillips, H., and Anderson, W. H. (1995). The bisindolylmaleimide GF 109203X, a selective inhibitor of protein kinase C, does not inhibit the potentiating effect of phorbol ester on ethanol-induced phospholipase C-mediated hydrolysis of phosphatidylethanolamine. *Biochim. Biophys. Acta* **1265,** 93–95.

Knight, Z. A., Schilling, B., Row, R. H., Kenski, D. M., Gibson, B. W., and Shokat, K. M. (2003). Phosphospecific proteolysis for mapping sites of protein phosphorylation. *Nat. Biotechnol.* **21,** 1047–1054.

Kobayashi, E., Nakano, H., Morimoto, M., and Tamaoki, T. (1989). Calphostin C (UCN-1028C), a novel microbial compound, is a highly potent and specific inhibitor of protein kinase C. *Biochem. Biophys. Res. Commun.* **159,** 548–553.

Ku, N. O., Liao, J., Chou, C. F., and Omary, M. B. (1996). Implications of intermediate filament protein phosphorylation. *Cancer Metastasis Rev.* **15,** 429–444.

Ku, N. O., Liao, J., and Omary, M. B. (1998). Phosphorylation of human keratin 18 serine 33 regulates binding to 14-3-3 proteins. *EMBO J.* **17,** 1892–1906.

Ku, N. O., Michie, S., Oshima, R. G., and Omary, M. B. (1995). Chronic hepatitis, hepatocyte fragility, and increased soluble phosphoglycokeratins in transgenic mice expressing a keratin 18 conserved arginine mutant. *J. Cell Biol.* **131,** 303–314.

Ku, N. O., and Omary, M. B. (1995). Identification and mutational analysis of the glycosylation sites of human keratin 18. *J. Biol. Chem.* **270,** 11820–11827.

Ku, N. O., and Omary, M. B. (2000). Keratins turn over by ubiquitination in a phosphorylation-modulated fashion. *J. Cell Biol.* **149,** 547–552.

Levitzki, A., Gazit, A., Osherov, N., Posner, I., and Gilon, C. (1991). Inhibition of protein-tyrosine kinases by tyrphostins. *Methods Enzymol.* **201,** 347–361.

Li, Y. M., and Casida, J. E. (1992). Cantharidin-binding protein: Identification as protein phosphatase 2A. *Proc. Natl. Acad. Sci. USA* **89,** 11867–11870.

Liao, J., Ku, N. O., and Omary, M. B. (1997). Stress, apoptosis, and mitosis induce phosphorylation of human keratin 8 at Ser-73 in tissues and cultured cells. *J. Biol. Chem.* **272,** 7565–7573.

Liao, J., and Omary, M. B. (1996). 14-3-3 proteins associate with phosphorylated simple epithelial keratins during cell cycle progression and act as a solubility cofactor. *J. Cell Biol.* **133,** 345–357.

Liao, P. C., Leykam, J., Andrews, P. C., Gage, D. A., and Allison, J. (1994). An approach to locate phosphorylation sites in a phosphoprotein: Mass mapping by combining specific enzymatic degradation with matrix-assisted laser desorption/ionization mass spectrometry. *Anal. Biochem.* **219,** 9–20.

Liu, J., Farmer, J. D., Jr., Lane, W. S., Friedman, J., Weissman, I., and Schreiber, S. L. (1991). Calcineurin is a common target of cyclophilin-cyclosporin A and FKBP-FK506 complexes. *Cell* **66,** 807–815.

Loboda, A. V., Krutchinsky, A. N., Bromirski, M., Ens, W., and Standing, K. G. (2000). A tandem quadrupole/time-of-flight mass spectrometer with a matrix-assisted laser desorption/ionization source: design and performance. *Rapid Commun. Mass Spectrom.* **14,** 1047–1057.

MacKintosh, C. (1993). *In* "Protein Phosphorylation, A Practical Approach" (D. G. Hardie, ed.), pp. 197–210. Oxford University Press, Inc., New York.

MacKintosh, C., and MacKintosh, R. W. (1994). Inhibitors of protein kinases and phosphatases. *Trends Biochem. Sci.* **19,** 444–448.

Mann, M., Hendrickson, R. C., and Pandey, A. (2001). Analysis of proteins and proteomes by mass spectrometry. *Annu. Rev. Biochem.* **70,** 437–473.

Mann, M., Ong, S. E., Gronborg, M., Steen, H., Jensen, O. N., and Pandey, A. (2002). Analysis of protein phosphorylation using mass spectrometry: Deciphering the phosphoproteome. *Trends Biotechnol.* **20,** 261–268.

Marin, P., Nastiuk, K. L., Daniel, N., Girault, J. A., Czernik, A. J., Glowinski, J., Nairn, A. C., and Premont, J. (1997). Glutamate-dependent phosphorylation of elongation factor-2 and inhibition of protein synthesis in neurons. *J. Neurosci.* **17,** 3445–3454.

Matsushima, R., Yoshizawa, S., Watanabe, M. F., Harada, K., Furusawa, M., Carmichael, W. W., and Fujiki, H. (1990). *In Vitro* and *in vivo* effects of protein phosphatase inhibitors, microcystins and nodularin, on mouse skin and fibroblasts. *Biochem. Biophys. Res. Commun.* **171,** 867–874.

McLachlin, D. T., and Chait, B. T. (2001). Analysis of phosphorylated proteins and peptides by mass spectrometry. *Curr. Opin. Chem. Biol.* **5,** 591–602.

Meijer, L., and Raymond, E. (2003). Roscovitine and other purines as kinase inhibitors. From starfish oocytes to clinical trials. *Acc. Chem. Res.* **36,** 417–425.

Mokhtari, A., Do, Khac, L., Tanfin, Z., and Harbon, S. (1985). Forskolin modulates cyclic AMP generation in the rat myometrium. Interactions with isoproterenol and prostaglandins E2 and I2. *J. Cyclic Nucleotide Protein Phosphor. Res.* **10,** 213–227.

Muid, R. E., Dale, M. M., Davis, P. D., Elliott, L. H., Hill, C. H., Kumar, H., Lawton, G., Twomey, B. M., Wadsworth, J., Wilkinson, S. E., *et al.* (1991). A novel conformationally restricted protein kinase C inhibitor, Ro 31-8425, inhibits human neutrophil superoxide generation by soluble, particulate and post-receptor stimuli. *FEBS Lett.* **293,** 169–172.

Neubauer, G., and Mann, M. (1999). Mapping of phosphorylation sites of gel-isolated proteins by nanoelectrospray tandem mass spectrometry: Potentials and limitations. *Anal. Chem.* **71,** 235–242.

Nishino, H., Kitagawa, K., Iwashima, A., Ito, M., Tanaka, T., and Hidaka, H. (1986). N-(6-phenylhexyl)-5-chloro-1-naphthalenesulfonamide is one of a new class of activators for Ca2+-activated, phospholipid-dependent protein kinase. *Biochim. Biophys. Acta* **889,** 236–239.

Nishio, H., Ikegami, Y., Segawa, T., and Nakata, Y. (1994). Stimulation of calcium sequestration by mezerein, a protein kinase C activator, in saponized rabbit platelets. *Gen. Pharmacol.* **25,** 413–416.

Nishizawa, K., Yano, T., Shibata, M., Ando, S., Saga, S., Takahashi, T., and Inagaki, M. (1991). Specific localization of phosphointermediate filament protein in the constricted area of dividing cells. *J. Biol. Chem.* **266,** 3074–3079.

Nuwaysir, L. M., and Stults, J. T. (1993). Electrospray ionization mass spectrometry of phosphopeptides isolated by on-line immobilized metal-ion affinity chromatography. *J. Am. Soc. Mass Spectrom.* **4,** 662–669.

Oda, Y., Nagasu, T., and Chait, B. T. (2001). Enrichment analysis of phosphorylated proteins as a tool for probing the phosphoproteome. *Nat. Biotechnol.* **19,** 379–382.

Ogawara, M., Inagaki, N., Tsujimura, K., Takai, Y., Sekimata, M., Ha, M. H., Imajoh-Ohmi, S., Hirai, S., Ohno, S., Sugiura, H., *et al.* (1995). Differential targeting of protein kinase C and CaM kinase II signalings to vimentin. *J. Cell Biol.* **131,** 1055–1066.

Roivainen, R., and Messing, R. O. (1993). The phorbol derivatives thymeleatoxin and 12-deoxyphorbol-13-O-phenylacetate-10-acetate cause translocation and down-regulation of multiple protein kinase C isozymes. *FEBS Lett.* **319,** 31–34.

Sahlgren, C. M., Mikhailov, A., Hellman, J., Chou, Y. H., Lendahl, U., Goldman, R. D., and Eriksson, J. E. (2001). Mitotic reorganization of the intermediate filament protein nestin involves phosphorylation by cdc2 kinase. *J. Biol. Chem.* **276,** 16456–16463.

Sahlgren, C. M., Mikhailov, A., Vaittinen, S., Pallari, H. M., Kalimo, H., Pant, H. C., and Eriksson, J. E. (2003). Cdk5 regulates the organization of Nestin and its association with p35. *Mol. Cell Biol.* **23,** 5090–5106.

Saitoh, M., Ishikawa, T., Matsushima, S., Naka, M., and Hidaka, H. (1987). Selective inhibition of catalytic activity of smooth muscle myosin light chain kinase. *J. Biol. Chem.* **262**, 7796–7801.

Sambrook, J., and Russell, D. W. (2001). "Molecular Cloning—A Laboratory Manual," Vol. 3. Cold Spring Harbor Laboratory Press, New York.

Seamon, K. B., and Daly, J. W. (1981). Forskolin: A unique diterpene activator of cyclic AMP-generating systems. *J. Cyclic Nucleotide Res.* **7**, 201–224.

Shetty, K. T., Link, W. T., and Pant, H. C. (1993). cdc2-like kinase from rat spinal cord specifically phosphorylates KSPXK motifs in neurofilament proteins: isolation and characterization. *Proc. Natl. Acad. Sci. USA* **90**, 6844–6848.

Simon, H. U., Yousefi, S., Dibbert, B., Hebestreit, H., Weber, M., Branch, D. R., Blaser, K., Levi-Schaffer, F., and Anderson, G. P. (1998). Role for tyrosine phosphorylation and Lyn tyrosine kinase in fas receptor-mediated apoptosis in eosinophils. *Blood* **92**, 547–557.

Spengler, B., Kirsch, D., Kaufmann, R., and Jaeger, E. (1992). Peptide sequencing by matrix-assisted laser-desorption mass spectrometry. *Rapid Commun. Mass Spectrom.* **6**, 105–108.

Sumi, M., Kiuchi, K., Ishikawa, T., Ishii, A., Hagiwara, M., Nagatsu, T., and Hidaka, H. (1991). The newly synthesized selective Ca2+/calmodulin dependent protein kinase II inhibitor KN-93 reduces dopamine contents in PC12h cells. *Biochem. Biophys. Res. Commun.* **181**, 968–975.

Takai, Y., Ogawara, M., Tomono, Y., Moritoh, C., Imajoh-Ohmi, S., Tsutsumi, O., Taketani, Y., and Inagaki, M. (1996). Mitosis-specific phosphorylation of vimentin by protein kinase C coupled with reorganization of intracellular membranes. *J. Cell Biol.* **133**, 141–149.

Tamaoki, T. (1991). Use and specificity of staurosporine, UCN-01, and calphostin C as protein kinase inhibitors. *Methods Enzymol.* **201**, 340–347.

Tepper, C. G., Jayadev, S., Liu, B., Bielawska, A., Wolff, R., Yonehara, S., Hannun, Y. A., and Seldin, M. F. (1995). Role for ceramide as an endogenous mediator of Fas-induced cytotoxicity. *Proc. Natl. Acad. Sci. USA* **92**, 8443–8447.

Toivola, D. M., Goldman, R. D., Garrod, D. R., and Eriksson, J. E. (1997). Protein phosphatases maintain the organization and structural interactions of hepatic keratin intermediate filaments. *J. Cell Sci.* **110(Pt 1)**, 23–33.

Tokutake, S., Hutchison, S. B., Pachter, J. S., and Liem, R. K. (1983). A batchwise purification procedure of neurofilament proteins. *Anal. Biochem.* **135**, 102–105.

Tsujimura, K., Ogawara, M., Takeuchi, Y., Imajoh-Ohmi, S., Ha, M. H., and Inagaki, M. (1994). Visualization and function of vimentin phosphorylation by cdc2 kinase during mitosis. *J. Biol. Chem.* **269**, 31097–31106.

Ui, M., Okada, T., Hazeki, K., and Hazeki, O. (1995). Wortmannin as a unique probe for an intracellular signalling protein, phosphoinositide 3-kinase. *Trends Biochem. Sci.* **20**, 303–307.

Umezawa, K., and Imoto, M. (1991). Use of erbstatin as protein-tyrosine kinase inhibitor. *Methods Enzymol.* **201**, 379–385.

Valgeirsdottir, S., Claesson-Welsh, L., Bongcam-Rudloff, E., Hellman, U., Westermark, B., and Heldin, C. H. (1998). PDGF induces reorganization of vimentin filaments. *J. Cell Sci.* **111(Pt 14)**, 1973–1980.

Veeranna, K., Shetty, T., Link, W. T., Jaffe, H., Wang, J., and Pant, H. C. (1995). Neuronal cyclin-dependent kinase-5 phosphorylation sites in neurofilament protein (NF-H) are dephosphorylated by protein phosphatase 2A. *J. Neurochem.* **64**, 2681–2690.

Villerbu, N., Gaben, A. M., Redeuilh, G., and Mester, J. (2002). Cellular effects of purvalanol A: A specific inhibitor of cyclin-dependent kinase activities. *Int. J. Cancer* **97**, 761–769.

Zhang, X., Herring, C. J., Romano, P. R., Szczepanowska, J., Brzeska, H., Hinnebusch, A. G., and Qin, J. (1998). Identification of phosphorylation sites in proteins separated by polyacrylamide gel electrophoresis. *Anal. Chem.* **70**, 2050–2059.

Zhou, W., Merrick, B. A., Khaledi, M. G., and Tomer, K. B. (2000). Detection and sequencing of phosphopeptides affinity bound to immobilized metal ion beads by matrix-assisted laser desorption/ionization mass spectrometry. *J. Am. Soc. Mass Spectrom.* **11**, 273–282.

PART III

Methods to Study Specific Intermediate Filament Proteins in Mammalian Systems

CHAPTER 15

Hair Keratins and Hair Follicle–Specific Epithelial Keratins

Lutz Langbein,[*] Herbert Spring,[†] Michael A. Rogers,[‡] Silke Praetzel,[*] and Juergen Schweizer[‡]

[*]Division of Cell Biology
German Cancer Research Center
69120 Heidelberg, Germany

[†]Division of Analytical Microscopy and Microinjection
German Cancer Research Center
69120 Heidelberg, Germany

[‡]Division of Normal and Neoplastic Epidermal Differentiation
German Cancer Research Center
69120 Heidelberg, Germany

METHODS IN CELL BIOLOGY, VOL. 78
Copyright 2004, Elsevier Inc. All rights reserved.
0091-679X/04 $35.00

I. Introduction

The large keratin multigene family is made up of the epithelial keratins (also designated "cytokeratins" or "soft keratins"), which are differentially expressed in the various types of epithelia, and the hair keratins ("hard keratins"), representing the major constituents of hard keratinized structures such as hairs, nails, and claws. Compared with epithelial keratins, hair keratins are generally characterized by their higher content of cysteine residues (7.6% vs. 2.9%; Yu *et al.*, 1993) within their head and tail domains. Epithelial and hair keratins are divided into acidic (type I) and basic/neutral (type II) members. Both together form the 10-nm intermediate filament (IF) cytoskeletal network of epithelial cells through an obligatory association of equimolar amounts of a type I and a type II keratin (for reviews, see e.g., Steinert and Roop [1988], Fuchs and Weber [1994], Mischke [1998]).

Earlier gel electrophoretic studies on native hair keratins of various mammals, including man, revealed the presence of four major type I (44–48 kDa) and an equal number of type II members (55–60 kDa) (Heid *et al.*, 1986a,b; Lynch *et al.*, 1986). On the basis of their two-dimensional patterns, the protein spots of each subfamily were numbered from 1 to 4 and further designated "H" for hair, plus an additional prefix "a" or "b", reflecting their nature as acidic (Ha) and basic/neutral (Hb) proteins, respectively. Besides these major Ha1–4 and Hb1–4 hair keratins, an additional, weakly detected "minor" protein pair was designated Hax/Hbx (Heid *et al.*, 1986a,b, 1988). Collectively, the hair keratin family was assumed to be composed of 8 to 10 individual members and therefore appeared considerably less complex than the epithelial keratin family with its 19 members known at that time.

Subsequent studies in our laboratories showed, that in reality, the human type I hair keratin gene subfamily alone had nine functional individual members clustered on chromosome 17q12–21, which could be subdivided into three groups. Genes of groups A (*hHa1, hHa3-I, hHa3-II, hHa4*) and B (*hHa7, hHa8*) each encoded structurally related hair keratins, whereas group C genes (*hHa2, hHa5, hHa6*) contained sequence information for structurally less related hair keratins (Rogers *et al.*, 1998). Similarly, the human type II hair keratin gene subfamily on chromosome 12q13 consisted of six functional individual members, which could be divided into only two groups. Once again, group A genes (*hHb1, hHb3, hHb6*) encoded structurally related hair keratins, whereas the products of the group C genes (*hHb2, hHb4, hHb5*) were rather distinct. The type II locus lacked functional counterparts of the type I group B genes (Rogers *et al.*, 2000). Because both hair keratin gene domains are flanked by epithelial keratin genes and no other hair keratin genes were found elsewhere in the human genome (Rogers *et al.*, 2000), we consider the human hair keratin family as being fully characterized.

Detailed expression studies at both the mRNA and protein level (Langbein *et al.*, 1999, 2001; Rogers *et al.*, 1997) showed that hair keratins hHa2/hHb2 and hHa5/hHb5 defined the early stages of hair differentiation, because hHa5/hHb5 were expressed in both the hair matrix and the lower hair cuticle, whereas the sequentially expressed hair keratins hHa2/hHb2 exhibited coexpression with hHa5/hHb5 only in hair cuticle cells (Fig. 1). Whereas the cuticular differentiation proceeded without the expression of further hair keratins, cells embarking on the inner matrix–cortex pathway sequentially expressed hair keratins hHa1 > hHa3-I > hHa3-II > hHa6 > hHa4 and hHb1 > hHb3 > hHb6, respectively. Furthermore, hHa8 was expressed heterogeneously in cortex cells. The expression of hHa7 was different from that of the other hair keratins in that it could not be

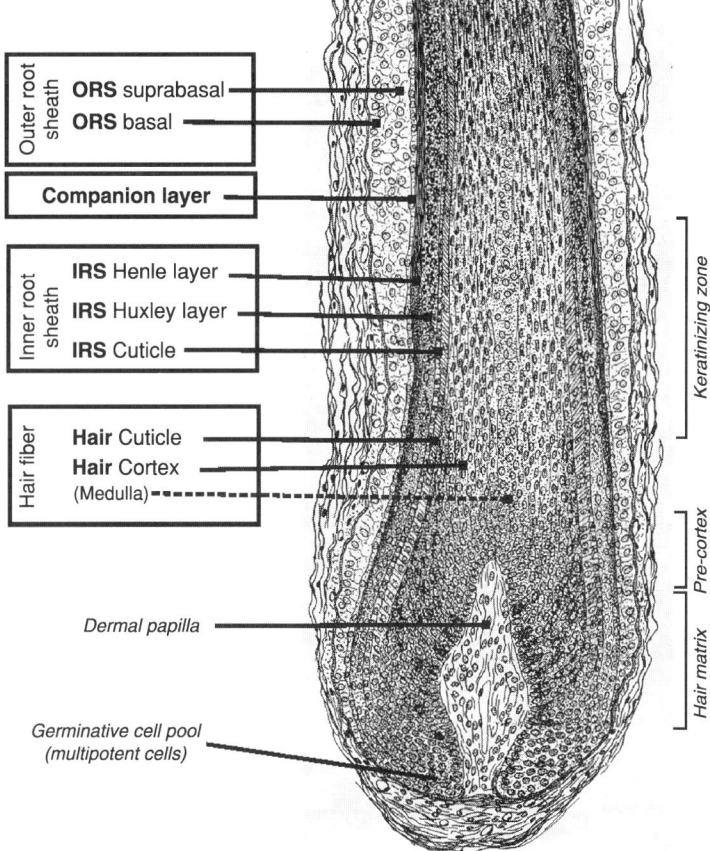

Fig. 1 The hair follicle. Designation of the tissue compartments and morphological description of the hair follicle. All compartments can be defined by special or specific combinations of distinct keratins. The hair forming unit is: (1) hair matrix/lower cortex (Ha5 and Hb5); (2) middle/upper cortex (Ha1, Ha3-I, Ha3-II, Ha4, Ha6, Ha7 and Ha8 as well as Hb1, Hb3, Hb4 and Hb6); (3) the medulla expressing a mixture of hair and epithelial keratins. The IRS is: (1) the IRS cuticle (K6irs3, K6irs2, with K6irs3 starting before K6irs2); (2) Huxley layer (K6irs4); (3) Henle layer (K6irs1, being the only type II keratin of this layer). The companion layer (K6hf). The ORS (K5 and K14). The figure was modified from the excellent and precise drawing of Bucher (1968).

detected in terminal scalp hair follicles but was present in central cortex cells of the rare and small vellus hair follicles of the scalp (compare Fig. 1). Surprisingly, hHb4 was undetectable in hair follicles but clearly demonstrable in the posterior compartment of the tongue filiform papillae. In addition to tissue localization, Western blotting was used to establish a two-dimensional catalog of human hair keratins (Langbein *et al.*, 1999, 2001).

The extension of the keratin analysis to the follicular compartments "outside" the hair-forming compartment, (i.e., the inner root sheath (IRS) [Henle layer, Huxley layer, IRS cuticle]) and the companion layer showed that virtually any of these tissue units expressed specific sets of epithelial keratins. Consistently, their genes were part of the known type I and type II keratin domains and often formed separate evolutionary branches. Keratin K6hf was characteristically expressed in the companion layer, a single-layered column of flat and vertically oriented cells between the cuboidal ORS (outer root sheath) and the IRS Henle cells, and thus defined this structure as an independent follicular compartment (Winter *et al.*, 1998). Remarkably, K6hf was also recently identified as a constituent of the medulla of the central hair-forming compartment (Wang *et al.*, 2003). In this structure, hair and epithelial keratins are obviously expressed as a "mixed" pattern. In addition, the IRS was characterized by distinct sets of keratins of its own. Whereas keratin K6irs1 was specifically expressed in all three compartments of the IRS (Langbein *et al.*, 2002a; Porter *et al.*, 2001), keratins K6irs2 and K6irs3 were both expressed in the IRS cuticle, but showed a different onset of expression in this compartment and starting with K6irs3 (Langbein *et al.*, 2002a). In contrast, keratin K6irs4 was specifically expressed in the Huxley layer and was ideally suited to confirm the occurrence of "*Flügelzellen*" (i.e., Huxley cells) forming horizontal cell extensions that pass through the Henle layer in its entire length (Langbein *et al.*, 2002a, 2003). Thus, together with the hair-forming compartment, all other tissue compartments of the hair follicle can now clearly be defined by single "marker" keratins or by specific patterns of keratins (Langbein *et al.*, 2003).

As members of a large multigene family, most keratin gene sequences and their encoded protein sequences display high homologies, or at least similarities within wide ranges. This is particularly true for the α-helical rod domains but also partially for the head and the tail domains (Parry, 1997; Steinert *et al.*, 1994). Therefore, there are considerable difficulties in selecting specific nucleotide sequences needed for the generation of RNA probes for *in situ* hybridization or peptide sequences used as specific immunogens. Moreover, the hair follicle, in particular the differentiated portion of the cuticle and the cortex, exhibits special morphological properties. Cortex cells accumulate an unprecedented number of up to 10, occasionally 11 different hair keratins (Langbein *et al.*, 2001), making their IF cytoskeleton extremely dense and compact (Hearle, 2000; Powell and Rogers, 1997; Swift, 1997). Moreover, the hair keratins are extremely rich in cysteine in their head and tail domains, by which they are cross-linked through a nearly uncountable number of keratin-associated proteins (KAPs) (Powell and Rogers, 1997; Rogers *et al.*, 2000, 2001, 2002, 2004).

This review is aimed at describing methods for the study of keratin expression in the hair follicle. These methods include: (1) *in situ* hybridization (ISH), (2) light microscopic immunohistochemistry (indirect immunofluorescence, IIF), (3) conventional electron microscopy (EM) and immunoelectron microscopy (IEM), and (4) one-dimensional (1-DE; Sodium dodecylsulfate-polyacrylamide gel electrophoresis [SDS-PAGE]) and two-dimensional (2-DE; IEF or NEPHGE) PAGE. Except for radioactive ISH, which will be described in detail, the description of the remaining and widely used methods will mainly focus in the framework of hair follicle studies on important or specific steps such as tissue processing, preparation of protein extracts, or selection of special tools. It will also relate to possible pitfalls and provide comments and advice regarding materials and instruments.

II. Procedures and Comments

A. *In Situ* Hybridization

1. Introduction

The type of tissue in which a given mRNA is expressed can be identified by Northern blot analysis or polymerase chain reaction (PCR) amplification. Both methods are, however, not suitable for the determination of the exact site of mRNA expression within the tissue of interest. To achieve this, ISH is the method of choice, in particular for cases in which antibodies are not (or not yet) available or if specific antibodies cannot be raised because of high homologies between the members of a multigene family. It should, however, be taken into consideration that ISH, as well as reverse transcriptase (RT) PCR or Northern blot procedures, do not provide definitive information on the translation and the existence of a functional protein.

The principle of ISH is based on the introduction of labeled nucleotides (rCTP, rGTP, rUTP, or rATP) into a cRNA probe by *in vitro* transcription of the respective cDNA, cloned into an appropriate plasmid (e.g., Bluescript), which contains RNA polymerase sites for sense and anti-sense transcription. The labeling can be done with either radioactively (^{35}S) or nonradioactively (e.g., Dig) modified ribonucleotides. As in most cases in our and others hands (Langbein *et al.*, 1993; 1999, 2000; Winter *et al.*, 1998; for a review see also, Wilkinson, 1992), the radioactive labeling is much more sensitive; this method will be described here.

2. Procedures

a. Solutions (should be prepared in advance)

1. *0.1% DEPC-H$_2$O* (6 L): 1 ml diethylpyrocarbonate (e.g., Sigma), add distilled water to 1 L.
2. *10 × PBS-DEPC* (1 L): 82 g NaCl, 2 g KCl, 14.4 g Na$_2$HPO$_4$, 2 g KH$_2$PO$_4$, add DEPC-H$_2$O to 1 L. Check that pH is 7.4, when diluted to 1× PBS-DEPC.

3. *20 × SSC* (1 L): Sterilize by autoclaving: 175 g (3 M) NaCl, 88.2 g (0.3 M) Na-citrate; adjust to pH 7 with HCl and add 0.1% DEPC-H_2O to 1 L.

4. *Deionized formamide*: Mix 100 ml formamide (e.g., Merck) with 5 g ion exchanger (e.g., Amberlite MB6, Serva; or Mixed Bed Resin, Sigma) and stir for 1 hour, filter, and store at room temperature, (RT).

5. *10 × salt solution*: 17.5 g (3M) NaCl, 10 ml (50 mM) EDTA, 25.5 ml 0.2 M NaH_2PO_4, 24.5 ml (0.2 M) Na_2HPO_4, 20 ml 1 M Tris-HCl, pH 6.8; add 0.1% DEPC-H_2O to 100 ml.

6. *50 × Denhardt's solution*: 5 g Ficoll 400, 5 g, polyvinylpyrrolidone, 5 g BSA, add H_2O to 500 ml. Filter through a sterile syringe using a sterile filter. Store at −20 °C.

7. *Hybridization buffer* (5 ml): 2.5 ml deionized formamide, 0.5 ml 10 × salt, 0.5 ml 50% dextransulfate, 100 μl 50× Denhardt's solution, 50 μl 1 M DTT, 250 μl yeast tRNA (e.g., Gibco) from a 10 mg/ml stock solution, 1.1 ml H_2O, 100 μg/ml denatured herring sperm DNA (e.g., Roche).

8. *TE-buffer*: 10 mM Tris-HCl, pH 7.4, 1 mM EDTA; sterilize by autoclaving. Store at RT.

9. *LB-buffer*: 500 mM Tris-HCl, pH 8, 100 mM $MgCl_2$, 50 mM DTT.

10. *Blocking solution* (0.1 M triethanolamine): Add 8 ml 98% triethanolamine (e.g., Sigma) to 600 ml PBS and bring pH to 8–8.5 by adding approximately 1 ml concentrated HCl.

b. Selection of Probes

The appropriate size of the ISH probe that is critical for the penetration into the tissue ranges from approximately 60–1000 Nt (Wilkinson, 1992). As a rule, smaller probes penetrate better but carry the risk of insufficient and/or unspecific labeling. Large probes might be hydrolyzed after labeling, using strong alkaline solutions, but this process is difficult to control and in most cases results in variable mixtures of smaller and larger fragments. The smaller fragments especially might lead to unsatisfactory results.

It is important that the selected nucleotide does not occur in clusters within the desired sequence, because this may lead to low transcription efficiency and possibly disintegration of the probe by radiation. Commonly, we use ^{35}S-rCTP. When only small probes are available, the choice of the labeled nucleotide might be critical, and the most abundant nucleotide should be selected to reach sufficient labeling of the probe. In the case of (large) multigene families such as hair and epithelial keratins, whose members usually exhibit a high degree of sequence conservation in the coding portions of their genes, the selection of 3'- or 5'-noncoding sequence motifs of a given gene is recommended (see Table I). Any of the cDNA sequences selected for *in vitro* transcription of cRNAs, including those for sense probes, has to be checked in DNA databases for specificity.

Table I
**cRNA Probes Used for Specific Human Keratins in the Hair Follicle with Cloning/
Restriction Sites, PCR Primers★**

Probe	Size (bp)	Region	Oligo-nucleotide sequences for PCR or DNA-fragment
hHa1	0.5 kb	3′-nc	PstI-XhoI cloned in BS-SK
hHa2	200	3′-nc	SphI/StyI-fragment cloned in BSII-SK
hHa3-I	250	3′-nc	PCR fragment cloned in EcoRV site of pMOS
hHa3-II	450	3′-nc	BamHI/XhoI-fragment cloned in BSII-KS
hHa4	154	3′-nc	HindIII cloned in BSII-KS
hHa5	300	3′-nc	PstI/ XhoI cloned in BS-SK
hHa6	170	5′-exon 1	taagagttggggctgctcagct tggagcgggagaacgcggagct
hHa7	252	3′-nc	cctgcctcctgtacttcttgtc tacagcttagaggcataggcag
hHa8	250	3′-nc	gttatacctttagaaaatctgg ttagaaacaacccaagaaaatg
ΨhHA	303	3′-nc	tcagcactcctagtccagcc caggaggcaacagaagagag
hHb1	373	3′-nc	PvuII/XhoI-fragment cloned in BSII-SK
hHb2	2541	3′-nc	Entire cDNA-BsrgI digest cloned in CR4,1
hHb3	334	3′-nc	SmaI/PstI-fragment cloned in BSII-KS
hHb4	469	3′-nc	agcagcgtctgtgccaccactg gcttagctggaactgctaatgg
hHb5	657	3′-nc	PvuII/XhoI-fragment cloned into EcoRV/XhoI site of BSII-KS
hHb6	242	3′-nc	PvuII/XhoI-fragment cloned into EcoRV/XhoI site in BSII-KS
K6hf	700	3′-nc	PCR product cloned in EcoRV
K6irs1	183	3′-nc	tggctgccagctttcctcctct gctagatgtggggtggggact
K6irs2	194	3′-nc	tgttttgcctgagccagtattg cccatctttctgcctccatc
K6irs3	263	3′-nc	acaatcccaatcagaagatgaa gatgcaaggagtccagtcag
K6irs4	254	3′-nc	ggaaatagatgctgccattctt ggctgtcaaagtcaccattct

*All subcloned DNA fragments are available from the authors. PCR, polymerase chain reaction.

c. Controls

The results of an ISH experiment have to be rigorously controlled. These controls include:

1. Hybridization without a labeled probe. Either the hybridization solution alone or RNAse digested antisense probes should be used to determine possible background labeling of the tissues (*negative control I*; compare Fig. 2C, which looks same as such a control).

2. Hybridization with the respective sense probe (*negative control II*; see also Fig. 2C).

3. Use of a probe for an appropriate gene whose expression profile and extent of expression in the tissue is known. This step allows the determination of possible mRNA degradation during tissue conservation and storage, but it is also indicative of the quality of the *in vitro* transcription procedure. For investigations on hair keratins on scalp sections, we normally use a keratin K14 probe (see also Fig. 2D). In this tissue, the K14 mRNA is specifically synthesized in the basal cell layer of the interfollicular epidermis, the outer root sheath, and the lining epithelia of the sebaceous and sweat glands. Because scalp sections also contain K14-negative areas (e.g., the dermis), the upper strata

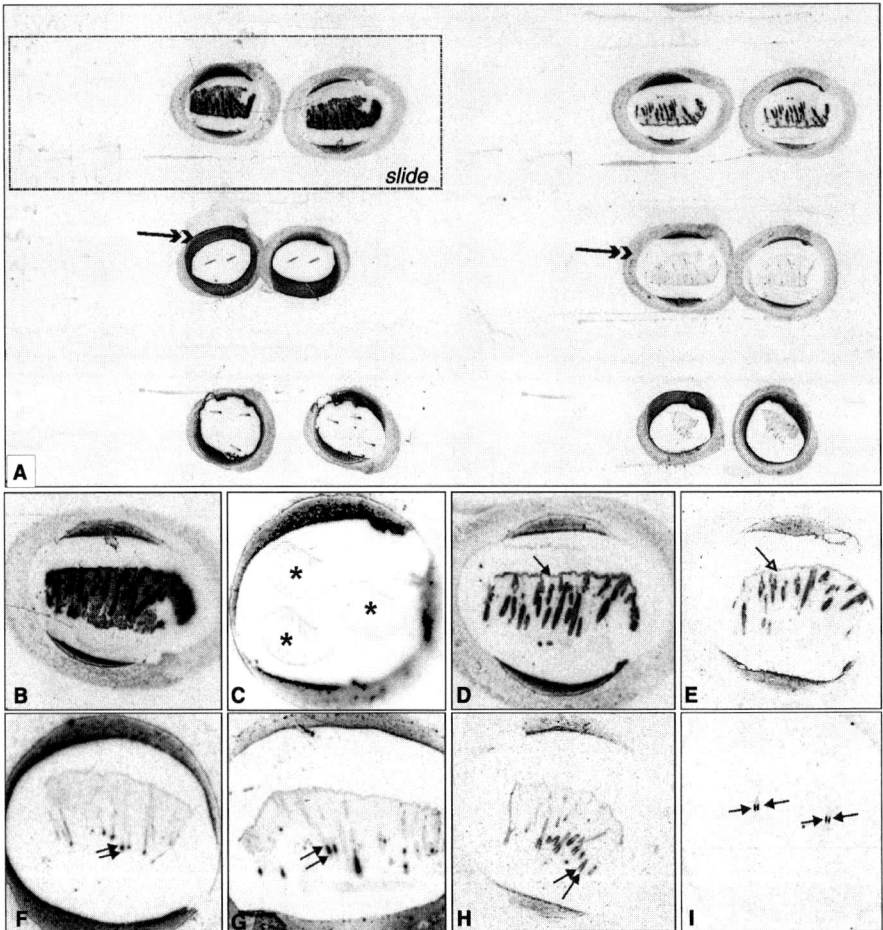

Fig. 2 Evaluation of an *in situ* hybridization (ISH) test x-ray film after 1 day of exposure (see, II.A.2.m). (A) Overview of test films obtained from six hybridized and washed slides. The contour of the upper left slide is outlined by a dotted line. (B) High background staining over the whole tissue section when RNase digestion has been left out or was incomplete. (C) Negative control. No labeling with the sense probe. (D, E) Positive control. Labeling of K14 mRNA (D, arrow) but not K16 mRNA (E, open arrow) in the interfollicular epidermis. Both mRNAs are, however, detectable in the ORS of the hair follicles. (F-I) Specific labeling of the mRNA of hHa5 (F; lowermost part of the follicle bulb, arrows), hHb5 (G; lower/middle part of the follicle, arrows), K6hf (H; two thin lines resulting from the labeling of the companion layer in the lower part of the follicle, arrows), and KAP 10.1 (I; two short parallel lines resulting from the labeling of the hair cuticle in the middle of the vertically oriented follicles, arrows). (B-H) Scalp skin sections. (I) Sections of plucked beard hairs.

of the interfollicular epidermis, the IRS, and the hair fiber, the probe is also useful for the evaluation of background staining (*positive control*; see also Fig. 2D,E).

d. Tissues

Throughout the ISH procedure, slide holders, scissors, forceps etc. used for tissue preparation must be extremely clean and sterile. Gloves are indispensable.

1. Clean instruments with ethanol.
2. Wash instruments in 3% (w/v) H_2O_2-solution for 10 minutes at RT and in DEPC/H_2O for 10 minutes at RT.
3. Wrap instruments in clean aluminium foil.

Instead of steps 1 and 2 above, aRNA-protecting system (e.g., RNase Erase System, ICN) might be used.

For cryoconservation or cryosectioning, tissue samples are immediately snap-frozen after surgical removal in liquid nitrogen–cooled isopentane and stored at $-80\,°C$. Single, freshly plucked beard hair follicles are placed and oriented horizontally on top of a drop of cryosectioning medium (e.g., Tissue Tek, Satura, Zoeterwoude, The Netherlands) on a small piece (approximately 1 cm × 1 cm) of Parafilm, carefully precooled on dry ice to a consistence that allows the follicle to be covered by TissueTek but prevents its further sinking into the medium. Subsequently, the whole sample is allowed to completely freeze on dry ice. The Parafilm can easily be removed, and the TissueTec-drop containing the follicle can be mounted onto the cryomicrotome. Usually, cryostate sections (nominally 5 μm) are prepared and mounted on commercially available adhesive glass slides (e.g., SuperFrost Plus; Menzel, Braunschweig, Germany) or glass slides treated with 3-aminopropyl-triethoxysilane (e.g., Sigma, Deisenhofen, Germany; see below). Sections must be air dried and can be stored at $-80\,°C$. Aldehyde-fixed material embedded in paraffin can also be used. Here, microwave treatment (see, II.B.2.b.2) is of advantage. In this case, prior assessment of mRNA conservation in the tissue by means of a positive control is, in particular, indispensable.

e. Silanization of Slides

1. Treat slides with 1 M HCl for 30 minutes under shaking and wash for 2×5 minutes in H_2O.
2. Dip slides in ethanol and let them dry.
3. Dip slides in 2% 3-aminopropyltrimethoxysilane (e.g., Fluka) in acetone, wash 2× in acetone, air-dry, and wrap in clean aluminium foil.
4. Autoclave slides.

f. Linearization of cDNA for Labeling of In Situ Hybridization Probes

1. Add 10 μg cDNA in 5 μl of a 10× endonuclease-buffer (depending on the enzyme of choice).
2. Add 50 U of the respective restriction endonuclease.

3. Add H_2O to 50 μl (Do not use DEPC-H_2O!).

4. Digest DNA for 2 hours at 37 °C and stop reaction by adding 2 μl 0.5 M EDTA.

5. Purify linearized cDNA with a commercial purification system (e.g., Qiaquick, Qiagen DNA purification columns) and elute in 40 μl TE.

6. Check linearization, purification, and concentration by loading 1 μl (should correspond to approximately 250 ng) on an agarose gel.

g. In Vitro *Transcription/Labeling of* In Situ *Hybridization Probes*

1. Add 1 μg linearized cDNA to 2 μl 10× transcriptions buffer.

2. Add 50 μCi ^{35}S-rCTP (or another labeled nucleotide of choice, see earlier).

3. Add 0.5 μl RNase inhibitor (e.g., Roche).

4. Add 2 μl of a mixture of rATP/rGTP/rUTP (5 mM each), when ^{35}S-rCTP was selected.

5. Add 0.5 μl T3- or T7-RNA polymerase (depending on the plasmid used and the direction of the cloned cDNA insert).

6. Bring to 20 μl with H_2O and incubate for 45 minutes at 37 °C.

7. Add 5 μl of 10× LB buffer, 1 μl (10U) DNase I (RNase-free; e.g., Roche) and 24 μl H_2O.

8. Incubate for 20 minutes at 37 °C.

9. Stop the reaction by adding 1 μl 0.5 M EDTA, 2 μl glycogen, 12.5 μl ice-cold 10 M NH_4-acetate, and 190 μl ice-cold ethanol.

10. Precipitate cRNA for at least 30 minutes at −70 °C or 2 hours at −20 °C, followed by centrifugation (13,000g) for 15 minutes at 4 °C.

11. Wash cRNA-pellet 2× using ice-cold 70% ethanol. Be careful not to lose the small and often only barely detectable pellet.

12. Let the pellet dry at approximately 35 °C for 15 minutes to evaporate EtOH, dissolve in 100 μl hybridization buffer for 15–20 minutes at 45–48 °C under shaking, and store at −20 °C.

13. Check the labeling efficiency by measuring 1 μl of the labeled probe in 5 ml scintillation solution (e.g., QuickScint, Zinsser).

h. Calculation of the Synthesized (Labeled) cRNA

1. 1 μCi of the labeled nucleotide is equivalent to 2.2×10^6 cpm, which means that 50 μCi of the nucleotide used at the beginning of the labeling process is 110×10^6 cpm (=theoretical labeling efficiency of 100%). Because only 1 μl out of 100 μl was measured in the counter, the determined activity value (B-value) has to be multiplied by 100 and then divided by 110×10^6.

2. As 50 μCi ^{35}S-rNTP is 0.05 nM, and 1 nmol of a nucleotide is 0.33 μg of synthesized cRNA, the "B value" $\times 0.05 \times 0.33 \times 4 \times 10,000$, results in the total yield in ng/ml.

For the ISH procedure, usually 1–3 ng cRNA diluted with hybridization buffer ("hybridization mix") is used per tissue section, depending on the size of the section. One *in vitro* labeling procedure yields cRNA sufficient for approximately 10–20 tissue sections, although a poor yield of labeled probe must not necessarily be a handicap for a good quality of labeling.

The labeled probe should be used immediately and should not be stored longer than 1 week.

i. Fixation and Blocking of Tissue Specimen

Use DEPC-H$_2$O and gloves throughout these procedures for protection against RNases.

1. Place 5 μm tissue cryosections on silanized slides, let dry on air, and store until use at $-80\,°$C.
2. Sections are placed on a clean tablet covered with fresh aluminium foil.
3. Encircle the sections with a silicon pen (e.g., DAKO-Pen) and place the slides into a clean, sterile cuvette.
4. Fix specimen with 4% formaldehyde/PBS (freshly prepared from *p*-formaldehyde) for 20 minutes at RT.
5. Wash specimen 2× 5 minutes with PBS.
6. Treat specimen with blocking solution to protect probes from unspecific binding to free aldehyde groups. Mix 200 ml blocking solution and 533 μl acetic-anhydride and incubate specimen under shaking at RT for 10 minutes.
7. Wash 2× 5 minutes with PBS. Leave slides in PBS until starting the prehybridization procedure.

j. Prehybridization and Hybridization

1. Clean a moist chamber with EtOH; cover its bottom with Whatman paper wetted with 2× SSC in 50% formamide.
2. Let the PBS drain off from the slides onto a Whatman paper and carefully remove the PBS outside the sections. To avoid direct contact of slides with the Whatman paper, use 1-mm thick plastic stripes ("spacers") onto which the slides are placed.
3. Cover each section completely with approximately 50 μl hybridization solution and incubate in an oven for 1 hour at 42 $°$C ("prehybridization").
4. Carefully remove the hybridization solution by means of a plastic pipette tip plugged into the flexible hose of a water jet vacuum pump.
5. Place approximately 25 μl hybridization-mix ("labeling mix": hybridization buffer plus 2–3 ng radiolabeled probe) onto the sections and cover them with a sterile coverslip. Try to avoid air bubbles.
6. Incubate for 5 minutes at 90 $°$C ("*denaturation*") on a heating plate.

7. Place slides back into the moist chamber, seal the chamber with Parafilm, and incubate at 42 °C overnight in an oven ("*hybridization*").

k. Posthybridization Washing

1. Bring a shaking water bath to 50 °C.
2. Remove the coverslips from the sections by immersing slides for 5–10 minutes in 2× SSC. The coverslips will come off easily.
3. Wash slides in the shaking water bath in 2× SSC/50% formamide/20 mM DTT for 30 minutes at 50 °C, then in 1× SSC/50% formamide/20 mM DTT for 30 minutes at 50 °C, and, finally, in 1× SSC/0.1% SDS for 5 minutes at RT.
4. Incubate the sections with 20 μg/ml RNaseA in 1× SSC for 30 minutes at 37 °C. The RNaseA digestion is very important in relation to background staining, which may often be prominent (see also Fig. 2B) if this step is omitted.
5. Wash the slides in the shaking water bath in 0.5× SSC/50% formamide/ 20 mM DTT at 50 °C for 30 minutes. In our hands, hybridization at a rather low temperature (i.e., 42 °C) is favorable for better binding of the antisense probe and reaches the high specificity by use of washing steps at higher temperature.
6. Dehydrate the samples through immersion in an ethanol series of 30%, 50%, 70%, and 100% EtOH, containing 0.3 M ammonium acetate, followed by immersion in 100% ethanol without ammonium acetate for 2–3 minutes and let them dry.

l. "Precheck" of In Situ Hybridization Labeling

For low-resolution evaluation of the quality and quantity of the ISH-labeling, we introduced a valuable "precheck" that uses exposure of the slides to an ordinary x-ray film. This method is also helpful for obtaining quick results during the establishment of optimizing conditions, thus resolving trouble-shooting problems or the validity of the probes.

1. Wrap an x-ray film cassette with Whatman paper, place the slides into the cassette, and fix them with adhesive tape.
2. Cover the slides with an x-ray film (e.g., Kodak, X-Omat AR) and develop after approximately 24 hours.

The quality and the strength of the ISH labeling and the final exposure time of the sections can now be evaluated (Fig. 2A). At this step, the possible background staining (Fig. 2B, background staining without RNAse digestion and Fig. 2D, E, positive control) and the failure of labeling of the sense probe (Fig. 2C) can also be controlled. The specificity of the probes used in seen in Fig. 2G (strong specific staining), Fig. 2F (weak specific staining), and Fig. 2H, I (specific staining in special or very restricted compartments of the follicle).

m. Detection of the **In Situ** *Hybridization Labeling on Tissue Sections*

To detect the radiolabeling on specimen, an autoradiography film emulsion (KODAK, NTB2) is used.

The following steps have to be done in a dark room.

1. Warm up the photoemulsion (e.g., 118 ml, per KODAK batch) to 42°C in a water bath and mix with 118 ml of 0.6 M prewarmed (42 °C) ammonium acetate. Subsequently in the dark (!), prepare aliquots of approximately 16 ml in scintillation vials. This volume depends on the size of the plastic cuvette used later (see later, II.C.2.n.3). We use a self-made 6.5 × 4 × 2 cm Plexiglas cuvette. For this purpose, we use a 25-ml plastic syringe through which a needle is sideward pushed so that the plunger automatically stops at the desired volume. Be careful to avoid air bubbles in the emulsion when pushing it out from the syringe. Store the aliquots at 4 °C in the dark.

2. Place a small bag of silica gel in a slide holder box. Warm a water bath up to 42 °C.

The following steps also have to done in the dark.

3. Warm the 16-ml film emulsion in the scintillation vials to 42 °C for 1 hour and fill into the plastic cuvette mentioned earlier (see II.C.2.n.1).

4. Dip a "test slide" (without a section) into the emulsion and check outside the dark room for the eventual formation of air bubbles and whether enough emulsion is in the cuvette.

5. Dip the section slides into the cuvette. A 16-ml aliquot is sufficient for at least 15 slides. Pour off the excess of the emulsion from the slides and let them dry for 2 hours on a rack.

6. Place the dried slides into the silica-gel–prepared slidebox. If varying exposure times are needed, use a separate box for each exposure time. Put the box(es) into the refrigerator at 4 °C. At this temperature the efficiency of autoradiography is higher than at RT. Expose the slides for the precalculated time(s) obtained from the results of x-ray test-film (normally 1–3 days).

n. Developing the Slides

1. Prepare a plastic box with ice.

2. Prepare conventional photo developer (diluted 1:1 with water; e.g., KODAK developer D19) and photo fixer (diluted 1:4 with water; e.g., KODAK fixer) in cuvettes. Several cuvettes with water will also be needed. In the dark, put slides in the right order in a slide-holder, develop the slides for 4 minutes, rinse with water for 1 minute, fix them for 5 minutes, and rinse again with water. Further rinsing steps can now be done outside the dark room.

3. The specimen should be counterstained for 3 minutes with Mayer's hematoxylin and washed 3× with water. Mount the slides in an appropriate embedding medium (e.g., Fluoromount).

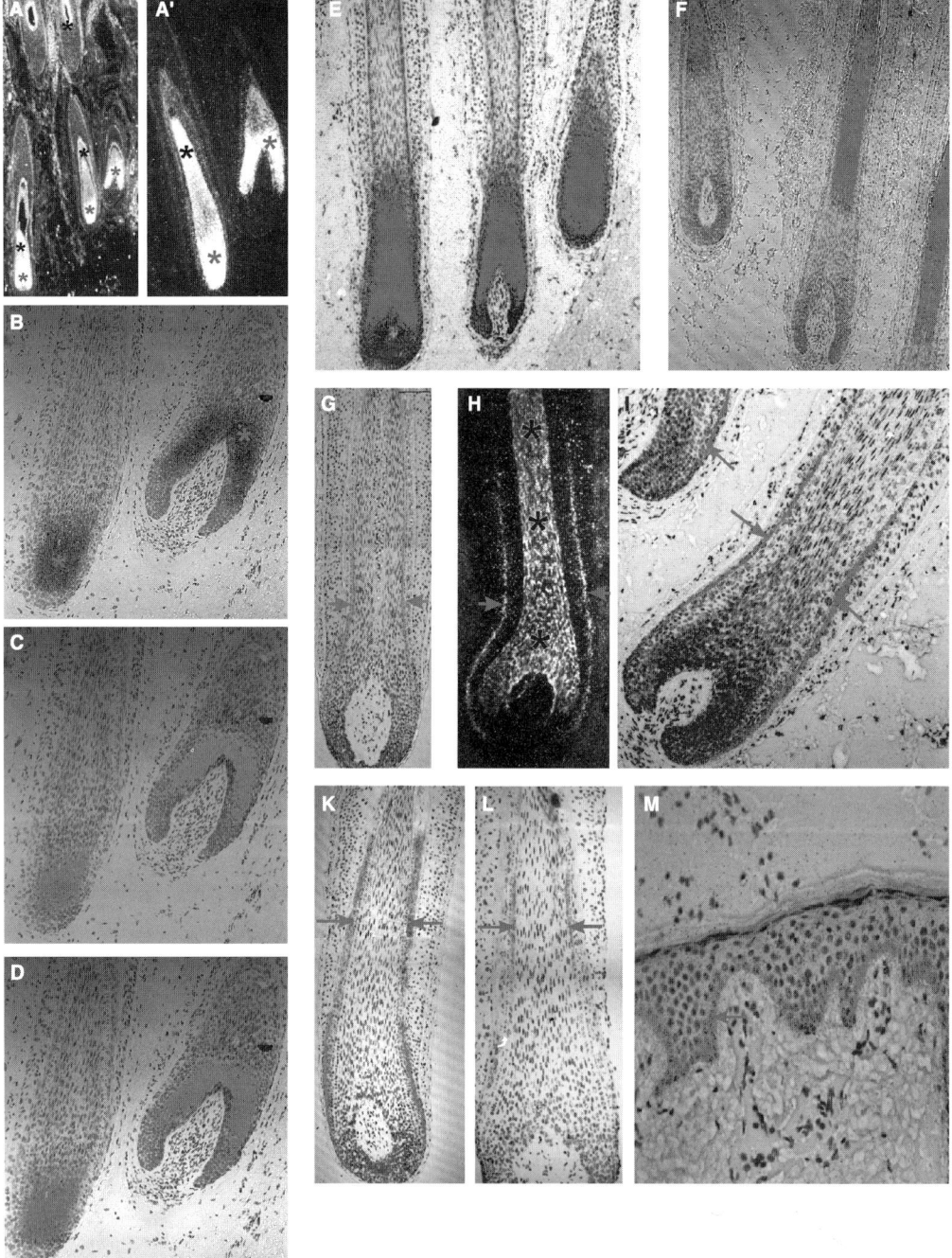

o. Analysis and Documentation

In principle, the labeling signals of radioactive ISH are based on the local appearance of metal silver grains in the photo emulsion as a result of the photochemical processing after radiation of these areas from the ^{35}S-labeled cRNA. Therefore, the microscopic methods used for the analysis of ISH results must be oriented according to this specific situation.

1. *Bright-field transmission light microscopy:* The use of this method is only applicable when an extremely strong signal (contrast) is achieved (Fig. 3B). This also holds true for phase-contrast microscopy.

2. *Dark-field microscopy:* For the detection of such metallic granular particles, as in the photo emulsion, this method is very convenient (see, Fig. 3A-A', H; see also, Langbein *et al.*, 1993). This type of microscopy is based on the illumination of fine structures from the side to observe them to be "lighted up" against a dark artificial background, in contrast to the bright-field technique against a bright one. Investigating the hair follicle and especially the hair fiber, which is extremely compact and hard from their densely packed masses of keratin filaments, these special dense protein structures (false "signal") are also seen very bright in dark-field microscopy (Fig. 3A-A', H). Therefore, a clear discrimination between the lightened protein structures and the lightened silver grains (ISH-signal) at such sites is difficult or impossible, although this method might be suitable for "soft" or nonkeratinized structures (see, Langbein *et al.*, 1993 and compare with Swensson *et al.*, 1998) Furthermore the dark-field imaging in principle does not allow us to concomitantly reveal the tissue morphology, and a precise correlation of ISH signals and expression site is difficult to obtain.

Fig. 3 Analysis and documentation of the ISH procedure. (A-A', H) The signals of the ISH labeling cannot be clearly identified by dark-field microscopy. Besides specific labeling of hHa5 mRNA (red asterisks in A-A') or K6hf mRNA (red arrows in H), false-positive images are detectable in nonlabeled areas of the hair cortex (black asterisks). (B) In the transmission image, only strong signals can be detected (red asterisks, hHa5 mRNA). (C, D) The sensitivity and specificity of the labeling in B can be strongly enhanced by reflection microscopy with a confocal laser scan microscope. The ISH signal, detected in false color, is given in red together with the green transmission image (C). The conversion of the green channel into a black-and-white image yields the best quality image, lacking background problems (D). (A-D) The hHa5 mRNA can be detected in the upper matrix and lower cortex including the hair cuticle (E, F; see also Fig. 2F) Specific mRNA staining of hHb5 (E; see also Fig. 2G) in the upper matrix up to the mid-cortex and hHa4 (F) in the upper cortex as the latest expressed hair keratin. (G) Specific staining of hHb2 in the hair cuticle. Note the high specificity and the clear identification of the labeling of the single-layered hair cuticle. (H, I) Comparison of the staining of K6hf mRNA in the companion layer using dark-field illumination (H) and reflection microscopy (I; see also Fig. 2H). Again, the high specificity and the clear identification of the labeling of the single cell layer are demonstrated. Note the problems regarding the identification of specific signals in H. (K-M) Specific labeling of K6irs1 mRNA (K) in the IRS, KAP10.1 mRNA (L; see also Fig. 2I) in the middle of the hair cuticle and K14 mRNA (M; positive control; see also Fig. 2D) in the basal cells of the interfollicular epidermis. (See Color Insert.)

3. *Reflection polarized contrast microscopy:* To overcome some of these problems and to enhance the signal-to-noise ratio, reflection polarized contrast microscopy (epipolarization microscopy) can be used, because it represents a suitable method for the detection of reflecting granular (e.g., metal) particles and allows the visualization of all the grains, regardless of their position relative to the focal plane (Halbhuber and Koenig, 2003; Speel *et al.*, 1993; van de Plas and Leunissen, 1992). Here the detectability of the silver particles can be strongly enhanced and the sensitivity of the method increased. However, the refractions from the light spectrum of the mercury lamp of the conventional microscope on the hair fiber protein structures cannot completely be eliminated and thus hinder the validation, in particular, of weak signals. Furthermore, the observation of the tissue morphology cannot be done and documented in parallel with this method.

To avoid the optical disturbances on the keratinized structures and to allow a parallel observation of histology and ISH signals, we use a confocal laser scanning microscope (e.g., LSM 510UV, Carl Zeiss, Germany) (Compare Montag *et al.*, 1988, Paddock, 2002). The instrument enables simultaneous visualization in epi-illumination reflection for the detection of hybridization signals and transmitted light in the bright field for hematoxylin staining. The images of the ISH signals are recorded with a He–Ne laser operating at a wavelength of 633 nm. Although the confocal properties of this microscope (e.g., small pinhole aperture) are not necessary, the use of the monochromatic long wave length light instead of the "full range spectrum" light of a conventional microscope (Hg-lamp for epi-illumination) does not only avoid any optical disturbances from the protein structures but also results in a clear image of a very high sensitivity of detection, which can be digitally loaded in one channel (e.g., red channel of a RGB-image). Furthermore, the transmission image of the tissue morphology can be loaded into a second (e.g., green) channel. The two image channels are then combined by an overlay in pseudocolor and seen simultaneously (Fig. 3C; ISH-signal in red, transmission image in green; see also, Rogers *et al.*, 1996). For better inspection and interpretation of the results, especially for a precise localization of the ISH signals on small structures, an electronic change of the green channel image into a black-and-white image represents a substantial improvement (Fig. 3D). This can be done with appropriate software for the computer, which adds the green (transmission image) into the red and the blue channel of the RGB format. Free download software (*Zeiss LSM Image Browser*, LSMib, Carl Zeiss; www.zeiss.com/de/micro/home.nsf) is easy to use for this process, and the results allow a precise and clear localization of the ISH signals, even in minor compartments of the hair follicle, without any disturbances (Fig. 3E–G, K–M; compare also Fig. 3H and I; see also, Langbein *et al.*, 1999, 2001, 2002a, 2003; Rogers *et al.*, 1997, 2004; Swensson *et al.*, 1998; Winter *et al.*, 1998, 2001; and compare also with Rogers *et al.*, 1996). It is noteworthy that this method, including polarized reflection microscopy by means of a normal light microscope, is also helpful when immunogold-labeling (with or without silver enhancement) in immunohistochemistry has to be assessed (see, Chapter II.C.2.c and Fig. 5; see also Speel *et al.*, 1993).

p. Double Labeling **In Situ** *Hybridization*

The radioactive ISH procedure described is normally suitable for the localization of a single mRNA species. Alternatively, double labeling with two different cRNA probes is also feasible (Fig. 4A). The double labeling procedure is applicable when the expression sites of the individual mRNAs are known and their synthesis does not overlap. For double labeling, the two probes have to be mixed before the hybridization procedure, whereas all other steps are as described for the single labeling protocol (see also, Trembleau *et al.*, 1993).

q. Combination of **In Situ** *Hybridization Labeling and Immunohistochemistry*

In principle, a combination of both methods can be done, although the results are usually of limited quality.

1. Both methods should be applied separately to check the intensity and localization of the respective signals (e.g., ISH silver grains for mRNA and IIF for protein). A prerequisite for such a combination is a rather strong IIF staining. An established positive control must be handled in parallel to check whether the respective antigen is accessible at the end of the ISH procedure, e.g., the fixation and incubation/washing steps.
2. The complete ISH process has to be done from steps II.A.2.a to II.A.2.m.
3. Follow IIF procedure II.B.2.c.1 (see below).
4. Treat tissue sections according to steps II.A.2.n to II.A.2.p.
5. Evaluation and documentation is carried out with fluorescence microscope and the method of choice for the IIF signal. Both can be done in parallel when a confocal laser scanning microscope equipped for IIF and reflection microscopy (Fig. 4B–B″) is available.

B. Light Microscopic Immunohistochemistry

1. Introduction

ISH provides information only about the occurrence and the localization of a given mRNA but not about the existence and stability of the respective protein. This information can only be obtained by immunohistochemistry, which is principally based on protein level. Because this method is widely used, this chapter will focus on protocols that are tailored for the detection of cytoskeletal proteins, in particular the keratins and their associated proteins of the hair follicle. Especially in the compact mid and upper cortex of the hair follicle, the detection of proteins is hampered by a poor accessibility of the antigens, which can, however, be overcome widely by the use of special treatment. As a rule, we prefer the indirect immunofluorescence technique (IIF) over immunoenzyme techniques (PAP-, APAAP-method) or the use of biotinylated antibodies (ABC-method), because this method is easier to handle, quicker, and avoids background problems from the enzyme reaction.

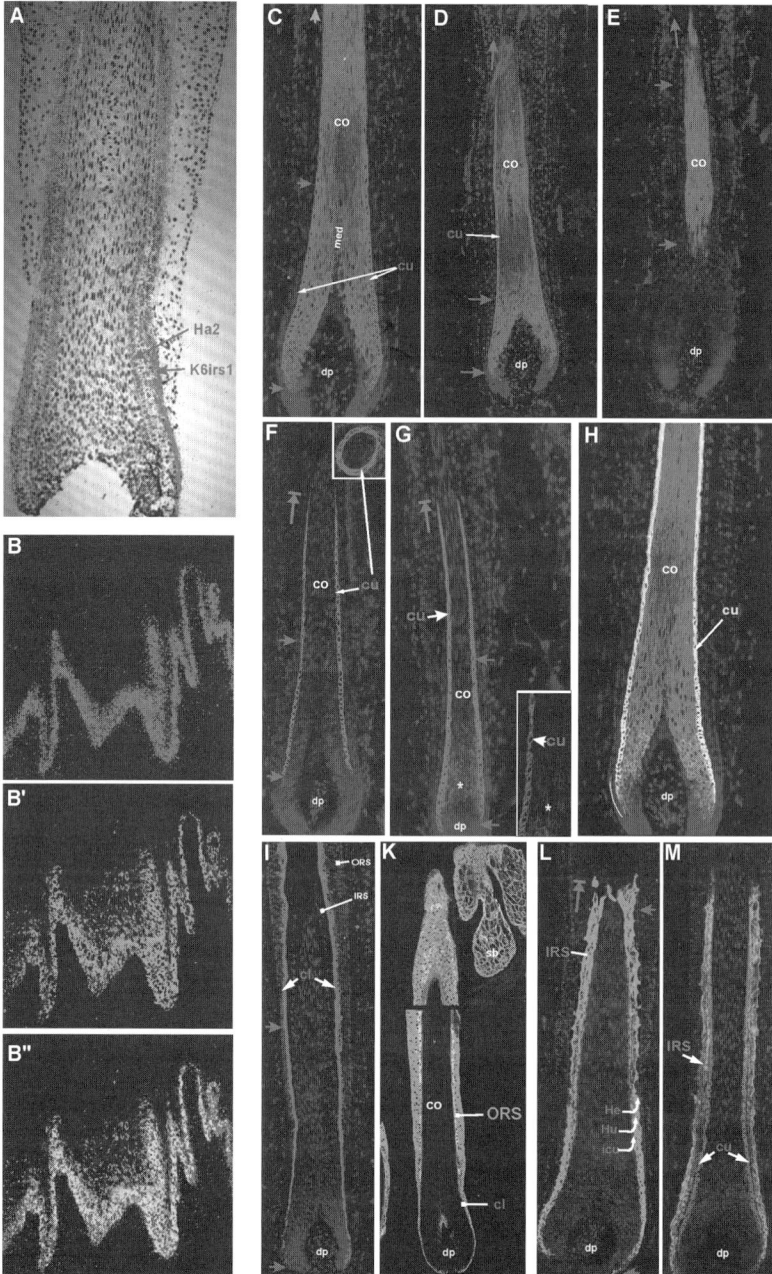

Fig. 4 Double-label ISH, combination of ISH/IIF and indirect immunofluorescence of various keratins in the hair follicle. (A) The specific staining of hHa2 mRNA in the hair cuticle and K6irs1 mRNA in the IRS can be clearly detected by double-label ISH. (B-B″) Although the signal for K14

2. Procedure

a. Selection of Peptides for Antibody Production

For obvious reasons, the selection of specific peptides (approximately 12–18 amino acids; compare Drenckhahn et al., 1993) for members of the keratin/hair keratin multigene family is mostly limited to the beginning of their head domains and the end of their tail domains, respectively (see, Tables II and III). Furthermore, the sequence should be of high antigenicity and contain a high extent of hydrophilic amino acids (compare Drenckhahn et al., 1993). All peptide sequences selected for antibody production, in particular smaller ones have to be carefully checked for specificity in protein databases. To improve antigenicity, the peptides should be coupled to a large protein (e.g., keyhole limpet protein). In cases in which no internal cysteine exists within the selected sequence, a cysteine residue has to be added at one end of the peptide for coupling to keyhole limpet protein (e.g., Peptide Specialty Laboratories, Heidelberg, Germany; www.peptid.de). If one end of a given peptide exhibits partial homologies/similarities to other keratin sequences, the cysteine should be added to this end. Surprisingly, occasionally even the use of a complete recombinant keratin protein for immunization may result in a monospecific antiserum.

In our hands, the immunization of guinea pigs yielded more specific antisera with lower background or unwanted staining problems compared with rabbits. Commonly, we use the following immunization scheme (see also, Langbein et al., 1999, 2001, 2002a, 2003; Winter et al., 1998, 2001):

1. Each animal is immunized by subcutaneous injections of 100 μg immunogen in 100 μl sterile PBS, supplemented with 100 μl complete Freunds adjuvant at multiple sites of the back skin. Prepare the immunogen with a hypersonic homogenizer to completely emulsify the mixture in an ice bath. After homogenization, it

---

mRNA (B; ISH labeling in red; cp. also Fig. 3M) is restricted to the basal cells of foot sole epidermis, the K14 protein is stable also in the lower to middle suprabasal layers (B' IIF staining in green). The yellow color in the merged images indicates the colocalization of K14 mRNA and protein (B'') in the basal cells of this epidermis. (C, D) Hair keratins hHb5 (C, see also Fig. 3G) and hHa5 (D, see also Fig. 3D) are both expressed in the hair cortex (co) and cuticle (cu). (E) hHa4 is the latest hair keratin and is expressed in the mid- to upper cortex. (F, G) Hair keratins hHb2 (F, see also Fig. 3G) and hHa2 (G) are specifically found in the hair cuticle. (H) By double-label IIF, the coexpression of hHa2 (green) and hHb5 (red) in the hair cuticle is clearly shown by the yellow (merged) color in this compartment. (I) Keratin K6hf is detectable throughout the companion layer (cl, see also Fig. 3I). (K) Keratin K6hf (red) is restricted to the cl, whereas keratin K14 (green, cp. also Fig. 3M) is found in the ORS and the interfollicular epidermis. (L) The keratin K6irs1 antibody stains the Henle and Huxley layers of the IRS, as well as the IRS-cuticle, see also Fig. 3K). (M) Double immunolabeling with K6irs1 (green) and hHa2 (red) shows that both cuticles are adjacent to each other. Red arrows in C–G, I, and L indicate the zones of the respective mRNA synthesis. Note that in contrast to the restricted occurrence of the mRNAs, the proteins are stably integrated in the cytoskeleton and detectable throughout the hair fiber. Compare the mRNA data (this figure) with the respective protein data (Fig. 3). The green stop-arrows in F, G, and L indicate that protein staining is no longer detectable in the terminally differentiated portions of the respective compartments. dp, dermal papilla; med, medulla. (See Color Insert.)

Table II

Human Type I and Type II Hair Keratins, Calculated Molecular Mass Values, Synthetic Peptides Used for the Generation of Antibodies/Antisera, Names of Antisera, Dilutions for ECL/IIF, and Source of Generation

Group/hair keratins	Calculated molecular mass (kDa)	Access.-Nr	N-/C-term	Oligopeptides used as antigens	Species/name of antiserum	Dilution ECL	Dilution IIF	Cross-reactive on mice[f]	Source
A-I: hHa1	47.2	Y16787	C	CAPRPRCGPCNSFVR	mou LH-Tric 1	1:2000	1:50	(+)	I.L.[c]
hHa1				Total recomb. protein	gp Ha1Prot.1	n.d.	1:5000	+	L.L.[d]
A-I: hHa 3-I/II[a]	46.2	Y16788 Y16789	C	EKPIGSCVTNPCGPRPSR	gp hHa3-II.2	1:5000	1:500	+	L.L.[d]
A-I: hHa4	44.7	Y16790	C	NSCGPCGTSQKGCCN	gp hH4.1	1:1000	1:300	+	L.L.[d]
B-I: hHa7	45.5	Y16793	C	PSCGPVTGGSPSGHGASMGR	gp hHa8.2	1:2000	1:1000	−	L.L.[d]
B-I: hHa8	46.4	Y16794	C	TTCGPTCGASTTGSRF	gp hHa9.1	1:80000	1:5000	+	L.L.[d]
B-I: cHaA[e]	48.3	AJ401055	N	SDHCSSLLSGQVSE	gp pseu a7.1	1:10000	1:2000	n.d.	L.L.[d]
C-I: hHa2	50.3	X90761	C	TCVPRTVGMPCSPCPQGRY	mou LH-Tric17	1:50	1:10	(+)	I.L.[c]
C-I: hHa5	47.2	Y16791	N	YSSSSCKLPSLSPVARS	gp hH5.2	1:30000	1:2500	+	L.L.[d]
C-I: hHa6	52.2	Y16792	C	APQVGTQIRTITEEI[b]	gp hHaRa.2	1:40000	1:1000	(+)	L.L.[d]
A-II: hHb1	54.8	Y19206	C	GSCGISSLGVGSCGSSCRKC	mou hHb1	1:500	1:50	(+)	I.L.[c]
A-II: hHb3	54.1	Y19208	C	CGFNSIGCGFRPGNF	gp hHb3.1	1:1000	1:500	+	L.L.[d]
A-II: hHb6	53.5	Y19211	C	APVVSTRVSSVPSNSNVVVGTTNA	gp hHb6co.2	1:1000	1:500	+	L.L.[d]
B-II: no functional gene in humans									
C-II: hHb2	56.6	Y19207	C	SIVGTGELYVPCEPQGLLSC	gp hHb8.2	1:2000	1:1000	+	L.L.[d]
C-II: hHb4	64.9	Y19209	N	IAAVGSRPIHCGVRF	gp hHb4/2am.2	1:2000	1:1500	+	L.L.[d]
hHb4			C	CGPSLGGARVAPATGDLLSTG	gp hHb2.2	1:2000	1:2000	n.d.	L.L.[d]
hHb4			N	SRPIHCGVRFGAGCGM	gp hHb4/am.1	1:2000	1:2000	n.d.	L.L.[d]
hHb4			C	GSMLISEACVPSVPC	gp hHb4/2co.1	1:2000	1:1500	n.d.	L.L.[d]
C-II: hHb5	55.7	Y19210	C	GRQITSGPSAIGGSITVVAPD-(C)	gp hHb5co.2	1:3000	1:1000	+	L.L.[d]

[a]Because of the high homology between hHa3-I and hHa3-II, this antibody recognizes both hair keratins.

[b]Derived from central portion of the tail domain.

[c]Mouse monoclonal antibody raised in the laboratory of Dr. I. M. Leigh, Centre for Cutaneous Research, Royal London Hospital, London, UK.

[d]Raised in the laboratory of Lutz Langbein (DKFZ Heidelberg) and available from PROGEN (Heidelberg, Germany, www.progen.de).

[e]cHaA is the chimpanzee functional gene of the human nonfunctional pseudogene ΨhHA.

[f]Sometimes the staining background on mice is higher than on human tissues.

gp, Guinea pig antiserum; mou, mouse monoclonal antibody.

Table III

Specific Human Keratins of the Hair Follicle, Calculated Molecular Mass Values, Synthetic Peptides Used for the Generation of Antibodies/Antisera, Names of Antisera, and Dilutions for ECL/IIF and Source of Generation

Keratins	Calc. mol. mass (kDa)	Acess.-Nr.	N-/C-term.	Peptides used as antigens	Species/name of antiserum	Dillution ECL	Dilution IIF	Cross-reactive on mice	Source
K6hf	59.5	Y19212	C	(C)-SGGHSLGAGLGGSGFSATSNRGL	gp Bax.1	1:40,000	1:2000	+	L.L.*
K6hf		Y17282		Total recomb. protein	gp K6hf-Prot.1	n.d.	1:10,000	+	L.L.
K6irs1	57.3	AJ308599	C	(C)-GGEGRSRGSANDYKDT	gp K6irs 1.2	1:2000	1:500–1000	+	L.L.*
K6irs2	55.9	AY033496	C	SYKTAADVKTKGSC	gp T3.2	1:20,000	1:2000	+	L.L.*
K6irs3	59.0	AJ508776	C	YSMLPGGCVTGSGN	gp T2.2	1:5000	1:2000	+	L.L.*
K6irs4	57.9	AJ508777	C	(C)-GKSTPASIPARKATR	gp T4.1	1:10,000	1:2000	+	L.L.*

gp, Guinea pig antiserum.

*Raised in the laboratory of Lutz Langbein (DKFZ Heidelberg) and available from PROGEN (Heidelberg, Germany; www.progen.de).

is recommended that the antigen mixture be aliquoted into an appropriate number of syringes and stored at −20 °C until use.

2. The following two or three booster injections are done at approximately 4-week intervals using incomplete Freunds adjuvant in the immunogen.

3. Test blood should be taken after the second and third booster injection to check the antibody titer and quality. If the titer is sufficiently high, take the blood 10 days after the last injection by heart puncture.

4. Prepare blood serum simply by centrifuging the clotted blood with blood centrifugation glass tubes with a paraffin buckler inside (e.g., BD Vacutainer SST tubes, 9.5 ml, Becton Dickinson). The undiluted antiserum should be supplemented with 1 mM NaN_3 and stored at −80 °C in adequate aliquots. Otherwise, a (highly) end-diluted antiserum should be supplemented with 5% normal goat serum and stored at 4 °C in the refrigerator or on ice during the IIF procedure.

5. Check the antisera for specificity by both immunohistochemistry on appropriate tissue sections and 1D or 2D SDS-PAGE and Western blot with adequate tissue extracts.

6. If necessary, antibodies can be affinity-purified with the antigen covalently bound to a carrier matrix (e.g., UltraLink system; Pierce) or to nitrocellulose powder (Schmidt *et al.*, 1997; for convenient and economic preparation of this powder, see Hammerl *et al.*, 1993) to which the antigen was bound. Briefly, 1 mg of the peptide or recombinant protein is bound to 0.4 ml wet (PBS) nitrocellulose powder, resuspended in a total volume of 1 ml PBS (2.7 mM KCl, 1.5 mM KH_2PO_4, 140 mM NaCl, 8.1 mM Na_2HPO_4, pH 7.4), and incubated for 2 hours at room temperature under shaking. After thoroughly rinsing the suspension with TBST (150 mM NaCl, 10 mM Tris-HCl, pH 8.0, 0.1% Triton X-100 [v/v]), the affinity powder is blocked for 30 minutes with 5% nonfat dry milk powder (NFMP) in TBST. This powder is exposed to 0.5 ml antiserum in a total volume of 2 ml TBST in an Eppendorf tube for 2 hours under shaking. The antigen-coated powder and bound antibodies are thoroughly (10 times) rinsed with 2 ml TBST each. The antibodies are repeatedly (two to three times) eluted with 1 ml 0.1 M glycine (pH 2.8 or a pH depending on the strength of antibody affinity) for 1 minute under shaking, centrifuged, and brought immediately to pH 7.5 by addition of 41 µl of 1 M Tris-HCl (pH 9.5). During all steps, the resuspended matrix material is pelleted by a short centrifugation step. After thoroughly rinsing with TBST, the affinity matrix can be reused or is stored at 4 °C on addition of 0.01% NaN_3. Eluted antibodies are supplemented with 0.01% NaN_3 and protease-free 5% bovine albumin (e.g., Serva, Germany) for stabilization and stored at −20 °C or −80 °C.

b. Preparation of Tissues

Fresh Tissue for Cryosections.   Routinely, human scalp skin obtained from surgical or cosmetical interventions is snap-frozen immediately after excision in isopentane and precooled in liquid nitrogen to approximately −140 °C (see

II.A.2.b; Langbein *et al.*, 1993, 1999, 2001, 2002a, 2003; Schmidt *et al.*, 1997). In parallel, we use a freshly plucked (beard) hair follicle (see, II.A.2.b). The use of silanized slides (e.g., SuperFrost Plus, Menzel, Germany) is recommended. The tissue sections are fixed in acetone at −20 °C for 10 minutes, dried on air, and stored, if necessary, at −80 °C.

Paraffin-Embedded Tissue Material. If only aldehyde-fixed and paraffin-embedded tissue material is available, one has to be aware that the antigenicity of keratins is often drastically reduced. Therefore, a microwave oven treatment is mandatory for antigen retrieval.

1. Deparaffinize paraffin sections by incubation in xylol (3× 5 minutes), 100% EtOH (2× 5 minutes), 95% EtOH, 80% EtOH, 70% EtOH, 50% EtOH (3 minutes each) and water.

2. Treat tissue sections in a microwave oven (e.g., MILESTONE, Diapath GmbH, Munich, Germany) for 1–2 minutes at 98 °C in sodium-citrate buffer, pH 6.0. Subsequently, the sections are cooled to room temperature. Depending on antigen/epitope, the use of alkaline conditions (0.1 M Tris-Cl, pH 9.5, 5% urea) might also be helpful instead of the citrate buffer and should be tested when the use of the acidic conditions does not give satisfying results. Rinse section in Tris-buffered saline (TBS, pH 7.4) at 4 °C.

3. Optionally, treat section with 0.001% trypsin in 0.05 M Tris-buffer (pH 7.4) for approximately 15 minutes at 37 °C, followed by rinsing with PBS. The concentration of the protease and the duration of digestion should be tested depending on the fixed tissue used.

c. Indirect Immunofluorescence Procedure
Single Label Indirect Immunofluorescence

1. After a brief wash in PBS, permeabilize sections with 0.1% Triton-X 100 for 2–3 minutes for better penetration of the antibodies.

2. Block unspecific binding of antibodies on sections with 5% normal serum of the species used for the generation of the secondary antibodies in PBS for 20 minutes at RT. Let the blocking normal serum drain off from slides.

3. Apply the primary antibody, appropriately diluted in PBS, for approximately 1 hour at RT in a moist chamber. Rinse sections 3× in PBS, 5 minutes each.

4. Incubate sections with the appropriately diluted secondary antibody for approximately 30 minutes at RT in a moist chamber. For nuclear counter-labeling, DNA is stained with DAPI (e.g., Serva, diluted 1:100,000), which can be added to the secondary antibody. Wash section 3× in PBS for 5 minutes each.

5. Rinse sections in water and in EtOH. Sections are air-dried and mounted (e.g., Fluoromount-G, Southern Biotechnology) (see Fig. 4C-G, I, L).

Double Label Indirect Immunofluorescence. For double labeling IIF, antibodies produced in different species have to be used. For an adequate documentation, their dilution should be separately adapted to a comparable staining intensity. Furthermore, be aware that the equipment of the microscope allows a clear separation of images from the different fluorochromes, which should not be detectable—even faintly—in the "wrong" channel. Narrow range filters are recommended.

1. For double labeling with appropriately diluted primary antisera raised in different species, these can either be mixed before incubation of the sections or reacted separately by consecutive incubation for 1 hour.

2. The same procedures can be applied for the different secondary antibodies with DAPI in one of the secondary antibodies (see Fig. 4H, K, M).

d. Special Indirect Immunofluorescence Conditions for the Demonstration of Keratin-Associated Proteins

The detection of keratin-associated proteins (KAPs) by immunohistochemistry raises several problems. KAPs are rather small molecules, approximately 5–40 kDa in size. Up to now, approximately 80 transcribed human KAP genes are known, which encode members of three chemically different KAP-gene super-families, the high sulfur (HS) KAPs, the ultrahigh sulfur (UHS) KAPs, and the high glycine-tyrosine (HGT) KAPs (Powell and Rogers, 1997; Rogers *et al.*, 2001, 2002, 2004). Because of their small size and their extremely high homology, virtually no KAP-protein-monospecific peptides can be derived and used as antigens. At most, "family-specific" antibodies can be raised (Shimomura *et al.*, 2003). Furthermore, KAPs are predominantly expressed in the highly differentiated portions of the hair follicle cortex and cuticle, where they cross-link the IFs formed from an extremely high number of hair keratins by use of disulfide bonds. The resulting compact structure of the mature hair fiber is barely penetrable for antibodies. To overcome this, at least in part, we elaborated a special IIF protocol (see Shimomura *et al.*, 2003), which is based on the application of reductive conditions to cleave the disulfide bonds and which includes the use of detergent for better accessibility and penetration of the antibodies.

1. Fix cryostat sections either in methanol ($-20\,°C$, 5 minutes) or use them unfixed.

2. Incubate sections for 2 hours at RT in "PBS-Ar/DTT." This reagent is prepared from PBS, supplemented with 10 mM DTT and 1 mM EDTA, and flowed through with argon gas for 2 hours.

3. Permeabilize sections by use of 0.1% Triton X-100 in PBS-Ar/DTT for 5 minutes.

4. Rinse sections three times with "PBS-Ar" (PBS-Ar/DTT without DTT) and apply the primary antiserum overnight at $4\,°C$. Rinse three times in PBS-Ar for 5 minutes each.

5. Follow steps II.B.2.c.1.ii.-v. as indicated previously.

6. In general, the staining is improved when unfixed cryostat sections are used, and the fixation is carried out subsequent to the application of the secondary antibodies.

e. Controls

For the IIF procedure, the results have to be checked by valid controls. These should include the following:

1. Incubation of specimen without the primary antibody (unspecific staining of the secondary antibody, *negative control* I).

2. In the case of new antibodies, use the primary antibody after absorption with the respective antigen in parallel (*negative control* II).

3. Use various tissues/cells known to be positive for the antigen (*positive control*).

4. Check that the fixation procedure used does not destroy the antigen or the antigen-binding site by use of cryosections and aldehyde-fixed tissue sections/cells and paraffin sections in parallel. Eventually antigen retrieval procedures (e.g., microwave or protease treatment) might be useful.

5. Check if a preincubation with a normal serum of the species of the secondary antibody (5% normal serum in PBS for 20 minutes) will be helpful and block unspecific binding of immunoglobulins.

6. In case of newly produced antisera, check their specificity on a dot-blot procedure for correct recognition of their antigen. Simply drop 5 ng and 50 ng of the antigen (recombinant protein or synthetic peptide *without* coupled KLH protein) in 10 μl of PBS on a sheet of nitrocellulose. Let the nitrocellulose dry for 2–3 hours, block the membrane with nonfat milk powder solution, and treat this blot according to conventional Western procedures for the detection of antigens.

7. In case of newly produced antisera, check their specificity on Western blots of 1DE- or 2DE-resolved reliable tissue extracts.

C. Conventional and Immunoelectron Microscopy

1. Introduction

For the investigation of morphology, conventional electron microscopy (EM; Fig. 5D, see also e.g. Ito, 1986, 1988, 1989; Langbein *et al.*, 2002a,b, 2003) and for the determination of proteins at the subcellular level immunoelectron microscopy (IEM; e.g., Langbein *et al.*, 2002a,b, 2003; Schmidt *et al.*, 1999) is indispensable. Because the detection of antigens is often rather poor in glutaraldehyde-fixed tissues and even more in a plastic (e.g., Epon) embedded tissue specimen, the use of the preembedding IEM is the method of choice in our hands. This holds particularly true with regard to the special conditions in the hair follicle. For this

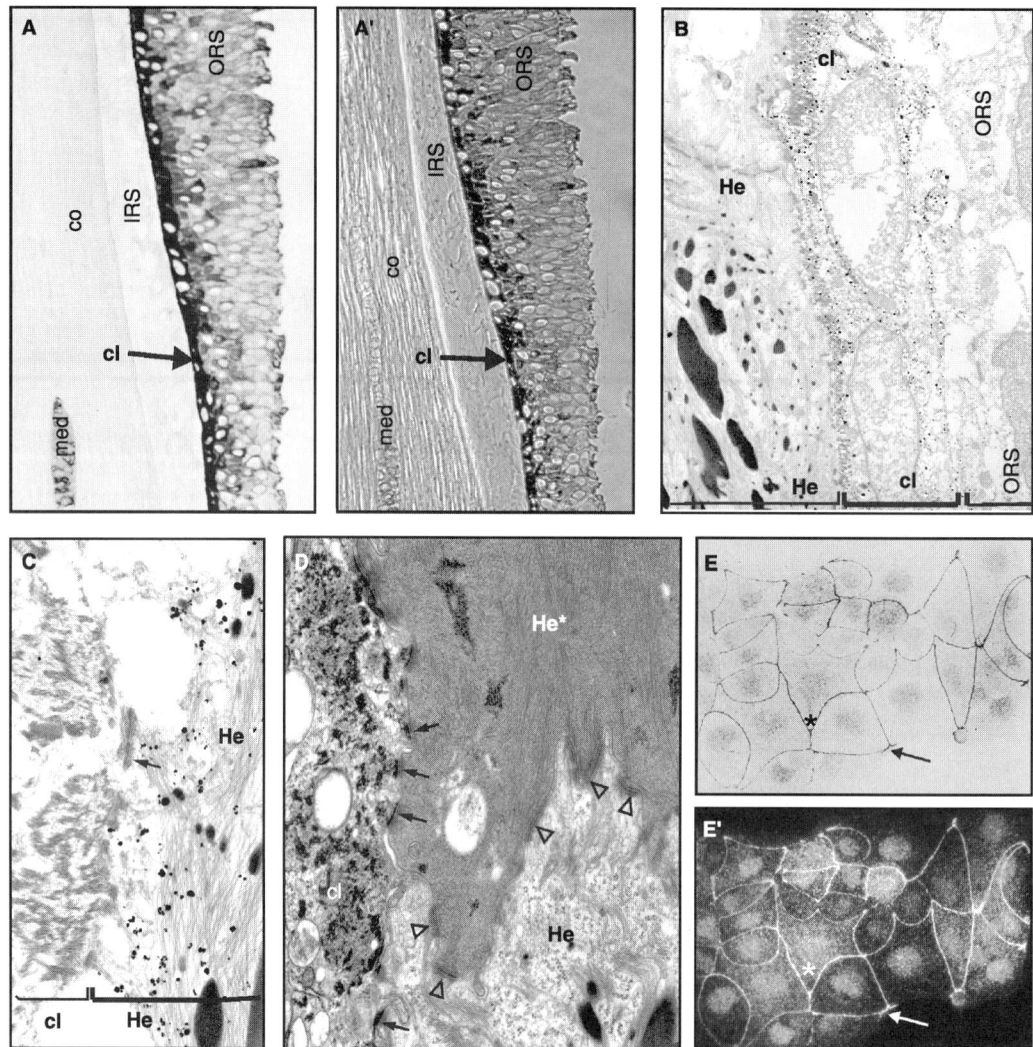

Fig. 5 Immunogold staining with silver enhancement at light and electron microscopic level. (A,B) The staining of K6hf is specifically restricted to the companion layer (cl) of the hair follicle. The pre-embedding immunogold labelling for IEM is controlled at the light microscopic level using transmitted light (A, dark label with arrow) or phase-contrast microscopy (A', dark label with arrow). Note the specific dark (in reality, dark brown to black) staining of keratin K6hf in the companion layer cells. With the electron microscope (B), the silver-enhanced gold particles decorate the companion layer cells between IRS Henle layer (He) and the ORS. (C) Keratin K6irs1 is detected in the Henle layer (the concomitant staining in the Huxley layer and the IRS cuticle is not shown in this figure). (D) Between Henle cells (triangles) and between Henle and companion layer cells (arrows), numerous desmosomes can be detected by conventional EM. Note the abrupt terminal differentiation of Henle cells (He, low-differentiated Henle cell; He*, terminally differentiated Henle cells containing tightly packed masses of keratin filaments. (E-E') Control of the specificity of pre-embedding immunogold labeling for desmoplakin in cells of a mammary carcinoma line, MCF-7, by bright-field (E) and reflection microscopy (E'). The arrows demarcate the specific labeling, whereas the asterisks mark some background staining best visible by reflection microscopy.

type of IEM, the immunolabelling has to be done on cryotissue sections comparable to the light microscopic immunohistochemistry. Fixed cryosections are incubated with the primary antibodies and gold-labeled secondary antibodies before embedding. Because of the situations in the hair follicle, the preembedding IEM method will be described in more detail, including special steps for treatment of the hair follicle tissue (compare Fig. 5B and C).

2. Solutions to be Prepared

Cacodylate buffer (0.5 M stock): Dissolve 107 g cacodylic acid (Na-salt) in 1000 ml distilled water and bring the pH to 7.2 with concentrated HCl. Add 0.5 M KCl and 25 mM $MgCl_2$. Use buffer at 10-fold (50 mM) dilution. Recheck pH after dilution.

3. Conventional Electron Microscopy Procedure

1. Cut tissue immediately after surgical removal into small pieces (approximately 3×3 mm) and fix in 2.5% glutaraldehyde in 50 mM cacodylate buffer for approximately 10 minutes at RT and for approximately 10 minutes on ice. The following steps should be done on ice or at 4 °C.
2. Wash tissue pieces 4× for 5 minutes each with 50 mM cacodylate buffer.
3. Postfix under the hood with freshly prepared 2% OsO_4 in 50 mM cacodylate buffer for 1–2 hours. Shortly rinse specimen twice in distilled water.
4. Contrast specimen in aqueous 0.5% urany acetate for 12–18 hours or overnight followed by rinsing in distilled water. If necessary, the transport of the specimen in distilled water to the local laboratory is possible.
5. Dehydrate specimen through an EtOH series of ice-cold 50%, 70%, 80%, 90%, 96%, and absolute EtOH at RT with 5–10 minutes for each step.
6. Further dehydrate specimen in a 2-ml glass Petri dish for 2× 20 minutes in absolute (water-free) EtOH at RT, in propyleneoxide for 2× 20 minutes (use the hood!). Specimen should not dry.
7. Incubate in propyleneoxide/Epon embedding medium (mixture of 1:1) overnight under gentle shaking. Add one half volume fresh Epon. Then follow step II.D.4.o. of the IEM protocol.

4. Preembedding Immunoelectron Microscopy Procedure

1. If possible, use freshly prepared tissue immediately snap-frozen in liquid nitrogen-cooled isopentane after excision (see, *ISH procedure*, II.A.2.d; *IIF procedure*, II.B.2.b.1).

2. Prepare cryosections of approximately 5-μm thick from the snap-frozen tissue. Clean silanized glass coverslips with petrolether, acetone, or ethanol to

make them absolutely fat free. Mount the cryosection on these cleaned coverslips and led sections shortly dry on air.

3. Fixation is done in 2% formaldehyde in PBS, freshly prepared from *p*-formaldehyde, for 7–10 minutes.

4. Quench residual free aldehyde groups by a washing with 50 mM NH_4Cl in PBS for 5 minutes.

5. Permeabilize specimens with 0.1% saponin in PBS for 5 minutes, or if necessary with 0.1% Triton X-100 in PBS, for 2 minutes. Duration and concentrations of this step should be tested, depending on antigen and type of tissue/structure to be detected.

6. Incubate specimen with the appropriately diluted primary antibodies for 2 hours, followed by thoroughly washing with PBS.

7. Incubate specimen for 2 hours or longer (overnight, for better penetration) with appropriately diluted secondary antibodies. In general, we use immunoglobulins coupled to 1.4-nm gold particles (Nanogold; Biotrend, Cologne, Germany). This type of gold particle is optimal for deep penetration into complex tissue structures. Wash samples 3× for 5 minutes with PBS.

8. Fix samples for 15 minutes with 2.5% glutaraldehyde and rinse with cacodylate buffer (see EM procedure, II.C.3.b.1).

9. Rinse samples 2× for 10 minutes with HEPES buffer (50 mM, pH 5.8) containing 200 mM sucrose to wash out the chloride ions.

10. For the detection of the very small (nano-) gold particles, these are enlarged by a silver-enhanced procedure with the HQ-Silver-Enhancement-Kit (Biotrend, see also manufacturer's protocol). The silver enhancement procedure has to be done in a dark box! The time of enhancement ranges from 3–6 minutes, and the procedure is stopped by washing the samples in 50 mM HEPES buffer containing 250 mM sodium thiosulfate. Rinse samples thoroughly (8–10×) with distilled water. To control the staining specificity, one slide is silver enhanced for 15 minutes and checked with different microscopic methods (e.g., bright-field [Fig. 5A], phase-contrast [Fig. 5A], or in particular reflection-contrast microscopy [Fig. 5E-E′; see also, ISH procedure II.A.2.p]).

11. Post-fix specimen with 0.2% OsO_4 in 50 mM cacodylate buffer for 30 minutes, followed by two washes with distilled water.

12. Dehydrate samples stepwise with EtOH and flat-embed them in Epon (see EM procedure, II.C.3.e).

13. Place the coverslips on glass slides covered with aluminium foil. Be aware that the tissue section is facing up. Put an Epon-filled gelatin capsule with its opening down onto the coverslip, which (partially) covers the tissue section, in particular the region of interest. Squeeze the capsule gently and release pressure immediately when the capsule is touching the coverslip.

14. Let the Epon polymerize over night at 60 °C in a conventional oven. Remove slide including the aluminium foil from the coverslip. Scrape off the

excess Epon from the bottom of the coverslips with a razor blade and cut off the coverslip protruding from the gelatin capsule with an old pair of scissors.

15. Replace capsule in the oven and heat to 60 °C and put the heated capsule directly into liquid nitrogen for 1 minute. Allow capsule to warm to room temperature. During this process, the coverslip detaches from the Epon and leaves the section in the Epon-filled capsule. As for conventional EM, the capsule with the specimen has to be trimmed for ultrathin sectioning with an ultramicrotome (e.g., Ultracut; Leica, Bensheim, Germany).

16. Electron micrographs are taken with an electron microscope (e.g., EM600 or EM900, Zeiss-LEO, Oberkochen, Germany) (Fig. 5B–D).

D. One- and Two-Dimensional Polyacrylamide Gel Electrophoresis and Western Blotting

1. Introduction

One-dimensional (1-DE; Fig. 6A, lane Coom) SDS-PAGE (Laemmli, 1970) or two-dimensional (2-DE; Fig. 6B) gel electrophoresis (isoelectric focusing, IEF or nonequilibrium pH gradient electrophoresis, NEPHGE; O'Farrell et al., 1975, 1977) both in combination with Western blots (Fig. 6A lanes 2–7, C-G) are the methods of choice to describe complex keratin patterns and to assess the quality and specificity of antibodies antisera raised against individual keratins. This section will mainly focus on the preparation of hair follicles and extracts thereof used for the analysis of keratins of the hair-forming compartment and the surrounding epithelial sheaths and layers (i.e., ORS, IRS, and the companion layer) (Langbein et al., 1999, 2001, 2002a, 2003; see also Heid et al., 1986a,b, 1988; Hofmann et al., 2002; Kitahara and Ogawa 1991; Stark et al., 1987; Steinert, 1975).

2. Procedures

a. Tissue Extraction and One-Dimensional Sodium Dodecylsulfate-Polyacrylamide Gel Electrophores

For obvious reasons, the use of total scalp skin as a source for electrophoretic studies on hair and epithelial keratin proteins that are specifically expressed in the human hair follicle is inappropriate in several aspects. Keratins that are expressed in "minor" compartments of the follicle (e.g., hair cuticle, Henle and/or Huxley layers, IRS cuticle, and companion layer) would be strongly "underrepresented" within extract of scalp skin, which contains large amounts (including the respective keratins) of epidermis, dermis, hair cortex, and ORS. Thus, inevitably, those studies must use freshly plucked hair follicles, with beard hairs being the most suitable starting material, because these possess particularly large bulbs that easily allow an evaluation even to the naked eye, when the follicle has been removed in toto (i.e., under preservation of the ORS and the dermal papilla). A further sign

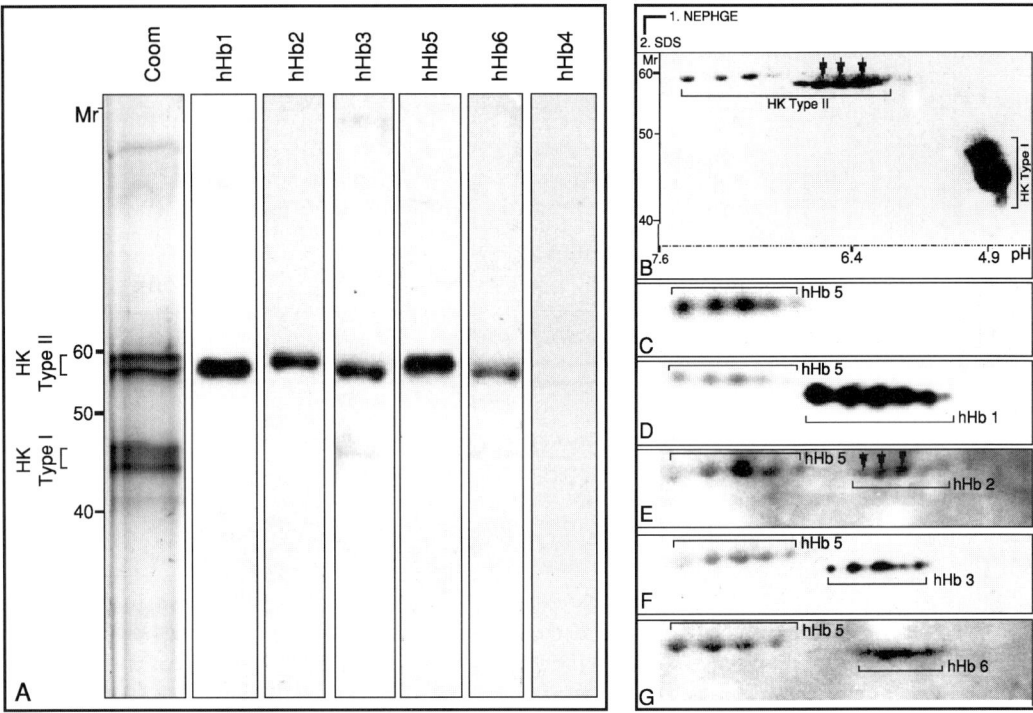

Fig. 6 Identification of human type II hair keratins by 1D and 2D PAGE and Western blot analysis. (A) One dimensional SDS-PAGE (10% polyacrylamide) of total keratin extracts from clipped hairs transferred to a nylon membrane. In lane "coom," the proteins are stained by Coomassie blue. The positions of the collective type II hair keratins (HK type II) and type I hair keratins (HK type I) are indicated. Western blots were performed with antisera against hHb1, hHb2, hHb3, hHb5, hHb6, and hHb4, each detecting a single protein band. Note absence of staining for hHb4, which does not occur in the hair follicle. The molecular weight ranges (Mr) between 40 and 60 kDa are indicated on the left hand side. (B–G) Two-dimensional gel electrophoresis of total keratin extracts from clipped hairs transferred to a nylon membrane. The proteins were separated by NEPHGE in the first dimension and by SDS-PAGE (SDS; 10% polyacrylamide) in the second dimension. (B) Coomassie staining of total (type II, type I) hair keratins (HK). Western blots were performed with the antiserum against hHb5 alone (C) or concomitantly with the antisera against hHb5 and hHb1 (D), hHb5 and hHb2 (E) with the arrows in E and B indicating hHb2, hHb5 and hHb3 (F), hHb5h and Hb6 (G).

for completely extracted follicles is bleeding after plucking. If necessary, plucked scalp hairs can also be used. On clipping of the lower, viable portion of the follicle, this material is homogenized in the presence of an appropriate extraction buffer for keratins (see later). For the homogenization of the follicles, the tool of choice is a microhomogenizer, completely made from borosilicate glass, including the pestle (e.g., size G1, 10–25 μl or G2, 25–100 μl; Carl Roth, Karlsruhe, Germany; www.carl-roth.de).

b. Hair Fiber and Hair Cuticle Keratins

1. Immediately after plucking, the hair follicles are freed from the adhering ORS, the companion layer, and the IRS simply by mechanically stripping these tissue compartments off by means of a clean sheet of cellulose paper.

2. Cut the plucked and stripped hair follicle approximately 2 mm above the follicle bulb, keep the follicles in sterile PBS, and analyze them under a microscope for the preservation of the cuticle.

3. Bring the follicles into 2× Laemmli sample buffer (125 mM Tris-HCl, pH 6.8, 2% SDS, 50 mM DTT or 20 mM mercaptoethanol, 10%, 10 mM EDTA; approximately 20 μl per follicle) and keep them on ice.

4. Thoroughly homogenize the follicles with the glass homogenizer mentioned previously for at least 10 minutes on ice. For a more vigorous homogenization, small amounts of sterile quartz sand can be added.

5. Add 2 μl of benzonase (DNAse, Merck, Darmstadt, Germany) and mix by pipetting according to the manufacturer's instructions. Centrifuge at 14,000 rpm with a laboratory centrifuge and prepare aliquots of the supernatant.

6. Briefly heat sample aliquots in a water bath at 98 °C, centrifuge them again for a complete sedimentation of nondissolved material, and analyze the supernatant on a 10% or 12% SDS-(check) mini-gel for optimal loading and running conditions (Fig. 6A, lane Coom).

c. ORS, Companion Layer and IRS Keratins

The following procedure is a partial extraction method and requires some experience.

1. Gently homogenize completely removed follicles in 2× Laemmli sample buffer by use of the glass homogenizer mentioned previously *above* (II.2.D.2.a) to dissolve the ORS, the companion layer, and the IRS. Add 2 μl of benzonase and mix slightly by pipetting.

2. After centrifugation and heating, aliquots of the supernatant are analyzed on a mini-test 10–12% SDS-PAGE Laemmli gel for optimal loading and running. It is recommended to homogenize the same number of follicles in the same amount of buffer for gradually extended times to evaluate the appropriate time of extraction. Contamination with small amounts of hair keratin cannot completely be avoided, but in our hands, this method leads to extracts in which the special epithelial keratins are strongly "enriched."

d. Tissue Extraction and Two-Dimensional Polyacrylamide Gel Electrophoresis

1. For 2-DE analysis, the keratin extracts dissolved in Laemmli buffer are first precipitated with chloroform/methanol (Wessel and Fluegge, 1984).

   a. Mix 1 vol. of Laemmli protein extract, with 3 vol. of MeOH, and vortex.

    b. Add 1 vol. of chloroform and vortex.

    c. Add 3 vol. of H_2O and vortex.

    d. Centrifuge for 1 minute at RT and 14,000 rpm and carefully discard the lower chloroform and the upper H_2O/methanol phase by use of a fine pipette. The middle (inter-) phase contains the precipitated proteins.

    e. Wash the precipitate in 3 vol. MeOH by vortexing and centrifuge for 1 minute RT at 14,000 rpm. Carefully discard the methanol, dry the protein precipitate on air, and store at −20 °C or directly dissolve it in "lysis buffer" (9.5 M urea, 2% NP40, 100 mM DTT, containing 2% ampholines either with IEF or NEPHGE, see below) for 2-D electrophoresis. For the loading of the 2-D gel tube for the first dimension, use an approximately 5–10-fold amount of the optimal extract used in the 1-D minigel.

2. Alternatively, the tissue can directly be homogenized in lysis buffer with a glass homogenizer mentioned previously. The homogenate is incubated overnight at 37 °C, heated at 60 °C for 3 minutes, centrifuged at 10,000 rpm for 5 minutes and loaded on the gel for the first dimension.

3. Type I (acidic) keratins are preferentially separated by isoelectric focusing (IEF) in the first dimension. The gels should contain the following mixture of ampholines: pH 5–7 (0.8% w/v), pH 4–6 (0.8% w/v), pH 3.5–10 (0.4% w/v) (e.g., BioRad, Munich, Germany).

4. Type II (basic to neutral) keratins are preferentially separated by nonequilibrium pH gradient electrophoresis (NEPHGE) (O'Farrell *et al.*, 1977). The gels should contain the following mixture of ampholines: pH 3.5–10 or 2–11 (2.5% w/v) (e.g., BioRad, Munich, Germany). In contrast to IIF, be aware that the anode is at the top of the electrophoresis chamber!

5. For resolution in the second dimension, use a 10% or 12% SDS-PAG (see earlier, II.D.2).

6. The gels are stained with Coomassie blue or directly used for Western blotting.

e. Western Blot Procedure

1. For Western blot procedures, transfer gels to PVDF membranes (e.g., Immobilon-P, Millipore, Eschborn, Germany) by conventional semidry or wet blotting procedure.

2. Stain membranes with 0.1% Coomassie blue R250 in 40% methanol and 1% acetic acid. Destain membranes in 50% methanol.

3. Block unspecific binding of immunoglobulins onto the membranes by incubation with 5% nonfat milk powder (NFMP) and 0.1% Tween 20 in Tris-buffered saline at pH 7.5 (TBST) for 1 hour.

4. Incubate membranes for 1 hour with the respective primary antibody at a suitable dilution (tested on 1-DE mini-check gel Western blot before). Wash 4× for 10 minutes each with TBST.

5. Apply horseradish peroxidase–coupled secondary antibodies (e.g., Dianova, diluted 1:10,000 in 5% NFMP in TBST) for 1 hour. Wash the membranes 3× for 10 minutes each.

6. Detect bound antibodies by use of the enhanced chemoluminiscence method (e.g., ECL-system; Amersham-Buchler, Braunschweig, Germany).

f. Controls

For an efficient control for the hair follicle, specific K6-related keratins, K6irs1-4 or K6hf, we use protein extracts from plucked hair follicles or scalp as *positive control*, whereas human foot sole epidermis extracts (Langbein *et al.*, 1993, 2002a, 2003; Winter *et al.*, 1998) positive for K6a were used as *negative control*. Considering that foot sole extracts contain a large number of other epithelial keratins, its use is also a convenient control for possible cross-reactions of antibodies against hair follicle keratins. Otherwise, the same holds true when hair follicle extracts are used to check epithelial keratin antibodies for a possible cross-reactivity with hair keratins or hair follicle–specific epithelial keratins. Keep in mind that an appropriate dilution of the primary antibodies and amount of the loaded proteins are prerequisites for a reliable Western staining.

III. Pitfalls and Problems

In this chapter, only the most important pitfalls are described. For more details, see also the hints given in the respective chapter of the review.

A. Tissues and Sections

1. *Handling of materials, tissues, and equipment*: (a) Avoid contamination with RNAses (ISH) or proteases (IIF). Therefore, tissue material from absolutely fresh tissue samples, immediately snap-frozen after surgical removal in liquid nitrogen-cooled isopentane and stored at −80 °C, should be used. Cryoconservation of material is superior over chemical fixation, including paraffin embedding, which often requires antigen retrieval methods (e.g., microwave treatment) (see II.B.2.b.2). (b) If necessary, sections can be stored at the same temperature (II.A.2.d). (c) For the ISH procedure, slide holders, scissors, forceps, etc. used for tissue preparation must be clean and sterile (II.A.2.d, i).

2. *Positive and negative controls* should be performed to check tissue integrity by use of an established ISH probe (see, II.A.c) or an established antibody with known expression sites for IIF (II.B.2.e).

3. *Loss of tissue sections*: Use commercially available or self-made silanized slides for optimal attachment of sections II.A.2.d, e).

B. *In Situ* Hybridization

1. *No or weak signals*: (a) Degradation of the probes or tissue mRNA by RNases and, therefore, use best quality tissues and avoid RNAse contamination during the procedure. (b) Check size and concentration of the probe and, in particular, the stringency of the hybridization/washing steps (II.A.2.f-g, k-l). (c) Avoid draining of the probe from the specimen. (d) Use DEPC water throughout, except for procedures involving enzymes (II.A.2.i) (e) Low labeling efficiency of the probe during *in vitro* transcription of cRNA may occur (II.A.2.b). (f) Check linearization, efficiency of *in vitro* transcription, and size of the ISH probes. Too large probes may lead to bad penetration, too small probes may produce background problems. (II.A.2.i, f). (g) Check nucleotides selected for labeling for their frequency in the transcript; if too small, switch to another nucleotide type (II.A.2.g). (h) Verify specificity of probes on appropriate control sections (II.A.2.b.). Be careful not to lose the cRNA pellet after labeling and do not "overdry" the pellet to avoid bad solubility (II.A.2.c). (i) Keep in mind that a low signal might be simply due to a low expression level of the gene. Use positive/negative controls.

2. *Distinction between low-level gene expression and background staining*: Use another probe of the same gene or a tissue section from another piece of tissue (controls!).

3. *Background over entire slide*: Avoid exposure to light during the procedure and check expiration date of the film emulsion. Check on a blank slide. (II.A.2.n-o).

4. *Signals arise outside of the hybridization target within the tissue*: This can only be handled by appropriate positive and negative controls (II.A.c).

5. *Uniform background over sections of distinct tissues*: Nonspecific binding of the probe on tissue may be caused by (a) an inappropriate dilution or size (toosmall) of the probe; (b) inappropriate hybridization/ washing temperatures (stringency) (II.A.2.k-l); (c) insufficient RNAse A treatment—check specificity of probes (II.A.2.b).

6. *Uniform background over sections of a distinct type of tissues*: Some tissues are more "sticky" than others. Use a different probe and/or try other hybridization/ washing conditions (II.A.2.k-l).

7. *Background on tissue section edges*: Occasionally, silver grains may occur preferentially at the margins of the tissue because of drying of the emulsion at the periphery of sections. Dry more slowly.

8. *Combination of ISH and IIF*: A prerequisite for such a combination is a rather strong IIF staining. An established positive control must be used in parallel

to check whether the respective antigen is accessible for IIF subsequent to the ISH procedure (II.A.2.r).

C. Antibody Production

Use preimmunserum for unwanted antibody reaction of the respective animal.

1. *No or low titer*: (a) Animals did not produce adequate amounts of antibodies because of a low antigenicity of the antigen. If possible, use another peptide sequence and check antigenicity with the respective software. Be aware that the peptide used is of good quality (mass spectroscopy controlled; II.B.2.a). (b) Check tissue samples by positive controls (II.B.2.e). (c) Store undiluted aliquots supplemented with 0.01% NaN_3 at $-80\,°C$. Do not freeze diluted antisera but store at $4\,°C$. (d) The immune response may vary from animal to animal. Use two animals for immunization with the same antigen. (e) Be aware that the detection system (e.g., IIF or Western) is correctly working.

2. *Cross-reaction with other antigens*: (a) Check databases for cross-reacting epitopes of the peptides in particular for antibodies against members of multigene families (II.B.2.a). (b) Purify antisera by affinity binding to the coupled antigen (without carrier protein, e.g., KLH; II.B.2.a.6).

D. Immunohistochemistry

In general, use adequate positive and negative controls to check the properties and quality of the primary and secondary antibodies, tissues, and detecting systems (immunoenzyme reaction, microscope equipment with the respective filters for the secondary antibodies).

1. *Weak or negative staining*: (a) Primary (or secondary or both) antibodies are inappropriately diluted. Use higher concentrations. (b) Tissue damage (antigens degraded) or inappropriately fixed (too long, false fixative for the respective antigen; II.B.2.b.1-2). This may happen when monoclonal antibodies are used, which only detect one epitope. Use antigen-retrieval protocols, mild protease digestion (trypsin or pepsin), or another type of fixation and detergent, respectively (TritonX-100 for IIF or saponin for IEM for better penetration of antibodies (II.B.2.b.1-2; II.B.2.c.3-4). (c) Wrong filters for the respective fluorescence labels. (d) Inappropriate storage of antibodies (see above). (e) Check correct species of the secondary antibody. (f) Incubate sections with antibodies for a longer time. (g) Keep in mind that the antigen may be present on only low amounts in the tissue selected for investigation.

2. *Very strong staining and/or heavy background*: (a) Primary (or secondary or both) antibodies are too concentrated. (b) Cross-reaction of antibodies with immunoglobulins of the species. Use antibodies cross-absorbed for the immunoglobulins of the respective species. (c) Inadequate washing after application of primary or secondary (or both) antibodies. (d) Inadequate blocking of the tissue sections. Block with normal serum of the species of the secondary antibody. No

quenching after aldehyde fixation (II.C.4.d). (e) Incubate sections with antibodies for a shorter time. (f) Strictly avoid drying of sections during the whole staining procedure. (g) For double-label IIF, use antibodies of different species not cross-reactive with each other and dilutions adapted to a comparably staining intensity. In every case, before double labeling, single labeling experiments with the selected antibodies have to be performed. (h) Use narrow-range fluorescence filters in the microscope for clear separation of images from the different fluorochromes (no detection—even faint—in the "wrong" channel). (i) Reduce time and temperature during silver enhancement (in the dark!) after immunogold labeling for immuno-EM (II.C.4.h).

E. Polyacrylamide Gel Electrophoresis and Western Procedures

In general, use adequate positive and negative controls to check the properties and quality of the primary and secondary antibodies, the enzyme reaction of the luminescence system, and the quality of tissue extracts. Use control extracts of tissues known to be positive or negative for the respective antigens.

1. *No or low signals*: (a) Check antibodies (primary and secondary) and dilutions by appropriate controls. (b) Use higher concentrated antibodies or load more material onto the gels. (c) Avoid loss of proteins by use of too high voltages and too long transfer of the gel to the membrane. (d) Be aware that the protein pellet is not lost during precipitation. Do not "overdry" the pellet. (II.D.2.b). (e) The proteins are not reliably separated during electrophoresis. Be sure that the polarity during electrophoresis is correct (e.g., for NEPHGE, the anode has to be on the top of the chamber; see II.D.2.b). (f) For the extraction of hair keratins, strong reducing conditions are required. Use sufficient amounts of DDT or mercaptoethanol in the probe buffers (II.D.2.a.1). (g) When hair keratins occurring only in small tissue compartments (i.e., the hair cuticle) have to be tested, be sure that these compartments are "enriched" in the extracts by selective preparation (e.g., remove ORS and IRS for cuticular hair keratins; compare II.D.1-2). (h) Homogenize tissues extensively. (i) Keep in mind that the protein of interest may not or may be present only in small amounts in your extract.

2. *Strong staining or background*: (a) Overall background by inappropriate blocking with non-fat milk powder solution or mishandling of ECL procedure. (b) Dilute primary and/or secondary antibodies to a higher extent. (c) Do not overload gels with the respective extract. If possible, use selectively prepared extracts (see above and II.D.2.a.2). (d) Use shorter exposure time for luminescence detection.

References

Bucher, O. (1968). Cytology, Histologie und mikroskopische Anatomie des Menschen. Huber, Bern.
Drenckhahn, D., Joens, T., and Schmitz, F. (1993). Production of polyclonal antibodies against proteins and peptides. *Meth. Cell Biol.* **37,** 8–57.

Fuchs, E., and Weber, K. (1994). Intermediate filaments: Structure, dynamics, function, and disease. *Annu. Rev. Biochem.* **63**, 345–382.

Halbhuber, K. J., and Koenig, K. (2003). Modern laser scanning microscopy in Biology, Biotechnology and Medicine. *Ann. Anat.* **185**, 1–20.

Hammerl, P., Hartl, A., and Thalhammer, J. (1993). Particulate nitrocellulose as a solid phase for protein immobilization in immuno-affinity chromatography. *J. Immunol. Meth.* **165**, 39–66.

Heid, H. W., Werner, E., and Franke, W. W. (1986a). The complement of native α-keratin polypeptides of hair-forming cells: A subset of eight polypeptides that differ from epithelial cytokeratins. *Differentiation* **32**, 101–109.

Heid, H. W., Moll, I., and Franke, W. W. (1986b). Patterns of expression of trichocytic and epithelial cytokeratins in mammalian tissues. I Human and bovine hair follicles. *Differentiation* **37**, 137–157.

Heid, H. W., Moll, I., and Franke, W. W. (1988). Pattern of expression of trichocytic and epithelial cytokeratins in mammalian tissues: I. Human and bovine hair follicies. *Differentiation* **37**, 137–158.

Hearle, J. W. S. (2000). A critical review of the structural mechanics of wool and hair fibres. *Int. J. Biol. Macromolec.* **27**, 123–138.

Hofmann, I., Winter, H., Muecke, N., Langowski, J., and Schweizer, J. (2002). The *in vitro* assembly of hair follicle keratins: Comparison of cortex and companion layer keratins. *Biol. Chem.* **383**, 1373–1381.

Ito, M. (1986). The innermost cell layer of the outer root sheath in anagen hair follicle: Light and electron microscopic study. *Arch Dermatol. Res.* **279**, 112–119.

Ito, M. (1988). Electron microscopic study on cell differentiation in anagen hair follicles in mice. *J. Invest. Dermatol.* **90**, 65–72.

Ito, M. (1989). Biological roles of the innermost cell layer of the outer root sheath in human anagen hair follicle: Further electron microscopic study. *Arch. Dermatol. Res.* **281**, 254–259.

Kitahara, T., and Ogawa, H. (1991). The extraction and characterization of human nail keratin. *J. Dermatol. Sci.* **2**, 402–406.

Laemmli, U. K. (1970). Cleavage of structural proteins during the assembly of the head of bacteriophage T4. *Nature* **227**, 680–685.

Langbein, L., Heid, H. W., Moll, I., and Franke, W. W. (1993). Molecular characterization of the body site-specific human cytokeratin 9-cDNA cloning, amino acid sequence, and tissue specificity of gene expression. *Differentiation* **55**, 57–73.

Langbein, L., Rogers, M. A., Winter, H., Praetzel, S., Beckhaus, U., Rackwitz, H. R., and Schweizer, J. (1999). The Catalog of Human Hair Keratins: I. Expression of the nine type I members in the hair follicle. *J. Biol. Chem.* **274**, 19874–19893.

Langbein, L., Rogers, M. A., Winter, H., Praetzel, S., and Schweizer, J. (2001). The catalog of human hair keratins: II. Expression of the six type II members in the hair follicle and the combined catalog of human type I and type II keratins. *J. Biol. Chem.* **276**, 35123–35132.

Langbein, L., Rogers, M. A., Praetzel, S., Aoki, N., Winter, H., and Schweizer, J. (2002a). A novel epithelial keratin, hK6irs1, is expressed differentially in all layers of the inner root sheath, including specialized Huxley cells ("*Flügelzellen*") of the human hair follicle *J. Invest. Dermatol.* **118**, 789–800.

Langbein, L., Grund, C., Kuhn, C., Praetzel, S., Kartenbeck, J., Brandner, J., Moll, I., and Franke, W. W. (2002b). Tight junctions and compositionally related junctional structures in mammalian stratified epithelia and cell cultures derived therefrom. *Eur. J. Cell Biol.* **81**, 419–435.

Langbein, L., Rogers, M. A., Praetzel, S., Winter, H., and Schweizer, J. (2003). K6irs1, K6irs 2, K6irs 3, and K6irs 4 represent the inner-root-sheath (IRS)-specific type II epithelial keratins of the human hair follicle. *J. Invest. Dermatol.* **120**, 512–522.

Lynch, M. H., O'Guin, W. M., Hardy, C., Mak, L., and Sun, T. T. (1986). Acidic and basic hair/nail ("hard") keratins: Their localization in upper cortical and cuticle cells of the human hair follicle and their relationship to "soft" keratins. *J. Cell. Biol.* **103**, 2593–2606.

Mischke, D. (1998). The complexity of gene families involved in epithelial differentiation: Keratin genes and the epidermal differentiation complex. *In* "Intermediate Filaments: Subcellular Biochemistry," (H. Herrmann and J. R. Harris, eds.), Vol. 31, pp. 71–95. Plenum Publishing Corp., New York.

Montag, M., Trendelenburg, M. F., and Spring, H. (1988). Video-microscopic image processing facilitates the evaluation of light microscopic autoradiography at high magnification. *J. Microscopy* **150,** 245–249.

O'Farrell, P. H. (1975). High resolution two-dimensional electrophoresis of proteins. *J. Biol. Chem.* **250,** 4007–4021.

O'Farrell, P. Z., Goodman, H. M., and O'Farrell, P. H. (1977). High resolution two-dimensional electrophoresis of basic as well as acidic proteins. *Cell* **12,** 1133–1142.

Paddock, S. (2002). Confocal reflection microscopy: The "other" confocal mode. *Biotechniques* **32,** 274–277.

Parry, D. A. D. (1997). Protein-chains in hair and epidermal keratin IF: Structural features and spatial arrangement. *In* "Formation and Structure of Human Hair," (P. Jollès, H. Zahn, and H. Hoecker, eds.), pp. 177–207. Birkhaeuser, Basel.

Powell, B. C., and Rogers, G. E. (1997). The role of keratin proteins and their genes in the growth, structure and properties of hair. *In* "Formation and Structure of Human Hair," (P. Jollès, H. Zahn, and H. Hoecker, eds.), pp. 59–148. Birkhaeuser, Basel.

Porter, R. M., Corden, L. D., Lunny, D. P., Smith, F. J., Lane, E. B., and McLean, W. H. (2001). Keratin K6irs is specific to the inner root sheath of hair follicle in mice and humans. *Br. J. Dermatol.* **145,** 558–568.

Rogers, M. A., Winter, H., Langbein, L., Krieg, T., and Schweizer, J. (1996). Genomic characterization of the human type I cuticular hair keratin hHa2 and identification of an adjacent novel type I hair keratin gene hHa5. *J. Invest. Dermatol.* **107,** 633–638.

Rogers, M. A., Langbein, L., Praetzel, S., Krieg, T., Winter, H., and Schweitzer, J. (1997). Sequences and differential expression of three novel human type-II hair keratins. *Differentiation* **61,** 357–362.

Rogers, M. A., Winter, H., Wolf, C., Heck, M., and Schweizer, J. (1998). Characterization of a 190-Kilobase pair domain of human type I hair keratin genes. *J. Biol. Chem.* **273,** 26683–26691.

Rogers, M. A., Winter, H., Langbein, L., Wolf, C., and Schweizer, J. (2000). Characterization of a 300–kilobase pair region of human DNA containing the type II hair kerath gene domain. *J. Invest. Dermatol.* **114,** 464–472.

Rogers, M. A., Langbein, L., Winter, H., Ehmann, C., Praetzel, S., Korn, B., and Schweizer, J. (2001). Characterization of a cluster of human high/ultrahigh sulfur keratin associated protein genes embedded in the type I keratin gene domain on chromosome 17q12-21. *J. Biol. Chem.* **276,** 19440–19451.

Rogers, M. A., Langbein, L., Winter, H., Praetzel, S., Ehmann, C., and Schweizer, J. (2002). Characterization of a first domain of human high glycine-tyrosine and high sulfur keratin associated protein (KAP) genes on chromosome 21q22.1. *J. Biol. Chem.* **277,** 48993–49002.

Rogers, M. A., Langbein, L., Winter, H., Beckmann, I., Praetzel, S., and Schweizer, J. (2004). Hair keratin associated proteins (KAPs): Characterization of a second high sulfur KAP gene domain on human chromosome 21. *J. Invest. Dermatol.* **120,** 147–158.

Schmidt, A., Langbein, L., Rode, M., Praetzel, S., Zimbelmann, R., and Franke, W. W. (1997). Plakophilins 1a and 1b: Widespread nuclear proteins recruited in specific epithelial cells as desmosomal plaque components. *Cell Tiss. Res.* **290,** 481–499.

Schmidt, A., Langbein, L., Praetzel, S., Rackwitz, H. R., and Franke, W. W. (1999). Plakophilin 3–A novel cell-type specific plaque protein. *Differentiation* **64,** 291–307.

Shimomura, Y., Aoki, N., Rogers, M. A., Langbein, L., Schweizer, J., and Ito, M. (2003). Characterization of the human keratin-associated protein 1 family members. *J. Invest. Dermatol. SP.* **8,** 96–99.

Speel, E. J. M., Kamps, M., Bonnet, J., Ramaekers, F. C. S., and Hopman, A. H. N. (1993). Multicolor preparations for *in situ* hybridization using precipitating enzyme cytochemistry in combination with reflection contrast microscopy. *Histochemistry* **100,** 357–366.

Stark, H. J., Breitkreutz, D., Limat, A., Bowden, P., and Fusenig, N. E. (1987). Keratins of the human hair follicle: "Hyperproliferative" keratins consistently expressed in outer root sheath cells *in vivo* and *in vitro Differentiation* **35,** 236–248.

Steinert, P. M. (1975). The extraction and characterization of bovine epidermal alpha keratin. *Biochem. J.* **149**, 39–48.

Steinert, P. M., and Roop, D. R. (1988). Molecular and cellular biology of intermediate filaments. *Annu. Rev Biochem.* **57**, 593–625.

Steinert, P. M., North, A. C. T., and Parry, D. A. D. (1994). Structural features of keratin intermediate filaments. *J. Invest. Dermatol.* **103**, 19S–24S.

Swensson, O., Langbein, L., McMillan, J. R., Churchill, L. J., Leigh, I. M., McLean, W. H. I., Lane, E. B., and Eady, R. A. J. (1998). Specialized keratin expression pattern in human ridged skin as an adaptation to high physical stress. *Br. J. Dermatol.* **139**, 767–775.

Swift, J. A. (1997). Morphology and histochemistry of human hair. *In* "Formation and Structure of Human Hair," (P. Jollès, H. Zahn, and H. Hoecker, eds.), pp. 149–175. Birkhaeuser, Basel.

Trembleau, A., Roche, D., and Calas, A. (1993). Combination of non-radioactive and radioactive *in situ* hybridization with immunohistochemistry: A new method allowing the simultaneous detection of two mRNAs and one antigen in the same brain tissue section. *J. Histochem. Cytochem.* **41**, 489–498.

Van de Plas, P. F. E. M., and Leunissen, J. L. M. (1992). Colloidal gold as a marker in molecular biology: The use of ultra-small gold particles. *In* "Non-radioactive Labelling and Detection of Biomolecules," (C. Kessler, ed.), pp. 116–126. Springer, Berlin.

Wang, Z., Wong, P., Langbein, L., Schweizer, J., and Coulombe, P. A. (2003). The type II epithelial keratin 6hf (K6hf), is, expressed in the companion layer, matrix and medulla in anagen-stage hair follicles. *J. Invest. Dermatol.* **121**, 1276–1282.

Wessel, D., and Fluegge, U. I. (1984). A method for the quantitative recovery of protein in dilute solution in the presence of detergents and lipids. *Anal. Biochem.* **138**, 141–143.

Wilkinson, D. G. (1992). "*In situ* hybridization. A practical approach," IRL press, Oxford, New York, Tokyo.

Winter, H., Langbein, L., Praetzel, S., Jacobs, M., Rogers, M. A., Leigh, I. M., Tidman, N., and Schweizer, J. (1998). A novel human type II cytokeratin, K6hf, specifically expressed in the companion layer of the hair follicle. *J. Invest. Dermatol.* **111**, 955–962.

Winter, H., Langbein, L., Krawczak, M., Cooper, D. N., Jave-Suarez, L. F., Rogers, M. A., Praetzel, S., Heidt, P., and Schweizer, J. (2001). Human type 1 hair keratin pseudogene ψhHaA has functional orthologs in the chimpanzee and gorilla: Evidence for recent inactivation of the human gene after the Pan-Homo divergence. *Hum. Genet.* **108**, 37–42.

Yu, J., Yu, D. W., Checkla, D. M., Freedberg, I. M., and Bertolino, A. P. (1993). Human hair keratins. *J. Invest. Dermatol.* **101**, 56S–59S.

CHAPTER 16

Skin: An Ideal Model System to Study Keratin Genes and Proteins

Kelsie M. Bernot,* Pierre A. Coulombe,*,[†] and Pauline Wong*

*Department of Biological Chemistry
The Johns Hopkins University School of Medicine
Baltimore, Maryland 21205

[†]Department of Dermatology
The Johns Hopkins University School of Medicine
Baltimore, Maryland 21205

METHODS IN CELL BIOLOGY, VOL. 78

I. Introduction

Keratins are among the most abundant proteins in epithelial cells, in which they occur as a cytoplasmic network of 10 to 12-nm-wide intermediate filaments (IFs). They are encoded by a large multigene family in mammals, with >50 individual genes partitioned into two major sequence types (Coulombe, 2003; Hesse, *et al.*, 2001; Moll *et al.*, 1982). A strict requirement for the heteropolymerization of type I and II keratin proteins during filament assembly underlies the pairwise transcriptional regulation of keratin genes. Regulation of keratin expression also depends on both the type and stage of differentiation in epithelia (Moll *et al.*, 1982). A major function fulfilled by keratin filaments is to act as a resilient yet pliable scaffold, which endows epithelial cells with the ability to sustain mechanical and nonmechanical stresses. Reflecting this crucial participation in mechanical support, inherited mutations altering the coding sequence of keratins underlie a large number of epithelial fragility disorders (Fuchs and Cleveland, 1998; Irvine and McLean, 1999; Omary, 2004). Keratins also have been found recently to significantly attenuate the cellular response to specific proapoptotic signals (Coulombe and Wong, 2004; Coulombe and Omary, 2002; Marceau *et al.*, 2001; Omary *et al.*, 2004; Oshima, 2002). So far, and like other recently identified functions, keratin-mediated modulation of apoptosis is manifested in a keratin sequence-dependent and context-dependent fashion. Skin represents an ideal system to study the properties of keratin proteins, the regulation of keratin genes, and the function of their encoded products. This chapter contains methods that have been particularly useful for the basic characterization of transgenic mouse models in which the expression of keratin genes has been manipulated.

A. The Family of Keratin Genes and Proteins

Keratins belong to the superfamily of IF proteins (Coulombe *et al.*, 2001; Fuchs, 1994; Steinert and Roop, 1988). They are heterogeneous in size (400–644 amino acid residues) and charge (pI ~4.7–8.4) and notoriously insoluble because of their primary structure. Keratin nomenclature originally was based on protein separation by both charge and size by means of two-dimensional electrophoresis (Moll *et al.*, 1982). Type I keratins tend to be smaller (40–64 kDa) and more acidic (pI ~4.7–6.1) compared with type II keratins, which are larger (52–68 kDa) and basic-neutral (pI ~5.4–8.4) in charge (Moll *et al.*, 1982). Comparison of either amino acid sequence (Fig. 1) or gene substructure (e.g., the number and position of introns) reveals two groups of approximately equal numbers of keratin sequences, designated type I (K9–K23; Ha1-Ha8; IRS1–4) and type II (K1–K8; Hb1–Hb6; K6irs1–4) IF proteins (Coulombe *et al.*, 2001; Fuchs, 1994; Hesse *et al.*, 2001; Langbein *et al.*, 2003; and references therein).

The coordinated regulation of the two distinct types of keratin-like genes, along with the obligatory heteropolymerization of their products to form 10- to 12-nm-wide filaments, is present as far back as primitive chordates (Karabinos *et al.*, 2004). In *homo sapiens*, functional type I and type II keratin genes are clustered on the long arms of chromosomes 17 and 12, respectively (Fig. 1). There is a notable exception to this principle, in that the type I keratin 18 (K18) locus is located next to K8 at the telomeric boundary of the type II cluster (Fig. 1). K8 and K18 exhibit a tight coregulation in both early embryonic and adult simple epithelia (Oshima, 1992) and likely are direct descendants of the ancestral pair of keratin genes. The key molecular features of individual keratin genes (e.g., number and position of introns) and those of the genomic clusters they form (e.g., transcriptional orientation, position relative to other family members) are nearly perfectly conserved in human and mouse (our own observations) and in many other species of mammals as well (Coulombe and Bernot, 2004; Hesse *et al.*, 2004). This conservation has direct implications for the evolution of keratin genes.

Comparing the primary structure of human keratins also reveals the existence of distinct subfamilies within each of the two keratin types (Fig. 1) and provides additional insight into the mechanisms presiding over their evolution. Groupings of individual genes within each of the two major clusters correspond to similar patterns of expression in specific types of epithelial cells or tissues. This situation applies for most genes expressed in simple epithelia (K7, K8, K18, K20), "soft" complex epithelia (K1–K6; K9–K17), "hard" epithelia such as hair shaft and nail (Ha1–Ha8; Hb1–Hb6), and even the highly restricted inner root sheath compartment of hair follicles (K6irs-1–4; IRS1–4). In addition to these intriguing features, the clustering of keratin genes is mirrored by a corresponding clustering of related protein products as revealed by phylogenetic analysis. This striking equivalence, highlighted by the color scheme used in Fig. 1, implies a hierarchical pattern of gene duplication and specialization during evolution.

Details about the structure of keratin proteins (Chapter 2), their assembly into 10- to 12-nm filaments (Chapter 1), and the regulation of their polymerized state (Chapters 13 and 14) are given elsewhere in this book. In the individual cells making up complex epithelia such as skin, keratin IFs are organized in a pancytoplasmic network extending from the surface of the nucleus to the cytoplasmic periphery, where they associate with membrane-spanning cell–cell or cell–matrix

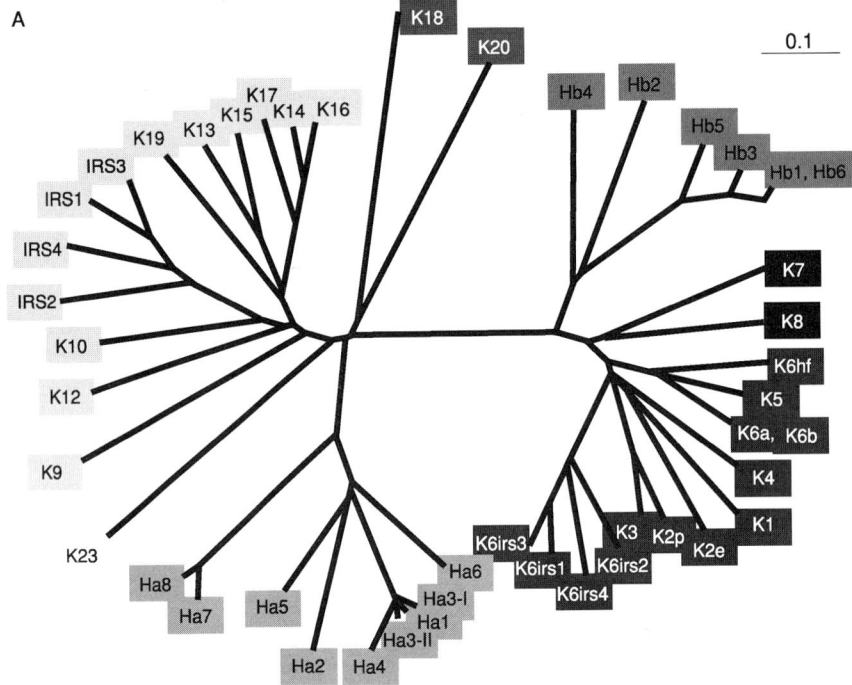

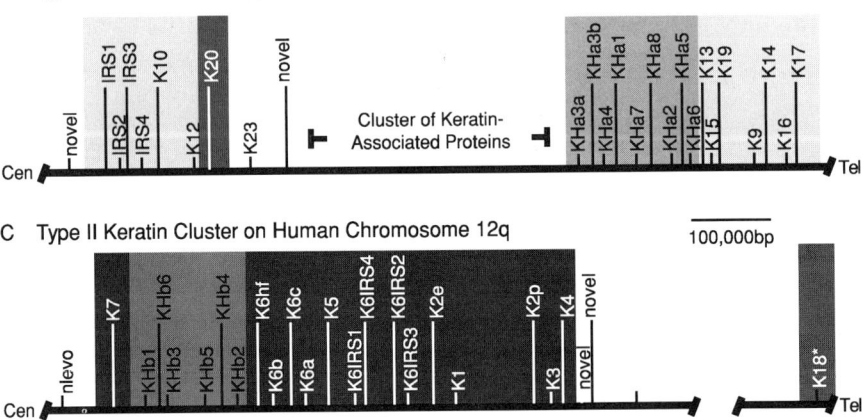

Fig. 1 The human keratin family *circa* 2003. (A) Comparison of the primary structure of human keratins with the publicly available ClustalW and TreeView software. Sequence relatedness is inversely correlated with the length of the lines connecting the various sequences and with the number and position of branch points. This comparison makes use of the sequences from the head and central rod domain for each keratin. Individual sequences were obtained from both the National Center for Biotechnology Information (www.ncbi.nlm.nih.gov) and the Human Genome Sequencing Consortium (www.ensembl.org) websites. A few keratins were left out for clarity purposes. Two major branches are

attachment complexes including desmosomes and hemidesmosomes (Fig. 2C) (Green and Gaudry, 2000; Jones, 1998). The reader can find further information about these processes in Chapters 25 through 28.

B. Differentiation-Dependent Regulation of Keratin Genes in Skin Epithelia

The tight relationship that has evolved between keratin gene regulation and epithelial cell differentiation is well established in skin tissue. The architectural complexity of adult skin epithelia, including the surface epidermis and a variety of epithelial appendages including hair, nail, and glands (Holbrook, 1993), is achieved through a temporally and spatially regulated sequence of simple decisions made by precursor cells during embryonic development (Coulombe and McGowan, 2000; Sengel, 1976). These sequential decisions result in the progressive restriction of fate determination, involving a variety of mesenchyme-derived and ectoderm-derived signals, and are reflected by specific changes in keratin gene regulation (Byrne, 1997; Coulombe and McGowan, 2000). In the end, more than half of all keratin genes are expressed in mature mammalian skin tissue (Fig. 2). Deviation from normal differentiation invariably results in altered keratin gene expression, such that "keratin profiling" is often exploited in the clinical setting to diagnose disease and determine its origin and/or course (Omary *et al.*, 2004; Osborn, 1983).

In the thin epidermis covering much of the body, mitotically active cells of the basal layer act as progenitors and consistently express K5 and K14 as their main keratin pair (Fuchs and Green, 1980; Nelson and Sun, 1983; Fig. 2F), along with lower levels of K15 (Lloyd *et al.*, 1995; Porter *et al.*, 2000). Onset of differentiation coincides with the appearance of the K1/K10 pair (Fuchs and Green, 1980; Woodcock-Mitchell *et al.*, 1982) through a robust transcriptional induction that occurs at the expense of K5/K14 (Byrne, 1997; Fuchs, 1995). Accordingly, K1/K10 proteins are readily detectable in the lowermost suprabasal layer of epidermis (Fig. 2G). The appearance of K1 and K10 coincides with a sudden

---

seen in this tree display, corresponding exactly to the known partitioning of keratin genes into type I and type II sequences. Beyond this dichotomy, each subtype is further segregated into major subgroupings, as indicated by the varied shadings. (B) Organization of functional type I keratin genes in the human genome. All functional type I keratin genes are clustered on the long arm of human chromosome 17. The only exception is K18 (asterisk), which is located at the telomeric (Tel) boundary of the type II gene cluster (Cen, centromere). Transcriptional orientation is from Tel to Cen for all genes. (C) All functional type II keratin genes are clustered on the long arm of human chromosome 12. The K8 and K18 genes are separated by 450,000 bp. Individual type I and type II keratin genes belonging to the same subgroup on the basis of the primary structure of their protein products (highlighted by the gray-scale boxes) tend to be clustered in the genome. Moreover, very highly related keratins are often encoded by neighboring genes (e.g., K5 and K6 paralogs; also, K14, K16, and K17), further emphasizing the primary role of gene duplication in generating keratin diversity. These features of the keratin family are virtually identical in mouse (not shown). (Adapted from Coulombe and Bernot [2004] with permission).

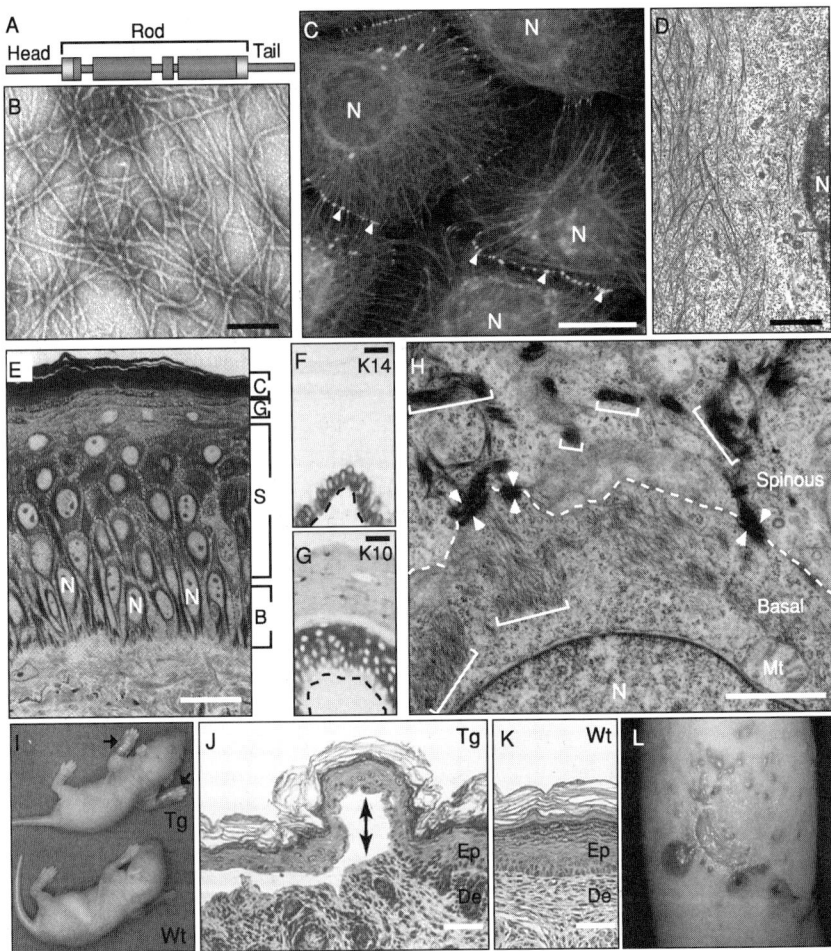

Fig. 2 Attributes, differential regulation, and disease association of keratin. (A) Schematic representation of the tripartite domain structure shared by all keratin and other intermediate filament (IF) proteins. A central α-helical "rod" domain acts as the major determinant of self-assembly and is flanked by nonhelical "head" and "tail" domains at the N-terminus and C-terminus, respectively. The ends of the rod domain contain 15–20 amino-acid regions (yellow), which are highly conserved among all IFs. (B) Visualization of filaments, reconstituted *in vitro* from purified K5 and K14, by negative staining and electron microscopy. Bar equals 125 nm. (C) Triple-labeling for keratin (red chromophore) and desmoplakin, a desmosome component (green chromophore) and DNA (blue chromophore) by indirect immunofluorescence of epidermal cells in culture. Keratin filaments are organized in a network that spans the entire cytoplasm and are attached to desmosomes at points of cell–cell contacts (arrowheads). Bar equals 30 μm. N, nucleus. (D) Ultrastructure of the cytoplasm of epidermal cells in primary culture as visualized by transmission electron microscopy. Keratin filaments are abundant and tend to be organized in large bundles of loosely packed filaments in the cytoplasm. Bar equals 5 μm. (E) Histological cross-section of resin-embedded human trunk epidermis, revealing the basal (B), spinous (S), granular (G), and cornified (C) compartments. Bar equals 50 μm. N, nucleus. (F and G) Differential distribution of keratin epitopes on human skin tissue cross-sections (similar to

and dramatic shift in the organization of cytoplasmic keratin filaments (Coulombe *et al.*, 1989), which exhibit significant bundling in early postmitotic keratinocytes (Fig. 2H). Another type II gene, K2e, is expressed at a later stage of differentiation (Collin *et al.*, 1992).

The epidermis of palm and sole skin is significantly thicker because of its specialization for resisting larger amounts of stress. This function is reflected in its architecture of alternating stripes of primary and secondary ridges (Swensson *et al.*, 1998) and specific changes in keratin expression. In the thick, stress-bearing, primary ridges, the major differentiation-specific type I K9 is presumed to foster a more resilient cytoskeleton (Knapp *et al.*, 1986; Langbein *et al.*, 1994; Swensson *et al.*, 1998). In the thinner secondary ridges, postmitotic cells preferentially express the type II K6a and type I K16 and K17. Relative to K1, K9, and K10, the properties of K6a, K16, and K17 are believed to foster greater cellular pliability (Bernot *et al.*, 2002; Wong and Coulombe, 2003), thus providing flexible "hinge" regions in between the more rigid, K1/K9-rich primary ridges (Swensson *et al.*, 1998). A similar principle likely underlies the substitution of K1/K10 with K6a, K6b, K16 and K17 in keratinocytes recruited from wound margins to participate in the restoration of the epidermal barrier after injury (Bernot *et al.*, 2002; Coulombe, 2003; Wong and Coulombe, 2003).

Consistent with its greater complexity, a significantly larger number of keratin genes are transcribed in the pilosebaceous unit (Langbein *et al.*, 1999, 2001, 2003). A mature hair follicle is composed of eight distinct epithelial layers organized in concentric circles along its main axis and segregated into three histological compartments. These compartments are, from inside out, the hair shaft proper; the inner root sheath (IRS); and the outer root sheath (ORS), an epithelium that wraps around the follicle and is contiguous with the epidermis (Holbrook and

---

frame E) as visualized by an antibody-based detection method. K14 occurs in the basal layer, where the epidermal progenitor cells reside (F). K10, on the other hand, is primarily concentrated in the differentiating suprabasal layers of epidermis (G). Dashed line indicates the basal lamina. Bar equals 100 μm. (H) Ultrastructure of the boundary between the basal and suprabasal cells in mouse trunk epidermis as seen by routine transmission electron microscopy. The sample, from which this micrograph was taken, is oriented in the same manner as (E). Organization of keratin filaments as loose bundles (brackets in basal cell) correlates with the expression of K5–K14 in basal cells, whereas the formation of much thicker and electron-dense filament bundles (brackets in spinous cell) reflects the onset of K1–K10 expression in early differentiating cells. Arrowheads point to desmosomes connecting the two cells. Bar equals 1 μm. N, nucleus (I) Newborn mouse littermates. The top mouse is transgenic (Tg) and expresses a mutated form of K14 in its epidermis. Unlike the control pup below (Wt), this transgenic newborn shows extensive blistering of its front paws (arrows). (J and K) Hematoxylin and eosin (H&E)–stained histological cross-section through paraffin-embedded newborn mouse skin similar to those shown in (I). Compared with the intact skin of a control littermate (K, Wt), the epidermis of the K14 mutant expressing transgenic pup (J, Tg) shows intraepidermal cleavage at the level of the basal layer, where the mutant keratin is expressed (opposing arrows). Bar equals 100 μm (L) Leg skin in a patient with the Dowling-Meara form of epidermolysis bullosa simplex. Characteristic of this severe variant of this disease, several skin blisters are grouped in a herpetiform fashion (Reproduced from Coulombe and Bernot [2004] with permission). (See Color Insert.)

Wolff 1993). The expression pattern of the ~27 keratin genes actively transcribed in hair follicles is detailed in Chapter 15 of this book, and only key features are highlighted here. The Ha1–Ha8 and Hb1–6 genes are unique to the hair shaft and related hard epithelia (e.g., nail). These genes exhibit a stage-specific regulation during the differentiation of matrix progenitors in the bulb region to hair cortex keratinocytes. The cysteine-rich head and tail domains of the "hard" keratin proteins form disulfide bridges that, along with the γ-glutamyl cross-links catalyzed by transglutaminases, foster the organization of large bundles of densely packed keratin IFs (Langbein, 1999; Langbein *et al.*, 2001, 2003). This organization undoubtedly endows hair and nail with its unique structural features, shape, and resilience. IRS1–4 and K6irs1–4 are found in the IRS compartment (Langbein *et al.*, 2003). By contrast, the ORS epithelium expresses several of the keratin genes expressed in the epidermis, with the notable exception of the differentiation-specific K1, K2e, and K10 (Stark *et al.*, 1987).

By the time an epidermal or hair cortex keratinocyte has completed differentiation, more than 80% of its protein content is keratin. This unique situation is reminiscent of globin gene expression in erythrocytes, another highly specialized cell type, and underscores the fundamental importance of keratin proteins to surface epithelia. Methods to study the mechanisms underlying the differential transcriptional regulation of keratin genes are outlined in Chapter 10.

C. Functions of Keratin Filaments and Their Involvement in Disease

The properties typifying keratin filaments collectively point to an important structural role. Evidence for this function came from transgenic mouse studies in which a mutated form of K14 capable of dominantly disrupting filament assembly and organization was targeted to the basal layer of epidermis *in vivo* (Coulombe *et al.*, 1991; Vassar *et al.*, 1991). Unlike controls, mutant-expressing mice exhibited massive intracellular lysis of basal cells in response to modest mechanical trauma to the skin (Fig. 2I–K). This mouse phenotype mimicked key aspects of a group of dominantly inherited epidermal disorders known as epidermolysis bullosa simplex (EBS, Fig. 2L), and it was shown shortly thereafter that this condition arises from mutations in either K14 (Bonifas *et al.*, 1991; Coulombe *et al.*, 1991) or K5 (Lane *et al.*, 1992). EBS turned out to be a paradigm that applies to the whole family of keratin genes and proteins and even nonkeratin IFs. Thus, keratin filaments fulfill a structural support function in all complex epithelia, ranging from the surface of the cornea to the upper gastrointestinal tract (Irvine and McLean, 1999; Omary *et al.*, 2004). *In vitro* biophysical assays showed that: (1) keratin IFs exhibit unique viscoelastic properties consistent with a major role of structural support in the cell, and (2) that disease-causing mutations significantly decrease the strength of keratin gels subject to mechanical strain (Ma *et al.*, 2001). Likewise, biophysical studies of live cells in culture showed that keratin IFs account for a substantial fraction of their viscoelastic properties and that disrupting keratin cytoplasmic organization causes cells to soften considerably (Beil *et al.*, 2003; Coulombe and Wong, 2004;

see Chapter 3 for details). Demonstration of a structural support role has been comparatively more difficult for simple epithelial tissues, but strong evidence to that effect now exists for liver hepatocytes and trophoblast giant cells (Hesse *et al.*, 2000; Ku *et al.*, 1995; Loranger *et al.*, 1997; Tamai *et al.*, 2000).

Transgenic mouse studies that focused on adult liver and extraembryonic tissue revealed that simple epithelial keratins 8/18 play an important cytoprotective role against chemical insults and in modulating the response to proapoptotic signals. In liver hepatocytes, interestingly, the function of cytoprotection requires phosphorylation of K8/K18. In regard to apoptosis, studies showed that K8/K18 filaments can bind to and either organize or sequester key intracellular effectors of the signaling machinery downstream from engagement of the tumor necrosis factor (TNF)-α and Fas receptors (Caulin *et al.*, 2000; Gilbert *et al.*, 2001; Inada *et al.*, 2001; Jaquemar *et al.*, 2003; Oshima, 2002). Of note, the presence of K17 is also required for the survival of matrix epithelial cells in mature hair foilicles (McGowan *et al.*, 2002). It seems likely that keratin IFs will be found to influence additional basic metabolic processes in the cell. In that regard, recent studies have shown that an intact K8/K18 filament network is required for the proper sorting of specific membrane proteins in polarized simple epithelial cells (Ameen *et al.*, 2001; Coulombe and Omary, 2002).

Finally, the role of keratin mutations in causing several epidermal, oral, ocular, and hair-related diseases is well established (Irvine and McLean, 1999; Omary *et al.*, 2004; see Chapter 6). These diseases are genetically well defined and typically inherited in an autosomal-dominant fashion, although recessive inheritance is occasionally seen. Most genetic lesions involved are missense mutations, although premature stop codons, small deletions, and mutations affecting mRNA splicing occur with some frequency. In general, disease severity correlates with the location and nature of the mutation within the keratin backbone. A keratin mutation database is being maintained at the University of Dundee and can be accessed online http://www.interfil.org (Cassidy *et al.*, 2002; see Chapter 6). For most of the keratin-based disorders affecting skin epithelia, mutations are causative for the disease and act through their ability to compromise the role of keratin IFs toward structural support (Omary *et al.*, 2004).

D. Selection of Methods for the Study of Keratins and Other Intermediate Filament Proteins in Skin

Of all the methods having been applied, it is the characterization of transgenic mouse models that has yielded the most innovative and definitive information about the function of cytoplasmic IF proteins in skin or any other tissue (Magin *et al.*, 2000; Takahashi *et al.*, 1999). Strategies and methodology toward the manipulation of genes in mice are described in Chapter 4 of this book. Here we focus on some of the basic methods needed as part of the initial characterization of a newly created transgenic mouse model, with an emphasis on the study of keratins in skin tissue. Unless their source is important for the success of an experiment, in which case a particular supplier is recommended, the reagents needed to perform the assays and methods described are listed without regard to source.

II. Collection of Mouse Skin Tissue for Analysis

A marked advantage of skin tissue is its accessibility for study. Accordingly, it is one of the best understood organs in the body. For the purpose of routine histology in mouse, dorsal back skin is traditionally studied. Because keratins are abundant and relatively insoluble proteins, the method of sacrifice should not interfere with the morphological analysis. Still, it is best to work quickly to prevent changes secondary to the loss of blood supply and to preserve the quality of morphology. Animals should be sacrificed by an institutionally approved method. In the United States, cervical dislocation or overdose with an approved anesthetic drug are common methods approved for mouse euthanasia.

A. Materials

Anesthetic; dissecting scissors; two forceps, preferably a curved and straight pair; Betadine; beaker with 70% ethanol for tools; razor blades; sterile bacterial dishes.

B. Adult Mouse Skin Samples

The amount of tissue that is required for an experiment may dictate whether one has to sacrifice the animal. For instance, general anesthesia can be performed to acquire a small skin sample (such as a 2 to 6-mm punch biopsy) from an adult mouse (for survival surgery). If the experiment performed calls for a large quantity of backskin, or skin tissue from body sites such as paws or whisker pads, the animal should be sacrificed before tissue harvesting.

To collect a small amount of tissue, anesthetize the animal and verify that it is completely unconscious by gently, but firmly, pressing against the footpad. If the animal responds, wait a few more minutes to allow the anesthesia to fully take effect. Small skin samples can be harvested quickly without much trauma. Institutional policies regarding survival surgeries should be followed.

In nonsurvival surgeries, animals should be sacrificed according to approved guidelines from the institution. For skin collection, use the tweezers to pinch the backskin near the tail, lift the skin from the rest of the body, and use the scissors to make a full-thickness incision through the fur and skin. Continue to cut along both sides of the midline to obtain a rectangular sample of backskin. With the curved portion of the scissors, gently separate away the skin from the rest of the body while trimming the facia when necessary and excise the desired amount of tissue.

The skin can be placed flat in a sterile Petri dish for further manipulation, such as trimming the sample before fixation or holding the sample temporarily while other tissues are being collected. However, if samples are to be used for protein or RNA extraction, they should be immediately snap-frozen by liquid nitrogen. See below for details on RNA and protein isolation from tissues.

C. Newborn Mouse Skin Samples

Because of their small size and the likelihood of rejection from their mothers after human manipulation, it is usually not feasible to subject mouse pups to survival surgery. The easiest method for sacrificing neonatal mice is decapitation either with surgical scissors or fresh razor blades. The tail and genital area, as well as the forelimbs and hindlimbs, are removed with scissors to facilitate the isolation of trunk skin. With tweezers and curved scissors, dorsal skin can be easily removed by cutting along the sides of the torso, gently prying away the skin from the main body.

The skin can be placed flat in a sterile Petri dish for further manipulation, such as trimming the sample before fixation or holding the sample temporarily while other tissues are being collected. However, if samples are to be used for protein or RNA extraction, they should be immediately frozen by liquid nitrogen. See below for details on RNA and protein isolation.

III. Preparation of Tissues for Morphological Studies

Histological analyses can be performed by a number of different methods, each with its own advantages and disadvantages. The most common histological method to view tissue morphology is to fix and process samples through paraffin embedding and stain sections with hematoxylin and eosin (H&E; Fig. 3A). This technique preserves morphology of the tissue and allows for visualization of most cells with the two counterstains. A potential limitation of the fixation and paraffin-embedding is that some epitopes desired for immunostaining can be masked during the processing of the tissue. Fresh-frozen sections offer an alternative to paraffin embedding. Tissues are either snap frozen in liquid nitrogen or fixed in paraformaldehyde before frozen embedding and sectioning. The fresh-frozen method offers the advantage of retaining epitope accessibility; however, morphology of the tissue sample is not as well preserved as in paraffin sections (compare Fig. 3B and C).

A. Materials

Paraffin processing: Bouin's fixative; 100%, 95%, 90%, 70%, and 50% EtOH; xylene; tissue marking dyes (Bradley Products, Inc., Bloomington, MN); paraplast paraffin (Tyco Healthcare/Kendall, Hampshire, UK). *Fresh frozen sectioning*:

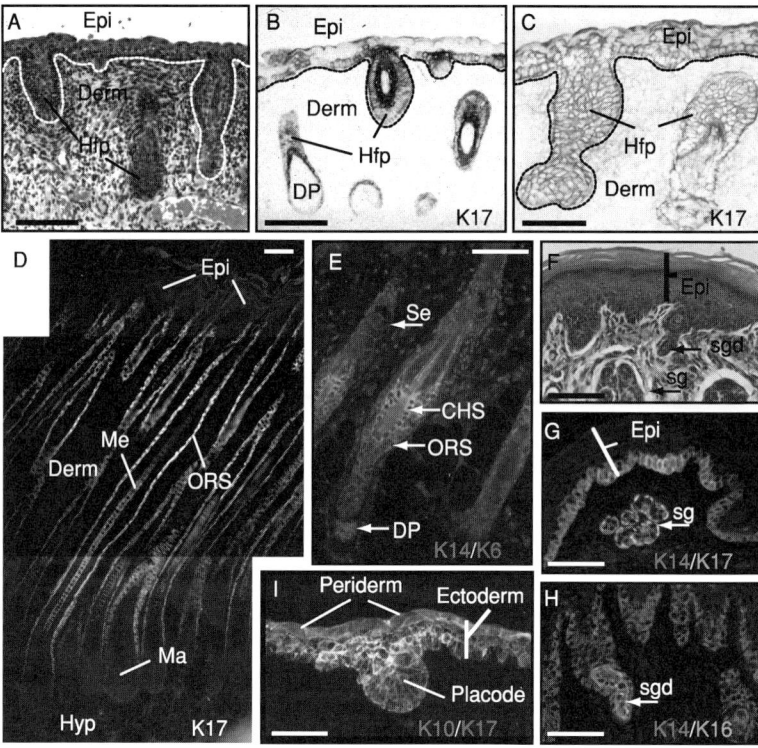

Fig. 3 Morphological analysis of mouse skin tissue. (A) Hematoxylin and eosin stain of whisker pads from a paraffin-embedded mouse embryo 16.5dpc, sectioned at 5 μm. Dotted line marks the boundary between the epidermis (Epi) and dermis (Derm). Hfp, hair follicle precurser. (B) Same sample as (A) but processed for immunohistochemistry of K17. DP, dermal papilla. (C) 5-μm section of an embryo 16.5 dpc fresh frozen and processed for immunohistochemistry of K17. Note the even staining of the hair follicle and developing epidermis and periderm compared with the uneven staining but better morphological preservation of (B). (D) Paraffin-embedded back skin from a 10-day-old mouse pup processed for immunohistochemistry of K17 during the anagen cycle of the hair. K17 antigen is present in the outer root sheath of the hair follicle and in the medulla of the hair shaft. Hyp, hypodermis; Me, medulla; Ma, matrix. (E) Paraffin-embedded back skin from an 18-day-old mouse pup was processed for indirect immunofluorescence of K14 (red, outer root sheath [ORS] and sebaceous gland [Se]), K6 (green, club hair sheath-CHS), and nuclei (blue). During the catagen phase of the hair cycle, the bulb of the hair undergoes a massive wave of apoptosis, whereas the dermal papilla (DP) migrates to the base of the club hair sheath. (F) H&E stain of a paraffin-embedded mouse foot pad. Note the much thicker epidermis that contains sweat glands but no hair follicles. (G) Fresh-frozen section of mouse foot pad processed for immunofluorescence of K14 (red) and K17 (green). Note the presence of these two keratins in the basal layer of the epidermis, as well as the myoepithelium of the sweat glands (sg). (H) Unlike K17, K16 is present only in sweat gland ducts (green, sgd) rather than in the myoepithelium. (I) Fresh-frozen section of e14.5 embryo depicting K17 immunogen (green) in the periderm and newly invaginating hair follicle. K17 is a very early marker of appendageal formation in the early stratifying epidermis compared with K10 (red), which is expressed only in suprabasal epidermis. Bars equal 50 μm. (See Color Insert.)

dry ice; liquid nitrogen; OCT compound (Sakura, Finetek, Torrance, CA); plastic embedding molds (Sakura, Finetek, Torrance, CA). *Paraformaldehyde*: DEPC-treated H_2O, PBS, and sucrose; paraformaldehyde; gluteraldehyde; methanol. *Transmission electron microscopy*: sodium cacodylate buffer; EM grade gluteraldehyde; paraformaldehyde; osmium tetroxide.

B. Preparation of Skin Tissue for Routine Morphological Study

The routine morphological analysis of skin is best carried out by use of sections prepared from paraffin-embedded tissue. For paraffin embedding, skin samples are fixed in Bouin's fixative overnight at 4 °C. This fixative functions well to retain the morphology of skin tissue. As a general rule, tissues should not be larger than $0.5–1\,cm^3$ for fixation, but size can vary depending on sample type or protocol need. Poor infiltration of fixative caused by excess sample size can yield inaccurate morphology and possible artifacts in the sample. Skin samples tend to roll up after removal from the torso. To prevent the skin from curling during fixation, the skin can be laid out flat in a weigh boat, with the epidermis facing up, and only then covered up with Bouin's fixative for 10–15 minutes before transferring to a closed container overnight. This initial step will help maintain a flat piece of tissue and allow for the proper face to be cut. After overnight fixation, samples are washed multiple times with 70% ethanol to remove excess fixative. At this point, tissues can be trimmed to expose the appropriate face for sectioning. In some cases, tissue-marking dyes (Bradley Products, Inc., Bloomington, MN) can be used to identify the face to be sectioned. In the mouse, hair follicles are oriented along the rostral-caudal axis; thus, a slice of skin parallel to this axis will result in a sectioning plane that contains the entire length of a single hair follicle along its long axis (Fig. 3D and E). Samples can be temporarily stored at 4°C in 70% ethanol before further processing for paraffin embedding.

Tissues are dehydrated with 30-minute washes of 50%, 75%, and 95% ethanol, two 15-minute washes of 100% ethanol, followed by two 10-minute washes with xylene. Samples are then infiltrated with molten paraffin at 60 °C while stirring in a beaker for 20–30 minutes with three consecutive rounds of fresh, molten paraffin. Finally, a small amount of fresh molten paraffin is added to the embedding mold. Carefully move mold over to some ice to slowly solidify it. As the paraffin cools, place the sample in the mold with correct orientation. The rest of the mold is then filled with molten paraffin to cover the sample and allowed to harden. The sample is now ready for sectioning onto slides with a microtome. A 5- to 10-μm thickness is desirable.

To prepare cut sections for immunostaining, slides are deparaffined with two washes of xylene, two washes of 100% ethanol, one wash each of 95%, 70%, and 50% ethanol, and a final wash in ddH_2O. All washes are 2.5 minutes each. If a horseradish peroxidase reaction is to be used for immunohistochemistry, endogenous peroxidase activity can be eliminated by treating slides with 0.3% H_2O_2 for 10 minutes at room temperature before immunostaining.

Occasionally, the processes of fixation and paraffin embedding mask the epitope of interest. In this case, an additional step to unmask the antigen before immunostaining will be required. After deparaffining slides, place samples in a heat-resistant slide rack and rinse three times in citrate buffer for 2 minutes at room temperature. During this time, boil ~500 ml of citrate buffer in a 1-L Pyrex beaker. Remove the boiling solution from heat, immediately add the slides to this solution in a heat-resistant rack, and allow it to incubate for 5–10 minutes. Rinse with phosphate-buffered saline (PBS) and begin the immunostaining protocol.

C. Fresh Frozen Sections

Chemical fixation followed by paraffin embedding often alters the reactivity of antibody epitopes on tissue sections. Alternatively, tissues can be snap-frozen with liquid N_2 without any prior fixation for the purpose of morphological studies, protein, and RNA extractions. Surgically harvested tissues should be quickly wrapped and labeled in foil, then plunged in liquid nitrogen. Large, multipurpose ice buckets are useful containers to hold liquid nitrogen for the purpose of fixation, because they are well insulated and allow for easy retrieval of specimens with forceps. Dry ice and 95% ethanol can also be used; however, one must label samples with something other ethanol-soluble ink.

For histological processing, tissues are embedded in OCT Compound (Sakura, FineTek, Torrance, CA), a water-soluble embedding medium. At room temperature, OCT remains a viscous liquid but quickly solidifies on dry ice. Fill disposable plastic embedding molds (Sakura, FineTek, Torrance, CA) with enough OCT to cover sample and place harvested tissue into molds with forceps. Care should be used to avoid introducing air bubbles into the OCT, because they will interfere with sectioning at a later point. To complete the process, carefully move the mold containing the tissue onto dry ice to harden the OCT. As the medium slowly solidifies, orientation of the tissue can be adjusted for the optimal cutting face. For sections that are already frozen, embedding is the same; however, the OCT will solidify faster. It is important not to let the sample thaw during the embedding procedure. OCT samples can be stored at −80 °C for several months. Sections of 5–10 μm should be cut and placed onto slides with a cryostat.

Because keratins are highly insoluble components of the cytoskeleton, they remain attached to the cells during immunostaining. However, some soluble proteins require fixation before immunostaining to prevent loss during subsequent washes and incubations. This potential problem can be solved by fixing slides with 4% paraformaldehyde for 3–5 minutes before staining.

D. Preparation of Skin Tissue for *In Situ* Hybridization

Paraformaldehyde is yet another method of fixation that allows for morphological and histological analyses, including *in situ* hybridization. Samples harvested for these types of experiments should be fixed immediately for optimal results. It is essential that tissues be handled quickly with RNAse-free instruments to protect against RNA degradation.

Samples should be immediately fixed in 4% paraformaldehyde that was freshly prepared in DEPC-treated H_2O. Fix tissues for 16–24 hours at 4 °C and rinse thoroughly with DEPC-treated PBS. For whole-mount or large tissue samples only, permeabilize with 50% methanol in DEPC-treated PBS briefly and then 100% methanol at −20 °C. Permeabilization time can range from 30 minutes to several months, depending on the size of tissue. Rehydrate by washing with 50% methanol in PBST-DEPC for 5 minutes on a rocker. Repeat with 30% methanol in PBST-DEPC with two final washes in PBST-DEPC for 5 minutes each. Transfer samples to 20% DEPC-treated sucrose, and incubate at 4 °C until tissues sink to the bottom. This time can range from 2 hours to overnight. Transfer samples to OCT compound and embed as described previously. If possible, cut sections per slide at 5–10 μm. The steps involved in the preparation of probes, and their application for the detection of specific keratin mRNAs in skin tissue sections, have been described elsewhere (Tong and Coulombe, 2004; Wang *et al.*, 2003).

E. Preparation of Skin Tissue for Transmission Electron Microscopy

Routine transmission electron microscopy (TEM) provides high-resolution imaging of individual cells and even cellular substructures, which is essential to the proper characterization of structural defects in skin epithelia. It is important to work quickly to get tissues into fixative when working with samples destined for electron microscopy; the slightest delay to fixation can result in subcellular damage (e.g., mitochondrial swelling) and blur the assessment of the impact of a genetic or other type of manipulation. Minced tissue pieces should be no more than ~1 mm^3 to ensure that penetration of fixative is both rapid and complete. Oftentimes, a buffered solution combining paraformaldehyde, a monofunctional aldehyde that penetrates tissues quickly, and glutaraldehyde, a bifunctional aldehyde with stronger fixative properties but slower penetration, is used as a primary fixative (Hayat, 2000).

All the fixative solutions should be freshly prepared and prechilled on ice. Harvest tissue and fix overnight at 4 °C with 2% TEM-grade gluteraldehyde, 1% paraformaldehyde in 0.1 M sodium cacodylate, pH 7.2. On the next day, rinse with 0.1 sodium cacodylate, pH 7.2, three times at room temperature. In a well-ventilated hood, postfix with 1% osmium tetroxide for 1 hour. It is best to work with osmium tetroxide in the hood with gloves and a mask, because it is very hazardous to humans. The samples will turn black as osmium infiltrates them. Collect osmium waste, and subsequent washes, in a separately labeled waste container and dispose according to institutional guidelines for hazardous disposal. Wash samples with sodium cacodylate, pH 7.2, three times. Rinse samples three additional times with double-distilled water and transfer tissues to 50% ethanol. Samples can be stored at 4 °C for a few days before proceeding through a standard TEM embedding protocol.

IV. Cell Culture Studies

Cell culture work offers a number of advantages compared with *in vivo* studies. Isolation of cells from tissues enriches for a certain population of cell types (e.g., keratinocytes, fibroblasts), which simplifies the experimental system. In addition, the environment can be easily manipulated to study effects of various treatments or growth conditions. Then cells are easily harvested and used for a large number of assays, such as cellular, protein, and RNA analyses. Yuspa, Hennings, and their colleagues have pioneered basic methods for the establishment of newborn mouse keratinocytes in primary culture, and their manipulation for the study of epidermal differentiation (e.g., Hennings and Hobrook, 1983; Hennings *et al.*, 1980; Yuspa *et al.*, 1989).

A. Materials

Two separate sets of tools—one for working with animals and one designated only for tissue culture (each set contains: two pairs of forceps—preferably one curved/one straight pair, dissecting scissors); Betadine antiseptic; 70% ethanol; sterile bacterial plates; Lymphoprep (Axis-Shield PoC AS, Oslo, Norway); 0.25% trypsin (Invitrogen, Cat#: 15050-065); cell scrapers (Sarstedt, Newton, NC); "mKer" keratinocyte medium (Rhouabhia *et al.*, 1992); AcuPunch, 4 mm (Acuderm, Ft. Lauderdale, FL); standard tissue culture materials are used.

B. Harvesting Keratinocytes from Newborn Mouse Skin

The protocol to isolate primary keratinocytes will take 2 days to complete. It is best to use pups that are between ages P0 (birth) and P4. Preparations from pups any older than P4 will yield a very low number of cells because of the developing hair follicles that invaginate deep into the dermis, preventing clean separation between the epidermis and dermis.

Step 1. On day 1, pups are sacrificed by decapitation with scissors and then submerged in Betadine for 3–5 minutes. Wash thoroughly with distilled water to remove excess Betadine and rinse three times with 70% ethanol. In the following steps, all work should be done in the tissue culture hood, and sterile technique should be used to avoid contamination. Surgical tools should be placed in a beaker containing 70% ethanol between uses.

Step 2. Manipulations can be done on the inner surface of sterile bacterial Petri dishes rather than the more expensive treated tissue-culture dishes. First, appendages must be removed. With forceps in one hand to hold a mouse pup and scissors in the other hand, remove the limbs at the second most distal joint, leaving a short stump. When removing the tail, it is important to remove the genital area as well, because this area is a potential source of bacterial contamination. Keep a piece of tissue for genotyping if necessary. Cut skin along the dorsal surface of the mouse starting at the location where the tail was removed and moving rostrally toward the head. With a second pair of tweezers, peel the skin away from the torso beginning near the shoulder region. Tease the skin away from the entire dorsal surface while leaving the ventral surface attached. Next remove the skin from the shoulders and forelimb stumps. At this point, the dorsal skin can be pulled around to the ventral side. Grasp both flaps of dorsal skin from the ventral side with one set of forceps. Grasp the deskinned shoulders with the other pair of forceps and pull the remaining skin over what is left of the hindlimbs, just like removing a pair of pants.

Step 3. Transfer the skin to a sterile Petri dish and place with the dermis side down. With the forceps, carefully spread skin out like a sheet such that the edges are not rolled under. Continue to work on the other littermates until all are at this stage. Skins can rest here up to ~45 minutes without any problem. If a longer holding period is needed (e.g., for a very large litter), Petri dishes should be placed on ice during this time.

Step 4. The next step is to float the skins on trypsin (0.25% Trypsin – Invitrogen Cat #: 15050-065) overnight. Place ~4 ml of trypsin into a sterile 60-mm dish. Peel the skin off of the Petri dish and gently float it on the trypsin such that the dermis side contacts the trypsin. Again make sure that the edges are not rolled under, because this will make separation more difficult. Place skins at 4 °C overnight; the optimal length of trypsin treatment is ~18 hours. Do not use trypsin with ethylenediamine tetraaceticacid (EDTA), because this is more damaging to the keratinocytes and will give rise to a lower cell yield.

Step 5. On the next day, again, all work should be performed in the hood while exercising sterile technique. With clean forceps soaked in 70% ethanol, gently place the skin, dorsal side down and dermis exposed, in a new sterile Petri dish as flat and extended as possible.

Step 6. Remove the dermis by gently pushing and teasing the dermis with tweezers toward the edges. Once it begins to peel off from the epidermis, remove the dermis completely. To isolate fibroblasts, retain the dermis and proceed as described in Section IV.D. To isolate keratinocytes, keep the Petri dish with the epidermis. Keratinocytes can be collected with a fresh sterile cell scraper (Sarstedt, Newton, NC) for each pup. Cells are scraped from the center of the skin out toward the edges and resuspended in 1 ml of mKer medium (Rouabhia *et al.*, 1992) or another media that contains serum to neutralize any residual trypsin, such as Dulbecco's modified Eagle's medium (DMEM) with 5% serum. It is important at this step to resuspend cells thoroughly, yet gently, to dissociate any keratinocyte clumps for more even growing in culture. Transfer resuspended cells to a 15-ml conical tube. Scrape back skin, rinse with media, and transfer cells two additional times with 1 ml each to retrieve any remaining cells and pool with rest of sample.

Step 7. Pellet cells at 300*g* for 10 minutes at 4 °C. Aspirate the media and resuspend cells in 5 ml of fresh media. Gently layer the suspension of cells on top of 8 ml of Lymphoprep (Axis-Shield PoC AS, Oslo, Norway). Do not rush the layering. A crisp interface between media and Lymphoprep will yield a better fraction of keratinocytes. Also do not combine cells from more than one pup on the

8 ml of Lymphoprep. The increased number of cells will overload the separation system, and the recovery of keratinocytes will be dramatically decreased.

Step 8. Spin at 800*g* for 30 minutes at 4 °C. After centrifugation, a white hazy layer can be seen at the interface. This is an enriched population of keratinocytes.

Step 9. Collect keratinocytes at interface with a fresh 10-ml pipette. Starting at the interface, collect ~8–9 ml of media and Lymphoprep. While aspirating, be careful to avoid the walls of the tube and the pellet of unwanted cells at the bottom. Transfer cells to fresh 15-ml conical tube, add 5 ml fresh media, mix gently, and pellet at 300*g* for 10 minutes at 4 °C. Remove media and resuspend cells in 1 ml of fresh mKer medium (Rouabhia *et al.*, 1992) to count cells. Plating cells at $20–30 \times 10^4$ cells/100-mm dish will yield approximately 80% confluency in 48–72 hours when grown in mKer medium. Change to fresh mKer media the following day to remove cellular debris and other noxious factors released from dying cells.

C. Keratinocyte Growth Conditions

For routine studies, primary keratinocytes are cultured in the presence of mKer medium (Rhouabhia *et al.*, 1992), which promotes both their proliferation and differentiation while deterring fibroblast growth. Keratinocytes prefer to grow as colonies. Over time, cells in the center of a colony are more likely to stratify and differentiate compared with cells located at the outer edges. Keratinocytes cultured in this standard mKer medium are depicted in Fig. 4A. Alternatively, the processes of keratinocyte proliferation and differentiation can be segregated through the use of media containing "low" vs. "high" concentrations of calcium ions (Hennings and Holbrook, 1983; Hennings *et al.*, 1980; Yuspa *et al.*, 1989). In the presence of low calcium (0.05 mM), keratinocytes proliferate without forming colonies or inducing differentiation-associated markers (Fig. 4B). The process of differentiation can be studied by switching the medium to a higher level of calcium (ranging from 0.12–1.2 mM). Such conditions allow for the formation of stable cell–cell contacts and for differentiation, and the pace of differentiation depends on how much calcium is added (Yuspa *et al.*, 1989). On switching to high calcium medium, morphological changes typically associated with differentiation, such as formation of tight cell–cell adhesions, begin within 15 minutes and are complete within 2 hours (Fig. 4C).

1. Standard Growth Conditions

Keratinocytes will readily grow on standard tissue culture dishes that have been treated with vacuum gas plasma to create an even hydrophilic surface; however, they will also grow on untreated glass coverslips, which is convenient when immunocytochemistry is to be performed on cultured cells. A good preparation of keratinocytes from a P3 mouse pup will generally yield a confluent, 10 cm plate within 48–72 hours after plating in mKer medium. If one wishes to count and plate equal numbers of cells, then $1–5 \times 10^4$ cells/35-mm dish is recommended. Cells

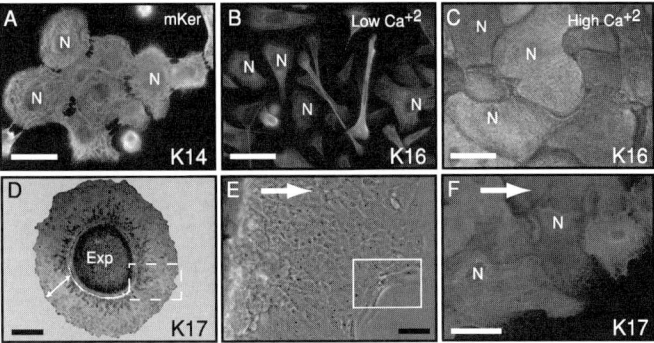

Fig. 4 Visualization of keratins in cultured keratinocytes. (A) Primary keratinocytes were cultured in mKer medium for 2 days followed by fixation and immunofluorescence visualization of K14. Note the dense filament networks and cell–cell contacts. Under mKer medium culture conditions, cells both proliferate and differentiate. Bars in (A–C) are 50 μm. N, nucleus. (B) 308 keratinocytes were cultured in the presence of "low" calcium (0.05 mM) for 48 hours, fixed, and visualized by indirect immunofluorescence for K16. Under low calcium conditions, cells proliferate, have a loosely packed keratin filament network, and very little cell–cell adhesion. (C) Duplicate cells from (B) were switched to 2 mM calcium for 24 hours followed by fixation and K16 immunostaining. On addition of high calcium medium, cells flatten and spread, synthesizing a more abundant and densely packed keratin filament network, and establish contiguous cell–cell contacts concomitant with entering the process of terminal differentiation. These morphological changes begin 15 minutes after addition of "high" calcium and are complete within 2 hours. (D) A 2-mm skin punch biopsy (Exp) was obtained from a 2-day-old pup and cultured as an explant in the presence of mKer medium. Throughout the next 8 days, keratinocytes and fibroblasts migrated and proliferated into the surrounding area of outgrowth (double-headed arrow). The explant and outgrowth were fixed and processed for immunohistochemistry for K17, which stains all of the keratinocytes in the outgrowth. (E) Phase-contrast image of live explant culture depicting a similar area to the dashed box in (D). Note that the cells migrate and proliferate as a sheet during outgrowth. Arrow denotes the direction of outgrowth. (F) Indirect immunofluorescence of K17 in cells at the leading edge of the explant, in a similar area to the box in (E).

typically adhere to the culture dish within several hours; however, some cells do not attach for various reasons and will float in the medium. It is desirable to change the medium the morning after plating to remove nonadherent cells and cellular debris. Thereafter, the medium should be changed every other day until cells are fixed or harvested. Unlike human skin keratinocytes, it is not possible to passage primary mouse skin keratinocytes under the conditions listed here; in general, it is best to complete experiments within 4–6 days after plating (for promising new findings, see Hager *et al.*, 1999).

2. Calcium Switch Protocol for Induction of Keratinocyte Differentiation

Because of the strict level of calcium required in this protocol, reagents must contain known levels of calcium. Calcium-free and magnesium-free phosphate buffer can be purchased from Cellgro (Mediatech, Herndon, VA). Fetal bovine serum must be chelated with chelex-100 resin (10 g resin/50 ml serum to be treated;

Biorad Laboratories, Inc., Hercules, CA). Resin is prepared by suspending 100 g of chelex-100 in 2.5 L of normal saline (21.75 g NaCl/2.5 L ddH$_2$O). The pH is adjusted to 7.4 while the solution is stirred for 10–15 minutes to verify the stability of the pH. The resin is then filtered to remove the saline without allowing the resin to dry out. It is useful to place several layers of Whatman filter paper on top of the filter so that layers can be removed consecutively as filtration slows with resin accumulation. The resin is scooped into a bottle containing fetal bovine serum (FBS) and stirred for 1 hour followed by filtration of the resin, this time discarding the resin. The FBS filtrate is then refiltered through a sterile 0.2 μm filter in the tissue culture hood. Aliquots of 40 ml are stored at −20 to −80 °C.

Media consists of EMEM without calcium (Biowhitaker, Walkersville, MD), 8% chelated serum, and 0.5% penicillin-streptomycin (Invitrogen, Carlsbad, CA). Residual calcium concentration in the medium should be verified with atomic absorption spectroscopy. Various amounts of 300 mM calcium chloride are added to this base medium to create low (0.05 mM), moderate (0.12 mM), and high (1.20–2.00 mM) calcium medium.

Keratinocyte culture under calcium switch conditions requires plating cells at a higher density than culture in mKer medium. To start, 5–10 × 10^5 cells/35-mm dish should be plated in moderate calcium (0.12–0.20 mM) overnight. Cells require this minimal amount of calcium to adhere to the plate. After overnight incubation at 37 °C, 5% CO$_2$, cells are washed with calcium-free PBS three times. Low calcium medium (0.05 mM calcium) is then added, and cells are allowed to proliferate for approximately 48 hours or until they are confluent. Cells may be fixed or harvested or induced to differentiate by adding either normal- (0.12–0.20 mM) or high- (1.2–2.0 mM) calcium medium. Cells are cultured an additional 24–48 hours before fixing or harvesting the differentiated cells.

D. Fixing Cells for Immunocytochemistry

Keratinocytes can be prepared for immunocytochemistry with a variety of standard fixation protocols depending on the antigen of interest. In general, cells are fixed in freshly prepared 3%–4% paraformaldehyde in PBS for 10–20 minutes at room temperature. Thereafter, cells are permeabilized with either 0.1% Triton X-100 in PBS for 10 minutes or 100% methanol for 5 minutes and washed three times with PBS. Permeabilization with either TritonX-100 or methanol works well when immunostaining for IFs or microtubules. To best reveal the F-actin networks with phalloidin, however, cells must be permeabilized with Triton X-100. After fixation, cells can be stored at 4 °C for a week or two before processing them for immunocytochemistry.

E. Fibroblast Cultures

The initial steps involved in the establishment of skin fibroblast in primary culture are identical to keratinocytes. After overnight incubation in trypsin and separation from epidermis, the dermis is placed in a sterile Petri dish and minced with sterile scissors. The pieces are then incubated in 0.35% collagenase type I (wt/vol) dissolved in DMEM for 30 minutes at 37 °C with a stirbar or shaking. Isolated fibroblasts are separated from the larger tissues by filtering the solution through a sterile gauze. Cells are pelleted by centrifugation at 200g for 5 minutes and resuspended and grown in DMEM with 10% FBS, penicillin, and streptomycin (Bridger *et al.*, 1993; Hager *et al.*, 1999).

F. *Ex Vivo* Skin Explant Cultures

Quantitating the contribution of keratinocytes to the wound epithelialization that ensues after acute injury to mouse skin is a very difficult task because of sectioning plane effects, the labile attachment of the newly formed epithelium to the wound bed, and the dominant contribution of dermis to wound closure (Mazzalupo *et al.*, 2002). As an alternative to experiments in skin tissue *in vivo*, we have developed an *ex vivo* skin explant culture assay that mimics the response of wound edge keratinocytes to full-thickness injury *in vivo*. This assay offers several advantages including quantitation, the possibility of manipulation (pharmacologically and otherwise), and the possibility of circumventing restrictive phenotypes (e.g., premature death of transgenic mice) (Wawersik and Coulombe, 2000; Wong and Coulombe, 2003). Detailed description of this assay and methods for quantitating keratinocyte outgrowth have been detailed by Mazzalupo *et al.* (2002).

Pups are sacrificed by decapitation, treated with antiseptic, and washed with 70% ethanol as described for primary keratinocyte preparation. Skins are removed and laid flat, dermis side down, in a sterile Petri dish. Full-thickness punches are taken from the back skin with a biopsy punch and transferred to a 24-well tissue culture dish with forceps soaked in 70% ethanol. Skin punches are placed flat in the center of each well, dermis side down, such that the edges are not rolled under. After 10 minutes to allow for adherence, 250 μl of mKer medium is added around the explant to moisten the edges but not cover the explant. Explants are placed in the tissue culture incubator overnight. On the following day, another 250 μl of mKer medium is added to submerge the explant. The media is changed every other day for the duration of the experiment, usually 8 days. Explants can be fixed in 4% paraformaldehyde and permeabilized with 100% methanol for IF immunostaining, or keratinocyte outgrowths can be harvested by cell scraping after removing the original punch biopsy.

V. Preparation and Analysis of Protein and RNA from Tissues and Cells

Protein and RNA isolation from tissues and cells is relatively simple; however, there are distinct advantages and disadvantages to each. Tissue samples harvested through animal surgery have the distinct advantage of a nearly intact *in vivo* context, which often cannot be duplicated in the tissue culture setting. However, such samples consist of a mixed population of cell types; skin represents a particular challenge at that level. Primary culture of keratinocytes isolated from skin consists of an enriched population of cells that can be easily manipulated. Cells in tissue culture can be subjected to a variety of growth conditions, and the protein or RNA collected from these cells represents their response. The protocols for collecting proteins and RNA have to be adjusted, depending on the source of the material.

A. Materials

Polytron hand-held homogenizer (Kinematica AG, Switzerland); TRIzol reagent (Invitrogen, Carlsbad, CA); metal mortal and pestle; hammer; liquid nitrogen; dry ice; Kimwipes; thermal protection gloves; disposable cell scrapers (Sarstedt, Newton, NC); and Seize X Protein A Immunoprecipitation Kit (Pierce Biotechnology, Rockford, IL) are used.

B. Preparation and Analysis of RNA and Proteins from Tissues

The initial procedure to extract RNA and proteins from skin is similar. Skin is a very robust tissue, and, therefore, vigorous mechanical disruption is needed to extract RNA and proteins. With tissues that have been snap-frozen, extraction begins with a metallic mortar and pestle approach. It is important to keep all tools and instruments as cold as possible to avoid alterations or degradation of proteins and RNA (Sambrook *et al.*, 1989). Before extraction, assemble samples, clean mortar and pestle with 70% ethanol, and chill spatulas on dry ice. DEPC-treated collection tubes for RNA isolation (Sambrook *et al.*, 1989) with appropriate buffer (see later) can be kept on wet ice.

With thermal gloves for protection, assemble the mortar and add a few milliliters of liquid nitrogen to further cool down the instrument. Next, carefully add the frozen skin sample and place the pestle in position. While securing the unit with one hand, use the hammer to pulverize the tissue with the pestle. Five to 10 strikes are usually sufficient to crush the skin enough for extraction. Tissue samples are then carefully transferred to individual tubes containing either RNA or protein extraction buffer with ice-cold spatulas. It is important to remove any residual tissue debris before beginning with the next sample. Thoroughly clean the mortal and pestle with 70% ethanol and Kimwipes, then allow to cool for a few minutes in the dry ice before proceeding with the next liquid nitrogen step.

The TRIzol reagent (Invitrogen, Carlsbad, CA) consistently and effectively isolates RNA from tissues and cells. To extract RNA, pulverized tissue is resuspended in TRIzol and further homogenized with a hand-held tissue disruptor for 2–3 minutes. Use 1 ml of TRIzol reagent for every 50–100 mg of tissue to avoid exceeding the capacity of the reagent. Extraction of RNA can be achieved by following the manufacturer's protocol.

For protein isolation, pulverized tissue should be resuspended in fresh protein extraction buffer containing 8 M urea (see Appendix). Tissue should be further disrupted by using a hand-held homogenizer unit. Transfer homogenate to microfuge tube and add β-mercaptoethanol to 5% of volume. Incubate samples at 37 °C for 1 hour with intermittent vortexing. Avoid longer incubations at this temperature to minimize urea-induced carbamylation of proteins. Pellet insoluble debris for 10 minutes at $14,000 \times g$ and transfer supernatant to fresh microfuge tube. A Bradford protein concentration assay (Bradford, 1976) can be performed to quantitate the amount of protein extracted.

C. Preparation and Analysis of RNA and Proteins from Cells

Isolation of RNA or proteins from tissue culture cells is a straightforward procedure. Primary keratinocytes destined for use in either RNA or protein assays should be grown to desired confluency. One must always keep in mind that once keratinocytes reach ~80% confluency, they will begin to stratify and alter their gene expression program, creating a more heterogeneous population of cells. Researchers should strive to obtain consistent keratinocyte confluence between experiments. To collect cells for RNA or protein work, aspirate the media and rinse with ice-cold PBS containing protease-inhibitors and phosphatase inhibitors to eliminate any residual media. It is important to work quickly to prevent any RNA or protein degradation and prevent the loss of protein modifications (e.g., phosphorylation), then collect cells with a cell scraper and transfer cells to a fresh microfuge tube. Rinse again with the ice-cold PBS and pool collected cells. Pellet cells at 4 °C for 10 minutes at $14,000g$ and remove PBS. Samples can be immediately frozen on dry ice and stored at -80 °C until further use or directly lysed for RNA or protein isolation.

As with RNA extraction from tissues, the TRIzol reagent is an effective and consistent means of RNA purification from cells. By following the manufacturer's protocol, extracting RNA from cell pellets from a 10-cm tissue culture plate usually requires only 1 ml of TRIzol. Although cells can be scraped and transferred into a microfuge tube before the addition of TRIzol, the manufacturer recommends lysing the cells directly on the tissue culture plate to avoid degradation of RNA during washes with PBS and scraping.

Protein extraction can be performed as a whole cell lysate, or lysates can be fractionated for enrichment of soluble and insoluble proteins (see Fig. 5A and B). Whole-cell lysates provide a quick and simple method for extracting proteins from cells. The whole-cell extraction buffer contains 8 M urea to dissociate protein–protein interactions and is conducive for sodium dodecylsulfate-polyacrylamic gel electrophoresis (SDS PAGE) analysis; however, it inhibits studies involving keratin-binding proteins and is not compatible with immunoprecipitation. Resuspend cell pellets from a 10-cm plate in 200–500 μl of cold, whole-cell extraction buffer and incubate for 15 minutes at 4 °C. Pellet insoluble material at $14,000g$ for 10 minutes. Remove supernatant and quantitate by use of the Bradford assay system.

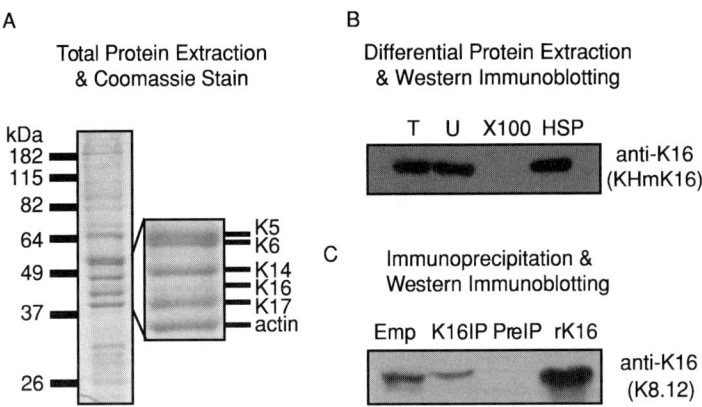

Fig. 5 Various analyses of skin keratin proteins. (A) Keratinocytes of the 308 cell line were cultured in mKer medium to 80% confluence. Proteins were extracted by resuspending the pelleted cells in 2× sodium dodecylsulfate-polyacrylamide gel electrophoresis (SDS-PAGE) buffer/β-mercaptoethanol and boiling for 5 minutes. A small amount was visualized by SDS-PAGE analysis. Individual keratin and actin bands are identified in the order of their migration in the magnified section to the right. Not shown is K15 (approximately the same size as K14). (B) Western blotting was performed with an antibody against K16 (KHmK16) on total protein lysates from SDS-PAGE sample buffer (T), 8 M urea (U), and Triton X-100 (X100). In addition, the final pellet from the high salt extraction was dissolved in 8 M urea and loaded onto the gel. To gauge relative amounts of keratin extraction proportionally, approximately 1% of a 10-cm plate of "SDS sample buffer" and "urea" samples was loaded compared to 25% for Triton X100 lysate sample and 0.1% for high salt pellet (HSP). Note that no keratin immunogen is visualized in the Triton supernatant, whereas urea and SDS-PAGE sample buffer are capable of extracting keratin protein. (C) 308 keratinocytes were cultured in mKer medium to 80% confluence, washed with PBS/protease inhibitors, and solubilized in 2% Empigen BB. Empigen lysate (Emp) was subjected to immunoprecipitation by incubation with KHmK16 antibody (K16IP) or preimmune sera (PreIP) bound to protein A–coated beads. Immunoprecipitation was carried out, and protein was released from the beads by boiling in the presence of SDS-PAGE sample buffer/β-mercaptoethanol. Samples were analyzed by SDS-PAGE analysis with recombinant K16 (rK16) loaded as a control. Western blotting with a different K16 antibody (K8.12) revealed immunogenic activity in the Empigen BB lysate, KHmK16 immunoprecipitation and recombinant K16 lanes, but no reactivity in the preimmune IP lane.

To enrich cell lysates for either the soluble or insoluble pools, high-salt fractionation steps can be used. Cells are initially lysed with PBS buffer containing 1% Triton-X100, 5 mM EDTA, protease inhibitors, and phosphatase inhibitors. Depending on the cell pellet size, 200–500 μl of lysis buffer can be used to lyse the cells. Incubate for 10 minutes at 4 °C and pellet the insoluble proteins at 14,000*g* for 10 minutes at 4 °C. Transfer the soluble fraction to a fresh tube and save as the "Triton-soluble fraction." Homogenize the remaining pellet with high-salt extraction buffer and shear with a 25-G syringe. Again, incubate for 10 minutes at 4 °C and pellet residual, insoluble proteins at 14,000*g* for 10 minutes at 4 °C. Transfer and save the high-salt supernatant in a new tube. Rinse pellet with PBS buffer to remove excess salts. The remaining pellet is enriched for keratin and other insoluble proteins. The buffer used to resuspend the insoluble pellet may depend on the next intended experiment. For instance, a buffer containing 8 M urea will solubilize most proteins and would be appropriate for SDS-PAGE analyses; however, if immunoprecipitation is a desired step, then cells can be resuspended in buffer containing 2% Empigen BB or 1% NP-40 with 5 mM EDTA and protease inhibitors in PBS. The limitation of these detergents is that they are not as effective at solubilization as a buffer containing 8M urea.

Immunoprecipitation is useful for studying IF-interacting proteins, as well as enriching for a particular protein. Given the highly insoluble nature of IFs, immunoprecipitation has been a challenge to develop; however, it is readily accomplished with the detergents Empigen BB and NP-40 (Lowthert *et al.*, 1995). These detergents are sufficient in maintaining some protein–protein interactions while solubilizing a partial fraction of the total keratin pool. Cell pellets collected from tissue culture can be lysed directly with 2% Empigen BB or 1% NP-40 extraction buffer with a 1-hour incubation at 4 °C. The insoluble proteins are removed by centrifugation at 14,000*g* for 10 minutes. The supernatant can be subjected to immunoprecipitation with the antibody of choice with the Seize X Protein A Immunoprecipitation Kit (Pierce Biotechnology, Rockford, IL) and following manufacturer's protocol. An example of keratin immunoprecipitation is depicted in Fig. 5C.

VI. Pearls and Pitfalls

A. Collection of Mouse Skin Tissue

Harvesting mouse skin is a straightforward procedure. The adequacy of the tissue sample for analysis depends on limiting the interval between death and preservation of the tissue—be it by fixation or snap-freezing. In adult mice, the thick fur coat often interferes with complete fixation and proper embedding. To circumvent this problem, the hair can be clipped very short before surgery with scissors or electrical clippers. Care should be used to avoid nicking the skin. Shaving or chemical depilation is not recommended because of the mild trauma it causes to the skin, which may complicate interpretations of the results.

B. Morphological Studies

Tissue orientation is key to successful morphological analysis of skin. Hair in particular requires very careful orientation during tissue collection, embedding, and sectioning. Hair follicles undergo a cyclic process of growth (anagen stage, Fig. 3D), regression (catagen stage, Fig. 3E), and rest (telogen stage). Details for studying these phases, including an excellent technique description of embedding samples for optimal hair follicle orientation, are provided by Müller-Röver *et al.* (2001). The same group provides detailed explanation for staging hair during embryonic hair morphogenesis (Paus *et al.*, 1999; see also Fig. 3I). Skin tissue obtained from areas other than the back (e.g., whisker pads, foot pads) can be accomplished in a similar manner to the techniques described for back skin. An example of glabrous footpad is depicted in Fig. 3F–H.

As always, the time between death of the animal and fixation or snap-freezing of the tissue sample should be limited as much as possible to reduce the possibility of poor morphology caused by postmortem changes. Incomplete fixative penetration or multiple freeze–thaw cycles can also result in poor morphology, lack of epitope accessibility, and degradation of protein and RNA. Immunohistochemistry requires a balance between morphological preservation (more fixation usually yields better morphology) and epitope preservation (less fixation usually yields better epitope accessibility). The protocols detailed here may need to be modified to suit individual antibodies; however, the breadth of protocols offered should lead to success in most instances. Of course, the best immunolocalization of antigens is usually obtained with sections prepared from frozen tissue. Otherwise, fixation with Bouin's rather than fresh paraformaldehyde is preferable when planning on immunolocalizing keratin epitopes on paraffin-embedded tissue sections.

A key issue during sample embedding is to maintain direct contact between the sample and the embedding medium. Air bubbles should be avoided, and care should be used to keep samples free from extraneous liquid (i.e., blood on the backskin) that can interfere with direct contact between sample and embedding medium. In the case of ultrastructural studies, particular care must be taken during fixation to avoid oxidative damage and osmotic stress (which may lead to cell shrinkage or swelling), as well as during the infiltration of the resin within the tissue sample. Consultation with a thorough textbook describing tissue preparation for ultrastructural studies is recommended.

C. Cell Culture Studies

Whereas most viable cells recovered from the epidermal sheet consist of basal keratinocytes, there is a sizable fraction of keratinocytes that originate from the upper segment of hair follicles (infundibulum and ORS in particular; see Kamimura *et al.*, 1997), along with melanocytes and dermal fibroblasts. One should be aware, therefore, of the heterogeneity of cell populations in these cultures. The use of mKer medium deters the growth of fibroblasts; when keratinocytes are not plated at a high-enough density, however, fibroblasts may take over the culture because of their faster proliferation rate. In addition, keratinocytes do not grow well when seeded at a low density, so plating efficiency is a key determinant of success with these cultures. Keeping track of calcium ions while preparing the medium to be used for keratinocyte culture is of paramount importance for reliable and reproducible results.

Keratinocytes in primary culture will stratify once the colony size exceeds 4–6 cell diameters in medium supplemented with moderate to high calcium, including mKer medium. Differentiating cells alter their gene expression profile, thus creating an even more heterogeneous population of cells. If one wishes to study the process of differentiation rather than the end-state, it is advisable to use differentiation-promoting medium containing "moderate" rather than "high" concentrations of calcium (Yuspa *et al.*, 1989). Culture under moderate-calcium conditions allows for the detection of increases in K1/K10 and other "early" markers of epidermal differentiation (Yuspa *et al.*, 1989). As the keratinocytes differentiate, they will begin to slough off into the supernatant, generally after only two to three layers of cells have formed; thus, complete, full-thickness epidermal stratification is not achieved under these culture conditions. For cell imaging studies, it is easiest to work with a monolayer of keratinocytes.

D. Isolation and Analysis of Protein and RNA from Skin Cells and Tissues

Protein and RNA extraction from skin is performed with similar methods to other organs; however, skin is a tougher tissue than internal organs such as liver because of the protective function of the hair and cornified layers of the skin. As with all tissue work, tissues should be processed quickly to limit the degradation of material. After homogenization, unless otherwise stated, extracts should be kept on ice while processing. Frequent freeze–thaw cycles lead to both RNA and protein degradation and should be avoided. RNA endonucleases are very stable and ubiquitous, so extreme care should be used to avoid contaminating any samples with RNAse. Ideally, experiments toward RNA isolation and analysis are conducted in a separate and dedicated area of the laboratory. At a minimum, solutions for RNA use should be kept separate from all others. Gloves should be worn at all times and changed frequently. Aerosol-barrier tips prevent contamination from pipettes. Total RNA quality can be roughly analyzed by running 1 μg of RNA on an agarose gel. The 28S and 18S ribosomal RNAs should be bright and strong with no "smearing." Ideally, the band intensity should be a 2:1 ratio of the larger (28S) to smaller (18S) band (Sambrook, 1989). If the larger (28S) band is weaker in intensity than the 18S band or if smearing is observed, then the RNA is partially degraded and should not be used for further analysis.

Protein degradation can be minimized by adding protease inhibitors during tissue washes and in all subsequent additions of solutions. These inhibitors should be added fresh each time a solution is used. Keratins are very abundant proteins in keratinocytes; they make up >50% of the total protein. The high salt extraction is a very good method for isolating total keratin protein, because almost no keratin (<1%) is solubilized by the Triton X-100 solution or the high-salt solution, so there is little loss of keratin sample. Also, there are not very many other proteins that remain insoluble throughout this procedure, so the keratin fraction is greatly enriched. If one does observe contaminating proteins remaining in the final pellet of the high-salt extraction, then the volume of extracting solutions should be increased. A good rule of thumb is to use 500 μl of extraction mix/10-cm plate followed by resuspension of the final pellet in 200 μl 8 M urea.

Keratins are very stable in heterotypic complexes *in vitro*. For example, although keratins are partially solubilized in Empigen BB, they remain in heterotypic complexes. Thus, an immunoprecipitation for K17 will also result in coimmunoprecipitation of K14, K5, and K6 from keratinocytes in culture. Keratins range from 40–60 kDa. The heavy chain of IgG is around 55 kDa. The Seize-X immunoprecipitation kit referenced here cross-links the immunoprecipitating antibody to protein A beads, so that when the precipitated material is eluted, the antibody remains attached to the beads. However, if one uses a generic protocol in which the antibody is released from the beads along with the immunoprecipitated material, SDS-PAGE analysis will reveal the presence of a very large band in the region where the epidermal keratins migrate, prohibiting identification of the keratins. Western blotting will circumvent this problem only if a different antibody developed in an alternate species is used. For example, a rabbit anti-K14 antibody could be used to immunoprecipitate followed by a mouse anti-K14 antibody for the Western blot.

VII. Conclusion

Mouse skin presents several advantages for the study of the mechanisms regulating the expression of keratin genes, and the properties, regulation, and function of their protein products. First, a very large fraction of known keratin genes are differentially expressed in skin epithelia, and other IF genes are expressed in nonepithelial compartments of the skin (notably, vimentin in dermal fibroblasts and in the endothelium). The compact nature of mouse keratin genes makes their manipulation through genetic engineering relatively straightforward. This said, well-known limitations of skin with regard to the study of keratin function in an *in vivo* setting include the clustered organization of type I and type II keratin genes, which prevents the generation of compound null mutations through matings, and the overlapping distribution of keratin proteins, which breeds functional redundancy (Coulombe and Omary, 2002).

Second, skin is visually accessible and readily amenable to survival surgery, and much is known about its biology. By virtue of its location at the interface with the outside world, skin, one of the largest organs in the body (in terms of both mass and surface area), fulfills a number of important functions, undergoes constant renewal, and enjoys significant clinical relevance, as well as social importance. Histologically, it features all major types of tissues with the exception of bone and cartilage and displays considerable variation between body sites. This heterogeneity is to be taken into account when extracting various types of macromolecules for analysis (e.g., DNA, RNA, proteins, and lipids) or, at a broader level, characterizing the impact of a genetic manipulation. Accessibility of the skin from the outside also means that it can be locally manipulated pharmacologically, although appropriate measures need to be taken in mice to circumvent the complications offered by a dense fur coat and very effective barrier function.

Third, it is possible to establish various types of skin cells, and even skin tissue explants, in primary culture. Cellular and subcellular analyses that may otherwise be difficult to carry out in skin tissue *in vivo* can be performed in an *ex vivo* setting. In the context of transgenic mouse studies, the findings stemming from the study of cells in primary culture can be readily related back to intact tissue physiology *in vivo*.

The methods and assays outlined in this chapter are typically needed during the initial analysis of mice in which the expression of a gene, keratin, or other is manipulated in skin tissue. Albeit simple and relatively unsophisticated, these methods have and will continue to be a mainstay in the characterization of the regulation and function of keratin genes in epithelial tissues.

Appendix

Bouin's fixative

72%	Saturated aqueous picric acid
5%	Glacial acetic acid
23%	Formaldehyde (37%)

Filter after mixing with 0.22-μm pore filter unit. Store at 4 °C.

Calcium switch media

EMEM without Ca^{+2}

8% Chelated FBS

0.5% Penicillin-streptomycin

Add desired amount of Ca^{+2} in the form of $CaCl_2$:

Low	0.05 mM
Moderate	0.12 mM
High	1.2–2.0 mM

0.1 M Citric acid stock ($C_6H_8O_7 \cdot H_2O$)	21.0 gr in 1 L
0.1 M Sodium citrate stock ($C_6H_5Na_3O_7 \cdot 2H_2O$)	29.4 gr in 1 L

Citrate buffer for antigen retrieval

For 500 ml of 0.01 M citrate buffer, pH 6.0, add 9 ml of citric acid, 41 ml of sodium citrate, and bring volume up to 500 ml.

Empigen BB extraction buffer

2% Empigen BB detergent [Calbiochem]

5 mM EDTA

1× PBS

Add 2 mM PMSF, protease inhibitors, and phosphatase inhibitors before use.

Fibroblast media

>DMEM
>
>10% FBS
>
>1% Penicillin-streptomycin

High salt extraction buffer

>10 mM Tris HCl, pH 7.6
>
>140 mM NaCl
>
>1.5 M KCl
>
>5 mM EDTA
>
>0.5% Triton-X100

Add protease inhibitors and phosphate inhibitors immediately before use.

mKer media

DMEM low glucose	3 parts
Ham's F-12	1 part
Fetal bovine serum	10%
Cholera toxin	1 nM
Insulin	5 μg/ml
EGF	10 ng/ml
Gentamicin	25 μg/ml
Hydrocortisone	400 ng/ml
Transferrin	5 μg/ml
3,3-5′ triiodo-L-thyroxine	2 nM
Penicillin	60 μg/ml

Store at 4 °C until use.

NP-40 extraction buffer

>1% NP-40 (also called IGEPAL CA-630)
>
>5 mM EDTA
>
>1× PBS

Add 2 mM PMSF, protease inhibitors, and phosphatase inhibitors before use.

Phenylmethylsulfonylflouride (PMSF)

Make 200 mM stock. Use at 1–2 mM concentration. (174.2 M.W.) Dissolve in isopropanol and store at −20 °C. Warm stock to room temperature to resuspend precipitate immediately before use, PMSF has a half-life of <1 hour in aqueous solutions. Use extreme caution, as PMSF is highly toxic.

Phosphatase inhibitors

	Stock	Use at
Sodium fluoride (MW 41.33)	1.0 M	50 mM
Sodium orthovanadate (MW 183.9)	0.2 M	0.2 mM
EDTA (MW 372.24)	0.5 M	2.0 mM

Protease inhibitor cocktail 1 1000× (PIC1)

2 mg/ml	Antipain
10 mg/ml	Aprotinin
10 mg/ml	Benzamidine
1 mg/ml	Leupeptin

Dissolve in sterile ddH$_2$O, aliquot, and store at −20 °C until use.

Protease inhibitor cocktail 2 1000× (PIC2)

1 mg/ml	Cymostatin
1 mg/ml	Pepstatin-A

Dissolve in DMSO, aliquot, and store at −20 °C until use.

Protein extraction buffer (pulverized skin)

8 M urea

50 mM Tris, pH 7.4

1 mM EGTA

2 mM DTT

2 mM PMSF (add immediately before use)

Triton cell lysis buffer

1% Triton X-100

5 mM EDTA

1× PBS

Add PMSF, protease inhibitors, and phosphates inhibitors immediately before use.

Whole-cell lysate extraction buffer (for tissue culture)

8 M urea

50 mM Tris, pH 7.4

1 mM EGTA

2 mM DTT

Add PMSF, protease inhibitors, and phosphates inhibitors immediately before use.

Acknowledgments

We would like to thank Dr. Kevin McGowan and Dr. Stacy Mazzalupo for images. Dr. Hsin-Yang Li contributed the fibroblast culture protocol. Efforts in P.A.C,;s laboratory are supported by grants AR44232 and AR42047 from the National Institutes of Health.

References

Ameen, N. A., Figueroa, Y., and Salas, P. J. (2001). Anomalous apical plasma membrane phenotype in CK8-deficient mice indicates a novel role for intermediate filaments in the polarization of simple epithelia. *J. Cell Sci.* **114,** 563–575.

Beil, M., Micoulet, A., von Wichert, G., Paschke, S., Walther, P., Omary, M. B., Van Veldhoven, P. P., Gern, U., Wolff-Hieber, E., Eggermann, J., *et al.* (2003). Sphingosylphosphorylcholine regulates keratin network architecture and visco-elastic properties of human cancer cells. *Nat. Cell Biol.* **5,** 803–811.

Bernot, K., McGowan, K., and Coulombe, P. A. (2002). Keratin 16 expression defines a subset of epithelial cells during skin morphogenesis and the hair cycle. *J. Invest. Dermatol.* **119,** 1137–1149.

Bonifas, J. M., Rothman, A. L., and Epstein, E. J. (1991). Epidermolysis bullosa simplex: Evidence in two families for keratin gene abnormalities [see comments]. *Science* **254,** 1202–1205.

Bradford, M. M. (1976). A rapid and sensitive method for the quantitation of microgram quantities of protein utilizing the principle of protein-dye binding. *Anal. Biochem.* **72,** 248–254.

Bridger, J. M., Kill, I. R., O'Farrell, M., and Hutchison, C. J. (1993). Internal lamin structures within G1 nuclei of human dermal fibroblasts. *J. Cell Sci.* **104** (Pt. 2), 297–306.

Byrne, C. (1997). Regulation of gene expression in developing epidermal epithelia. *Bioessays* **19,** 691–698.

Cassidy, A. J., Lane, E. B., Irvine, A. D., and McLean, W. H. I. (2002). "The Human Intermediate Filament Mutation Database."http://www.interfil.org.

Caulin, C., Ware, C. F., Magin, T. M., and Oshima, R. G. (2000). Keratin-dependent, epithelial resistance to tumor necrosis factor-induced apoptosis. *J. Cell Biol.* **149,** 17–22.

Collin, C., Moll, R., Kubicka, S., Ouhayoun, J. P., and Franke, W. W. (1992). Characterization of human cytokeratin 2, an epidermal cytoskeletal protein synthesized late during differentiation. *Exp. Cell Res.* **202,** 132–141.

Coulombe, P. A. (2003). Wound epithelialization: Accelerating the pace of discovery. *J. Invest. Dermatol.* **121,** 219–230.

Coulombe, P. A., and Bernot, K. M. (2004). Keratins and the skin. *In* "Encyclopedia of biological chemistry," (W. Lennarz and M. D. Lane, eds.), vol. 2 pp. 497–504. Elsevier, Oxford.

Coulombe, P. A., Hutton, M. E., Letai, A., Hebert, A., Paller, A. S., and Fuchs, E. (1991). Point mutations in human keratin 14 genes of epidermolysis bullosa simplex patients: Genetic and functional analyses. *Cell* **66,** 1301–1311.

Coulombe, P. A., Hutton, M. E., Vassar, R., and Fuchs, E. (1991). A function for keratins and a common thread among different types of epidermolysis bullosa simplex diseases. *J. Cell Biol.* **115,** 1661–1674.

Coulombe, P. A., Kopan, R., and Fuchs, E. (1989). Expression of keratin K14 in the epidermis and hair follicle: Insights into complex programs of differentiation. *J. Cell Biol.* **109,** 2295–2312.

Coulombe, P. A., Ma, L., Yamada, S., and Wawersik, M. (2001). Intermediate filaments at a glance. *J. Cell Sci.* **114,** 4345–4347.

Coulombe, P. A., and McGowan, K. (2000). Morphogenesis of skin epithelia. *In* "Cell polarity," (D. G. Drubin, ed.), pp. 285–313. Oxford University Press, Oxford.

Coulombe, P. A., and Omary, M. B. (2002). 'Hard' and 'soft' principles defining the structure, function and regulation of keratin intermediate filaments. *Curr. Opin. Cell Biol.* **14,** 110–122.

Coulombe, P. A., and Omary, M. B. (2002). Hard and soft principles defining the structure, function and regulation of keratin intermediate filaments. *Curr. Opin. Cell Biol.* **14,** 110–122.

Coulombe, P. A., and Wong, P. (2004). Cytoplasmic intermediate filaments revealed as dynamic and multipurpose scaffolds. *Nat. Cell Biol.* **6,** 699–706.

Fuchs, E. (1994). Intermediate filaments and disease: Mutations that cripple cell strength. [review]. *J. Cell Biol.* **125,** 511–516.

Fuchs, E. (1995). Keratins and the skin. *Ann. Rev. Cell Dev. Biol.* **11,** 123–153.

Fuchs, E., and Cleveland, D. W. (1998). A structural scaffolding of intermediate filaments in health and disease. *Science* **279,** 514–519.

Fuchs, E., and Green, H. (1980). Changes in keratin gene expression during terminal differentiation of the keratinocyte. *Cell* **19,** 1033–1042.

Gilbert, S., Loranger, A., Daigle, N., and Marceau, N. (2001). Simple epithelium keratins 8 and 18 provide resistance to Fas-mediated apoptosis. The protection occurs through a receptor-targeting modulation. *J. Cell Biol.* **154,** 763–774.

Green, K. J., and Gaudry, C. A. (2000). Are desmosomes more than tethers for intermediate filaments? *Nat. Rev. Mol. Cell Biol.* **1,** 208–216.

Hager, B., Bickenbach, J. R., and Fleckman, P. (1999). Long-term culture of murine epidermal keratinocytes. *J. Invest. Dermat.* **112,** 971–976.

Hayat, M. A. (2000). Principles and techniques of electron microscopy: Biological applications. Cambridge University Press, Cambridge, UK

Hennings, H., and Holbrook, K. A. (1983). Calcium regulation of cell-cell contact and differentiation of epidermal cells in culture. *Exp. Cell Res.* **143,** 127–142.

Hennings, H., Michael, D., Cheng, C., Steinert, P., Holbrook, K., and Yuspa, S. H. (1980). Calcium regulation of growth and differentiation of mouse epidermal cells in culture. *Cell* **19,** 245–254.

Hesse, M., Franz, T., Tamai, Y., Taketo, M. M., and Magin, T. M. (2000). Targeted deletion of keratins 18 and 19 leads to trophoblast fragility and early embryonic lethality. *EMBO J.* **19,** 5060–5070.

Hesse, M., Magin, T. M., and Weber, K. (2001). Genes for intermediate filament proteins and the draft sequence of the human genome: Novel keratin genes and a surprisingly high number of pseudogenes related to keratin genes 8 and 18. *J. Cell Sci.* **114,** 2569–2575.

Hesse, M., Zimek, A., Weber, K., and Magin, T. M. (2004). Comprehensive analysis of keratin gene clusters in humans and rodents. *Eur. J. Cell Biol.* **83,** 19–26.

Holbrook, K. A., and Wolff, K. (1993). The structure and development of skin. *In* "Dermatology in general medicine." (T. B. Fitzpatrick, A. Z. Eisen, K. Wolff, I. M. Freedberg, and M. D. Austen, eds.), pp. 97–144. McGraw-Hill, New York.

Inada, H., Izawa, I., Nishizawa, M., Fujita, E., Kiyono, T., Takahashi, T., Momoi, T., and Inagaki, M. (2001). Keratin attenuates tumor necrosis factor-induced cytotoxicity through association with TRADD. *J. Cell Biol.* **155,** 415–426.

Irvine, A. D., and McLean, W. H. (1999). Human keratin diseases: The increasing spectrum of disease and subtlety of the phenotype-genotype correlation. *Br. J. Dermatol.* **140,** 815–828.

Jaquemar, D., Kupriyanov, S., Wankell, M., Avis, J., Benirschke, K., Baribault, H., and Oshima, R. G. (2003). Keratin 8 protection of placental barrier function. *J. Cell Biol.* **161,** 749–756.

Jones, L. C. R., Hopkinson, S. B., and Goldfinger, L. E. (1998). Structure and assembly of hemidesmosomes. *Bioessays* **20,** 488–494.

Kamimura, J., Lee, D., Baden, H. P., Brissette, J., and Dotto, G. P. (1997). Primary mouse keratinocyte cultures contain hair follicle progenitor cells with multiple differentiation potential. *J. Invest. Dermatol.* **109,** 534–540.

Karabinos, A., Zimek, A., and Weber, K. (2004). The genome of the early chordate Ciona intestinalis encodes only five cytoplasmic intermediate filament proteins including a single type I and type II keratin and a unique IF-annexin fusion protein. *Gene* **326,** 123–129.

Knapp, A. C., Franke, W. W., Heid, H., Hatzfeld, M., Jorcano, J. L., and Moll, R. (1986). Cytokeratin No. 9, an epidermal type I keratin characteristic of a special program of keratinocyte differentiation displaying body site specificity. *J. Cell Biol.* **103,** 657–667.

Ku, N. O., Michie, S., Oshima, R. G., and Omary, M. B. (1995). Chronic hepatitis, hepatocyte fragility, and increased soluble phosphoglycokeratins in transgenic mice expressing a keratin 18 conserved arginine mutant. *J. Cell Biol.* **131,** 1303–1314.

Lane, E. B., Rugg, E. L., Navsaria, H., Leigh, I. M., Heagerty, A. H., Ishida, Y. A., and Eady, R. A. (1992). A mutation in the conserved helix termination peptide of keratin 5 in hereditary skin blistering. *Nature* **356,** 244–246.

Langbein, L., Heid, H. W., Moll, I., and Franke, W. W. (1994). Molecular characterization of the body site-specific human epidermal cytokeratin 9: CDNA cloning, amino acid sequence, and tissue specificity of gene expression. *Differentiation* **55,** 164.

Langbein, L., Rogers, M. A., Praetzel, S., Winter, H., and Schweizer, J. (2003). K6irs1, K6irs2, K6irs3, and K6irs4 represent the inner-root-sheath-specific type II epithelial keratins of the human hair follicle. *J. Invest. Dermatol.* **120,** 512–522.

Langbein, L., Rogers, M. A., Winter, H., Praetzel, S., and Schweizer, J. (2001). The catalog of human hair keratins. II. Expression of the six type II members in the hair follicle and the combined catalog of human type I and II keratins. *J. Biol. Chem.* **276,** 35123–35132.

Langbein, L., Rogers, M. A., Winter, H., Praetzel, S., Beckhaus, U., Rackwitz, H. R., and Schweizer, J. (1999). The catalog of human hair keratins. I. Expression of the nine type I members in the hair follicle. *J. Biol. Chem.* **274,** 19874–19884.

Lloyd, C., Yu, Q. C., Cheng, J., Turksen, K., Degenstein, L., Hutton, E., and Fuchs, E. (1995). The basal keratin network of stratified squamous epithelia: Defining K15 function in the absence of K14. *J. Cell Biol.* **129,** 1329–1344.

Loranger, A., Duclos, S., Grenier, A., Price, J., Wilson-Heiner, M., Baribault, H., and Marceau, N. (1997). Simple epithelium keratins are required for maintenance of hepatocyte integrity. *Am. J. Pathol.* **151,** 1673–1683.

Lowthert, L. A., Ku, N. O., Liao, J., Coulombe, P. A., and Omary, M. B. (1995). Empigen BB: A useful detergent for solubilization and biochemical analysis of keratins. *Biochem. Biophys. Res. Commun.* **206**(1), 370–379.

Ma, L., Yamada, S., Wirtz, D., and Coulombe, P. A. (2001). A 'hot-spot' mutation alters the mechanical properties of keratin filament networks. *Nat. Cell Biol.* **3,** 503–506.

Magin, T. M., Hesse, M., and Schroder, R. (2000). Novel insights into interemdiate filament function from studies of transgenic and knockout mouse. *Protoplasma* **211,** 140–150.

Marceau, N., Loranger, A., Gilbert, S., Daigle, N., and Champetier, S. (2001). Keratin-mediated resistance to stress and apoptosis in simple epithelial cells in relation to health and disease. *Biochem. Cell Biol.* **79,** 543–555.

Mazzalupo, S., Wawersik, M. J., and Coulombe, P. A. (2002). An *ex vivo* assay to assess the potential of skin keratinocytes for wound epithelialization. *J. Invest. Dermatol.* **118**(5), 866–870.

McGowan, K. M., Tong, X., Colucci-Guyon, E., Langa, F., Babinet, C., and Coulombe, P. A. (2002). Keratin 17 null mice exhibit age- and strain-dependent alopecia. *Genes Dev.* **16,** 1412–1422.

Moll, R., Franke, W. W., Schiller, D. L., Geiger, B., and Krepler, R. (1982). The catalog of human cytokeratins: Patterns of expression in normal epithelia, tumors and cultured cells. *Cell* **31,** 11–24.

Müller-Rover, S., Handjiski, B., van der Veen, C., Eichmuller, S., Foitzik, K., McKay, I. A., Stenn, K. S., and Paus, R. (2001). A comprehensive guide for the accurate classification of murine hair follicles in distinct hair cycle stages. *J. Invest. Dermatol.* **117,** 3–15.

Nelson, W. G., and Sun, T. T. (1983). The 50- and 58-kdalton keratin classes as molecular markers for stratified squamous epithelia: Cell culture studies. *J. Cell Biol.* **97,** 244–251.

Omary, M. B., Coulombe, P. A., and McLean, W. H. I. (2004). Intermediate filaments and their associated diseases. *N. Engl. J. Med.* Invited review. In press.

Osborn, M., and Weber, K. (1983). Tumor diagnosis by intermediate filaments typing: A novel tool for surgical pathology. *Lab. Invest.* **48,** 372–393.

Oshima, R. G. (1992). Intermediate filament molecular biology. *Curr. Opin. Cell Biol.* **4,** 110–116.

Oshima, R. G. (2002). Apoptosis and keratin intermediate filaments. *Cell Death Differ.* **9,** 486–492.

Paus, R., Muller-Rover, S., Van Der Veen, C., Maurer, M., Eichmuller, S., Ling, G., Hofmann, U., Foitzik, K., Mecklenburg, L., and Handjiski, B. (1999). A comprehensive guide for the recognition and classification of distinct stages of hair follicle morphogenesis. *J. Invest. Dermatol.* **113,** 523–532.

Porter, R. M., Lunny, D. P., Ogden, P. H., Morley, S. M., McLean, W. H., Evans, A., Harrison, D. L., Rugg, E. L., and Lane, E. B. (2000). K15 expression implies lateral differentiation within stratified epithelial basal cells. *Lab. Invest.* **80,** 1701–1710.

Rouabhia, M., Germain, L., Belanger, F., Guignard, R., and Auger, F. A. (1992). Optimization of murine keratinocyte culture for the production of graftable epidermal sheets. *J. Dermatol.* **19,** 325–334.

Sambrook, J., Fritsch, E. F., and Maniatis, T. (1989). "Molecular cloning: A laboratory manual." Cold Spring Harbor Laboratory, Cold Spring Harbor, N. Y.

Sengel, P. (1976). "The morphogenesis of skin." Cambridge University Press, Cambridge.

Stark, H. J., Breikreutz, D., Limat, A., Bowden, P., and Fusenig, N. E. (1987). Keratins of the human hair follicle: "Hyperproliferative" keratins consistently expressed in outer rrot sheat cells *in vivo* and *in vitro Differentiation* **35,** 236–248.

Steinert, P. M., and Roop, D. R. (1988). Molecular and cellular biology of intermediate filaments. *Annu. Rev. Biochem.* **57,** 593–625.

Swensson, O., Langbein, L., McMillan, J. R., Stevens, H. P., Leigh, I. M., McLean, W. H., Lane, E. B., and Eady, R. A. (1998). Specialized keratin expression pattern in human ridged skin as an adaptation to high physical stress. *Br. J. Dermatol.* **139,** 767–775.

Takahashi, K., Coulombe, P. A., and Miyachi, Y. (1999). Using transgenic models to study the pathogenesis of keratin-based inherited skin disease. *J. Dermatol. Sci.* **21,** 73–95.

Tamai, Y., Ishikawa, T., Bosl, M. R., Mori, M., Nozaki, M., Baribault, H., Oshima, R. G., and Taketo, M. M. (2000). Cytokeratins 8 and 19 in the mouse placental development. *J. Cell Biol.* **151,** 563–572.

Tong, X., and Coulombe, P. A. (2004). A novel mouse type I intermediate filament gene, keratin 17n (K17n), exhibits preferred expression in nail tissue. *J. Invest. Dermatol.* **122,** 965–970.

Vassar, R., Coulombe, P. A., Degenstein, L., Albers, K., and Fuchs, E. (1991). Mutant keratin expression in transgenic mice causes marked abnormalities resembling a human genetic skin disease. *Cell* **64,** 365–380.

Wang, Z., Wong, P., Langbein, L., Schweizer, J., and Coulombe, P. A. (2003). Type II epithelial keratin 6hf (K6hf) is expressed in the companion layer, matrix, and medulla in anagen-stage hair follicles. *J. Invest. Dermatol.* **121,** 1276–1282.

Wawersik, M., and Coulombe, P. A. (2000). Forced expression of keratin 16 alters the adhesion, differentiation, and migration of mouse skin keratinocytes. *Mol. Biol. Cell* **11,** 3315–3327.

Wong, P., and Coulombe, P. A. (2003). Loss of keratin 6 (K6) proteins reveals a function for intermediate filaments during wound repair. *J. Cell Biol.* **163,** 327–337.

Woodcock-Mitchell, J., Eichner, R., Nelson, W. G., and Sun, T. T. (1982). Immunolocalization of keratin polypeptides in human epidermis using monoclonal antibodies. *J. Cell Biol.* **95,** 580–588.

Yuspa, S. H., Kilkenny, A. E., Steinert, P. M., and Roop, D. R. (1989). Expression of murine epidermal differentiation markers is tightly regulated by restricted extracellular calcium concentrations *in vitro*. *J. Cell Biol.* **109,** 1207–1217.

CHAPTER 17

Studying Simple Epithelial Keratins in Cells and Tissues

Nam-On Ku,★ Diana M. Toivola,★ Qin Zhou,★ Guo–Zhong Tao,★ Bihui Zhong,† and M. Bishr Omary★

★Department of Medicine
Palo Alto VA Medical Center and Stanford University
Palo Alto, California 94304

†Division of Gastroenterology
The First Affiliated Hospital of Sun Yet-sen University
Guangzhou 510080, China

I. Introduction

Intermediate filament (IF) proteins consist of a large family of nuclear (i.e., lamins) and cytoplasmic (e.g., keratins, neurofilaments, vimentin, desmin) cytoskeletal proteins (Fuchs and Weber, 1994; Herrmann *et al.*, 2003). They are expressed in a tissue-preferential manner such as keratins (K) in epithelial cells, neurofilaments in neuronal cells, vimentin in mesenchymal cells, and desmin in muscle. The major epithelial keratins (excluding the "hard" keratins found in epidermal appendages; see Chapter 15) include type I (K9–K20) and type II (K1–K8) IF proteins. This list of 20 keratins does not account for additional members (e.g., K6a-f) that are expressed in keratinocytes, particularly during wound healing (see Chapter 16). All epithelial cells express at least one type I and one type II keratin and, regardless of their number in an epithelial cell, are typically found as noncovalent heteropolymers in a 1:1 type I to II molar ratio (Coulombe and Omary, 2002; Fuchs and Weber, 1994). The major keratins of simple-type epithelia (i.e., epithelia consisting of a single layer of cells) are K7 and K8 (type II) and K18, K19, and K20 (type I). However, some of these keratins may be found in other epithelia (e.g., K7 and K19 in stratified epithelia) (Moll, 1998; Smith *et al.*, 2002).

The relative solubility of simple epithelial keratins, compared with the minimal solubility of epidermal and other nonsimple-epithelial keratins, has helped facilitate the characterization of their regulation (Omary *et al.*, 1998). This includes the identification of several phosphorylation and glycosylation sites of K8/K18/K19, defining their degradation by caspases during apoptosis and identifying several keratin-associated proteins such as heat shock protein 70 (hsp70) and 14-3-3 proteins (Fig. 1), glucose-regulated protein (Grp78), tumor necrosis factor (TNF) receptor type 2, and protein kinases including 40-kDa PKCε-related kinase, p38, and c-Jun *N*-terminal kinases (Coulombe and Omary, 2002).

Although progress has been achieved in the biochemical studies of K8/K18/ K19/K20 by use of *in vitro* cell culture systems, studies in transgenic mice have added significantly to our understanding of K8/K18 function of protecting cells from necrotic and apoptotic forms of injury induced by mechanical and nonmechanical stresses. For example, studies of mice that lack K8 or that overexpress K18 Arg89→Cys demonstrated the importance of an intact K8/K18 network in protecting hepatocytes from a variety of stresses, including liver perfusion, partial hepatectomy, griseofulvin, or microcystin-LR toxicity, and apoptosis (Omary *et al.*, 2002). Simple epithelial keratins are also essential for providing structural integrity during embryonic development. For example, although K18-null or K19-null mice have a normal life span, K8/K19 or K18/K19 double-null mice manifest embryonic lethality caused by placental defects with trophoblast giant cell cytolysis and abnormalities in the trophoblast layer (Hesse *et al.*, 2000; Tamai *et al.*, 2000).

Studies in transgenic mice that overexpress phosphorylation-mutant K18 demonstrated the *in vivo* functional significance for keratin phosphorylation. In these studies, the function of the two major K18 phosphorylation sites, Ser52 and Ser33, was examined by overexpressing their corresponding K18 Ser→Ala mutants. In human cultured cells and tissues, Ser52 accounts for most of K18 phosphorylation during interphase, and this phosphorylation increases during mitosis and a variety of stresses. Although there was no significant impact on

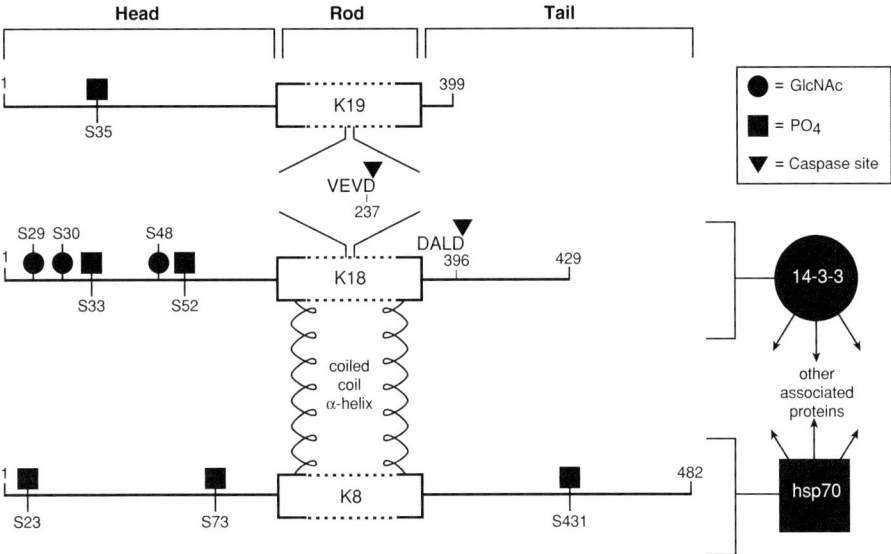

Fig. 1 Posttranslational modifications of K8/K18/K19. Phosphorylation and glycosylation occur at the head and/or tail domains of keratins (as is the case for all other intermediate filament [IF] proteins). Caspase-mediated cleavage occurs in the rod and tail domains of K18 and the rod domain of K19 and other IF proteins. VEVD or similar consensus sequences are found within the rod domain of many IF proteins, whereas DALD is a unique caspase site that is found only in the tail domain of K18. The sizes of the head, rod, and tail are not drawn to scale but consist, respectively, of 87, 315, and 80 amino acids (aa) for K8; 82, 308, and 39 aa for K18; and 71, 315, and 13 aa for K19. The schematic also highlights the association of K18 with 14-3-3 proteins and K8 with heat shock protein 70 (hsp70); 14-3-3 and hsp70 also interact with other proteins.

filament organization when K18 Ser52→Ala (S52A) was overexpressed in transgenic mice, these mice were significantly more predisposed to hepatotoxic injury compared with mice that overexpress wild-type (WT) K18 (Ku *et al.*, 1998b). This provided direct evidence for the importance of keratin phosphorylation in cytoprotection. With regard to K18 Ser33, it becomes hyperphosphorylated during mitosis, with consequent K18 association with the 14–3–3 protein family (Ku *et al.*, 1998a; Liao and Omary, 1996). Findings in transgenic mice that overexpress K18 Ser33→Ala (S33A) demonstrated that K18 Ser33 phosphorylation plays an important role in: (1) filament organization in the liver after mitosis and in the pancreas under basal conditions, (2) regulation of hepatocyte nuclear 14–3–3 redistribution during mitosis, and (3) modulation of hepatocyte mitotic progression, in a limited fashion, after partial hepatectomy (Ku *et al.*, 2002b).

This chapter focuses on describing techniques used to study simple epithelial keratins in cell culture systems and in mice. The methods that we will focus on include: (1) isolation of keratins, (2) keratin posttranslational modifications,

(3) keratin staining in tissues and cells, and (4) simple epithelial keratins in organ-specific injury models.

II. Materials and Methods

A. Isolation of Keratins

1. High Salt Extraction of Keratins

Although simple epithelial keratins are significantly more soluble than epidermal keratins, most (~95%) cellular K8/K18 is insoluble. In general, IF proteins can be solubilized in solutions containing a high concentration of urea or guanidinium hydrochloride, which produce soluble tetrameric or oligomeric subunits, or in solutions containing denaturing detergents such as sodium dodecylsulfate (SDS), which generate solubilized denatured monomers. A highly enriched keratin fraction can also be obtained with high salt extraction (HSE) (Fig. 2A) (Achtstaetter *et al.*, 1986; Chou *et al.*, 1993), which is a very useful tool for the isolation of most of the keratin content in tissues, primary cultured cells and cell

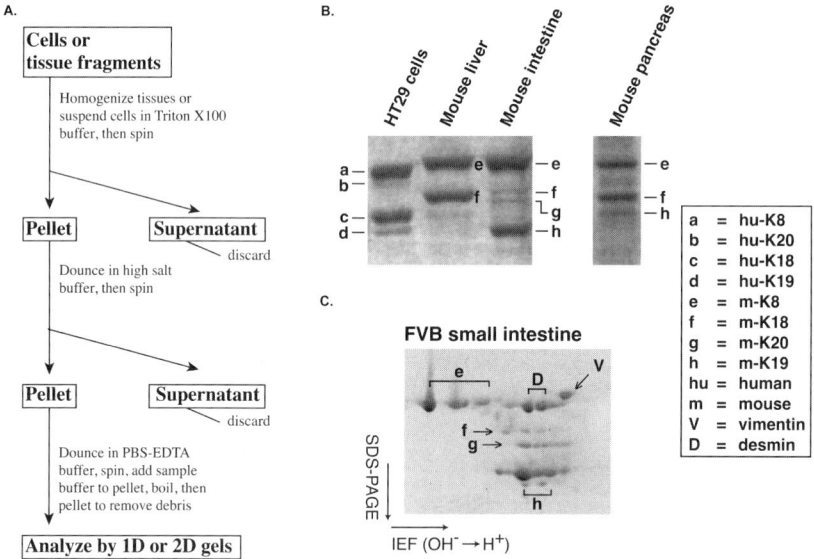

Fig. 2 (A) Flowchart of the high salt extraction method. (B) Example of keratins isolated by high salt extraction (HSE) then analyzed by sodium dodecylsulfate-polyacrylamide gel electrophoresis (SDS-PAGE). Keratins were isolated from the human colonic cell line HT29 or from the indicated mouse tissues. (C) Two-dimensional gel of HSE of the small intestine of nontransgenic FVB mice (Zhou *et al.*, 2003). The boxed enclosure summarizes the abbreviations used in B and C.

lines (e.g., Fig. 2B). Of note, all vendors listed in this chapter are based in the United States.

a. Reagents

- Phosphate-buffered saline (PBS, pH 7.4): 36 g NaCl plus 80 ml of 5 M Na_2HPO_4 (titrated to pH 7.4) brought up to 4 L with distilled (d) H_2O.
- High salt buffer (HSB): 10 mM Tris-HCl, pH 7.6, 140 mM NaCl, 1.5 M KCl, 5 mM ethylenediaminetetraacetic acid (EDTA), 0.5% Triton X100, plus a protease inhibitor mix consisting of 1 mM phenylmethylsulfonyl fluoride (PMSF) added fresh (from a 1 M stock solution prepared in methanol) and a premixed protease inhibitor cocktail consisting of 10 μM leupeptin, 10 μM pepstatin, and 25 μg/ml aprotinin (added fresh). The premixed protease inhibitor cocktail is prepared as a 1000× stock (e.g., 10 mM leupeptin) and stored in aliquots (at $-20\,^\circ$C).
- Triton X100 buffer: 1% Triton X100 plus 5 mM EDTA in PBS (pH 7.4), plus PMSF and the protease inhibitor cocktail as above.
- 4× Nonreducing sample buffer (4× NRSB): 10 ml 2× stacking gel buffer, 16 ml glycerol, 6.4 ml 0.1% bromophenol blue, and 3.2 g SDS brought to a total volume of 40 ml with dH_2O.
- 2× Stacking gel buffer: 1.75 ml concentrated phosphoric acid (or 25.6 ml of 1 M), and 5.7 g Tris base brought to 50 ml with dH_2O.

b. Steps

1. Cells grown in suspension are harvested by centrifugation (5 minutes, 1000 rpm), whereas adherent cells are scraped into an Eppendorf tube. Cells are washed 2× with PBS (22 $^\circ$C). The cell pellet is resuspended in 1 ml of Triton X100 buffer, mixed, and incubated over ice (2 minutes), then repelleted (10 minutes; 14,000 rpm; 4 $^\circ$C) and processed as in step "3" below.

2. For tissue fragments, add 1 ml of ice cold Triton X100 buffer per ~0.5 cm^3 of isolated tissue, homogenize by douncing 50 strokes—or by using a PowerGen 125 tissue homogenizer (Fisher Scientific) (for tissues difficult to dounce such as the intestine) for 1 minute (keep over ice during mechanical homogenization), then spin (10 min; 14,000 rpm; 4 $^\circ$C) and discard the supernatant.

3. The pellet, which consists mainly of nuclei and nonsolubilized membrane and cytoskeletal proteins, is resuspended in 1 ml of HSB, dounced (100 strokes over ice), mixed (30 minutes) with a rotator placed in a cold room, then repelleted (20 minutes; 14,000 rpm; 4 $^\circ$C).

4. Discard the supernatant, then add 1 ml PBS with 5 mM EDTA, dounce 100 strokes (as a pellet washing step), then spin 10 minutes as previously. Add 200–400 μl of 2× NRSB, vortex to dissolve the pellet as much as possible, heat to 95 $^\circ$C (2–4 minutes), and spin (2 minutes; 14,000 rpm; 22 $^\circ$C).

5. Load 5–10 μl onto SDS-polyacrylamide gel electrophoresis (PAGE) gel and store the remaining sample (which may be aliquoted depending on the extent of anticipated freeze-thawing) at −20 °C. An example of HSE keratin profiles, on isolation from cells and tissues, is shown in Fig. 2B. If two-dimensional (2D) gel analysis (isoelectric focusing then SDS-PAGE) is performed after HSE, 2D sample buffer (which contains 9.5 M urea and ampholyte) is added to the pellet at the last step of the procedure followed by vortexing and heating to 37 °C (urea-containing samples should not be boiled). An example of 2D gel analysis of keratins is shown in Fig. 2C.

2. Keratin Immunoprecipitation

Immunoprecipitation has been successfully used as an aid to study K8/K18/K19/K20 biochemical properties, such as their posttranslational modifications and associated proteins. These keratins are more amenable to such studies, compared with epidermal keratins, because of their relative solubility in detergents that maintain antigen (keratin)–antibody binding. These studies include the characterization of keratin posttranslational modifications (Ku and Omary, 1994, 1995, 1997; Liao *et al.*, 1997) and associated proteins (Ku *et al.*, 1998a; Liao and Omary, 1996; Liao *et al.*, 1995a). For example, heat shock protein 70 (hsp70) and 14-3-3 protein (Fig. 3) were identified as keratin-associated proteins after coimmunoprecipitation then microsequencing of individual coprecipitated protein bands (after in-gel protease digestion and isolation of released peptides). The following protocol offers a general guideline for keratin immunoprecipitation, although optimization may be required for other specific IF proteins and antibodies. A list of monoclonal antibodies (mAb) and other immunologic reagents useful for studying simple epithelial keratins are listed in Table I.

a. Materials
- Cell line: HT29 cells, an adherent human colonic carcinoma cell line (American Type Culture Collection), were cultured at 37 °C in RPMI 1640 medium supplemented with 10% fetal calf serum (FCS), 100 units/ml penicillin, 100 μg/ml streptomycin, and 2 mM L-glutamine.
- Beads: protein A Sepharose 4B (Cat. No. 10-1041) or protein G Sepharose 4B (Cat. No. 10-1242) (Zymed).
- Antibody: monoclonal mouse anti-human K18 antibody (L2A1) (Table I) (Chou and Omary, 1993).
- Lysis buffer: 1% Nonidet P-40 (NP-40) or 1% Empigen BB (Calbiochem) in PBS (pH 7.4) containing 5 mM EDTA, 0.1 mM PMSF, and the protease inhibitor cocktail used in the HSE Section.
- Washing buffer: 0.1% NP-40 or 0.1% Empigen in PBS (pH 7.4) containing 5 mM EDTA.

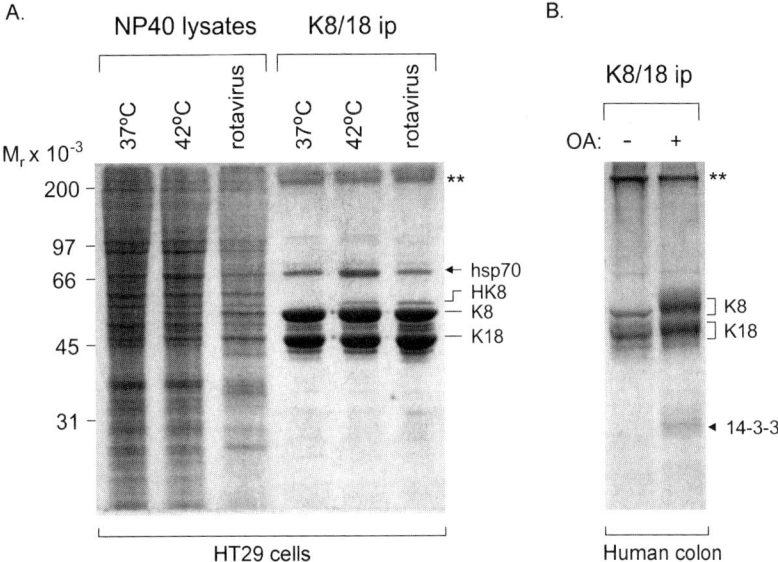

Fig. 3 Coimmunoprecipitation of K8/K18 and their associated proteins. (A) HT29 cells were cultured at 37 °C (16 hours) as normal control, at 42 °C (16 hours), or in the presence of rotavirus (12 hours), respectively (Liao *et al.*, 1995c). Rotavirus infection was performed by addition of trypsin-activated rotavirus into serum-free medium (2.8×10^7 plague forming units/dish) and incubated for 1 hour. The virus-containing medium was then removed followed by incubation in normal medium for 12 hours. Cells were solubilized with NP-40 lysis buffer, followed by precipitation with monoclonal antibody (mAb) L2A1. The immunoprecipitates (ip) were analyzed by 10% SDS-PAGE followed by Coomassie staining. Note that heat shock protein (hsp)70 coimmunoprecipitated with K8/K18, and a protein band of hyperphosphorylated K8 (HK8) was induced after heat treatment or rotavirus infection. (B) Freshly isolated human colon biopsies were cultured at 37 °C with or without okadaic acid (OA; 1 μg/ml; 2 hours), homogenized with NP-40 lysis buffer, followed by immunoprecipitation with mAb L2A1. Note K8/K18 and 14-3-3 protein association (confirmed by microsequencing of the isolated bands and by immunoblotting using antibodies to 14-3-3 proteins, not shown) after OA treatment, which induces K18 S33 phosphorylation and 14-3-3 binding with K18 (Liao and Omary, 1996; Ku *et al.*, 1998a). OA also induces generalized K8/K18 hyperphosphorylation, with retardation of their gel migration. **Highlights the immunoglobulins.

b. Preparation of NP-40 Cell Lysate

1. Rinse cells in a tissue culture dish (10 cm, at 80%–90% confluence) with 5 ml of prewarmed (37 °C) PBS-EDTA, then scrape the cells into 1 ml of prewarmed PBS-EDTA buffer.

2. Pellet the cells into an Eppendorf tube by a brief centrifugation (15 seconds; 10,000 rpm).

3. Add 1.3 ml of prechilled lysis buffer and vortex, then gently shake on a rotator (2 hours, 4 °C).

Table I
Antibodies Directed to Simple-Type Epithelial Keratins

Keratin	Antibody name	Host	Species reactivity*	Applications	Vendor
K7	RCK-105	Mouse	h, m	ICC,WB	ICN Biochemicals, Progen
K8	M20	Mouse	h	ICC, WB	Labvision
	Troma I	Rat	m	ICC, WB	DSHB
	E3264	Rabbit	h	ICC, WB	Spring Bioscience
	GT3†	Rabbit	h	ICC, WB, P	Epitomics
K18	DC10	Mouse	h	ICC, WB	Labvision
	L2A1	Mouse	h	ICC, WB, P	Labvision
	GT6	Rabbit	h	ICC, WB, P	Epitomics
K19	4.62	Mouse	h	ICC, WB	Sigma
	KA4	Mouse	h	ICC, WB	Sigma
	E2634	Rabbit	h	ICC, WB	Spring Bioscience
K20	Ks20.8	Mouse	h, m	ICC, WB	Labvision
	ITKs20.10	Mouse	h	ICC, WB	Progen
	Q6	Mouse	h	ICC‡ WB, P	Labvision
Pan keratin	8.13§	Mouse	h	ICC, WB, P	Sigma
K8 phospho-S73 (h)	LJ4	Mouse	h, m	ICC, WB, P	Labvision
K8 phospho-S431 (h)	2D6	Mouse	h, m	ICC, WB, P	Labvision
	5B3	Mouse	h, m	ICC, WB	Labvision
K18 phospho-S33 (h)	IB4	Mouse	h, m	ICC, WB, P	Labvision
K18 43-kDa fragment (cleaved at D396)	M30	Mouse	h, m	ICC, WB	Roche Applied Science

Antibodies for keratins with applications for immunocytochemistry (ICC), Western blotting (WB) and immunoprecipitation (P) are shown. ICC indicates that the antibody works on acetone-fixed frozen sections unless otherwise stated. Websites for the companies from which the antibodies may be purchased are DSHB, www.uiowa.edu/~dshbwww, Epitomics, www.epitomics.com, ICN, www.mpbio.com, Labvision (Neomarkers), www.labvision.com, Progen, www.progen.de, Roche Applied Science, www.roche-applied-science.com, Sigma, www.sigma.com, Spring Bioscience, www.springbio.com
‡PFA fixation only for mAb Q6.
*h, human; m, mouse.
†Recognizes human non-phospho-K8 S73–containing epitope selectively (Tao *et al.*, manuscript in preparation).
§Reacts with type II keratins K1/K5/K6/K7/K8 and type I keratins K10/K11/K18.

    4. Pellet the cell lysate (14,000 rpm; 20 minutes; 4 °C), then carefully transfer the supernatant of solubilized material (NP-40 cell lysate), without disturbing the pellet, to a clean tube.

c. Cell Lysate Preclearing

    1. Transfer 50 μl of the protein A (or protein G) beads slurry to an Eppendorf tube and add 500 μl cold lysis buffer.

    2. Spin at 10,000 rpm for 30 seconds and wash one additional time with 500 μl of cold lysis buffer.

3. Add 600 μl of NP-40 cell lysate to the Eppendorf tube and shake at 4 °C for 30–60 minutes.

4. Spin (10,000 rpm; 10 minutes; 4 °C) then transfer the cleared lysate to a new tube.

d. Immunoprecipitation

1. Add 5–10 μg of antibody to the Eppendorf tube containing the cold precleared NP-40 lysate. Incubate at 4 °C for 1 hour, then add 50 μl of prewashed protein A slurry and incubate for 2 hours at 4 °C on a rotator.

2. Spin (5000 rpm; 30 seconds; 4 °C) then carefully remove the supernatant and wash the beads three times with 1.0 ml of washing buffer.

3. Add 50 μl of 2× nonreducing Laemmli sample buffer. Vortex then heat (95 °C, 2 minutes).

4. Spin (14,000 rpm; 1 minute) and collect the supernatant, analyze a fraction (typically 10–30 μl) by SDS-PAGE, then use Coomassie staining or transfer the gels of interest to membranes for immunoblotting.

B. Posttranslational Modifications of Keratins

Keratins are highly dynamic and reorganize in response to many stimuli such as mitosis and apoptosis. Potential mechanisms that may regulate keratin filament organization include keratin-associated proteins and posttranslational modifications, such as phosphorylation, glycosylation, and proteolysis. The current state of knowledge of the posttranslational modifications of K8/K18/K19 (summarized in Fig. 1) was made possible by the approaches and some of the techniques described in this section.

General information regarding overall keratin phosphorylation can be obtained by gel analysis of the *in vivo* or *in vitro* labeled keratins. The 2D gel analysis also provides information regarding the relationship of the charged phosphoisoforms to the neutral O-GlcNAc–containing species that may or may not be phosphorylated (Liao *et al.*, 1996). K8/K18/K19/K20 isoforms are separated in the first dimension by use of gels containing 1.6% Bio-Lyte 5/7 ampholyte and 0.4% Bio-Lyte 3/10 ampholyte (Bio-Rad), as detailed further in manufacturer's instructions. K8/K18 phosphorylation is also detected by immunostaining (see Section II.C) or immunoblotting with site-specific phospho-antibodies (Table I). In addition to the K8/K18/K19 serine phosphorylation sites known to date (Fig. 1), K8/K19 pervanadate-mediated tyrosine phosphorylation was demonstrated by immunoblotting with anti-phosphotyrosine antibodies and phosphoamino acid (PAA) analysis of K8 and K19 isolated from metabolically labeled cells, although the modified K8/K19 phosphotyrosine residues remain to be identified (Feng *et al.*, 1999).

Several methods have been used to identify *in vivo* phosphorylation sites, including the microsequencing of chemically or protease-generated peptides

(Steinert, 1988). Alternatively, *in vivo* or *in vitro* labeled tryptic or chymotryptic keratin peptides, generated from purified keratins and isolated after peptide mapping (electrophoresis then chromatography; van der Geer and Hunter, 1994), can be subjected to manual Edman degradation to identify the position of the radiolabeled residue(s) within the isolated radiolabeled peptide (Ku and Omary, 1994, 1995). The peptide mapping technique can be applied to both phosphorylation and glycosylation sites, which are predicted on the basis of the location of the radiolabeled residue and the likely modified keratin peptide (predicted from known keratin sequences). The predicted modification sites can then be confirmed by mutation of the potentially involved residue(s), transfection of the wild type (WT) and predicted modification-mutant cDNAs, *in vivo* labeling, then peptide mapping to confirm the absence of a radiolabeled peptide in the maps generated from the mutant vs the WT constructs (Ku and Omary, 1994, 1995). Mass spectrometry techniques (see Chapter 14) may also be used to identify O-GlcNAc or phosphorylation sites, as done for the characterization of neurofilament phosphorylation (Betts *et al.*, 1997) and the glycosylation of a variety of proteins (Dell and Morris, 2001). Improvements in the sensitivity of these techniques could make the characterization of IF posttranslational modifications less tedious.

In addition, a reverse immunologic approach can be used to identify phosphorylation residues (Omary *et al.*, 1998). This approach entails generating monoclonal antibodies against hyperphosphorylated keratins purified from cells treated with phosphatase inhibitors, such as okadaic acid. Antibodies are screened for specific binding to phosphorylated keratins. Potential target phosphorylation sites are predicted and then confirmed by combining the information generated from: (1) binding of the antiphosphokeratin antibodies to keratins phosphorylated *in vitro* with a panel of kinases and (2) the consensus sequence of the kinase(s) and the known keratin sequences. Alternatively, the phosphorylation sites recognized by the generated antibodies can be identified by epitope mapping of the antibodies of interest with synthetic peptides and confirmed by generating site-specific keratin mutants and then immunoblotting (using a dot blot technique) with the test antibodies. This approach was used to identify phosphorylation at K8 Ser-73, which plays a significant role in keratin filament organization in response to stress, apoptosis, and mitosis (Liao *et al.*, 1997).

Type I keratins, including K18 and K19, undergo caspase-mediated proteolysis during apoptosis, whereas their type II partner, K8, manifests remarkable resistance to apoptosis-associated degradation. The methods used to identify the *in vivo* cleavage sites have relied on microsequencing of keratin fragments obtained from apoptotic cells (Ku *et al.*, 1997), by testing keratin fragmentation with purified recombinant caspases (Caulin *et al.*, 1997), by epitope mapping of an antibody (called M30, Table I) that recognizes K18 only after caspase cleavage (Leers *et al.*, 1999). The potential caspase cleavage sites are confirmed by mutation (e.g., K18 D237E) followed by transfection of WT and mutant cDNAs then testing for caspase-mediated cleavage after inducing apoptosis in transfected cells (Caulin *et al.*, 1997; Ku and Omary, 2001).

In this section, we will focus on the *in vivo* and *in vitro* detection of keratin phosphorylation, glycosylation, and proteolysis. For each posttranslational modification, this section is subdivided into: "Reagents," "*In Vivo* modification," and "*In Vitro* modification."

1. Phosphorylation

In vivo metabolic labeling of cells is performed with $^{32}PO_4$ in phosphate-free medium, supplemented with 0.1% (v/v) normal medium to prevent phosphate depletion during labeling and to facilitate uptake of the labeled phosphate. After *in vivo* labeling, the immunoprecipitated keratins can be used for PAA analysis to confirm the presence of one or more of the radiolabeled PAA residues phospho-serine, phosphotyrosine, and phosphothreonine. The PAA analysis is carried out by acid hydrolysis of the ^{32}P-labeled keratins, followed the separation of PAA by one-dimensional or two-dimensional thin-layer electrophoresis (van der Geer and Hunter, 1994). Alternatively, global changes in keratin phosphorylation can be detected by immunoblotting with commercially available antibodies against phosphoserine, phosphotyrosine, or phosphothreonine (Feng *et al.*, 1999).

For *in vitro* phosphorylation studies, the amount of purified kinase and components of the kinase buffer are variable, depending on the kinase and its cofactor requirements. We describe, as an example, *in vitro* human K8 phosphorylation by p42 MAP kinase (Erk2) (Fig. 4A), which is a likely physiological kinase for human K8 Ser431 (Ku *et al.*, 2002a).

a. Reagents

[^{32}P]-orthophosphoric acid and [γ-^{32}P]-ATP (3,000 Ci/mmol) (DuPont New England Nuclear); phosphate-free Dulbecco's modified Eagles medium (DMEM) and dialyzed FCS (Invitrogen); yeast adenosine triphosphate (ATP) (A2383, Sigma); p42 MAP kinase (New England Biolabs); $1\times$ *in vitro* kinase buffer (50 mM Tris-HCl [pH 7.5], 10 mM $MgCl_2$, 1 mM EGTA, 2 mM DTT, and 0.01% Brij 35). The kinase buffer may vary.

b. In Vivo *Phosphorylation*

1. Incubate cells (in 100-mm dishes) with 5 ml of phosphate-free DMEM containing 10% dialyzed FCS 2 mM, L-glutamine, 100 units/ml penicillin, and 100 μg/ml streptomycin (30 minutes, 37 °C).
2. Add 50 μl of normal medium and 250–500 μCi/ml [^{32}P]-orthophosphoric acid and incubate for 5 hours at 37 °C. The labeling time can vary, depending on the nature of the experiment, with a 3- to 5-hour labeling period typically providing steady-state incorporation.
3. Wash cells with PBS three times.
4. Isolate K8/K18 as described in Section II.A, and then analyze by SDS-PAGE and autoradiography.

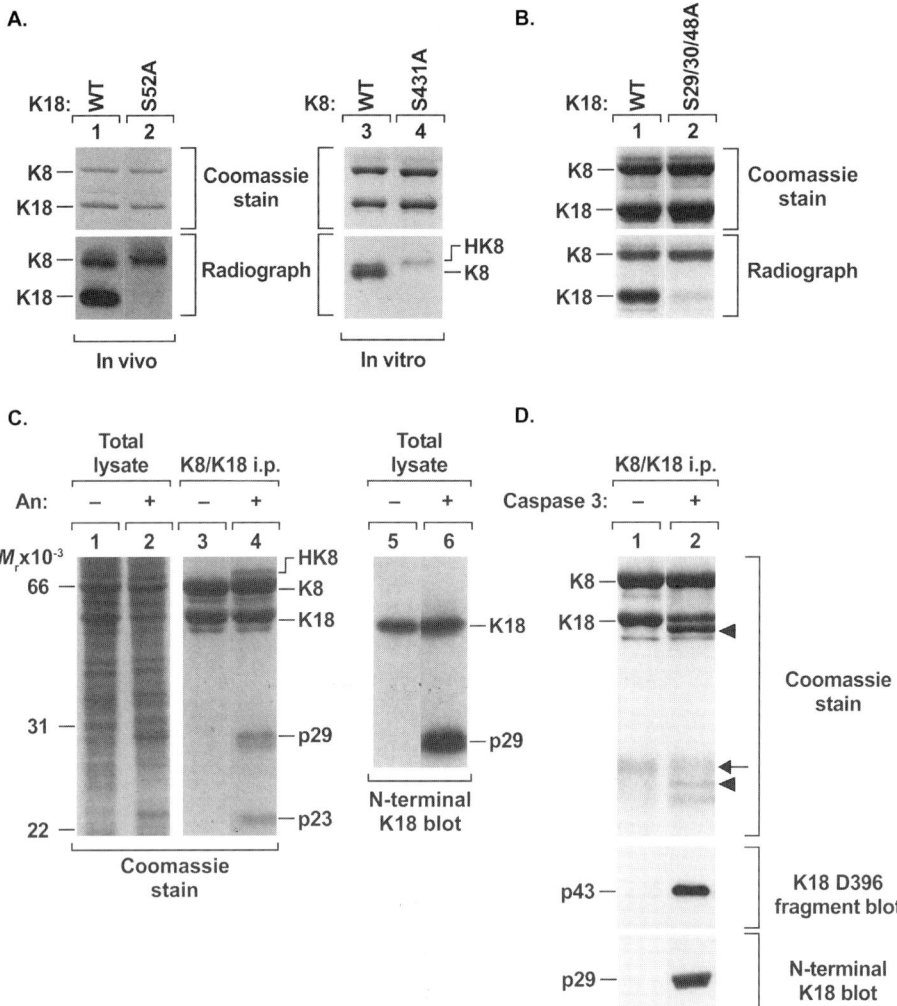

Fig. 4 *In vivo* and *in vitro* posttranslational modifications of K8/K18. (A) *In vivo* and *in vitro* phosphorylation of K8/K18. For *in vivo* phosphorylation, cultured cells that express WT K8 and WT K18, or WT K8 with Ser52→Ala K18, were labeled with $^{32}PO_4$ followed by solubilization, immunoprecipitation of K8/K18, SDS-PAGE, then Coomassie staining (lanes 1, 2). Note that mutation of K18 Ser52, the major K18 phosphorylation site, results in near total abolishment of K18 phosphorylation. For *in vitro* phosphorylation, BHK-21 cells were cotransfected with WT K18 and one of the K8 constructs, WT or S431A. After 3 days, K8/K18 immunoprecipitates were prepared then used for *in vitro* phosphorylation by p42 MAP kinase (Erk2), boiled, and then analyzed by SDS-PAGE and autoradiography (lanes 3, 4). Note that mutation of K8 Ser431 results in abolishment of K8 phosphorylation, but minor phosphorylation is detected in a distinct hyperphosphorylated K8 (HK8) caused by K8 Ser73 phosphorylation (Ku *et al.*, 2002a). (B) *In vitro* galactosylation of K8/K18. K8/K18 were immunoprecipitated from cells that express WT K8/K18 or WT K8 and K18 S29/30/48A. The immunoprecipitates were *in vitro* labeled with UDP-[^{3}H]galactose and galactosyltransferase,

c. In Vitro *Phosphorylation*

1. Prepare K8/K18 immunoprecipitates as described in Section II.A.2.

2. For 10 μl of immune complex–associated beads (total bead volume), add 15 μl of 1× kinase buffer and heat for 2 minutes (90 °C) to inactivate endogenous kinase activity.

3. After cooling to room temperature, add 5 μCi of [γ-^{32}P]-ATP, 20 μM ATP, and 1 unit of p42 MAP kinase (total reaction volume = 25 μl) and incubate 15 minutes at 22 °C.

4. Add 25 μl of 2× Laemmli sample buffer containing 4% SDS and 20% glycerol and boil for 2 minutes.

5. Pellet then analyze the labeled K8/K18 immunoprecipitates by SDS-PAGE and autoradiography.

2. Glycosylation

Detection and confirmation of keratin glycosylation is performed by use of a combination of techniques, including *in vivo* labeling and *in vitro* galactosylation (Chou *et al.*, 1992; Greis *et al.*, 1996; Hart, 2003; King and Hounsell, 1989; Ku and Omary, 1995). Metabolic labeling of cells is performed with ^{3}H-glucosamine, and the nature of the attached saccharides is assessed after acid hydrolysis of the purified protein. *In vitro* galactsylation is carried out with immunopurified keratins, UDP^{3}H-galactose, and galactosyltransferase. This adds a radiolabeled galactose to accessible terminal GlcNAc on the protein backbone (Fig. 4B) and provides an indirect assessment of a protein's O-GlcNAc modification.

a. Reagents

D-[6-^{3}H]-glucosamine (40 Ci/mmol) and uridine diphosphate (UDP)-[4,5-^{3}H]-galactose (36.7 Ci/mmol) (DuPont New England Nuclear); glucose-free RPMI 1640 medium (Invitrogen); Amplify fluorographic reagent (Amersham

---

followed by SDS-PAGE, Coomassie staining, and fluorography. Mutation of K18 Ser29/30/48 (to Ala), the major glycosylation sites of K18, abolishes most of K18 glycosylation. (C) *In vivo* K18 fragmentation during anisomycin-induced apoptosis. HT29 cells were treated with anisomycin (10 μg/ml) for 24 hours. The cells were solubilized in 1× Laemmli sample buffer (total lysate) or in 1% NP-40 followed by K8/K18 immunoprecipitation (ip). Total lysates and K8/18 ip were analyzed by SDS-PAGE and Commassie staining or were immunoblotted with anti-K18 (N-terminal epitope) antibody. p29 represents an N-terminal K18 fragment that is generated after K18 cleavage at Asp237. (D) *In vitro* K18 fragmentation by caspase-3. K8/K18 ip were incubated with buffer alone or with buffer containing recombinant caspase-3 for 3 hours. The K8/K18 ip were separated by SDS-PAGE, then analyzed by Commassie staining, or by blotting with Ab DC10 that recognizes K18 (aa 1-237) or Ab M30 that specifically recognizes the exposed Asp396 of K18 (i.e., on release of the K18 tail fragment, aa 397–429). Both K18 Asp396 and Asp237 are caspase-3 *in vitro* digestion sites that generate p43 and p29 (arrowheads), respectively, after sequential cleavage. Arrow = nonspecific bond.

Biosciences); galactosyltransferase (G5507, Sigma); 1× galactosyltransferase buffer (20 mM MnCl$_2$, 100 mM sodium cacodylate [pH 6.5]; 1× washing buffer
(0.5% NP-40 in PBS [pH 7.4]).

b. In Vivo *Glycosylation*

1. For confluent cells in one 100-mm dish, dry 500 μCi of D-[6-^{3}H]-glucosamine
   with a SpeedVac concentrator and resuspend with 50 μl of normal medium.

2. Incubate the cells (one 100-mm dish) in 5 ml of glucose-free RPMI 1640
   medium containing 10% dialyzed FCS 2 mM L-glutamine, 100 units/ml
   penicillin, and 100 μg/ml streptomycin for 30 minutes at 37 °C.

3. Add 50 μl of the radioactive-containing medium (final concentration;
   100 μCi/ml [^{3}H]-glucosamine) and incubate for 12 hours at 37 °C.

4. Wash cells with PBS three times.

5. Isolate K8/K18 as described in Section II.A and then analyze by SDS-PAGE
   and fluorography (for this, we use Amplify fluorographic reagent as recommended by the supplier).

c. In Vitro *Galactosylation*

1. Prepare K8/K18 immunoprecipitiates as described in Section II.A.2.

2. For 10 μl of immune complex–associated beads, dry 0.6 μCi of UDP-[^{3}H]galactose
   in a SpeedVac concentrator and resuspend with 20 μl of 1× galactosyltransferase
   buffer.

3. Add 25 μm of galactosyltransferase and incubate for 2 hours at 37 °C.

4. Wash the immunoprecipitates with 0.4 ml of 1× washing buffer two times.

5. Boil beads in 50 μl of 2× Laemmli sample buffer for 1 minute.

6. Analyze the labeled samples by SDS-PAGE and fluorography.

3. Cleavages by Caspases

We describe *in vivo* K18 fragmentation during anisomycin-mediated apoptosis
and *in vitro* K18 cleavage by purified recombinant caspase 3 (Fig. 4C,D)
(Ku and Omary, 2001; Ku *et al.*, 1997). Similar findings were reported in other
K8/K18-expressing systems with etoposide (125–250 μg/ml, >24 hours) or dauno-
mycin (2–6 μg/ml, >24 hours) treatment of cultured cells to induce apoptosis or by
use of caspases 3, 6, or 7 for *in vitro* K18 cleavage (Caulin *et al.*, 1997). Notably,
K18 and K19 degradation can be easily detected in BHK hamster kidney cells that
are transfected with K8/K18 or K8/K19 cDNA constructs with LipofectAMINE
(Invitrogen), without adding any proapoptotic reagents (Ku *et al.*, 1997). K18
fragmentation is also observed in transgenic mice after intraperitoneal injection of
Fas antibody (0.15 μg antibody/g body weight) or injection of TNF-α (15 ng/g)
with actinomycin D (1 μg/g) (Ku *et al.*, 2003b) (see also Section II.D.1).

a. Reagents

Anisomycin (CalBiochem); recombinant caspase-3 (Upstate Biotechnology Inc.); 1× caspase buffer (25 mM Hepes, 1 mM dithiothreitol [pH 7.5]).

b. **In Vivo** *Keratin Fragmentation*

1. Incubate ~80%–90% confluent cells in 10 ml of normal culture medium containing 10 μg/ml anisomycin for 16–24 hours at 37 °C.
2. Wash cells with PBS three times (need to include apoptotic floater cells).
3. Isolate K8/K18 as described in Section II.A and then analyze by SDS-PAGE and Coomassie staining or by immunoblotting.

c. **In Vitro** *Fragmentation*

1. Prepare K8/K18 immunoprecipitates as described in Section II.A.2.
2. For 10 μl of immune complex–associated beads, add 20 μl of 1× caspase buffer with 1 unit of recombinant caspase-3 and incubate for 3 hours at 37 °C.
3. Add 25 μl of 2× Laemmli sample buffer containing 4% SDS and 20% glycerol and boil for 3 minutes.
4. Analyze the fragmented K8/K18 by SDS-PAGE and immunoblotting.

C. Keratin Staining in Tissues and Cells

Immunofluorescence staining combined with confocal microscopy are indispensable tools in studying keratin filament organization, disassembly, degradation, phosphorylation, and interaction with other proteins. Because keratins are very abundant proteins composed of filament networks that undergo dynamic alterations during physiological and stress conditions, they are easy to visualize. Antibodies to posttranslationally modified keratins, such as the phospho-specific keratin antibodies and the M30 antibody to the caspase degraded K18 (Table I and Fig. 5) (Ku and Omary, 1997; Ku *et al.*, 1998a; Leers *et al.*, 1999; Liao *et al.*, 1995b, 1997) are helpful specific tools to aid our understanding of simple epithelial keratin dynamics and regulation.

1. Reagents

O.C.T. compound (optimum cutting compound) and Vinyl Tissue-Tek Cryomolds (Sakura); sterile Lab-Tek Chamber slides with cover (Nalgen Nunc Int.); gelatin (Kodak); chrom alum (chromium potassium sulfate, 12-hydrate; EMD Chemicals); charged superfrost/plus microscope slides and premium cover glasses thickness 1 (17 mm; Fisher scientific); 100% acetone or methanol stored at −20 °C; freshly made 0.5% para-formaldehyde (PFA) in PBS (heat PFA powder in H_2O to 60 °C and add NaOH until dissolved, cool to 22 °C, add 5× strength solution of

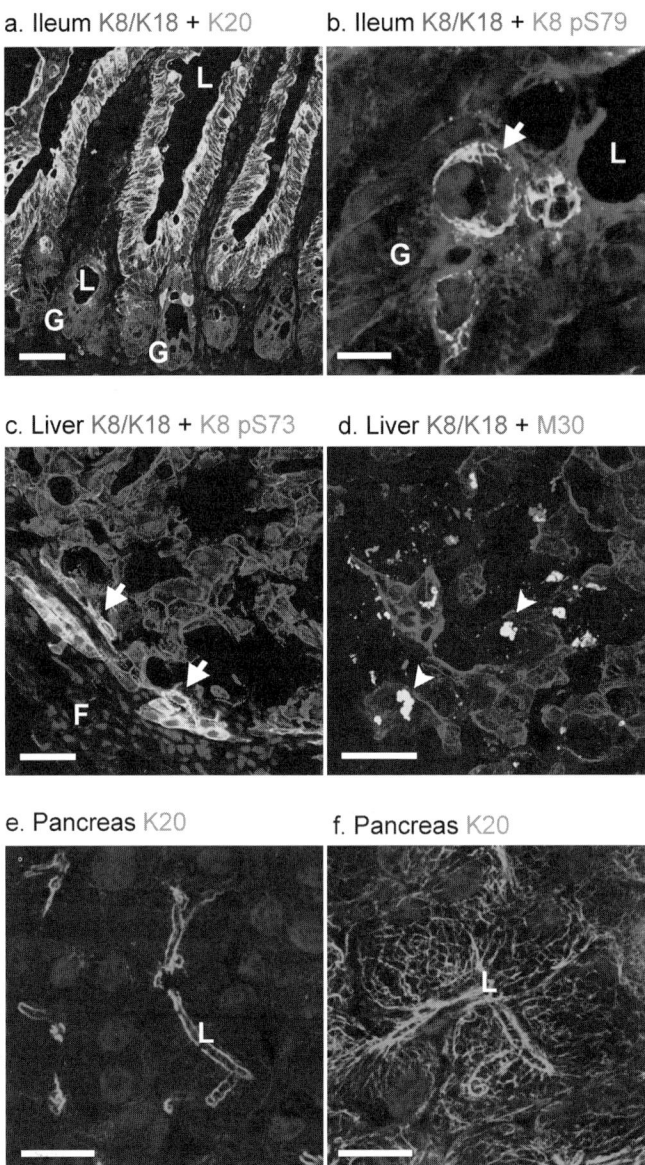

Fig. 5 Analysis of keratins in mouse and human tissues by fluorescence microscopy. Sections of normal mouse ileum (a,b), human cirrhotic liver (c,d), and mouse pancreas (e,f) treated with saline (e) or 48 hours after 7 hourly injections of 50 μg/kg caerulein (f) were double- or triple-stained for keratins (± nuclear staining) then viewed with a BioRad MRC1024 confocal microscope. K8/K18 (red) in (a–d) were detected with a rabbit anti-keratin antibody and a CY5-conjugated goat anti-rabbit antibody and further pseudo-colored red with the Adobe photoshop program. K20 (green) was detected with Ab Ks20.8 Ab, phospho-K8 (green) in (b) and (c) was probed with mAb LJ4, and the K18

PBS and water to a final 1× PBS solution, pH 7.4); buffer A (2.5% [w/v] BSA [bovine serum albumin, Sigma] in PBS [pH 7.4]; buffer B (2% normal serum [that matches the host species of the secondary antibody] in buffer A); buffer C (0.2 mg/ml ribonuclease A [Sigma] in buffer B); fluorescent-labeled secondary antibodies diluted in buffer B; nuclear fluorescent probes (Toto-3 iodide, Yo-Pro-1 iodide) prepared in buffer C, and ProLong antifade mounting media (Molecular Probes).

2. Steps

a. Staining of Frozen Tissue Sections

1. Sacrifice test animals by CO_2 inhalation and immediately excise desired tissues, immerse (with orientation if needed, such as for intestinal tissues) in an O.C.T.–filled cryomold (placed over a block of dry ice), then allow to quickly freeze on dry ice. Store blocks at −80 °C.

2. Precoat microscope slides with gelatin: add 5 g gelatin to 1 L of water (55 °C) and dissolve, then add 0.5 g chrom alum. Filter the warm solution with Whatman filter paper and store at 4 °C. Wipe slides free of dust, dip in the coating solution, drain off excess solution, stand vertical to dry, and store at 22 °C.

3. Cut thin sections (6 μm) with a cryostat and place on gelatin-coated cover slides.

- For acetone fixation: allow sections to air dry (1 hour), then fix in −20 °C acetone (10 minutes), and air dry (1 hour). Slides can be stored at −20 °C and when used refixed immediately (while frozen) in −20 °C acetone for 10 minutes, air dried (1 hour), or dried with drie-rite desiccant (22 °C; W A Hammond Drierite Co Ltd).

- For PFA fixation: air dry sections (5 minutes) and fix in 0.5% PFA (for labeling that requires PFA, such as FITC-phalloidin) for 20 minutes (22 °C), then wash slides with PBS.

4. For subsequent incubations, place slides in a moisture chamber (a box with lid with the bottom covered with wet paper towels).

5. Wash 3× with PBS (for PFA fixation: permeabilize cell membranes with 0.2% NP-40 in PBS, 5 minutes [22 °C], then wash 3× with PBS).

6. Block with buffer A for 10 minutes, then with buffer B for 10 minutes.

---

apoptotic caspase-generated 43-kDa fragment (green) in (d) was detected with mAb M30 (Table I). Nuclei in (a–f) were stained with propidium iodide and pseudo-colored blue. Note that although K8/K18 (red, a) are expressed throughout the ileal epithelium, K20 (green) is present in the villus cells (yellow when colocalized with K8/K18) but is absent in the basal glands. Arrows in (b) and (c) show phosphorylated K8 (S79 in mouse and S73 in human) in a mitotic cell in a mouse ileal gland and in hepatocytes of a human cirrhotic liver, respectively. Arrowheads in (d) point to keratin and M30-positive aggregates indicative of apoptosis. Note dramatic induction of filaments (panel f). F, fibrosis; G, intestinal glands; L, lumen. Scale bars: a, c, d = 50 μm, b = 10 μm, e, f = 20 μm. (See Color Insert.)

7. Incubate with primary antibodies 45–60 minutes (e.g., mAb L2A1 and rabbit antibody E2634, Table I) diluted in buffer B. In general, use 1:50–1:100 dilutions of mouse mAb ascites and 1:500–1:2000 of rabbit polyclonal antibodies. Depending on nature and extent of experiments, antibody titration may be necessary.

8. Wash by incubating with PBS 3× (5 minutes/incubation), then block with buffer C (10 minutes).

9. Incubate with the fluorescently labeled secondary antibodies (1:50–1:200) and the nuclear marker (e.g., Toto-3; for double or triple staining) diluted in buffer C for 30 minutes, then wash 3× with PBS. For example, we use Texas red-conjugated goat anti-rabbit (Molecular Probes, T-2767) and FITC-conjugated goat anti-mouse antibody (BioSource, AMI4408). Preabsorbed secondary antibodies should be purchased when double/triple staining is carried out.

10. Mount sections under coverslips with fresh Prolong anti-fade mounting media, then allow to dry.

11. View in microscope with appropriate filters and lasers for the fluorophores used (see Fig. 5 for tissue staining examples). Stained slides can be stored at 4 °C for several weeks.

b. Staining of Cells

1. Sterilize coverslips by soaking them in 99% ethanol for 5 minutes, dry one at a time in an open flame, and place in a sterile 2-ml tissue culture dish or use sterile plastic chamber slides.

2. Plate cells onto the coverslips or plastic chamber slides at $\sim 10^4$ cells/ml and culture until cells are attached and dividing.

3. If applicable, treat cells with stimulatory agent of choice (e.g., phosphatase inhibitor). Remove media and wash cells 3× with PBS.

4. If treatment generates floater (detached) cells, collect and centrifuge the medium at 1000 rpm, 5 minutes, further wash cells 3× with PBS by centrifugation as earlier. Incubate floater cells in all further steps in Eppendorf tubes. Alternatively, transfer cells to slides with a cytospin centrifuge (e.g., Shandon Cytospin, Thermo Electron Corp).

5. Fix cells with methanol (−20 °C, 10 minutes), then air dry. For PFA fixation, use 0.5% PFA for 20 minutes (22 °C).

6. Stain as described for the frozen tissues.

D. Simple Epithelial Keratins in Organ-Specific Injury Models

Simple-type glandular epithelia express K8 (primarily) and K7 (at low levels, typically in ductal cells) as the type II keratins and variable levels of type I keratins (K18, K19, or K20), depending on the cell type, as found in several digestive organs, including hepatocytes (K8 and K18), gallbladder (K8 and K19>K18), pancreas (K8 and K18>K19/K20), small intestine and colon (K8 and K19>K18/K20) (Fig. 2B). This section addresses a variety of techniques that can be used to examine organ-specific injury models (Table II). Their usefulness is in comparing the phenotype of transgenic mouse lines and in testing the effect of a keratin transgene or a keratin-null in a tissue-specific fashion. General principles of animal experiments to keep in mind, independent of organs, include the following.

- Pilot experiments need to be carried out to optimize dosing to avoid potential rapid lethality caused by variables related to mouse genotypes and strain differences.
- Sex- and age-matched mice ($\sim$2- to 3-month-old) need to be used.
- Organs and blood are collected after euthanizing the mice by CO_2 inhalation. Blood is collected by intracardiac puncture (0.5–1.0 ml with a 22-G needle) followed immediately by organ harvesting.
- Dissected organs are generally divided into several pieces, depending on the experiment and used for: (1) fixation with 10% formalin followed by paraffin embedding, sectioning, then hematoxylin/eosin staining; (2) snap freezing in optimum cutting-temperature compound, sectioning, then fixing in cold acetone or PFA for subsequent immunofluorescence staining (see Section II.C); and (3) snap freezing in liquid nitrogen for subsequent biochemical analysis (see Section II.A).

1. Liver Injury and Regeneration Models

The drug doses described herein are based on studies with K8 or K18 transgenic mouse models in an FVB strain background.

a. Toxin Administration

1. Griseofulvin: Mice are fed a powdered Lab Diet (PMI Feeds, Inc) with 1.25% griseofulvin (Sigma) for 7–17 days. The amount of griseofulvin was chosen on the basis of preliminary experiments that showed that this dosing and duration lead to fatality within 10 days in the K18 Arg 89→Cys overexpressing transgenic mice that are highly susceptible to drug-induced liver injury (Ku et al., 1996). This injury model is different than the longer duration feeding regimen, which induces Mallory body formation (see Chapter 8).

2. Acetaminophen (McNeil): Mice are deprived of solid food for 3 hours followed by administration of acetaminophen by gavage (400 μg/g body wt, diluted with H_2O to a final volume of 0.4–0.5 ml), then switching the water source to 5% dextrose in water. After 2 hours, the mice are allowed to feed ad lib and observed intermittently for 72 hours, followed by tissue harvesting and analysis (Ku et al., 1996).

3. Microcystin-LR (MLR) (Alexis Corp): MLR is a hepatotoxin that inhibits type 1 and type 2A serine/threonine phosphatases and causes extensive intrahepatic hemorrhage (see also Chapter 14). MLR is prepared in dimethylsulfoxide as a

Table II
Simple Epithelial Keratin-Related Disease Animal Models

Organ	Model/Toxin	Disease	Specifics	References
Liver	Griseofulvin	Acute/chronic hepatitis	Feed mice for 7–17 days (1.25% griseofulvin in food)	Ku et al. (1996)
	Acetaminophen	Acute hepatitis	Gavage, 400 μg/g*	Ku et al. (1996)
	Microcystin-LR	Acute hepatitis	Single IP[†]30 ng/g	Toivola et al. (1998) Ku et al. (1998)
	Fas antibody	Acute hepatitis	Single IP 0.15 μg/g	Gilbert et al. (2001) Ku et al. (2003b)
	TNF-α (with Act. D[†])	Acute hepatitis	Single IP 15 ng/g TNF-α (with 1 μg/g Act. D[‡]) Single IV[§]30 μg/g	Ku et al. (2003b)
Gallbladder and liver	Concanavalin A	Acute hepatitis		Caulin et al. (2000)
	Lithogenic diet	Gallstones and steatohepatitis	Feed mice for 5 weeks	Tao et al. (2003)
Pancreas	Caerulein	Acute pancreatitis (generally mild)	Hourly seven IP injections; no age or sex preference	Toivola et al. (2000a,b) Zhong et al. (2004) Zhong and Omary (2004)
	Choline-deficient, ethionine-supplemented diet (CD diet)	Acute pancreatitis (generally severe)	Feed mice for up to 3 days; more effective in females; causes ∼50% lethality in 16–20 g weight mice	Toivola et al. (2000a,b) Zhong et al. (2004) Zhong and Omary (2004)
Small Intestine	Gamma irradiation	Stem cell apoptosis	Whole-body irradiation at 8 Gy	Martin et al. (1998)
	Rotavirus infection	Diarrhea and crypt cell injury	Oral or intestinal loop infection	Burns et al. (1995) Ludert et al. (1996)
Colon	Dextran sulfate sodium	Colitis	Feed mice for up to 12 days (2% dextran sulfate sodium in water)	Siegmund et al. (2002) Ku et al. (unpublished results)

*Reagents are administrated per gram of mouse body weight.
[†]IP, Intraperitoneal.
[‡]Act. D, actinomycin D.

1 mg/ml stock solution and then diluted in PBS (pH 7.4). Mice are fasted overnight then given 30 ng/g mouse wt intraperitoneally (IP), followed by processing of the livers after 3 hours (Ku *et al.*, 1998b; Toivola *et al.*, 1998).

4. Proapoptotic reagents: Apoptosis is induced by IP injection of Fas antibody (0.15 μg/g body wt) (PharMingen) to overnight-fasted mice. Alternatively, purified TNF-α (15 ng/g) (Sigma) with actinomycin D (1 μ/g) (Sigma) is injected. Mice are killed by CO_2 inhalation 4 hours after injection (Ku *et al.*, 2003b). Concanavalin A (Caulin *et al.*, 2000) or Fas antibody (Gilbert *et al.*, 2001) can also induce apoptosis-mediated liver injury as tested in K8-null mice.

b. Partial Hepatectomy

Partial hepatectomy is performed by sequentially removing the lateral, left, and right median lobes from mice (Wilson *et al.*, 1953; Yokoyama *et al.*, 1953). Mice are euthanized 1, 2, 3, 7, or 10 days afterwards, depending on the nature of the experiment. For control sham-hepatectomy, mice undergo anesthesia, abdominal wall and peritoneal incision, liver exposure, and then closure of the incision (Ku *et al.*, 2002b). Peak mitotic activity is noted 2–3 days after liver resection.

2. Gallbladder and Liver: High-Fat Lithogenic Diet-induced Injury

Male mice (4–6 weeks old) are fed a lithogenic diet (Cat. No. 960393, ICN Bio-medicals) containing 1.23 g cholesterol, 0.48 g sodium cholate, 17.84 g butter fat, 0.98 g corn oil, 48.33 g sucrose, 19.33 g casein, and essential vitamins and minerals for 5 weeks (Tao *et al.*, 2003). All mice are allowed free access to water. This diet induces gallbladder stone formation and causes liver injury from fat deposition. A dissection microscope needs to be used to optimally visualize stone or stone-debris formation.

3. Pancreatitis Models

Pancreatic injury can be induced by various mouse models (Table II) (Banerjee *et al.*, 1994; Lerch and Adler, 1994) with subsequent effects on keratin expression, degradation, phosphorylation, and filament organization (Toivola *et al.*, 2000a,b; Zhong and Omary, 2004; Zhong *et al.*, 2004). Caerulein, an analog of the secreta-gogue cholecystokinin, causes pancreatic autolysis and mild edematous, nonlethal pancreatitis (Lerch and Adler, 1994). The choline-deficient, ethionine-supplemented diet (CD diet) causes severe acute hemorrhagic pancreatitis with fat necrosis and results in significant fatality in young female mice, which are most susceptible to this injury (Lerch and Adler, 1994). Pancreatitis can also be induced by other models such as the coxsackievirus B4, which causes acute or chronic pancreatitis depending on the virus strain (Caggana *et al.*, 1993; Toivola *et al.*, unpublished observations). The pancreatitis models that have been used to study keratin biology in the pancreas (Table II) will be described.

a. Caerulein–Induced Acute Pancreatitis

1. Starve mice from solid food overnight, but allow water intake ad libitum. Susceptibility to injury in this model seems to be sex and age independent, but there are strain differences (Zhong and Omary, 2004).

2. Inject IP 0.9% saline (in control mice) or 50 μg/kg of mouse wt caerulein diluted in 0.9% NaCl (Research Plus; or #C-9026, Sigma) once every hour for 6 hours (seven total injections).

3. Sacrifice mice 1–240 hours after the first injection and collect blood by cardiac puncture, then quickly excise and process the pancreas. In this model, caerulein induces reversible keratin degradation, hyperphosphorylation, and overexpression. An example of K20 overexpression with dramatic formation of K20-containing cytoplasmic filaments is shown in Fig. 5e,f.

b. Choline–Deficient, Ethionine Supplemented Diet Induced Acute Pancreatitis

1. Starve young female mice (14–19 g, typically ~4–5 weeks old) from solid food for 12–16 hours, but allow water intake ad libitum.

2. Mix CD powdered chow (#TD90262, Harlan Teklad), thoroughly with 0.5% DL-ethionine (#E-5139, Sigma).

3. Allow mice to feed with the CD diet for 3 days, then switch to a normal diet for 1–7 days. The diet is placed in cups at a corner of the cage and refurbished daily. Note that DL-ethionine is carcinogenic; thus, excess food and bedding need to be discarded as hazardous waste.

4. Depending on the age of the mice, CD diet induced lethality, when present, typically occurs in 1–2 days, but no later than 5 days, after discontinuation of the diet. Complete disruption of keratin filament networks together with vacuole formation and cell death is most prominent 60 hours after the onset of the diet (Toivola *et al.*, 2000a,b; Zhong and Omary, 2004; Zhong *et al.*, 2004).

4. Small Intestine Injury Models

a. Irradiation

Two-month-old male mice are examined 4, 24, 48, 72 hours after whole-body gamma irradiation, given as an acute 8-Gy dose. Cell injury and apoptosis occur primarily in the crypt region of the epithelium (Martin *et al.*, 1998).

b. Rotavirus–Induced Diarrhea Model in Suckling Mice

Five-day-old mice are orally inoculated with 10^4 diarrhea dose 50 (DD_{50}) of a virulent wild-type murine rotavirus strain ECw. Presence of diarrhea in the suckling mice is confirmed by gently pressing the lower abdomen of the mice and resolves 10 days after infection (Burns *et al.*, 1995).

c. **In Vivo** *Small Bowl Loop Rotavirus Infection*

Mice are given ketamine and xylazine for anesthesia. A ventral abdominal incision is made in 10-day-old suckling mice or 4- to 6-week-old mice. Two loops of small bowel (jejunum or ileum) approximately 0.5 to 1 cm in length are isolated with fine suture material. The loops are injected with virus by use of a Hamilton syringe, and the abdomen is closed with suture or glue. After 4–10 hours, injected mice are euthanized followed by harvesting of the isolated loops for imaging and biochemical studies. Analgesia (0.005–0.05 mg/kg of bupronorphine) and hydration (0.5 ml saline) are administrated IP postoperatively (Ludert *et al.*, 1996).

5. Colitis Injury Model

Dextran sulfate sodium (DSS) causes colitis by interfering with the intestinal epithelial cell barrier with subsequent induction of cytokines and other inflammatory mediators (Blumberg *et al.*, 1999). For lethality experiments, mice are given 2% DSS (ICN Pharmaceuticals, Inc) that is dissolved in the drinking water. DSS is administered for up to 12 days followed by switching to normal drinking water. Mice are assessed daily for up to 4 weeks (Ku *et al.*, unpublished observation; Siegmund *et al.*, 2002). Lethality typically begins toward the end of the DSS administration period. For nonlethality studies, mice are given 2% DSS for 10 days or less followed by removal of the colons for histology, immunofluorescence, and biochemical studies.

III. Pearls and Pitfalls

A. Isolation of Keratins

1. High-Salt Extraction

1. Although the HSE method retains most of the keratins (>90%) in the cell, a small portion of soluble keratins are discarded as part of the supernatant fractions.

2. Occasionally, nonkeratin proteins and nucleic acid genomic material may coprecipitate with the keratins because of insufficient homogenization and douncing or inadequate solubilization buffer to cell/tissue pellet volume ratio. This will result in a high background after gel analysis. Also, keratin degradation may occur during the isolation procedure, so protease inhibitors are added in the buffer (EDTA is essential to minimize keratin degradation during HSE and immunoprecipitation), and the procedures should be performed as efficiently as possible.

3. In some occasions, keratin analysis of individually isolated fractions including cytosolic (detergent free) membrane or "loosely" associated cytoskeleton, cytoskeletal, or remaining insoluble fractions may need to be carried out. For this, we previously described a sequential extraction method (see Omary *et al.*, 1998 for details).

2. Immunoprecipitation

1. We routinely test the anti-keratin antibodies by immunoprecipitation with both protein A and G, and then select what works better. In addition, we routinely covalently conjugate our antibodies to protein A or G Sepharose beads (with a kit purchased from Pierce) to minimize the amount of antibody release from the beads after adding the sample buffer.

2. Nonspecific protein binding can be a problem, which is generally overcome by the preclearing step or washing with salt or detergent-containing buffers. Inclusion of a control immunoprecipitation where the test antibody is replaced by a nonrelevant immunoglobulin is important.

3. Immunoprecipitation of NP-40 (or other nonionic detergent) solubilized cells can isolate, if quantitative, 10%–15% of the total cellular keratins, whereas Empigen extracts ∼50% of the total keratin pool. However, these quantities can be significantly less, depending on antibody affinity and experimental conditions (e.g., interphase vs. mitotic cells; incubation of cells with phosphatase inhibitors that increase keratin solubility).

4. When testing keratin-associated proteins by coimmunoprecipitation, it is essential to use two or more antikeratin antibodies to avoid "pseudo-association" by means of molecular mimicry of antibody binding (e.g., see Zhou *et al.*, 2000).

5. We typically use nonreducing sample buffer for gel analysis, particularly for immunoprecipitate analysis of K8/K18/K19, which do not contain any cysteines.

B. Posttranslational Modification of Keratins

1. [^{32}P]-orthophosphoric acid usually is sold in 1-ml volume, independent of the purchased millicuries, hence, it may be necessary to custom request the volume as needed to ensure high specific activity.

2. D-[6-^{3}H]-glucosamine and uridine diphosphate (UDP)-[4,5-^{3}H]-galactose are dried, just before application, by use of a SpeedVac concentrator to remove the stabilizing organic solvent, typically 90% ethanol.

C. Keratin Staining in Tissues and Cells

1. Most antibodies for keratins and other IF proteins work best when the cells or tissues are fixed with water-free −20 °C methanol or acetone (10 minutes), whereas some antibodies require formalin fixation or the use of antigen retrieval protocols (Table I). Double labeling of a specimen with antibodies or reagents, which requires acetone (e.g., keratin antibodies), or PFA (e.g., fluorescent-phalloidin, which binds to F-actin), can be done by use of

mild PFA fixation (0.5% for 20 minutes), which gives reasonable staining results.

2. Isotype-specific secondary antibodies or fluorophore-conjugated antibodies can be used to double-label with two mouse monoclonal antibodies simultaneously. Online resources for fluorophores and help with selection of secondary antibodies can be found at http://fluorescence.bio-rad.com/, http://www.probes.com/handbook/, and http://www.jacksonimmuno.com/.

D. Simple Epithelial Keratins in Organ–Specific Injury Models

1. In assessing injury models in mice, there is significant variability in the extent of injury. This variability needs to be taken into consideration so that adequate numbers of mice are analyzed, because comparison among genotypes may be missed or overinterpreted if not enough mice are examined.

2. For the CD diet that induces pancreatitis, the extent of injury becomes limited if old (the 4- to 5-week age window is critical for lethality-type experiments) mice are used.

IV. Concluding Remarks

We have described tools to study simple epithelial keratins in cell culture and in mice. The biochemical approaches we described in Section II.A and II.B can be used to expand on our current knowledge of the regulation of K7, K8, K18, K19, and K20 and are routinely used in our laboratory. In addition, such approaches can be extended to other keratins and IF proteins (see also Chapters 5, 13, and 14) for which analysis of their posttranslational modifications and regulation by associated proteins has been somewhat limited. Although the lower solubility of nonsimple epithelial keratins can be challenging and may hinder the usefulness of such approaches, we believe that they can still be applicable.

In addition to the biochemical approaches, analysis of keratin organization at the light microscope level as highlighted in Section II.C is critical in understanding the relationship of keratin organization to its regulation. It is clear from cell culture systems, and more importantly from animal model studies, that mutation at a single residue (e.g., K18 Arg89→Cys) can have a tremendous effect on keratin function, organization, and posttranslational modifications (Ku *et al.*, 2003b). Hence, assessment of keratin organization in tissues and cells, as described in Section II.C, provides a powerful tool, particularly with the increasing ability to generate specialized antibody probes. For example, probes to specific phosphorylation sites of K8 and K18 (Table I), and to K19 and K20 (unpublished results), and to the K18 caspase-generated cleavage site (Table I) provide essential and highly sensitive adjuncts to study keratin regulation at the single cell level.

As K8 and K18 mutations become recognized as a likely risk factor for a variety of human digestive diseases (Omary *et al.*, 2004), the usefulness of the injury models described in Section II.D will become even more essential. The relationship of K8 and K18 to human disease includes liver disease (K8 and K18 mutations: Ku *et al.*, 2001, 2003a), inflammatory bowel disease (K8 mutations: Owens *et al.*, 2004), and chronic pancreatitis (K8 mutations: Cavestro *et al.*, 2003). It remains to be determined whether mutations in K7, K19, and K20 will cause or predispose patients to digestive or nondigestive (e.g., in the case of K19) diseases that relate to their cell preferential expression. Regardless, understanding the pathogenesis of the currently known K8 and K18 mutations and the function of simple epithelial keratins should be facilitated by the use of approaches described in this chapter in combination with animal and tissue culture models of disease-related mutations.

Acknowledgments

We thank Kris Morrow for his assistance with figure preparation. Our work is supported by NIH grants DK52951 and DK47918 and Department of Veterans Affairs Merit Award (M.B.O.). We are also grateful to VA Research Enhancement Award Program (Q.Z.) and Crohn's & Colitis Foundation of America Research Award support (G-Z.T.).

References

Achtstaetter, T., Hatzfeld, M., Quinlan, R. A., Parmelee, D. C., and Franke, W. W. (1986). Separation of cytokeratin polypeptides by gel electrophoretic and chromatographic techniques and their identification by immunoblotting. *Methods Enzymol.* **134,** 355–371.

Banerjee, A. K., Galloway, S. W., and Kingsnorth, A. N. (1994). Experimental models of acute pancreatitis. *Br. J. Surg.* **81,** 1096–1103.

Betts, J. C., Blackstock, W. P., Ward, M. A., and Anderton, B. H. (1997). Identification of phosphorylation sites on neurofilament proteins by nanoelectrospray mass spectrometry. *J. Biol. Chem.* **272,** 12922–12927.

Blumberg, R. S., Saubermann, L. J., and Strober, W. (1999). Animal models of mucosal inflammation and their relation to human inflammatory bowel disease. *Curr. Opin. Immunol.* **11,** 648–656.

Burns, J. W., Krishnaney, A. A., Vo, P. T., Rouse, R. V., Anderson, L. J., and Greenberg, H. B. (1995). Analyses of homologous rotavirus infection in the mouse model. *Virology* **207,** 143–153.

Caggana, M., Chan, P., and Ramsingh, A. (1993). Identification of a single amino acid residue in the capsid protein VP1 of coxsackievirus B4 that determines the virulent phenotype. *J. Virol.* **67,** 4797–4803.

Caulin, C., Salvesen, G. S., and Oshima, R. G. (1997). Caspase cleavage of keratin 18 and reorganization of intermediate filaments during epithelial cell apoptosis. *J. Cell Biol.* **138,** 1379–1394.

Caulin, C., Ware, C. F., Magin, T. M., and Oshima, R. G. (2000). Keratin-dependent, epithelial resistance to tumor necrosis factor-induced apoptosis. *J. Cell Biol.* **149,** 17–22.

Cavestro, G. M., Frulloni, L., Nouvenne, A., Neri, T. M., Calore, B., Ferri, B., Bovo, P., Okolicsanyi, L., Di Mario, F., and Cavallini, G. (2003). Association of keratin 8 gene mutation with chronic pancreatitis. *Dig. Liver Dis.* **35,** 416–420.

Chou, C. F., and Omary, M. B. (1993). Mitotic arrest-associated enhancement of O-linked glycosylation and phosphorylation of human keratins 8 and 18. *J. Biol. Chem.* **268,** 4465–4472.

Chou, C. F., Riopel, C. L., Rott, L. S., and Omary, M. B. (1993). A significant soluble keratin fraction in 'simple' epithelial cells. Lack of an apparent phosphorylation and glycosylation role in keratin solubility. *J. Cell Sci.* **105,** 433–444.

Chou, C. F., Smith, A. J., and Omary, M. B. (1992). Characterization and dynamics of O-linked glycosylation of human cytokeratin 8 and 18. *J. Biol. Chem.* **267,** 3901–3906.

Coulombe, P. A., and Omary, M. B. (2002). "Hard" and "Soft" principles defining the structure, function and regulation of keratin intermediate filaments. *Curr. Opin. Cell Biol.* **14,** 110–122.

Dell, A., and Morris, H. R. (2001). Glycoprotein structure determination by mass spectrometry. *Science* **291,** 2351–2356.

Feng, L., Zhou, X., Liao, J., and Omary, M. B. (1999). Pervanadate-mediated tyrosine phosphorylation of keratins 8 and 19 via a p38 mitogen-activated protein kinase-dependent pathway. *J. Cell Sci.* **112,** 2081–2090.

Fuchs, E., and Weber, K. (1994). Intermediate filaments: Structure, dynamics, function, and disease. *Annu. Rev. Biochem.* **63,** 345–382.

Gilbert, S., Loranger, A., Daigle, N., and Marceau, N. (2001). Simple epithelium keratins 8 and 18 provide resistance to Fas-mediated apoptosis. The protection occurs through a receptor-targeting modulation. *J. Cell Biol.* **154,** 763–773.

Greis, K. D., Hayes, B. K., Comer, F. I., Kirk, M., Barnes, S., Lowary, T. L., and Hart, G. W. (1996). Selective detection and site-analysis of O-GlcNAc-modified glycopeptides by beta-elimination and tandem electrospray mass spectrometry. *Anal. Biochem.* **234,** 38–49.

Hart, G. W. (2003). Structural and functional diversity of glycoconjugates: A formidable challenge to the glycoanalyst. *Methods Mol. Biol.* **213,** 3–24.

Herrmann, H., Hesse, M., Reichenzeller, M., Aebi, U., and Magin, T. M. (2003). Functional complexity of intermediate filament cytoskeletons: From structure to assembly to gene ablation. *Int. Rev. Cytol.* **223,** 83–175.

Hesse, M., Franz, T., Tamai, Y., Taketo, M. M., and Magin, T. M. (2000). Targeted deletion of keratins 18 and 19 leads to trophoblast fragility and early embryonic lethality. *EMBO J.* **19,** 5060–5070.

King, I. A., and Hounsell, E. F. (1989). Cytokeratin 13 contains O-glycosidically linked N-acetylglucosamine residues. *J. Biol. Chem.* **264,** 14022–14028.

Ku, N. O., Azhar, S., and Omary, M. B. (2002a). Keratin 8 phosphorylation by p38 kinase regulates cellular keratin filament reorganization: Modulation by a keratin 1-like disease causing mutation. *J. Biol. Chem.* **277,** 10775–10782.

Ku, N. O., Darling, J. M., Krams, S. M., Esquivel, C. O., Keeffe, E. B., Sibley, R. K., Lee, Y. M., Wright, T. L., and Omary, M. B. (2003a). Keratin 8 and 18 mutations are risk factors for developing liver disease of multiple etiologies. *Proc. Natl. Acad. Sci. USA* **100,** 6063–6068.

Ku, N. O., Gish, R., Wright, T. L., and Omary, M. B. (2001). Keratin 8 mutations in patients with cryptogenic liver disease. *N. Engl. J. Med.* **344,** 1580–1587.

Ku, N. O., Liao, J., and Omary, M. B. (1997). Apoptosis generates stable fragments of human type I keratins. *J. Biol. Chem.* **272,** 33197–33203.

Ku, N. O., Liao, J., and Omary, M. B. (1998a). Phosphorylation of human keratin 18 serine 33 regulates binding to 14-3-3 proteins. *EMBO J.* **17,** 1892–1906.

Ku, N. O., Michie, S., Resurreccion, E. Z., Broome, R. L., and Omary, M. B. (2002b). Keratin binding to 14-3-3 proteins modulates keratin filaments and hepatocyte mitotic progression. *Proc. Natl. Acad. Sci. USA* **99,** 4373–4378.

Ku, N. O., Michie, S. A., Soetikno, R. M., Resurreccion, E. Z., Broome, R. L., and Omary, M. B. (1998b). Mutation of a major keratin phosphorylation site predisposes to hepatotoxic injury in transgenic mice. *J. Cell Biol.* **143,** 2023–2032.

Ku, N. O., Michie, S. A., Soetikno, R. M., Resurreccion, E. Z., Broome, R. L., Oshima, R. G., and Omary, M. B. (1996). Susceptibility to hepatotoxicity in transgenic mice that express a dominant-negative human keratin 18 mutant. *J. Clin. Invest.* **98,** 1034–1046.

Ku, N. O., and Omary, M. B. (1994). Identification of the major physiologic phosphorylation site of human keratin 18: Potential kinases and a role in filament reorganization. *J. Cell Biol.* **127,** 161–171.

Ku, N. O., and Omary, M. B. (1995). Identification and mutational analysis of the glycosylation sites of human keratin 18. *J. Biol. Chem.* **270,** 11820–11827.

Ku, N. O., and Omary, M. B. (1997). Phosphorylation of human keratin 8 *in vivo* at conserved head domain serine 23 and at epidermal growth factor-stimulated tail domain serine 431. *J. Biol. Chem.* **272**, 7556–7564.

Ku, N. O., and Omary, M. B. (2001). Effect of mutation and phosphorylation of type I keratins on their caspase-mediated degradation. *J. Biol. Chem.* **276**, 26792–26798.

Ku, N. O., Soetikno, R. M., and Omary, M. B. (2003b). Keratin mutation in transgenic mice predisposes to Fas but not TNF-induced apoptosis and massive liver injury. *Hepatology* **37**, 1006–1014.

Leers, M. P., Kolgen, W., Bjorklund, V., Bergman, T., Tribbick, G., Persson, B., Bjorklund, P., Ramaekers, F. C., Bjorklund, B., Nap, M., Jornvall, H., and Schutte, B. (1999). Immunocytochemical detection and mapping of a cytokeratin 18 neo-epitope exposed during early apoptosis. *J. Pathol.* **187**, 567–572.

Lerch, M. M., and Adler, G. (1994). Experimental animal models of acute pancreatitis. *Int. J. Pancreatol.* **15**, 159–170.

Liao, J., Ku, N. O., and Omary, M. B. (1996). Two-dimensional gel analysis of glandular keratin intermediate filament phosphorylation. *Electrophoresis* **17**, 1671–1676.

Liao, J., Ku, N. O., and Omary, M. B. (1997). Stress, apoptosis, and mitosis induce phosphorylation of human keratin 8 at Ser-73 in tissues and cultured cells. *J. Biol. Chem.* **272**, 17565–17573.

Liao, J., Lowthert, L. A., Ghori, N., and Omary, M. B. (1995a). The 70-kDa heat shock proteins associate with glandular intermediate filaments in an ATP-dependent manner. *J. Biol. Chem.* **270**, 915–922.

Liao, J., Lowthert, L. A., Ku, N. O., Fernandez, R., and Omary, M. B. (1995b). Dynamics of human keratin 18 phosphorylation: Polarized distribution of phosphorylated keratins in simple epithelial tissues. *J. Cell Biol.* **131**, 1291–1301.

Liao, J., Lowthert, L. A., and Omary, M. B. (1995c). Heat stress or rotavirus infection of human epithelial cells generates a distinct hyperphosphorylated form of keratin 8. *Exp. Cell Res.* **219**, 348–357.

Liao, J., and Omary, M. B. (1996). 14-3-3 proteins associate with phosphorylated simple epithelial keratins during cell cycle progression and act as a solubility cofactor. *J. Cell Biol.* **133**, 345–357.

Ludert, J. E., Feng, N., Yu, J. H., Broome, R. L., Hoshino, Y., and Greenberg, H. B. (1996). Genetic mapping indicates that VP4 is the rotavirus cell attachment protein *in vitro* and *in vivo*. *J. Virol.* **70**, 487–493.

Martin, K., Kirkwood, T. B., and Potten, C. S. (1998). Age changes in stem cells of murine small intestinal crypts. *Exp. Cell Res.* **241**, 316–323.

Moll, R. (1998). Cytokeratins as markers of differentiation in the diagnosis of epithelial tumors. *Subcell. Biochem.* **31**, 205–262.

Omary, M. B., Coulombe, P. A., and McLean, W. H. I. (2004). Intermediate filament proteins and their associated diseases. *N. Engl. J. Med.* in press.

Omary, M. B., Ku, N. O., Liao, J., and Price, D. (1998). Keratin modifications and solubility properties in epithelial cells and *in vitro*. *Subcell. Biochem.* **31**, 105–140.

Omary, M. B., Ku, N. O., and Toivola, D. M. (2002). Keratins: Guardians of the liver. *Hepatology* **35**, 251–257.

Owens, D. W., Wilson, N. J., Hill, A. J., Rugg, E. L., Porter, R. M., Hutcheson, A. M., Quinlan, R. A., Van Heel, D., Parkes, M., Jewell, D. P., Campbell, S. S., Ghosh, S., Satsangi, J., and Lane, E. B. (2004). Human keratin 8 mutations that disturb filament assembly observed in inflammatory bowel disease patients. *J. Cell Sci.* **117**, 1989–1999.

Siegmund, B., Lehr, H. A., and Fantuzzi, G. (2002). Leptin: A pivotal mediator of intestinal inflammation in mice. *Gastroenterology* **122**, 2011–2025.

Smith, F. J., Porter, R. M., Corden, L. D., Lunny, D. P., Lane, E. B., and McLean, W. H. (2002). Cloning of human, murine, and marsupial keratin 7 and a survey of K7 expression in the mouse. *Biochem. Biophys. Res. Commun.* **297**, 818–827.

Steinert, P. M. (1988). The dynamic phosphorylation of the human intermediate filament keratin 1 chain. *J. Biol. Chem.* **263,** 13333–13339.

Tamai, Y., Ishikawa, T., Bosl, M. R., Mori, M., Nozaki, M., Baribault, H., Oshima, R. G., and Taketo, M. M. (2000). Cytokeratins 8 and 19 in the mouse placental development. *J. Cell Biol.* **151,** 563–572.

Tao, G. Z., Toivola, D. M., Zhong, B., Michie, S. A., Resurreccion, E. Z., Tamai, Y., Taketo, M. M., and Omary, M. B. (2003). Keratin-8 null mice have different gallbladder and liver susceptibility to lithogenic diet-induced injury. *J. Cell Sci.* **116,** 4629–4638.

Toivola, D. M., Baribault, H., Magin, T., Michie, S. A., and Omary, M. B. (2000a). Simple epithelial keratins are dispensable for cytoprotection in two pancreatitis models. *Am. J. Physiol. Gastrointest. Liver Physiol.* **279,** G1343–G1354.

Toivola, D. M., Ku, N. O., Ghori, N., Lowe, A. W., Michie, S. A., and Omary, M. B. (2000b). Effects of keratin filament disruption on exocrine pancreas-stimulated secretion and susceptibility to injury. *Exp. Cell Res.* **255,** 156–170.

Toivola, D. M., Omary, M. B., Ku, N. O., Peltola, O., Baribault, H., and Eriksson, J. E. (1998). Protein phosphatase inhibition in normal and keratin 8/18 assembly-incompetent mouse strains supports a functional role of keratin intermediate filaments in preserving hepatocyte integrity. *Hepatology* **28,** 116–128.

van der Geer, P., and Hunter, T. (1994). Phosphopeptide mapping and phosphoamino acid analysis by electrophoresis and chromatography on thin-layer cellulose plates. *Electrophoresis* **15,** 544–554.

Wilson, M. E., Stowell, R. E., Yokoyama, H. O., and Tsuboi, K. K. (1953). Cytological changes in regenerating mouse liver. *Cancer Res.* **13,** 86–92.

Yokoyama, H. O., Wilson, M. E., Tsuboi, K. K., and Stowell, R. E. (1953). Regeneration of mouse liver after partial hepatectomy. *Cancer Res.* **13,** 80–85.

Zhong, B., and Omary, M. B. (2004). Actin overexpression parallels severity of pancreatic injury. *Exp. Cell Res.* **299,** 404–414.

Zhong, B., Zhou, Q., Toivola, D. M., Tao, G. Z., Resurreccion, E. Z., and Omary, M. B. (2004). Organ-specific stress induces mouse pancreatic keratin overexpression in association with NF-kappaB activation. *J. Cell Sci.* **117,** 1709–1719.

Zhou, Q., Toivola, D. M., Feng, N., Greenberg, H. B., Franke, W. W., and Omary, M. B. (2003). Keratin 20 helps maintain intermediate filament organization in intestinal epithelia. *Mol. Biol. Cell* **14,** 2959–2971.

Zhou, X., Liao, J., Meyerdierks, A., Feng, L., Naumovski, L., Bottger, E. C., and Omary, M. B. (2000). Interferon-alpha induces nmi-IFP35 heterodimeric complex formation that is affected by the phosphorylation of IFP35. *J. Biol. Chem.* **275,** 21364–21371.

CHAPTER 18

Muscle Intermediate Filament Proteins

Richard M. Robson, ★ **Ted W. Huiatt,** ★ **and Robert M. Bellin** †

★Muscle Biology Group, Departments of Biochemistry, Biophysics, and
Molecular Biology and of Animal Science
Iowa State University
Ames, Iowa 50011

†Department of Biology
College of the Holy Cross
Worcester, Massachusetts 01610

I. Introduction

The intermediate filaments (IF) in adult striated muscle cells encircle myofibrils at all of their Z-lines and link all adjacent myofibrils within the cell. The transverse IFs also help tie the peripheral layer of myofibrils to the outer cell membrane at sites called costameres (Fig. 1). A small number of longitudinally oriented IFs located along the surface of myofibrils appear to connect adjacent Z-lines in the same myofibril (Tokuyasu *et al.*, 1983a,b). Furthermore, the IFs also link the

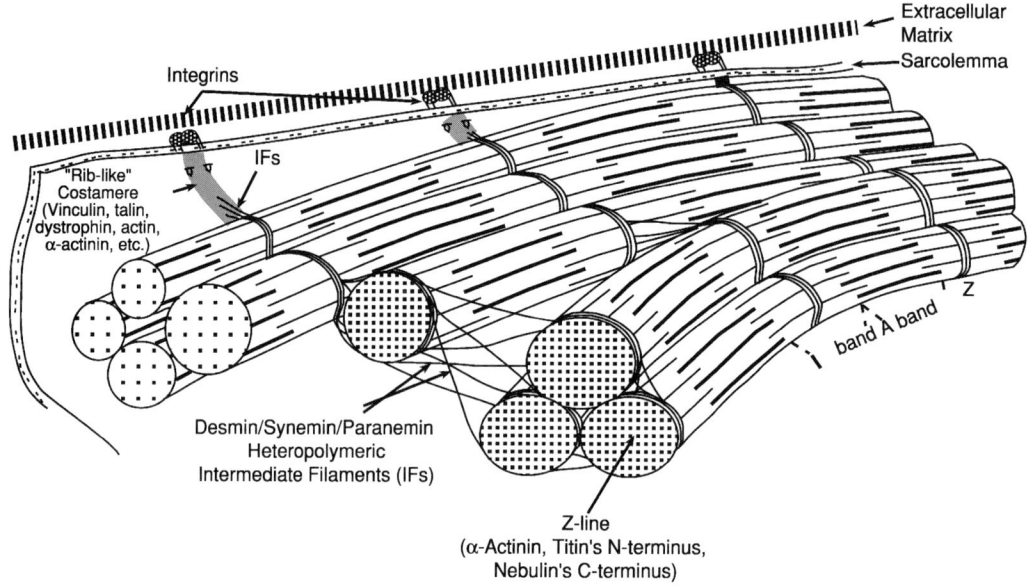

Fig. 1 Schematic depicting the primary location of desmin/synemin/paranemin heteropolymeric intermediate filaments (IFs) at the Z-lines of adult striated muscle cells (Modified from Bellin *et al.*, 2001; with permission of The American Society for Biochemistry and Molecular Biology, Inc.). Also depicted are the attachment points of IFs to structures (myofibrillar Z-lines and costameres) containing proteins that have been found to interact with synemin. Paranemin would be present at most in very minor amounts in skeletal muscle cells, with more present in cardiac muscle cells (see text). For a recent review on costameres, see Ervasti (2003).

myofibrillar Z-lines to nearby mitochondria located between adjacent myofibrils (Reipert *et al.*, 1999; Yang and Makita, 1997), and to nuclei aligned along and near the outer cell membrane (Carlsson *et al.*, 2000). Thus, IFs appear to function as mechanical integrators of cellular space (Lazarides, 1980) and thereby to provide the overall cytoskeletal integrity and strength, as well as the organization, necessary for supporting contraction. Although many studies have been reported, the role(s) of IFs within early developing muscle cells remain(s) less clear.

The five muscle cell IF proteins include desmin, vimentin, synemin, paranemin[1] (avian)/nestin (mammalian ortholog), and syncoilin. The major IF protein in adult skeletal muscle cells is desmin (53 kDa), a type III IF protein. Early stage skeletal muscle myoblasts initially express vimentin (54 kDa) (Sejersen and Lendahl, 1993), also a type III IF protein, with desmin expression initiated in committed myoblasts before fusion into multinucleated myotubes (Bennett *et al.*, 1979;

---

[1]Avian paranemin has also been referred to as IFAPa-400, EAP-300, and transition (see Darenfed *et al.* [2001]; Hemken *et al.* [1997]; Napier *et al.* [1999]).

Capetanaki *et al.*, 1997). The ratio of vimentin to desmin in skeletal muscle tissue decreases during muscle development. For example, in mammalian (porcine) skeletal muscle, the ratio is 3:1 at 45 days gestation, 1:6 at birth, and 1:20 in adult (30 mo) (Bilak *et al.*, 1987). Similar qualitative results were found in avian (chicken) skeletal muscle by Tokuyasu *et al.* (1984) who reported that the distributions of desmin and vimentin were coincidental during myotube development, with the amount of vimentin decreasing with myotube maturity, and essentially undetectable at hatching. The desmin and vimentin observed in early stages of development are present as heteropolymeric IFs containing both proteins (Tokuyasu *et al.*, 1985). The very small amount of vimentin that can be observed in samples of adult skeletal muscle by Western blotting or two-dimensional gel electrophoresis may be accounted for by the presence of non-skeletal muscle cells.

The unique, very large IF proteins, synemin (avian, 183 kDa) and nestin (human, 177 kDa), are variably classified as type VI (Coulombe *et al.*, 2001), type IV (Hesse *et al.*, 2001), or type III IF proteins (Herrmann *et al.*, 2003). Type VI will be used herein. The type VI IF proteins have very short N-terminal head domains (e.g., avian and human synemins, 10 residues; avian paranemin, 16 residues; human nestin, 7 residues) and extremely long C-terminal tail domains. The type VI IF proteins also differ in expression pattern. Both avian synemin (Granger and Lazarides, 1980) and paranemin (Breckler and Lazarides, 1982) were originally identified and referred to as IF-associated proteins (Price and Lazarides, 1983; Robson, 1989). Synemin (Becker *et al.*, 1995; Bellin *et al.*, 1999) and paranemin (Hemken *et al.*, 1997) were subsequently both shown to be IF proteins. Although avian paranemin and mammalian nestin are considered orthologs, sharing ~50% sequence identity in their rod domains, their tail domains share only ~22% identity. Synemin is present in both developing and mature avian and mammalian skeletal muscle cells (Bilak *et al.*, 1998; Price and Lazarides, 1983), but nestin, which colocalizes at the Z-lines in early postnatal mammalian skeletal muscle cells, is downregulated and essentially absent in adult muscle cells (Sejersen and Lendahl, 1993; Sjöberg *et al.*, 1994). All IFs in muscle cells appear to contain at least two of the five muscle IF proteins (e.g., desmin/ synemin heteropolymeric IFs in adult muscle cells). In some developmental stages, three or even four of the IF proteins (with the possible exception of syncoilin) may be present within the same heteropolymeric IFs (Bilak *et al.*, 1998; Carlsson *et al.*, 2002; Hemken *et al.*, 1997). Whereas highly purified desmin (Huiatt *et al.*, 1980) or vimentin (Ip *et al.*, 1985a) will form synthetic IFs under *in vitro* IF-forming conditions, neither synemin (Bilak *et al.*, 1998) nor nestin (Steinert *et al.*, 1999) alone forms IFs *in vitro*. Both expressed paranemin (Hemken *et al.*, 1997) and synemin (Bellin *et al.*, 1999) will form heteropolymeric IFs with the vimentin present in SW13.C1 vim$^+$ cells. Provided the molar ratio of synemin to desmin is kept at or below ~1:25, heteropolymeric IFs, albeit not beautiful, will form *in vitro* (Hirako *et al.*, 2003). Likewise, provided the molar ratio of nestin to vimentin does not exceed ~1:4, heteropolymeric IFs will form *in vitro* (Steinert *et al.*, 1999). Interestingly, expression of the type III IF proteins vimentin, glial fibrillary acidic protein (GFAP), or peripherin in IF-free SW13.C2 vim – cells, but not of the type III IF protein desmin, resulted in the formation of extended IF networks (Schweitzer *et al.*, 2001). The latter studies indicate that desmin filaments clearly differ in organizational properties from other type III IF proteins. Of the type VI IF proteins, synemin, paranemin, and nestin, only coexpression of para-nemin with the desmin resulted in formation of an extended, desmin-containing IF network. The ability of syncoilin to form heteropolymeric IFs remains unclear. This recently discovered, novel type IV IF protein (predicted molecular mass of 53.6 kDa) has a 156-amino acid residue N-terminal head but only a very short 18-residue C-terminal tail (Newey *et al.*, 2001). Both synemin (Mizuno *et al.*, 2001) and syncoilin (Newey *et al.*, 2001; Poon *et al.*, 2002) interact with α-dystrobrevin, a component of the dystrophin-associated protein complex (DAPC) located at the sarcolemma of skeletal and cardiac muscle cells. Syncoilin also interacts with desmin, but it has not yet been shown able to assemble into heteropolymeric IFs (Poon *et al.*, 2002). Syncoilin is concentrated in skeletal muscle cells at the sarcolemma and neuromuscular junction, and slight labeling with syncoilin anti-bodies occurs at the myofibrillar Z-lines. As has been shown for the cytoplasmic type III IF protein peripherin (Landon *et al.*, 1989), the cytoplasmic type VI IF protein synemin is expressed as multiple splice variants. These include a larger (~180 kDa) α-synemin and smaller (~150 kDa) β-synemin, which is miss-ing 312 residues near the end of the long C-terminal tail domain present in human α-synemin (Titeux *et al.*, 2001). β-synemin is the same protein referred to as desmuslin, which was shown to interact with α-dystrobrevin (Mizuno *et al.*, 2001), a component of the DAPC. The expression patterns of mammalian (human) α- and β-synemins differ, with the smaller β-synemin the major isoform expressed in striated muscle cells, and both isoforms expressed in similar amounts in smooth muscle cells (Titeux *et al.*, 2001). Recent studies reveal that the mouse synemin gene encodes three isoforms by means of alternative splicing (Xue *et al.*, 2004). The largest two, synemin H and synemin M, are similar to the human orthologs, α- and β-synemin. The third isoform, synemin L, represents a new form with a much shorter tail. The latter small homolog, L-synemin, has also been found in humans.[2]

Synemin appears to have a very important cytoskeletal cross-linking function. The rod domains of desmin, vimentin, synemin, and paranemin[3] all have the ability to interact with each other (Bellin *et al.*, 1999, 2001; Duval *et al.*, 1995; Hemken *et al.*, 1997). It has been suggested that, in mixed polymers of type III and type VI IF proteins, the type VI proteins occupy positions at the periphery of the IFs (Hermann and Aebi, 2000; Steinert *et al.*, 1999). Thus, the long tail domain of synemin likely extends from the surface of desmin/synemin heteropolymeric IFs in muscle cells. The C-terminal part (~300 amino acids) of the very long tail domain of avian synemin interacts with the α-actinin head and rod domains, vinculin tail

---

[2]D. Paulin, personal communication.
[3]S. A. Lex, T. W. Huiatt, and R. M. Robson, unpublished observations.

domain, and dystrophin[4] (see Bellin *et al.*, 2001, and Fig. 1 herein). An additional, important way in which the IFs are able to bind to myofibrillar Z-lines (Hijikata *et al.*, 1999; Schröder *et al.*, 1999) and to the costameric sites located along the sarcolemma (Schröder *et al.*, 1999), and thereby fulfill their cytoskeletal cross-linking role(s), is the presence of the cytolinker protein plectin (Steinbock and Wiche, 1999; Wiche, 1998), (see Chapter by Gerhard Wiche on plectin herein). Plectin, which binds to both desmin (Reipert *et al.*, 1999) and synemin,[5] also plays a major role in linking the myofibrils to the nearby mitochondria (Reipert *et al.*, 1999). A central portion of the rod domain of desmin has been shown to interact with the Z-line end (C-terminal region) of nebulin (Bang *et al.*, 2002), providing yet another way in which the myofibrillar Z-line/IF linkage may be strengthened. It has been recently reported that linkage of the peripheral layer of myofibrils to costameres may also involve specific cytokeratins (O'Neill *et al.*, 2002). The multiple mechanisms by which the myofibrillar Z-line/IF interaction and the myofibrillar/costamere interaction are fortified add further support for the critical functional importance and role of these key cytoskeletal linkages in striated muscle cells.

Nestin is always expressed with another IF protein (e.g., neurofilaments in developing neurons, vimentin and/or desmin in developing muscle cells) (Sjöberg *et al.*, 1994). Recent studies have also provided a possible, important function for nestin. Sahlgren *et al.* (2001) have shown that nestin is phosphorylated by cdc2 kinase, and the nestin-containing IFs undergo partial disassembly and reorganization during mitosis. In addition, nestin appears to influence the phosphorylation-dependent disassembly of the vimentin-containing IFs during mitosis (Chou *et al.*, 2003).

The importance of desmin in the striated muscle cell cytoskeleton has been demonstrated by several laboratories that used mice without desmin (Balogh *et al.*, 2002; Boriek *et al.*, 2001; Haubold *et al.*, 2003; Li *et al.*, 1996, 1997; Milner *et al.*, 1996; Thornell *et al.*, 1997). The most significant effects, including disorganized sarcomeres, myofibrils, and overall muscle cell architecture, are evident in muscles undergoing repeated cycles of contraction/relaxation, such as the heart and diaphragm (Li *et al.*, 1996; Milner *et al.*, 1996). The location of plectin at the myofibrillar Z-lines and along the sarcolemma in skeletal muscle cells is relatively unaffected in mice without desmin (Carlsson *et al.*, 2000). However, the type VI IF proteins are essentially absent at the Z-line striations/connections between adjacent myofibrils normally present in wild-type mice, but are still present and located at the neuromuscular and myotendinous junctions (Carlsson *et al.*, 2000). Significant abnormalities have been found in mitochondria within desmin-null mice (Milner *et al.*, 2000), indicating the desmin-containing heteropolymeric IFs also have an important role in the mitochondrial location/position and respiratory function in skeletal and cardiac muscle cells.

[4]R. C. Bhosle, D. E. Michele, K. P. Campbell, D. Paulin, Z. Li, and R. M. Robson, unpublished observations.

[5]D. L. Walker, G. Wiche, and R. M. Robson, unpublished observations.

Many desmin IF myopathies, which exhibit structural changes in the myofibrillar Z-line regions, have been described (e.g., Carlsson and Thornell, 2001; Carlsson *et al.*, 2002). On the basis of the expression and cellular localization of synemin in several muscle diseases and myofibrillar myopathies, it has recently been suggested that synemin also participates in these disorders (Olive *et al.*, 2003). Synemin is an extremely calpain-labile substrate (Bilak *et al.*, 1998), even much more so than desmin (O'Shea *et al.*, 1979). This suggests that any muscle disease in which calpain proteolytic activity is elevated will likely exhibit degradation of synemin, and thereby cause significant damage to the overall muscle cell cytoskeleton.

The assembly/disassembly state of muscle IF proteins is generally considered to be regulated by posttranslational covalent modifications, especially the phosphorylation/dephosphorylation state of desmin (Inagaki *et al.*, 1988, 1996; also see Robson, 1989, and the chapter herein by Masako Inagaki). An additional mechanism for controlling the assembly state of desmin is a rather unique modification involving reversible mono-ABP-ribosylation ribosylation of arginyl residues within the desmin N-terminal head domain (Yuan *et al.*, 1999; Zhou *et al.*, 1996).

II. Materials

A. Commercially Available Muscle Intermediate Filament Proteins

Of the proteins discussed in this chapter, only vimentin is currently available as a purified sample from a commercial source. Cytoskeleton Inc. sells purified recombinant hamster vimentin at a concentration of 5 mg/ml (Cat. No. V01). Cytoskeleton Inc. also supplies purified recombinant vimentin as part of a "Biochem Kit" that includes the buffer reagents necessary for inducing assembly of the vimentin into IFs.

B. Antibodies for the Study of Muscle Intermediate Filament Proteins

A large number of suppliers have antibodies available that recognize vimentin from a range of tissue sources. The Developmental Studies Hybridoma Bank (University of Iowa) has available the AMF-17b mouse monoclonal antibody produced against chicken vimentin. This antibody has also been found to recognize all mammalian forms of vimentin except those from rodents. We have successfully used this antibody for both tissue culture staining and Western blot analysis of human vimentin. Santa Cruz Biotechnology Inc. has available several anti-vimentin antibodies. Catalog numbers SC-7557, SC-5565 and SC-7558 are all polyclonal antibodies that recognize mouse and human vimentin. Santa Cruz also sells three different phosphovimentin-specific antibodies that recognize vimentin that has been phosphorylated at Ser39, Ser72, or Ser83. Sigma Immunochemicals produces three anti-vimentin monoclonal antibodies made in goat (Cat. No. V4630) or mouse (Cat. No. V6630 and Cat. No. V5255). In addition to those specifically listed, there are many other commercial sources of vimentin-specific antibodies, including Abcam, ABR-Affinity Biosciences, BD Biosciences

Pharmingen, Biodesign International, Chemicon, Novus Biologicals, Serotec, Stressgen, U. S. Biological, and Zymed.

Similar to the list supplied for vimentin, a large number of suppliers have antibodies available that recognize desmin. The Developmental Studies Hybridoma Bank has available the D-3 and D-76 mouse monoclonal antibodies produced against chicken desmin. In agreement with the information listed by the Hybridoma Bank for these two antibodies, we have successfully used the D-3 antibody for tissue staining and the D-76 antibody for Western blot analysis of avian desmin. Santa Cruz Biotechnology Inc. has available several anti-desmin antibodies. Catalog numbers SC-7559, SC-7556, and SC-14026 are all polyclonal antibodies that recognize mouse and human desmin. Santa Cruz also sells a mouse monoclonal against desmin (Cat. No. SC-23879) that is reported to recognize desmin from all species tested. Sigma Immunochemicals produces an anti-desmin monoclonal antibody made in mouse (Cat. No. D1033) and a rabbit polyclonal antibody (Cat. No. D8281). In addition to those specifically listed, there are many other commercial sources of desmin-specific antibodies (including Abcam, BD Biosciences Pharmingen, Beckman Coulter, Chemicon, Novus Biologicals, Serotec, U. S. Biological, and Zymed).

For the orthologous proteins paranemin and nestin, antibodies are available from several sources for these proteins. The Developmental Studies Hybridoma Bank has available a mouse monoclonal antibody named Rat-401 produced against rat nestin. The Hybridoma Bank also stocks two mouse monoclonals (EAP3 and A2B11) against chicken transitin, a protein that has been shown to be highly homologous, if not identical, to paranemin. Santa Cruz Biotechnology Inc. has available a rabbit polyclonal antibody against nestin (Cat. No. SC-20978). In addition to those specifically listed, there are many other commercial sources of nestin-specific antibodies (including Abcam, BD Biosciences Pharmingen, Chemicon, Novus Biologicals, and U. S. Biological). No other commercial antibodies are available for paranemin at this time.

Although antibodies for both syncoilin and synemin have been described in the literature, no antibodies against either of these proteins are currently commercially available.

III. Procedures

A. Purification of Intermediate Filament Proteins from Muscle

1. Desmin

Two major sources of muscle have been used in our laboratory for preparation of desmin, namely avian (turkey) smooth muscle (gizzard) and mammalian skeletal muscle (porcine semitendinosis and biceps femoris). Either fresh or quick-frozen stored muscle can be used. Procedures we first developed for preparation of avian smooth muscle desmin incorporated an initial step modified from

Sobieszek and Bremel (1975) in which we first prepared "washed" contractile elements. Turkey gizzards are an excellent source of desmin, which makes up ~1% of total muscle protein (Huiatt *et al.*, 1980). All muscle types (skeletal, cardiac, and smooth) within a given species express the same desmin molecule. Thus, if species is not important in a given study, and one wishes to conduct experiments with tissue-purified desmin, turkey gizzards are strongly preferred as starting material.

We subsequently adapted the procedure described in Huiatt *et al.* (1980) for purification of desmin from mammalian (porcine) skeletal muscle (see flowchart in Fig. 2). In this procedure, we incorporated an important step described in Small and Sobieszek (1977), which involves extraction/solubilization of desmin in acetic acid from a crude desmin-containing fraction. The major desmin-containing fractions resulting from this procedure are shown in Fig. 3. The final yield of purified desmin obtained from "washed myofibrils" prepared initially from 600 g of ground porcine skeletal muscle is ~24 mg, which represented ~0.05% of the myofibrillar/cytoskeletal protein fraction (O'Shea *et al.*, 1981).

Skeletal muscle samples (~1 kg) are often available shortly after death from a nearby hog slaughterhouse, provided permission is obtained. The muscle must be immersed in ice and used as soon as possible to reduce postmortem proteolysis of the desmin by the calpains. Alternatively, the muscle sample can be cut into ~2 to 3-cm cubes, packed in dry ice for transport, and then stored at or below $-20\,°C$ for several months or more.

The individual purification steps are described in further detail for porcine skeletal muscle desmin in O'Shea *et al.* (1981) The major steps in that procedure follow. All steps are done at 0–4 °C.

1. The porcine skeletal muscle (semitendinosis and biceps femoris) is coarsely ground in a precooled meat grinder.

2. 600 g of the ground muscle is homogenized in 6 vol of standard salt solution (100 mM KCl, 1 mM NaN_3, 2 mM EGTA, 2 mM $MgCl_2$, 20 mM potassium phosphate, pH 6.8) in a Waring blender and centrifuged at 2000*g* for 10 minutes.

3. The sediment is resuspended in 5 vol of standard salt solution also containing 1% Triton X-100 by homogenization in a Waring blender and centrifuged at 2000*g* for 10 minutes.

4. The sediment is resuspended in 5 vol 50 mM Tris-HCl, pH 7.6, 5 mM ethylenediamine tetraacetic acid (EDTA), 1% Triton X-100 by homogenization in a Waring blender and centrifuged at 2000*g* for 10 minutes.

5. The sediment is resuspended in 5 vol 150 mM KCl, 5 mM EDTA, 1% Triton X-100 by homogenization in a Waring blender and centrifuged at 2000*g*.

6. The sediment (referred to as washed myofibrils) is resuspended in 5 vol 500 mM NaCl, 5 mM EDTA, 5 mM adenosine triphosphate (ATP), 1 mM cysteine, 40 mM imidazole-HCl, pH 7.1, by homogenization in a Waring blender, stirred on a magnetic stirrer for 30 minutes, and centrifuged at 13,800*g* for 30 minutes.

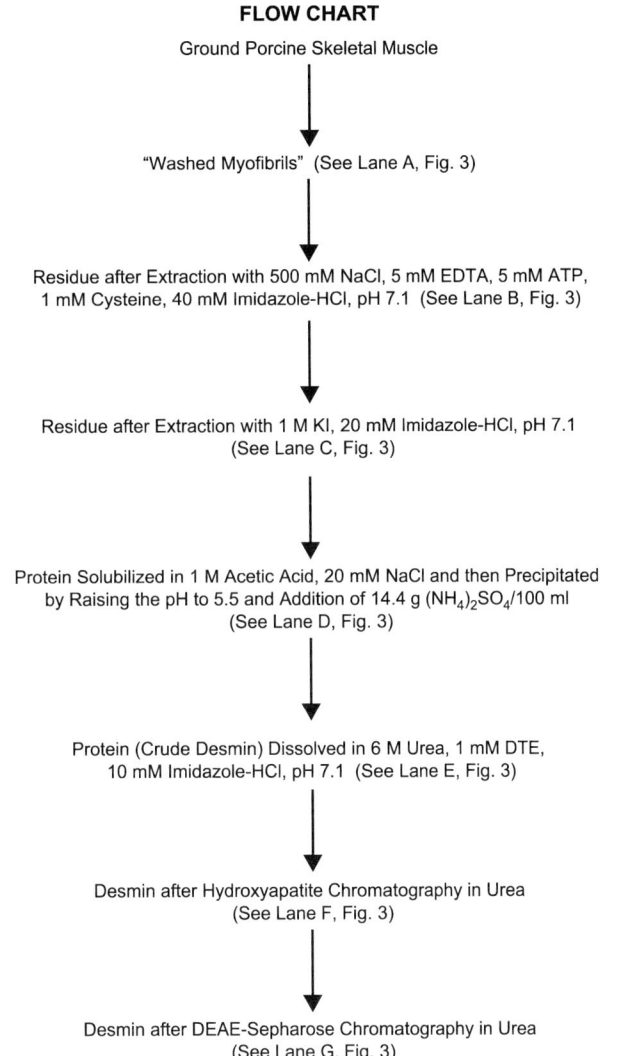

FLOW CHART

Ground Porcine Skeletal Muscle

"Washed Myofibrils" (See Lane A, Fig. 3)

Residue after Extraction with 500 mM NaCl, 5 mM EDTA, 5 mM ATP,
1 mM Cysteine, 40 mM Imidazole-HCl, pH 7.1 (See Lane B, Fig. 3)

Residue after Extraction with 1 M KI, 20 mM Imidazole-HCl, pH 7.1
(See Lane C, Fig. 3)

Protein Solubilized in 1 M Acetic Acid, 20 mM NaCl and then Precipitated
by Raising the pH to 5.5 and Addition of 14.4 g $(NH_4)_2SO_4$/100 ml
(See Lane D, Fig. 3)

Protein (Crude Desmin) Dissolved in 6 M Urea, 1 mM DTE,
10 mM Imidazole-HCl, pH 7.1 (See Lane E, Fig. 3)

Desmin after Hydroxyapatite Chromatography in Urea
(See Lane F, Fig. 3)

Desmin after DEAE-Sepharose Chromatography in Urea
(See Lane G, Fig. 3)

Fig. 2 Flowchart showing summary of major steps in purification of mammalian skeletal muscle desmin (From Robson *et al.*, 1984; with permission of Food and Nutrition Press, Inc., Trumbull, CT 06611). See O'Shea *et al.* (1981) for a detailed description of the procedure.

7. The previous step is repeated, but the stirring step is increased to 4 hours. The resulting sediment can be stored overnight at $-25\,°C$ or used immediately for the following step.

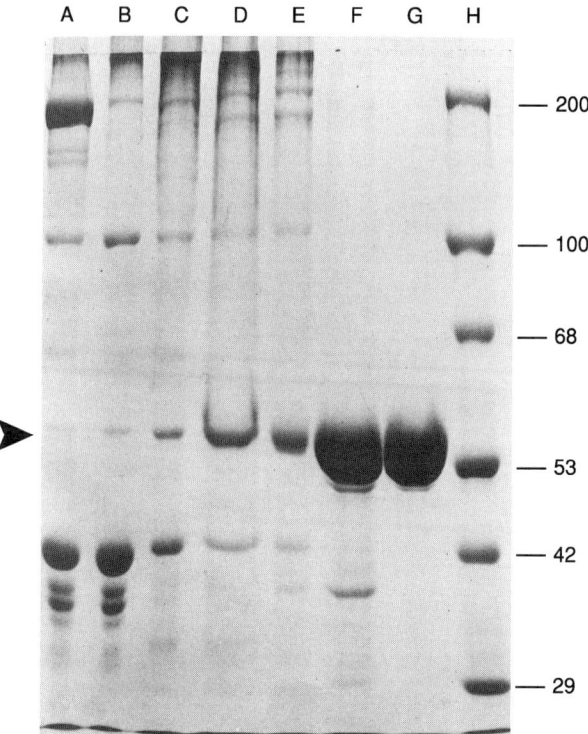

Fig. 3 Sodium dodecyl sulfate–polyacrylamide gel slab showing major fractions obtained during purification of porcine skeletal muscle desmin (From Robson *et al.*, 1984; with permission of Food and Nutrition Press, Inc., Trumbull, CT 06611). See flowchart in Fig. 2. Lane A, Washed myofibrils; lane B, residue remaining after extraction with 500 mM NaCl-containing solution; lane C, residue remaining after extraction with 1 M KI-containing solution; lane D, protein solubilized in 1 M acetic acid, 20 mM NaCl and then precipitated by adjustment of pH to 5.5 and addition of ammonium sulfate; lane E, protein (crude desmin) dissolved in 6 M urea-containing solution; lane F, hydroxyapatite-purified desmin; lane G, DEAE-Sepharose–purified desmin; lane H, polypeptide molecular weight standards. The arrowhead on left of figure marks the position of the skeletal muscle desmin. The electrophoretic system used (Laemmli 1970) and details concerning preparation of samples for electrophoresis are described in O'Shea *et al.* (1981). The amount of protein loaded was 20 μg in A, B, C, F, and G and 15 μg in D and E.

8. The sediment is resuspended in 5 vol 1 M KI, 20 mM imidazole-HCl, pH 7.1, by homogenization in a Waring blender, stirred on a magnetic stirrer for 2 hours, and centrifuged at 13,800g for 30 minutes.

9. The sediment is again resuspended in 5 vol 1 M KI, 20 mM imidazole-HCl, pH 7.1, by homogenization in a Waring blender, stirred on a magnetic stirrer for 2 hours, and centrifuged at 13,800g for 30 minutes.

10. The sediment is resuspended in 5 vol deionized H_2O by homogenization in a Waring blender and centrifuged at 13,800g for 15 minutes.

11. Step "10" is repeated.

12. The sediment is resuspended in 3.3 vol of 1 M acetic acid, 20 mM NaCl by homogenization in a Waring blender, stirred gently for ~10 hours (or overnight) at 4 °C, and then centrifuged at 13,800g for 1 hour.

13. The sediment is discarded, and the supernatant is filtered through Whatman 541 paper, then adjusted to pH 5.5 by adding 10 M NaOH. Ammonium sulfate is slowly added (14.4 g/100 ml of supernatant), the suspension is left standing for 30 minutes, and then centrifuged at 13,800g for 15 minutes.

14. The sediment is dissolved in ~50 ml of 6 M urea, 1 mM dithioerythritol (DTE), 10 mM imidazole-HCl, pH 7.1, by homogenization in a Dounce tissue grinder, dialyzed ~10 hours (overnight) against another batch of the same solution, but at pH 7.2, and centrifuged at 95,500g for 2 hours. The supernatant (crude desmin; see Fig. 3, lane E) is saved, and the sediment is discarded.

15. The crude desmin in 6 M urea, 1 mM DTE, 10 mM imidazole-HCl, pH 7.2, is subjected to hydroxyapatite column chromatography in the same solution and eluted with a linear 0–60 mM sodium phosphate gradient.

16. Fractions are analyzed by sodium dodecylsulfate-polyacrylamide gel electrophoresis (SDS-PAGE).

17. The fractions enriched in desmin are collected (partially purified desmin; see Fig. 3, lane F), dialyzed against 6 M urea, 1 mM DTE, 10 mM imidazole-HCl, pH 7.2, and subjected to DEAE-Sepharose-FF column chromatography in the same solution, eluted with a linear 0–250 mM NaCl gradient.

18. The fractions are subjected to SDS-PAGE, and those fractions containing desmin, without contaminants, are pooled as appropriate. The purified desmin contains about 1–2% of a 50-kDa desmin degradation product (see Fig. 3, lane G). To ensure desmin is correctly identified, Western blotting should be done with desmin antibodies.

19. The DEAE-Sepharose-FF-purified desmin is dialyzed extensively against 10 mM Tris-acetate, pH 8.5, 1 mM DTE. The sample is then clarified by centrifugation at 134,000 g for 1 hour. The purified soluble desmin is then aliquoted into vials and can be stored for more than 2 weeks under nitrogen and toluene vapor at 0–4 °C, or for years in liquid nitrogen.

2. Vimentin

Mammalian vimentin can be purified in reasonable yield from porcine aorta. Smooth muscle contains large amounts of IFs, but unlike most smooth muscle, the aorta close to the heart expresses primarily vimentin rather than desmin (Frank and Warren, 1981). A yield of 60–70 mg of highly purified vimentin is obtained from 400 g of ground aorta.[6] The purification procedure is similar to that for preparation of desmin from avian smooth muscle (turkey gizzard) and involves the following major steps.

1. A washed myofibril fraction is prepared by homogenization of the ground aorta in 10 vol of 100 mM KCl, 2 mM $MgCl_2$, 2 mM ethyleneglycol-bis-(β-aminoethylether) N,N,N^1,N^1-tetraacetic acid (EGTA), 1 mM NaN_3, 0.01% (w/v) penicillin, 0.01% streptomycin, 20 mM K-phosphate, pH 6.8, with a Polytron homogenizer (Brinkmann). Triton X-100 is added to a concentration of 1% (v/v), and the suspension is centrifuged at 2000*g* for 10 minutes. The myofibril sediment is washed a total of four additional times by homogenizing the sediment in the same buffer followed by centrifugation. Triton X-100 (1%) is included in all but the final wash.

2. Actomyosin is removed by high-ionic strength extraction. The myofibril pellet is suspended by homogenization with the Polytron in 10 vol of 500 mM NaCl, 5 mM EDTA, 5 mM ATP, 1 mM cysteine, 40 mM imidazole-HCl, pH 7.0. The suspension is stirred for 1 hour at 4 °C, and the insoluble IF fraction is collected by centrifugation at 14,000*g* for 20 minutes. The sediment is resuspended and extracted a second time with stirring for 2 hours and centrifuged to sediment the IF fraction.

3. The sediment remaining after extraction of actomyosin is then extracted with urea to solubilize the vimentin by homogenization in 5 vol of 8 M urea, 0.1% 2-mercaptoethanol (MCE), 10 mM imidazole-HCl, pH 7.0, stirring for 1 hour at 4 °C, and centrifugation at 14,000*g* for 30 minutes. The supernatant is saved, the sediment is extracted again with the same solution, and the two extracts are combined.

4. The combined extract is subjected to batch chromatography on diethyl-aminoethyl (DEAE)-cellulose (Whatman DE-52). The combined extract is mixed with 300 ml (packed vol) of DE-52 previously equilibrated with buffer A (6 M urea, 0.1% MCE, 20 mM imidazole, pH 7.0) and stirred for 1 hour. The DEAE cellulose is collected by centrifugation (14,000*g* for 15 minutes) and washed twice with buffer A. Bound protein is eluted with two 2-L washes of buffer A + 200 mM NaCl, and the combined eluates are concentrated by ultrafiltration, yielding a crude, DEAE-cellulose purified vimentin fraction.

5. The DEAE-cellulose–purified vimentin is dialyzed against buffer A, clarified at 143,000*g* for 1 hour, and then subjected to chromatography on hydroxyapatite (HA-Ultragel; BioSepra) eluted with a 12–50 mM Na-phosphate gradient in buffer A.

6. The hydroxyapatite-purified vimentin is dialyzed overnight against buffer A containing 50 mM NaCl, clarified, and chromatographed on DEAE-Sepharose FF, eluted with a linear gradient of 50–250 mM NaCl.

---

[6]M. K. Hartzer and R. M. Robson, unpublished observations.

7. The final, purified vimentin is dialyzed against 1 mM DTE, 10 mM Tris-acetate, pH 8.5, followed by dialysis against 10 mM Tris-acetate, pH 8.5, to remove urea.

3. Synemin

An excellent source of muscle for preparation of purified synemin is fresh (or harvested and quick frozen) turkey gizzard smooth muscle. The procedure is based on those described in Sandoval *et al.* (1983) and Bilak *et al.* (1998). A yield of 9–15 mg of purified synemin was obtained from nine preparations starting with 500 g of trimmed turkey gizzard by Bilak *et al.* (1998). The major steps in the procedure are:

1. The muscle is minced/ground and homogenized in 10 mM Tris, 140 mM KCl, 2 mM EGTA, 0.1% MCE, 0.2 mM phenylmethylsulfonyl fluoride (PMSF), pH 7.5 (buffer A), and centrifuged at 13,000*g* for 20 minutes. The sediment is then subjected to a series of sequential extractions with:

   a. Buffer A (two times)

   b. 10 mM Tris, 10 mM EGTA, 0.1% MCE, 0.2 mM PMSF, pH 7.5 (low ionic strength buffer)

   c. 0.6 M KI, 10 mM $Na_2S_2O_3$, 0.1% MCE, 0.2 mM PMSF, pH 7.5 (high ionic strength buffer)

   d. Three washes with cold H_2O containing 0.1% MCE

   e. Two washes with cold acetone and then air dried to yield an acetone powder

2. The synemin is extracted from the acetone powder ($\sim$12 g) by homogenization in 15 vol (w/v) of 6 M urea, 2 mM dithiothreitol (DTT), 5 mM Na *p*-tosyl-L-arginine methyl ester (TAME), 0.2 mM PMSF, 10 mM Tris-HCl, pH 7.5 (buffer B), stirred for 12 hours at 0–4 °C, and centrifuged at 14,000*g* for 30 minutes. The supernatant is saved. The sediment is resuspended in 10 vol of buffer B, stirred for 8 hours and centrifuged again at 14,000*g* for 30 minutes. The supernatant is combined with the previous one. The combined urea extract in buffer B is clarified at 183,000*g* for 1 hour.

3. The supernatant (960 mg) (see Fig. 4, lane U; from Bilak *et al.*, 1998) is loaded on to a 2.5 × 41 cm column of hydroxyapatite (HA-Ultrogel; BioSepra) that has been equilibrated in buffer B.

4. The column is eluted sequentially with

   a. 200 ml of buffer B

   b. 400 ml of a 10–35 mM linear phosphate gradient in 6 M urea, 2 mM DTT, 5 mM TAME, pH 7.5 (solution C)

   c. 200 ml of solution C containing 35 mM potassium phosphate

   d. 400 ml of a 35–200 mM linear phosphate gradient in solution C

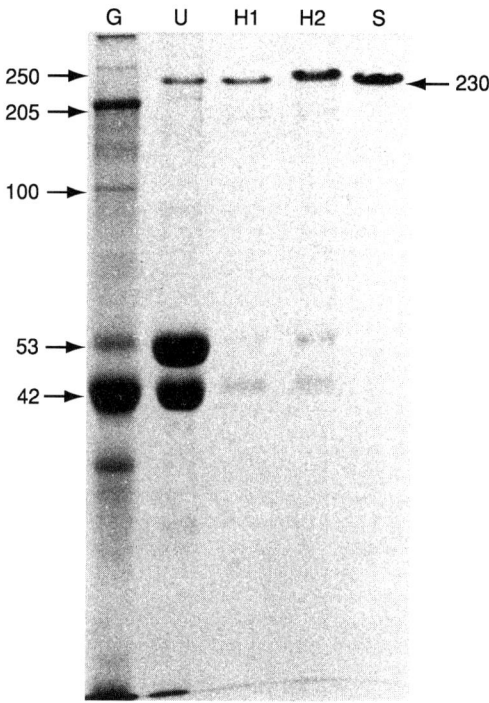

Fig. 4 Sodium dodecylsulfate–polyacrylamide gel electrophoresis analysis of major fractions obtained during purification of avian synemin (Reprinted from Bilak *et al.*, 1998; with permission from Elsevier). Lane G, Turkey gizzard standard with approximate migration distances of filamin (250 kDa), myosin heavy chain (205 kDa), α-actinin (100 kDa), desmin (53 kDa), and actin (43 kDa) marked at the left side, migration position of synemin (230 kDa) is marked at the right side; lane U, urea extract loaded onto the first hydroxyapatite column; lane H1, pooled protein sample collected from the first hydroxyapatite column; lane H2, pooled protein sample collected from the second hydroxyapatite column; lane S, DEAE–Sephacel purified synemin.

5. The synemin-containing fractions, determined by SDS-PAGE, eluting between 115 and 140 mM phosphate, are combined (see Fig. 4, lane H1), dialyzed against buffer B containing 1 mM TAME, subjected to a second hydroxyapatite (Bio-Rad) column chromatography step, and eluted as described for the previous column chromatographic step.

6. The fractions containing synemin elute between 120 and 140 mM phosphate and are pooled (see Fig. 4, lane H2).

7. The hydroxyapatite-purified synemin (∼25 mg) is dialyzed against buffer I (6 M urea, 2 mM EGTA, 2 mM DTT, 1 mM TAME, 20 mM Tris-HCl, pH 7.5) (Sandoval *et al.*, 1983) and subjected to chromatography on a 1.6 × 40 cm

DEAE-Sephacel (Sigma) column equilibrated in buffer I. The column is eluted with a 0–200 mM linear NaCl gradient in buffer I. The synemin-containing fractions eluting between 135 and 160 mM NaCl are pooled and dialyzed against 6 M urea, 0.1% (v/v) MCE, 10 mM Tris-HCl (see Fig. 4, lane S). Aliquots of the purified synemin can be stored at −70 °C or lower for at least a year.

A procedure for purification of synemin from mammalian (rabbit) stomach smooth muscle has recently been published (Hirako *et al.*, 2003). The procedures are generally similar to those used for avian synemin, except the "acetone powder" step is not used. The yield of purified synemin was approximately 1.4 mg from 3.5 g of the rabbit stomach smooth muscle layer. Two bands were identified by SDS-PAGE at 200 and 180 kDa, corresponding to the α-synemin and β-synemin splice variants described in human smooth muscle by Titeux *et al.* (2001). Bands migrating at ∼115 and 140 kDa below the purified synemin doublet were not identified.

4. Paranemin/Nestin

The major source of muscle used for preparation of paranemin is embryonic chick skeletal muscle. An average yield (five preparations) of 1.7 mg of purified paranemin was obtained from 100 g of embryonic skeletal muscle (Hemken *et al.*, 1997). The procedures are described in Hemken *et al.* (1997) in which the initial steps were based on those in Breckler and Lazarides (1982).

1. Breast and thigh muscles are dissected from 12 dozen 14-day-old chick embryos.

2. The muscle (∼100 g wet weight) is homogenized in 235 ml of 130 mM EDTA, 20 mM Tris-HCl, pH 7.5 (see Fig. 5, lane 2; paranemin is a minor band migrating at ∼280 kDa) and is centrifuged at 20,700g for 30 minutes.

3. The supernatant (2–4 °C) is filtered through glass wool and centrifuged for 90 minutes at 125,000g. The crude high-speed supernatant (see Fig. 5, lane 3, containing more of the 280-kDa paranemin) is subjected to chromatography on a 2.6-cm × 110-cm Bio-Gel A-5m (Bio-Rad) column previously equilibrated with 100 mM NaCl, 0.1 mM EDTA, 20 mM Tris-HCl, pH 7.5. The column is eluted with equilibration buffer at 24 ml/h. Paranemin-enriched fractions elute near the column void volume and are collected, pooled, and dialyzed against 6 M urea, 1 mM DTT, 0.2 mM PMSF, 10 mM Tris-HCl, pH 7.5.

4. The dialysate is chromatographed on a 2.6 cm × 33 cm column of HA-Ultrogel (Sepracor) previously equilibrated in the dialysis buffer. The column is eluted with the dialysis buffer containing a linear 0–500 mM potassium phosphate gradient, pH 7.5, at a flow rate of 24 ml/h.

5. The fractions containing paranemin are eluted at ∼200 mM phosphate and are pooled, dialyzed vs. 6 M urea, 1 mM DTT, 5 mM EDTA, 0.2 mM PMSF, 20 mM Tris-HCl, pH 7.5, and then chromatographed on a 2.6 cm × 20 cm DEAE-cellulose (Whatman, DE-52) column previously equilibrated with dialysis buffer.

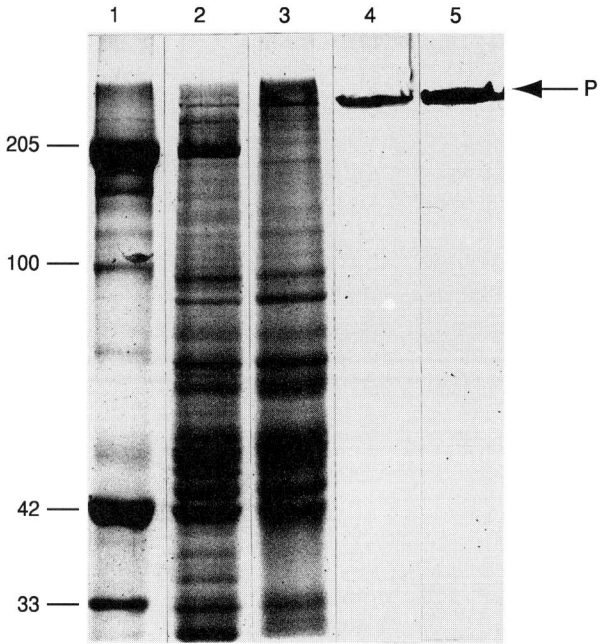

Fig. 5 Purification of paranemin from 14-day-old embryonic (E14) chick skeletal muscle. Adult chicken cardiac myofibril protein markers (lane 1) indicated in kilodaltons are myosin heavy chain (205), α-actinin (100), actin (42), and tropomyosin (33). Lane 2, whole muscle homogenate from E14 chick skeletal muscle; lane 3, crude high-speed supernatant before loading onto gel filtration column; lane 4, purified paranemin from DEAE–cellulose column; lane 5, Western blot of purified paranemin using monoclonal antibody 4D3. P, paranemin.

SDS-PAGE of purified paranemin is shown in lane 4, and a Western blot, which uses monoclonal antibody 4D3, of a duplicate sample of purified paranemin is shown in lane 5.

Nestin, the mammalian ortholog of paranemin, has been purified from baby hamster kidney (BHK-21) cells (Steinert *et al.*, 1999). The IFs in BHK-21 cells contain primarily vimentin and desmin but also high molecular weight proteins that co-cycle with vimentin and desmin during *in vitro* assembly/disassembly (see refs. in Steinert *et al.*, 1999). Starting with BHK-21 cells grown in 850-cm^2 roller bottles, Steinert *et al.* (1999) succeeded in preparing ∼1 to 2 μg of highly purified nestin per bottle.

5. Syncoilin

Tissue purification of syncoilin has not been reported.

B. Expression of Muscle Intermediate Filament Proteins and Domains

Muscle cell IF proteins and protein subdomains have been successfully expressed in bacteria by the use of standard molecular biological techniques. Once purified from bacterial lysates, the expressed proteins and domains have been used in a variety of biochemical assays ranging from IF assembly studies to *in vitro* protein interaction assays (Bellin *et al.*, 1999, 2001), as described in the following sections.

Essentially any standard protein expression vector can be used for the bacterial expression of muscle cell IF proteins; however, there are certain advantages and disadvantages to various systems that should be noted. The use of 6× His tag systems, such as the pProEX HT vectors from Life Technologies, is quite convenient when expressing IF proteins or subdomains that will be used in protein interaction studies. This is because the small six-amino-acid purification tag used in these systems seems to have no adverse effect on these types of assays. However, it should be noted that this small fusion tag does not aid in increasing the solubility of the attached IF protein or protein subdomain, often resulting in highly insoluble expressed protein. As a result, proteins produced this way must often be solubilized in a denaturing solution, such as 6 M urea, before purification on a chelating column, such as Ni-NTA, which can be used in the presence of high urea concentrations. After purification, the denaturant can often be successfully removed while retaining solubility of the expressed IF protein by dialysis of the protein sample into a low ionic strength solution buffered at a pH of ∼8.5.

One possible way to avoid the solubility problems during expression is by the use of a larger fusion protein, which often acts to increase the solubility of the attached IF protein. Possibilities for expressing these types of fusions include the pGEX family of vectors (Amersham Pharmacia Biotech), which cause expression of the protein of interest as a fusion with glutathione-S-transferase (GST), or the pMAL family of vectors (New England Biolabs), which cause expression as a fusion with maltose-binding protein (MBP). GST-fusion and MBP-fusion proteins are often quite soluble and can be purified from bacterial lysates often in one step by the use of a conjugated glutathione column for GST fusions or a conjugated amylose column for MBP fusions. It should be noted that these relatively large purification tags often do interfere with downstream uses of the expressed proteins. This is especially important to consider in the case of GST-fusion for use in protein interaction studies, because GST has been noted to interact with itself (Niedziela-Majka *et al.*, 1998). As a result, expression of IF proteins as fusions with GST or MBP normally requires proteolytic removal of the purification tag before use in assembly or interaction assays. Most commercially available vectors for expression of these types of fusions also incorporate a convenient proteolysis site between the purification tag and the protein of interest to make tag removal fairly easy.

C. Solubility and Assembly of Muscle Intermediate Filament Proteins

One of the major properties of muscle IF proteins is their ability to assemble into IFs *in vitro*, either alone, for desmin and vimentin, or in combination with other IF proteins. Furthermore, because studies that use these proteins require knowledge of the properties of the purified proteins to be able to choose reasonable conditions for biochemical experiments, this brief section on the solubility and assembly characteristics of the major muscle IF proteins is included. Additional information on the methods used to examine assembly of IF proteins can be found in the chapter by Herrmann in this volume.

1. Methods to Measure Muscle Intermediate Filament Protein Assembly

Assembly of IF proteins can be easily quantitated by sedimentation. To determine the percentage of total protein that is assembled into IFs under a specific set of conditions, the protein is dialyzed or diluted into the appropriate assembly buffer, and the solution is centrifuged at high speed (we routinely use 183,000g for 1 hour) to sediment the insoluble IFs. The supernatant is carefully removed, and the protein concentration of the supernatant is measured either by the Folin-Lowry (Lowry *et al.*, 1951) or BCA (Pierce) methods. The percentage of protein sedimented can then be calculated, or the insoluble sediment can be dissolved in SDS buffer to measure the amount of protein in the sediment fraction. One can also prepare samples of the supernatants and sediments for analysis by SDS-PAGE, followed by quantitation of the gels by densitometry. Analysis by SDS-PAGE is also useful for examination of the coassembly of two IF proteins, or for analysis of interactions, as described in Section III.E.3 of this chapter. Centrifugation at lower g forces (e.g., 17,000g for 30 minutes) can provide information about the assembly process by sedimenting only large filaments and leaving short or partially assembled filaments in the supernatant. Data from typical experiments using high-speed and low-speed centrifugation to analyze the solubility and assembly, respectively, of desmin and vimentin are shown in Fig. 6.

Qualitative measurement of IF assembly, as well as information on the structure of the assembled filaments, can be obtained by examination in the transmission electron microscope (TEM) of samples that have been negatively stained with uranyl acetate. For this procedure, a drop of the protein sample, at a concentration of ~0.1 mg/ml, is placed on a 400-mesh copper grid covered with a carbon or carbon-coated formvar support film and allowed to sit undisturbed for 30 seconds to 1 minute. The grid is then washed with 3–5 drops of distilled water, and a drop of 2% (w/v) aqueous uranyl acetate is added. Excess stain is removed by touching the grid with the torn edge of a piece of filter paper, and the grid is allowed to air dry before examination in the TEM. Significantly better spreading of the sample and stain is achieved if the grids have been glow discharged for 1–2 minutes in a vacuum evaporator before adding the protein.

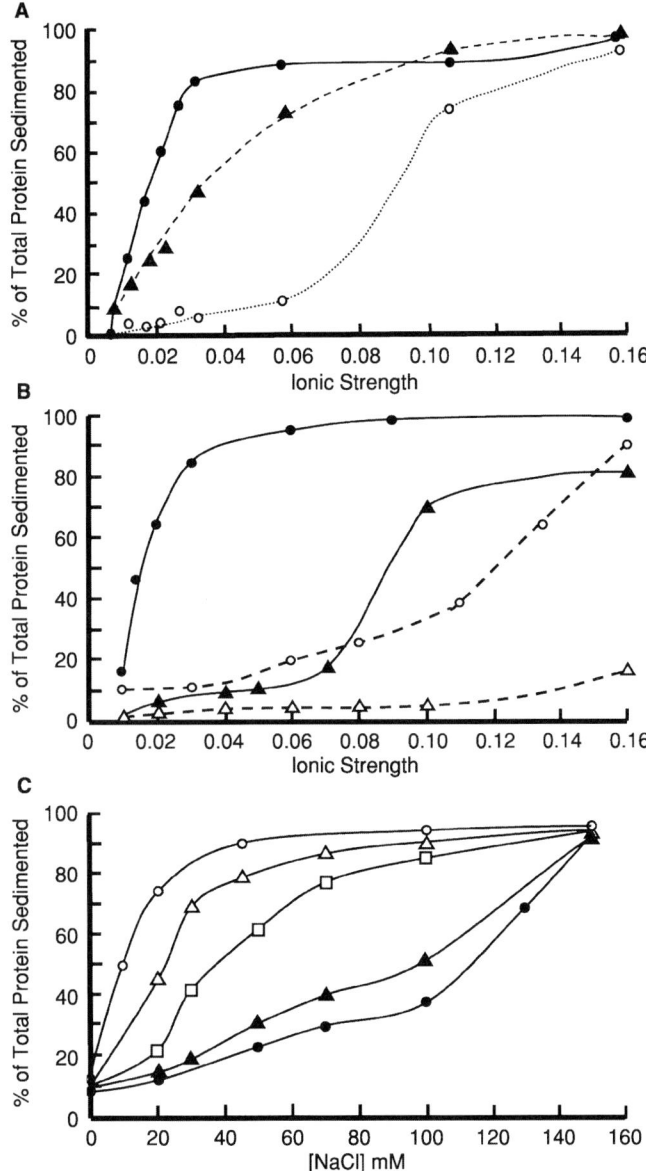

Fig. 6 Effect of ionic strength on the solubility and assembly of tissue-purified desmin and vimentin. Samples of avian desmin (1 mg/ml) (A), porcine desmin, or vimentin (0.6 mg/ml) (B) or mixtures of porcine desmin and vimentin (total protein concentration of 0.6 mg/ml) (C) in 10 mM Tris-acetate, pH 8.5, were dialyzed overnight at 4°C against a series of solutions of increasing NaCl concentration, buffered with either 10 mM imidazole-HCl, pH 7.0, or 10 mM Tris-HCl, pH 8.5. After dialysis, samples were centrifuged at either high speed (183,000*g* for 1 hour) to measure overall solubility or low speed (17,000*g* for 30 minutes) to measure assembly. The supernatants were carefully removed, the

2. Desmin and Vimentin

Desmin and vimentin are both type III IF proteins that have similar solubility and assembly properties. We have purified avian (turkey) and mammalian (porcine) desmin and porcine vimentin in our laboratory, as described earlier in Sections III.A and III.B, and characterized their solubility and assembly by centrifugation coupled with electron microscope examination. Both desmin and vimentin are purified by extraction and chromatography in solutions containing 6–8 M urea. The purified proteins are each soluble in buffers of very low ionic strength (e.g., 10 mM Tris-acetate) and pH of 8 or higher. Urea is removed by overnight dialysis of the purified desmin or vimentin against several changes of 10 mM Tris-acetate, pH 8.5, 5 mM MCE, followed by dialysis against 10 mM Tris-acetate, pH 8.5, and finally centrifugation at 183,000g to remove any aggregates. Desmin remains soluble at concentrations up to approximately 8 mg/ml; at higher concentrations, the protein forms a gel-like aggregate. The soluble desmin (Geisler *et al.*, 1992; Ip *et al.*, 1985b), or vimentin (Herrmann *et al.*, 1996; Strelkov *et al.*, 2003) is a tetramer of the 53-kDa (desmin) or 54-kDa (vimentin) subunit polypeptides, a structure usually referred to as a protofilament. Assembly of the soluble protofilaments to form 10-nm diameter IFs is induced by lowering the pH and/or increasing the ionic strength, either by dialysis or dilution.

Measurement of the solubility of avian desmin by high-speed centrifugation shows that, in low ionic strength buffers (10 mM Tris or imidazole), the solubility decreases below pH 8. Desmin is ~50% sedimented at pH 7.4, 90% sedimented at pH 7.0, and forms an insoluble precipitate at pH values between 4.5 and 6.5 (data not shown). Analysis of the effect of ionic strength, by use of increasing concentrations of NaCl, on the solubility and assembly of avian desmin at pH 7.0 and 8.5 is shown in Fig. 6A. Under these conditions, a small amount of desmin remains soluble after high-speed centrifugation, which represents the critical concentration for assembly, equivalent to the concentration of unassembled protein in equilibrium with the assembled polymer. At pH 7.0, desmin is maximally insoluble at all ionic strengths above 0.02, as determined by high-speed centrifugation, and thus data for high-speed centrifugation at pH 7.0 is not included in this graph. Low-speed centrifugation shows that, at pH 7.0, assembly increases with increasing

---

amount of protein remaining in the supernatant was determined by the Folin–Lowry method, and the percentage of the total protein sedimented was calculated. (A) Solubility and assembly of tissue-purified avian desmin at pH 8.5 or pH 7.0. Closed circles, pH 8.5, high-speed centrifugation; open circles, low-speed centrifugation, pH 8.5; triangles, low-speed centrifugation, pH 7.0. (B) Assembly of tissue-purified porcine desmin and vimentin at pH 8.5 and 7.0, determined by low-speed centrifugation. Closed circles, desmin at pH 7.0; open circles, vimentin at pH 7.0; closed triangles, desmin at pH 8.5; open triangles, vimentin at pH 8.5. Desmin and vimentin were purified from porcine cardiac muscle and porcine aortic smooth muscle, respectively. (C) Coassembly of porcine desmin and vimentin at pH 7.0 determined by low-speed centrifugation. Ratios of desmin to vimentin (w/w) were: open circles, 100:0 (desmin alone); open triangles, 75:25; open squares, 50:50; closed triangles, 25:75, closed circles, 0:100 (vimentin alone).

ionic strength, reaching maximal assembly at an ionic strength of ∼0.15 (Fig. 6A, solid triangles). Electron microscopy (data not shown) confirmed that the desmin in the supernatant remaining after low-speed centrifugation consists of oligomers and short filaments and that the effect of increasing ionic strength is formation of more and longer filaments. At pH 8.5, the solubility, determined by high-speed centrifugation (closed circles, Fig. 6A) also decreases rapidly with increasing NaCl concentration, reaching a plateau at about 85% sedimented at an ionic strength between 0.03 and 0.04. Thus, the critical concentration is higher at pH 8.5 than at pH 7.0. Measurement of assembly at pH 8.5 by low-speed centrifugation (open circles, Fig. 6A) shows that assembly is less complete at pH 8.5 than at pH 7.0, at all ionic strengths up to 0.15. The effect of KCl on solubility is identical to that of NaCl, but divalent cations are much more effective, with 2 mM $MgCl_2$ yielding minimum solubility at either pH 7.0 or 8.5. Phosphate buffers are also much more effective at inducing assembly than typical monovalent buffer ions. Electron microscopy of negatively stained desmin shows that the insoluble desmin under all these conditions exists as filaments approximately 10 nm in diameter. The structure of the filaments appears to vary under different conditions, particularly at low or high pH, with the most regular filaments formed at pH ∼7 and ionic strengths of 0.16 or greater.

Avian desmin and mammalian (porcine) desmin have similar solubility properties. However, desmin and vimentin differ in their solubility. Comparison of the solubility of mammalian desmin with mammalian vimentin indicates that vimentin requires lower pH and higher ionic strength than desmin for complete assembly. For example, measurement of solubility by high-speed centrifugation shows that, at low ionic strength (10 mM buffers), desmin is 50% soluble at pH 7.4, whereas the 50% solubility point for vimentin is pH 6.5.[6] A comparison of the effect of ionic strength on the assembly of mammalian desmin and vimentin is shown in Fig. 6B. At pH 7.0, the percentage of protein sedimented was lower for vimentin (Fig. 6B, open circles) than for desmin (closed circles) at all ionic strengths up to 0.15. The difference is more marked at pH 8.5. At pH 8.5, more than 80% of the vimentin (open triangles, Fig. 6B) remains in the low-speed supernatant even at an ionic strength of 0.16, where desmin is largely in the sediment (closed triangles, Fig. 6B).

Purified mammalian desmin and vimentin are capable of coassembly *in vitro*, as shown by measurement of the effect of increasing ionic strength, at pH 7.0, on the assembly of mixtures of desmin and vimentin (Fig. 6C). If the two proteins do not coassemble, then the assembly curve for a 50:50 mixture of desmin and vimentin should be biphasic, with a rapid increase in assembly at lower ionic strengths because of the assembly of desmin, a plateau at about 60% sedimented at an ionic strength of 0.3, and then a gradual increase in assembly with increasing ionic strength reflecting vimentin assembly. The actual data from mixtures of desmin and vimentin at ratios (w/w) of 75:25, 50:50, and 25:75, as well as with the individual proteins, shows a smooth curve at all ratios, reflecting the assembly of desmin–vimentin homopolymers.

Thus, these two type III IF proteins have similar, but not identical, assembly properties, and they are capable of coassembling into heteropolymeric filaments. In combination, the data shown here should provide information useful in examining the assembly and interactions of these proteins. For both proteins, the most consistent assembly is obtained at or near physiological conditions. The typical assembly buffer used to examine the effect of factors such as protein covalent modification or of other proteins on assembly consists of 10–20 mM Tris, imidazole, or other similar buffer at pH 7.0–7.5; 100–150 mM KCl or NaCl; and 2–5 mM $MgCl_2$ (Yuan *et al.*, 1999). These conditions yield the most consistent and regular filaments.

3. Synemin and Paranemin/Nestin

These unusual, large IF proteins do not form IFs by themselves *in vitro* but are able to form heteropolymeric filaments along with desmin or vimentin. Avian synemin has been purified and characterized in our laboratory (Bilak *et al.*, 1998). After purification, avian synemin is normally in 6 M urea, 0.1% MCE, 10 mM Tris-HCl, pH 7.5. Like desmin, the purified synemin remains soluble when dialyzed against 10 mM Tris-HCl, 0.1% MCE, pH 8.5, followed by dialysis against 10 mM Tris-HCl, pH 8.5, with no MCE. Cross-linking in 10 mM triethanolamine-HCl, pH 8.5, shows that the soluble synemin is a dimer of 230-kDa subunits. Avian synemin becomes insoluble when dialyzed against solutions with a pH lower than 8.0, as determined by high-speed centrifugation, and is completely insoluble between pH 6.0 and 4.5. The solubility of synemin also decreases with increasing ionic strength at either pH 8.5 or 7.5. However, low-speed centrifugation demonstrated that synemin does not form filaments, and this was clearly shown by EM of negatively stained samples. In 10 mM Tris-HCl, pH 7.27, the largely soluble synemin consists of globular particles ~11 nm in diameter, but under conditions in which desmin assembles into filaments, the avian synemin is seen as globular aggregates that vary in diameter from 15–25 nm. Later studies using expression in SW13 cells to examine interactions (Bellin *et al.*, 1999) demonstrated that avian synemin can coassemble with vimentin into IF networks within a cellular environment.

Rabbit smooth muscle synemin, also purified in urea, exhibits solubility properties similar to avian desmin (Hirako *et al.*, 2003). In addition, these investigators demonstrated that synemin and desmin, at a molar ratio of 1:25, can coassemble *in vitro* to form filaments similar in appearance to desmin filaments, as seen by TEM examination of rotary shadowed samples. When the ratio of synemin to desmin is increased to 6:25, the filaments become shorter and more branched. In this study, filament assembly was done by dialysis of the proteins, initially in urea, into 50 mM Tris-HCl, pH 7.5, 1 mM $MgCl_2$, 1% bovine serum albumin (BSA), and 1 mM DTT.

Paranemin has been purified from embryonic chick skeletal muscle (Hemken *et al.*, 1997), but the *in vitro* assembly properties of this protein have not been characterized. Expression in SW13 cells (Hemken *et al.*, 1997) showed that para-nemin can form IFs with vimentin and that coexpression of paranemin and desmin leads to the formation of filament networks in SW13 cells (Schweitzer *et al.*, 2001). Nestin, the mammalian ortholog of paranemin, has been purified from BHK-21 cells and the assembly *in vitro* characterized by Steinert *et al.* (1999). Results demonstrated that nestin alone does not form filaments but can coassemble with vimentin or the type IV IF protein α-internexin to form heterodimers and IFs, but only if the amount of nestin does not exceed 25% of the total protein. Higher amounts of nestin appear to inhibit IF formation.

4. Syncoilin

Little is known about the assembly properties of syncoilin, because it has not been purified other than as a thioredoxin fusion protein that was used to make antibodies (Newey *et al.*, 2001). Poon *et al.* (2002) demonstrated that syncoilin interacts with desmin by yeast two-hybrid analysis and that syncoilin is colocalized with desmin in mouse muscle. These investigators (Poon *et al.*, 2002) also examined the interaction of syncoilin and desmin in mouse muscle cells by immunoprecipitation. In extracts of mouse muscle in 150 mM NaCl, 1% (v/v) Nonidet P-40, 0.05% (w/v) SDS, and 50 mM Tris, pH 7.4, syncoilin was found mostly in the supernatant after centrifugation at 30,000 rpm for 45 minutes, whereas desmin was found mostly in the insoluble pellet fraction as expected, but a small amount of desmin remained in the soluble fraction. Desmin and syncoilin could be coimmunoprecipitated from this soluble fraction with antibodies to either desmin or syncoilin, again demonstrating an interaction between these two proteins. However, cotransfection of COS-7 cells with desmin and/or syncoilin demonstrated that syncoilin did not form IFs either alone or with desmin, but instead disrupted desmin assembly. The ability of purified syncoilin to form IFs *in vitro* has not been examined, but the experiments of Poon *et al.* (2002) suggest that syncoilin does not assemble into IFs by itself.

D. Cellular Localization of Muscle Intermediate Filament Proteins

A substantial amount of data concerning the organization, structure, function, and interactions of muscle IFs and IF proteins has been generated over the past ~25 years by light and electron microscope localization studies. Thus, it is difficult to present a compendium of methods for study of muscle IF proteins without including methods for IF protein localization. However, the antibody localization methods used in our laboratory and in many other laboratories are not specific to muscle IFs and muscle cells but, instead, are commonly used methods that can be found in a wide variety of protocol manuals. Thus, this section will be limited to a brief overview, with appropriate references, of the procedures that we or others have used successfully for cellular localization of muscle IF proteins. Commercially available antibodies are listed herein in Section II.B.

The basic methods for indirect immunofluorescence localization in cell cultures are described in detail by Osborn (1994). Cells are usually grown on glass coverslips, although fixation and labeling in cell culture plates can also be done. For cells such as embryonic chick muscle cells, muscle cell lines (C2C12), or SW13 cells, we and our colleagues have used three common fixation methods, namely: (1) fixation with methanol at $-20\,^\circ\mathrm{C}$ for 6 minutes (Bellin *et al.*, 1999; Hemken *et al.*, 1997); (2) fixation with 70% acetone and 30% methanol for 10 minutes at $-20\,^\circ\mathrm{C}$ (Schweitzer *et al.*, 2001); and (3) fixation with 3.7% formaldehyde in PBS for 10 minutes followed by permeabilization by treatment with 0.2% Triton X-100 for 1 minute. All three fixation methods yield equivalent structural preservation. The choice depends on the antibody used, because some antibodies recognize epitopes that are altered by a specific fixation method. We have successfully used all of these fixation methods with antibodies to desmin, synemin, paranemin, and vimentin. For secondary antibodies, we recommend the use of the Alexa dye-conjugated antibodies from Molecular Probes, Inc. rather than the more common rhodamine and fluorescein conjugates, because the Alexa conjugates are brighter and much less susceptible to fading.

A number of studies have been done with antibody localization at the electron microscope level to study muscle IFs. The earlier studies of Tokuyasu and coworkers (Tokuyasu *et al.*, 1983a,b, 1985) used immunogold labeling on ultrathin frozen sections to localize desmin in chicken skeletal (Tokuyasu *et al.*, 1983a) and cardiac muscle (Tokuyasu *et al.*, 1983b) and to analyze the organization of vimentin and desmin in developing chicken skeletal muscle (Tokuyasu *et al.*, 1985). More recently, our laboratory (Chou *et al.*, 1992, 1994) reported the use of post-embedding immunogold labeling to characterize the organization of desmin IFs and the relative distribution of desmin and the Z-line protein α-actinin in the developing embryonic chick gizzard. Finally, Schröder *et al.* (1999) described a method for single or double-immunogold labeling of frozen sections with antibodies to plectin and desmin, followed by glutaraldehyde fixation, epon embedding, and sectioning for TEM. This novel procedure was used to demonstrate colocalization of plectin and desmin in human skeletal muscle. Each of these EM localization methods is described in detail in the cited references.

E. Muscle Intermediate Filament Protein Interactions

Because muscle cell IF proteins are involved in maintaining the structural integrity of muscle cells, many of the studies on these proteins have focused on the physical linkages that hold these IF proteins together and link them to other cellular structures. A wide range of methods can be used to determine protein interactions, and most of them have at least some application in the study of muscle cell IF proteins. We will comment on several of these methods that have proven useful for the study of IF proteins; however, other published methods not detailed here, such as coimmunoprecipitation and GST-pulldowns, may also be useful when exploring the uncharacterized interactions of proteins with IFs. In addition, because all protein interaction assays have strengths and weaknesses, it is advisable to use multiple methods when characterizing new protein–protein interaction pairs to minimize the potential for false-positive or false-negative results.

1. Blot Overlays

This method is essentially a modification of a standard Western blot procedure to serve as a simple assay for *in vitro* protein interaction. It has been used to show interactions of muscle IF proteins with each other and with other muscle cell proteins (Bellin *et al.*, 1999, 2001). The advantage of this system is that assays can be completed fairly rapidly, and a wide range of buffer conditions can be tested (e.g., a divalent cation–containing buffer can be used in place of the Tris-buffered saline described later). However, the method does require purified proteins and is entirely *in vitro* in nature. The steps in the procedure are as follows.

1. Two samples of a purified protein and/or lysate of interest are subjected to SDS-PAGE and transferred to nitrocellulose membranes, as is done for typical Western blotting, to produce two identical blots. It is also a good idea to include other proteins or lysates of interest on each blot that are known not to interact with the overlay protein. These serve as negative controls to demonstrate specificity in overlay protein binding.

2. The blots are then each blocked with 5% nonfat dry milk powder (Blotto) in Tris-buffered saline-Tween 20 (TBS-T) (10 mM Tris-base, 100 mM NaCl, 0.1% Tween 20, pH 7.5) for 1 hour on a rocking plate mixer at room temperature.

3. One blot, the "overlay blot," is then incubated with a possible interacting purified protein of interest, normally at ~10 ug/ml in 1% Blotto in TBS-T. The second blot, the "control blot," is incubated in 1% Blotto in TBS-T containing no purified protein. The incubations are conducted for 2 hours at room temperature on a rocking plate mixer. The incubation step can also be carried out overnight at 4 °C.

4. The blots are each then separately washed three times for 10 minutes each with 1% Blotto in TBS-T on a rocking plate mixer at room temperature.

5. The blots are then both incubated with a primary antibody against the overlaid protein. The primary antibody should be diluted with 5% Blotto in TBS-T, and this incubation is normally done for 1 hour at room temperature on a rocking plate mixer. Although precise antibody dilution will need to be determined empirically, a dilution that works well in a typical Western blot will tend to be appropriate for this overlay procedure.

6. The blots are then each separately washed three times for 10 minutes each with 5% Blotto in TBS-T on a rocking plate mixer at room temperature.

7. The blots are then both incubated with an enzyme (e.g., horseradish peroxidase [HRP]) coupled with secondary antibody that will recognize the primary antibody used in step 5. The secondary antibody should be diluted with 5% Blotto in TBS-T, and this incubation is normally done for 1 hour at room temperature on a rocking plate mixer.

8. The blots are then each separately washed three times for 10 minutes each with TBS-T (no Blotto) on a rocking plate mixer at room temperature.

9. The blots are subjected to a detection procedure to visualize the binding of the secondary antibody. With HRP-conjugated secondary antibodies, enhanced chemiluminescence (ECL) detection methods, such as the ECL kit from Amersham Pharmacia Biotech, are quite effective for visualization. Regardless of the method used, both blots should be subjected to the detection method and any subsequent treatments (e.g., exposure to film, development of film) at the same time to ensure identical treatment.

10. Results are interpreted by comparing the overlay blot with the control blot. Bands appearing on the overlay blot but not the control blot indicate some level of interaction between the overlaid protein and that specific protein band on the blot. To demonstrate the interaction with increased confidence, it is desirable to conduct a reciprocal overlay experiment, where overlay protein is now on the blot, and the blot protein is used in the overlay.

2. SW13 Cell Transfection

The SW13 vimentin$^+$ (SW13.C1 vim$^+$) and vimentin$^-$ (SW13.C2 vim$^-$) clonal lines (Sarria *et al.*, 1990, 1994; Schweitzer *et al.*, 2001) established in the laboratory of Robert Evans provide a convenient way to study coassembly of a protein of interest with vimentin in a cellular context. It should be noted that the IF protein synemin has been demonstrated to be endogenously expressed, along with vimentin in SW13.C1 vim$^+$ cells, but is lacking, along with vimentin, from SW13.C2 vim$^-$ cells (Bellin *et al.*, 1999). This technique provides a useful cell-based counterpart to the cosedimentation assay described later in this chapter.

1. Use of SW13 cells for coassembly studies requires a cDNA expression vector designed to produce the protein of interest in mammalian cells. Both the pRC/RSV (Invitrogen) and pCDNA3 vectors have been found to be well suited for this method.

2. Separate cultures of SW13.C1 vim$^+$ and SW13.C2 vim$^-$ cells are maintained in DMEM–F12 media with 5% fetal calf serum. Cells are passaged into 35-mm diameter dishes containing a sterile coverslip and grown to ~70% confluency.

3. The cDNA construct is separately transfected into the SW13.C1 vim$^+$ and SW13.C2 vim$^-$ cells with FuGENE 6 reagent (Boehringer Mannheim). Although other transfection agents are also acceptable for this purpose, FuGENE 6 seems to yield consistently good transfection levels of the SW13 clonal lines, which are fairly difficult to transfect. Exact ratios of reagent to DNA should be determined empirically; however, a combination of 3 μl transfection reagent to 1 μg DNA is often suitable for a 35-mm diameter dish.

4. Protein expression should be visualized 24–48 hours after transfection by standard immunocytochemistry methods. Double-label immunofluorescence that

uses a rabbit polyclonal antibody against the protein of interest and the anti-vimentin monoclonal antibody AMF-17b (Developmental Studies Hybridoma Bank), along with Alexa 488–coupled goat anti-rabbit and Alexa 546–coupled goat anti-mouse secondary antibodies (Molecular Probes) has proven to be a useful approach.

5. Analysis of the results is straightforward. Proteins that will copolymerize or interact with vimentin filaments are easily noted by colabeling with the vimentin network in the SW13.C1 vim$^+$ cells. Proteins that are able to form IFs in the absence of vimentin are likely to do so in the SW13.C2 vim$^-$ cells, whereas assembly incompetent proteins often form punctate arrays in the cell cytoplasm.

3. Cosedimentation

By taking advantage of the self-assembly properties of the muscle cell IF proteins desmin or vimentin, it is possible to determine IF protein interactions by a cosedimentation assay used in Bellin *et al.* (1999). The method described focuses on a cosedimentation experiment that uses desmin purified by methods described earlier.

1. The interaction of a purified protein of interest with purified desmin is tested by use of three different sets of conditions, each based on controlling the solubility of desmin. Three assay tubes should be set up on the basis of the following descriptions: (1) *soluble desmin*: desmin and the protein of interest are mixed in 10 mM Tris-HCl, pH 8.5; (2) *filament-forming conditions*: desmin and the protein of interest are first mixed in 10 mM Tris-HCl, pH 8.5, and then the mixture is adjusted to IF-forming conditions by titrating the pH to 7.0 with addition of 2 M imidazole-HCl, pH 6.0, and by addition of MgCl$_2$ and NaCl to 1 mM and 100 mM, respectively; (3) *preformed filaments*: desmin by itself is first assembled into filaments in 100 mM NaCl, 1 mM MgCl$_2$, 10 mM imidizole-HCl, pH 7.0, and then the desmin filaments and the protein of interest are mixed. For each of the three sets of conditions, an equal amount by weight (normally ~25 μg) of purified protein of interest and purified desmin should be used in each sample. Bovine serum albumin (BSA) (Sigma) can also be added (normally ~10 μg) to each sample as an internal control.

2. All reaction tubes are then subjected to high-speed centrifugation conditions (100,000g for 20 minutes), which will sediment desmin filaments and any associated protein(s).

3. The resulting supernatants and pellets are separated by SDS-PAGE and stained with Coomassie Blue R-250 by use of standard methods.

4. Individual samples of desmin and of the protein of interest, in each of the three sets of conditions, should also be subjected to the same high-speed centrifugation to test their sedimentation behavior in the absence of the other protein.

5. By comparing the presence of the protein of interest in the supernatant and pellet fractions of each mixed assembly assay tube with the results from the protein of interest assayed in the absence of desmin, the presence of an interaction with desmin may be demonstrated. In addition, it is possible to assess whether the interaction with desmin is during filament formation, as would be the case in the formation of heteropolymeric filaments, or after filament formation, such as would be expected for a filament cross-linking protein.

4. Yeast Two-Hybrid Assays

Two-hybrid protein interaction assays have become fairly commonplace methods for detecting protein interactions between two specific proteins or for screening for unknown proteins. Several studies have been published that use the *Gal4* yeast two-hybrid system to test protein interactions between binary pairs of proteins, including IF proteins or protein domains (Bellin *et al.*, 2001; Carpenter and Ip, 1996; Meng *et al.*, 1997). For any of these methods, specialized DNA constructs encoding the proteins of interest must be produced. Several commercial kits are available for construction and screening of two-hybrid interactions, but many IF protein interactions have been assayed with the pPC86 and pPC97 yeast two-hybrid vectors originally described by Chevray and Nathans (1992). Whereas there is not ample space in this chapter to include a complete explanation of the yeast two-hybrid system, one basic protocol for conducting semiquantitative interactions with the pPC86 and pPC97 vectors with commercial transfection and assay kits is described in this section.

1. Plan the yeast cotransformations necessary to test the protein interactions of interest. Each co-transformation should include one construct made in the pPC86 vector and one in the pPC97 vector. Interactions between the fusion proteins produced from these vectors will bring the DNA-binding and transactivation domains of Gal4 close enough together to result in the transcription of β-galactosidase. For proper interpretation of results, it is essential to include a solid battery of controls for each experiment. Every construct should be cotransfected with the opposite empty vector alone (e.g., pPC86-desmin with pPC97-no insert) to be sure it will not activate β-galactosidase expression without an interaction partner. The specificity of interactions should be confirmed by performing vector-swap experiments for each protein pair (e.g., if pPC86-desmin gave a positive result with pPC97-protein of interest, then pPC97-desmin should give the same result when cotransfected with pPC86-protein of interest). Finally, at least one positive control should be included with each set of cotransformations (e.g., desmin dimerization: pPC86-desmin and pPC97-desmin).

2. Cotransform the PCY2 strain of yeast (Chevray and Nathans, 1992) with each of the vector pairs described in step 1 with the Alkali-Cation Yeast Transformation kit (Bio 101) by following the protocol provided with the kit. Although the optimum amount of each plasmid DNA should be determined empirically, ~400 ng of plasmid DNA for each construct normally works well.

3. Grow 2-ml yeast cultures of each double-transformant in double dropout Leu⁻ Trip⁻ media, 180 rpm at 30 °C for 72 hours.

4. Induce protein expression from the vectors by changing the yeast to growth in galactose containing Leu⁻ Trip⁻ media and growing for an additional 3 hours, 180 rpm at 30 °C.

5. Record an A_{600} value for each culture for later use in assay normalization.

6. Prepare yeast lysates by pelleting and freezing the yeast from each 2-ml culture in liquid nitrogen for 5 minutes followed by resuspending in 100 μl of lysate buffer (1 mM MgCl₂, 0.1% NaN₃, 10 mM sodium phosphate, pH 7.0).

7. Carry out semiquantitative determinations of the β-galactosidase activity levels in the lyastes with the FluorAce β-galactosidase Reporter Assay Kit (BioRad). This kit can be modified for use with yeast cells as follows: mix 20 μl of yeast lyaste with 50 μl of the kit's 1× reaction buffer, including 4-MUG and MCE, directly in wells of a black microtiter plate kept on ice. Incubate the plate for 3 hours at 37 °C, and then add of 150 μl of the kit's 1× stop buffer to terminate the detection reaction. It is best to set up three replicate wells for each lysate.

8. Read the assay plate on a fluorescence microplate reader with excitation at ~355 nm and emission at ~460 nm.

9. Determine the average of fluorescence readings for each of the three replicate wells and divide this number by the A_{600} value. These results represent the normalized fluorescence averages for each interaction pair. To calculate relative interaction affinities, divide each of these normalized averages by the value for a well-characterized protein interaction (e.g., the dimerization of desmin).

IV. Pearls and Pitfalls

It is important to develop a good understanding of the biochemical characteristics of muscle cell IF proteins to be successful in conducting experiments involving these proteins. Most muscle cell IF proteins, especially the large proteins synemin and paranemin/nestin, are very susceptible to proteolysis during preparation and storage. As a result, protease inhibitor cocktails should be used routinely, and all preparation steps should be conducted in a cold room. Extraction and chromatography of the muscle IF proteins is done in urea-containing solutions. It is important to deionize the concentrated (8 M) urea stock solution by passing it over a mixed-bed ion-exchange column to remove traces of cyanate, which can modify proteins. We have noted also that the Bradford protein assay (Bradford, 1976) available commercially as the Bio-Rad Protein Assay, gives unreliable results with desmin, because the protein precipitates in the reagent. Thus, we recommend the use of the Folin-Lowry (Lowry et al., 1951) or BCA (Pierce) protein assays for measuring the concentration of desmin. The large type VI IF proteins synemin and paranemin/nestin only compose a very small percentage of total muscle cell protein. Any method focused on identifying endogenous expression (e.g., Western blotting) or purification of these proteins from tissue sources requires a fairly large amount of starting material to detect or to obtain a useful amount of protein. As noted earlier herein, the solubility of all the muscle IF proteins greatly depends on the ionic strength and pH of storage and assay buffers, so these factors should be considered when planning experiments.

There are two important points to be considered when using SDS-PAGE to characterize samples containing the muscle IF proteins. First, the molecular masses of desmin (53 kDa) and vimentin (54 kDa) are similar. Use of appropriate molecular mass markers/standards in this range is advised for SDS-PAGE of samples that could contain both, such as embryonic muscle samples. Western blotting with commercially available antibodies is strongly recommended for unambiguous protein identification. Second, the large type VI IF proteins synemin and paranemin/nestin migrate much more slowly by SDS-PAGE than expected from the molecular masses predicted from their cDNA sequence. Avian synemin has a predicted mass of 182 kDa but migrates by SDS-PAGE with an estimated mass of ~230 kDa. Avian paranemin has a predicted molecular mass of 193 kDa but migrates by SDS-PAGE with an estimated mass of ~280 kDa.

V. Concluding Remarks

Many unanswered questions remain regarding the precise mechanisms underlying how the large IF proteins, synemin and paranemin/nestin, function within muscle cells. The detailed manner in which they are structurally packed and arranged with respect to the type III IF proteins in heteropolymeric IFs also remains unclear. A basic question that many have considered is simply whether type III IF proteins can function within any cells without the presence of a type IV or VI protein. Although some progress is being made in identifying proteins within other cytoskeletal structures with which muscle IF proteins interact, it is likely, or at least possible, that we have not yet identified many of the most important ones. Syncoilin is an intriguing muscle IF protein. That it does not seem to, or at least has not yet been shown to, form heteropolymeric IFs may put it in a functional class of its own. Considering that much of the IF protein research published these past three decades has been focused on the complexities of the keratins and neurofilament triplet proteins, we hope that this chapter has provided some background information and methods that will be useful for those relatively unfamiliar with muscle IF proteins.

Acknowledgments

The research from our laboratories included within this chapter was supported by grants from the USDA National Research Initiative Competitive Grants Program, Awards 89-37265-4441, 92-37206-8051, 96-35206-3744, 96-35206-3857, 99-35206-8676, 00-35206-9381 and 2003-35206-12823; and from the Muscular Dystrophy Association. We thank Dr. Denise Paulin for sharing unpublished information with us. We thank Lynn Newbold for help with the manuscript and Mary Sue Mayes for help with preparation of the figures.

References

Balogh, J., Merisckay, M., Li, Z., Paulin, D., and Arner, A. (2002). Hearts from mice lacking desmin have a myopathy with impaired active force generation and unaltered wall compliance. *Cardiovasc. Res.* **53,** 439–450.

Bang, M. L., Gregorio, C., and Labeit, S. (2002). Molecular dissection of the interaction of desmin with the C-terminal region of nebulin. *J. Struct. Biol.* **137,** 119–127.

Becker, B., Bellin, R. M., Sernett, S. W., Huiatt, T. W., and Robson, R. M. (1995). Synemin contains the rod domain of intermediate filaments. *Biochem. Biophys. Res. Commun.* **213,** 796–802.

Bellin, R. M., Sernett, S. W., Becker, B., Ip, W., Huiatt, T. W., and Robson, R. M. (1999). Molecular characteristics and interactions of the intermediate filament protein synemin. Interactions with alpha-actinin may anchor synemin-containing heterofilaments. *J. Biol. Chem.* **274,** 29493–29499.

Bellin, R. M., Huiatt, T. W., Critchley, D. R., and Robson, R. M. (2001). Synemin may function to directly link muscle cell intermediate filaments to both myofibrillar Z-lines and costameres. *J. Biol. Chem.* **276,** 32330–32337.

Bennett, G. S., Fellini, S. A., Toyama, Y., and Holtzer, H. (1979). Redistribution of intermediate filament subunits during skeletal myogenesis and maturation *in vitro. J. Cell. Biol.* **82,** 577–584.

Bilak, S. R., Bremner, E. M., and Robson, R. M. (1987). Composition of intermediate filament subunit proteins in embryonic, neonatal and postnatal porcine skeletal muscle. *J. Anim. Sci.* **64,** 601–606.

Bilak, S. R., Sernett, S. W., Bilak, M. M., Bellin, R. M., Stromer, M. H., Huiatt, T. W., and Robson, R. M. (1998). Properties of the novel intermediate filament protein synemin and its identification in mammalian muscle. *Arch. Biochem. Biophys.* **355,** 63–76.

Boriek, A. M., Capetanaki, Y., Hwang, W., Officer, T., Badshah, M., Rodarte, J., and Tidball, J. G. (2001). Desmin integrates the three-dimensional mechanical properties of muscles. *Am. J. Physiol. Cell Physiol.* **280,** C46–C52.

Bradford, M. M. (1976). A rapid and sensitive method for the quantitation of microgram quantities of protein utilizing the principle of protein-dye binding. *Anal. Biochem.* **72,** 248–254.

Breckler, J., and Lazarides, E. (1982). Isolation of a new high molecular weight protein associated with desmin and vimentin filaments from avian embryonic skeletal muscle. *J. Cell Biol.* **92,** 795–806.

Capetanaki, Y., Milner, D. J., and Weitzer, G. (1997). Desmin in muscle formation and maintenance: Knockouts and consequences. *Cell Struct. Funct.* **22,** 103–116.

Carlsson, L., Li, Z. L., Paulin, D., Price, M. G., Breckler, J., Robson, R. M., Wiche, G., and Thornell, L. E. (2000). Differences in the distribution of synemin, paranemin, and plectin in skeletal muscles of wild-type and desmin knock-out mice. *Histochem. Cell Biol.* **114,** 39–47.

Carlsson, L., and Thornell, L. E. (2001). Desmin-related myopathies in mice and man. *Acta Physiol. Scand.* **171,** 341–348.

Carlsson, L., Fischer, C., Sjöberg, G., Robson, R. M., Sejersen, T., and Thornell, L. E. (2002). Cytoskeletal derangements in hereditary myopathy with a desmin L345P mutation. *Acta Neuropathol. (Berl.)* **104,** 493–504.

Carpenter, D. A., and Ip, W. (1996). Neurofilament triplet protein interactions: Evidence for the preferred formation of NF-L-containing dimers and a putative function for the end domains. *J. Cell Sci.* **109,** 2493–2498.

Chevray, P. M., and Nathans, D. (1992). Protein interaction cloning in yeast: Identification of mammalian proteins that react with the leucine zipper of Jun. *Proc. Natl. Acad. Sci. USA* **89,** 5789–5793.

Chou, R.-G. R., Stromer, M. H., Robson, R. M., and Huiatt, T. W. (1992). Assembly of contractile and cytoskeletal elements in developing smooth muscle cells. *Dev. Biol.* **149,** 339–348.

Chou, R.-G. R., Stromer, M. H., Robson, R. M., and Huiatt, T. W. (1994). Substructure of cytoplasmic dense bodies and changes in distribution of desmin and alpha-actinin in developing smooth muscle cells. *Cell. Motil. Cytoskeleton* **29**, 204–214.

Chou, Y. H., Khuon, S., Herrmann, H., and Goldman, R. D. (2003). Nestin promotes the phosphorylation-dependent disassembly of vimentin intermediate filaments during mitosis. *Mol. Biol. Cell* **14**, 1468–1478.

Coulombe, P. A., Ma, L., Yamada, S., and Wawersik, M. (2001). Intermediate filaments at a glance. *J. Cell Sci.* **114**, 4345–4347.

Darenfed, H., Ma, X., Davis, L., Juge, N., Savard, P. E., Cole, G. J., and Vincent, M. (2001). Molecular polymorphism of the intermediate filament protein transitin. *Histochem. Cell Biol.* **116**, 397–409.

Duval, M., Ma, X., Valet, J.-P., and Vincent, M. (1995). Purification of developmentally regulated avian 400-kDa intermediate filament associated protein. Molecular interactions with intermediate filament proteins and other cytoskeleton components. *Biochem. Cell Biol.* **73**, 651–657.

Ervasti, J. M. (2003). Costameres: The Achilles' heel of Herculean muscle. *J. Biol. Chem.* **278**, 13591–13594.

Frank, E. D., and Warren, L. (1981). Aortic smooth muscle cells contain vimentin instead of desmin. *Proc. Natl. Acad. Sci. USA* **78**, 3020–3024.

Geisler, N., Schunemann, J., and Weber, K. (1992). Chemical cross-linking indicates a staggered and antiparallel protofilament of desmin intermediate filaments and characterizes one higher-level complex between protofilaments. *Eur. J. Biochem.* **206**, 841–852.

Granger, B. L., and Lazarides, E. (1980). Synemin: A new high molecular weight protein associated with desmin and vimentin filaments in muscle. *Cell* **22**, 727–738.

Haubold, K. W., Allen, D. L., Capetanaki, Y., and Leinwand, L. A. (2003). Loss of desmin leads to impaired voluntary wheel running and treadmill exercise performance. *J. Appl. Physiol.* **95**, 1617–1622.

Hemken, P. M., Bellin, R. M., Sernett, S. W., Becker, B., Huiatt, T. W., and Robson, R. M. (1997). Molecular characteristics of the novel intermediate filament protein paranemin. Sequence reveals EAP-300 and IFAPa-400 are highly homologous to paranemin. *J. Biol. Chem.* **272**, 32489–32499.

Herrmann, H., and Aebi, U. (2000). Intermediate filaments and their associates: Multi-talented structural elements specifying cytoarchitecture and cytodynamics. *Curr. Opin. Cell Biol.* **12**, 79–90.

Herrmann, H., Haner, M., Brettel, M., Muller, S. A., Goldie, K. N., Fedtke, B., Lustig, A., Franke, W. W., and Aebi, U. (1996). Structure and assembly properties of the intermediate filament protein vimentin: The role of its head, rod and tail domains. *J. Mol. Biol.* **264**, 933–953.

Herrmann, H., Hesse, M., Reichenzeller, M., Aebi, U., and Magin, T. M. (2003). Functional complexity of intermediate filament cytoskeletons: From structure to assembly to gene ablation. *Int. Rev. Cytol.* **223**, 83–175.

Hesse, M., Magin, T. M., and Weber, K. (2001). Genes for intermediate filament proteins and the draft sequence of the human genome: Novel keratin genes and a surprisingly high number of pseudogenes related to keratin genes 8 and 18. *J. Cell Sci.* **114**, 2569–2575.

Hijikata, T., Murakami, T., Imamura, M., Fujimaki, N., and Ishikawa, H. (1999). Plectin is a linker of intermediate filaments to Z-discs in skeletal muscle fibers. *J. Cell Sci.* **112**, 867–876.

Hirako, Y., Yamakawa, H., Tsujimura, Y., Nishizawa, Y., Okumura, M., Usukura, J., Matsumoto, H., Jackson, K. W., Owaribe, K., and Ohara, O. (2003). Characterization of mammalian synemin, an intermediate filament protein present in all four classes of muscle cells and some neuroglial cells: Co-localization and interaction with type III intermediate filament proteins and keratins. *Cell Tissue Res.* **313**, 195–207.

Huiatt, T. W., Robson, R. M., Arakawa, N., and Stromer, M. H. (1980). Desmin from avian smooth muscle. Purification and partial characterization. *J. Biol. Chem.* **255**, 6981–6989.

Inagaki, M., Gonda, Y., Matsuyama, M., Nishizawa, K., Nishi, Y., and Sato, C. (1988). Intermediate filament reconstitution *in vitro*. The role of phosphorylation on the assembly-disassembly of desmin. *J. Biol. Chem.* **263**, 5970–5978.

Inagaki, M., Matsuoka, Y., Tsujimura, K., Ando, S., Tokui, T., Takahashi, T., and Inagaki, N. (1996). Dynamic property of intermediate filaments: Regulation by phosphorylation. *BioEssays* **18,** 481–487.

Ip, W., Hartzer, M. K., Pang, Y.-Y. S., and Robson, R. M. (1985a). Assembly and vimentin *in vitro* and its implications concerning the structure of intermediate filaments. *J. Mol. Biol.* **183,** 365–375.

Ip, W., Heuser, J. E., Pang, Y.-Y. S., Hartzer, M. K., and Robson, R. M. (1985b). Subunit structure of desmin and vimentin protofilaments and how they assemble into intermediate filaments. *Ann. N. Y. Acad. Sci.* **455,** 185–199.

Landon, F., Lemonnier, M., Benarous, R., Hue, C., Fiszman, M., Gros, F., and Portier, M. M. (1989). Multiple mRNAs encode peripherin, an neuronal intermediate filament protein. *EMBO J.* **8,** 1719–1726.

Lazarides, E. (1980). Intermediate filaments as mechanical integrators of cellular space. *Nature* **283,** 249–256.

Li, Z., Colucci-Guyon, E., Pincon-Raymond, M., Mericskay, M., Pournin, S., Paulin, D., and Babinet, C. (1996). Cardiovascular lesions and skeletal myopathy in mice lacking desmin. *Dev. Biol.* **175,** 362–366.

Li, Z., Mericskay, M., Agbulut, O., Butler-Browne, G., Carlsson, L., Thornell, L. E., Babinet, C., and Paulin, D. (1997). Desmin is essential for the tensile strength and integrity of myofibrils but not for myogenic commitment, differentiation, and fusion of skeletal muscle. *J. Cell Biol.* **139,** 129–144.

Lowry, O. H., Rosebrough, N. J., Farr, A. L., and Randall, R. J. (1951). Protein measurement with the Folin phenol reagent. *J. Biol. Chem.* **193,** 265–275.

Meng, J. J., Bornslaeger, E. A., Green, K. J., Steinert, P. M., and Ip, W. (1997). Two-hybrid analysis reveals fundamental differences in direct interactions between desmoplakin and cell type-specific intermediate filaments. *J. Biol. Chem.* **272,** 21495–21503.

Milner, D. J., Weitzer, G., Tran, D., Bradley, A., and Capetanaki, Y. (1996). Disruption of muscle architecture and myocardial degeneration in mice lacking desmin. *J. Cell Biol.* **134,** 1255–1270.

Milner, D. J., Mavroidis, M., Weisleder, N., and Capetanaki, Y. (2000). Desmin cytoskeleton linked to muscle mitochondrial distribution and respiratory function. *J. Cell Biol.* **150,** 1283–1298.

Mizuno, Y., Thompson, T. G., Guyon, J. R., Lidov, H. G., Brosius, M., Imamura, M., Ozawa, E., Watkins, S. C., and Kunkel, L. M. (2001). Desmuslin, an intermediate filament protein that interacts with alpha-dystrobrevin and desmin. *Proc. Natl. Acad. Sci. USA* **98,** 6156–6161.

Napier, A., Yuan, A., and Cole, G. J. (1999). Characterization of the chicken transitin gene reveals a strong relationship to the nestin intermediate filament class. *J. Mol. Neurosci.* **12,** 11–22.

Newey, S. E., Howman, E. V., Ponting, C. P., Benson, M. A., Nawrotzki, R., Loh, N. Y., Davies, K. E., and Blake, D. J. (2001). Syncoilin, a novel member of the intermediate filament superfamily that interacts with alpha-dystrobrevin in skeletal muscle. *J. Biol. Chem.* **276,** 6645–6655.

Niedziela-Majka, A., Rymarczyk, G., Kochman, M., and Ozyhar, A. (1998). GST-Induced dimerization of DNA-binding domains alters characteristics of their interaction with DNA. *Protein Expr. Purif.* **14,** 208–220.

Olive, M., Goldfarb, L., Dagvadorj, A., Sambuughin, N., Paulin, D., Li, Z., Goudeau, B., Vicart, P., and Ferrer, I. (2003). Expression of the intermediate filament protein synemin in myofibrillar myopathies and other muscle diseases. *Acta Neuropathol. (Berl.)* **106,** 1–7.

O'Neill, A., Williams, M. W., Resneck, W. G., Milner, D. J., Capetanaki, Y., and Bloch, R. J. (2002). Sarcolemmal organization in skeletal muscle lacking desmin: Evidence for cytokeratins associated with the membrane skeleton at costameres. *Mol. Biol. Cell* **13,** 2347–2359.

Osborn, M. (1994). Immunofluorescence microscopy of cultured cells. *In* "Cell Biology: A Laboratory Handbook" (J. E. Celis, ed.), Vol. 2, pp. 347–354. Academic Press, San Diego.

O'Shea, J. M., Robson, R. M., Huiatt, T. W., Hartzer, M. K., and Stromer, M. H. (1979). Purified desmin from adult mammalian skeletal muscle: A peptide mapping comparison with desmins from adult mammalian and avian smooth muscle. *Biochem. Biophys. Res. Commun.* **89,** 972–980.

O'Shea, J. M., Robson, R. M., Hartzer, M. K., Huiatt, T. W., Rathbun, W. E., and Stromer, M. H. (1981). Purification of desmin from adult mammalian skeletal muscle. *Biochem. J.* **195,** 345–356.

Poon, E., Howman, E. V., Newey, S. E., and Davies, K. E. (2002). Association of syncoilin and desmin: Linking intermediate filament proteins to the dystrophin-associated protein complex. *J. Biol. Chem.* **277**, 3433–3439.

Price, M. G., and Lazarides, E. (1983). Expression of intermediate filament-associated proteins paranemin and synemin in chicken development. *J. Cell Biol.* **97**, 1860–1874.

Reipert, S., Steinbock, F., Fischer, I., Bittner, R. E., Zeold, A., and Wiche, G. (1999). Association of mitochondria with plectin and desmin intermediate filaments in striated muscle. *Exp. Cell Res.* **252**, 479–491.

Robson, R. M. (1989). Intermediate filaments. *Curr. Opin. Cell Biol.* **1**, 36–43.

Robson, R. M., Shea, J. M., Hartzer, M. K., Rathbun, W. E., LaSalle, F., Schreiner, P. J., Kasang, L. E., Stromer, M. H., Lusby, M. L., Ridpath, J. F., Pang, Y.-Y., Evans, R. R., Zeece, M. G., Parrish, F. C., and Huiatt, T. W. (1984). Role of new cytoskeletal elements in maintenance of muscle integrity. *J. Food Biochem.* **8**, 1–24.

Sahlgren, C. M., Mikhailov, A., Hellman, J., Chou, Y. H., Lendahl, U., Goldman, R. D., and Eriksson, J. E. (2001). Mitotic reorganization of the intermediate filament protein nestin involves phosphorylation by cdc2 kinase. *J. Biol. Chem.* **276**, 16456–16463.

Sandoval, I. V., Colaco, C. A. L. S., and Lazarides, E. (1983). Purification of the intermediate filament-associated protein, synemin, from chicken smooth muscle. Studies on its physicochemical properties, interaction with desmin, and phosphorylation. *J. Biol. Chem.* **258**, 2568–2576.

Sarria, A. J., Nordeen, S. K., and Evans, R. M. (1990). Regulated expression of vimentin cDNA in cells in the presence and absence of a preexisting vimentin filament network. *J. Cell Biol.* **111**, 553–565.

Sarria, A. J., Lieber, J. G., Nordeen, S. K., and Evans, R. M. (1994). The presence or absence of a vimentin-type intermediate filament network affects the shape of the nucleus in human SW-13 cells. *J. Cell Sci.* **107**, 1593–1607.

Schröder, R., Warlo, I., Herrmann, H., van der Ven, P. F., Klasen, C., Blumcke, I., Mundegar, R. R., Furst, D. O., Goebel, H. H., and Magin, T. M. (1999). Immunogold EM reveals a close association of plectin and the desmin cytoskeleton in human skeletal muscle. *Eur. J. Cell Biol.* **78**, 288–295.

Schweitzer, S. C., Klymkowsky, M. W., Bellin, R. M., Robson, R. M., Capetanaki, Y., and Evans, R. M. (2001). Paranemin and the organization of desmin filament networks. *J. Cell Sci.* **114**, 1079–1089.

Sejersen, T., and Lendahl, U. (1993). Transient expression of the intermediate filament nestin during skeletal muscle development. *J. Cell Sci.* **106**, 1291–1300.

Sjöberg, G., Jiang, W. Q., Ringertz, N. R., Lendahl, U., and Sejersen, T. (1994). Colocalization of nestin and vimentin/desmin in skeletal muscle cells demonstrated by three-dimensional fluorescence digital imaging microscopy. *Exp. Cell Res.* **214**, 447–458.

Small, J. V., and Sobieszek, A. (1977). Studies on the function and composition of the 10-nm(100-Å) filaments of vertebrate smooth muscle. *J. Cell Sci.* **23**, 243–268.

Sobieszek, A., and Bremel, R. D. (1975). Preparation and properties of vertebrate smooth-muscle myofibrils and actomyosin. *Eur. J. Biochem.* **55**, 49–60.

Steinbock, F. A., and Wiche, G. (1999). Plectin: A cytolinker by design. *Biol. Chem.* **380**, 151–158.

Steinert, P. M., Chou, Y. H., Prahlad, V., Parry, D. A., Marekov, L. N., Wu, K. C., Jang, S. I., and Goldman, R. D. (1999). A high molecular weight intermediate filament-associated protein in BHK-21 cells is nestin, a type VI intermediate filament protein. Limited co-assembly *in vitro* to form heteropolymers with type III vimentin and type IV alpha-internexin. *J. Biol. Chem.* **274**, 9881–9890.

Strelkov, S. V., Herrmann, H., and Aebi, U. (2003). Molecular architecture of intermediate filaments. *Bioessays* **25**, 243–251.

Thornell, L., Carlsson, L., Li, Z., Mericskay, M., and Paulin, D. (1997). Null mutation in the desmin gene gives rise to a cardiomyopathy. *J. Mol. Cell. Cardiol.* **29**, 2107–2124.

Titeux, M., Brocheriou, V., Xue, Z., Gao, J., Pellissier, J. F., Guicheney, P., Paulin, D., and Li, Z. (2001). Human synemin gene generates splice variants encoding two distinct intermediate filament proteins. *Eur. J. Biochem.* **268**, 6435–6449.

Tokuyasu, K. T., Dutton, A. H., and Singer, S. J. (1983a). Immunoelectron microscopic studies of desmin (skeletin) localization and intermediate filament organization in chicken cardiac muscle. *J Cell Biol.* **96,** 1727–1735.

Tokuyasu, K. T., Dutton, A. H., and Singer, S. J. (1983b). Immunoelectron microscopic studies of desmin (skeletin) localization and intermediate filament organization in chicken cardiac muscle. *J. Cell Biol.* **96,** 1736–1742.

Tokuyasu, K. T., Maher, P. A., and Singer, S. J. (1984). Distributions of vimentin and desmin in developing chick myotubes *in vivo*. I. Immunofluorescence study. *J. Cell Biol.* **98,** 1961–1972.

Tokuyasu, K. T., Maher, P. A., and Singer, S. J. (1985). Distributions of vimentin and desmin in developing chick myotubes *in vivo*. II. Immunoelectron microscopic study. *J. Cell Biol.* **100,** 1157–1166.

Wiche, G. (1998). Role of plectin in cytoskeleton organization and dynamics. *J. Cell Sci.* **111**(Pt. 17), 2477–2486.

Xue, Z. G., Cheraud, Y., Brocheriou, V., Izmiryan, A., Titeux, M., Paulin, D., and Li, Z. (2004). The mouse synemin gene encodes three intermediate filament proteins generated by alternative exon usage and different open reading frames. *Exp. Cell Res.* **298,** 431–444.

Yang, Y.-G., and Makita, T. (1997). Immunocytochemical localization of desmin filaments in skeletal muscle of neonatal swine. *Acta Hiscochem. Cytochem.* **30,** 157–164.

Yuan, J., Huiatt, T. W., Liao, C. X., Robson, R. M., and Graves, D. J. (1999). The effects of mono-ADP-ribosylation on desmin assembly-disassembly. *Arch. Biochem. Biophys.* **363,** 314–322.

Zhou, H., Huiatt, T. W., Robson, R. M., Sernett, S. W., and Graves, D. J. (1996). Characterization of ADP-ribosylation sites on desmin and restoration of desmin intermediate filament assembly by de-ADP-ribosylation. *Arch. Biochem. Biophys.* **334,** 214–222.

CHAPTER 19

[^{35}S]Methionine Metabolic Labeling to Study Axonal Transport of Neuronal Intermediate Filament Proteins *In Vivo*

Stéphanie Millecamps and Jean-Pierre Julien

Research Center of CHUL and Department of Anatomy and Physiology
Laval University
Quebec, G1V 4G2, QC Canada

I. Introduction

In mammals, mature axons are characterized by an extensive cross-linked network of three structural components with characteristic diameter: actin microfilaments (7 nm), intermediate filaments (IFs) (8–10 nm), and microtubules (24 nm) (Hirokawa *et al.*, 1984). These cytoskeletal proteins are synthesized in neuronal cell body and transported down the axon toward the nerve ending. Defects in axonal transport can cause motor neuron diseases and neuropathies as shown by the discovery of gene mutations/dysfunctions for components of the microtubule-based transport in subsets of mouse models and human cases (Bommel *et al.*, 2002; Bomont *et al.*, 2000; Ding *et al.*, 2002; Hafezparast *et al.*, 2003; LaMonte *et al.*, 2002; Martin *et al.*, 2002; Xia *et al.*, 2003; Zhao *et al.*, 2001). In experimental animal models, one method to measure the rate of transported cytoskeletal proteins is to analyze segments of the sciatic nerve after injection of radioactive amino acid precursors into the spinal cord. This *in vivo* approach has been especially suitable to investigate the axonal transport of neurofilament (NF) proteins, the major IF proteins in large myelinated axons. Here we will describe this pulse-chase radiolabeling procedure in mice and its applications in detecting axonal transport defects.

NFs are the most abundant and widely expressed IF proteins in the adult nervous system. NFs are composed of three polypeptides termed the NF light (NF-L, 61 kDa), medium (NF-M, 90 kDa), and heavy (NF-H, 115 kDa) subunits (Ching and Liem, 1993; Lee *et al.*, 1993). The posttranslation modifications confer them apparent molecular weights on sodium dodecyl sulfate (SDS) gels of 68 kDa (NF-L), 150 kDa (NF-M), and 200 kDa (NF-H). Two other types of IF proteins, α-internexin and peripherin, are also expressed in subsets of adult neurons. These neuronal IFs proteins appear at different developmental stages and remain expressed in distinctive subsets of adult neuronal populations. α-Internexin (66 kDa) is expressed at much lower levels than the three NF subunits in large motor neurons of the spinal cord and the cranial nerve nuclei, but it is the dominant IF in small-caliber axons interneurons and cerebellar granule cells (Fliegner *et al.*, 1994). It is expressed at high levels in embryonic neurons but at lower levels after birth (Fliegner *et al.*, 1994; Kaplan *et al.*, 1990). Peripherin (57 kDa) is another IF protein that is expressed early in development with the outgrowth of specific central nervous system (CNS) and peripheral nervous system (PNS) neuronal populations. Its expression gradually disappears as NF subunits are synthesized in most CNS neurons. It persists in adult cytoskeleton mainly in autonomic nerves and in sensory neurons (Parysek *et al.*, 1988). Peripherin expression is increased in spinal motor neurons after nerve injury of the sciatic nerve (Oblinger *et al.*, 1989; Terao *et al.*, 2000; Troy *et al.*, 1990), whereas expression of all NF subunits is reduced (Muma *et al.*, 1990; Portier *et al.*, 1982). It is also increased after cerebral lesions or ischemia (Beaulieu *et al.*, 2002). In contrast to NFs, α-internexin and peripherin can self-assemble into homopolymers (Beaulieu *et al.*, 1999; Cui *et al.*, 1995; Ho *et al.*, 1995). They also interact with each of the three NF subunits *in vitro* (Athlan and Mushynski, 1997; Beaulieu *et al.*, 1999; Ching and Liem, 1993) or *in vivo* (Parysek *et al.*, 1991).

Like all members of the IF family, neuronal IFs contain a highly conserved central α-helical domain called the rod domain, which is involved in the formation of coiled-coil dimers that line up to form an antiparallel tetramer. Tetramers link end-to-end to form protofilaments (2- to 3-nm diameter) that laterally associate to form protofibrils (4- to 5-nm diameter). Four protofibrils intertwine to produce a ropelike IF structure (10-nm diameter). Unlike most IFs, the assembly of NFs requires at minimum the heteropolymerization of NF-L with either NF-M or NF-H (Ching and Liem, 1993; Lee *et al.*, 1993). The NF-M and NF-H subunits project their carboxy-terminal domains at the periphery of the filaments to form cross-bridges (Hirokawa *et al.*, 1984; Julien *et al.*, 1983). After their synthesis in the perikarya, NF proteins assemble into filaments (Black *et al.*, 1986; Nixon *et al.*, 1989). In the axons, NF proteins are recovered almost exclusively with the Triton-insoluble proteins (Black *et al.*, 1986; Morris and Lasek, 1982; Nixon *et al.*, 1989). The turnover of axonal cytoskeleton requires active transport of newly synthesized proteins from the cell body to the neuronal processes and nerve endings. Pulse-labeling experiments with [^{35}S]methionine revealed that components of antero-grade transport (from the cell body to the axon tip) can be distinguished by both their speed and the cargoes transported. The fast component is the move-ment of membrane-bound organelles along the microtubules track. It moves at about 200–400 mm/day. The slow axonal transport conveys cytoskeletal pro-teins and is divided in two rate components. The slow component a (SCa) carries (0.1–1 mm/d) microtubules, NFs, α-internexin, and peripherin in axons (Chadan *et al.*, 1994; Filliatreau *et al.*, 1988; Kaplan *et al.*, 1990). The slow component b (SCb), which is slightly faster (2–8 mm/d), carries actin (microfilaments), spec-trin, and other cytoplasmic proteins (Hoffman and Lasek, 1975; Nixon and Logvinenko, 1986).

How IF proteins move into axons has long been controversial. Two hypotheses, the "assembled polymer" and the "unassembled subunit" transport models, have been debated. The first theory was that NF proteins are assembled in the cell body immediately after synthesis and are transported as intact polymers by sliding mechanism on the microtubule network (Lasek, 1986; Lasek *et al.*, 1984; Tytell *et al.*, 1981). This theory was then challenged by the subunit transport model that states that cytoskeletal polymers are stationary in the axon and that NFs are conveyed as free subunits or small oligomers along microtubules (Okabe *et al.*, 1993; Terada *et al.*, 1996). It implies that the transported proteins exchange with those in the stationary NFs. In support of this model, virally expressed NF-M proteins were transported at slow rates without NF-L or NF-H into axons of transgenic mice with reduced NF content in the axons (Terada *et al.*, 1996). However, the former hypothesis is supported by recent studies based on the direct visualization of green fluorescent protein–tagged NFs moving bidirectionally in growing axons of cultured embryonic neurons (Ackerley *et al.*, 2003; Roy *et al.*, 2000; Wang *et al.*, 2000). The movement of the fluorescent polymers was bidirec-tional but mainly anterograde and consisted of intermittent rapid movements (0.7 μm/s) interspersed by long pauses (Roy *et al.*, 2000; Wang *et al.*, 2000). The direct observation of moving microtubules in living cells demonstrated that these cytoskeletal components are also conveyed along axons by a similar mechanism (Wang and Brown, 2002).

The slow transport wave observed *in vivo* by the pulsed radiolabeled techniques is the sum of individual NF movements spending the most of their time pausing (Brown, 2000). The movement of the polymers is fast (few μm/sec), but the resultant transport is slow (few μm/min) because of long pauses that might result from dissociation of NFs with the motor protein. Recent evidence suggests that NFs interact with several fast motor proteins, including kinesin (Yabe et al., 1999), dynein (Shah et al., 2000), and myosin (Brown, 2003; Rao et al., 2002a). Posttranslational modifications of NF proteins by kinases and phosphatases are involved in regulation of subunit assembly (Bennett and DiLullo, 1985; Nixon and Lewis, 1986; Sternberger and Sternberger, 1983), and in controlling the transition between moving and stationary NF pools (Lewis and Nixon, 1988). This is supported by studies demonstrating that transport is slowed when NFs are phosphorylated (Archer et al., 1994; Jung et al., 2000a,b; Nixon et al., 1987; Watson et al., 1989). The role of NF-H side arm phosphorylation as a regulator of axonal transport was recently visualized in cultured cortical neurons. Fluorescence-tagged NF-H constructs that were mutated within KSP repeats moved significantly faster than the nonmutated one (Ackerley et al., 2003). Phosphorylation may promote detachment of NFs from the anterograde motor and subsequent transfer to the retrograde motor (Miller et al., 2002).

Explaining the molecular requirements for motility of neuronal IF proteins may provide new insight on pathogenic mechanisms associated with neurodegenerative disorders. It is now recognized that IF disorganization can provoke disease. For example, NF-L mutations found in subsets of Charcot-Marie-Tooth (CMT) disease disrupt assembly and axonal transport of NFs in cultured neurons (Brownlees et al., 2002; Jordanova et al., 2003; Perez-Olle et al., 2002; Zuchner et al., 2004). Yet, the exact molecular mechanism of neurodegeneration by NF-L mutations remains to be explained. Similarly, a small number of patients with sporadic amyotrophic lateral sclerosis (ALS) ($\sim$1% cases) exhibit codon deletions or insertion in the KSP phosphorylation domain of the NF-H gene (Al-Chalabi et al., 1999; Figlewicz et al., 1994; Tomkins et al., 1998), but how these variants might contribute to motor neuron degeneration remains unknown. Finally, variant peripherin forms may also exist in some ALS cases, but their neurotoxicity is not understood (Robertson et al., 2003).

Defects in axonal transport of NF proteins have been detected with pulse-chase labeling with [^{35}S]methionine in various animal models of pathological conditions as listed in Table I. The first reports of transport impairment of NFs in neurons came from studies on intoxication-induced neuropathies. These impairments of axonal transport are selective for NFs (Bizzi et al., 1984; Griffin et al., 1985; Monaco et al., 1985, 1989; Pappolla et al., 1987; Parhad et al., 1986), whereas in other models of proximal peripheral neuropathies, such as those associated with diabetes or hypothyroidism, perturbations of axonal transport affect components of both SCa and SCb (Medori et al., 1985, 1988; Stein et al., 1991). Transport abnormalities of NFs were subsequently reported in various mouse models of neurological disorders, including wobbler mice (Mitsumoto and Gambetti, 1986), myelin-deficient mice (de Waegh and Brady, 1990; Kirkpatrick et al., 2001) and mice expressing mutant forms of superoxide dismutase (SOD1) linked to ALS (Williamson and Cleveland, 1999; Zhang et al., 1997). More recently, in vivo pulse

Table I

Models with Defects in Axonal Transport of ^{35}S-Radiolabeled Intermediate Filament Proteins

Models	Rate of slow component A transport	References
Overexpression of NF proteins		
hNF-H	Slowed (NF-L/NF-M/NF-H /tubulin/actin)	Collard *et al.*, 1995
mNF-H	Slowed (NF-L/NF-M/NF-H)	Marszalek *et al.*, 1996
Knock out of IF genes		
α-Internexin −/−	Normal (NF-M)	Yuan *et al.*, 2003
NF-L −/−	Normal (NF-M)	Yuan *et al.*, 2003
α-Internexin −/−; NF-L −/−	Abolished (NF-M)	Yuan *et al.*, 2003
NF-M −/−	Increased (NF-L/NF-H)	Jacomy *et al.*, 1999
NF-H −/−	Increased (NF-L/NF-M)	Zhu *et al.*, 1998
	Normal (NF-L/NF-M)	Rao *et al.*, 2002b
NF-M −/−; NF-H −/−	Very low (NF-L)	Jacomy *et al.*, 1999
	No transport (NF-L)	Yuan *et al.*, 2003
NF-H tail −/−	Normal (NF-L/NF-M)	Rao *et al.*, 2002b
Neurodegenerative models		
SOD1^{G37R}/SOD1^{G85R} transgenic	Slowed (NFs/tubulin)	Williamson and Cleveland, 1999
Wobbler	Slowed (NFs/tubulin/actin)	Mitsumoto and Gambetti, 1986
Trembler mutant (PNS myelin-deficient mice)	Slowed (NFs) Increased (tubulin)	de Waegh and Brady, 1990
Shiverer mutant (CNS myelin-deficient mice)	Increased (NF-M/tubulin/Sca)	Kirkpatrick *et al.*, 2001
Neuropathies		
ββ′–iminodipropionitrile	Slowed (NFs)	Griffin *et al.*, 1985; Parhad *et al.*, 1986
Aluminium salts	Slowed (NFs/peripherin)	Bizzi *et al.*, 1984
Carbon disulfide	Increased (NFs)	Pappolla *et al.*, 1987
2,5-Hexanedione	Increased (NFs)	Monaco *et al.*, 1985
Hypothyroidism	Slowed (NFs/tubulin)	Stein *et al.*, 1991
Diabetes	Slowed (fastC/SCa/SCb)	Medori *et al.*, 1985; Medori *et al.*, 1988

radiolabeling analyses of spinal motor neurons or retinal ganglion cell (RGC) neurons in IF-deficient mice have provided important insights into the role of individual proteins in assembly and transport of IF proteins into axons. For example, by use of double knockout mice for NF-M and NF-H, two studies have provided evidence that NF-L alone is incapable of efficient transport in sciatic or optic nerve without the other NF protein partners (Jacomy *et al.*, 1999; Yuan *et al.*,

2003). In contrast, NF-M did not require other NF proteins to be transported in optic axons if it contained α-internexin protein (Yuan *et al.*, 2003). The concomitant disruptions of α-internexin and NF-L genes abolished transport of NF-M in optic axons. Interestingly, the loss of NF-M or NF-H proteins in mice increased the transport rate of the remaining NF proteins in the sciatic nerve (Jacomy *et al.*, 1999; Zhu *et al.*, 1998) but not in optic axons (Rao *et al.*, 2002b). Thus, these distinct axon populations may exhibit different requirements for axonal transport of IF proteins. The molecular mechanisms responsible for these differences remain to be explained. Perhaps, this reflects variations in the content of molecular motors and/or IF's interacting proteins.

Pulse-chase labeling with [^{35}S]methionine constitutes the standard procedure to analyze the *in vivo* transport of cytoskeletal proteins down the sciatic nerve or the optic nerve of mice. After injection of this radioactive amino acid into cell bodies of spinal motor neurons or RGC neurons, the isotope emission allows the monitoring of newly synthesized IF proteins as they travel within axons. From the analysis of the distance covered by radiolabeled proteins at different time intervals, the rate of their transport can be measured. Here we describe the experimental procedure for analysis of IF protein transport into motor axons after injection of [^{35}S]methionine into spinal cord of the mouse.

II. Materials and Instrumentation

A. Reagents

The reagents needed for the injection include L-[^{35}S]methionine (1175 Ci/mmol, Perkin Elmer, Life Sciences, Boston, MA), xylazine (CDMV, St-Hyacinthe, QC, Canada), ketamine (CDMV, St-Hyacinthe, QC, Canada), buprenorphine (Reckitt & Colman Pharmaceuticals, Inc. Richmond, VA), chloral hydrate (Fisher Scientific, Nepean, ON, Canada), and povidone iodine (Laboratoire Atlas, Montreal, QC, Canada).

B. Apparatus

The equipment needed for the surgery includes a heating pad (model TP-22G Harvard Apparatus), a Gaymar T/pump (model TP-500, Gaymar Industries, Orchard Park, NY), a small electric drill with engraving cutter number 105 (Dremel MultiPro, Rotary Tool 395T6 model, Racime, WI), a syringe pump model 100 (KD Scientific, New Hope, PA), polyethylene tubing number 801000 (A-M Systems Inc., Carlsborg, WA), Hamilton syringe (Fisher Scientific, Nepean, ON, Canada), needle FSSP 9715835 model (Fisher Scientific, Nepean, ON, Canada), Polytron PT1200C (Kinematica AG, Switzerland), and microdissecting scissors and tweezers (Harvard Apparatus, Saint-Laurent, QC, Canada).

III. Procedures

A. Injection of [^{35}S]Methionine into Spinal Cord

1. Preparation of [^{35}S]Methionine

   a. Steps

   1. Dry out the [^{35}S]methionine by lyophilization (Speedvac evaporator).
   2. Dissolve the residual concentrate in a phosphate-buffered saline (PBS) solution to obtain a concentration of 500 μCi in 2 μl.
   3. Prepare samples of 2 μl and freeze them at -80 °C until they are injected.

2. Preparation of the Injection Device

   a. Steps

   1. Connect a Hamilton syringe to the tubing attached to needle.
   2. Fill the syringe and the tubing with a PBS solution.
   3. Place the syringe against the piston and adjust the settings of the pump to inject 1 μl at a rate of 0.5 μl/min. Check these settings by measuring the volume of PBS that flows out of the needle in 2 minutes.

3. Surgery

   a. Steps

   1. Deeply anesthetize the mice with a cocktail of xylazine (75 mg/kg)/ketamine (10 mg/kg) diluted in PBS administered intraperitoneally.

   2. Keep the animals on a heating pad during surgery to maintain the body temperature at 37 °C.

   3. Shave the back skin and disinfect with povidone iodine solution. All operating instruments must be sterilized.

   4. Incise the skin with a razor blade and grab the skin on each side of the spinal column with a pair of hemostatic forceps.

   5. The muscles must be carefully incised with surgical scissors and microdissecting tweezers to expose the laminae. Under the operating microscope, drill a 1- to 2-mm^2 window in the right laminae of the vertebra at the level of the L5 spinal cord with a small electric drill to expose the right spinal cord without damage.

   6. Evacuate 5 μl of PBS from the needle, aspirate 2 μl of air. Place the needle in the [^{35}S]methionine solution and carefully aspirate 2 μl. The air volume maintains a space between the PBS solution and the [^{35}S]methionine solution. Place the Hamilton syringe against the piston of the pump.

   7. Under the operating microscope, position the needle into the anterior horn area: 1-mm deep from the dorsal surface and 1 mm from the middle groove.

A smooth and constant flow (1 μl at a rate of 0.5 μl/min) is delivered by the syringe pump. The needle must be kept in place for 2 minutes after the end of injection to avoid flow back of liquid during the removal. Remove the needle slowly.

8. Close the muscles in layers with silk sutures.

9. Apply wound suture clips to close the skin incision. Cover the wound with topical antibiotics.

10. Maintain the mouse warmth until it wakes up. One hour after waking up, the mice are administered a subcutaneous injection of buprenorphine (0.1 mg/kg).

B. Monitoring the Radioactivity

It is recommended that all the surgery and dissection procedures be performed in an isolated location. A fume hood minimizes the risk of inhaling the volatile [^{35}S]methionine. Appropriate precautions have to be followed during handling and housing of injected animals. Litter, food, and carcasses of animals have to be placed and evacuated in a waste container specific for this radioisotope. Similarly, all solutions used for gel electrophoresis have to be collected into a waste disposal specific for this radioisotope. All equipment, apparatus, and surfaces (benches, walls, floors, cages, speedvac) must be monitored for possible [^{35}S]methionine contamination with a "wipe test" procedure. Suspected areas can be wiped with a moist filter paper that is placed in a vial containing scintillation fluid and counted in a liquid scintillation counter for beta-contamination. Any count rate >0.5 B/cm^2 must be decontaminated.

C. Dissection of the Sciatic Nerve and Roots

a. Steps

1. One to 4 weeks after injection, the mice are killed with an overdose of chloral hydrate (1 g/kg) and decapitated.

2. Firmly pin the mouse body on a board and completely remove the skin from the back and the legs.

3. Remove the lamina of vertebra one by one, starting at the cervical (rostral) end to the sacrum (caudal), and carefully remove the arachnoid membrane to expose the spinal cord.

4. Dissect out the right sciatic nerve and L5 ventral and dorsal roots. The sciatic nerve is the main nerve trunk running from the spinal cord down to the leg. Begin the dissection by cutting the leg muscles. The peroneal branch of the sciatic nerve can be found between the two large calf muscles, the peroneus and the gastrocnemius muscles. Horizontally, incise the right leg at the level of these muscles to expose the nerve. Free the sciatic nerve from surrounding tissue by carefully cutting the tissue parallel to the nerve, then by firmly cutting the *ilium* bone. Carefully cut the articular process of L5 vertebra to expose the roots.

5. Grab the muscular extremity of the nerve between forceps and detach the nerve throughout its course up to the spinal cord by also collecting L5 ventral root and L5 dorsal root ganglia (DRG). Also remove the injected region of the spinal cord.

6. Carefully place the entire sciatic nerve and roots onto a plastic plate with a ruler. The ventral root and DRG are pooled into one fraction corresponding to an axonal length of 12 mm (fraction 1). Cut the sciatic nerve into 3-mm consecutive segments starting from the L5 dorsal root ganglion to the muscular extremity (fractions 2–8). Freeze the different fractions in liquid nitrogen and keep them at −80 °C until homogenization.

D. Protein Homogenization and Fractionation

a. Steps

1. Homogenize each segment with a Polytron in 100 μl of 1% Triton X-100 extraction buffer made in 10 mM Tris-HCl, pH 7.5, 150 mM NaCl, 1 mM EDTA, and 10 μg/ml of each protease inhibitor (aprotinin, leupeptin, pepstatin).

2. Centrifuge the mixture at 13,000 rpm for 10 minutes. Collect the Triton X-100–soluble fraction.

3. To the pellet, add 100 μl of a 0.5% SDS buffer containing 8 M urea and 2% 2-mercaptoethanol with protease inhibitors. Mechanically dissociate the pellet. Centrifuge at 13,000 rpm for 10 minutes and collect the supernatant (Triton X-100–insoluble fraction). Measure the protein concentration of the Triton X-100–insoluble fraction with the standard Bradford assay (Bio-Rad, Mississauga, ON, Canada).

4. Add to samples (20 μg of proteins prepared from each nerve segment) SDS sample buffer at final concentration of 62.5 mM Tris, pH 6.8, 10% glycerol, 2% SDS, 5% 2-mercaptoethanol, and 0.001% bromophenol blue. Boil the samples at 100 °C for 5 minutes and load them into wells of 4–8% polyacrylamide gels (1.5-mm thick).

5. Electrophoresis is carried out at a constant current of 30 mA until the bromophenol blue dye has run to the bottom of the gel.

6. Place the polyacrylamide gel in a plastic box and stain it with Coomassie Brilliant blue solution (1% Coomassie blue, 40% methanol, 10% glacial acetic acid) for 30 minutes.

7. Wash the gel in a standard Coomassie destaining and fixative solution (30% methanol, 10% glacial acetic acid) for 1 hour. Pour off the fixing solution in radioactive disposal. Check the quality of the migration; the intensity of NF proteins at 68, 150, and 200 kDa (for NF-L, NF-M and NF-H, respectively); and the similar concentration of proteins loaded in each well.

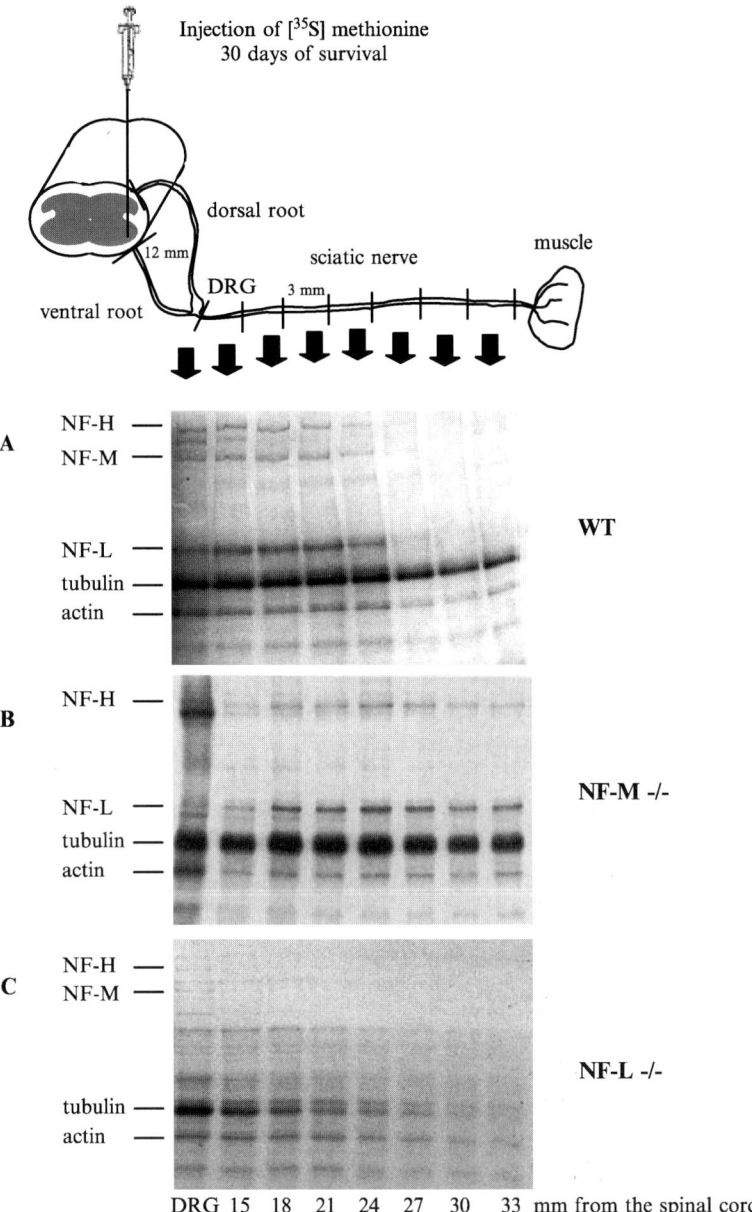

Fig. 1 Axonal transport of neurofilament (NF) proteins in sciatic nerve of normal mice and of mice knockout for NF-M or NF-L. Fluorographs of slow-transport profiles in motor axons of the sciatic nerves from normal (WT) mouse (A), NF-M knockout (B), and NF-L knockout (C) mice 30 days after intraspinal injection of [^{35}S]methionine. L5 ventral root, dorsal root, and dorsal root ganglia (DRG) were pooled in one fraction (DRG), and protein extract was loaded on the first lane (left). Consecutive 3-mm segments of the sciatic nerve were loaded in each successive lane of the gel extending distally to

E. Fluorography

Fluorography is a procedure that increases the sensitivity of detection of ^{35}S-labeled proteins. It is recommended instead of autoradiography. This is achieved through infusion of organic scintillant into the gel. The fluorographic scintillant converts the emitted energy of the isotope to visible light and increases the signal detected by X-ray film.

a. Steps

1. For fluorographic enhancement of the signal, soak the fixed gel in Amplify reagent (Amersham Biosciences, Baie d'Urfe, QC, Canada) with agitation for 30 minutes.

2. Then place the gel on a sheet of Whatman 3 MM filter paper, cover it with a plastic wrap, and dry the gel completely at 60–80 °C for 30–90 minutes under vacuum with a conventional gel dryer.

3. Expose the gel to Kodak Biomax film in a tightly closed cassette at −80 °C. The first exposure time ranges from 5 weeks to 2 months, depending on the signal intensity.

IV. Comments

Alternatively, *in vivo* axonal transport of IF proteins can be done with measurement of pulse radiolabeling analyses in RGCs after intravitreal injection of [^{35}S]methionine (Nixon and Logvinenko, 1986). The primary optic pathway includes the optic nerve, the optic chiasm, and part of the optic tract extending to the lateral geniculate nucleus. In this system, the analysis of axonal transport is usually carried out at 3 and 7 days after radiolabel injection. Basically, the optic pathways from groups of three mice are dissected and cut into consecutive 1-mm segments on dry ice. The Triton X-100–insoluble preparations from each segment are then subjected to SDS-polycrylamide gel electrophoresis and fluorography as described previously.

The rate of NF transport is ~0.25 mm/d in optic axons compared with ~1 mm/d in motor axons. Moreover, be aware that the two systems, spinal motor neurons and RGCs, might yield different results. For example, NF-M requires NF-L for transport in axons of spinal motor neurons (Fig. 1), whereas it is transported at

the right. Positions of NF-H (200 kDa), NF-M (150 kDa), NF-L (68 kDa), tubulin (55 kDa), and actin (45 kDa) are indicated. (A) Slowly transported [^{35}S]-labeled proteins in the Triton X100–insoluble fraction of a wild-type animal. The peak of radioactivity for the three NF subunits occurred 18 mm from the injection site with a leading edge of radioactive protein at 24 mm. In mice lacking NF-M (B), the peak and leading edge of radiolabeled NF-L and NF-M proteins were detected more distally, at 24 and 33 mm, respectively. The velocity of NF-L and NF-H transport was increased in absence of NF-M subunit. As shown in (C), NF-L is required for transport of other NF subunits. In NF-L knockout mice, no radiolabeled NF-M and NF-H proteins were detected in segments of the sciatic nerve.

normal rates in optic axons in the absence of NF-L (Yuan *et al.*, 2003). Another discrepancy is that disruption of the NF-H gene increased the rate of transport of NF-L and NF-M proteins in motor axons (Jacomy *et al.*, 1999), whereas it had no effects on the rate of NF proteins in optic axons (Rao *et al.*, 2002b).

V. Pitfalls

Mortality during or after surgical procedures may occur. This could result from anesthetic overdose, dehydration of the animal, hypothermia, or pain. All these parameters can be easily monitored. Alternatively, mercaptoethanol contained in the initial [^{35}S]methionine solution was not sufficiently evaporated. Because mercaptoethanol is highly toxic, it must be entirely eliminated before injection of [^{35}S]methionine solution.

To avoid cracking the gel during drying, soak it in 7% acetic acid, 7% methanol, and 10% glycerol for 5 minutes before immersing it in Amplify fluorographic reagent.

Darken film. Either the gel was not enough dried or it was overexposed. Reexpose the gel for a shorter time, but avoid developing the film too fast for the first exposure, because the faint signals will be difficult to detect past the half-life of [^{35}S]methionine (87.5 days). Be careful of any leak during the long storage of the film.

There is no signal on the film. Perhaps the needle was blocked during injection. Make sure between two injections that the needle is not clogged by blood or tissue.

To reduce the variation in amount of radiolabeled amino acid incorporated into axonal proteins, corresponding nerve segments from three different animals can be pooled into one fraction.

Bands are diffused. This may be due to a long immersion in scintillator solution or to a poor contact between the gel and the film. Use a good-quality cassette to maintain an even pressure over all film surface.

References

Ackerley, S., Thornhill, P., Grierson, A. J., Brownlees, J., Anderton, B. H., Leigh, P. N., Shaw, C. E., and Miller, C. C. (2003). Neurofilament heavy chain side arm phosphorylation regulates axonal transport of neurofilaments. *J. Cell Biol.* **161,** 489–495.

Al-Chalabi, A., Andersen, P. M., Nilsson, P., Chioza, B., Andersson, J. L., Russ, C., Shaw, C. E., Powell, J. F., and Leigh, P. N. (1999). Deletions of the heavy neurofilament subunit tail in amyotrophic lateral sclerosis. *Hum. Mol. Genet.* **8,** 157–164.

Archer, D. R., Watson, D. F., and Griffin, J. W. (1994). Phosphorylation-dependent immunoreactivity of neurofilaments and the rate of slow axonal transport in the central and peripheral axons of the rat dorsal root ganglion. *J. Neurochem.* **62,** 1119–1125.

Athlan, E. S., and Mushynski, W. E. (1997). Heterodimeric associations between neuronal intermediate filament proteins. *J. Biol. Chem.* **272,** 31073–31078.

Beaulieu, J. M., Kriz, J., and Julien, J. P. (2002). Induction of peripherin expression in subsets of brain neurons after lesion injury or cerebral ischemia. *Brain Res.* **946,** 153–161.

Beaulieu, J. M., Robertson, J., and Julien, J. P. (1999). Interactions between peripherin and neurofilaments in cultured cells: Disruption of peripherin assembly by the NF-M and NF-H subunits. *Biochem. Cell Biol.* **77,** 41–45.

Bennett, G. S., and DiLullo, C. (1985). Slow posttranslational modification of a neurofilament protein. *J. Cell Biol.* **100,** 1799–1804.

Bizzi, A., Crane, R. C., Autilio-Gambetti, L., and Gambetti, P. (1984). Aluminum effect on slow axonal transport: A novel impairment of neurofilament transport. *J. Neurosci.* **4,** 722–731.

Black, M. M., Keyser, P., and Sobel, E. (1986). Interval between the synthesis and assembly of cytoskeletal proteins in cultured neurons. *J. Neurosci.* **6,** 1004–1012.

Bommel, H., Xie, G., Rossoll, W., Wiese, S., Jablonka, S., Boehm, T., and Sendtner, M. (2002). Missense mutation in the tubulin-specific chaperone E (Tbce) gene in the mouse mutant progressive motor neuronopathy, a model of human motoneuron disease. *J. Cell Biol.* **159,** 563–569.

Bomont, P., Cavalier, L., Blondeau, F., Ben Hamida, C., Belal, S., Tazir, M., Demir, E., Topaloglu, H., Korinthenberg, R., Tuysuz, B., Landrieu, P., Hentati, F., and Koenig, M. (2000). The gene encoding gigaxonin, a new member of the cytoskeletal BTB/kelch repeat family, is mutated in giant axonal neuropathy. *Nat. Genet.* **26,** 370–374.

Brown, A. (2000). Slow axonal transport: Stop and go traffic in the axon. *Nat. Rev. Mol. Cell. Biol.* **1,** 153–156.

Brown, A. (2003). Axonal transport of membranous and nonmembranous cargoes: A unified perspective. *J. Cell Biol.* **160,** 817–821.

Brownlees, J., Ackerley, S., Grierson, A. J., Jacobsen, N. J., Shea, K., Anderton, B. H., Leigh, P. N., Shaw, C. E., and Miller, C. C. (2002). Charcot-Marie-Tooth disease neurofilament mutations disrupt neurofilament assembly and axonal transport. *Hum. Mol. Genet.* **11,** 2837–2844.

Chadan, S., Le Gall, J. Y., Di Giamberardino, L., and Filliatreau, G. (1994). Axonal transport of type III intermediate filament protein peripherin in intact and regenerating motor axons of the rat sciatic nerve. *J. Neurosci. Res.* **39,** 127–139.

Ching, G. Y., and Liem, R. K. (1993). Assembly of type IV neuronal intermediate filaments in nonneuronal cells in the absence of preexisting cytoplasmic intermediate filaments. *J. Cell Biol.* **122,** 1323–1335.

Collard, J. F., Cote, F., and Julien, J. P. (1995). Defective axonal transport in a transgenic mouse model of amyotrophic lateral sclerosis. *Nature* **375,** 61–64.

Cui, C., Stambrook, P. J., and Parysek, L. M. (1995). Peripherin assembles into homopolymers in SW13 cells. *J. Cell Sci.* **108**(Pt. 10), 3279–3284.

de Waegh, S., and Brady, S. T. (1990). Altered slow axonal transport and regeneration in a myelin-deficient mutant mouse: The trembler as an *in vivo* model for Schwann cell-axon interactions. *J. Neurosci.* **10,** 1855–1865.

Ding, J., Liu, J. J., Kowal, A. S., Nardine, T., Bhattacharya, P., Lee, A., and Yang, Y. (2002). Microtubule-associated protein 1B: A neuronal binding partner for gigaxonin. *J. Cell Biol.* **158,** 427–433.

Figlewicz, D. A., Krizus, A., Martinoli, M. G., Meininger, V., Dib, M., Rouleau, G. A., and Julien, J. P. (1994). Variants of the heavy neurofilament subunit are associated with the development of amyotrophic lateral sclerosis. *Hum. Mol. Genet.* **3,** 1757–1761.

Filliatreau, G., Denoulet, P., de Nechaud, B., and Di Giamberardino, L. (1988). Stable and metastable cytoskeletal polymers carried by slow axonal transport. *J. Neurosci.* **8,** 2227–2233.

Fliegner, K. H., Kaplan, M. P., Wood, T. L., Pintar, J. E., and Liem, R. K. (1994). Expression of the gene for the neuronal intermediate filament protein alpha-internexin coincides with the onset of neuronal differentiation in the developing rat nervous system. *J. Comp. Neurol.* **342,** 161–173.

Griffin, J. W., Parhad, I., Gold, B., Price, D. L., Hoffman, P. N., and Fahnestock, K. (1985). Axonal transport of neurofilament proteins in IDPN neurotoxicity. *Neurotoxicology* **6,** 43–53.

Hafezparast, M., Klocke, R., Ruhrberg, C., Marquardt, A., Ahmad-Annuar, A., Bowen, S., Lalli, G., Witherden, A. S., Hummerich, H., Nicholson, S., Morgan, P. J., Oozageer, R., Priestley, J. V., Averill, S., King, V. R., Ball, S., Peters, J., Toda, T., Yamamoto, A., Hiraoka, Y., Augustin, M.,

Korthaus, D., Wattler, S., Wabnitz, P., Dickneite, C., Lampel, S., Boehme, F., Peraus, G., Popp, A., Rudelius, M., Schlegel, J., Fuchs, H., Hrabe de Angelis, M., Schiavo, G., Shima, D. T., Russ, A. P., Stumm, G., Martin, J. E., and Fisher, E. M. (2003). Mutations in dynein link motor neuron degeneration to defects in retrograde transport. *Science* **300,** 808–812.

Hirokawa, N., Glicksman, M. A., and Willard, M. B. (1984). Organization of mammalian neurofilament polypeptides within the neuronal cytoskeleton. *J. Cell Biol.* **98,** 1523–1536.

Ho, C. L., Chin, S. S., Carnevale, K., and Liem, R. K. (1995). Translation initiation and assembly of peripherin in cultured cells. *Eur. J. Cell Biol.* **68,** 103–112.

Hoffman, P. N., and Lasek, R. J. (1975). The slow component of axonal transport. Identification of major structural polypeptides of the axon and their generality among mammalian neurons. *J. Cell Biol.* **66,** 351–366.

Jacomy, H., Zhu, Q., Couillard-Despres, S., Beaulieu, J. M., and Julien, J. P. (1999). Disruption of type IV intermediate filament network in mice lacking the neurofilament medium and heavy subunits. *J. Neurochem.* **73,** 972–984.

Jordanova, A., De Jonghe, P., Boerkoel, C. F., Takashima, H., De Vriendt, E., Ceuterick, C., Martin, J. J., Butler, I. J., Mancias, P., Papasozomenos, S., Terespolsky, D., Potocki, L., Brown, C. W., Shy, M., Rita, D. A., Tournev, I., Kremensky, I., Lupski, J. R., and Timmerman, V. (2003). Mutations in the neurofilament light chain gene (NEFL) cause early onset severe Charcot-Marie-Tooth disease. *Brain* **126,** 590–597.

Julien, J. P., Smoluk, G. D., and Mushynski, W. E. (1983). Characteristics of the protein kinase activity associated with rat neurofilament preparations. *Biochim. Biophys. Acta* **755,** 25–31.

Jung, C., Yabe, J. T., Lee, S., and Shea, T. B. (2000a). Hypophosphorylated neurofilament subunits undergo axonal transport more rapidly than more extensively phosphorylated subunits *in situ*. *Cell. Motil. Cytoskeleton* **47,** 120–129.

Jung, C., Yabe, J. T., and Shea, T. B. (2000b). C-terminal phosphorylation of the high molecular weight neurofilament subunit correlates with decreased neurofilament axonal transport velocity. *Brain Res.* **856,** 12–19.

Kaplan, M. P., Chin, S. S., Fliegner, K. H., and Liem, R. K. (1990). Alpha-internexin, a novel neuronal intermediate filament protein, precedes the low molecular weight neurofilament protein (NF-L) in the developing rat brain. *J. Neurosci.* **10,** 2735–2748.

Kirkpatrick, L. L., Witt, A. S., Payne, H. R., Shine, H. D., and Brady, S. T. (2001). Changes in microtubule stability and density in myelin-deficient shiverer mouse CNS axons. *J. Neurosci.* **21,** 2288–2297.

LaMonte, B. H., Wallace, K. E., Holloway, B. A., Shelly, S. S., Ascano, J., Tokito, M., Van Winkle, T., Howland, D. S., and Holzbaur, E. L. (2002). Disruption of dynein/dynactin inhibits axonal transport in motor neurons causing late-onset progressive degeneration. *Neuron* **34,** 715–727.

Lasek, R. J. (1986). Polymer sliding in axons. *J. Cell Sci. Suppl.* **5,** 161–179.

Lasek, R. J., Garner, J. A., and Brady, S. T. (1984). Axonal transport of the cytoplasmic matrix. *J. Cell Biol.* **99,** 212s–221s.

Lee, M. K., Xu, Z., Wong, P. C., and Cleveland, D. W. (1993). Neurofilaments are obligate heteropolymers *in vivo*. *J. Cell Biol.* **122,** 1337–1350.

Lewis, S. E., and Nixon, R. A. (1988). Multiple phosphorylated variants of the high molecular mass subunit of neurofilaments in axons of retinal cell neurons: Characterization and evidence for their differential association with stationary and moving neurofilaments. *J. Cell Biol.* **107,** 2689–2701.

Marszalek, J. R., Williamson, T. L., Lee, M. K., Xu, Z., Hoffman, P. N., Becher, M. W., Crawford, T. O., and Cleveland, D. W. (1996). Neurofilament subunit NF-H modulates axonal diameter by selectively slowing neurofilament transport. *J. Cell Biol.* **135,** 711–724.

Martin, N., Jaubert, J., Gounon, P., Salido, E., Haase, G., Szatanik, M., and Guenet, J. L. (2002). A missense mutation in Tbce causes progressive motor neuronopathy in mice. *Nat. Genet.* **32,** 443–447.

Medori, R., Autilio-Gambetti, L., Monaco, S., and Gambetti, P. (1985). Experimental diabetic neuropathy: impairment of slow transport with changes in axon cross-sectional area. *Proc. Natl. Acad. Sci. USA* **82,** 7716–7720.

Medori, R., Jenich, H., Autilio-Gambetti, L., and Gambetti, P. (1988). Experimental diabetic neuropathy: Similar changes of slow axonal transport and axonal size in different animal models. *J. Neurosci.* **8,** 1814–1821.

Miller, C. C., Ackerley, S., Brownlees, J., Grierson, A. J., Jacobsen, N. J., and Thornhill, P. (2002). Axonal transport of neurofilaments in normal and disease states. *Cell Mol. Life Sci.* **59,** 323–330.

Mitsumoto, H., and Gambetti, P. (1986). Impaired slow axonal transport in wobbler mouse motor neuron disease. *Ann. Neurol.* **19,** 36–43.

Monaco, S., Autilio-Gambetti, L., Lasek, R. J., Katz, M. J., and Gambetti, P. (1989). Experimental increase of neurofilament transport rate: Decreases in neurofilament number and in axon diameter. *J. Neuropathol. Exp. Neurol.* **48,** 23–32.

Monaco, S., Autilio-Gambetti, L., Zabel, D., and Gambetti, P. (1985). Giant axonal neuropathy: Acceleration of neurofilament transport in optic axons. *Proc. Natl. Acad. Sci. USA* **82,** 920–924.

Morris, J. R., and Lasek, R. J. (1982). Stable polymers of the axonal cytoskeleton: the axoplasmic ghost. *J. Cell Biol.* **92,** 192–198.

Muma, N. A., Hoffman, P. N., Slunt, H. H., Applegate, M. D., Lieberburg, I., and Price, D. L. (1990). Alterations in levels of mRNAs coding for neurofilament protein subunits during regeneration. *Exp. Neurol.* **107,** 230–235.

Nixon, R. A., and Lewis, S. E. (1986). Differential turnover of phosphate groups on neurofilament subunits in mammalian neurons *in vivo. J. Biol Chem.* **261,** 16298–16301.

Nixon, R. A., Lewis, S. E., Dahl, D., Marotta, C. A., and Drager, U. C. (1989). Early posttranslational modifications of the three neurofilament subunits in mouse retinal ganglion cells: Neuronal sites and time course in relation to subunit polymerization and axonal transport. *Brain Res. Mol Brain Res.* **5,** 93–108.

Nixon, R. A., Lewis, S. E., and Marotta, C. A. (1987). Posttranslational modification of neurofilament proteins by phosphate during axoplasmic transport in retinal ganglion cell neurons. *J. Neurosci.* **7,** 1145–1158.

Nixon, R. A., and Logvinenko, K. B. (1986). Multiple fates of newly synthesized neurofilament proteins: Evidence for a stationary neurofilament network distributed nonuniformly along axons of retinal ganglion cell neurons. *J. Cell Biol.* **102,** 647–659.

Oblinger, M. M., Wong, J., and Parysek, L. M. (1989). Axotomy-induced changes in the expression of a type III neuronal intermediate filament gene. *J. Neurosci.* **9,** 3766–3775.

Okabe, S., Miyasaka, H., and Hirokawa, N. (1993). Dynamics of the neuronal intermediate filaments. *J. Cell Biol.* **121,** 375–386.

Pappolla, M., Penton, R., Weiss, H. S., Miller, C. H., Jr., Sahenk, Z., Autilio-Gambetti, L., and Gambetti, P. (1987). Carbon disulfide axonopathy. Another experimental model characterized by acceleration of neurofilament transport and distinct changes of axonal size. *Brain Res.* **424,** 272–280.

Parhad, I. M., Griffin, J. W., Hoffman, P. N., and Koves, J. F. (1986). Selective interruption of axonal transport of neurofilament proteins in the visual system by beta,beta'-iminodipropionitrile (IDPN) intoxication. *Brain Res.* **363,** 315–324.

Parysek, L. M., Chisholm, R. L., Ley, C. A., and Goldman, R. D. (1988). A type III intermediate filament gene is expressed in mature neurons. *Neuron* **1,** 395–401.

Parysek, L. M., McReynolds, M. A., Goldman, R. D., and Ley, C. A. (1991). Some neural intermediate filaments contain both peripherin and the neurofilament proteins. *J. Neurosci. Res.* **30,** 80–91.

Perez-Olle, R., Leung, C. L., and Liem, R. K. (2002). Effects of Charcot-Marie-Tooth-linked mutations of the neurofilament light subunit on intermediate filament formation. *J. Cell Sci.* **115,** 4937–4946.

Portier, M. M., Croizat, B., and Gros, F. (1982). A sequence of changes in cytoskeletal components during neuroblastoma differentiation. *FEBS Lett.* **146,** 283–288.

Rao, M. V., Engle, L. J., Mohan, P. S., Yuan, A., Qui, D., Cataldo, A., Hassinger, L., Jacobsen, S., Lee, V. M., Andreadis, A., Julien, J. P., Bridgman, P. C., and Nixon, R. A. (2002a). Myosin Va binding to neurofilaments is essential for correct myosin Va distribution and transport and neurofilament density. *J. Cell Biol.* **159,** 279–290.

Rao, M. V., Garcia, M. L., Miyazaki, Y., Gotow, T., Yuan, A., Mattina, S., Ward, C. M., Calcutt, N. A., Uchiyama, Y., Nixon, R. A., and Cleveland, D. W. (2002b). Gene replacement in mice reveals that the heavily phosphorylated tail of neurofilament heavy subunit does not affect axonal caliber or the transit of cargoes in slow axonal transport. *J. Cell Biol.* **158,** 681–693.

Robertson, J., Doroudchi, M. M., Nguyen, M. D., Durham, H. D., Strong, M. J., Shaw, G., Julien, J. P., and Mushynski, W. E. (2003). A neurotoxic peripherin splice variant in a mouse model of ALS. *J. Cell Biol.* **160,** 939–949.

Roy, S., Coffee, P., Smith, G., Liem, R. K., Brady, S. T., and Black, M. M. (2000). Neurofilaments are transported rapidly but intermittently in axons: Implications for slow axonal transport. *J. Neurosci.* **20,** 6849–6861.

Shah, J. V., Flanagan, L. A., Janmey, P. A., and Leterrier, J. F. (2000). Bidirectional translocation of neurofilaments along microtubules mediated in part by dynein/dynactin. *Mol. Biol. Cell* **11,** 3495–3508.

Stein, S. A., Kirkpatrick, L. L., Shanklin, D. R., Adams, P. M., and Brady, S. T. (1991). Hypothyroidism reduces the rate of slow component A (SCa) axonal transport and the amount of transported tubulin in the hyt/hyt mouse optic nerve. *J. Neurosci. Res.* **28,** 121–133.

Sternberger, L. A., and Sternberger, N. H. (1983). Monoclonal antibodies distinguish phosphorylated and nonphosphorylated forms of neurofilaments *in situ. Proc. Natl. Acad. Sci. USA* **80,** 6126–6130.

Terada, S., Nakata, T., Peterson, A. C., and Hirokawa, N. (1996). Visualization of slow axonal transport *in vivo. Science* **273,** 784–788.

Terao, E., Janssens, S., van den Bosch de Aguilar, P., Portier, M., and Klosen, P. (2000). *In vivo* expression of the intermediate filament peripherin in rat motoneurons: Modulation by inhibitory and stimulatory signals. *Neuroscience* **101,** 679–688.

Tomkins, J., Usher, P., Slade, J. Y., Ince, P. G., Curtis, A., Bushby, K., and Shaw, P. J. (1998). Novel insertion in the KSP region of the neurofilament heavy gene in amyotrophic lateral sclerosis (ALS). *Neuroreport* **9,** 3967–3970.

Troy, C. M., Muma, N. A., Greene, L. A., Price, D. L., and Shelanski, M. L. (1990). Regulation of peripherin and neurofilament expression in regenerating rat motor neurons. *Brain Res.* **529,** 232–238.

Tytell, M., Black, M. M., Garner, J. A., and Lasek, R. J. (1981). Axonal transport: Each major rate component reflects the movement of distinct macromolecular complexes. *Science* **214,** 179–181.

Wang, L., and Brown, A. (2002). Rapid movement of microtubules in axons. *Curr. Biol.* **12,** 1496–1501.

Wang, L., Ho, C. L., Sun, D., Liem, R. K., and Brown, A. (2000). Rapid movement of axonal neurofilaments interrupted by prolonged pauses. *Nat. Cell Biol.* **2,** 137–141.

Watson, D. F., Griffin, J. W., Fittro, K. P., and Hoffman, P. N. (1989). Phosphorylation-dependent immunoreactivity of neurofilaments increases during axonal maturation and beta,beta′-iminodipropionitrile intoxication. *J. Neurochem.* **53,** 1818–1829.

Williamson, T. L., and Cleveland, D. W. (1999). Slowing of axonal transport is a very early event in the toxicity of ALS-linked SOD1 mutants to motor neurons. *Nat. Neurosci.* **2,** 50–56.

Xia, C. H., Roberts, E. A., Her, L. S., Liu, X., Williams, D. S., Cleveland, D. W., and Goldstein, L. S. (2003). Abnormal neurofilament transport caused by targeted disruption of neuronal kinesin heavy chain KIF5A. *J. Cell Biol.* **161,** 55–66.

Yabe, J. T., Pimenta, A., and Shea, T. B. (1999). Kinesin-mediated transport of neurofilament protein oligomers in growing axons. *J. Cell Sci.* **112**(Pt. 21), 3799–3814.

Yuan, A., Rao, M. V., Kumar, A., Julien, J. P., and Nixon, R. A. (2003). Neurofilament transport *in vivo* minimally requires hetero-oligomer formation. *J. Neurosci.* **23,** 9452–9458.

Zhang, B., Tu, P., Abtahian, F., Trojanowski, J. Q., and Lee, V. M. (1997). Neurofilaments and orthograde transport are reduced in ventral root axons of transgenic mice that express human SOD1 with a G93A mutation. *J. Cell Biol.* **139,** 1307–1315.

Zhao, C., Takita, J., Tanaka, Y., Setou, M., Nakagawa, T., Takeda, S., Yang, H. W., Terada, S., Nakata, T., Takei, Y., Saito, M., Tsuji, S., Hayashi, Y., and Hirokawa, N. (2001). Charcot-Marie-Tooth disease type 2A caused by mutation in a microtubule motor KIF1Bbeta. *Cell* **105,** 587–597.

Zhu, Q., Lindenbaum, M., Levavasseur, F., Jacomy, H., and Julien, J. P. (1998). Disruption of the NF-H gene increases axonal microtubule content and velocity of neurofilament transport: Relief of axonopathy resulting from the toxin beta, beta′-iminodipropionitrile. *J. Cell Biol.* **143,** 183–193.

Zuchner, S., Vorgerd, M., Sindern, E., and Schroder, J. M. (2004). The novel neurofilament light (NEFL) mutation Glu397Lys is associated with a clinically and morphologically heterogeneous type of Charcot-Marie-Tooth neuropathy. *Neuromuscul. Disord.* **14,** 147–157.

CHAPTER 20

Lamins

Georg Krohne

Division of Electron Microscopy
Biocenter of the University of Würzburg
Am Hubland
D-97074 Würzburg, Germany

METHODS IN CELL BIOLOGY, VOL. 78
0091-679X/04 $35.00

I. Introduction

The aim of this article is to describe the biochemical properties, some methods, and tools allowing us to enrich, purify, or immunoprecipitate lamins. Therefore, the chapter focuses primarily on structural aspects of lamins that are relevant for these topics. The molecular architecture of the nuclear envelope, the interaction of lamina proteins, the evolution of lamins, dynamics of nuclear envelope proteins during mitosis, the structure of lamin polymers *in vivo*, and diseases caused by mutations of lamins and other nuclear envelope proteins have been reviewed recently (Cohen *et al.*, 2001; Dechat *et al.*, 2000a; Goldman *et al.*, 2002; Gruenbaum *et al.*, 2003; Krohne, 1998).

Lamins are type V intermediate filament (IF) proteins that are localized subjacent to the inner nuclear membrane and are major structural components of the nuclear lamina. Each lamin molecule is composed of three main structural domains (Fig. 1A). These are the short aminoterminal head domain of 26–55 amino acids in length followed by the central α-helical rod domain (354 amino acids in length; for exceptions for the *C. elegans* lamin; Karabinos *et al.*, 2003; Riemer *et al.*, 1993) and the carboxyterminal tail. The α-helical segments (1A, 1B, 2; Fig. 1A) of the rod domain contain heptad repeats with hydrophobic amino acids at the first and fourth position, characteristic of coiled coil-forming proteins. The carboxyterminal tail has in most lamins a length of 160–260 amino acids (for exceptions see Karabinos *et al.*, 2003). The central 115 amino acids of the carboxyterminal tail are organized into an immunoglobulin (Ig) fold (Krimm *et al.*, 2002) that can be seen as spherical particle by electron microscopy (for exceptions see Karabinos *et al.*, 2003). The carboxyterminal tail contains two sequence elements that are not present in cytoplasmic IF proteins. These are the nuclear localization

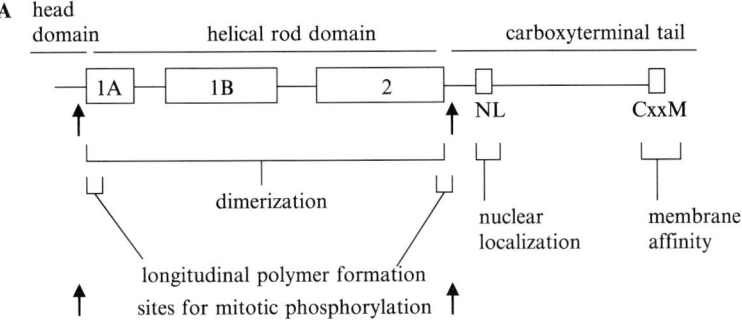

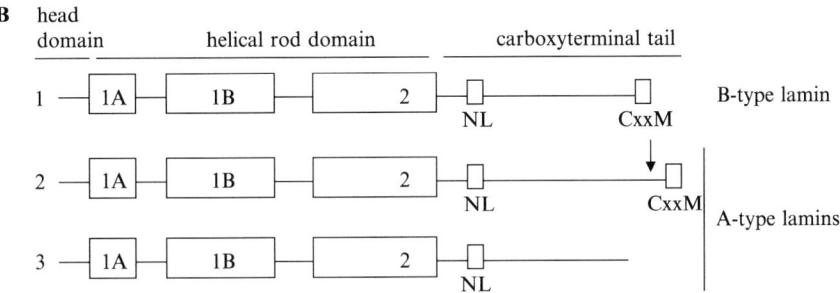

Fig. 1 Schematic drawings of a lamin molecule (A) and lamin isotypes (B) of vertebrates and invertebrates. (A) The three major domains (head domain, helical rod domain, carboxyterminal tail) are marked. The helical rod domain contains three α-helical segments with heptad repeats (1A, 1B, 2). Domains of the lamin molecule involved in nuclear localization (NL), the interaction with the inner nuclear membrane (CxxM motif; membrane affinity; C, cysteine; x, any amino acid; M, methionine), in individual steps of polymer formation (dimerization, longitudinal polymer formation) and in the disassembly of the polymer (sites for mitotic phoshorylation; arrows) are marked. (B) B-type (1) and A-type lamins (2, 3) of vertebrates and invertebrates. B-type lamins of vertebrates include lamins B1, B2, and in addition LIII in amphibians and fishes (Hofemeister *et al.*, 2002). In the analyzed invertebrates, one B-lamin has been identified (Erber *et al.*, 1999; Karabinos *et al.*, 2003). A-type lamins of vertebrates contain a CxxM motif in their primary amino acid sequence (2) except for the mammalian lamin C (3). During maturation of lamin A, the CxxM motif including the carboxyterminal 18 (mammals; Weber *et al.*, 1989) or 20 (chicken; Hennekes and Nigg, 1994) amino acids are proteolytically removed (arrow). The lamin C of *Drosophila melanogaster* does not contain a CxxM motif in the primary amino acid sequence (3). Germ line–specific splice variants of the mammalian lamin A gene and the lamin B2 gene are not shown.

signal (NL) and the CxxM motif (CxxM; C, cysteine; x, any amino acid; M, methionine) at the direct carboxyterminus. The CxxM motif is posttranslationally processed by the farnesylation of the cysteine, followed by proteolytic removal of the last three amino acids and carboxyl methylation of the cysteine. These modifications confer a higher hydrophobicity to the lamin carboxyterminus and are essential for the targeting to the inner nuclear membrane (Maske *et al.*, 2003).

Lamins are present in cells in two major forms. Immediately after synthesis, they form parallel aligned dimers, and they are also present as dimers in mitotic cells. Lamin polymers are formed by the association of the evolutionary conserved end domains of the α-helical rod (Fig. 1A), resulting in the formation of head–tail polymers. It is not known how many of these head–tail polymers are aggregated laterally in a staggered form in the lamina (Krohne, 1998). During mitosis, lamins are phosphorylated at evolutionary conserved serines flanking the rod domain on both sides (Fig. 1A), resulting in the disassembly of the polymer into dimers that are either soluble (A-type lamins) or remain associated with membranes (B-type lamins).

Two major lamin isotypes can be distinguished according to the presence or absence of the modified cysteine of the CxxM motif. The farnesylated and carboxyl methylated cysteine of the CxxM motif remains permanently attached to B-type lamins (Fig. 1B), whereas it is proteolytically removed from A-type lamins during the final maturation of the carboxyterminus. In two characterized A-type lamins, no CxxM motif is present in the primary sequence (Fig. 1B). Further major differences between A-type and B-type lamins are their behavior during mitosis (see earlier) and their expression during embryonic development. B-type lamins are expressed in the embryo and all somatic cells, whereas the expression of A lamins starts late during embryonic development in vertebrates and insects (for review see Gruenbaum *et al.*, [2003]).

II. Materials

A. Stock Solutions Required for Most Preparations

MgCl$_2$: 1 M MgCl$_2$ in H$_2$O (stored at −20 °C).

Dithiothreitol (DTT): 1 M in H$_2$O, store at −20 °C.

Phenylmethylsulfonylchloride (PMSF): stock solution (stored at 4 °C) is 200 mM in methanol; final concentration in the buffer is 0.2 mM.

Trypsin inhibitor: stock solution (stored at −20 °C) is 100 mg/ml H$_2$O; final concentration in the buffer: 10 μg/ml.

DNase (50 mg/ml): 25 mg of DNase (DNAse I, AppliChem, Darmstadt, Germany) is solubilized in 250 μl Tris-buffer (100 mM NaCl, 2 mM MgCl$_2$, 20 mM Tris, pH 7.5), then 250 μl 100% glycerol is added. The stock solution (50 mg DNase/ml) is stored at −20 °C.

RNase (10 mg/ml): 10 mg of RNase (RNAse A, Sigma R5503; Sigma Munich, Germany) is solubilized in 500 μl Tris-buffer (100 mM NaCl, 2 mM MgCl$_2$, 20 mM Tris, pH 7.5) and heated for 10 minutes to 80 °C to inactivate proteases. Then 500 μl 100% glycerol is added. The stock solution (10 mg RNase/ml) is stored at −20 °C.

B. Stock Solutions, Buffers, and Reagents for the Production of Lamins in Bacteria and Lamin Isolation

Ampicillin: (100 mg/ml H_2O), sterile filtered, store at $-20\,°C$.

IPTG (isopropyl-1-thio-β-D-galactopyranoside): 1 M in H_2O, store at $-20\,°C$.

LB-medium (1000 ml): 10 g Bacto Tryptone, 5 g Bacto Yeast Extract, 10 g NaCl, pH 7.4, adjust with NaOH.

LB-plates: LB-medium is supplemented with 1.5% Bacto Agar, autoclaved, cooled to 50 °C, ampicillin is added (150 μg ampicillin/ml), and plates are poured. Bacto Tryptone, Bacto Yeast Extract, and Bacto Agar are from Becton Dickinson (Le Pont de Claix, France).

Bacterial lysis buffer: 1 mM ethylenediamine tetraacetic acid (EDTA) 25% sucrose, 50 mM Tris, pH 8.0.

Bacterial detergent buffer: 0.2 M NaCl, 1% (w/v) sodium deoxycholic acid, 1.5% (w/v) Nonidet P-40, 2 mM EDTA, 20 mM Tris, pH 7.5

Bacterial Triton buffer: 0.75% Triton X-100, 2 mM EDTA; pH 7.5, adjusted with NaOH.

Bio-Rad Poly-Prep column (10 ml volume; Bio-Rad, Hercules, CA)

DEAE-Sepharose (fast flow; Pharmacia, Freiburg, Germany)

CM-Sepharose (fast flow; Pharmacia, Freiburg, Germany)

C. Buffers for the Isolation of Lamins from Sf9 Cells

Sf9-phosphate buffer: 100 mM NaCl, 1 mM EGTA, 1 mM EDTA, 10 mM Na_2HPO_4; trypsin inhibitor: 10 μg/ml; PMSF: 0.2 mM; pH 7.2.

Sf9-phosphate/Triton buffer: 0.5% Triton X-100, 50 mM NaCl, 1 mM EDTA, 1 mM EGTA, 10 mM Na_2HPO_4; trypsin inhibitor: 10 μg/ml; PMSF: 0.2 mM; pH 7.2.

DNase II buffer: 2 mM $MgCl_2$, 10 mM Tris, pH 7.5.

D. Buffers for the Isolation and Extraction of Liver Nuclei

Medium A: 0.44 M sucrose, 2 mM $MgCl_2$, 70 mM KCl, 10 mM Tris, pH 7.4.

2.1 M sucrose: 2.1 M sucrose, 2 mM $MgCl_2$, 10 mM Tris, pH 7.4.

2.6 M sucrose: 2.6 M sucrose, 2 mM $MgCl_2$, 10 mM Tris, pH 7.4.

DNase I buffer: 2 mM $MgCl_2$, 10 mM Tris, pH 8.5.

DNase II buffer: 2 mM $MgCl_2$, 10 mM Tris, pH 7.5.

5 mM Tris: 5 mM Tris, pH 7.5.

10 mM Tris: 10 mM Tris, pH 7.5.

High salt buffer: 500 mM NaCl, 1 mM DTT, 10 mM Tris, pH 7.5.

Triton high salt buffer: 2% Triton X-100, 500 mM KCl, 1 mM DTT, 20 mM Tris, pH 9.0.

E. Buffers for the Enrichment and Solubilization of Lamins from HeLa and *Xenopus* A6 Cells

Phosphate-buffered saline (PBS): 137 mM NaCl, 3 mM KCl, 7 mM Na_2HPO_4, 1.5 mM KH_2PO_4, pH 7.2–7.4.

10 mM Tris: 10 mM Tris, pH 7.5.

DNase I buffer: 2 mM $MgCl_2$, 10 mM Tris, pH 8.5.

DNase II buffer: 2 mM $MgCl_2$, 10 mM Tris, pH 7.5.

5 mM Tris: 5 mM Tris, pH 7.5.

High salt buffer: 500 mM NaCl, 1 mM DTT, 10 mM Tris, pH 7.5.

HeLa-Triton/pH 9.0 buffer: 1% Triton X-100, 150 mM NaCl, 50 mM Tris, pH 9.0.

Xenopus-Triton/pH 7.4 buffer: 1% Triton X100, 150 mM NaCl, 50 mM Tris, pH 7.4.

F. Buffers for Lamin Immunoprecipitations

Human, *Xenopus, Drosophila* Cells: see Section II.G

IP-buffer: 1% Triton X100, 150 mM NaCl, 50 mM Tris, pH 7.4.

G. Cell Culture Media

HeLa cells (human epitheloid carcinoma cells): Dulbecco's modified Eagle's medium (DMEM) (GIBCO/Invitrogen, Karlsruhe, Germany) containing 10% fetal calf serum (FCS), 100 U/ml penicillin, 100 IU/ml streptomycin, and 2 mM l-glutamine; 37 °C, 5% CO_2.

Xenopus A6 cells (kidney epithelial cells of Xenopus laevis): 18.5 ml H_2O, 15 ml FCS, 100 IU/ml penicillin, 100 IU/ml streptomycin, 2 mM L-glutamine, add DMEM (GIBCO) to a final volume of 150 ml; 27 °C, 5% CO_2.

Sf9 cells (butterfly Spodoptera frugiperda): TC100 medium (GIBCO) supplemented with 10% FCS, 100 IU/ml penicillin, 100 IU streptomycin, and 2 mM L-glutamine; 27 °C, no CO_2. Cells are routinely grown in 50-ml Falcon flasks (Labor Schubert, Schwandorf/Germany) containing 5 ml medium.

Drosophila S2 and Kc167 cells: Schneider's *Drosophila* medium (GIBCO) containing 10% FCS, 50 IU/ml penicillin, 50 IU/ml streptomycin, and 2 mM L-glutamine; 27 °C, no CO_2. Cells are routinely grown in 50-ml Falcon flasks (Labor Schubert) containing 5 ml medium.

III. Purification of Lamins Produced in Bacteria

If larger quantities of a specific lamin are required for *in vitro* binding studies, the generation of specific antibodies, or the affinity purification of polyclonal lamin antibodies, it is very helpful to produce the specific lamin in bacteria. In my laboratory, we have cloned the complete coding sequence of a lamin cDNA including the stop codon into the bacterial expression vectors pET17 or pET21 (Calbiochem-Novabiochem, Schwalbach, Germany) using, if possible, the restriction enzymes NdeI (5′ end) and XhoI (3′ end). We are routinely using the bacterial strain *Escherichia coli* BL21-CodonPlus (Stratagene, Amsterdam, The Netherlands) for the transformation with pET vectors. *E. coli* BL21-CodonPlus bacteria containing a pET lamin vector can be induced to synthesize this lamin by the addition of IPTG to the culture medium.

A. Steps for the Identification of Lamin-Producing Bacterial Colonies

Day 1: Transform *E. coli* BL21-CodonPlus cells with the pET-lamin plasmid, plate the cells on LB plates containing 150 μg ampicillin/ml, and grow them overnight at 37 °C. We grow this bacterial strain routinely at 37 °C; all liquid cultures are grown under shaking in LB-medium containing 100 μg ampicillin/ml.

Day 2: Bacteria of individual colonies are grown overnight in 2–5 ml LB medium.

Day 3: Inoculate 10 ml of LB medium with 0.5–1 ml of an overnight culture, measure the cell density (OD_{600}), grow the cells to a density of 0.5–0.8 (OD_{600}). Take 1 ml of the cells, transfer them into an 1.5-ml Eppendorf tube (Eppendorf, Hamburg, Germany), spin the cells down (3 minutes at 3000g in a tabletop centrifuge), and freeze the pellet (control; noninduced bacteria). Induce the protein synthesis by addition of IPTG (1 mM final concentration) to the remaining 9 ml of the culture. Let the bacteria cells grow for another 5 hours, take 1 ml of the culture, pellet, and freeze the cells as described previously for the control.

Day 4: Proteins of bacterial pellets were separated by sodium dodecylsulfate-polyacrylamide gel electrophoresis (SDS-PAGE) (Laemmli, 1970), transferred to nitrocellulose (Kyhse-Andersen, 1984), and incubated with commercially available lamin antibodies (Table I). Antibodies bound to the filter are detected with an enhanced chemiluminescence system (Amersham Buchler, Braunschweig, Germany) by exposure to x-ray film (X-Omat AR; Eastman Kodak C., Rochester, NY). All lamins analyzed in my laboratory (*Xenopus* lamins A, B1; *Drosophila* lamins Dm0, C; 1mn-1 of *C. elegans*) were abundant in the bacteria after induction with IPTG and could be easily visualized in SDS-PAGE after staining with Coomassie blue (Fig. 2A; see Table II for molecular weights of lamins). A bacterial colony that expresses the desired lamin is then inoculated overnight in 50 ml LB medium.

Table I
Commercially Available Mouse Monoclonal Lamin Antibodies

Lamin antibody	Isotype; recognized species	Supplier
R27	Lamin A/C; A-type lamins from mammals to fish	Progen
X223	Lamin B2; from mammals to *Xenopus*	Progen
X67	Lamins A/C, B1; from mammals to *Xenopus*, negative on mouse and rat	Progen
X155	Lamin B2; *Xenopus*, trout, *Danio*; lamin LIII in *Xenopus*	Progen
X167	Lamins A/C, B1, B2; human to trout, weak on mouse and rat	Progen
IQ168	Lamin A; human, rat, mouse, bovine, dog	ImmuQuest
IQ169	Lamin B1; rat human, mouse, bovine, rabbit dog, sheep	ImmuQuest
IQ170	Lamin B2; human, mouse, hamster, *Xenopus*	ImmuQuest
ADL195	Lamin Dm0; *Drosophila*	DSHB
ADL84	Lamin Dm0; *Drosophila*	DSHB
ADL67	Lamin Dm0; *Drosophila*	DSHB
LC28	Lamin C; *Drosophila*	DSHB

The antibodies can be purchased from the following companies. ProgenBiotechnik GmbH, Maaβstraβe 30, 69123 Heidelberg, Germany, tel: +49 6221 8278-0, fax: +49 6221 827824, www.progen.de; ImmuQuest Ltd, 18 Houghton Banks, Ingleby Barwick, Cleveland, TS17 5AL, UK, tel: +44(0) 1642 308 831, fax: +44(0) 1642 769 146, Email: cutomerservice@immuquest.com; and Developmental Studies Hybridoma Bank (DSHB), Department of Biological Sciences, The University of Iowa, 28 Biology Building East, Iowa City, IA 52242, tel: (319)335-3826, fax: (319)335-2077.

Day 5: Inoculate 2×200 ml LB medium with 5–10 ml of the overnight culture, measure the cell density (OD_{600}), grow the cells to a density of 0.5–0.8 (OD_{600}), induce the protein synthesis by addition of IPTG (1 mM final concentration), and let the bacteria cells grow for another 5 hours. Pellet the bacteria (10 minutes, $4000g$) of a 200-ml culture stepwise in two sterile centrifuge tubes with a screw cap (50-ml volume; Hartenstein, Würzburg, Germany), remove the supernatant, and fill each tube again with 50 ml bacteria culture, repeat the centrifugation, and freeze the wet pellets (each tube contains the bacterial pellet of 100 ml culture) at $-70\,°C$ or proceed directly with the extraction of lamins. My experience is that A-type lamins are more susceptible to degradation and should, therefore, be extracted directly without storage of the bacteria.

B. Purification of Lamins from Bacterial Inclusion Bodies

All lamins that we have analyzed in my laboratory form insoluble aggregates in bacteria and are present in inclusion bodies. Therefore, in the first step, inclusion bodies are purified. All steps are performed at $4\,°C$ if not indicated otherwise. The pellets of 200 ml bacterial culture are completely resuspended with a loose-fitting Dounce homogenizer in 8 ml of bacterial lysis buffer. Then 4 ml bacterial lysis buffer containing 50 mg lysozyme per milliliter is added and incubated on ice for 30 minutes. Subsequently, 200 μl of 1.0 M $MgCl_2$, 200 μl of 0.2 M PMSF, 200 μg

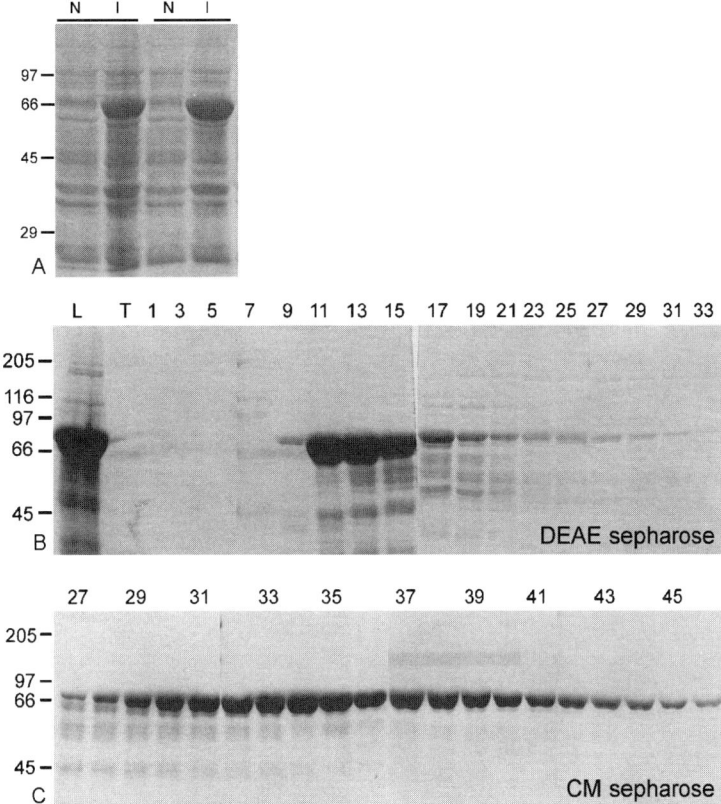

Fig. 2 Expression of lamins in bacteria and lamin purification by ion exchange chromatography. Sodium dodecyl sulfate-polyacrylamide gel electrophoresis (SDS-PAGE) (11% acrylamide) that have been stained with Coomassie blue are shown. Molecular masses of reference proteins (in kDa) are marked. (A) Total protein of two bacterial colonies before (N = non induced) and after the induction (I, induced) of the synthesis of the *C. elegans* lamin lmn-1. The prominent polypeptide band (lane I) with a slightly higher mobility than the 66-kDa marker represents lmn-1. (B, C) Purification of bacterially expressed lamin Dm0 of *Drosophila melanogaster* by anion (B, DEAE sepharose) and cation (C, CM sepharose) exchange chromatography. The prominent polypeptide band at 66 kDa represents lamin Dm0. Lane L (in B), proteins of bacterial inclusion bodies solubilized in 8 M "basic" urea buffer that had been used for the loading of the DEAE column shown. Lane T (in B), proteins not bound to the column. Every second fraction of the DEAE sepharose column (B, fractions 1–33) was analyzed. Proteins contained in fractions 10–16 (B) were pooled, dialyzed against the "acidic" urea buffer, and fractionated on a CM sepharose column (C). Proteins contained in fractions 27–46 (C) are shown. Lamin Dm0 contained in fractions 34–38 is highly concentrated (approximately 0.5 mg/ml) and pure enough for *in vitro* studies.

RNase, and 500 μg DNase are added, and the incubation is continued on ice for 20 minutes. Finally, 20 ml of bacterial detergent buffer is added, and lysed bacteria are fractionated by centrifugation (6000*g*, 10 minutes, 4 °C). The resulting

Table II

Theoretical Isoelectric Points (pI) and Theoretical Molecular Weights (mw) of A- and B-Type Lamins Deduced from Amino Acid Sequences Encoded by Published Lamin cDNAs (Accession Numbers in Brackets) Using the Program ExPASy-Compute pI/Mw tool (http://us.expasy.org/tools/pi_tool.html)

Species	A-type lamins	B-type lamins
	pI/mw (accession number)	pI/mw (accession number)
Human	A: 6.57/74139 (P02545)	B1: 5.11/66408 (NM_005573)
	C: 6.40/65134 (BC003162)	B2: 5.29/67688 (BC006551)
Mouse	A: 6.54/74237 (BC015302)	B1: 5.11/66884 (NM_010721)
	C: 6.22/52651 (NM_019390)	B2: 5.44/67029 (X54098)
Chicken	A: 6.50/73164 (X16879)	B1: 5.15/66530 (X16878)
		B2: 5.31/67940 (X16880)
Xenopus laevis	A: 7.07/74919 (X06345)	B1: 4.99/66450 (X06344)
		B2: 5.63/71470 (X54099)
		LIII: 5.97/67300 (X13169)
Danio rerio	A: 6.56/73732 (AF397016)	B1: 5.15/66704 (AJ250201)
		B2: 5.18/65905 (AJ005936)
		LIII: 5.54/67054 (AF397015)
Drosophila melanogaster	C: 6.42/69855 (NM_079018)	Dm0: 6.32/71249 (P08928)
Caenorhabditis elegans		lmn-1: 5.43/64084 (NP_492371)

Complement of the molecular characterized lamins expressed in somatic cells and oocytes of selected vertebrates and invertebrates. Lamin LIII is expressed only in oocytes, Sertoli cells (Hofemeister *et al.*, 2002), and in *Xenopus* in a few other specialized cells (Benavente *et al.*, 1985). Splice variants of the mouse and rat lamin A (lamin C2) and lamin B2 gene (lamin B3) that are expressed during spermatogenesis have not been listed (for review see Gruenbaum *et al.*, 2003). For details on lamins of other invertebrates see Erber *et al.* (1999), Karabinos *et al.* (2003), and Gruenbaum *et al.* (2003).

pellets are washed three times with bacterial Triton buffer and pelleted after each washing step as described previously. Lamins are highly enriched in the final pellet containing inclusion bodies.

Triton-washed pellets are resuspended in 8 ml of 8.0 M "basic" urea solution at room temperature and centrifuged at 15 °C for 50 minutes at 100,000g. The supernatant enriched in the solubilized lamin (Fig. 2B, lane L) is stored at -70 °C. The required pH value of the "basic" urea buffer can be calculated from the isoelectric points of lamins (Table II), and the required pH value of the "basic" urea buffer is listed in Table III.

C. Column Fractionation (1. Column: DEAE–Sepharose)

The column fractionation is performed at room temperature. A Bio-Rad Poly-Prep column (diameter, 0.8 cm; length, 10 cm) is connected with a tube pump, thus allowing a constant flow (1–2 drops/s). The column is filled with 5 ml of DEAE-sepharose and washed with 50 ml of the same "basic" urea buffer that

Table III

Recommended Buffers for the Purification of Bacterially Expressed Lamins by Ion Exchange Chromatography

Lamin/species	"Basic" urea buffer	"Acidic" urea buffer
Lamin A, C/human, mouse lamin A/chicken, *Danio rerio*	20 mM Tris-HCl, pH 7.5	50 mM sodium acetate, pH 5.6
Lamin A/*Xenopus*	20 mM Tris-HCl, pH 7.8	20 mM sodium phosphate, pH 6.4
Lamins B1, B2/human, mouse, chicken, *Xenopus, Danio rerio*	20 mM sodium phosphate, pH 6.2	50 mM sodium acetate, pH 4.5
Lamins Dm0, C/*Drosophila melanogaster*	20 mM Tris-HCl, pH 7.5	50 mM sodium acetate, pH 5.7
lmn-1/*Caenorhabditis elegans*	50 mM sodium acetate, pH 6.5	50 mM sodium acetate, pH 4.6

All buffers contained 8.0 M urea, 2 mM dithiothreitol [DTT], and 0.2 mM phenylmethylsulfonylfluoride (PMSF). These reagents are therefore not listed in the table. The pH of each buffer should be approximately 0.6–1.2 pH units above ("basic" buffer) or below ("acidic" buffer) the pI of the specific lamin (see Table II for pI values).

had been used for the extraction of a specific lamin from the inclusion bodies (Table III). The solubilized lamins in "basic" urea buffer are loaded onto the column (Fig. 2B, lane L), nonbound proteins are removed by washing the column with 40–50 ml of 8.0 M "basic" urea buffer, and bound proteins are eluted with a linear salt gradient (0–250 mM NaCl in 8.0 M "basic" urea solution). The salt gradient is generated with a gradient mixer by filling 40 ml of 8.0 M "basic" urea buffer in the first reservoir that is directly connected with the column. The second reservoir is filled with 40 ml of 8.0 M "basic" urea buffer that contains 250 mM NaCl. The salt gradient is formed in the first reservoir with a magnetic stirrer. Fractions of 1.5 ml are collected. To assay for lamins, 10 μl of each fraction is mixed with fourfold concentrated SDS sample buffer (Laemmli, 1970), and polypeptides are analyzed by SDS-PAGE and staining of the polyacrylamide gels with Coomassie blue (Fig. 2B, fractions 1–33). Fractions enriched for the specific lamin (Fig. 2B, fractions 10–16) are pooled and stored at −70 °C or processed for dialysis.

D. Dialysis of Lamins

Dialysis tubings (10- to 15-mm diameter) are prepared by boiling for 10 minutes in 2 mM EDTA, washed with distilled water, and stored at 4 °C in 10% EtOH. Fractions enriched in the specific lamin (Fig. 2B, fractions 10–16) are dialyzed for a total of 5 hours at 20 °C against an "acidic" urea buffer (see Table III) with a pH that is approximately 0.6–1.2 pH unit below the isoelectric point of the selected lamin. The dialysis buffer is changed once.

E. Column Fractionation (2. Column: CM–Sepharose)

A column is filled with 5 ml of CM-sepharose as described previously for the DEAE-sepharose. Each column is washed with 30–40 ml of the 8.0 M "acidic" urea buffer that has been used for dialysis. The dialyzed proteins are loaded onto the column. The column is washed, and bound proteins are eluted with a linear salt gradient (0–250 mM NaCl in 8.0 M "acidic" urea buffer specific for the selected lamin) exactly as described for the DEAE-sepharose column. Fractions containing the purified lamins are identified as described previously (Fig. 2C, fractions 27–46). The lamin concentration in these fractions is in the range of 0.1–1.0 mg/ml. Fractions were stored at −70 °C. The lamin concentration can be quantitated by SDS-PAGE and Coomassie blue staining compared with bovine serum albumin (BSA) (Sigma, Munich, Germany) standards run in the same gel (0.2–4 μg BSA per lane). In the experiment shown (Fig. 2C), fractions 34–38 are pure enough for *in vitro* studies (see III.G). The lamin concentration in these fractions is approximately 0.5 mg/ml.

F. Chloroform/Methanol Precipitation of Lamins

This method can be routinely used to precipitate lamins from urea solutions. Because lamins do not lose their assembly properties after chloroform/methanol precipitation, this method is ideal to concentrate and desalt lamins without dialysis.

Procedure: Four volumes of methanol, 1 volume of chloroform, and 3 volumes of distilled water are added to 1 volume of the protein in the urea solution, mixed, and centrifuged for 2 minutes at 11,000*g* in a tabletop centrifuge. The denatured protein forms a white precipitate at the interface between the chloroform layer (bottom) and the water–methanol layer (top). The supernatant consisting of water and methanol is discarded, 3 volumes of methanol are added, and the precipitated protein is pelleted for 2 minutes at 11,000*g*. The pellet is briefly air dried and then dissolved at a protein concentration of 0.5–1.0 mg/ml in 8.0 M urea buffered with 20 mM sodium phosphate set at the desired pH and stored at −70 °C.

G. Use of the Purified Lamins for *In Vitro* Studies

Solubility of dialyzed lamins: For *in vitro* binding studies, 20 μl of lamin in urea buffer (8.0 M urea buffered with 20 mM sodium phosphate) are dialyzed with UM membranes (Millipore Corp., Bedford, MA), against 500–1000 ml of dialysis buffer A (for A-type lamins; 20 mM sodium phosphate, 300 mM KCl, pH 8.2) or dialysis buffer B (for B-type lamins; 20 mM sodium phosphate, 500 mM KCl, pH 8.4) for 1 hour at room temperature. The elevated salt concentration and high pH largely prevent the polymerization of lamins during dialysis. Under these dialysis conditions, renatured lamins had predominantly formed parallel aligned dimers and short longitudinal aggregates. This buffer is tolerated when microinjected into cells. Lamins reconstituted in dialysis buffers A or B can be added to

Xenopus egg extracts competent for nuclear reconstitution without destroying this *in vitro* system. Lamins present in dialysis buffers A or B should be centrifuged (3 minutes, 11,000*g*) after dialysis to remove aggregates (for experimental details on *Xenopus* egg extracts see Gant *et al.*, 1999; Lourim and Krohne, 1993; Newmeyer and Wilson, 1991; Spann *et al.*, 1997).

Quantification of lamins: We have used bacterially expressed lamins for the quantification of *Xenopus* lamins (Lourim *et al.*, 1996). In addition, lamins or lamin domains expressed in bacteria can be used to study *in vitro* individual steps of polymer formation (for review see Stuurman *et al.*, 1998) and to solve the atomic structure of distinct lamin domains (Krimm *et al.*, 2002).

Solid-phase binding studies: In addition, we have used bacterially expressed *Drosophila* lamins Dm0 and C to study their *in vitro* binding to proteins of the inner nuclear membrane (Wagner *et al.*, 2004). For this purpose, chloroform/methanol–precipitated lamins are solubilized at a final concentration of 1 μg/μl in PBS containing 4 M urea. Of this stock solution, 0.5 μl is mixed with 19.5 μl PBS. Wells of a 96-well plate (Greiner BIO-ONE GmbH, Frickenhausen) are coated with the desired protein (0.5 μg protein in 20 μl PBS/well) by incubation for 2 hours at 18 °C, followed by an incubation with PBS containing 0.5% BSA for 2 hours at 18 °C. Wells are then washed three times with PBS and incubated with the [^{35}S] methionine-labeled protein (3 μl *in vitro* translated protein + 17 μl PBS/BSA per well) for 2 hours at 18 °C. [^{35}S] methionine-labeled proteins are *in vitro* synthesized in rabbit reticulocyte lysates by coupled *in vitro* transcription/translation with the T7-RNA polymerase in the TNT system (Promega, Madison, WI) as recommended by the supplier. Wells are finally washed three times with PBS, and bound proteins are solubilized in sample buffer for SDS-PAGE. Control wells are coated with BSA. To detect [^{35}S]methionine-labeled proteins after separation by SDS-PAGE, gels are incubated for 1 hour in dimethylsulfoxide, followed by an incubation for 3 hours in Rotifluoreszint D (Roth, Karlsruhe, Germany), and finally for 1 hour in H$_2$O. Gels are dried and exposed to x-ray films. For densitometric analysis of x-ray films, we used Scion Image for Windows (Scion Corporation, MD).

Lamins expressed in bacteria have also been used after separation by SDS-PAGE and transfered to nitrocellulose for blot overlay assays (Dechat *et al.*, 2000b).

Labeling of lamins with fluorescein: The labeling of purified lamins with the fluorescent dye 5-iodoacetamidofluorescein (5-IAF) and their use for the study of lamin dynamics in microinjected vertebrate cells has been described in detail previously (Benavente and Krohne, 1998; Schmidt and Krohne, 1995; Schmidt *et al.*, 1994).

H. Other Advantages of pET–Lamin Constructs

The pET plasmids have the advantage that they can also be used for the *in vitro* synthesis of [^{35}S] methionine-labeled lamins in rabbit reticulocyte lysates by coupled *in vitro* transcription/translation with the T7-RNA polymerase in the TNT system. For radioactive labeling of proteins during synthesis, 0.5 μg of plasmid DNA and 40 μCi [^{35}S] methionine (Amersham) were used per experiment. The [^{35}S] methionine-labeled lamins are ideal for transport studies in microinjected amphibian oocytes (Krohne et al., 1989) and blot overlay assays (Krohne et al., 1987). The cotranslation of a lamin and a putative binding partner in the rabbit reticulocyte lysate allows us to study complex formation between these proteins. Complex formation can be verified by immunoprecipitation (Wagner et al., 2004; see also Section VIII).

I. Comments

Bacterial colonies producing lamins can be stored for 1–2 weeks at 4 °C on replica plates. A reduced protein production was noted when older colonies were used.

IV. Production of Lamins in Insect Cells Using the Baculovirus System

Lamins synthesized in eucaryotic cells are posttranslationally processed and modified (see Introduction). If modified lamin are required in larger quantities, they can be expressed in Sf9 cells of *Spodoptera frugiperda* by use of the Baculogold Baculovirus Expression Vector System of Pharmingen (BD Biosciences Pharmingen, San Diego, CA).

A. Lamin Constructs

All our lamin cDNAs used for cloning into the transfer plasmids pVL 1392, pVL1393, and pAcSG2 contained a ribosomal binding site (RBS) in front of the initiating ATG (Gieffers and Krohne, 1991; Klapper et al., 1997; Krohne et al., 1998). This RBS together with the coding region of the specific lamin was cloned into the transfer plasmids. In several cases, the bacterial expression vectors pET17 or pET21 containing the coding region of a specific lamin (see earlier) could be restricted with XbaI and NotI to obtain the required lamin DNA together with the RBS. For transfection experiments, the plasmid DNA is purified with QIAGEN Plasmid Midi Kit (QIAGEN, Hilden, Germany). The purified DNA is resuspended in distilled H_2O, sterile filtered (Millipore filters, pore size 0.22 μm; Hartenstein, Würzburg, Germany), and stored at −20 °C. The quotient of the optical density of the DNA at 260 nm/280 nm should be 1.7–2.0, and its concentration 0.3–1 μg/μl H_2O.

B. Culture of Sf9 Cells

Sf9 cells of the butterfly *Spodoptera frugiperda* are grown in the absence of CO_2 at 27 °C in TC100 medium (Gibco, Eggenstein, Germany) supplemented with 10% FCS, 100 IU/ml penicillin, 100 IU streptomycin, and 2 mM L-glutamine. Cells are routinely grown in 50-ml Falcon flasks (Labor Schubert, Schwandorf, Germany) containing 5 ml medium. All transfections are performed in flasks of this size.

C. Transfection and Amplification of Recombinant Viruses

Step 1: Approximately 3–4 hours before transfection, 2×10^6 Sf9 cells growing at a logarithmic rate are resuspended in 5 ml of complete TC 100 medium and cultured for at least 2 hours until they had attached to the bottom of the flask.

Step 2: The transfection solutions are prepared in polystyrol Falcon centrifuge tubes (Sarstedt, Nümbrecht, Germany) as follows.
  • One tube contains 1–2 μg of the transfer plasmid and 100–500 ng of linearized BaculoGold-DNA (Pharmingen) in a final volume of 12 μl H_2O.
  • A second tube contains 6 μl Lipofectin (Gibco, Eggenstein, Germany) and 6 μl H_2O.
  • The solutions of both tubes are combined and gently mixed, avoiding the shearing of the viral DNA, and subsequently incubated for 15 minutes at room temperature.

Step 3: When the DNA is ready for transfection, all attached Sf9 cells are washed two times each with 5 ml of incomplete TC100 medium (medium without FCS, glutamine, and antibiotics), avoiding their resuspension.
  • 2.5 ml of incomplete TC100 medium is added per flask, followed by 24 μl of the transfection solution.
  • The Sf9 cells are then incubated for 2 hours at 27 °C, after which 2.5 ml of complete TC100 medium is added to each flask.

Step 4: The cells are cultured for 7 days without feeding. The medium containing recombinant viruses is cleaned from cell debris by centrifugation (10 minutes, 2000g) and stored at 4 °C.

Step 5: Recombinant baculoviruses contained in the culture medium are amplified by growing 2×10^6 Sf9 cells for 7 days in 3 ml of complete TC100 medium supplemented with 2 ml of culture medium containing the recombinant baculoviruses. The medium containing recombinant viruses is cleaned from cell debris by centrifugation (10 minutes, 2000g).

Step 6: The amplification needs to be repeated a second time to ensure a high virus titer. The virus containing culture medium that has been cleaned of cell debris by centrifugation (2000g; 10 minutes) is stored at either 4 °C or at -196 °C in liquid nitrogen.

D. Production of Lamins in Sf9 Cells

Approximately 2×10^6 Sf9 cells per flask contained in 5 ml supplemented TC 100 medium (see earlier) are infected with the amplified recombinant viruses (1 ml virus containing medium per flask), harvested 2–3 days later by centrifugation

(1000g, 5 minutes), washed twice with incomplete TC100 medium, and then once with Sf9-phosphate buffer (100 mM NaCl, 1 mM EDTA, 1 mM EGTA, 10 mM Na$_2$HPO$_4$; trypsin inhibitor: 10 μg/ml; PMSF: 0.2 mM; pH 7.2). After each washing step, cells are pelleted (1000g, 5 minutes).

All further steps are performed at 4°C. Cells harvested from one flask are resuspended by pipetting in Sf9-phosphate/Triton buffer (0.5% Triton X-100, 50 mM NaCl, 1 mM EDTA, 1 mM EGTA, 10 mM Na$_2$HPO$_4$; trypsin inhibitor: 10 μg/ml; PMSF: 0.2 mM; final pH 7.2), incubated for 15 minutes, and then pelleted by centrifugation (3000g, 10 minutes). The supernatant containing solubilized proteins is removed, the pellet is washed once with DNase II buffer (2 mM MgCl$_2$, 10 mM Tris, pH 7.5), pelleted (3000g, 10 minutes), resuspended in 2 ml DNase II buffer supplemented with DNase (100 μg) and RNase (100 μg), and incubated for 15 minutes at 20°C. Cell remnants are pelleted (3000g, 10 minutes), resuspended once in 2 ml Sf9-phosphate buffer, and pelleted again (3000g, 10 minutes). Lamins can be solubilized in urea buffer and fractionated by ion exchange chromatography (see Section III.C and Tables II and III).

E. Comments

In my laboratory we have expressed lamins A and B1 of *Xenopus laevis*, lamins DmO and C of *Drosophila melanogaster*, and deletion mutants of these lamins in Sf9 cells. Coomassie blue–stained SDS-PAGE after separation of total cellular proteins of infected Sf9 revealed that each tested lamin was the most abundant cellular protein and visible as a prominent polypetide band (Klapper *et al.*, 1997; Krohne *et al.*, 1998).

V. Selection of Cultured Cells and Tissues for Lamin Enrichment

The lamin complement of a nucleus is different depending on the cell line, the tissue, and the species. Cell lines, isolated cells, and organs are available that express only B-type lamins, whereas A-type lamins are only expressed in combination with B lamins. A selection of suitable cells and tissues is summarized in Table IV. In the following chapters, the enrichment of lamins from cells and tissues by cell fractionation and immunoprecipitations is described. Commercially available monoclonal lamin antibodies suitable for immunoblotting, immunoprecipitations, and immunofluorescence are listed in Table I. In my laboratory, the antibodies X223, R27, X155, X67, and X167 have been generated.

VI. Enrichment of Lamins from Isolated Mammalian Liver Nuclei

The preparation of highly purified nuclei is the prerequisite for the isolation of lamins. We have used bovine, rat, and mouse liver successfully. Mouse and rat liver have the advantage that these organs contain less connective tissue than

Table IV
Suitable Somatic Cells and Tissues for the Study of Lamins

Species	Cell/tissue	Lamin complement
Human	HeLa	A, C, B1, B2
	Myeloid leukemia cell line	B2 strong,
	HL-60 (nondifferentiated) [1]	B1 weak, traces of A and C
Mouse	Liver/hepatocytes	A, C, B1, B2
	F9 embryonal carcinoma cell [2]	B1, B2
Rat	Liver/hepatocytes	A, C, B1, B2
Chicken	Fibroblasts [3]	A, B1, B2
	14-day-old embryo: brain [4]	B1, B2
Xenopus laevis	Cultured kidney epithelial cells (A6)	A, B1, B2
Drosophila melanogaster	Schneider S2 cells	Dm0
	Kc167 cells	C, Dm0

The enrichment of lamins from more specialized cells (amphibian oocytes, nucleated erythrocytes of amphibian and chicken) has been described (Krohne and Franke, 1983). Meiotic cells of the male germ line of mouse and rat have been used for the biochemical characterization of splice variants of the mammalian lamin A (lamin C2; Alsheimer *et al.*, 1999) and the lamin B2 gene (lamin B3; Furukawa and Hotta, 1993; for review see Gruenbaum *et al.*, 2003; Krohne, 1998). For further details on cell lines and tissues marked by numbers in brackets see the following references: [1] Olins *et al.*, 2001. [2] Lebel *et al.*, 1987; Collard and Raymond, 1990. [3] Lehner *et al.*, 1986. [4] Lehner *et al.*, 1987.

bovine liver. We have either prepared nuclei from fresh liver or alternatively have cut fresh liver with scissors into small pieces (size, 5–10 mm) and transferred them immediately into liquid nitrogen. Frozen tissue has been stored up to 7 days at $-70\,^{\circ}$C. Nuclei prepared from fresh and frozen tissue had the same quality in respect to lamins.

A. Isolation of Nuclei and Digestion with Nucleases

The whole preparation should be carried out at $4\,^{\circ}$C, and the solutions contain 10 μg trypsin inhibitor/ml and 0.2 mM PMSF/ml.

The preparation is described for 50–70 g of rat liver (livers from 6–7 young rats; Sprague-Dawley). The liver is minced with scissors in 200 ml medium A and washed once with fresh medium A. The minced tissue pieces are forced with a press through a fine metal sieve (pore size, 05–1 mm). Most of the connective tissue is held back in the sieve. To the homogenate, medium A is added to a final volume of 300 ml. Portions of 30–40 ml are then homogenized with the aid of a motor-driven Potter-Elvehjem glass-Teflon homogenizer (setting, highest rotation speed) to further disrupt cells and tissue pieces. The homogenate is then filtered immediately through 4–6 layers of nylon gauze (No. 7–200; Swiss Silk Bolting Cloth Mfg, Zurich, Switzerland) or cheesecloth. The filtrate is then centrifuged in a rotor with swinging buckets at 900g for 12 minutes. The supernatant is carefully decanted, and the loose pellet containing nuclei, intact cells, and endoplasmic reticulum is resuspended with a loose-fitting Dounce homogenizer in a final volume of 30 ml medium A. The homogenate is adjusted with 2.6 M sucrose to 2.0 M (final volume, 90–100 ml); 16–17 ml of the homogenate in 2.0 M sucrose is then layered on top of a 19-ml cushion of 2.1 M sucrose and centrifuged for 60 minutes in six swinging buckets (110,000g; SW28, Beckman centrifuge). The purified nuclei are contained in the white pellet, and the sheet of material at the top of the centrifuge tube contains cells, tissue debris, and cytoplasmic membranes and is enriched in cytokeratins.

The six nuclear pellets are resuspended with the aid of a loose-fitting Dounce homogenizer in 30 ml medium A, pelleted again (10 minutes, 1000g), and resuspended with a Dounce homogenizer in 30–40 ml of 5 mM Tris (pH 7.5). The nuclei are incubated for 5 minutes to allow their swelling, pelleted (3000g, 10 minutes), and resuspended in 20 ml of DNase buffer I (Dounce homogenizer) as described previously and incubated with 150 μg DNase and 150 μg RNase for 15 minutes at 20 °C. Nuclei are pelleted (3000g, 10 minutes), resuspended a second time in 5 mM Tris (pH 7.5), pelleted (3000g, 10 minutes) after an incubation for 5 minutes, and resuspended in 20 ml DNase buffer II (Dounce homogenizer). Nuclei are digested a second time with nucleases (15 minutes at 20 °C; 100 μg DNase, 100 μg RNase), pelleted (3000g, 10 minutes), washed once with 10 mM Tris, pH 7.5, and pelleted again. The wet pellet can be stored frozen at −70 °C. These extracted nuclei are highly enriched in lamins and contain in addition the integral membrane proteins of the inner nuclear membrane, the pore complex membrane, and peripheral pore complex proteins (nucleoporins). Remnants of the chromatin are also present in this fraction.

B. Isolation of Nuclear Envelopes and Lamins

Nuclear envelope fraction: Nuclei that have been digested the second time with nucleases (in DNase buffer II for 15 minutes at 20 °C; 100 μg DNase, 100 μg RNase) are pelleted and resuspended (Dounce homogenizer) in 10 ml high salt buffer (10 mM Tris pH, 7.5, 500 mM NaCl, 1 mM DTT), incubated for 10 minutes at 4 °C, and pelleted (4000g, 10 minutes). The pellet is washed once with 10 mM Tris (pH 7.5). By this treatment nearly all of the chromatin has been solubilized, resulting in a fraction highly enriched in nuclear envelopes.

Lamin purification

1. If lamins need to be separated from integral membrane proteins, the following extraction is recommended. The nuclear envelope pellet is solubilized in 5 ml of 8 M "basic" urea buffer (pH 7.5; see Table III) and pelleted (100,000g, 60 minutes, 15 °C). Lamins are recovered in the supernatant, whereas membranes and integral membrane proteins are contained in the pellet. If a further subfractionation of lamins by ion exchange chromatography (see Section III.C, E; Tables II and III) is required, the following urea buffers are recommended.

Rat liver lamins A/C can be enriched as follows. The nuclear envelope pellet is solubilized in 5 ml of 8 M "acidic" urea buffer with a pH close to the pI of lamin B1 (see Table II; 50 mM sodium acetate, pH 5.1; lamin B1 is the predominant

B-type lamin of hepatocytes), pelleted (100,000*g*, 60 minutes, 15 °C), and the supernatant is passed over a CM-sepharose column (1–2 ml bed volume), and bound lamins are eluted with a salt gradient (see Section III.C). At this pH, lamins B1 and B2 are bound very weakly to the ion exchange column.

Rat liver lamins B1/B2 can be enriched as follows. The nuclear envelope pellet is solubilized in 5 ml of 8 M "basic" urea buffer with a pH close to the pI of lamins A/C (see Table II; 50 mM sodium phosphate, pH 6.5), pelleted (100,000*g*, 60 minutes, 15 °C), and the supernatant is passed over a DEAE-sepharose column (1–2 ml bed volume), and bound lamins are eluted with a salt gradient (see Section III.C). At this pH, lamins A/C are bound very weakly to the ion exchange column.

2. Gerace and coworkers have fractionated urea-extracted lamins (8 M urea, 1 mM DTT, 20 mM Tris, pH 8.8) by passing the urea extract over DEAE-cellulose (De-52; Whatman Inc. Clifton, NJ; 0.5–1.0 ml bed volume) and phosphocellulose columns (P11; Whatman Inc.; 0.5–1.0 ml bed volume) in sequence (Aebi *et al.*, 1986; Foisner and Gerace, 1993; Glass and Gerace, 1990). Lamins A/C were eluted from the phosphocellulose column, and lamins B1/B2 with a salt gradient from the DEAE column.

3. Alternatively, lamins can be solubilized together with integral membrane proteins as follows. The nuclear envelope fraction is resuspended by homogenization in 4 ml Triton high salt buffer (2% Triton X-100, 500 mM KCl, 1 mM DTT, 20 mM Tris, pH 9.0; Aebi *et al.*, 1986), extracted for 20 minutes at 4 °C by end-over-end rotation, and centrifuged (200,000*g*, 40 minutes). Nearly all lamins are recovered in the supernatant.

C. Comments

Especially for bovine liver, a small meat-mincing machine can also be used for the initial homogenization. Minced mouse liver can directly be homogenized with a motor-driven Potter-Elvehjem glass–Teflon homogenizer, because it contains less connective tissue than rat or bovine liver. A mouse liver weighs approximately 1 g. We noted that the nuclear fraction is more contaminated by connective tissue if more than 70 g of liver is taken for the preparation. If the preparation is scaled down, volumes of buffers should be reduced proportionally. The preparation of nuclei is a combination of the methods described by Krohne and Franke (1983) and Kaufmann *et al.* (1983).

VII. Enrichment of Lamins from Cultured Cells (Tested for *Xenopus* A6 Cells and HeLa Cells)

All preparation should be carried out at 4 °C if not indicated otherwise, and the solutions contain a final concentration of 10 μg trypsin inhibitor/ml and 0.2 mM PMSF/ml. Five Petri dishes (10-cm diameter) with cells grown to near confluency are used for one preparation.

A. Preparation by Nuclease Digestion

Cells are washed two times with PBS, each Petri dish is filled with 2–3 ml PBS, cells are scraped of the Petri dishes with a rubber policeman, and pelleted (900g, 10 minutes). Cells are resuspended with a Dounce homogenizer in 20–30 ml swelling buffer (10 mM Tris, pH 7.5), incubated for 10 minutes, and pelleted (3000g, 10 minutes).

The pellet is resupended in 5 ml DNase I buffer and incubated for 15 minutes at 20 °C with DNase (50 μg) and RNase (100 μg). Nuclease digested cell remnants are pelleted (3000g, 10 minutes), resupended (Dounce homogenizer) in 20 ml swelling buffer (10 mM Tris, pH 7.5), incubated for 5 minutes, and pelleted again. The pellet is resuspended in 5 ml DNase buffer II, incubated with nucleases exactly as described previously, and pelleted. The pellet is resuspended in 3 ml high salt buffer (500 mM NaCl, 1 mM DTT, 10 mM Tris, pH 7.5), incubated for 10 minutes at 4 °C, pelleted (10,000g, 10 minutes), washed once with 10 mM Tris (pH 7.5), and the final pellet can be stored at −70 °C. Lamins can be further enriched from high salt extracted nuclei by their solubilization in urea buffer followed by ion exchange chromatography (see Sections III.C and VI.B)

B. Lamin Solubilization without Prior Nuclease Digestion

Cells are washed two times with PBS, each Petri dish is filled with 2–3 ml PBS, cells are scraped off the Petri dishes with a rubber policeman, and pelleted (900g, 10 minutes). Cells are resuspended with a Dounce homogenizer in 20 ml swelling buffer (10 mM Tris, pH 7.5), incubated for 10 minutes, and pelleted (3000g, 10 minutes). The pellet is homogenized in 4 ml HeLa-Triton/pH 9.0 buffer (1% Triton X-100, 150 mM NaCl, 50 mM Tris, pH 9.0) and extracted for 20 minutes by end-over-end rotation. Insoluble remnants are pelleted (100,000g, 60 minutes). When HeLa cells are used, approximately 50% of total lamins are recovered in the supernatant. If the salt concentration in the Triton/pH 9.0 buffer is increased to 300 mM, the chromatin is released from nuclei, and nonsoluble cellular remnants are no longer pelletable.

For *Xenopus* A6 cells, a Triton buffer with a lower pH is sufficient to obtain the same result (*Xenopus*-Triton/pH 7.4; buffer: 1% Triton X100, 150 mM NaCl, 50 mM Tris, pH 7.4).

C. Comments

Dounce homogenization of cells in the swelling buffer (10 mM Tris pH 7.5) followed by centrifugation of the homogenate (3000g, 10 minutes) results in supernatant that contains a large proportion of the total cytoplasmic IF proteins, whereas more than 90% of the lamins are recovered in the pellet fraction. Nevertheless, nuclei from cultured cells that have been isolated by hypotonic shock are more contaminated with cytoplasmic IF proteins than nuclei isolated from mouse or rat liver. In addition, it is less expensive to isolate milligram amounts of lamins from liver than from cultured cells.

VIII. Immunoprecipitation of Lamin from Cell Extracts

A. Preparation of Cell Extracts

To study lamin containing protein complexes by immunoprecipitation in my laboratory, we have prepared cell extracts as follows. All protein preparations are performed at 4 °C, and the lysis buffers/immunoprecipitation buffer (IP-buffer: 1% Triton X100, 50 mM Tris, pH 7.4, and 150 mM NaCl) is complemented before use with 0.2 mM PMSF and 10 μg trypsin inhibitor/per milliliter.

Cells grown to near confluency in 10-cm Petri dishes (*Xenopus* A6 cells; HeLa cells) are washed twice with PBS, cells are then scraped off the Petri dishes with a rubber policeman, and pelleted (900*g*, 10 minutes). The cell pellet is washed once more with PBS, pelleted (900*g*, 10 minutes), resuspended by pipetting in IP buffer (500 μl IP buffer/cell pellet obtained from one Petri dish), and incubated for 20 minutes by end-over-end rotation. Cell lysates are fractionated by centrifugation at 11,000*g*. Supernatants are shock frozen in liquid nitrogen and stored at −70 °C.

Schneider S2 and Kc167 cells are grown in flasks and are loosely attached to the plastic. Cells are detached by pipetting, pelleted (600*g*, 10 minutes), washed twice with culture medium without FCS and antibiotics, and washed once with PBS. Cells are pelleted after each washing step (600*g*, 10 minutes) and are finally extracted with IP buffer as described for HeLa and A6 cells.

Supernatants that have been stored frozen are centrifuged after thawing (11,000 *g*, 15 minutes).

B. Preabsorption of the Extract

All steps are performed at 4 °C. For each immunoprecipitation, 10 mg of swollen protein-A-sepharose (Pharmacia Biotech, Uppsala, Sweden) is used and for the preadsorption 20 mg sepharose/ml extract is used. The required protein-A-sepharose is swollen for 20 minutes in 10 ml IP buffer, pelleted (500*g*, 5 minutes), washed once more with IP buffer, pelleted again, and resuspended in IP buffer at a concentration of 10 mg sepharose/500 μl.

For the preadsorption, 20 mg of sepharose is pelleted (500*g*, 5 minutes), mixed with 1 ml of the 11,000*g* supernatant of extracted cells, and incubated by end-over-end rotation for 20 minutes. The sepharose is pelleted (500*g*, 5 minutes), the preabsorbed supernatant is taken for immunoprecipitations, and the sepharose pellet is discarded.

C. Coupling of Antibodies to Protein-A-Sepharose

For all immunoprecipitations performed with mouse monoclonal antibodies first, 1.25 μg of affinity-purified rabbit antibodies against mouse IgG (RAM; Dianova, Hamburg, Germany) is bound in 50–100 μl PBS to 10 mg of swollen protein-A-sepharose. The sepharose is incubated with the RAM for 45–60 minutes and resuspended occasionally. The protein-A-sepharose is washed twice with IP buffer and pelleted (500*g*, 5 minutes). In the next step, 1–2 μg of a mouse monoclonal lamin antibody (Table I) is bound the RAM-protein-A-sepharose as described previously.

D. Immunoprecipitation

One milliliter of the preabsorbed 11,000g supernatant is mixed with 10 mg of swollen protein-A-sepharose that had been coupled to mouse monoclonal lamin antibodies. After incubation for 2 hours by end-over-end rotation, the protein-A-sepharose is pelleted (500g, 5 minutes) and washed three times with IP buffer, and finally once with PBS. After each washing step, the sepharose is pelleted at 500g and at 1500g (5 minutes) after the final wash with PBS. The supernatant is removed, SDS-sample buffer is added, and the samples are boiled for 5 minutes at 96 °C before separation by SDS-PAGE.

To detect whether a specific nuclear envelope protein is coimmunoprecipitated with lamins, guinea pig antibodies raised against this protein are used for immunoblotting. Because guinea pig antibodies of the IgG class are binding directly to protein-A, immunprecipitations in my laboratory are often performed with guinea pig antibodies specific for a distinct nuclear envelope protein, and mouse monoclonal lamin antibodies are used to detect coimmunoprecipitated lamins.

E. Comments

I noted significant differences with respect to the solubilization of lamins by IP buffer when HeLa cells, *Xenopus* A6 cells, and *Drosophila* cells (Schneider S2 cells, Kc167 cells) were compared. At least 50% of lamins and integral membrane proteins of the inner nuclear membrane could be recovered in the 11,000g supernatant of A6 and Schneider S2 cells, whereas only 10% of HeLa lamins are solubilized under the same conditions.

Immunprecipitations with antibodies against lamins, lamin-associated polypeptide 2 (LAP2), and the *Drosophila* lamin B receptor performed in my laboratory revealed that extracts of A6, S2, and Kc167 cells possess significant amounts of protein complexes that contain lamins and integral membrane proteins of the inner nuclear membrane (Lang and Krohne, 2003; Wagner *et al.*, 2004). Some of these protein complexes disintegrate when the salt concentration in the IP buffer is elevated to 300 mM NaCl (Lang and Krohne, 2003; for other immunoprecipitation protocols see Dechat *et al.*, 2000b).

References

Aebi, U., Cohn, J., Buhle, L., and Gerace, L. (1986). The nuclear lamina is a meshwork of intermediate-type filaments. *Nature (London)* **323**, 560–564.

Alsheimer, M., von Glasenapp, E., Hock, R., and Benavente, R. (1999). Architecture of the nuclear periphery of rat pachytene spermatocytes: Distribution of nuclear envelope proteins in relation to synaptonemal complex attachment sites. *Mol. Biol. Cell* **10**, 1235–1245.

Benavente, R., and Krohne, G. (1998). *In vivo* systems to study the dynamics of nuclear lamins. *In* "Methods in Cell Biology" (M. Berrios, ed.), Vol 53, pp. 591–602. Academic Press, San Diego.

Benavente, R., Krohne, G., and Franke, W. W. (1985). Cell type-specific expression of nuclear lamina proteins during development of *Xenopus laevis*. *Cell* **41,** 177–190.

Cohen, M., Lee, K. K., Wilson, K. L., and Gruenbaum, Y. (2001). Transcriptional repression, apoptosis, human disease and the functional evolution of the nuclear lamina. *Trends Biochem. Sci.* **26,** 41–47.

Collard, J. F., and Raymond, Y. (1990). Transfection of human lamins A and C into mouse embryonal carcinoma cells possessing only lamin B. *Exp. Cell Res.* **186,** 182–187.

Dechat, T., Vleck, S., and Foisner, R. (2000a). Lamina-associated polypeptide 2 isoforms and related proteins in cell cycle-dependent nuclear structure dynamics. *J. Struct. Biol.* **129,** 335–345.

Dechat, T., Korbei, B., Vaughan, O. A., Vlcek, S., Hutchison, C. J., and Foisner, R. (2000b). Lamina-associated polypeptide 2α binds intranuclear A-type lamins. *J. Cell Sci.* **113,** 3473–3484.

Erber, A., Riemer, D., Hofemeister, H., Bovenschulte, M., Stick, R., Panopoulou, G., Lehrach, H., and Weber, K. (1999). Characterization of the *Hydra* lamin and its gene: a molecular phylogeny of metazoan lamins. *J. Mol. Evol.* **49,** 260–271.

Foisner, R., and Gerace, L. (1993). Integral membrane proteins of the nuclear envelope interact with lamins and chromosomes, and binding is modulated by mitotic phosphorylation. *Cell* **73,** 1267–1279.

Furukawa, K., and Hotta, Y. (1993). cDNA cloning of a germ line specific lamin B3 from mouse spermatocytes and analysis of its function by ectopic expression in somatic cells. *EMBO J.* **12,** 97–106.

Gant, T. M., Harris, C. A., and Wilson, K. L. (1999). Roles of LAP2 proteins in nuclear assembly and DNA replication: Truncated LAP2β proteins alter lamina assembly, envelope formation, nuclear size, and DNA replication efficiency in *Xenopus laevis* extracts. *J. Cell Biol.* **144,** 1083–1096.

Gieffers, C., and Krohne, G. (1991). *In vitro* reconstitution of recombinant lamin A and a lamin A mutant lacking the carboxy-terminal tail. *Eur. J. Cell Biol.* **55,** 191–199.

Glass, J. R., and Gerace, L. (1990). Lamins A and C bind and assemble at the surface of mitotic chromosomes. *J. Cell Biol.* **111,** 1047–1057.

Goldman, R. D., Gruenbaum, Y., Moir, R. D., Shumaker, D. K., and Spann, T. P. (2002). Nuclear lamins: Building blocks of nuclear architecture. *Genes Dev.* **16,** 533–547.

Gruenbaum, Y., Goldman, R. D., Meyuhas, R., Mills, E., Margalit, A., Fridkin, A., Dayani, Y., Prokocimer, M., and Enosh, A. (2003). The nuclear lamina and its functions in the nucleus. *Int. Rev. Cytol.* **226,** 1–62.

Hennekes, H., and Nigg, E. (1994). The role of isoprenylation in membrane attachment of nuclear lamins,. *J. Cell Sci.* **107,** 1019–1029.

Hofemeister, H., Kuhn, C., Franke, W. W., Weber, K., and Stick, R. (2002). Conservation of the gene structure and membrane targeting signals of germ cell specific lamin LIII in amphibians and fish. *Eur. J. Cell Biol.* **81,** 51–60.

Karabinos, A., Schunemann, J., Meyer, M., Aebi, U., and Weber, K. (2003). The single nuclear lamin of *Caenorhabditis elegans* forms *in vitro* stable intermediate filaments and paracrystals with a reduced axial periodicity. *J. Mol. Biol.* **325,** 241–247.

Kaufmann, S. H., Gibson, W., and Shaper, J. H. (1983). Characterization of the major polypeptide of the rat liver nuclear envelope. *J. Biol. Chem.* **258,** 2710–2719.

Klapper, M., Exner, K., Kempf, A., Gehrig, C., Stuurman, N., Fisher, P. A., and Krohne, G. (1997). Assembly of A- and B-type lamins studied *in vivo* with the baculovirus system. *J. Cell Sci.* **110,** 2519–2532.

Krimm, I., Ostlund, C., Gilquin, B., Couprie, J., Hossenlopp, P., Mornon, J. P., Bonne, G., Courvalin, J. C., Worman, H. J., and Zinn-Justin, S. (2002). The Ig-like structure of the C-terminal domain of lamin A/C, mutated in muscular dystrophies, cardiomyopathy, and partial lipodystrophy. *Structure (Camb)* **10,** 811–823.

Krohne, G. (1998). Lamin assembly *in vivo*. *In* "Subcellular Biochemistry: Intermediate Filaments" (H. Herrmann and J. Robin Harris, eds.), pp. 563–586. Plenum Publishing, London, UK.

Krohne, G., and Franke, W. W. (1983). Proteins of pore complex-lamina structures from nuclei and membranes. *Methods Enzymol.* **96J**, 597–608.

Krohne, G., Stuurman, N., and Kempf, A. (1998). Assembly of *Drosophila* lamin DmO and C mutant proteins studied with the baculovirus system. *Eur. J. Cell Biol.* **77**, 276–283.

Krohne, G., Waizenegger, I., and Höger, T. H. (1989). The conserved carboxy-terminal cysteine of nuclear lamins is essential for lamin association vith the nuclear envelope. *J. Cell Biol.* **109**, 2003–2011.

Krohne, G., Wolin, S., McKeon, F., Franke, W., and Kirschner, M. (1987). Nuclear lamin L₁ of *Xenopus* laevis: cDNA cloning, amino acid sequence and binding specificity of a member of the lamin B subfamily. *EMBO J.* **6**, 3801–3808.

Kyhse-Andersen, J. (1984). Electroblotting of multiple gels: A simple apparatus without tank for rapid transfer of proteins form polyacrylamide to nitrocellulose. *J. Biochem. Biophys. Methods* **10**, 203–210.

Laemmli, U. (1970). Cleavage of structural proteins during assembly of the head of the bacteriophage T4. *Nature (London)* **227**, 680–685.

Lang, C., and Krohne, G. (2003). Lamina-associated polypeptide 2β (LAP2β) is contained in a protein complex together with A- and B-type lamins. *Eur. J. Cell Biol.* **82**, 143–153.

Lebel, S., Lampron, C., Royal, A., and Raymond, Y. (1987). Lamins A and C appear during retinoic acid-induced differentiation of mouse embryonal carcinoma cells. *J. Cell Biol.* **105**, 1099–1104.

Lehner, C. F., Kurer, V., Eppenberger, H. M., and Nigg, E. A. (1986). The nuclear lamin protein family in higher vertebrates. Identification of quantitatively minor lamin proteins by monoclonal antibodies. *J. Biol. Chem.* **261**, 13293–13301.

Lehner, C. F., Stick, R., Eppenberger, H. M., and Nigg, E. A. (1987). Differential expression of nuclear lamin proteins during chicken development. *J. Cell Biol.* **105**, 577–587.

Lourim, D., Kempf, A., and Krohne, G. (1996). Characterization and quantitation of three B-type lamins in *Xenopus* oocytes and eggs: Increase of lamin L₁ protein synthesis during meiotic maturation. *J. Cell Sci.* **109**, 1775–1785.

Lourim, D., and Krohne, G. (1993). Membrane-associated lamins in *Xenopus* egg extracts: Identification of two vesicle populations. *J. Cell Biol.* **123**, 501–512.

Maske, C. P., Hollinshead, M. S., Higbee, N. C., Bergo, M. O., Young, S. G., and Vaux, D. J. (2003). A carboxyl-terminal interaction of lamin B1 is dependent on the CAAX endoprotease Rce1 and carboxymethylation. *J. Cell Biol.* **162**, 1223–1232.

Newmeyer, D., and Wilson, K. (1991). Egg extracts for nuclear import and nuclear assembly reactions. *Methods Cell Biol.* **36**, 607–635.

Olins, A. L., Herrmann, H., Lichter, P., Kratzmeier, M., Doenecke, D., and Olins, D. E. (2001). Nuclear envelope and chromatin compositional differences comparing undifferentiated and retinoic acid- and phorbol ester-treated HL-60 cells. *Exp. Cell Res.* **268**, 115–127.

Riemer, D., Dodemont, H., and Weber, K. (1993). A nuclear lamin of the nematode *Caenorhabditis elegans* with unusual structural features; cDNA cloning and gene organization. *Eur. J. Cell Biol.* **62**, 214–223.

Schmidt, M., and Krohne, G. (1995). *In vivo* assembly kinetics of fluorescently labeled *Xenopus* lamin A mutants. *Eur. J. Cell Biol.* **68**, 345–354.

Schmidt, M., Tschödrich-Rotter, M., Peters, R., and Krohne, G. (1994). Properties of fluorescently labeled *Xenopus* lamin A *in vivo*. *Eur. J. Cell Biol.* **65**, 70–81.

Spann, T. P., Moir, R. D., Goldman, A. E., Stick, R., and Goldman, R. D. (1997). Disruption of nuclear lamin organization alters the distribution of replication factors and inhibits DNA synthesis. *J. Cell Biol.* **136**, 1201–1212.

Stuurman, N., Heins, S., and Aebi, U. (1998). Nuclear lamins: Their structure, assembly, and interactions. *J. Struct. Biol.* **122**, 42–66.

Wagner, N., Weber, D., Seitz, S., and Krohne, G. (2004). The lamin B receptor of *Drosophila melanogaster*. *J. Cell Sci.* **117**, 2015–2028.

Weber, K., Plessmann, U., and Traub, P. (1989). Maturation of nuclear lamin A involves a specific carboxy-terminal trimming, which removes the polyisoprenylation site from the precursor; implications for the structure of the nuclear lamina,. *FEBS Lett.* **257**, 411–414.

CHAPTER 21

The Intermediate Filament Systems in the Eye Lens

Ming Der Perng,[*] **Aileen Sandilands,**[†] **Jer Kuszak,**[‡]
Ralf Dahm,[§] **Alfred Wegener,**[¶] **Alan R. Prescott,**[#] **and**
Roy A. Quinlan[*]

[*]School of Biological and Biomedical Sciences
The University of Durham
Durham DH1 3LE, UK

[†]Department of Molecular and Cellular Pathology
Ninewells Hospital, The University of Dundee
Dundee DD1 9SY, UK

[‡]Departments of Ophthalmology and Pathology
Rush-Presbyterian-St Luke's Medical Centre
Chicago, Illinois 60612

[§]Project Manager ZF-MODELS IP
Max Planck Institute for Developmental Biology
D-72076 Tübingen, Germany

[¶]Experimental Ophthalomology
University of Bonn
Bonn, 53105, Germany

[#]CHIPs
School of Life Sciences
The University of Dundee
Dundee DD1 5EH, UK

I. Introduction

Over the past 3 years there have been many exciting developments in the study of lens intermediate filaments (IFs). As for the other IF proteins, mutations in one of the eye lens-specific IF proteins, CP49, have been found to be the cause of inherited cataract. This has been followed by the generation of gene knockouts and the exciting discovery that the lens-specific IFs are exquisitely linked to one of the most important features of an eye lens—its optical properties. A major emphasis of lens research is the process of cataractogenesis, because this is the prime pathological state that affects function. Our most recent discovery has helped refocus the emphasis onto the much more fundamental question of lens transparency and the role of the cytoskeleton in determining lens function. Indeed, it is now clear that the lens-specific cytoskeleton is a key element in organizing lens fiber cell architecture. The cytoskeleton helps organize the precise lens fiber cell arrangement in the lens, which is essential for the minimization of light scatter and the ultimate function of the lens—forming an image on the retina. This naturally draws experimentation toward the generation of new knockout and transgenic models and away from the biochemistry and cell biology of the lens. To fully appreciate the role of the cytoskeleton in lens function, however, it will be necessary to map posttranslational modifications, protein processing, and assembly characteristics of the lenticular IF proteins to the altered protein associations and changes in lens cell architecture. This chapter complements the knockout approach by suggesting protocols to purify and analyze lens IF proteins and their associated chaperones from the different cellular compartments and general assistance in the analysis of the IF networks of the lens.

The first lens IF protein to be purified from the lens was vimentin (Geisler and Weber, 1981) and, indeed, bovine eye lenses are still a useful source for this protein. At that time, CP49 and filensin were not described as IF proteins, although the unique lens filaments composing these two proteins were described and named *beaded filaments* (Maisel and Perry, 1972). When filensin [(Gounari *et al.*, 1993), also called CP115 (FitzGerald, 1988) or Beaded Filament Structural Protein 1 (Bfsp1)], and CP49 [(Hess *et al.*, 1993), also called phakinin, (Merdes *et al.*, 1993) or Bfsp2 (Jakobs *et al.*, 2000)] were cloned and sequenced, their IF origins became obvious. Both proteins are expressed only in the lens (Gounari *et al.*, 1993; Hess *et al.*, 1993; Merdes *et al.*, 1993) and form typical 10-nm filaments when coassembled *in vitro* (Carter *et al.*, 1995a; Goulielmos *et al.*, 1996). CP49 and filensin serve as good markers for lens epithelial cell differentiation into lens fiber cells (Ireland *et al.*, 2000; Wigle *et al.*, 1999) although keratins (Kasper and Viebahn, 1992), vimentin (Ellis *et al.*, 1984b; Kasper and Viebahn, 1992; Kasper *et al.*, 1988; Ramaekers *et al.*, 1982b; Sandilands *et al*, 1995a), nestin (Yang *et al.*, 2000), and synemin (Tawk *et al.*, 2003) are also expressed in lens cells during development and differentiation. Filensin and CP49, however, are present throughout all stages of bovine, human, and chicken lens fiber cell differentiation (FitzGerald, 1988; Ireland *et al.*, 2000a; Quinlan, 1991; Sandilands *et al.*, 1995a). Both proteins are also present in cortical and nuclear fiber cells of the developing mouse lens (Blankenship *et al.*, 2001) and, as shown from the analysis of the CP49 knockout mouse, CP49 and filensin expression is tightly linked (Sandilands *et al.*, 2003).

Normally, CP49 and filensin form a separate and distinct filament network from the vimentin filaments in lens fiber cells as judged by immunofluorescence microscopy (Sandilands *et al.*, 1995a). *In vitro* studies confirm that filensin and CP49 do not coassemble either with keratins (Merdes *et al.*, 1993) or with vimentin (Carter *et al.*, 1995b; Merdes *et al.*, 1993). *In vitro* studies have also shown that neither filensin nor CP49 can assemble into 10-nm filaments on their own (Carter *et al.*, 1995b; Merdes *et al.*, 1993). The fact that vimentin is lost from lens fiber cells at late stages of differentiation (Ellis *et al.*, 1984a; Ramaekers *et al.*, 1980; Ramaekers *et al.*, 1982a), whereas the CP49/filensin network is maintained (Sandilands *et al.*, 1995a), is also evidence that the two networks are separate. Only recently have data been presented that filensin could associate with vimentin-containing filament material in some special circumstances, for instance, in the lens fiber cells of CP49 knockout animals (Sandilands *et al.*, 2004). Clearly some pathological situations might therefore change normal filensin associations.

Coassembly of filensin with CP49 produces 10-nm filaments *in vitro* (Carter *et al.*, 1995a), and evidence has been presented to show that the C-terminal tail domain of filensin contributes to the beaded nature of the filament (Goulielmos *et al.*, 1996). Filensin and CP49 are beaded filament components, but there are other proteins that contribute to this specific structural feature of the beaded filament, and these are the lens chaperones, αA-crystallin and αB-crystallin. Coassembly of filensin, CP49, and α-crystallins produces structures that look more similar to native beaded filaments than just CP49 and filensin on their own (Carter *et al.*, 1995a). It has been demonstrated that these protein chaperones are needed for IF network formation (Perng *et al.*, 2004), and perhaps this is one of the functions of this association in the lens. The CP49/filensin/α-crystallin complex is certainly robust, surviving extraction in buffers containing 1.5 M KCl and 1% Triton X-100 (Nicholl and Quinlan, 1994).

The differentiation status of the lens fiber cells also correlates with the apparent subcellular distribution of CP49/filensin. The lens epithelia cells cover just the anterior half the adult lens, whereas the lens fiber cells compose the bulk of the lens (Fig. 1). The lens fiber cells are formed from lens epithelial cells at the lens equator as part of their differentiation program.

The lens capsule (area 1) is covered only on the anterior half by epithelial cells (area 2). In the most recently differentiated fiber cells (area 3) of normal bovine and human lenses, filensin and CP49 are predominantly plasma membrane associated (Ireland *et al.*, 2000; Sandilands *et al.*, 1995a). More mature cells (i.e., those deeper into the lens [areas 5–7]), show both cytoplasmic and plasma

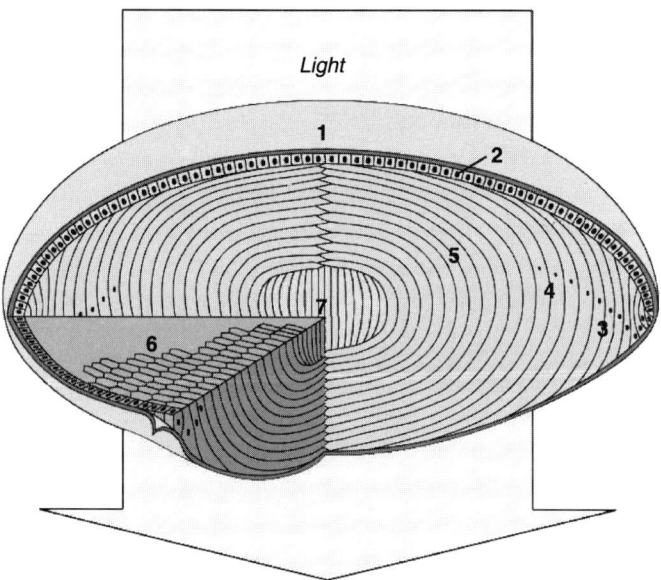

Fig. 1 Schematic of the eye lens showing the major features and regions. The lens of the eye is enclosed by the lens capsule (1). The anterior surface of the lens (top) consists of a single layer of epithelial cells (2). Epithelial cells near the equator of the lens form a reservoir from which the lens grows throughout development and throughout the life of the individual. Initially, the cuboidal epithelial cells elongate (3) in two directions until their ends reach the two lens poles (top and bottom) and are joined at the lens sutures. At this stage, the fiber cells degrade their organelles (4), for example, the nuclei, symbolized here as black dots. The bulk of the lens thus consists of long, ribbonlike fiber cells devoid of cytoplasmic organelles (5). When cross-sectioned, cortical fiber cells display a characteristic hexagonal profile (6). The very core of the lens (lens nucleus) is composed of the primary fiber cells (7). These cells form shortly after the closure of the lens vesicle by linear elongation of the epithelial cells covering the posterior surface of the hollow lens vesicle. (See Color Insert.)

membrane staining (Sandilands *et al.*, 1995a). In the 6 month bovine lens, the transition zone occurs approximately 200 μm deep into the fiber cell mass (area 4). The mechanisms for these changes have not yet been characterized, but could involve proteolytic processing of filensin (Sandilands *et al.*, 1995b) or phosphorylation, which has been linked to increased insolubilization of CP49 (Ireland *et al.*, 1993). This transition coincides with other major changes in lens fiber cells such as transcriptional shutdown (Dahm *et al.*, 1998; Gribbon *et al.*, 2002), the loss of nuclei and mitochondria (Bassnett, 1997; Bassnett and Beebe, 1992; Kuwabara and Imaizumi, 1974; Sandilands *et al.*, 1995b). The use of peptide-specific antibodies show that filensin is proteolytically processed during lens fiber cell differentiation (Sandilands *et al.*, 1995b). This processing of filensin coincides with different subcellular distributions for the C-terminal tail domain compared with the N-terminal IF protein domain (Sandilands *et al.*, 1995b) and reveal this to be an important post-translational modification whose significance has yet to be determined. The functional importance of such changes has yet to be determined.

The overexpression (Capetanaki *et al.*, 1989) or the inappropriate expression of IF proteins (Dunia *et al.*, 1990) in the lens by transgenesis causes fiber cell plasma membrane disruption, which results in cataract. Normal lens cell differentiation is disrupted, with fiber cell nuclei being retained (Capetanaki *et al.*, 1989). Also, CP49 mutations are the cause of lens cataracts in humans (Conley *et al.*, 2000; Jakobs *et al.*, 2000). In contrast, the loss of vimentin (Colucci-Guyon *et al.*, 1994), CP49 (Alizadeh *et al.*, 2002; Sandilands *et al.*, 2003), or filensin (Alizadeh *et al.*, 2003) does not induce cataracts in the knockout mouse. Careful analyses of the CP49 knockout demonstrated an increase in light scatter that was due to dramatically altered plasma membrane organization and the loss of beaded filament structure (Sandilands *et al.*, 2003). It has been proposed that filensin and CP49 comprise a specialized fiber cell plasma membrane–associated cytoskeleton within the lens (Georgatos *et al.*, 1994), a proposal that concurs with the initial results of the CP49 knockout (Sandilands *et al.*, 2003). Enucleated fiber cells, however, have both membrane and cytoplasmic networks of filensin-CP49, suggesting other roles for beaded filaments in lens fiber cells besides organizing the plasma membrane compartment, such as facilitating accommodation.

Comparative studies reveal some species differences in expression patterns for lens epithelial and fiber cells. For instance, both CP49 and filensin are exclusively expressed in lens fiber cells throughout embryogenesis in the mouse (Blankenship *et al.*, 2001; Ireland *et al.*, 2000), as indicated by the lens-specific nature of promoters; the role of transcription factors and enhancer sequences in this lens fiber cell specific expression have been identified (Masaki and Quinlan, 1997; Masaki *et al.*, 1998, 2002; Sandilands *et al.*, 2003). This is not the same for all mammals, however, because during human embryogenesis CP49 is expressed in epithelial cells at week 4/5 (Ireland *et al.*, 2000; Quinlan and Prescott 2004), the time of lens vesicle closure. By week 9 in the human, CP49 and filensin have become restricted to lens fiber cells. The lens epithelial cells in some mouse strains can, however, express other IF proteins, glial fibrillatory acidic protein

(GFAP), and desmin. For instance, in BalbC mice, the lens epithelial cells, but not the fiber cells, express GFAP from E18 onward (Boyer *et al.*, 1990; Hatfield *et al.*, 1985). Then the transforming growth factor (TGF) β-induced wound healing response of rat lens initiates desmin expression in the epithelial cells (Gordon-Thomson *et al.*, 1998). So lens cells can potentially express a wide range of IF proteins; which is influenced by species, strain, and cell environment.

Species differences constantly feature in the study of lenticular IF proteins and range from the mildly annoying to the very significant. The most important differences to be aware of are:

1. The apparent molecular weight of filensin on sodium dodecysulfate–polyacrylamide gel electrophoresis (SDS-PAGE) can vary between mammals. The bovine filensin is 115 kDa, whereas rat is closer to 94 kDa, which is most likely due to different C-terminal sequences (Masaki and Quinlan, 1997).

2. Birds express a CP49 splice variant that is 49 amino acids longer in helix I (Wallace *et al.*, 1998).

3. Some fish express forms of CP49 [Zebra fish: ENSDARG00000011998; Puffer fish: ENSDARG00000015055; and Trout (Binkley *et al.*, 2002)] that have shorter rod domains compared with mammalian homologues, posing interesting structural questions associated with heterodimer formation (Wallace *et al.*, 1998). The trout CP49 even has a C-terminal tail domain (Binkley *et al.*, 2002), which is not the case for all mammalian and most fish homologues, making it more like the lens IF protein found in squid and octopus (Tomarev *et al.*, 1993).

4. During development in the mouse lens, CP49 and filensin are expressed first in the more differentiated fiber cells (Ireland *et al.*, 2000) and then seem to concentrate at the anterior ends of the fiber cells (Blankenship *et al.*, 2001). In the human lens, CP49 and filensin are expressed in both lens epithelial and lens fiber cells (Ireland *et al.*, 2000), although some concentration at the posterior end of the lens fiber cells is observed early in development (Ireland *et al.*, 2000). In the mouse, expression starts at E13, well after lens vesicle closure, but in the human, the earliest detected expression is at the time of lens vesicle closure at week 4 (Ireland *et al.*, 2000).

Then, perhaps, the most significant for mouse transgenic and knockout studies is the following difference:

5. Some mouse strains contain a mutation in CP49 that effectively produces a knockout through an altered splice site sequence (Sandilands *et al.*, 2004). This has implications for the analysis of other lens-specific knockouts, especially when a strain dependence in phenotype has already been observed (Gong *et al.*, 1999).

Like other inbred strains, mice too are prone to ophthalmological problems, and this CP49 mutation is one example. In the 129X1 strain, the fiber cell cytoskeleton is disrupted, but this mouse does not present with a cataract, rather there is increased light scattering (Sandilands *et al.*, 2003). This can be detected in the animal with a slit lamp (Sandilands *et al.*, 2003). Other inbred strains can present with other problems such as microophthalmia as reported for C57B16

(Robinson *et al.*, 1993; Smith *et al.*, 1994), making it necessary to monitor animal stocks continually for pathological conditions (Smith and Sundberg, 1996) to prevent confusion when classifying knockout/transgenic phenotypes. The presence of the CP49 mutation now adds a genetic factor to consider when considering phenotype.

The discovery of the mouse mutation in the CP49 gene was made by the lack of antibody staining for CP49 in these lenses (Sandilands *et al.*, 2004). Clearly, antibodies are important reagents for the study of lens IFs, but there are few commercially available antibodies to filensin and CP49. One clone, 7B10, that originated from the Quinlan laboratory recognizes human, but not mouse, filensin. Our studies have developed two polyclonal antibodies specific to filensin (Sandilands *et al.*, 2004) and CP49 (Sandilands *et al.*, 1995a), and these cross-react with human, bovine, rat, mouse, and chicken. Other polyclonal antibodies have been made by the Fitzgerald laboratory (Blankenship *et al.*, 2001), which recognize mouse proteins, and the Ireland laboratory (Ireland and Maisel, 1984; Ireland and Maisel, 1989). Monoclonal antibodies are also available, but these too are prone to limited species cross-reactivities; for instance, R2D2 is an excellent antibody that works well with bovine (FitzGerald, 1988), but not human or mouse proteins in our hands. Likewise, clone MA9 is an excellent antibody for chicken filensin (Ireland *et al.*, 2000). The problem of poor mouse cross-reactivity is not unique to the beaded filament proteins, because it also exists for several popular vimentin monoclonals (Bohn *et al.*, 1992), and for this reason we developed a rabbit polyclonal antibody 3052 (Sandilands *et al.*, 1995a), an approach also taken by the Fitzgerald laboratory (Blankenship *et al.*, 2001). These immunological tools are then sufficient for analysis of beaded filament protein distributions and associations in the lens. This chapter includes the basic methods to track the distribution of IF proteins in the eye lens, as well as some microscopy techniques for looking at morphology, so that when a novel mutation causes an eye phenotype in a transgenic or knockout mouse, these protocols provide the basics to begin the analysis.

II. Materials and Instrumentation

This section emphasizes materials and some specialized instrumentation to monitor lens optical properties. In addition to the column matrices used in the purification of lens IF proteins and their associated protein chaperones, the α-crystallins, details of the primer sets needed to identify the natural CP49 mutation present in some mouse strains and a list of CP49, filensin and vimentin antibodies for the study of the IFs are presented. These antibody tools underpin the light and electron microscopy studies, and protocols are detailed in section III. Last, details of the equipment used to monitor lens optical properties are presented.

A. Column Matrices for Purification of Lens Intermediate Filament Proteins

All the protocols presented here can be applied to standard low-pressure chromatography systems. The separations use the following media:

- Fractogel EMD TMAE 650S—anion exchanger (Merck, Darmstadt, Germany)
- Fractogel EMD-COO⁻ 650S—cation exchanger (Merck, Darmstadt, Germany)
- Bio-Gel HT gel—hydroxyapetite media (BioRad, UK)

B. Detection of the Mouse CP49 Mutation—Primer Details

For genomic DNA, the different alleles can be detected with the following primers:

- Exon2 forward (129X1/SvJ) 5′ CAG TCA TGT GGT TCT GGA AGC3′
- Exon2 forward (C57BL/6J) 5′ AAG TTT CAC CAC ATT CTC CAG C3′
- Exon2 reverse 5′ CCG TGG TCT GGA GTC TGG 3′

These primers generate a 335-bp product for the mutant *Bfsp2* (e.g., 129X1/SvJ) and a 203-bp product for the wild-type allele (e.g., C57BL/6J). For a standard polymerase chain reaction (PCR), 100 ng of mouse genomic DNA is amplified in a reaction volume of 50 μl/s. Typically, 1.5 mM $MgCl_2$ is used; 0.2 U of *Taq* polymerase (Bioline) is used per PCR reaction.

If cDNA is available, the following primers can be used to distinguish the two products

CP49-For	5′ TCA CCT GGA GAG CAA GGC 3′
CP49-Rev	5′ ACT TAC TTC CTC TTC CGC TGC 3′

In the case of the wild-type allele, a 207-bp product is amplified (e.g., C3H), but for the mutant allele a 124-bp product is made (e.g., 129X1/SvJ). For the PCR primers described previously, the typical conditions used for PCR are as follows: An initial template denaturation step of 94 °C for 3 minutes is followed by 40 cycles of 94 °C for 30 seconds; 58 °C for 30 seconds; 72 °C for 1 minute, and a final extension step for 5 minutes at 72 °C. PCR products are visualized by gel electrophoresis on a 2% (w/v) agarose gel.

C. Antibodies against Beaded Filament Proteins

Antibodies avaiiable in the scientific community for the study of the lens intermediate filaments are found in Table I.

Table I
Antibodies Available to CP49, Filensin, and Vimentin

Antibody	Antigen	Species reactivity	Reference
Rabbit polyclonal – 2981	Bovine CP49 – purified by ion exchange chromatography	Bovine, human, mouse, rat, chicken	Sandilands et al. (1995a)
Rabbit polyclonal	Recombinant mouse CP49	Mouse	Blankenship et al. (2001)
Rabbit polyclonal (IMRGC_899)	Chicken CP49 – gel purified	Chicken	Ireland and Maisel (1984)
Mouse monoclonal	Bovine lens material	Bovine, mouse, chicken, and tiger salamander	FitzGerald (1988)
Rabbit polyclonal – 3241	53-kDa fragment containing the rod domain of filensin	Bovine, human, mouse, chicken	Sandilands et al. (2004)
Rabbit polyclonal–URNOT-76	Recombinant bovine filensin	Mouse	Blankenship et al. (2001)
Mouse monoclonal – R2D2	Bovine lens material	Bovine	FitzGerald (1988)
Mouse monoclonal – MA9	Chicken lens material	Chicken	Ireland et al. (2000)
Mouse monoclonal – 7B10	Bovine and human lens material	Bovine, human and sheep, but not chicken, rat, or mouse	Sandilands et al. (1995b)
Rabbit polyclonal-3052	Bovine lens vimentin	Bovine, human, mouse, rat and chicken	Sandilands et al. (1995a)
Mouse monoclonal–Vim 3B4	Recombinant vimentin	Mouse	Blankenship et al. (2001)

D. Measuring Lens Optical Properties

Crystalline lenses are a principal component in the process of dynamic focusing or accommodation. To perform its role in vision, a lens must be transparent. Thus, it is commonly held that whereas a transparent lens is normal, a lens with either a reduced level of transparency or a localized opacity is abnormal and/or pathological. A lens that is transparent, however, to the naked eye is not necessarily optically efficient. Biological lenses have different levels of optical quality that are directly related to the structure of the lens. Therefore, when assessing the effects of experimental or naturally occurring lens pathological conditions, it is important to recognize that lens function (sharpness of focus) can be significantly altered even in the absence of compromised lens clarity.

The optical quality of lenses (sharpness of focus, i.e., spherical aberration and intensity of scatter) can be measured with a low-power helium–neon laser scanner designed and developed by Sivak and associates (Sivak *et al.*, 1986) at the University of Waterloo, School of Optometry in Ontario, Canada (contact Dr. Sivak for equipment supply (jsivak@sciborg.uwaterloo.ca); see details of the ScanTox system at http://www.optometry.uwaterloo.ca/~sivakgrp/scantox_brochure.htm). The laser scanner consists of a low-power (2 mW) helium–neon laser mounted on a computer-controlled X-Y table and a television camera with a video frame digitizer (Fig. 2). The scanning laser beam is used, so that a programmed series of beams pass sequentially at precise increments through the lens. The location of the beams is directed through different parts of the lens (e.g., on sutures, near sutures, off sutures; see Fig. 8 later). Back vertex distance (BVD; spherical aberration) for each laser beam is recorded from the posterior surface of the lens to the focal point. Repeated measurements of BVDs indicate instrument reproducibility

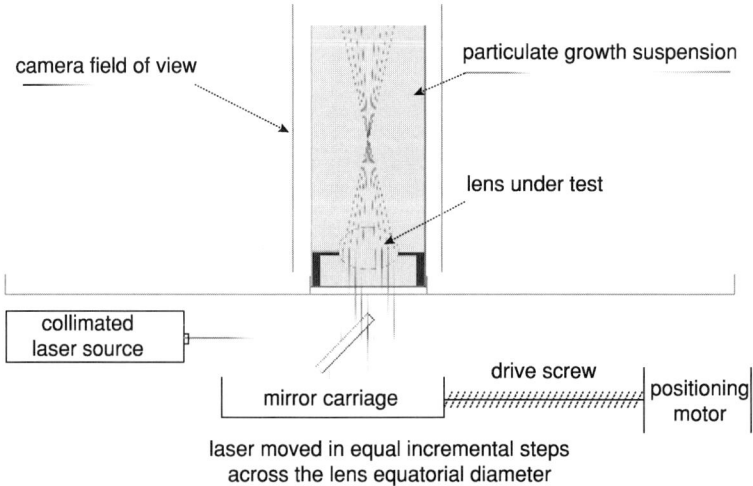

Fig. 2 Instrument for measuring lens optical properties.

within 0.32%. Changes in these distances with beam position (sharpness of focus) are influenced by the degree of longitudinal spherical aberration. In addition, morphological irregularities of the lens can misdirect the laser beam, causing a more scattered appearance of the focal point. With this type of analysis, sharpness of focus and/or the loss of focus, whatever the cause, referred to as variability in BVD, is defined as the standard error of the mean of the lens BVD (mm) with interanimal variance expressed as the standard error of the mean.

III. Procedures

In this section, the protocols needed to purify the individual IF proteins and protocols for studying lens IF function are described. The purifications detail the methods we have used to purify filensin, the 53-kDa fragment of filensin, CP49, vimentin, and αA-crystallin and αB-crystallin from bovine lens material. Recombinant expression of these individual proteins can achieve similar goals, and this is especially valuable when biomass is limiting, such as in the case of mouse or human lenses. Nevertheless, posttranslational modifications are clearly very important in lens filament function (Ireland *et al.*, 1993; Sandilands *et al.*, 1995b), and as we seek the identity of binding partners and the details of filament function in the lens, purification protocols for native proteins from the lens will be needed.

Methods detailing the immunochemical and histological characterization of the lens are also described, as well as PCR protocols, to detect the naturally occurring mutation in CP49 that is present in some mouse strains, notably the 129 strains that are used to generate knockouts and therefore could significantly influence lens phenotypes. Last, we detail the optical methods needed to characterize lens optical properties, an essential tool in the analysis of this tissue.

A. Preparation of Lens Fractions

Between 25 and 50 fresh bovine lenses were obtained from the abattoir and processed within an hour. Lenses were isolated, decapsulated, and then stirred on ice for 30 minutes in extraction buffer (see later for details). The ratio of extraction buffer wet to weight of lenses was 2:1 (approximately 50 to 100 ml for 25 to 50 lenses), and this volume was used at all other extraction stages. The suspension contained the outer cortical lens layers, which was poured off and then homogenized in a loose-fitting (B pestle) Dounce homogenizer. This fraction was centrifuged for 20 minutes at 30,000g at 4 °C to produce a supernatant and pellet fraction, both of which are retained.

The pellet from this initial fractionation was extracted twice more with extraction buffer. At each of these steps, the supernatants were also retained. The final pellet fraction can then be extracted with buffers containing detergents or high salt to further enrich for the cytoskeletal proteins.

The retained supernatants can be used to isolate another cytoskeletal fraction called the plasma membrane cytoskeleton complex (PMCC). This is achieved by layering the supernatants onto a 0.85 M sucrose cushion in isolation buffer and centrifuged for 1 hour at 80,000*g* at 4 °C. The resulting pellet is termed the lens fiber cell PMCC. Although this can also be used to purify lens cytoskeletal proteins (Carter *et al.*, 1995a), it is only a fraction of the total and it is better to use the pellet fraction described earlier for large-scale protein purification.

B. Purification of Lens Intermediate Filament Proteins

1. Purification of CP49

A major problem in purifying CP49 is contamination by its 40 kDa breakdown product. To obtain sufficient full-length CP49 protein for assembly studies, we developed a protocol that uses a weaker anion exchanger (EMD-TMAE 650S) followed by further purification on a cation exchanger (EMD-COO$^-$650S) to obtain CP49 free from its proteolytic contaminant. This particular protocol requires a pump and gradient controller–based system, because a nonlinear gradient is used for the separations.

Lenses were extracted with the method described in (A), with the following modifications. The extraction buffer used was 10 mM sodium phosphate, pH7.4, 100 mM KCl, 5 mM EDTA, and 0.5 mM DTT. The final pellet was extracted twice more with extraction buffer containing 1M KCl before finally washing once more in extraction buffer. The pellet was then frozen at $-20\,^\circ$C until required.

The starting material was approximately 20 g of accumulated pellets from the extraction of 100 lenses. This is a typical preparation. This pellet material was extracted with an equal volume (20 ml) of 8 M urea, 20 mM Tris-HCl pH8.0, 2 mM EDTA, 50 mM 2-mercaptoethanol at room temperature before centrifugation at 13,000 rpm in a Beckman JS13 swing out rotor. After centrifugation, the supernatant was removed, and the pellet was once again extracted for 30 minutes at room temperature, but this time with 4 M urea, 10 mM Tris-HCl pH8.0, 1 mM EDTA, 25 mM 2-mercaptoethanol. This step was included to ensure that all the cytoskeletal material was extracted from the starting material. The centrifugation step was repeated, and then the resulting supernatants were centrifuged for 2 hours at 28,000 rpm in a Beckman SW28 rotor at 4 °C. The supernatant from this spin was diluted 1:1 with 8 M urea, 20 mM Tris-HCl pH8.0, 2 mM EDTA, 50 mM 2-mercaptoethanol, and then loaded onto a Fractogel EMD-TMAE 650S column, 150 × 26 mm at a flow rate of between 2 and 4 ml/min^{-1}. The protein concentration of the extract was $\sim$16 mg/ml^{-1}, and approximately 1.6 g of protein was loaded onto the column. The column was washed in column buffer A (8 M urea, 20 mM Tris-HCl pH8.0, 2 mM EDTA, 50 mM 2-mercaptoethanol) until the baseline absorbance returned to near baseline, which took between 10 and 15 column volumes. The proteins were eluted with column buffer A containing 1 M guanidine hydrochloride (8 M urea, 20 mM Tris-HCl pH8.0, 2 mM EDTA, 1 M guanidine hydrochloride, 50 mM 2-mercaptoethanol) at a flow rate of 4.5 ml/min^{-1} using the gradient profile 0%–40% column buffer B for 80 minutes followed by 40%–100% column buffer for 30 minutes. One minute fractions were collected throughout the elution, and the fractions were monitored by SDS-PAGE, as shown in Fig. 3A.

Fractions 17–24 (Fig. 3A) enriched in CP49 were selected for further purification on a 15–2.5-cm Fractogel EMD-COO$^-$ 650S column. These fractions containing approximately 120 mg protein were dialyzed overnight against 7 M urea, 20 mM sodium formate pH4.0, 2 mM EDTA, and 1 mM DTT and loaded onto the column. Proteins bound to the column were eluted with a linear gradient. The column buffer for this stage (column buffer B) contained 7 M urea, 20 mM sodium formate, pH 4.0, 2 mM EDTA, and 50 mM 2-mercaptoethanol, and the gradient was developed with column buffer B containing 0.3 M guanidine hydrochloride. A flow rate of 20 ml/h^{-1} was used and 3.75 ml fraction collected, and the total gradient was 600 ml. Fraction composition was monitored by SDS-PAGE (Fig. 3B), and fractions 45–47 were pooled and stored at −80°C. The yield of CP49 from this procedure was 8 mg.

2. Purification of Vimentin, Filensin, and its 53-kDa Fragment

Fractions enriched in vimentin, filensin, and the 53-kDa fragment of filensin (Fractions 30–17 in Fig. 3A) were dialyzed against hydroxyapatite column buffer (7 M urea, 10 mM sodium phosphate, pH 7.5, 1 mM DTT, 0.2 mM PMSF) before being loaded onto a hydroxyapatite column (2.6 × 10 cm). The preparation of vimentin, filensin, and the 53-kDa fragment is based on previously published procedures (Quinlan *et al.*, 1992). As shown in Fig. 3C, vimentin is recovered in the flow-through and subsequent 50-ml wash from the hydroyapatite column. Filensin and its 53-kDa fragment are eluted from the column with a 200-ml linear gradient of 10–100 mM sodium phosphate in hydroxyapatite column buffer. Protein purity was checked by SDS-PAGE. Fractions containing purified vimentin, filensin and the 53-Kda fragment of filensin can be stored at −80°C until required. This part of the purification procedure for lens IF proteins can be scaled down to suit the availability of lens material.

C. Preparing Lenses for Immunofluorescence Microscopy

Adult bovine lenses were dissected out of the eye with a scalpel blade to open the eye around its middle. The cut is made at right angles to the optical axis to complete a two-thirds section. The cut should not be complete, because pushing on the outside of the cornea with the index finger will displace the lens, which can be released from the eye by severing the suspensory ligaments with the tip of the scalpel blade. Dissected lenses were then immediately plunge-frozen into liquid nitrogen. Sections, 30–40-μm thick, were cut in a Reichert-Jung cryocut 1800

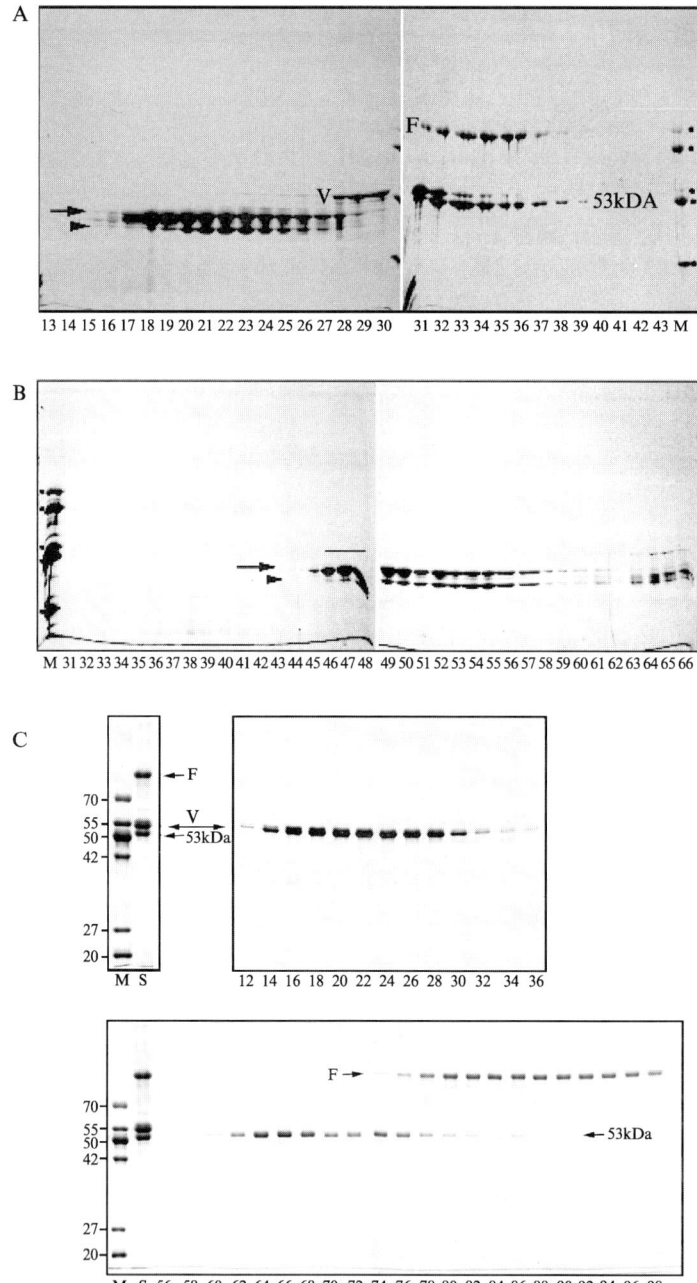

Fig. 3 Purification of CP49 from bovine lenses. (A). Purification of CP49-enriched fractions on Fractogel EMD 650S TMAE; 20 g of lens membranes were extracted as described in section IIIB2, and

cryostat (Leica, Vienna, Austria) at $-20\,°C$, mounting the lens in mounting fluid so that sections parallel to the optical axis are cut. The lens cryosections were fixed in a solution of 4% w/v paraformaldehyde in phosphate-buffered saline (PBS) or in a methanol-acetone mixture (1:1 v/v) stored at $-20\,°C$. The sections were then washed with PBS (3×), permeabilized with 1% v/v Nonidet P-40 (VWR, UK) in PBS and then blocked with 10% v/v normal goat serum (Sigma, UK) for 20 minutes before incubation with the primary antibodies. After 1 hour incubation at room temperature, the sections were washed again with PBS (3×) before incubating in the secondary antibody solution (1 hour at room temperature). In some cases, the DNA stains propidium iodide (Sigma, UK), Sytox (Molecular Probes, The Netherlands), DAPI (4′,6-diamidino-2-phenylindole; Sigma), or Hoechst 33528 (bis-benzimide; Sigma) were included in the secondary antibody solution at a final concentration of 1 μg/ml. All solutions were made up in 0.3% w/v bovine serum albumin (BSA) (Sigma) and 0.02% w/v sodium azide (Sigma, UK) in PBS. An example of a bovine lens stained with IF antibodies is shown in Fig. 4.

A similar procedure is used to process mouse and rat lenses, but rather than removing the lens from the eye, it is left intact to aid orientation of the smaller lens. The cornea and optic nerve serve as good orientation features. Sections are cut using the procedure described earlier, although section thickness is reduced to $\sim$15 μm. The eye lens can be easily damaged when being removed from the animal or the eye itself and therefore the dissection needs to be done carefully.

1.6 g of protein was loaded onto a 2.5 w × 15 cm Fractogel EMD-TMAE 650S column in column buffer A (8 M urea, 20 mM Tris-HCl pH 8.0, 2 mM EDTA, 50 mM 2-mercaptoethanol). After extensive washing with column buffer, the proteins were eluted with a nonlinear gradient from 0–1 M guanidine hydrochloride. Fractions 17–24 enriched in CP49 (arrow) and its breakdown product (arrowhead). Vimentin (V), filensin (F), and its 53 kDa breakdown product (53 kDa) all elute later than CP49. These fractions can be taken for hydroxyapatite purification of vimentin and filensin as described in the following. (B) Purification of CP49 from its breakdown product by cation exchange chromatography. Fractions 17–24 were dialyzed overnight into column buffer B (7 M urea, 20 mM sodium formate, pH 4.0, 2 mM EDTA, and 50 mM 2-mercaptoethanol). Approximately 120 mg of protein was loaded onto a 2.5-× 10-cm column of Fractogel EMD-C00⁻ 650S, and bound proteins were eluted by a linear 600-ml gradient from 0–0.3 M guanidine hydrochloride. Full-length CP49 (arrow) eluted slightly ahead of the major breakdown product (arrowhead). Fractions indicated by the horizontal line were pooled and stored at $-80\,°C$ until required. (C) Pooled fractions containing most of the vimentin and a few contaminants (track S) were further purified by hydroxyapatitie column chromatography. After dialysis of the selected fractions against hydroxyapatite column buffer, the sample was loaded onto the column. Vimentin was recovered in the flowthrough and wash fractions (14–30). Filensin and its 53 kDa fragment were eluted with a linear gradient of 10–100 mM sodium phosphate in column buffer. Note that under these conditions, most of the 53 kDa fragment was separated from filensin, which eluted later in the gradient. The position of vimentin, 53 kDa fragment, and filensin are indicated by arrows and appropriate labels. Molecular weight markers (track M, kDa × 10^{-3}) were HSC70 (70 kDa), vimentin (55 kDa), GFAP (50 kDa), actin (42 kDa), HSP27 (27 kDa), and αB-crystallin (20 kDa).

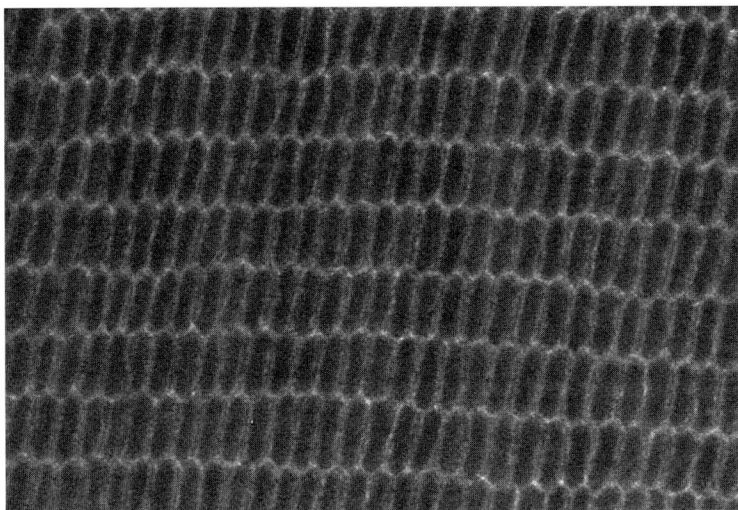

Fig. 4 Immunofluorescence staining of a bovine lens section with filensin polyclonal antibodies. By use of polyclonal antibodies to filensin, the plasma membrane distribution in these fiber cells can be seen. They are equivalent to cells seen in region 3 in Fig. 1. The regular hexagonal distribution of the lens fiber cells is very obvious, and the obtuse sectioning angle gives the appearance of double lines along the long axes of the fiber cells.

D. Preparation of Eye Sections for Histochemistry

Whole rat or mouse eyes were fixed for more than 48 hours in a solution of 4% w/v paraformaldehyde (Sigma, UK) in PBS (Sigma). The eyes were then dehydrated twice in 100% ethanol, processed through two changes of 100% xylene (5 minutes each) and embedded in paraffin wax overnight. 7- to 10-μm-thick sections were cut in a Reichert-Jung 2035 benchtop microtome (Leica, Vienna, Austria) and collected on APES-coated glass slides (APES; aminopropylethoxy-silane, Sigma). For immunohistochemistry, the sections were dewaxed in two changes of 100% xylene and rehydrated to H_2O through a series of ethanol changes (two times 100% and once 50%).

Staining of rehydrated sections was performed in Mayer's Haemalum (hematoxylin) solution (Gurr/BDH/Merck, Poole, UK) of the acidophilic nuclear stain hematoxylin for 5 minutes. The stained sections were subsequently dehydrated in a series of ethanol (50% v/v ethanol in water for 5 minutes; 100% ethanol twice for 5 minutes), cleared in xylene (100% xylene twice for 5 minutes), and mounted in DPX mounting medium (Agar Scientific, Stanstead, UK).

E. Freeze Fracture Electron Microscopy

Dissected lenses were fixed immediately in five times their volume of Peter's fixative (2.5% v/v glutaraldehyde, 1.0% w/v paraformaldehyde, and 0.02 M $CaCl_2$ in a 0.08 M cacodylate buffer; all Sigma) for at least 24 hours. In large lenses, the outer cortical regions of the lenses were then dissected into cubes of approximately 1 mm^3 corresponding to the areas of interest and cryoprotected in 2.3 M sucrose (Sigma) in PBS (Sigma) overnight. The lens pieces were plunge frozen in liquid propane in a Reichert KF 80 and subsequently fractured and replicated with platinum (2 nm) and carbon support (15 nm) in a Cressington CFE50C (Cressington Scientific Instruments Ltd., Watford, UK) or a Balzers BAF 300 (Bal-Tec AG, Balzers, Lichtenstein) at 160 K and 10^{-5} Pa. The replicas were cleaned by floating them on 5% v/v hypochloric acid for 24 hours, washed in distilled water, transferred onto grids, and examined by transmission

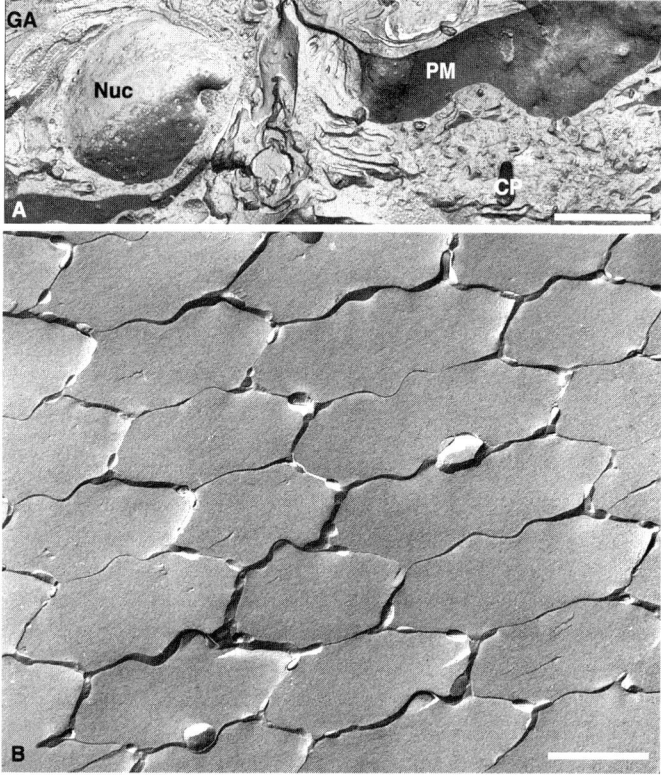

Fig. 5 Freeze fracture replicas of bovine lens epithelial and mature fiber cells. (A) The cytoplasm of epithelial cells (CP) contains numerous organelles, including a prominent nucleus (Nuc) with adjacent Golgi apparatus (GA). PM; plasma membrane between two epithelial cells. (B) Typical hexagonal profiles of cross-fractured fiber cells in the lens cortex. The cytoplasm of mature fiber cells is completely uniform because of the absence of intracellular organelles and consists of an optically homogenous solution largely composed of crystallin proteins. Note the highly ordered arrangement of the fiber cells in the lens. Scale bars: A, 2 μm; B, 5 μm.

electron microscopy (TEM). An example showing lens fiber cells is presented in Fig. 5.

F. Scanning Electron Microscopy

Adult lenses were fixed as described previously (section E). Large bovine lenses were dehydrated through a modified series of ethanol ($2\times$ 50%, $1\times$ 60%, $1\times$ 70%, $1\times$ 80%, $1\times$ 90%, $2\times$ 100%; 24 hours each). The ethanol was subsequently replaced by acetone through two changes of ethanolacetone 1:1 (v/v) and two changes of 100% acetone (24 hours each). The lenses were then inserted into a Polaron E3000 Series II critical point drying apparatus (Polaron, Watford, UK), the liquid acetone replaced with liquid CO_2, and the lenses were allowed to impregnate with CO_2. After 1 hour, the chamber was flushed with fresh CO_2. The temperature was then slowly raised to $37\,°C$ (critical point of CO_2 is $32\,°C$), and the gaseous CO_2 was vented from the chamber. The critical point dried lenses were manually fractured to reveal the layers of fiber cells and mounted onto aluminium stubs with silver paint. Specimens were then sputter coated for 5 minutes in a Polaron E5100 Series "cool stage" sputter coater fitted with an Au/Pd target (coating thickness approximately 100 nm). The specimens were examined in a Jeol JSM-35 scanning electron microscope (Jeol UK Ltd.) or Hitachi 4900 at an accelerating voltage of 15 kV.

G. Fixing Lenses for Electron Microscopy

After removal of the lens from eye, it was added directly into fixative solution (1.25% v/v glutaraldehyde, 1% w/v paraformaldehyde in 0.08 M cacodylate buffer, pH 7.2, + 0.02% w/v $CaCl_2$). Depending on the size of the lens, lens pieces, or lens fraction, they were left for as short as 1 hour to 24–48 hours at room temperature. Larger lenses were divided into smaller pieces to facilitate fixation. After this incubation, the lens samples were then washed twice in cacodylate buffer and stored at $4\,°C$ until processing could be undertaken. The lenses were then postfixed in 1% v/v OsO_4 in 1.5% w/v sodium ferrocyanide in cacodylate buffer for 1 hour before dehydrating in an alcohol series from 50% to 100% for 10 minutes in each concentration. Samples were then moved into 100% propylene oxide and incubated for 20 minutes, changing the solution after 10 minutes and reducing the propylene oxide concentration stepwise while at the same time increasing the Durcopan concentration. The samples were then changed into a 1:1 solution of propylene oxide Durcupan resin and left for 2 hours, changing the solution after 1 hour. Last, the samples were left in 100% v/v Durcupan overnight, changing into fresh Durcupan in EM capsules (Agar Aids, UK) and polymerizing at $60\,°C$ overnight. Sections were cut on ultramicrotome at 70–100 nm thickness and stained with 3% w/v uranyl acetate followed by Reynolds lead citrate (1.33 g lead nitrate + 1.76 g Na citrate + 8 ml 1 M NaOH in 50 ml water) before viewing in the electron microscope.

H. Antigen Retrieval from Samples Prepared for Electron Microscopy

Antigen retrieval from these EM samples is possible by use of the following technique. Durcupan sections were processed for immunogold labeling by first etching the resin for 8 minutes with 1% v/v periodic acid. The sections were washed quickly in distilled water before being treated for 10 minutes with 2% w/v sodium periodate to remove the OsO_4. The sections were once again washed quickly with distilled water and with PBS (3 × 2 minutes) before incubating in 0.5% w/v fish skin gelatin for 30 minutes. The sections were then ready to probe with antibodies. The sections were incubated in primary antibody diluted in fish skin gelatin for 90 minutes, washed in PBS (2 × 10 minutes), and then incubated in protein A gold for 120 minutes. The sections were washed extensively in water (10 × 1 minute) and then dried in a grid box under a desk lamp. Finally, the sections were exposed to osmium vapor (4% v/v solution) for 30 seconds and then stained with 3% w/v uranyl acetate for 15 minutes, washed in water (4 ×) and last stained with Reynold's lead citrate (1.33 g lead nitrate + 1.76 g Na citrate + 8 ml 1 M NaOH in 50 ml water) for 15 minutes, washed (4 ×) and then dried. An example is shown in Fig. 6.

I. Mouse Strain Differences

A list of those strains so far tested for the mutant allele are given in Table II.

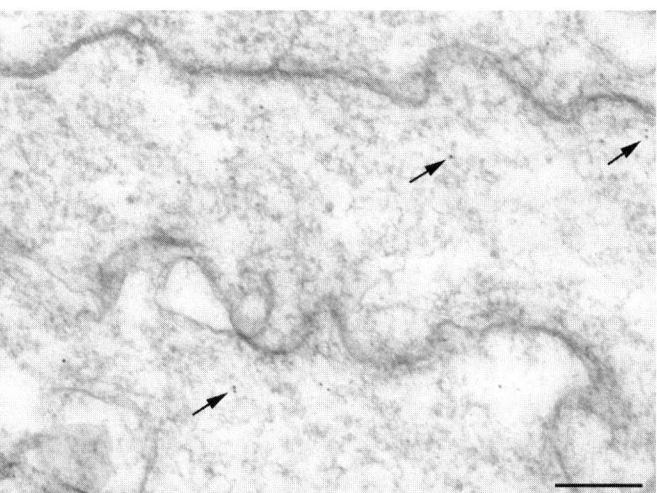

Fig. 6 Antigen retrieval of vimentin epitopes from Durcopan-embedded lenses. Immunogold labeling of vimentin filaments as described (Sandilands *et al.*, 2004) with polyclonal antibodies to vimentin (3052) that are then detected by protein A–labeled gold. Notice the two types of filament in this section (beaded and smooth) and that the vimentin antibodies as detected by the immunogold particles (arrows) detect the smooth 10-nm filaments. Bar = 200 nm.

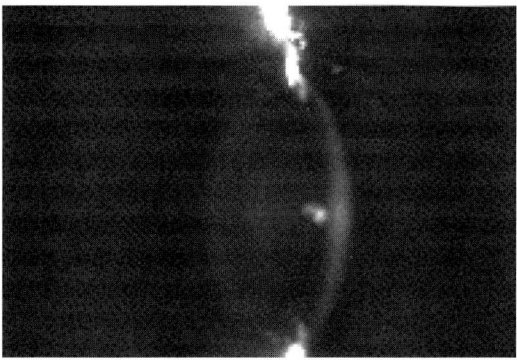

Fig. 7   Slit-lamp image of a lens in the eye of a 4-month-old mouse, strain C3H. Notice the lens is clear showing no obvious aberrations.

Table II
Mouse Strain Expression of *Bfsp2* Alleles

Mouse strains with wild-type *Bfsp2*	Mouse strains with mutant *Bfsp2*
AKR	129X1/SvJ
BalbC	129S1/ImJ
C3HEI	129S2/SvPas
C57B16J	129S4/SvJae
DBA	101
NMR1	
SEC	
SWR	
102E1	

J. Slit-Lamp Examination

Just as for humans, the eye lens in animal (mouse) eyes can be examined with a slit lamp. Commercial systems supplied by Nikon or Zeiss can be used for animal examinations. A topical application of 2 drops ($\sim$60 μl) of atropine (0.5%) were put into the eyes of the animals 20 minutes before examination. The animal was then positioned at the slit lamp, so that the lens could be visualized to confirm the presence or absence of opacities or other light-scattering regions in the lens. Each eye takes approximately 0.5–1 minute to examine for the skilled operator. The animals need not be anesthetised. Recovery from the atropine treatment takes approximately 1 hour. Examination and recovery should occur in a darkened room. As an animal procedure, slit-lamp examination will require authorization. An example of a normal mouse lens is shown in Fig. 7.

K. Measurement of Lens Optical Properties

The following is a description of how the function of most crystalline lenses can be easily and reproducibly quantified.

1. Lens Optical Quality Analysis

Lenses were prepared for optical quality analysis as follows: Within 2 minutes of death, eyes were carefully excised, and the lens was removed and suspended on a beveled washer designed to support its equatorial rim. The lens was then placed in a specially designed two-chambered cell made of glass and silicon rubber in the instrument described in Section II.D. Both lens surfaces were bathed in a culture medium (25 ml) consisting of M199 (Invitrogen) with Earle's salts and 8% v/v FBS. Previous work involving periodic measurements of medium osmolarity and pH indicates little or no change over the duration of the experiment. Lenses were positioned so that the initial series of scans will pass, as closely as possible, directly along a defined suture branch (see Kuszak, 1995; Kuszak and Al-Ghoul, 2002; Kuszak *et al.*, 1991, 1994, 2000; Sivak *et al.*, 1994; Fig. 8). Data are normally collected for 10 scanning positions per series for small lenses (mice) and for as many as 18 scanning positions for larger lenses, such as chicken or human. Then a mechanical linkage makes it possible to rotate the lens in defined increments (15 degrees) through a total of 360 degrees ensuring comprehensive examination of lens optic as it relates to lens structure. Thus, the number of laser beams routinely passed through lenses varies from a total of 120 scanning positions per small lens (mouse) to 216 scanning positions per large lens (human). In the case of a lens with a line suture (e.g., frog and rabbit), this means that 36 scans are transmitted along or "on" suture planes, 72 "near" suture planes, and 108 "off" suture planes. Similarly, in the case of a mouse lens with a Y suture, this means that 30 scans are transmitted "on" suture planes, 60 "near" suture planes, and 30 "off" suture planes. It is important to differentiate between scans that pass on or near sutures as opposed to scans that pass off sutures, because suture planes are naturally occurring sites of structural disorder oriented directly along the visual axis (Fig. 8). These regions of the lens are inferior in optical quality by comparison to the highly ordered radial cell columns (see Kuszak *et al.*, 2004 for a full discussion).

Statistical calculations to determine the significance of differences between average BVD and variability in BVD are carried out by use of a one-way analysis of variance (ANOVA) or the Kruskal–Wallis ANOVA when dealing with nonnormal populations. A probability value of at least $P < 0.05$ is considered significant.

2. Quantifiable Data from Lens Laser Scan Analysis

Each series of laser beams passed through a manufactured glass lens produces a scan profile (Fig. 9A), which is less than perfect, given the spherical aberration and broad range of focus. In contrast, a natural lens taken from a fish has none of

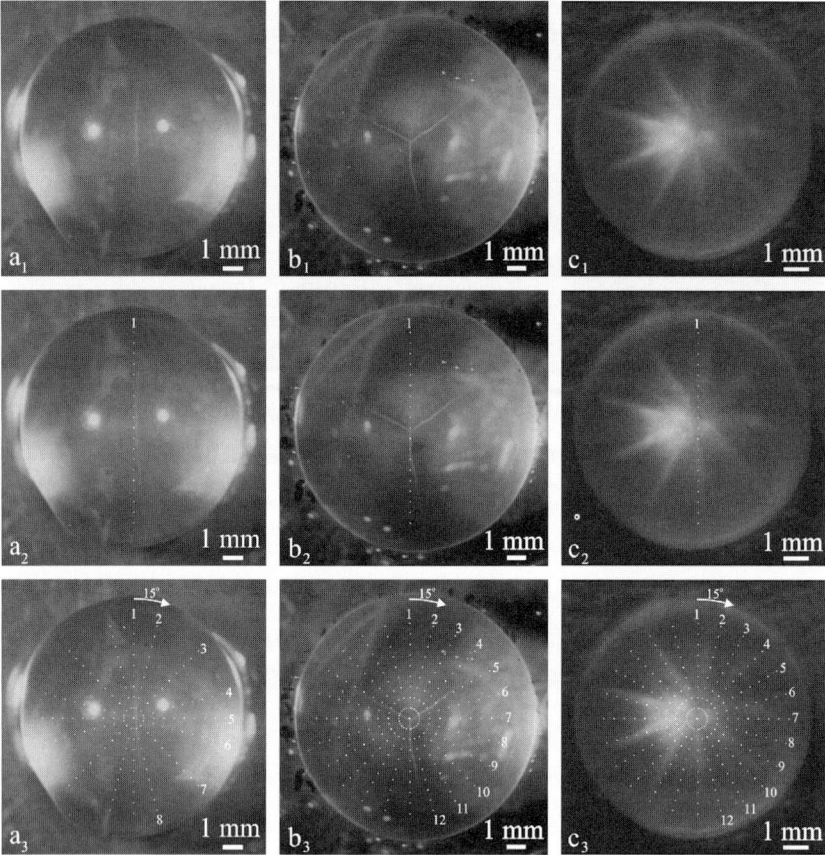

Fig. 8 Determining the optical properties of the eye lens. These examples show the importance of lens morphology to optical analysis. The three examples show the anterior face of three different lenses that have three different suture arrangements. The sutures, as areas of optical imperfection, will influence the laser analysis of the optical properties of the lens. A rabbit lens (a_1–a_3) has a typical line suture (a_1), and the first measurements have been made along this suture (a_2–1). In the next measurement (a_3), the lens is rotated by 15 degrees (a_3–2) to make the second series of measurements. Six other measurements are made (a_3–3–8) to complete the analysis of the lens. The scan series 1 is, therefore, an on-suture plane, whereas 2, 4, 6, and 8 are near-suture with scans on lines 3 and 7 being off-suture. Plane 5 is also on-suture, because the line suture on the posterior side of the lens is positioned at 90 degrees to the suture on the anterior surface—a further consideration when conducting this analysis of the lens. In the case of a Y-suture lens, such as the mouse or the bovine lens shown here (b_1–b_3), the lens is again positioned so that the first set of scans align with one of the lens sutures on the anterior face, which can be clearly seen as a "Y" shape in the middle of the lens (b_1). The sutures on the posterior surface are displaced by 60 degrees, and, therefore, this first scan series will be completely on suture (b_2–1). Of all the subsequent scans (b_3–2–12), only scans on lines 3, 7, and 11 will be off-suture. In the last example, a primate lens is shown, and here the suture arrangements are quite different again, now showing a starlike appearance (c_1). From the first (c_2–1) to all subsequent scans (c_3–1), none will be "off-suture" because of the suture arrangement in such lenses.

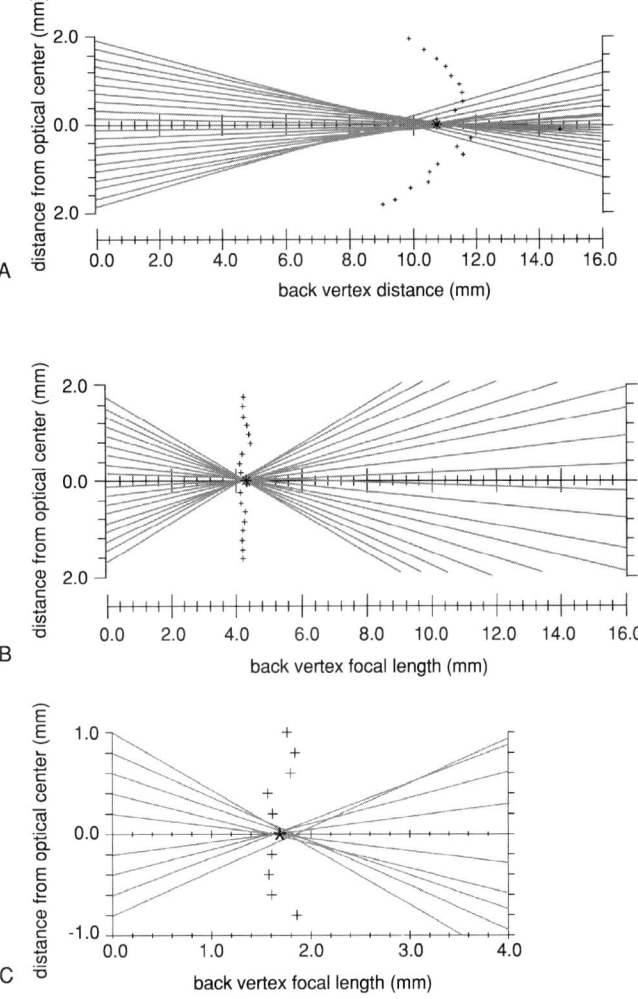

Fig. 9 Measuring optical properties of lenses. In the first example (A), the properties of a glass bead are shown. The average BVD (*) of all the laser beams is plotted on the graph shown. Sharpness of focus, or variability in average BVD, is plotted (+) for each beam relative to the average BVD. Note that this ground-glass bead exhibits longitudinal spherical aberration and a broad range of focus as seen by the speard in (Adapted from Jacob G. Sivak, PhD, OD, University of Waterloo, Waterloo, Ontario, Canada). In contrast, a natural lens from a fish (B) of essentially the same size and shape as the previous ground-glass lens, exhibits minimal longitudinal spherical aberration and a sharp focus evident by the near vertical plotting of crosses (Adapted from Jacob G. Sivak, PhD, OD, University of Waterloo, Waterloo, Ontario, Canada) for an example of a series of off-suture scans. Last, an example of a mouse lens (C) is shown for off-suture scans illustrating. By comparison with the fish lens, the relatively poor optical properties of the mouse lens is seen by the scatter in the BVD values (+) for each beam relative to the average BVD for the whole lens.

these problems (Fig. 9B). Last, a mouse lens profile is shown (Fig. 9C) demonstrating how the lenses in this mammal do not match the optical properties of the fish lens. The laser scan system is not only useful for studying the compromised function of cataractous lenses, but also for studying the optical properties of "clear" lenses, which has proven to be so important for examining the effects of the CP49 knockout (Sandilands *et al.*, 2003).

IV. Pearls and Pitfalls

This chapter includes the basic methods to purify important lens fractions, the key IF proteins, and their associated proteins. As we consolidate the animal models that have been developed in the past few years, lens fractionation and protein complex purification will once again become important, and the methods presented here give a number of critical protocols and useful assays. One of the pitfalls in the purification of lens proteins is proteolysis, especially when urea buffers are used, but the methods here describe the effective purification of CP49, filensin, and vimentin for use *in vitro* assembly and structural studies. Both CP49 and filensin are extensively posttranslationally modified, and accessing different purified fractions will assist in identifying these modifications and their role in filament assembly. The list of antibodies is a good starting point for the immunological characterization of the lens, and because many are not commercially available, it provides an essential resource. A method is described to allow the researcher to recover antigenicity from lens material previously prepared for electron microscopy. This makes it possible to use archival material, as well as obtaining lens sections with good cell preservation. The description of primer sets to identify known mutations in some commonly used mouse strains will also enable the researcher to screen for the potential influence of CP49 in lens phenotypes caused by knockout or transgenic approaches. Last, but not least, the description of methods for the preparation, sectioning, and examination of eye lenses are presented as a starting point for the researcher beginning to explore lens function.

V. Discussion and Concluding Remarks

It is a very exciting time to be involved in lens research. The IF proteins of the lens help determine its optical properties. Loss of lens optical properties need not translate into the most common lenticular disease state, namely cataract. Instead, changing the CP49 and filensin networks offers a way to modulate the optical properties of the lens. We hope to discover those filament interactions that are needed to preserve lens function. Of course, the animal models will also help us understand mechanisms of cataractogensis, and in combination with the optical studies, we can dream of the futuristic possibility of regenerating superlenses for combating cataract.

References

Alizadeh, A., Clark, J., Seeberger, T., Hess, J., Blankenship, T., and FitzGerald, P. G. (2003). Targeted deletion of the lens fiber cell-specific intermediate filament protein filensin. *Invest. Ophthalmol. Vis. Sci.* **44,** 5252–5258.

Alizadeh, A., Clark, J. I., Seeberger, T., Hess, J., Blankenship, T., Spicer, A., and FitzGerald, P. G. (2002). Targeted genomic deletion of the lens-specific intermediate filament protein CP49. *Invest. Ophthalmol. Vis. Sci.* **43,** 3722–3727.

Bassnett, S. (1997). Fiber cell denucleation in the primate lens. *Invest. Ophthalmol. Vis. Sci.* **38,** 1678–1687.

Bassnett, S., and Beebe, D. C. (1992). Coincident loss of mitochondria and nuclei during lens fibre cell differentiation. *Dev. Dynam.* **194,** 85–93.

Binkley, P. A., Hess, J., Casselman, J., and FitzGerald, P. (2002). Unexpected variation in unique features of the lens-specific type I cytokeratin CP49. *Invest. Ophthalmol. Vis. Sci.* **43,** 225–235.

Blankenship, T. N., Hess, J. F., and FitzGerald, P. G. (2001). Development- and differentiation-dependent reorganization of intermediate filaments in fiber cells. *Invest. Ophthalmol. Vis. Sci.* **42,** 735–742.

Bohn, W., Wiegers, W., Beuttenmuller, M., and Traub, P. (1992). Species-specific recognition patterns of monoclonal antibodies directed against vimentin. *Exp. Cell Res.* **201,** 1–7.

Boyer, S., Maunoury, R., Gomes, D., Hill, N. B. de, A. M., and Dupouey, P. (1990). Expression of glial fibrillary acidic protein and vimentin in mouse lens epithelial cells during development *in vivo* and during proliferation and differentiation *in vitro*: Comparison with the developmental appearance of GFAP in the mouse central nervous system. *J. Neurosci. Res.* **27,** 55–64.

Capetanaki, Y., Smith, S., and Heath, J. P. (1989). Overexpression of the vimentin gene in transgenic mice inhibits normal lens cell differentiation. *J. Cell Biol.* **109,** 1653–1664.

Carter, J. M., Hutcheson, A. M., and Quinlan, R. A. (1995a). *In vitro* studies on the assembly properties of the lens beaded filament proteins: Co-assembly with α-crystallin but not with vimentin. *Exp. Eye Res.* **60,** 181–192.

Carter, J. M., Hutcheson, A. M., and Quinlan, R. A. (1995b). *In vitro* studies on the assembly properties of the lens proteins CP49, CP115: Coassembly with alpha-crystallin but not with vimentin. *Exp. Eye Res.* **60,** 181–192.

Colucci-Guyon, E., Portier, M.-M., Dunia, I., Paulin, D., Pournin, S., and Babinet, C. (1994). Mice lacking vimentin develop and reproduce without an obvious phenotype. *Cell* **79,** 679–694.

Conley, Y. P., Erturk, D., Keverline, A., Mah, T. S., Keravala, A., Barnes, L. R., Bruchis, A., Hess, J. P., FitzGerald, P. G., Weeks, D. E., Ferrell, R. E., and Gorin, M. B. (2000). A juvenile-onset, progressive cataract locus on chromosome 3q21-q22 is associated with a missense mutation in the beaded filament structural protein-2. *Am. J. Hum. Genet.* **66,** 1426–1431.

Dahm, R., Gribbon, C., Quinlan, R. A., and Prescott, A. R. (1998). Changes in the nucleolar and coiled body compartments precede lamina and chromatin reorganization during fibre cell denucleation in the bovine lens. *Eur. J. Cell Biol.* **75,** 237–246.

Dunia, I., Pieper, F., Manenti, S., van de Kemp, A., Devilliers, G., Benedetti, E. L., and Bloemendal, H. (1990). Plasma membrane-cytoskeleton damage in eye lenses of transgenic mice expressing desmin. *Eur. J. Cell Biol.* **53,** 59–74.

Ellis, M., Alousi, S., Lawniczak, J., Maisel, H., and Welsh, M. (1984a). Studies on lens vimentin. *Exp. Eye Res.* **38,** 195–202.

Ellis, M., Alousi, S., Lawniczak, J., Maisel, H., and Welsh, M. (1984b). Studies on lens vimentin. *Exp. Eye Res.* **38,** 195–202.

FitzGerald, P. G. (1988). Immunochemical characterization of a Mr 115 lens fiber cell-specific extrinsic membrane protein. *Curr. Eye Res.* **7,** 1243–1253.

Geisler, N., and Weber, K. (1981). Isolation of polymerization-competent vimentin from porcine eye lens tissue. *FEBS Lett.* **125,** 253–256.

Georgatos, S. D., Gounari, F., and Remington, S. (1994). The beaded intermediate filaments and their potential functions in eye lens. *Bioessays* **16,** 413–418.

Gong, X., Agopian, K., Kumar, N. M., and Gilula, N. B. (1999). Genetic factors influence cataract formation in alpha 3 connexin knockout mice. *Dev. Genet.* **24,** 27–32.

Gordon-Thomson, C., de Iongh, R. U., Hales, A. M., Chamberlain, C. G., and McAvoy, J. W. (1998). Differential cataractogenic potency of TGF-beta1,-beta2, and-beta3 and their expression in the postnatal rat eye. *Invest. Ophthalmol. Vis. Sci.* **39,** 1399–1409.

Goulielmos, G., Remington, S., Schwesinger, F., Georgatos, S. D., and Gounari, F. (1996). Contributions of the structural domains of filensin in polymer formation and filament distribution. *J. Cell Sci.* **109,** 447–456.

Gounari, F., Merdes, A., Quinlan, R., Hess, J., FitzGerald, P. G., Ouzounis, C. A., and Georgatos, S. D. (1993). Bovine filensin possesses primary and secondary structure similarity to intermediate filament proteins. *J. Cell Biol.* **121,** 847–853.

Gribbon, C., Dahm, R., Prescott, A. R., and Quinlan, R. A. (2002). Association of the nuclear matrix component NuMA with the Cajal body and nuclear speckle compartments during transitions in transcriptional activity in lens cell differentiation. *Eur. J. Cell Biol.* **81,** 557–566.

Hatfield, J. S., Skoff, R. P., Maisel, H., Eng, L., and Bigner, D. D. (1985). The lens epithelium contains glial fibrillary acidic protein (GFAP). *J. Neuroimmunol.* **8,** 347–357.

Hess, J. F., Casselman, J. T., and FitzGerald, P. G. (1993). cDNA analysis of the 49 kDa lens fiber cell cytoskeletal protein: A new, lens-specific member of the intermediate filament family? *Curr. Eye Res.* **12,** 77–88.

Ireland, M., and Maisel, H. (1984). A cytoskeletal protein unique to lens fibre cell differentiation. *Exp. Eye Res.* **38,** 637–645.

Ireland, M., and Maisel, H. (1989). A family of lens fiber cell specific proteins. *Lens Eye Toxic. Res.* **6,** 623–638.

Ireland, M. E., Klettner, C., and Nunlee, W. (1993). Cyclic AMP-mediated phosphorylation and insolubilization of a 49-kDa cytoskeletal marker protein of lens fiber terminal differentiation. *Exp. Eye Res.* **56,** 453–461.

Ireland, M. E., Wallace, P., Sandilands, A., Poosch, M., Kasper, M., Graw, J., Liu, A., Maisel, H., Prescott, A. R., Hutcheson, A. M., Goebel, D., and Quinlan, R. A. (2000). Up-regulation of novel intermediate filament proteins in primary fiber cells: An indicator of all vertebrate lens fiber differentiation? *Anat. Rec.* **258,** 25–33.

Jakobs, P. M., Hess, J. F., FitzGerald, P. G., Kramer, P., Weleber, R. G., and Litt, M. (2000). Autosomal-dominant congenital cataract associated with a deletion mutation in the human beaded filament protein gene BFSP2. *Am. J. Hum. Genet.* **66,** 1432–1436.

Kasper, M., Moll, R., Stosiek, P., and Karsten, U. (1988). Patterns of cytokeratin and vimentin expression in the human eye. *Histochemistry* **89,** 369–377.

Kasper, M., and Viebahn, C. (1992). Cytokeratin expression and early lens development. *Anat. Embryol. Berl.* **186,** 285–290.

Kuszak, J. R. (1995). The ultrastructure of epithelial and fiber cells in the crystalline lens. *Int. Rev. Cytol. Surv. Cell Biol.* **163,** 305–350.

Kuszak, J. R., and Al-Ghoul, K. J. (2002). A quantitative analysis of sutural contributions to variability in back vertex distance and transmittance in rabbit lenses as a function of development, growth, and age. *Optom. Vis. Sci.* **79,** 193–204.

Kuszak, J. R., Peterson, K. L., Sivak, J. G., and Herbert, K. L. (1994). The interrelationship of lens anatomy and optical quality. II. Primate lenses. *Exp. Eye Res.* **59,** 521–535.

Kuszak, J. R., Sivak, J. G., Herbert, K. L., Scheib, S., Garner, W., and Graff, G. (2000). The relationship between rabbit lens optical quality and sutural anatomy after vitrectomy. *Exp. Eye Res.* **71,** 267–281.

Kuszak, J. R., Sivak, J. G., and Weerheim, J. A. (1991). Lens optical quality is a direct function of lens sutural architecture. *Invest. Ophthalmol. Vis. Sci.* **32,** 2119–2129.

Kuszak, J. R., Zoltoski, R. K., and Sivertson, C. (2004). Fibre cell organization in crystalline lenses. *Exp. Eye Res.* **78,** 673–687.

Kuwabara, T., and Imaizumi, M. (1974). Denucleation process in the lens. *Invest. Ophtahlmol. Vis. Sci.* **13,** 973–981.

Maisel, H., and Perry, M. M. (1972). Electron microscope observations on some structural proteins of the chick lens. *Exp. Eye Res.* **14,** 7–12.

Masaki, S., Kamachi, Y., Quinlan, R. A., Yonezawa, S., and Kondoh, H. (1998). Identification and functional analysis of the mouse lens filensin gene promoter. *Gene* **214,** 77–86.

Masaki, S., and Quinlan, R. A. (1997). Gene structure and sequence comparisons of the eye lens specific protein, filensin, from rat and mouse: Implications for protein classification and assembly. *Gene* **201,** 11–20.

Masaki, S., Yonezawa, S., and Quinlan, R. (2002). Localization of two conserved cis-acting enhancer regions for the filensin gene promoter that direct lens-specific expression. *Exp. Eye Res.* **75,** 295–305.

Merdes, A., Gounari, F., and Georgatos, S. D. (1993). The 47-kD lens-specific protein phakinin is a tailless intermediate filament protein and an assembly partner of filensin. *J. Cell Biol.* **123,** 1507–1516.

Nicholl, I. D., and Quinlan, R. A. (1994). Chaperone activity of α-crystallins modulates intermediate filament assembly. *EMBO J.* **13,** 945–953.

Perng, M. D., Wen, S. F., Van Den, I. P., Prescott, A. R., and Quinlan, R. A. (2004). Desmin aggregate formation by R120G {alpha}B-crystallin is caused by altered filament interactions and is dependent upon network status in cells. *Mol. Biol. Cell* **15,** 2335–2346.

Quinlan, R. A. (1991). The soluble plasma membrane-cytoskeleton complexes and aging in the lens. *In* "Eye Lens Membranes and Aging" (G. F. J. M. Vrensen and J. Clauwaert, eds.), Vol. 15, pp. 171–184. Eurage, Leiden.

Quinlan, R., and Prescott, A (2004). Lens cell cytoskeleton. *In* "Development of the Ocular Lens" (F. J. Lovicu and M. L. Robinson, eds.), pp. 597–624. Cambridge University Press. Cambridge, UK.

Quinlan, R. A., Carter, J. M., Hutcheson, A. M., and Campbell, D. G. (1992). The 53 kDa polypeptide component of the bovine fibre cell cytoskeleton is derived from the 115 kDa beaded filament protein: Evidence for a fibre cell specific intermediate filament protein. *Curr. Eye Res.* **11,** 909–921.

Ramaekers, F. C., Osborn, M., Schmid, E., Weber, K., Bloemendal, H., and Franke, W. W. (1980a). Identification of the cytoskeletal proteins in lens-forming cells, a special epitheloid cell type. *Exp. Cell Res.* **127,** 309–327.

Ramaekers, F. C., Poels, L. G., Jap, P. H., and Bloemendal, H. (1982a). Simultaneous demonstration of microfilaments and intermediate-sized filaments in the lens by double immunofluorescence. *Exp. Eye Res.* **35,** 363–369.

Ramaekers, F. C. S., Dunia, I., Dodemot, H. J., Bendetti, E. L., and Bloemendal, H. (1982b). Lenticular intermediate-sized filaments: Biosynthesis and interaction with plasma membrane. *Proc. Natl. Acad. Sci. USA* **79,** 3208–3212.

Robinson, M. L., Holmgren, A., and Dewey, M. J. (1993). Genetic control of ocular morphogenesis: Defective lens development associated with ocular anomalies in C57BL/6 mice. *Exp. Eye Res.* **56,** 7–16.

Sandilands, A., Prescott, A. R., Carter, J. M., Hutcheson, A. M., Quinlan, R. A., Richards, J., and FitzGerald, P. G. (1995a). Vimentin and CP49/Filensin form distinct networks in the lens which are independantly modulated during lens fibre cell differentiation. *J. Cell Sci.* **108,** 1397–1406.

Sandilands, A., Prescott, A. R., Hutcheson, A. M., Quinlan, R. A., Casselman, J. T., and FitzGerald, P. G. (1995b). Filensin is proteolytically processed during lens fiber cell differentiation by multiple independant pathways. *Eur. J. Cell Biol.* **67,** 238–253.

Sandilands, A., Prescott, A. R., Wegener, A., Zoltoski, R. K., Hutcheson, A. M., Masaki, S., Kuszak, J. R., and Quinlan, R. A. (2003). Knockout of the intermediate filament protein CP49 destabilises the lens fibre cell cytoskeleton and decreases lens optical quality, but does not induce cataract. *Exp. Eye Res.* **76,** 385–391.

Sandilands, A., Wang, X., Hutcheson, A. M., James, J., Prescott, A. R., Wegener, A., Pekny, M., Gong, X., and Quinlan, R. A. (2004). Bfsp2 mutation found in mouse 129 strains causes the loss of CP49 and induces vimentin-dependent changes in the lens fibre cell cytoskeleton. *Exp. Eye Res.* **78,** 109–123.

Sivak, J. G., Gershon, D., Dovrat, A., and Weerheim, J. (1986). Computer assisted scanning laser monitor of optical quality of the excised crystalline lens. *Vis. Res.* **26,** 1873–1879.

Sivak, J. G., Herbert, K. L., Peterson, K. L., and Kuszak, J. R. (1994). The interrelationship of lens anatomy and optical quality. I. Non-primate lenses. *Exp. Eye Res.* **59,** 505–520.

Smith, R. S., Roderick, T. H., and Sundberg, J. P. (1994). Microphthalmia and associated abnormalities in inbred black mice. *Lab. Anim. Sci.* **44,** 551–560.

Smith, R. S., and Sundberg, J. P. (1996). Ophthalmic abnormalities in inbred mice. *In* "Pathobiology of the Gaeing Mouse" (U. Mohr, ed.), Vol. 2, pp. 117–123. ILSI press, Washington DC.

Tawk, M., Titeux, M., Fallet, C., Li, Z., Daumas-Duport, C., Cavalcante, L. A., Paulin, D., and Moura-Neto, V. (2003). Synemin expression in developing normal and pathological human retina and lens. *Exp. Neurol.* **183,** 499–507.

Tomarev, S. I., Zinovieva, R. D., and Piatigorsky, J. (1993). Primary structure and lens-specific expression of genes for an intermediate filament protein and a beta-tubulin in cephalopods. *Biochim. Biophys. Acta* **1216,** 245–254.

Wallace, P., Signer, E., Paton, I. R., Burt, D., and Quinlan, R. (1998). The chicken CP49 gene contains an extra exon compared to the human CP49 gene which identifies an important step in the evolution of the eye lens intermediate filament proteins. *Gene* **211,** 19–27.

Wigle, J. T., Chowdhury, K., Gruss, P., and Oliver, G. (1999). Prox1 function is crucial for mouse lens-fibre elongation. *Nat. Genet.* **21,** 318–322.

Yang, J., Bian, W., Gao, X., Chen, L., and Jing, N. (2000). Nestin expression during mouse eye and lens development. *Mech. Dev.* **94,** 287–291.

PART IV

Methods to Study Specific
Intermediate Filament Proteins in
Non–Mammalian Systems

CHAPTER 22

Fish Keratins

Michael Schaffeld and Jürgen Markl

Institute of Zoology
Johannes Gutenberg University
55099 Mainz, Germany

I. Introduction
II. Materials, Procedures, and Pitfalls
   A. Preparation of Fish Tissues and Cytoskeletal Proteins
   B. Two-Dimensional Polyacrylamide Gel Electrophoresis of Fish Keratins
   C. Immunoblotting and Complementary Keratin Blot Binding Assay
   D. Indirect Immunofluorescence Microscopy of Fish Frozen Tissue Sections
   E. Preparation of Total RNA and mRNA from Fish Tissues
   F. Cloning and Sequencing of cDNA Encoding Fish Keratins
   G. Peptide Mass Fingerprinting of Fish Keratins using Matrix-Assisted
      Laser-Desorption/Ionization Time of Flight Mass Spectrometry
   H. Multiple Sequence Alignment and Phylogenetic Tree Construction
III. Discussion and Concluding Remarks
   References

I. Introduction

Intermediate filaments (IF) represent one of the three major cytoskeletal protein filament systems characteristic for most vertebrate cells. IF networks extend from the cell nucleus to the plasma membrane, attach to desmosomes, and interact with a variety of cell structures, thereby contributing to the tensile strength and shape of the cell. Recent data suggest that IF proteins also play a role in central cellular processes such as apoptosis and signal transduction (reviews: Kirfel *et al.*, 2003; Paramio and Jorcano, 2002). IF proteins are a complex multigene family made up of, in man, 65 different members (Hesse *et al.*, 2001). They are expressed in cell-specific and tissue-specific patterns that can, in addition, vary with the developmental stage of the animal. According to their primary structures, six IF protein

METHODS IN CELL BIOLOGY, VOL. 78
Copyright 2004, Elsevier Inc. All rights reserved.
0091-679X/04 $35.00

types have been defined. The α-cytokeratins, here further simply referred to as "keratins," represent type I and type II and encompass most of the IF protein diversity. Keratin filaments are always assembled from type I/II heterodimers; in tetrapods (i.e., "terrestrial" vertebrates such as mammals and amphibians), sharks, and lampreys, keratins are almost exclusively restricted to and are typical for epithelial cells. However, in teleost fishes, keratins are additionally present in a variety of mesenchymal cell types (Conrad *et al.*, 1998; Groff *et al.*, 1997; Markl and Franke, 1988; Markl *et al.*, 1989); in the nonteleost vertebrates, mesenchymally derived cells usually do not express keratins but a type III IF protein termed vimentin (Herrmann *et al.*, 1989; Lazarides, 1982; Schaffeld *et al.*, 2001a; Schultess, 2001; Schultess *et al.*, 2001). Vimentin in turn is structurally closely related to desmin, another type III protein, which is typical for muscle cells. Thus, if we trace the evolution and expression of keratins within the Vertebrata, we will constantly also stumble over vimentin and desmin; therefore, these two IF proteins will also be briefly discussed in this article.

IF proteins resembling vertebrate types I, II, and III have been detected in lower chordates such as an ascidian and two lancelets (e.g., Erber *et al.*, 1998; Karabinos *et al.*, 2000, 2002, 2004; Luke and Holland, 1999; Riemer and Weber, 1998; Riemer *et al.*, 1998; Wang *et al.*, 2000), but true orthologs of vimentin, desmin, or individual human keratins (e.g., the most abundant and presumably most "ancient" keratins, K8 and K18) have not been detected outside of the Vertebrata. In all comprehensively studied vertebrates, including a lamprey, a shark, several teleosts, and various tetrapods, "E" keratins expressed in stratified epithelia (notably the epidermis) can be distinguished from "S" keratins appearing in simple epithelia (review: Markl and Schechter, 1998). Therefore, four major keratin subtypes have been defined: IE, IIE, IS, IIS. Of the typical human IS/IIS keratin pair K8 and K18, K8 is clearly discernible in the freshwater lamprey, which belongs to the most ancient group of living vertebrates, the Agnatha ("jawless vertebrates": lampreys and hagfishes), as well as in all studied Gnathostomata ("jawed vertebrates") such as shark, trout, frog, and man. However, a true ortholog of K18 is not determinable in lamprey at the sequence level but exists in all hitherto investigated gnathostomians. In contrast to K18, the panel of IE keratins detected and sequenced in trout and zebrafish is not directly comparable to the panel of IE keratins earlier identified in human or *Xenopus*; apparently, the two panels stem from a common root, but result from independent radiations (Schaffeld *et al.*, 2002a,b, 2003). Correspondingly, apart from K8, there is no teleost type II keratin detected that is directly orthologous to a certain keratin of frog or man. This hypothesis of an independent radiation of the keratins in teleosts and tetrapods is also supported by the recent analysis of the keratin genes in the teleost fish *Fugu rubripes* (Zimek *et al.*, 2003) compared with the human keratin genes (Hesse *et al.*, 2001). As exemplified in the shark *Scyliorhinus stellaris*, in elasmobranchs a very simple but highly specialized system of IE/IIE keratins exists, which is not directly comparable to the IE/IIE keratins sequenced in bony animals (Schaffeld *et al.*, 1998, 2004), and the keratin system of the freshwater lamprey is even more remote (Schultess, 2001; Schultess *et al.*, 2001). Each vertebrate class evolved a specific design for a highly rigid but flexible skin; this might have triggered evolution of different panels of α-keratins to establish tensile strength of the respective type of stratified epidermis. In contrast, the situation in the simple epithelia is much more uniform, and, indeed, in the studied Gnathostomata, they invariantly express K8/K18 orthologs, and in lamprey a comparable keratin pair exists, which in phylogenetic analyses can serve as useful outgroup.

It is clear that keratins are not only complex in terms of their expression patterns within a given animal species but also with respect to their evolution within the Vertebrata. On the other hand, keratins are excellent molecular markers in ontogeny and pathology, which is relevant not only in man but also in fish, if one considers the commercial applications in trout or salmon aquaculture, as well as basic research on genetic model organisms such as the fugu or the zebrafish. Tracing keratin evolution prevents false interpretation of fragmentary sequence or antibody labeling data, for which we give two examples: (1) Several authors claimed to have sequenced zebrafish K8 (Chua and Lim, 2000; Gong *et al.*, 2002; Imboden *et al.*, 1997; Ju *et al.*, 1999), but we have recently shown by phylogenetic analysis that they had, indeed, sequenced several zebrafish "E" keratins (Schaffeld *et al.*, 2003). Starting from K8/18 candidates previously biochemically identified (Conrad *et al.*, 1998) and supported by EST library mining, we have cDNA cloned and sequenced the authentic orthologs of human K8 and K18 in zebrafish. (2) By sequencing vimentin and desmin in various lower vertebrates, including shark, we detected that a published sequence of a type III protein from another elasmobranch, the electric ray *Torpedo californica* (Frail *et al.*, 1990; EMBL accession no X51533) is indeed a vimentin. This protein had previously been considered as a specialized electric organ IF protein rather than a vimentin or desmin (for review, see Markl and Schechter, 1998).

We believe that tracing IF protein evolution might help to refine the phylogenetic tree of the Vertebrata at certain deep furcations that so far resist evolutionary analyses, as for example the "Actinopterygii problem" (phylogenetic relationship of bichirs, gars, bowfins, sturgeons, and teleosts). It might also unravel clues of specific functions of individual IF protein subtypes, which in turn would help to understand the biological role of their diversity in man. One example are the mesenchymal keratins in teleosts that probably reinforce these cells to better resist osmotical stress. Another is the fish optic nerve keratins, which might be correlated with the ability of this nerve to regenerate (Druger *et al.*, 1992; 1994; Fuchs *et al.*, 1994; Giordano *et al.*, 1989; 1990; Markl and Schechter, 1998). A third example is that the regular *in vitro* assembly of vimentin is affected by temperature and works best at the physiological temperature of the animal in its natural environment (Cerda *et al.*, 1998; Herrmann *et al.*, 1993, 1996a,b; Schaffeld *et al.*, 2001a).

To analyze fish keratins, several methods originally developed by others in studies of human and frog keratins have been adapted. In most cases, we apply antibodies developed against frog or human keratins. In any given fish species, we follow a specific strategy:

1. We identify the type I and type II keratins at the protein level (by two-dimensional polyacrylamide gel electrophores [2D-PAGE] and complementary keratin blot binding [CKBB] assays on isolated cytoskeletal proteins).

2. We analyze their tissue-specific expression patterns, thereby defining "E" and "S" keratins (by immunoblotting of cytoskeletal proteins to check antibody specificities and immunofluorescence microscopy of frozen tissue sections using the defined antibodies).

3. We clone and sequence cDNAs encoding keratins, vimentin, and desmin, followed by phylogenetic tree analysis.

4. We correlate the sequences to their respective protein spots in 2D-PAGE by matrix-assisted laser-desorption/ionization time of flight mass spectrometry (MALDI-MS) peptide mass fingerprinting.

With respect to vimentin and desmin, we apply the same methods, except for the CKBB assay.

II. Materials, Procedures, and Pitfalls

We have modified and adapted a panel of methods originally developed in studies of human and frog keratins. Within the scope of this chapter, we want to explain their application and relevance within the context of investigations of keratins, vimentin, and desmin in fish. At the same time, we will concentrate on specific problems that arise when studying fish keratins.

A. Preparation of Fish Tissues and Cytoskeletal Proteins

1. Preparation of Fish Tissues

To decontaminate the dissecting instruments (especially from ribonucleases), autoclave them at 120 °C for 30 minutes. Alternatively, soak them in 0.5 M NaOH for 30 minutes and afterwards rinse them thoroughly in sterile distilled water. Finally, wash the instruments in 100% ethanol and let them air dry. Large fish such as lamprey, shark, bichir, sturgeon, lungfish, carp, goldfish, and trout may be killed after MS 222 narcosis (0.5–1.0 g/L) by blood loss through cutting their tail artery. Small fish such as zebrafish may be directly killed by a cut through their neck region.

Dissection has to be performed permanently on ice. We recommend immediately excising and using (or snap freezing) those organs that are more sensitive to internal degradation processes such as liver, spleen, stomach, pancreas, intestine, kidney, gill, and brain. In contrast, skin, fin, eye, gonads, and muscles (including cardiac muscle) seem to be much more resistant to degradation. If necessary, cut the tissue into smaller pieces (approximately 0.125–1.0 cm^3) before the freezing process.

Tissue samples selected for preparation of cytoskeletal proteins or extraction of RNA should be either immediately used or, alternatively, shock-frozen by dropping them into liquid nitrogen. Tissue samples for immunofluorescence microscopy should, however, be shock-frozen by dropping them into isopentane that has been precooled (down to about approximately $-130\,°C$) in liquid nitrogen. If you wish a tissue block in a defined orientation, you may stick it, before freezing, onto a drop of cryo-embedding compound resting on a small piece of parafilm sheet. The parafilm will disintegrate completely during the freezing process. Freezing the tissue in liquid nitrogen–cooled isopentane should prevent formation of ice crystals within the cells, thereby maintaining their structural integrity. Under no circumstances should the tissue be allowed to partly defrost before sectioning. Use plastic tweezers to handle the frozen samples and a container with dry ice for their transport to the cryomicrotome. The frozen tissue samples can be stored at $-70\,°C$.

2. Preparation of Cytoskeletal Proteins

To purify IF proteins from fish tissues, you should perform a standard low/high salt extraction. IFs are insoluble in physiological buffers and will not dissolve even when incubated with high concentrations of salt such as 1.5 M KCl. In many cases, we found that the keratins were remarkably stable during extraction; however, in certain animal species and tissue types, we observed a considerable degree of keratin degradation. Therefore, we advise the addition of PMSF plus a special cocktail of other protease inhibitors (leupeptin, pepstatin, and benzamidin) to all extraction buffers. If you use frozen tissue samples, do not allow them to thaw before homogenization. Moreover, all extraction steps should be performed on ice. To avoid contamination with human keratins (especially K9), wear gloves in each step and protect the equipment from dust. Note that several antibodies that have a high affinity for fish keratins strongly react to K9 (e.g., monoclonal antibodies KL1 and AE3). The following solutions are required:

- Low salt buffer (pH 7.2): 150 mM NaCl, 10 mM Tris/HCl, 5 mM ethylene diamine tetraacetic acid (EDTA), 1% Triton × 100
- High salt buffer (pH 7.2): 1.5 M KCl, 10 mM Tris/HCl, 5 mM EDTA, 1% Triton × 100
- TE = Tris/EDTA buffer (pH 7.2): 10 mM Tris/HCl, 5 mM EDTA
- Protease inhibitor stock solutions (except benzamidin each in ethanol p.a.): 0.38 M polymethyl sulfonyl fluoride (PMSF), 10.5 mM leupeptin, 2 M benzamidin, 1.5 mM pepstatin A; stock solutions of leupeptin, benzamidin, and pepstatin may be stored at $-20\,°C$. The PMSF stock solution has to be freshly prepared each time and should then be kept at room temperature.

Precool the three buffers on ice. Shortly before buffer use, add protease inhibitor stock solutions: 0.8 μl of leupeptin per ml buffer (final concentration 8.4 μM), 0.5 μl of benzamidin (final concentration 1 mM), 0.7 μl of pepstatin (final concentration 1 μM), and 7.5 μl PMSF (final concentration 2.8 mM). Take care that you continuously agitate the extraction buffers when adding the protease inhibitor solutions to avoid precipitation (especially important for PMSF, which is poorly soluble in water). Thoroughly homogenize the tissue in a fourfold volume of ice-cooled low salt buffer using an Ultra Turrax® (multiple times with maximum speed!). Subsequently centrifuge the homogenate for 10 minutes with at least 10,000g at 4 °C (or 15,000 rpm in a Sorvall refrigerated centrifuge RC5B Rotor SS34) and discard the supernatant. Resuspend the pellet in an equivalent volume of ice-cooled high salt buffer and homogenize as before. Centrifuge as described previously and again discard the supernatant. Depending on the prepared tissue type, this high salt extraction has to be repeated several times to remove most of the soluble proteins. Fish muscle preparations will still contain high amounts of actin if not extracted several times under high salt conditions. Extractions from fish skin contain large amounts of keratin but often also much collagen; if not removed sufficiently, the collagen may interfere with subsequent procedures (e.g., PAGE). The final pellet should then be thoroughly resuspended in an adequate volume of precooled TE buffer (use vortex or if necessary pipette tip). Transfer the suspension into one or more 1.5-ml reaction caps, centrifuge at 12,000g at 4 °C in a refrigerated bench centrifuge and remove the supernatant. To remove the salts and the detergent, repeat this washing step at least twice by adding 1 ml TE buffer to each cap and resuspending the precipitate carefully. If the pellet is very stiff and difficult to resuspend, you may also use a truncated pipette tip. After the last washing step, the pellet may be used directly for PAGE or stored at −70 °C.

B. Two-Dimensional Polyacrylamide Gel Electrophoresis of Fish Keratins

An excellent method to analyze the complex patterns of IF proteins (especially keratins) in fish is 2D-PAGE. An example is given in Fig. 1a. The protein spots so obtained can then be processed either by blotting (e.g., for immunoassays and overlay assays, Fig. 1a′–a‴) or by excising them from the gel (e.g., for peptide mass fingerprinting; see Section II.G).

For 2D-PAGE, in the first dimension separation, we use either isoelectric focusing (IEF) or nonequilibrium pH gradient gel electrophoresis (NEPHGE), as described by O'Farrell (1975, 1977) and Achtstätter *et al.* (1986). The NEPHGE system is preferable when acidic to neutral keratins and "basic" keratins (those with isoelectric points above pH 7) are to be separated in the same gel, which is often necessary with samples from mammals. The IEF system is more useful for separating keratins with isoelectric points below pH 7.0. Because in fish basic keratins are usually lacking (as deduced from our studies in lamprey, shark, trout, goldfish, carp, zebrafish, bichir, sturgeon, and lungfish), we routinely apply only the IEF system. In the second dimension of both systems, sodium dodecyl sulfate (SDS) PAGE is applied, which separates the proteins according to their different molecular masses. We usually perform SDS-PAGE in a low salt buffer system according to Laemmli (1970).

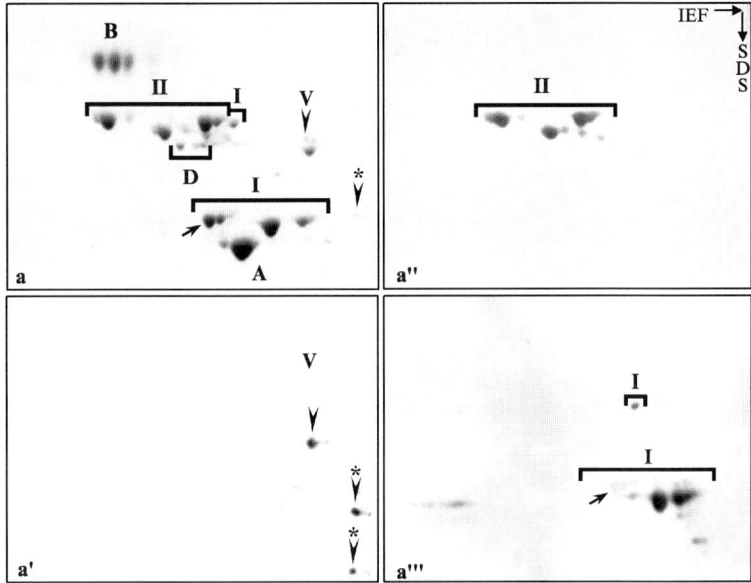

Fig. 1 (a) Two-dimensional polyacrylamide gel electrophoresis (Coomassie Blue stained) of cytoskeletal proteins extracted from bichir tongue, with isoelectric focusing in the first dimension and sodium dodecylsulfate-polyacrylamide gel electrophoresis in the second dimension. Actin (A) and bovine serum albumin (B) served as marker proteins. (a') Immunoblot of (a) using monoclonal anti-vimentin antibody VIM 14.13 to identify vimentin (V, labeled with arrowheads; * marks vimentin degradation products). (a'') complementary keratin blot binding (CKBB) assay of (a) using biotinylated human keratin K18 to identify the type II keratins. (a''') CKBB assay of (a) using biotinylated human keratin K8 to identify the type I keratins. Note that one bichir keratin (arrow) is only weakly detected in this test. However, it is intensively stained with several monoclonal anti-keratin antibodies such as AE3 (not shown). Unpublished data obtained in our laboratory by Miriam Bremer.

In proteomics studies with hundreds or thousands of protein spots, systems that use immobilized pH gradients are certainly superior to the system with carrier ampholytes described here (Tables I and II), but in case of the keratin complement of a given species, the latter system yields excellent, highly reproducible separations and is much cheaper. In our laboratory, we use the vertical electrophoresis system PHERO-vert-180 available from Biotec Fischer (Reiskirchen, Germany), including an insert for 2D round gel electrophoresis that is available as an accessory. Ampholines can be purchased from Amersham Biosciences (Uppsala, Sweden) or another suitable supplier.

1. Sample Preparation

Sample preparation is a key factor for successful 2D-PAGE analysis. It requires an appropriate sample buffer that is used not only for electrophoretic separation of the proteins but also for their denaturation, thereby ensuring that their

Table I

Buffers and Solutions (Final Concentrations) for Two-Dimensional Polyacrylamide Gel Electrophoresis of Fish Cytoskeletal Proteins

Chemicals and stock solutions	Buffers					
	Lysis-1	Lysis-2	Lysis-A	Lysis-K$_{NEP}$	Lysis-K$_{IEF}$	Equilibration
Urea	9.5 M	9.5 M	9.5 M	6 M	6 M	—
SDS	0.5%	—	—	—	—	2%
NP40	—	5% (w/v)	2% (w/v)*	5% (w/v)	—	—
0.5 M Tris/HCL (pH 6.8)	—	—	—	—	—	60 mM
DTT	25 mM	25 mM	25 mM	—	—	20 mM
Glycerol	—	—	—	—	—	10% (w/v)
Ampholines pH 3–10	2% (w/v)	2% (w/v)	0.4% (w/v)	1% (w/v)	0.2% (w/v)	—
Ampholines pH 4–6	—	—	0.8% (w/v)	—	0.4% (w/v)	—
Ampholines pH 5–7	—	—	0.8% (w/v)	—	0.4% (w/v)	—

Aliquot lysis 1, 2, A, and K and store at −25°C.
*For shark cytoskeletal preparations from skin, we increased the NP40 concentration up to 6%.

Table II

Gel Compositions (Final Concentrations) for Two-Dimensional Polyacrylamide Gel Electrophoresis of Fish Cytoskeletal Proteins

Chemicals and stock solutions	First dimension*		Second dimension[†]	
	NEPHGE	IEF	Stacking gel	Separation gel
Acrylamide:bisacryamide				
28.38%:1.62%	4% (w/v) [1.33 ml]	4% (w/v) [1.33 ml]	—	—
30.0%:0.80%	—	—	3.9% (w/v) [1.3 ml]	10% (w/v) [10 ml]
Urea	9.5 M [5.7 g]	9.5 M [5.7 g]	—	—
NP40 10% (w/v)	2% (w/v) [2 ml]	2% (w/v) [2 ml]	—	—
SDS 20% (w/v)	—	—	0.1% (w/v) [0.05 ml]	0.1% (w/v) [0.15 ml]
0.5 M Tris/HCl (pH 6.8)	—	—	125 mM [2.5 ml]	—
1.5 M Tris/HCl (pH 8.8)	—	—	—	375 mM [7.5 ml]
Ampholines pH 3–5	5% [0.5 ml]	1% [0.1 ml]	—	—
Ampholines pH 4–6	—	2% [0.2 ml]	—	—
Ampholines pH 5–7	—	2% [0.2 ml]	—	—
Deionized water	[2 ml]	[2 ml]	[6.1 ml]	[12.4 ml]
APS 10% (w/v)	0.15% (v/v) [15 μl]	0.1% (v/v) [10 μl]	1% (v/v) [0.1 ml]	1% (v/v) [0.3 ml]
TEMED	0.1% (v/v) [10 μl]	0.07% (v/v) [7 μl]	0.05% (v/v) [5 μl]	0.05% (v/v) [15 μl]

*Values in square brackets are to make 10 ml of gel mixture (10–12 focusing gels).

[†]Values in square brackets are to make 10 ml (30 ml) of stacking (separation) gel mixture (about 1 slab gel). The catholyte for IEF and NEPHGE is 20 mM NaOH, the anolyte 10 mM H_3PO_4; the catholyte, as well as anolyte, for the second dimension is composed of 23 mM Tris, 190 mM glycine, and 0.2% (w/v) SDS.

migration behavior in IEF or NEPHGE is reproducible. To reach the goal of a complete solubilization and denaturation of the keratins, several components are required in the sample buffer: a chaotrophe, a nonionic detergent, a reductant, ampholines, and eventually protease inhibitors.

Urea is the chaotrophe of choice, because it disrupts hydrogen bonds (very effective) and hydrophobic interactions (less effective) without influencing the intrinsic charge of the protein. Warning! Never heat urea-containing solutions above 37 °C, because in solution urea forms an equilibrium with ammonium cyanate that can react with amino groups, leading to artifactual charge heterogeneity of the separated proteins. Furthermore, always use fresh urea solutions or freeze solutions for later use and only buy good-quality urea. To increase disruption of hydrophobic interactions, a nonionic and nonlinear detergent such as NP40 is additionally required in the sample buffer. To break disulfide bridges by converting them into sulfhydryl groups, a comparatively mild reducing agent such as DTT is sufficient. Apart from their major role, the addition of ampholines (usually up to 2%) is beneficial in several aspects: they support protein solubilization, scavenge cyanate ions, and during centrifugation help to precipitate contaminating nucleic acids.

In solutions containing 9.5 M urea, most mammalian keratins are solubilized to monomers (Franke *et al.*, 1983), and this agrees with our observations in fish. However, several keratins require 10.0 M urea to be completely released from their heterotypic complexes. Depending on the intended first-dimension separation, dissolve the cytoskeleton pellet in an adequate volume of lysis-A solution (for IEF) or lysis-1 solution (for NEPHGE). The optimal volume of the sample buffer for solubilizing the proteins at a maximal concentration must be determined experimentally for the particular tissue or material. To guarantee a sufficient concentration of urea while resuspending the precipitate, add a spatula tip of crystalline urea to the sample buffer. Resuspend the pellet by repeated agitation with a pipette tip and allow enough time (at least 15–30 minutes at room temperature) for the sample buffer to work, because the reduction of proteins with DTT in particular normally takes some time. The lysis-1 buffer that is used to prepare the samples for NEPHGE contains 0.5% (w/v) SDS to increase the initial solubilization, especially of the alkaline proteins. Before electrophoresis, the SDS must be removed from the proteins by competitive displacement with NP40, which is achieved by adding an identical volume of lysis-2 solution (containing an excess of NP40) to the sample. After centrifugation of the solubilized samples for 10 minutes at 10,000g at room temperature, 10–80 μl of the protein sample can be loaded onto the focusing gel. The appropriate amount of protein applied to the gel should be experimentally determined. It depends on the complexity of the polypeptide mixture and on the desired resolution. If you intend to stain the gels with Coomassie blue after electrophoresis or transfer the protein pattern to nitrocellulose, add 3 μg of α-actin from rabbit muscle and 3 μg of bovine serum albumin as internal markers to each focusing gel (loading sample).

After centrifugation, you may also store the samples at −20 °C. After thawing, take care that the urea is redissolved completely before electrophoresis. Do not put the samples on ice afterwards, because the urea in the buffer might precipitate. Be aware of potential protease activity. In some cytoskeletal preparations, we observed progressing degradation, even after solubilization in sample buffer. In such cases, add protease inhibitors to the sample.

2. First-Dimension Electrophoresis

We perform the first-dimension separation in tube gels as described by O'Farrell (1975, 1977); the tube gels are cast to a height of 125 mm in 130-mm glass tubes of 3.0-mm inside diameter (you need a little less than 1 ml of gel mixture per tube). Before casting the tube gels, thoroughly clean the tubes with acetone, rinse them several times with deionized water, and then dry them by exposing them to compressed air. The gel and buffer modifications for IEF and NEPHGE that we use in our laboratory are summarized in Table I and II. We do not degas any of the required gel solutions, and we use a Pasteur pipette with an elongated tip to cast the gels. Instead of an 8-M urea solution, we use deionized water to overlay the focusing gels after casting. Then the top of the tube is sealed with parafilm. In this condition, the gels can be stored in the dark at warm room temperature (at least 27 °C) for about 4 days. Allow the gels to polymerize for at least 4 hours. Immediately before use, remove the parafilm seals from the glass tube. Then use a water filled plastic syringe to carefully press the gel about 3–5 mm out of the end of the glass tube and cover this end with a double-layered piece of gauze bandage (held in place by an appropriate section of silicon or latex tubing) to prevent the gels from escaping from the tube during electrophoresis. Thoroughly rinse the gel surface with deionized water and get rid of the excess fluid with a facial tissue. Avoid touching the gel. In contrast to the IEF procedure described by O'Farrell, we do not prerun the tube gels. For IEF and for NEPHGE, directly apply the samples (10–80 μl) onto the tube gel and overlay them with 20 μl of the appropriate lysis-K solution (see Table I). Then fill the tubes with 20 mM NaOH for IEF and 10 mM H_3PO_4 for NEPHGE. After filling the respective tanks with catholyte and anolyte IEF, we start at 200 V for 15–30 minutes and then continue at 300 V for 30 minutes. From then, IEF gels are usually exposed to a constant voltage of 400 V for 16–18 hours. For focusing under NEPHGE conditions, the running conditions are 10 min/200 V, 10 min/300 V, and 5–7 h/400 V. Remember that in NEPHGE analyzes the polarity is reverse compared with IEF. In NEPHGE, the cathode is at the bottom and the anode on the top of the gel. After the run, the tube gels are carefully forced out of the glass tubes by compressed air with an appropriate plastic syringe. You probably have to connect the syringe to a truncated pipette tip (use parafilm) for a better fit to the glass tube's inner diameter. Incubate the tube gels in small plastic Petri dishes (diameter 5 cm) in equilibration buffer by gently shaking them for 20 minutes at room temperature (mark the basic and the acidic end of the focusing gel on the Petri dish bottom). After removing the buffer, you may either use the tube gels directly for the second-dimension separation or store them in the Petri dish at $-20\,^{\circ}\mathrm{C}$. They are stable for several months.

3. Second-Dimension Electrophoresis

We perform the second-dimension separation in slab gels containing a separation gel and a stacking gel. The method is as described by O'Farrell (1975), but we applied minor variations in gel and buffer composition (see Tables I and II) and gel dimensions (see below). The separation gel is $12.5 \times 15 \times 0.15$ cm, and the stacking gel is $2 \times 15 \times 0.15$ cm in size. The front gel plates that we use possess a special notch (grinding) for the reception of the tube gels. After pouring the separation gel, we overlay it with 100% ethanol instead of water. Wait at least 1 hour, then cast the stacking gel; to form a flat surface on its top, apply a thin layer of deionized water. After polymerization of the stacking gel (takes about 30 minutes), envelope the casted slab gels with water-soaked paper tissues and store them in an airtight container at $4\,^{\circ}\mathrm{C}$. Allow the gels to rest overnight before use to guarantee complete polymerization. To load the tube gel onto the slab gel, simply put it carefully into the notch of the front plate and then quickly underlay it with melted agarose solution (1% agarose in equilibration buffer plus 0.05%–0.1% [w/v] bromphenol blue) with a Pasteur pipette. (We always put the tube gel in the same orientation onto the SDS gel; basic end to the left, acidic end to the right). Allow 5 minutes for the agarose to gelatinize. Gels are usually run with a constant current of 25 mA in the stacking and 35 mA in the separation gel until the bromphenol blue front reaches the bottom of the gel (after approximately 5 hours). To achieve a better resolution in the second dimension, you may extend the running time, maximally another hour.

a. Pitfalls

• The quality of all chemicals should be at least of analytical grade.

• The sample buffer (lysis buffer) has to be prepared freshly. Alternatively, make small portions (1 ml) and store frozen in test tubes at $-20\,^{\circ}\mathrm{C}$.

• If protein samples have initially been solubilized in SDS-containing buffers, you can use them for IEF or NEPHGE if you dilute them at least 1:8 to 1:10 in lysis buffer (lysis-2 for NEPHGE).

• If there are problems forcing the tube gels out of the glass tubes, treat the inner wall of the tubes with acryl glide plate coating.

• When preparing the gel mixture for the first dimension, the urea sometimes hardly dissolves. In this case, you may first mix the deionized water, acrylamide, and NP40, and thereafter solubilize the required amount of urea in this solution by heating it under slight stirring to $35\,^{\circ}\mathrm{C}$ maximally. Let the solution cool down somewhat, and then add the heat-sensitive ampholytes, APS and TEMED; then pour the gels.

• Take care that your laboratory is not too cold when casting the gels to avoid the urea coming out of solution during the casting process.

- If air bubbles are trapped at the bottom of the glass tube, they may be removed by carefully tapping the long side of the glass tube against the laboratory bench.

- Clean plates with SDS solution after use to get rid of all keratin contaminants. If you have severe problems with human keratin K9 contamination, you may prerun the SDS gels for 30–60 minutes, at 25 mA per gel.

- Ammonium persulfate solution should always be prepared freshly. A 10% solution of ammonium persulfate may be used for 2–3 days if stored in a refrigerator.

- Keratins dispersed in solution (e.g., in samples from a *in vitro* translation) may be prepared by adding the same volume of a lysis buffer that contains all ingredients except urea in the twofold concentration; then add sufficient amounts of solid urea.

- Coelectrophoresis, in which cytoskeletal preparations from two different tissues are combined, has proved to be a very powerful approach in fish. In many cases, it gave a hint as to whether or not two spots from different tissues but with similar electrophoretic migration behavior represent the same keratin. For each tissue, 1/2 to 3/4 of the protein amount optimal for the respective single preparations should be applied.

- For separations of shark cytoskeletal preparations from skin, we increased the NP-40 concentration in the lysis-A solution up to 6% (w/v) to achieve a better denaturation of keratins for the first-dimension separation.

4. Coomassie Blue R250 Staining of Gels

Coomassie blue is a very efficient staining dye for detection of the major proteins separated by 2D-PAGE. In a single spot, it will detect 0.2–15 μg of protein (depending on its properties). The gel may at first be stained with Coomassie blue to show the most abundant keratins and subsequently can be restained with a silver-based method to reveal, in addition, the minor protein components. The Coomassie blue staining protocol provided here stains and fixes the proteins simultaneously by use of a staining solution composed of 0.1% (w/v) Coomassie blue dissolved in a 7.5% (v/v) acetic acid/40% (v/v) methanol solution. The methanol and acetic acid precipitate the proteins, thereby fixing them in the gel. To prepare the Coomassie blue staining solution, dissolve the dye in the desired volume of methanol by stirring overnight at room temperature. Subsequently, add the appropriate volume of glacial acetic acid and fill up to the final volume with deionized water. Afterwards, filtrate the solution through a middle-fine fluted filter (e.g., filter 595 1/2 available from Schleicher and Schüll, Germany) and store in a dark bottle.

Cover the gel with staining solution in a container with a loosely fitting lid to prevent evaporation. Stain at room temperature with gentle shaking for at least 1 hour (1.0-mm thick gels) and 3 hours (1.5-mm thick gels), or better overnight. Then suck off the staining solution and destain the gel with destaining solution (20% [v/v] isopropanol/7.5% [v/v] acetic acid) under gentle shaking. Change it several times until the protein spots are visible with minimal background. For thick gels, this may take several hours. Store the gels in 7.5% (v/v) acetic acid, which in time will destain the background entirely. You may also shrink-wrap the Coomassie blue–stained gels together with a small portion of 7.5% (v/v) acetic acid and store them in the refrigerator. If the acetic acid is refilled regularly, the gels are stable for years and are still usable for peptide mass fingerprinting. During the entire procedure, handling of the gels should be minimized, and gloves should be worn at all times.

a. Pitfalls

• Before staining, you may first equilibrate gels for 5–10 minutes in 20% isopropanol to remove most of the SDS that otherwise would form precipitates with the Coomassie blue. This step is not obligatory, but according to our experience it significantly prevents deterioration of the staining solution and allows us to use it repeatedly for a longer period.

• Staining of multiple gels in one container may adversely affect the staining process, sometimes resulting in an inhomogenous staining.

C. Immunoblotting and Complementary Keratin Blot Binding Assay

1. Transfer of Proteins from Polyacrylamide Gels to Nitrocellulose Sheets

a. Transfer Procedure

Nitrocellulose sheets (0.45-μm pore size) must be handled wearing gloves to avoid protein contamination, notably with keratins, from your fingers. For the protein transfer, we use a buffer system described by Herrmann and Wiche (1987), mainly consisting of boric acid titrated to high pH (8.8). This buffer ensures a very efficient transfer even of very high molecular weight proteins such as plectin (Mr 550,000), especially when performed overnight. To reduce any reformation of disulphide bridges, you may additionally add DTT to the transfer buffer, but this is not obligatory (transfer or "blotting" buffer: 25 mM boric acid pH 8,8; 2 mM EDTA; 1 mM DTT). The tranfer is performed in a PHERO-blot wet blotting tank from Biotec Fischer (Reiskirchen, Germany).

After 2D-PAGE, remove one of the glass plates and let the gel rest on the other one. Remove the stacking gel with one of the spacers, and label the orientation of the gel by cutting off one of its corners. Transfer the gel carefully from the glass plate into a container filled with blotting buffer and equilibate it for 10–20 minutes by gentle shaking at room temperature. Use wet gloves to handle the gels to avoid damage. Meanwhile, prepare the required filter papers (Whatman 3MM) and nitrocellulose membranes and soak them in transfer buffer. Cut the nitrocellulose pieces to the same size as the gel. After preincubation of the gel is finished, mount it on a gel holder as follows:

• Gel holder 1
• Blotting sponge

- Two layers of Whatman 3MM filters
- Nitrocellulose sheet
- Gel
- Two layers of Whatman 3MM filters
- Blotting sponge
- Gel holder 2

Avoid air bubbles between the layers (especially the gel and the nitrocellulose membrane) when preparing this "sandwich." Press the layers together with a rod, for example, the long side of a 10-ml pipette. Then put the gel holder into the transfer tank, which contains a magnetic stir. Watch out for correct orientation of the gel holder (nitrocellulose membrane to the anode, gel to the cathode)! Now fill the tank with transfer buffer and take care that the gel holder electrodes are completely submersed in transfer buffer, and then connect the power supply. (Before turning on the power supply, again verify that the gel holder is positioned such that the proteins will transfer in a negative to positive direction!) Start the transfer at 100 mA and raise the current every 5 to 10 minutes by 100 mA steps until 500 mA is reached. Under continous stirring, continue the transfer at room temperature for 1–2 hours at 500 mA. If you want to perform the transfer overnight in the same way, raise the current to only 300 mA and run the equipment at 4–8°C. After completion, remove the gel holder from the transfer tank. Dismantle the blot and peel the nitrocellulose sheet from the gel by taking great care not to tear or crease it. If you want to check the efficiency of the protein transfer, you should keep the gel and stain it with Coomassie blue or other protein dye.

b. Reversible Staining of Transferred Proteins with Ponceau S

The quality of the protein transfer to the membrane may be monitored by staining the nitrocellulose sheet with Ponceau S. After dismantling the blot, briefly rinse the membrane in deionized water and then incubate it for 5 minutes at room temperature in Ponceau S stock solution (Sigma-Aldrich) diluted 1:10 in deionized water. Then destain the filters in deionized water until the ratio of background to protein stain is as desired. Afterwards let the membranes air-dry on Whatman 3 MM paper. For later identification of labeled protein spots, we copy the protein pattern and the margins of the dried nitrocellulose sheet onto a clean transparency (use a lightbox to better recognize the spots!). The dried membranes may be directly used for further analyses or stored at room temperature between two sheets of Whatman 3 MM paper.

2. Immunoblotting

In immunoblotting, proteins that have been transferred from a polyacrylamide gel to an immobilizing membrane may be detected by specific antibodies. After the nonspecific protein-binding sites have been blocked with a detergent and/or protein-based blocking agent, the membrane is incubated with primary antibodies (monoclonal or polyclonal) specific for the target antigen. Secondary antibodies, conjugated to an enzyme such as alkaline phosphatase, that are specific for the primary antibodies are used to localize the target antigen on the membrane. The resulting antigen/antibody complexes are detected by developing the membrane in an enzyme substrate reaction mixture that produces an insoluble colored reaction product. Ensure that all described incubations are performed with the protein side of the membrane up. The following solutions are required.

- TBST (pH 8.0): 10 mM Tris-Cl, 140 mM NaCl, 0.1% Tween 20
- Washing buffer: 0.5 M NaCl, 0.1% Triton × 100 in TBST
- AP buffer (pH 9.5): 100 mM Tris/Cl, 100 mM NaCl, 50 mM MgCl$_2$
- NBT solution: 50 mg/ml p-Nitrobluetetrazoliumchloride in 70% dimethylform-amide (store in the dark at $-20\,^{\circ}$C)
- BCIP solution: 50 mg/ml 5-brom-4-chlor-3-indolylphosphat-p-toluidin in 100% dimethylformamide (store in the dark at $-20\,^{\circ}$C)

c. Membrane Blocking

Before blocking, rehydrate the nitrocellulose membrane in deionized water. Then transfer the sheets to Petri dishes filled with blocking solution and incubate for 1–2 hours at room temperature under gentle shaking. Depending on the applied primary antibody, use as blocking solution either TBST, 10% goat serum (or serum that comes from the animal in which the secondary antibodies have been raised) in TBST, or 1%–10% low-fat milk powder in TBST. The optimal blocking condition has to be determined experimentally for each primary antibody. Most of our tested antibodies provided good results with 10% goat serum in TBST. If the background is too high, use low-fat milk powder instead, but under these conditions, some antibodies may no longer detect their epitope. For example, in zebrafish, antibody V9 specifically detected vimentin (Cerda *et al.*, 1998) when the membrane was blocked with 10% goat serum but barely reacted when blocking was performed with different concentrations of milk powder.

d. Primary Antibody Incubation

To use minimal primary antibody for incubation, we generally seal the nitrocellulose membranes in a plastic bag with an appropriate volume of antibody solution (a minimum of 40 μl/cm^2 is required) and incubate for at least 2 hours at room temperature under gentle motion; for this, the bag is taped to a slowly rotating plate. The optimal working dilution has to be determined experimentally for each primary antibody. Usually one tenth of the dilution found to be optimal for immunofluorescence microscopy is a good start. For the dilution of the antibodies, use TBST. Because some monoclonal antibodies barely recognize their specific epitope when the corresponding keratin is complexed with SDS, we usually perform the primary antibody incubation overnight at 4–8 $^{\circ}$C to improve refolding of the polypeptide chain. After incubation, wash the membranes three to six times for 15 minutes in TBST or washing buffer at room temperature.

e. Secondary Antibody Incubation

For detection of the primary antibodies, we use alkaline phosphatase–conjugated secondary antibodies from Dianova (Hamburg, Germany) raised in goat against the species that developed the primary antibody. After the last washing step after primary antibody incubation, overlay the membrane with a sufficient volume of conjugate that has been diluted 1:7500 in TBST, and under gentle shaking incubate for 60 minutes at room temperature (to save antibody solution you may incubate several membranes in a single Petri dish). Then rinse the membrane again three to six times as described previously in TBST or washing buffer.

f. Colorimetric Detection

During the last washing step, freshly prepare a sufficient amount of detection solution by adding 3.3 μl BCIP and 6.6 μl NBT per milliliter AP buffer and store at room temperature. After the final wash, briefly rinse the membranes in AP buffer and afterwards remove the excess fluid by gently tapping the blot onto Whatman 3 MM blotting paper (do not allow the blot to dry). Then pour a sufficient amount of detection solution onto each blot and develop until the spots have reached the desired intensity. A 7×9 cm membrane requires about 5 ml of substrate solution. Protect the solution from strong light. Reactive spots will turn purple within 1–15 minutes. Weak reactions may be developed longer (up to 60 minutes), but this will also increase background staining. Stop the color reaction by washing the membrane in deionized water for several minutes and change the water at least once. Air-dry the developed nitrocellulose membranes and store them between two sheets of Whatman 3 MM paper. Protect the membranes from light.

g. Pitfalls

• It is essential to demonstrate by a negative control (with primary antibody lacking and with known negative primary antibody raised in the same species) that a positive reaction is due to the presence of a specific antigen.

• You may save the primary antibody solution for repeated use by passing it through a sterile filter syringe and storing it at 4 °C. You should add 0.02% sodium azide as a preservative. We recommend a maximum of two repeats.

• If necessary, use washing buffer instead of TBST to lower the background reaction.

• Handle the nitrocellulose membranes with care, because the dye may also precipitate at scratches.

• To save on antibody solution, peripheral regions are usually cut away from the membrane before incubation. Take care, however, that nothing is removed that is necessary for a direct comparison with the original gel.

• All incubation steps should be performed with the protein face of the membrane up.

• In several cases, we have observed a strong cross-reaction of anti-vimentin antibodies with fish type II keratins and/or desmin. Frequently, anti-desmin antibodies also recognize vimentin (especially if polyclonals are applied). Antibodies raised against keratins from man or frog in fish often show a different specifty (e.g., a type I specific antibody labels type II keratins or even other IF proteins) or are completely unreactive. Antibodies that have been positively tested by immunofluorescence microscopy may be negative in immunoblotting experiments. An overview of the antibodies we used in fish and their reactions is given in Fig. 2.

3. Complementary Keratin Blot Binding Assay

This method is based on the characteristic property of keratins to form obligate type I/II heterodimers as the first step in filament assembly. Although *in vivo* keratins are expressed in specific pairs, *in vitro* each type I and type II keratin will combine to form heterodimers, even if they stem from very distant vertebrate species (see Fig. 1 and Conrad *et al.*, 1998; Markl *et al.*, 1989; Schaffeld *et al.*, 1998). This enables one to identify the type I and type II keratins in blots after 2D-PAGE of cytoskeletal proteins by exposing the latter to labeled human keratin K8 and K18, respectively. The following solutions are required (only use fresh urea buffers!).

• 2x-urea buffer (pH 8.0): 20 mM Tris/Cl, 8 M urea, 6% BSA, 0.05%–0.1% Tween 20
• 1x-urea buffer (pH 8.0): 10 mM Tris/Cl, 4 M urea, 3% BSA, 0.05%–0.1% Tween 20
• TBST (pH 8.0): 10 mM Tris-Cl, 150 mM NaCl, 0.1% Tween 20
• High-salt extraction buffer (pH 8.0): 10 mM Tris/Cl, 1.5 M NaCl, 30 mM KCl, 0.5% Triton × 100, 0.1% Tween 20
• TBS (pH 7.6): 20 mM Tris/Cl, 137 mM NaCl
• AP buffer (pH 9.5): 100 mM Tris/Cl, 100 mM NaCl, 50 mM $MgCl_2$
• BNHS solution: 1 mg/ml biotinyl-N-hydroxysuccinimid in dimethylformamide (store at $-20\,°C$)
• NBT solution: 50 mg/ml Nitro Blue tetrazolium chloride in 70% dimethylformamide (store in the dark at $-20\,°C$)
• BCIP solution: 50 mg/ml 5-brom-4-chlor-3-indolylphosphate-p-toluidin in 100% dimethylformamide (store in the dark at $-20\,°C$)

h. *Production of Biotinylated Keratins by* In Vitro *Transcription/Translation*

To reduce the chance of RNase contamination, wear gloves throughout the procedure and use microcentrifuge tubes and pipette tips that have been autoclaved and handled only with gloves. The *in vitro* translation of a keratin cDNA

Monoclonal antibodies to detect intermediate filaments (mainly keratins) in fish

antibody	subclass	antigen	supplier	lamprey		shark		sturgeon		bichir		zebrafish		carp		trout	
				F	W	F	W	F	W	F	W	F	W	F	W	F	W
Ks pan 1-8	IgG2a	human K1-8	Progen, Germany	+	n	+	+	+	+	+	+	+	−	+	−	n	n
RGE53	IgG1	human K18	Progen, Germany	n	n	n	n	n	n	n	n	n	n	n	n	+	+
K8.60	IgG1	human K1/10/11	Progen, Germany	+	+	+	−	+	n	+	n	−	n	+	−	+	+
AE1	IgG1	type I keratins	Progen, Germany	−	n	−	n	+	n	+	+	+	n	+	+	+	+
AE3	IgG1	type II keratins	Progen, Germany	+	+	−	n	+	n	+	+	−	n	−	−	+	+
Ks19.2	IgG2b	human K19	Progen,, Germany	n	n	n	n	n	n	−	n	n	n	n	n	+	+
Ks18.04*	IgG1	human K18	Progen, Germany	+	+	−	n	+	n	+	n	+	+	+	+	+	+
K8.13	IgG2a	human K1/5/6/7/8/10/11/18	Sigma-Aldrich	−	n	+	+	+	n	+	n	n	n	n	n	n	n
KL1	IgG1	human keratins broad range	Immunotech, Marseille, France	+	+	−	n	+	+	+	n	+	+	+	+	+	+
A45-B/B3	IgG1	human K 8/18/19	Micromet, Martinsried, Germany	+	+	−	n	+	n	+	n	+	n	+	n	+	+
LU5	IgG1	mammalian keratins	Chemicon Int. Ltd	+	n	n	n	+	n	+	n	n	n	n	n	n	n
C10	IgG1	K8	gift from J.Bartek	+	+	n	n	n	n	n	n	n	n	+	+	+	+
LE63	IgG	keratins kangaroo rat	gift from E. B. Lane	−	n	n	n	n	n	n	n	n	n	n	n	+	+
LE64	IgG2a	K19 kangaroo rat	gift from E. B. Lane	n	n	−	n	n	n	+	+	+	+	n	n	n	n
LE65	IgG2a	K18 kangaroo rat	gift from E. B. Lane	n	n	+	n	n	n	+	n	n	n	n	n	n	n
68.4	IgG1	K18* Xenopus laevis	gift from B. Fouquet	+	n	+	+	+	−	+	n	+	−	+	+	+	−
46.6	IgM	K1/8 Xenopus laevis	gift from B. Fouquet	−	n	+	+	+	+	+	n	n	n	n	n	+	+
1.3.1	IgM	K2/18 Xenopus laevis	gift from B. Fouquet	−	n	−	n	+	+	+	n	n	n	n	n	+	+
164.4	IgG1	K1/8 Xenopus laevis	gift from B. Fouquet	+	n	+	+	+	+	+	+	+	+	+	+	+	+
79.14	IgG1	K1/8 Xenopus laevis	gift from B. Fouquet	−	n	+	+	+	+	+	+	+	+	+	+	+	+
26.8	IgG1	K1/8 Xenopus laevis	gift from B. Fouquet	−	n	+	+	+	+	+	n	n	n	n	n	+	+
F1F2	IgG1	simple type I keratins trout	our group	n	n	n	n	n	n	n	n	n	n	n	n	+	+
VIM 14.13	IgG1	vimentin Xenopus laevis	gift from B. Fouquet	+	n	+	+	+	+	+	+	+	+	+	+	+	+
V9	IgG1	vimentin mammals	Beckman Coulter	n	n	−	n	n	n	+	+	+	+	n	n	n	n
VIM 3B4	IgG2a	vimentin	Chemicon Int. Ltd	+	+	−	n	n	n	n	n	−	n	−	n	−	−
VIM 13.2	IgM	vimentin	Sigma-Aldrich	+	+	+	+	n	n	n	n	n	n	n	n	n	n
H5	IgG1	chicken vimentin	DSHB, University of Iowa	+	n	+	+	n	n	n	n	n	n	n	n	n	n
DE-B-5	IgG1	human desmin	Chemicon Int. Ltd	−	+	+	+	+	−	−	n	−	−	n	n	n	n
D33	IgG 1	human desmin	DakoCytomation, Denmark	+	+	n	n	+	n	+	+	n	n	n	n	n	n
IFA	IgG 1	IF proteins	gift from Harald Herrmann	+	+	+	+	+	+	+	+	n	n	n	n	n	n
69.16	IgG 1	mitochondria Xenopus	gift from B. Fouquet	−	n	+	+	+	n	n	n	n	n	−	n	n	n

Fig. 2 Monoclonal antibodies reactive to fish intermediate filaments. F, reaction in indirect immunofluorescence microscopy on cryostat sections; W, reaction in Western blots (immunoblots); +, positive reaction; −, negative reaction; n, not tested. Note that several antibodies show cross-reactions with intermediate filament protein types other than those against which they have been raised. *Ks18.04 is also refered to as C04; in trout, zebrafish, goldfish, and carp, C04 exclusively labels keratin K18, whereas C10 specifically recognizes K8 (shown in trout and zebrafish: Conrad *et al.*, 1998; Markl *et al.*, 1989; Schaffeld *et al.*, 2002a,b, 2003). In trout, keratin specific antibody 1.3.1 specifically labels mitochondrial antigens, whereas mitochondria-specific antibody 69.16 in shark specifically labels K8 (see Markl and Schechter, 1998; Schaffeld *et al.*, 2004). In addition to the listed monoclonal antibodies, we used polyclonal antibodies (GPpoly) that we raised against keratins isolated from RTG2 cells and specifically recognize a broad range of keratins in different tissues of all investigated fish species, including lamprey, shark, bichir, sturgeon, and all teleosts (Conrad *et al.*, 1998; Markl *et al.*, 1989; Schaffeld *et al.*, 1998). DSHB, Developmental Studies Hybridoma Bank.

clone may be performed with the TNT Coupled Reticulocyte Lysate System in combination with the Transcend Non-Radioactive Translation Detection System (Promega). This is a one-tube transcription/translation system, accomplishing the incorporation of biotinylated lysine into nascent polypeptides during translation. This biotinylated lysine is added to the translation reaction as a precharged, ε-labeled biotinylated lysine-tRNA complex (Transcend tRNA) rather than a free amino acid. You may directly use circular plasmid cDNA constructs for the reaction. Thaw the Transcend tRNA on ice, whereas directly before use the reticulocyte lysate has to be thawed rapidly in your hand and then put on ice. According to the manufacturer instructions, incubate 1 μg of the respective circular DNA construct with an adequate TNT RNA polymerase (T3, T7, or SP6; depending on the construct and direction of cDNA cloning) and reticulocyte lysate in the presence of Transcend tRNA as follows:

- TNT rabbit reticulocyte lysate, 25 μl
- TNT reaction buffer, 2 μl
- TNT RNA polymerase (T3, T7, SP6), 1 μl
- Complete amino acid mix 1 mM, 1 μl (alternatively add 0.5 μl each of two different minus amino acid mixes)
- RNasin 40 U/μl, 1 μl
- DNA (1 μg), x μl
- H$_2$O (RNase free), make up to 49 μl

Mix the reaction gently using a pipette (avoid air bubbles) and subsequently add 1 μl of Transcend tRNA. Again mix the sample, centrifuge for 1 minute in a microcentrifuge, and incubate for 75–90 minutes at 30 °C (use a water bath or a Thermocycler). In this way, prepare six reactions of 50 μl each and pool them after incubation. Separate 3 μl as control for gel analysis. To further purify the biotinylated keratins, add an equal volume of 2× urea buffer to the sample and put it on a PD10 Sephadex G25 column (Amersham Biosciences, Uppsala, Sweden) that has previously been equilibrated with 25 ml 1× urea buffer. Elute the column with 3.5 ml 1× urea buffer and immediately collect in 300-μl fractions. Pick 3 μl of each fraction and prepare it for SDS electrophoresis in a 10% acrylamide gel (5 × 8 × 0.1 cm). After PAGE, transfer the proteins on nitrocellulose sheets and detect the biotinylated proteins as described in the following. Pool the fractions containing the labeled keratin (it usually occurs in fractions 5–8, with the strongest signal in 6 and 7). Use the fractions directly for CKKB assays or store them in 100-μl aliquots at −20 °C. If the yield of labeled keratin seems to be too low, you may increase the applied Transcend tRNA up to twice the standard amount to improve labeling.

Biotinylated proteins can also be removed from the translation reaction with biotin binding resins such as SoftLink soft release avidin resin. However, nascent proteins synthesized in the presence of Transcend tRNA bind strongly to SoftLink resin and cannot be eluted by use of "soft-release" nondenaturing conditions (for more information, see instruction manual of the supplier).

i. Production of Biotinylated Keratins by Directly Biotinylating Recombinant Keratins

Dilute the recombinant keratin to a concentration of 1 mg/ml with TBS buffer. Add 50 μl BNHS per ml of protein solution, and after mixing, incubate at room temperature for 1 hour. Dialyze the reaction overnight at 8 °C against TBS. Check the biotinylated product by 1D- or 2D-PAGE. You may store the sample at −20 °C until use. Do not repeatedly freeze and thaw the sample.

j. Incubation of IF Proteins Transferred to Nitrocellulose with Biotinylated Keratin Probes

Incubate all membranes separately and process them protein side up. Block the nitrocellulose sheets for at least 1 hour at room temperature in TBST. Meanwhile, prepare the binding solution by adding 100 μl of the *in vitro* translated keratin probe or 10–20 μg of the labeled recombinant keratin to 1 ml of 1× urea buffer. After blocking, shrink wrap each filter with 1–3 ml of the binding solution and incubate them at least for 4 hours at room temperature (or better) overnight, at 8 °C on a plate that rotates the blots continuously. Subsequently, wash the filters four to six times for 10–15 minutes in high-salt extraction buffer at room temperature with gentle shaking. Then the biotinylated keratins may either be visualized by binding streptavidin-alkaline phosphatase (streptavidin-AP) or streptavidin-horseraddish peroxidase (streptavidin-HRP) followed either by colorimetric or chemiluminescent detection. The latter may further enhance the sensitivity of the detection and is preferable if only very low amounts of cytoskeletal preparations are available and/or the signals yielded by steptavidin-AP are too weak. A simple protocol for the streptavidin-AP detection is given in the following. As negative controls for the CKKB assay, incubate blots with 1× urea buffer only. This is important to exclude signals on the basis of biotinylated proteins that may be present in the cytoskeletal preparations, although so far, we did not observe this problem in our experiments.

k. Colorimetric (BCIP/NBT) Detection of Biotinylated Proteins

Process the filters protein side up. Do not allow the membranes to fully dry during any of the subsequent steps. Perform all the washing and incubation steps at room temperature with gentle shaking. Use a shallow container that is slightly larger than the membrane. For streptavidin-AP incubation and the color development reaction, use just enough solution to submerge the membrane. Before starting the detection process, briefly rinse the nitrocellulose membranes in TBST. Then block the membranes for 1 hour at room temperature in 10% milk powder dissolved in TBST. Again rinse the filter in TBST. Subsequently, incubate the membrane for 1 hour at room temperature with streptavidin-AP (Promega, Mannheim, Germany) at a dilution of 1:5000 in 5% milk powder/TBST. Then wash four times for 15 minutes in TBST. During the last washing step, freshly prepare a sufficient amount of detection solution by adding 3.3 μl BCIP and 6.6 μl NBT per ml AP buffer and store at room temperature. After the final washing, further treat the blots as described in "*colorimetric detection*" of immunoblots.

D. Indirect Immunofluorescence Microscopy of Fish Frozen Tissue Sections

Immunofluorescence microscopy of cytoskeletal proteins has the advantage that the labeled filaments glow like neon lamps in total darkness (see Fig. 3). Consequently, details are detectable that otherwise would completely escape light microscopy, for example, when horseraddish peroxidase staining is used. In our IF protein studies, we exclusively use frozen tissue sections, because we encountered serious artifact problems when we applied other methods such as paraffin sectioning. The standard procedure elaborated on mammalian tissues also works perfect with fish tissues (Markl and Franke, 1988). A required standard solution is PBS (140 mM NaCl, 2.7 mM KCl, 1.5 mM KH_2PO_4, 8.1 mM $Na_2HPO_4 \times 2H_2O$). For incubation of sections it is not necessary to further adjust the pH of the PBS.

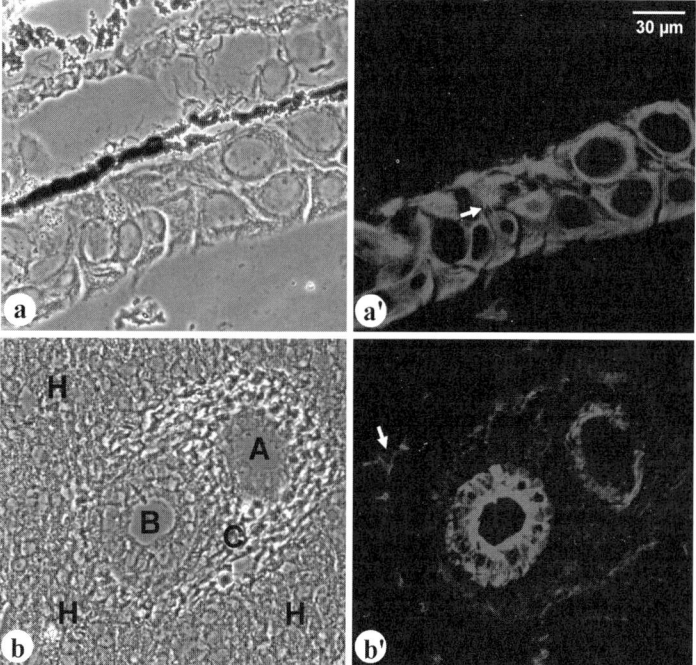

Fig. 3 Immunofluorescence microscopy of 5-μm cryostat sections of lungfish skin (young specimen; a, a′) and trout liver (b, b′) with monoclonal anti-keratin antibodies KL1 (a, a′) and 68.4 (b, b′), respectively. The same fields are presented using phase-contrast optics (a, b) and epifluorescence optics (a′, b′); note the morphological details visible because of the immunofluorescence stain. (a, a′) On lungfish skin, antibody KL1 specifically recognizes the epidermal keratinocytes but is completely unreactive to mesenchymal cells such as the pigment cell layer or dermal fibroblasts. To withstand mechanical stress, keratinocytes are packed with keratin filaments (e.g., arrow in a′). (b, b′) Antibody 68.4 on trout liver tissue, specifically staining a hepatic artery (A), bile duct (B), and the surrounding interstitial cells, as well as the lining of bile canaliculi (arrow) formed by hepatocytes (H). (See Color Insert.)

1. Production of Mounting Medium (Mowiol) for Cryostat Sectioning

Adjust the pH of the PBS buffer to 8.0 and autoclave it at 120 °C for 30 minutes. Then dissolve 20 g of Mowiol 488 (Höchst AG, Frankfurt, Germany) in 80 ml PBS by stirring 24 hours at room temperature. Subsequently, add 40 ml of water-free glycerol and stir overnight. Afterwards centrifuge the mixture with 10,000g (12,000 rpm; Sorvall refrigerating centrifuge RC5B; Rotor SS34) at room temperature for 15 minutes. Then fill the mounting medium into 2-ml reaction caps and store it at −20 °C until use. After thawing, store it at 4 °C but do not freeze it again for storage.

2. Production of Cryostat Frozen Tissue Sections

Freeze the tissue sample onto the cutting block of your cryostat with an adequate cryoembedding compound and cut 5-μm sections. Tissues with even densities usually work best. It is important to avoid thicker sections to obtain single cell layers. The optimal orientation of the tissue block has to be determined experimentally. Often, better results are achieved if the large side of the cut section is at 90 degrees to the blade. Depending on the tissue, the box temperature and the cutting block temperature should range between −15 °C and −30 °C. The optimal cutting temperature has to be determined individually for each tissue and organism. Usually, more fatty tissues (such as liver) require higher cutting temperatures. If the sections wrinkle while sectioning and look mushy afterwards, try using a decreased temperature; if they look brittle and contain breaks, switch to a higher temperature. To cut sections, first trim the tissue block by producing 10- to 15-μm sections to obtain a good cutting surface. Then adjust the section thickness to 5 μm and start producing collectable sections (depending on the cryostate, it usually will take several sections until the preset thickness is achieved). It is important to clean the blade and the antiroll plate with acetone after each section. Make sure that the acetone has evaporated entirely before the next sectioning. This should prevent sticking of the section, or part of it, to the blade. While sectioning, the antiroll plate should be parallel to the knife edge and only protrude very slightly over the edge. Its optimal orientation has to be determined very carefully. Pick up sections on a microscope slide (before use, carefully clean all slides by dipping them in petroleum benzene or acetone and letting them air dry at room temperature) by holding it over but not touching the cut sections and allow them to air dry overnight. Because of the temperature difference between blade and slide, the section should immediately hop onto the slide. Depending on the size of the sections, you may collect up to four specimens on one slide. Fix the sections for 10 minutes in acetone at −20 °C and let them air dry again. Sections can be stored at −20 °C if they are wrapped in aluminium foil, but sometimes they yield poor results after such storage; therefore, we generally advise the use of tissue sections for immunofluorescence microscopy immediately after drying from acetone fixation.

3. Incubation with Antibodies

All incubation steps should be performed in a humid chamber that can be easily constructed from a gel-staining box by putting a water-soaked Whatman 3 MM paper at the bottom. Antibodies should be diluted in PBS. Under no circumstances should the sections be allowed to dry up during the incubation steps, because this would create serious artifacts.

a. Primary Antibodies

Most of the antibodies available for interaction with fish IF proteins were initially raised against mammalian or *Xenopus* IFs. Figure 2 summarizes monoclonal antibodies against keratins, vimentin, and desmin that we found to work well on fish tissues. How much to dilute such antibodies to work "nicely" on a given fish tissue has to be determined empirically, but the following values are a good guide. If supplied as cell culture supernatant, a monoclonal antibody may be used from dilution 1:10 to undiluted, whereas a polyclonal antibody might work at dilution 1:200 to 1:100. Monoclonal antibodies supplied as ascites fluid can be diluted in the range of 1:100 to 1:1000. However, the commercially available antibodies should first be applied according to the data sheet of the supplier, even if it does not mention fish tissues. Encircle each section with a PAP pen for immunostaining (e.g., liquid blocker super PAP pen from Sigma-Aldrich) to separate them from each other. This allows the testing of different antibodies on a single slide. Incubate at least two sections with the same antibody. Apply 20–40 μl of antibody solution on each section, and cover it entirely by use of the side of the pipette tip. Be careful not to touch the sections. Place the slides horizontally into the humid chamber and let them incubate at room temperature for 1 hour. As a negative control, use PBS or antibody medium instead of the primary antibody solution; if available, you can additionally apply an antibody if previous work determined that the tissue is negative. You might also include a positive control by using a tissue that you know will react. After incubation, remove most of the excess fluid by gently knocking the long side of the slide onto a sheet of facial tissue, but make sure that the sections do not dry up. To remove the remaining antibody solution, rinse the slide by submerging it briefly in PBS. Collect up to 10 slides in a holder and rinse them three times for 5 minutes in PBS. Do not shake holder or tank, because this could wash away the sections.

b. Secondary Antibodies

These antibodies are raised against antibodies from the species used to develop the primary antibody. We usually purchase them from Dianova (Hamburg, Germany). Use Texas Red (exitation 596 nm/emission 620 nm; color red) or Indocarbocyanin (Cy3, excitation 552 nm/emission 568 nm; color red/orange) as antibody conjugate at a dilution of 1:200 in PBS. These dyes are less prone to fading under ultraviolet (UV) light exposure than, for example, FITC (fluoresceinisothiocyanate; excitation 495 nm/emission 525 nm; color green), and give bright and distinct reactions. If you want to simultaneously stain the cell nuclei, just add 1 μl of 1 mg/ml Höchst 33258 DNA stain (2p-(4-hydroxyphenyl)-5-(4-methyl-l-piperazinyl)-2,5p-bi-benzimidazole in 0.9 % [w/v] NaCl) stock solution (exitation 360 nm/emission 470 nm; color blue) to 1 ml of secondary antibody solution. Be careful, DNA dyes are extremely poisonous! After the last rinse following the primary antibody incubation, again remove the excess fluid by wiping the empty areas of the slide with a facial tissue. You will have to work very quickly to avoid the sections drying out. Then cautiously add 20–40 μl of secondary antibody solution to the PBS buffer that remained on each section. Incubate for 1 hour at room temperature in the humid chamber. Thereafter, rinse the slides three times in PBS as described previously. After the last rinse, briefly submerge the slides in distilled water to remove the salts. As the final step, dehydrate the sections by submerging the slides for 5 minutes in 100% ethanol. Subsequently, let the sections air-dry until the ethanol has completely evaporated. The slides are now ready for mounting.

c. Double Antibody Staining

This works nicely if you combine, in the first incubation step, a monolonal antibody raised in mouse and polyclonal antibodies raised, for example, in guinea pig; simply use a mixture but maintain the original concentrations. The secondary antibody can then be a mixture of goat anti-guinea pig immunoglobulin Texas Red conjugate, which will give a red signal and goat anti-mouse immunoglobulin FITC conjugate, which will give a green color. We found this technique to be very successful when comparing the expression of "E" keratins versus "S" keratins, keratin versus vimentin, or keratin versus collagen (Markl and Franke, 1988). The last combination has been particularly useful to understand tissue organization in fish.

4. Mounting of Sections

Mowiol is a highly viscous fluid, and you have to be very careful to prevent formation of air bubbles while handling it. Cover all sections with a drop of Mowiol with a Pasteur pipette or a slightly truncated pipette tip. Then carefully add the coverslip. After 10 minutes, cautiously correct the position of the coverslip, put the slides in a light-protected folder, and let them dry overnight in a horizontal position (storing them vertically might allow the cover slips to move before the Mowiol is completely dry). When examining the slides with a fluorescence microscope (with the appropriate filters), doing it within 24 hours will probably yield the best results. Start your observations with the negative and positive controls. Take pictures immediately. Prolonged storage often increases background fluorescence and/or decreases the specific fluorescence. Some fluorochromes, especially FITC, are prone to fading or quenching during exposure to

UV light, so minimize viewing time. Because fluorochromes are light sensitive, perform unusually long incubations in the dark, and store your slides in the dark.

5. Staining of Fish Cell Lines

The staining of cells grown on microsope slides works in the same way as described previously for tissue sections with only one alteration. At first, incubate the cells for 10 minutes in methanol at $-20\,°C$ before putting them directly into the $-20\,°C$ cold acetone for further fixation. Before incubation with antibodies, encircle each well with a PAP pen.

6. Pitfalls

• When applying anti-keratin antibodies to fish tissue sections, a truly positive reaction is so brilliant that it can be easily distinguished from background staining. On the other hand, we found that for most fish tissues, nonspecific primary antibodies yield a faint immunofluorescence; its strength varies with the antibody applied.

• Some tissues emit autofluorescence when exposed to UV light. For example, in most of the fish species tested, skeletal muscle and liver tissue showed comparatively strong background staining. (In certain shark digestive tract tissues, we even observed nonspecific binding of several primary antibodies to the mucous.)

• We often have immunofluorescence microscopic series in which a positive reaction is lacking, and in those cases, inexperienced candidates tend to claim "positives" that merely come from background staining. This is even more so when they take digital images instead of photomicrography, because the software easily transforms weak background fluorescence into brilliant stain. We therefore have the rule in our laboratory that nobody working with the immunofluorescence microscope is allowed to perform, during data collection, digital contrast enhancement.

• Moreover, in each series adequate negative *and* positive controls must be included. This sounds trivial but is very often ignored, even in our own laboratory. Especially for beginners, but also for those who are unsure whether the applied antibodies are still working, it is strongly recommended that a whole series of positive controls be run before studying an unexplored fish species or tissue. For this purpose, we always use tissues from rainbow trout, because their reactions are particularly well documented.

• Antisera that we raised in guinea pig against cytoskeletal preparations from rainbow trout tissues mostly show strong cross-reactions with collagen or other extracellular material (see Markl and Franke, 1988), whereas antisera raised against cytoskeletal preparations from the rainbow trout cell line RTG2 that do not express any collagens are highly specific for keratins even in evolutionary distant species such as shark (Conrad *et al.*, 1998; Markl *et al.*, 1989; Schaffeld *et al.*, 1998).

E. Preparation of Total RNA and mRNA from Fish Tissues

1. Total RNA Preparation

Very crucial for experimental procedures such as RT-PCR or cDNA library construction is how RNA isolation is performed. Successful isolation of RNA depends on suppression of endogenous RNases during cell/tissue lysis and avoiding contamination with exogenous RNases during the isolation process. We generally advise that tissues be removed from the freshly killed animal as fast as possible and the RNA isolation is started immediately with thorough homogenization of the tissue in the required lysis buffer. Alternately, you may snap-freeze the excised tissue in liquid nitrogen and store it at $-70\,°C$ for later use. However, this may result in a lower yield and quality of the isolated RNA, especially if the tissue is stored for a longer period of time. If you use frozen tissue for RNA isolation, ensure that the tissue does not thaw before starting homogenization in lysis buffer.

In our experiments, the yield of intact total RNA from fish tissues was usually low when we applied commercially available standard kits that are based on guanidinium thiocyanate (GTC) extraction followed by acidic phenol/chloroform purification or specific binding of RNA to a silicagel matrix. Furthermore, these preparations were still contaminated with considerable amounts of genomic DNA that sometimes disturbed the following experiments. Usually we obtained much better results when we applied a protocol modified from Chirgwin *et al.* (1979), which includes GTC extraction and subsequent sedimentation of RNA by ultracentrifugation through a dense cushion of cesium chloride. The following solutions are required

- GTC (pH 7.0): 4 M guanidinium thiocyanate, 25 mM sodium citrate, 20 mM DTT
- CsCl: 5.7 M CsCl, 25 mM sodium acetate
- TE/SDS (pH 7.5): 10 mM Tris/Cl, 1 mM EDTA, 0.5% SDS
- Phenol (pH 7.5–8.8): Roti-Phenol, equilibrated in TE buffer (Roth, Karlsruhe, Germany)
- Phenol/chloroform/isoamyl alcohol: Roti-Phenol/Choroform 25:24:1, equilibrated with TE buffer (Roth, Karlsruhe, Germany)
- Chloroform/isoamyl alcohol (24:1)
- 3 M sodium acetate (pH 5.5)
- 100% ethanol
- 70% (v/v) ethanol

Filtrate the GTC and CsCl solutions through a sterile filter (0.45 μm) and store them at $4\,°C$.

Perform all homogenization steps on ice. Thoroughly homogenize the tissue in an appropriate volume (as large as possible) of precooled GTC buffer with an Ultra Turrax. Subsequently, further homogenize the suspension with the help of a hand-held glass homogenator (Dounce) by at least 20 strokes with the loose-fitting pestle, followed by 10–20 strokes with the tight-fitting pestle. It is important that all cells are ruptured during this step. If necessary, subsequently sediment remaining particulate material, such as debris from the fish scales, by centrifugation for 10 minutes at 1000–1500g and 4 °C. After centrifugation, you may either freeze the supernatant in liquid nitrogen for later use or directly continue by loading it onto the CsCl cushions. To prepare the GTC homogenate for ultracentrifugation, adjust it to the required volume as indicated in Table III (use GTC buffer).

Incubate the ultracentrifugation tubes for 30 minutes in 0.5 M NaOH before use. Afterwards, clean them with sterile water and 100% ethanol and let them air-dry. Fill CsCl solution into the tubes, and carefully overlay the CsCl cushion up to the top of the tube with the tissue homogenate in GTC buffer (for volumes see Table III; let the homogenate pour down the wall of the tube to overlay the CsCl cushion; use a glass pipette). Centrifuge in a Beckmann SW 40 or SW 28 rotor with either of the conditions shown in Table III. After ultracentrifugation, proceed with one tube after another as follows.

Carefully remove the GTC buffer down to the CsCl cushion with a Pasteur pipette that is connected to a water jet vacuum pump (do not disturb the cushion). Subsequently, with a hot scalpel blade cut off the tube directly at the upper boarderline of the CsCl cushion. By quickly flipping the shortened tube upside down, discard the CsCl solution and remove the excess fluid by resting it upside down on a facial tissue. Do not allow the RNA pellet to air dry. Avoid contamination of the RNA pellet with remaining GTC buffer or excess CsCl solution from the tube wall. Thoroughly resuspend each RNA pellet in prewarmed (65 °C) 250–1000 μl TE/SDS until the RNA is solubilized; quickly transfer the solution to 2-ml reaction vials for further treatment. To avoid degradation immediately, add 1 volume of phenol to each sample and mix thoroughly (use a vortex). Centrifuge for 3 minutes at room temperature and 10,000g. After centrifugation, transfer the upper (aqueous and RNA containing) phase to a fresh reaction cap and repeat this phenol extraction three to four times. Afterwards, extract the solution in this way three to five

Table III
Conditions for Ultracentrifugation of the Guanidium Thiocyanate (GTC) Homogenate

Rotor	CsCl volume (ml) for each tube	GTC homogenate total volume (ml) for 6 tubes	Conditions (at 20 °C) (h/rpm)
SW 40	2	40	16.5/32,000
SW 28	5.5	190	24/20,000
SW 28	5.5	190	20/24,000

For Beckmann SW40-Rotor, use 1–3 g; for SW28-Rotor use 3–10 g tissue.

times with phenol/chloroform and one to two times with chloroform. To increase the RNA yield, you may rewash the interphase with 200–500 μl TE/SDS after the first extraction with phenol. After the last extraction step, add 1/10 volume of 3 M sodium acetate, pH 5.5, to the supernatant, mix thoroughly, and precipitate the sample RNA by mixing it with 2–3 volumes of 100% ethanol and let it incubate overnight at $-20\,°C$. The next day, centrifuge the suspension at $10,000g$, at $4\,°C$, then discard the supernatant and wash the pellet twice with 70% (v/v) ethanol. Subsequently, air dry the precipitate (do not dry at high temperature or too long) and resuspend it in sterilized water. Then take an aliquot of the solubilized RNA for photometric determination of its concentration and contamination with proteins (deduced from the OD_{260}/OD_{280} ratio). Furthermore, check the quality (integrity) of the RNA on an analytical agarose gel [(for protocols see, e.g., Sambrook *et al.* (1989) and Sambrook and Russell (2001)]. The remaining solubilized RNA should be precipitated immediately as described previously and stored at $-20\,°C$.

a. Pitfalls
- Create an absolutely RNase-free environment.
- Use only double deionized and sterilized water.
- RNA from liver should be precipitated with 1 volume of 8 M LiCl to avoid the precipitation of large amounts of glycogen.
- The number of phenol/chloroform extractions performed should depend on the amount of interphase material that is formed after each centrifugation. Extract until the interphase is almost clear.

2. mRNA Isolation

For efficient isolation of high-quality mRNA (polyA + RNA) from total RNA, we successfully applied the PolyATract mRNA isolation system from Promega that uses their MagneSphere technology. Furthermore, we tested the PolyATtract System 1000 that isolates messenger RNA directly from crude cell or tissue lysates, thereby eliminating the often time-consuming need for total RNA isolation. This system is remarkably quick and worked well for fish (we tested it for zebrafish and lungfish), but it is presently much more expensive than the first method. For detailed protocols, refer to the supplier's instruction manuals.

F. Cloning and Sequencing of cDNA Encoding Fish Keratins

1. cDNA Libraries

We usually construct cDNA libraries with the ZAP Express cDNA synthesis kit and ZAP Express cDNA Gigapack III Gold Cloning Kit from Stratagene according to the instruction manual. This system allows directional cloning of the cDNA into the λ-ZAP expression vector that can be screened with either cDNA or antibody probes. By rapid *in vivo* excision of the pBK-CMV phagemid, the isolated clones are then accessible for standard characterization in a plasmid system (for details see instruction manual). We achieved best results with 3–5 μg of mRNA for constructing the cDNA libraries. However, in several cases, good cDNA libraries have also been derived from significantly lower amounts of mRNA (1–2 μg). To increase the yield of the cDNA library construction, we generally perform all required ligation and precipitation steps at least overnight at 4 °C and −20 °C, respectively. For broad-range antibody screening for keratin coding cDNAs, we usually use antisera raised against keratins extracted from the rainbow trout cell line RTG2. For similar purposes, we also apply monoclonal antibody IFA specific for a common motif shared by a variety of different IF proteins (Pruss *et al.*, 1981). However, screening fish cDNA libraries with antibody IFA mainly provided cDNA clones coding for type II keratins and the type III proteins vimentin and desmin, whereas clones encoding type I keratins have rarely been obtained. We furthermore screen the libraries with digoxigenin (DIG)-labeled probes derived from already available cDNA clones of the same or other species or from cDNA fragments that we previously obtained from RT-PCR experiments. A standard PCR that uses DIG-labeled nucleotide mix (PCR DIG Probe Synthesis Kit available from Roche) is a very quick and efficient way to construct such cDNA probes. We usually purify the labeling reactions by electrophoresis in an agarose gel followed by gel extraction of the respective DNA fragment. To test the efficiency of the labeling process, we produce different probe dilutions (1:10–1:10000) and bind and detect 1 μl of each dilution on a nylon sheet (dot blot analysis). The stringency of the following screening procedure depends on the applied conditions for the hybridization and the "hot" washing steps (time/temperature/salt) and has to be adjusted individually for each combination of cDNA probe and desired target sequence. For details concerning screening of cDNA libraries with antibodies or DIG-labeled cDNA probes, see the picoBlue immunoscreening kit instruction manual from Stratagene, the DIG application manual for filter hybridization from Roche, and the publications of Sambrook *et al.* (1989) and Sambrook and Russell (2001).

If isolated cDNA clones do not represent the complete sequence desired, we usually construct DIG-labeled probes from their 5′ end (at least 300 bp in length) and use them to rescreen the library under stringent conditions. Alternatively, we perform a PCR by directly applying the phage suspension of the cDNA library as matrix DNA by use of a vector sequence (e.g., T3 promoter) as upstream primer in combination with a cDNA-specific downstream primer. Apply at least one fold of the complexity (primary titer) of the cDNA library to such a PCR run. Before PCR, you may lyse the phages by initially heating the reaction (without the enzyme) up to 95 °C for 5 minutes. If those approaches are not successful, we complete the cDNA sequence by RACE-PCR (see later).

2. Reverse Transcriptase-Polymerase Chain Reaction

Usually, we have good results with either the Quiagen One Step RT-PCR Kit, or the Expand RT-PCR Kit from Roche followed by standard PCR, using

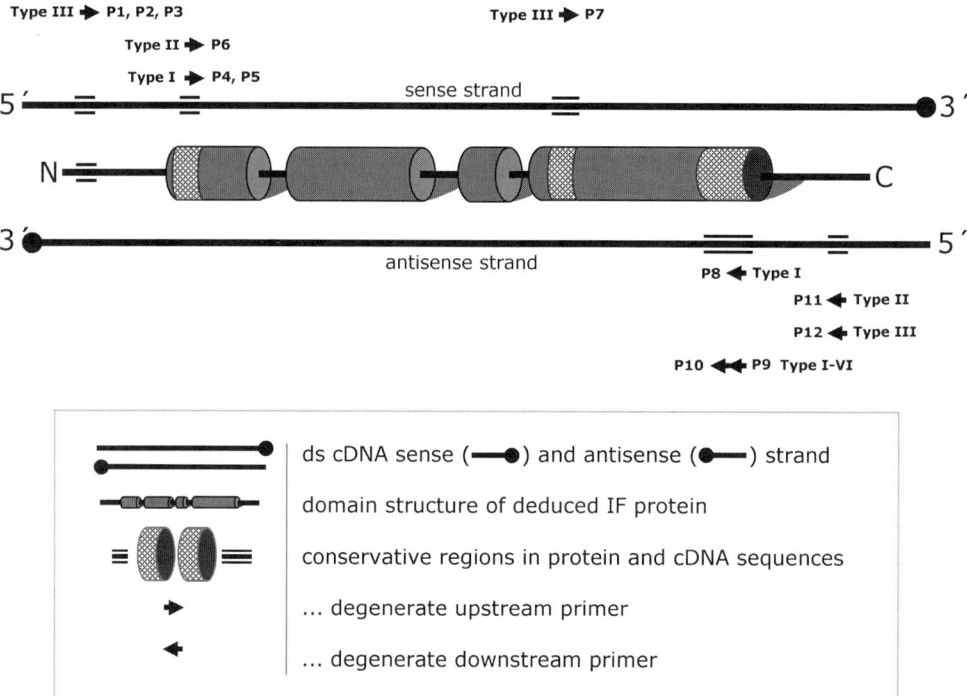

Fig. 4 Potential primer binding sites in conservative regions of intermediate filament proteins as deduced from a comparison of sequences from different evolutionary distant vertebrates. For cross reference to the primer sequences (P1–P12) see Table IV. The figure has been modified from Jan Schultess (2001).

Taq polymerase. By comparing the cDNA sequences and their deduced amino acid sequences from different vertebrates, we have identified several sites that seem eligible for the construction of degenerate primers to amplify IF proteins from evolutionary distant groups (Fig. 4). For example, primer combinations that successfully amplified almost the entire rod domain of several type I keratins from evolutionary distant species such as lamprey, shark, and birchir are CYTGAAYSASCGYYTGGC or CCTSAACGACCGYCTGGC as degenerate forward and GTAKGTBGCWATYTCMRCHTCSAG or TCYTCBCCYTC-CAGSAGYTTCCTGTA as degenerate reverse primers. Further possible degenerate primers to amplify fragments of type I–III IF proteins from different fish species are included in Table IV, and their general positions are illustrated in Fig. 4.

3. RACE-Polymerase Chain Reaction

For 5' Race, we use the GeneRacer Kit (Invitrogen, Karlsruhe, Germany) according to the supplier's instruction manual. By dephosphorylation of non-mRNA and truncated mRNA, the GeneRacer Kit ensures amplification of full-length

Table IV
Possible Degenerate Primer Sequences to Amplify Type I–III Coding cDNA Fragments from Evolutionary Distant Vertebrates

Name	Target*	US/DS[†]	Sequence in 5′ => 3′ orientation
P1	Desmin (head)	US	CAG TCC TAC ACC TGC GAG ATH GA
P2	Vimentin (head)	US	TTT GCT GAC CTS TCN GAR GCT GC
P3	Desmin (head)	US	TCC TCY TAC CGC CGC ACY TTY GG
P4	Type I keratin (1A rodstart)	US	CYT GAA YSA SCG YYT GGC
P5	K18 (1A rodstart)	US	CCT SAA CGA CCG YCT GGC
P6	Type II keratin (1A rodstart)	US	ATY GCH GAD GCH GAG SAR CGY GG
P7	Type III keratin (2B rod)	US	GAR GAV TGG TAY AAR TCH AAR
P8	Type I (2B enrod)	DS	GTA KGT BGC WAT YTC MRC HTC SAG
P9	Type I–VI (2B endrod)	DS	GTA RGT GGC RAT YTC VAY RTC MAG
P10	Type I–VI (2B endrod)	DS	TCY TCB CCY TCC AGS AGY TTC CTG TA
P11	K8 (tail)	DS	CCRTCBTKRGTYTCVAYSKTCTT
P12	Desmin (tail)	DS	CTC RCT SAC SAC CTC NCC ATC

*As by-products you may also obtain cDNA fragments coding for other intermediate filament proteins.
[†]US, upstream; DS, downstream.

transcripts only (for details see the supplier's instruction manual; for an example see Schaffeld *et al.*, 2004).

4. Cloning of Polymerase Chain Reaction Products

To perform rapid and efficient cloning of a PCR product into a plasmid vector, we usually use the TOPO XL or the TOPO TA PCR Cloning Kit from Invitrogen life technologies according to the instruction manual. After only 5 minutes at room temperature, the ligation is complete and ready for transformation into *E. coli*. The linearized TOPO vector contains a topisomerase that is covalently attached to its protruding 3′-deoxy-thymidine (T), optimizing the cloning into the plasmid DNA (for more details see instruction manual).

5. Pitfalls

• In a given fish species, different type I or type II keratin sequences may show an overall sequence identity up to 90% even at the DNA level. In such cases, stretches made up of up to 100 identical base pairs (sometimes even more) may be found scattered throughout the entire sequence alignment of the respective keratins, including their more variable head or tail domain and even extending to their 3′ UTR (see Fig. 5). This situation may complicate many experimental approaches. In particular, it has to be taken into account for the following intentions: amplification of a specific target sequence by PCR or RT-PCR (amplified fragments should be long enough to reveal characteristic differences; verify all

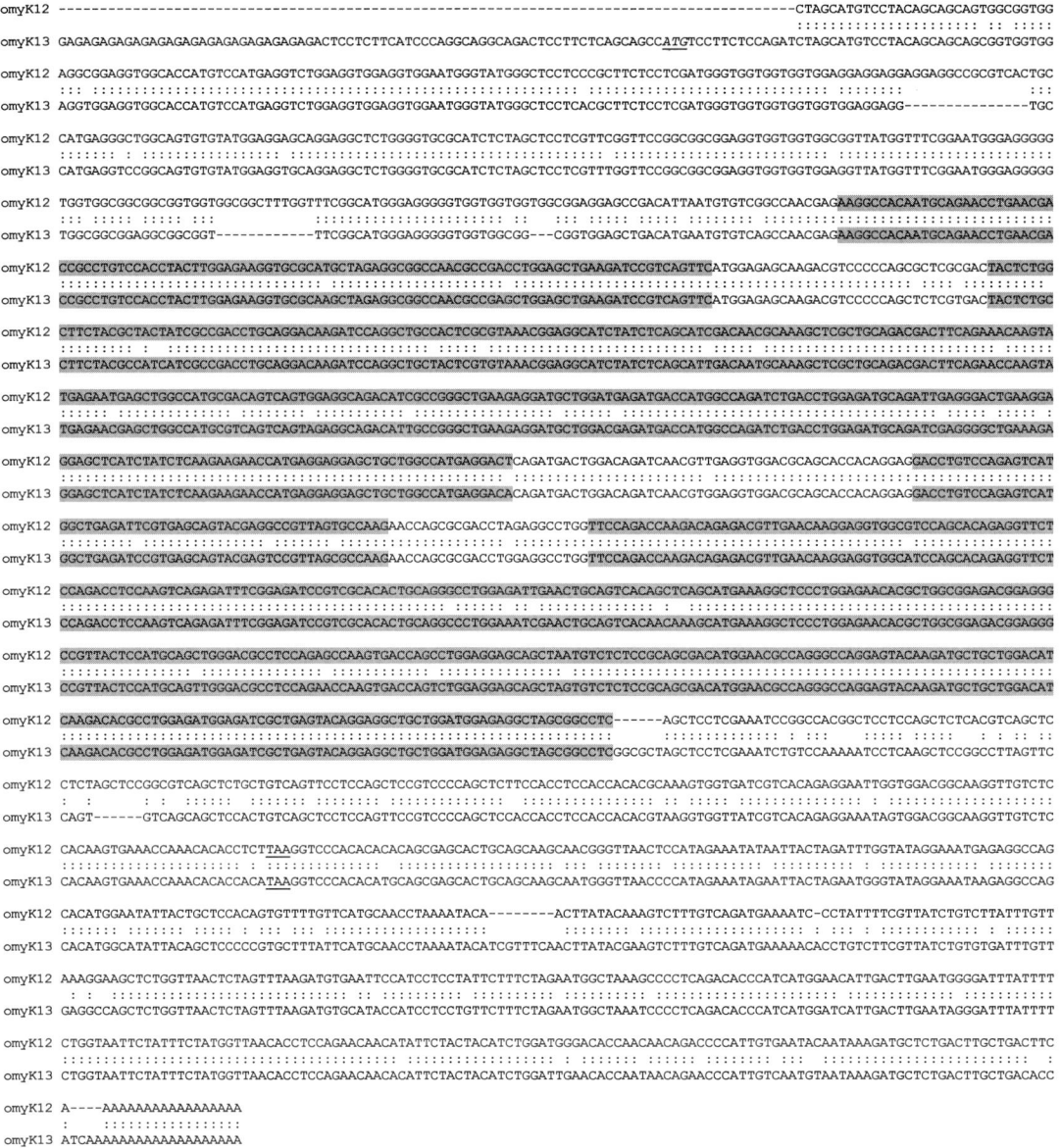

Fig. 5 Alignment of cDNA clones coding for the rainbow trout type I keratins OmyK12 and OmyK13 (see Schaffeld *et al.*, 2002a), showing the high identity that both sequences share even within their nonhelical head/tail encoding regions and 3′ untranslated regions. Start and stop codons are underlined; the sections encoding the helical parts of the rod are shadowed in grey. Note that the cDNA sequence encoding OmyK12 lacks several bp from its 5′ end.

products by sequencing); RACE-PCR to recover lacking fragments of cDNA clones (choose overlapping regions exceeding 250 bp); tracing expression of a keratin by *in situ* hybridization or Northern blot experiments (choose very stringent conditions); or assignment of cDNA clones to proteins extracted from tissues by PMF with MALDI mass spectrometry (see below Section II.G).

• Heterolog screening of cDNA libraries with probes derived from keratin coding cDNA clones as a by-product frequently also yielded the isolation of genes encoding vimentin and/or desmin or vice versa, even if stringent conditions have been applied. Furthermore, we frequently isolated myosin encoding cDNA clones.

G. Peptide Mass Fingerprinting of Fish Keratins with Matrix–Assisted Laser–Desorption/Ionization Time of Flight Mass Spectrometry

Peptide mass fingerprinting (PMF) with matrix-assisted laser-desorption/ ionization time of flight (MALDI-TOF) mass spectrometry is a powerful tool to assign obtained amino acid sequences of IF proteins to the polypeptides that have been extracted from tissue. This technique includes trypsinolysis of cytoskeletal proteins separated by 2D-PAGE, elution of the obtained fragments from the gel, determination of their molecular masses by MALDI-TOF mass spectrometry, and comparison of such values to a theoretical digest on the basis of the amino acid sequence. The mass of each peptide detected by PMF will be the sum of the amino acids present, including any modifications that those amino acids might have undergone. Oxidation of methionine residues to methionine sulfoxide was the most frequent modification we observed in our investigations of fish keratins. In this case, 16 Da have to be added to the theoretical mass of the fragment for each modified methionine. Consequently, this also has to be considered if the cysteine residues in the protein sample are modified before 2D-PAGE (e.g., by treatment with iodoacetamide or iodoacetic acid to permanently block the sulphide groups). Generally, proteins for analyses may be provided to commercial PMF services in polyacrylamide gels stained with Coomassie blue. In combination with PMF by MALDI-TOF mass spectrometry, only a few silver-staining protocols provide satisfactory results [see, e.g., Moertz *et al.* (2001) and Shevchenko *et al.* (1996); and citations therein]. When sufficient material is available, strong protein spots rather than weak ones should be provided, because the former clearly improve data quality. An important point is the quality of the electrophoresis in that sharp, well-focused spots yield better results than diffuse ones, even though both may contain the same amount of protein.

The obtained PMF spectrum of each spot contains the mass/charge ratio (m/z) of the detected fragments. In MALDI-TOF mass spectrometry, the peptides are generally ionized by protonation, resulting in a single charge on each fragment. Thus, the m/z values given in a PMF spectrum represent the masses of the detected fragments. Peaks that most likely represent peptides of the digested protein spot are labeled automatically by certain computer programs (done by the commercial services). It is also possible that the commercial service just provides the list of obtained mass values. Note that the given values refer to the monoisotopic mass (the first peak of the isotope distribution) of the protonized (mass +1) fragments. Do not expect that all of the masses you receive will be fragments from your investigated protein! Several might stem from contaminants.

A theoretical tryptic digest of the deduced amino acid sequence may be performed on http://www.expasy.org/tools/peptide-mass.html with the settings shown in Fig. 6. As a result, you will receive a table consisting of five columns showing: (1) the masses of the resulting peptides; (2) their position within the amino acid sequence; (3) the number of missed cleavages in the respective peptide; (4) the position of possibly oxidized methionines in this fragment and its modified mass; and (5) the sequence of the fragment. Compare the masses with those from your PMF spectra. This may be done by computational Internet tools, but in our laboratory, we prefer doing it by hand, because sometimes not all hits are automatically found. In our experiments, one factor complicated the assignment of new fish keratin sequences to their protein spots, namely, the great similarity observed between certain keratins of a given fish species (see III.F). Therefore, assignment of obtained primary structures to polypeptides extracted from tissues should be based on at least the following four criteria: (1) The total number of matching peptides should be high enough. (An overall sequence coverage of at least 30% should be achieved.) (2) Peptides (masses) that are characteristic for the studied sequence should be present. (At least three masses should be absent in simulated tryptic digests of the other putative keratin sequences. If from the studied animal species only a few keratin sequences are known, identification will remain uncertain.) (3) A maximal mass tolerance of 150 ppm should be allowed. (This means that only fragments showing a maximal mass difference of ±0.15 if compared with the theoretical value should be considered for identification. Usually, the average mass tolerance of our PMF analyses is less than ±0.05 (50 ppm).) (4) The maximum number of allowed missed cleavages should be 1. (This allows all concatenations of two adjoining masses to be added to the list of theoretical peptides under consideration but excludes combinations of more than two masses.)

For phosphorylated variants of a genuine keratin, you will receive an almost identical PMF spectrum. Generally, this modification is said to suppress the ionization of the respective fragment, and it has also been suggested that these groups may be removed from the peptide during the mass spectrometry [see e.g., McLachlin and Chait (2001) and citations therein]. Thus, phosphorylated peptides will rarely be easily detected in MALDI-TOF experiments. For examples of our PMF analyses, see Schaffeld et al. (1998, 2002a,b, 2003, 2004).

Useful websites for MALDI-TOF

http://www.aber.ac.uk/parasitology/Proteome/MS_Tut.html
http://kss.kribb.re.kr/PMF/PMFhelp.html

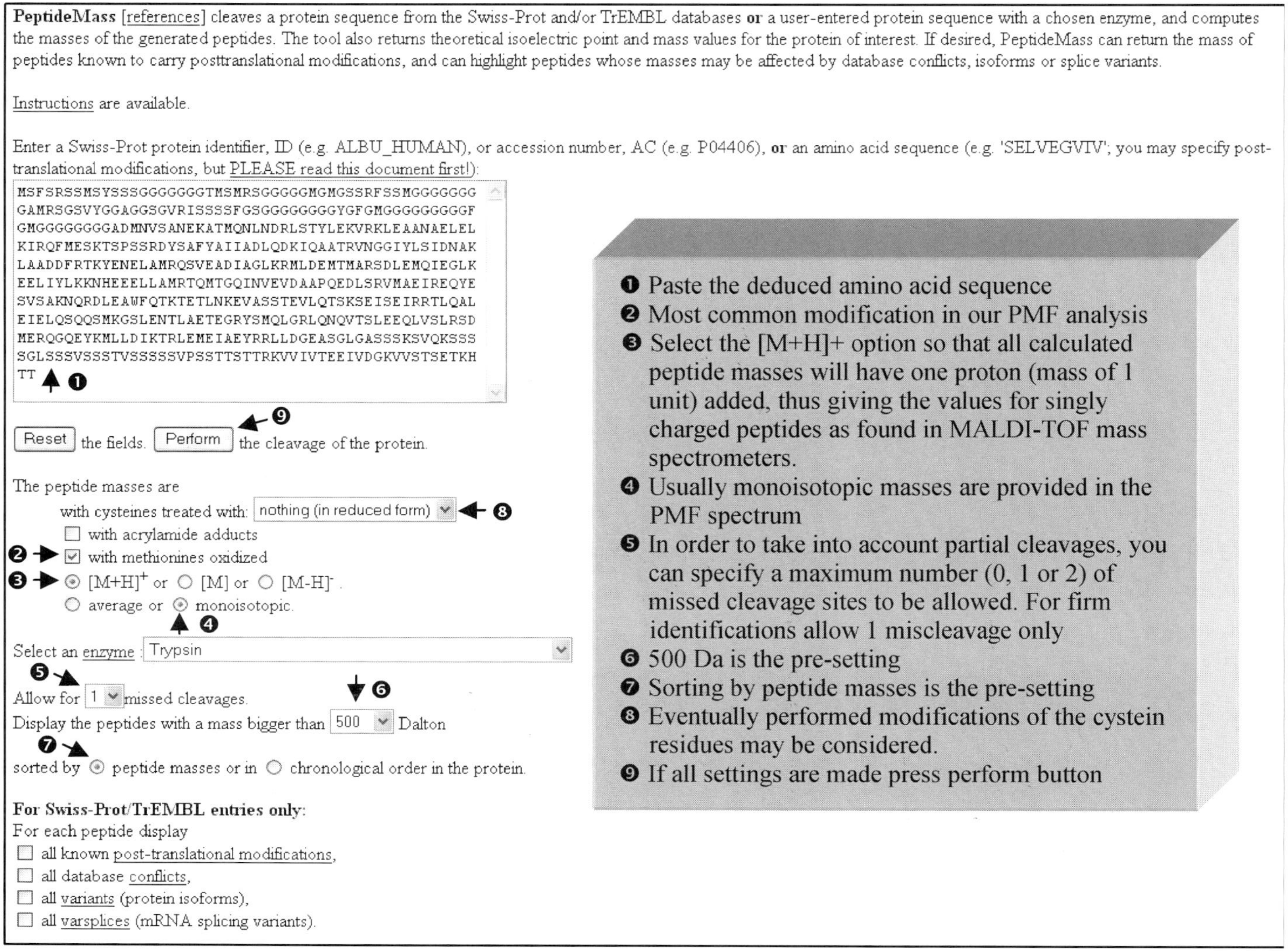

Fig. 6 Internet tool as available under http://www.expasy.org/tools/peptide-mass.html to perform a theoretical typical digest of a deduced amino acid sequence. Follow the steps and adjust the settings as described in 1 to 9. For further explanations see text (Section II.G).

http://www.matrixscience.com/search_form_select.html

http://prospector.ucsf.edu/mshome4.0.htm

http://prospector.ucsf.edu/ucsfhtml4.0/msfit.htm

http://www.expasy.org/tools/

H. Multiple Sequence Alignment and Phylogenetic Tree Construction

We perform multiple alignments of the deduced amino acid sequences by the computer program CLUSTAL_X (version 1.81; Higgins *et al.*, 1996; Thompson *et al.*, 1994, 1997) that is available for free download under ftp://ftp-igbmc.u-strasbg.fr/pub/ClustalX/. You will find online help under http://www-igbmc.u-strasbg.fr/BioInfo/ClustalX/Top.html. Then alignments are further edited by hand. For phylogenetic analyses of the deduced amino acid sequences, we use the program package PHYLIP (Felsenstein, 2000; actual version 3.6 b, free download and online help at http://evolution.genetics.washington.edu/phylip.html). Usually we calculate the distances between pairs of protein sequences according to the Jones Taylor Thornton substitution model (Jones *et al.*, 1992) and construct the trees by the Neighbor Joining method (Saitou and Nei, 1987). Then we test the reliability of the trees by bootstrap analysis (Felsenstein, 1985) with at least 100 replications (SEQBOOT program from the PHYLIP package). To display the calculated trees on a Windows PC, we use the TREEVIEW software program (Page, 1996; free download and online help at http://taxonomy.zoology.gla.ac.uk/rod/treeview.html). For a more profound understanding and a better statistical support of the evolutionary relationships of IF proteins, it might be relevant to perform further analysis with Maximum-Likelihood and/or Bayesian approaches. However, in case of the keratins, the trees we constructed so far by these more sophisticated methods did not differ significantly from those calculated by the simpler Neighbor Joining method.

In most cases, an accurate alignment of the head and tail domain of keratins is not possible if keratins from several evolutionary very distant species are compared. In such cases, we usually calculate their phylogenetic relationships on the basis of an alignment of their central helical rod domains. However, the head and tail sequences of keratins also contain motifs of significant phylogenetic information, and sometimes the presence or lack of such a motif may provide important hints for their evolution (e.g., see discussion for identification of K8 orthologs in different species). Interestingly, almost all type II keratins from lamprey to man show a highly conserved motif of approximately 20 amino acids within their head domains that directly precedes the central rod.

In phylogenetic trees derived from multiple sequence alignments, K18 candidates form a discrete branch that makes it easy to classify them as orthologs. In contrast, the various K8s branch off from their respective taxonomic group at a basal position (see also Schaffeld *et al.*, 1998, 2002a,b), probably because radiation of IIE keratins repeatedly started from K8-like progenitors. How are they then identified as orthologs? For this, we have previously proposed three criteria

(Schaffeld *et al.*, 2002b): (1) coexpression with K18 in simple epithelia and absence in keratinocytes; (2) basal (sister group) position in the phylogenetic tree compared with the other type II keratins of the same species; and (3) a typical motif "VxKxxETxDGxxVSESSxV" in the tail domain, (e.g., in trout:-NKKSVVIK-MIETKDGK-VVSESSEVVDD-). Although this motif exists in various other IF proteins, it is absent in all other type II keratins sequenced so far.

III. Discussion and Concluding Remarks

Our results as documented in Schaffeld *et al.* (1998, 2001a, 2002a,b, 2003, 2004) are examples for the successful experimental combination of 2D-PAGE, CKBB assay, immunoblotting, cDNA sequencing, and peptide mass fingerprinting by MALDI-TOF mass spectometry for the analysis of complex keratin systems. 2D-PAGE reveals, in the tissues studied, the entire range of cytoskeletal polypeptides that are possible keratin candidates. In CKBB assays, keratins are clearly identified (even more efficiently than by immunoblotting), and, moreover, they can be immediately classified into type I and type II polypeptides. Immunoblotting links the biochemically identified polypeptides to their expression patterns observed by immunofluorescence microscopy. And ultimately, peptide mass fingerprinting allows us to group the polypeptides separated by 2D-PAGE into clusters, each representing a genuine keratin and its modifications. With respect to cDNA sequencing, peptide mass fingerprinting not only provides identification of the sequenced protein but also confirms, with the theoretical cleavage products calculated from the sequence corresponding to the experimentally observed peptide pattern, that the derived protein sequence must be correct.

Our biochemical and immunohistological investigations of lamprey, trout, and shark have demonstrated that fishes express complex systems of keratins. Moreover, we were able to show that throughout all presently investigated vertebrates, including amphibians and mammals, keratins are consistently expressed in epithelial cells, and that "E" keratins expressed in the epidermis and other stratified epithelia can be distinguished from "S" keratins appearing in simple epithelia (review: Markl and Schechter, 1998). But are the keratins that we observe in the different fish species directly comparable to those expressed in frog or man? Our sequence analyses of keratins from lamprey, shark, trout, and zebrafish (Schaffeld *et al.*, 1998, 2002a,b, 2003, 2004; Schultess, 2001; Schultess *et al.*, 2001) clearly revealed that identification of keratin counterparts in evolutionary distant species is not achieved by simply comparing the new sequences to those already in the data banks, but requires a comprehensive phylogenetic analysis. This, in turn, calls for keratin sequences from a variety of phylogenetically distant species that we and others are attempting to provide; in this context, we try to improve the trees by analyzing the keratin complement of a given species as completely as possible. Of course, for this, an ultimate approach is to use the data from a genome project (Hesse *et al.*, 2001; Zimek *et al.*, 2003) if available.

The phylogenetic tree in Fig. 7 illustrates the range of keratin sequences known to date (Tables V–VII). This tree has been calculated from an alignment of most keratins completely sequenced so far. It also includes, for the first time, various sequences from the freshwater lamprey, the bichir, the sturgeon, and the lungfish

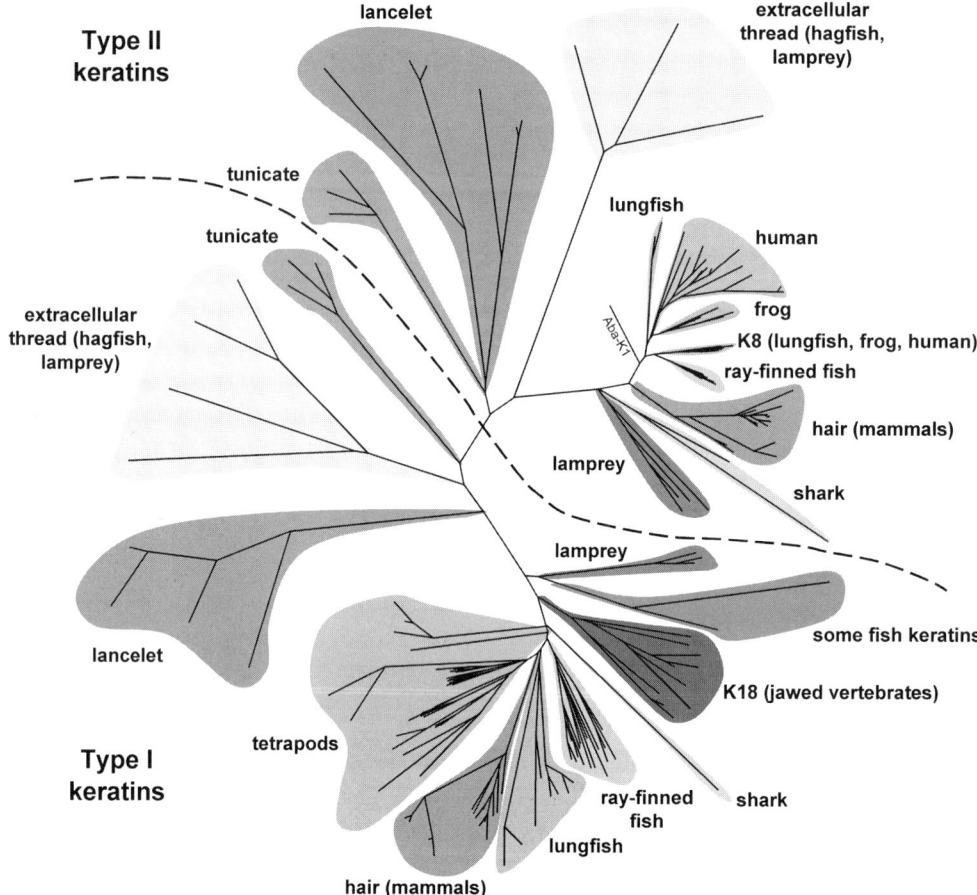

Fig. 7 Unrooted phylogenetic tree of most currently available chordate keratin sequences (see Tables V–VII) generated from multiple sequence alignment of their *rod domains* (we exclusively included sequences encoding at least the entire rod domain; partial sequences have been excluded from the alignment). We also included the *Branchiostoma* C sequences, which according to Karabinos *et al.* (2002) show keratin-like features (we discarded the X sequence, which showed a very long branch near the base of the E2/D2 origin). Distances were calculated according to the Jones Taylor Thornton matrix (Jones *et al.*, 1992), and the tree was constructed by the Neighbor Joining method. This tree shows that the keratins of lower chordates, as well as the Agnatha thread keratins, are distinct from the type I and II keratins found in vertebrates. For further discussion see text (Section III). (See Color Insert.)

Table V

European Molecular Biology Laboratory Accession Numbers of Most to Date Available Vertebrate Type I Keratin Sequences

Name	Accession	Species	Name	Accession	Species
Aba-K10	AJ493255	*Acipenser baeri*	Lfl-K18	AJ308118	*Lampetra fluviatilis*
Aba-K11	AJ493256		Lfl-K!1	AJ289860	
Aba-K12	AJ493257		Lfl-K!2	AJ308111	
Aba-K13	AJ493258		Lfl-K!3	AJ308112	
Aba-K14	AJ493259		Mmu-K10	L00193	*Mus musculus*
Aba-K15	AJ493260		Mmu-K12	U02880	
Aba-K18	AJ493261		Mmu-K13	U13921	
Bta-K6B	X02870	*Bos taurus*	Mmu-K14	BC011074	
Bta-K19	X04152		Mmu-K15	D16313	
Cau-K18	L09744	*Carassius auratus*	Mmu-K16	AF053235	
Cau-K49	L09743		Mmu-K17	AB013608	
Cau-K50	M86918		Mmu-K18	M22832	
Cfa-K9	AF000949	*Canis familiaris*	Mmu-K19	M28698	
Dre-K18	AJ493269		Mmu-K20	BC057172	
Dre-cyt1	AF084461		Mmu-Kc29	AB013607	
Dre-zfCKI	AF197880		Mmu-KRik473	AK014642	
Est-Kγ	U20546	*Eptatretus stoutii*	Mmu-K1M1	M27734	
Gga-K19	AB016281	*Gallus gallus*	Mmu-KHA2	X75649	
Ggo-K_GhaA	AJ401055	*Gorilla gorilla*	Nvi-K17	AY008292	
Hsa-K9	X75015	*Homo sapiens*	Oar-K1M1_8c1	AF227758	*Ovis aries*
Hsa-K10	BC034697		Oar-K1M2	M23912	
Hsa-K12	D78367		Ocu-K12	X77665	*Oryctolagus cuniculus*
Hsa-K13	X14640		Omy-K10	AJ272372	*Oncorhynchus mykiss*
Hsa-K14	BC042437		Omy-K11	AJ272371	
Hsa-K15	X07696		Omy-K12	AJ427868	
Hsa-K16	AF061812		Omy-K13	AJ427867	
Hsa-K17	Z19574		Omy-K18	Y14289	
Hsa-K18	X12881		Ptr-K19	X82579	*Potorous tridactylus*
Hsa-K19	Y00503		Ptro-K_GHaA	AJ401054	*Pan troglodytes*
Has-K20	X73501		Rno-K21	M63665	*Rattus norvegicus*
Hsa-K_HA1	Y16787		Sst-K10	AJ623268	*Scyliorhinus stellaris*
Hsa-K_HA2	X90761		Sst-K18	Y14647	
Hsa-K_HA3b	Y16789		Xla-K18	BC054993	*Xenopus laevis*
Hsa-K_HA4	Y16790		Xla-K1C0	Y00968	
Hsa-K_HA5	Y16791		Xla-K1C1	X04804	
Hsa-K_HA6	Y16792		Xla-K1C4	M18155/	
Hsa-K_HA7	Y16793			X04807	
Hsa-K_HA8	Y16793		Xla-K1C5	M11032	
Lfl-K10	AJ308116	*Lampetra fluviatilis*	Xla-KXAKa	AB045600	
Lfl-K11	AJ308117		Xla-KXAKb	AB045601	

Note that data from EST and GENOME projects are mostly not included. The sequences of bichir and lungfish that stem from our laboratory are not yet submitted to the EMBL databank. Hair keratin sequences are shadowed in grey. We also included the agnathan thread keratins alpha and gamma.

Table VI

European Molecular Biology Laboratory Accession Numbers of Most to Date Available Vertebrate Type II Keratin Sequences

Name	Accession	Species	Name	Accession	Species
Aba-K1	AJ493262	*Acipenser baeri*	Mmu-K1	M10937	*Mus musculus*
Aba-K2	AJ493263		Mmu-K5	AF306785	
Cau-K8	P18520*	*Carassius auratus*	Mmu-K6	K02108	
Cfa-K2	AF293846	*Canis familiaris*	Mmu-K8	D90360	
Dre-K1	AF197909	*Danio rerio*	Mmu-K_HB	AY028606	
Dre-K4	AF134850		Mmu-K_HB2	AK009099	
Dre-K8	AJ460000		Mmu-K_HB_f	M92088	
Dre-zf-K8	—**		Mmu-K_HB_k	AY028607	
Est-Kα	U11865	*Eptatretus stoutii*	Nvi-KII	AJ001295	*Notophthalmus viridescens*
Gga-K_oto	AF072698	*Gallus gallus*	Nvi-K8	AJ001296	
Hsa-K1	AF237621	*Homo sapiens*	Oar-K_HB	X72379	*Ovis aries*
Hsa-K2	AF019084		Oar-K_HB5	P25691*	
Hsa-K3	P12035*		Oar-K_HB7	P15241*	
Hsa-K4	AY043326		Oar-K_HB9	X62509	
Hsa-K5	AF274874		Ocu-K3	X74371	*Oryctolagus cuniculus*
Hsa-K6a	BC008807		Omy-K1	AJ272369	*Oncorhynchus mykiss*
Hsa-K7	BC002700		Omy-K2	AJ272370	
Hsa-K8	BC000654		Omy-K3	AJ315933	
Hsa-K_HB1	Y13621		Omy-K8a	AJ272373	
Hsa-K_HB2	Y19207		Omy-K8b	X92522	
Hsa-K_HB3	Y19208		Ptr-K8	X70987	*Potorous tridactylus*
Hsa-K_HB4	Y19209		Rca-K8	AB056480	*Rana catesbeiana*
Hsa-K_HB5	Y19210		Rca-K_ie	AF229168	
Hsa-K_HB6	AJ000263		Rca-K_rlk	AB050956	
Lfl-K1	AJ401159	*Lampetra fluviatilis*	Rno-K8	M63482	*Rattus norvegicus*
Lfl-K2	AJ308115		Sst-K1	Y17257	*Scyliorhinus stellaris*
Lfl-K3	AJ308113		Sst-K8	Y14648	
Lfl-K8	AJ308114		Xla-K2C1	X02894	*Xenopus laevis*
Lfl-Kα1	AJ289859		Xla-K2C2	X02895	
Lfl-Kα2	AJ308120		Xla-K55	X14427	
Lfl-Kα3	AJ308112		Xla-K8	M13811	
Mmu-KII	AK019521	*Mus musculus*	Xla-K_xlk	AB045599	

Note that data from EST and GENOME projects are mostly not included. The sequences of bichir and lungfigh that stem from our laboratory are not yet submitted to the EMBL databank. Hair keratin sequences are shadowed in grey.We also included the agnathan thread keratins alpha and gamma.

—** Incomplete amino acid sequence published by Imboden *et al.* (1997) that is not available as a EMBL or GenBank submission.

Table VII
European Molecular Biology Laboratory Accession Numbers of Lower Chordate Keratin Sequences

Type I			Type II		
Name	Accession	Species	Name	Accession	Species
Bfl-IF1	AF108192	*Branchiostoma floridae*	Bfl-C1	AJ223577	*Branchiostoma floridae*
Bfl-K1	AJ245432		Bfl-C2	AJ245429	
			Bfl-D1	AJ223581	
			Bfl-E2	AJ245431	
Bla-E1	AJ010294	*Branchiostoma lanceolatum*	Bla-C2	AJ223578	*Branchiostoma lanceolatum*
Bla-K1	AJ245426		Bla-D1	AJ223579	
Bla-X1	AJ245427		Bla-E2	AJ010293	
Bla-Y1	AJ245428				
Cin-D	AJ298331	*Ciona intestinalis*	Cin-C	AJ298332	*Ciona intestinalis*
Moc-D	AJ417905	*Molgula oculata*	Moc-C	AJ417906	*Molgula oculata*
Scl-D	AJ271145	*Styela clava*	Scl-C	AJ271146	*Styela clava*

We also included the *Branchiostoma* C and X sequences, because they show keratin-like features and in phylogenetic trees are grouped with *Branchiostoma* keratins D1 and E2.

that we obtained recently (Schaffeld *et al.*, 2001b; Schultess *et al.*, 2001). Our previously published trees strongly suggest that all the "E" keratins so far sequenced, plus those "S" keratins that are neither K8 nor K18 orthologs, diversified independently in elasmobranchs, teleosts, and tetrapods, respectively, whereas true K8 and K18 orthologs are present in all studied jawed vertebrates or Gnathostomata (Schaffeld *et al.*, 1998, 2002a,b, 2003, 2004). The present tree shows, in addition, that the keratins from the lamprey (a member of the ancient Agnatha) also originate from an independent radiation event. In this case, orthologs of K8 and K18 are not immediately obvious, but putative candidates have also been identified. Separate branches are formed by IF proteins from the lancelet *Branchiostoma* (a cephalochordate) and the tunicate *Ciona* (an urochordate), functionally identified as type I and type II keratins (Karabinos *et al.*, 1998, 2000, 2002; Wang *et al.*, 2000), by mammalian hair keratins and by so-called thread keratins. The latter have first been described in hagfish (Downing *et al.*, 1984; Koch *et al.*, 1994, 1995; Spitzer *et al.*, 1984, 1988), but recently also in lamprey (Schultess, 2001; Schultess *et al.*, 2001) and are components of the slime secreted from skin glands (not shown for lamprey yet). The major vertebrate groups have very different skin types, and their evolution might have stimulated radiation of all these different subsets of "E" keratins.

Because nothing in biology makes sense but in the light of evolution, we are convinced that we need detailed knowledge of the keratin system in diverse species to ultimately understand the biological functions of these prominent cytoskeletal elements in man. Most of the methods described in this chapter have been obtained by us from Werner Franke's laboratory in Heidelberg and have been routinely used and adapted to fish keratins in our laboratory for the past 10 years. Rather than studying specific aspects of IFs by single techniques, we prefer a broad survey of the keratin complement of a given fish species by use of the whole range of methods described here. This not only allows us to trace the evolution of the keratin multigene family and the Vertebrata, respectively, but also provides a firm platform from which more specific details such as developmental expression patterns, expression control, associated proteins, and structure-function relationships of keratins can be investigated.

Acknowledgments

We thank Robin Harris for careful proofreading of the manuscript, Miriam Bremer for providing the data for Tables V–VII and Fig. 1, and Jens Limbarth for the data shown in Fig. 3.

References

Achtstätter, T., Hatzfeld, M., Quinlan, R. A., Parmelee, D. C., and Franke, W. W. (1986). Separation of cytokeratin polypeptides by gel electrophoretic and chromatographic techniques and their identification by immunoblotting. *Methods Enzymol.* **134**, 355–371.

Cerda, J., Conrad, M., Markl, J., Brand, M., and Herrmann, H. (1998). Zebrafish vimentin: molecular characterization, assembly properties and developmental expression. *Eur. J. Cell Biol.* **77**, 1–13.

Chirgwin, J. M., Przybyla, A. E., MacDonald, R. J., and Rutter, W. J. (1979). Isolation of biologically active ribonucleic acid from sources enriched in ribonuclease. *Biochemistry* **18**, 5294–5299.

Chua, K. L., and Lim, T. M. (2000). Type I and type II cytokeratin cDNAs from the zebrafish (*Danio rerio*) and expression patterns during early development. *Differentiation* **66**, 31–41.

Conrad, M., Lemb, K., Schubert, T., and Markl, J. (1998). Biochemical identification and tissue-specific expression patterns of keratins in the zebrafish. *Danio rerio. Cell Tissue Res.* **293**, 195–205.

Downing, S. W., Spitzer, R. H., Koch, E. A., and Salo, W. L. (1984). The hagfish slime gland thread cell. 1. A unique cellular system for the study of intermediate filaments and intermediate filament-microtubule interactions. *J. Cell Biol.* **98**, 653–669.

Druger, R. K., Glasgow, E., Fuchs, C., Levine, E. M., Matthews, J. P., Park, C. Y., and Schechter, N. (1994). Complex expression of keratins in goldfish optic nerve. *J. Comp. Neurol.* **340**, 269–280.

Druger, R. K., Levine, E. M., Glasgow, E., Jones, P. S., and Schechter, N. (1992). Cloning of a type I keratin from goldfish optic nerve: Differential expression of keratins during regeneration. *Differentiation* **52**, 33–43.

Erber, A., Riemer, D., Bovenschulte, M., and Weber, K. (1998). Molecular phylogeny of metazoan intermediate filament proteins. *J. Mol. Evol.* **47**, 751–762.

Felsenstein, J. (1985). Confidence limits on phylogenies: An approach using the bootstrap. *Evolution* **39**, 783–791.

Felsenstein, J. (2000). PHYLIP (phylogeny inference package). Version 3.6 alpha. Distributed by the author, Department of Genetics, University of Washington, Seattle.

Frail, D. E., Mudd, J., and Merlie, J. P. (1990). Nucleotide sequence of an intermediate filament cDNA from *Torpedo californica. Nucleic Acids Res.* **8**, 7.

Franke, W. W., Schiller, D. L., Hatzfeld, M., and Winter, S. (1983). Protein complexes of intermediate-sized filaments. *Proc. Natl. Acad. Sci. USA* **80**, 7113–7117.

Fuchs, C., Druger, R. K., Glasgow, E., and Schechter, N. (1994). Differential expression of keratins in goldfish optic nerve during regeneration. *J. Comp. Neurol.* **343**, 332–340.

Giordano, S., Glasgow, E., Tesser, P., and Schechter, N. (1989). A type II keratin is the major intermediate filament protein expressed in glial cells of the goldfish visual pathway: Molecular cloning and sequence analysis. *Neuron* **2**, 1507–1516.

Giordano, S., Hall, C., Quitschke, W., Glasgow, E., and Schechter, N. (1990). Keratin 8 of simple epithelia is expressed in glia of the goldfish nervous system. *Differentiation* **44**, 163–172.

Gong, Z., Ju, B., Wang, X., He, J., Wan, H., Sudha, P. M., and Yan, T. (2002). Green fluorescent protein expression in germ-line transmitted transgenic zebrafish under a stratified epithelial promoter from Keratin8. *Dev. Dyn.* **223**, 204–215.

Groff, J. M., Nayan, D. K., and Higgins, R. J. (1997). Cytokeratin-filament expression in epithelial and non-epithelial tissues of the common carp (*Cyprinus carpio*). *Cell Tissue Res.* **287**, 375–384.

Herrmann, H., Eckelt, A., Brettel, M., Grund, C., and Franke, W. W. (1993). Temperature-sensitive intermediate filament assembly: Alternative structures of *Xenopus laevis* vimentin *in vitro* and *in vivo*. *J. Mol. Biol.* **234**, 99–113.

Herrmann, H., Fouquet, B., and Franke, W. W. (1989). Expression of intermediate filament proteins during development of *Xenopus laevis*. I. cDNA clones encoding different forms of vimentin. *Development* **105**, 279–298.

Herrmann, H., Häner, M., Brettel, M., Müller, S., Goldie, K., Fedtke, B., Lustig, A., Franke, W. W., and Aebi, U. (1996a). Structure and assembly properties of the intermediate filament protein vimentin: The role of its head, rod and tail domains. *J. Mol. Biol.* **264**, 933–953.

Herrmann, H., Münick, M. D., Brettel, M., Fouquet, B., and Markl, J. (1996b). Vimentin in a cold-water fish, the rainbow trout: Highly conserved primary structure but unique assembly properties. *J. Cell Sci.* **109**, 569–578.

Herrmann, H., and Wiche, G. (1987). Plectin and IFAP-300K are homologous proteins binding to microtubule-associated proteins 1 and 2 and to the 240-kilodalton subunit of spectrin. *J. Biol. Chem.* **262**, 1320–1325.

Hesse, M., Magin, T. M., and Weber, K. (2001). Genes for intermediate filament proteins and the draft sequence of the human genome: Novel keratin genes and a surprisingly high number of pseudogenes related to keratin genes 8 and 18. *J. Cell Sci.* **114**, 2569–2575.

Higgins, D. G., Thompson, J. D., and Gibson, T. J. (1996). Using CLUSTAL for multiple sequence alignments. *Methods Enzymol.* **266**, 383–402.

Imboden, M., Goblet, C., Korn, H., and Vriz, S. (1997). Cytokeratin 8 is a suitable epidermal marker during zebrafish development. *C. R. Acad. Sci. III* **320**, 689–700.

Jones, D. T., Taylor, W. R., and Thornton, J. M. (1992). Therapid generation of mutation data matrices from protein sequences. *Comput. Appl. Biosci.* **8**, 275–282.

Ju, B., Xu, Y., He, J., Liao, J., Yan, T., Hew, C. L., Lam, T. J., and Gong, Z. (1999). Faithful expression of green fluorescent protein (GFP) in transgenic zebrafish embryos under control of zebrafish gene promoters. *Dev. Genet.* **25**, 158–167.

Karabinos, A., Riemer, D., Erber, A., and Weber, K. (1998). Homologues of vertebrate type I, II and II intermediate filament (IF) proteins in an invertebrate: The IF multigene family of the cephalochordate *Branchiostoma*. *FEBS Letters* **437**, 15–18.

Karabinos, A., Riemer, D., Panopoulou, G., Lehrach, H., and Weber, K. (2000). Characterisation and tissue-specific expression of the two keratin subfamilies of intermediate filament proteins in the cephalochordate Branchiostoma. *Eur. J. Cell. Biol.* **79**, 17–26.

Karabinos, A., Schunemann, J., Parry, D. A., and Weber, K. (2002). Tissue-specific co-expression and *in vitro* heteropolymer formation of the two small *Branchiostoma* intermediate filament proteins A3 and B2. *J. Mol. Biol.* **316**, 127–137.

Karabinos, A., Zimek, A., and Weber, K. (2004). The genome of the early chordate *Ciona intestinalis* encodes only five cytoplasmic intermediate filament proteins including a single type I and type II keratin and a unique IF-annexin fusion protein. *Genetics* **326**, 123–129.

Kirfel, J., Magin, T. M., and Reichelt, J. (2003). Keratins: A structural scaffold with emerging functions. *Cell Mol. Life Sci.* **60**, 56–71.

Koch, E. A., Spitzer, R. H., Pithawalla, R. B., Castillos, F. A., and Parry, D. A. D. (1995). Hagfish biopolymer: A type I/type II homologue of epidermal keratin filaments. *Int. J. Biol. Macromol.* **17**, 283–292.

Koch, E. A., Spitzer, R. H., Pithawalla, R. B., and Parry, D. A. D. (1994). An unusual filament subunit from the cytoskeletal biopolymer released extracellularly into seawater by the primitve hagfish (*Eptatretus stouti*). *J. Cell. Sci.* **107,** 3133–3140.

Laemmli, U. K. (1970). Cleavage of structural proteins during the assembly of the head of bacteriophage T4. *Nature* **277,** 680–685.

Lazarides, E. (1982). Intermediate filaments: A chemically heterogeneous, developmentally regulated class of proteins. *Annu. Rev. Biochem.* **51,** 219–250.

Luke, G. N., and Holland, P. W. (1999). Amphioxus type I keratin cDNA and the evolution of intermediate filament genes. *J. Exp. Zool.* **285,** 50–56.

Markl, J., and Franke, W. W. (1988). Localization of cytokeratins in tissues of the rainbow trout: Fundamental differences in expression pattern between fish and higher vertebrates. *Differentiation* **39,** 97–122.

Markl, J., and Schechter, N. (1998). Fish intermediate filament proteins in structure, function and evolution. *In* "Intermediate Filaments" (H. Herrmann and J. R. Harris, eds.), Vol. 31, pp. 1–33. Plenum Press, New York.

Markl, J., Winter, S., and Franke, W. W. (1989). The catalog and the expression complexity of cytokeratins in a teleost fish, the rainbow trout. *Eur. J. Cell. Biol.* **50,** 1–16.

McLachlin, D. T., and Chait, B. T. (2001). Analysis of phosphorylated proteins and peptides by mass spectrometry. *Curr. Opin. Chem. Biol.* **5,** 591–602.

Moertz, E., Krogh, T. N., Vorum, H., and Gorg, A. (2001). Improved silver staining protocols for high sensitivity protein identification using matrix-assisted laser desorption/ionization-time of flight analysis. *Proteomics* **1,** 1359–1363.

O'Farrell, P. H. (1975). High resolution of two dimensional electrophoresis. *J. Biol. Chem.* **250,** 4007–4021.

O'Farrell, P. Z., Googman, H. M., and O'Farrell, P. H. (1977). High resolution two dimensional gelelectrophoresis of basic as well as acidic proteins. *Cell* **12,** 1133–1142.

Paramio, J. M., and Jorcano, J. L. (2002). Beyond structure: do intermediate filaments modulate cell signalling? *Bioessays* **24,** 836–844.

Pruss, R. M., Mirsky, R., Raff, M. C., Thorpe, R., Dowding, A. J., and Anderton, B. H. (1981). All classes of intermediate filaments share a common antigenic determinant defined by a monoclonal antibody. *Cell* **27,** 419–428.

Riemer, D., Karabinos, A., and Weber, K. (1998). Analysis of eight cDNAs and six genes for intermediate filament (IF) proteins in the cephalochordate *Branchiostoma* reveals differences in the IF multigene families of lower chordates and the vertebrates. *Gene* **211,** 361–373.

Riemer, D., and Weber, K. (1998). Common and variant properties of intermediate filament proteins from lower chordates and vertebrates; two proteins from the tunicate *Styela* and the identification of a type III homologue. *J. Cell. Sci.* **111,** 2967–2675.

Saitou, N., and Nei, M. (1987). The neighbor-joining method: A new method for reconstructing phylogenetic trees. *Mol. Biol. Evol.* **6,** 1406–1425.

Sambrook, J., Fritsch, E. F., and Maniatis, T. (1989). "Molecular Cloning: A Laboratory Handbook" Cold Spring Harbor Laboratory Press, Cold Spring Harbor, New York.

Sambrook, J., and Russell, D. W. (2001). "Molecular Cloning: A Laboratory Manual." Cold Spring Harbor Laboratory Press, Cold Spring Harbor, New York.

Schaffeld, M., Löbbecke, A., Lieb, B., and Markl, J. (1998). Tracing keratin evolution: Catalog, expression patterns and primary structure of shark (*Scyliorhinus stellaris*) keratins. *Eur. J. Cell Biol.* **77,** 69–80.

Schaffeld, M., Herrmann, H., Schultess, J., and Markl, J. (2001a). Vimentin and desmin of a cartilaginous fish, the shark *Scyliorhinus stellaris*: Sequence, expression patterns and *in vitro* assembly. *Eur. J. Cell Biol.* **80,** 692–702.

Schaffeld, M., Schultess, J., Haberkamp, M., Bremer, M., and Markl, J. (2001b). Intermediate filament protein evolution in fish: Sequences from lamprey, shark, bichir, sturgeon and trout. *First Joint French-German Congress on Cell Biology, Biol. Cell* **93,** 235.

Schaffeld, M., Höffling, S., Haberkamp, M., Conrad, M., and Markl, J. (2002a). Type I keratin cDNAs from the rainbow trout: Independent radiation of keratins in fish. *Differentiation* **70,** 292–299.

Schaffeld, M., Haberkamp, M., Braziulis, E., Lieb, B., and Markl, J. (2002b). Type II keratin cDNAs from the rainbow trout: Implications for keratin evolution. *Differentiation* **70,** 282–291.

Schaffeld, M., Knappe, M., Hunzinger, C., and Markl, J. (2003). cDNA sequences of the authentic keratins 8 and 18 in zebrafish. *Differentiation* **71,** 73–82.

Schaffeld, M., Höffling, S., and Markl, J. (2004). Differential expression of "E" and "S" keratins in the shark *Scyliorhinus stellaris,* and cDNA sequence of a novel epidermal type I keratin. *Eur. J. Cell Biol.* **83,** 359–368.

Schultess, J. (2001). Molekulare Evolution der Intermediärfilament-Proteine des Flussneunauges *Lampetra fluviatilis,* pp. 1–149. Thesis. Faculty of Biology, University of Mainz, Mainz, Germany.

Schultess, J., Schaffeld, M., and Markl, J. (2001). Intermediate filament protein sequences of the cyclostome *Lampetra fluviatilis. First Joint French-German Congress on Cell Biology, Biol. Cell* **93,** 235.

Shevchenko, A., Wilm, M., Vorm, O., and Mann, M. (1996). Mass spectrometric sequencing of proteins from silver-stained polyacrylamide gels. *Anal. Chem.* **68,** 850–858.

Spitzer, R. H., Downing, S. W., Koch, E. A., Salo, W. L., and Saidel, L. J. (1984). Hagfish slime gland thread cells. II. Isolation and characterization of intermediate filament components associated with the thread. *J. Cell Biol.* **98,** 670–677.

Spitzer, R. H., Koch, E. A., and Downing, S. W. (1988). Maturation of hagfish gland thread cells: composition and characterization of intermediate filament polypeptides. *Cell Motil. Cytoskeleton* **11,** 31–45.

Thompson, J. D., Higgins, D. G., and Gibson, T. J. (1994). CLUSTAL W: Improving the sensivity of progressive multiple sequence alignment through sequence weighting, position-specific gap penalties and weight matrix choice. *Nucleic Acids Res.* **22,** 4673–4680.

Thompson, J. D., Gibson, T. J., Plewniak, F., Jeanmougin, F., and Higgins, D. G. (1997). The CLUSTAL_X windows interface: Flexible strategies for multiple sequence alignment aided by quality analysis tools. *Nucleic Acids Res.* **25,** 4876–4882.

Wang, J., Karabinos, A., Schunemann, J., Riemer, D., and Weber, K. (2000). The epidermal intermediate filament proteins of tunicates are distant keratins; A polymerisation-competent hetero coiled coil of the *Styela* D protein and *Xenopus* keratin 8. *Eur. J. Cell Biol.* **79,** 478–487.

Zimek, A., Stick, R., and Weber, K. (2003). Genes coding for intermediate filament proteins: Common features and unexpected differences in the genomes of humans and the teleost fish *Fugu rubripes. J. Cell Sci.* **116,** 2295–2302.

CHAPTER 23

Performing Functional Studies of *Xenopus laevis* Intermediate Filament Proteins Through Injection of Macromolecules into Early Embryos

Christine Gervasi and Ben G. Szaro

Department of Biological Sciences and the Center for Neuroscience Research
University at Albany, State University of New York
Albany, New York 12222

METHODS IN CELL BIOLOGY, VOL. 78

I. Introduction

A. Intermediate Filament Heterogeneity Suggests Complex Functions

One of the great puzzles of intermediate filaments (IFs) is that despite their status as the most abundant cytoskeletal component, their precise function has been elusive; because unlike microfilaments and microtubules, IFs are mostly dispensable for the basic cellular processes that are essential for life. Arguably, this very dispensability may have freed IFs to evolve into the most diverse family of cytoskeletal proteins. Analysis of the initial draft of the human genome reported 127 IF gene candidates (Lander *et al.*, 2001), 65 of which are considered functional (Hesse *et al.*, 2001). The expression of each member of this family is remarkably cell-type specific and highly regulated during development, further suggesting that IFs subserve a variety of functions supporting the diverse morphologies and subtle behaviors that define unique cell types.

In neurons, IF composition evolves with each successive phase of axon outgrowth; in representative organisms of all the major vertebrate classes (e.g., *Xenopus* [Gervasi *et al.*, 2000; Szaro *et al.*, 1989], zebrafish [Asch *et al.*, 1998; Canger *et al.*, 1998; Leake *et al.*, 1999], chick [Bennett *et al.*, 1988; Tapscott *et al.*, 1981], and rodents [Fliegner *et al.*, 1994; Pachter and Liem, 1984; Shaw and Weber, 1982; Shaw *et al.*, 1985]), these expression patterns are largely conserved. As axons develop, the composition of neuronal IFs (nIFs) evolves from one that resembles other cell types to one that is uniquely neuronal. At axon initiation, type III nIFs predominate. In the mammalian central nervous system (CNS), this protein is vimentin (Boyne *et al.*, 1996; Cochard and Paulin, 1984; Shea *et al.*, 1993), whereas in *Xenopus* CNS, it is peripherin (Gervasi *et al.*, 2000; Goldstone and Sharpe, 1998; Undamatla and Szaro, 2001); both of these IFs are found in nonneuronal cell types as well. The α-internexin–like nIF proteins and NF-M emerge later during axon elongation (Charnas *et al.*, 1992; Fliegner *et al.*, 1994; Lin and Szaro, 1994; Szaro *et al.*, 1989; Zhao and Szaro, 1997); in *Xenopus* spinal cord, NF-L and NF-H expression are both delayed until after axons reach their targets (Charnas *et al.*, 1992; Lin and Szaro, *et al.*, 1994; Szaro *et al.*, 1989). In all vertebrates, the levels of NF-L, NF-M, and NF-H expression and the phosphorylation of NF-M and NF-H tail domains increase still further after synaptogenesis (Carden *et al.*, 1987; de Waegh *et al.*, 1992; Sánchez *et al.*, 1996; Schwartz *et al.*, 1990, 1994; Szaro *et al.*, 1989; Undamatla and Szaro, 2001). Conversely, nIF head domain phosphorylation, which destabilizes nIFs and stimulates subunit exchange (Giasson and Mushynski, 1998; Inagaki *et al.*, 1996), increases at axonal branch points, implicating nIF destabilization in collateral branching (Landmesser and Swain, 1992).

These developmentally regulated changes must be linked in some way to axon outgrowth, because they are recapitulated in successfully regenerating axons (Asch *et al.*, 1998; Glasgow *et al.*, 1994; Goldstein *et al.*, 1987, 1988; Wong and Oblinger, 1990; Zhao and Szaro, 1994), are abnormal in injured axons that cannot regenerate (Jiang *et al.*, 1994; Mikucki and Oblinger, 1991), and depend on external cues encountered by regrowing axons (Liuzzi and Tedeschi, 1992; Zhao and Szaro, 1995, 1997). Although this remarkably well-conserved, highly orchestrated procession of nIF protein expression during axon outgrowth argues strongly that individual nIFs play important roles in axon development, their precise contributions remain unclear.

Studies of the function of other IFs suggest some possibilities. Genetic studies relating IFs to skin diseases (Fuchs and Cleveland, 1998) have demonstrated that they are essential for maintaining the mechanical integrity of cells, forming an integrated network with microtubules and microfilaments through IF-associated proteins. Thus, because IFs are the major mechanical stabilizers of the cytoplasm (Chou *et al.*, 2001; Goldman *et al.*, 1996, 1999), the nIFs might provide structural support to the growing axon, which must maintain enough plasticity to permit growth while simultaneously consolidating structure and branching. IFs also provide a stable scaffolding for localizing other cellular proteins and organelles, which in turn enables these components to function efficiently (Faigle *et al.*, 2000).

Studies in transgenic mice have now demonstrated examples of both classes of function for nIF proteins, at least in adult mice. For example, nIFs are essential for the expansion of axon caliber, which occurs during the maturation of myelinated axons, after synaptogenesis (Eyer and Peterson, 1994; Hoffman *et al.*, 1987; Sakaguchi *et al.*, 1993; Williamson *et al.*, 1996; Zhao *et al.*, 1995). This expansion is attributed to an increase in the expression of NF-L and NF-M (Garcia *et al.*, 2003; Rao *et al.*, 2002, 2003). NF-M is also essential to localize dopamine D1 receptors properly, which affects desensitization (Kim *et al.*, 2002).

Although transgenic mice have been unequivocally successful as a model for studying the role nIFs play in the adult nervous system and in neuropathologies, they have unfortunately revealed little about their roles earlier in development (Beaulieu *et al.*, 2000; Elder *et al.*, 1998a,b; Eyer and Peterson, 1994; Jacomy *et al.*, 1999; Levavasseur *et al.*, 1999; Rao *et al.*, 1998; Zhu *et al.*, 1998). Mice lacking the various nIF subunits, either individually or in combination, are typically born with only relatively mild phenotypes, with the most significant being a 34% loss of sensory neurons in peripherin knockouts (Lariviere *et al.*, 2002).

Despite these relatively mild phenotypes at birth, closer examination suggests that nIFs do influence earlier development. For example, loss of NF-L in mice and birds reduces axon numbers at birth by as much as 20% (Julien, 1999; Yamasaki *et al.*, 1991) and slows rates of peripheral nerve regeneration (Jiang *et al.*, 1996; Zhu *et al.*, 1997). Moreover, birds lacking NF-L are born with tremors. Because nearly all the observations on transgenic mice to date were made postnatally, deficits arising *in utero* would be hidden from view and may be compensated for by the time of birth. Moreover, because nIFs are not essential for life, their functions are likely to be subtle and thus require more direct observation for detection. Such subtlety of function has been observed for other IFs. For example, although initially no phenotype was observed for vimentin knockout mice (Colucci-Guyon *et al.*, 1994), subsequent studies of vimentin null fibroblasts in culture revealed deficits in cell motility (Eckes *et al.*, 1998). Whereas such subtle defects may not affect survival of an organism in a controlled laboratory environment, they may nonetheless influence its success in its natural environment. The high degree of conservation of nIF protein expression patterns during development suggests that they play important roles in mediating processes critical for producing a competitive organism in the wild. Observing axon development directly may therefore better reveal these functions. The South African claw-toed frog, *Xenopus laevis*, is ideally suited for such studies.

B. *Xenopus laevis* as a Vertebrate Model for Studying Intermediate Filament Involvement in Development

The power of simple systems to study IF function is now well demonstrated for invertebrates, such as *C. elegans* (Karabinos *et al.*, 2001). To study IF function in vertebrate neural development, our laboratory has been using *Xenopus laevis* embryos. Over the past 60 years, *Xenopus* has become a standard laboratory organism for studies in cell, molecular, and developmental biology. Four characteristics have helped make this happen (Gurdon, 1996). (1) Its lifestyle is permanently aquatic, making it possible to maintain *Xenopus* in small aquaria. (2) It is highly resistant to disease, most likely because its natural habitat in Africa is small stagnant ponds. (3) Hundreds of fertile embryos can be obtained on demand, either through artificial fertilization, or through natural spawnings stimulated by injecting human chorionic gonadotropin. (4) Under proper rearing conditions, *Xenopus laevis* can be raised from eggs to sexually mature adults in about a year, which, although too long for genetic experiments, is short enough for easily maintaining viable stocks. More recently, another species of *Xenopus*, *Xenopus tropicalis* (formerly called *Silurana tropicalis*), has been used to complement *X. laevis* for genetic experiments, chiefly because of its ancestrally diploid genome (the *X. laevis* genome duplicated some 35–40 million years ago) and because it can be reared to adulthood in as little as 3 to 4 months.

Indeed, the small size of *X. laevis* embryos (1 mm in diameter) and their external development make it easy to study early developmental stages in this system. Moreover, during the first few days of development, which encompass developmental stages spanning from cleavage through early swimming tadpoles, IF compositions may be manipulated through the injection of antibodies, RNAs, DNAs, and, most recently, antisense morpholino oligonucleotides into *Xenopus* embryonic blastomeres. Because these molecules, for the most part, remain intact and undiluted within descendants of the injected blastomere (Jacobson and Hirose, 1978; Vize *et al.*, 1991), injections at the two-cell stage yield animals with one side serving as an internal control for the other. Thus, directly comparing the two sides permits the characterization of subtle effects of altering IF composition between developing sister cells. Moreover, well-defined methods of culturing newly differentiating *Xenopus* spinal cord neurons (Tabti and Poo, 1991) and performing immunocytochemistry (Dent *et al.*, 1989) and *in situ* hybridization on embryos (Shain and Zuber, 1996) in whole mounts makes it easier to assay the effects of altering IF composition both in culture and in the intact animal.

The history of injecting macromolecules into *Xenopus* blastomeres to study development began with the injection of horseradish peroxidase to study cell lineage (Hirose and Jacobson, 1979; Jacobson and Hirose, 1978), a method that was pioneered in the leech (Weisblat *et al.*, 1978). Fortuitously, horseradish peroxidase activity remains in descendants of the injected blastomeres for several days, through early swimming tadpole stages. Other compounds, such as fluorescent dextrans, were soon found to be equally effective as lineage tracers and provided the additional advantage that they can be visualized in living embryos (Gimlich and Gerhart, 1984). Investigators then began experimenting with injecting other macromolecules, not only to trace cell lineages but also to alter gene function. Taking a cue from studies that used intracellularly injected antibodies to disrupt cytokeratins in epithelial cells (Klymkowsky *et al.*, 1983), Warner *et al.* (1984) were the first to inject antibodies into *Xenopus* embryos, in their case, to disrupt gap junctions. We have injected antibodies against *Xenopus* NF-M to disrupt nIFs during the first few days of neural development. As described more extensively in the "Discussion", these antibodies blocked transport of type IV nIFs into the axon and altered axonal growth dynamics (Lin and Szaro, 1995; Szaro *et al.*, 1991; Walker *et al.*, 2001). Klymkowsky *et al.* (1992) microinjected antibodies to cytokeratins to disrupt the deep cytokeratin network, which inhibited gastrulation movements.

Conversely, *in vitro* transcribed mRNAs may be injected to effectively increase expression of molecules during development, as first described for N-CAM (Kintner, 1988). This technique complements the ability to disrupt gene function with the ability to augment gene expression. Injected mRNAs have also been used to express truncated mRNAs encoding dominant negative forms of proteins to disrupt the function of various genes, including vimentin (Christian *et al.*, 1990). We have also used this method to disrupt nIFs during development (Lin and Szaro, 1996), which complemented our studies with antibodies. However, the use of dominant negatives to disrupt IFs is generally less specific than the use of antibodies, because they tend to disrupt multiple IFs. For example, dominant negative NF-M disrupts not only nIFs but also desmin filaments of muscle cells descended from the same injected blastomere.

Antisense methods have only recently begun to live up to their early promise as a more generally applicable method of inhibiting gene function than antibodies and dominant negatives. At first, the efficacy of full-length antisense RNAs seemed to be reduced by an endogenous heteroduplex RNA unwinding activity present in early cleavage stages (Rebagliati and Melton, 1987). Despite these early disappointing reports, injection of antisense RNA has nonetheless been used successfully to suppress expression of genes that act later in development, including *goosecoid, BMP-4* (Steinbeisser *et al.*, 1995), *FGF* (Lombardo and Slack, 1997), and *slug* (Carl *et al.*, 1999). In general, this method is most successful for relatively rare RNAs translated between gastrulation and neurulation, properties that generally preclude its use for studying nIFs. Although RNAi has also shown some promise as a tool for suppressing gene expression in *Xenopus* embryos (Nakano *et al.*, 2001), it, too, seems to work best with genes expressed at lower levels than are most IFs.

Encoding antisense mRNAs into plasmids with strong promoters has overcome some of the developmental stage limitations of antisense RNA. This method was first used to demonstrate a role for the membrane protein 4.1 in *Xenopus* retinal development (Giebelhaus *et al.*, 1988). Although this method lengthens the period of development that can be studied with antisense methods, its general utility is reduced because injected DNAs are typically expressed by only a minority of the cells descended from the injected blastomere (Vize *et al.*, 1991). Nevertheless, injected plasmids encoding truncated, dominant negative desmin and vimentin have been used to disrupt the endogenous desmin network present at intersomitic junctions, producing severe structural defects (Cary and Klymkowsky, 1995). The prospects of using still stronger promoters, such as that of the adenovirus VAI gene, which is read by RNA polymerase III (Nichols *et al.*, 1995), to obtain sufficiently high levels of antisense expression, or of using inducible promoters, such as that of *hsp70* (Wheeler *et al.*, 2000), may make this method more useful for inhibiting gene expression during development if the problem of mosaic expression were overcome.

Injected antisense oligonucleotides, which had been supremely successful in suppressing protein translation in unfertilized oocytes (Sumikawa and Miledi, 1988), have generally been unsuccessful after fertilization, at least in their unmodified form. This is most likely because of their exceptionally short half-lives in fertilized embryos. Nevertheless, they have been a valuable tool for disrupting maternally expressed genes active in early development. Torpey *et al.* (1992), for example, injected antisense oligonucleotides into unfertilized *Xenopus* oocytes to suppress the expression of maternally expressed cytokeratins. These oocytes were then transplanted into a surrogate mother so that they could be laid and fertilized to develop as embryos. Suppressing maternally expressed cytokeratins disrupted the superficial embryonic keratin network, which compromised the mechanical strength of the embryo and inhibited wound healing. Both of these properties aid in the survival of an externally developing embryo in the wild.

Fortunately, chemically modifying oligonucleotides has greatly improved their efficacy in suppressing gene expression after fertilization. For example, phosphothiorate antisense oligonucleotides have been used successfully to study the role of *engrailed* in retinotectal map formation (Retaux *et al.*, 1996), but because these oligos also inhibited bFGF binding, such oligonucleotides are considered prone to nonspecific effects. More recently, antisense morpholino oligonucleotides have become a powerful new tool for suppressing specific expression of mRNAs in *Xenopus* more specifically (Nutt *et al.*, 2001). First used in zebrafish (Nasevicius and Ekker, 2000), these oligos are rapidly becoming the method of choice for suppressing gene expression during the first days of *Xenopus* development, and we have successfully used one to suppress expression of peripherin through stage 42

(3 days of development). The remainder of this chapter describes the methods we have used to study nIF function during *Xenopus* development, as well as some of the results obtained with them.

II. Instrumentation and Materials

A. Useful *Xenopus* References

Because *X. laevis* is widely used as a model system for development, there are many valuable resources available for the investigator. The following lists some of those we have found most useful. The normal developmental table of Nieuwkoop and Faber (1994) is indispensable for staging embryos. Although the times to reach a given developmental stage given in this book are only accurate at 22 °C, the effects of temperature are linear and readily predictable across the temperature range for normal development (16–27 °C). Developmental times for other temperatures may be estimated by multiplying the time given in Nieuwkoop and Faber by the following correction factor (R), where T is the temperature in °C (Lin & Szaro, 1994 ID: 351): R = 0.095T −1.104.

Another useful reference is Volume 36 of this series (Kay and Peng, 1991). Chapter 1 (Wu and Gearhart, 1991) describes how to raise *Xenopus* in the laboratory. Chapter 5 (Keller, 1991) discusses early development, and Chapter 11 (Heasman *et al.*, 1991) tells how to use antisense oligonucleotides to suppress expression of maternally expressed messages. Use of fluorescent tracers to follow cell lineage is outlined in Chapter 15 (Gimlich, 1991), and general methods for studying altered gene function in *Xenopus* are discussed in Chapter 20 (Vize *et al.*, 1991). A useful method for whole mount immunostaining of *Xenopus* embryos is given in Chapter 22 (Klymkowsky and Hanken, 1991), and whole mount *in situ* hybridization methods are presented in Chapter 23 (O'Keefe *et al.*, 1991) and in Appendix G (Harland, 1991). Further useful improvements to these methods are described by Shain and Zuber (1996). Chapter 26 (Peng *et al.*, 1991) describes preparing cultures from *Xenopus* embryonic neural tube. Another useful reference for this technique is by Tabti and Poo (1991). There is also an excellent book on the biology of *Xenopus* edited by Tinsley and Kobel (1996).

B. Instrumentation

1. Injection Micropipettes

Injection micropipettes are prepared from 1.5-mm OD glass capillary tubes (Kwik-Fil Borosilicate Glass Capillaries, TFW-150; World Precision Instruments, Sarasota, FL). To pull the micropipettes, we use a Narishige Model PC-10, which may be used in either single or double pull mode to prepare pipettes for either microinjection or patch clamp, respectively. Some experimentation is required to determine the optimum settings (we use the puller in single mode, with heater #1 set to 0 and heater #2 set to 58).

Once pipettes are pulled, we bevel them on a Narishige EG-44 micro-grinder. The motor is set to a speed of 90, and the pipette is held at an angle of 35 degrees. While observing the tip through a dissecting microscope, we lower the pipette to the surface of the grinding wheel, keeping its surface wet with a moistened kimwipe. We then touch the tip of the needle to the grinding wheel for several seconds. The amount of water drawn into the needle is usually a good indicator of the size of the opening. Afterward, we inspect and measure the diameter of the tip under a microscope. The ideal tip is approximately 5–6 μm in its outer diameter, with a smooth opening (Fig. 1A). Tips smaller than this tend to clog. If the tips are too large, they will damage the embryo and promote backflow. Tips within the appropriate size range should yield appropriately sized injection volumes with a pulse duration of 150–400 ms at 16 psi. Significantly shorter pulse lengths than this damage the embryo. Again, some practice is required to achieve good tips.

2. Injection Apparatus

The injection setup consists of the following (Fig. 1D):

1. A good dissecting microscope (e.g., Leica/Wild M3) with an ocular micrometer in the eyepiece for measuring drop size and a fiberoptic light source (e.g., Fostec).

2. A joystick micromanipulator (e.g., Narishige MN-151) mounted onto a metal baseplate (e.g., Narishige IP) with a magnetic stand (e.g., Narishige GJ-1).

3. A stable table. We use one designed for holding a balance.

4. A device for delivering controlled air pulses capable of ejecting calibrated amounts of solution from the injection micropipette. We use a Picospritzer II by General Valve Corporation (Fairfield, NJ). This may be connected either to a tank of dry nitrogen or to the house air supply through a filter, which prevents contaminating the air line with oil and moisture.

5. A dish for holding the embryos in place while they are injected (Fig. 1C). We fill a 60-mm Petri dish with melted beeswax that has been blackened with a few grams of bone black stirred into it before it is poured. Once the wax has solidified, the small end of a borosilicate pasteur pipette that has been sealed by fire polishing is heated in an alcohol lamp to make small indentations in the surface. With a little practice, it is possible to lay out a grid work of holes that are just the right size for holding the embryos in place. Another alternative is to make a mesh-coated injection dish. To make this dish, cut a piece of 800-μm polypropylene mesh to fit into the bottom of a polystyrene Petri dish. A drop of methylene chloride is used to melt the mesh to the bottom of the dish.

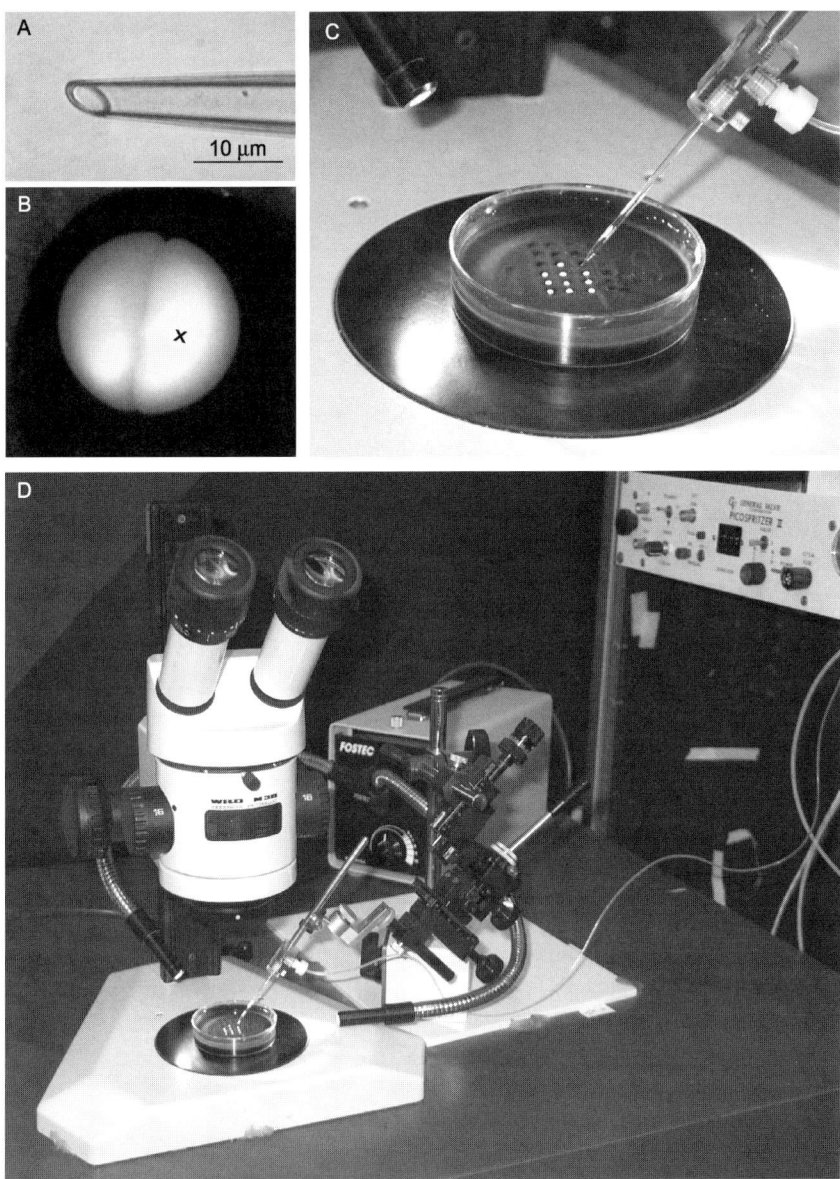

Fig. 1   (A) The magnified tip of a properly beveled injection micropipette. (B) A two-cell stage albino *Xenopus* embryo. The "X" indicates the target site for injections. (C) Embryos lined up and ready for injection. The embryos are placed into indentations made on the surface of a wax-filled Petri dish, which is filled with 5% Ficoll in HBS. The injection micropipette (right), mounted in its holder, is being positioned above the embryos. (D) An overview of the injection setup, which includes a dissecting microscope, a joystick micromanipulator on its baseplate, and a fiberoptic illuminator sitting on a balance table. The microscope and baseplate are held in position on the table with tacky wax. Air pressure for the injections is provided by the house air line and is regulated by a rack-mounted Picospritzer II (upper right).

6. A fluorescence dissecting microscope with appropriate filters for visualizing the distribution of microinjected fluorescent dyes and expressed enhanced green fluorescent protein (eGFP) in the embryos.

C. Solutions

1. Hepes Buffered Steinberg's Solution (HBS)

> 58.22 mM NaCl
> 0.67 mM KCl
> 0.34 mM Ca(NO$_3$)$_2$
> 0.83 mM MgSO$_4$
> 10 mM Hepes (Sigma)
> Titrate to pH 7.6 with NaOH
> Autoclave or filter through a 0.22-μm filter
> Add 500 μl pen/strep per liter of solution

2. Dejellying Solution

> 0.39 g dithiothreitol
> 3.1 g Tris base (Sigma)
> 500 ml dechlorinated tap water

3. 5% Ficoll in HBS

> 50 g Ficoll (Type 400, Sigma) dissolved in 800 ml HBS.
> After it has dissolved, bring the volume to 1 L with HBS.
> Filter sterilize through a 0.22-μm filter.
> Add 500 μl pen/strep solution and store at 4 °C.

4. Human Chorionic Gonatotropin Stock

> Chorulon (Intervet Inc., Millsboro, DE)
> Dissolve lyophilized powder with sterile distilled water to 3300 U/ml, then aliquot and store them at −20 °C.

5. Pen/Strep Solution

> 10,000 units penicillin-G, sodium salt (Sigma)
> 7500 units streptomycin sulfate (Sigma)

10 ml sterile distilled water

Sterilize by filtration through at 0.22-μm filter

D. Macromolecular Reagents for Altering Intermediate Filaments

As described in Section I, antibodies, RNAs, plasmid DNAs, and antisense morpholino oligonucleotides have been the most successful reagents for studying IF function in *Xenopus*. Each of these must be prepared following stringent standards, such as dissolving them in water of the highest quality (e.g., that produced by the Barnstead Thermolyne Easypure UV/UF system, Dubuque, IA).

1. Antibodies

Antibodies must be purified. We use monoclonal antibodies purified from ascites fluid with the MapsII kit (Biorad, Richmond, CA). After purification, antibodies are dialyzed extensively against HBS and then concentrated in HBS to a final concentration of 7–10 μg/μl with Centricon (Millipore, Billerica, MA) filters. The solutions are then split into aliquots and stored at $-80\,°C$. Just before injection, the antibodies are thawed and mixed with an equal volume of fluorescent dextrans, which serve as a fluorescent tracer dye, and then spun in a microfuge at 4 °C for 10 minutes to remove any particles that might clog the injection pipettes. Several fluorescent lysinated dextrans are available (Molecular Probes, Eugene, OR) in a variety of colors that work well in *Xenopus*: Oregon Green 488 (injected at 7.5 mg/ml), FITC (injected at 7.5 mg/ml), rhodamine (injected at 1.2 mg/ml), and Cascade Blue (injected at 1.2 mg/ml).

Both the specificity and the efficacy of each new antibody for blocking IF function must be determined experimentally. For antibodies that do not cause deleterious side effects, doses ranging from 100–200 ng per embryo are usually well tolerated, and their effects on IFs are both long lasting and dramatic. For example, injected anti–NF-M blocks transport not only of NF-M but also that of the other nIFs that copolymerize along with it, yielding neurites that are largely devoid of type IV nIFs through the end of the third day of development (equivalent to stage 42). The two anti–NF-M antibodies that worked successfully for us, XC10C6 (Szaro and Gainer, 1988) and RM0270 (Wetzel *et al.*, 1989), target epitopes in the head-rod and the C-terminus of NF-M, respectively. However, several others that targeted dephosphorylated epitopes within the NF-M tail domain did not work well. These latter antibodies yielded severe developmental abnormalities at doses as low as 7 ng/embryo, which began well before endogenous NF-M is expressed. When used on tissue sections at higher concentrations than those normally used for immunocytochemistry, these latter NF-M antibodies also stained nuclei. Thus, at the high concentrations used for injections, it is likely that they targeted other proteins.

Because antibodies vary, several controls should be performed when a new antibody is being used for the first time. We used the following to help confirm that effects obtained with anti–NF-M were specific: (1) Two separate NF-M monoclonal antibodies, which targeted distinct epitopes, produced similar effects, reducing the possibility that antibodies caused effects by binding to other molecules. (2) Both Fab fragments and whole IgGs yielded similar effects, thereby reducing the possibility that cross-linked antigen–antibody complexes had non-specifically blocked other cellular components from the axon. (3) Injection of a nonfunction blocking antibody to *Xenopus* β-tubulin (Chu and Klymkowsky, 1987), which is available from the Developmental Studies Hybridoma Bank (Iowa City, IA), as well as antibodies to rat (but not frog) NF-M, rat neurophysins, and bacterial β-galactosidase (Promega, Madison, WI), had no effects. These controls demonstrated that antibodies themselves, even when bound to other cellular components, are well tolerated. (4) Anti–NF-M also failed to block transport of actin, β-tubulin, or mitochondria, thereby further showing that these essential proteins and organelles moved into the axon. (5) Effects of anti–NF-M were limited to those nIF subunits that colocalize with NF-M and began only after NF-M was expressed, further strengthening arguments for the antibody's specificity. The investigator should consider these, as well as appropriate additional controls relevant to the particular IF being studied, to confirm the specificity of any antibody used to disrupt IFs in *Xenopus*.

2. RNAs

RNAs encoding full-length proteins may be injected to obtain increased expression of IFs, or truncated RNAs may be used as dominant negatives to block IF function. Injected RNAs are useful for the first few days of development. They are detectable in the embryo for at least 1 day of development (Fig. 2A), and their translated proteins often persist even longer (Fig. 2B-D).

Precautions must be followed to guard against RNases, including using RNase-free water (DEPC-treated water cannot be used, because it is toxic to embryos), RNase-plastic ware, RNase-barrier pipette tips, and RNase-glassware. Glassware may be made RNase free by rinsing it in either 50% nitric acid or RNase Zap (Ambion, Austin, TX).

RNAs are transcribed with the Message Machine kit (Ambion) under conditions that add a 5′ methyl guanosine cap, which is necessary to stabilize the RNA once it is injected into the embryo. In addition, our RNAs contain the 3′ untranslated region (UTR) of rabbit β-globin instead of that of SV40, because the globin UTR has a cytoplasmic polyadenylation signal. This ensures that the RNA will become polyadenylated in the embryo, which both stabilizes it and increases its translation. In addition, the coding sequences should be cloned into the plasmid at a site close to the RNA polymerase promoter, because intervening sequences may reduce translation. We have had good success with a modified pGem3Z (Promega, Madison, WI) as a vector. Into its *Hind* III site, which is just downstream of the

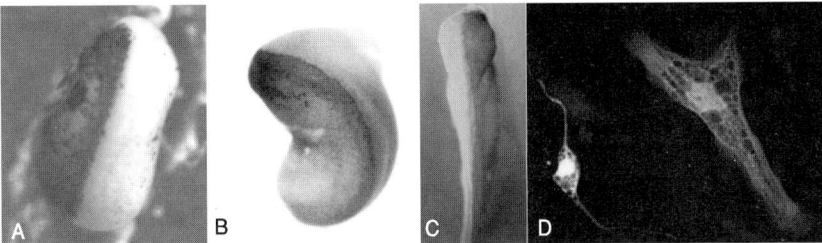

Fig. 2 (A) Distribution in a 1-day-old embryo (stage 24) of peripherin antisense RNA injected at the two-cell stage, visualized by *in situ* hybridization. (B) Distribution at stage 24 of XNIF protein translated from synthetic RNA injected at the two-cell stage, visualized by immunocytochemistry. (C) Expression in a 2 1/2-day-old embryo (stage 35/36) of exogenous β-gal RNA injected at the two-cell stage, visualized by histochemical staining. (D) Exogenous eGFP expression in a neuron (lower left) and a muscle cell (upper right) in culture 2 days after injection of RNA at the two-cell stage and 21 hours after plating. (A-C) Dorsal views of unilaterally labeled embryos. Rostral is at the top.

Sp6 promoter, we cloned a *Not* I restriction site immediately upstream of the 3'UTR of rabbit β-globin. Protein coding sequences are then cloned into this *Not* I site. The vector may be linearized within the 3'UTR at a *Xho* I site, which is downstream of the internal polyadenylation signal.

To act as a fluorescent tracer, synthetic mRNA encoding eGFP (1–3 ng/embryo) is mixed with the test RNA. RNAs used for injection are dissolved directly in RNase-free water (EasyPure UV/UF, Barnstead Thermolyne) rather than in Tris EDTA, which is toxic. Embryos can generally tolerate at least 6–10 ng of RNA injected at the two-cell stage before deleterious, nonspecific side effects arise from overloading the cells with RNA.

3. Plasmid DNAs

Plasmid DNA should be prepared either by cesium chloride density ultracentrifugation or by use of kits that produce high-quality DNA (e.g., Qiagen, Valencia, CA). When such kits are used, care should be taken to use *Escherichia coli* strains that preferentially yield supercoiled rather than nicked DNA (e.g., DH5 α or XL-1 Blue). Any of several promoters may be used to drive expression ubiquitously, including the *Xenopus borealis* cytoskeletal actin promoter and a cytomegalovirus (CMV) promoter. Typically, 100 pg of DNA are injected into each embryo.

4. Antisense Morpholino Oligonucleotides

Antisense morpholino oligonucleotides are excellent for suppressing gene expression in injected *Xenopus* embryos. Morpholino oligonucleotides consist of a nucleic acid base, a morpholine ring, and a nonionic phosphorodiamidate intersubunit linkage. These modifications preserve the spacing between bases, which is needed for Watson-Crick base pairing, and render the oligos highly resistant to enzymatic degradation within the cell. Morpholino oligos are custom made by Gene Tools, LLC (Philomath, OR), which has a website (www.gene-tools.com) that details how to design an effective antisense oligonucleotide. As a general starting point for blocking translation, they recommend 25-mers that target the 5' UTR extending as far as 22 nucleotides into the coding domain of the targeted mRNA. We typically inject 20 ng per embryo at the two-cell stage.

III. Procedures

A. Maintaining *Xenopus* Adults

In our experiments, we use adults primarily from the strain of periodic albinos (Hoperskaya, 1975; Tompkins, 1977), because these animals lack dense pigment, which interferes with whole mount assays. These frogs may be obtained commercially from a variety of sources, including Nasco (Ft. Atkinson, WI), *Xenopus* I (Dexter, MI), and *Xenopus* Express (Plant City, FL; and Vernassal, France).

We maintain our colony in autoclavable polycarbonate hamster (two to four adults) and rodent cages (one adult), using approximately 2 L of dechlorinated municipal tap water per adult. Summertime algal blooms in our municipal water supply have caused us some difficulty. We solved this problem by filtering the water through a carbon charcoal tank (US Filter, Warrendale, PA), which removes chlorine and many organic contaminants, then passing it through an ultraviolet light sterilizing unit and a 0.22-μm filter. The water is then piped into 30-gallon holding tanks to allow it to equilibrate to room temperature. Adults are kept on a 12/12 day–night cycle and fed "moist salmon diet" (Rangen Inc., Buhl, ID), approximately 10 pellets per frog, three times a week. At least once a week, their water is changed and the aquaria wiped clean with a soap-free nylon scrubber.

B. Obtaining and Dejellying Embryos

Although many laboratories prefer to use artificial fertilization to obtain embryos, which requires sacrificing the males, we prefer instead to stimulate natural spawnings. This has the combined advantages of conserving the males and of allowing eggs to be laid at a steady pace throughout the morning. Sexually mature, gravid females are selected for their red cloaca, and males for their black nuptial pads, which are on the inside surface of their forelimbs.

1. Using a 1-ml tuberculin syringe and a 27-gauge needle, inject males twice with 333 units of human chorionic gonadotropin (HCG)—the first time, 2 days before, and the second time, on the evening of the spawning. Inject females the evening of the spawning with 500–700 units of HCG. For injection, wrap frogs in a wet paper towel to cover their heads and calm them and expose the injection site on the dorsal thigh. Insert the needle under the skin, on the dorsal surface of the

thigh, immediately caudal to the lateral line, taking care not to penetrate the muscle. Then push the tip of the needle gently across the lateral line into the dorsal lymph sac, keeping the tip high, just under the skin. Once the tip is inside the lymph sac, inject the hormone.

2. After injections, place the mating pair into a small aquarium (we use hamster cages) filled with dechlorinated tap water to a level that will just keep them completely covered when clasped. Cover the aquarium with a heavy metal lid to prevent the animals from escaping, and drape it with a white towel or laboratory coat. Place the covered tank within site of a window to allow the morning sun to hit the tank. To trigger spawning, the water temperature should rise above 22 °C (to help trigger spawning behavior, some investigators place a small aquarium heater into the tank with a timer set to warm the water to 24 °C early in the morning). In our hands, spawning typically begins soon after sunup, approximately 12 hours after injecting the adults, and continues uninterrupted for several hours through the morning.

3. Beginning 15–20 minutes after they are laid, collect the fertilized eggs. This may be done by drawing them into a 10-ml plastic serological pipette (made by breaking off the end of the pipette and attaching a large rubber bulb to draw the water). If careful, the mating pair may be gently lifted and placed into a second aquarium without separating them, which allows the spawning to continue. Eggs may then be more easily gathered by gently dislodging them from the bottom of the tank, using the edge of your flattened hand and then swirling the water in a circle. The eggs will gather in the center of the vortex.

4. For injections, several dozen newly fertilized eggs are collected at once in a finger bowl filled with dechlorinated tap water. The tap water is then poured off and replaced with dejellying solution. Gently swirl the eggs until the jelly comes loose, usually within 1–3 minutes (leaving the eggs in dejellying solution any longer than this will make them too soft to inject). Pour off the dejellying solution and rinse the eggs 10 times in dechlorinated tap water, discarding as many of the dead eggs (the large, flaccid white ones) as possible. After the last rinse, transfer the eggs into 20% HBS, supplemented with 0.5 mg/L gentamycin. The embryos are more fragile after dejellying than before, so treat them gently. Embryos are then parceled into 60-mm polystyrene Petri dishes filled with 20% HBS, 0.5 mg/L gentamycin (Sigma). Keep several dishes of embryos as uninjected controls to monitor the vitality of the spawning (>90% normal development is good). The rest of the embryos are reared to the two-cell stage for injection. At room temperature, first cleavage begins approximately 90 minutes after fertilization, and subsequent cleavages occur approximately every 20–30 minutes thereafter.

C. Injecting Embryos

1. Before injecting the eggs, prepare the workplace. Solutions for injection are spun in a microcentrifuge for 10 minutes and stored on ice. Also, prepare a clean glass microscope slide by wrapping it tightly with a single layer of Parafilm. After

turning on the air supply to the Picospritzer, set its regulator to 16 psi and the pulse duration to an initial value of 10 msec.

2. To draw the injection solution into the injection micropipette, mount it into one end of a piece of 1.5-mm ID Tygon tubing. The other end of the tubing is placed onto a 18-gauge needle mounted onto a 5-ml syringe. Pipette 20 μl of clean mineral oil onto the Parafilm-covered glass slide, and then, depending on the number of eggs to be injected, pipette 1.5–3 μl of injection solution into the mineral oil. While observing the drop through the microscope, draw this solution into the micropipette with the syringe. Remove the injection micropipette from the tubing and insert it into the halter of the micropipette holder mounted on the micromanipulator.

3. To calibrate the injection pipette, insert its tip into the mineral oil droplet again while observing it under the dissecting microscope. With an initial pulse duration of 10 msec, inject a drop into the mineral oil. With the ocular micrometer, measure the diameter of the droplet and calculate its volume. Adjust the pulse duration, which should fall between 150 and 400 msec (see Section II.B.1), until the desired volume is attained. This must be done each time a new micropipette and injection solution are used. We routinely choose volumes between 8 and 15 nl for injection. Injection volumes of more than 25 nl produce abnormal embryos, whereas volumes less than 5 nl may disperse poorly through the embryo.

4. Select two-cell stage embryos for injection and with a plastic transfer pipette (e.g., Samco #202, San Fernando, CA), and transfer them to the injection dish containing 5% Ficoll in HBS, which collapses the vitelline membrane. The Ficoll solution should be changed after each round of injection to avoid diluting it too much and to prevent contamination. Position the embryos into the wax wells and orient them, dorsal side up. With the injection dish and embryos in view under the microscope, bring the tip of the injection pipette into the field of view. While observing the tip of the pipette, inject several pulses into the Ficoll solution to clear any backfilled solution from the pipette tip and ensure that the tip has not clogged. Then, placing the needle just above the targeted embryo, pulse solution through the tip once again to clear backfill. Immediately afterward, lower the micropipette along its axis and gently insert the tip into the embryo. We generally orient the needle at approximately a 45-degree angle (Fig. 1C) and target a point 20–30 degrees south of the pole, midway across the egg from the plane defined by the cleavage furrow (Fig. 1B). Once the tip has penetrated, pulse the injection solution once and then withdraw the tip. Immediately afterward, pulse solution through the tip again to remove any yolk that may have clogged the tip.

5. Transfer the injected embryos into a 60-mm polystyrene Petri dish containing 5% Ficoll in HBS. Be sure to use untreated Petri dishes and not tissue culture dishes, whose treated plastic will damage the embryos. Place no more than 30 embryos into each dish and space them evenly. Leave them in 5% Ficoll/HBS for several hours to allow the injection wound to heal, but transfer them before

gastrulation begins. Transfer embryos into 100% HBS, then successively for 1–5 minutes each through 70% and 50% HBS. Finally, place them into 20% HBS for rearing at 20–23 °C to the desired stage for analysis. During transfer, avoid carrying solutions from one dish to the next by letting the embryos sink to the tip of the pipette after drawing them up. The embryos will drop from the tip into the next solution when the transfer pipette breaks the surface. If you are having trouble with wound healing, doubling the amount of calcium nitrate in the HBS often helps.

IV. Pearls and Pitfalls

A. Troubleshooting

Early monitoring of injected embryos and isolating those that were successfully injected and healthy from those that are damaged or unlabeled is critical for success. One advantage of coinjecting fluorescent tracer dyes or eGFP is that the embryos may be rapidly screened under the fluorescence dissecting microscope to determine whether or not the injections were successful. By the time of neurulation, successfully injected embryos will be labeled on one side of the midline and will appear free of major morphological defects. Table I lists some of the major symptoms of trouble and the remedies we have found most helpful.

B. Ensuring Healthy Spawnings

Healthy spawnings are a key element of success, and keeping your breeding colony well fed and free of disease are essential. Avoid introducing new adults into your colony until you have quarantined them and are certain that they bring no new pathogens into the colony. In the absence of any new animals, an outbreak of disease is often a sign that the animals are stressed. Identifying the source of this stress can be difficult. Check the water temperature, which should be maintained between 18° and 22 °C. Also, check for biological contaminants in the water; green algae, black mold, or flocculent white growths in the tanks, water lines, or filters are indications that the carbon filter and ultraviolet sterilization unit are no longer functioning properly. Change them regularly as recommended by the manufacturer. Also, to avoid placing the rest of the colony at risk, it is better to sacrifice chronically sick animals than to try to nurse them back to health. Healthy animals that become sluggish or stop eating should be isolated and watched carefully. Often, these can be brought back to good health by adding NaCl to a concentration of 0.1 M or a teaspoon of sulfadiazine to the water. When an animal dies, dispose of it promptly and sterilize the aquarium.

If the entire colony stops breeding, double check both the water temperature and the length of the day/night cycle. In its natural habitat, during the summer,

Table I
Troubleshooting Chart

Symptoms	Probable cause	Suggested remedy
Uninjected embryos are frayed and dying	1. HBS is of the wrong pH or composition	1. Check the pH and composition of the HBS
	2. Bacterial contamination	2. Sterilize solutions and tools
	3. Too many embryos crowded into the rearing dish	3. Reduce the density of embryos to <30 per dish and remove the dead embryos
	4. Unhealthy spawning	4. Try another spawning
	5. Over dejellying	5. Dejelly for less time
Injected embryos are extruding large amounts of yolk	1. Injection tip is too large or broken	1. Use a smaller tip
	2. Poor wound healing	2. Raise the calcium in the HBS
	3. Unhealthy spawnings	3. Try another spawning
	4. Over dejellying	4. Reduce the dejellying time
Injected embryos are deformed and dying	1. Borderline healthy spawning	1. Try another spawning
	2. Too high an injection pressure	2. Lower the injection pressure
	3. Toxic injection solution	3. Repurify the injection solution or reduce the dose of the injection solution
Injected blastomeres fail to cleave	Toxic injection solutions	Reduce the concentration of the fluorescent tracer or the reagent or repurify the injection solution
Injected embryos are unlabeled	Clogged injection pipette	Pulse solution through the pipette before each injection. If the tip is visibly clogged and cannot be cleared, change the pipette
Large variations in the intensity of labeling among injected embryos	Backfilling of Ficoll into the injection pipette between injections, or partial clogging	Pulse injection solution through the pipette until fresh solution can be seen leaving the tip
Injected label is confined to small domains within the embryo	1. Injection volumes were too small	1. Increase the injection volume
	2. Injections were done too near the 4-cell stage	2. Inject slightly earlier
	3. Partially clogged tip	3. Clear the injection pipette with pulses of solution
Embryos are labeled on both sides	1. Injections were done too early	1. Delay injections
	2. Injection site was too close to the plane of cleavage	2. Move the injection site away from the plane of cleavage
	3. Injection pressure was too high or the needle tip was too large	3. Lower the injection pressure or the needle diameter
Incomplete gastrulation	Embryos transferred to 20% HBS too late	Transfer embryos to 20% HBS by stage 9, before gastrulation begins
Injected solution forms pools just under the cortex of the blastomere	The injection tip was inserted too shallow	Insert the tip further into the embryo

X. laevis burrows into the mud to estivate as its pools dry up. As long as the water temperature and day/night cycles are regulated, a healthy colony should remain productive throughout the year.

Xenopus are most fecund between the ages of 18 months and 5 years, although we have had many that produced healthy spawnings for even longer. During their productive years, mating pairs can deliver healthy spawnings repeatedly. We generally like to give the females at least 2 months and the males 3–4 weeks rest between spawnings. A female that has not been spawned in many months often produces large numbers of dead eggs. Generally, her eggs will improve if she is spawned again within a few months. However, sometimes a frog fails repeatedly to spawn healthy embryos. If so, they may have simply grown too old or else be unfit, and it is best to retire them.

V. Discussion and Concluding Remarks

A. Functional Studies of Neuronal Intermediate Filaments in *Xenopus*

We have successfully used antibodies to NF-M (both whole IgGs and Fabs) and mRNAs encoding truncated, dominant negative NF-M mRNAs to study the involvement of type IV nIFs during axonal outgrowth (Lin and Szaro, 1995, 1996; Szaro *et al.*, 1991; Walker *et al.*, 2001). These reagents effectively block the transport of NF-M, together with the α-internexin homolog, XNIF (Fig. 3B–D), which is coexpressed with NF-M in developing axons. The loss of these two type IV nIFs significantly slowed axon outgrowth in intact animals (Fig. 3E,F) and in culture. Thus, although nIFs are not essential for axon growth, they nonetheless facilitate it.

Through time-lapse video microscopy, we quantitatively measured axonal growth parameters in dissociated spinal cord cultures made from antibody-injected embryos (Walker *et al.*, 2001). In these cultures, normal axonal outgrowth begins with an initial outburst of neuritic outgrowth. After this outburst, neurites typically settle into a second phase of slower growth, which is characterized by long repeated pauses; in *Xenopus*, this phase typically precedes expression of both NF-M and XNIF (Undamatla and Szaro, 2001). Then, as expression of these two nIFs begins, the average rate of axonal growth accelerates, because the periods of extension lengthen and pausing behavior diminishes. Regardless of whether axons are in the slower or more rapid phase of growth, the velocity of extension during active growth itself remains the same, at 35–40 μm/h. Blocking transport of NF-M and XNIF had no effect on the timing of outgrowth, the length of time axons were able to maintain their growth, or the velocity of extension during active growth. It did, however, reduce the average rate of growth during the phase of rapid elongation, and it did this by increasing pausing and retraction behavior at the expense of extension. The persistence of this slower mode of growth, which is more characteristic of young axons, led us to conclude that NF-M and XNIF

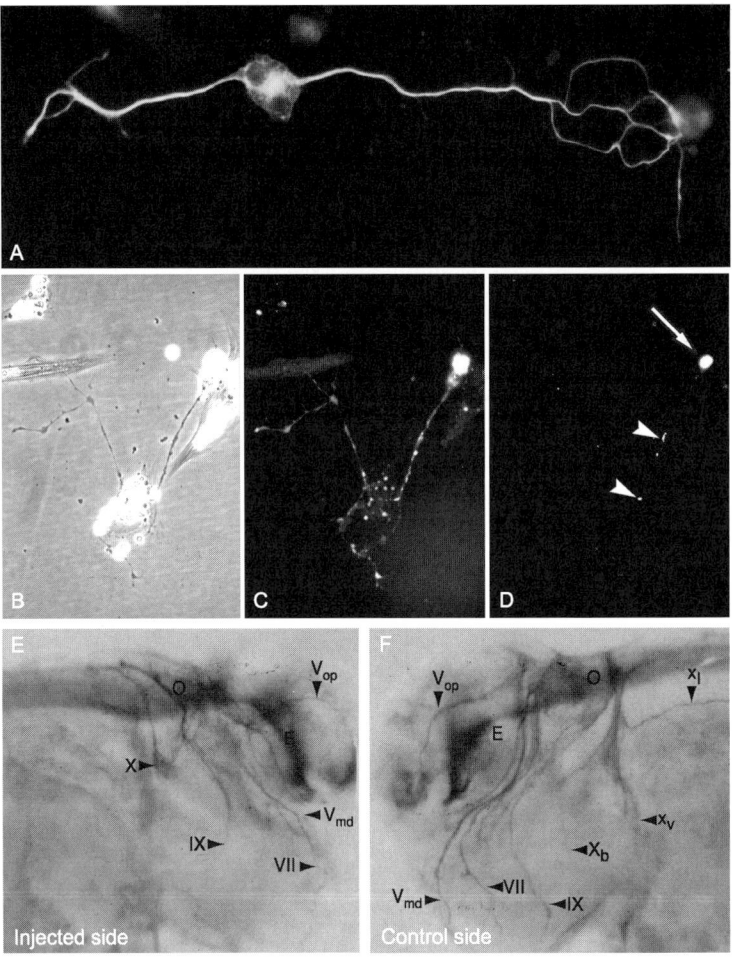

Fig. 3 (A) Immunofluorescence staining for NF-M illustrating its normal distribution in a cultured neuron 24 hours after plating (2 days after fertilization). (B–D) Transport of the α-internexin–like subunit, XNIF, is blocked by the injected NF–M antibody through 42 hours of outgrowth in culture (3 days after fertilization). (B) Phase-contrast image. (C) Oregon Green dextran 488 fills the neuritic processes within the same cell as that shown in (B). (D) Immunostaining this same cell for XNIF shows that this protein is largely confined to the cell body (arrow) and a few scattered places along the neurite (arrowheads point to examples). (E–F) Inhibition of cranial nerve development by a truncated dominant negative NF–M. The embryo was injected unilaterally at the two-cell stage with RNA and then immunostained as a whole mount at stage 37/38 (3 days old) with an antibody to MAP-1 to stain axons. Cranial nerve outgrowth was retarded on the injected side of the embryo (E) compared with the contralateral, uninjected side (F). Roman numerals designate branches of the cranial nerves. l, lateral branch; md, mandibular branch; op, opthalmic branch; b, brachial branch; v, ventral branch; E, eye; O, otic vesicle.

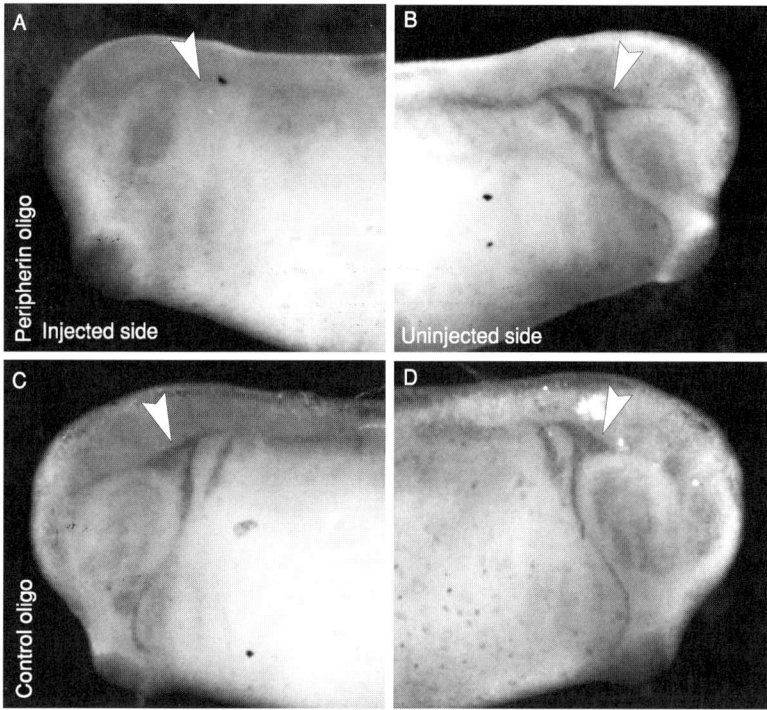

Fig. 4 A peripherin antisense morpholino oligonucleotide suppresses peripherin expression. Embryos were injected at the two-cell stage with 20 ng of peripherin morpholino antisense oligonucleotide (A, B) or with 20 ng of a generic control morpholino oligonucleotide (C, D). At stage 29/30–31, embryos were immunostained in whole mount for peripherin. Peripherin expression was undetectable on the side injected with the peripherin oligo (A), whereas it filled the CNS, trigeminal ganglia (arrowhead), and other cranial nerves on the uninjected side of experimental embryos (B) and on both sides of embryos injected with the control oligo (C, D).

constitute a rate-limiting component needed for the longer bouts of extension that characterize the more rapid average elongation of long axons.

We have now begun to study the specific contribution of peripherin to axonal outgrowth by suppressing its expression with an antisense morpholino oligonucleotide. As determined by whole-mount immunocytochemistry (Fig. 4), injection of as much as 20 ng of peripherin oligo per embryo has had no overt effects on general development, whereas it specifically suppresses peripherin protein expression through the third day of development (stage 42). We are now analyzing axonal outgrowth and the transport of the other nIFs in these embryos in detail. Thus, this method provides a powerful adjunct to antibody injection for disrupting IFs, both individually and in combination.

B. Interpreting Molecular Lesions

Because IFs exist as complexes, interpreting the phenotypes generated by molecular lesion studies should be considered carefully. Comparing results among different types of lesions can be useful in these considerations, because they yield different information. For example, because IFs form complex integrated structural networks, suppressing expression of a single IF protein may release the other components that bind to it to interact abnormally with the rest of the cell. Alternatively, disrupting the entire complex with an antibody might yield different phenotypes. Similarly, because IFs are copolymers, severely disrupting their stoichiometries by suppressing or overexpressing one of the subunits generally produces aggregates of the remaining IF proteins (Beaulieu *et al.*, 1999; Wong *et al.*, 1995). The size and cellular location of these aggregates often determine whether or not they are deleterious in themselves (Beaulieu *et al.*, 2000). In addition, because mRNAs and proteins can influence the transcription and translation of other genes through complex, interdependent genetic regulatory networks, different compensatory effects may arise depending on the molecular lesion. For example, because some regulatory networks sense levels of primary transcript in the nucleus, suppressing protein translation with an antisense morpholino oligonucleotide or disrupting IF formation with an antibody in the cytoplasm might result in different compensatory effects than would a targeted gene knockout, which would remove even the newly transcribed primary transcript. For these reasons, the various means of producing molecular lesions should be viewed as complementary tools for parsing function from phenotype. Because several of these methods can be applied at once in *Xenopus*, it may prove useful in these considerations.

C. Concluding Remarks

Injection of macromolecules into *Xenopus* embryos has proven to be an effective method of manipulating IFs to study their function during vertebrate development. To date, injection of antibodies to specific IFs and expression of truncated, dominant negative IF proteins by mRNA injection have been the most exploited of these reagents. They have demonstrated that although IFs are generally dispensable for vertebrate development to proceed, they nonetheless enhance many of its essential processes, such as gastrulation, muscle organization, and axonal outgrowth. Helping these processes to occur effectively and in proper synchrony is likely to be important for producing normal and highly competitive individuals. This is especially true for externally developing embryos, such as *Xenopus*, which must survive in the wild even before development is complete. Thus, although IFs may not be essential for the survival of individual cells, they are arguably indispensable for the survival of the species. This feature alone might account for their diversity and persistence in evolution.

The application of antisense morpholino oligonucleotides for suppressing expression of specific IF proteins during development creates even more possibilities for exploring IF function during development. It is now also relatively straightforward to create transgenic *Xenopus* embryos by fertilizing oocytes with sperm nuclei that have been incubated with transgenes (Amaya and Kroll, 1999). To date, this technique has only been used with IFs to study the properties of their gene promoters (Roosa *et al.*, 2000). As the relevant promoters needed to obtain targeted and inducible expression of transgenes are applied to IFs, this method will undoubtedly open new frontiers in studies of IF function as well.

The ability to alter IF composition in an organism where development can be observed directly makes *Xenopus* a powerful system. By using it appropriately with other model systems, we may yet learn how this complex family of genes, with its exquisitely regulated patterns of expression, contributes to the cell and to the organism. Thus, *Xenopus* is likely to remain an important model system for IF studies for some time to come.

Acknowledgments

The writing of this chapter was supported by the National Science Foundation (IBN0236886). We thank Andrew Smith for supplying photographs.

References

Amaya, E., and Kroll, K. L. (1999). A method for generating transgenic frog embryos. *Methods Mol. Biol.* **97,** 393–414.

Asch, W. S., Leake, D., Canger, A. K., Passini, M. A., Argenton, F., and Schechter, N. (1998). Cloning of zebrafish neurofilament cDNAs for plasticin and gefiltin: Increased mRNA expression in ganglion cells after optic nerve injury. *J. Neurochem.* **71,** 20–32.

Beaulieu, J.-M., Jacomy, H., and Julien, J.-P. (2000). Formation of intermediate filament protein aggregates with disparate effects in two transgenic mouse models lacking the neurofilament light subunit. *J. Neurosci.* **20,** 5321–5328.

Beaulieu, J.-M., Robertson, J., and Julien, J.-P. (1999). Interactions between peripherin and neurofilaments in cultured cells: Disruption of peripherin assembly by the NF-M and NF-H subunits. *Biochem. Cell Biol.* **77,** 41–45.

Bennett, G. S., Hollander, B. A., and Laskowska, D. (1988). Expression and phosphorylation of the mid-sized neurofilament protein NF-M during chick spinal cord neurogenesis. *J. Neurosci. Res.* **21,** 376–390.

Boyne, L. J., Fischer, I., and Shea, T. B. (1996). Role of vimentin in early stages of neuritogenesis in cultured hippocampal neurons. *Int. J. Dev. Neurosci.* **14,** 739–748.

Canger, A. K., Passini, M. A., Asch, W. S., Leake, D., Zafonte, B. T., Glasgow, E., and Schechter, N. (1998). Restricted expression of the neuronal intermediate filament protein plasticin during zebrafish development. *J. Comp. Neurol.* **399,** 561–572.

Carden, M. J., Trojanowski, J. Q., Schlaepfer, W. W., and Lee, V. M. Y. (1987). Two-stage expression of neurofilament polypeptides during rat neurogenesis with early establishment of adult phosphorylation patterns. *J. Neurosci.* **7,** 3489–3504.

Carl, T. F., Dufton, C., Hanken, J., and Klymkowsky, M. W. (1999). Inhibition of neural crest migration in *Xenopus* using antisense slug RNA. *Dev. Biol.* **213,** 101–115.

Cary, R. B., and Klymkowsky, M. W. (1995). Disruption of intermediate filament organization leads to structural defects at the intersomite junction in *Xenopus* myotomal muscle. *Development* **121,** 1041–1052.

Charnas, L. R., Szaro, B. G., and Gainer, H. (1992). Identification and developmental expression of a novel low molecular weight neuronal intermediate filament protein in *Xenopus laevis. J. Neurosci.* **12,** 3010–3024.

Chou, Y.-H., Helfand, B. T., and Goldman, R. D. (2001). New horizons in cytoskeletal dynamics: Transport of intermediate filaments along microtubule tracks. *Curr. Opin. Cell Biol.* **13,** 106–109.

Christian, J. L., Edelstein, N. G., and Moon, R. T. (1990). Overexpression of wild-type and dominant negative mutant vimentin subunits in developing *Xenopus* embryos. *The New Biologist.* **2,** 700–711.

Chu, D. T. W., and Klymkowsky, M. W. (1987). Experimental analysis of cytoskeletal function in early *Xenopus laevis* embryos. *First Intntl. Symp. Cytoskel. Dev.* **8,** 140–142.

Cochard, P., and Paulin, D. (1984). Initial expression of neurofilaments and vimentin in the central and peripheral nervous system of the mouse embryo *in vivo. J. Neurosci.* **4,** 2080–2094.

Colucci-Guyon, E., Portier, M.-M., Dunia, I., Paulin, D., Pournin, S., and Babinet, C. (1994). Mice lacking vimentin develop and reproduce without an obvious phenotype. *Cell* **79,** 679–694.

de Waegh, S., Lee, V. M. Y., and Brady, S. T. (1992). Local modulation of neurofilament phosphorylation, axonal caliber, and slow axonal transport by myelinating Schwann cells. *Cell* **68,** 451–463.

Dent, J. A., Polson, A. G., and Klymkowsky, M. W. (1989). A wholemount immunocytochemical analysis of the expression of the intermediate filament protein vimentin in *Xenopus. Development* **105,** 61–74.

Eckes, B., Dogic, D., Colucci-Guyon, E., Wang, N., Maniotis, A., Ingber, D., Merckling, A., Langa, F., Aumailley, M., Delouvèe, A., Koteliansky, V., Babinet, C., and Krieg, T. (1998). Impaired mechanical stability, migration and contractile capacity in vimentin deficient fibroblasts. *J. Cell Sci.* **111,** 1897–1907.

Elder, G. A., Friedrich, V. L., Jr., Bosco, P., Kang, C., Giourov, A., Tu, P.-H., Lee, V. M. Y., and Lazzarini, R. A. (1998a). Absence of the mid-sized neurofilament subunit decreases axonal calibers, levels of light neurofilament (NF-L) and neurofilament content. *J. Cell Biol.* **141,** 727–739.

Elder, G. A., Friedrich, V. L., Jr., Kang, C., Bosco, P., Gourov, A., Tu, P.-H., Zhang, B., Lee, V. M. Y., and Lazzarini, R. A. (1998b). Requirement of heavy neurofilament subunit in the development of axons with large calibers. *J. Cell Biol.* **143,** 195–205.

Eyer, J., and Peterson, A. (1994). Neurofilament-deficient axons and perikaryal aggregates in viable transgenic mice expressing a neurofilament-beta-galacotsidase fusion protein. *Neuron* **12,** 389–405.

Faigle, W., Colucci-Guyon, E., Louvard, D., Amigorena, S., and Galli, T. (2000). Vimentin filaments in fibroblasts are a reservoir for SNAP23, a component of the membrane fusion machinery. *Mol. Biol. Cell.* **11,** 3485–3494.

Fliegner, K. H., Kaplan, M. P., Wood, T. L., Pintar, J. E., and Liem, R. K. H. (1994). Expression of the gene for the neuronal intermediate filament protein alpha-internexin coincides with the onset of neuronal differentiation in the developing rat nervous system. *J. Comp. Neurol.* **342,** 161–173.

Fuchs, E., and Cleveland, D. W. (1998). A structural scaffolding of intermediate filaments in health and disease. *Science* **279,** 514–519.

Garcia, M. L., Lobsiger, C. S., Shah, S. B., Deerinck, T. J., Crum, J., Young, D., Ward, C. M., Crawford, T. O., Gotow, T., Uchiyama, Y., Ellisman, M. H., Calcutt, N. A., and Cleveland, D. W. (2003). NF-M is an essential target for the myelin-directed "outside-in" signaling cascade that mediates radial axonal growth *J. Cell Biol.* **163,** 1011–1020.

Gervasi, C., Stewart, C.-B., and Szaro, B. G. (2000). *Xenopus laevis* peripherin (XIF3) is expressed in radial glia and proliferating neural epithelial cells as well as in neurons. *J. Comp. Neurol.* **423,** 512–531.

Giasson, B. I., and Mushynski, W. E. (1998). Intermediate filament disassembly in cultured dorsal root ganglion neurons is associated with amino-terminal head domain phosphorylation of specific subunits. *J. Neurochem.* **70,** 1869–1875.

Giebelhaus, D. H., Eib, D. W., and Moon, R. T. (1988). Antisense RNA inhibits expression of membrane skeleton protein 4.1 during embryonic development of *Xenopus. Cell* **53,** 601–615.

Gimlich, R. L. (1991). Fluorescent dextran clonal markers. *In* "Methods in cell biology" (B. K. Kay and H. B. Peng, eds.), Vol. 36, pp. 285–298. Academic Press, San Diego.

Gimlich, R. L., and Gerhart, J. C. (1984). Early cellular interactions promote embryonic axis formation in *Xenopus laevis*. *Dev. Biol.* **104,** 117–130.

Glasgow, E., Druger, R. K., Fuchs, C., Lane, W. S., and Schechter, N. (1994). Molecular cloning of gefiltin (ON1): Serial expression of two new neurofilament mRNAs during optic nerve regeneration. *EMBO J.* **13,** 297–305.

Goldman, R. D., Chao, Y.-H., Prahlad, V., and Yoon, M. (1999). Intermediate filaments: Dynamic processes regulating their assembly, motility, and interactions with other cytoskeletal systems. *FASEB J.* **13,** S261–S265.

Goldman, R. D., Khuon, S., Chou, Y. H., Opal, P., and Steinert, P. M. (1996). The function of intermediate filaments in cell shape and cytoskeletal integrity. *J. Cell Biol.* **134,** 971–983.

Goldstein, M. E., Cooper, H. S., Bruce, J., Carden, M. J., Lee, V. M. Y., and Schlaepfer, W. W. (1987). Phosphorylation of neurofilament proteins and chromatolysis following transection of rat sciatic nerve. *J. Neurosci.* **7,** 1586–1594.

Goldstein, M. E., Weiss, S. R., Lazzarini, R. A., Shneidman, P. S., Lees, J. F., and Schlaepfer, W. W. (1988). mRNA levels of all three neurofilament proteins decline following nerve transection. *Mol. Brain Res.* **3,** 287–292.

Goldstone, K., and Sharpe, C. R. (1998). The expression of *Xif3* in undifferentiated anterior neuroectoderm, but not in primary neurons, is induced by the neuralizing agent noggin. *Int. J. Dev. Biol.* **42,** 757–762.

Gurdon, J. B., (1996). *In* "Introductory comments: *Xenopus* as a laboratory organism" (R. C. Tinsley and H. R. Kobel, eds.), pp. 3–6. Oxford University Press, Oxford.

Harland, R. M. (1991). *In situ* hybridization: An improved whole mount method for *Xenopus* embryos. *In* "Methods in cell biology" (B. K. Kay and H. B. Peng, eds.), Vol. 36, pp. 685–696. Academic Press, San Diego.

Heasman, J., Holwill, S., and Wylie, C. C. (1991). Fertilization of cultured *Xenopus* oocytes and use in studies of maternally inherited molecules. *In* "Methods in cell biology" (B. K. Kay and H. B. Peng, eds.), Vol. 36, pp. 214–231. Academic Press, San Diego.

Hesse, M., Magin, T. M., and Weber, K. (2001). Genes for intermediate filament proteins and the draft sequence of the human genome: Novel keratin genes and a surprisingly high number of pseudogenes related to keratin genes 8 and 18. *J. Cell Sci.* **114,** 2569–2575.

Hirose, G., and Jacobson, M. (1979). Clonal organization of the central nervous system of the frog: I. Clones stemming from individual blastomeres of the 16-cell and earlier stages. *Develop. Biol.* **71,** 191–202.

Hoffman, P. N., Cleveland, D. W., Griffin, J. W., Landes, P. W., Cowan, N. J., and Price, D. L. (1987). Neurofilament gene expression: A major determinant of axonal caliber. *Proc. Natl. Acad. Sci. USA* **84,** 3472–3476.

Hoperskaya, O. A. (1975). The development of animals homozygous for a mutation causing periodic albinism (a[p]) in *Xenopus laevis*. *J. Embryol. Exp. Morphol.* **34,** 253–264.

Inagaki, M., Matsuoka, Y., Tsujimura, K., Ando, S., Tokui, T., Takahashi, T., and Inagaki, N. (1996). Dynamic property of intermediate filaments: Regulation by phosphorylation. *BioEssays* **18,** 481–487.

Jacobson, M., and Hirose, G. (1978). Origin of the retina from both sides of the embryonic brain: A contribution to the problem of crossing at the optic chiasma. *Science* **202,** 637–639.

Jacomy, H., Zhu, Q., Beaulieu, J.-M., and Julien, J.-P. (1999). Disruption of Type IV intermediate filament network in mice lacking the neurofilament medium and heavy subunits. *J. Neurochem.* **73,** 972–984.

Jiang, T. Q., Pickett, J., and Oblinger, M. M. (1994). Comparison of changes in Beta-tubulin and NF gene expression in rat DRG neurons under regeneration-permissive and regeneration-prohibitive conditions. *Brain Res.* **637,** 233–241.

Jiang, X. M., Zhao, J. X., Ohnishi, A., Itakura, C., Mizutani, M., Yamamoto, T., Murai, Y., and Ikeda, M. (1996). Regeneration of myelinated fiber after crush injury is retarded in sciatic nerves of mutant Japanese quails deficient in neurofilaments. *Acta Neuropathol. (Berl.).* **92,** 467–472.

Julien, J. P. (1999). Neurofilament functions in health and disease. *Curr. Opin. Cell Biol.* **9,** 554–560.

Karabinos, A., Schmidt, H., Harborth, J., and Weber, K. (2001). Essential roles for four cytoplasmic intermediate filament proteins in *Caenorhabditis elegans* development. *Proc. Natl. Acad. Sci. USA* **98,** 7863–7868.

Kay, B. K., and Peng, H. B. (1991). "Methods in cell biology, Vol. 36, *Xenopus laevis*: Practical uses in cell and molecular biology." Academic Press, San Diego.

Keller, R. (1991). Early embryonic development in *Xenopus laevis. In* "Methods in cell biology" (B. K. Kay and H. B. Peng, eds.), Vol. 36, pp. 62–116. Academic Press, San Diego.

Kim, O.-K., Ariano, M. A., Lazzarini, R. A., Levine, M. S., and Sibley, D. R. (2002). Neurofilament-M interacts with the D1 dopamine receptor to regulate cell surface expression and desensitization. *J. Neurosci.* **22,** 5920–5930.

Kintner, C. (1988). Effects of altered expression of the neural cell adhesion molecule, N-CAM, on early neural development in *Xenopus* embryos. *Neuron* **1,** 545–555.

Klymkowsky, M. W., and Hanken, J. (1991). Whole-mount staining of *Xenopus* and other vertebrates. *In* "Methods in cell biology" (B. K. Kay and H. B. Peng, eds.), Vol. 36, pp. 420–443. Academic Press, San Diego.

Klymkowsky, M. W., Miller, R. H., and Lane, E. B. (1983). Morphology, behavior, and interaction of cultured epithelial cells after the antibody-induced disruption of keratin filament organization. *J. Cell Biol.* **96,** 494–509.

Klymkowsky, M. W., Shook, D. R., and Maynell, L. A. (1992). Evidence that the deep keratin filament systems of the *Xenopus* embryo act to ensure normal gastrulation. *Proc. Natl. Acad. Sci. USA* **89,** 8736–8740.

Lander, E. S., Linton, L. M., and others (2001). Initial sequencing and analysis of the human genome. *Nature* **409,** 860–921.

Landmesser, L., and Swain, S. (1992). Temporal and spatial modulation of a cytoskeletal antigen during peripheral axonal pathfinding. *Neuron* **8,** 291–305.

Lariviere, R. C., Nguyen, M. D., Ribera-da-Silva, A., and Julien, J.-P. (2002). Reduced number of unmyelinated sensory axons in peripherin null mice. *J. Neurochem.* **81,** 525–532.

Leake, D., Asch, W. S., Canger, A. K., and Schechter, N. (1999). Gefiltin in zebrafish embryos: Sequential gene expression of two neurofilament proteins in retinal ganglion cells. *Diff.* **65,** 181–189.

Levavasseur, F., Zhu, Q., and Julien, J.-P. (1999). No requirement of alpha-internexin for nervous system development and for radial growth of axons. *Mol. Brain Res.* **69,** 104–112.

Lin, W., and Szaro, B. G. (1994). Maturation of neurites in mixed cultures of spinal cord neurons and muscle cells from *Xenopus laevis* embryos followed with antibodies to neurofilament proteins. *J. Neurobiol.* **25,** 1235–1248.

Lin, W., and Szaro, B. G. (1995). Neurofilaments help maintain normal morphologies and support elongation of neurites in *Xenopus laevis* cultured embryonic spinal cord neurons. *J. Neurosci.* **15,** 8331–8344.

Lin, W., and Szaro, B. G. (1996). Effects of intermediate filament disruption on the early development of the peripheral nervous system of *Xenopus laevis. Dev. Biol.* **179,** 197–211.

Liuzzi, F. J., and Tedeschi, B. (1992). Axo-glial interactions at the dorsal root transitional zone regulate neurofilament protein synthesis in axotomized sensory neurons. *J. Neurosci.* **12,** 4783–4792.

Lombardo, A., and Slack, J. M. W. (1997). Inhibition of *eFGF* expression in *Xenopus* embryos by antisense mRNA. *Dev. Dyn.* **208,** 162–169.

Mikucki, S. A., and Oblinger, M. M. (1991). Corticospinal neurons exhibit a novel pattern of cytoskeletal gene expression after injury. *J. Neurosci. Res.* **30,** 213–225.

Nakano, H., Amemiya, S., Shiokawa, K., and Taira, M. (2001). RNA interference for the organizer-specific gene Xlim-1 in *Xenopus* embryos. *Biochem. Biophys. Res. Commun.* **274,** 434–439.

Nasevicius, A., and Ekker, S. C. (2000). Effective targeted gene 'knockdown' in zebrafish. *Nature Genetics.* **26**, 129–130.

Nichols, A., Rungger-Brändle, E., Muster, L., and Rungger, D. (1995). Inhibition of *Xhox1A* gene expression in *Xenopus* embryos by antisense RNA produced from an expression vector read by RNA polymerase III. *Mech. Dev.* **52**, 37–49.

Nieuwkoop, P. D., and Faber, J. (1994). "Normal table of *Xenopus laevis* (Daudin)." Garland Publishing, New York.

Nutt, S. L., Bronchain, O. J., Hartley, K. O., and Amaya, E. (2001). Comparison of morpholino based translational inhibition during the development of *Xenopus laevis* and *Xenopus tropicalis. Genesis* **30**, 110–113.

O'Keefe, H. P., Melton, D. A., Ferreiro, B., and Kintner, C. (1991). *In situ* hybridization. *In* "Methods in cell biology" (B. K. Kay and H. B. Peng, eds.), Vol. 36, pp. 443–695. Academic Press, San Diego.

Pachter, J. S., and Liem, R. K. H. (1984). The differential appearance of neurofilament triplet polypeptides in the developing rat optic nerve. *Dev. Biol.* **103**, 200–210.

Peng, H. B., Baker, L. P., and Chen, Q. (1991). Tissue culture of *Xenopus* neurons and muscle cells as a model for studying synaptic induction. *In* "Methods in cell biology" (B. K. Kay and H. B. Peng, eds.), Vol. 36, pp. 511–526. Academic Press, San Diego.

Rao, M. V., Campbell, J., Yuan, A., Kumar, A., Gotow, T., Uchiyama, Y., and Nixon, R. A. (2003). The neurofilament middle molecular mass subunit carboxyl-terminal tail domains is essential for the radial growth and cytoskeletal architecture of axons but not for regulating neurofilament transport rate. *J. Cell Biol.* **163**, 1021–1031.

Rao, M. V., Garica, M. L., Miyazaki, Y., Gotow, T., Yuan, A., Mattina, S., Ward, C. M., Calcutt, N. A., Uchiyama, Y., Nixon, R. A., and Cleveland, D. W. (2002). Gene replacement in mice reveals that the heavily phosphorylated tail of neurofilament heavy subunit does not affect axonal caliber or the transit of cargoes in slow axonal transport. *J. Cell Biol.* **158**, 681–693.

Rao, M. V., Houseweart, M. K., Williamson, T. L., Crawford, T. O., and Folmer, J. (1998). Neurofilament-dependent radial growth of motor axons and axonal organization of neurofilaments does not require the neurofilament heavy subunit (NF-H) or its phosphorylation. *J. Cell Biol.* **143**, 171–181.

Rebagliati, M. R., and Melton, D. A. (1987). Antisense RNA injections in fertilized frog eggs reveal an RNA duplex unwinding activity. *Cell* **48**, 599–605.

Retaux, S., McNeill, L., and Harris, W. A. (1996). Engrailed, retinotectal targeting, and axonal patterning in the midbrain during *Xenopus* development: An antisense study. *Neuron* **16**, 63–75.

Roosa, J. R., Gervasi, C., and Szaro, B. G. (2000). Structure, biological activity of the upstream regulatory sequence, and conserved domains of a middle molecular mass neurofilament gene of *Xenopus laevis. Mol. Brain Res.* **82**, 35–51.

Sánchez, I., Hassinger, L., Paskevich, P. A., Shine, H. D., and Nixon, R. A. (1996). Oligodendroglia regulate the regional expansion of axon caliber and local accumulation of neurofilaments during development independently of myelin formation. *J. Neurosci.* **16**, 5095–5105.

Sakaguchi, T., Okada, M., Kitamura, T., and Kawasaki, K. (1993). Reduced diameter and conduction velocity of myelinated fibers in the sciatic nerve of a neurofilament-deficient mutant quail. *Neurosci. Lett.* **153**, 65–68.

Schwartz, M. L., Shneidman, P. S., Bruce, J., and Schlaepfer, W. W. (1990). Axonal dependency of the postnatal upregulation in neurofilament expression. *J. Neurosci. Res.* **27**, 193–201.

Schwartz, M. L., Shneidman, P. S., Bruce, J., and Schlaepfer, W. W. (1994). Stabilization of neurofilament transcripts during postnatal development. *Mol. Brain Res.* **27**, 215–220.

Shain, D. H., and Zuber, M. X. (1996). Sodium dodecyl sulfate (SDS)-based whole-mount *in situ* hybridization of *Xenopus laevis* embryos. *J. Biochem. Biophys. Meth.* **31**, 185–188.

Shaw, G., Banker, G. A., and Weber, K. (1985). An immunofluorescence study of neurofilament protein expression by developing hippocampal neurons in tissue culture. *Eur. J. Cell Biol.* **39**, 205–216.

Shaw, G., and Weber, K. (1982). Differential expression of neurofilament triplet proteins in brain development. *Nature* **298**, 277–279.

Shea, T. B., Beerman, M. L., and Fischer, I. (1993). Transient requirement for vimentin in neuritogenesis: Intracellular delivery of anti-vimentin antibodies and antisense oligonucleotides inhibit neurite initiation, but not elongation of exiting neurites in neuroblastoma. *J. Neurosci. Res.* **36**, 66–76.

Steinbeisser, H., Fainsod, A., Niehrs, C., Sasai, Y., and De Robertis, E. M. (1995). The role of *gsc* and *BMP-4* in dorsal-ventral patterning of the marginal zone in *Xenopus*: A loss-of-function study using antisense RNA. *EMBO J.* **14**, 5230–5243.

Sumikawa, K., and Miledi, R. (1988). Repression of nicotinic acetylcholine receptor expression by antisense RNAs and an oligonucleotide. *Proc. Natl. Acad. Sci. USA* **85**, 1302–1306.

Szaro, B. G., and Gainer, H. (1988). Identities, antigenic determinants, and topographic distributions of neurofilament proteins in the nervous systems of adult frogs and tadpoles of *Xenopus laevis*. *J. Comp. Neurol.* **273**, 344–358.

Szaro, B. G., Grant, P., Lee, V. M. Y., and Gainer, H. (1991). Inhibition of axonal development after injection of neurofilament antibodies into a *Xenopus laevis* embryo. *J. Comp. Neurol.* **308**, 576–585.

Szaro, B. G., Lee, V. M. Y., and Gainer, H. (1989). Spatial and temporal expression of phosphorylated and non-phosphorylated forms of neurofilament proteins in the developing nervous system of *Xenopus laevis*. *Dev. Brain Res.* **48**, 87–103.

Tabti, N., and Poo, M.-M. (1991). Culturing spinal neurons and muscle cells from *Xenopus* embryos. *In* (G. Banker and K. Goslin, eds.), pp. 137–154. MIT Press, Cambridge.

Tapscott, S. J., Bennett, G. S., Toyama, Y., Kleinbart, F., and Holtzer, H. (1981). Intermediate filament proteins in the developing chick spinal cord. *Dev. Biol.* **86**, 40–54.

Tinsley, R. C., and Kobel, H. R. (1996). "The biology of *Xenopus*." Oxford University Press, Oxford.

Tompkins, R. (1977). Grafting analysis of the periodic albino mutant of *Xenopus laevis*. *Proc. Natl. Acad. Sci. USA* **76**, 4350–4354.

Torpey, N., Wylie, C. C., and Heasman, J. (1992). Function of maternal cytokeratin in *Xenopus* development. *Nature* **357**, 413–415.

Undamatla, J., and Szaro, B. G. (2001). Differential expression and localization of neuronal intermediate filament proteins within newly developing neurites in dissociated cultures of *Xenopus laevis* embryonic spinal cord. *Cell Motil. Cytoskel.* **49**, 16–32.

Vize, P. D., Hemmati-Brivanlou, A., Harland, R. M., and Melton, D. A. (1991). Assays for gene function in developing *Xenopus* embryos. *In* "Methods in cell biology" (B. K. Kay and H. B. Peng, eds.), Vol. 36, pp. 368–388. Academic Press, San Diego.

Walker, K. L., Yoo, H.-K., Undamatla, J., and Szaro, B. G. (2001). Loss of neurofilaments alters axonal growth dynamics. *J. Neurosci.* **21**, 9655–9666.

Warner, A. E., Guthrie, S. C., and Gilula, N. B. (1984). Antibodies to gap junctional protein selectively disrupt junctional communication in the early amphibian embryo. *Nature* **311**, 127–131.

Weisblat, D. A., Sawyer, R. T., and Stent, G. S. (1978). Cell lineage analysis by intracellular injection of a tracer enzyme. *Science* **202**, 1295–1298.

Wetzel, D. M., Lee, V. M. Y., and Erulkar, S. D. (1989). Long term cultures of neurons from adult frog brain express GABA and glutamate-activated channels. *J. Neurobiol.* **20**, 255–270.

Wheeler, G. N., Hamilton, F. S., and Hoppler, S. (2000). Inducible gene expression in transgenic *Xenopus* embryos. *Curr. Biol.* **10**, 849–852.

Williamson, T. L., Marszalek, J. R., Vechio, J. D., Bruijn, L. I., Lee, M. K., Xu, Z., Brown, R. H., Jr., and Cleveland, D. W. (1996). Neurofilaments, radial growth of axons, and mechanisms of motor neuron disease. *Cold Spring Harbor Symp. Quant. Biol.* **61**, 709–723.

Wong, J., and Oblinger, M. M. (1990). Differential regulation of peripherin and neurofilament gene expression in regenerating rat DRG neurons. *J. Neurosci. Res.* **27**, 332–341.

Wong, P. C., Marszalek, J., Crawford, T. O., Xu, Z. S., Hsieh, S. T., Griffin, J. W., and Cleveland, D. W. (1995). Increasing neurofilament subunit NF-M expression reduces axonal NF-H, inhibits radial growth, and results in neurofilamentous accumulation in motor neurons. *J. Cell Biol.* **130,** 1413–1422.

Wu, M., and Gearhart, J. (1991). Raising *Xenopus* in the laboratory. *Methods Cell Biol.* **36,** 3–18.

Yamasaki, H., Itakura, C., and Mizutani, M. (1991). Hereditary hypotrophic axonopathy with neurofilament deficiency in a mutant strain of the Japanese quail. *Acta Neuropathol.* **82,** 427–434.

Zhao, J. X., Ohnishi, A., Itakura, C., Mizutani, M., Yamamoto, T., Hojo, T., and Murai, Y. (1995). Smaller axon and unaltered numbers of microtubules per axon in relation to number of myelin lamellae of myelinated fibers in the mutant quail deficient in neurofilaments. *Acta Neuropathol.* **89,** 305–312.

Zhao, Y., and Szaro, B. G. (1994). The return of phosphorylated and nonphosphorylated epitopes of neurofilament proteins to the regenerating optic nerve of *Xenopus laevis*. *J. Comp. Neurol.* **343,** 158–172.

Zhao, Y., and Szaro, B. G. (1995). The optic tract and tectal ablation influence the composition of neurofilaments in regenerating optic axons of *Xenopus laevis*. *J. Neurosci.* **15,** 4629–4640.

Zhao, Y., and Szaro, B. G. (1997). Xefiltin, a *Xenopus laevis* neuronal intermediate filament protein, is expressed in actively growing optic axons during development and regeneration. *J. Neurobiol.* **33,** 811–824.

Zhu, Q., Couillard-Despres, S., and Julien, J.-P. (1997). Delayed maturation of regenerating myelinated axons in mice lacking neurofilaments. *Exp. Neurol.* **148,** 299–316.

Zhu, Q., Lindenbaum, M., Levavasseur, F., Jacomy, H., and Julien, J.-P. (1998). Disruption of the NF-H gene increases axonal microtubule content and velocity of neurofilament transport: Relief of axonopathy resulting from the toxin β,β'-iminodipropionitrile. *J. Cell Biol.* **143,** 183–193.

CHAPTER 24

Intermediate Filaments in
Caenorhabditis elegans

Alexandra Fridkin,[*] Anton Karabinos,[†] and Yosef Gruenbaum[*]

[*]Department of Genetics
The Institute of Life Sciences
The Hebrew University of Jerusalem
Jerusalem, 91904, Israel

[†]Max Planck Institute for Biophysical Chemistry
Department of Biochemistry
37077 Goettingen, Germany

METHODS IN CELL BIOLOGY, VOL. 78
Copyright 2004, Elsevier Inc. All rights reserved.
0091-679X/04 $35.00

I. Introduction

A. *C. elegans*:

The free-living, soil nematode *Caenorhabditis elegans* is an excellent model organism for studying intermediate filament (IF) structure and function. The adult is only about 1-mm long and feeds on bacteria. There are two sexes: males, which produce sperm, and hermaphrodites, which produce both oocytes and sperm and can self-fertilize. *C. elegans* has a short life cycle of ~3.5 days and a simple body containing only 959 somatic cells, which includes all major cell types (neurons, muscle, intestine, hypodermis, germ cells, and others). Fertilization in *C. elegans* occurs inside the hermaphrodite, where embryo development starts, eggs are laid, and the embryo develops until it hatches ~13.5 h after the first cleavage. The animal undergoes four larval stages (L1–L4) and becomes a mature adult. The adult hermaphrodite lays about 300 eggs during the first 4 days of reproductive life. There are several major advantages for using *C. elegans* to study IFs. The entire cell lineage is known, enabling real-time microscopic analysis of development and differentiation, and the animal remains transparent throughout its life cycle. The entire genome has been sequenced revealing only 11 cytoplasmic IF genes and one nuclear lamin gene. This number is significantly lower than >65 cytoplasmic IF genes and 3 lamin genes in mammals. In addition, transgenic techniques, as well as the RNA interference technique, sophisticated genetic analyses, and the production of a large collection of mutant lines all make *C. elegans* especially attractive for studying the functions of evolutionarily conserved genes including IFs. In this chapter, we will include a short review of our current knowledge of IFs in *C. elegans* and will summarize useful techniques for their analyses.

II. Intermediate Filaments in *C. elegans*

A. Cytoplasmic Intermediate Filaments in *C. elegans*

The 11 cytoplasmic IF sequences in the *C. elegans* database were sorted according to their evolutionary subgroup and named A (A1–A4), B (B1–B2), C (C1–C2), D (D1–D2), and E (E1) (Dodemont *et al.*, 1994; Karabinos *et al.*, 2001). All these IFs have a head domain, a rod domain containing a lamin-like length of the coil 1b, and a tail domain. Six of these proteins (A1–A4 and B1–B2) also have a globular (lamin-homology) segment in their tail domain. Green Fluorescent protein (GFP) promoter reporters and specific antibodies were used to analyze the spatial and temporal pattern of expression of the 11 cytoplasmic IF proteins. These analyses revealed that A1 and B1 produce two alternatively spliced variants (A1a, A1b, B1a, and B1b, respectively) and that one or both splice variants of the B1 gene are always coexpressed in a tissue-specific manner with at least one member of the A family in hypodermis, pharynx, pharyngeal-intestinal valve, excretory cells, uterus, vulva, and rectum (Karabinos *et al.*, 2001, 2003a). Antibodies showed that C2 is first expressed in the late embryo and continues to be expressed in the cytoplasm and desmosomes of intestinal cells and in the pharynx desmosomes (Karabinos *et al.*, 2002), whereas strong intestinal expression was observed for proteins B2, C1, D1, D2, and E1 (A. Karabinos and K. Weber, unpublished results). The coexpression pattern of the A and B proteins described previously relates to their *in vitro* polymerization properties. In blot overlay assays of all recombinant IF proteins, B1 binds strongly to A1–A4, while mixing it with with A1a, A1b, A2, or A3 yielded heteropolymeric IF in filament assembly assays (Karabinos *et al.*, 2003a) (see Fig. 1A). To determine the roles of cytoplasmic IFs, RNA interference (RNAi; Fire *et al.*, 1998) on the 11 *C. elegans* genes was performed by standard microinjection of the double-stranded RNA (dsRNA). This study revealed that A1, A2, A3, and B1 genes are essential for nematode development (Karabinos *et al.*, 2001). These results show that single IF genes are essential. RNAi silencing of either A3 or B1 caused late embryonic lethality, whereas downregulating of A1 or A2 and the genetic A2 knockout (Hapiak *et al.*, 2003) induced larval lethality. Because the efficiency of the downregulation of the IF proteins was not analyzed, it was possible that genetic knockout or another form of dsRNA delivery could be more effective than microinjection in particular developmental stages and could perhaps also reveal essential function for some of the other six IF genes. This has been observed with C2, whereas C2 RNAi delivery by microinjection caused only a mild dumpy phenotype of adults

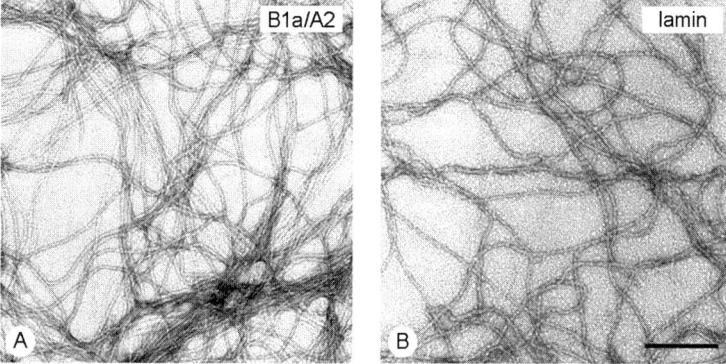

Fig. 1 Electron micrographs of intermediate filaments (IFs) assembled *in vitro* from Ce-lamin and from equal amounts of the Bla/A2 proteins. (A) Equal amounts (0.6 mg/ml) of the recombinant IF proteins B1a and A2 in urea buffer were mixed together and dialyzed for 20 h against the 10 mM Tris-HCl, pH = 7.2, buffer containing 1 mM β-mercaptoethanol. (B) *C. elegans* lamin (0.1 mg/ml) in urea buffer was dialyzed for 4 h against 2 mM Tris-HCl, pH = 9, and then 16 h against 15 mM Tris-HCl, pH = 7.4. Both buffers contained 1 mM DTT. In both (A) and (B), the samples were negatively stained with 2% uranyl acetate and viewed with an electron microscope. Scale bar represents 200 nm. Data reprinted by permission of *J. Mol. Biol.* from Karabinos *et al.* (2003a) and from Karabinos *et al.* (2003b).

(Karabinos *et al.*, 2001). Lethality in about 3% of the adult progeny occurred on application of dsRNA by feeding (Karabinos *et al.*, 2003a).

The phenotypes were always observed in tissues where the IF proteins are expressed. For example, partial downregulation of B1 causes paralysis of late larva, detachment of hypodermis from cuticle, displacement of body muscles, and uneven excretory canals, which all essentially match the expression pattern of B1 in larva (Karabinos *et al.*, 2003a). On the cellular level, the downregulation of B1 also caused a striking structural defect in the filament bundles of the pharyngeal marginal cells of live larvae as visualized by the A1-GFP fluorescence (Karabinos *et al.*, 2003a). These B1 phenotypes partially overlap with those of A2 and A3, which is in agreement with the B1/A coexpression pattern and the observation of an obligatory heteropolymeric B1/A IF formation *in vitro* (see earlier).

With fluorescence recovery after photobleaching (FRAP) analysis in anesthetized *C. elegans*, the properties of the GFP-tagged IF proteins have been analyzed for the first time in a living organism (Karabinos *et al.*, 2003a). Ala-GFP containing filaments in the marginal cells of the pharynx showed no recovery after 4 h, and similar results were obtained for the filament bundles visualized by the B1a-GFP. In both experiments, the first signs of fluorescence recovery were visible approximately 5 h after photobleaching. These results indicate that in nematode cells there is an equilibrium between unpolymerized IF subunits and the polymerized IF of the tonofibrils. Moreover, a similar rate of fluorescence recovery for A1 and B1 proteins is in line with the obligatory heteropolymeric nature of the polymerized IF (see earlier).

B. Nuclear Intermediate Filaments in *C. elegans*

The *C. elegans* genome contains a single lamin gene, termed *lmn-1*, encoding a single lamin protein, termed Ce-lamin (Liu *et al.*, 2000). The structure of Ce-lamin is similar to that of *Drosophila* and vertebrate lamins with two exceptions: it lacks 14 residues (2 heptads) at the amino terminal part of coil 2b and it does not contain a cdk1 site in front of coil 1a (Riemer *et al.*, 1993). Like other lamins it forms dimers and head-to-tail polymers. Interestingly, by use of low ionic strength buffers, which are known to yield *in vitro* 10-nm filaments with invertebrate cytoplasmic IF proteins (Herrmann *et al.*, 2003), the vertebrate and *Drosophila* lamins formed only paracrystals with a 24- to 25-nm axial repeat, which is half the length of the lamin rod domain (Stuurman *et al.*, 1998). In contrast, Ce-lamin formed IFs *in vitro* with an average diameter of 10.2 ($\pm$0.7) nm, and no paracrystals were observed in these experiments (Karabinos *et al.*, 2003b) (Fig. 1B). These results demonstrate the ability of lamins to form IFs under certain conditions, which fits the observation that the cytoplasmic IF vimentin, fused to a nuclear localization signal, can form IF in mammalian cell nuclei (Dreger *et al.*, 2002) and that such IF-like structures were also observed in the nuclear lamina of *Xenopus* oocytes (Aebi *et al.*, 1986). Nevertheless, the IF-like structures were so far not found in nuclei of wild-type *C. elegans* or mammals (Cohen *et al.*, 2002).

Ce-lamin is expressed ubiquitously during *C. elegans* development and is present both at the nuclear periphery and nucleoplasm (Liu *et al.*, 2000). During mitosis, it remains in a spindle envelope until mid-late anaphase, similar to inner nuclear membrane (INM) proteins and in contrast to nucleoporins (Lee *et al.*, 2000). RNAi downregulation of Ce-lamin (*lmn-1 (RNAi)*) caused an early embryonic arrest, indicating that lamins are essential components of the nuclear envelope (Liu *et al.*, 2000). Ce-lamin is required for maintaining nuclear shape, because at the two-cell stage embryo the *lmn-1 (RNAi)* nuclei showed rapid changes in nuclear shape. Despite the gross defects in nuclear morphology, nuclear divisions still occurred in these embryos. Downregulation of Ce-lamin also caused mitotic arrest with chromatin anaphase bridges and unequal distribution of chromatin into daughter nuclei (Fig. 3). In addition, most nuclei of the *lmn-1 (RNAi)* embryos showed an abnormal clustering of nuclear pore complexes (Liu *et al.*, 2000). Most animals that hatched from eggs laid outside the most potent window of RNAi activity had no germ cells or reduced amounts and defective germ cells. A fraction of these germ cells had multiple nuclei, large nuclei, or spermatocyte-like nuclei with condensed chromatin. A small fraction of the F2 embryos from the semisterile animals developed into fertile adults with a high incidence of males (12.25%). These results again support the idea that lamins are required for normal chromatin segregation.

C. Proteins of the Inner Nuclear Membrane that Interact with Ce-Lamin

Several characterized INM proteins interact with Ce-lamin and/or require Ce-lamin for their INM localization. These include Ce-emerin (Gruenbaum *et al.*, 2002), Ce-MAN1 (Liu *et al.*, 2003), UNC-84 (Lee *et al.*, 2002), and matefin (Fridkin *et al.*, 2004). In addition, many of the recently discovered 67 mammalian INM proteins (Schirmer *et al.*, 2003) have homologs in *C. elegans* and are potentially lamin interacting proteins.

III. General Methodology

Studying *C. elegans* IF genes uses a large number of different conventional *C. elegans* techniques, which can be found in books (for example see Epstein and Shakes (1995) and on the *C. elegans* Web site (http://elegans.swmed.edu/). A useful Web site to follow different *C. elegans* strains and available deletions in specific genes is http://www.wormbase.org/. In this chapter, we will describe methods that were either developed or adjusted for studying IF genes in *C. elegans*. We will describe how *C. elegans* IFs can be assembled *in vitro*, detected *in vivo* by indirect immunofluorescence or immunogold electron microscopy (EM), processed for thin-section EM, and live imaging to follow IFs in worms.

IV. Forming Intermediate Filaments *In Vitro*

A. Protein Expression and Purification

To make filaments, the IF proteins are first bacterially expressed by use of any of the available bacterial expression systems and purified to near homogeneity. We routinely use the pET20 vector (Novagen) and the *E. coli* strain BL21(DE3)pLysS. However, there are no constraints on which bacterial expression system to use. Recombinant IF proteins are always found in the inclusion body preparations.

To purify the protein from pET expression plasmid, we frequently use the His_6 tag (Goldberg *et al.*, 1999; Liu *et al.*, 2003), which, depending on the specific pET vector, is fused either at the amino or at the carboxy end of the expressed protein, and Ni-NTA agarose beads that bind to the His_6 tag. For a small-scale purification, a 100-ml bacterial culture is resuspended in 10 ml lysis buffer (50 mM Tris-HCl, pH 8.0, 0.3 M NaCl, 1 mM PMSF, 8 M urea) and kept overnight at $-20\,^{\circ}C$. The bacteria are then sonicated on ice four times, 30 s each, with a Sonic sonicator (Heat Systems Ultrasonic Inc. NY) with 3-mm tip at 37% power and centrifuged at 12,000*g* for 15 min at $4\,^{\circ}C$. The supernatant is transferred to a tube containing 3 ml Ni-NTA agarose beads (Quiagen, Cat. #30210) in buffer containing 8 M urea, 50 mM Tris-HCl, pH 8.0, 0.5 M NaCl, and incubated for 1 h at $4\,^{\circ}C$ under constant mixing. The Ni-NTA beads are then used to make a small column (0.8-cm diameter), and the column is washed three times, each with 15 ml of buffer containing 50 mM Tris-HCl, pH 8.0, 0.5 M NaCl, 8 M urea, 20 mM imidazole (Merck, Germany Cat. # 104716). The column is further washed until the OD_{280} is less than 0.05. The bound protein is then eluted with 15 ml buffer containing 20 mM Tris-HCl, pH 7.5, 100 mM NaCl, 8 M urea, 200 mM imidazole, and 1 ml fractions are collected. The fractions with the highest OD_{280} are pooled. When necessary, the His_6-tag is removed by exoproteolytic digestion with the TAGZyme system from QUIAGEN (Valencia, CA): http://www1.qiagen.com/literature/handbooks/INT/ProteinPurification.aspx.

Comments:

1. In general, bacterially expressed *C. elegans* IFs are stable and not toxic to bacteria.
2. Freshly transfected bacteria give higher levels of expression.

B. Making Ce–Lamin Intermediate Filaments *In Vitro*

Bacterially expressed and purified *C. elegans* lamin (0.1 mg/ml) in urea buffer is dialyzed at room temperature for 4 h against a buffer containing 2 mM Tris-HCl, pH 9 and then for up to 16 h against a buffer containing 15 mM Tris-HCl, pH 7.4. Both buffers contain 1 mM of DTT and are made in Milli-Q water. The protein sample is placed on a glow-discharged and carbon-coated copper grid, washed on drop of distilled water, negatively stained with 2% uranyl acetate, and viewed with an EM. Under these conditions, Ce-lamin makes stable 10-nm IFs (Fig. 1B; Karabinos *et al.*, 2003b). Paracrystals of Ce-lamin with an axial repeat length of 44.7 nm are obtained when the low-salt buffers contain 20–30 mM $CaCl_2$.

Comments:

1. Use only highly purified water. We use the Milli Q water (Millipore, Billerica Massachusetts).

C. Making *C. elegans* Cytoplasmic Intermediate Filaments *In Vitro*

Equimolar amounts of recombinant proteins from the A and B IF subgroups are dialyzed at room temperature for 20 h on dialysis filters against either the 5 mM Tris-HCl buffer, pH 7.2, containing 15 mM NaCl, or the 10 mM Tris-HCl buffer, pH 7.2. Both buffers contain 1 mM 2-mercaptoethanol. The sample is then negatively stained with 2% uranyl acetate and viewed with an EM.

V. Indirect Immunofluorescence Analysis

A. Preparation of Poly-Lysine–Coated Slides

Slides (10–2066 Cell-Line/Erie Scientific CO) are washed with distilled water and soap, then washed with double distilled water (DDW), dried in oven at 65 °C, and cooled to room temperature. A 150-μl drop of poly-Lysine (Sigma P-1524 poly-Lysine 0.1–0.2% in DDW) is placed in the middle of the slide and left for 30 min at room temperature. The excess poly-Lysine is removed with a Gilson pipette and the slide is placed in a dry oven for 3 h at 65 °C. The slides can be stored for few weeks in a clean box.

B. Antibody Staining of Larvae and Adults

Wild-type, mutant, or RNAi-treated larvae or adult *C. elegans* are collected and washed in DDW or placed on an empty NGM plate for couple of min to get rid of bacteria. NGM plates are prepared by adding 3 g NaCl, 2.5 g bacto-peptone, and 17 g agar to 975 ml DDW, autoclaving and then adding 1 ml of cholesterol (5g/L in ethanol), 1 ml of 1 M $CaCl_2$, 1 ml of 1 M $MgSO_4$, and 25 ml of 1 M KH_2PO_4. A 9-μl drop of DDW is pipetted on the poly-Lysine–covered area, and the *C. elegans* are placed in this drop. It is not recommended to mix animals from different larval stages or larvae and adults. Place a 40- (or longer) × 20-mm coverslip on top of the animals perpendicular to the slide and tap gently on it until the larva is somewhat flattened or eggs are coming out of the adult vulva. The slide is then placed either for >10 min on dry ice or the area of the slide containing the worms is immersed for few seconds in liquid N_2, and then the coverslip is quickly removed. For methanol/acetone fixation, the slide is immediately incubated in methanol for 20 min at $-20\,^{\circ}$C, then in acetone for 10 min at $-20\,^{\circ}$C, and then in phosphate-buffered saline (PBS) containing 0.1% Tween-20 (PBST) for 15 min at $22\,^{\circ}$C. For formaldehyde fixation, the slide is immediately incubated in methanol for 4 min at $-20\,^{\circ}$C, and then in a freshly prepared PBST containing 3.7% formaldehyde and incubated for 20 min at $22\,^{\circ}$C. Unless otherwise mentioned, all further steps are performed at $22\,^{\circ}$C. For both fixation methods, the slides are washed twice, 3 min each, with PBST and incubated for 10 min with PBST containing either 5% normal goat serum or 10% low-fat milk (blocker solution) to block nonspecific epitopes. After a quick wash with PBST, a drop of 50- to 100-μl primary antibody at the proper dilution in the blocker solution is placed on top of the worms, the worms are covered with 22×22-mm coverslip, and incubated for 3 h at $22\,^{\circ}$C or overnight at $4\,^{\circ}$C on a shaker. The coverslip is then removed by dipping the slide in PBST, and the slides are then washed three times, 30 min each, in PBST. A 50- to 100-μl drop of secondary antibody at the proper dilution in PBST is placed on top of the worms and covered with 22×22-mm coverslip and incubated for 2 h at $22\,^{\circ}$C. The slides are then washed three times, 30 min each, in PBST and incubated for 10 min in PBS containing 1 μg/ml DAPI to stain the DNA. After a single wash for 5 min in PBS, a drop of 20 μl glycerol containing 2% *N*-propyl gallate or any of the commercial mounting solutions is placed on top of the worms. The worms are then covered with a coverslip, and the edges are sealed with a commercial nail hardener.

Comments:

1. If animals do not stick well to slide: (1) further remove the bacteria from the animals by dipping them quickly in DDW before placing them on the slide and (2) make fresh poly-Lysine solution.

C. Antibody Staining of Embryos

For a large-scale antibody staining of embryos, adult hermaphrodites are grown on three 9-cm NGM plates. Embryos are collected from NGM plates by use of the Clorox/M9 technique. Each plate is washed with 3 ml M9 buffer (3 g KH_2PO_4, 6 g Na_2HPO_4, 5 g NaCl, 1 ml 1 M $MgSO_4$, per liter), and a rubber policeman is used to detach the worms and embryos from the plate. Each plate is washed again with 2 ml M9, and all washes are collected in a 15-ml conical tube. The tube is centrifuged at 3000g for 1 min, and the pellet is resuspended and washed two times in 5 ml DDW. A solution containing 5 ml DDW, 0.4 ml of 10 N NaOH and 0.66 ml of 3% sodium hypochlorite (which dissolve the adults while leaving intact embryos), is added to the packed embryos, and the tube is occasionally vortexed until adults are not present anymore. If, after 10 min, adults are still present in the solution, an additional 0.3 ml of 3% sodium hypochlorite is added, and the tube is incubated for additional 3–10 min. The tube is then centrifuged at 3000*g* for 1 min, and the pelleted embryos are washed three times with DDW. The embryos are then transferred to a 2-ml Eppendorf tube, the volume is adjusted with DDW to 610 μl, and several solutions are added in the following order: (1) 140 μl buffer containing 30 mM Na-PIPES, pH 7.4 ((piperazine-1,4-bis (2-ethanesulfonic acid) disodium salt)), 160 mM KCl, 40 mM NaCl, 20 mM Na_2EGTA, 10 mM spermidine-HCl, in 50% methanol; (2) 500 μl of 4% formaldehyde in DDW; (3) 250 μl 50 mM of Na_2EGTA in 90% methanol. The tube is vortexed and incubated for 15 min on ice with occasional mixing and frozen in liquid N_2. The tube is then thawed under tap water and left on a shaker for 15–20 min at room temperature. The embryos are then washed once with a buffer containing 100 mM Tris-HCl, pH 7.4, 1% Triton X-100, 1 mM EDTA and once with PBS containing 0.1% bovine serum albumin (BSA), 0.5% Triton X-100, 5 mM sodium azide, and 1 mM EDTA (PBST-B), incubated for 15 min at room temperature, and resuspended in 1 ml PBST-B. The fixed embryos can be stored for several weeks at 4 °C.

Antibody staining is performed essentially as described (Chen *et al.*, 2000). Embryo suspension (100 μl) is centrifuged for 1 min at 3000 rpm in a microfuge, and the supernatant is removed; 20 μl of primary antibodies diluted in PBS containing 0.45% Triton X-100, 1 mM EDTA, 1% BSA, and 5 mM sodium azide (PBST-A) are added, and the embryos are incubated overnight at 22 °C on a shaker. Embryos are then washed four times, 25 min each, with PBST-A; 20 μl of secondary antibodies diluted in PBST-A are added, and the tube is incubated for 2 h at 22 °C with occasional mixing. Embryos are then washed four times, 25 min each, with PBST-B, incubated for 10 min with PBS containing 1 μg/ml DAPI, washed once with PBS, and placed on a slide with a mounting medium (see earlier) for viewing.

Comments:

1. Many normal sera contain antibodies against epitopes in *C. elegans*. Affinity purification of these antibodies should get rid of these epitopes.

VI. Immunogold–Electron Microscopy Preembedding Staining of *C. elegans* Embryos

The protocol that we present here is based on Cohen *et al.* (2002). *C. elegans* embryos are collected from three 9-cm NGM plates with the Clorox/M9 technique as described earlier. Embryos are resuspended in 400 μl DDW and fixed by adding the embryos to a 1-ml solution containing 80 mM KCl, 20 mM NaCl, 1.3 mM Na_2EGTA, 3.2 mM spermidine, 7.5 mM sodium HEPES, pH 6.5, 25% methanol, 2% paraformaldehyde, and 1.5% glutaraldehyde, and immediately immersed in liquid N_2. The embryos are thawed under tap water and incubated at room temperature for 45 min in the same fixative. Fixed embryos are washed once in

1 ml of 100 mM Tris-HCl, pH 7.5 and three times, 15 min each, in PBS containing 0.1% BSA and 1 mM EDTA (PBSB). Embryos are then pelleted and incubated overnight in primary antibodies diluted in PBS containing 1 mM EDTA and 1% BSA. Excess antibodies are removed by four washes, 1 h each, in PBSB. Embryos are then incubated overnight with 6-nm gold-conjugated secondary antibodies at the dilution recommended by the manufacturer. Embryos are pelleted and fixed for 1 h at room temperature with 2.5% glutaraldehyde. After washing in PBS and centrifugation at 2000g, the embryo pellet is mounted in 15 μl of 3.4% low-melting agarose. The agarose block is transferred to PBS and washed four times, 30 min each, in PBS. Embryos are then postfixed in 1% buffered OsO_4 containing 1.5% reduced $K_4Fe(CN)_6$ for 1 h at room temperature, dehydrated in graded series of ethanol (50%, 70%, 100%), 10 min each, followed by three additional incubations, 10 min each, in 100% ethanol and subjected to two changes of propylene oxide. The sample is then incubated in a series of graded Epon/ethanol solutions, 2 h each, at room temperature with Epon/ethanol (1:2), Epon/ethanol (1:1), Epon/ethanol (2:1), overnight in fresh 100% Epon at 4 °C in a desiccating chamber, and three more times, 2 h each, at room temperature with 100% Epon. The Epon is left to polymerize for 2 days at 60 °C in a dry oven. The Epon block is then sectioned with a Diatome diamond knife to give 80- to 90-nm thin sections. The sections are picked up on 200-mesh thin bar copper grids, stained with uranyl acetate and lead citrate, and viewed with a transmission EM. A similar protocol was used to immuno-EM localize lamins in *C. elegans* (Fig. 2) (Cohen *et al.*, 2002).

Near-complete removal of membranes results in better antibody penetration. In these experiments, the washed embryos are spun down and resuspended in 1.25 ml of fixation solution containing 0.4% freshly made formaldehyde, 18% methanol, and 10 mM Na_2EGTA. After 20 min incubation on ice, embryos are subjected to two rounds of freezing in liquid N_2 and thawing under tap water and then strongly agitated on a shaker for 20 min at room temperature. Embryos are then washed once with 100 mM Tris-HCl, pH 7.6, containing 1 mM EDTA and 0.5% Triton

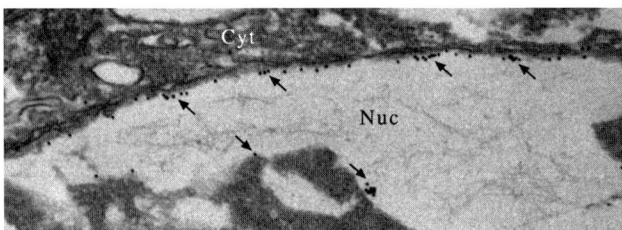

Fig. 2 Immunogold transmission electron microscope labeling of Ce-lamin in *C. elegans* embryos. Rabbit antibodies against the rod and tail regions of Ce-lamin were used as primary antibodies and 12-nm gold-conjugated goat-anti–rabbit IgG were used as secondary antibodies to localize Ce-lamin in embryos. The positions of the gold particles near the inner nuclear membrane and in the nuclear interior are marked by arrows. Nuc, Nucleus; Cyt, Cytoplasm; Data reprinted by permission of *J. Struct. Biol.* from Cohen *et al.* (2002).

X-100 and once with PBS containing 1 mM EDTA and 0.5% Triton X-100 (PBS-TR) and incubated for 15 min at room temperature with PBS-TR. Embryos are then incubated overnight at room temperature with primary antibodies diluted in PBS containing 0.5% Triton X-100 and 0.1% BSA (PBS-TRB). After four washes, 30 min each, with PBS-TR, embryos are incubated with gold-conjugated secondary antibodies as earlier. Embryos are washed four times, 30 min each, with PBST and fixed for 1 h at room temperature in 2.5% glutaraldehyde. Immunogold labeling, postfixation with OsO_4, Epon embedding, sectioning, staining with uranyl acetate and lead citrate, and viewing are performed as described previously.

VII. Sample Preparation for Transmission Electron Microscopy Analysis of *C. elegans* Embryos

A. Embryos

Embryos are collected by the standard Clorox/M9 technique as described previously. A ~22- × 22-mm square at or near the center of a Falcon 3001 Petri dish (35 × 10 mm) is marked. One drop (100–200 μl) of 0.1% poly-Lysine (Sigma P8920) is placed on this 22- × 22-mm square and left incubated for 5 min at room temperature. Excess poly-Lysine is removed with a Gilson pipette. The Petri dish is glued with double-stick tape to the lid of a 50-ml conical tube (Petri dish facing up), and ~100 μl embryos are placed on the poly-Lysine–treated area of the Petri dish. The tubes are centrifuged at 1000g for 1 min in a swinging bucket. This step ensures that the embryos remain on the poly-Lysine for the rest of the procedure. The Petri dish is removed from the tube and immediately placed on ice. One milliliter of fixative containing 1% freshly made paraformaldehyde, 2.5% EM grade glutaraldehyde (Agar Scientific LTD, Essex, UK), 0.1 M sodium Hepes, pH 7.0, is placed on top of the embryos, the plate is sealed with parafilm and immediately microwaved (Shef exclusive microwave oven SF-1875E at 70% of microwave power. Pelco-brand microwave, which provides a continuous power of 550 W, was also used successfully) on ice for 20 sec *on*, 10 sec *off*, and then again 20 sec *on*. The sample is then cooled on ice for 5 min in a fume hood, and the microwave cycle is repeated once more. Fixation is completed by further incubation of the sample for 1 h at room temperature. Excess fixative solution is then removed, and the embryos are washed three times, 10 min each, with 1 ml of 0.2 M sodium Hepes, pH 5.6. One milliliter of solution containing 1% OsO_4, 0.5% reduced $K_4Fe(CN)_6$, 0.1 M sodium Hepes, pH 5.6, is added, and the sample is incubated in the dark for 1 h. Dehydration in graded series of ethanol, Epon infiltration, embedding, and polymerization are performed as described previously. The Epon block is then separated from the Petri dish by soaking in liquid N_2. Epon sectioning, staining, and viewing are also performed as described previously. An example of wild-type and lamin downregulated embryos is shown in Fig. 3.

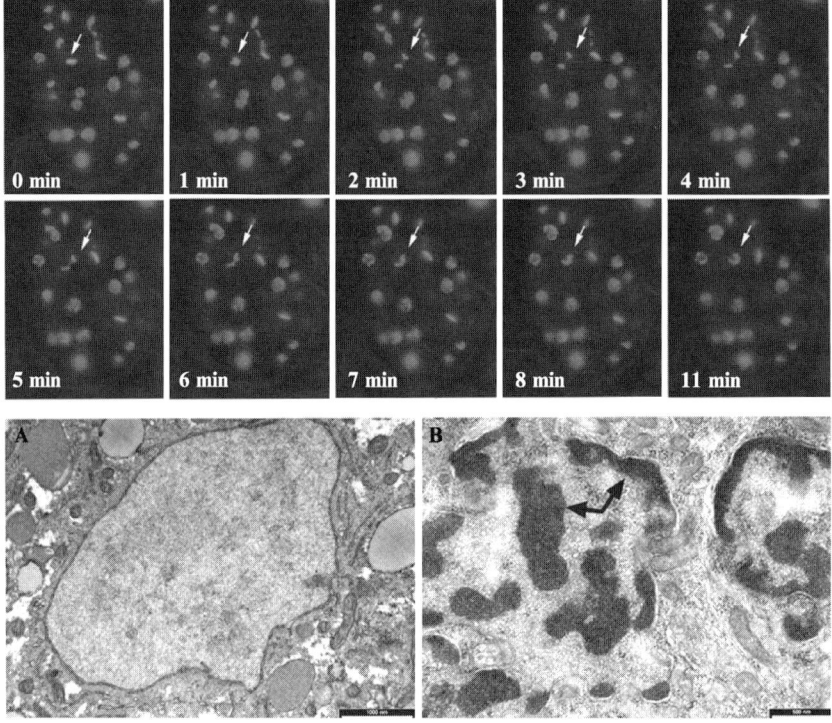

Fig. 3 Nuclear morphology of wild-type embryos and embryos with RNAi-depleted Ce-lamin. *Upper panels*, Time-lapse microscopy of a live *C. elegans lmn-1 (RNAi)* embryo expressing histone H2B fused to GFP (Praitis *et al.*, 2001). Most nuclei in the embryo look abnormal. The arrow points toward a metaphase nucleus, which forms an anaphase bridge that was not resolved. The embryo was viewed with a Zeiss Axioplan II microscope equipped for fluorescence. An Axiocam CCD camera and the Axio Vision Image Analysis package were used to record time-lapse data every minute. *Lower panels*, Wild-type and *lmn-1 (RNAi)* embryos were visualized by thin section transmission electron microscopy. (A) A nucleus from a control animal had normal nuclear shape and chromatin distribution and typical nuclear envelope with two membranes and a uniform lumen. (B) A nucleus in *lmn-1 (RNAi)* embryos, in which Ce-lamin is significantly downregulated. This nucleus had drastic change in shape and condensation of chromatin (arrows). Bars represent 1.0 μm in panel A and 0.6 μm in panel B. The EM data reprinted by permission of *J. Struct. Biol.* from Cohen *et al.* (2002).

VIII. Live Imaging of *C. elegans*

A. Sample Preparation

Live imaging of *C. elegans* requires optimal sample preparation, keeping the worms from moving during the imaging period, and minimal photo damage to the worms. Embryos, which hardly move until about the 1.5-fold stage, are dissected out by cutting adults in depression slides under the dissecting microscope with two sharp needles as scissors in egg salts solution (118 mM NaCl, 40 mM KCl, 3.4 mM CaCl$_2$ and 5 mM sodium Hepes, pH 7.4) (Edgar, 1995). Dissected eggs are transferred to an agar pad (Sulston and Horvitz, 1977), which is prepared by adding a drop of 3–5% agar at 60 °C with the warm Pasteur pipette to a clean slide placed parallel between two "spacer" slides that are raised by one or two layers of adhesive tape. The agar drop is flattened with a second slide, which is coated with silicon in a perpendicular orientation such that the slide is supported by the "spacer" slide, while making sure that bubbles do not form. After the agar is hard, the perpendicular top slide is carefully removed, and a small drop of 2 μl of egg salts solution with the embryos is added. Older animals, which tend to move, are placed on agar pads in M9 buffer containing 1 mM levamisole (Sigma cat. #L9756), which stops movement of larvae and adults. An apparatus for temperature-controlled microscopy can also be used to anesthetize the worms, using temperatures between 4–8 °C (Rabin and Podbilewicz, 2000; Rieder and Cole, 1998). Sodium azide (3–10 mM; Sigma S8032) also anesthetizes the worms but slowly kills the animals. Enough M9 or egg salt solution is then added on top of the sample to avoid drying, and the sample is finally covered with a coverslip sealed by vacuum grease. The sample slide is placed as quickly as possible on the microscope stage, and data are collected. For an extensive period of observation of living larval specimens, it is possible to coat the center of the coverslip with a thin layer of *E. coli* OP50. The larvae stay in the area with the bacterial lawn and feed. In addition, worms can be observed live with Nomarski optics and then removed for immunostaining or EM.

Comments:

1. Excess liquid can result in movement of animals during imaging. Use a Kimwipe to remove some of the liquid.
2. Put a thin layer of bacteria at the center of the coverslip to avoid the animal walking outside the field of view.

B. Sample Observation

After the worms are localized with low magnification lens (4× or 10×), maximum resolution and magnification are obtained with a high-magnification lens (63× or 100×) with a high numerical aperture (1.3 NA or higher). Plan Apochromat oil immersion objectives are used to follow cell fate and to follow intracellular and intercellular events. Plan or Planfluor objectives are used for Differential Interference Contrast (DIC) observations and Planachromat or Planfluor objectives for fluorescence analyses. Leica, Zeiss, Olympus, and Nikon research microscopes have equivalent optics appropriate for *C. elegans* live microscopy. Nikon oil immersion objectives (NA = 1.4) have the longest working distance permitting deeper optical sectioning with confocal microscopy. Four-dimensional microscopic observations can be obtained with many commercially available systems, including Zeiss, Nikon, Leica, Olympus, PerkinElmer, Deltavision, BioRad, and others.

Comments:

1. Keep the room temperature uniform.
2. Commercially available deconvolution programs can be used to remove the out-of-focus information.

C. Fluorescence Recovery after Photobleaching Analyses

For fluorescence recovery after photobleaching (FRAP) experiments, we use the Zeiss confocal laser scanning microscope LSM 510 with software version 2.02. Worms expressing GFP-tagged proteins are anesthetised for 20 min by the method of Kirby *et al.* (Kirby *et al.*, 1990) in 0.1% tricaine and 0.01% tetramisole in water. This procedure stops movement of analyzed worms and provides the opportunity to photobleach a specific region of live GFP-expressing animals. Photobleaching is performed with a laser wavelength of 488 nm (serial laser scanning with manually adjusted z-coordinates) on worms mounted in water on slides with an agar pad (see earlier). The bleached animals are placed on NGM plates seeded with *E. coli* OP50, and recovery of fluorescence is monitored at 30-min intervals with conventional fluorescence microscopy (see earlier). The anesthetized animals usually resumed a normal mode of locomotion 3 h after bleaching. Analyzing the lateral average fluorescence of the bleached and the nonbleached areas and subtracting the average background fluorescence allows fluorescence quantification by the laser scanning microscopy.

Acknowledgments

We thank Yonatan Tzur, Ayelet Margalit, Ester Neufeld, Dale Shumaker, Jürgen Schünemann, and Klaus Weber for critical review of this manuscript. We would also like to thank Benjamin Podbilewicz for his contribution to the live imaging part. This work was funded by grants from the Israel Science Foundation (ISF) and the National Institutes of Health (GM64535) and the Binational Science Foundation USA-Israel (BSF).

References

Aebi, U., Cohn, J., Buhle, L., and Gerace, L. (1986). The nuclear lamina is a meshwork of intermediate-type filaments. *Nature* **323,** 560–564.

Chen, F., Hersh, B. M., Conradt, B., Zhou, Z., Riemer, D., Gruenbaum, Y., and Horvitz, H. R. (2000). Translocation of *C. elegans* CED-4 to nuclear membranes during programmed cell death. *Science* **287,** 1485–1489.

Cohen, M., Tzur, Y. B., Neufeld, E., Feinstein, N., Delannoy, M. R., Wilson, K. L., and Gruenbaum, Y. (2002). Transmission electron microscope studies of the nuclear envelope in *Caenorhabditis elegans* embryos. *J. Struct. Biol.* **140,** 232–240.

Dodemont, H., Riemer, D., Ledger, N., and Weber, K. (1994). Eight genes and alternative RNA processing pathways generate an unexpectedly large diversity of cytoplasmic intermediate filament proteins in the nematode *Caenorhabditis elegans*. *EMBO J.* **13,** 2625–2638.

Dreger, C. K., Konig, A. R., Spring, H., Lichter, P., and Herrmann, H. (2002). Investigation of nuclear architecture with a domain-presenting expression system. *J. Struct. Biol.* **140,** 100–115.

Edgar, L. G. (1995). Blastomere culture and analysis. *In* "C. elegans: Modern biological analysis of an organism" (H. F. Epstein and D. C. Shakes, eds.), pp. 303–320. Academic Press, San Diego.

Epstein, H. F., and Shakes, D. C. (1995). "*Caenorhabditis elegans*: Modern biological analysis of an organism." Academic Press, San Diego.

Fire, A., Xu, S., Montgomery, M. K., Kostas, S. A., Driver, S. E., and Mello, C. C. (1998). Potent and specific genetic interference by double-stranded RNA in *Caenorhabditis elegans. Nature* **391,** 806–811.

Fridkin, A., Mills, E., Margalit, A., Neufeld, E., Lee, K. K., Feinstein, N., Cohen, M., Wilson, K. L., and Gruenbaum, Y. (2004). Matefin, a *C. elegans* germ-line specific SUN-domain nuclear membrane protein, is essential for early embryonic and germ cell development. *Proc. Natl. Acad. Sci. USA.* **101,** 6987–6992.

Goldberg, M., Harel, A., Brandeis, M., Rechsteiner, T., Richmond, T. J., Weiss, A. M., and Gruenbaum, Y. (1999). The tail domain of lamin Dm0 binds histones H2A and H2B. *Proc. Natl. Acad. Sci. USA* **96,** 2852–2857.

Gruenbaum, Y., Lee, K. K., Liu, J., Cohen, M., and Wilson, K. L. (2002). The expression, lamin-dependent localization and RNAi depletion phenotype for emerin in *C. elegans. J. Cell Sci.* **115,** 923–929.

Hapiak, V., Hresko, M. C., Schriefer, L. A., Saiyasisongkhram, K., Bercher, M., and Plenefisch, J. (2003). mua-6, a gene required for tissue integrity in *Caenorhabditis elegans* encodes a cytoplasmic intermediate filament. *Dev. Biol.* **263,** 330–342.

Herrmann, H., Hesse, M., Reichenzeller, M., Aebi, U., and Magin, T. M. (2003). Functional complexity of intermediate filament cytoskeletons: From structure to assembly to gene ablation. *Int. Rev. Cytol.* **223,** 83–175.

Karabinos, A., Schmidt, H., Harborth, J., Schnabel, R., and Weber, K. (2001). Essential roles for four cytoplasmic intermediate filament proteins in *Caenorhabditis elegans* development. *Proc. Natl. Acad. Sci. USA* **98,** 7863–7868.

Karabinos, A., Schulze, E., Klisch, T., Wang, J., and Weber, K. (2002). Expression profiles of the essential intermediate filament (IF) protein A2 and the IF protein C2 in the nematode *Caenorhabditis elegans. Mech. Dev.* **117,** 311–314.

Karabinos, A., Schulze, E., Schunemann, J., Parry, D. A., and Weber, K. (2003a). *In vivo* and *in vitro* evidence that the four essential intermediate filament (IF) proteins A1, A2, A3 and B1 of the nematode *Caenorhabditis elegans* form an obligate heteropolymeric IF system. *J. Mol. Biol.* **333,** 307–319.

Karabinos, A., Schunemann, J., Meyer, M., Aebi, U., and Weber, K. (2003b). The single nuclear lamin of *Caenorhabditis elegans* forms *in vitro* stable intermediate filaments and paracrystals with a reduced axial periodicity. *J. Mol. Biol.* **325,** 241–247.

Kirby, C., Kusch, M., and Kemphues, K. (1990). Specification of male development in *Caenorhabditis elegans*: The fem genes. *Dev. Biol.* **105,** 234–239.

Lee, K. K., Gruenbaum, Y., Spann, P., Liu, J., and Wilson, K. L. (2000). *C. elegans* nuclear envelope proteins emerin, MAN1, lamin, and nucleoporins reveal unique timing of nuclear envelope breakdown during mitosis. *Mol. Biol. Cell* **11,** 3089–3099.

Lee, K. K., Starr, D., Liu, J., Cohen, M., Han, M., Wilson, K., and Gruenbaum, Y. (2002). Lamin-dependent localization of UNC-84, a protein required for nuclear migration in *C. elegans. Mol. Biol. Cell* **13,** 892–901.

Liu, J., Lee, K. K., Segura-Totten, M., Neufeld, E., Wilson, K. L., and Gruenbaum, Y. (2003). MAN1 and emerin have overlapping function(s) essential for chromosome segregation and cell division in *C. elegans. Proc. Natl. Acad. Sci. USA* **100,** 4598–4603.

Liu, J., Rolef-Ben Shahar, T., Riemer, D., Spann, P., Treinin, M., Weber, K., Fire, A., and Gruenbaum, Y. (2000). Essential roles for *Caenorhabditis elegans* lamin gene in nuclear organization, cell cycle progression, and spatial organization of nuclear pore complexes. *Mol. Biol. Cell* **11**, 3937–3947.

Praitis, V., Casey, E., Collar, D., and Austin, J. (2001). Creation of low-copy integrated transgenic lines in *Caenorhabditis elegans*. *Genetics* **157**, 1217–1226.

Rabin, Y., and Podbilewicz, B. (2000). Temperature-controlled microscopy for imaging living cells: Apparatus, thermal analysis, and temperature dependency of embryonic elongation in *C. elegans*. *J. Microsc.* **199**, 214–223.

Rieder, C., and Cole, R. (1998). Perfusion chambers for high-Resolution video light microscopic studies of vertebrate cell monolayers: Some considerations and design. *Meth. Cell Biol.* **56**, 253–275.

Riemer, D., Dodemont, H., and Weber, K. (1993). A nuclear lamin of the nematode *Caenorhabditis elegans* with unusual structural features; cDNA cloning and gene organization. *Eur. J. Cell Biol.* **62**, 214–223.

Schirmer, E. C., Florens, L., Guan, T., Yates, J. R., and Gerace, L. (2003). Nuclear membrane proteins with potential disease links found bysubtractive proteomics. *Science* **531**, 1380–1382.

Stuurman, N., Heins, S., and Aebi, U. (1998). Nuclear lamins: Their structure, assembly, and interactions. *J. Struct. Biol.* **122**, 42–66.

Sulston, J. E., and Horvitz, H. R. (1977). Postembryonic cell lineages of the nematode *Caenorhabditis elegans*. *Dev. Biol.* **56**, 110–156.

PART V

Methods to Study Intermediate Filament Associated Proteins

CHAPTER 25

Plectin

Günther A. Rezniczek, Lubomír Janda, and Gerhard Wiche

Institute of Biochemistry and Molecular Cell Biology
University of Vienna
Max F. Perutz Laboratories
Vienna Biocenter
A-1030 Vienna, Austria

══════════   **I. Introduction**

A. Historical Perspective

The cytoskeleton, a complex network of protein filaments and associated proteins, provides eukaryotic cells with both mechanical strength and dynamic abilities, such as the positioning and trafficking of vesicles, organelles, and proteins; the alignment and segregation of chromosomes during mitosis; coordinated and directed movement; muscle contraction; and change of shape. In recent years, it has been realized that the different cytoskeletal networks (i.e., actin fibers, intermediate filaments [IFs], and microtubules) perform many of their tasks in cooperation. This interplay relies on an important family of structurally and, in part, functionally related proteins referred to as plakins or cytolinkers (Fuchs and Karakesisoglou, 2001; Fuchs and Wiche, 2004; Leung *et al.*, 2002; Wiche, 1998) that are capable of interlinking different elements of the cytoskeleton. The >500-kDa protein plectin is one of the best characterized cytolinkers and the most versatile one. In fact, when plectin was first isolated nearly 25 years ago (Pytela and Wiche, 1980), it was postulated that it might be involved in the organization and network formation of the cytoskeleton, based on the identification of IFs as interacting partners, its presumptive structural relationship to high molecular mass microtubule-associated proteins, and because of its immunolocalization within dense cytoplasmic network arrays of cultured cells. Presumptive network formation was the reason for giving plectin its name, based on the Greek word πλεκτή, meaning net or mesh (Wiche and Baker, 1982; Wiche *et al.*, 1982).

However, a better understanding of the exact roles of plectin molecules and their mechanisms of action evolved only slowly. It was more than a decade after the original isolation of plectin when the cloning and sequencing of plectin cDNA, first reported for rat (Wiche *et al.*, 1991), opened the door for molecular genetic studies and the analyses of gene structure and regulation. In the past 6–8 years, great progress was made. Of particular importance were the characterization of the exon–intron organization and the chromosomal localization of the human gene, both first reported by Liu *et al.* (1996) and independently confirmed by McLean *et al.* (1996), and the finding that the hereditary disease epidermolysis bullosa simplex muscular dystrophy (EBS-MD) is caused by defects in the plectin gene. The important role of plectin for the mechanical stability of living cells was further confirmed by defining the molecular interactions of plectin with a number of cytoskeletal proteins. These included components of all three major cytoskeletal filament systems and of several junctional complexes. The generation of plectin (−/−) animals by targeted gene disruption in mice (Andrä *et al.*, 1997) provided insights into additional novel roles of plectin as regulator of cellular signaling processes. Recent studies revealed that the molecular diversity of plectin is further increased by an unusual 5′ complexity of plectin transcripts, giving rise to at least 18 putative protein isoforms, all presumably carrying out distinct functions.

B. Structural and Molecular Properties

Full-length plectin has a deduced molecular mass ranging from 499–533 kDa, depending on the particular plectin isoform and based on mouse sequences (Rezniczek et al., 2003). Rotary shadowing electron microscopy (EM) of purified plectin molecules (Foisner and Wiche, 1987) revealed a dumbbell-like structure made up of a central 200-nm-long rod domain flanked by large globular domains (Fig. 1A). Secondary structure predictions based on cDNA and deduced amino acid sequences confirmed the multidomain structure of plectin. They revealed that the rod domain is characterized by long stretches of heptad repeats, distinctive of α-helical coiled coils (McLachlan and Stewart, 1975), and exhibits a staggered strict period of 10.4 for acidic and basic residues, suggesting an energetically most favorable parallel arrangement of a plectin dimer (Wiche et al., 1991). This is supported by both the microscopic dimensions of plectin and gel permeation high-performance liquid chromatography (HPLC) data, indicating a molecular mass of plectin molecules in solution of slightly over 1.1×10^6 (Foisner and Wiche, 1987; Weitzer and Wiche, 1987). The structure of the 214-kDa carboxy-terminal globular domain, encoded by a single very large (>6 kb) exon, is dominated by six highly homologous ~300 amino acid residues–long repeat domains (Wiche et al., 1991), which also occur in other cytolinker (plakin) protein family members, such as desmoplakin (Green et al., 1990), the epithelial and neuronal isoforms of BPAG1/dystonin (Sawamura et al., 1991), envoplakin (Ruhrberg et al., 1996), and epiplakin (Fujiwara et al., 2001; Spazierer et al., 2003). On the basis of computer modeling, the strongly conserved central core regions of plectin repeat domains were postulated to contain five ankyrin-like repeats with a predicted tertiary structure of a solenoid, where strategic cysteine residues play an important role (Janda et al., 2001). The recently reported crystal structure of two of desmoplakin's three repeat domains (Choi et al., 2002) confirmed the predicted ankyrin-like repeat substructure of the core regions. As was deduced from its cDNA sequence (Elliott et al., 1997; Liu et al., 1996; McLean et al., 1996), the amino-terminal domain of plectin harbors a conventional actin-binding domain (ABD) comprising two tandemly arranged calponin homology (CH) domains (Stradal et al., 1998), preceded by alternative sequences encoding alternative first exons. The amino-terminal domain further contains a so-called plakin domain. The various subdomains of the molecule are depicted in Fig. 1B. For a recent review on the whole family of plakin proteins, see Röper et al. (2002).

C. Gene Organization and Isoform Diversity

Plectin was first cloned and sequenced from rats (Wiche et al., 1991). The sequence of the human gene and its exon–intron organization and localization (q24) on chromosome 8 were reported 5 years later (Liu et al., 1996; McLean et al., 1996). A more detailed analysis of the rat gene locus followed (Elliott et al., 1997). The most detailed analysis of a plectin gene locus, however, has been carried out in mouse (Fuchs et al., 1999). Analysis of the genomic exon–intron organization of

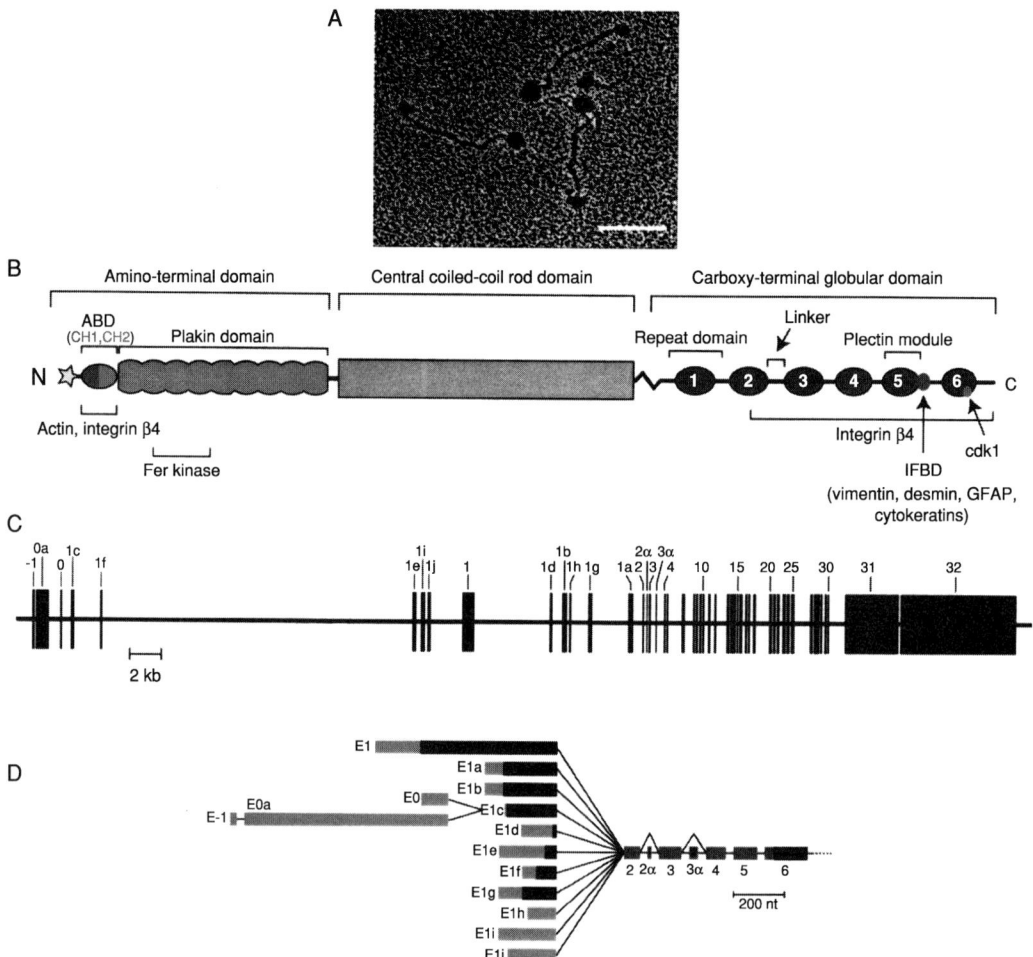

Fig. 1 Protein and gene structure of plectin. (A) Electron microscopy of plectin molecules after rotary shadowing with platinum-carbon. A dumbbell like structure is revealed. Bar, 100 nm. (B) Schematic map of the protein. The tripartite structure of plectin molecules is made up of a central rod flanked by amino-terminal and carboxy-terminal (globular) domains. The amino-terminal domain contains an actin-binding domain, consisting of two calponin homology domains (dark and light red), and a region called plakin domain (light green). The carboxy-terminal globular domain contains six highly homologous plectin repeat domains (blue), each consisting of a plectin module and a linker region, one of which harbors an intermediate filament binding domain (IFBD; light blue). The carboxy-terminal domain also contains a unique cell cycle kinase cdk1 phosphorylation site (Malecz *et al.*, 1996) shown in red. Alternative splicing of plectin transcripts leads to expression of protein isoforms with different amino termini (yellow star). Regions responsible for direct binding to various proteins are indicated below schematics. (C) Genomic organization of the murine plectin gene. Exons are represented as black boxes, introns by lines. Plectin (exons −1 to 32) extends over ∼62 kb; exon 32 terminates at the location of the polyadenylation consensus sequence. (D) Schematic representation of alternative plectin transcripts. Alternative splicing of the 5′-end of the plectin gene gives rise to at least

the murine plectin gene, which contains well over 40 exons and spans over 62 kb on chromosome 15, revealed an unusual 5′ transcript complexity of plectin isoforms (Fig. 1C and D). In total, 16 alternatively spliced exons have been identified so far, 11 of them (1–1j) directly splicing into a common exon 2 (which is the first exon to encode plectin's highly conserved ABD), three (−1, 0a, 0) into exon 1c, and two additional exons (2α and 3α) are optionally spliced within the exons encoding the ABD (Fig. 1D). In a recent study, the human and rat gene loci were reanalyzed on the basis of newly available genome sequences and compared with the mouse gene. In the rat, all 11 alternative first exons identified in mice were found, whereas in the human gene so far, the presence of 8 of these exons has been confirmed (Zhang *et al.*, 2004). Furthermore, isoforms lacking exon 31 (encoding the rod domain) were identified in rats (Elliott *et al.*, 1997), mice (unpublished data), and humans (Brown *et al.*, 2001). A rodless human plectin variant, as judged by molecular size, was also identified on the protein level (Brown *et al.*, 2001; Schröder *et al.*, 2000). It is conceivable that heterodimers containing full-length plectin isoforms with distinct amino termini encoded by alternative first coding exons could be formed. This may lead to increased functional diversity and potential for differential regulation of plectin.

D. Expression and Subcellular Localization

Plectin is a widespread if not ubiquitous protein of mammalian cells. By use of antisera and a panel of monoclonal antibodies (Foisner *et al.*, 1994; Wiche and Baker, 1982), it was shown that plectin is expressed in a variety of tissues and mammalian cell lines. It is particularly prominent in various types of muscle, stratified and simple epithelia, and cells forming the blood–brain barrier (Errante *et al.*, 1994; Wiche *et al.*, 1983, 1984, 1989). At the cellular level, plectin codistributes with different types of IFs and is located at plasma membrane attachment sites of IFs and microfilaments, such as hemidesmosomes (Rezniczek *et al.*, 1998; Wiche *et al.*, 1984), desmosomes (Eger *et al.*, 1997), Z-line structures and dense plaques of striated and smooth muscle, intercalated disks of cardiac muscle, and focal contacts (Seifert *et al.*, 1992; Wiche *et al.*, 1983). Prominent plectin expression in cells forming tissue layers at the interface between tissues and fluid-filled cavities was found at the surfaces of kidney glomeruli, liver bile canaliculi, bladder urothelium, gut villi, ependymal layers lining the cavities of brain and spinal cord,

---

16 different transcripts: 11 alternative first exons splicing into exon 2 are shown. Eight of them are first coding exons (1–1g), whereas three (1h, 1i, 1j) are noncoding. Three additional noncoding exons (E0 and E0a, splicing into E1c, and E−1, splicing into E0a) and two optionally spliced exons (2α and 3α, inserted between exons 2 and 3, and 3 and 4, respectively) are also shown. Exons are shown as boxes and splice events as lines connecting individual boxes; noncoding regions are orange, regions coding in all cases are black, regions coding only in conjunction with a first coding exon are brown, and the optionally spliced exons 2α and 3α are green. Note that although there is an in-frame ATG in exon 4, translation initiation for isoforms with noncoding first exons occurs only in exon 6. (See Color Insert.)

and endothelial cells of blood vessels (Errante *et al.*, 1994; Wiche *et al.*, 1983; Yaoita *et al.*, 1996). RNase protection experiments performed on a panel of mouse tissues and cell types with antisense riboprobes specific for the various alternative first exons of plectin provided clear evidence for the occurrence of tissue-specific or dominant plectin isoforms (Andrä *et al.*, 2003; Fuchs *et al.*, 1999). For example, plectin transcripts containing exon 1d were exclusively found in skeletal and heart muscle, whereas exon 1a–containing transcripts were dominant in organs rich in epithelial cell types, such as lung, small intestine, and skin. Tissue-specific expression was also characteristic of the two optionally spliced exons 2α and 3α, with isoforms containing exon 2α being expressed in brain, heart, and skeletal muscle, and exon 3α being brain-specific (Fuchs *et al.*, 1999). In a recent study, the influence of the alternative first exons on the stability of gene products, initiation of translation, and subcellular localization of isoforms was investigated systematically. It was demonstrated that, although all plectin variants were expressed in transfected cells as intact full-length plectin species (Fig. 2A), their subcellular localization was highly dependent on their short alternative amino-terminal sequence encoded by the different first exons (Rezniczek *et al.*, 2003). For instance, plectin 1a was specifically associated with hemidesmosome-like structures in keratinocytes, plectin 1b was found to be associated exclusively with mitochondria, and plectin 1f was concentrated in vinculin-positive structures at actin stress fiber ends. Concurring with the notion of plectin isoforms having different and specific functions, the hemidesmosomal defect observed in plectin (−/−) keratinocytes could be rescued by overexpression of plectin-1a, but not plectin-1c, which colocalized with microtubuli (Andrä *et al.*, 2003; see Fig. 2B and C).

E. Molecular Interactions of Plectin

Consistent with its varied cellular localization, plectin has been shown to interact with a variety of cytoskeletal structures and proteins on the molecular level. Plectin was originally identified as an IF-binding protein, and direct interaction was first demonstrated with vimentin IFs. Plectin's IF-binding site was mapped to a ~50-amino-acid-long sequence, including a short stretch of amino acids with characteristics of a bipartite nuclear localization signal, located between the highly conserved core regions of the carboxy-terminal repeats 5 and 6 (Nikolic *et al.*, 1996). Apart from vimentin, several other types of IF subunit proteins have been shown to interact with plectin, including desmin (Reipert *et al.*, 1999), glial fibrillary acidic protein (GFAP) (Foisner *et al.*, 1988), the nuclear IF protein lamin B (Foisner *et al.*, 1991b), and type I and type II cytokeratins (Geerts *et al.*, 1999; Steinböck *et al.*, 2000).

Plectin contains a highly conserved ABD close to its amino terminus, and analysis of plectin (−/−) cells demonstrated a novel important role of plectin as regulator of actin cytoskeleton dynamics (Andrä *et al.*, 1998). Splicing of the optional exons 2α and 3α within the ABD was shown to modulate its actin binding activity (Fuchs *et al.*, 1999). The presence of exon 2α in recombinant

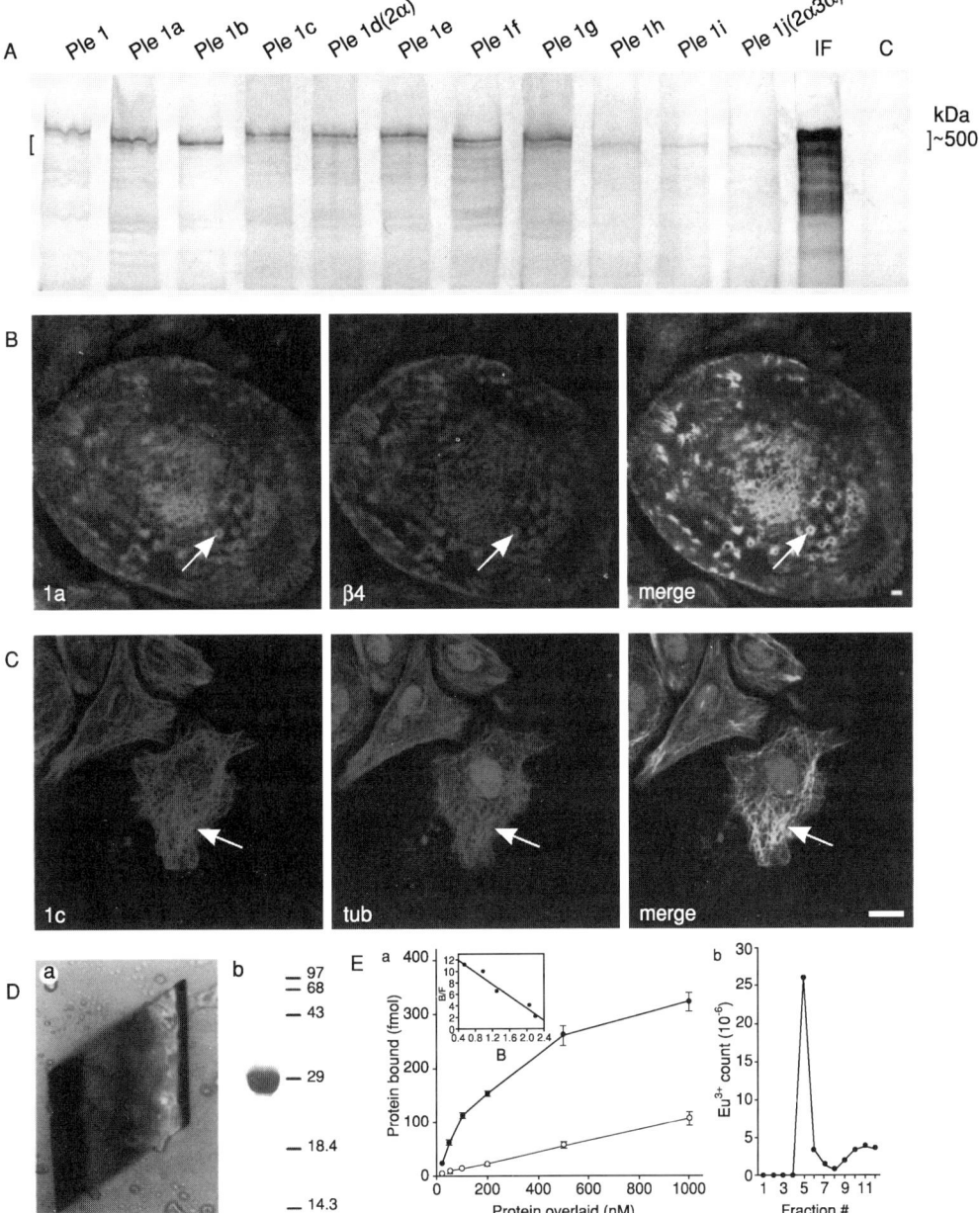

Fig. 2 Various examples of plectin isoform analyses. (A) Expression of recombinant plectin isoforms in full length. Mouse isoform cDNA constructs containing 11 alternative first exons (1–1j, in some cases with exons 2α and 3α, as indicated) were cloned into an EGFP-fusion expression vector and proteins expressed in plectin-deficient mouse fibroblasts. Cells were lysed and expressed proteins

ABD proteins, as contained in the muscle-specific isoform of plectin, increased binding to actin *in vitro*, suggesting that this "fine-tuning" mechanism based on alternative splicing is likely to optimize the proposed biological role of plectin as a cytolinker opposing intense mechanical forces in tissues like striated muscle. The ABD of plectin has recently been found to be multifunctional, binding not only to actin but also to the integrin subunit β4 (Geerts *et al.*, 1999; Rezniczek *et al.*, 1998) and vimentin (Sevcik *et al.*, 2004). In addition, the signaling molecule phosphatidylinositol-4,5-bisphosphate (PIP$_2$), regulating binding of plectin to actin, binds within the ABD (Andrä *et al.*, 1998). Binding of integrin β4 and actin to plectin's ABD has been shown to be mutually exclusive (Garcia-Alvarez *et al.*, 2003; Geerts *et al.*, 1999; Rezniczek *et al.*, 2003). For plectin transcripts containing the noncoding first exons 1h, 1i, or 1j, it was demonstrated that initiation of translation occurs not at the first possible start codon in exon 4, but only in exon 6. As a consequence, the resulting protein species lack the entire first CH domain of the ABD and fail to bind to integrin β4 and actin (Fuchs *et al.*, 1999; Rezniczek *et al.*, 2003). Additional binding sites for integrin β4 have been shown in the plakin and carboxy-terminal domains of plectin (Koster *et al.*, 2004; Rezniczek *et al.*, 1998).

Direct interaction of plectin with the membrane skeleton proteins fodrin and α-spectrin (Herrmann and Wiche, 1987), the desmosomal protein desmoplakin (Eger *et al.*, 1997), and the nonreceptor tyrosine kinase Fer (Lunter and Wiche, 2002) have also been demonstrated. In a recent yeast two-hybrid screen, the receptor for activated C kinase 1 (RACK1) was identified as a plectin binding partner, and it was shown that RACK1 is sequestered to the cytoskeleton through plectin during initial stages of cell adhesion (Osmanagic-Myers and Wiche, 2004). Furthermore, whole mount EM was used to demonstrate that plectin is capable of

---

analyzed by immunoblotting (SDS-5%-polyacrylamide gel) with anti-plectin serum 9. Plectin from a rat glioma C6 cell IF preparation served as positive control and size marker (IF), and a lysate from mock-transfected cells as negative control (C). The bracket indicates a molecular mass range of 498–562 kDa. (B) Localization of plectin 1a at hemidesmosome-like stable anchoring complexes (SAC). Immortalized keratinocytes were processed for double immunofluorescence microscopy with antibodies specific for plectin isoform 1a (1a) and antibodies to integrin β4 (β4). Note that plectin 1a colocalizes with integrin β4 at SACs, forming the typical Swiss cheese–like pattern (arrow). Bar, 10 μm. (C) Colocalization of plectin 1c with microtubules in mouse keratinocytes. Double immunofluorescence microscopy was carried out with plectin isoform 1c–specific antibodies (1c) and monoclonal antibodies to tubulin (tub). Note colocalization of plectin 1c and microtubules (arrows). Bar, 10 μm. (D) Crystal (a) and SDS-PAGE (b) of a plectin ABD isoform including five amino acid residues encoded by tissue specifically spliced exon 2α. The protein was expressed in *E. coli*, purified to homogeneity, and crystallized by use of the hanging-drop method. (E) Europium overlay assay. (a) Binding of plectin's ABD (MBP-fusion protein; closed circles) to the cytoplasmic domain of integrin β4 (MBP negative control, open circles). Scatchard transformation of the binding data is shown in the inset. (b) Typical elution profile of a Eu^{3+}-labeled protein from a Bio-Rad P6 column, showing baseline-separation of the protein (fraction 5) from the unreacted labeling reagent starting at fraction 9. A and E (panel a) reproduced from Rezniczek *et al.* (2003), ©2003 Oxford University Press. B and C reproduced from Andrä *et al.* (2003), ©2003 The Society for Investigative Dermatology, Inc. (See Color Insert.)

physically linking IFs to microtubules (Svitkina *et al.*, 1996), and *in vitro* interaction of purified plectin with microtubule-associated proteins (MAP2 and MAP1 subtypes) from the brain has been reported (Herrmann and Wiche, 1987). Specific binding domains that have been mapped on the molecule are indicated in Fig. 1B.

F. Plectin and Human Diseases

In 1996, a number of groups reported that patients with EBS-MD, an autosomal-recessive severe skin-blistering disease combined with muscular dystrophy, lack plectin expression in skin and muscle tissues because of defects in the plectin gene (Chavanas *et al.*, 1996; Gache *et al.*, 1996; McLean *et al.*, 1996; Pulkkinen *et al.*, 1996; Smith *et al.*, 1996). On the basis of plectin's prominence at plasma membrane junctional sites of IFs, this had been anticipated, and the direct interaction of plectin with the hemidesmosomal integrin subunit $\beta 4$ generating a linkage between the IF cytoskeleton and the extracellular matrix provided a molecular model for the skin-blistering phenotype (Rezniczek *et al.*, 1998). In most cases of EBS-MD, homozygous mutations in the plectin gene led to premature stop codons within the >3 kb-long exon encoding the rod domain of the protein. In addition, several cases with mutations outside of exon 31 have been described: a 9-bp deletion in an exon preceding the rod (Pulkkinen *et al.*, 1996), a compound heterozygous nonsense/one-amino acid-insertion (close to plectin's ABD) mutation (Bauer *et al.*, 2001), a homozygous 16-bp insertion mutation close to the IF binding site (Schröder *et al.*, 2002), and a 14-bp deletion in an exon encoding part of the plakin domain (Charlesworth *et al.*, 2003). Furthermore, an autosomal-dominant form of hereditary EBS (EBS-Ogna) without muscular dystrophy is also linked to a mutation in the plectin gene (Koss-Harnes *et al.*, 1997, 2002). The skin and muscle phenotypes of humans, as far as documented, were also found in plectin-deficient mice (Andrä *et al.*, 1997). Further analysis of these mice and of mice with specifically inactivated plectin isoforms will open new perspectives regarding plectin's involvement in diseases other than those known as EBS-MD.

G. Main Considerations

Plectin presents a number of challenges to the researcher. Its molecular size alone makes plectin quite difficult to handle, and seemingly simple techniques become demanding tasks. Although some of the experimental limitations imposed by plectin's size can be overcome by carefully selected conditions and protocols, protein fragments of smaller size will usually have to be used. But size is only part of the problem. Although the vast diversity of differentially spliced plectin transcripts—varying only subtly in their 5'-end structure—unfolded new and exciting perspectives regarding functional variability, it also added a tremendous amount of complexity. Because there is increasing evidence that different plectin isoforms fulfill distinct and specific roles within a cell (Andrä *et al.*, 2003; Rezniczek *et al.*,

2003; see Fig. 2B and C for examples), a major question arising when investigating the function(s) of plectin is which isoform one is dealing with. In the following discussion, we will provide strategies and information about available tools to study plectin in the light of this pertinent question.

1. How to Distinguish Isoforms?

Plectin has been characterized as an extremely versatile and multifunctional protein, and there is evidence that its alternative amino termini mediate interaction with yet unknown proteins or alternatively modulate the ability of other parts of the molecule to interact with binding partners. Thus, it will be important to distinguish plectin isoforms from each other. One way to achieve this (e.g., after successful coimmunoprecipitation) might be to separate the captured isoform(s) on high-resolution sodium dodecylsulfate-polyacrylamide gel electrophoresis (SDS-PAGE) alongside isoform size markers. As has recently been shown with full-length cDNA constructs, intact plectin can be overexpressed in cultured cells and individual isoforms resolved sufficiently to clearly distinguish them (Rezniczek *et al.*, 2003; Fig. 2A). Together with the possibility to determine which isoforms are expressed in a given cell (e.g., by means of isoform-specific RT-PCR or RNase protection assay), this may be a viable option. Alternatively, if sufficient and pure enough protein can be obtained, amino-terminal protein sequencing and/ or mass spectroscopic techniques may give the desired answers.

Clearly, the most desirable tools to identify isoforms *in situ* are isoform-specific antibodies. Although raising of such antibodies has been successful in a few cases (plectin 1a and 1c; Andrä *et al.*, 2003; Rezniczek *et al.*, 1998), antibodies for all the different sequences encoded by alternative coding exons are not yet available. For some sequences, like those encoded by exons 1d and 2α, both only five amino acids long, it may prove impossible to obtain specific antibodies. The problem is even more severe for plectin 1h, 1i, and 1j, the isoforms starting with noncoding first exons. The putative single protein species that results from translation of these different transcripts starting at the first ATG within exon 6 (Rezniczek *et al.*, 2003) has no known features that would enable their distinction among each other or from other isoforms (except size).

Given the lack of isoform-specific antibodies in most cases, a way to study a certain isoform is the use of tagged cDNA expression constructs. As has been shown, ectopic expression of tagged full-length plectin in cultured cells is possible. The use of fluorescent tags (e.g., green fluorescent protein) allows studying of plectin dynamics *in vivo*. For isoforms 1h, 1i, and 1j, this seems to be the only option currently available.

2. How to Study Isoform Functions?

As already stressed in the previous section, one should always bear in mind that the "ubiquitous plectin" is the combination of a multitude of distinctly behaving isoforms. Many *in vitro* and *in vivo* approaches that have been used to study plectin will be discussed in the following sections. In Section II.A, we will first describe how to isolate and characterize plectin from tissues and cultured cells. This includes a discussion of plectin antibodies. Because the use of purified cellular plectin with a molecular mass of >500 kDa is not feasible for many *in vitro* experiments, and it is often difficult to obtain sufficiently large quantities of protein, we will discuss the recombinant expression and purification of smaller plectin fragments in systems such as *E. coli* (Section II.B). To understand the function of a protein, it is important to identify binding partners. We will comment on several *in vitro* and *in vivo* assays that have been used to this end, including blot and microtiter plate overlay assays, yeast two-hybrid assays, cosedimentation and coimmunoprecipitation (Section II.C), and the expression of (full-length) plectin in cultured mammalian cells to study protein interaction and subcellular localization (Section II.D). Another important tool for the study of plectin function is the use of knockout animals, which we will discuss in Section II.E. Last, we will address the topic of protein crystallization, a difficult but powerful tool to understand molecular mechanisms involving large proteins, such as plectin, at the atomic level (Section II.F). Table I provides an extensive methodological literature review.

II. Materials and Methods

A. Purification of Cellular Plectin and Identification of Plectin in Cells

1. Purification from Rat Glioma C6 Cells

a. Materials and Instrumentation

Rat glioma C6 cells (ATCC CCL-107) cultured in Dulbecco's modified eagle's medium (DMEM) with 10% fetal calf serum; ultracentrifuge; Dounce homogenizer; glass/Teflon homogenizer; Sepharose CL-4B (Sigma-Aldrich, CL-4B-200) gel permeation column.

b. Procedure

Rinse confluent cultures of C6 cells with ice-cold 6 mM Na_2HPO_4/KH_2PO_4, pH 7.4, 171 mM NaCl, 3 mM KCl, 10 mM 6-amino-hexanic acid, 0.2 mM EGTA, 0.2 mM PMSF (solution A) containing 1% glucose. Scrape off the cells in the same solution, harvest cells (centrifuge at 300g), and wash them twice. Resuspend cell pellets at 1 g (wet weight) per 5 ml of solution A supplemented with 10 mM $MgCl_2$, 600 mM KCl, 5 mM 2-mercaptoethanol, 3 mM PMSF, 1% (v/v) Triton X-100, 20 μg/ml RNase A, 50 μg/ml DNase I, and disrupt the cells with several strokes in a Dounce homogenizer at 4 °C over 5 minutes. To obtain IF pellets, sediment the insoluble cell residues at 30,000g for 20 minutes at 4 °C and wash them three times in solution A containing 5 mM 2-mercaptoethanol, 3 mM PMSF, 0.2% Triton X-100, 600 mM KCl. Disperse pellets in 2 ml of 34 mM PIPES, pH 7.5, 7 M urea, 1.4 mM EGTA, 1.4 mM $MgCl_2$, 5 mM 2-mercaptoethanol

Table I
Methodological Literature Survey*

Affinity chromatography
61
Antibodies (generation, characterization, purification)
3, 10, 12, 17, 18, 22, 24, 25, 28, 29, 30, 35, 37, 42, 46, 48, 50, 55, 57, 58, 59, 60, 68, 70, 73
Bacterial expression
His-tag: 2, 19, 20, 23, 30, 42, 47, 53, 55, 56, 61, 64, 65
GST-fusion: 9, 20, 24, 31, 40
MBP-fusion: 20, 21, 24, 31, 42, 43
Blot overlay
14, 16, 15, 27, 30, 55, 61
CD (circular dichroism)
13
cDNA constructs
1, 2, 3, 4, 9, 11, 20, 21, 24, 30, 34, 36, 40, 41, 42, 43, 45, 46, 47, 48, 55, 56, 57, 64, 71
Cell culture
2, 3, 5, 7, 10, 12, 13, 14, 17, 18, 19, 20, 22, 24, 25, 26, 29, 30, 33, 34, 35, 36, 37, 38, 39, 43, 45, 46, 47, 48, 49, 52, 53, 54, 55, 56, 59, 64, 66, 67, 68, 70, 72, 73
Cell synchronization
19, 43
Coimmunoprecipation
10, 16, 17, 19, 26, 30, 36, 42, 43, 49, 50, 58, 70
Cosedimentation (Actin)
20, 23, 24, 40
Crystallization
23, 61, 65
Dot blot
15, 24
DSC (differential scanning calorimetry)
23
Electron microscopy
4, 6, 9, 13, 14, 15, 18, 20, 28, 29, 30, 33, 37, 47, 48, 52, 54, 55, 58, 59, 62, 67, 70, 72
Europium microtiter plate overlay assay
2, 4, 21, 42, 47, 53, 55, 56, 64
FACS (fluorescent activated cell sorting)
19, 46
Fluorescence microscopy
2, 3, 4, 5, 7, 10, 12, 13, 14, 17, 18, 19, 20, 22, 24, 25, 26, 28, 29, 30, 33, 34, 35, 36, 37, 38, 39, 43, 45, 46, 47, 48, 49, 51, 53, 55, 56, 57, 58, 59, 60, 64, 66, 67, 68, 70, 72, 73

Histology
1, 8, 12, 29, 33, 39, 54, 57, 59, 62, 66
HPLC (high performance liquid chromatography)
13, 67
Immunoblotting
1, 3, 4, 5, 7, 8, 10, 12, 14, 16, 15, 17, 19, 22, 24, 28, 29, 30, 32, 37, 38, 39, 40, 44, 45, 46, 48, 50, 55, 56, 57, 58, 62, 67, 70, 72
ITC (isothermal titration calorimetry)
23
Knock-out (targeted gene inactivation)
1
Microinjection (antibodies)
18
MS (mass spectrometry, MALDI)
38, 65
PCR (phenotyping, RACE)
1, 2, 4, 7, 8, 11, 21, 24, 30, 33, 36, 40, 41, 42, 45, 47, 64
Peptide mapping (proteolytic cleavage)
5, 14, 15, 19, 26, 27, 38, 44 (*CNBr digestion*), 48 (*Edman degradation*), 63, 69
Phosphorylation
19, 38, 42, 43, 49
Pull down
10, 20, 40
Purification (cellular plectin)
10, 13, 14, 15, 17, 18, 19, 22, 25, 26, 32, 37, 40, 43, 44, 46, 48, 49, 52, 55, 67, 68, 69, 70
Radioactive labeling
10, 16, 19, 20, 26, 27, 55
RNase protection
3, 4, 11, 21, 56, 58, 59, 71
Transfection (cultured cells)
2, 3, 5, 7, 10, 12, 13, 14, 17, 18, 19, 20, 22, 24, 25, 26, 29, 30, 33, 34, 35, 36, 37, 38, 39, 43, 45, 46, 47, 49, 53, 55, 56, 59, 64, 66, 72
Ultracentrifugation
14
Yeast two-hybrid system
20, 24, 34, 35, 36, 40, 46, 49

Notes: This table represents an alphabetized index of methods used to study plectin with references to relevant published articles.
References: [1–3]Andrä *et al.*, 1997, 1998, 2003; [4]Bauer *et al.*, 2001; [5]Beil *et al.*, 2002; [6]Bohn *et al.*, 1996; [7]Brown *et al.*, 2001; [8]Charlesworth *et al.*, 2003; [9]Clubb *et al.*, 2000; [10]Eger *et al.*, 1997; [11]Elliott *et al.*, 1997; [12]Errante *et al.*, 1994; [13]Foisner and Wiche, 1987; [14–19]Foisner *et al.*, 1988, 1991a,b, 1994, 1995, 1996; [20]Fontao *et al.*, 2001; [21]Fuchs *et al.*, 1999; [22]Gache *et al.*, 1996; [23]Garcia-Alvarez *et al.*, 2003; [24]Geerts *et al.*, 1999; [25]Gonzales *et al.*, 2001; [26,27]Herrmann and Wiche, 1983, 1987;

(solution B) per 10 g of cells at 4 °C and homogenize in a glass/Teflon homogenizer for 30 minutes. Centrifuge the solution at 200,000g for 30 minutes at 4 °C and load the supernatant on a Sepharose CL-4B gel permeation column equilibrated in solution B. Analyze eluted fractions by SDS-PAGE and store plectin-containing fraction at −70 °C. Alternatively, the high salt/Triton X-100–resistant IF pellets can be homogenized in 0.3 mM Na_2PO_4/KH_2PO_4, pH 8.0, 5 mM EGTA, 0.5% sodium N-lauroyl-sarcosinate (Sigma-Aldrich, L9150) (solution C) and separated on a Sepharose CL-4B column in the same solution. For most experiments, plectin samples in solutions B or C need to be desalted into solutions of low ionic strength (e.g., on BioGel P10 columns [Bio-Rad, Richmond, CA]) by use of 2 mM Tris-HCl (pH 8.0), 2 mM ammonium acetate (pH 8.0), or 5 mM borate (pH 8.5) supplemented with 5 mM 2-mercaptoethanol and/or 3 mM PMSF, if desired.

2. Isolation from Bovine Lens

a. Materials and Instrumentation

Fresh cattle eyes; forceps; scalpel; liquid nitrogen; porcelain mortar; glass/Teflon homogenizer; Dounce homogenizer; Sephacryl S-500 (Amersham Biosciences).

b. Procedure

Dissect cattle eyes shortly after slaughter. Remove the lens capsule with forceps and separate epithelial and cortical cell layers from the nuclear lens region with a scalpel and freeze them by immersion in liquid nitrogen. Grind the frozen lens tissue pieces in a mortar to yield a fine powder and suspend it in 10 volumes of 6 mM sodium phosphate, pH 7.1, 771 mM NaCl, 3 mM KCl, 100 mM $MgCl_2$, 8 mM 2-mercaptoethanol, 2 mM PMSF, and 1% Triton X-100. Homogenize in a glass/Teflon homogenizer for 15 minutes in the presence of 50 μg/ml DNase I and 20 μg/ml RNase A. Perform this and all subsequent steps at 0–4 °C. Spin down insoluble cell residues in a centrifuge for 20 minutes at 46,000g and wash by repeated (2×) suspension in 10 volumes 100 mM EGTA (pH 7.4) and subsequent centrifugation. The washed pellets can either be stored at −20 °C or dissolved at once in 5 volumes of 0.3 mM sodium phosphate, pH 8.9, 2 mM PMSF, 8 mM 2-mercaptoethanol, 0.1 mM EGTA, 1% sodium-N-lauroyl-sarcosinate (SLS), and homogenized in a Dounce homogenizer for 10 minutes. Centrifuge the resulting solution for 15 minutes at 46,000g and load 15-ml aliquots of the clear supernatants onto a 245-ml Sephacryl S-500 gel permeation column (2.6 × 50 cm)

[28]Hieda et al., 1992; [29,30]Hijikata et al., 1999, 2003; [31]House et al., 2003; [32]Kalman and Szabo, 2000; [33]Koss-Harnes et al., 2002; [34–36]Koster et al., 2001, 2003, 2004; [37]Koszka et al., 1985; [38]Larsen et al., 2002; [39]Lie et al., 1998; [40]Litjens et al., 2003; [41]Liu et al., 1996; [42]Lunter and Wiche, 2002; [43]Malecz et al., 1996; [44]Muenchbach et al., 1998; [45]Niessen et al., 1997; [46]Nievers et al., 2000; [47]Nikolic et al., 1996; [48]Okumura et al., 1999; [49]Osmanagic-Myers and Wiche, 2004; [50]Proby et al., 1999; [51]Pulkkinen et al., 1996; [52]Pytela and Wiche, 1980; [53,54]Reipert et al., 1999, 2004; [55,56]Rezniczek et al., 1998, 2003; [57–59]Schröder et al., 1997, 1999, 2002; [60]Seifert et al., 1992; [61]Sevcik et al., 2004; [62]Smith et al., 1996; [63]Stegh et al., 2000; [64]Steinböck et al., 2000; [65]Urbanikova et al., 2002; [66]Vita et al., 2003; [67]Weitzer and Wiche, 1987; [68]Wiche and Baker, 1982; [69–72]Wiche et al., 1982, 1983, 1991, 1993; [73]Zernig and Wiche, 1985.

equilibrated at room temperature in 0.3 mM sodium phosphate, pH 8.5, 5 mM EGTA, 1 mM PMSF, 0.25% SLS. Collect 9-ml fractions and analyze by SDS-PAGE. Up to 50% of the total plectin is usually recovered free of contaminants in early fractions. Fractions not used immediately for further experiments can be stored at −20 °C.

3. Sodium Dodecylsulfate–Polyacrylamide Gel Electrophoresis and Immunoblotting

a. Materials and Instrumentation

Standard protein gel electrophoresis assembly (e.g., Bio-Rad Mini-Protein III), tank blot (Bio-Rad Mini Protean III), or semi-dry blotting apparatus (Bio-Rad Transblot SD). Sample buffer (prepare as 2×; final concentrations 50 mM Tris-HCl, pH 6.8, 100 mM DTT, 2% SDS, 10% glycerol, 0.1% bromophenol blue); acrylamide mix (29:1 ratio of acrylamide versus bis-acrylamide); separating gel (5% acrylamide mix, 375 mM Tris-HCl, pH 8.8, 0.1% SDS, 0.1% ammonium persulfate, 0.08% TEMED); stacking gel (4% acrylamide mix, 125 mM Tris-HCl, pH 6.8, 0.1% SDS, 0.1% ammonium persulfate, 0.1% TEMED); running buffer (25 mM Tris, 250 mM glycine, 0.1% SDS); tank-blot transfer solution (3.05 g/L Tris, 14.4 g/L glycine); semi-dry transfer solution (5.81 g/L Tris, 2.22 g/L glycine, 3.7 g/L SDS, 20% methanol).

b. Procedure

Prepare SDS-5% polyacrylamide gels with 4% stacking gels. Prepare the samples by mixing them with 2× sample buffer and heating to 95 °C for 3 minutes before loading. Run the gels at 20 mA per gel until the bromophenol blue front reaches the bottom. Stain gels with any staining protocol or continue with blotting.

Equilibrate the gel and filter paper, pads, and membrane in transfer solution and assemble the gel/membrane sandwich. Tank-blot overnight at 35 V at 4 °C, or for 1 hour at 100 V with an ice-pack in a Bio-Rad Mini Protean chamber; alternatively, use a semi-dry transfer cell at 160 mA (max. 25 V) for 1 hour. After blotting, transferred proteins can be reversibly stained with Ponceau-S solution (0.5% Ponceau-S, 1% acetic acid). Carry out standard immunodetection with suitable antibodies.

4. Immunofluorescence Microscopy

Standard immunofluorescence techniques are used to visualize plectin. To study the subcellular localization of plectin, the use of confocal laser scanning microscopy is highly recommended. Adherent cells grown in culture can be fixed with either ice-cold methanol for 90 seconds or with 4% paraformaldehyde in phosphate-buffered saline (PBS) for 15 minutes with subsequent permeabilization (5 minutes, 0.05% Triton X-100 in PBS). After fixation, apply blocking solution (optional) and antibodies. For tissue sections of frozen or paraffin-embedded material, the monoclonal antibodies usually perform better than the sera, with the limitation that they lack broader species cross-reactivity.

5. Antibodies

Although many antibodies against plectin have been raised by a number of laboratories, only few are commercially available. We have compiled a table of plectin-specific antibodies and their uses as cited in the literature (Table II).

6. RNase Protection Assay

a. Materials and Instrumentation

RNA transcription plasmid (e.g., pSP64; Promega, Madison, WI); $[\alpha\text{-}^{32}P]GTP$; heat-adjustable water bath; sequencing gel assembly (e.g., Bio-Rad Sequi-Gen GT); InstantImager (PerkinElmer, Boston, MA) or other similar instrumentation for quantitation of radioactive signals. All buffers and solutions must be RNase-free.

b. Preparation of Radiolabeled Antisense Probes

Insert cDNA fragments corresponding to the RNA probes into a transcription vector so that transcription will give rise to the corresponding antisense RNAs. Probes should be designed $\sim$80–120 bp in length. Linearize 20 μg of the riboprobe plasmid with a suitable restriction enzyme with a recognition sequence immediately downstream of the cDNA fragment that will be transcribed. Verify complete digestion by analyzing a small part of the reaction with agarose gel electrophoresis. Extract the restriction mix with phenol/chloroform/isoamyl alcohol (PCI; 25:24:1) and with chloroform. Precipitate DNA (sodium acetate/ethanol), wash with 75% ethanol, air-dry, and redissolve in H_2O at 1 μg/μl. Prepare the labeling mix consisting of 8.5 μl H_2O, 2 μl 10 × SP6 buffer, 1 μl BSA (2 mg/ml), 2 μl rNTP (5 mM each), 1 μl RNAsin (40 U/μl), 0.5 μl SP6 polymerase (20 U/μl), linearized riboprobe plasmid (1 μg/μl), 4 μl $[\alpha\text{-}^{32}P]GTP$ (800 Ci/mmol, 1 Ci/ml) and incubate for 1 hour at 40 °C. Add stop solution (1 μl of 10 mg/ml yeast tRNA, 0.5 μl of 0.5 M EDTA, 28.5 μl H_2O) and extract with PCI (2×) and with chloroform. Precipitate nucleic acids with 0.1 volumes of 3 M sodium acetate and 3 volumes of ethanol at −20 °C for 30 minutes and centrifuge at full speed in an Eppendorf tabletop centrifuge for 15 minutes. Wash the pellets with 100 μl 75% ethanol, dissolve in 10 μl urea blue juice (7 M urea, 0.4% bromophenol blue, 0.4% xylene cyanole), and run on a polyacrylamide gel (see later). Determine the exact positions of the labeled probes in the gel by exposing X-ray–sensitive film to the gel for 1 minute and excise the corresponding gel pieces. With a pipette tip, mince the gel pieces in 300 μl cracking solution (10 mM Tris-HCl, pH 7.5, 0.3 M NaCl, 0.1% SDS, 0.1 mM EDTA) and shake vigorously for 2 hours in a thermomixer at room temperature. Centrifuge for 5 minutes at full speed and transfer the supernatants to fresh tubes. Extract with PCI and chloroform. Take small aliquots of the eluted probes and dilute to 100 μl in H_2O and count in a scintillation counter. Activities of 20,000–40,000 cpm/μl are

Table II
Survey of Plectin Antibodies

Antibody*	Type[†], Host	Immunogen, epitope	Reactivity	Source[‡]/Methods[§]
Commercially available antibodies				
7A8 (14)	mAb, M (IgG$_1$)	Plectin (R), rod	R, B, Ha	Source (SA, #P-9318); EM (23, 43, 44, 53); IB (5, 13, 12, 14, 16, 18, 23, 24, 29, 31, 36, 46, 60); IH (50); IF (5, 14, 15, 19, 23, 24, 25, 26, 29, 31, 43, 53, 62); IP (16)
10F6 (14)	mAb, M (IgG$_1$)	Plectin (R), rod	H, M, R, B, Ha	Source (BMS, #BMS165); EM (9); IB (9, 12, 14, 17); IF (1, 9, 26, 34, 49, 62); IP (9, 40)
Clone 31	mAb, M	Plectin (H), C-terminal domain (aa 3062–3184)	H	Source (BD, #611348)
P1 (47)	pAb, GP	Plectin (H), C-terminal domain (aa 4367–4684)	H	Source (PG, #GP20); IB (47); IF (47)
P2 (47)	pAb, GP	Plectin (H), C-terminal domain (aa 4367–4684)	H, M, B	Source (PG, #GP21); EM (47); IB (6, 47, 52); IF (6, 33, 47, 48, 52)
N19	pAb, G (IgG)	Plectin (H), N-terminal domain	H, M, R	Source (SCBT, #sc-7573); IB; IH
C20	pAb, G (IgG)	Plectin (H), C-terminal domain	H, M, R	Source (SCBT, #sc-7572); IB (6, 29); IH; IP; IF (6, 29)
Other antibodies described in the literature				
1A2 (14)	mAb, M	Plectin (R), rod	R, B	EM (12); IB (12, 14); IH (10); IF (49, 62)
1D8 (14)	mAb, M	Plectin (R), rod	R, B, Ha	EM (15); IB (10, 12, 14, 15); IH (10); IF (10, 15)
4C4 (14)	mAb, M	Plectin (R), rod	R, B	IB (14)
4C10 (14)	mAb, M	Plectin (R), rod	R	IB (14)
5B3 (14)	mAb, M (IgG$_1$)	Plectin (R), rod	H, R, B	IB (4, 14, 17, 44); IF (4, 17, 26, 44)
5C6 (14)	mAb, M	Plectin (R), rod	R, B	IB (14); IF (62)
5C10 (14)	mAb, M (IgM)	Plectin (R), rod	H, M, R, B, Ha	IB (14); IF (26)
5D9 (14)	mAb, M	Plectin (R), rod	R, Ha	IB (14)
6B8 (14)	mAb, M	Plectin (R), rod	R	EM (12, 15); IB (12, 14); IF (15, 49, 62)
6C6 (14)	mAb, M (IgG$_2$)	Plectin (R), rod	H, M, R, B, Ha	EM (50); IB (14, 17); IF (26)
121 (39)	mAb, M (IgG$_1$)	Plectin (B)	H, R, B	EM (22, 47, 50); IB (17, 22, 29, 32, 39, 41, 46); IH (32); IF (7, 17, 18, 22, 27, 28, 29, 32, 37, 38, 39, 42, 46, 54); IP (41)

(continues)

Table II *(continued)*

Antibody*	Type[†], Host	Immunogen, epitope	Reactivity	Source[‡]/Methods[§]
417D (8)	mAb, M	Plectin (H), N-terminal domain (aa 478–487)	H, Ha	EM (8); IB (8)
9 (3)	AS, rabbit	Plectin (R), N-terminal domain (E419–V541)	M, R	IB (3, 34, 40, 45, 51); IF (3)
21, 46 (56)	AS, rabbit	Plectin (R)	M, R	EM (11, 15, 30, 49, 58); IB (16, 20, 21, 30, 55, 56, 58, 59); IF (2, 3, 9, 10, 11, 16, 20, 40, 49, 56, 57, 58, 59, 61, 62); IP (16, 35)
123C (3)	AS, rabbit	Plectin (R), C-terminal repeat domain 5	M, R	IB (3)
E1a (44)	AS, rabbit	Plectin (M) peptide, plectin 1a	M, R	EM (44); IB (3); IF (3)
E1c (3)	AS, rabbit	Plectin (M) (M1-S17), plectin 1c	M	IB (3); IF (3)
F3 (8)	AS, GP	Plectin (H), N-terminal domain (aa 478–487)	H, R	IB (8)
D16 (18)	AS, rabbit	Plectin (H), ABD (aa 1-339)	H, R	IB (29); IF (18, 29)
Plectin N (52)	AS, GP	Plectin (H), N-terminal domain (aa 587–601)	H	IB (52)
(24)	AS, rabbit	Plectin (R), rod	H, R	EM (24); IB (24); IF (24); IP (24)

Abbreviations used: B, bovine; G, goat; GP, guinea pig; H, human; Ha, hamster; M, mouse; R, rat.

*Numbers in parenthesis given here are the original references.

[†]mAb, monoclonal antibody; AS, antiserum.

[‡]Commercial sources: SA, Sigma-Aldrich; BMS, Bender MedSystems, Vienna, Austria; BD, BD Biosciences Pharmingen; PG, ProGen, Heidelberg, Germany; SCBT, Santa Cruz Biotechnology, Santa Cruz, CA; catalog numbers (#).

[§]Methods: EM, electron microscopy; IB, immunoblotting; IF, immunofluorescence; IH, immunohistochemistry; IP, immunoprecipitation; MI, microinjection. Numbers in parenthesis, references as specified below.

References: [1–3]Andrä *et al.*, 1997, 1998, 2003; [4]Bauer *et al.*, 2001; [5]Beil *et al.*, 2002; [6]Brown *et al.*, 2001; [7]Charlesworth *et al.*, 2003; [8]Clubb *et al.*, 2000; [9]Eger *et al.*, 1997; [10]Errante *et al.*, 1994; [11–16]Foisner *et al.*, 1988, 1991a,b, 1994, 1995, 1996; [17]Gache *et al.*, 1996; [18]Geerts *et al.*, 1999; [19]Gonzales *et al.*, 2001; [20,21]Herrmann and Wiche, 1983, 1987; [22]Hieda *et al.*, 1992; [23,24]Hijikata *et al.*, 1999, 2003; [25]Kalman and Szabo, 2000; [26]Koss-Harnes *et al.*, 2002; [27–29]Koster *et al.*, 2001, 2003, 2004; [30]Koszka *et al.*, 1985; [31]Larsen *et al.*, 2002; [32]Lie *et al.*, 1998; [33]Litjens *et al.*, 2003; [34]Lunter and Wiche, 2002; [35]Malecz *et al.*, 1996; [36]Muenchbach *et al.*, 1998; [37]Niessen *et al.*, 1997; [38]Nievers *et al.*, 2000; [39]Okumura *et al.*, 1999; [40]Osmanagic-Myers and Wiche, 2004; [41]Proby *et al.*, 1999; [42]Pulkkinen *et al.*, 1996; [43]Reipert *et al.*, 1999; [44,45]Rezniczek *et al.*, 1998, 2003; [46–48]Schröder *et al.*, 1997, 1999, 2002; [49]Seifert *et al.*, 1992; [50]Smith *et al.*, 1996; [51]Spazierer *et al.*, 2003; [52]Stegh *et al.*, 2000; [53]Svitkina *et al.*, 1996; [54]Vita *et al.*, 2003; [55]Weitzer and Wiche, 1987; [56]Wiche and Baker, 1982; [57–61]Wiche *et al.*, 1982, 1983, 1984, 1991, 1993; [62]Yaoita *et al.*, 1996.

usually achieved. Use the probes immediately or store them at −20 °C for up to 48 hours.

c. Hybridization and RNase Digestion

Mix 100,000 cpm of the corresponding probes with 1 μg of the mRNA preparations and coprecipitate with 0.1 volumes of 3 M sodium acetate and 3 volumes of ethanol at −20 °C for 30 minutes. After 15 minutes centrifugation, wash pellets with 75% ethanol, air-dry briefly, and dissolve in 10 μl hybridization solution (80% formamide, 0.4 M NaCl, 40 mM PIPES, 1 mM EDTA, pH 7.5). After 16 hours at 60 °C, cool on ice and add 350 μl RNase solution (10 mM Tris-HCl, pH 7.5, 300 mM NaCl, 5 mM EDTA, 5 μg/ml RNase A, 100 U/ml RNase T1). Incubate at 37 °C for 1 hour, and then stop the reaction with 10 μl of 20% SDS and 5 μl of 20 mg/ml proteinase K. After 15 minutes at 37 °C, extract the samples with PCI and chloroform. Add 1 μg of 10 mg/ml yeast tRNA and precipitate RNA as before. After centrifugation (15 minutes, full speed), carefully remove the supernatant and wash the pellets with 150 μl of 75% ethanol. Dissolve the air-dried pellets in 10 μl of urea blue juice.

d. Preparation of Labeled Length Markers

Restriction fragments resulting from *Hpa*II-digestions of pUC18 plasmid are suitable length markers (501, 404, 353, 242, 190, 147, 110, 89, 67, 34, 34, 26 bp). Digest 2 μg pUC18 with 1 U *Hpa*II overnight in 20 μl and inactivate the enzyme at 70 °C for 10 minutes. Label the ends with the Klenow DNA-polymerase fragment (5 μl digested pUC18; 5 μl of 10 × Klenow buffer; 1 μl of 5 U/μl Klenow fragment, Promega; 2 μl [α-^{32}P]dCTP; 37 μl H$_2$O) for 30 minutes at 37 °C. Extract with PCI, precipitate and dissolve in 100 μl of urea blue juice, and store at −20 °C for up to 1 month. Measure a small aliquot in a scintillation counter, and use 20,000 cpm per slot and run.

e. Polyacrylamide Gel Electrophoresis

To prepare a 8% polyacrylamide gel (0.4-mm spacers), mix 27.6 g urea, 12 ml 5 × TBE, 16 ml of a 29% acrylamide/1% bis-acrylamide solution, and adjust to a final volume of 60 ml. Start polymerization with 300 μl ammonium persulfate (10% stock solution) and 50 μl TEMED. Prerun the gel at 45 W for 30 minutes. To denature RNA double strands, heat the samples to 90 °C for 3 minutes before loading. Run the gels at 45 W for 1–2 hours (gel temperature during separation should be ~45 °C). Dry gels at 80 °C and expose to x-ray film or quantify directly with an InstantImager or PhosphorImager.

7. Electron Microscopy

Negative staining and rotary shadowing EM have extensively been used by our group to visualize plectin purified from cells (e.g., Foisner and Wiche, 1987) (see Fig. 1A) and to characterize the molecular interaction of plectin's carboxy-terminal repeat domain 5 (expressed in *E. coli*) with vimentin and keratin 5/14 IFs (Nikolic *et al.*, 1996; Steinböck *et al.*, 2000).

Immunogold EM of plectin in cryosections or lowicryl-embedded tissue sections (Reipert *et al.*, 1999; Rezniczek *et al.*, 1998), or in cells grown on coverslips (Svitkina *et al.*, 1996) with mouse monoclonal anti-plectin antibody 7A8 have been described. This antibody gives excellent results when used in purified form, but because of lack of species cross-reactivity, it only works well on rat tissue. For other antibodies used in EM refer to Table II.

B. Expression and Purification of Recombinant Plectin Fragments in *Escherichia coli*

Here, we provide a basic protocol for the purification of His-tagged proteins under both denaturing (urea; isolation of inclusion bodies) and nondenaturing conditions. We rely on the pET vector system (EMD/Merck Biosciences, formerly Novagen), driven by the T7*lac* promoter, and use for expression bacterial strains BL21 (DE3) or BL21-CodonPlus (DE3)-RILP (both from Stratagene, La Jolla, CA; the latter producing an increased supply of rare *E. coli* tRNAs that correspond to codons used more frequently by other organisms).

1. Materials and Instrumentation

Expression plasmids, compatible host strain; sonicator (Sonopuls HD70, Bandelin, Berlin, Germany); activated nickel-chelate column (His•Bind, Novagen).

2. Preparation of Bacterial Stock

Freshly transform the expression plasmid into BL21(DE3) with any standard transformation method and grow colonies on LB plates overnight. Use several colonies to inoculate 100 ml TB medium (12 g peptone, 24 g yeast extract, 4 ml glycerol in 900 ml H_2O; autoclave; add 100 ml potassium phosphate: 23.1g/L KH_2PO_4, 125.4g/L K_2HPO_4) containing 1% glucose, 100 μg/ml carbenicillin (in contrast to ampicillin, carbenicillin is not degraded by the released β-lactamase, thus preventing plasmid loss), and 34 μg/ml chloramphenicol and grow overnight. Harvest the cells under sterile conditions (Sorvall GSA rotor, 4000 rpm, 10 minutes, 4 °C) and resuspend the cell pellet in 4 ml sterile 10% glycerol. Aliquot cell suspension (200 μl), shock freeze in liquid nitrogen, and store at −80 °C.

3. Induction

Use ∼100 μl of bacterial stock to inoculate 100 ml TB medium containing 1% glucose and 100 μg/ml ampicillin. Grow cells at 37 °C until an OD_{600} of 0.5–0.6 is reached (approximately 2 hours). Add IPTG (from a 1 M stock solution) to a final concentration of 0.1–1 mM; induce for 3–5 hours at 30 °C. Cool cells on ice (∼15 minutes) and harvest cells (GSA, 4000 rpm, 10 minutes, 4 °C).

4. Nondenaturing (Native) Purification

Resuspend the pellet in 5 ml sonication solution (50 mM Tris-HCl, pH 9.0, 2 mM EDTA, 1% Triton X-100). An optional freeze–thaw cycle assists in cell lysis. Thaw frozen cells and put them on ice. Disrupt by sonication (80% power, 10–20 seconds continuous pulse on a Bandelin Sonopuls HD70) with brief cooling steps (put the tube into liquid nitrogen) in between sonication cycles (5–10). Separate insoluble matter by centrifugation, resuspend the pellet in 5 ml of sonication solution, and repeat sonication (2–3 pulses) and centrifugation. Load combined supernatants after filtration (0.45 μm) onto a His•Bind column ($\sim$5-ml bed volume) prepared by washing with 3 column volumes (cv) of H_2O, charging with 5 cv of 50 mM $NiSO_4$, and equilibrating with 3 cv of column buffer (50 mM Tris-HCl, pH 9.0, 20% glycerol). Wash with 3 cv of column buffer and elute the protein with 50 mM Tris-HCl, pH 9.0, 20% glycerol, 250 mM L-histidine.

5. Purification from Inclusion Bodies under Denaturing Conditions

Resuspend the pellet in 5 ml of 20 mM Tris-HCl, pH 7.9, 500 mM NaCl. An optional freeze–thaw cycle assists in cell lysis. Thaw frozen cells and incubate with lysozyme (0.1 mg/ml) and 0.1% Triton X-100 for 15 minutes at 30 °C. Sonicate on ice (10 cycles of 30 seconds continuous pulse, 90% power, 30-second pauses for cooling) and centrifuge (20,000g, 15 minutes, 4 °C). Resuspend the pellet (mainly inclusion bodies) thoroughly in 2–5 ml (1/20–1/50 culture volume) binding solution (20 mM Tris-HCl, pH 7.9, 500 mM NaCl, 6 M urea) containing 5 mM imidazole. Remove insoluble matter by centrifugation (39,000g, 20 minutes, 4 °C). Prepare the His•Bind column ($\sim$5-ml bed volume) by washing with 3 cv of H_2O, charging with 5 cv of 50 mM $NiSO_4$, and equilibrate with 3 cv of binding solution containing 5 mM imidazole. Filter the supernatant through a 0.45-μm filter and load onto the column. Wash with 3 cv of binding solution containing 5 mM imidazole and 5 cv of binding solution containing 20 mM imidazole. Elute the protein with 250–500 mM imidazole in binding solution.

C. Binding Assays

1. Microtiter–Plate Overlay Assay Using Eu^{3+}-Labeled Proteins

a. Materials and Instrumentation

DELFIA Eu^{3+}-labeling kit (PerkinElmer); 96-well microtiter plates (e.g., Nunc MaxiSorp); fluorometer capable of time-resolved measurement (e.g., PerkinElmer VICTOR^2 D); gel filtration columns for desalting.

b. Procedure

Dialyze plectin or the putative binding partner or controls to be labeled into 50 mM $NaHCO_3$ (pH 8.5). The protein concentration should be in the range of 0.5–1.5 mg/ml. Other solutions can be used for labeling as long as they do not contain any free amines or sodium azide. Add the labeling reagent (follow the manufacturer's instructions regarding the amount to be used) and let the labeling reaction continue for 24–72 hours at room temperature. Clarify the solutions (100,000g, 10 minutes, 4 °C) and apply them to small gel filtration columns (e.g., Bio-Rad P6) to separate proteins from unreacted labeling reagent. Characterize the labeled protein solutions by measuring protein content and amounts of bound Eu^{3+} ions. Typical labeling degrees range from 0.2–2 Eu^{3+} ions per protein molecule.

Prepare solutions of plectin or candidate proteins to be coated in 25 mM $Na_2B_4O_7$ (pH 9.3) at concentrations of 100–200 nM. Apply 100 μl per well of an adsorbent 96-well microtiter plate in triplicate, cover the plate, and incubate overnight at 4 °C. Discard the coating solutions and wash twice with PBS. Block the wells with 100 μl of 4% BSA in PBS for 1 hour at room temperature. Discard the blocking solution and add 100 μl of the Eu^{3+}-labeled proteins at the desired concentrations in overlay solution (PBS containing 1 mM EGTA, 2 mM $MgCl_2$, 1 mM DTT, 0.1% Tween-20). The composition of the overlay solution may need to be adapted to the specific assay. Incubate for 1.5 hours at room temperature. Wash the plates excessively (about six times) with overlay solution and bring the plates to near-dryness by banging them on stacks of paper towels. Add 100 μl DELFIA Enhancement Solution to each well and shake on a plate shaker for a minimum of 5 minutes. For each plate, also include 100 μl of 1 nM Eu^{3+}-standard. Measure the plates in a time-resolved fluorometer and convert the fluorescence values to concentrations by comparing with the Eu^{3+}-standard.

2. Cosedimentation of Maltose–Binding Protein Fusions of Plectin with Actin

a. Materials

Purified plectin ABD (expressed as a maltose-binding protein [MBP] fusion protein; see above), purified rabbit skeletal muscle α-actin (Cytoskeleton, Denver, CO, is an excellent source).

b. Procedure

Proteins should be dialyzed into 2 mM Tris-HCl, pH 8.0, 0.2 mM $CaCl_2$, 0.5 mM ATP (actin G-solution, AGS) and kept on ice. Clarify all purified proteins by centrifugation at 100,000g for 1 hour at 4 °C before use. Premix actin (2.5 μM) with the plectin MBP-ABD fusion protein (or MBP alone as a control) and initiate actin polymerization by the addition of 0.1 volume of 2 mM Tris-HCl, pH 8.0, 20 mM $MgCl_2$, 1 M KCl, 5 mM ATP (initiation mix, IM). After incubation at room temperature for 1 hour, pellet actin filaments by centrifugation at 100,000 g for 1 hour at 20 °C. Wash pellets with actin polymerization solution (APS; AGS with the addition of 0.1 volumes IM). Analyze equal amounts of pellet and supernatant by SDS-PAGE.

To assess actin-bundling activity of plectin, perform low-speed sedimentation. Polymerize actin at 24 μM for 3 hours at room temperature, then dilute to 4.5 μM in APS containing various amounts of MBP-plectin ABD or MBP alone. After

1-hour incubation at room temperature, centrifuge at 14,000*g* for 1 minute, and analyze pellets and supernatants as before.

3. Coimmunoprecipitation of Plectin and Bound Proteins from Cell Lysates

We provide two basic protocols that have been used successfully to precipitate signaling complexes associated with plectin from cultured cells (Osmanagic-Myers and Wiche, 2004).

a. Protocol 1 (Total Lysate, More Stringent)

Grow cells to the desired density, wash twice with ice-cold PBS, and collect cells in 1 ml (per 66-cm^2-plate) ice-cold 50 mM HEPES, pH 7.0, 5 mM MgCl$_2$, 1 mM EGTA, 100 mM NaCl, 0.5% Triton X-100, 0.1 mM DTT, 50 μg/ml DNase I, 20 μg/ml RNase A, 1 mM PMSF, 10 mM benzamidine, 10 μg/ml approtinin, 10 μg/ml pepstatin, 10 μg/ml leupeptin (lysis solution) with a rubber policeman and transfer the cell suspension to 1.5-ml reaction tubes. To ensure optimal activity of DNase and RNase, incubate the tubes at room temperature for 10 minutes. (Breakup of nucleic acids with enzymes is much gentler than the widespread use of large-gauge needles.) Then, add Triton X-100 to a final concentration of 1%. Optionally, SDS can be added to 0.1% at this point. Leave at room temperature for another 5 minutes before removing insoluble matter by centrifugation at full speed (14,000 rpm) in an Eppendorf microfuge for 1 minute. Transfer the supernatant to a fresh tube. A 30-minute preclearing step using protein A (or G) sepharose (35 μl of a 10% solution per 500 μl supernatant) equilibrated in lysis buffer is essential to prevent high background. After removal of the protein A (or G) beads (2 minutes, full speed, Eppendorf microfuge), incubate the lysates with antibodies by rotation either for 3 hours at room temperature or overnight at 4 °C. To recover antibody complexes, add 50 μl of a 50% protein A (or G) sepharose slurry (per 500 μl lysate) and incubate by rotation for 3–6 hours at 4 °C. Recover the beads (centrifuge for 1 minute with a slow increase from 0–3000 rpm over 30 seconds) and wash three times with 1 ml of 50 mM HEPES, pH 7.0, 5 mM MgCl$_2$, 1 mM EGTA, 100 mM NaCl, 1% Triton X-100, at 4 °C.

b. Protocol 2 (Cellular Fractionation, Less Stringent)

Perform all steps at 4 °C or on ice. Quantities are for one 60-cm^2-plate seeded with 8×10^5 mouse fibroblasts grown overnight. Wash cells with PBS and collect them in PBS with a rubber policeman. Sediment cells by centrifugation at 350*g* for 3 minutes and gently resuspend them in 300 μl of 10 mM PIPES, pH 6.8, 0.01% digitonin, 300 mM sucrose, 100 mM NaCl, 3 mM MgCl$_2$, 5 mM EDTA, 0.2 mM DTT, 1 mM PMSF, 10 μg/ml approtinin, 10 μg/ml leupeptin, 10 μg/ml pepstatin, 1 mM benzamidine, 1 mM Na$_3$VO$_4$ (lysis solution 1) and incubate with end-over-end rotation for 10 minutes. Centrifuge for 1 minute at 15,800*g* and transfer the supernatant (cytosolic fraction) to a fresh tube. Resuspend the pellet in 300 μl of lysis solution 2 (10 mM PIPES, pH 7.4, 0.5% Triton X-100, 300 mM sucrose, 100 mM NaCl, 3 mM MgCl$_2$, 3 mM EDTA, 0.2 mM DTT, and protease/phosphatase inhibitors as in lysis solution 1) and agitate (end-over-end rotation) for 20 minutes. Centrifuge at 15,800g for 1 minute to obtain the membrane (supernatant) and cytoskeleton (pellet) fractions. Before adding primary antibodies, an optional preclearing step can be performed. Add primary antibodies and incubate 3 hours to overnight with end-over-end rotation. Add protein A (or G) sepharose and incubate for 3 hours to overnight (incubation times need to be individually balanced between specificity and efficiency of precipitation). Remove the beads by careful centrifugation (slowly accelerate within 30 seconds from 0–3000 rpm in an Eppendorf microfuge and leave at 3000 rpm for 1 minute) and wash three to four times with the respective lysis solution (inhibitors are not necessary).

D. Ectopic Expression of Plectin in Mammalian Cells

For cloning of plectin fragments or full-length plectin cDNAs (insert sizes of up to 14.2 kb) and transfection into cells, standard techniques can be used. Adequate efficiency of transfection of full-length plectin expression plasmids into a number of different cell types was successfully achieved with common liposomal and nonliposomal transfection agents such as Lipofectamine (Life Technologies, San Diego, CA) or FuGENE 6 (Roche Diagnostics, Basel, Switzerland).

E. Targeted Gene Inactivation in Mice and Culture of Plectin–Deficient Cells

Mice lacking plectin have been created through targeted inactivation of the plectin gene by disruption of exons 2–4 or the rod-encoding exon 31 (Andrä et al., 1997). In both cases, plectin expression was lost. Plectin-deficient mice have severe skin blistering caused by rupture of basal keratinocytes without preventing hemidesmosome or desmosome formation and die within a few days after birth. Furthermore, myofibril integrity is severely impaired in skeletal and heart muscle. Thus, these animals exhibit the same skin and muscle phenotypes that were described for patients with EBS-MD as far as documented.

To perform detailed analyses on the cellular level, isolation and in vitro culture of cells derived from plectin-deficient mice is an important tool. Usually, primary cell lines isolated from tissues are preferred but not suitable for all cell types and analyses. To establish immortalized plectin-deficient cell lines, we have intercrossed plectin (+/−) and p53 (−/−) mice to obtain plectin (−/−)/p53 (−/−) double knockout mice (Andrä et al., 2003). This method generates a defined mutation (absence of p53) that facilitates immortalization of cells that otherwise are difficult to immortalize (Metz et al., 1995).

F. Crystallization of the Plectin Actin–Binding Domain

Due to the large size plectin, it cannot be crystallized in its entirety but only in parts. The first domain of plectin whose structure has been solved was the ABD located close to the amino terminus of the protein. In fact, the structure of two plectin ABDs have independently been solved, one from human and the other from mouse plectin (Garcia-Alvarez et al., 2003; Sevcik et al., 2004).

1. To Prepare ABD Crystals

A cDNA expression vector derived from pET15b (Novagen) encoding the mouse plectin ABD version including exon 2α-derived sequences with an amino-terminal His-tag was engineered for overexpression in *Escherichia coli*. After removal of the His-tag by thrombin digestion, the final protein was 237 residues long, containing the plectin ABD(2α) (amino acids 181–417; accession no. AF 188008) flanked by six amino-terminal (GSHMEF) and two carboxy-terminal (EF) residues (added as cloning requirements). The protein was dialyzed against ammonium carbonate and lyophilized. Two crystalline forms (I and II) were obtained by the hanging-drop vapor diffusion method.

Monoclinic crystals (form I), belonging to the P2₁ space group with two mole-cules in the asymmetrical unit, were grown from 4-μl drops containing equivalent amounts of protein and precipitant solutions. The protein solution was prepared by dissolving the lyophilized plectin ABD in 50 mM Tris-HCl, pH 9.0, to a concentration of 20 mg/ml. The precipitant solution contained 10% PEG 8000, 2% dioxane (v/v), and 0.1 M Tris-HCl, pH 8.5–9.0 (Urbanikova *et al.*, 2002).

Orthorhombic crystals in the P2₁2₁2₁ space group with only one molecule in the asymmetrical unit (form II) were prepared by the same method combined with seeding. Drops (6 μl) were prepared by mixing equal volumes of protein solution (see earlier) and precipitant solution (10% PEG 8000, 0.1 M cacodylate buffer, pH 6.5, 0.2 M calcium acetate). Drops were collected after 24 hours of equilibration, and the precipitate was removed by centrifugation. The supernatants were used to form new drops, in which microcrystals had grown. To obtain better crystals, the procedure was repeated with solutions with lower concentrations of protein (10 mg/ml) and precipitant (8% PEG 8000). The precipitant solution was enriched with dioxane (2%, v/v) to reduce the twinning tendency of crystals. The previously obtained microcrystals were used as seeds. Crystals reached dimensions of up to 0.8 mm after 1–2 days (Sevcik *et al.*, 2004; see Fig. 2D).

The orthorhombic crystal form of human ABD (also including the exon-2α encoded amino acids) was prepared by Garcia-Alvarez *et al.* (2003). Crystals were obtained at room temperature by vapor diffusion after mixing equal volumes of protein solution (20 mg/ml in 20 mM Tris-HCl, pH 7.0, 150 mM NaCl) and precipitant solution (0.1 M Tris-HCl, pH 7.0, 15% propanol).

III. Pearls and Pitfalls

A. Purification of Cellular Plectin and Identification of Plectin in Cells

1. Isolation from Cells and Tissue

In some cell lines such as CHO and rat glioma C6 cells, plectin is relat-ively abundant, accounting for up to 1% of the total protein (Herrmann and Wiche, 1983). A typical Triton X-100/high-salt–resistant IF preparation contains approximately 35% of the total cellular protein and 75% of cellular plectin (Herrmann and Wiche, 1983) with plectin and vimentin as its major polypeptide components and a plectin/vimentin mass ratio of 0.2–0.3 in C6 cells (Foisner and Wiche, 1987). Thus, C6 cells were originally chosen as a source for plectin mainly because sufficient amounts of the protein can easily be purified from them, but other cells containing plectin and vimentin can be used as well. However, for economic reasons, isolation of plectin from tissues rather than from cultured cells is preferable. Bovine lens tissues, especially epithelial and cortical fiber cells, are a relatively rich source of plectin. The overall yield of purified plectin can be expected to be ~300 mg per 100 lenses (~200 g). The amount of plectin in epithelial and cortical cell layers of lenses again comes close to 1% of the total protein content (Weitzer and Wiche, 1987). In both cases, gel permeation chromatography of urea or sodium *N*-laurosylsarcosine–solubilized IF preparations yields purified plectin fractions that elute with the void volume of the column or shortly afterward.

2. Detection of Plectin Protein and Transcripts

For the resolution of plectin proteins on gels, standard SDS-PAGE protocols are used. We found 5%–6% polyacrylamide gels suitable for good resolution of plectin molecules while guaranteeing easy handling; 4% and lower gels may provide better resolution but can be very tedious to handle. For blotting, we recommend a tank-blot system and transferring overnight. No pH adjustments are necessary for the running and transfer solutions. If you need to adjust the pH, never use hydrochloric acid, because the chlorid ions will interfere with separation/transfer! Instead, use a concentrated glycin solution.

It is interesting to note that because of its α-helical nature and surface charge, the rod domain is highly antigenic. Thus, when the whole molecule was used as an immunogen to raise monoclonal antibodies, most of them were reactive with epitopes residing within the rod domain, as determined by epitope mapping by use of a variety of methods (Foisner *et al.*, 1994; Wiche *et al.*, 1991). To raise new antibodies against specific domains of plectin, it would be preferable to immunize animals with recombinant proteins or peptides. In the light of the multiplicity of plectin variants, the main focus of plectin antibody generation will be to raise antibodies specific for a particular isoform. We have used isoform-specific peptides to generate such antibodies in the past; however, except for an antibody specific for plectin isoform 1a (Rezniczek *et al.*, 1998), the results were less than satisfactory. We found an approach more useful in which short glutathion-S-transferase (GST)–tagged target sequences serve as immunogen in rabbits, and MBP fusions of the same sequences immobilized on CNBr-Sepharose are used for subsequent affinity purification of the crude sera. An antibody specific for plectin 1c was obtained in this way (Andrä *et al.*, 2003).

RNase protection assays have proven extremely useful to determine the tissue and cell type distribution of plectin transcripts containing alternative exons.

Antisense riboprobes are fairly simple to obtain, and besides providing qualitative results, this assay is also suitable for quantitation of RNA expression (Fuchs *et al.*, 1999; Rezniczek *et al.*, 2003).

B. Expression and Purification of Recombinant Plectin Fragments in *Escherichia coli*

Many *in vitro* assays rely on the availability of more or less large quantities of pure proteins. However, purification of proteins or domains of proteins serving as building blocks of the cytoskeleton frequently brings a number of problems, and expression of full-length plectin in bacteria is not possible because of its size. Smaller fragments representing functional and structural subdomains of plectin have been expressed on a wide scale. These recombinant proteins often accumulate and aggregate in the intracellular space of bacteria forming inclusion bodies and lack biological activity. Although the presence of inclusion bodies can make preliminary isolation steps very simple, this often leads to difficulties with protein refolding, correct formation of disulfide bonds, and thus recovery of full biological activity. Therefore, the use of physiological ("native") purification conditions is preferable. By fine-tuning the expression conditions (concentration of inducer, stringent reducing conditions, or temperature of induction), otherwise insoluble plectin fragments can be obtained in a soluble form and with much better chances of retaining their activity.

Purification tags that have been used for the expression of plectin fragments are the His-tag (binding to nickel-chelating affinity columns), GST (binding to reduced glutathion immobilized on a column), and MBP (binding to amylose resins). The His-tag is usually the favored tag because it is small, and purification under both denaturing and physiological conditions is rapid and robust. The large GST and MBP tags are usually only used when either very short plectin sequences are expressed (e.g., as antigens for antibody production [GST fusions]) or when the fragments exhibit very low solubility. For instance, plectin's ABD, especially versions containing the exon 2α and 3α-encoded sequences, had low solubility under conditions suitable for interaction with actin or integrin β4 in overlay and cosedimentation assays. This problem could be solved by use of MBP fusions of the ABD with comparatively much higher solubility (Fuchs *et al.*, 1999; Geerts *et al.*, 1999). However, crystallization of the plectin ABD without any tag was successful after removal of the His-tag after purification (Garcia-Alvarez *et al.*, 2003; Sevcik *et al.*, 2004; Urbanikova *et al.*, 2002).

C. Binding Assays

1. Blot Overlay Assays

A number of plectin-binding partners have been identified with blot overlay assays. One of the first interactions described was the binding of plectin to spectrin/fodrin, high molecular weight microtubule–associated proteins, and a number of IF proteins, including vimentin and lamin B with either radiolabeled proteins or immunological detection (Foisner *et al.*, 1988, 1991b; Herrmann and Wiche, 1987). Interaction of plectin with integrin β4 was mapped to multiple molecular sites with a blot overlay assay with purified recombinant proteins (Rezniczek *et al.*, 1998), and Geerts *et al.* (1999) used a dot blot assay to characterize the interaction of a radiolabeled plectin ABD with actin and integrin β4.

2. Microtiter–Plate Overlay Assays

Besides blot overlay assays, immobilization of proteins on microtiter plates and overlay of (labeled) candidate binding partners is a common method to quantitatively assess binding. Usually, the soluble protein is either labeled radioactively or detected with labeled antibodies after binding. We have adapted the dissociation-enhanced lanthanide fluorescence immunoassay (DELFIA) system (PerkinElmer, formerly Wallac) to measure direct plectin–protein interactions and found it an excellent alternative to radioactive labeling of proteins. Fluorescence is quantitatively measured by use of time-resolved fluorometry, so that short-lived background fluorescence and interference from the samples and environment are not a concern. Apart from being a nonradioactive procedure, the use of this assay system has the advantage of being highly sensitive with a wide dynamic range and even allows multiplexing (Europium, Samarium and Terbium can be used together as they have mutually distinct fluorescence emission spectra). This assay system has been used to characterize binding of plectin to integrin β4 (Rezniczek *et al.*, 1998, 2003), actin (Andrä *et al.*, 1998; Fuchs *et al.*, 1999), vimentin (Nikolic *et al.*, 1996; Steinböck *et al.*, 2000), desmin (Reipert *et al.*, 1999; Steinböck *et al.*, 2000), and keratins (Steinböck *et al.*, 2000) and to determine and explain the effects of a compound heterozygous plectin mutation occurring in a patient (Bauer *et al.*, 2001). Plectin or plectin fragments are either labeled and overlaid onto immobilized candidate proteins or vice versa (see Fig. 2E for exemplary results obtained with this assay system).

3. Yeast Two–Hybrid Assay

Yeast two-hybrid assays are convenient tools to screen for new interaction partners of a protein (domain) and to map and characterize known interactions. The group of A. Sonnenberg has used yeast two-hybrid assays extensively to study interactions of plectin with integrin β4 and actin (Geerts *et al.*, 1999; Litjens *et al.*, 2003), keratins, vimentin, and GFAP (Geerts *et al.*, 1999), and BP180 (Koster *et al.*, 2003). On the basis of our own experience, however, we find that two-hybrid assays, while being very powerful, have serious limitations in credibility of results (false-negative and false-positive results). Also, use of the yeast two-hybrid assay for quantitation of interactions is to be viewed with caution. In the case of plectin interactions, microtiter plate binding and cosedimentation assays have proven better means to obtain quantitative results. The general pitfalls of two-hybrid assays have been discussed in detail elsewhere (e.g., Stephens and Banting, 2000).

4. Cosedimentation Assays

Because plectin was originally identified as an IF-associated protein, it was shown early to cosediment with filamentous vimentin and other IF preparations. Later, plectin was also found associated with focal contacts and actin stress fibers (Seifert *et al.*, 1992), but attempts to cosediment plectin purified from C6 cells with actin *in vitro* failed. After the plectin sequence was known and the ABD at its amino terminus had been identified, cosedimentation of this domain with actin has been reported numerous times. In some studies (Fontao *et al.*, 2001; Geerts *et al.*, 1999; Litjens *et al.*, 2003), versions of the ABD fused to MBP (amino-terminal) were used, while Garcia-Alvarez *et al.* (2003) worked with His-tagged proteins. Fontao *et al.* (2001) reported an apparent K_d of $\sim$0.3 μM for plectin-actin binding, which was in excellent agreement with the K_d reported previously for a plectin fragment encoded by exons 1–24, which was measured with an Eu^{3+}-overlay assay (Andrä *et al.*, 1998). The lower K_d of 22 μM observed by Garcia-Alvarez *et al.* (2003) was likely due to the absence of additional sequences in the amino-terminal of the ABD sample used. This is also in line with the finding that stress fiber association of plectin 1d (exon 1d encodes only five amino acids) was less pronounced than that of other isoforms in transfected cells (Rezniczek *et al.*, 2003).

5. Coimmunoprecipitation and Pull-Down Assays

The main difficulty with plectin immunoprecipitation lies, again, in the size of the protein and its relative insolubility because of its numerous associations with cytoskeletal and membrane structures. The right balance between detergent solubilization and resulting stringency of the conditions and conservation of labile complexes must be found. When working with mouse cells or tissues, ideally a negative control from a plectin-deficient animal is included in the analysis.

D. Ectopic Expression of Plectin in Mammalian Cells

Ectopic (over)expression of proteins or protein domains in mammalian cells is a commonly used method to obtain information about their subcellular localization and function and has frequently been used to study plectin. For example, transfection experiments with truncated forms of plectin indicated a role of plectin's carboxy-terminal globular domain in IF association (Wiche *et al.*, 1993), which was then further pinpointed to the fifth plectin repeat domain (Nikolic *et al.*,

1996). Overexpressed repeat 5 not only decorates IFs but eventually leads to their bundling and finally to a complete collapse of the IF network.

Although potential pitfalls of ectopic overexpression, such as changes in the cell's biochemistry in response to the transfection agents, nonphysiological (too high) concentrations of the expressed protein, and aggregation have to be considered, the method has several advantages, particularly in the case of extraordinary large proteins such as plectin. Individual domains and mutant forms of the protein can be studied, and with the inclusion of sequences encoding epitope tags (e.g., myc, HA, His) the expression products can be easily visualized or isolated. When fluorescent tags, such as EGFP and DsRed (BD Biosciences Clontech) or their variants, are used, the expressed fusion protein can be visualized *in vivo*. For expression of plectin, mostly myc, HA, and EGFP-tags have been used, predominantly located at the carboxy terminus. When carboxy-terminal tags and amino-terminal fragments of plectin were used, it was found that in some cases the fidelity of protein translation (i.e., the recognition of the correct start codon) was suboptimal. This was much less the case when the respective 5'-UTR sequences were included (Rezniczek *et al.*, 2003).

Interesting possibilities arose with the availability of plectin-deficient cells, either from patients with a gene defect in plectin leading to loss of plectin expression (e.g., EBS-MD keratinocytes; Geerts *et al.*, 1999) or from plectin knockout mice (Andrä *et al.*, 1997). For example, in plectin-deficient keratinocytes, a specific rescue effect of particular plectin isoforms on hemidesmosomal defects could be observed after ectopic expression (Andrä *et al.*, 2003).

Mammalian cells have also successfully been used as "test tubes" to perform binding assays (e.g., mapping of plectin-integrin $\beta4$ binding; Rezniczek *et al.*, 1998). Differently tagged fragments of both proteins were overexpressed in PtK2 cells, and interacting fragments colocalized in distinct patterns not observed when expressed individually.

E. Targeted Gene Inactivation in Mice

The early death of plectin knockout mice forestalls the opportunity to analyze possible disorders developing only later, such as neuronal defects, that have been described in cases of EBS-MD (Lie *et al.*, 1998; Smith *et al.*, 1996). Therefore, it is highly desirable to obtain animal models of adult EBS-MD. This could be achieved by performing a conditional knockout through crossing mice with inducible and/or tissue-specific Cre or Flp recombinase expression with mice whose plectin gene has been modified to include the respective recognition sites (loxP or FRT) flanking the regions to be eliminated. Another approach to overcome the limitation of early death could be the targeting of individual plectin isoforms. A much milder phenotype is to be expected here, and, in addition, it might reveal specific functions of these isoforms. Both approaches are currently under way, and the knockout of individual isoforms has been successful in overcoming early lethality (unpublished).

IV. Concluding Remarks

After almost 25 years since its first description by Pytela and Wiche (1980), plectin has fulfilled much of the initial promise as an organizer of the cytoskeleton. Recent discoveries revealed plectin's involvement in regulatory processes, creating a new perspective on functional potentials of plectin that extend far beyond mere structural responsibilities. Plectin's widespread distribution, strategic localization, and multitude of known (and more probably still unrevealed) binding partners make it ideally fit to act as a scaffolding platform, bringing together the right molecules at the right time and location, to perform their tasks. This role is further supported by an elaborate fine-tuning mechanism of plectin provided by the alternative splicing of its gene, leading to subtle, yet functionally significant, variations in the expressed isoforms.

The future focus of plectin research will have to be put on the understanding of the roles of the individual isoforms. Pinpointing specific isoform functions (as, for instance, plectin 1a defects which are most likely responsible for the skin blistering in patients with EBS-MD) might also allow us to tailor practicable gene therapy approaches for patients with plectin gene mutations. We hope that this chapter will spur new interest in plectin and provide a methodical starting point for researchers new to plectin.

Acknowledgments

We thank Selma Osmanagic-Myers and Michael Zörer for providing the immunoprecipitation and RNase protection assay protocols, and Kerstin Andrä for panels B and C of Fig. 2. Studies carried out in the authors' laboratory were supported by grants of the Austrian Science Research Fund.

References

Andrä, K., Lassmann, H., Bittner, R., Shorny, S., Fässler, R., Propst, F., and Wiche, G. (1997). Targeted inactivation of plectin reveals essential function in maintaining the integrity of skin, muscle, and heart cytoarchitecture. *Genes Dev.* **11,** 3143–3156.

Andrä, K., Nikolic, B., Stöcher, M., Drenckhahn, D., and Wiche, G. (1998). Not just scaffolding: Plectin regulates actin dynamics in cultured cells. *Genes Dev.* **12,** 3442–3451.

Andrä, K., Kornacker, I., Jörgl, A., Zörer, M., Spazierer, D., Fuchs, P., Fischer, I., and Wiche, G. (2003). Plectin-isoform-specific rescue of hemidesmosomal defects in plectin (−/−) keratinocytes. *J. Invest. Dermatol.* **120,** 189–197.

Bauer, J. W., Rouan, F., Kofler, B., Rezniczek, G. A., Kornacker, I., Muss, W., Hametner, R., Klausegger, A., Huber, A., Pohla-Gubo, G., *et al.* (2001). A compound heterozygous one amino-acid insertion/nonsense mutation in the plectin gene causes epidermolysis bullosa simplex with plectin deficiency. *Am. J. Pathol.* **158,** 617–625.

Beil, M., Leser, J., Lutz, M. P., Gukovskaya, A., Seufferlein, T., Lynch, G., Pandol, S. J., and Adler, G. (2002). Caspase 8-mediated cleavage of plectin precedes F-actin breakdown in acinar cells during pancreatitis. *Am. J. Physiol. Gastrointest. Liver Physiol.* **282,** G450–460.

Bohn, W., Etzrodt, D., Foisner, R., Wiche, G., and Traub, P. (1996). Cytoskeleton architecture of C6 rat glioma cell subclones whole mount electron microscopy and immunogold labeling. *Scanning Microsc. Suppl.* **10,** 285–293.

Brown, M. J., Hallam, J. A., Liu, Y., Yamada, K. M., and Shaw, S. (2001). Cutting edge: Integration of human T lymphocyte cytoskeleton by the cytolinker plectin. *J. Immunol.* **167,** 641–645.

Charlesworth, A., Gagnoux-Palacios, L., Bonduelle, M., Ortonne, J. P., De Raeve, L., and Meneguzzi, G. (2003). Identification of a lethal form of epidermolysis bullosa simplex associated with a homozygous genetic mutation in plectin. *J. Invest. Dermatol.* **121,** 1344–1348.

Chavanas, S., Pulkkinen, L., Gache, Y., Smith, F. J., McLean, W. H., Uitto, J., Ortonne, J. P., and Meneguzzi, G. (1996). A homozygous nonsense mutation in the PLEC1 gene in patients with epidermolysis bullosa simplex with muscular dystrophy. *J. Clin. Invest.* **98,** 2196–2200.

Choi, H. J., Park-Snyder, S., Pascoe, L. T., Green, K. J., and Weis, W. I. (2002). Structures of two intermediate filament-binding fragments of desmoplakin reveal a unique repeat motif structure. *Nat. Struct. Biol.* **9,** 612–620.

Clubb, B. H., Chou, Y. H., Herrmann, H., Svitkina, T. M., Borisy, G. G., and Goldman, R. D. (2000). The 300-kDa intermediate filament-associated protein (IFAP300) is a hamster plectin ortholog. *Biochem. Biophys. Res. Commun.* **273,** 183–187.

Eger, A., Stockinger, A., Wiche, G., and Foisner, R. (1997). Polarisation-dependent association of plectin with desmoplakin and the lateral submembrane skeleton in MDCK cells. *J. Cell Sci.* **110,** 1307–1316.

Elliott, C. E., Becker, B., Oehler, S., Castanon, M. J., Hauptmann, R., and Wiche, G. (1997). Plectin transcript diversity: Identification and tissue distribution of variants with distinct first coding exons and rodless isoforms. *Genomics* **42,** 115–125.

Errante, L. D., Wiche, G., and Shaw, G. (1994). Distribution of plectin, an intermediate filament-associated protein, in the adult rat central nervous system. *J. Neurosci. Res.* **37,** 515–528.

Foisner, R., and Wiche, G. (1987). Structure and hydrodynamic properties of plectin molecules. *J. Mol. Biol.* **198,** 515–531.

Foisner, R., Leichtfried, F. E., Herrmann, H., Small, J. V., Lawson, D., and Wiche, G. (1988). Cytoskeleton-associated plectin: *In situ* localization, *in vitro* reconstitution, and binding to immobilized intermediate filament proteins. *J. Cell Biol.* **106,** 723–733.

Foisner, R., Feldman, B., Sander, L., and Wiche, G. (1991a). Monoclonal antibody mapping of structural and functional plectin epitopes. *J. Cell Biol.* **112,** 397–405.

Foisner, R., Traub, P., and Wiche, G. (1991b). Protein kinase A- and protein kinase C-regulated interaction of plectin with lamin B and vimentin. *Proc. Natl. Acad. Sci. USA* **88,** 3812–3816.

Foisner, R., Feldman, B., Sander, L., Seifert, G., Artlieb, U., and Wiche, G. (1994). A panel of monoclonal antibodies to rat plectin: Distinction by epitope mapping and immunoreactivity with different tissues and cell lines. *Acta Histochem.* **96,** 421–438.

Foisner, R., Bohn, W., Mannweiler, K., and Wiche, G. (1995). Distribution and ultrastructure of plectin arrays in subclones of rat glioma C6 cells differing in intermediate filament protein (vimentin) expression. *J. Struct. Biol.* **115,** 304–317.

Foisner, R., Malecz, N., Dressel, N., Stadler, C., and Wiche, G. (1996). M-phase-specific phosphorylation and structural rearrangement of the cytoplasmic cross-linking protein plectin involve p34cdc2 kinase. *Mol. Biol. Cell* **7,** 273–288.

Fontao, L., Geerts, D., Kuikman, I., Koster, J., Kramer, D., and Sonnenberg, A. (2001). The interaction of plectin with actin: Evidence for cross-linking of actin filaments by dimerization of the actin-binding domain of plectin. *J. Cell Sci.* **114,** 2065–2076.

Fuchs, E., and Karakesisoglou, I. (2001). Bridging cytoskeletal intersections. *Genes Dev.* **15,** 1–14.

Fuchs, P., Zörer, M., Rezniczek, G. A., Spazierer, D., Oehler, S., Castanon, M. J., Hauptmann, R., and Wiche, G. (1999). Unusual 5′ transcript complexity of plectin isoforms: Novel tissue-specific exons modulate actin binding activity. *Hum. Mol. Genet.* **8,** 2461–2472.

Fuchs, P., and Wiche, G. (2004). Intermediate filament linker proteins: Plectin and BPAG1. *In* "Encyclopedia of Biological Chemistry" (W. J. Lennartz and M. D. Lane, eds.). Elsevier, New York, In press.

Fujiwara, S., Takeo, N., Otani, Y., Parry, D. A., Kunimatsu, M., Lu, R., Sasaki, M., Matsuo, N., Khaleduzzaman, M., and Yoshioka, H. (2001). Epiplakin, a novel member of the Plakin family originally identified as a 450-kDa human epidermal autoantigen. Structure and tissue localization. *J. Biol. Chem.* **276,** 13340–13347.

Gache, Y., Chavanas, S., Lacour, J. P., Wiche, G., Owaribe, K., Meneguzzi, G., and Ortonne, J. P. (1996). Defective expression of plectin/HD1 in epidermolysis bullosa simplex with muscular dystrophy. *J. Clin. Invest.* **97,** 2289–2298.

Garcia-Alvarez, B., Bobkov, A., Sonnenberg, A., and de Pereda, J. M. (2003). Structural and functional analysis of the actin binding domain of plectin suggests alternative mechanisms for binding to F-actin and integrin beta4. *Structure* **11,** 615–625.

Geerts, D., Fontao, L., Nievers, M. G., Schaapveld, R. Q., Purkis, P. E., Wheeler, G. N., Lane, E. B., Leigh, I. M., and Sonnenberg, A. (1999). Binding of integrin alpha6beta4 to plectin prevents plectin association with F-actin but does not interfere with intermediate filament binding. *J. Cell Biol.* **147,** 417–434.

Gonzales, M., Weksler, B., Tsuruta, D., Goldman, R. D., Yoon, K. J., Hopkinson, S. B., Flitney, F. W., and Jones, J. C. (2001). Structure and function of a vimentin-associated matrix adhesion in endothelial cells. *Mol. Biol. Cell* **12,** 85–100.

Green, K. J., Parry, D. A., Steinert, P. M., Virata, M. L., Wagner, R. M., Angst, B. D., and Nilles, L. A. (1990). Structure of the human desmoplakins. Implications for function in the desmosomal plaque. *J. Biol. Chem.* **265,** 2603–2612.

Herrmann, H., and Wiche, G. (1983). Specific *in situ* phosphorylation of plectin in detergent-resistant cytoskeletons from cultured Chinese hamster ovary cells. *J. Biol. Chem.* **258,** 14610–14618.

Herrmann, H., and Wiche, G. (1987). Plectin and IFAP-300K are homologous proteins binding to microtubule-associated proteins 1 and 2 and to the 240-kilodalton subunit of spectrin. *J. Biol. Chem.* **262,** 1320–1325.

Hieda, Y., Nishizawa, Y., Uematsu, J., and Owaribe, K. (1992). Identification of a new hemidesmosomal protein, HD1: A major, high molecular mass component of isolated hemidesmosomes. *J. Cell Biol.* **116,** 1497–1506.

Hijikata, T., Murakami, T., Imamura, M., Fujimaki, N., and Ishikawa, H. (1999). Plectin is a linker of intermediate filaments to Z-discs in skeletal muscle fibers. *J. Cell Sci.* **112,** 867–876.

Hijikata, T., Murakami, T., Ishikawa, H., and Yorifuji, H. (2003). Plectin tethers desmin intermediate filaments onto subsarcolemmal dense plaques containing dystrophin and vinculin. *Histochem. Cell Biol.* **119,** 109–123.

House, C. M., Frew, I. J., Huang, H. L., Wiche, G., Traficante, N., Nice, E., Catimel, B., and Bowtell, D. D. (2003). A binding motif for Siah ubiquitin ligase. *Proc. Natl. Acad. Sci. USA.* **100,** 3101–3106.

Janda, L., Damborsky, J., Rezniczek, G. A., and Wiche, G. (2001). Plectin repeats and modules: Strategic cysteines and their presumed impact on cytolinker functions. *Bioessays* **23,** 1064–1069.

Kalman, M., and Szabo, A. (2000). Plectin immunopositivity appears in the astrocytes in the white matter but not in the gray matter after stab wounds. *Brain Res.* **857,** 291–294.

Koss-Harnes, D., Jahnsen, F. L., Wiche, G., Soyland, E., Brandtzaeg, P., and Gedde-Dahl, T., Jr. (1997). Plectin abnormality in epidermolysis bullosa simplex Ogna: Non-responsiveness of basal keratinocytes to some anti-rat plectin antibodies. *Exp. Dermatol.* **6,** 41–48.

Koss-Harnes, D., Hoyheim, B., Anton-Lamprecht, I., Gjesti, A., Jorgensen, R. S., Jahnsen, F. L., Olaisen, B., Wiche, G., and Gedde-Dahl, T., Jr. (2002). A site-specific plectin mutation causes dominant epidermolysis bullosa simplex Ogna: Two identical *de novo* mutations. *J. Invest. Dermatol.* **118,** 87–93.

Koster, J., Kuikman, I., Kreft, M., and Sonnenberg, A. (2001). Two different mutations in the cytoplasmic domain of the integrin beta 4 subunit in nonlethal forms of epidermolysis bullosa prevent interaction of beta 4 with plectin. *J. Invest. Dermatol.* **117,** 1405–1411.

Koster, J., Geerts, D., Favre, B., Borradori, L., and Sonnenberg, A. (2003). Analysis of the interactions between BP180, BP230, plectin and the integrin alpha6beta4 important for hemidesmosome assembly. *J. Cell Sci.* **116,** 387–399.

Koster, J., Van Wilpe, S., Kuikman, I., Litjens, S. H., and Sonnenberg, A. (2004). Role of binding of plectin to the integrin {beta}4 subunit in the assembly of hemidesmosomes. *Mol. Biol. Cell* **15,** 1211–1223.

Koszka, C., Leichtfried, F. E., and Wiche, G. (1985). Identification and spatial arrangement of high molecular weight proteins (M_r 300 000-330 000) co-assembling with microtubules from a cultured cell line (rat glioma C6). *Eur. J. Cell Biol.* **38,** 149–156.

Larsen, A. K., Moller, M. T., Blankson, H., Samari, H. R., Holden, L., and Seglen, P. O. (2002). Naringin-sensitive phosphorylation of plectin, a cytoskeletal cross-linking protein, in isolated rat hepatocytes. *J. Biol. Chem.* **277,** 34826–34835.

Leung, C. L., Green, K. J., and Liem, R. K. (2002). Plakins: A family of versatile cytolinker proteins. *Trends Cell Biol.* **12,** 37–45.

Lie, A. A., Schröder, R., Blümcke, I., Magin, T. M., Wiestler, O. D., and Elger, C. E. (1998). Plectin in the human central nervous system: Predominant expression at pia/glia and endothelia/glia interfaces. *Acta Neuropathol.* **96,** 215–221.

Litjens, S. H., Koster, J., Kuikman, I., van Wilpe, S., de Pereda, J. M., and Sonnenberg, A. (2003). Specificity of binding of the plectin actin-binding domain to beta4 integrin. *Mol. Biol. Cell* **14,** 4039–4050.

Liu, C. G., Maercker, C., Castanon, M. J., Hauptmann, R., and Wiche, G. (1996). Human plectin: Organization of the gene, sequence analysis, and chromosome localization (8q24). *Proc. Natl. Acad. Sci. USA.* **93,** 4278–4283.

Lunter, P. C., and Wiche, G. (2002). Direct binding of plectin to Fer kinase and negative regulation of its catalytic activity. *Biochem. Biophys. Res. Commun.* **296,** 904–910.

Malecz, N., Foisner, R., Stadler, C., and Wiche, G. (1996). Identification of plectin as a substrate of p34cdc2 kinase and mapping of a single phosphorylation site. *J. Biol. Chem.* **271,** 8203–8208.

McLachlan, A. D., and Stewart, M. (1975). Tropomyosin coiled-coil interactions: Evidence for an unstaggered structure. *J. Mol. Biol.* **98,** 293–304.

McLean, W. H., Pulkkinen, L., Smith, F. J., Rugg, E. L., Lane, E. B., Bullrich, F., Burgeson, R. E., Amano, S., Hudson, D. L., Owaribe, K., *et al.* (1996). Loss of plectin causes epidermolysis bullosa with muscular dystrophy: cDNA cloning and genomic organization. *Genes Dev.* **10,** 1724–1735.

Metz, T., Harris, A. W., and Adams, J. M. (1995). Absence of p53 allows direct immortalization of hematopoietic cells by the myc and raf oncogenes. *Cell* **82,** 29–36.

Muenchbach, M., Dell'Ambrogio, M., and Gazzotti, P. (1998). Proteolysis of liver plectin by mu-calpain. *Biochem. Biophys. Res. Commun.* **249,** 304–306.

Niessen, C. M., Hulsman, E. H., Oomen, L. C., Kuikman, I., and Sonnenberg, A. (1997). A minimal region on the integrin beta4 subunit that is critical to its localization in hemidesmosomes regulates the distribution of HD1/plectin in COS-7 cells. *J. Cell Sci.* **110,** 1705–1716.

Nievers, M. G., Kuikman, I., Geerts, D., Leigh, I. M., and Sonnenberg, A. (2000). Formation of hemidesmosome-like structures in the absence of ligand binding by the (alpha)6(beta)4 integrin requires binding of HD1/plectin to the cytoplasmic domain of the (beta)4 integrin subunit. *J. Cell Sci.* **113,** 963–973.

Nikolic, B., Mac Nulty, E., Mir, B., and Wiche, G. (1996). Basic amino acid residue cluster within nuclear targeting sequence motif is essential for cytoplasmic plectin-vimentin network junctions. *J. Cell Biol.* **134,** 1455–1467.

Okumura, M., Uematsu, J., Hirako, Y., Nishizawa, Y., Shimizu, H., Kido, N., and Owaribe, K. (1999). Identification of the hemidesmosomal 500 kDa protein (HD1) as plectin. *J. Biochem. (Tokyo)* **126,** 1144–1150.

Osmanagic-Myers, S., and Wiche, G. (2004). Plectin-RACK1 (Receptor for Activated C Kinase 1) Scaffolding: A Novel Mechanism to Regulate Protein Kinase C Activity. *J. Biol. Chem.* **279,** 18701–18710.

Proby, C., Fujii, Y., Owaribe, K., Nishikawa, T., and Amagai, M. (1999). Human autoantibodies against HD1/plectin in paraneoplastic pemphigus. *J. Invest. Dermatol.* **112**, 153–156.

Pulkkinen, L., Smith, F. J., Shimizu, H., Murata, S., Yaoita, H., Hachisuka, H., Nishikawa, T., McLean, W. H., and Uitto, J. (1996). Homozygous deletion mutations in the plectin gene (PLEC1) in patients with epidermolysis bullosa simplex associated with late-onset muscular dystrophy. *Hum. Mol. Genet.* **5**, 1539–1546.

Pytela, R., and Wiche, G. (1980). High molecular weight polypeptides (270,000–340,000) from cultured cells are related to hog brain microtubule-associated proteins but copurify with intermediate filaments. *Proc. Natl. Acad. Sci. USA* **77**, 4808–4812.

Reipert, S., Steinböck, F., Fischer, I., Bittner, R. E., Zeöld, A., and Wiche, G. (1999). Association of mitochondria with plectin and desmin intermediate filaments in striated muscle. *Exp. Cell Res.* **252**, 479–491.

Reipert, S., Fischer, I., and Wiche, G. (2004). High-pressure freezing of epithelial cells on sapphire coverslips. *J. Microsc.* **213**, 81–85.

Rezniczek, G. A., de Pereda, J. M., Reipert, S., and Wiche, G. (1998). Linking integrin alpha6beta4-based cell adhesion to the intermediate filament cytoskeleton: Direct interaction between the beta4 subunit and plectin at multiple molecular sites. *J. Cell Biol.* **141**, 209–225.

Rezniczek, G. A., Abrahamsberg, C., Fuchs, P., Spazierer, D., and Wiche, G. (2003). Plectin 5′-transcript diversity: Short alternative sequences determine stability of gene products, initiation of translation and subcellular localization of isoforms. *Hum. Mol. Genet.* **12**, 3181–3194.

Röper, K., Gregory, S. L., and Brown, N. H. (2002). The 'spectraplakins': Cytoskeletal giants with characteristics of both spectrin and plakin families. *J. Cell Sci.* **115**, 4215–4225.

Ruhrberg, C., Hajibagheri, M. A., Simon, M., Dooley, T. P., and Watt, F. M. (1996). Envoplakin, a novel precursor of the cornified envelope that has homology to desmoplakin. *J. Cell Biol.* **134**, 715–729.

Sawamura, D., Li, K., Chu, M. L., and Uitto, J. (1991). Human bullous pemphigoid antigen (BPAG1). Amino acid sequences deduced from cloned cDNAs predict biologically important peptide segments and protein domains. *J. Biol. Chem.* **266**, 17784–17790.

Schröder, R., Mundegar, R. R., Treusch, M., Schlegel, U., Blümcke, I., Owaribe, K., and Magin, T. M. (1997). Altered distribution of plectin/HD1 in dystrophinopathies. *Eur. J. Cell Biol.* **74**, 165–171.

Schröder, R., Warlo, I., Herrmann, H., van der Ven, P. F., Klasen, C., Blümcke, I., Mundegar, R. R., Fürst, D. O., Goebel, H. H., and Magin, T. M. (1999). Immunogold EM reveals a close association of plectin and the desmin cytoskeleton in human skeletal muscle. *Eur. J. Cell Biol.* **78**, 288–295.

Schröder, R., Fürst, D. O., Klasen, C., Reimann, J., Herrmann, H., and van der Ven, P. F. (2000). Association of plectin with Z-discs is a prerequisite for the formation of the intermyofibrillar desmin cytoskeleton. *Lab. Invest.* **80**, 455–464.

Schröder, R., Kunz, W. S., Rouan, F., Pfendner, E., Tolksdorf, K., Kappes-Horn, K., Altenschmidt-Mehring, M., Knoblich, R., van der Ven, P. F., Reimann, J., *et al.* (2002). Disorganization of the desmin cytoskeleton and mitochondrial dysfunction in plectin-related epidermolysis bullosa simplex with muscular dystrophy. *J. Neuropathol. Exp. Neurol.* **61**, 520–530.

Seifert, G. J., Lawson, D., and Wiche, G. (1992). Immunolocalization of the intermediate filament-associated protein plectin at focal contacts and actin stress fibers. *Eur. J. Cell Biol.* **59**, 138–147.

Sevcik, J., Urbanikova, L., Kostan, J., Janda, L., and Wiche, G. (2004). Actin-binding domain of mouse plectin: Crystal structure and binding to vimentin. *Eur. J. Biochem.* **271**, 1873–1884.

Smith, F. J., Eady, R. A., Leigh, I. M., McMillan, J. R., Rugg, E. L., Kelsell, D. P., Bryant, S. P., Spurr, N. K., Geddes, J. F., Kirtschig, G., *et al.* (1996). Plectin deficiency results in muscular dystrophy with epidermolysis bullosa. *Nat. Genet.* **13**, 450–457.

Spazierer, D., Fuchs, P., Pröll, V., Janda, L., Oehler, S., Fischer, I., Hauptmann, R., and Wiche, G. (2003). Epiplakin gene analysis in mouse reveals a single exon encoding a 725 kDa protein with expression restricted to epithelial tissues. *J. Biol. Chem.* **278**, 31657–31666.

Stegh, A. H., Herrmann, H., Lampel, S., Weisenberger, D., Andrä, K., Seper, M., Wiche, G., Krammer, P. H., and Peter, M. E. (2000). Identification of the cytolinker plectin as a major early *in vivo* substrate for caspase 8 during CD95- and tumor necrosis factor receptor-mediated apoptosis. *Mol. Cell. Biol.* **20**, 5665–5679.

Steinböck, F. A., Nikolic, B., Coulombe, P. A., Fuchs, E., Traub, P., and Wiche, G. (2000). Dose-dependent linkage, assembly inhibition and disassembly of vimentin and cytokeratin 5/14 filaments through plectin's intermediate filament-binding domain. *J. Cell Sci.* **113**, 483–491.

Stephens, D. J., and Banting, G. (2000). The use of yeast two-hybrid screens in studies of protein:Protein interactions involved in trafficking. *Traffic* **1**, 763–768.

Stradal, T., Kranewitter, W., Winder, S. J., and Gimona, M. (1998). CH domains revisited. *FEBS Lett.* **431**, 134–137.

Svitkina, T. M., Verkhovsky, A. B., and Borisy, G. G. (1996). Plectin sidearms mediate interaction of intermediate filaments with microtubules and other components of the cytoskeleton. *J. Cell Biol.* **135**, 991–1007.

Urbanikova, L., Janda, L., Popov, A., Wiche, G., and Sevcik, J. (2002). Purification, crystallization and preliminary X-ray analysis of the plectin actin-binding domain. *Acta Crystallogr. D Biol. Crystallogr.* **58**, 1368–1370.

Vita, G., Monici, M. C., Owaribe, K., and Messina, C. (2003). Expression of plectin in muscle fibers with cytoarchitectural abnormalities. *Neuromuscul. Disord.* **13**, 485–492.

Weitzer, G., and Wiche, G. (1987). Plectin from bovine lenses. Chemical properties, structural analysis and initial identification of interaction partners. *Eur. J. Biochem.* **169**, 41–52.

Wiche, G., and Baker, M. A. (1982). Cytoplasmic network arrays demonstrated by immunolocalization using antibodies to a high molecular weight protein present in cytoskeletal preparations from cultured cells. *Exp. Cell Res.* **138**, 15–29.

Wiche, G., Herrmann, H., Leichtfried, F., and Pytela, R. (1982). Plectin: A high-molecular-weight cytoskeletal polypeptide component that copurifies with intermediate filaments of the vimentin type. *Cold Spring Harb. Symp. Quant. Biol.* **46**, 475–482.

Wiche, G., Krepler, R., Artlieb, U., Pytela, R., and Denk, H. (1983). Occurrence and immunolocalization of plectin in tissues. *J. Cell Biol.* **97**, 887–901.

Wiche, G., Krepler, R., Artlieb, U., Pytela, R., and Aberer, W. (1984). Identification of plectin in different human cell types and immunolocalization at epithelial basal cell surface membranes. *Exp. Cell Res.* **155**, 43–49.

Wiche, G. (1989). Plectin: General overview and appraisal of its potential role as a subunit protein of the cytomatrix. *Crit. Rev. Biochem. Mol. Biol.* **24**, 41–67.

Wiche, G., Becker, B., Luber, K., Weitzer, G., Castanon, M. J., Hauptmann, R., Stratowa, C., and Stewart, M. (1991). Cloning and sequencing of rat plectin indicates a 466-kD polypeptide chain with a three-domain structure based on a central alpha-helical coiled coil. *J. Cell Biol.* **114**, 83–99.

Wiche, G., Gromov, D., Donovan, A., Castanon, M. J., and Fuchs, E. (1993). Expression of plectin mutant cDNA in cultured cells indicates a role of COOH-terminal domain in intermediate filament association. *J. Cell Biol.* **121**, 607–619.

Wiche, G. (1998). Role of plectin in cytoskeleton organization and dynamics. *J. Cell Sci.* **111**, 2477–2486.

Yaoita, E., Wiche, G., Yamamoto, T., Kawasaki, K., and Kihara, I. (1996). Perinuclear distribution of plectin characterizes visceral epithelial cells of rat glomeruli. *Am. J. Pathol.* **149**, 319–327.

Zernig, G., and Wiche, G. (1985). Morphological integrity of single adult cardiac myocytes isolated by collagenase treatment: Immunolocalization of tubulin, microtubule-associated proteins 1 and 2, plectin, vimentin, and vinculin. *Eur. J. Cell Biol.* **38**, 113–122.

Zhang, T., Haws, P., and Wu, Q. (2004). Multiple variable first exons: a mechanism for cell- and tissue-specific gene regulation. *Genome Res.* **14**, 79–89.

CHAPTER 26

In Vitro Methods for Investigating Desmoplakin–Intermediate Filament Interactions and Their Role in Adhesive Strength

Tracie Y. Hudson,[*] Lionel Fontao,[†] Lisa M. Godsel,[*] Hee-Jung Choi,[‡] Arthur C. Huen,[*] Luca Borradori,[†] William I. Weis,[‡] and Kathleen J. Green[*]

[*]Departments of Pathology and Dermatology
Northwestern University Feinberg School of Medicine
Chicago, Illinois 60611

[†]Department of Dermatology
University Hospital
Geneva, Switzerland CH-1211

[‡]Departments of Structural Biology and Molecular and Cellular Physiology
Stanford University School of Medicine
Stanford, California 94305

I. Introduction

Desmosomes are intercellular junctions that play critical roles in the organization and integrity of tissues within multicellular organisms (reviewed in Getsios *et al.*, 2004; Huber, 2003). Frequently described as "spot welds," desmosomes anchor intermediate filaments (IF) to plasma membrane–associated desmosomal cadherins through a highly organized and complex protein scaffolding that includes, but is not limited to, members of the armadillo gene family (plakoglobin and plakophilins) and the IF-associated protein desmoplakin. Loss of desmosome integrity because of genetic mutation leads to tissue fragility, notably in the heart and complex tissues such as epidermis, that is manifested in response to mechanical stress (McGrath, 1999; McMillan and Shimizu, 2001).

Desmoplakins occur as two major spliced forms, desmoplakins I and II (DPI and II), with predicted molecular weights (MWs) of 332 and 259 kDa, respectively, and are responsible for linking the desmosomal complex to the cytoplasmic IF network (Green and Bornslaeger, 1999). Desmoplakins are members of the plakin family of cytolinkers, which are large modular proteins containing various combinations of functional molecules that interact with IF, actin, and microtubules. As a group, the plakins are involved in linking various cytoskeletal elements to membrane structures and also cross-linking cytoskeletal elements within the cytoplasm (Fuchs and Yang, 1999; Leung *et al.*, 2002; Ruhrberg and Watt, 1997).

Desmoplakins were first identified as one of the major components of enriched fractions of desmosomes isolated from bovine tissues (Skerrow and Matoltsy, 1974 and reviewed in Hertzberg *et al.*, 1992). DPI was subsequently identified with immunohistochemical techniques as a constitutive component of the cytoplasmic desmosome plaque in epithelia, myocardial and Purkinje fiber cells of the heart, arachnoidal cells of the meninges, and the dendritic reticulum cells of the germinal centers of lymph nodes (reviewed in Schmidt *et al.*, 1994). DPII is biochemically and immunologically similar to DPI and is now known to be derived from the same gene (Mueller and Franke, 1983; Virata *et al.*, 1992). This smaller form has a more limited distribution, being expressed at lower levels in some simple epithelia and absent in heart tissue (Angst *et al.*, 1990). Purified and rotary shadowed desmoplakin appears as a rod-shaped protein with globular head and tail, a structure that is consistent with predictions made from the cDNA sequence published around the same time (Green *et al.*, 1990; O'Keefe *et al.*, 1989).

DP is built from three major protein domains—a 1056 amino acid N-terminal domain (containing the so-called plakin domain found in all but one family member), followed by an 889 amino acid central rod domain, and a C-terminal 926 residue IF binding domain (Green and Bornslaeger, 1999). DPII lacks two thirds of the central rod, but its N-terminus and C-terminus seem to be identical to DPI (Virata *et al.*, 1992). The DP N-terminus contains the binding sites for armadillo family members plakoglobin and plakophilins and thus is required for targeting the DP–IF complex to the desmosomal cadherin tails, whereas the central rod facilitates dimerization and, possibly, oligomerization of DP (Bonne *et al.*, 2003; Chen *et al.*, 2002; Hatzfeld *et al.*, 2000; Kowalczyk *et al.*, 1997; Meng *et al.*, 1997).

The carboxy-terminal domain of desmoplakin (DPCT) mediates DP binding to IF in desmosomes (Kouklis *et al.*, 1994; Stappenbeck and Green, 1992; Stappenbeck *et al.*, 1993a), and it is this interaction that will be the focus of this chapter. IFs are the predominant load-bearing elements of the cytoskeleton, and desmosomes are found in large abundance in tissues that are exposed to mechanical stress such as the heart and epidermis. The importance of DP during early embryogenesis, subsequent morphogenesis of epithelial and vascular tissues, and integrity of adult tissues is underscored by the existence of mouse models (Gallicano *et al.*, 1998, 2001; Vasioukhin *et al.*, 2001), as well as patients with mutations in DP, who exhibit cardiomyopathies and cutaneous keratodermas (Godsel *et al.*, in press; McGrath, 1999; McMillan and Shimizu, 2001). As the primary, and required, IF-binding protein in all desmosome-bearing tissues, DP exhibits functional flexibility in its capacity to associate with the type III IF proteins vimentin and desmin as well as simple and complex epithelial keratins. However, the specific sequences within DP's C-terminal 926 residues involved in these associations seems to differ depending on IF type (Fontao *et al.*, 2003; Meng *et al.*, 1997; Stappenbeck *et al.*, 1993a), and recent research has focused on defining the requirements for binding to, and regulating, the association between DP and cell type–specific IFs.

The first two methods in this chapter describe strategies for examining IF–DP interactions—*in vitro* purification/cosedimentation and yeast three-hybrid (Y3H) analyses. The third and final method section will be devoted to a description of functional assays for determining the mechanical properties of epithelial cell sheets harboring mutant DPs that compromise the IF–desmosome connection. Together these experimental strategies provide an opportunity to examine how engineered and naturally occurring DP mutations affect IF binding and adhesive strength of epithelial cell sheets.

II. Materials and Methods

A. Generation of Recombinant Carboxy-Terminal Domain of Desmoplakin and Cosedimentation Assays

Although the literature is in general agreement that IF binding is mediated by the DP C-terminus, how this domain interacts with different IFs is still poorly understood. The DP C-terminus can be broken down into three subdomains, denoted A, B, and C, each connected by intervening sequences (Fig. 1). Each subdomain includes 4.5 copies of a 38-residue repeating motif called the "plakin repeat" (PR) and is thus referred to as a "plakin repeat domain" (PRD), and the

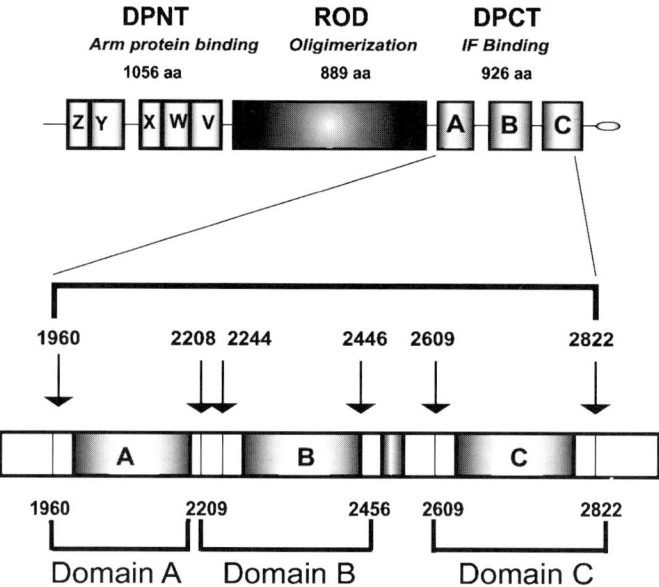

Fig. 1   Domain structure of full length desmoplakin (DP) is shown at the top of the figure, with an enlargement of the C-terminal intermediate filament binding domain below, showing the three plakin repeat domains A, B, and C, with the amino acid positions of the chymotryptic cleavage sites on the top. The boundaries of the carboxyl-terminal region of desmoplakin (DPCT) used for purification and cosedimentation assays are 1960–2822. The recombinant protein lacks the final 49 residues of DP.

linker between B and C harbors an additional PR-like sequence. This repeat was first predicted on the basis of analyses of the primary amino acid sequence (Green *et al.*, 1990), and more recently refined on the basis of proteolytic mapping analysis and high-resolution structural studies of crystallized protein domains (Choi *et al.*, 2002). In addition, a 68-residue tail following the final repeat harbors sequences which seem to fine tune the regulation of IF–DP binding, possibly in a IF-specific manner (Fontao *et al.*, 2003; Meng *et al.*, 1997; Stappenbeck *et al.*, 1993a).

The relative roles of the PRDs and intervening and terminal sequences in IF binding are yet to be fully appreciated. Recent structural studies identified a basic groove on the surface of the C PRD that could possibly represent an IF binding site (Choi *et al.*, 2002). Consistent with this idea, all the PRDs seemed to be able to interact individually with vimentin in cosedimentation assays, albeit weakly. Inclusion of B plus C or all three PRDs in the recombinant protein increased association with vimentin (Choi *et al.*, 2002). Other studies, including Y3H analysis, which will be described in this chapter, have highlighted the importance of the linker domain between B and C in both DP and in other plakin family members (Fontao *et al.*, 2003; Nikolic *et al.*, 1996). Inclusion of this linker sequence at the terminus of the DP B PRD did not seem to alter cosedimentation with vimentin, however. High-resolution structural analysis of cocrystals of IF peptides and DP may be required to sort out these issues. In the meantime, point mutations engineered in the context of the DP C-terminus promise to provide insight into the importance of surface residues in the basic groove in DP–IF interactions.

In this section, we provide a detailed description of the methods for purification of bacterially produced recombinant DP C-terminus (DPCT), which can be used for *in vitro* binding studies such as the DPCT/vimentin cosedimentation assay. The procedure for the DPCT/vimentin cosedimentation assay will also be detailed in this section.

1. Materials and Instrumentation

a. Protein Expression and Purification

The vector used for expressing DPCT in the protocol described here is pPROEX-hTc (Cat. No. 10711-018; Invitrogen). Although this vector is no longer commercially available, pTrcHis A, B, C (Cat. No. V360-20; http://www. invitrogen.com) can be used as an alternative. The DPCT plasmid can be propagated in One Shot Top 10 Competent DH5α (Cat. No. C4040-1; Invitrogen). The transformants are grown in Luria broth (10 g tryptone, 5 g yeast extract, and 10 g NaCl/L) with 0.1 mg/ml ampicillin (Cat. No. A9393; Sigma). Isopropyl-β-D thiogalactoside (IPTG) (Cat. No. BP1755-10; Fisher Biotech) is used for induction of DPCT expression. Other equipment required includes a spectrophometer and a refrigerated centrifuge (e.g., Sorvall RC-5B floor model).

For cell lysis, 10× BugBuster (Cat. No. 70921-4; Novagen; http://www. emdbiosciences.com/html/NVG/home.html) is diluted to 1× in Ni-A buffer (50 mM Tris-Cl pH 8.0, 0.1 M NaCl, 5 mM imidazole, 5 mM β-mercaptoethanol) containing complete EDTA-free protease inhibitor tablets (Cat. No. 1873580; Roche). 75KU Lysozyme (Cat. No. 71110-3; Novagen), and 250 U Benzoase Nuclease (Cat. No. 70746-3) are added to facilitate lysis and reduce viscosity. Other laboratory materials required include an orbital platform shaker and a refrigerated ultracentrifuge (e.g., Beckman TL-100 tabletop model). DPCT is purified from the bacterial supernatant obtained in the preceding steps by loading it on an Econo Column (Cat. No. 737-1511; Biorad) containing Ni-NTA agarose (Cat. No. 30230; Qiagen). Several buffers are needed for the column purification step: Ni-A, Ni-B, Ni-C, and Ni-E. Ni-A (50 mM Tris-Cl pH 8.0, 0.1 M NaCl, 5 mM imidazole, 5 mM β-mercaptoethanol) is used to equilibrate the column. Ni-B (50 mM Tris-Cl pH 8.0, 0.8 M NaCl, 5 mM β-mercaptoethanol) is used to wash the column. Ni-C (50 mM Tris-Cl pH 8.0, 0.1 M NaCl, 10 mM imidazole, 5 mM β-mercaptoethanol) is also used to wash the column. Ni-E (50 mM Tris-Cl pH 8.0, 0.1 M NaCl, 200 mM imidazole, 5 mM β-mercaptoethanol) is used to elute the His-tagged protein from the column. The flow of the various buffers through the column can be controlled with a two-way stopcock (Cat. No. 7328102; Biorad). An anti-His tag antibody used to identify the recombinant DPCT is available from Oncogene Res. Prod. (now part of Calbiochem;

www.emdbiosciences.com/html/CBC/home.html; cat #OB05-100 μg His tag [Ab-1] monoclonal mouse IgG), and is used at a 1:100 dilution of the 100 μg/ml stock for immunoblotting. Recombinant TEV (Cat. No. 10127–017; Invitrogen) is used to remove the His tag from DPCT. Recombinant TEV (rTEV) storage buffer (50 mM Tris-Cl, 1 mM EDTA, 5 mM DTT, 50% [v/v] glycerol, 0.1% [w/v] Triton X-100), as well as DTT (0.1 M) and 20X rTEV cleavage buffer (1 M Tris-HCl pH 8.0, 10 mM EDTA) are supplied with the enzyme. Before running the TEV-treated DPCT through the second Ni-NTA agarose column, it must be dialyzed into Ni-F (25 mM Tris-Cl pH 8.0, 0.1 M NaCl, 5 mM imidizole, and 2 mM β-mercaptoethanol) buffer by use of dialysis tubing (Cat. No. 68100; Pierce).

Ion exchange chromatography is used to further purify the full-length protein. A 5-ml HiTrap Q HP column (Cat. No. 17115401; Amersham Biosciences) along with a 20-ml syringe is used for this purpose. QA solution (50 mM Tris-Cl, pH 8.0, 2 mM β-mercaptoethanol, 0.5 mM PMSF) is used to equilibrate the column and to prepare solutions for an NaCl gradient (100 mM–250 mM). Column fractions containing full-length DPCT can be pooled, concentrated, and transferred into QF buffer (20 mM HEPES [pH 7.5] 0.16 M NaCl, 2 mM β-mercaptoethanol, 0.5 mM EDTA) by use of Centricon Plus-20 columns (Cat. No. UFC2LTK08). Complete Protease Inhibitor Tablets (Cat. No. 1697498; Roche) can be added during the final dialysis and concentration.

b. Cosedimentation Assay

Recombinant Syrian hamster vimentin was purchased in lyophilized form from Cytoskeleton (http://www.cytoskeleton.com). It can be purchased as a kit containing resuspension and polymerization buffers (Cat. No. BK020). The individual components can also be purchased separately. These components include vimentin, which is sold in 5 × 50-μg aliquots (Cat. No. V01; Cytoskeleton), vimentin subunit buffer (5 mM PIPES, pH 7.0, 1 mM DTT) (Cat. No. VSB01-001; Cytoskeleton), and vimentin polymerization buffer (5 mM PIPES, pH 7.0, 1 mM DTT, 3.4 mM NaCl) (Cat. No. VPB01-001; Cytoskeleton). Other laboratory equipment needed includes an ultracentrifuge.

2. Procedures

a. Protein Expression

pPROEX-hTc vector by Invitrogen Life Technologies, or a similar vector, can be used for expression of DPCT in *Escherichia coli*. The protein is expressed as an amino terminal 6 histidine fusion. The histidine sequence has a strong affinity for Ni-NTA resin, which facilitates purification of the protein. Also encoded within the vector is the sequence for the Tobacco Etch Virus (TEV) protease cleavage site, which can be used to remove the histidine tag by use of recombinant TEV. It is recommended that clones expressing the highest levels of intact protein first be identified with small-scale expression experiments. Western blotting can be used to visualize the results of the small-scale experiment if the results are not easily observed by Coomassie brilliant blue or other similar colorimetric detection methods. In large-scale preparations, the yield of DPCT is sufficient to be detected with Coomassie.

Once suitable clones have been selected, large-scale purification can be performed. The first step involves lysis of the bacteria. A wide variety of methods are available for large-scale lysis of bacteria such as a French press, sonication, or chemical lysis with commercially prepared protein extraction reagents. The method detailed in the following describes lysis with the chemical agent Bugbuster, which is purchased as a $10\times$ stock and diluted in the appropriate buffer. Once the cells have been lysed and the crude DPCT preparation obtained, Ni-NTA resin is used to separate DPCT from other contaminating proteins.

Expression steps (small scale)

1. Transform DH5α competent cells with the pPROEX-DPCT DNA.
2. Inoculate 2 ml of Luria broth (LB) media containing 0.1 mg/ml ampicillin with a single pPROEX-DPCT colony. Incubate the culture overnight, shaking at 37 °C.
3. The next day inoculate 50 ml of LB containing 0.1 mg/ml ampicillin with 1 ml of the overnight culture. Grow at 37 °C with shaking for approximately 2.5 hours. Periodically check the OD_{600} of the culture during this time.
4. When the culture reaches OD_{600} of 0.6–0.7 (typically 2–2.5 hours), remove 100 μl and set aside. Induce the remaining culture with 0.5 mM IPTG. Continue the incubation another 3 hours and remove a second 100 μl sample.
5. Take the 100-μl samples and add 100 μL 3$\times$ Laemmli buffer, boil 15 minutes, run 15 μl on a 10% SDS-PAGE gel.
6. Perform Western blot analysis to check for DPCT expression, with an anti-His tag antibody.

Expression steps (large scale)

1. Transform DH5α competent cells with the pPROEX-DPCT DNA.
2. Inoculate 50 ml of Luria broth (LB) media containing 0.1 mg/ml ampicillin with a single colony from stock previously tested in the small-scale experiment detailed earlier. Incubate the culture overnight, shaking at 37 °C.
3. Use 20 ml of the overnight culture per 1 L of LB plus 0.1 mg/ml ampicillin. Incubate the culture at 37 °C with shaking. Check the OD_{600} periodically.
4. When the culture reaches an OD_{600} of 0.6–0.7 (typically 2–2.5 hours), induce the culture with 0.5 mM IPTG. Continue incubation for 3 hours.
5. Centrifuge the bacteria at 2600g for 15 minutes, freeze pellet(s) overnight at −20 °C to facilitate lysis. Two liters of culture should result in 6–8 g (wet weight) of bacteria.

b. Purification of Carboxyl-Terminal Domain of Desmoplakin
Lysis steps

1. Thaw cells in room temperature water. Prepare 1× Bugbuster Protein Extraction reagent in Ni-A buffer.
2. Dissolve one complete EDTA-free protease inhibitor tablet in 50 ml of 1× bugbuster in Ni-A buffer.
3. Resuspend pellet(s), 6–8 mg wet weight, in 20 ml 1× Bugbuster, 10 μL benzonase nuclease (25 U/μL). Shake slowly for 40 minuts and 2.5 μL rLysozyme (30 KU/μl), shake slowly for 40 minutes at room temperature.
4. Spin the sample at 100,000*g* in an ultracentrifuge for 1 hour.
5. Separate the supernatant from the pellet. Store the supernatant on ice.
6. If the pellet still contains unlysed bacteria, repeat steps 3 and 4 with the pellet obtained in step 5 until the pellet has a glassy appearance, which indicates efficient cell lysis.
7. Combine the supernatants and place on ice.

Ni-column purification steps

1. Add 6 ml of Ni-NTA agarose beads per 2 L of culture to a 10-cm by 1.5-cm Econo column.
2. Wash with 5 column volumes of deionized water.
3. Equilibrate the column with 5 column volumes of Ni-A buffer.
4. Adjust the flow rate of the column to 1 ml/min.
5. Add the sample to the column (save flow-through and all washes as a precaution; also take small 50-μl samples at each step, which will be used to monitor the progress of the purification by gel electrophoresis).
6. Wash in 10 column volumes of Ni-A buffer.
7. Wash in 2 column volumes of Ni-B buffer.
8. Wash in 1 column volume of Ni-C buffer.
9. Elute in 4 column volumes of Ni-E buffer, collect the eluate in 3-ml fractions.
10. Strip column with 2 column volumes of 0.1 M EDTA.
11. Check progress by gel electrophoresis and Coomassie brilliant blue staining. Most of the DPCT should elute in the first 3–4 elution fractions.
12. Combine the desired eluates.

Cleavage of his-tag with TEV protease (optional)

1. Add 1× TEV buffer to pooled fractions from above.
2. Add 0.1 M DTT to a final concentration of 1 μM
3. Add 1 U per 3 μg protein of TEV protease (10 U/μl; 1 U cleaves > 85% of 3 μg control substrate in 1 hour at 37 °C).

4. Rotate overnight at 4 °C.

5. Check cleavage by Western blotting of untreated and TEV-treated samples.

Ni–NTA column (2) purification steps (if TEV cleavage was performed)

1. Dialyze the sample into Ni–F buffer with dialysis tubing.

2. Prepare a Ni–NTA column as detailed in section 2b.

3. Wash in 5 volumes of Ni–F buffer.

4. Apply sample to column; DPCT should now be present in the flow-through.

c. Removal of Carboxy–Terminal Domain of Desmoplakin Degradation Products from the Protein Preparation

At this point, the protein sample will contain largely DPCT and low molecular weight degradation products of the full-length DPCT (98 kDa). Removal of some of the lower molecular weight break-down products can be accomplished with ion exchange chromatography. If further purification is desired, size exclusion chromotagraphy can also be performed after the ion exchange purification. The procedure for DPCT purification with HiTrap Q HP 5-ml purification columns is detailed in the following. A Mono Q anion exchange column can also be used; however, it is more expensive than HiTrap.

Anion exchange purification steps (save all flow-through and washes to check for the presence of DPCT)

1. Dialyze sample into QA buffer using dialysis tubing.

2. Fill a 20-ml syringe with 5 column volumes of QA buffer.

3. Remove the twist off end from the column.

4. Wash the column with 5 column volumes of QA at a flow rate of 5 ml/min.

5. Wash the column with 5 column volumes of 1 M NaCl in QA buffer.

6. Wash the column with 5 column volumes of QA buffer.

7. Apply the dialyzed sample to the column with a 20-ml syringe at flow rate of 5 ml/min.

8. Wash the column with at least 3 column volumes of QA.

9. Elute DPCT with a 150-ml stepwise gradient of NaCl (100 mM–250 mM). Each "step" in the gradient should consist of 15 ml of NaCl solution. Collect 5-ml fractions.

10. After completion, strip and regenerate the column by washing with 5 column volumes of 1 M NaCl followed by 5 volumes of QA buffer.

11. Run a small amount of each wash and elution fraction on a 10% SDS-PAGE gel and stain by Coomassie brilliant blue or the staining method of your choice.

12. DPCT at 98 kDa should elute in the range of 180–190 mM NaCl. Pool the desired fractions.

13. Dialyze and concentrate to 200 µl with a Centricon Plus-20 centrifugal filter device and buffer QF.

14. Divide the 200 µl into small aliquots and freeze at −80 °C.

d. Cosedimentation Assay

DPCT purified as described previously can be used for *in vitro* binding assays with IFs such as vimentin or keratin. One such assay is cosedimentation. In this assay, DPCT is mixed with polymerized IFs, incubated, and spun at a 100,000g to pellet polymerized IF and associated proteins.

Cosedimentation assay steps

1. Resuspend 50 µg lyophilized vimentin in 100-µl vimentin subunit buffer to a final concentration of 0.5 µg/µl.

2. Before polymerization, aliquot 20 µL vimentin per tube.

3. Add NaCl to a final concentration of 150 mM to each aliquot.

4. Polymerize at 37 °C for 1 hour.

5. Add DPCT in a 1:1 or 1:5 ratio and incubate at room temperature for 1 hour.

6. Centrifuge at 20 °C for 30 minutes at 100,000g.

7. *Immediately* separate supernatant from pellet.

8. Suspend the pellet in 40 µl of 1× Laemmli sample buffer.

9. Add Laemmli sample buffer to the supernatant.

10. Run 20 µl of the supernatant and pellet on a 10% SDS-PAGE gel.

3. Comments

The amount of pure DPCT obtained from the previous procedure ranges from 0.75–2 mg/2L bacterial culture. Protease inhibitors can be added to the buffer used in the final dialysis and concentration step to minimize protein degradation. Regarding the cosedimentation assay, DPCT should be found in the pellet when polymerized vimentin is present. In the absence of polymerized vimentin, DPCT should remain in the supernatant.

B. Yeast Three-Hybrid Assays

Defining the interactions between DP and other plakin family members has also been approached with a modification of the popular yeast two-hybrid reporter system for protein–protein interactions. In the variation of the technique described here, a yeast three-hybrid approach was developed to facilitate examination of interactions between DP and keratin IF; the latter are obligate heteropolymers (Fuchs and Weber, 1994; Parry and Steinert, 1999). The strategy allows the concomitant expression of three fusion proteins, one plakin and two

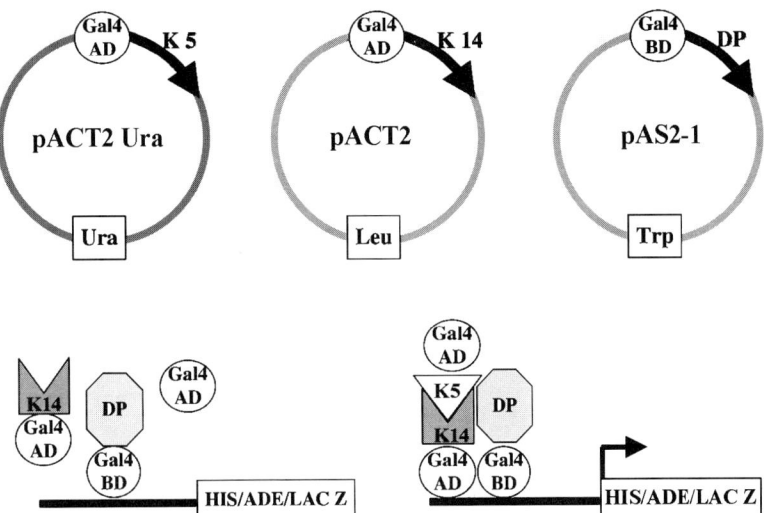

Fig. 2 The yeast three-hybrid system is adapted from the Clontech Matchmaker yeast two-hybrid system. Three different plasmids each containing a selectable marker, Ura, Leu, and Trp, are used. cDNA fragments for distinct keratin heterodimers (such as K5 and K14) are subcloned in frame with the activation domain (AD) of Gal4, whereas desmoplakin (DP) deletion mutants are cloned in frame with the DNA binding domain (DB) of Gal4. When expressed in yeast, Gal4 fusion proteins are exported to the nucleus. When no interaction occurs between a Gal4-AD fusion and a Gal4-BD fusion, as is the case for the K14 monomer and DP fusions (left bottom panel), the reporter genes are not activated, and yeast cannot grow on media lacking either histine or adenine, and no β-galactosidase activity is observed. When Gal4-DB fused to DP and Gal4-AD fused to K5 and K14 are expressed, dimerized K5/K14 associates with DP. Gal4-AD and Gal4-BD are consequently brought in close proximity, resulting in restoration of the transcriptional activity of Gal4. Reporter genes are activated, and yeast can grow on selective markers.

complementary keratins. Previous studies have demonstrated that IF proteins are capable of strong interaction in yeast (Leung and Liem, 1996; Meng *et al.*, 1996; Schnabel *et al.*, 1998). In this system, the LEU2 selective marker in the plasmid pACT2 has been replaced with URA3 to generate pACT2-URA, thus facilitating the simultaneous expression of three plasmids: a GAL4-BD fusion-protein (such as in the pAS2-1 plasmid) and two GAL4-AD fusion-protein (such as in pACT2 or pACT2-URA plasmids), each encoding a different selection marker, TRP1, LEU2, or URA3, respectively (Fig. 2). By introducing two complementary type I and type II keratins that are expected to properly dimerize (Leung and Liem, 1996; Meng *et al.*, 1996; Schnabel *et al.*, 1998), each fused to GAL4-AD, one in pACT2 and the other in pACT2-URA (Fig. 2), the effect of of keratin heterodimerization on interactions with the plakin IF-binding domain fused to GAL4-BD in pAS2-1 can be studied in yeast. The PJ69-4A strain is used because triple transformants can be easily selected on medium lacking tryptophane, leucine, and uracyl.

1. Materials and Instrumentation

Yeast strain: The *Saccharomyces cerevisiae* strain PJ69-4A (Dr. P. James, Department of Biomolecular Chemistry, University of Wisconsin, Madison, WI) contains the genetic markers trp1-901, leu2-3, ura3-52, his3-200, gal4Δ, gal80Δ, LYS2::-GAL1-HIS3, GAL2-ADE2, and met2::GAL7-lacZ (James *et al.*, 1996). It contains three reporter genes, HIS3, ADE2, and Lac-Z each driven by a different promoter under the control of GAL4 transcriptional factor. PJ69-4A yeasts must be restreaked every 3 weeks on YPD plates. A glycerol stock can be used for long-term storage but should not be used directly to inoculate liquid YPD. Indeed, spontaneous revertants of the ADE-2 gene exist, which will grow as white colonies on YPD plate and should not be used. Fresh pink colonies should be used to inoculate YPD medium.

YPD plates: 20 g/L bacto peptones (Difco, no. 211677; www.difco.com), 10 g/L bacto yeast extracts (Difco, no. 212750), 20 g/L bacto agar, add H_2O to 950 ml, adjust to pH 5.8, autoclave, and cool down to 55 °C. Add glucose to 2% (50 ml of a filter sterilized 40% solution). Pour plates, about 20 ml per 8-cm-diameter Petri dishes.

YPD medium: 20 g/L bacto peptones (Difco no. 211677), 10 g/L bacto yeast extracts (Difco, no. 212750), add H_2O to 950 ml, adjust to pH 5.8, autoclave, and cool down to 55 °C. Add glucose to 2% (50 ml of a filter sterilized 40% solution).

YNB: Yeast nitrogen base (Difco no. 233520)

Synthetic Complete (SC) dropout medium: SC-LWU, SC-LWU-His +3AT, SC-LWU-Ade plates and liquid (Table I)

• All amino acids are from Fluka. Stock solutions are made with double-distilled water and filter sterilized.

• 3-amino-1,2,4 triazole. (3AT) (Fluka no. 09540) stock solution is prepared with double-distilled water, filter sterilized, and stored at −20 °C in 1-ml aliquots.

• Glutamic and aspartic acid stocks have to be buffered to pH 7 with NaOH to be completely solubilized.

Liquid media:

• Media are prepared from amino acid stock solution and powder (YNB, glucose). After adding water, adjust the pH to 5.4.

Solid medium:

• A liquid medium about two times concentrated is first prepared and buffered to pH 5.4, then autoclaved. Melted (microwaved) agarose is added, and plates are poured (about 20 ml per 8-cm Petri dishes). Plates are then dried under a laminar flow for about 30 minutes. This step allows faster drying in a subsequent step (c 3).

Carrier DNA: Herring testes carrier DNA at 10 mg/ml (Eppendorf; www.eppendorf.com). Just before use, denature DNA by placing at 95 °C for 15 minutes and immediately cool on ice.

TE buffer (10×): 0.1 M tris-HCl, 10 mM EDTA, pH 7.5. Autoclaved.

Table I

Synthetic Complete (SC) Dropout Medium Recipes

	SC-L-W-U		SC-L-W-U-His +3AT		SC-L-W-U-Ade	
	1 L	40 Plates	1 L	40 Plates	1 L	40 Plates
Adenine, 0.2%	10 ml	10 ml	10 ml	10 ml	0	0
Uracyl, 0.2%	0	0	0	0	0	0
L-histidine, 1%	2 ml	2 ml	0	0	2 ml	2 ml
L-arginine, 1%	2 ml	2 ml	2 ml	2 ml	2 ml	2 ml
L-methionine, 1%	2 ml	2 ml	2 ml	2 ml	2 ml	2 ml
L-tyrosine, 0.044%	68 ml	68 ml	68 ml	68 ml	68 ml	68 ml
L-isoleucine, 1%	3 ml	3 ml	3 ml	3 ml	3 ml	3 ml
L-lysine, 1%	3 ml	3 ml	3 ml	3 ml	3 ml	3 ml
L-phenylalanine	5 ml	5 ml	5 ml	5 ml	5 ml	5 ml
L-glutamic acid, 1%	10 ml	10 ml	10 ml	10 ml	10 ml	10 ml
L-aspatic acid, 1%	10 ml	10 ml	10 ml	10 ml	10 ml	10 ml
L-valine, 3%	5 ml	5 ml	5 ml	5 ml	5 ml	5 ml
L-threonine, 4%	5 ml	5 ml	5 ml	5 ml	5 ml	5 ml
L-serine, 8%	5 ml	5 ml	5 ml	5 ml	5 ml	5 ml
3 AT (1 M)	0	0	2 ml	2 ml	0	0
Yeast nitrogen base	6.7 g	6.7 g	6.7 g	6.7 g	6.7 g	6.7 g
Glucose	20 g	20 g	20 g	20 g	20 g	20 g
H_2O	870	370	866	366	880	380
Agar, 4%	0	500 ml	0	500 ml	0	500 ml

LiAc (10×): 1 M lithium acetate. Adjust to pH 7.5 with acetic acid and autoclave.

50% PEG 4000: Polyethylene glycol, (average molecular weight, 3350). Autoclaved.

TE/LiAc (1×): 1 ml LiAc (10×) 1 ml TE buffer (10×), 8 ml water.

PEG/LiAc (1×): 1 ml LiAc (10×) 1 ml TE buffer (10×), 8 ml of 50% PEG 4000.

DMSO.

Z buffer (10×): Na_2HPO_4 600 mM, NaH_2PO_4 400 mM, KCl 100 mM, $MgSO_4$ 10 mM, pH 7.0, autoclaved, or sterilized by filtration.

ONPG (o-nitrophenyl β-D-galactopyranoside): Sigma (no. N1127).

Na_2CO_3: 1 M sterilized by filtration.

TCA 100%.

TCA buffer: 20 mM Tris, pH 8, 50 mM ammonium acetate, 2 mM EDTA, 10 μg/ml leupeptin, 10 μg/ml pepstatin, 10 μg/ml aprotinin.

Glass beads: 0.5-mm diameter.

Mini-BeadBeater-1 (Biospec Products Cat. # 3110BX; biospec.com) with tubes (Cat. No. 1083).

Gal4-antibodies: Santa-Cruz biotechnology, Gal4 (DBD) RK5C1 (no. SC 510), Gal4 TA, C-10 (no. SC 1663).

2. Procedures

Yeast are cotransformed with three defined constructs in pAS2-1, pACT2 (Clontech), or pACT2-URA (a derivative of pACT2, see Fontao *et al.*, 2003). Triple transformants are selected on agar synthetic complete (SC) medium lacking leucine, tryptophan, and uracil, respectively (SC-L-W-U). Then selected yeast clones are analyzed by use of the plate assay to test the activity of HIS and ADE reporter genes or with β-galactosidase assays to analyze the activity of Lac-Z. The major steps are summarized in Fig. 3.

a. Steps in the Preparation of Competent Yeast

During this procedure, yeast are grown in rich medium; therefore, cultures are very susceptible to contamination. It is advisable to grow a control tube of YPD without yeast inoculation. During the last step of this procedure (day 2) yeast must be in mid log phase (OD_{600} 0.4–0.6) to become competent. The protocol listed in the following is sufficient to perform approximately 25 transformations.

1. Use fresh colony. Inoculate one colony of PJ69-4A from YPD plate in 1 ml YPD, vortex vigorously, and add 4 ml YPD. Grow overnight at 30 °C.
2. Transfer the 5-ml culture to 45 ml YPD, vortex, and grow overnight at 30 °C.
3. Vigorously vortex the overnight culture and dilute 1/10 with YPD and grow the yeast for 3–4 hours at 30 °C to have an OD_{600} between 0.4–0.6.
4. Spin 10 minutes at 1000g at room temperature.
5. Wash the pellet twice with 25 ml of double-distilled water.
6. Spin 10 minutes at 1000g and resuspend the yeast pellet in 2.5 ml TE/LiAc. Store on ice.

b. Steps in Yeast Transformation

It is recommended that positive and negative controls be performed for each series of tests. Negative controls should include a transactivation test (the bait protein only with an empty prey plasmid or vice versa) and a prey plasmid tested against a bait plasmid that produces a noninteracting protein. In this context, some keratin constructs were observed to have a tendency to transactivate when tested against empty plasmid but did not transactivate when tested against plakin mutants that do not bind to keratins.

1. In one tube, mix 3 plasmid DNAs (about 0.8 μg of each pACT2, pACT2-URA, pAS2-1 encoding prey and bait to be tested) and 10 μl denatured carrier DNA.
2. Add 0.1 ml competent yeast and vortex.
3. Add 0.6 ml PEG/LiAc and vortex.
4. Incubate at 30 °C for 30 minutes.
5. Add 75 μl of DMSO and mix gently.
6. Heat 15 minutes at 42 °C and chill on ice.

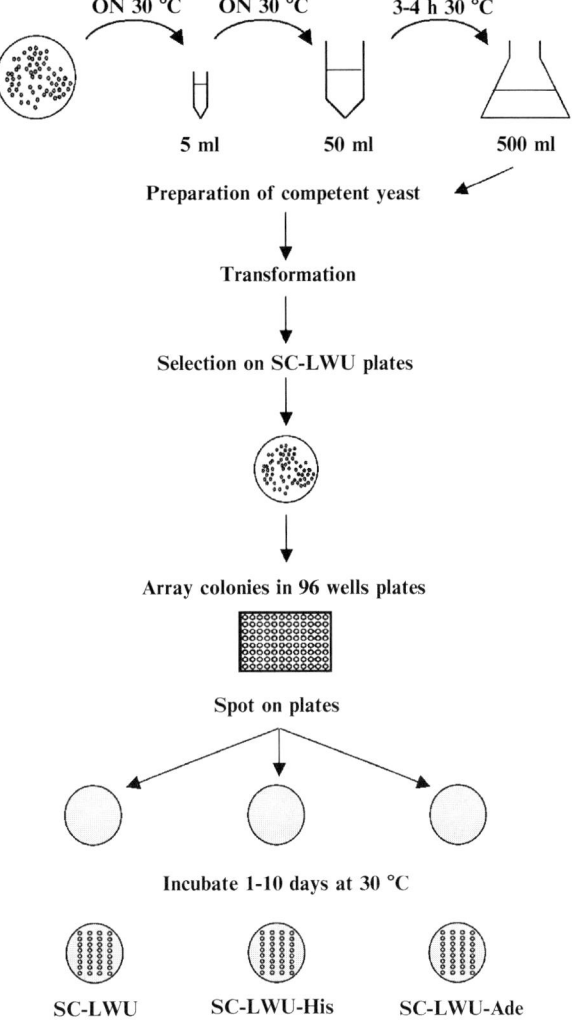

Fig. 3 Diagram of the major steps of the yeast three-hybrid procedure.

7. Spin 5 minutes at 1000g, remove the supernatant, and resuspend the pellet in 0.5 ml TE (1×) buffer.

8. Plate 200 μl per plate (SC-LWU) and allow plate to dry.

Grow yeast at 30 °C for 4–6 days until colonies reach about 1 mm in diameter. The number of colonies present after 6 days is variable and likely depends on the quality of the plasmid DNA and the toxicity of the prey and bait fusion proteins.

Keep in mind that in this system the Gal4 fusion proteins are constitutively expressed.

c. Yeast Three–Hybrid Interaction Analysis
Steps in plate assays

1. Array eight colonies from each transformation in 96-well plates. Each colony is resuspended in 100–200 μl of water. The volume is adjusted according to the size of the colony to maintain about the same cell density in each well.
2. Spot 3 μl of each resuspended colony with a multichannel pipette on SC-LWU (positive control), SC-L-W-U-His +3AT, SC-L-W-U-Ade plates. Colonies are typically spotted along four vertical columns consisting of eight spots each per 8-cm Petri dish.
3. Allow drops to dry and incubate the plates at 30 °C for 1–10 days. Positive interactions will be identified as large white colonies on SC-LWU, SC-LWU-His +3AT, SC-LWU-Ade plates. Although in general all eight colonies grow on these three different plates, variable results have been observed (see Pitfalls).

Steps in β-galactosidose assays

1. Resuspend an entire colony from SC-LWU plate obtained in step C.3 in 10 ml liquid SC-LWU. Analyze approximately three to five different colonies individually to perform a statistical analysis of the data.
2. Incubate cultures overnight at 30 °C with shaking.
3. Vortex each culture and measure OD_{600} with SC-LWU as blank. In general, after an overnight incubation, the OD will be $0.3 < OD_{600} < 0.8$.
4. Spin down 1 OD in a 2-ml Eppendorf tube and resuspend the yeast pellet in 1 ml Z buffer ($1\times$).
5. Measure the OD_{600} of a 100-μl aliquot and record this value; it will be used subsequently to calculate the β-galactosidase activity. It is recommended that the OD be measured quickly, because yeast has a tendency to settle at the bottom of the cuvette.
6. To the 900 μl that remains, add 10 μl of 0.1% SDS (prepared in Z buffer $1\times$) and 50 μl of chloroform and vortex for 10 seconds. This step permeabilizes yeast plasma membranes and cell walls.
7. Add 200 μl of ONPG (4 mg/ml in Z buffer $1\times$). Note the time of the addition and incubate at 30 °C until a coloration of pale yellow develops.
8. Stop the reaction by adding 500 μl of 1 M Na_2CO_3. Note the time of the addition.
9. Spin down cell debris at 10,000g for 5 minutes and measure the OD_{450} of the supernatant. We measure the OD in duplicate by arraying supernatants in 96-well plates. β-galactosidase activity $= 1000\times (OD_{450}/(OD_{600} \times$ time of incubation (in minutes))).

d. Steps in GAL4-Fusion Protein Expression Analysis

Different methods exist to extract proteins from yeast. We have tested several, and the TCA method is the only one that gave reliable results in our experience. It is recommended that several colonies be analyzed for each construct.

1. Inoculate 5 ml of SC-LWU medium with one colony from SC-LWU plates (Step C.3).
2. Dilute the overnight culture with 15 ml of SC-LWU and grow cells until OD_{600} reaches 0.4–0.5 (note the OD value).
3. Centrifuge at 1000g at 4 °C for 5 minutes and resuspend the pellet in 10 ml of ice-cold water.
4. Centrifuge at 1000g at 4 °C for 5 minutes, resuspend cells in 1 ml of ice-cold water, and transfer to tubes for Beadbeater.
5. Centrifuge at 1000g at 4 °C for 5 minutes, remove the water, and immediately freeze the cells in liquid N_2.
6. Resuspend cell pellet in 100 μl of TCA buffer per 7.5 OD units; immediately add the same volume of 20% ice-cold TCA and vortex.
7. Add about the same volume of glass beads. Disrupt the cells in Beadbeater (30 seconds at 5000 rpm), keep on ice, and proceed to the next tube.
8. Perform a second 30-second run of Beadbeater for each tube, keep on ice.
9. Pipette out the supernatant above glass beads and transfer to a clean Eppendorf tube.
10. Wash beads with 500 μl of a buffer containing 50% TCA buffer and 50% of TCA 20%. Perform another run of 30 seconds with Beadbeater. Pipette out the supernatant and transfer it to the corresponding first extract.
11. Centrifuge the supernatant from step 10 for 10 minutes at 14,000g at 4 °C. Remove the supernatant carefully (pellet should be almost dry) and dry the pellet for 5–10 minutes in speed vac. This step reduces the amount of TCA in the pellet that contains cell debris and proteins.
12. Resuspend the pellet in 1× SDS buffer. Use 20 μl of SB per 1 OD of cell. The color should be blue; if not, neutralize dropwise with 1 M Tris pH 8.
13. Resuspend the pellet by sonication, boil the samples for 10 minutes at 100 °C.
14. Centrifuge and load 10–25 μl of supernatant per lane in minigel for Western blot analysis.

C. Assays to Investigate Intercellular Adhesive Strength

Assays to test the consequences of engineered and naturally occurring mutations in junctional molecules have frequently been limited to a microscopic examination of junction morphology and assembly state. Less emphasis has been placed on developing methods to study the strength or fragility of cell sheets formed from cells harboring mutations in these proteins. Here we discuss an adaptation of an assay that we find to be simple, reproducible, and informative (Calautti *et al.*, 1998). It uses an enzyme called dispase, which, unlike trypsin, specifically cleaves extracellular matrix molecules that provide attachment to the underlying substrate. This protease can therefore be used to lift intact epithelial cell sheets from the underlying substrate without damaging the cell–cell contacts or the cell membranes. The strength of cell–cell attachment can be quantified by counting the number of dissociated fragments released from the monolayer after dispase treatment and mild mechanical stress. Cell dissociation can be viewed with time-lapse imaging for direct observation of the cells as the monolayer separates into fragments. In the following discussion, we have used this assay to quantify differences in adhesive strength exhibited by cultured cell monolayers stably expressing inducible forms of the plaque protein DP by use of the Tet On system. For further reference to generation of the cell lines and induction of the inducible promoter, refer to Huen *et al.* (2002). This procedure can also be applied to other types of cells expressing different mutant proteins or treated with blocking antibodies or pharmacological agents.

1. Materials and Instrumentation

a. Growth of Monolayers, Protein Induction, and Dispase Treatment

Cells are grown in 60-mm culture dishes (cat. # 353002; BD Falcon) in Dulbecco's modified of Eagle medium (DMEM; cat # 10-017-CV) containing 10% fetal bovine serum (FBS; cat. # 3H 30070.03; HyClone) and 100 μg/ml penicillin/streptomycin (cat. # 30-002-C1; Mediatech, Inc.), 400 μg/ml G418 (cat. # 61-234-RG; Mediatech), and 1 μg/ml puromycin (cat. # 540411; Calbiochem); 2–4 ug/ml doxycycline (cat. # D9891; Sigma-Aldrich) is used to induce protein production 24–48 hours before confluence of the monolayer. Cell monolayers are washed in Dulbecco's phosphate-buffered saline with calcium and magnesium (DPBS+; cat. # 21-030-CV; Mediatech, Inc.), and 2 ml dispase (cat. # 210455; Roche Diagnostics) at 2.4 U/ml is added for extracellular matrix digestion (ECM).

b. Placement of Mechanical Stress on the Extracellular Matrix–Released Monolayers

Cell monolayers are placed in 5 ml DPBS+ in 15-ml conical tubes (cat. # 14-959-53A; BD Falcon) on a tabletop rocker. After inversion, the monolayers and fragments are transferred to a fresh 60-mm dish for quantification.

c. Cytotoxicity Determination: LDH Assay

The Cytotoxicity Detection Kit (cat. #: 1 644 793; Roche Diagnostics) is a colorimetric assay for quantification of cell lysis by measurement of lactate dehydrogenase (LDH) activity in the supernatant. This assay is performed with 96-well plates and a microplate reader with a 490 to 492-nm filter; 1% Triton X-100 buffer (1% Triton X-100, 145 mM NaCl, 10 mM Tris-HCl, pH 7.4, 5 mM EDTA, 2 mM

EGTA, 1 mM PMSF) is used to lyse a monolayer as a positive control for 100% LDH activity.

d. Quantification of the Cell–Cell Attachment Strength

Monolayer fragments are counted by use of a dissecting microscope (Leica MZ6). To aid in the counting of the fragments a glass Pasteur pipette (cat. # 13-678-20A; Fisher Scientific) can be fashioned as a tool (see later). The pipette can be sealed with heat to melt the tip and then closed off with tweezers.

e. Imaging Modification of Dispase Assay: Growth of Monolayers

Cells are grown as described previously (Section 1.a) with the exception that they are plated on 40-mm diameter glass coverslips (cat. # 40-1313-0319; Bioptechs, Inc.; www.bioptechs.com) contained within 60-mm dishes; 100% ethanol and an alcohol burner are used to sterilize the coverslips after being cleaned with Ross Lens tissue (SPI Supplies).

f. Imaging Modification of Dispase Assay: Wounding of Monolayers

Monolayers are wounded with the edge of a clean fresh razor blade or the tip of a 30-gauge hypodermic needle.

g. Imaging Modification of Dispase Assay: Assembly of the Imaging Chamber

Coverslips are mounted into a Focht Chamber System 2 (FCS2), a closed system live-cell chamber (cat. # 060319-2-03; Bioptechs, Inc.). This chamber contains a chamber top with perfusion ports and dovetailed base, a 0.25 mm or 0.5 mm lower gasket, a 0.75-mm upper gasket with perfusion clearance holes (gasket set; cat. # 060319-2-0719; Bioptechs, Inc.), and a microaqueduct coverslip (cat. # 030119-5; Bioptechs, Inc.) with perfusion clearance holes (Fig. 4). The chamber is clamped into the custom FCS2 stage adapter (cat. # 060319-2-12; Bioptechs, Inc.) of a Leica DMIRBE inverted microscope. The DMIRBE is fitted with a $40 \times$ oil immersion objective and equipped for phase and fluorescence with halogen and mercury illumination.

h. Imaging Modification of Dispase Assay: Treatment of Monolayer with Dispase and Imaging of Cells during Treatment

Alcohol-resistant pump tubing (cat. # 126-014-51; Fisher Scientific) is attached to either side of the chamber top on the perfusion ports to allow for laminar flow of media over the monolayer. One length of tubing runs to a beaker or flask to collect waste. The other length of tubing is attached to a peristaltic pump (Rainin) to transport DPBS+ containing 48 U/ml of dispase over the monolayer. Monolayers can be imaged with an Orca 100 Hamamatsu digital camera (cat. # C4742-95) or equivalent and analyzed using OpenLab software (Improvision, Inc.; www.improvision.com) or other equivalent software. Movies of the time-lapse recording can be generated with Adobe Premier software.

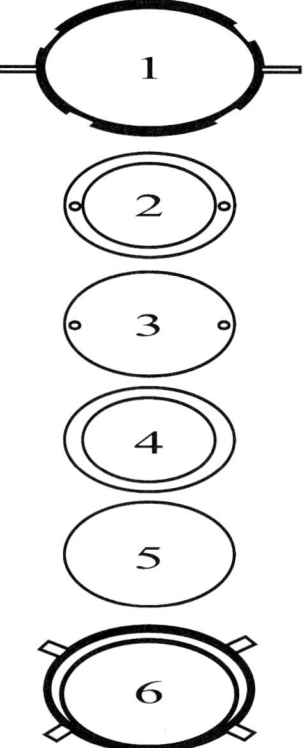

Fig. 4 Schematic for the closed imaging chamber. The imaging chamber consists of a chamber top (1), which contains perfusion ports for media flow to and from the enclosed chamber. An upper gasket (2) is placed between the top and the microaqueaduct slide (3), both of which contain perfusion clearance holes. A lower gasket (4) without perfusion holes is placed between the microaqueaduct slide and the coverslip (5) with the monolayer of cells. Finally, the chamber bottom (6) is placed over the rest of the components and slowly clamped down to hold the chamber together.

i. Imaging Modification of Dispase Assay: Measurement of Cell Distortion

Cells from the time-lapse recordings are chosen for measurement and analyzed with OpenLab (Improvision, Inc.) or equivalent and Microsoft Excel software to measure the aspect ratio (diameter along the axis of stretch/diameter orthogonal to the axis of stretch).

2. Procedures

a. Steps in the Growth of Monolayers and Dispase Treatment

1. Cells are cultured in 60-mm dishes in the appropriate media until the monolayer is completely confluent. Cells that require induction for protein expression should be treated with 2–4 μg/ml doxycycline for at least 24 hours before confluency.

2. Allow the cells to grow an additional 24 hours. Monolayers should be completely confluent, or the cells will be released from the dish as fragments rather than as a single intact monolayer.

3. Rinse cells twice with 5 ml DPBS+ to wash off media.

4. Add 2 ml of dispase at 2.4 U/ml in DPBS+.

5. Incubate the monolayer without rocking for 30–45 minutes at 37 °C/5% CO_2.

b. Steps in the Placement of Mechanical Stress on the Extracellular Matrix–Released Monolayers

1. Monolayers will now be separated from the extracellular matrix and will be curled up in the bottom of the dish. The monolayer should be handled with great care from this point to avoid fragmentation.

2. Transfer the monolayers to 15-ml conical tubes with wide-bore pipettes. Alternatively, wide-bore P1000 pipette tips or P1000 tips that have had the tips cut off can be used for transfer.

3. Carefully rinse the monolayers with 5 ml of DPBS+.

4. Aspirate to remove the DPBS+ wash. Leave approximately 1 ml of the wash above the monolayer to avoid any disturbance to the cells.

5. Repeat steps 3 and 4.

6. Add a final 5 ml of DPBS+ to the 15-ml conical tube and cap tightly.

7. Secure the conical tubes to a tabletop rocker with the rocking arm set to the widest possible arc.

8. Subject the monolayer to 25–50 inversion cycles. The number and strength of inversions can be adjusted for cell types with different adhesive strengths.

9. Transfer the contents of the conical tube into a fresh 60-mm culture dish.

c. Steps in Cytotoxicity/Cytolysis Determination: LDH Assay

1. 500 μl of the supernatant from the dispase-treated, mechanically manipulated samples is removed from each of the 60-mm dishes and placed in 1.5-ml tubes.

2. The supernatant is gently spun at 200–400g for 5 minutes to remove debris.

3. 50 μl of the spun supernatant is placed in a well of a 96-well plate. Three wells are used for each sample in the dispase assay (assayed in triplicate).

4. As a negative control for LDH release, test 500 μl taken from the dispase-released monolayer before the application of stress.

5. The positive control for the LDH release is to lyse a confluent monolayer in 1% Triton-X-100 buffer. When using this control for the assay, a number of dilutions should be performed so that the result is within the detectable limits of the LDH assay.

d. Steps in Quantification of the Cell–Cell Attachment Strength

1. Place the 60-mm dish on a dissecting microscope and count the fragments with $0.63\times$ magnification.

2. Melt the end of a glass Pasteur pipette and close it off with a tweezers. Use this tool to move fragments in the dish from one side to the other as they are counted.

3. To achieve statistical significance, the assay is done three separate times, in triplicate. Statistical analysis and charting of the data by bar graphs can be done with Excel or any other statistics program of choice.

4. A photograph of the monolayers in the dish that uses a digital camera can be a useful complement to the graphed data.

e. Steps in Imaging Modification of Dispase Assay: Growth of Monolayers

1. Cells are cultured on 40-mm diameter sterile coverslips that are placed in 60-mm dishes in the appropriate media until the monolayer is completely confluent. Cells that require induction for protein expression should be treated with 2–$4\,\mu g/ml$ doxycycline for at least 24 hours before confluency.

2. Allow the cells to grow an additional 24 hours. Monolayers should be completely confluent, or the cells will be released from the dish as fragments rather than as a single intact monolayer.

f. Steps in Imaging Modification of Dispase Assay: Wounding of Monolayers

1. The coverslip is then taken from the incubator and rinsed twice with 5 ml of DPBS+.

2. Make sure that the chamber is ready for assembly and the microscope is set up for time-lapse imaging (see section g, which follows).

3. Bring the coverslip to the microscope and wound the monolayer in the center with a fresh, sharp razor blade or a fresh 30-gauge hypodermic needle with light pressure.

g. Steps in Imaging Modification of Dispase Assay: Assembly of the Imaging Chamber and Placement on the Microscope

1. Directly after wounding of the monolayer, assemble the imaging chamber by connecting the tubing to the perfusion ports of the chamber top. One length of tubing should be connected to a peristaltic pump or syringe. The other length of tubing should run to a waste receptacle, such as a flask or beaker. Refer to Fig. 4 for a schematic of the imaging chamber components. It is best to put the chamber together upside down starting with the chamber top held upside down.

2. Pump imaging media or DPBS+ through the tubing and be sure there are no bubbles in the tubing leading to the chamber top. A bubble in the chamber can make imaging difficult.

3. Place the microaqueduct slide onto the top of the chamber. Line up the holes in the slide with the perfusion ports on the chamber top.

4. Place the top gasket over the microaqueduct slide, again aligning the holes in the gasket to the perfusion ports on the chamber top.

5. Pump some media or DPBS+ onto the microaqueduct slide.

6. Place the coverslip, monolayer side toward the microaqueduct slide.

7. Add oil to the bottom of the coverslip and put the 40× oil objective in place so that the monolayer is in focus.

8. Flow DPBS+ over the coverslip by laminar flow through the perfusion ports with the peristaltic rabbit pump set at the lowest setting so as not to disturb the monolayer.

9. Again be sure that the monolayer is in focus and that the software and digital camera for time-lapse data collection are ready to begin imaging. Focus on one side of the wound area. Imaging the edge of wound is best, because one or the other wound edge is typically where stress on the monolayer is greatest. The gasket over the edge of the monolayer will hold the monolayer in place as the dispase begins to cleave the extracellular matrix.

h. Steps in Imaging Modification of Dispase Assay, Treatment of Monolayer with Dispase, and Imaging of Cells during Treatment

1. Change the media flowing over the monolayer to DPBS+ containing 48 U/ml dispase and immediately begin imaging.

2. Image the monolayer every 4–8 seconds with several Z stacks of approximately 2- to 4-μm apart at each time point. The program for data collection can be designed to take phase and/or fluorescence images. Phase images are best for determining the amount of stress placed on the cells and can be used for quantification (see section i. following regarding measurement of cell distortion). Fluorescence is very useful for demonstrating the expression and localization of tagged proteins (GFP, RFP, YFP, CFP, BFP) in the cells that will be used for quantifying cell distortion.

3. The monolayer will split at the wound edge, and the split will move beyond the field of view within 5–40 minutes; therefore, monitor the field of view during recording.

4. The imaging software can be used to compile the individual time-lapse images into a movie format such as Quicktime or AVI. Alternatively, TIFF images can be compiled into a movie format with Adobe Premiere. For an example of video clips comparing control cells and cells expressing the dominant negative mutant DPNTP, which uncouples IF from the desmosome, refer to Huen *et al.* (2002) and http://www.jcb.org/ cgi/content/full/ jcb.200206098/DC1.

i. Steps in Imaging Modification of Dispase Assay, Measurement of Cell Distortion

1. Imaging software, such as OpenLab (Improvision) or MetaMorph (Universal Imaging; http://www.image1.com/products/metamorph/) can be used to measure the aspect ratio of the cells. Determine the diameter along the axis of stretch and divide that value by the diameter orthogonal to the axis of stretch. Do this for each time point captured in the time-lapse recording.

2. The aspect ratios can be charted for each cell over time and then averages and standard deviations of a population of cells can be determined and charted with Microsoft Excel or any other statistics software.

3. Comments

The time-lapse imaging modification of the dispase protocol was described for the Bioptechs, Inc. FCS2 chamber. However, any chamber built for perfusion that contains a gasket that is clamped down over the entire monolayer edge should be acceptable. The success of this procedure is dependent on the monolayer is clamped down around its edge, thus allowing generation of force at the wound edges in the center of the monolayer.

The cytotoxicity detection assay is important to demonstrate that the cells are not undergoing lysis caused by mechanical manipulations that are too stringent. Although we have described an LDH activity assay to detect cell fragility, other assays could be used for the study, including traditional methods such as [3H]-thymidine or alkaline phosphatase release. After testing a number of different epithelial lines and primary cells with 50 or more inversion cycles on a tabletop rocker, we found no detectable cell lysis in the presence of 1.8–2.0 mM calcium.

III. Pitfalls and Pearls

A. Generation and Purification of Recombinant Carboxy-Terminal Domain of Desmoplakin and Cosedimentation Assay

1. DPCT Expression and Purification

1. In our hands, expression of DPCT from the pPROEX-hTc vector is leaky. Equal amounts of DPCT were obtained with or without induction by IPTG in DH5α and BL21 cells, and IPTG induction resulted in an increased level of degradation without increasing the level of expression.

2. Shake bacterial cultures at 250 rpm with plenty of aeration to encourage robust bacterial growth and protein production.

3. As mentioned earlier, after the protein has been extracted from the bacteria, it must be kept cold during purification to minimize degradation. Protease inhibitors can also be added at various stages of the purification. Avoid freeze/thawing the sample between purification steps. Frequent freeze/thaw cycles

increase degradation and may result in protein precipitating. Protein can be stored on ice at 4 °C for up to a week. At the end of the purification procedure divide the DPCT into small aliquots and store at −80 °C.

4. Loss of protein occurs at each stage of the purification. The amount lost from start to finish is about 90%. If significant amounts of DPCT are needed for subsequent assays, more than 2 L of bacterial culture should be used.

5. When comparing the amount of protein loss during dialysis, we found that the use of dialysis cassettes resulted in more protein loss than traditional dialysis tubing.

6. Depending on how the purified recombinant DPCT will be used, removal of the His-Tag may not be necessary. If it is necessary to remove the tag, several small reactions (less than $600 \, \mu l$/reaction) may result in more efficient cleavage than a large reaction. A custom TEV buffer containing 10 mM Tris-Cl, pH 8.0, 0.1% NP40, 150 mM NaCl, 0.5 mM EDTA (final concentration) results in more efficient cleavage in our hands than the buffer supplied with the enzyme (L. Goldfinger, personal communication).

2. Cosedimentation Assays with Carboxy-Terminal Domain of Desmoplakin and Vimentin

1. The manner in which the NaCl is added to the unpolymerized vimentin can influence the length and quality of the polymerized vimentin filaments. Adding the salt all at once at the start of the 1-hour incubation favors short filaments. Adding the salt in small increments every 20 minutes during the 1-hour incubation favors long filaments that sediment more efficiently during the high-speed spin.

2. Keep the final volume of DPCT:vimentin as small as possible to optimize protein–protein interactions.

3. If DPCT or unpolymerized vimentin is found in the pellet fraction after the 100,000g spin, preclear the unpolymerized vimentin and DPCT sample at 100,000g before using it in the assay. Separate the supernatant from the pellet, and use the precleared supernatant in the assay.

4. Separate the supernatant and pellet immediately after the high-speed spin. This reduces the chances that some of the pellet will get carried over into the supernatant when the two are separated.

5. Use a negative protein control such as bovine serum albumin (BSA) when assaying for DPCT's ability to bind IFs such as vimentin. BSA should not cosediment with IFs under the conditions outlined in section A.2.d.1.

B. Yeast Three-Hybrid Assays

1. A significant number of keratin and plakin constructs, despite a correct DNA sequence, did not produce Gal4-fusion proteins. Although there is no clear explanation for this observation, it is conceivable that mutations outside of the

inserted cDNA affect protein expression. In general, this problem is resolved by testing another DNA miniprep. Therefore, it is important to confirm expression of the Gal4-fusion protein by Western blot of yeast extracts at least once.

2. In β-galactosidase assays, variations among series of experiments were observed. Therefore, if samples from different series are to be compared, always include the same positive control in each series and use it to adjust the obtained results. As for the plate test, negative or transactivation controls must be conducted in parallel to appreciate the baseline of β-galactosidase activity.

3. In plate assays, it is recommended that plates be checked every day, because growth rates are not always identical. This is particularly important for constructs that have a tendency for transactivation. In some cases, after 2 days at 30 °C, positive interactions are visible, whereas transactivation is not visible. However, in this case, it is critical that cell density between tested proteins and transactivation controls is comparable. Waiting for longer periods may mask true positives that cannot be discriminated from the transactivation control plates. If the results are equivocal, it is recommended that a β-galactosidase assay be performed to better assess the strength of the interactions.

4. In some plate assays, colonies grow only on plates lacking histidine but not on plates lacking adenine. This can probably be explained by the fact that the ADE reporter gene is less sensitive than the HIS reporter gene.

5. In some plate assays, only half of the colonies grow on SC-LWU-His +3AT and/or SC-LWU-Ade. In this case, repeat the assay by analyzing eight new colonies. If similar results are obtained, it is likely that the detected interactions are weak. In such a situation, it is recommended that a β-galactosidase assay be performed.

6. Some Gal4 fusion proteins are not detectable by Western blot analysis, although they give positive interaction in yeast assays. This usually happens with small proteins (15–20 kDa) that are in fusion with Ga14 AD. There is no clear explanation for this observation. It is likely that the protein is either toxic or that its expression is downregulated in yeast.

C. Assays to Investigate Intercellular Adhesive Strength

1. For optimal results in the dispase assay, the number of inversions and the strength of force placed on the monolayer may need to be varied depending on cell type analyzed and conditions in which the cells are grown. For example, primary keratinocytes in 1.8 mM calcium may require even more inversions to observe measurable differences in fragmentation between subject groups.

2. Cells maintained in a low-calcium environment will fragment on treatment with dispase without any mechanical stress placed on the monolayer. For example, human primary keratinocytes cultured from 0.05 mM–0.1 mM calcium will fragment on dispase treatment and additional mechanical stress should not be used under these conditions. At 0.2 mM calcium, an intact monolayer will be

present after incubation in dispase, and a reduced number of inversions can be administered.

3. Making the time-lapse movies can be challenging, and it may take a number of attempts to obtain usable data. Taking images of multiple Z planes at each time point is important, because the focal plane of the cells may change over time as the monolayer lifts off the coverslip. The use of the slowest possible speed or turning off the peristaltic pump once the dispase-containing buffer has filled the chamber can help avoid disturbing the lifting monolayer and consequently keep the field of view in focus.

4. Fluorescence is a useful way to monitor which cells in the monolayer are expressing a protein of interest. For example, cells expressing a dominant negative form of a cytoplasmic desmosomal protein were identifiable because of a C-terminal fluorescent protein (GFP) tag (see Huen *et al.* [2002]). In this way, only cells expressing the dominant negative mutant were selected for quantification of stress with aspect ratio determinations from time-lapse images.

5. If one is determining the effect of expression of an ectopic molecule on adhesive strength, this assay requires homogeneous expression of mutant molecules throughout the monolayer to observe consistent, reproducible results.

IV. Discussion

In this chapter, we have highlighted two very different approaches for investigating IF interactions with one of their major binding partners, the plakin family member DP. The use of bacterially expressed purified DP protein in cosedimentation assays has been described for polymerized forms of the type III intermediate filament vimentin. However, by controlling conditions under which the proteins are combined, the approach could be adapted to studying the dependence of the interaction on other oligomeric forms and types of IFs.

Because of their relative insolubility and the fact that they exist as obligate heterodimers, keratins present additional challenges. These challenges can be met in a number of other ways, including solid-phase binding assays, which were not discussed here but have been used for this purpose (Kouklis *et al.*, 1994; Meng *et al.*, 1997). Such assays have potential pitfalls of background because of non-specific charged interactions. Another approach was highlighted here: i.e., an adaptation of the Y2H technique in which three selectable markers are engineered into the system thus allowing co-expression of three proteins, including type I and type II keratins and DP.

The various approaches that have been used to address the question of IF–DP interactions have provided data that are frequently in agreement but differ in certain areas. For instance, the importance of the flexible C-terminal tail (following the C PRD) in DP has been confirmed in multiple studies (Fontao *et al.*, 2003; Meng *et al.*, 1997; Stappenbeck *et al.*, 1993b) as has the importance of DP serine

2849 contained therein (Fontao *et al.*, 2003; Stappenbeck *et al.*, 1994). However, discrepancies do exist. In certain assays, a type II epidermal keratin seems to be sufficient in providing a binding site for DP (Kouklis *et al.*, 1994; Meng *et al.*, 1997), whereas other studies support the requirement for a heterodimer (Fontao *et al.*, 2003). The importance of the linker region downstream of the DP "B" subdomain is also not clear (Choi *et al.*, 2002; Fontao *et al.*, 2003). These uncertainties emphasize the importance of the use of variety of complementary approaches in defining these interactions. It should also be kept in mind that, regardless of the approach, the use of deletions may unmask interactions that do not normally occur *in vivo*. Cocrystallization of IF peptides with DP will be a challenge for the future that may help to unravel how DP–IF interactions occur.

The value of developing functional assays to test the result of engineered and naturally occurring mutations in junctional molecules has become clear. Here we have described one assay that was adapted to studying the effects of engineered mutations and cells derived from patients with naturally occurring truncations of DP (Huen *et al.*, 2002). This assay can also be used to study the effects of pharmacological agents, blocking antibodies, and mutations in other plaque and transmembrane proteins on adhesive strength. Another assay which can be used in conjunction with the dispase assay is called the hanging drop assay. Although not discussed here, it can be used to assess adhesive strength of cells allowed to aggregate in suspension, without ever having had the opportunity to establish contacts with the substrate (Huen *et al.*, 2002; Thoreson *et al.*, 2000).

Together, the methods described in this chapter provide a toolbox with which to understand both the molecular requirements for establishing an IF–desmosome connection and for analyzing the functional effects of perturbing this interaction on the integrity of epithelial cell sheets.

Acknowledgments

Work in the authors' laboratories is funded by NIH grants R01AR41836, R01AR43380, and project #4 of P01 DE12328 (KJG), NIH R01 GM56169 (WIW), and the Swiss National Foundation for Scientific Research 3100-067860/1 (LB). L. M. Godsel was supported by NIH T32 AR07593-05 and H.-J. Choi was supported by fellowships from the Korea Science and Engineering Foundation and the American Heart Association.

References

Angst, B. D., Nilles, L. A., and Green, K. J. (1990). Desmoplakin II expression is not restricted to stratified epithelia. *J. Cell Sci.* **97,** 247–257.

Bonne, S., Gilbert, B., Hatzfeld, M., Chen, X., Green, K. J., and van Roy, F. (2003). Defining desmosomal plakophilin-3 interactions. *J. Cell Biol.* **161,** 403–416.

Calautti, E., Cabodi, S., Stein, P. L., Hatzfeld, M., Kedersha, N., and Dotto, G. P. (1998). Tyrosine phosphorylation and src family kinases control keratinocyte cell-cell adhesion. *J. Biol. Chem.* **141,** 1449–1465.

Chen, X., Bonne, S., Hatzfeld, M., van Roy, F., and Green, K. J. (2002). Protein binding and functional characterization of plakophilin 2. Evidence for its diverse roles in desmosomes and beta-catenin signaling. *J. Biol. Chem.* **277,** 10512–10522.

Choi, H. J., Park-Snyder, S., Pascoe, L. T., Green, K. J., and Weis, W. I. (2002). Structures of two intermediate filament-binding fragments of desmoplakin reveal a unique repeat motif structure. *Nat. Struct. Biol.* **9,** 612–620.

Fontao, L., Favre, B., Riou, S., Geerts, D., Jaunin, F., Saurat, J. H., Green, K. J., Sonnenberg, A., and Borradori, L. (2003). Interaction of the bullous pemphigoid antigen 1 (BP230) and desmoplakin with intermediate filaments is mediated by distinct sequences within their COOH terminus. *Mol. Biol. Cell* **14,** 1978–1992.

Fuchs, E., and Weber, K. (1994). Intermediate filaments: Structure, dynamics, function and disease. *Ann. Rev. Biochem.* **63,** 345–382.

Fuchs, E., and Yang, Y. (1999). Crossroads on cytoskeletal pathways. *Cell* **98,** 547–550.

Gallicano, G. I., Bauer, C., and Fuchs, E. (2001). Rescuing desmoplakin function in extra-embryonic ectoderm reveals the importance of this protein in embryonic heart, neuroepithelium, skin and vasculature. *Development* **128,** 929–941.

Gallicano, G. I., Kouklis, P., Bauer, C., Yin, M., Vasioukhin, V., Degenstein, L., and Fuchs, E. (1998). Desmoplakin is required early in development for assembly of desmosomes and cytoskeletal linkage. *J. Biol. Chem.* **143,** 2009–2022.

Getsios, S., Huen, A. C., and Green, K. J. (2004). Working out the strength and flexibility of desmosomes. *Nat. Rev. Mol. Cell Biol.* **5,** 271–281.

Green, K. J., and Bornslaeger, E. A. (1999). Desmoplakin. *In* "Guidebook to the extracellular matrix, anchor, and adhesion proteins," (T. Kreis and R. Vale, eds.), pp. 103–105. Oxford University Press, Oxford.

Green, K. J., Parry, D. A. D., Steinert, P. M., Virata, M. L. A., Wagner, R. M., Angst, B. D., and Nilles, L. A. (1990). Structure of the human desmoplakins: Implications for function in the desmosomal plaque. *J. Biol. Chem.* **265,** 2603–2612.

Hatzfeld, M., Haffner, C., Schulze, K., and Vinzens, U. (2000). The function of plakophilin 1 in desmosome assembly and actin filament organization. *J. Cell Biol.* **149,** 209–222.

Hertzberg, E., Tsukita, S., Green, K. J., and Stevenson, B. (1992). Isolation of intercellular junctions by cell fractionation. *In* "Cell-cell interations: A practical approach," (W. Gallin, D. Paul, and B. Stevenson, eds.), pp. 111–142. Oxford University Press, Oxford.

Huber, O. (2003). Structure and function of desmosomal proteins and their role in development and disease. *Cell Mol. Life Sci.* **60,** 1872–1890.

Huen, A. C., Park, J. K., Godsel, L. M., Chen, X., Bannon, L. J., Amargo, E. V., Hudson, T. Y., Mongiu, A. K., Leigh, I. M., and Kelsell, D. P. (2002). Intermediate filament-membrane attachments function synergistically with actin-dependent contacts to regulate intercellular adhesive strength. *J. Cell Biol.* **159,** 1005–1017.

James, P., Halladay, J., and Craig, E. A. (1996). Genomic libraries and a host strain designed for highly efficient two-hybrid selection in yeast. *Genetics* **144,** 1425–1436.

Kouklis, P. D., Hutton, E., and Fuchs, E. (1994). Making a connection: Direct binding between keratin intermediate filaments and desmosomal proteins. *J. Biol. Chem.* **127,** 1049–1060.

Kowalczyk, A. P., Bornslaeger, E. A., Borgwardt, J. E., Palka, H. L., Dhaliwal, A. S., Corcoran, C. M., Denning, M. F., and Green, K. J. (1997). The amino-terminal domain of desmoplakin binds to plakoglobin and clusters desmosomal cadherin-plakoglobin complexes. *J. Biol. Chem.* **139,** 773–784.

Leung, C. L., Green, K. J., and Liem, R. K. (2002). Plakins: A family of versatile cytolinker proteins. *Trends Cell Biol.* **12,** 37–45.

Leung, C. L., and Liem, R. K. (1996). Characterization of interactions between the neurofilament triplet proteins by the yeast two-hybrid system. *J. Biol. Chem.* **271,** 14041–14044.

McGrath, J. A. (1999). Hereditary diseases of desmosomes. *J. Derm. Sci.* **20,** 85–91.

McMillan, J. R., and Shimizu, H. (2001). Desmosomes: Structure and function in normal and diseased epidermis. *J. Dermatol.* **28,** 291–298.

Meng, J.-J., Bornslaeger, E. A., Green, K. J., Steinert, P. M., and Ip, W. (1997). Two hybrid analysis reveals fundamental differences in direct interactions between desmoplakin and cell type specific intermediate filaments. *J. Biol. Chem.* **272,** 21495–21503.

Meng, J. J., Khan, S., and Ip, W. (1996). Intermediate filament protein domain interactions as revealed by two-hybrid screens. *J. Biol. Chem.* **271**, 1599–1604.

Mueller, H., and Franke, W. W. (1983). Biochemical and immunological characterization of desmoplakins I and II, the major polypeptides of the desmosomal plaque. *J. Mol. Biol.* **163**, 647–671.

Nikolic, B., MacNulty, E., Mir, B., and Wiche, G. (1996). Basic amino acid residue cluster within nuclear targeting sequence motif is essential for cytoplasmic plectin-vimentin network junctions. *J. Cell Biol.* **134**, 1455–1467.

O'Keefe, E. J., Erickson, H. P., and Bennett, V. (1989). Desmoplakin I and desmoplakin II: Purification and characterization. *J. Biol. Chem.* **264**, 8310–8318.

Parry, D. A., and Steinert, P. M. (1999). Intermediate filaments: Molecular architecture, assembly, dynamics and polymorphism. *Q. Rev. Biophys.* **32**, 99–187.

Ruhrberg, C., and Watt, F. M. (1997). The plakin family: Versatile organisers of cytoskeletal architecture. *Curr. Opin. Genet. Dev.* **7**, 392–397.

Schmidt, A., Heid, H. W., Schafer, S., Nuber, U. A., Zimbelmann, R., and Franke, W. W. (1994). Desmosomes and cytoskeletal architecture in epithelial differentiation: Cell type-specific plaque components and intermediate filament anchorage. *Eur. J. Cell Biol.* **65**, 229–245.

Schnabel, J., Weber, K., and Hatzfeld, M. (1998). Protein-protein interactions between keratin polypeptides expressed in the yeast two-hybrid system. *Biochim. Biophys. Acta* **1403**, 158–168.

Skerrow, C. J., and Matoltsy, A. G. (1974). Chemical characterization of isolated epidermal desmosomes. *J. Biol. Chem.* **63**, 524–530.

Stappenbeck, T. S., Bornslaeger, E., Virata, M. L. A., Corcoran, C. M., and Green, K. J. (1993a). Sequences in desmoplakin required for interaction with intermediate filaments and incorporation into desmosomes. *J. Invest. Derm.* **100**, 523.

Stappenbeck, T. S., Bornslaeger, E. A., Corcoran, C. M., Luu, H. H., Virata, M. L. A., and Green, K. J. (1993b). Functional analysis of desmoplakin domains: Specification of the interaction with keratin versus vimentin intermediate filament networks. *J. Biol. Chem.* **123**, 691–705.

Stappenbeck, T. S., and Green, K. J. (1992). The desmoplakin carboxyl terminus coaligns with and specifically disrupts intermediate filament networks when expressed in cultured cells. *J. Biol. Chem.* **116**, 1197–1209.

Stappenbeck, T. S., Lamb, J. A., Corcoran, C. M., and Green, K. J. (1994). Phosphorylation of the desmoplakin COOH terminus negatively regulates its interaction with keratin intermediate filament networks. *J. Biol. Chem.* **269**, 29351–29354.

Thoreson, M. A., Anastasiadis, P. Z., Daniel, J. M., Ireton, R. C., Wheelock, M. J., Johnson, K. R., Hummingbird, D. K., and Reynolds, A. B. (2000). Selective uncoupling of p120(ctn) from E-cadherin disrupts strong adhesion. *J. Cell Biol.* **148**, 189–202.

Vasioukhin, V., Bowers, E., Bauer, C., Degenstein, L., and Fuchs, E. (2001). Desmoplakin is essential in epidermal sheet formation. *Nat. Cell Biol.* **3**, 1076–1085.

Virata, M. L. A., Wagner, R. M., Parry, D. A. D., and Green, K. J. (1992). Molecular structure of the human desmoplakin I and II amino terminus. *Proc. Natl. Acad. Sci. USA* **89**, 544–548.

CHAPTER 27

Studying Cytolinker Proteins

Dmitry Goryunov, Conrad L. Leung, and Ronald K.H. Liem

Columbia University College of Physicians and Surgeons
New York, New York 10032

I. Introduction

The cytoskeleton of most eukaryotic cells is composed of three major filamentous networks: microfilaments (MFs), microtubules (MTs), and intermediate filaments (IFs). The cytoskeleton was once thought of as a mere scaffold for cell structure, but it is now known to be a highly dynamic system of interconnected filamentous networks. The cytoskeleton plays vital roles in a variety of cellular processes, including motility, adhesion, cytokinesis, morphological changes, intracellular trafficking and signaling, and structural integrity. A large number of

METHODS IN CELL BIOLOGY, VOL. 78
Copyright 2004, Elsevier Inc. All rights reserved.
0091-679X/04 $35.00

proteins have been identified that can associate with the different cytoskeletal networks. The cytoskeleton is involved in the cellular processes just listed through these associated proteins. Studying cytoskeletal-associated proteins is therefore likely to yield insights not only into the regulation of the cytoskeleton but also into the mechanisms of cell function in general.

Over the past decade, it has become increasingly clear that the individual cytoskeletal subsystems are organized and function in an interdependent manner (Herrmann and Aebi, 2000; Leung *et al.*, 2002; Schaerer-Brodbeck and Riezman, 2000). Proteins have been identified that can link different types of filaments to one another, as well as to various kinds of intercellular junctions and compartments. Some of the best-studied cytolinker proteins belong to the plakin protein family (Leung *et al.*, 2002; Ruhrberg and Watt, 1997; Wiche, 1998). This family includes plectin, microtubule-actin cross-linking factor (MACF), envoplakin, periplakin, desmoplakin, epiplakin, and bullous pemphigoid antigen 1 (BPAG1). Many of these proteins were initially characterized as IF-binding proteins and subsequently shown to be able to bind other cytoskeletal components.

Most of these proteins are large α-helical coiled-coil molecules that have binding domains for one or more of the three cytoskeletal networks. They can also associate with components of various junctional complexes. The widely expressed plectin has been shown to associate with MTs, MFs, and IFs (Svitkina *et al.*, 1996). Plectin is essential for the mechanical integrity of skin, skeletal muscle, and heart. It has also been implicated in processes involving MF rearrangement in cultured cells (Andra *et al.*, 1998). The *BPAG1* gene gives rise to multiple alternatively spliced isoforms that contain binding domains for all three cytoskeletal subsystems (Leung *et al.*, 2001). Consistent with the presence of major BPAG1 isoforms in neurons and skeletal muscle, *BPAG1*-null mice exhibit severe neurological and motor abnormalities (Dalpe *et al.*, 1999; Guo *et al.*, 1995). MACF, formerly known as ACF7, is another plakin capable of integrating different components of the cytoskeleton. It contains an N-terminal actin-binding domain (ABD) and a C-terminal MT-binding domain (MTBD) and has been shown to cross-link MFs and MTs in transfected cells (Leung *et al.*, 1999a). The presence of a plakin domain, the signature feature of the plakin protein family, in MACF makes it likely that MACF also interacts with integrins or other membrane proteins involved in cell–cell or cell–substratum interactions. Recent data suggest that it may be involved in MT guidance along MFs, as well as in MT capture at the plasma membrane (Kodama *et al.*, 2003). Consistent with a role for MACF-like proteins in anchoring the cytoskeleton to the plasma membrane, mutations in the *Drosophila* ortholog of *MACF, kakapo/short stop*, lead to epidermal blistering resulting from detachment of the basal membrane from the underlying muscle cells (Gregory and Brown, 1998). In addition, Shot, the protein product of *kakapo/short stop*, seems to be necessary for certain stages of neurogenesis, because mutations in the gene also result in defects in axonal extension (Lee *et al.*, 2000). Shot has recently been shown to recruit EB1/APC and promote MT assembly at the muscle-tendon junction (Subramanian *et al.*, 2003).

The GAS2 family of proteins was identified relatively recently and represents another class of potential MT-actin cytolinkers (Brancolini *et al.*, 1992; Zuchman-Rossi *et al.*, 1996). It currently includes three members, GAS2, GAR17, and GAR22 (Goriounov et al., 2003). Their shared characteristics are the presence of an N-terminal ABD of the calponin homology (CH) family and a C-terminal MTBD called the GAS2-related (GAR) domain. *GAS2* was identified in a screen for genes induced in murine fibroblasts by growth arrest—hence its name, growth arrest–specific gene 2 (Schneider *et al.*, 1988). GAS2 protein has been shown to associate with MFs in cultured cells, presumably through its CH domain (Brancolini *et al.*, 1992). Its GAR domain colocalizes with MTs in transfected COS7 cells (Sun *et al.*, 2001). Human *GAS2*-related gene on chromosome 22 (*GAR22*) was originally identified in a search for putative tumor suppressors. It is closely related to *GAS2*. The primary transcript of *GAR22* undergoes alternative splicing, resulting in two mRNA splice-forms (Zucman-Rossi *et al.*, 1996). *GAR17* (*GAS2*-related on chromosome 17) was identified and cloned based on sequence similarity to *GAR22*. It also undergoes alternative splicing giving rise to two isoforms. The cytoskeletal-binding properties of the corresponding GAR17 and GAR22 isoforms are practically identical despite some differences in domain composition (Goriounov *et al.*, 2003).

II. Methods for Studying Cytolinker Proteins

Various methods have been used to determine whether a protein is capable of linking different cytoskeletal components. These include immunofluorescent visualization of MT, IF, and actin networks in transfected cells, *in vitro* binding assays based on protein cosedimentation with polymerized filaments, *in vitro* overlay ("Far-Western") assays, two-hybrid assays showing interactions with monomeric cytoskeletal proteins, and treatments of transfected cells with cytoskeleton-depolymerizing drugs. An initial indication that a protein might function as a cytolinker can be obtained from the analysis of its primary structure showing the presence of domains previously reported to interact with different cytoskeletal subsystems. As a first step, we usually transfect cultured cells with a plasmid expressing the protein in question and determine the intracellular distribution of the transfected protein by means of indirect immunofluorescent microscopy. A potential cytolinker should show colocalization with more than one of the cytoskeletal filamentous networks, if to varying degrees. With triple-staining techniques, it might also be possible to show that the protein induces enhanced coalignment of one type of filament with another. This coalignment suggests that actual bridging of the cytoskeleton, rather than independent binding to its different components, has occurred.

However, the results of such immunostaining should not be considered evidence of direct binding of the protein to the different types of filaments. This conclusion usually relies on the demonstration of *in vitro* binding in cosedimentation/overlay assays, as well as corresponding protein interactions in two-hybrid experiments. The most conclusive evidence of a protein cross-linking various types of filaments is obtained by electron microscopy.

A. Triple Staining of Transfected Cells

When a candidate cytolinker protein has been shown to associate or colocalize with two individual cytoskeletal components, one can perform triple staining of cells transfected with a protein-encoding plasmid. Because transient transfections generally result in protein overexpression, enhanced coalignment of the two cytoskeletal components may be observed if the candidate protein can indeed bridge the cytoskeleton *in vivo*. This kind of assay provides strong, albeit indirect, evidence that the protein is an actual cytolinker. Described in the following is a procedure for triple staining of COS7 cells transfected with a putative actin-MT linking protein, GAR22β.

GAR22β protein contains both an ABD and an MTBD and can associate with either type of filament in transfected cells (Fig. 1A; Goriounov *et al.*, 2003). These findings suggest that this protein may be capable of physically linking MFs and MTs. To explore this possibility, we transfected FLAG-tagged GAR22β into COS7 cells and stained the cells with a polyclonal anti-tubulin antibody, a monoclonal anti-FLAG antibody, and fluorescently labeled phalloidin. GAR22α, whose MTBD is apparently masked, was used as a negative control (Goriounov *et al.*, 2003).

Reagents and Equipment

- FLAG-tagged GAR22α and GAR22β cDNAs in pcDNA3 vector (pFLAG-GAR22α and pFLAG-GAR22β).
- LipofectAMINE PLUS reagent (Invitrogen #18324-012 and #11514-015).
- 4% paraformaldehyde in PBS, pH 7.4 (add 4 g of paraformaldehyde powder to 70 ml of PBS, heat to 50–60 °C with constant stirring, add 1 N NaOH drop by drop until the powder is dissolved, then adjust the pH to 7.4 with hydrochloric acid).
- Phosphate-buffered saline (PBS), pH 7.4.
- Normal goat serum.
- Parafilm.
- 100- or 150-mm Petri dishes.
- Monoclonal FLAG M2 antibody (Sigma #F3165), polyclonal anti-tubulin antibody (Sigma #T3526).
- AlexaFluor-594–conjugated goat anti-rabbit antibody (Molecular Probes #A11012), AlexaFluor-488–conjugated goat anti-mouse antibody (Molecular Probes #A11001), phalloidin-AlexaFluor-647 (Molecular Probes #A22287).
- Aquamount slide-mounting medium (Lerner Laboratories).
- Glass slides and coverslips (Fisher #12–548-A).

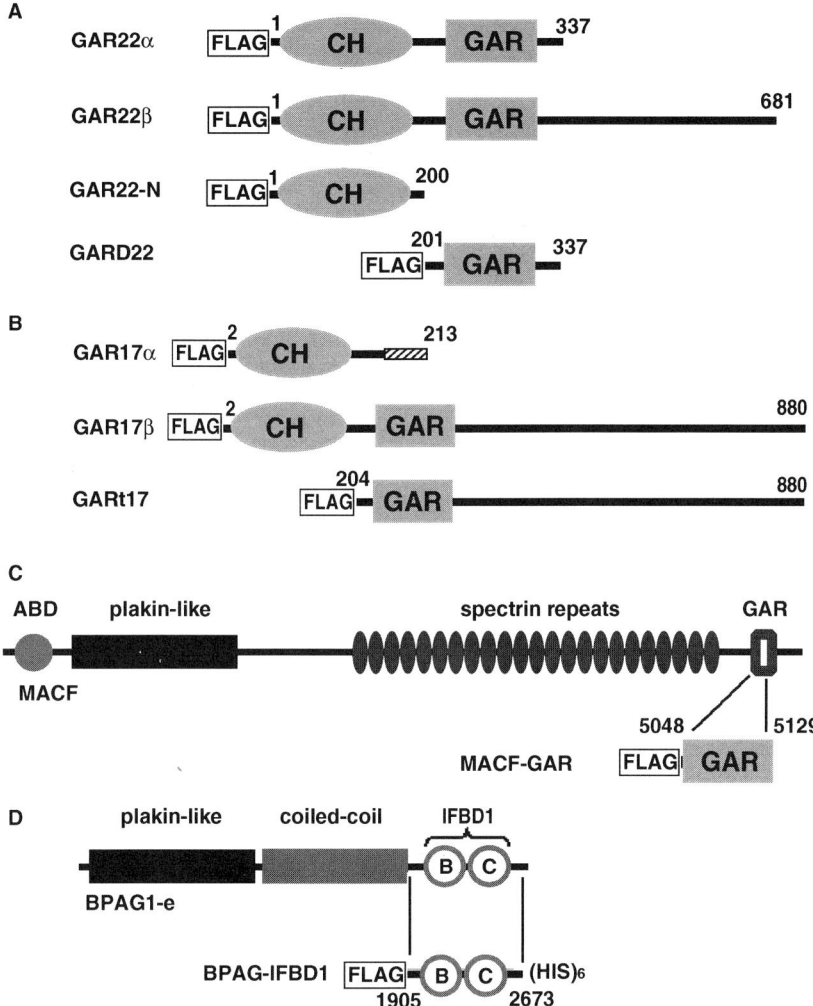

Fig. 1 Cytolinker protein constructs used in the experiments described in this review. (A) GAR22 constructs. Both full-length isoforms of human GAR22 (α and β) contain a calponin homology (CH) domain and a GAS2-related (GAR) domain. The first 337 amino acids of GAR22β are identical to GAR22α. The additional C-terminal sequence of hGAR22β does not contain any known domains and seems to be highly degenerate and unstructured. GAR22-N is a truncated version of GAR22 that retains only the actin-binding CH domain. GARD22 consists of the microtubule (MT)-binding GAR domain only. All constructs contain a FLAG epitope tag at their N-termini. The amino acids of GAR22 making up each construct are indicated. (B) GAR17 constructs. Human GAR17α contains only a CH domain; human GAR17β contains both a CH domain and a GAR domain. The last 27 amino acids of GAR17α (hatched box) are different from the corresponding sequence of GAR17β because of a frameshift introduced into the α splice-form by alternative splicing. GARt17 consists of the GAR domain and the C-terminal sequence of GAR17β. The latter does not contain any known domains, similar to the C-terminus of GAR22β. The amino acids making up each construct are indicated. (C) MACF is a giant protein containing domains

- Fluorescent microscope (e.g., Nikon Eclipse E800) with a mounted digital camera (e.g., SPOT, Diagnostic Instruments, Inc.).
- Photoshop 6.0 (or similar) image-processing program.

Procedure

1. Seed COS7 cells at ~30% confluency on coverslips sterilized in burning ethanol (100%).

2. Next day, transfect the cells with pFLAG-GAR22β using LipofectAMINE PLUS reagent as suggested by the manufacturer. Incubate at 37 °C for 24–48 hours.

3. Wash the cells with PBS three times.

4. Fix the cells for 10 minutes with freshly prepared ice-cold 4% paraformaldehyde (in PBS, pH 7.4). The paraformaldehyde solution can be stored at 4 °C for up to a month.

5. Wash three times with PBS.

6. Place the coverslips face up on a flat piece of Parafilm in a Petri dish.

7. Pipette 300 μl of 5% (in PBS) normal goat serum on each coverslip. Cover the Petri dish and incubate for 30 minutes at room temperature to block nonspecific binding sites.

8. Aspirate the blocking solution and add 300 μl of PBS containing monoclonal FLAG M2 antibody (1:500), polyclonal tubulin antibody (1:50), and phalloidin-AlexaFluor-647 (1:1000). Incubate for 1 hour at room temperature.

9. Wash the coverslips three times with PBS, 5 minutes each.

10. Add 300 μl of PBS containing AlexaFluor-594–conjugated goat anti-rabbit antibody and AlexaFluor-488–conjugated goat anti-mouse antibody (both at 1:1000). Incubate for 1 hour at room temperature.

11. Wash three times with PBS, 5 minutes each.

12. Aspirate the PBS, carefully blot off excess liquid by holding the edge of the coverslip against a Kimwipe tissue, and place it face down onto a drop of Aquamount medium on a glass slide.

---

capable of binding actin (actin-binding doman, ABD) and MTs (GAR domain). It also contains a plakin domain and a number of spectrin repeats. MACF-GAR is a construct composed of the GAR domain of MACF that is FLAG-tagged at its N-terminus. The amino acids of MACF comprising the GAR domain are indicated. (D) BPAG1 has multiple alternative splice-forms combining various cytoskeletal-binding domains. Shown here is the epidermally expressed BPAG1-e isoform, which consists of a plakin domain, a coiled-coil rod, and an intermediate-filament binding domain (IFBD1). In turn, IFBD1 is composed of two plakin repeat domains (PRDs, B and C). BPAG-IFBD1 is a construct that consists of IFBD1 with a FLAG-tag at its N-terminus and a His-tag at its C-terminus. The BPAG1-e amino acids comprising the IFBD1 domain are indicated. (Modified from Goriounov *et al.*, 2003; Leung *et al.*, 1999a, 2001; Sun *et al.*, 2001.)

13. Let the mounting medium dry for 40 minutes, then rinse the slides briefly in distilled water to remove residual salt. Air-dry the slides overnight at room temperature.

14. Take pictures of the cells at wavelengths corresponding to all three labels and superimpose them to produce a red-green-blue (RGB) image. In the resulting image, areas of triple-color overlay, which would correspond to triple colocalization and likely cross-linking, are represented in white.

An example of GAR22-actin-MT triple staining is shown in Fig. 2. The long isoform of GAR22, GAR22β, seems capable of bridging MTs and MFs in transfected COS7 cells. GAR22α, the short isoform, does not cause enhanced MT-MF coalignment (not shown).

B. *In Vitro* Binding Assays

In vitro binding assays are generally used to show that a potential cytoskeletal binding protein is capable of associating with certain types of filaments. In such an assay, preassembled filaments are incubated with the protein under study, and the reaction mixture is centrifuged at high speed to pellet the filaments. The presence of the protein in the pellet is indicative of filament binding, whereas its presence in the supernatant shows lack of binding. The protein can be labeled radioactively, which increases the sensitivity of the assay. Purified recombinant protein can also be used, in which case a more rigorous conclusion can be drawn as to the direct nature of its interaction with the filaments.

We prepare radioactively labeled proteins with the STP3-T7 system from Novagen. The Single Tube Protein System 3 (STP3) is designed for efficient *in vitro* synthesis of proteins directly from supercoiled or linear DNA templates containing a T7 or Sp6 promoter. The system is based on a two-step protocol in which transcription with T7 or Sp6 polymerase is directly followed by translation in an optimized rabbit reticulocyte lysate. We use pcDNA3 or its derivatives for cloning cDNAs to be transcribed/translated. Other vectors containing T7 or Sp6 promoters can also be used. We have found linearization to be inessential for efficient transcription in this system.

1. Preparation of Radioactively Labeled Proteins

Reagents and Equipment

- STP3/T7 *in vitro* transcription/translation system (Novagen #70192).
- ^{35}S-methionine (NEN Life Sciences Products #NEG-009A).
- 30 °C waterbath.
- pcDNA3 vectors encoding proteins to be studied.
- Strong-wall centrifuge tubes (e.g., Beckman #343778, to be used with a TLA45 rotor).
- Refrigerated ultracentrifuge (e.g., Beckman model TL-100).

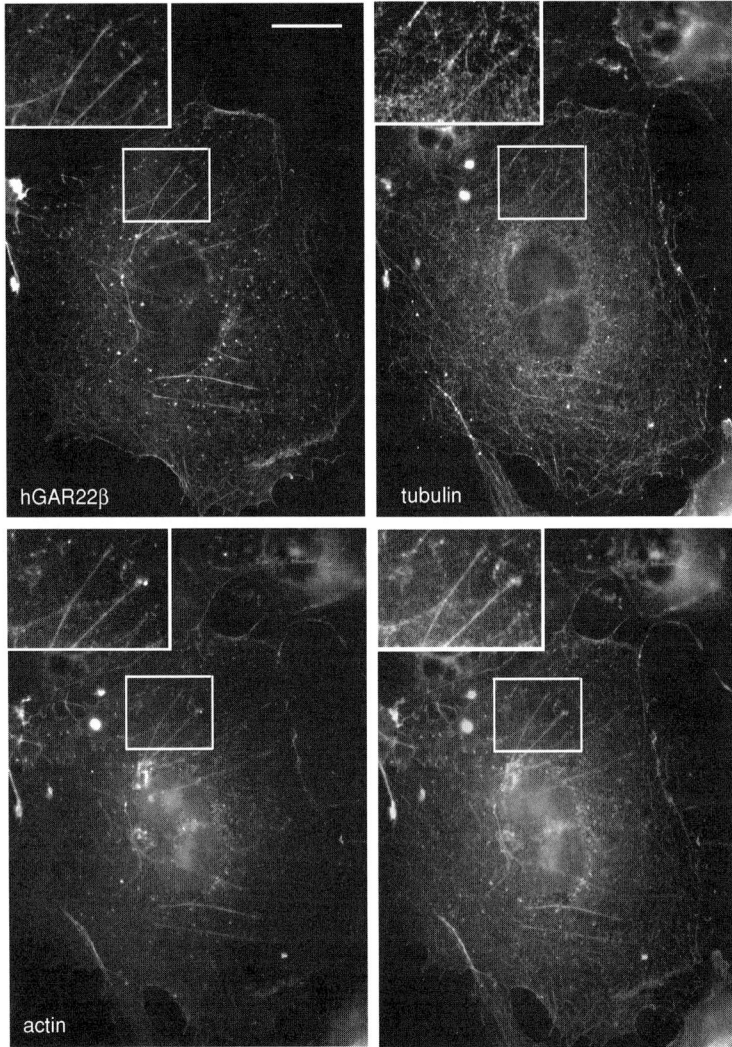

Fig. 2 GAR22β bridges MFs and MTs in transfected cells. COS7 cells were transiently transfected with pFLAG-GAR22β and triple-stained with a monoclonal anti-FLAG antibody (A), a polyclonal anti-tubulin antibody (B), and fluorescently labeled phalloidin (C). (D) Superimposed image, with the FLAG staining in blue, the tubulin staining in red, and the actin staining in green. Triple-stained filaments, likely resulting from cross-linking by GAR22β, are white-colored. Bar, 10 μm. (From Goriounov *et al.*, 2003.) (See Color Insert.)

Procedure

1. Add 0.5 μg of protein-encoding plasmid to 8 μl of transcription mix; adjust the volume with water to 10 μl.

2. Incubate the transcription reaction at 30 °C for 15 minutes.

3. Add 30 μl of translation mix and 4 μl (40 μCi) of ^{35}S-methionine; adjust the volume with water to a total of 50 μl.

4. Incubate the translation reaction at 30 °C for 60 minutes. At this point the reaction can be frozen at $-$ 80 °C.

5. Just before use, spin the protein at 150,000g for 15 minutes at 4 °C. Collect the supernatant (your radioactively labeled protein) and use it for *in vitro* binding.

2. Actin-Binding Assay

To test labeled proteins for the ability to bind actin filaments, we use the Actin Binding Protein Spin-Down Biochem Kit from Cytoskeleton, Inc. This kit provides an easy and convenient way to prepare F-actin and then use it to cosediment any binding proteins. The procedure consists of actin polymerization followed by *in vitro* protein binding and sedimentation by high-speed centrifugation.

Reagents and Equipment

- Actin Binding Protein Spin-Down Biochem Kit (Cytoskeleton #BK001).
- 5× SDS sample buffer.
- SDS-PAGE apparatus.
- Gel-dryer (e.g., BioRad model 583) with a vacuum pump.
- X-ray film (e.g., BioMax MS, Kodak).
- Strong-wall centrifuge tubes (e.g., Beckman #343778, to be used with a TLA45 rotor).
- Refrigerated ultracentrifuge (e.g., Beckman model TL-100).
- Gel-fixing solution (40% methanol, 7% acetic acid, 0.5% Coomassie); destaining solution (40% methanol, 7% acetic acid).
- Whatman paper.

F-actin Preparation

1. Thaw a 250-μg aliquot of monomeric actin on ice (kit component AKL99).

2. Add 250 μl of General Actin Buffer (BSA01) and pipette up and down five times. The resulting actin concentration is 1 mg/ml.

3. Incubate on ice for 30 minutes.

4. Add 25 μl of Actin Polymerization Buffer (BSA02) and incubate at room temperature for 1 hour. This is your F-actin stock (23 μM). It can be stored at 4 °C for up to a week.

Protein Binding and Sedimentation

1. For each protein to be tested, set up two microfuge tubes on ice.

2. Pipette 40 μl of F-actin stock into one tube and 40 μl of F-actin buffer (BSA01:BSA02 10:1) into the other. The latter will provide a negative spin-down control for your protein.

3. Add 10 μl of labeled protein (prespun at 150,000g for 20 minutes at 4 °C) to each tube, mix by stirring gently with the pipette tip. Use α-actinin (AT01) or BSA to serve as positive and negative binding controls, respectively (final concentration, 2 μM).

4. Incubate all tubes at room temperature for 30 minutes.

5. Layer the binding reactions on top of 100 μl of Cushion Buffer (BSA03) in strong-wall centrifuge tubes. The use of cushion will reduce nonspecific binding.

6. Spin the tubes at 150,000g for 1.5 hours at room temperature.

7. Collect 40 μl of supernatant from each tube and add 10 μl of 5× SDS sample buffer. Remove the remainder of the supernatant with a pipette tip and discard as liquid radioactive waste. Resuspend the actin pellet in 50 μl of 1× sample buffer. Boil the samples for 5 minutes and run 10 μl of each on an SDS-PAGE gel.

8. Fix and stain the gel in fixing solution with constant shaking for 1 hour at room temperature; destain in destaining solution overnight. Placing Kimwipe tissues or small pieces of sponge in the container will accelerate destaining, because the released dye will be absorbed by these materials.

9. Dry the gel in a vacuum dryer on a sheet of Whatman paper. To prevent gel cracking, wet the paper thoroughly in destaining solution before placing the gel on it. Cover the gel with thin plastic film (Saran-Wrap or similar). Set the dryer for 80 °C and 1–2 hours (depending on gel size).

10. Remove the plastic wrap and expose the gel to x-ray film for 24–72 hours at −80 °C. Use an intensifying screen.

Examples of results obtained from this actin-binding assay are shown in Fig. 3A. The GARD22 protein, which consists of the GAR domain of GAR22 (Fig. 1A), remains in the supernatant both in the absence and in the presence of F-actin. It is therefore incapable of binding actin *in vitro*. In contrast, the GAR22-N protein, which contains a CH domain, remains in the supernatant in the absence of filaments but is found in the pellet in the presence of F-actin. GAR22-N can therefore bind actin filaments *in vitro*, although this binding could still be mediated by components of the *in vitro* transcription/translation mixture.

3. Microtubule–Binding Assay

Similar approaches can be used to test whether a protein is capable of binding MTs. As in the actin-binding assays, radioactively labeled proteins provide the highest sensitivity. However, purified recombinant proteins should be used to

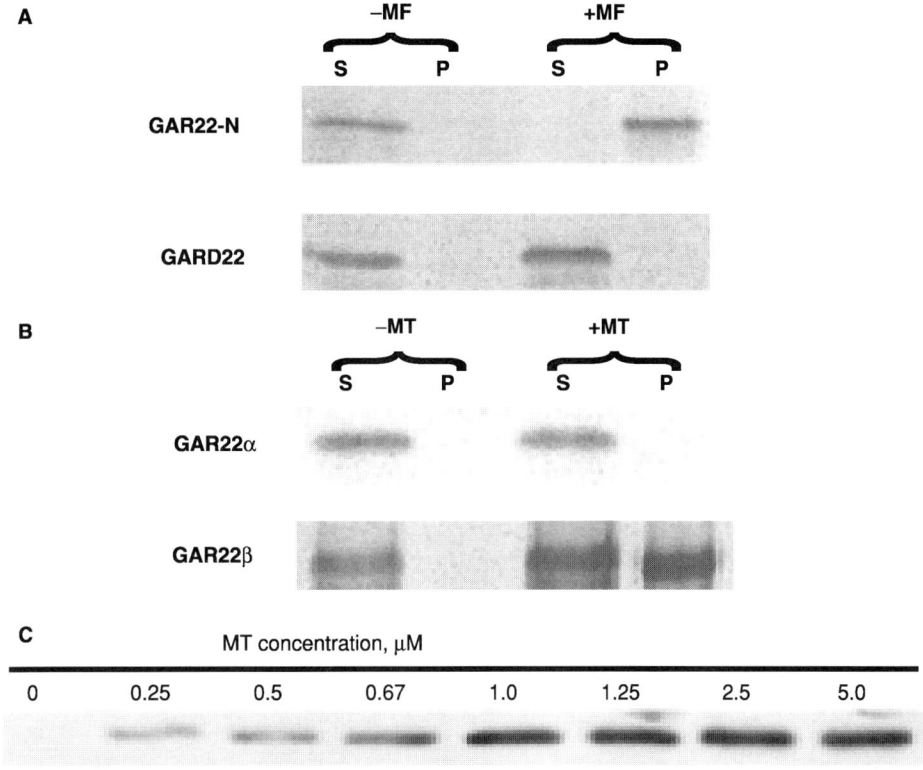

Fig. 3 *In vitro* binding assays. (A, B) GAR22 proteins labeled with ^{35}S-methionine were incubated with preassembled MFs (+MF; A) or MTs (+MTs; B), the reaction mixtures were centrifuged at high speed, and the resulting supernatant (S) and pellet (P) fractions were run on an SDS-PAGE gel. The gels were dried and the radioactively labeled proteins visualized by autoradiography. The presence of a protein in a pellet fraction indicates binding to the corresponding filaments. Reactions without filaments (−MF, −MT) were run to control for nonspecific protein aggregation. (Modified from Goriounov *et al.*, 2003.) (C) Determination of the K_D of binding of MACF-GAR to MTs. *In vitro* binding assays that used purified MACF-GAR protein were performed with indicated concentrations of MTs. Pellet fractions were subjected to SDS-PAGE and MACF-GAR protein visualized with Western blotting with FLAG M2 monoclonal antibody. (Modified from Leung *et al.*, 1999b.)

determine whether the interaction with MTs is direct. We usually use radioactively labeled proteins with the Microtubule Associated Protein Spin-Down Kit from Cytoskeleton, Inc. ^{35}S-labeled proteins are prepared as described previously. The basic assay includes incubation of the protein of interest with preassembled MTs, high-speed centrifugation to spin down the MTs and associated proteins through a buffered cushion, and analysis of the pellet and supernatant fractions by SDS-PAGE followed by autoradiography.

Reagents and Equipment

- Microtubule Associated Protein Spin-Down Kit (Cytoskeleton #BK029).
- 5× SDS sample buffer.
- SDS-PAGE apparatus.
- Gel-dryer (e.g., BioRad model 583) with a vacuum pump.
- X-ray film (e.g., BioMax MS, Kodak).
- Strong-wall centrifuge tubes (e.g., Beckman #343778, to be used with a TLA45 rotor).
- Refrigerated ultracentrifuge (e.g., Beckman model TL-100).
- Gel-fixing solution (40% methanol, 7% acetic acid, 0.5% Coomassie); destaining solution (40% methanol, 7% acetic acid).
- Whatman paper.
- 35 °C waterbath.

Microtubule Preparation

1. Thaw a 20-μl aliquot of tubulin (5 mg/ml) on ice.
2. Add 2.5 μl of MT Cushion Buffer and incuabte at 35 °C for 20 minutes.
3. In the meantime, add taxol to 180 μl of Tubulin Dilution Buffer (TDB) to a final concentration of 40 μM. Prewarm the resulting solution at 35 °C.
4. Add the TDB/taxol solution to the MTs and mix thoroughly but gently with a pipette tip. Keep the MTs at room temperature.

You now have a population of MTs that are 5- to 10-μm in length at a concentration of ~10 nM (equivalent to 50 μM tubulin). The MTs will remain stable for up to a week when stored at 4 °C.

Microtubule Binding and Sedimentation

1. Mix 10 μl of radioactively labeled protein with 20 μl of MTs and 20 μl of TDB/taxol solution (final volume 50 μl) in a microfuge tube. In a control reaction, mix 10 μl of protein with 40 μl of TDB/taxol (negative spin-down control). Use 2 μg of MAP2 or BSA as positive and negative binding controls, respectively.

2. Leave the reactions at room temperature for 20 minutes for binding to attain equilibrium.

3. Layer each reaction on top of 100 μl of Cushion Buffer (supplemented with 20 μM taxol) in strong-wall centrifuge tubes.

4. Centrifuge at 100,000g for 40 minutes at room temperature.

5. Take 40 μl of the supernatant and mix with 10 μl of 5× SDS sample buffer. Boil for 5 minutes. This is your supernatant fraction.

6. Layer 100 μl of TDB on the remaining supernatant and carefully aspirate the mixture with a pipette tip from the *bottom* of the tube, so that the freshly added TDB is removed last and thus washes the pellet.

7. Resuspend the pellet in 50 μl of 1× SDS sample buffer, boil for 5 minutes. This is your pellet fraction.

8. Run 10 μl of each fraction on an SDS-PAGE gel.

9. Fix and stain the gel in fixing solution with constant shaking for 1 hour at room temperature; destain in destaining solution overnight. Placing Kimwipe tissues or small pieces of sponge in the container will accelerate destaining, because the released dye will be absorbed by these materials.

10. Dry the gel in a vaccum dryer on a sheet of Whatman paper. To prevent gel cracking, wet the paper thoroughly in destaining solution before placing the gel on it. Cover the gel with thin plastic film (Saran-Wrap or similar).

11. Visualize the labeled protein by autoradiography. MAP2 and BSA are visualized by Coomassie staining.

Figure 3B represents an example of the MT-binding assay described previously. GAR22α, the shorter isoform of GAR22 (Fig. 1A), remains in the supernatant in reactions with or without MTs. Therefore, GAR22α is unable to bind MTs, at least *in vitro*. The longer isoform, GAR22β, is present in the pellet when incubated with MTs. This result indicates that it binds MTs *in vitro*. Note that some of the protein still remains in the supernatant. In our hands, this is a common result for MT-binding proteins and may be due to an excess of radioactively labeled protein relative to the MTs.

4. Microtubule–Binding Assay Using Purified Proteins

In some cases, the use of purified recombinant proteins is the only option available to the investigator. For example, it is indispensable for the quantitative determination of the affinity of binding, the inverse measure of which is known as the dissociation constant, K_D. Described in the following is an assay that we used to determine the K_D of binding of the GAR domain of MACF to MTs (Fig. 1C; Sun *et al.*, 2001).

Reagents and Equipment

- pET protein expression system (Novagen #70757 or similar).
- pET-MACF-GAR plasmid encoding FLAG- and His-tagged MACF-GAR protein.
- BL21 competent cells.
- LB medium.
- IPTG stock solution (1 M in water; Fisher #327280–050).
- Column charge solution (50 mM $NiSO_4$).
- Binding buffer (20 mM Tris-HCl, pH 7.9; 0.5 M NaCl, 5 mM imidazole).
- Wash buffer (20 mM Tris-HCl, pH 7.9; 0.5 M NaCl, 60 mM imidazole).
- Elution buffer (20 mM Tris-HCl, pH 7.9; 0.5 M NaCl, 1 M imidazole).
- Microtubule Associated Protein Spin-Down Kit (Cytoskeleton #BK029).

- 35 °C waterbath.
- Strong-wall centrifuge tubes (e.g., Beckman #343778, to be used with a TLA45 rotor).
- Refrigerated ultracentrifuge (e.g., Beckman model TL-100).
- PEM buffer (80 mM PIPES, pH 6.8; 1 mM $MgCl_2$, 1 mM EGTA).
- Ponceau S stain solution (Sigma #P7170).
- PBS-T (0.05% Tween-20 in PBS, pH 7.4).
- Immobilon-P PVDF membrane (Millipore).
- 5× SDS sample buffer.
- SDS-PAGE apparatus.
- Sonicator (e.g., Branson Sonifier 250).
- Electroblotting apparatus.
- Fat-free dry milk.
- FLAG M2 monoclonal antibody (Sigma #F3165).
- Goat anti-mouse IgG conjugated with horseradish peroxidase (Sigma).
- Rocking platform.
- Chemiluminescence detection kit (ECL; Amersham-Phramacia).
- X-ray film (e.g., BioMax MS, Kodak).
- Scanner.
- NIH Image program (or similar).

Microtubule-Actin Crosslinking Factor-GAR Protein Production and Purification

1. Transform BL21 competent cells with pET-MACF-GAR.
2. The following day, pick a colony to set up an overnight miniculture (2–3 ml).
3. Add 1–2 ml of the overnight culture to 200 ml of LB containing 50 μg/ml ampicillin; grow to an OD of 0.4–0.6 (monitor OD at 20-minute intervals).
4. Add IPTG to 1–2 mM to induce protein expression; incubate for 2 hours.
5. Collect the bacteria by centrifugation (5000*g* for 5 minutes) and resuspend in 20 ml of ice-cold binding buffer.
6. Sonicate on ice in short bursts (10–15 seconds) until the sample is no longer viscous (usually a total of 60–80 seconds).
7. Clear the lysate by centrifugation at 20,000*g* for 20 minutes at 4 °C; filter the supernatant through a 0.45-μm membrane (syringe-end filter).
8. Load the supernatant onto a Ni^{2+}-charged column.
9. Wash with 10 volumes of binding buffer, then with 6 volumes of wash buffer.
10. Elute the bound protein with elution buffer.
11. Dialyze the purified protein against PEM buffer at 4 °C overnight with three buffer changes. Store the protein in aliquots at -80 °C. It is important to note that proteins tend to lose solubility when stored at -80 °C for a long period of time.

Microtubule Binding and Western Blotting

1. Prepare MTs as described previously (Section II.B).
2. Set up binding reactions with purified MACF-GAR and serially diluted MTs (0–5 μM). At the highest concentration, MTs should be present in significant molar excess relative to MACF-GAR.
3. Process the reaction mixtures as in a standard MT-binding assay (Section II.B). Dissolve the pellet in 50 μl of $1\times$ sample buffer and boil for 5 minutes.
4. Run the samples on an SDS-PAGE gel, loading equal volumes.
5. Transfer the proteins onto a PVDF membrane for 1 hour at 300 mA.
6. Stain the membrane with Ponceau stain to verify transfer.
7. Destain with PBS-T for 10 minutes on a rocking platform.
8. Block the membrane with 5% milk in PBS-T for 1 hour at room temperature.
9. Incubate the membrane with FLAG M2 monoclonal antibody (1:2000 in PBS-T) for 1 hour at room temperature.
10. Wash three times with PBS-T for 5 minutes each.
11. Incubate the membrane with HRP-conjugated goat anti-mouse IgG (1:2000 in PBS-T) for 1 hour at room temperature.
12. Wash three times with PBS-T for 5 minutes each.
13. Visualize the MACF-GAR protein by ECL.
14. Scan the film and determine the relative amounts of bound MACF-GAR protein with NIH Image program.

Figure 3C shows the results of MT-MACF-GAR binding assays using serial dilutions of MTs. The K_D is determined as the MT concentration at which half of the MACF-GAR protein cosediments with MTs. In the example shown, the K_D of binding is estimated to be $\sim$0.6 μM. This value is comparable to those for other MT-binding proteins, such as MAP2 (Coffey and Purich, 1995) and XMAP230 (Andersen *et al.*, 1994).

C. Two-Hybrid Assay

The yeast two-hybrid assay was developed by Stanley Fields and coworkers (Fields and Song, 1989). It is a sensitive *in vivo* method for identifying proteins that interact with a protein of interest. It can also be used for studying interactions between any two given proteins. The assay is based on reconstitution of a functional GAL4 transcriptional activator and consequent activation of a reporter gene under the control of a GAL4-responsive promoter. The target (or one of the two proteins under study) is cloned into a special vector such that a fusion between the target protein and the GAL4 DNA-binding domain (GBD) is generated. The library (or the second of the two proteins) is cloned into another vector such that fusions between the library proteins and the GAL4 activation domain (GAD) are generated. The two types of hybrid plasmids are then cotransformed into a yeast host strain such as SFY526, and the cells are plated on synthetic medium lacking leucine and tryptophan to select those transformants that contain both plasmids. If the two proteins interact, a functional GAL4 activator is reconstituted, and the expression of the reporter gene (lacZ in the case of SFY526) is induced. Reporter activation can be detected by a β-galactosidase activity assay.

The yeast two-hybrid system is a useful tool to characterize *in vivo* association of cytolinkers with various filament proteins. Our laboratory has used this system to study interactions between the intermediate filament binding domain 1 (IFBD1) of BPAG1 and neuronal IF proteins (Leung *et al.*, 1999b). As an example, we describe the assay used to test the interactions of BPAG-IFBD1 (Fig. 1D) with peripherin and vimentin. BPAG-IFBD1 cDNA was fused with GBD in vector pGBT9. Peripherin and vimentin cDNAs were fused with GAD in vector pGAD424. The plasmids were contransformed into yeast strain SFY526 by the lithium-acetate method, and interaction between the proteins was determined by the qualitative β-galactosidase filter lift assay.

Reagents and Equipment

- pGBT9-IFBD1, pGAD424-peripherin, and pGAD424-vimentin plasmids.
- *Saccharomyces cerevisae* strain SFY526.
- TE buffer (10 mM Tris-HCl, pH 8.0; 1 mM EDTA).
- YPD medium (20 g/L Difco peptone, 10 g/L yeast extract; adjust pH to 5.8, autoclave; 15 g/L agar—for plates only).
- Dextrose solution (40% in water; autoclave).
- Z buffer ($Na_2HPO_4 \times 7H_2O$ – 16.1g/L, $NaH_2PO_4 \times H_2O$ – 5.5 g/L, KCl – 0.75 g/L, $MgSO_4 \times 7H_2O$ – 0.246 g/L; autoclave).
- 10X LiAc (1 M lithium acetate, Sigma #L6883; adjust pH to 7.5 with dilute acetic acid, autoclave).
- PEG/LiAc solution (40% PEG4000, 1× TE buffer, 1× LiAc; prepare fresh just before use).
- PEG stock solution (polyethylene glycol, average molecular weight 3350, Sigma #P3640; 50% in water, autoclave).
- X-gal stock solution (20 mg/ml 5-bromo-4-chloro-3-indolyl-β-D-galactopyranoside (Fisher #BP1615100) in DMF; store in the dark at $-20\,°C$).
- Z buffer/X-gal solution (100 ml Z buffer, 0.27 ml β-mercaptoethanol, 1.67 ml X-gal stock solution).
- Whatman #1 paper filters (VWR#28310-026 for use with 100-mm plates, #28310-106 for 150-mm plates).
- 100- or 150-mm Petri dishes.
- Beckman TJ-6 (or similar model) centrifuge.
- 30 °C shaker.
- Liquid nitrogen.

- Salmon sperm DNA (Sigma #D1626; dissolve in TE and sonicate to reduce viscosity, extract DNA with phenol-chloroform, precipitate with ethanol, resuspend in TE at 10 mg/ml; boil for 20 minutes and cool immediately on ice; store in aliquots at $-20\,°C$).
- Synthetic Trp$^-$Leu$^-$ selection medium (300 mg/L L-isoleucine (Sigma #I7383), 1500 mg/L L-valine (#V0500), 200 mg/L L-adenine hemisulfate (#A9126), 200 mg/L L-arginine HCl (#A5131), 200 mg/L L-histidine HCl monohydrate (#H9511), 300 mg/L L-lysine HCl (#L1262), 200 mg/L L-methionine (#M9625), 500 mg/L L-phenylalanine (#P5030), 2000 mg/L L-threonine (#T8625), 300 mg/l L-tyrosine (#T3754), 200 mg/L L-uracil (#U0750)).

Preparation of Competent Yeast Cells

1. Inoculate SFY526 yeast cells in 20 ml of YPD containing 2% dextrose (1 ml of 40% stock solution). Incubate overnight at 30 °C with continuous shaking (230–250 rpm).
2. Transfer 10–20 ml of the overnight culture to 300 ml of YPD (add 15 ml of 40% dextrose) in a 1-L flask to produce an OD_{600} of 0.2. Incubate at 30 °C with shaking for 3 hrs
3. Centrifuge the cells in a Beckman TJ-6 centrifuge at 2200 rpm (1000g) for 5 minutes at room temperature.
4. Resuspend the pellet in 35 ml of sterile H_2O or TE buffer.
5. Spin as in step 3.
6. Resuspend the pellet in 1.5 ml of sterile 1× TE/LiAc.
7. Store on ice until transformation.

Cotransformation of Competent Yeast Cells

1. Add 0.1 μl of each plasmid, together with 100 μl of salmon sperm DNA, to a microfuge tube.
2. Add 100 μl of competent yeast cells, mix well.
3. Add 0.6 ml of sterile PEG/LiAc solution, vortex to mix.
4. Incubate at 30 °C with shaking (200 rpm) for 30 minutes.
5. Add 70 μl (10% final concentration) of DMSO, mix gently.
6. Heat-shock for 15 minutes at 42 °C, swirling occasionally.
7. Chill on ice briefly.
8. To pellet the cells, centrifuge for 5 seconds at 14,000 rpm in a microcentrifuge.
9. Resususpend the pellet in 0.5 ml of TE.
10. Spread 100 μl of the transformation mixture onto a 100-mm plate containing synthetic selection medium.
11. Incubate the plates at 30 °C (face down) for 2–3 days (until colonies appear).

Filter Assay of β-Galactosidase

1. Streak individual colonies (or spread them in small patches) directly onto filters layered over agar plates containing selection medium. Incubate at 30 °C for 1–2 days.

2. Presoak a sterile Whatman #1 filter in Z buffer/X-Gal solution.

3. Place another filter on the surface of the agar plate containing the streaked patches. Poke holes through the filter into the agar to orient the filter.

4. Lift the filter from the agar with forceps and transfer it (colonies facing up) to liquid nitrogen. Submerge completely for 10 seconds.

5. Thaw the filter at room temperature and place it (colonies facing up) on top of the presoaked filter.

6. Incubate at 30 °C for 0.5–8 hours until blue staining develops.

7. Positives can be identified by aligning the filter to the agar plates by use of the marker holes.

In the assay just described, cotransformation of yeast cells with pGBT9-IFBD1 and pGAD424-peripherin resulted in β-galactosidase–positive colonies. Cotransformations of pGBT9-IFBD1 and pGAD424 or pGBT9 and pGAD424-peripherin did not yield positive colonies. These results indicate that peripherin and BPAG-IFBD1 interact *in vivo*. This interaction is most likely direct and identifies peripherin as a probable filamentous target for BPAG1 binding via its IFBD1. In contrast, cotransformation of yeast with pGBT9-IFBD1 and pGAD424-vimentin did not result in any β-galactosidase–positive colonies, suggesting that vimentin does not interact with BPAG-IFBD1 *in vivo*.

D. Overlay Binding Assay ("Far-Western")

Overlay binding assays represent another approach to test whether a potential cytolinker interacts with filament proteins. The principle of the assay is similar to that of Western blotting, hence the alternative informal name "Far-Western." One of the proteins under study is resolved on an SDS-PAGE gel and transferred onto a membrane, whereas the other potential interactor is used to probe the membrane, much like the primary antibody in Western.

Overlay binding assays have been used to study the interactions of plectin and desmoplakin with IF proteins (Meng *et al.*, 1997; Nikolic *et al.*, 1996). Our laboratory has examined the interactions between BPAG-IFBD1 and IF proteins with such an assay (Leung *et al.*, 1999b). Bacterially expressed IF proteins and BPAG-IFBD1 (Fig. 1D) were purified by column chromatography. Purified IF proteins were resolved by SDS-PAGE and transferred onto a nylon membrane. After the membrane was washed to remove the SDS and partially renature the transferred proteins, it was incubated with purified BPAG-IFBD1. The bound IFBD1 protein was then detected with an anti-FLAG monoclonal antibody.

Reagents and Equipment

- pET protein expression system (Novagen #70757 or similar).
- pET-NF-L, pET-NF-M, pET-NF-H, and pET-peripherin plasmids encoding His-tagged proteins.
- pET-BPAG-IFBD1 plasmid encoding FLAG- and His-tagged BPAG-IFBD1.
- BL21 competent cells.
- LB medium.
- PBS-T (0.05% Tween-20 in PBS, pH 7.4).
- Column charge solution (50 mM $NiSO_4$).
- Binding buffer (20 mM Tris-HCl, pH 7.9; 0.5 M NaCl, 5 mM imidazole).
- Wash buffer (20 mM Tris-HCl, pH 7.9; 0.5 M NaCl, 60 mM imidazole).
- Elution buffer (20 mM Tris-HCl, pH 7.9; 0.5 M NaCl, 1 M imidazole).
- Ponceau S stain solution (Sigma #P7170).
- Nylon transfer membrane (Millipore).
- IPTG stock solution (1 M in water; Fisher #327280-050).
- Chemiluminescence detection kit (ECL; Amersham-Phramacia).
- 5× SDS sample buffer.
- SDS-PAGE apparatus.
- Sonicator (e.g., Branson Sonifier 250).
- Electroblotting apparatus.
- FLAG M2 monoclonal antibody (Sigma #F3165).
- Goat anti-mouse IgG-conjugated with horseradish peroxidase (Sigma).
- Rocking platform.
- X-ray film (e.g., BioMax MS, Kodak).

Procedure

1. Purify His-tagged IFBD1 and IF proteins as described in Section II.C.
2. Run 20 pmol of purified IF proteins on an SDS-PAGE gel.
3. Transfer the proteins onto a nylon membrane for 1 hour at 300 mÅ.
4. Stain the membrane with Ponceau stain to verify transfer.
5. Destain with PBS-T for 10 minutes on a rocking platform.
6. Block the membrane with 5% BSA in PBS-T overnight at 4 °C.
7. Incubate the membrane with 50 μg of purified IFBD1 protein in 5–10 ml of PBS-T for 1 hour at room temperature.
8. Wash three times with PBS-T for 5 minutes each.
9. Incubate the membrane with FLAG-M2 monoclonal antibody in PBS-T for 1 hour at room temperature.
10. Wash three times with PBS-T for 5 minutes each.

11. Incubate the membrane with HRP-conjugated goat anti-mouse IgG (1:1000) for 1 hour at room temperature.

12. Wash three times with PBS-T for 5 minutes each.

13. Visualize the bound IFBD1 protein by ECL.

As shown in Fig. 4B, BPAG-IFBD1 associated with peripherin but not with neurofilament triplet proteins. This result identifies peripherin as the likely interaction partner of BPAG-IFBD1 *in vivo*, whereas neurofilaments seem unlikely to interact with BPAG-IFBD1.

E. Nocodazole–Protection Assay

Nocodazole is a synthetic compound first identified in a screen for antihelminthic agents (De Brabander *et al.*, 1975). It causes microtubule depolymerization by binding to free tubulin dimers and preventing them from incorporating

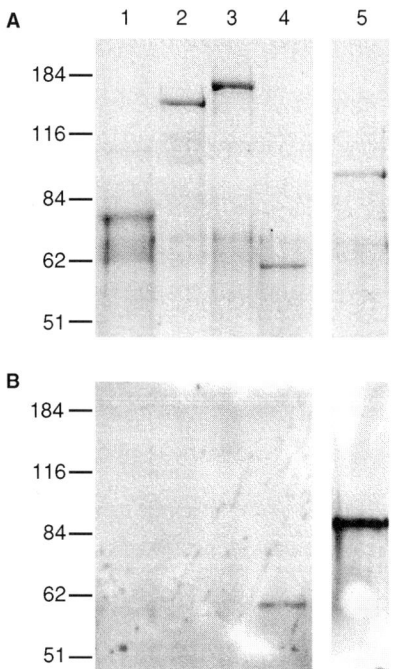

Fig. 4 BPAG-IFBD1 interacts specifically with peripherin in the overlay binding assay. (A) Coomassie staining of purified NF-L (lane 1), NF-M (lane 2), NF-H (lane 3), peripherin (lane 4), and FLAG/His-tagged BPAG-IFBDI (lane 5). (B) Equal amounts of purified NF-L (lane 1), NF-M (lane 2), NF-H (lane 3), and peripherin (lane 4) were resolved by SDS-PAGE and transferred onto a nylon membrane. The membrane was probed with purified BPAG-IFBD1 and the bound protein detected with FLAG M2 monoclonal antibody. Lane 5 is a FLAG M2 Western blot of purified BPAG-IFBD1. (Modified from Leung *et al.*, 1999b.)

into MTs (Hoebeke *et al.*, 1976). This compound has been widely used for the study of the MT cytoskeleton and its associated proteins. In particular, the ability of a protein to stabilize MTs and protect them from nocodazole-induced depolymerization has been used as an indication of high-affinity binding. We routinely use nocodazole treatment to assess the stringency of the association between MTs and potential cytolinker proteins or their individual domains. The procedure is based on a standard immunostaining protocol of transfected cells (see previously). The only significant modification is that the cells are treated with nocodazole just before fixation. Also, as in most other cases in which no phalloidin staining is performed, we prefer to fix the cells with methanol. Described in the following is a nocodazole treatment experiment comparing the MT-stabilizing properties of GAR17β and its truncated version, GARt17 (Fig. 1B; Goriounov *et al.*, 2003).

Reagents and Equipment

- pFLAG-GAR17β and pFLAG-GARt17 plasmids.
- LipofectAMINE PLUS reagent (Invitrogen #18324-012 and #11514-015).
- Methanol (precool at $-20\,^\circ$C just before use).
- Nocodazole stock solution (2 mg/ml in DMSO; Sigma #M1404).
- PBS, pH 7.4.
- Normal goat serum.
- Parafilm.
- 100- or 150-mm Petri dishes.
- Monoclonal FLAG M2 antibody (Sigma, F-3165), polyclonal anti-tubulin antibody (Sigma #T3526).
- AlexaFluor-594-conjugated goat anti-rabbit antibody (Molecular Probes #A11012), AlexaFluor-488-conjugated goat-anti-mouse antibody (Molecular Probes #11001).
- Aquamount slide-mounting medium (Lerner Laboratories).
- Glass slides and coverslips (Fisher #12-548-A).
- Fluorescent microscope (e.g., Nikon Eclipse E800) with a mounted digital camera (e.g., SPOT; Diagnostic Instruments, Inc.).
- Photoshop 6.0 (or similar) image-processing program.

Procedure

The procedure is similar to that described in Section I, with the following modifications.

1. Before fixing the cells, add nocodazole to the culture medium (10 μM final concentration). Incubate at $37\,^\circ$C for 1.5 hours. This will result in complete depolymerization of the MTs.
2. Fix the cells with methanol for 5 minutes at $-20\,^\circ$C.
3. Exclude phalloidin from the primary antibody incubation.

As shown in Fig. 5, transfected full-length GAR17β binds MTs but is unable to protect them from nocodazole. In contrast, GARt17, which lacks the CH domain, not only binds to but also bundles MTs. Furthermore, GARt17 also protects the MTs from nocodazole-induced depolymerization. These results show that the MT-stabilizing ability of GARt17 is enhanced compared with the full-length protein. The reason is unclear but could be due to the elimination of a masking effect by the actin-binding CH domain.

F. Latrunculin-Protection Assay

MF-depolymerizing drugs have been used extensively to study actin dynamics and MF involvement in cellular processes. Cytochalasins are a family of compounds excreted by various molds. Their principal mode of action is to bind to the fast growing plus ends of actin filaments, thereby preventing the addition of actin monomers at these ends. This leads to depolymerization of the filaments by way of loss of monomers at their minus ends (Sampath and Pollard, 1991). The action of cytochalasins on F-actin is accompanied by numerous side effects that can compromise the interpretation of an experiment. Another class of actin-depolymerizing drugs, called latrunculins, has therefore gained significant popularity in recent years.

Latrunculins A and B are natural marine products that bind actin and disrupt its organization. They were first isolated from the Red Sea sponge *Latrunculia magnifica* and later found in unrelated Pacific sponges (Spector *et al.*, 1983). The mechanism of action is a highly specific sequestration of monomeric actin, mimicking the activity of the G-actin–sequestering proteins β-thymosins (Coue *et al.*, 1987). Latrunculin A binds actin more strongly and is a more potent inhibitor of actin polymerization than Latrunculin B, even though the structures of the two compounds are very similar (Spector *et al.*, 1999). The effect of Latrunculin A on the MF network is phenomenologically similar to that of the cytochalasins, but it acts at significantly lower concentrations (starting from 0.1 μM, one to two orders of magnitude less than the effective concentrations of the cytochalasins). This may account for the relative specificity of Latrunculin A action compared with the cytochalasins.

Latrunculins have been extensively used to study the role of the actin cytoskeleton in various cellular processes. They can also be used to characterize the nature of the interactions between MFs and cytolinker proteins. Some actin-binding proteins will stabilize MFs and protect them from depolymerization by latrunculins. Described in the following is an experiment in which we tested the ability of GAR17 and GAR22 proteins and their CH domains (Fig. 1A,B) to protect MFs from Latrunculin A–induced depolymerization.

Reagents and Equipment

- pFLAG-GAR17α and pFLAG-GAR22-N plasmids.
- LipofectAMINE PLUS reagent (Invitrogen #18324-012 and #11514-015).

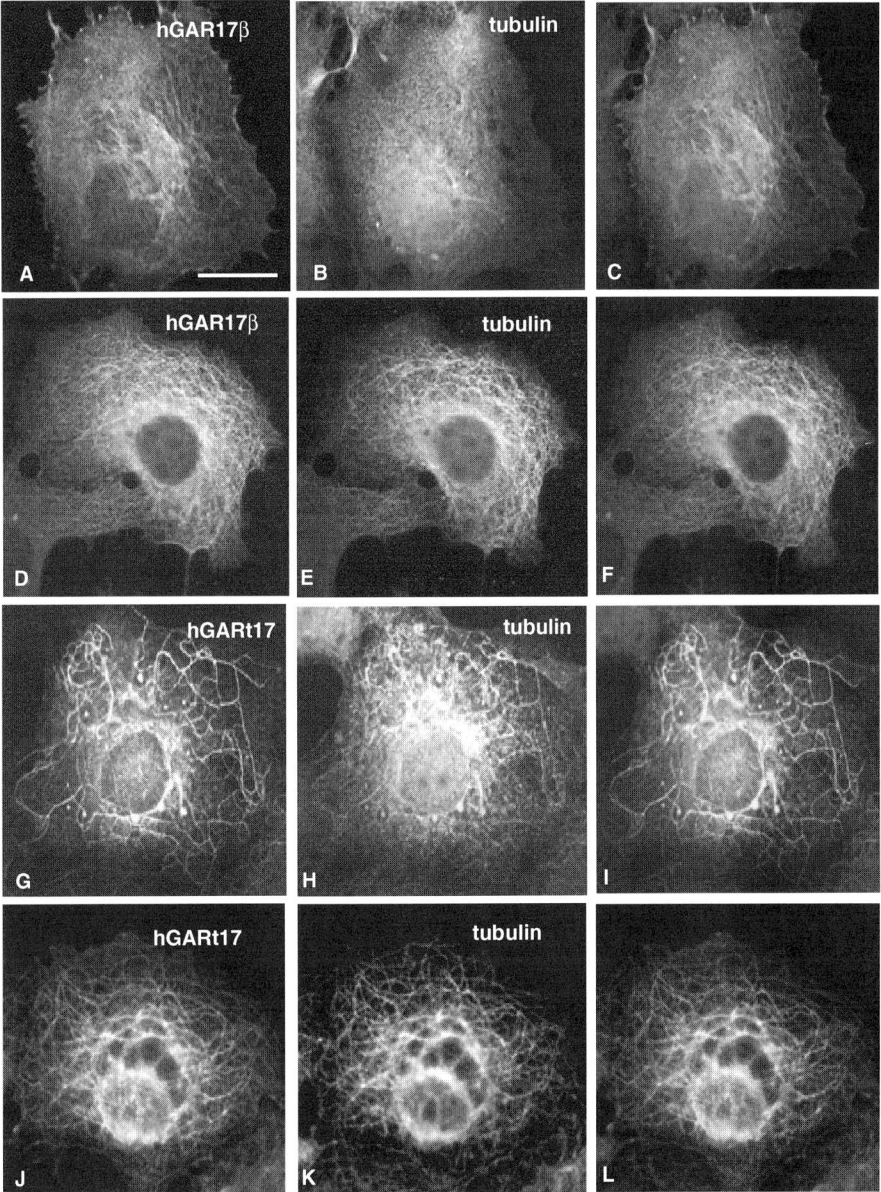

Fig. 5 GARt17 but not GAR17β protects MTs from nocodazole. COS7 cells were transiently transfected with pFLAG-GAR17β (A–F) or pFLAG-GARt17 (G–L), treated with nocodazole (A–C, G–I), and stained with monoclonal FLAG M2 antibody (A, D, G, J) and polyclonal anti-tubulin antibody (B, E, H, K). (C, F, I, L) Superimposed images, with the FLAG staining in green and the tubulin staining in red. Both GAR17β and GARt17 bound MTs in untreated cells (F, L). After nocodazole treatment, MTs were preserved only in the cells expressing GARt17 (I). The preserved MTs appear bundled, as do the MTs in untreated GARt17-transfected cells. Bar, 15 μm. (See Color Insert.)

- 4% paraformaldehyde in PBS, pH 7.4 (add 4 g of paraformaldehyde powder to 70 ml of PBS, heat to 50–60 °C with constant stirring, add 1 N NaOH drop by drop until the powder is dissolved; then adjust the pH to 7.4 with hydrochloric acid).
- Latrunculin A stock solution (Molecular Probes #L12370).
- PBS, pH 7.4.
- Normal goat serum.
- Parafilm.
- 100- or 150-mm Petri dishes.
- Monoclonal FLAG M2 antibody (Sigma #F-3165), AlexaFluor-488–conjugated goat anti-mouse antibody (Molecular Probes #11001), AlexaFluor-594-phalloidin (Molecular Probes # A12381).
- Aquamount slide-mounting medium (Lerner Laboratories).
- Glass slides and coverslips (Fisher #12-548-A).
- Fluorescent microscope (e.g., Nikon Eclipse E800) with a mounted digital camera (e.g., SPOT; Diagnostic Instruments, Inc.).
- Photoshop 6.0 (or similar) image-processing program.

Procedure

The procedure is similar to that described in Section I, with the following modifications.

1. Before fixing the cells, add Latrunculin A to the culture medium (final concentration 0.5, 5, or 25 μM). Incubate at 37 °C for 1 hour. This will result in complete depolymerization of the MFs in untransfected cells.
2. Use monoclonal FLAG M2 antibody (1:500) and AlexaFluor-594-phalloidin (1:500) in the primary antibody incubation.
3. Use only AlexaFluor-488-conjugated goat anti-mouse antibody (1:1000) in the secondary antibody incubation.

CH Domains of GAR17 and GAR22 Protect Microfilaments from Latrunculin A

Both GAR17α (Fig. 6A,B) and GAR22-N (Fig. 6C,D) were able to protect a considerable portion of the MFs from depolymerization by 25 μM Latrunculin A. The resulting actin structures appeared as thick spiky bundles. No individual filaments or stress fibers could be observed. In most cells, transfected and untransfected, the nuclei were displaced from the cell center. The plasma membrane showed multiple protrusions and blebbing, probably resulting from the disintegration of the cortical MF network. Similar results were obtained with Latrunculin A concentrations of 5 μM (data not shown) and 0.5 μM (Fig. 6E,F). At 0.5 μM, even though the actin cytoskeleton of the untransfected cells remained largely intact (Fig. 6F, cell at the bottom), spiky actin bundles still formed in most transfected cells. Neither full-length isoform of GAR22 nor GAR17β as able to protect MFs to any considerable degree. The GAR domain of GAR22 was also ineffective (data not shown).

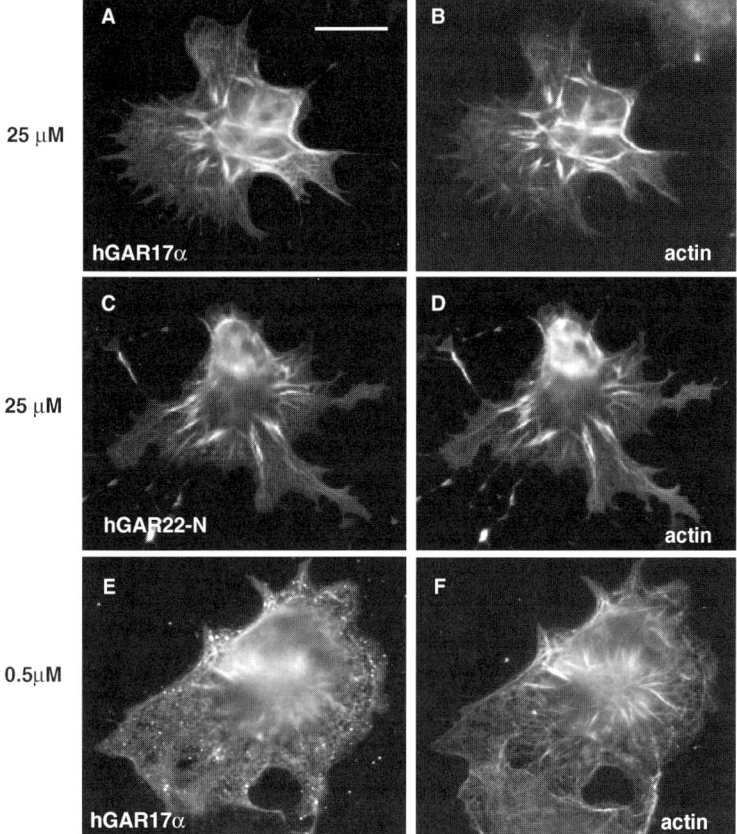

Fig. 6 GAR22-N and GAR17α protect MFs from Latrunculin A. COS7 cells were transfected with pFLAG-GAR17α (A, B, E, F) or pFLAG-GAR22-N (C, D), treated with 25 μM (A–D) or 0.5 μM (E, F) Latrunculin A for 1 hour, and stained with monoclonal FLAG M2 antibody (A, C, E) and fluorescently labeled phalloidin (B, D, F). Both proteins protected MFs from Latrunculin A, resulting in the formation of spiky actin bundles. Bar, 15 μm.

III. Pearls and Pitfalls

A. Triple Staining of Transfected Cells

In our hands, the presence of a FLAG epitope tag seldom results in any differences in the cytoskeletal-binding profile of a protein. To rule out any tag-dependent effects, one should transfect cells with an untagged version of the protein and use protein-specific antibodies for staining. Transient transfection followed by immunocytochemical analyses can be used to determine whether a protein associates with certain cytoskeletal components. This method, however, cannot distinguish between direct and indirect interactions.

Data obtained with this technique can be meaningful only when compared with a single-filament binding control (e.g., a truncated version of the protein lacking its MTBD or ABD or an unrelated MT or actin-binding protein). In the example shown in Fig. 2, GAR22α serves as such a control. Statistical analysis involving multiple cells should be conducted to find out whether there is a significant increase in the number of triple-stained filaments. We used the χ^2 criterion for 2×2 contingency tables with the Yates discontinuity correction factor (Bailey, 1981). Thirty cells transfected with each construct (α or β) were selected randomly, and their pictures were taken to generate a triple image. A cell was scored as positive if it contained at least one triple-stained (white) filament equal to or exceeding one fifth of the length of the cell or at least two triple-stained filaments each equal to or exceeding one tenth of the length of the cell. The following formula was used to calculate the χ^2:

$$\chi^2 = n(/ad - bc/ - 0.5\,n)^2/(a+b)(c+d)(a+c)(b+d)$$

where

    a = number of negative β-transfected cells (8)
    b = number of positive β-transfected cells (22)
    c = number of negative α-transfected cells (27)
    d = number of positive α-transfected cells (3)
    n = total number of cells (60)

For the given data, $\chi^2 = 22.22$. If $\chi^2 > 7.88$ (discontinuity factor $= 1$), then $P < 0.005$. Because $22.22 > 7.88$, the difference between the numbers of positive α-transfected and β-transfected cells is considered statistically significant ($P < 0.005$).

B. *In Vitro* Binding Assays

MT or actin binding in the sedimentation assays described in Sections II.B.1 and II.B.2 does not prove that the radioactively labeled protein interacts with the filaments directly. The binding may be mediated by some components of the *in vitro* transcription/translation mix present in the protein preparation. To rule out this possibility, one should repeat the assay with recombinant purified proteins and use Western blotting for protein visualization. The most common problem with this approach is the insolubility of recombinant proteins, which results in sedimentation regardless of binding. The addition of small amounts of detergents such as SDS may help overcome this obstacle but could also interfere with binding.

To demonstrate that the association of a given protein with the filaments is direct, one can use an assay similar to that described in Section II.B.3 and used for K_D determination. Because purified protein is used in this assay, cosedimentation with the filaments is interpreted as evidence of direct binding. If purified protein is soluble in PEM buffer and can be produced in sufficient quantities, one can measure the amount of bound protein directly by Coomassie staining. This will eliminate the need for Western blotting and may result in a more accurate K_D. If filament binding requires posttranslational modification of the protein under study, it should be purified from a eukaryotic expression system, such as the baculovirus system.

C. Two-Hybrid Assay

When interpreting two-hybrid data, one has to keep in mind that the assay characterizes interactions of monomeric proteins rather than filaments. It is possible that the interaction of a given protein with a cytoskeletal filament protein in the two-hybrid assay will be different from its interaction with actual filaments in cells.

D. Overlay Binding Assay

This assay detects interactions between monomeric proteins. Therefore, one has to exercise caution when extrapolating overlay assay results to the *in vivo* interactions with filaments. A good confirmatory *in vitro* assay is a slot-blot assay that uses polymerized filaments. In this assay, *in vitro* polymerized filaments are immobilized on a membrane, which is then probed with the protein under study, similar to the standard overlay assay described in Section II.D (Leung *et al.*, 1999b).

E. Nocodazole-Protection Assay

It is important to emphasize that this assay does not provide evidence of direct binding to MTs. Both the association and the MT-stabilizing effect may be indirect. Yet, combined with *in vitro* data that use purified recombinant proteins, the assay can serve as confirmation of the ability of a protein to interact with MTs *in vivo* and provide a measure of the strength of this association. MT bundling is frequently observed with proteins that bind MTs with high affinity/avidity. Although the mechanism of bundling is unknown, such proteins often (but not always) turn out to be capable of MT stabilization in the nocodazole-protection assay.

F. Latrunculin-Protection Assay

To the best of our knowledge, our data in Section II.F represent the first report of any protein capable of protecting MFs in mammalian cells from depolymerization induced by Latrunculin A. The only similar published data have been reported by Danninger and Gimona (2000), who observed protection of MFs from Latrunculin B by h1 calponin. They used the less potent of the two compounds at a relatively low concentration (0.2 μM). In that study, the preserved actin filaments were described as stress fibers, although they appeared to contain clusters of spiky filaments near their ends. We used concentrations of the more potent Latrunculin

A as high as 25 μM and observed similar degrees of MF protection in all cases. The preserved actin structures had the appearance of thick, short, and characteristically spiky bundles located predominantly in the central region of the cell.

The authors of the aforementioned study did not attribute the actin-stabilizing ability of h1 calponin to its CH domain but rather to its C-terminal repeats and the spacer sequence between the CH domain and the repeats. However, our results point to the CH domain as responsible for MF protection. The sequences flanking the CH domain in either GAR22-N or GAR17α exhibit no similarity to the corresponding sequences of calponin. It is, therefore, likely that these proteins and h1 calponin stabilize MFs by means of their CH domains.

The lack of protection by both full-length GAR22 isoforms or GAR17β suggests that the actin-stabilizing ability of the CH domains of GAR17 and GAR22 is negatively regulated by their C-terminal sequences, possibly in a manner similar to that of h2 calponin (Danninger and Gimona, 2000). Whether this actin-stabilizing capacity plays any physiological role is unclear. We have found that GAR22 does not undergo apoptosis-induced cleavage (unpublished observations), arguing against the possibility of the release of CH-domain–containing fragments as a result of apoptotic protease cleavage. The mechanism of the observed MF stabilization and bundling is unknown. Because GAR22-N and GAR17α do not seem to bundle MFs in untreated cells (Goriounov *et al.*, 2003), the mechanism likely impinges on a depolymerization-dependent event or intermediate.

IV. Concluding Remarks

We have described a number of techniques that are commonly used to study cytolinker proteins. We focused on methods that are used to demonstrate that a given protein can interact with various filamentous networks or their monomeric components, thus justifying the name "cytolinker." Most of the techniques presented are indirect (immunofluorescent staining of transfected cells, *in vitro* binding assays using radioactively labeled proteins, nocodazole- and latrunculin-protection assays) or involve the monomeric forms of the filament protein (two-hybrid and overlay assays). These factors represent intrinsic limitations that the investigator should bear in mind when interpreting data obtained by these techniques. Positive triple staining or cosedimentation of unpurified proteins in a spin-down assay do not prove that the protein associates directly with the filaments. Similarly, lack of interactions observed in two-hybrid or standard overlay assays does not rule out the possibility that the protein under study interacts with filaments *in vivo*. Spin-down binding assays that use purified proteins or overlay assays with polymerized filaments provide evidence of direct interaction *in vitro*. Yet they still cannot be relied on as indicators of direct binding *in vivo*. Electron microscopy of cells or tissues showing a protein directly bridging different filamentous systems provides perhaps the most rigorous evidence of such *in vivo* associations (Svitkina *et al.*, 1996).

Finally, the demonstration of cytoskeletal cross-linking by a protein in an *in vitro* or *in vivo* system still leaves open the question of its physiological relevance. Functional studies should be performed that test the significance of the linkage in various cell types and at different developmental stages under various conditions. Cytoskeleton-related disruption of tissue integrity or cell structure as well as defects in development and differentiation in mutant organisms would validate a crucial filament-organizing role of the protein. Only then can one conclude that the protein under study is indeed a functional cytolinker.

Acknowledgments

This work was supported by grants from the National Institutes of Health and the Muscular Dystrophy Association.

References

Andersen, S., Buendia, B., Dominguez, J., Sawyer, A., and Karsenti, E. (1994). Effect on microtubule dynamics of XMAP230, a microtubule-associated protein present in *Xenopus laevis* eggs and dividing cells. *J. Cell Biol.* **127,** 1289–1299.

Andra, K., Nikolic, B., Stocher, M., Drenckhahn, D., and Wiche, G. (1998). Not just scaffolding: Plectin regulates actin dynamics in cultured cells. *Genes Dev.* **12,** 3442–3451.

Bailey, N. T. J. (1981). *In* "Statistical Methods in Biology," p. 216. Halsted Press, New York.

Brancolini, C., Bottega, S., and Schneider, C. (1992). Gas2, a growth arrest-specific protein, is a component of the microfilament network system. *J. Cell Biol.* **117,** 1251–1261.

Coffey, R., and Purich, D. (1995). Non-cooperative binding of the MAP-2 microtubule-binding region to microtubules. *J. Biol. Chem.* **270,** 1035–1040.

Coue, M., Brenner, S. L., Spector, I., and Korn, E. D. (1987). Inhibition of actin polymerization by Latrunculin A. *FEBS Lett.* **213,** 316–318.

Dalpe, G., Mathieu, M., Comtois, A., Zhu, E., Wasiak, S., De Repentigny, Y., Leclerc, N., and Kothary, R. (1999). Dystonin-deficient mice exhibit an intrinsic muscle weakness and an instability of skeletal muscle cytoarchitecture. *Dev. Biol.* **210,** 367–380.

Danninger, C., and Gimona, M. (2000). Live dynamics of GFP-calponin: Isoform-specific modulation of the actin cytoskeleton and autoregulation by C-terminal sequences. *J. Cell Sci.* **113,** 3725–3736.

De Brabander, M., Van de Veire, R., Aerts, F., Geuens, G., Borgers, M., Desplenter, L., and De Cree, J. (1975). *In* "Microtubules and Microtubule Inhibitors" (M. Borgeres and M. De Brabander, eds.), pp. 509–521. North-Holland Publishing Company, Amsterdam.

Fields, S., and Song, O. (1989). A novel genetic system to detect protein-protein interactions. *Nature* **340,** 245–247.

Goriounov, D., Leung, C. L., and Liem, R. K. (2003). Protein products of human Gas2-related genes on chromosomes 17 and 22 (hGAR17 and hGAR22) associate with both microfilaments and microtubules. *J. Cell Sci.* **116,** 1045–1058.

Gregory, S. L., and Brown, N. H. (1998). *Kakapo*, a gene required for adhesion between and within cell layers in *Drosophila*, encodes a large cytoskeletal linker protein related to plectin and dystrophin. *J. Cell Biol.* **143,** 1271–1282.

Guo, L., Degenstein, L., Dowling, J., Yu, Q. C., Wollman, R., Perman, B., and Fuchs, E. (1995). Gene targeting of BPAG1: Abnormalities in mechanical strength and cell migration in stratified epithelia and neurological degeneration. *Cell* **81,** 233–243.

Herrmann, H., and Aebi, U. (2000). Intermediate filaments and their associates: Multi-talented structural elements specifying cytoarchitecture and cytodynamics. *Curr. Opin. Cell Biol.* **12,** 79–90.

Hoebeke, J., Nijen, G., and De Brabander, M. (1976). Interaction of oncodazole (R17934), a new anti-tumoral drug, with rat brain tubulin. *Biochem. Biophys. Res. Commun.* **69,** 319–324.

Kodama, A., Karakesisoglou, I., Wong, E., Vaezi, A., and Fuchs, E. (2003). ACF7: An essential integrator of microtubule dynamics. *Cell* **115,** 343–354.

Lee, S., Harris, K. L., Whitington, P. M., and Kolodziej, P. A. (2000). Short stop is allelic to kakapo, and encodes rod-like cytoskeletal-associated proteins required for axon extension. *J. Neurosci.* **20,** 1096–1108.

Leung, C. L., Sun, D., Zheng, M., Knowles, D. R., and Liem, R. K. H. (1999a). Microtubule actin cross-linking factor (MACF): A hybrid of dystonin and dystrophin that can interact with the actin and microtubule cytoskeletons. *J. Cell Biol.* **147,** 1275–1285.

Leung, C. L., Sun, D., and Liem, R. K. (1999b). The intermediate filament protein peripherin is the specific interaction partner of mouse BPAG1-n (dystonin) in neurons. *J. Cell Biol.* **144,** 435–446.

Leung, C. L., Zheng, M., Prater, S. M., and Liem, R. K. H. (2001). The BPAG-1 locus: Alternative splicing produces multiple isoforms with distinct cytoskeletal linker domains, including predominant isoforms in neurons and muscles. *J. Cell Biol.* **154,** 691–697.

Leung, C. L., Green, K. J., and Liem, R. K. (2002). Plakins: A family of versatile cytolinker proteins. *Trends Cell Biol.* **12,** 37–45.

Meng, J., Bornslaeger, E., Green, K., Steinert, P., and Ip, W. (1997). Two-hybrid analysis reveals fundamental differences in direct interactions between desmoplakin and cell type-specific intermediate filaments. *J. Biol. Chem.* **272,** 21495–21503.

Nikolic, B., Nutly, E., Mir, D., and Wiche, G. (1996). Basic amino acid residue cluster within nuclear targeting sequence motif is essential for cytoplasmic plectin-vimentin network junctions. *J. Cell Biol.* **134,** 1455–1467.

Ruhrberg, C., and Watt, F. (1997). The plakin family: Versatile organizers of cytoskeletal architecture. *Curr. Opin. Genet. Dev.* **7,** 392–397.

Sampath, P., and Pollard, T. D. (1991). Effects of cytochalasins, phalloidin and pH on the elongation of actin filaments. *Biochemistry* **30,** 1973–1980.

Schaerer-Brodbeck, C., and Riezman, H. (2000). Interdependence of filamentous actin and microtubules for asymmetric cell division. *Biol. Chem.* **381,** 815–825.

Schneider, C., King, R. M., and Philipson, L. (1988). Genes specifically expressed at growth arrest of mammalian cells. *Cell* **54,** 783–793.

Spector, I., Shochet, N. R., Kashman, Y., and Groweiss, A. (1983). Latrunculins: Novel marine toxins that disrupt microfilament organization in cultured cells. *Science* **219,** 493–495.

Spector, I., Braet, F., Shochet, N. R., and Bubb, M. R. (1999). New anti-actin in the study of the organization and function of the actin cytoskeleton. *Microsc. Res. Tech.* **47,** 18–37.

Subramanian, A., Prokop, A., Yamamoto, M., Sugimura, K., Uemura, T., Betschinger, J., Knoblich, J., and Volk, T. (2003). Shortstop recruits EB1/APC1 and promotes microtubule assembly at the muscle-tendon junction. *Curr. Biol.* **13,** 1086–1095.

Sun, D., Leung, C. L., and Liem, R. K. (2001). Characterization of the microtubule-binding domain of microtubule actin cross-linking factor (MACF): Identification of a novel group of microtubule-associated proteins. *J. Cell Sci.* **114,** 161–172.

Svitkina, T. M., Verkhovsky, A. B., and Borisy, G. G. (1996). Plectin sidearms mediate interaction of intermediate filaments with microtubules and other components of the cytoskeleton. *J. Cell Biol.* **135,** 991–1007.

Wiche, G. (1998). Role of plectin in cytoskeleton organization and dynamics. *J. Cell Sci.* **111,** 2477–2486.

Zucman-Rossi, J., Legoix, P., and Thomas, G. (1996). Identification of new members of the Gas2 and Ras families in the 22q12 chromosome region. *Genomics* **38,** 247–254.

CHAPTER 28

Keratin Bundling Proteins

Pawel Listwan and Joseph A. Rothnagel

Department of Biochemistry and Molecular Biology and
The Centre for Functional and Applied Genomics
University of Queensland
Brisbane, Queensland 4072, Australia

I. Introduction

Keratin bundling proteins are a subcategory of intermediate filament associated proteins (IFAPs) that are defined by their ability to organize keratin filaments into macrofibrilar arrays. Although there are a growing number of proteins that can be considered IFAPs (Coulombe and Omary, 2002), only filaggrin and a related protein have been shown unequivocally to bundle keratin and other intermediate filaments into parallel ropelike structures. The pioneering approach that led to the characterization of filaggrin as an IFAP with bundling ability, by Beverly Dale and Peter Steinert and their colleagues, remains the most effective way to determine this property (Dale *et al.*, 1978; Mack *et al.*, 1993; Steinert *et al.*, 1981). Since the identification of filaggrin, several other putative keratin bundling proteins have been identified such as trichohyalin (Rothnagel and Rogers, 1986), repetin (Krieg *et al.*, 1997), and hornerin (Makino *et al.*, 2001), but their ability to bundle filaments has not yet been proven.

A. Filaggrin—The Archetypal Keratin Bundling Protein

Filaggrin was so-named because of its filament aggregating properties (Steinert *et al.*, 1981) and became the first IFAP to be systematically studied. The early biochemical studies focused on a 25- to 35-kDa protein, which seemed to be a major component of mammalian epidermal extracts, but it later became evident that these peptides were, in fact, specific breakdown products of a much larger precursor (reviewed in Presland and Dale, 2000). The precursor, termed profilaggrin, accumulates within cytoplasmic bodies known as keratohyalin granules, which are a prominent feature of granular layer keratinocytes (Manabe *et al.*, 1991; Stevens *et al.*, 1990). Keratohyalin granules are storage sites for several different proteins in addition to profilaggrin, where they are sequestered from other cellular constituents until they are required. Profilaggrin is a short-lived, high-molecular weight (>300 kDa), phosphorylated polyprotein (Lonsdale-Eccles *et al.*, 1984; Ramsden *et al.*, 1983; Scott and Harding, 1981) that contains multiple repeating units of filaggrin. There are up to 20 or more filaggrin repeats in rodents (Rothnagel and Steinert, 1990) and 10–12 repeats in humans (Markova *et al.*, 1993; Presland *et al.*, 1992). The release of individual filaggrin subunits from profilaggrin requires dephosphorylation by protein phosphatase PP2A and the activity of three different proteases, including profilaggrin proteinase 1, calpain, and furin (Presland and Dale, 2000). Importantly, only filaggrin and not the phosphorylated precursor is able to bundle intermediate filaments (Harding and Scott, 1983; Lonsdale-Eccles *et al.*, 1982; Steinert *et al.*, 1981). The processing of filaggrin continues in the stratum corneum, where it is digested into free amino acids that provide the high osmolarity necessary for the retention of water and maintenance of tissue flexibility (Horii *et al.*, 1983; Scott *et al.*, 1982).

B. Filament Bundling Assays—The "Gold Standard"

Although several methods are available for showing interactions between proteins such as immunochemical staining, immunoprecipitation, various reiterations of the two-hybrid assay, biosensor experiments, photobleaching of fluorescently tagged proteins, and fluorescent resonance energy transfer, none of these assays is able to show filament bundling per se. As stated previously, the most convincing approach for determining the ability of a protein to bundle intermediate filaments is the biochemical method first used to characterize filaggrin (Dale *et al.*, 1978; Steinert *et al.*, 1981). These workers used simple solution biochemistry to measure the degree of bundling spectrophotometrically and then visualized the quality of the macrofibrils produced by electron microscopy. This procedure requires the isolation of relatively pure fractions of keratin or other intermediate filaments and the protein to be tested. Intermediate filaments can be isolated from tissues such as bovine tongue or mouse epidermis or from appropriate cultured cell lines. Similarly, the protein to be tested can be purified from these same sources or, if possible, recombinant approaches can be used. These often have the advantage of producing near-pure proteins with fewer purification steps. An important finding from the studies on filaggrin was that it was able to bundle other intermediate filaments such as desmin, vimentin, and neurofilaments but not with actin microfilaments or microtubules, illustrating a high degree of specificity (Mack *et al.*, 1993; Steinert *et al.*, 1981). Later it was shown that the amino acid composition and charge distribution of the interacting peptide were important parameters for filament bundling. For filaggrin it was determined that the minimum requirement for filament bundling was a repeating motif of 14 residues containing 3 β-turns (Mack *et al.*, 1993). This observation led to the formation of the ionic zipper hypothesis (Mack *et al.*, 1993), in which charged residues within filaggrin interact with charges along the rod domains of intermediate filaments and so draw filaments together into a parallel array.

C. Filaggrin-2—A New Filament Bundling Protein

We have recently identified a novel protein (termed Filaggrin-2; Flg-2) that is expressed late in epidermal differentiation and is similar to profilaggrin in both structure and function (Listwan *et al.*, in preparation; Rothnagel *et al.*, 2002). A polyclonal antibody that recognizes the repeat sequences localized filaggrin-2 to the granular (in keratohyalin granules) and cornified layers of the epidermis. The mouse *Flg-2* gene encodes a 260-kDa protein of 2362 amino acids, which has the same structural organization as profilaggrin—a N-terminal calcium-binding domain, followed by a unique sequence that corresponds to the B-domain of profilaggrin, a repetitive sequence domain, and a unique C-terminal domain (Fig. 1).

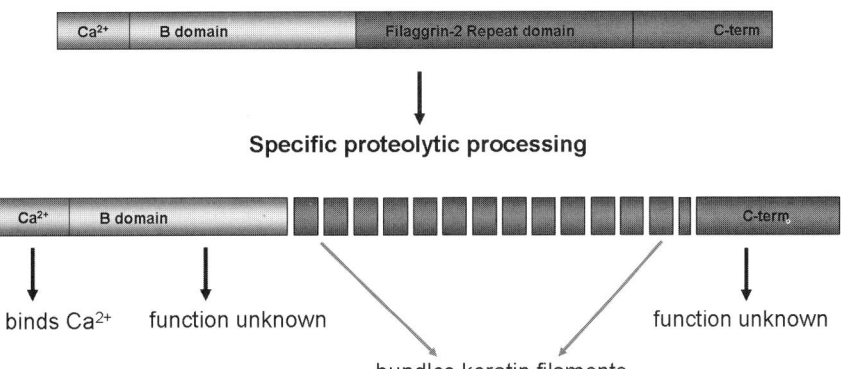

Fig. 1 Schematic showing the domain structure of mouse filaggrin-2. The proprotein consists of a N-terminal calcium binding domain, a B-domain, a repetitive domain, and a C-terminal tail as indicated and is processed into smaller peptides. The repetitive domain consists of 13 complete repeats and 1 partial repeat at the C-terminal end. These repeats vary in size between 74 and 80 residues and are capable of bundling keratin filaments. (See Color Insert.)

A potential recognition sequence for a furinlike protease is present at the end of the B-domain, suggesting that filaggrin-2 is processed in a manner similar to profilaggrin (Presland *et al.*, 1997). In addition, Western blot analysis shows both high (>300 kDa) and low (<30 kDa) molecular weight products in epidermal extracts, consistent with processing of a polypeptide precursor as seen for profilaggrin. The repetitive domain consists of several tandem repeats that vary in size between 74 and 80 residues. The filaggrin-2 repeat has a basic pI and is rich in proline, glycine, serine, glutamine, arginine, and histidine. At the end of each repeat is a short stretch of hydrophobic residues that may be functionally equivalent to the linker region found between filaggrin repeats (Resing *et al.*, 1989; Rothnagel and Steinert, 1990).

The pI and charge distribution of filaggrin-2 repeats are similar to filaggrin and suggested to us that these repeat peptides would be capable of binding to keratin filaments. We, therefore, produced a recombinant peptide containing a complete repeat of filaggrin-2 and added it to keratin filaments extracted from bovine tongue, as had been done previously for filaggrin (Dale *et al.*, 1978; Steinert *et al.*, 1981). We initially analyzed this interaction in a turbidity assay and observed a marked increase in absorbance. This was not seen for the control protein, bovine serum albumin, which had no effect on absorbance, whereas poly-L-arginine produced an intermediate change in turbidity. Analysis of the same mixtures by transmission electron microscopy revealed that only the filaggrin-2 repeat protein was able to order the keratin filaments into almost parallel bundles that were similar to those produced by profilaggrin (Fig. 2).

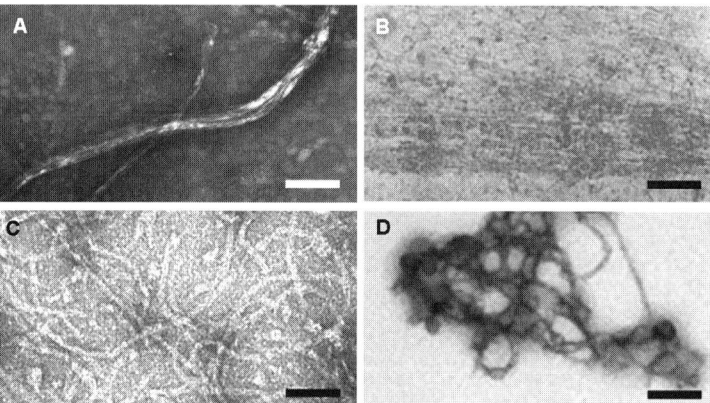

Fig. 2 Formation of keratin filament macrofibrils on the addition of recombinant mouse filaggrin-2 repeat peptides to bovine keratin filaments. (A) Mixing of filaggrin-2 with keratin filaments resulted in the formation of condensed macrofibrils. (B) In some areas, multiple linear arrays were observed on the addition of filaggrin-2. (C) Bovine keratin filaments did not show aggregation in the presence of bovine serum albumin. (D) Poly-L-arginine produced some bundling but no ordered arrays. Scale bars represent 100 nm in (A), (C), and (D) and 200 nm in (B).

D. The Other Contenders—Trichohyalin, Repetin, and Hornerin

Like profilaggrin, trichohyalin, repetin, and hornerin are members of the S100-fused class of proteins, and their genes are found within the epidermal differentiation cluster on lq21 and the syntenic region on mouse chromosome 3 (Ridinger *et al.*, 1998). These proteins consist of a N-terminal calcium-binding domain with a central repetitive domain that is flanked by unique sequences (Feitz *et al.*, 1993; Kreig *et al.*, 1997; Lee *et al.*, 1993; Makino *et al.*, 2001). In addition, these proteins are all expressed at varying levels in late differentiated cells of squamous stratified epithelia.

Trichohyalin is a 190- to 220-kDa protein (depending on the species) that is predominantly found in the inner root sheath and medulla of hair follicles (Rothnagel and Rogers, 1986). Lesser amounts are present within the epithelial layers of the tongue, esophagus, forestomach, and interfollicular epidermis (Hamilton *et al.*, 1991, 1992; O'Guin *et al.*, 1992). In wool and hair follicles, trichohyalin first accumulates in the trichohyalin granules of the inner root sheath and medulla and is later dispersed from these granules during differentiation of these structures. These granules are considered to be functionally equivalent to the keratohyalin granules of the interfollicular epidermis. In squamous stratified epithelia, trichohyalin has been shown to coexist with filaggrin in composite keratohyalin granules (Manabe and O'Guin, 1992, 1994). An examination of the primary sequence predicts that trichohyalin has an extended alpha-helical conformation (Fietz *et al.*, 1993; Lee *et al.*, 1993) and a charge distribution that would enable it to sandwich between individual filaments (Lee *et al.*, 1993; Steinert *et al.*, 2003). Although trichohyalin consists of a series of repetitive elements, there is no evidence of proteolytic processing as occurs in profilaggrin. However, immunoelectron microscopy has revealed that trichohyalin leaves the granules and interacts with the keratin filaments concomitant with their differentiation into condensed, hardened, and ordered arrays (Rothnagel and Rogers, 1986). Subsequent studies established that trichohyalin is cross-linked to keratin filaments and several other proteins by transglutaminases in inner root sheath cells of the hair follicle (Steinert *et al.*, 2003). On this evidence it has been suggested that trichohyalin acts as a filament-bundling protein, but this has not been shown biochemically.

Repetin is a 130-kDa protein that is highly expressed in the tongue and forestomach of mice with lower levels found the epidermis (Krieg *et al.*, 1997), where it seems to localize to keratohyalin granules (Koch *et al.*, 2000). The central region of repetin consists of 49 tandem repeats of 12 amino acids each. Each repeat consists of four glutamines, two glycines, two serines, and single lysine, arginine, aspartic acid and histidine residues, which are found in the same relative position in most of the repeating peptides. Although these repeats are similar to those found in trichohyalin, there is no evidence that repetin can bundle filaments or, indeed, binds to keratin filaments. The role of repetin is not yet known, but it has been found cross-linked to trichohyalin and is currently thought to be a component of the cornified envelope (Jarnik *et al.*, 2002; Steinert *et al.*, 2003).

Hornerin is a 260-kDa protein that is expressed in the granular layer of the epidermis, tongue, esophagus, and forestomach of the mouse, where it has been shown to localize to keratohyalin granules (Makino *et al.*, 2001, 2003). The repetitive domain consists of 12 repeats that are between 170 and 177 residues long and have a pI of 11.02. The repeats are rich in glycine, serine, glutamine, arginine, and histidine residues, an amino acid composition that is most similar to filaggrin. In addition, there is evidence of extensive proteolytic breakdown of hornerin in epidermal extracts, which suggests processing of the type seen for profilaggrin and filaggrin-2. Notwithstanding the similarity with filaggrin, it remains to be determined whether hornerin has filament-bundling properties.

II. Methods

Keratin-bundling experiments are relatively straightforward and easy methods that involve mixing appropriate amounts of isolated keratin filaments with the recombinant protein that is under investigation. There are two methods that are used together to monitor macrofibril formation. These are spectrophotometric monitoring of turbidity caused by filament aggregation and transmission electron microscopy (TEM) that allows visual examination of the overall appearance and organization of the filaments on mixing with the protein of interest. Both methods use the same preparation of isolated keratin filaments and that of the protein to be tested for filament bundling. The turbidity assay is used for a quick assessment of filament bundling and to determine the molar quantities of each component necessary for optimal interactions. Examination by electron microscopy is essential, because the turbidity assay does not give any indication of the quality of the macrofibrils produced. For instance, it has been observed that metal ions such as Ca^{2+}, Zn^{2+}, Eu^{3+}, and U^{4+} produce a large increase in turbidity caused by aggregation of a few filaments but do not produce extensive or ordered macrofibrilar arrays (Mack *et al.*, 1993).

A. Isolation of Keratin Intermediate Filaments

One of the richest sources of the keratin filaments is bovine tongue epithelium. The protocol involves the separation of the tongue mucosa from the muscles by forceps and soaking in $1\times$ PBS containing 20 mM EDTA and the protease inhibitors leupeptin, pepstatin, phenylmethylsulfonylfluroride (PMSF), aprotinin, and EDTA-Na_2. Surface epithelium is then separated from the underlying connective tissue and minced with the scalpel blade in a disassembly buffer containing 50 mM Tris-HCL, pH 9.0, 8 M urea, 0.2% β-mercaptoethanol supplemented with the protease inhibitors listed previously, at room temperature (RT) for 2 hours. The preparation is then clarified by centrifugation at 100,000g at RT for 30 minutes. The Bradford assay can be used to estimate protein concentration of the supernatant, which is subsequently diluted in a filament disassembly buffer to a final protein concentration of 0.5 mg/ml. The filament preparation is then dialyzed twice against assembly buffer containing 10 mM Tris-HCl, pH 7.4, 0.2% β-mercaptoethanol, and the protease inhibitors previously listed overnight at 4 °C. The dialyzed preparation is again centrifuged at 100,000g at 4 °C for 30 minutes, and the pellet containing the polymerized intermediate filaments is resuspended in the assembly buffer for the macrofibril formation experiments. An aliquot of the preparation can be assessed for purity by sodium dodecyl sulfate-polyacrylamide gel electrophoresis (Fig. 3). We find that these preparations contain less than 5% of nonintermediate filament proteins that copurify with the filaments and are thought to be IFAPs (see Steinert *et al.*, 1976). As can be seen from control experiments, the presence of these proteins does not interfere with the bundling assay (Fig. 2). The isolated bovine intermediate filaments can be stored at a concentration of 1 mg/ml in 20% glycerol at −80 °C.

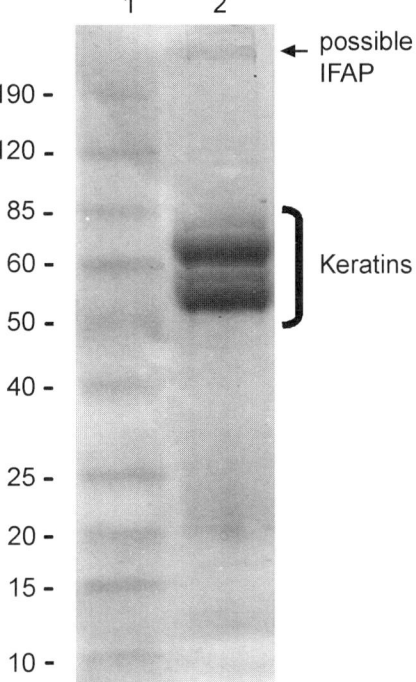

Fig. 3 A Coomassie-blue–stained sodium dodecylsulfate-polyacrylamide gel electrophoresis gel of a keratin intermediate filament preparation isolated from bovine tongue epithelium. Lane 1 contains the Benchmark prestained protein ladder (Invitrogen); lane 2, isolated filament preparation. Five intermediate filament proteins are present in the preparation: K14 (50 kDa), K13 (52 kDa), K10 (56 kDa), K5 (58 kDa), and K1 (61 kDa).

B. Preparation of the Filament Bundling Protein

The protein to be tested for bundling properties can be isolated from tissues or cell culture sources, but it is often faster and easier to use recombinant approaches such as the pGEX bacterial expression system (Smith and Johnson, 1988) or a baculovirus expression system (Smith et al., 1983). To avoid potential nonspecific or spurious interactions between the recombinant protein and the keratin filaments, any tags that were used in the purification of the recombinant protein should be removed before bundling experiments. We expressed the filaggrin-2 repeat peptide in pGEX-T2 (Amersham Pharmacia Biotech) as a GST fusion protein and used thrombin to cleave the peptide from GST. The peptide open-reading frame was cloned into the *Bam*HI site, which is on the boundary of the GST sequences, so that no exogenous amino acids are present in the purified peptide after thrombin cleavage. The GST fusion protein expression is induced by adding 1 mM isopropyl-β-D-thiogalactopyranoside (IPTG) to the bacterial culture, which is then grown for a further 3 hours. The bacterial culture is then centrifuged at 7700g for 20 minutes at 4 °C, and the cell pellet is resuspended in 50 μl of ice-cold phosphate-buffered saline (PBS) containing 2.5 mM PMSF. The resuspended cells are sonicated on ice and then mixed gently in 1% Triton X-100 and 2.5 mM PMSF for 30 minutes to aid the solublization of the fusion protein. The insoluble material is removed by centrifugation at 12,000g for 30 minutes at 4 °C and discarded. The fusion protein can be purified by use of a glutathione Sepharose 4B column (Amersham Pharmacia Biotech). The bacterial sonicate is applied to the column and washed three times by the addition of 10 bed volumes of PBS and allowed to drain. The matrix is removed from the column and resuspended in thrombin cleavage buffer (50 mM Tris-HCL, pH 8.5, 150 mM NaCl, 2.5 mM CaCl$_2$ and 5 mM 1 M dithiothreitol), and then incubated for 30 minutes at RT on a rotary wheel. The matrix is washed three times with 4 volumes of thrombin cleavage buffer before adding bovine thrombin to a final concentration of approximately 5 U/mg of fusion protein and then incubated for 2 hours at 37 °C on a rotary wheel. The released fusion protein can then be collected into a fresh tube after centrifugation at 12,000g. For the filaggrin-2 peptide, the supernatant was passed through an FPLC to separate the fusion protein from thrombin, and the purified peptide was stored at a concentration of 0.3 mg/ml in 50 mM Tris-HCL, pH 8.5, 150 mM NaCl, 1.0 mM EDTA at −80 °C. We routinely snap freeze our protein samples in liquid nitrogen to avoid the formation of ice crystals within the sample.

C. Macrofibril Formation Assay

We found that a ratio of 1:2 of mouse filaggrin-2 recombinant protein (0.2 mg/ml in assembly buffer) to the keratin filament preparation gave the highest level of bundling. For controls we used bovine serum albumin (BSA) as a negative control and poly-L-arginine as a positive control, both at a concentration of 1 mg/ml (in assembly buffer) as used in the earlier studies (Mack et al., 1993). The extent of macrofibril formation was determined by monitoring absorbance at 300 nm every

10 seconds for 3 minutes at RT (Steinert *et al.*, 1981). The same conditions were used for the TEM analysis. Filaggrin-2 (or BSA or poly-L-arginine) and the filament preparation were mixed together, and after 5 minutes of incubation at RT, 2 μl of mixtures were placed on carbon-coated grids, negatively stained with 4% uranyl acetate, dried, and examined in Philips Tecnai 12 transmission electron microscope (Fig. 2). It should be noted that this assay is not limited to keratin filaments and can also be used for other intermediate filaments such as vimentin, desmin, and neurofilaments as has already been demonstrated by others (Mack *et al.*, 1993).

III. Pearls and Pitfalls

Bovine tongue epithelium is a rich source of keratin intermediate filaments, which can be isolated with a high degree of purity. The protocol is not challenging and involves commonly used reagents and equipment. One key step that must not be circumvented or rushed is the dialysis, which should be maintained for at least 16 hours with two changes of the assembly buffer. The addition of protease inhibitors is also essential to maintain the integrity of the isolated filaments. The isolated intermediate filaments are invariably of high purity with only minor amounts of proteins that copurify and probably represent IFAPs. Low levels of these contaminating proteins are unlikely to compromise filament bundling experiments.

These assays require that the protein to be tested is soluble in the assembly buffer. Solubility can be problematic for bacterial expressed proteins and may require the use of other expression systems such as baculovirus expression in insect cells. Alternatively, it may be preferable to isolate the protein directly from tissue or cultured cells. The expression and purification of adequate quantities of the protein to be tested will be far more challenging than the preparation of the intermediate filaments themselves. However, once both components are available, the bundling assays are not only easy to set up, but the results are available relatively quickly (minutes to a few hours).

IV. Concluding Remarks

To date, only two proteins, filaggrin and filaggrin-2, have been shown unequivocally to function as keratin-bundling proteins. It remains to be seen whether this exclusive group will acquire new members, with trichohyalin, hornerin, and repetin all requiring further testing. Although this review has focused on keratin-bundling proteins, it is probable that novel or known IFAPs will be identified that bundle other intermediate filament types such as vimentin and desmin. In addition, the remarkably linear arrays of neurofilaments observed in some axons may be mediated by a specific bundling protein, and the role of neural IFAPs in this activity would seem worthy of further investigation.

References

Coulombe, P. A., and Omary, M. B. (2002). 'Hard' and 'soft' principles defining the structure, function and regulation of keratin intermediate filaments. *Curr. Opin. Cell Biol.* **14,** 110–122.

Dale, B. A., Holbrook, K. A., and Steinert, P. M. (1978). Assembly of stratum corneum basic protein and keratin filaments in macrofibrils. *Nature* **276,** 729–731.

Fietz, M. J., McLaughlan, C. J., Campbell, M. T., and Rogers, G. E. (1993). Analysis of the sheep trichohyalin gene: Potential structural and calcium-binding roles of trichohyalin in the hair follicle. *J. Cell Biol.* **121,** 855–865.

Hamilton, E. H., Payne, R. E., Jr., and O'Keefe, E. J. (1991). Trichohyalin: Presence in the granular layer and stratum corneum of normal human epidermis. *J. Invest. Dermatol.* **96,** 666–672.

Hamilton, E. H., Sealock, R., Wallace, N. R., and O'Keefe, E. J. (1992). Trichohyalin: Purification from porcine tongue epithelium and characterization of the native protein. *J. Invest. Dermatol.* **98,** 881–889.

Harding, C. R., and Scott, I. R. (1983). Histidine-rich proteins (filaggrins): Structural and functional heterogeneity during epidermal differentiation. *J. Mol. Biol.* **170,** 651–673.

Horii, I., Kawasaki, K., Koyama, J., Nakayama, Y., Nakajima, K., Okazaki, K., and Seiji, M. (1983). Histidine-rich protein as a possible origin of free amino acids of stratum corneum. *J. Dermatol.* **10,** 25–33.

Jarnik, M., de Viragh, P. A., Scharer, E., Bundman, D., Simon, M. N., Roop, D. R., and Steven, A. C. (2002). Quasi-normal cornified cell envelopes in loricrin knockout mice imply the existence of a loricrin backup system. *J. Invest. Dermatol.* **118,** 102–109.

Koch, P. J., de Viragh, P. A., Scharer, E., Bundman, D., Longley, M. A., Bickenbach, J., Kawachi, Y., Suga, Y., Zhou, Z., Huber, M., Hohl, D., Kartasova, T., Jarnik, M., Steven, A. C., and Roop, D. R. (2000). Lessons from loricrin-deficient mice: Compensatory mechanisms maintaining skin barrier function in the absence of a major cornified envelope protein. *J. Cell Biol.* **151,** 389–400.

Krieg, P., Schuppler, M., Koesters, R., Mincheva, A., Lichter, P., and Marks, F. (1997). Repetin (Rptn), a new member of the "fused gene" subgroup within the S100 gene family encoding a murine epidermal differentiation protein *Genomics* **43,** 339–348.

Lee, S. C., Kim, I. G., Marekov, I. N., O'Keefe, E. J., Parry, D. A., and Steinert, P. M. (1993). The structure of human trichohyalin. Potential multiple roles as a functional EF-hand-like calcium-binding protein, a cornified cell envelope precursor, and an intermediate filament-associated (cross-linking) protein. *J. Biol. Chem.* **268,** 12164–12176.

Lonsdale-Eccles, J. D., Teller, D. C., and Dale, B. A. (1982). Characterization of a phosphorylated form of the intermediate filament-aggregating protein filaggrin. *Biochemistry* **21,** 5940–5948.

Lonsdale-Eccles, J. D., Resing, K. A., Meek, R. L., and Dale, B. A. (1984). High-molecular-weight precursor of epidermal filaggrin and hypothesis for its tandem repeating structure. *Biochemistry* **23,** 1239–1245.

Mack, J. W., Steven, A. C., and Steinert, P. M. (1993). The mechanism of interaction of filaggrin with intermediate filaments. The ionic zipper hypothesis. *J. Mol. Biol.* **232,** 50–66.

Makino, T., Takaishi, M., Morohashi, M., and Huh, N. H. (2001). Hornerin, a novel profilaggrin-like protein and differentiation-specific marker isolated from mouse skin. *J. Biol. Chem.* **276,** 47445–47452.

Makino, T., Takaishi, M., Toyoda, M., Morohashi, M., and Huh, N. H. (2003). Expression of hornerin in stratified squamous epithelium in the mouse: A comparative analysis with profilaggrin. *J. Histochem. Cytochem.* **51,** 485–492.

Manabe, M., Sanchez, M., Sun, T. T., and Dale, B. A. (1991). Interaction of filaggrin with keratin filaments during advanced stages of normal human epidermal differentiation and in ichthyosis vulgaris. *Differentiation* **48,** 43–50.

Manabe, M., and O'Guin, W. M. (1992). Keratohyalin, trichohyalin and keratohyalin-trichohyalin hybrid granules: An overview. *J. Dermatol.* **19,** 749–755.

Manabe, M., and O'Guin, W. M. (1994). Existence of trichohyalin-keratohyalin hybrid granules: Co-localization of two major intermediate filament-associated proteins in non-follicular epithelia. *Differentiation* **58,** 65–75.

Markova, N. G., Marekov, L. N., Chipev, C. C., Gan, S. Q., Idler, W. W., and Steinert, P. M. (1993). Profilaggrin is a major epidermal calcium-binding protein. *Mol. Cell Biol.* **13,** 613–625.

O'Guin, W. M., Sun, T. T., and Manabe, M. (1992). Interaction of trichohyalin with intermediate filaments: Three immunologically defined stages of trichohyalin maturation. *J. Invest. Dermatol.* **98,** 24–32.

Presland, R. B., and Dale, B. A. (2000). Epithelial structural proteins of the skin and oral cavity: Function in health and disease. *Crit. Rev. Oral Biol. Med.* **11,** 383–408.

Presland, R. B., Haydock, P. V., Fleckman, P., Nirunsuksiri, W., and Dale, B. A. (1992). Characterization of the human epidermal profilaggrin gene. Genomic organization and identification of an S-100-like calcium binding domain at the amino terminus. *J. Biol. Chem.* **267,** 23772–23781.

Presland, R. B., Kimball, J. R., Kautsky, M. B., Lewis, S. P., Lo, C. Y., and Dale, B. A. (1997). Evidence for specific proteolytic cleavage of the N-terminal domain of human profilaggrin during epidermal differentiation. *J. Invest. Dermatol.* **108,** 170–178.

Ramsden, M., Loehren, D., and Balmain, A. (1983). Identification of a rapidly labelled 350 K histidine-rich protein in neonatal mouse epidermis. *Differentiation* **23,** 243–249.

Resing, K. A., Walsh, K. A., Haugen-Scofield, J., and Dale, B. A. (1989). Identification of proteolytic cleavage sites in the conversion of profilaggrin to filaggrin in mammalian epidermis. *J. Biol. Chem.* **264,** 1837–1845.

Ridinger, K., Ilg, E. C., Niggli, F. K., Heizmann, C. W., and Schafer, B. W. (1998). Clustered organization of S100 genes in human and mouse. *Biochim. Biophys. Acta* **1448,** 254–263.

Rothnagel, J. A., and Rogers, G. E. (1986). Trichohyalin, an intermediate filament-associated protein of the hair follicle. *J. Cell Biol.* **102,** 1419–1429.

Rothnagel, J. A., and Steinert, P. M. (1990). The structure of the gene for mouse filaggrin and a comparison of the repeating units. *J. Biol. Chem.* **265,** 1862–1865.

Rothnagel, J. A., Listwan, P., Karunaratne, S., and Zhang, D. (2002). Characterization of a novel gene with filaggrin-like repeats that is expressed late in epidermal differentiation [abstract]. *J. Invest. Dermatol.* **119,** 271.

Scott, I. R., and Harding, C. R. (1981). Studies on the synthesis and degradation of a high molecular weight, histidine-rich phosphoprotein from mammalian epidermis. *Biochim. Biophys. Acta* **669,** 65–78.

Scott, I. R., Harding, C. R., and Barrett, J. G. (1982). Histidine-rich protein of the keratohyalin granules. Source of the free amino acids, urocanic acid and pyrrolidone carboxylic acid in the stratum corneum. *Biochim. Biophys. Acta* **719,** 110–117.

Smith, D. B., and Johnson, K. S. (1988). Single-step purification of polypeptides expressed in Escherichia coli as fusions with glutathione S-transferase. *Gene* **67,** 31–40.

Smith, G. E., Summers, M. D., and Fraser, M. J. (1983). Production of human beta interferon in insect cells infected with a baculovirus expression vector. *Mol. Cell Biol.* **12,** 2156–2165.

Steinert, P. M., Cantieri, J. S., Teller, D. C., Lonsdale-Eccles, J. D., and Dale, B. A. (1981). Characterization of a class of cationic proteins that specifically interact with intermediate filaments. *Proc. Natl. Acad. Sci. USA* **78,** 4097–4101.

Steinert, P. M., Idler, W. W., and Zimmerman, S. B. (1976). Self-assembly of bovine epidermal keratin filaments *in vitro. J. Mol. Biol.* **108,** 547–567.

Steinert, P. M., Parry, D. A., and Marekov, L. N. (2003). Trichohyalin mechanically strengthens the hair follicle: Multiple cross-bridging roles in the inner root sheath. *J. Biol. Chem.* **278,** 41409–41419.

Steven, A. C., Bisher, M. E., Roop, D. P., and Steinert, P. M. (1990). Biosynthetic pathways of filaggrin and loricin–two major proteins expressed by terminally differentiated epidermal keratinocytes. *J. Struct. Biol.* **104,** 105–162.

CHAPTER 29

Lamin-Associated Proteins

Cecilia Östlund and Howard J. Worman

Departments of Medicine and of Anatomy and Cell Biology
College of Physicians and Surgeons
Columbia University
New York, New York 10032

I. Introduction

A. The Nuclear Envelope and Lamina

The nuclear envelope surrounds the cell nucleus in eukaryotic interphase cells but disassembles during mitosis in higher eukaryotes. It is composed of the nuclear lamina, nuclear pore complexes (NPCs), and nuclear membranes (Fig. 1) (Burke and Stewart, 2002; Goldman et al., 2002; Holaska et al., 2002; Worman and

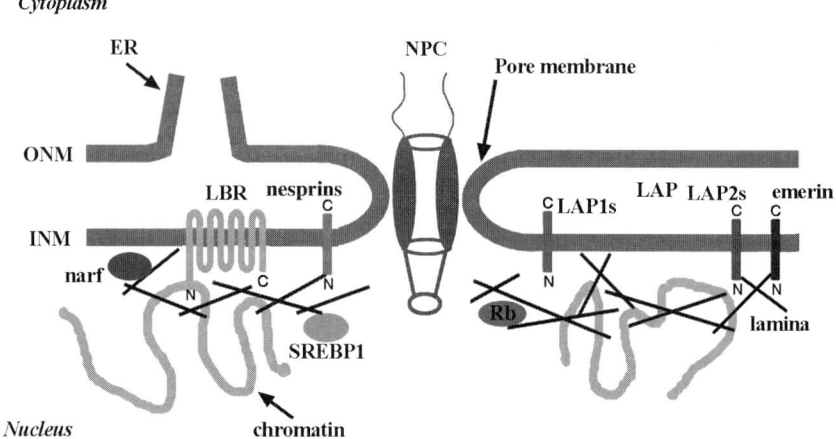

Fig. 1 Schematic diagram of the nuclear envelope showing the nuclear lamina and associated proteins. The nuclear membranes consist of the outer nuclear membrane (ONM), continuous with the ER, the pore membrane, and the inner nuclear membrane (INM). The NPC is associated with the pore membrane. The lamina, chromatin, and some well-characterized lamin-binding proteins of nucleus and INM are also shown. Modified from Östlund *et al.* (2002) with permission from Landes Bioscience. (See Color Insert.)

Courvalin, 2000). The nuclear membranes consist of three distinct, but interconnected, parts: the outer nuclear membrane, which is directly continuous with the endoplasmic reticulum (ER) membrane, the pore membrane, and the inner nuclear membrane. The lamina is an intermediate filament network, which lines the inner aspect of the inner nuclear membrane in higher eukaryotic cells. The main components of the lamina are the lamin proteins, which can be divided into two major subgroups, the A-type lamins, which have a neutral isoelectric point and the acidic B-type lamins (reviewed by Stuurman *et al.*, 1998). In mammals, lamin A and lamin C are the major A-type lamins and are different splice forms of the *LMNA* gene (Lin and Worman, 1993). The major forms of B-type lamins are lamin B1 and lamin B2, encoded by *LMNB1* and *LMNB2*, respectively (Biamonti *et al.*, 1992; Höger *et al.*, 1990; Lin and Worman, 1995). All lamins, like other intermediate filament proteins, have a central, α-helical coiled–coiled rod domain flanked by amino-terminal "head" and carboxyl-terminal "tail" domains. The tail domains of lamins A, B1, and B2 possess a CaaX-box motif, which is subject to farnesylation. Although the farnesyl group is attached to the B-type lamins throughout their lifespan, it is removed from lamin A during the processing from pre-lamin A to mature lamin A.

During the past few years, there has been an increased interest in the lamins as genetic studies have identified several diseases, sometimes called "laminopathies," which are caused by mutations in *LMNA*. Among these diseases are autosomal-dominant Emery-Dreifuss muscular dystrophy (AD-EDMD), dilated cardiomyopathy and conduction defect 1, limb girdle muscular dystrophy type 1B, Dunnigan-type familial partial lipodystrophy, autosomal-recessive Charcot-Marie-Tooth disorder type 2B1, and Hutchinson-Gilford progeria (reviewed by Burke and Stewart, 2002; Östlund and Worman, 2003; Worman and Courvalin, 2002, 2004).

Several proteins have been shown to interact with lamins (Fig. 1, Table I). These include not only integral proteins of the inner nuclear membrane, most of which have been shown to interact with lamins, but also several nucleoplasmic proteins, including some transcription factors. Although the exact functions of many of the lamin-associated proteins are not well understood, this suggests a dual role for lamins, both as structural proteins of the nuclear envelope and as regulators of transcription and chromatin organization. This review will focus primarily on the transmembrane proteins that associate with lamins.

Before the lamins had been shown to be intermediate filament proteins and their cDNA sequences were known (Aebi *et al.*, 1986; Fisher *et al.*, 1986; McKeon *et al.*, 1986), the lamina was identified as a biochemical fraction that could be purified together with the NPCs from rat liver nuclei (Dwyer and Blobel, 1976). Purified nuclei (Blobel and Potter, 1966) were treated with DNAse to remove chromatin, yielding nuclear envelopes. The nuclear envelopes were then treated with 2% Triton X-100 to solubilize the membranes and washed again with 500 mM KCl to remove residual chromatin, yielding a pore complex–lamina fraction (Dwyer and Blobel, 1976). Because proteins tightly associated with lamins copurify with lamins whereas others are washed away during these extraction conditions, this purification method and variants have been instrumental in the identification of lamin-associated proteins, as well as in verifying interactions between lamins and putative binding partners.

B. The Lamin B Receptor

1. Identification of Lamin B Receptor

In 1988, Worman *et al.* showed that B-type, but not A-type, lamins bound in a saturable fashion to turkey erythrocyte nuclear envelopes that had been extracted with 8 M urea to remove endogenous lamins (Worman *et al.*, 1988b). (Turkey erythrocytes, as other nonmammalian erythrocytes, contain a nucleus.) This suggested that B-type lamins were associated with the inner nuclear membrane by means of a transmembrane "receptor" functioning as an attachment site for the nuclear lamina. In an initial overlay binding assay experiment to identify such a receptor in the inner nuclear membrane, lamin-depleted nuclear envelopes were electrophoretically separated and transferred to nitrocellulose sheets. A dilute solution of ^{125}I-labeled B-type lamins was then incubated with the blot, where they bound to an abundant protein with an apparent molecular mass of 58 kDa, which was named the "lamin B receptor (LBR)." Antibodies were obtained by immunizing guinea pigs with purified LBR. The antibodies were then used in

Table I
Lamins and Some Well-Characterized Lamin-Binding Proteins

Protein	MW (kDa)	Integral membrane protein	Suggested function(s)	Commercial antibody sources*
Emerin	29	Yes	Mutated in X-EDMD.	Abcam, ImmuQuest, Novocastra, St Cruz
Lamin A	72	No	Structural support of the nuclear envelope. Regulation of transcription and chromatin organization. Lamins A and C are mutated in several diseases.	Abcam, BD Biosciences, Chemicon, ImmuQuest, Novocastra, St Cruz, USB
Lamin B1	65	No		Abcam, Chemicon, ImmuQuest, Novocastra, St Cruz, USB
Lamin B2	68	No		Abcam, Chemicon, ImmuQuest, Novocastra, St Cruz, USB
Lamin C	62	No		Abcam, BD Biosciences, Chemicon, ImmuQuest, Novocastra, St Cruz, USB
LAP 1A	75	Yes	Lamin targeting after mitosis.	Abcam
LAP 1B	68	Yes		
LAP 1C	55	Yes		
LAP 2α	75	No	Lamin targeting after mitosis.	Abcam, ImmuQuest,
LAP 2β	53	Yes		Abcam
LAP 2γ	39	Yes		

LBR	58	Yes	Preservation of chromatin structure, sterol reductase, mutated in Pelger-Huët anomaly and HEM/Greenberg dysplasia.	ImmuQuest
Narf	52	No	Putative role in pre-lamin A processing.	
Nesprin 1α	131	Yes	Maintenance of nuclear organization and structural integrity.	Abcam, ImmuQuest
Nesprin 2α	61	Yes		
Otefin (*Drosophila*)	45	No	Attachment of membrane vesicles to chromatin.	
Rb	110	No	Tumor suppression, cell cycle regulation.	BD Biosciences, Novocastra, St Cruz
SREBP1	68	No	Adipocyte differentiation, cholesterol biosynthesis.	Abcam, BD Biosciences, St Cruz
UNC-84 (*C. elegans*)	99 and 125	Yes	Nuclear migration and positioning.	
YA (*Drosophila*)	91	No	*Drosophila* development.	

*Commercial antibody sources identified through an Internet search on April 29, 2004, using Google.com and Biocompare.com. For details on the antibodies, see abcam.com, bdbiosciences.com, chemicon.com, immuquest.com, novocastra.co.uk, scbt.com, and usbweb.com.

Western blots of cellular fractions, and LBR was shown to be resistant to alkali extraction (0.10 M Na_2CO_3, pH 11.5) from membranes and was soluble only in a mixture of strong detergents and salt (2% Triton X-100 and 2 M KCl), indicating that it was an integral membrane protein. Indirect immunofluorescence microscopy with these antibodies yielded a smooth nuclear rim staining. These data strongly suggested that the protein was a lamin-binding integral membrane protein of the nuclear envelope in turkey erythrocytes (Worman et al., 1988b). A later study confirmed that LBR was exclusively localized to the nuclear membrane in chicken liver cells (Worman et al., 1990), and its localization to the inner nuclear membrane was confirmed by immunoelectron microscopy with a specific monoclonal antibody from a battery of antibodies raised against abundant nuclear proteins from chicken (Bailer et al., 1991).

2. Cloning of the Lamin B Receptor cDNA and Gene

The LBR protein was electroeluted from SDS-PAGE gels, and the amino acid sequences of several fragments were determined by microsequencing. Partially degenerate polymerase chain reaction (PCR) primers were then synthesized and used to amplify an oligomere used to design a probe for screening of a lambda Zap chicken embryonal cDNA library (Worman et al., 1990). Sequencing of several overlapping clones revealed that chicken LBR is a protein of 637 amino acids with a nucleoplasmic, amino-terminal domain of 204 amino acids without a cleavable signal sequence, followed by a hydrophobic domain with eight putative transmembrane segments as determined by a Kyte-Doolittle hydropathy plot (Worman et al., 1990). The cDNA for human LBR was isolated from a HeLa cell cDNA library with ^{32}P-labeled chicken LBR as a probe (Ye and Worman, 1994). Human LBR has 615 amino acids, and its amino acid sequence is 68% identical to that of chicken LBR, and has the same topological organization.

The gene for human LBR was isolated by screening of genomic libraries with human LBR cDNA as a probe (Schuler et al., 1994). It contains 13 protein-coding exons. The first four exons encode the nucleoplasmic amino-terminal domain, and exons 5–13 encode the hydrophobic domain. Exons 4 and 5 are separated by a relatively large intron of more than 10 kb, suggesting that the LBR gene may have evolved from recombination of two genes, one encoding a nuclear protein, which became the first four coding exons, and one encoding an integral membrane protein similar to sterol reductases (see later), which became the last nine exons (Schuler et al., 1994).

3. Interactions between Lamin B Receptor and the Lamina and Chromatin

As described previously, LBR was identified as a lamin B–binding protein. To further characterize this interaction, rat liver nuclear envelopes were extracted with 8 M urea to yield a solution enriched in lamins (Ye and Worman, 1994). This solution was dialyzed against phosphate-buffered saline (PBS) and incubated with the complete amino-terminal region of LBR or parts thereof fused to glutathione-S-transferase (GST) attached to glutathione-sepharose beads. Immunoblotting of unbound and bound fractions demonstrated that the complete amino-terminal region of LBR binds to lamin B1, whereas shorter parts of it do not.

LBR has also been shown to interact with chromatin. In a yeast two-hybrid screen, two human heterochromatin proteins, HP1$^{Hs\alpha}$ and HP1$^{Hs\gamma}$, homologous to *Drosophila* HP1, bound to the amino-terminal domain of LBR (Ye and Worman, 1996). The interactions were confirmed in several different *in vitro* binding and coimmunoprecipitation assays (Ye and Worman, 1996). Several other studies have also shown that LBR can associate with chromatin and DNA. *In vivo*, LBR-containing membrane vesicles are targeted to chromosomes early during mitotic nuclear envelope reassembly (Buendia and Courvalin, 1997; Chaudhary and Courvalin, 1993). The amino-terminal domain of LBR binds directly to double-stranded DNA *in vitro*, preferentially to the nucleosomal linker (Duband-Goulet and Courvalin, 2000; Ye and Worman, 1994).

4. Targeting of Lamin B Receptor to the Inner Nuclear Membrane

Several studies have addressed the question of how LBR and other integral proteins are targeted to the inner nuclear membrane (Ellenberg *et al.*, 1997; Östlund *et al.*, 1999; Rolls *et al.*, 1999; Soullam and Worman, 1993, 1995; Wu *et al.*, 2002). Their targeting is different than targeting of soluble proteins to the nucleus, which occurs through the central channel of the NPCs and is dependent on nuclear localization sequences, often composed of one or two short stretches of basic amino acids (Nakielny and Dreyfuss, 1999). The targeting of integral proteins to the inner nuclear membrane and their retention there often requires several regions of the protein. The results of most targeting studies of such proteins are consistent with a diffusion-retention model (Soullam and Worman, 1995). In this model, proteins are synthesized on and cotranslationally inserted into the rough ER membrane. Because the ER, outer nuclear membrane, pore membrane, and inner nuclear membrane are continuous with each other, proteins can potentially move between all these domains by lateral diffusion. To reach the inner nuclear membrane, they have to pass through the pore membrane, where the NPCs are situated. Ultrastructural studies have shown the NPCs to contain eight lateral channels adjacent to the membrane (Akey and Radermacher, 1993; Hinshaw *et al.*, 1992), and integral membrane proteins have been suggested to reach the inner nuclear membrane by diffusion through these channels (Soullam and Worman, 1995). They are then retained and immobilized in the inner nuclear membrane by binding to other proteins or structures, such as the nuclear lamina and chromatin. Early immunofluorescence microscopy studies that used tagged full-length or truncated forms of LBR showed that at least two different targeting/retention signals are independently sufficient to localize the protein to the inner nuclear membrane. One signal is situated in the amino-terminal region of LBR and the other in the first transmembrane spanning region (Smith and Blobel, 1993; Soullam and Worman, 1993, 1995).

If integral proteins reach the inner nuclear membrane by lateral diffusion, they must then somehow be retained there or they may diffuse back to the ER. Fluorescent recovery after photobleaching (FRAP) and fluorescence loss in photobleaching (FLIP) experiments with the amino-terminal domain and the first transmembrane-spanning region of LBR fused to green fluorescent protein (GFP) have shown that LBR-GFP diffuses rapidly and freely in the ER but becomes immobilized when it reaches the inner nuclear membrane (Ellenberg *et al.*, 1997). This decrease in diffusion in the inner nuclear membrane indicates binding to fixed structural components, such as the lamina and/or chromatin. Studies of other inner nuclear membrane proteins have yielded similar results (Östlund *et al.*, 1999; Rolls *et al.*, 1999; Wu *et al.*, 2002). (For a more detailed description of immunofluorescence microscopy, FRAP, and FLIP, see "Procedures".)

5. Putative Functions of Lamin B Receptor

The exact function(s) of LBR is not understood. As described previously, LBR interacts with B-type lamins, and this may be involved in attaching the lamina to the inner nuclear membrane *in vivo*. However, other interactions, such as between B-type lamins and other proteins such as LAP2β (see later), between different lamin types, and between the farnesyl-groups of B-type lamins and the nuclear membrane may also be involved in this process. LBR also interacts with chromatin through both protein–protein and protein–DNA interactions. In addition to attaching the nuclear lamina, LBR may also be involved in the association of chromatin with the inner nuclear membrane. Binding of LBR to chromatin may also be responsible for the targeting of membranes to decondensing chromosomes in anaphase and initiating nuclear envelope reassembly during mitosis (Chaudhary and Courvalin, 1993; Collas *et al.*, 1996). A role for LBR in chromatin organization has also been suggested by the finding that heterozygous mutations in LBR cause Pelger-Huët anomaly, an autosomal-dominant disorder characterized by abnormal nuclear shape and chromatin organization in blood granulocytes (Hoffmann *et al.*, 2002). Rare homozygous individuals with Pelger-Huët anomaly have varying degrees of developmental delays, epilepsy, and skeletal abnormalities (see later).

The carboxyl-terminal, hydrophobic domain of LBR is homologous to yeast sterol reductases (Schuler *et al.*, 1994). BLAST searches of The Institute for Genomic Research EST database and GenBank identified several overlapping expressed sequence tags from human cDNA clones of various sizes. The clones were obtained and found to encode two human paralogues of LBR, SR-1 (TM7SF2) and SR-2 (DHCR7) (Holmer *et al.*, 1998). The homology between these proteins and LBR was restricted to the hydrophobic region of LBR, with SR-1 and SR-2 lacking a hydrophilic amino-terminal region. Immunofluorescence microscopy showed that SR-1 and SR-2 were localized to the ER (Holmer *et al.*, 1998). Simultaneous work from other groups showed that SR-2/DHCR7 is a 7-dehydrocholesterol reductase mutated in patients with Smith-Lemli-Optiz syndrome (Fitzky *et al.*, 1998; Moebius *et al.*, 1998). Although LBR demonstrated sterol Δ^{14}-reductase activity when expressed in yeast (Silve *et al.*, 1998), a role in human cholesterol biosynthesis was not expected, because all enzymes identified previously in this pathway had been localized to the ER membrane, not the inner nuclear membrane. However, such a role for LBR was recently suggested when a mutation in LBR was identified in a fetus with autosomal-recessive HEM/Greenberg skeletal dysplasia (Waterham *et al.*, 2003). Gas chromatography/mass spectrometry of lipids from skin fibroblasts from this fetus showed elevated levels of cholesta-8, 14-dien-3β-ol, which is a substrate for sterol Δ^{14}-reductase in the cholesterol biosynthesis pathway. The role of the nuclear envelope in the cholesterol biosynthetic pathway may, therefore, need to be reevaluated. The heterozygous mother of the fetus with HEM/Greenberg skeletal dysplasia also had hypolobulated granulocytes (Waterham *et al.*, 2003), suggesting that HEM/Greenberg skeletal dysplasia and homozygous Pelger-Huët anomaly are allelic disorders with a wide clinical spectrum. The different characteristics of the diseases, caused by heterozygous and homozygous mutations, suggest that LBR has at least two different physiological functions, to preserve chromatin structure and to function as a sterol Δ^{14}-reductase.

C. Lamina-Associated Polypeptides

1. Identification of the Lamina–Associated Polypeptide 1 Proteins

The lamina-associated polypeptides (LAPs) were first identified by use of monoclonal antibodies raised against a pore complex–lamina fraction isolated from rat liver nuclear envelopes by sequential extraction with 1 M NaCl and 2% Triton X-100 (Senior and Gerace, 1988). One of the antibodies (RL13) labeled three proteins, a major polypeptide of 75 kDa and two minor components of 68 kDa and 55 kDa, on immunoblots of rat liver nuclear envelopes. These proteins were later named LAP1A, LAP1B, and LAP1C, respectively (Foisner and Gerace, 1993). The three proteins were resistant to extraction with 0.1 M NaOH, 4 M guanidine-HCl, and urea, showing them to be integral membrane proteins (Foisner and Gerace, 1993; Senior and Gerace, 1988). When nuclear envelopes are incubated with buffer containing 2% Triton X-100 and 50 mM KCl, the nuclear membranes are solubilized, whereas NPCs and lamina remain intact (Dwyer and Blobel, 1976). All of LAP1A and 1B, and most of LAP1C, remained associated with the pore complex–lamina fraction during this treatment, and LAP1A and LAP1B were also resistant to 2% Triton X-100 and 500 mM KCl, which will solubilize many pore complex proteins as well, leaving a fraction highly enriched in the lamina (Senior and Gerace, 1988). These data suggested LAP1A and 1B had strong direct or indirect associations with the lamina, and that LAP1C may also be attached to the lamins, but more weakly. Immunofluorescence microscopy and electron microscopy with indirect immunogold labeling showed the RL13 antigens to be localized to the inner nuclear membrane, with the RL13 epitope facing the nucleoplasm, supporting the biochemical studies (Senior and Gerace, 1988). Interestingly, the expression of different LAP1 isoforms seems to be developmentally regulated. Although all three isoforms were seen on immunoblots of total cell homogenates from rat liver, spleen, brain, and kidney, only LAP1C was seen in extracts from a number of cultured rat cell lines (Senior and Gerace, 1988). Differential expression of different isotypes was also seen in a variety of mouse cell lines (Martin *et al.*, 1995).

2. Cloning of Lamina-Associated 1 Polypeptide Proteins

Despite the LAP1 proteins being among the first lamin-binding proteins identified, LAP1C cDNA from rat was not cloned until 1995 (Martin *et al.*, 1995) and human LAP1B in 2002 (Kondo *et al.*, 2002). Cloning of full-length LAP1A cDNA has not yet been reported. LAP1C cDNA was identified by the screening of rat liver cDNA expression libraries with antibody RL13. Analysis of the cDNA sequence confirmed the presence of a transmembrane spanning region and determined that the amino-terminal region contained the RL13 epitope, showing LAP1C to be a type II integral membrane protein (Martin *et al.*, 1995). Despite several attempts in the same study, no full-length clones of LAP1A or 1B were isolated. However, partial cDNAs identified suggested that these isoforms contain insertions after nucleotides 713 and 1059 in the LAP1C cDNA, suggesting that the three isoforms are the result of alternative RNA splicing (Martin *et al.*, 1995). Human LAP1B, which was isolated from an expression library from HeLa cells, shows 73.6% amino acid identity with the predicted sequence from rat (Kondo *et al.*, 2002).

Analysis of the subcellular localization of different LAP1B deletion mutants with GFP fused to the amino-termini showed that the transmembrane segment (amino acids 339–351) together with the second half of the amino-terminal, nucleoplasmic region (amino acids 183–339) were sufficient for correct targeting of the protein to the nuclear rim (Kondo *et al.*, 2002). The first 189 amino acids were sufficient for a nucleoplasmic localization, without any accumulation at the nuclear rim. The transmembrane segment together with the luminal carboxyl-terminus was localized throughout the ER. These data suggest retention of the proteins in the nucleus by interaction with other nuclear components, such as the lamina and/or chromatin, combined with a need for the transmembrane segment for an insertion in the nuclear membrane localization, and are similar to data from several other nuclear proteins. Constructs containing the complete nucleoplasmic region, with or without the transmembrane domain, were more resistant to extraction with 0.5% Triton X-100 than constructs containing parts of the nucleoplasmic region, suggesting that full activity of retention may require the entire length of the nucleoplasmic domain (Kondo *et al.*, 2002). The role of the luminal carboxyl-terminus is not clear. No sequence similarity has been identified between

LAP1 and other known proteins, and the function of the LAP1 isoforms is not understood. Like other lamin-binding proteins of the nuclear envelope, they may help attach lamins to the inner nuclear membrane in interphase and target them to reforming nuclear envelopes at the end of mitosis.

3. Identification of the Lamina–Associated Polypeptide 2 Proteins and Binding of the Lamina–Associated Polypeptide Proteins to Lamins

An additional protein of 53 kDa was identified with another monoclonal antibody (RL29) prepared against proteins of a pore complex–lamina fraction from rat liver and was named LAP2 (later renamed LAP2β). As with the LAP1 antigens, LAP2 was localized to the nuclear envelope as shown by immunofluorescence microscopy (Foisner and Gerace, 1993). It was also resistant to extraction with 8 M urea, as well as extraction with 0.1 M NaOH and 4 M guanidine-HCl, showing it to be an integral membrane protein (Foisner and Gerace, 1993). An interaction with the lamins was shown by resistance to extraction with 250 mM NaCl and 1% Triton X-100. To investigate the binding of LAPs 1 and 2 to the lamins in further detail, Foisner and Gerace (1993) incubated the LAP proteins extracted from lamin-depleted nuclear membranes with 1% Triton X-100 and 500 mM NaCl with purified and assembled lamins A, C, and B1, and with vimentin as a control. LAP1A and 1B bound to all three lamins, whereas LAP1C did not bind any lamin. LAP2 bound only to lamin B1. None of the proteins bound to vimentin. The same results were obtained with LAP proteins purified by immunoaffinity with immobilized monoclonal antibodies, suggesting that the interactions between the LAPs and the lamins are direct (Foisner and Gerace, 1993).

Simultaneously with the identification of LAP2, cDNAs for three proteins called thymopoietins (TP) were cloned and shown to be alternative spice forms of one human gene localized to chromosome 12q22. They were named TPα (75 kDa), TPβ (51 kDa), and TPγ (39 kDa) (Harris *et al.*, 1994, 1995). Thymopoietin had previously been identified as a 5-kDa protein in bovine thymus, but this molecule was shown to be a degradation product of these longer proteins (Harris *et al.*, 1995). Comparisons of the sequences from human TPβ and rat LAP2 showed them to be orthologes (Furukawa *et al.*, 1995; Harris *et al.*, 1995), and the TP proteins are now generally known as LAP2α, LAP2β, and LAP2γ. These three proteins share the first 187 amino acids. LAP2α then has a specific domain (amino acid 187–693). LAP2β and LAP2γ are more closely related, being identical except for 109 amino acids specific to LAP2β inserted at amino acid 220. Human LAP2β has a total of 453 amino acids and LAP2γ 344 (Harris *et al.*, 1994). The carboxyl-terminal region common to LAP2β and LAP2γ contains a transmembrane spanning region (amino acid 410–433 in LAP2β), whereas LAP2α is predicted to be a nonmembrane protein. In mouse, four additional LAP2 splice forms (β', ε, δ, and ζ) have been identified (Berger *et al.*, 1996). They are all closely related, shorter forms of LAP2β.

Immunofluorescence microscopy studies of HeLa cells transfected with hemagglutinin (HA)-tagged LAP2β deletion mutants showed that a region between amino acid 298 and 373 is important for nuclear envelope targeting, but that regions flanking this area are necessary for a complete Triton X-100–resistant integration at the nuclear rim (Furukawa et al., 1995). Binding experiments with the yeast two-hybrid system showed that the nuclear targeting region (amino acid 298–373) was responsible for binding to lamin B1 and lamin B2 (Furukawa et al., 1998). When this fragment was microinjected into metaphase HeLa cells, it did not affect the reassembly of the nucleus, but strongly inhibited nuclear volume increase in interphase. Injection into G1 cells also affected nuclear growth and inhibited entry into S phase (Yang et al., 1997). These data suggested that one function of LAP2β is to regulate lamina dynamics involved in nuclear growth.

4. Binding of Lamina–Associated Polypeptide Proteins to Chromatin

Experiments in which immunoaffinity purified LAP1A, LAP1B, and LAP2β were incubated with mitotic chromosomes showed a direct binding between LAP2β (but not LAP1A or 1B) and chromosomes (Foisner and Gerace, 1993). The interaction between LAP2β and chromosomes, as well as the interaction between LAP2β and lamin B1, was abolished by mitosis-specific phosphorylation of LAP2β. Several regions of LAP2β have been implicated in the interaction with chromatin. Amino acids 244–296 of LAP2β fused to GST bound to double-stranded and single-stranded DNA-cellulose in vitro (Furukawa et al., 1997), whereas the first 85 amino acids were sufficient for binding to mitotic chromosomes in digitonin-permeabilized rat kidney cells (Furukawa et al., 1998). A two-hybrid assay showed amino acids 67–137 of LAP2β to interact with barrier-to-autointegration factor (BAF) (Furukawa, 1999). BAF prevents autointegration of retroviral DNA (Chen and Engelman, 1998; Lee and Craigie, 1998). Its normal cellular function is not clear, but it binds double-stranded DNA nonspecifically, and thereby bridges DNA molecules to large nucleoprotein complexes. The binding between LAP2β and BAF was confirmed by in vitro studies in which the formation of protein complexes between purified recombinant LAP2 fragments (from the region common to all LAP2 proteins) and purified recombinant BAF were detected on native gels (Shumaker et al., 2001). Analysis with mutant LAP2 proteins confirmed that this binding occurs mainly, although maybe not completely, through the LEM-domain, which encompasses amino acids 103–159. The LEM-domain is present in several inner nuclear membrane proteins (Cohen et al., 2001; Lin et al., 2000; Schirmer et al., 2003). Structural analysis of this domain by nuclear magnetic resource has shown it to be mainly composed of two large, parallel α-helices and possibly involved in protein–protein interactions (Cai et al., 2001; Laguri et al., 2001; Wolff et al., 2001). A similar "LEM-like" domain is present in the LAP2 DNA–binding region between amino acids 1 and 50 (Cai et al., 2001; Laguri et al., 2001; Lin et al., 2000).

Immunofluorescence microscopy of NRK cells showed that LAP2β localizes to the surface of chromosomes during late anaphase and early telophase, when the lamins are still predominantely cytoplasmic, suggesting a role for LAP2β in the early steps of nuclear envelope reassembly after mitosis, and a possible role in the assembly of the lamins (Foisner and Gerace, 1993). The interaction between the LAP2 proteins and the chromatin-bound BAF may be important in this process, but the physiological role of the LEM-BAF binding, as well as possible other functions of the LEM-domain, remain to be conclusively investigated.

5. Other Interactions

LAP2β has also been shown to bind the mouse ortholog of the *Drosophila* germ–cell-less protein (mGCL) in a two-hybrid screen (Nili *et al.*, 2001). The binding was confirmed in *in vitro* binding assays with GST-mGCL and ^{35}S-labeled LAP2β (the shorter LAP2ζ was shown not to bind mGCL). mGCL interacts with the DP3α component of the E2F transcription factor, translocates this transcription factor to the nuclear envelope, and reduces the transcriptional activity of E2F-DP heterodimer (de la Luna *et al.*, 1999). When LAP2β was coexpressed with E2F5 or DP3α in H1299 cells, LAP2β was shown to reduce their transcriptional activity measured by a luciferase reporter gene under the control of a promoter with E2F binding sites (Nili *et al.*, 2001).

LAP2α shares the first chromatin-binding domain and the LEM-domain with other LAP2 proteins but lacks the transmembrane spanning regions and the region binding to B-type lamins. Immunofluorescence microscopy showed a partial colocalization between A-type lamins, specifically intranuclear lamin A/C structures, and LAP2α (Dechat *et al.*, 2000). Coimmunoprecipitation studies with anti-LAP2α antibodies showed that a fraction of lamin A/C, but not B-type lamins, bound to LAP2α in interphase cells. This binding was abolished in mitotic cells. Blot overlay studies in which ^{35}S-labeled lamins were incubated with transblotted recombinant LAP2α fragments confirmed that LAP2α binds to A-type lamins rather than B-type and showed that this interaction occurs between amino acids 319 and 566 of lamin A/C and the unique, terminal 78 amino acids of LAP2α (Dechat *et al.*, 2000).

D. Emerin

1. Identification, Cloning, and Characterization of Emerin

In 1994, Bione and colleagues identified emerin as the protein that, when mutated, caused X-linked EDMD (Bione *et al.*, 1994). This disease had previously been linked to distal chromosome Xq28 by positional cloning (Yates *et al.*, 1993). Sequencing of the cDNA of eight transcripts in this region with high levels of expression in brain and/or muscle showed that all subjects with EDMD had mutations in the *STA* gene, now called *EMD* (Bione *et al.*, 1994). Northern blot analysis of emerin mRNA showed a ubiquitous tissue distribution, with the highest expression in skeletal muscle and heart. The mRNA was shown to encode a protein of 254 amino acids, with a hydrophobic region forming a putative membrane anchoring 11 amino acids from its carboxyl-terminus. Two regions of homology with LAP2β were found, one close to the amino-terminus (the LEM-domain, see earlier) and one by the membrane-spanning domain. Although these regions of homology had been identified, it came as a surprise when two groups in 1996 showed emerin to be a nuclear envelope protein (Manilal *et al.*, 1996; Nagano *et al.*, 1996), because no muscular dystrophy had previously been linked to abnormalities in this subcellular compartment. Immunohistochemical studies with monoclonal antibodies generated against the first 188 amino acids of emerin and polyclonal antibodies generated against synthetic peptides showed that the protein is localized to the nuclear envelope. Emerin also localized to the nuclear fraction in biochemical fractionation of skeletal muscle, brain, and liver tissue. Inner nuclear membrane localization of emerin was later confirmed by immunogold electron microscopy (Cartegni *et al.*, 1997) and by fluorescence microscopy studies of cells selectively permeabilized with the detergent digitonin (Fig. 2; see also "Procedures"), which permeabilizes the plasma membrane but leaves the nuclear membranes intact (Adam *et al.*, 1990; Dabauvalle *et al.*, 1999; Östlund *et al.*, 1999). Emerin was absent from skeletal and cardiac muscle biopsy specimens from one patient with EDMD and later shown to be absent or severely diminished in most patients as determined by Western blotting (Manilal *et al.*, 1998; Nagano *et al.*, 1996). Because reverse transcriptase (RT) PCR showed that emerin mRNA levels were normal, ruling out mRNA instability, this suggested that truncated forms of the emerin protein are unstable.

2. Interactions between Emerin, A-Type Lamins, and Barrier-to-Autointegration Factor

A functionally important interaction between emerin and A-type lamins is suggested by the finding that AD-EDMD, phenotypically identical to the X-linked form of the disease, is caused by mutations in the *LMNA* gene, which encodes the A-type lamins (Bonne *et al.*, 1999). Several lines of experimental evidence support this interaction. In *in situ* matrix preparations in which cells were treated with detergents, DNAse, and 2M NaCl, emerin remained bound to the nuclear lamina (Squarzoni *et al.*, 1998). In cells of a knockout mouse lacking A-type lamins, emerin is partly mislocalized to the ER, whereas LAP2β localization is unaffected (Sullivan *et al.*, 1999). Emerin is also mislocalized in HeLa cells, where A-type lamins are depleted by RNAi and in cell lines expressing low levels of A-type lamins (Harborth *et al.*, 2001; Vaughan *et al.*, 2001). Expression in transfected cells of some lamin A and lamin C mutants that cause muscle disease caused a redistribution of emerin to the ER in some cells (Östlund *et al.*, 2001; Raharjo *et al.*, 2001). A direct interaction between A-type lamins and emerin has been shown by biomolecular interaction analysis with a BIAcore biosensor (Clements *et al.*, 2000) and by affinity column chromatography, immunoprecipitation, and the yeast

Triton X-100

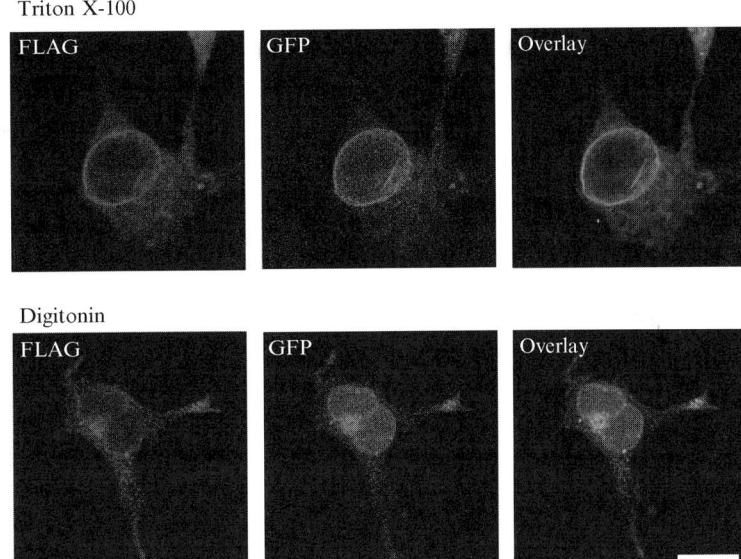

Fig. 2 Emerin localizes to the inner nuclear membrane. Immunofluorescence studies of COS-7 cells that were transfected with plasmids expressing emerin tagged with a FLAG-tag and fused to GFP. The cells were permeabilized with Triton X-100 (upper panels) or digitonin (lower panels) and were then labeled with anti-FLAG monoclonal antibodies recognized by rhodamine-conjugated secondary antibodies (left panels). The middle panels show fluorescence from GFP and the panels to the right an overlay of signals from the two channels. Although fluorescence from GFP is seen at the nuclear rim under both permeabilization conditions, the anti-FLAG antibodies do not label the nuclear rim in digitonin-permeabilized cells, indicating that the epitope is inside the nucleus. Bar, 10 μm. Reprinted from Östlund et al. (1999) with permission from The Company of Biologists, Limited. (See Color Insert.)

two-hybrid assay (Sakaki et al., 2001). In vitro studies have shown that the region between amino acids 70 and 178 of the amino-terminal region of emerin interacts with the tail region common to lamins A and C (amino acid 384–566). Lysates from bacteria expressing wild-type or mutant emerin, where amino acids conserved between human emerin and LAP2β were substituted by alanines, were resolved by sodium dodecylsulfate-polyacrylamide gel electrophoresis (SDS-PAGE) and transferred to membranes. The blots were then probed with ^{35}S-labeled lamin A, which bound to wild-type emerin and emerin with mutations outside the region between amino acid 70 and 178, but failed to bind proteins with mutations in this region (Lee et al., 2001). This lamin-binding region overlaps with a region important for the nuclear envelope targeting of emerin (Östlund et al., 1999; Tsuchiya et al., 1999) (see later). Experiments that used the yeast two-hybrid system showed that the region between amino acids 384 and 566 of lamin A is involved in this interaction with emerin (Sakaki et al., 2001).

Blot experiments in which emerin with conserved residues mutated to alanines was probed with ^{35}S-labeled BAF, as described previously for lamins, showed that emerin, like LAP2, binds to the DNA-bridging protein BAF by means of the LEM-domain. The significance of BAF binding to LEM-domain proteins is not well understood, but BAF is suggested to target emerin to the chromatin after mitosis, something corroborated by the fact that fluorescence microscopy showed that GFP-tagged emerin with mutations in the LEM-domain did not localize correctly in telophase (Lee *et al.*, 2001).

3. Targeting and Dynamics of Emerin

Several regions of the amino-terminal, nucleoplasmic domain of emerin preceding the transmembrane segment (amino acids 1–219) are involved in its inner nuclear membrane targeting and retention. Immunofluorescence miscroscopy of cells transfected with full-length and truncated forms of emerin showed that the region between amino acids 117 and 170 localizes to the nucleus but is not sufficient for targeting of an unrelated transmembrane segment from a nonnuclear protein to the nuclear envelope (Östlund *et al.*, 1999; Tsuchiya *et al.*, 1999). A portion of emerin containing amino acids 3–169 is sufficient for targeting of a transmembrane-spanning protein to the inner nuclear membrane (Cartegni *et al.*, 1997; Östlund *et al.*, 1999). Together, these experiments indicate a role for amino acids 3–116 in targeting of emerin to the inner nuclear membrane (Östlund *et al.*, 1999). Several regions of emerin may, however, play a role in the efficient targeting of the protein to the inner nuclear membrane, because a chimeric protein with GFP fused to amino acids 107–254, containing parts of the amino-terminal domain and the transmembrane and luminal regions, also targets to the nuclear rim (Tsuchiya *et al.*, 1999). FRAP experiments with cells transiently transfected with emerin fused to GFP showed that the diffusional mobility of emerin was decreased in the inner nuclear envelope compared with the ER (although to a lesser extent than shown with LBR), consistent with an interaction with nuclear proteins such as A-type lamins and BAF (Östlund *et al.*, 1999). FLIP experiments showed that there was very little back flow of protein from the inner nuclear membrane to the ER (Fig. 3).

4. The Function of Emerin

The function of emerin is not clear. Cells in which emerin has been depleted with RNAi exhibit no obvious abnormal phenotype (Harborth *et al.*, 2001). However, mutations leading to an absence or severe reduction of emerin lead to a muscular dystrophy and cardiomyopathy indistinguishable from that caused by some lamin A/C mutations. That a lack of emerin causes no obvious effect on the lamins, whereas a lack of A-type lamins causes a mislocalization of emerin (Harborth *et al.*, 2001; Sullivan *et al.*, 1999; Vaughan *et al.*, 2001) suggests that emerin may have important functions in muscle other than anchoring the lamina in the nuclear envelope.

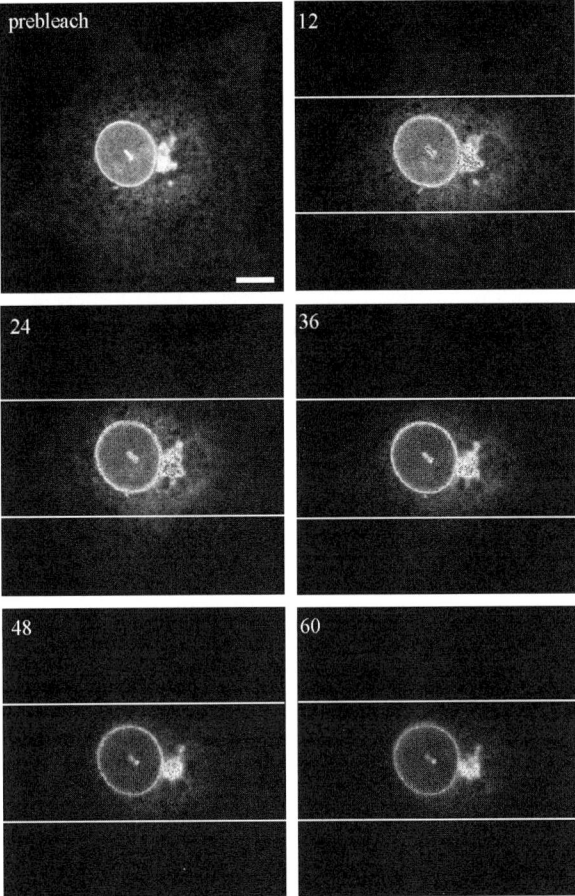

Fig. 3   FLIP experiment on COS-7 cells expressing emerin fused to GFP to study the continuity of nuclear envelope and ER membranes by repeated bleaching of parts of the cytoplasm. Areas above the upper line and below the lower line were bleached every 80 seconds. The figure shows confocal micrographs taken before the first bleach and after 12, 24, 36, 48, and 60 rounds of bleaching. Although the fluorescent signal in the ER disappears, the fluorescence in the nuclear envelope remains relatively unchanged. Bar, 5 μm. Reprinted from Östlund *et al.* (1999) with copyright permission from The Company of Biologists, Limited.

E. Other Lamin-Associated Proteins

Several other proteins may be directly or indirectly associated with lamins. The *Caenorhabditis elegans* protein UNC-84 colocalizes with Ce-lamin, the only lamin present in this organism, during interphase but is not detected at the nuclear envelope in embryos rendered lamin-deficient by RNAi (Lee *et al.*, 2002). Mutations in the *unc-84* gene cause defects in nuclear migration and positioning during *C. elegans* development. An orthologous protein has also been identified in a proteomic screen identifying nuclear envelope proteins in mouse cells (Dreger *et al.*, 2001). Immunofluorescence microscopy of tissue from worms with *unc-84* mutations showed that the nuclear envelope proteins UNC-83 and ANC-1, which are involved in nuclear migration and positioning, respectively, are mislocalized (Starr and Han, 2002; Starr *et al.*, 2001). ANC-1 contains 8546 amino acids and binds to actin. It has been suggested that ANC-1 is localized in the outer nuclear membrane and somehow is retained there through interaction with an unidentified binding partner, mainly present in the lumen between the inner and outer nuclear membrane, which in turn binds to UNC-84 (Starr and Han, 2003). ANC-1 could therefore tether the nuclei to the cytoplasmic actin cytoskeleton. Shorter forms of two mammalian homologs to ANC-1, nesprin-1α and nesprin-2α (also called syne-1 and 2 or myne 1 and 2), are also apparently localized to the inner nuclear membrane (Mislow *et al.*, 2002b; Zhang *et al.*, 2001). Blot overlay experiments showed nesprin-1α interacts with both emerin and lamin A (Mislow *et al.*, 2002a). Nesprin-1α, like emerin, is also mislocalized to the ER in fibroblasts from a human subject homozygous for a lamin A/C Y259X mutation, where A-type lamins were absent (Muchir *et al.*, 2003). Transfection of wild-type lamin A or C cDNA into these cells restored correct localization of emerin and nesprin-1α to the nuclear envelope, showing that A-type lamins are required for correct anchorage of these proteins.

Several proteins have been shown to interact with *Drosophila* lamin Dm. Otefin, a peripheral protein of the inner nuclear membrane, which contains a LEM-domain and a hydrophobic domain that may function as a membrane-anchor (Ashery-Padan *et al.*, 1997; Harel *et al.*, 1989), interacts with lamin Dm in the yeast two-hybrid system, and the two proteins can be coimmunoprecipitated from *Drosophila* embryo membrane vesicle extracts (Goldberg *et al.*, 1998). Lamin Dm also interacted with young arrest (YA), a protein essential for *Drosophila* development, in a two-hybrid screen (Goldberg *et al.*, 1998). Other proteins reported to bind to lamins include Narf, a protein identified in a two-hybrid screen with a HeLa cell cDNA library as specifically interacting with the prelamin A precursor form but not with mature lamin A (Barton and Worman, 1999), and the retinoblastoma (Rb) protein, a tumor suppressor and cell cycle regulator (Mancini *et al.*, 1994). Interestingly, the A-type lamins bound only the hypophosphorylated, active form of Rb in an *in vitro* binding assay, suggesting a role of the lamina in cell cycle progression and tumorigenesis. Lamin A has also been shown to bind sterol response element binding protein 1 (SREBP1) in a yeast two-hybrid screen, *in vitro* GST pull-down assays, and *in vivo* coimmunoprecipitation experiments (Lloyd *et al.*, 2002). The physiological significance of this interaction is not yet clear, but because SREBP1 is an adipocyte differentiation factor, it may be relevant for the role of lamins in familial partial lipodystrophy.

New lamin interaction partners are being identified regularly; however, the physiological significance of the interactions is generally not understood. A recent proteomics study that used multidimensional protein identification technology (MudPIT) identified 337 uncharacterized open-reading frames unique to the nuclear envelope fraction (Schirmer *et al.*, 2003). Most of these were resistant to treatment with salt and detergent, suggesting that they are associated with lamins. Although all of these proteins may not be true lamin-binding proteins, this study clearly shows that a lot remains to be learned about the nuclear lamina and its associated proteins.

II. Materials

A. Purification of Lamin Fraction from Rat Liver

Routine chemicals and protease inhibitors can be purchased from Sigma-Aldrich (St. Louis, MO): aprotinin (A1153), chymostatin (C7260), dithiothreitol (DTT; D0632), DNAse I (D4527), ethylenediaminetetraacetic acid (EDTA; E9884), KCl (P5405), leupeptin (L2023), MgCl$_2$ (M8266), pepstatin A (P5318), phenylmethyl-sulfonyl fluoride (PMSF; P7626), sucrose (S8501), triethanolamine (TEA; T1502), Tris(hydroxymethyl)aminomethane hydrochloride (Tris-HCl; T3253), urea (U5128). Sprague-Dawley rats can be purchased from Charles River laboratories (Wilmington, MA). Alternatively, frozen rat livers can be purchased from Pel-Freeze Biological (Rogers, AR). SW28 centrifuge tubes can be purchased from Beckman Coulter (Fullerton, CA).

B. Immunofluorescence Microscopy Studies of Lamin-Associated Proteins

The following routine chemicals and protease inhibitors can be purchased from Sigma-Aldrich: bovine serum albumin (BSA; A6003), digitonin (D5628), methanol (M1775), paraformaldehyde (P6148), Triton X-100 (T9284), Tween-20 (P7949). SlowFade Light Antifade Kit (S-7461) was purchased from Molecular Probes (Eugene, OR); Lab-Tek Chamber slides from Nalge Nunc International (Naperville, IL). Antibodies that recognize some of the lamin-associated proteins are available from commercial sources (Table I).

C. Photobleaching Studies of Integral, Lamin-Associated Proteins

Enhanced green fluorescent protein (EGFP) and related vectors in the living color series are available from BD Biosciences Clontech (Palo Alto, CA). Lab-Tek Chambered Coverglass Systems are from Nalge Nunc International.

III. Procedures

A. Purification of Lamin Fraction from Rat Liver

Livers from newly killed young Sprague-Dawley rats may yield better results, but if one wants to avoid handling and killing rats, frozen rat liver can be purchased from Pel-freeze biological. It is crucial to keep liver tissue cold by

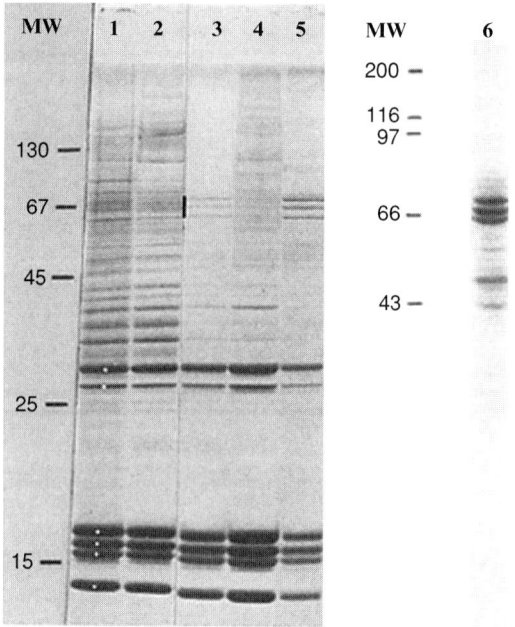

Fig. 4 Proteins separated by SDS-polyacrylamide gel electrophoresis and stained by Coomassie Blue from purified rat liver nuclei (*lane 1*, procedure 1, step 17), supernatant (*lane 2*), and pellet (*lane 3*) from nuclei treated with DNAse I at pH 8.5 (procedure 2, step 6), supernatant (*lane 4*), and pellet (nuclear envelopes; *lane 5*) from nuclei treated with DNAse I at pH 7.5 (procedure 2, step 10). Lane 6 shows 8 M urea extracts of salt-washed nuclear envelopes (procedure 3, step 4). The vertical bar indicates lamins A, B1, and C. White dots indicate histone bands. Lanes 1 to 5 taken from Dwyer and Blobel (1976). (Reproduced from *The Journal of Cell Biology*, 1976, vol. 70, 581–591 by copyright permission from The Rockefeller University Press). Lane 6 taken from Worman *et al.* (1988a). (Reproduced from *The Journal Biological Chemistry*, 1988, vol. 263, 12135-12141 by copyright permission from The American Society for Biochemistry & Molecular Biology).

working in a cold room. The protocol described in the following, modified from several sources (Aaronson and Blobel, 1975; Blobel and Potter, 1966; Dwyer and Blobel, 1976) is one of many. For another description of purification of nuclei and nuclear envelopes from rat liver, as well as preparation of nuclei from COS cells, see Kihlmark and Hallberg (1998). For examples of nuclear proteins and nuclear subfractions separated by SDS-PAGE, see Fig. 4.

1. Purification of Rat Liver Nuclei

a. Solutions and biological material

1. Solution 1: 0.25 M sucrose, 50 mM TEA-HCl, pH 7.5, 25 mM KCl, 5 mM MgCl$_2$. Make 1.5 L.

2. Solution 2: 2.3 M sucrose, 50 mM TEA-HCl, pH 7.5, 25 mM KCl, 5 mM MgCl$_2$. Make 0.7 L.

3. 1 M DTT, 0.1 M PMSF (optional: 1 mg/ml aprotinin, 10 mg/ml leupeptin, 10 mg/ml pepstatin, 10 mg/ml chymostatin). Add 1 mM DTT and 0.1 mM PMSF to solutions 1 and 2 in all steps; in steps 1 and 3, add 0.5 mM PMSF and protease inhibitors as indicated.

4. Biological material: Sprague-Dawley rats (male or female), range 48×120 g to 24×360 g or approximately 24 frozen livers.

b. Steps

1. Kill overnight-fasted rats by methods established by the American Veterinary Medical Association Panel on Euthanasia, extract livers, and place in ice-cold solution 1 with 0.5 mM PMSF and 10 μg/ml aprotinin, or let frozen livers thaw in this solution.

2. Rinse livers and mince over ice with razors.

3. Homogenize 4–8 livers at a time in a 300-ml Potter-Elvehjem homogenizer at 100 motor setting (15 strokes) in solution 1 (about 100 ml to fill vessel) with 0.5 mM PMSF and 1 μg/ml each of leupeptin, pepstatin, and chymostatin.

4. Pour homogenate over cheesecloth (two double layers) attached with rubber bands over large beaker, move with spoon to filter through.

5. Repeat step 4.
Repeat steps 3–5 until sufficient amounts of homogenate are ready.

6. Pour filtrate into centrifuge tubes for low-speed spin, dilute with solution 1 to about 400 ml.

7. Centrifuge at 1200g for 10 minutes at 4 °C.

8. Remove tubes carefully from centrifuge, decant supernatant slowly retaining the soft pellet; with spatula transfer the pellet to clean Potter-Elvehjem homogenizer without adding any buffer.

9. Homogenize low-speed pellet in Potter-Elvehjem homogenizer (five strokes) and pour in large cylinder.

10. Measure volume of pellets and add exactly 2 volumes of solution 2; cover cylinder tightly with parafilm and mix thoroughly.

11. Dispense this mix into SW28 tubes allowing for cushion; underlay 5 ml of solution 2 as cushion (mix slightly the interface to create a gradual rather than abrupt increase in density).

12. Centrifuge 1 hour at 27,000 rpm or 1 hour 15 minutes at 24,000 rpm in a SW28 rotor at 4 °C.

13. Take tubes out of centrifuge/rotor, break semisolid upper layer, and invert quickly to decant supernatant; leave inverted over paper towels for a few minutes. Wipe clean inside walls of tubes with paper without touching pellets.

14. Take out nuclear pellets with spatula and place into a 50-ml Dounce homogenizer with less than half its capacity of solution 1; wash tubes and spatula; homogenize manually with care and adjust to 100 ml in cylinder.

15. Measure optical density at 260 nm of a 1:100 dilution in solution 1; dispense adequate amounts into tubes.

16. Centrifuge at 1200g for 10 minutes at 4 °C.

17. Decant supernatant; freeze pellets in liquid N_2 (vortex first), and store at −70 °C.

2. Preparation of Nuclear Envelopes from Rat Liver Nuclei

a. Steps

The following volumes are for 500 OD at 260-nm nuclei from section 1.

1. Vortex nuclear pellet.

2. Add 5 ml of 0.1 mM $MgCl_2$, 1 mM DTT, and 0.5 mM PMSF while vortexing.

3. Add 20 ml 10% sucrose, 20 mM TEA-HCL, pH 8.5, 0.1 mM $MgCl_2$, 1 mM DTT, and 0.5 mM PMSF.

4. Add 25 μl of 2 mg/ml DNAse I made fresh. Invert to mix, without making foam, and incubate at room temperature on a rocker for 15 minutes.

5. Underlay with 30% sucrose, 20 mM TEA-HCl, pH 8.5, 0.1 mM $MgCl_2$, 1 mM DTT, and 0.5 mM PMSF.

6. Centrifuge in a swing-out rotor at 15,000g for 20 minutes at 4 °C.

7. Remove supernatant and vortex pellet well.

8. Add 5 ml of 10% sucrose, 20 mM TEA-HCl, pH 7.5, 0.1 mM $MgCl_2$, 1 mM DTT, and 0.5 mM PMSF while vortexing. Resuspend with a Pasteur pipette if necessary.

9. Add 25 μl of 2 mg/ml DNAse I. Mix and incubate as in step 4.

10. Underlay and spin as in steps 5 and 6, except that TEA-HCL should be at pH 7.5.

11. Remove supernatants.

12. The pellet is the nuclear envelopes.

3. Purification of Lamina and Lamin–Associated Proteins from Nuclear Envelopes

a. Steps

1. Resuspend pellet from section 2 in 1 M KCl, 20 mM Tris-HCl, pH 7.6, 1 mM DTT, and 0.5 mM PMSF.

2. Centrifuge at 14,000 rpm for 20 minutes in an Eppendorf centrifuge (supernatant is KCl-wash).

3. Resuspend pellet (lamin-enriched fraction) in 600 μl 8 M urea, 1 mM EDTA, 40 mM Tris-HCl, pH 8.0, and 0.2 mM PMSF.

4. Centrifuge at 365,000g for 70 minutes.

5. The supernatant is a lamin-rich urea extract; the pellet is relatively enriched in nuclear envelope membranes and membrane proteins (e.g., LBR).

B. Immunofluorescence Microscopy Studies of Lamin-Associated Proteins

For antibodies to have access to proteins in the inner nuclear envelope, a detergent that permeabilizes the nuclear membranes such as Triton X-100 must be used. To be able to distinguish proteins in the outer and inner nuclear membranes, digitonin, which selectively permeabilizes the plasma membrane but leaves the nuclear membrane intact at appropriate concentration and under certain conditions, can be used (Adam *et al.*, 1990). Antibodies will then be unable to reach epitopes inside the nucleus. For transfected cells, it is crucial to prove that a studied cell expresses a nuclear protein and that lack of staining is really due to inaccessibility of the epitope, not lack of protein expression. This can be accomplished by use of chimeric proteins tagged with GFP. Nuclear GFP fluorescence will be detectable even in digitonin permeabilized cells (see also Fig. 2).

1. Studies of Cells Permeabilized with Triton X-100

a. Solutions

1. Solution A: 0.1% Tween-20 in PBS.
2. Solution B: 0.1% Tween-20 and 2% BSA in PBS.

b. Steps

1. Grow cells in LAB-TEK glass chamberslides until desired cell density is reached.
2. Wash cells three times with PBS.
3. Fix cell with ice-cold methanol for 6 minutes at $-20\,°C$.
4. Permeabilize cells with 0.5% Triton X-100 in PBS for 2 minutes at room temperature.
5. Wash three times 3 minutes with solution A.
6. Incubate with primary antibodies in solution B for 1 hour.
7. Wash four times 3 minutes with solution A.
8. Incubate with secondary antibodies in solution B for 1 hour.
9. Wash four times 3 minutes with solution A.
10. Wash three times 5 minutes with PBS.
11. Remove the chambers from slides and dip in methanol.
12. Let slides air dry.
13. Mount slides with mounting media, (e.g., SlowFade and seal with clear nail polish).

2. Studies of Digitonin-Permeabilized Cells

a. Steps:

1. Grow cells in glass chamberslides until desired cell density is reached.
2. Wash cells three times with PBS.

3. Fix cells with 2% paraformaldehyde in PBS for 30 minutes on ice.

4. Wash cells three times with PBS.

5. Incubate with precooled 40 μg/ml digitonin in PBS for 10 minutes on ice.

6. Washes and incubations are as in steps 5–13 in the protocol for Triton X-100 permeabilized cells, except that Tween-20 is excluded from the buffers, and all steps are performed on ice. Be very careful, because the cells are less well attached to the slide in this protocol.

C. Photobleaching Studies of Integral, Lamin-Associated Proteins

EGFP is the fluorescent protein most commonly used in photobleaching studies, but recently yellow and blue variants of EGFP are also available from BD Biosciences Clontech's living color series. Stably or transiently transfected cells are grown in Lab-Tek Chambered Coverglass System until the desired confluency is reached. It is preferable to use a growth media without phenol red shortly before and during the experiments, because this compound may cause background fluorescence. Both FRAP and FLIP have been successfully used to study lamin-associated proteins (Ellenberg et al., 1997; Östlund et al., 1999; Rolls et al., 1999; Wu et al., 2002). As these techniques have become increasingly popular during the last years, new microscope systems (e.g., Zeiss LSM 510 from Carl Zeiss, Inc.) are ready to use for these types of experiments, and the detailed design of an experiment will depend on the type of microscope available. It is optimal if the experiments can be performed on a 37 °C stage. A lens providing enough depth to bleach through the z-axis of the cell should be used (e.g., 40 × 0.9 NA).

Choose a representative cell and define bleach and scan regions. Collect one image prebleach, then bleach at 100% laser power with the 488-nm line for EGFP. The number of bleach iterations will have to be optimized for the protein being studied. Immediately afterwards, collect images to monitor the recovery in the bleached area with attenuated laser emission until the recovery reaches a plateau. The frequency with which the recovery (1–10 seconds) is monitored will depend on the mobility of the protein.

The total cellular fluorescence is determined for each image and compared with the initial total fluorescence to determine the amount of signal lost during bleach and imaging. The fluorescence signal measured in the region of interest normalized to the change in total fluorescence is determined as $I_{rel} = (T_0 I_t)/(T_t I_0)$, where T_0 is total cellular intensity during prebleach, T_t is the total cellular intensity at time-point t, I_0 is the average intensity in the region of interest during prebleach, and I_t is the average intensity in the region of interest at timepoint t (Phair and Misteli, 2000). The diffusion constant D can be determined by fitting the experimental data to an empirical formula for one-dimensional diffusion (Ellenberg et al., 1997) and verified by a numerical computer simulation of free diffusion on qualitative and quantitative data sets (Sciaky et al., 1997). However, for many purposes, $t_{1/2}$, the time when the recovery has reached a 50% value, is a sufficient measure of protein mobility. (For data analysis methods, see also The National Resource for Cell Analysis and Modeling website: www.nrcam.uchc.edu).

FLIP (Cole *et al.*, 1996) uses cycles of photobleaching of chosen areas at full laser illumination and a single whole-cell image with attenuated emission to monitor loss of fluorescence outside the photobleached area (see Fig. 3). This routine is repeated (in the case of emerin every 80 seconds for 80 minutes; Östlund *et al.*, 1999) until fluorescence is completely lost from the bleached region. The loss from other areas of the cell is a measure of the mobility of the protein between the different areas of the cell.

IV. Pearls and Pitfalls

A pearl of advice: when dealing with protein–protein interactions involving nuclear lamins, be extremely careful. It is crucial to verify that the interaction between a lamin and another protein is specific. Because different members of the intermediate filament family show regions of high homology, care has to be taken to verify that an interaction with a lamin does not occur with any intermediate filament protein. Cytoplasmic intermediate filament proteins should be used as negative controls. Domain-specific interactions between portions of the lamin not shared with other intermediate filament proteins could also address this specificity issue.

Like all other intermediate filament proteins, lamins polymerize and are difficult to study because of their insolubility in aqueous solutions. Biochemical experiments, such as coimmunoprecipitations or GST pull-down assays, have to be well controlled, because the lamins precipitate by themselves. Studies that use domains of lamins that are more soluble in aqueous solutions may be of value in performing GST pull-down assays. As with all protein–protein interactions, pull-down assays should be carried out under conditions of increasing stringency, such as increasing ionic strength or detergent concentrations, to obtain some idea about affinity. Affinities may also be established by biomolecular interaction analysis with a BIAcore biosensor, as done by Clements *et al.* (2000) to examine the interaction between lamin A and emerin.

To demonstrate *in vivo* significance, any putative lamin-associated protein identified with *in vitro* binding assays should be studied with the immunofluorescence microscopic and photobleaching methods described in this chapter. *In vivo* colocalization is critical to confirm an interaction in cells. Ideally, localization of the putative binding partner is affected by the expression of a mutant lamin or lack of lamins, as has been shown for emerin and A-type lamins (Östlund *et al.*, 2001; Raharjo *et al.*, 2001; Sullivan *et al.*, 1999). Similarly, dependence of the diffusional mobility or sequestration of a putative lamin-associated protein on expression of the lamin *in vivo* can be demonstrated with FRAP and FLIP. Other methods, such a fluorescence resonance energy transfer, could also potentially help show an interaction *in vivo*.

Remember: when dealing with protein–protein interactions involving nuclear lamins, be extremely careful. Strict attention to details and controls will likely lead to rewarding results. Carelessness could lead to the identification of false-positive interactions or low-affinity *in vitro* ones with no cell biological significance.

V. Discussion and Concluding Remarks

Although many proteins have been shown to interact with the lamins, the *in vivo* functions of these interactions are often not clear. Working with genetically easily modified organisms, such as *Drosophila* and *C. elegans*, may help determine some of the functions of the lamins and their associated proteins, as will studies on genetically modified mice. The identification of diseases caused by mutations in these proteins may also help explain their functions, because the disease phenotype can be informative. With the number of nuclear envelope proteins increasing rapidly because of progress in the field of proteomics (Schirmer *et al.*, 2003), methods such as those described here will be important to verify lamin association of newly identified proteins. It will also be crucial to develop reliable methods to study the role of these proteins in the structural integrity of the nucleus, as well as in transcriptional regulation.

References

Adam, S. A., Sterne-Marr, R., and Gerace, L. (1990). Nuclear protein import in permeabilized mammalian cells requires soluble cytoplasmic factors. *J. Cell Biol.* **111,** 807–816.

Aebi, U., Cohn, J., Buhle, L., and Gerace, L. (1986). The nuclear lamina is a meshwork of intermediate-type filaments. *Nature* **323,** 560–564.

Akey, C. W., and Radermacher, M. (1993). Architecture of the *Xenopus* nuclear pore complex revealed by three-dimensional cryo-electron microscopy. *J. Cell Biol.* **122,** 1–19.

Aaronson, R. P., and Blobel, G. (1975). Isolation of nuclear pore complexes in association with a lamina. *Proc. Natl. Acad. Sci. USA* **72,** 1007–1011.

Ashery-Padan, R., Weiss, A. M., Feinstein, N., and Gruenbaum, Y. (1997). Distinct regions specify the targeting of otefin to the nucleoplasmic side of the nuclear envelope. *J. Biol. Chem.* **272,** 2493–2499.

Bailer, S. M., Eppenberger, H. M., Griffiths, G., and Nigg, E. A. (1991). Characterization of a 54-kD protein of the inner nuclear membrane: Evidence for cell cycle-dependent interaction with the nuclear lamina. *J. Cell Biol.* **114,** 389–400.

Barton, R. M., and Worman, H. J. (1999). Prenylated prelamin A interacts with Narf, a novel nuclear protein. *J. Biol. Chem.* **274,** 30008–30018.

Berger, R., Theodor, L., Shoham, J., Gokkel, E., Brok-Simoni, F., Avraham, K. B., Copeland, N. G., Jenkins, N. A., Rechavi, G., and Simon, A. J. (1996). The characterization and localization of the mouse thymopoietin/lamina-associated polypeptide 2 gene and its alternatively spliced products. *Genome Res.* **6,** 361–370.

Biamonti, G., Giacca, M., Perini, G., Contreas, G., Zentilin, L., Weighardt, F., Guerra, M., Della Valle, G., Saccone, S., Riva, S., and Falaschi, A. (1992). The gene for a novel human lamin maps at a highly transcribed locus of chromosome 19 which replicates at the onset of S-phase. *Mol. Cell. Biol.* **12,** 3499–3506.

Bione, S., Maestrini, E., Rivella, S., Mancini, M., Regis, S., Romeo, G., and Toniolo, D. (1994). Identification of a novel X-linked gene responsible for Emery-Dreifuss muscular dystrophy. *Nat. Genet.* **8**, 323–327.

Blobel, G., and Potter, V. R. (1966). Nuclei from rat liver: Isolation method that combines purity with high yield. *Science* **154**, 1662–1665.

Bonne, G., Raffaele di Barletta, M., Varnous, S., Bécane, H.-M., Hammouda, E.-H., Merlini, L., Muntoni, F., Greenberg, C. R., Gary, F., Urtizberea, J.-A., Duboc, D., Fardeau, M., Toniolo, D., and Schwartz, K. (1999). Mutations in the gene encoding lamin A/C cause autosomal dominant Emery-Dreifuss muscular dystrophy. *Nat. Genet.* **21**, 285–288.

Buendia, B., and Courvalin, J.-C. (1997). Domain-specific disassembly and reassembly of nuclear membranes during mitosis. *Exp. Cell Res.* **230**, 133–144.

Burke, B., and Stewart, C. L. (2002). Life at the edge: The nuclear envelope and human disease. *Nat. Rev. Mol. Cell Biol.* **3**, 575–585.

Cai, M., Huang, Y., Ghirlando, R., Wilson, K. L., Craigie, R., and Clore, G. M. (2001). Solution structure of the constant region of nuclear envelope protein LAP2 reveals two LEM-domain structures: One binds BAF and the other binds DNA. *EMBO J.* **20**, 4399–4407.

Cartegni, L., Raffaele di Barletta, M., Barresi, R., Squarzoni, S., Sabatelli, P., Maraldi, N., Mora, M., Di Blasi, C., Cornelio, F., Merlini, L., Villa, A., Cobianchi, F., and Toniolo, D. (1997). Heart-specific localization of emerin: New insights into Emery-Dreifuss muscular dystrophy. *Hum. Mol. Genet.* **6**, 2257–2264.

Chaudhary, N., and Courvalin, J.-C. (1993). Stepwise reassembly of the nuclear envelope at the end of mitosis. *J. Cell Biol.* **122**, 295–306.

Chen, H., and Engelman, A. (1998). The barrier-to-autointegration protein is a host factor for HIV type 1 integration. *Proc. Natl. Acad. Sci. USA* **95**, 15270–15274.

Clements, L., Manilal, S., Love, D. R., and Morris, G. E. (2000). Direct interaction between emerin and lamin A. *Biochem. Biophys. Res. Commun.* **267**, 709–714.

Cohen, M., Lee, K. K., Wilson, K. L., and Gruenbaum, Y. (2001). Transcriptional repression, apoptosis, human disease and the functional evolution of the nuclear lamina. *Trends Biochem. Sci.* **26**, 41–47.

Cole, N. B., Smith, C. L., Sciaky, N., Terasaki, M., Edidin, M., and Lippincott-Schwartz, J. (1996). Diffusional mobility of Golgi proteins in membranes of living cells. *Science* **273**, 797–801.

Collas, P., Courvalin, J.-C., and Poccia, D. (1996). Targeting of membranes to sea urchin sperm chromatin is mediated by a lamin B receptor-like integral membrane protein. *J. Cell Biol.* **135**, 1715–1725.

Dabauvalle, M.-C., Müller, E., Ewald, A., Kress, W., Krohne, G., and Müller, C. R. (1999). Distribution of emerin during the cell cycle. *Eur. J. Cell Biol.* **78**, 749–756.

Dechat, T., Korbei, B., Vaughan, O. A., Vlcek, S., Hutchison, C. J., and Foisner, R. (2000). Lamina-associated polypeptide 2α binds to intranuclear A-type lamins. *J. Cell Sci.* **113**, 3473–3484.

de la Luna, S., Allen, K. E., Mason, S. L., and La Thangue, N. B. (1999). Integration of a growth-suppressing BTB/POZ domain protein with the DP component of the E2F transcription factor. *EMBO J.* **18**, 212–228.

Dreger, M., Bengtsson, L., Schöneberg, T., Otto, H., and Hucho, F. (2001). Nuclear envelope proteomics: Novel integral membrane proteins of the inner nuclear membrane. *Proc. Natl. Acad. Sci. USA* **98**, 11943–11948.

Duband-Goulet, I., and Courvalin, J.-C. (2000). Inner nuclear membrane protein LBR preferentially interacts with DNA secondary structures and nucleosomal linker. *Biochemistry* **39**, 6483–6488.

Dwyer, N., and Blobel, G. (1976). A modified procedure for the isolation of a pore complex-lamina fraction from rat liver nuclei. *J. Cell Biol.* **70**, 581–591.

Ellenberg, J., Siggia, E. D., Moreira, J. E., Smith, C. L., Presley, J. F., Worman, H. J., and Lippincott-Schwartz, J. (1997). Nuclear membrane dynamics and reassembly in living cells: Targeting of an inner nuclear membrane protein in interphase and mitosis. *J. Cell Biol.* **138**, 1193–1206.

Fisher, D. Z., Chaudhary, N., and Blobel, G. (1986). cDNA sequencing of nuclear lamins A and C reveals primary and secondary structural homology to intermediate filament proteins. *Proc. Natl. Acad. Sci. USA* **83**, 6450–6454.

Fitzky, B. U., Witsch-Baumgartner, M., Erdel, M., Lee, J. N., Paik, Y. K., Glossmann, H., Utermann, G., and Moebius, F. F. (1998). Mutations in the Δ7-sterol reductase gene in patients with Smith-Lemli-Opitz syndrome. *Proc. Natl. Acad. Sci. USA* **95**, 8181–8186.

Foisner, R., and Gerace, L. (1993). Integral membrane proteins of the nuclear envelope interact with lamins and chromosomes, and binding is modulated by mitotic phosphorylation. *Cell* **73**, 1267–1279.

Furukawa, K. (1999). LAP2 binding protein 1 (L2BP1/BAF) is a candidate mediator of LAP2-chromatin interaction. *J. Cell Sci.* **112**, 2485–2492.

Furukawa, K., Panté, N., Aebi, U., and Gerace, L. (1995). Cloning of a cDNA for lamina-associated polypeptide 2 (LAP2) and identification of regions that specify targeting to the nuclear envelope. *EMBO J.* **14**, 1626–1636.

Furukawa, K., Glass, C., and Kondo, T. (1997). Characterization of the chromatin binding activity of lamina-associated polypeptide (LAP) 2. *Biochem. Biophys. Res. Commun.* **238**, 240–246.

Furukawa, K., Fritze, C. E., and Gerace, L. (1998). The major nuclear envelope domain of LAP2 coincides with its lamin binding region but is distinct from its chromatin interaction domain. *J. Biol. Chem.* **273**, 4213–4219.

Goldberg, M., Lu, H., Stuurman, N., Ashery-Padan, R., Weiss, A. M., Yu, J., Bhattacharyya, D., Fisher, P. A., Gruenbaum, Y., and Wolfner, M. F. (1998). Interaction among Drosophila nuclear envelope proteins lamin, otefin, and YA. *Mol. Cell Biol.* **18**, 4315–4323.

Goldman, R. D., Gruenbaum, Y., Moir, R. D., Shumaker, D. K., and Spann, T. P. (2002). Nuclear lamins: Building blocks of nuclear architecture. *Genes Dev.* **16**, 533–547.

Harborth, J., Elbashir, S. M., Bechert, K., Tuschl, T., and Weber, K. (2001). Identification of essential genes in cultured mammalian cells using small interfering RNAs. *J. Cell Sci.* **114**, 4557–4565.

Harel, A., Zlotkin, E., Nainudel-Epszteyn, S., Feinstein, N., Fisher, P. A., and Gruenbaum, Y. (1989). Persistence of major nuclear envelope antigens in an envelope-like structure during mitosis in *Drosophila melanogaster* embryos. *J. Cell Sci.* **94**, 463–470.

Harris, C. A., Andryuk, P. J., Cline, S., Chan, H. K., Natarajan, A., Siekierka, J. J., and Goldstein, G. (1994). Three distinct human thymopoietins are derived from alternatively spliced mRNAs. *Proc. Natl. Acad. Sci. USA* **91**, 6283–6287.

Harris, C. A., Andryuk, P. J., Cline, S. W., Mathew, S., Siekierka, J. J., and Goldstein, G. (1995). Structure and mapping of the human thymopoietin (TMPO) gene and relationship of human TMPO β to rat lamin-associated polypeptide 2. *Genomics* **28**, 198–205.

Hinshaw, J. E., Carragher, B. O., and Milligan, R. A. (1992). Architecture and design of the nuclear pore complex. *Cell* **69**, 1133–1141.

Hoffmann, K., Dreger, C. K., Olins, A. L., Olins, D. E., Shultz, L. D., Lucke, B., Karl, H., Kaps, R., Müller, D., Vayá, A., Aznar, J., Ware, R. E., Sotelo Cruz, N., Lindner, T. H., Herrmann, H., Reis, A., and Sperling, K. (2002). Mutations in the gene encoding the lamin B receptor produce an altered nuclear morphology in granulocytes (Pelger-Hüet anomaly). *Nat. Genet.* **31**, 410–414.

Höger, T. H., Zatloukal, K., Waizenegger, I., and Krohne, G. (1990). Characterization of a second highly conserved B-type lamin present in cells previously thought to contain only a single B-type lamin. *Chromosoma* **99**, 379–390.

Holaska, J. M., Wilson, K. L., and Mansharamani, M. (2002). The nuclear envelope, lamins and nuclear assembly. *Curr. Opin. Cell Biol.* **14**, 357–364.

Holmer, L., Pezhman, A., and Worman, H. J. (1998). The human LBR/sterol reductase multigene family. *Genomics* **54**, 469–476.

Kihlmark, M., and Hallberg, E. (1998). Preparation of nuclei and nuclear envelopes. *In* "Cell biology: A laboratory handbook" (J. E. Celis, ed.), Vol. 2, pp. 152–158. Academic Press, San Diego.

Kondo, Y., Kondoh, J., Hayashi, D., Ban, T., Takagi, M., Kamei, Y., Tsuji, L., Kim, J., and Yoneda, Y. (2002). Molecular cloning of one isotype of human lamina-associated polypeptide 1s and a topological analysis using its deletion mutants. *Biochem. Biophys. Res. Commun.* **294**, 770–778.

Laguri, C., Gilquin, B., Wolff, N., Romi-Lebrun, R., Courchay, K., Callebaut, I., Worman, H. J., and Zinn-Justin, S. (2001). Structural characterization of the LEM motif common to three human inner nuclear membrane proteins. *Structure* **9**, 503–511.

Lee, K. K., Haraguchi, T., Lee, R. S., Koujin, T., Hiraoka, Y., and Wilson, K. L. (2001). Distinct functional domains in emerin bind to lamin A and DNA-bridging protein BAF. *J. Cell Sci.* **114**, 4567–4573.

Lee, K. K., Starr, D., Cohen, M., Liu, J., Han, M., Wilson, K. L., and Gruenbaum, Y. (2002). Lamin-dependent localization of UNC-84, a protein required for nuclear migration in *Caenorhabditis elegans*. *Mol. Biol. Cell* **13**, 892–901.

Lee, M. S., and Craige, R. (1998). A previously unidentified host protein protects retroviral DNA from autointegration. *Proc. Natl. Sci. USA* **95**, 1528–1533.

Lin, F., and Worman, H. J. (1993). Structural organization of the human gene encoding nuclear lamin A and nuclear lamin C. *J. Biol. Chem.* **268**, 16321–16326.

Lin, F., and Worman, H. J. (1995). Structural organization of the human gene (*LMNB1*) encoding nuclear lamin B1. *Genomics* **27**, 230–236.

Lin, F., Blake, D. L., Callebaut, I., Skerjanc, I. S., Holmer, L., McBurney, M. W., Paulin-Levasseur, M., and Worman, H. J. (2000). MAN1, an inner nuclear membrane protein that shares the LEM domain with lamina-associated polypeptide 2 and emerin. *J. Biol. Chem.* **275**, 4840–4847.

Lloyd, D. J., Trembath, R. C., and Shackleton, S. (2002). A novel interaction between lamin A and SREBP1: Implications for partial lipodystrophy and other laminopathies. *Hum. Mol. Genet.* **11**, 769–777.

Mancini, M. A., Shan, B., Nickerson, J. A., Penman, S., and Lee, W.-H. (1994). The retinoblastoma gene product is a cell cycle-dependent, nuclear matrix-associated protein. *Proc. Natl. Acad. Sci. USA* **91**, 418–422.

Manilal, S., Nguyen thi Man, Sewry, C. A., and Morris, G. E. (1996). The Emery-Dreifuss muscular dystrophy protein, emerin, is a nuclear membrane protein. *Hum. Mol. Genet.* **5**, 801–808.

Manilal, S., Recan, D., Sewry, C. A., Hoeltzenbein, M., Llense, S., Leturcq, F., Deburgrave, N., Barbot, J., Nguyen thi Man, Muntoni, F., Wehnert, M., Kaplan, J., and Morris, G. E. (1998). Mutations in Emery-Dreifuss muscular dystrophy and their effects on emerin protein expression. *Hum. Mol. Genet.* **7**, 855–864.

Martin, L., Crimaudo, C., and Gerace, L. (1995). cDNA cloning and characterization of lamina-associated polypeptide 1C (LAP1C), an integral protein of the inner nuclear membrane. *J. Biol. Chem.* **270**, 8822–8828.

McKeon, F. D., Kirschner, M. W., and Caput, D. (1986). Homologies in both primary and secondary structure between nuclear envelope and intermediate filament proteins. *Nature* **319**, 463–468.

Mislow, J. M. K., Holaska, J. M., Kim, M. S., Lee, K. K., Segura-Totten, M., Wilson, K. L., and McNally, E. M. (2002a). Nesprin-1α self-associates and binds directly to emerin and lamin A *in vitro*. *FEBS Lett.* **525**, 135–140.

Mislow, J. M. K., Kim, M. S., Davis, D. B., and McNally, E. M. (2002b). Myne-1, a spectrin repeat transmembrane protein of the myocyte inner nuclear membrane, interacts with lamin A/C. *J. Cell Sci.* **115**, 61–70.

Moebius, F. F., Fitzky, B. U., Lee, J. N., Paik, Y. K., and Glossmann, H. (1998). Molecular cloning and expression of the human Δ7-sterol reductase. *Proc. Natl. Acad. Sci. USA* **95**, 1899–1902.

Muchir, A., van Engelen, B. G., Lammens, M., Mislow, J. M., McNally, E., Schwartz, K., and Bonne, G. (2003). Nuclear envelope alterations in fibroblasts from LMGD1B patients carrying nonsense Y259X heterozygous or homozygous mutation in lamin A/C gene. *Exp. Cell Res.* **291**, 352–362.

Nagano, A., Koga, R., Ogawa, M., Kurano, Y., Kawada, J., Okada, R., Hayashi, Y. K., Tsukahara, T., and Arahata, K. (1996). Emerin deficiency at the nuclear membrane in patients with Emery-Dreifuss muscular dystrophy. *Nat. Genet.* **12**, 254–259.

Nakielny, S., and Dreyfuss, G. (1999). Transport of proteins and RNAs in and out of the nucleus. *Cell* **99**, 677–690.

Nili, E., Cojocaru, G. S., Kalma, Y., Ginsberg, D., Copeland, N. G., Gilbert, D. J., Jenkins, N. A., Berger, R., Shaklai, S., Amariglio, N., Brok-Simoni, F., Simon, A. J., and Rechavi, G. (2001). Nuclear membrane protein LAP2β mediates transcriptional repression alone and together with its binding partner GCL (germ-cell-less). *J. Cell Sci.* **114,** 3297–3307.

Östlund, C., and Worman, H. J. (2003). Nuclear envelope proteins and neuromuscular diseases. *Muscle Nerve* **27,** 393–406.

Östlund, C., Ellenberg, J., Hallberg, E., Lippincott-Schwartz, J., and Worman, H. J. (1999). Intracellular trafficking of emerin, the Emery-Dreifuss muscular dystrophy protein. *J. Cell Sci.* **112,** 1709–1719.

Östlund, C., Bonne, G., Schwartz, K., and Worman, H. J. (2001). Properties of lamin A mutants found in Emery-Dreifuss muscular dystrophy, cardiomyopathy and Dunnigan-type partial lipodystrophy. *J. Cell Sci.* **114,** 4435–4445.

Östlund, C., Wu, W., and Worman, H. J. (2002). Targeting and retention of proteins in the inner and pore membranes of the nuclear envelope. *In* "Nuclear envelope dynamics in embryos and somatic cells" (P. Collas, ed.), pp. 29–41. Landes Bioscience, Georgetown, TX.

Phair, R. D., and Misteli, T. (2000). High mobility of proteins in the mammalian cell nucleus. *Nature* **404,** 604–609.

Raharjo, W. H., Enarson, P., Sullivan, T., Stewart, C., and Burke, B. (2001). Nuclear envelope defects associated with *LMNA* mutations causing dilated cardiomyopathy and Emery-Dreifuss muscular dystrophy. *J. Cell Sci.* **114,** 4447–4457.

Rolls, M. M., Stein, P. A., Taylor, S. S., Ha, E., McKeon, F., and Rapoport, T. A. (1999). A visual screen of a GFP-fusion library identifies a new type of nuclear envelope membrane protein. *J. Cell Biol.* **146,** 29–43.

Sakaki, M., Koike, H., Takahashi, N., Sasagawa, N., Tomioka, S., Arahata, K., and Ishiura, S. (2001). Interaction between emerin and nuclear lamins. *J. Biochem.* **129,** 321–327.

Schirmer, E. C., Florens, L., Guan, T., Yates, J. R., 3rd, and Gerace, L. (2003). Nuclear membrane proteins with potential disease links found by subtractive proteomics. *Science* **301,** 1380–1382.

Schuler, E., Lin, F., and Worman, H. J. (1994). Characterization of the human gene encoding LBR, an integral protein of the nuclear envelope inner membrane. *J. Biol. Chem.* **269,** 11312–11317.

Sciaky, N., Presley, J., Smith, C., Zaal, K. J. M., Cole, N., Moreira, J. E., Terasaki, M., Siggia, E., and Lippincott-Schwartz, J. (1997). Golgi tubule traffic and the effects of brefeldin A visualized in living cells. *J. Cell Biol.* **139,** 1137–1155.

Senior, A., and Gerace, L. (1988). Integral membrane proteins specific to the inner nuclear membrane and associated with the nuclear lamina. *J. Cell Biol.* **107,** 2029–2036.

Shumaker, D. K., Lee, K. K., Tanhehco, Y. C., Craigie, R., and Wilson, K. L. (2001). LAP2 binds to BAF-DNA complexes: requirement for the LEM domain and modulation by variable regions. *EMBO J.* **20,** 1754–1764.

Silve, S., Dupuy, P. H., Ferrara, P., and Loison, G. (1998). Human lamin B receptor exhibits sterol C14-reductase activity in *Saccharomyces cerevisiae. Biochim. Biophys. Acta.* **1392,** 233–244.

Smith, S., and Blobel, G. (1993). The first membrane spanning region of the lamin B receptor is sufficient for sorting to the inner nuclear membrane. *J. Cell Biol.* **120,** 631–637.

Soullam, B., and Worman, H. J. (1993). The amino-terminal domain of the lamin B receptor is a nuclear envelope targeting signal. *J. Cell Biol.* **120,** 1093–1100.

Soullam, B., and Worman, H. J. (1995). Signals and structural features involved in integral membrane protein targeting to the inner nuclear membrane. *J. Cell Biol.* **130,** 15–27.

Squarzoni, S., Sabatelli, P., Ognibene, A., Toniolo, D., Cartegni, L., Cobianchi, F., Petrini, S., Merlini, L., and Maraldi, N. M. (1998). Immunocytochemical detection of emerin within the nuclear matrix. *Neuromuscul. Disord.* **8,** 338–344.

Starr, D. A., and Han, M. (2002). Role of ANC-1 in tethering nuclei to the actin cytoskeleton. *Science* **298,** 406–409.

Starr, D. A., and Han, M. (2003). ANChors away: An actin based mechanism of nuclear positioning. *J. Cell Sci.* **116,** 211–216.

Starr, D. A., Hermann, G. J., Malone, C. J., Fixsen, W., Priess, J. R., Horvitz, H. R., and Han, M. (2001). *unc-83* encodes a novel component of the nuclear envelope and is essential for proper nuclear migration. *Development* **128,** 5039–5050.

Stuurman, N., Heins, S., and Aebi, U. (1998). Nuclear lamins: Their structure, assembly, and interactions. *J. Struct. Biol.* **122,** 42–66.

Sullivan, T., Escalante-Alcalde, D., Bhatt, H., Anver, M., Bhat, N., Nagashima, K., Stewart, C. L., and Burke, B. (1999). Loss of A-type lamin expression compromises nuclear envelope integrity leading to muscular dystrophy. *J. Cell Biol.* **147,** 913–919.

Tsuchiya, Y., Hase, A., Ogawa, M., Yorifuji, H., and Arahata, K. (1999). Distinct regions specify the nuclear membrane targeting of emerin, the responsible protein for Emery-Dreifuss muscular dystrophy. *Eur. J. Biochem.* **259,** 859–865.

Vaughan, O. A., Alvarez-Reyes, M., Bridger, J. M., Broers, J. L. V., Ramaekers, F. C. S., Wehnert, M., Morris, G. E., Whitfield, W. G. F., and Hutchison, C. J. (2001). Both emerin and lamin C depend on lamin A for localization at the nuclear envelope. *J. Cell Sci.* **114,** 2577–2590.

Waterham, H. R., Koster, J., Mooyer, P., van Noort, G., Kelley, R. I., Wilcox, W. R., Wanders, R. J. A., Hennekam, R. C. M., and Oosterwijk, J. C. (2003). Autosomal recessive HEM/Greenberg skeletal dysplasia is caused by 3β-hydroxysterol Δ^{14}-reductase deficiency due to mutations in the lamin B receptor gene. *Am. J. Hum. Genet.* **72,** 1013–1017.

Wolff, N., Gilquin, B., Courchay, K., Callebaut, I., Worman, H. J., and Zinn-Justin, S. (2001). Structural analysis of emerin, an inner nuclear membrane protein mutated in X-linked Emery-Dreifuss muscular dystrophy. *FEBS Lett.* **501,** 171–176.

Worman, H. J., and Courvalin, J.-C. (2000). The inner nuclear membrane. *J. Membr. Biol.* **177,** 1–11.

Worman, H. J., and Courvalin, J.-C. (2002). The nuclear lamina and inherited disease. *Trends Cell Biol.* **12,** 591–598.

Worman, H. J., and Courvalin, J.-C. (2004). How do mutations in lamins A and C cause disease? *J. Clin. Invest.* **113,** 349–351.

Worman, H. J., Lazaridis, I., and Georgatos, S. D. (1988a). Nuclear lamina heterogeneity in mammalian cells. *J. Biol. Chem.* **263,** 12135–12141.

Worman, H. J., Yuan, J., Blobel, G., and Georgatos, S. D. (1988b). A lamin B receptor in the nuclear envelope. *Proc. Natl. Acad. Sci. USA* **85,** 8531–8534.

Worman, H. J., Evans, C. D., and Blobel, G. (1990). The lamin B receptor of the nuclear envelope inner membrane: A polytopic protein with eight potential transmembrane domains. *J. Cell Biol.* **111,** 1535–1542.

Wu, W., Lin, F., and Worman, H. J. (2002). Intracellular trafficking of MAN1, an integral protein of the nuclear envelope inner membrane. *J. Cell Sci.* **115,** 1361–1372.

Yang, L., Guan, T., and Gerace, L. (1997). Lamin-binding fragment of LAP2 inhibits increase in nuclear volume during the cell cycle and progression into S phase. *J. Cell Biol.* **139,** 1077–1087.

Yates, J. R., Warner, J. P., Smith, J. A., Deymeer, F., Azulay, J. P., Hausmanowa-Petrusewicz, I., Zaremba, J., Borkowska, J., Affara, N. A., and Ferguson-Smith, M. A. (1993). Emery-Dreifuss muscular dystrophy: Linkage to markers in distal Xq28. *J. Med. Genet.* **30,** 108–111.

Ye, Q., and Worman, H. J. (1994). Primary structure analysis and lamin B and DNA binding of human LBR, an integral protein of the nuclear envelope inner membrane. *J. Biol. Chem.* **269,** 11306–11311.

Ye, Q., and Worman, H. J. (1996). Interaction between an integral protein of the nuclear envelope inner membrane and human chromodomain proteins homologous to *Drosophila* HP1. *J. Biol. Chem.* **271,** 14653–14656.

Zhang, Q., Skepper, J. N., Yang, F., Davies, J. D., Hegyi, L., Roberts, R. G., Weissberg, P. L., Ellis, J. A., and Shanahan, C. M. (2001). Nesprins: a novel family of spectrin-repeat-containing proteins that localize to the nuclear membrane in multiple tissues. *J. Cell Sci.* **114,** 4485–4498.

INDEX

VOLUMES IN SERIES

Founding Series Editor
DAVID M. PRESCOTT

Volume 1 (1964)
Methods in Cell Physiology
Edited by David M. Prescott

Volume 2 (1966)
Methods in Cell Physiology
Edited by David M. Prescott

Volume 3 (1968)
Methods in Cell Physiology
Edited by David M. Prescott

Volume 4 (1970)
Methods in Cell Physiology
Edited by David M. Prescott

Volume 5 (1972)
Methods in Cell Physiology
Edited by David M. Prescott

Volume 6 (1973)
Methods in Cell Physiology
Edited by David M. Prescott

Volume 7 (1973)
Methods in Cell Biology
Edited by David M. Prescott

Volume 8 (1974)
Methods in Cell Biology
Edited by David M. Prescott

Volume 9 (1975)
Methods in Cell Biology
Edited by David M. Prescott

Volume 10 (1975)
Methods in Cell Biology
Edited by David M. Prescott

Volume 11 (1975)
Yeast Cells
Edited by David M. Prescott

Volume 12 (1975)
Yeast Cells
Edited by David M. Prescott

Volume 13 (1976)
Methods in Cell Biology
Edited by David M. Prescott

Volume 14 (1976)
Methods in Cell Biology
Edited by David M. Prescott

Volume 15 (1977)
Methods in Cell Biology
Edited by David M. Prescott

Volume 16 (1977)
Chromatin and Chromosomal Protein Research I
Edited by Gary Stein, Janet Stein, and Lewis J. Kleinsmith

Volume 17 (1978)
Chromatin and Chromosomal Protein Research II
Edited by Gary Stein, Janet Stein, and Lewis J. Kleinsmith

Volume 18 (1978)
Chromatin and Chromosomal Protein Research III
Edited by Gary Stein, Janet Stein, and Lewis J. Kleinsmith

Volume 19 (1978)
Chromatin and Chromosomal Protein Research IV
Edited by Gary Stein, Janet Stein, and Lewis J. Kleinsmith

Volume 20 (1978)
Methods in Cell Biology
Edited by David M. Prescott

Advisory Board Chairman
KEITH R. PORTER

Volume 21A (1980)
Normal Human Tissue and Cell Culture, Part A: Respiratory, Cardiovascular, and Integumentary Systems
Edited by Curtis C. Harris, Benjamin F. Trump, and Gary D. Stoner

Volume 21B (1980)
Normal Human Tissue and Cell Culture, Part B: Endocrine, Urogenital, and Gastrointestinal Systems
Edited by Curtis C. Harris, Benjamin F. Trump, and Gray D. Stoner

Volume 22 (1981)
Three-Dimensional Ultrastructure in Biology
Edited by James N. Turner

Volume 23 (1981)
Basic Mechanisms of Cellular Secretion
Edited by Arthur R. Hand and Constance Oliver

Volume 24 (1982)
The Cytoskeleton, Part A: Cytoskeletal Proteins, Isolation and Characterization
Edited by Leslie Wilson

Volume 25 (1982)
The Cytoskeleton, Part B: Biological Systems and *in Vitro* Models
Edited by Leslie Wilson

Volume 26 (1982)
Prenatal Diagnosis: Cell Biological Approaches
Edited by Samuel A. Latt and Gretchen J. Darlington

Series Editor
LESLIE WILSON

Volume 27 (1986)
Echinoderm Gametes and Embryos
Edited by Thomas E. Schroeder

Volume 28 (1987)
***Dictyostelium discoideum:* Molecular Approaches to Cell Biology**
Edited by James A. Spudich

Volume 29 (1989)
Fluorescence Microscopy of Living Cells in Culture, Part A: Fluorescent Analogs, Labeling Cells, and Basic Microscopy
Edited by Yu-Li Wang and D. Lansing Taylor

Volume 30 (1989)
Fluorescence Microscopy of Living Cells in Culture, Part B: Quantitative Fluorescence Microscopy—Imaging and Spectroscopy
Edited by D. Lansing Taylor and Yu-Li Wang

Volume 31 (1989)
Vesicular Transport, Part A
Edited by Alan M. Tartakoff

Volume 32 (1989)
Vesicular Transport, Part B
Edited by Alan M. Tartakoff

Volume 33 (1990)
Flow Cytometry
Edited by Zbigniew Darzynkiewicz and Harry A. Crissman

Volume 34 (1991)
Vectorial Transport of Proteins into and across Membranes
Edited by Alan M. Tartakoff

Selected from Volumes 31, 32, and 34 (1991)
Laboratory Methods for Vesicular and Vectorial Transport
Edited by Alan M. Tartakoff

Volume 35 (1991)
Functional Organization of the Nucleus: A Laboratory Guide
Edited by Barbara A. Hamkalo and Sarah C. R. Elgin

Volume 36 (1991)
***Xenopus laevis:* Practical Uses in Cell and Molecular Biology**
Edited by Brian K. Kay and H. Benjamin Peng

Series Editors
LESLIE WILSON AND PAUL MATSUDAIRA

Volume 37 (1993)
Antibodies in Cell Biology
Edited by David J. Asai

Volume 38 (1993)
Cell Biological Applications of Confocal Microscopy
Edited by Brian Matsumoto

Volume 39 (1993)
Motility Assays for Motor Proteins
Edited by Jonathan M. Scholey

Volume 40 (1994)
A Practical Guide to the Study of Calcium in Living Cells
Edited by Richard Nuccitelli

Volume 41 (1994)
Flow Cytometry, Second Edition, Part A
Edited by Zbigniew Darzynkiewicz, J. Paul Robinson, and Harry A. Crissman

Volume 42 (1994)
Flow Cytometry, Second Edition, Part B
Edited by Zbigniew Darzynkiewicz, J. Paul Robinson, and Harry A. Crissman

Volume 43 (1994)
Protein Expression in Animal Cells
Edited by Michael G. Roth

Volume 44 (1994)
***Drosophila melanogaster:* Practical Uses in Cell and Molecular Biology**
Edited by Lawrence S. B. Goldstein and Eric A. Fyrberg

Volume 45 (1994)
Microbes as Tools for Cell Biology
Edited by David G. Russell

Volume 46 (1995) (in preparation)
Cell Death
Edited by Lawrence M. Schwartz and Barbara A. Osborne

Volume 47 (1995)
Cilia and Flagella
Edited by William Dentler and George Witman

Volume 48 (1995)
***Caenorhabditis elegans:* Modern Biological Analysis of an Organism**
Edited by Henry F. Epstein and Diane C. Shakes

Volume 49 (1995)
Methods in Plant Cell Biology, Part A
Edited by David W. Galbraith, Hans J. Bohnert, and Don P. Bourque

Volume 50 (1995)
Methods in Plant Cell Biology, Part B
Edited by David W. Galbraith, Don P. Bourque, and Hans J. Bohnert

Volume 51 (1996)
Methods in Avian Embryology
Edited by Marianne Bronner-Fraser

Volume 52 (1997)
Methods in Muscle Biology
Edited by Charles P. Emerson, Jr. and H. Lee Sweeney

Volume 53 (1997)
Nuclear Structure and Function
Edited by Miguel Berrios

Volume 54 (1997)
Cumulative Index

Volume 55 (1997)
Laser Tweezers in Cell Biology
Edited by Michael P. Sheez

Volume 56 (1998)
Video Microscopy
Edited by Greenfield Sluder and David E. Wolf

Volume 57 (1998)
Animal Cell Culture Methods
Edited by Jennie P. Mather and David Barnes

Volume 58 (1998)
Green Fluorescent Protein
Edited by Kevin F. Sullivan and Steve A. Kay

Volume 59 (1998)
The Zebrafish: Biology
Edited by H. William Detrich III, Monte Westerfield, and Leonard I. Zon

Volume 60 (1998)
The Zebrafish: Genetics and Genomics
Edited by H. William Detrich III, Monte Westerfield, and Leonard I. Zon

Volume 61 (1998)
Mitosis and Meiosis
Edited by Conly L. Rieder

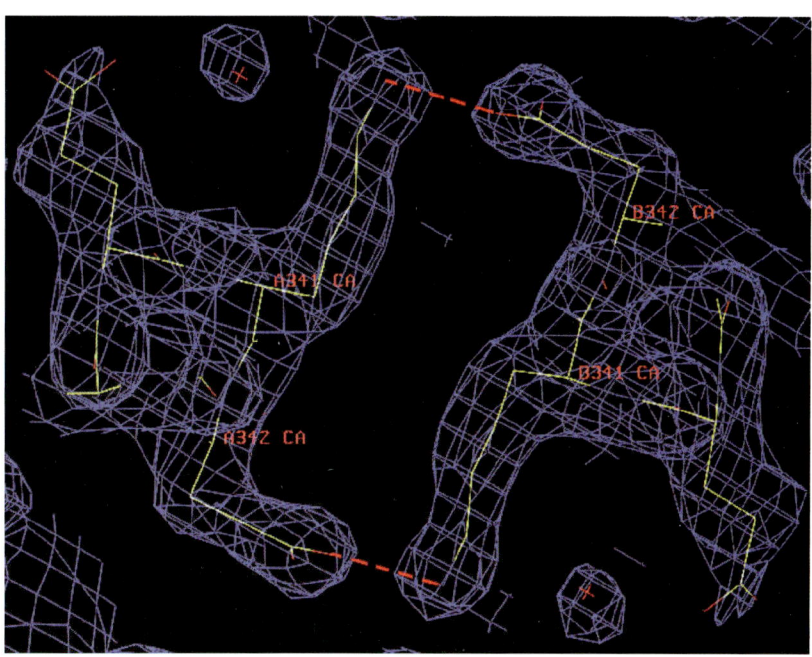

Chapter 2, Fig. 3 Part of the atomic structure of the lamin A coil 2B fragment overlaid with the 2.3
Å-resolution electron density map ($2F_{obs}$-F_{calc}, 1.5σ level). The interchain salt bridges formed between
residues Lys341 and Glu342 are shown with dashed lines.

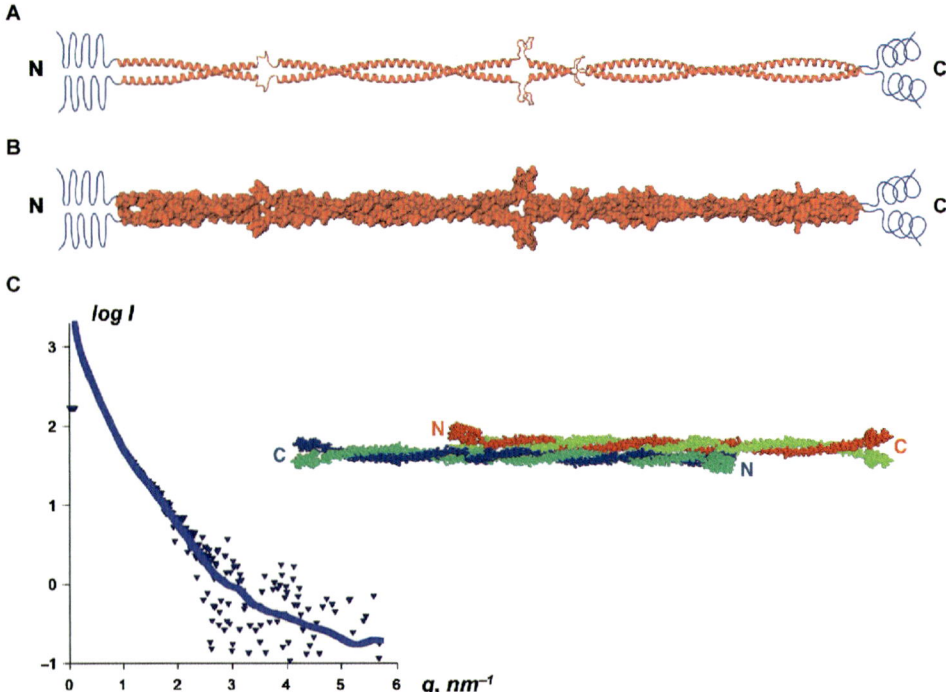

Chapter 2, Fig. 4 (A) Model of the vimentin dimer shown as a ribbon diagram. The coiled-coil rod and the terminal domains are shown in red and blue, respectively. (B) Same model with the coiled-coil rod shown as surface. (C) Model of the vimentin tetramer based on SAXS data. (C) Experimental SAXS data obtained from the solution of vimentin tetramers (dark blue triangles), as well as the scattering curve computed from the above tetramer model (light blue line). Shown is the dependence of the scattered intensity (arbitrary units, log scale) on $q = 4\pi \sin \theta / \lambda$, where 2θ is the scattering angle and λ is the wavelength.

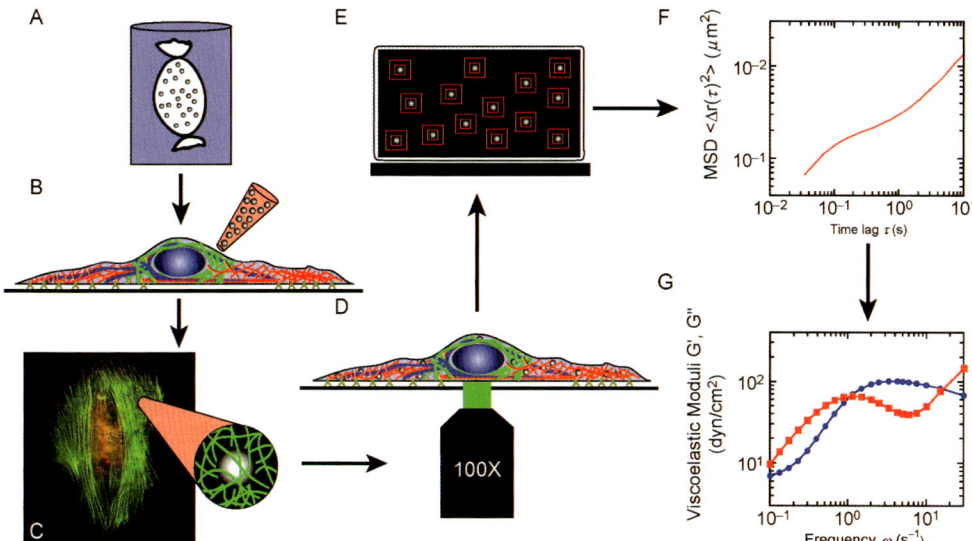

Chapter 3, Fig. 2 Schematic of intracellular microrheology (ICM). (A) Particles are first prepared by dialysis against microinjection buffer. (B) Dialyzed particles are then microinjected into the cytoplasm of adherent cells and become embedded in the cytoskeletal network (C). (D) The Brownian displacements of particles injected into the cell are monitored by video-based particle tracking with high-magnification fluorescence microscopy. (E) The time-dependent x and y coordinates of recorded particles are tracked by measurement of intensity-weighted centroid displacements. (F) Mean-squared displacements are calculated for each particle and used to evaluate the frequency-dependent viscoelastic moduli of the cell (G).

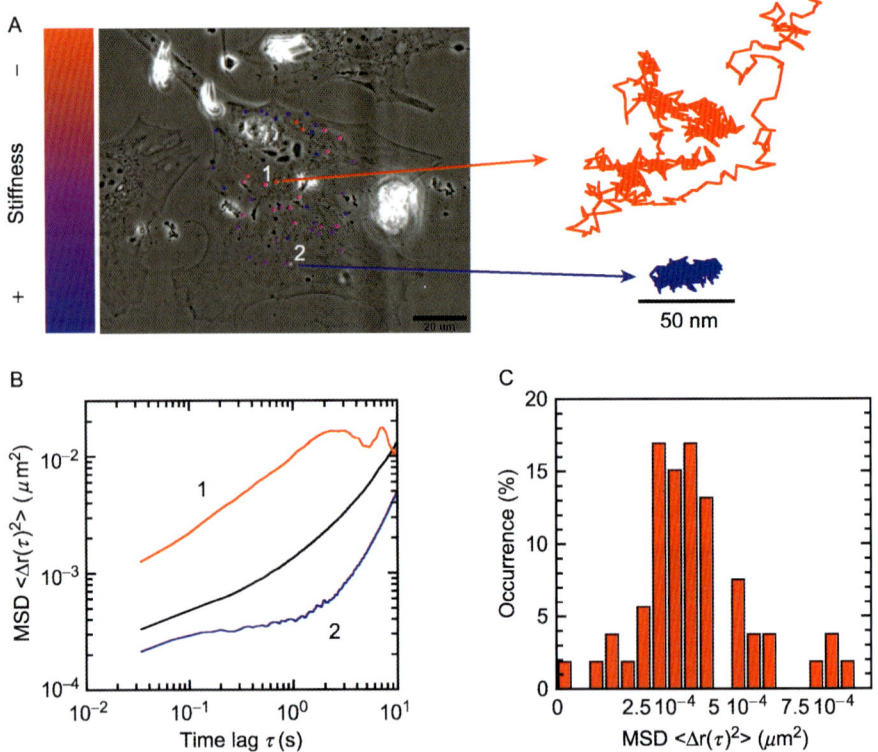

Chapter 3, Fig. 3 Application of intracellular microrheology. (A) Local mapping of the mechanical properties of a migrating cell at the edge of a wounded fibroblast monolayer. Here we only show tracked particles and have neglected aggregates that violate assumptions made in the generalized Stokes–Einstein relationship. Each particle position was color coded, corresponding to the local value of the creep compliance at that position in the cell. The color indicators at each particle position do not reflect the size of the particle (100 nm). Indicator size was increased to aid visual presentation. The time-dependent trajectory of a particle close to the nucleus (1) is much more confined than a particle toward the leading edge (2). (B) The mean-squared displacements (MSDs) of particles depicted in (A) together with the ensemble averaged MSD of all particles within the cell. The MSD of the particle close to the nucleus (1) is much larger than a particle toward the leading edge (2), suggesting that the leading edge of migrating cells is much stiffer than the perinuclear region. (C) Distribution of the MSDs of particles embedded within the cytoplasm of the cell shown in (A). The mechanical properties of the cell are highly heterogeneous, thus illustrating the necessity of local measurements within the cell.

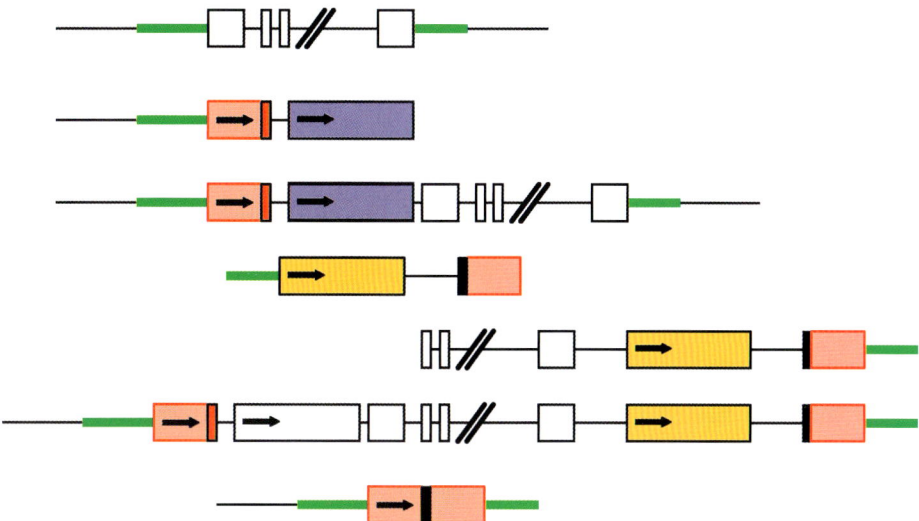

Chapter 4, Fig. 7 Generation of large-scale genomic deletions based on an HPRT-deficient ES cell line. Top, Target locus. Below, Targeting vector for 5′ flank of targeting vector, consisting of a nonfunctional half-HPRT-minigene (red arrow indicates minigene promoter), followed by a loxP site (red box) and a neomycin resistance gene (blue box, arrow marking promoter). Below, After homologous recombination, G418-resistant ES cell clones are identified by PCR. Below, Targeting vector for 3′ end, consisting of a puromycin resistance gene (yellow box) and the 3′ half of a nonfunctional HPRT-minigene (red), preceded by a loxP site. Below, After homologous recombination, puromycin-resistant ES cell clones are identified by PCR. Complex rearrangements can occur after Cre activity, depending on the cell cycle phase in which Cre is active and the chromosomal integration of the two targeting vectors (refer to Yu and Bradley, 2001). Below, After Cre activity, the marker genes and the genomic sequences between the two loxP sites are deleted, and surviving correctly targeted ES cell clones are resistant against HAT and sensitive against G418 and puromycin.

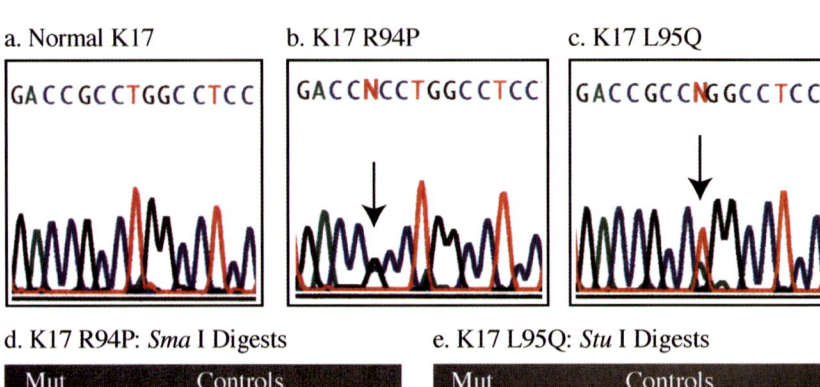

a. Normal K17

GACCGCCTGGCCTCC

b. K17 R94P

GACCNCCTGGCCTCC

c. K17 L95Q

GACCGCCNGGCCTCC

d. K17 R94P: *Sma* I Digests

Mut   Controls

e. K17 L95Q: *Stu* I Digests

Mut   Controls

Chapter 6, Fig. 5 (a–c) Mutations in K17 revealed by direct DNA sequencing; and (d and e), confirmation of mutations by restriction digestion. (a) Normal K17 sequence in exon 1, corresponding to codons 93–97, base numbers 277–291, inclusive. (b) The equivalent sequence shown in (a) from an affected individual with pachyonychia congenita type 2 (PC-2), showing heterozygous missense mutation 281G > C (arrow), which predicts the amino acid substitution R94P in K17. (c) The equivalent sequence shown in (a) derived from another PC-2 patient showing missense mutation 284T > A (arrow) leading to amino acid change L95Q. (d) Confirmation of mutation R94P by *Sma*I digestion. Lane 1; digested polymerase chain reaction (PCR) product from the affected individual (marked "Mut") gives an additional band because of the introduction of a *Sma*I site in one primer, which depends on the mutation. Lanes 2–5 are digested PCR products from normal unrelated controls. (e) Confirmation of mutation L95Q by *Stu*I digestion. Lane 1; digestion of PCR product from the affected individual (marked "Mut") shows an additional band caused by a new *Stu*I site created by the mutation. Lanes 2–5 show digested PCR products from normal unrelated controls.

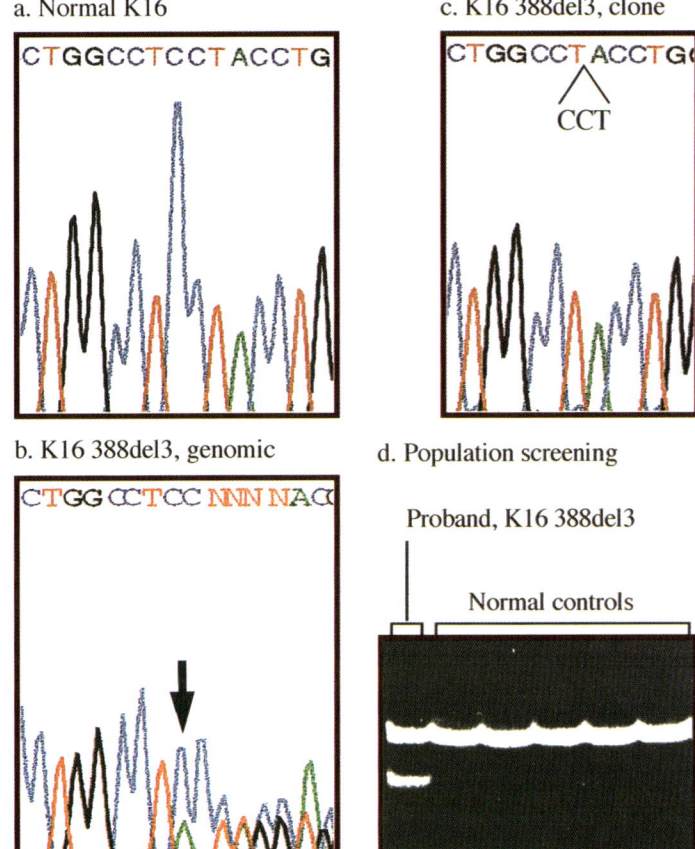

a. Normal K16

c. K16 388del3, clone

b. K16 388del3, genomic

d. Population screening

Proband, K16 388del3

Normal controls

Chapter 6, Fig. 6 K16 mutation detection and confirmation. (a) K16 genomic sequence derived from a normal individual by direct sequencing of polymerase chain reaction (PCR) products. Coding sequence base numbers 382–396, encompassing codons 128–132, are shown. (b) The equivalent K16 sequence as shown in (a), derived from an individual with pachyonychia congenita type 1 (PC-1). Arrow indicates the start of sequence overlap caused by the heterozygous 3-bp deletion, 388del3. (c) The equivalent K16 sequence as shown in (a) derived from the mutant allele cloned into pCR2.1 vector. This clarifies that the mutation is a 3-bp deletion: removal of a CCT repeat, as indicated. (d) Confirmation of mutation 388del3 and exclusion from normal control individuals. A 140-bp PCR fragment spanning the mutation was amplified from genomic DNA from the proband and 50 unaffected, unrelated individuals (five of which are shown here). PCR products were separated on 6% sequencing gels. Note that the smaller mutant band is much weaker than the normal one, because the PCR used here amplifies both the functional K16 gene and its pseudogenes.

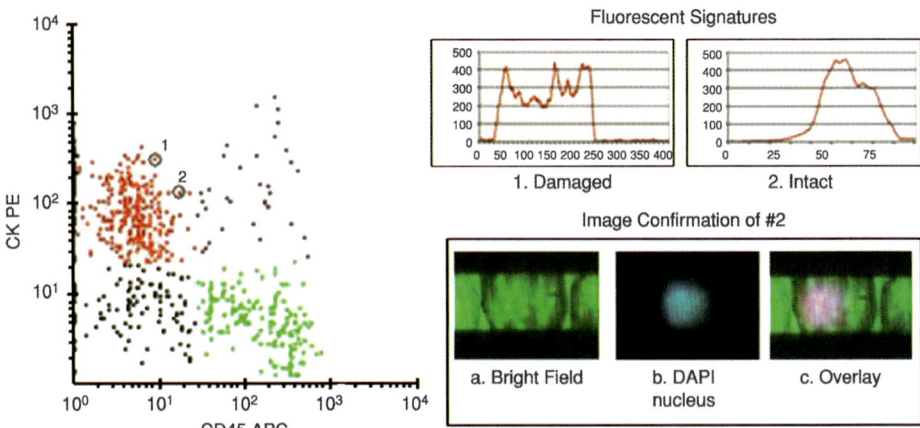

Chapter 7, Fig. 1 The dotplot on the left shows the leukocyte staining (CD45-APC) and target cell staining (CK-rPE) of the separated immunomagnetic labeled cells in the CellTracks system. The fluorescent signature of the target candidate no. 1 is displayed in the histograms (upper right), whereas the bright field image of the cytoplasm and the fluorescence signal of DAPI (DNA staining of the nucleus) are displayed in the figure right below. (source: http://www.immunicon.com)

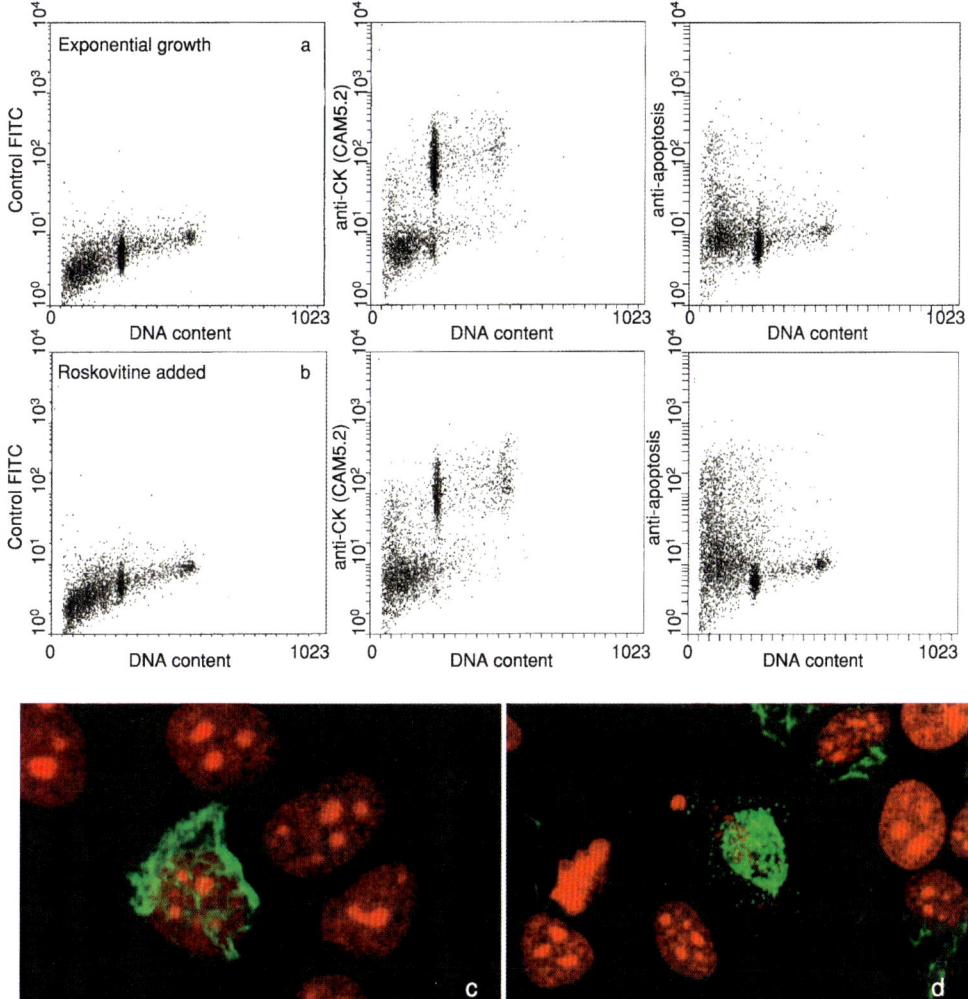

Chapter 7, Fig. 4   Multiparameter flow cytometric analysis of apoptosis in exponentially growing MR65 cells (a) and after 8 hours of treatment with roskovitine (b). In the left dotplots, the negative control is depicted (DNA (x-axis) vs. mouse-Ig-FITC); the middle dotplots show the cytokeratin expression (labeled by the pan-CK antibody (CAM5.2) vs. DNA content, whereas in the right dotplots the expression of caspase cleaved CK18 (y-axis) is depicted vs. DNA content (x-axis). Panels (c) and (d) show the expression of this caspase-cleaved CK18 as visualized by confocal scanning laser microscopy after immunostaining with the M30-cytodeath antibody (green) combined with DNA staining with propidium iodide (red).

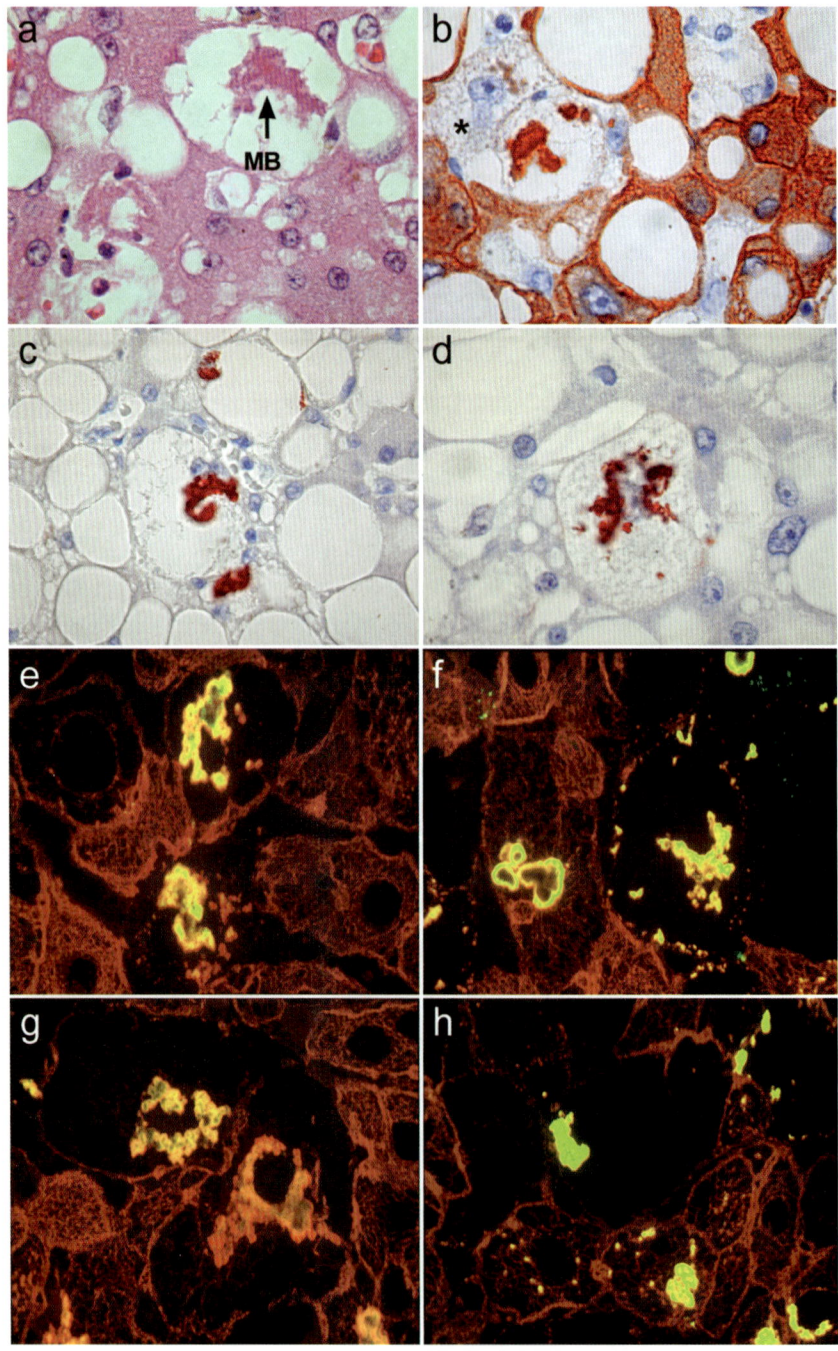

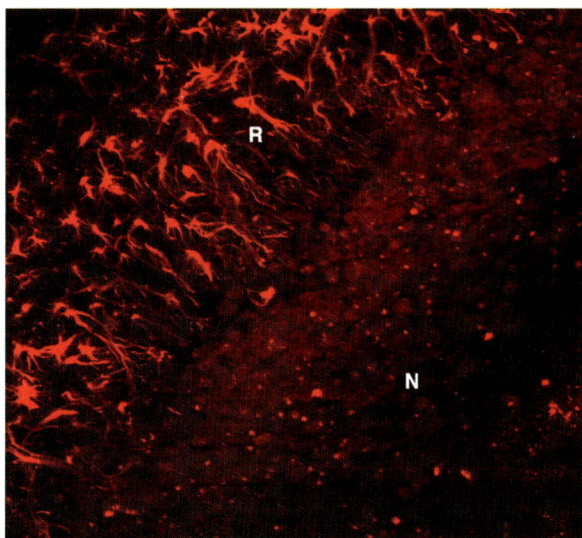

Chapter 9, Fig. 2   Reactive gliosis with characteristic hypertrophy of astrocyte processes around an ischemic lesion in the brain. Astrocytes are visualized by use of antibodies against glial fibrillary acidic protein.

Chapter 8, Fig. 1   Mallory bodies (MBs) in a human liver with alcoholic steatohepatitis (a–d). (a) Hematoxylin-eosin–stained section (MB in a ballooned hepatocytes is indicated by arrow). (b) Immunohistochemistry on paraffin-embedded tissue with the monoclonal keratin antibodies to K8 and K18 (K8.8 and DC-10, NeoMarkers). (c) Immunohistochemistry with antibody p62CT. (d) Immunohistochemistry with a polyclonal antibody to ubiquitin (Dako). Note that MBs are present in ballooned hepatocytes with a loosened or even undetectable keratin intermediate filament (IF) network (ballooned hepatocyte with no detectable keratin IF is indicated by asterisk in (b). Double-label immunofluorescence microscopy on frozen mouse livers after 2 months of 3,4-diethoxycarbonyl-1,4-dihydrocollidine (DDC) intoxication (e–h) with combinations of the polyclonal antibody to K8/K18 (50K160, in red) and antibodies to nonkeratin MB components (in green). (e) antibody to ubiquitin; (f) antibody to p62; (g) antibody to SMI 31; (h) antibody M_M120–1.

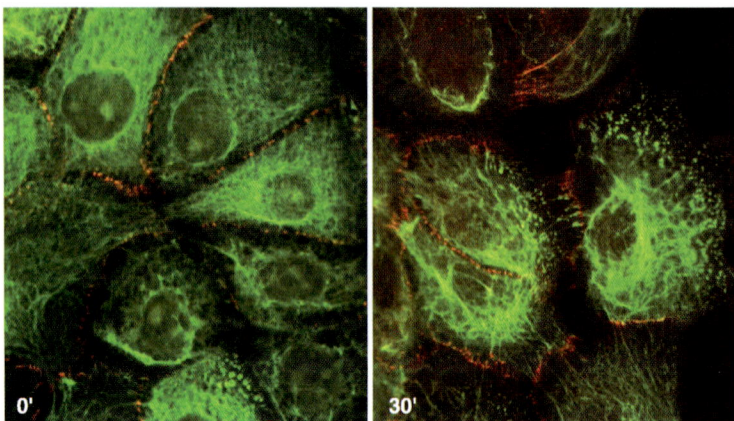

Chapter 9, Fig. 3 Mechanical stress can be used to reveal cytoskeleton weaknesses, as demonstrated by the effect of stretch on mutant keratin filaments in cells from patients with EBS (Russell, 2004; Russell *et al.*, 2004). The KEB-7 cells shown here carry the K14 mutation R125P, typical of severe EBS (Morley *et al.*, 2003) and causing basal keratinocyte fragility and epidermal blistering. Before stretch (0') keratin staining reveals occasional aggregates. After 30 minutes of an oscillating stretch (30'), the keratin filament network is severely disrupted. Wild-type cell keratin filaments do not break down under this stress (not shown). Keratin filaments (green) visualized by immunofluorescence with polyclonal antiserum to keratin 5 and fluorescein-conjugated second antibody; desmosomes (red) shown by staining with mouse monoclonal antibody 11-5F to desmoplakin and Texas Red-conjugated second antibody. (Picture courtesy of David Russell.)

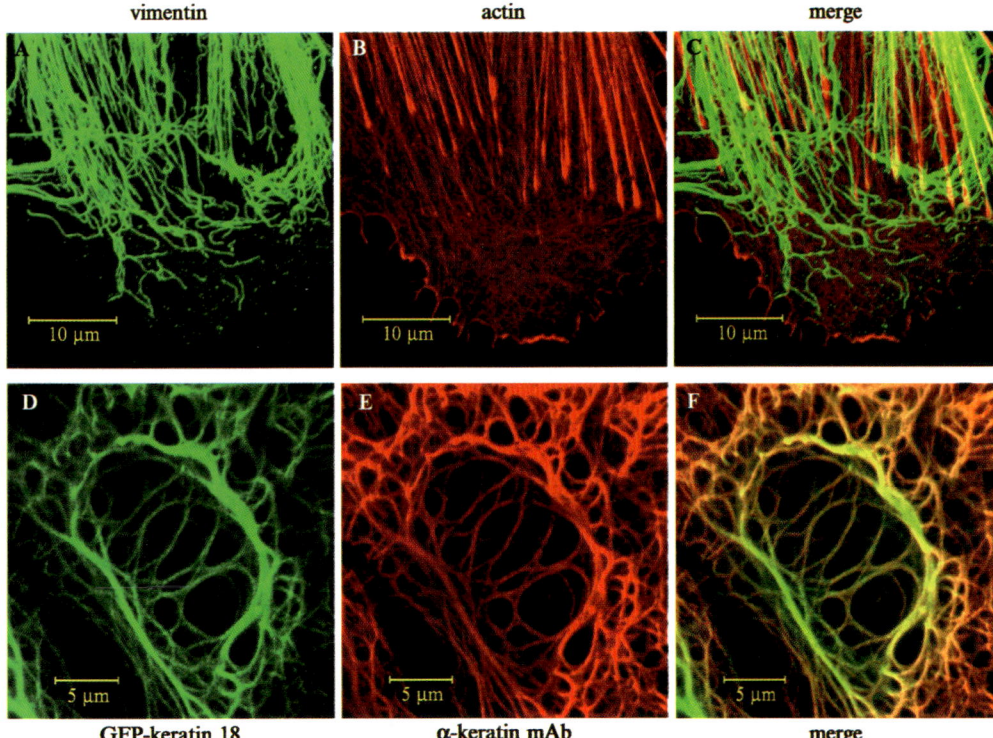

vimentin   **actin**   **merge**

GFP-keratin 18   **α-keratin mAb**   **merge**

Chapter 11, Fig. 2 Human endothelial cells fixed in formaldehyde and stained with a mouse monoclonal antibody to vimentin (V-9 clone) followed by a goat anti-mouse-Alexa 488 secondary antibody, (A) together with phalloidin-Alexa 568 to show F-actin (B). Overlay (C) reveals colocalization (yellow) between vimentin IF and the tips of actin stress fibers. Transformed lung alveolar epithelial cell (A549 line) expressing green fluovescent protein (GFP)-tagged keratin 18 (D) immunostained with a mouse monoclonal antibody to keratin 18 followed by a goat anti-mouse-Alexa 568 (red) conjugated secondary antibody (E). Overlay shows that the staining patterns are coincident and that GFP-keratin is fully incorporated into the endogenous network (F).

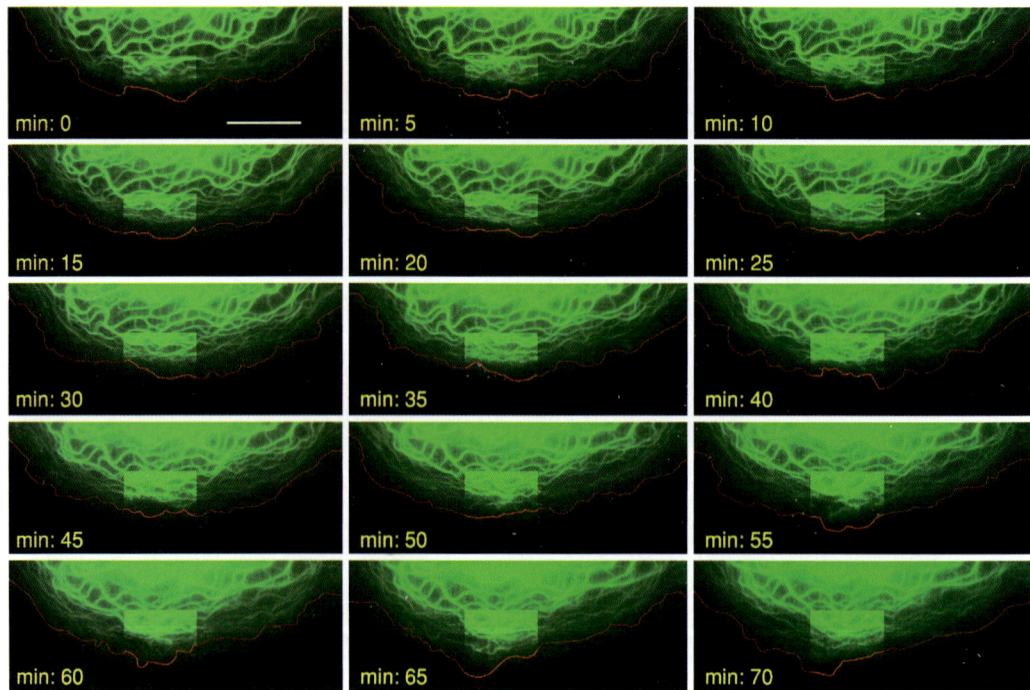

Chapter 12, Fig. 3 Epifluorescence images depicting a peripheral region of a living hepatocellular carcinoma–derived PK18-5 cell synthesizing fluorescent keratin hybrid HK18-YFP. Note that changes in keratin distribution are difficult to discern from the series of still images, but that they are easily recognized in the corresponding time-lapse series that is provided as movie 1, revealing a continuous inward flow of fluorescent material. The look-up table was adjusted in the square region in the middle to enhance low fluorescent structures. The cell edge is demarcated by a red line. Its position was determined from phase-contrast images that were recorded in parallel. Bar, 5 μm.

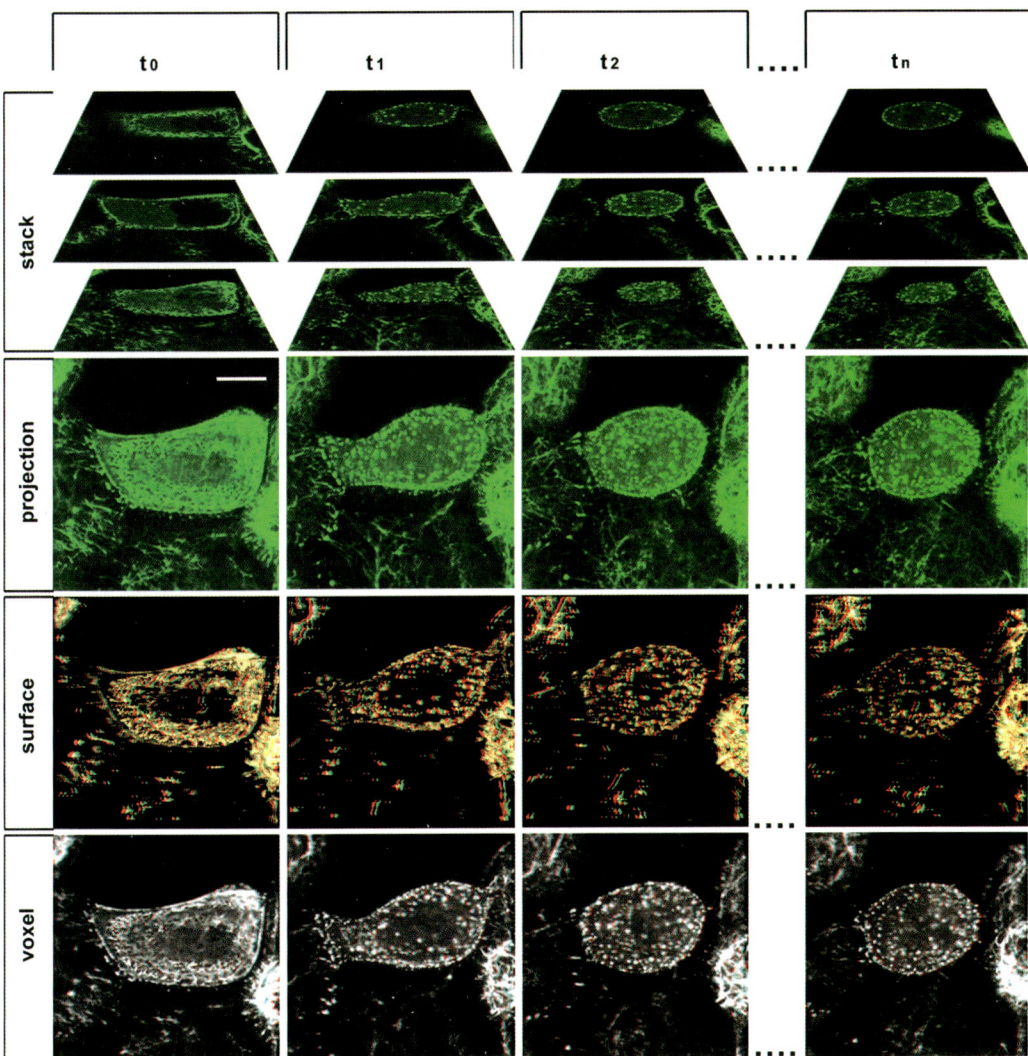

Chapter 12, Fig. 5 Different types of presentation of a time series of z-stacks of confocal laser scan fluorescence images that were recorded in a mitotic AK13-1 cell producing HK13-EGFP. The top part of the panel (*stack*) depicts three confocal fluorescence micrographs of a z-stack at four different time points. The corresponding movie 3 was assembled from a single focal plane (recording intervals, 2.5 minutes). The next row of pictures presents projection views of the entire z-stacks at the different time points. The complete series of projection images was combined into movie 4. Below, stacks were reconstructed into 3D anaglyph pictures that should be viewed with red-green glasses for complete 3D visualization. For comparison, surface and voxel reconstructions are shown. The complete time series of the 3D images are provided as movies 5 and 6, respectively. Bar, 5 μm.

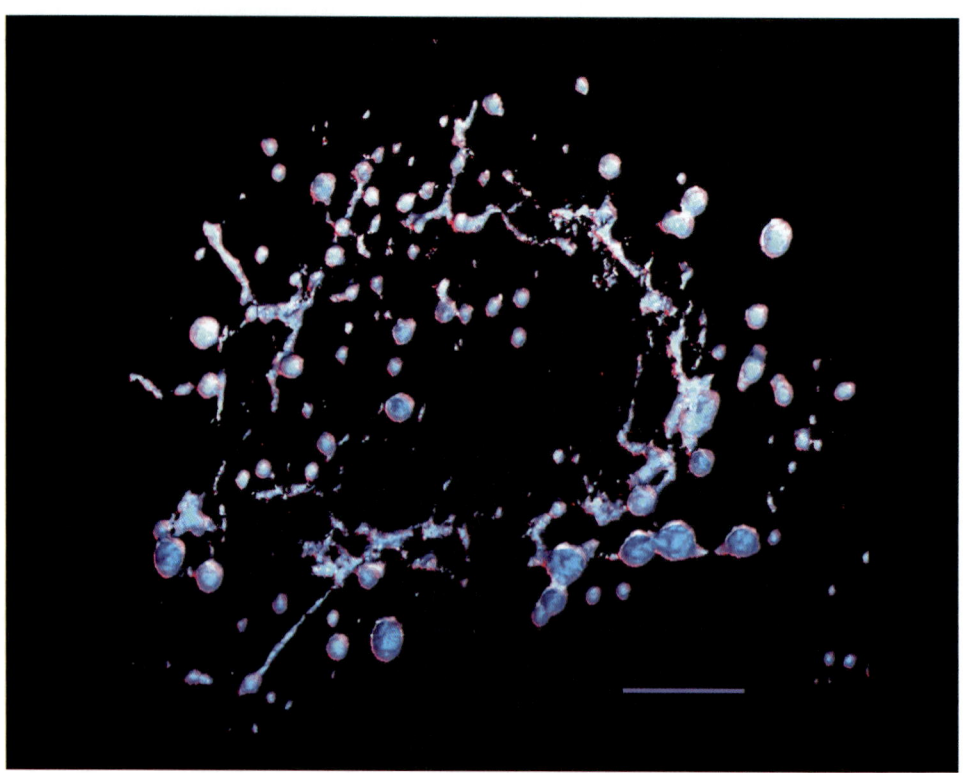

Chapter 12, Fig. 6 Three-dimensional reconstruction (surface view) of fluorescence images recorded in 32 confocal planes at 1024 × 1024 pixel of a single AK13-1 cell producing fluorescent HK13-EGFP fusion proteins. The cell was treated with 0.1 µg/ml okadaic acid for 4 hours and was fixed with methanol/acetone before imaging. Keratin granules are seen together with residual perinuclear KF aggregates. Bar, 5 µm.

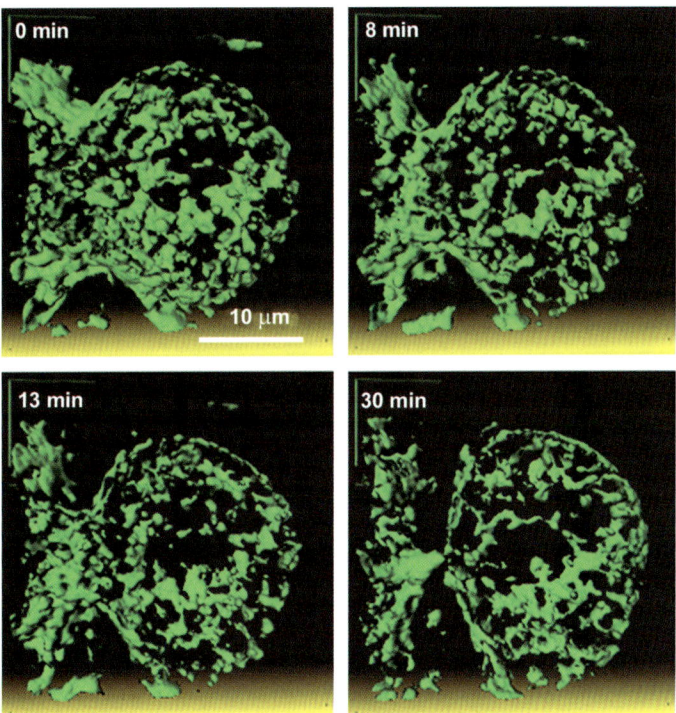

Chapter 12, Fig. 9 Four-dimensional analysis of the keratin fluorescence in a dividing AK13-1 cell producing HK13-EGFP. Images were recorded in multiple focal planes with a confocal laser scan microscope every 1.6 minutes and were used to generate the 3D reconstructions (surface view). Four consecutive reconstructions are depicted highlighting redistribution of granular keratin. The complete time series is provided as movie 7.

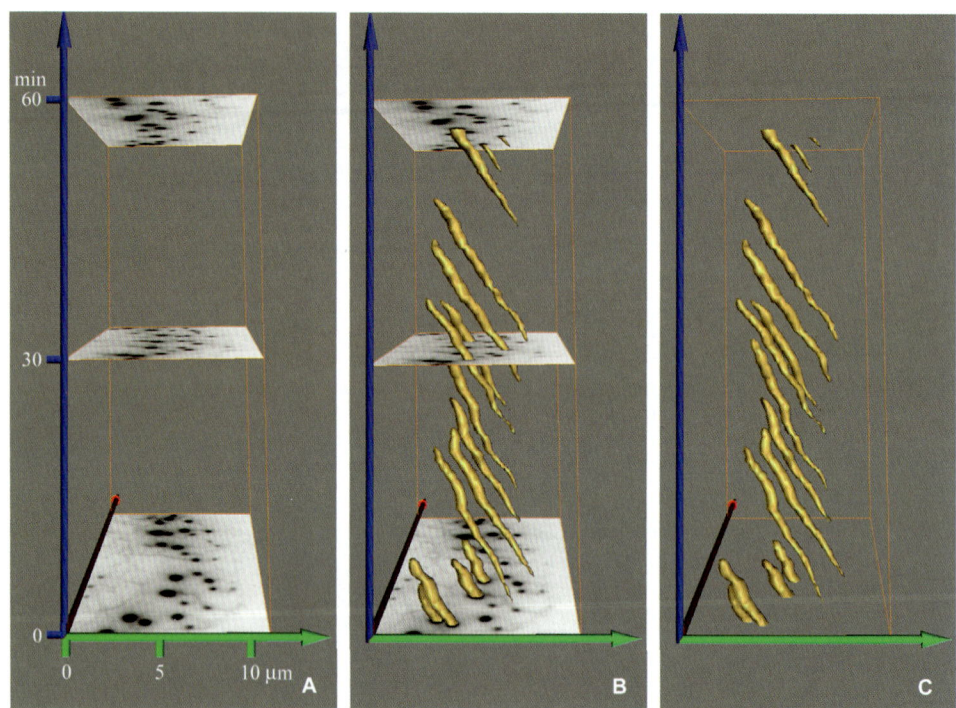

Chapter 12, Fig. 10 Preparation of time-space diagrams. (A) Three epifluorescence images (inverse presentation) that are taken from a time-lapse recording of a small region in the cell periphery of an AK13-1 cell synthesizing HK13-EGFP after a short treatment with 10 mM orthovanadate. Surface views were prepared from the time series as indicated in (B). For clarity only a few representative traces are depicted. These surface views (C) reveal a lifetime of ~20 minutes and a consistent inward-directed movement from the subplasmalemmal region at right to the cell interior at left of ~250 nm/min.

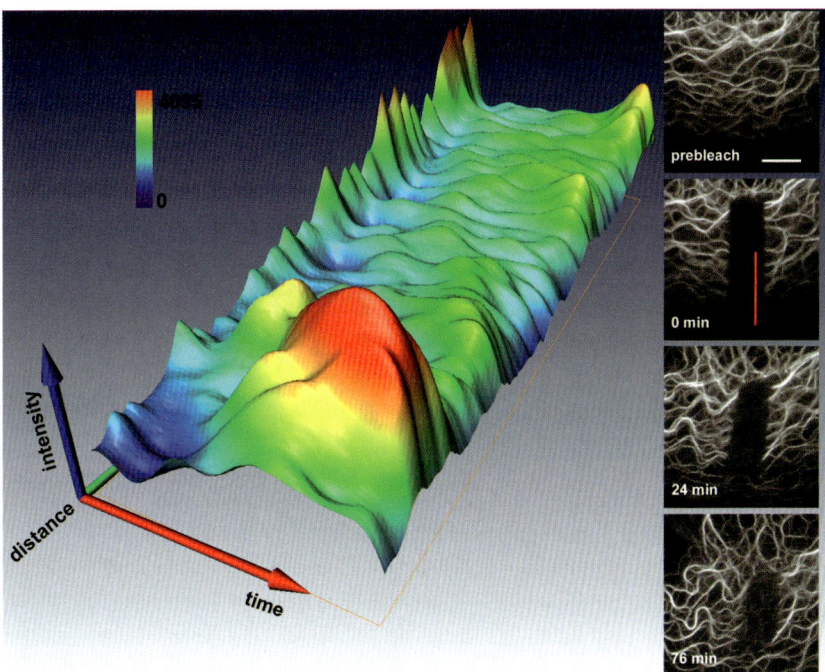

Chapter 12, Fig. 11 Time-intensity diagram depicting fluorescence alterations that were observed in a fluorescence recovery after a photobleaching experiment of a PK18-5 cell synthesizing fluorescent HK18-YFP chimeras. The diagram was derived from the image series presented in movie 8. Four images from the movie are shown at right presenting the prebleach keratin cytoskeleton, the location of the bleached area together with the line (in red) that was selected for intensity measurements, and two time points during fluorescence recovery (24 minutes and 76 minutes postbleach). The diagram at left displays the values of fluorescence intensity that were measured along the red line (*distance*) in relation to time. Note that the most fluorescence recurs in the cell periphery originating beneath the plasma membrane that moved gradually outside the imaged area during recording. Bar, 5 μm.

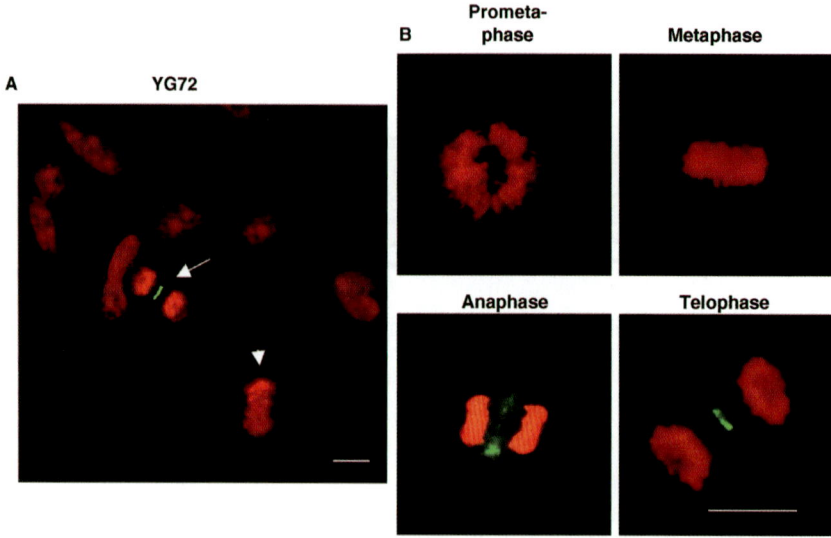

Chapter 13, Fig. 8   Immunocytochemistry with YG72. (A) and (B) Fluorescent photomicrographs of U251 glioma cells stained with YG72 (green). Chromosomes were stained with propidium iodide (red). An arrow and an arrowhead in (A) show a telophase and metaphase cell, respectively. Bars represent 10 μm. Modified with permission from Yasui *et al.* (2001).

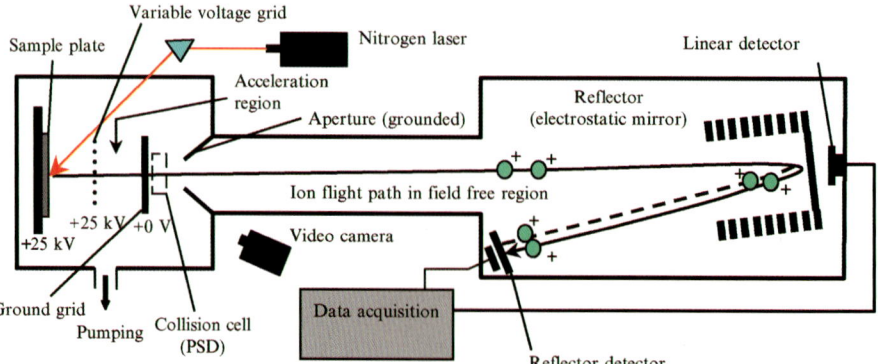

Chapter 14, Fig. 1   The principle of MALDI-TOF MS (see text for details).

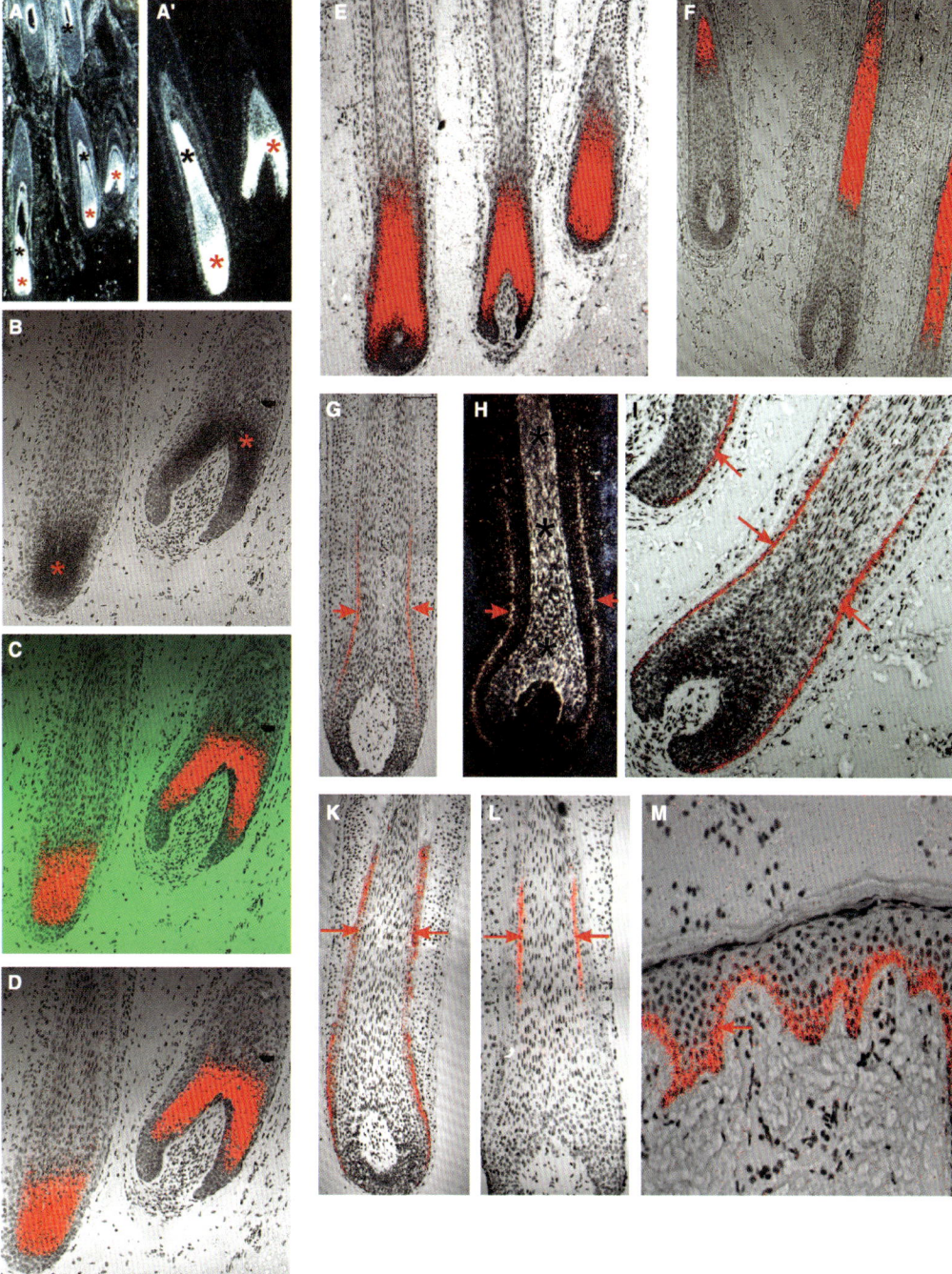

Chapter 15, Fig. 3 Analysis and documentation of the ISH procedure. (A-A′, H) The signals of the ISH labeling cannot be clearly identified by dark-field microscopy. Besides specific labeling of hHa5 mRNA (red asterisks in A-A′) or K6hf mRNA (red arrows in H), false-positive images are detectable

(*Continued*)

in nonlabeled areas of the hair cortex (black asterisks). (B) In the transmission image, only strong signals can be detected (red asterisks, hHa5 mRNA). (C, D) The sensitivity and specificity of the labeling in B can be strongly enhanced by reflection microscopy with a confocal laser scan microscope. The ISH signal, detected in false color, is given in red together with the green transmission image (C). The conversion of the green channel into a black- and-white image yields the best quality image, lacking background problems (D). (A-D) The hHa5 mRNA can be detected in the upper matrix and lower cortex including the hair cuticle (E, F; see also Fig. 2F) Specific mRNA staining of hHb5 (E; see also Fig. 2G) in the upper matrix up to the mid-cortex and hHa4 (F) in the upper cortex as the latest expressed hair keratin. (G) Specific staining of hHb2 in the hair cuticle. Note the high specificity and the clear identification of the labeling of the single-layered hair cuticle. (H, I) Comparison of the staining of K6hf mRNA in the companion layer using dark-field illumination (H) and reflection microscopy (I; see also Fig. 2H). Again, the high specificity and the clear identification of the labeling of the single cell layer are demonstrated. Note the problems regarding the identification of specific signals in H. (K-M) Specific labeling of K6irs1 mRNA (K) in the IRS, KAP10.1 mRNA (L; see also Fig. 2I) in the middle of the hair cuticle and K14 mRNA (M; positive control; see also Fig. 2D) in the basal cells of the interfollicular epidermis.

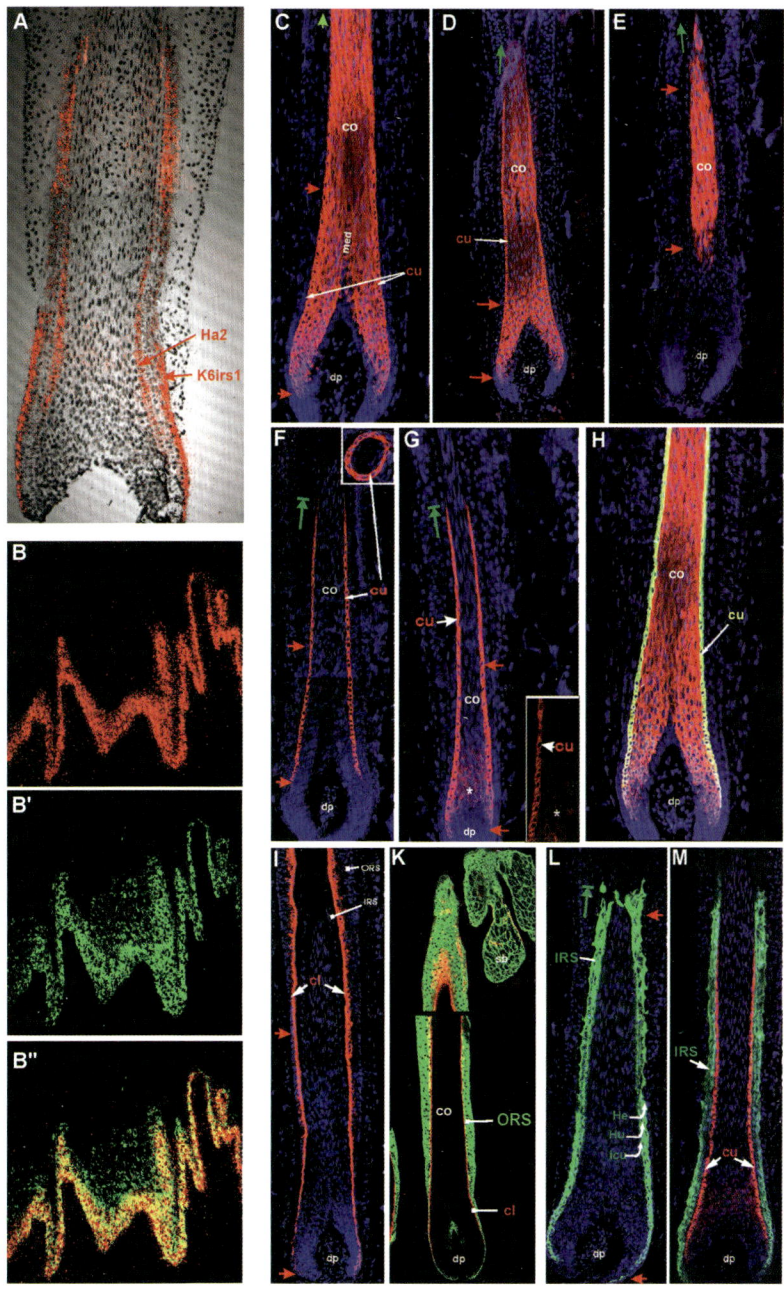

Chapter 15, Fig. 4 Double-label ISH, combination of ISH/IIF and indirect immunofluorescence of various keratins in the hair follicle. (A) The specific staining of hHa2 mRNA in the hair cuticle and K6irs1 mRNA in the IRS can be clearly detected by double-label ISH. (B–B″) Although the signal for

(*Continued*)

K14 mRNA (B; ISH labeling in red; cp. also Fig. 3M) is restricted to the basal cells of foot sole epidermis, the K14 protein is stable also in the lower to middle suprabasal layers (B′ IIF staining in green). The yellow color in the merged images indicates the colocalization of K14 mRNA and protein (B″) in the basal cells of this epidermis. (C, D) Hair keratins hHb5 (C, see also Fig. 3G) and hHa5 (D, see also Fig. 3D) are both expressed in the hair cortex (co) and cuticle (cu). (E) hHa4 is the latest hair keratin and is expressed in the mid- to upper cortex. (F, G) Hair keratins hHb2 (F, see also Fig. 3G) and hHa2 (G) are specifically found in the hair cuticle. (H) By double-label IIF, the coexpression of hHa2 (green) and hHb5 (red) in the hair cuticle is clearly shown by the yellow (merged) color in this compartment. (I) Keratin K6hf is detectable throughout the companion layer (cl, see also Fig. 3I). (K) Keratin K6hf (red) is restricted to the cl, whereas keratin K14 (green, cp. also Fig. 3M) is found in the ORS and the interfollicular epidermis. (L) The keratin K6irs1 antibody stains the Henle and Huxley layers of the IRS, as well as the IRS-cuticle, see also Fig. 3K). (M) Double immunolabeling with K6irs1 (green) and hHa2 (red) shows that both cuticles are adjacent to each other. Red arrows in C–G, I, and L indicate the zones of the respective mRNA synthesis. Note that in contrast to the restricted occurrence of the mRNAs, the proteins are stably integrated in the cytoskeleton and detectable throughout the hair fiber. Compare the mRNA data (this figure) with the respective protein data (Fig. 3). The green stop-arrows in F, G, and L indicate that protein staining is no longer detectable in the terminally differentiated portions of the respective compartments. dp, dermal papilla; med, medulla.

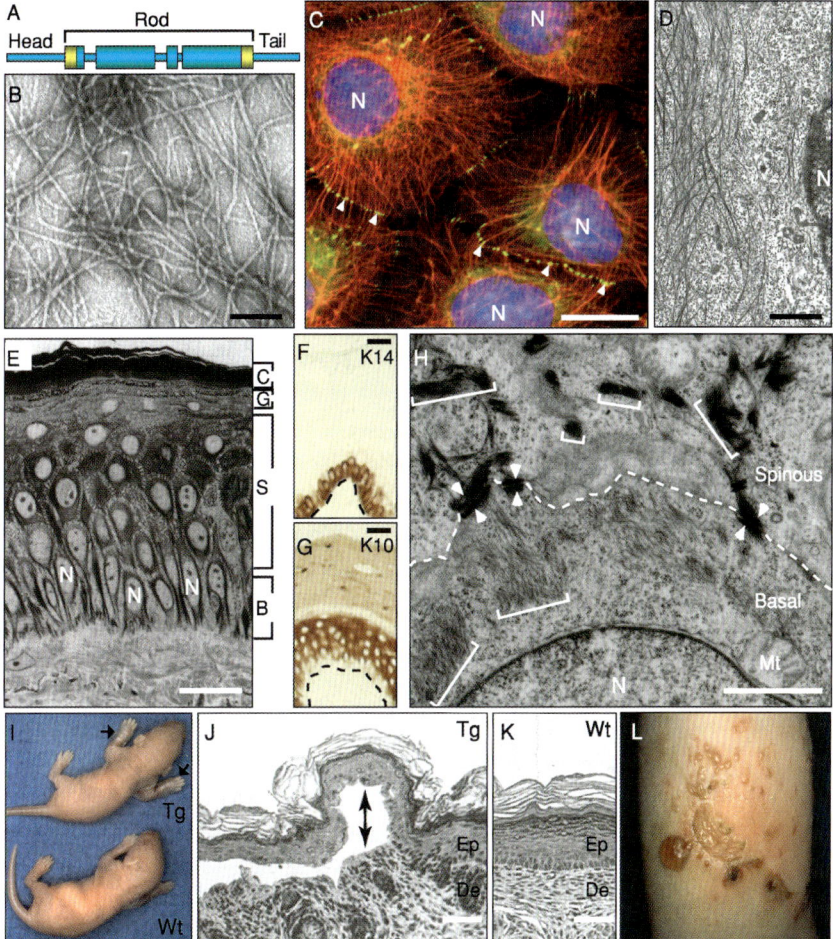

Chapter 16, Fig. 2 Attributes, differential regulation, and disease association of keratin. (A) Schematic representation of the tripartite domain structure shared by all keratin and other intermediate filament (IF) proteins. A central α-helical "rod" domain acts as the major determinant of self-assembly and is flanked by nonhelical "head" and "tail" domains at the N-terminus and C-terminus, respectively. The ends of the rod domain contain 15–20 amino-acid regions (yellow), which are highly conserved among all IFs. (B) Visualization of filaments, reconstituted *in vitro* from purified K5 and K14, by negative staining and electron microscopy. Bar equals 125 nm. (C) Triple-labeling for keratin (red chromophore) and desmoplakin, a desmosome component (green chromophore) and DNA (blue chromophore) by indirect immunofluorescence of epidermal cells in culture. Keratin filaments are organized in a network that spans the entire cytoplasm and are attached to desmosomes at points of cell–cell contacts (arrowheads). Bar equals 30 μm. N, nucleus. (D) Ultrastructure of the cytoplasm of epidermal cells in primary culture as visualized by transmission electron microscopy. Keratin filaments are abundant and tend to be organized in large bundles of loosely packed filaments in the cytoplasm. Bar equals 5 μm. (E) Histological cross-section of resin-embedded human trunk epidermis, revealing the basal (B), spinous (S), granular (G), and cornified (C) compartments. Bar equals 50 μm. N, nucleus. (F and G) Differential distribution of keratin

(Continued)

epitopes on human skin tissue cross-sections (similar to frame E) as visualized by an antibody-based detection method. K14 occurs in the basal layer, where the epidermal progenitor cells reside (F). K10, on the other hand, is primarily concentrated in the differentiating suprabasal layers of epidermis (G). Dashed line indicates the basal lamina. Bar equals 100 μm. (H) Ultrastructure of the boundary between the basal and suprabasal cells in mouse trunk epidermis as seen by routine transmission electron microscopy. The sample, from which this micrograph was taken, is oriented in the same manner as (E). Organization of keratin filaments as loose bundles (brackets in basal cell) correlates with the expression of K5–K14 in basal cells, whereas the formation of much thicker and electron-dense filament bundles (brackets in spinous cell) reflects the onset of K1–K10 expression in early differentiating cells. Arrowheads point to desmosomes connecting the two cells. Bar equals 1 μm. N, nucleus (I) Newborn mouse littermates. The top mouse is transgenic (Tg) and expresses a mutated form of K14 in its epidermis. Unlike the control pup below (Wt), this transgenic newborn shows extensive blistering of its front paws (arrows). (J and K) Hematoxylin and eosin (H&E)–stained histological cross-section through paraffin-embedded newborn mouse skin similar to those shown in (I). Compared with the intact skin of a control littermate (K, Wt), the epidermis of the K14 mutant expressing transgenic pup (J, Tg) shows intraepidermal cleavage at the level of the basal layer, where the mutant keratin is expressed (opposing arrows). Bar equals 100 μm (L) Leg skin in a patient with the Dowling-Meara form of epidermolysis bullosa simplex. Characteristic of this severe variant of this disease, several skin blisters are grouped in a herpetiform fashion (Reproduced from Coulombe and Bernot [2004] with permission).

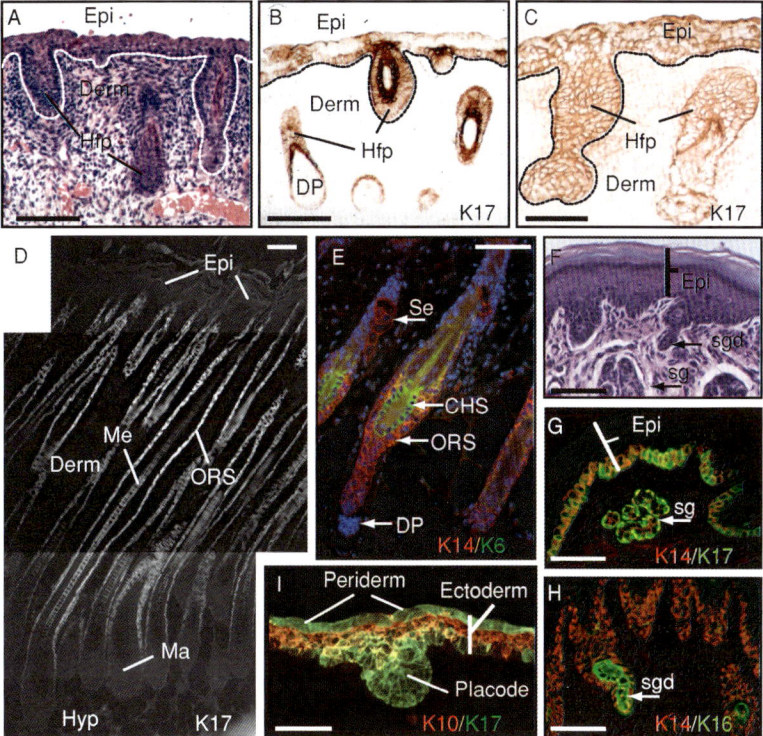

Chapter 16, Fig. 3 Morphological analysis of mouse skin tissue. (A) Hematoxylin and eosin stain of whisker pads from a paraffin-embedded mouse embryo 16.5dpc, sectioned at 5 μm. Dotted line marks the boundary between the epidermis (Epi) and dermis (Derm). Hfp, hair follicle precurser. (B) Same sample as (A) but processed for immunohistochemistry of K17. DP, dermal papilla. (C) 5-μm section of an embryo 16.5 dpc fresh frozen and processed for immunohistochemistry of K17. Note the even staining of the hair follicle and developing epidermis and periderm compared with the uneven staining but better morphological preservation of (B). (D) Paraffin-embedded back skin from a 10-day-old mouse pup processed for immunohistochemistry of K17 during the anagen cycle of the hair. K17 antigen is present in the outer root sheath of the hair follicle and in the medulla of the hair shaft. Hyp, hypodermis; Me, medulla; Ma, matrix. (E) Paraffin-embedded back skin from an 18-day-old mouse pup was processed for indirect immunofluorescence of K14 (red, outer root sheath [ORS] and sebaceous gland [Se]), K6 (green, club hair sheath-CHS), and nuclei (blue). During the catagen phase of the hair cycle, the bulb of the hair undergoes a massive wave of apoptosis, whereas the dermal papilla (DP) migrates to the base of the club hair sheath. (F) H&E stain of a paraffin-embedded mouse foot pad. Note the much thicker epidermis that contains sweat glands but no hair follicles. (G) Fresh-frozen section of mouse foot pad processed for immunofluorescence of K14 (red) and K17 (green). Note the presence of these two keratins in the basal layer of the epidermis, as well as the myoepithelium of the sweat glands (sg). (H) Unlike K17, K16 is present only in sweat gland ducts (green, sgd) rather than in the myoepithelium. (I) Fresh-frozen section of e14.5 embryo depicting K17 immunogen (green) in the periderm and newly invaginating hair follicle. K17 is a very early marker of appendageal formation in the early stratifying epidermis compared with K10 (red), which is expressed only in suprabasal epidermis. Bars equal 50 μm.

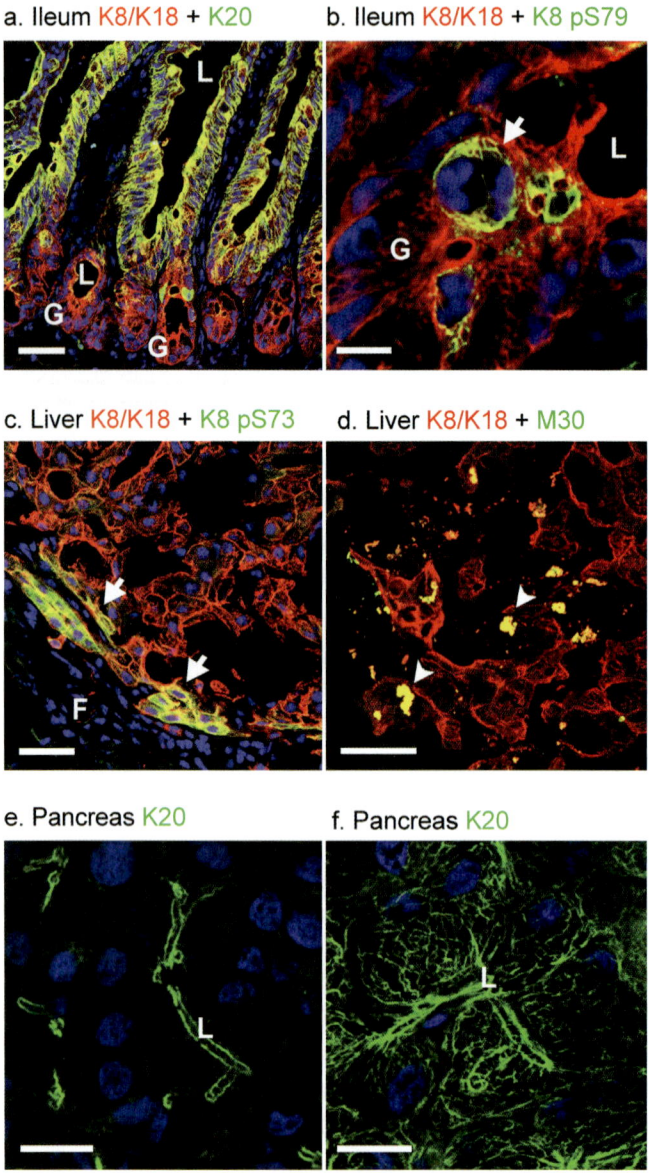

Chapter 17, Fig. 5 Analysis of keratins in mouse and human tissues by fluorescence microscopy. Sections of normal mouse ileum (a,b), human cirrhotic liver (c,d), and mouse pancreas (e,f) treated with saline (e) or 48 hours after 7 hourly injections of 50 μg/kg caerulein (f) were double- or triple-stained for keratins ($\pm$ nuclear staining) then viewed with a BioRad MRC1024 confocal microscope. K8/K18 (red) in (a–d) were detected with a rabbit anti-keratin antibody and a CY5-conjugated goat anti-rabbit antibody and further pseudo-colored red with the Adobe photoshop program. K20 (green) was detected with Ab Ks20.8 Ab, phospho-K8 (green) in (b) and (c) was probed with mAb LJ4, and the K18 apoptotic caspase-generated 43-kDa fragment (green) in (d) was detected with mAb M30

(Continued)

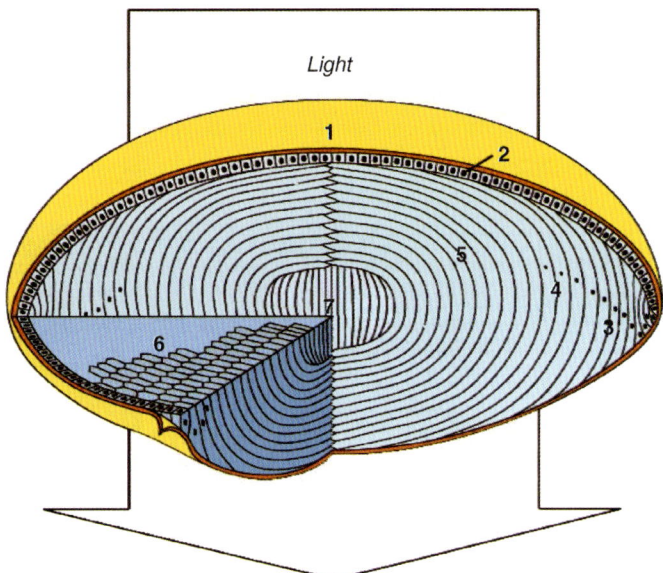

Chapter 21, Fig. 1 Schematic of the eye lens showing the major features and regions. The lens of the eye is enclosed by the lens capsule (1). The anterior surface of the lens (top) consists of a single layer of epithelial cells (2). Epithelial cells near the equator of the lens form a reservoir from which the lens grows throughout development and throughout the life of the individual. Initially, the cuboidal epithelial cells elongate (3) in two directions until their ends reach the two lens poles (top and bottom) and are joined at the lens sutures. At this stage, the fiber cells degrade their organelles (4), for example, the nuclei, symbolized here as black dots. The bulk of the lens thus consists of long, ribbonlike fiber cells devoid of cytoplasmic organelles (5). When cross-sectioned, cortical fiber cells display a characteristic hexagonal profile (6). The very core of the lens (lens nucleus) is composed of the primary fiber cells (7). These cells form shortly after the closure of the lens vesicle by linear elongation of the epithelial cells covering the posterior surface of the hollow lens vesicle.

(Table I). Nuclei in (a–f) were stained with propidium iodide and pseudo-colored blue. Note that although K8/K18 (red, a) are expressed throughout the ileal epithelium, K20 (green) is present in the villus cells (yellow when colocalized with K8/K18) but is absent in the basal glands. Arrows in (b) and (c) show phosphorylated K8 (S79 in mouse and S73 in human) in a mitotic cell in a mouse ileal gland and in hepatocytes of a human cirrhotic liver, respectively. Arrowheads in (d) point to keratin and M30-positive aggregates indicative of apoptosis. Note dramatic induction of filaments (panel f). F, fibrosis; G, intestinal glands; L, lumen. Scale bars: a, c, d $= 50\,\mu$m, b $= 10\,\mu$m, e, f $= 20\,\mu$m.

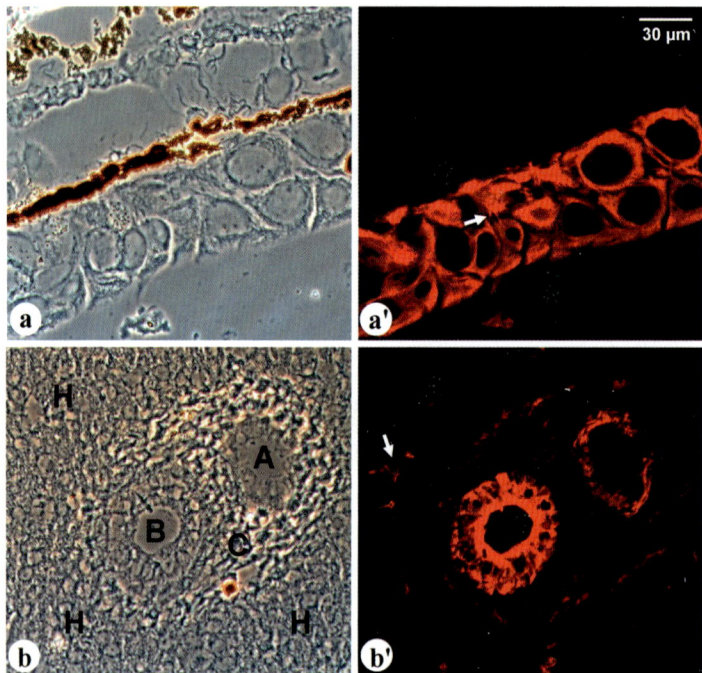

Chapter 22, Fig. 3 Immunofluorescence microscopy of 5-μm cryostat sections of lungfish skin (young specimen; a, a′) and trout liver (b, b′) with monoclonal anti-keratin antibodies KL1 (a, a′) and 68.4 (b, b′), respectively. The same fields are presented using phase-contrast optics (a, b) and epifluorescence optics (a′, b′); note the morphological details visible because of the immunofluorescence stain. (a, a′) On lungfish skin, antibody KL1 specifically recognizes the epidermal keratinocytes but is completely unreactive to mesenchymal cells such as the pigment cell layer or dermal fibroblasts. To withstand mechanical stress, keratinocytes are packed with keratin filaments (e.g., arrow in a′). (b, b′) Antibody 68.4 on trout liver tissue, specifically staining a hepatic artery (A), bile duct (B), and the surrounding interstitial cells, as well as the lining of bile canaliculi (arrow) formed by hepatocytes (H).

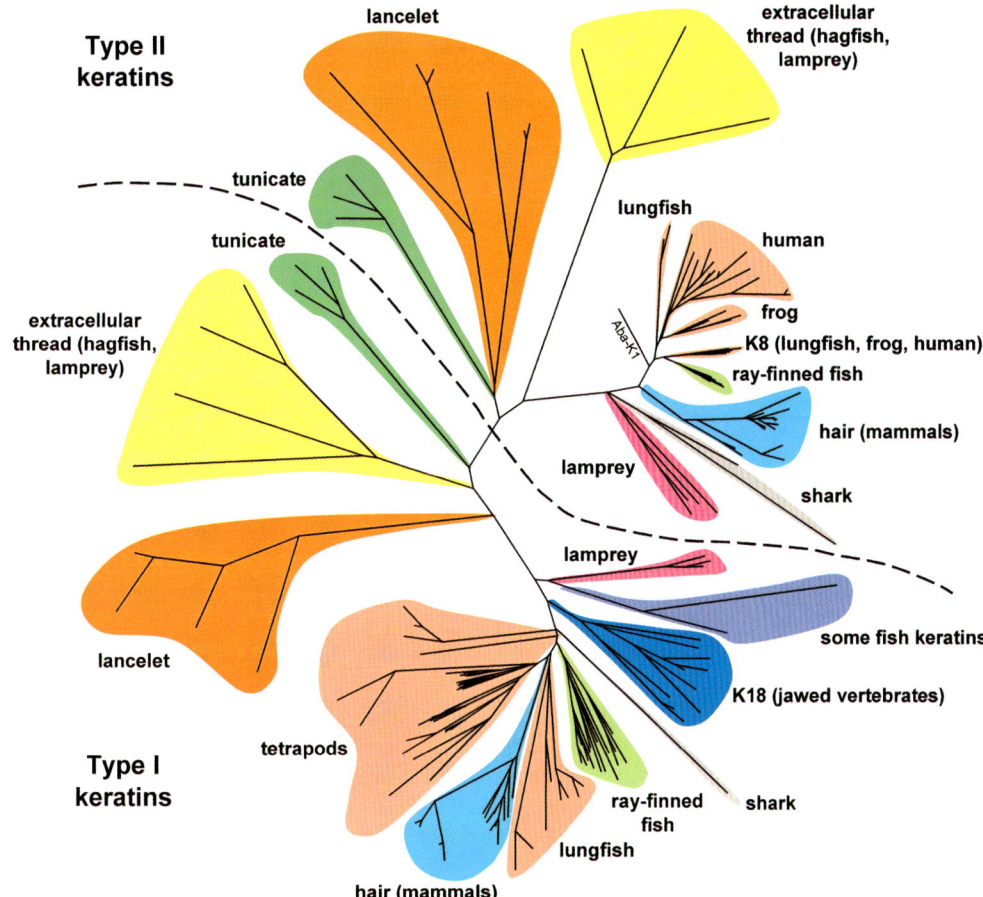

Type II keratins

lancelet

extracellular thread (hagfish, lamprey)

tunicate

tunicate

lungfish

human

frog

K8 (lungfish, frog, human)

ray-finned fish

hair (mammals)

shark

Aba-K1

lamprey

lamprey

some fish keratins

extracellular thread (hagfish, lamprey)

lancelet

K18 (jawed vertebrates)

tetrapods

ray-finned fish

shark

Type I keratins

hair (mammals)

lungfish

Chapter 22, Fig. 7 Unrooted phylogenetic tree of most currently available chordate keratin sequences (see Tables V–VII) generated from multiple sequence alignment of their *rod domains* (we exclusively included sequences encoding at least the entire rod domain; partial sequences have been excluded from the alignment). We also included the *Branchiostoma* C sequences, which according to Karabinos *et al.* (2002) show keratin-like features (we discarded the X sequence, which showed a very long branch near the base of the E2/D2 origin). Distances were calculated according to the Jones Taylor Thornton matrix (Jones *et al.*, 1992), and the tree was constructed by the Neighbor Joining method. This tree shows that the keratins of lower chordates, as well as the Agnatha thread keratins, are distinct from the type I and II keratins found in vertebrates. For further discussion see text (Section III).

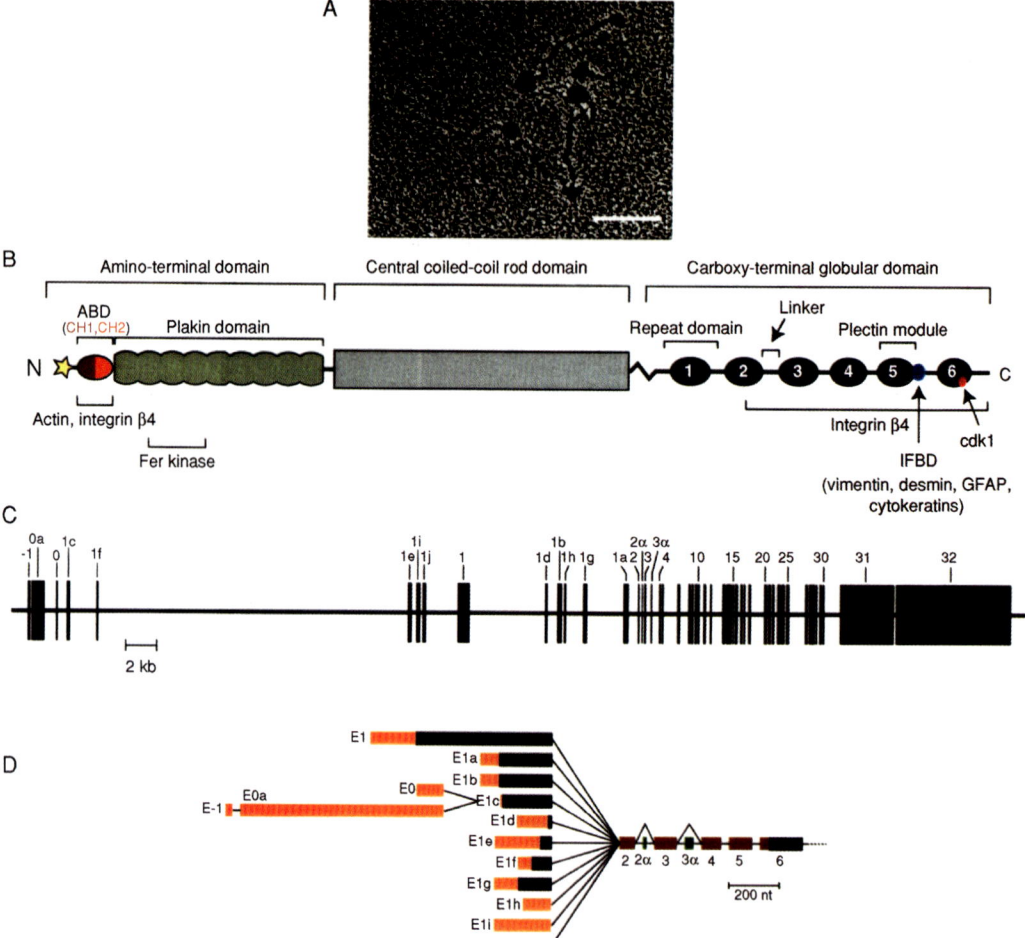

Chapter 25, Fig. 1 Protein and gene structure of plectin. (A) Electron microscopy of plectin molecules after rotary shadowing with platinum-carbon. A dumb bell like structure is revealed. Bar, 100 nm. (B) Schematic map of the protein. The tripartite structure of plectin molecules is made up of a central rod flanked by amino-terminal and carboxy-terminal (globular) domains. The amino-terminal domain contains an actin-binding domain, consisting of two calponin homology domains (dark and light red), and a region called plakin domain (light green). The carboxy-terminal globular domain contains six highly homologous plectin repeat domains (blue), each consisting of a plectin module and a linker region, one of which harbors an intermediate filament binding domain (IFBD; light blue). The carboxy-terminal domain also contains a unique cell cycle kinase cdk1 phosphorylation site (Malecz *et al.*, 1996) shown in red. Alternative splicing of plectin transcripts leads to expression of protein isoforms with different amino termini (yellow star). Regions responsible for direct binding to various proteins are indicated below schematics. (C) Genomic organization of the murine plectin gene. Exons are represented as black boxes, introns by lines. Plectin (exons −1 to 32) extends over ∼62 kb; exon 32 terminates at the location of the polyadenylation consensus sequence. (D) Schematic representation of alternative plectin transcripts. Alternative splicing of the 5′-end of the plectin gene gives rise to at least 16 different transcripts: 11 alternative first exons splicing into exon 2 are shown. Eight of them are first coding exons (1–1g), whereas three (1h, 1i, 1j) are noncoding. Three additional noncoding exons (E0 and E0a, splicing into E1c, and E − 1, splicing into E0a) and two optionally spliced exons (2α and 3α, inserted between exons 2 and 3, and 3 and 4, respectively) are also shown. Exons are shown as boxes and splice events as lines connecting individual boxes; noncoding regions are orange, regions coding in all cases are black, regions coding only in conjunction with a first coding exon are brown, and the optionally spliced exons 2α and 3α are green. Note that although there is an in-frame ATG in exon 4, translation initiation for isoforms with noncoding first exons occurs only in exon 6.

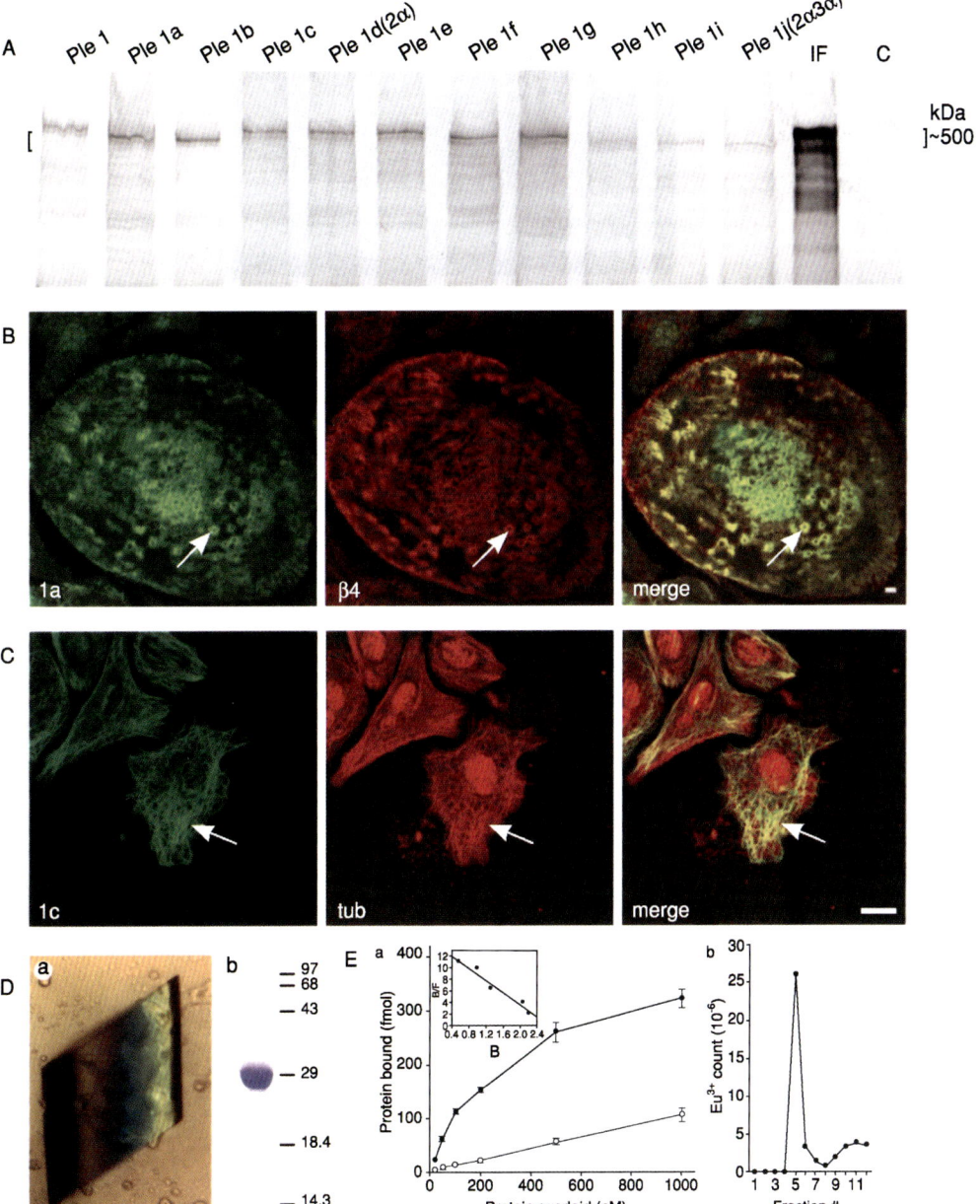

Chapter 25, Fig. 2 Various examples of plectin isoform analyses. (A) Expression of recombinant plectin isoforms in full length. Mouse isoform cDNA constructs containing 11 alternative first exons (1–1j, in some cases with exons 2α and 3α, as indicated) were cloned into an EGFP-fusion expression vector and proteins expressed in plectin-deficient mouse fibroblasts. Cells were lysed and expressed

(*Continued*)

proteins analyzed by immunoblotting (SDS-5%-polyacrylamide gel) with anti-plectin serum 9. Plectin from a rat glioma C6 cell IF preparation served as positive control and size marker (IF), and a lysate from mock-transfected cells as negative control (C). The bracket indicates a molecular mass range of 498–562 kDa. (B) Localization of plectin 1a at hemidesmosome-like stable anchoring complexes (SAC). Immortalized keratinocytes were processed for double immunofluorescence microscopy with antibodies specific for plectin isoform 1a (1a) and antibodies to integrin $\beta4$ ($\beta4$). Note that plectin 1a colocalizes with integrin $\beta4$ at SACs, forming the typical Swiss cheese–like pattern (arrow). Bar, 10 μm. (C) Colocalization of plectin 1c with microtubules in mouse keratinocytes. Double immunofluorescence microscopy was carried out with plectin isoform 1c–specific antibodies (1c) and monoclonal antibodies to tubulin (tub). Note colocalization of plectin 1c and microtubules (arrows). Bar, 10 μm. (D) Crystal (a) and SDS-PAGE (b) of a plectin ABD isoform including five amino acid residues encoded by tissue specifically spliced exon 2α. The protein was expressed in *E. coli*, purified to homogeneity, and crystallized by use of the hanging-drop method. (E) Europium overlay assay. (a) Binding of plectin's ABD (MBP-fusion protein; closed circles) to the cytoplasmic domain of integrin $\beta4$ (MBP negative control, open circles). Scatchard transformation of the binding data is shown in the inset. (b) Typical elution profile of a Eu^{3+}-labeled protein from a Bio-Rad P6 column, showing baseline-separation of the protein (fraction 5) from the unreacted labeling reagent starting at fraction 9. A and E (panel a) reproduced from Rezniczek *et al.* (2003), ©2003 Oxford University Press. B and C reproduced from Andrä *et al.* (2003), ©2003 The Society for Investigative Dermatology, Inc.

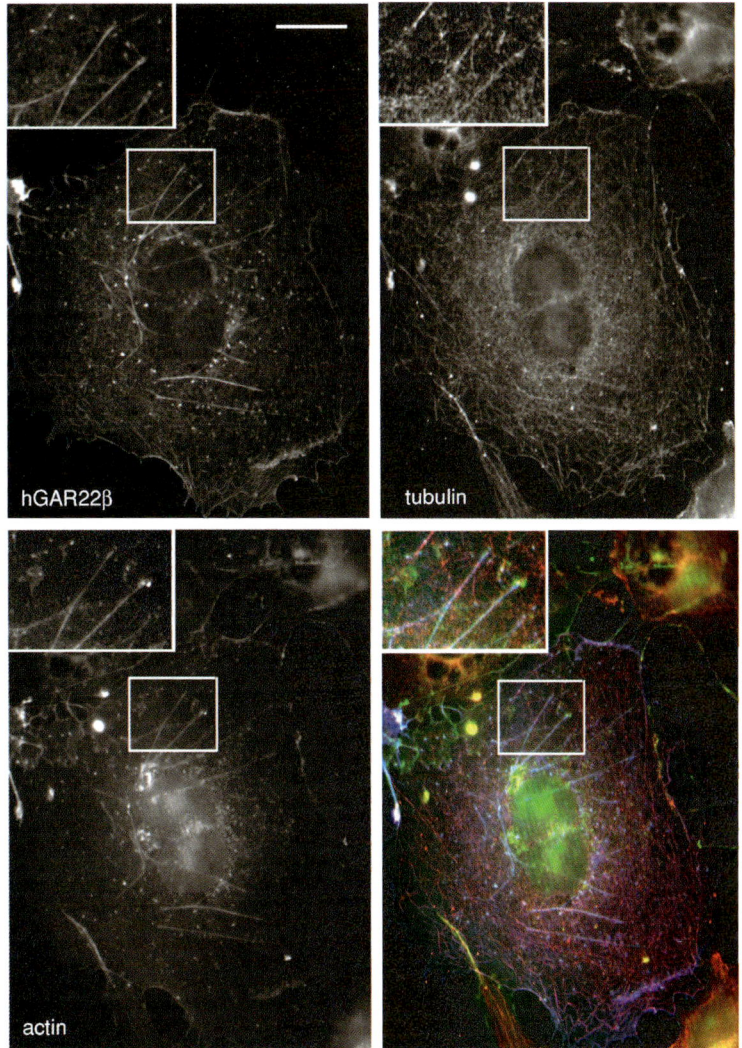

Chapter 27, Fig. 2 GAR22β bridges MFs and MTs in transfected cells. COS7 cells were transiently transfected with pFLAG-GAR22β and triple-stained with a monoclonal anti-FLAG antibody (A), a polyclonal anti-tubulin antibody (B), and fluorescently labeled phalloidin (C). (D) Superimposed image, with the FLAG staining in blue, the tubulin staining in red, and the actin staining in green. Triple-stained filaments, likely resulting from cross-linking by GAR22β, are white-colored. Bar, 10 μm. (From Goriounov *et al.*, 2003.)

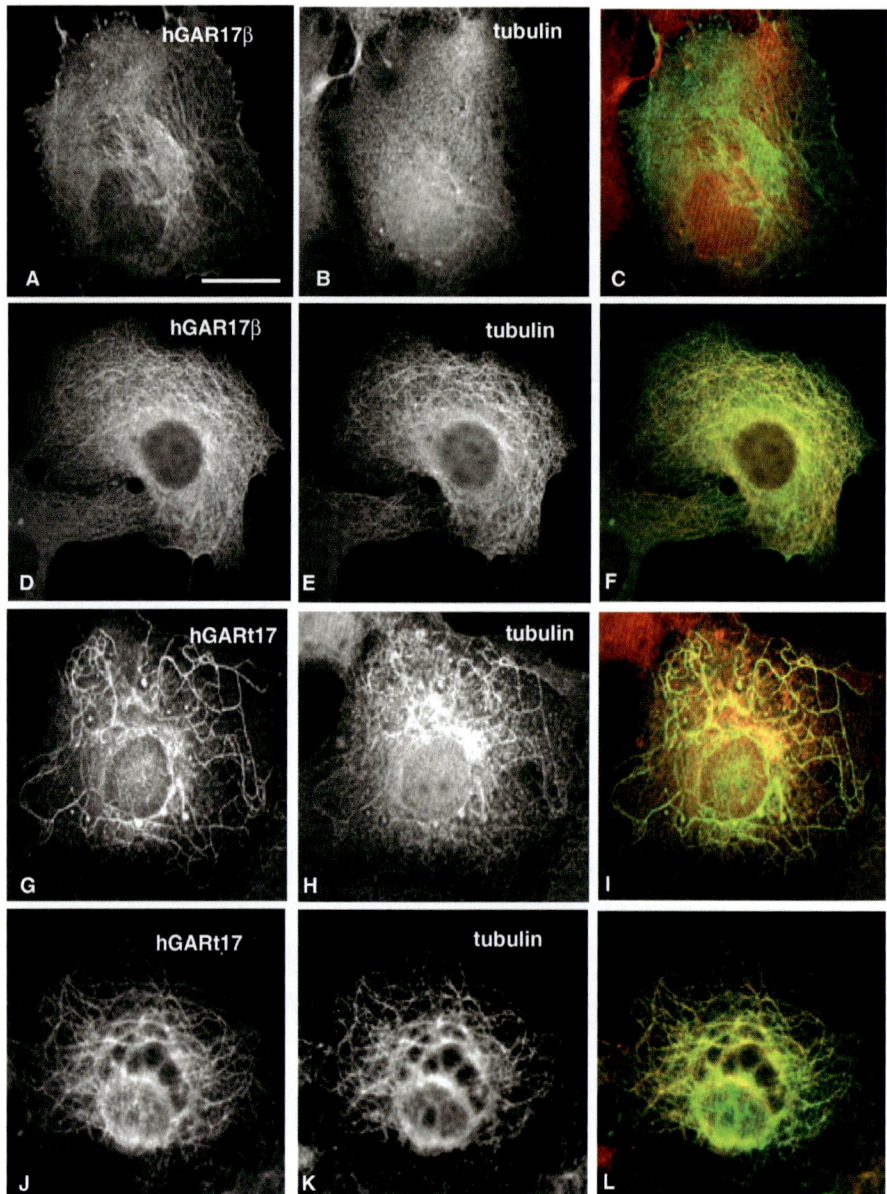

Chapter 27, Fig. 5 GARt17 but not GAR17β protects MTs from nocodazole. COS7 cells were transiently transfected with pFLAG-GAR17β (A–F) or pFLAG-GARt17 (G–L), treated with nocodazole (A–C, G–I), and stained with monoclonal FLAG M2 antibody (A, D, G, J) and polyclonal anti-tubulin antibody (B, E, H, K). (C, F, I, L) Superimposed images, with the FLAG staining in green and the tubulin staining in red. Both GAR17β and GARt17 bound MTs in untreated cells (F, L). After nocodazole treatment, MTs were preserved only in the cells expressing GARt17 (I). The preserved MTs appear bundled, as do the MTs in untreated GARt17-transfected cells. Bar, 15 μm.

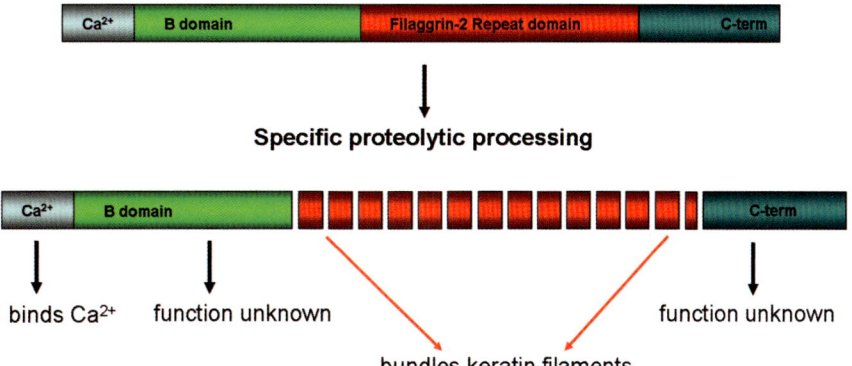

Specific proteolytic processing

binds Ca²⁺   function unknown   function unknown

bundles keratin filaments

Chapter 28, Fig. 1   Schematic showing the domain structure of mouse filaggrin-2. The proprotein consists of a N-terminal calcium binding domain, a B-domain, a repetitive domain, and a C-terminal tail as indicated and is processed into smaller peptides. The repetitive domain consists of 13 complete repeats and 1 partial repeat at the C-terminal end. These repeats vary in size between 74 and 80 residues and are capable of bundling keratin filaments.

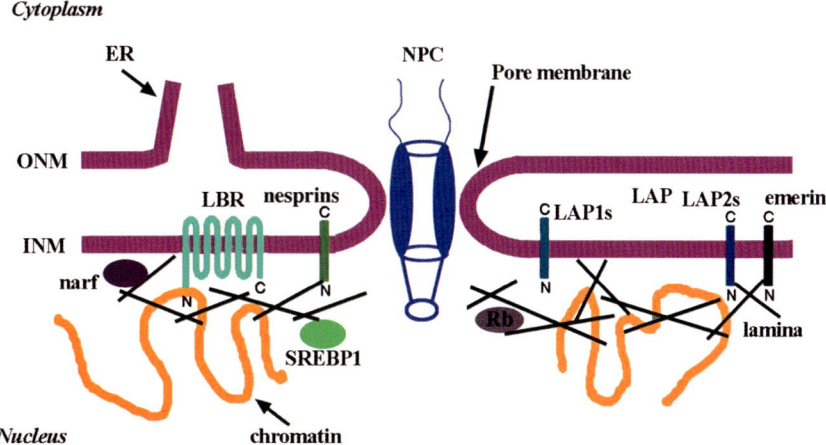

Chapter 29, Fig. 1   Schematic diagram of the nuclear envelope showing the nuclear lamina and associated proteins. The nuclear membranes consist of the outer nuclear membrane (ONM), continuous with the ER, the pore membrane, and the inner nuclear membrane (INM). The NPC is associated with the pore membrane. The lamina, chromatin, and some well-characterized lamin-binding proteins of nucleus and INM are also shown. Modified from Östlund *et al.* (2002) with permission from Landes Bioscience.

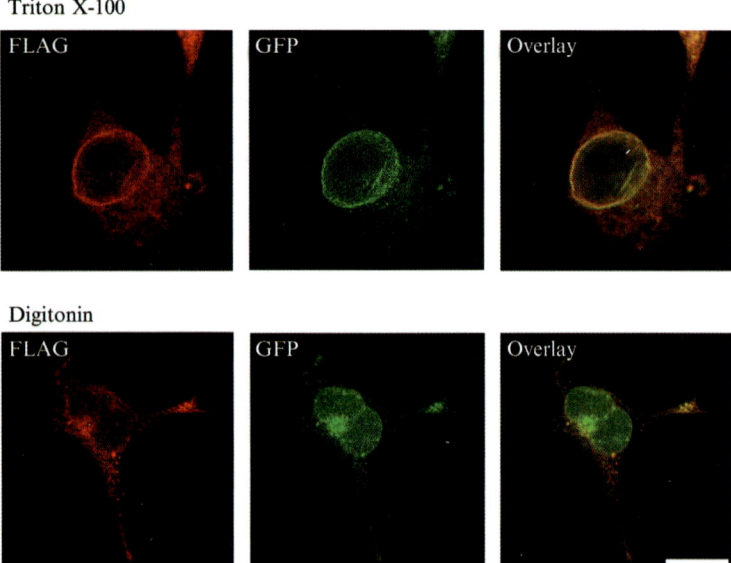

Chapter 29, Fig. 2 Emerin localizes to the inner nuclear membrane. Immunofluorescence studies of COS-7 cells that were transfected with plasmids expressing emerin tagged with a FLAG-tag and fused to GFP. The cells were permeabilized with Triton X-100 (upper panels) or digitonin (lower panels) and were then labeled with anti-FLAG monoclonal antibodies recognized by rhodamine-conjugated secondary antibodies (left panels). The middle panels show fluorescence from GFP and the panels to the right an overlay of signals from the two channels. Although fluorescence from GFP is seen at the nuclear rim under both permeabilization conditions, the anti-FLAG antibodies do not label the nuclear rim in digitonin-permeabilized cells, indicating that the epitope is inside the nucleus. Bar, 10 μm. Reprinted from Östlund *et al.* (1999) with permission from The Company of Biologists, Limited.